PERIODIC TABLE OF THE ELEMENTS

Metals
Nonmetals
Metalloids

	IA (1)	IIA (2)	IIIB (3)	IVB (4)	VB (5)	VIB (6)	VIIB (7)	VIIIB (8)	VIIIB (9)	VIIIB (10)	IB (11)	IIB (12)	IIIA (13)	IVA (14)	VA (15)	VIA (16)	VIIA (17)	0 (18)
1	1 **H** 1.0079																1 **H** 1.0079	2 **He** 4.0026
2	3 **Li** 6.941	4 **Be** 9.0122											5 **B** 10.811	6 **C** 12.011	7 **N** 14.0067	8 **O** 15.9994	9 **F** 18.9984	10 **Ne** 20.1797
3	11 **Na** 22.9898	12 **Mg** 24.3050											13 **Al** 26.9815	14 **Si** 28.0855	15 **P** 30.9738	16 **S** 32.066	17 **Cl** 35.4527	18 **Ar** 39.948
4	19 **K** 39.0983	20 **Ca** 40.078	21 **Sc** 44.9559	22 **Ti** 47.88	23 **V** 50.9415	24 **Cr** 51.9961	25 **Mn** 54.9380	26 **Fe** 55.847	27 **Co** 58.9332	28 **Ni** 58.69	29 **Cu** 63.546	30 **Zn** 65.39	31 **Ga** 69.723	32 **Ge** 72.61	33 **As** 74.9216	34 **Se** 78.96	35 **Br** 79.904	36 **Kr** 83.80
5	37 **Rb** 85.4678	38 **Sr** 87.62	39 **Y** 88.9059	40 **Zr** 91.224	41 **Nb** 92.9064	42 **Mo** 95.94	43 **Tc** (98)	44 **Ru** 101.07	45 **Rh** 102.9055	46 **Pd** 106.42	47 **Ag** 107.8682	48 **Cd** 112.411	49 **In** 114.82	50 **Sn** 118.710	51 **Sb** 121.75	52 **Te** 127.60	53 **I** 126.9045	54 **Xe** 131.29
6	55 **Cs** 132.9054	56 **Ba** 137.327	57 **La** 138.9055 *	72 **Hf** 178.49	73 **Ta** 180.9479	74 **W** 183.85	75 **Re** 186.207	76 **Os** 190.2	77 **Ir** 192.22	78 **Pt** 195.08	79 **Au** 196.9665	80 **Hg** 200.59	81 **Tl** 204.3833	82 **Pb** 207.2	83 **Bi** 208.9804	84 **Po** (210)	85 **At** (210)	86 **Rn** (221)
7	87 **Fr** (223)	88 **Ra** (226)	89 **Ac** (227) **	104 **Unq** (261)	105 **Unp** (262)	106 **Unh** (263)	107 **Uns** (262)	108 **Uno**	109 **Une**									

*Lanthanide Series

58 **Ce** 140.115	59 **Pr** 140.9076	60 **Nd** 144.24	61 **Pm** (145)	62 **Sm** 150.36	63 **Eu** 151.965	64 **Gd** 157.25	65 **Tb** 158.9253	66 **Dy** 162.50	67 **Ho** 164.9303	68 **Er** 167.26	69 **Tm** 168.9342	70 **Yb** 173.04	71 **Lu** 174.967

** Actinide Series

90 **Th** 232.0381	91 **Pa** 231.0359	92 **U** 238.0289	93 **Np** (237)	94 **Pu** (244)	95 **Am** (243)	96 **Cm** (247)	97 **Bk** (247)	98 **Cf** (251)	99 **Es** (252)	100 **Fm** (257)	101 **Md** (258)	102 **No** (259)	103 **Lr** (262)

Note: Atomic masses are IUPAC values (up to four decimal places). More accurate values for some elements are given in the table inside the back cover.

Place
Stamp
Here

Angus McDonald
Saunders College Publishing
Public Ledger Building, Suite 1250
150 South Independence Mall West
Philadelphia, PA 19106-3412

Fundamentals of ANALYTICAL CHEMISTRY

Seventh Edition

Douglas A. Skoog
Stanford University

Donald M. West
San Jose State University

F. James Holler
University of Kentucky

SAUNDERS COLLEGE PUBLISHING
Harcourt Brace College Publishers

Fort Worth Philadelphia San Diego
New York Orlando Austin San Antonio Toronto
Montreal London Sydney Tokyo

Text Typeface: Times Roman
Compositor: Progressive Information Technologies
Vice President; Publisher: John Vondeling
Associate Editor: Jennifer Bortel
Managing Editor: Carol Field
Project Editor: Beth Ahrens
Manager of Art and Design: Carol Bleistine
Art Director: Caroline McGowan
Text Designers: Caroline McGowan/Chazz Bjanes
Art & Design Coordinator: Kathleen Flanagan
Cover Designer: Chazz Bjanes
Text Artwork: Vantage Art, Inc.
Vice President, Director of EDP: Tim Frelick
Production Manager: Alicia Jackson
Vice President, Marketing: Marjorie Waldron
Product Manager: Angus McDonald

Cover Credit: © Tony Stone Images/Andrew Brookes

Printed in the United States of America

FUNDAMENTALS OF ANALYTICAL CHEMISTRY, Seventh Edition
0-03-005938-0

Library of Congress Catalog Card Number: 95-067683

78901234 069 10 98765

Preface

The seventh edition of *Fundamentals of Analytical Chemistry,* like its predecessors, is an introductory textbook designed primarily for a one- or two-semester course for chemistry majors. Since the publication of the sixth edition, the scope of analytical chemistry has continued to expand and includes applications in biology, medicine, materials science, ecology, and other related fields. The content of courses in analytical chemistry varies from institution to institution and depends upon the available facilities, time allocated to analytical chemistry in the chemistry curriculum, and desires of individual instructors. It is with these thoughts in mind that we have designed the seventh edition of *Fundamentals of Analytical Chemistry* to form a versatile and modular basis for courses in analytical chemistry.

OBJECTIVES

A major objective of this text is to provide a rigorous background in those chemical principles that are particularly important to analytical chemistry. A second goal is to develop an appreciation for the difficult task of judging the accuracy and precision of experimental data and to show how these judgments can be sharpened by the application of statistical methods. A third aim is to introduce a wide range of techniques of modern analytical chemistry. A final goal is to teach those laboratory skills that will give students confidence in their ability to obtain high-quality analytical data.

COVERAGE

The material in this text covers both fundamental and practical aspects of chemical analysis. Following a brief introduction in Chapter 1, Chapters 2, 3, and 4 present topics in statistics and data analysis that are important in analytical chemistry. An overview of classical gravimetric and titrimetric analysis is then presented in Chapters 5 and 6. The theory of aqueous solutions, activities, and chemical equilibria is detailed in Chapters 7–9. In Chapters 10–17, we consider the theory and practice of various titrimetric methods of analysis. Chapters 18–21 cover various electrical methods such as potentiometry, electrogravimetry, coulometry, and voltammetry. Chapters 22–26 on various spectroscopic methods are followed by Chapter 27, which is concerned with kinetic aspects of analytical chemistry. The various types of chromatography are then considered in Chapters

28–30. Practical facets of the preparation and analysis of real samples are detailed in Chapters 31–34. Finally, Chapters 35 and 36 provide discussions of the equipment and practice of chemical analysis and many detailed procedures for specific analyses.

Organization

Fundamentals of Analytical Chemistry, Seventh Edition, provides a gradual progression from the theoretical to the practical topics of analytical chemistry. Each chapter is organized as a learning unit; many chapters can stand alone, thus affording instructors the option of including or omitting topics as time and resources permit.

Important Equations

Equations that we feel are most important have been highlighted with a color screen for emphasis and ease of review.

Mathematical Level

Most of the development of the theory of chemical analysis requires a working knowledge of college algebra. A few concepts presented require basic differential and integral calculus.

Worked Examples

A large number of worked examples serve as aids in understanding the concepts of analytical chemistry. As in the sixth edition, we follow the practice of including units in chemical calculations and using dimensional analysis as a check on their correctness. The examples are also models for the solution of problems found at the end of most of the chapters.

Questions and Problems

An extensive set of questions and problems is included at the end of most chapters. Answers to approximately half of the problems are given at the end of the book.

Appendixes and Endpapers

Included in the appendixes are important references of analytical chemistry, tables of chemical constants, designations for filtering media, a section on the use of logarithms and exponential numbers, a section on normality and equivalents (terms that are not used in the text itself), a list of compounds recommended for the preparation of standard materials, and derivations of the expressions for propagation of measurement uncertainties.

The endpapers provide a full-color chart of chemical indicators, a periodic table of the elements, a table of atomic numbers and masses, a table of common molar masses, and a list of cross references between this book and *Mathcad® Applications for Analytical Chemistry.*

CHANGES IN THE SEVENTH EDITION

Users of the sixth edition will find numerous changes in the seventh edition not only in content by also in style and format.

Content

In Chapter 3 on statistical treatment of data, we have added an entirely new section on quality assurance in which we illustrate the use of control charts for judging the quality of an instrument performance and the quality of a manufactured product.

In Section 18D-7 on ion-selective electrodes, we have introduced a description of ion-selective field effect transistors, which have just appeared on the market for the determination of pH.

In Chapter 33, which deals with decomposing and dissolving samples, we have introduced a new section on the use of microwave ovens and furnaces for treatment of difficult samples of various kinds.

In Chapter 36, we have introduced a new experiment on the determination of magnesium in milk of magnesia preparations by ion exchange.

New to this edition is a Glossary, which appears at the end of the text.

In response to the request of many students, we have increased the number of margin notes that contain definitions, points of particular importance, historical notes, line drawings and photographs of modern analytical equipment, and thought provoking questions. All important definitions are highlighted in color. In addition, we have added margin notes that refer to related exercises in the workbook *Mathcad Applications for Analytical Chemistry.* These margin notes are marked with an icon.

We have also added several new boxed and highlighted *Features*. These Features consist of derivations of equations that the instructor may or may not choose to assign, historical notes, and illustrations of applications of analytical chemistry in the modern world.

Format

Style

We have continued to make style and format changes that make the text more readable and student-friendly. Throughout, we have endeavored to use shorter sentences, simpler words, and the active voice. We have also shortened chapters and increased their number so that students are not so overwhelmed by the amount of material to be mastered in one sitting. Although a certain amount of vocabulary must be introduced when study of a new field is undertaken, we have attempted to avoid jargon whenever possible.

Illustrations

All illustrations have been redrawn for greater clarity, and many photographs have been added to this edition. We have also provided full-color inserts to show in dramatic fashion various important color changes, chemical phenomena, and pieces of equipment.

ANCILLARIES

Instructor's Manual

The instructor's manual is a complete set of worked-out solutions to the problems in the text.

Overhead Transparencies

Approximately 100 transparencies are available for overhead projection. Many of the transparencies have multiple figures from the text.

Mathcad® Applications for Analytical Chemistry[1]

This ancillary introduces students to the command structure of the *Mathcad* gradually and introduces program syntax as needed to perform statistical calculations, solve systems of equilibrium equations, carry out least-squares analysis, analyze multicomponent mixtures using multiple linear regression, and accomplish many other computational and graphical tasks. *Mathcad* is a commercially available mathematical notebook, computational tool, and equation solver that allows students to perform many of the calculations of analytical chemistry quickly and accurately. Results can be easily plotted, printed, and formatted to enhance student understanding of the concepts presented in *Fundamentals of Analytical Chemistry,* Seventh Edition. *Mathcad* is available in versions for both IBM-PC running under DOS and Windows and Macintosh computers.

See *Mathcad Applications for Analytical Chemistry,* **pp. 15–18.**

We have introduced margin notes throughout the text similar to the one shown in the margin here that provide cross-references to relevant sections of *Mathcad Applications for Analytical Chemistry.* In addition, we have provided a comprehensive listing of cross-references in the endpages in the back of this book.

ACKNOWLEDGMENTS

We wish to acknowledge with thanks the comments and suggestions of the following who have reviewed the manuscript at various stages in its production: Professor T. Carter Gilmer, The University of Michigan—Dearborn; Professor Frederick C. Sauls, King's College; Professor Joseph J. Topping, Towson State University; and Professor Timothy A. Nieman, University of Illinois.

Our thanks also to Professor C. Marvin Lang and Gary Shulfer of the University of Wisconsin, Stevens Point, for providing color slides of the postage stamps that appear in the color plates and in the marginal notes. In addition we would like to thank Professor Mike McClure of Hopkinsville Community College, Hopkinsville, KY, for providing details of the analysis described in the Feature in Chapter 1 and Ms. Susan Spencer of Autodesk, Inc., for providing Hyperchem™, which was used to generate a number of the molecular models depicted in the book.

---

[1]F. J. Holler, *Mathcad® Applications for Analytical Chemistry.* Philadelphia: Saunders College Publishing, 1994.

Our writing team enjoys the services of a superb technical reference librarian, Ms. Maggie Johnson, head of the University of Kentucky Chemistry/Physics Library. She assisted us in many ways in the production of this book, including checking references, performing literature searches, and providing background information for many of the Features. We are grateful for her competence, enthusiasm, and good humor.

Finally, we want to acknowledge with thanks the various people at Saunders College Publishing for their friendly assistance in completing this project in record time. Among these are Project Editor Beth Ahrens, Copy Editor Andrew Potter, Art Director Caroline McGowan, and Production Manager Alicia Jackson. Our thanks also to Publisher John Vondeling and Associate Editor Jennifer Bortel.

Douglas A. Skoog
Donald M. West
F. James Holler

Contents Overview

Contents

CHAPTER

1

Introduction

Analytical chemistry involves separating, identifying, and determining the relative amounts of the components in a sample of matter. Qualitative analysis reveals the chemical identity of the species in the sample. Quantitative analysis establishes the relative amount of one or more of these species, or *analytes,* in numerical terms. Qualitative information is required before a quantitative analysis can be undertaken. A separation step is usually a necessary part of both a qualitative and a quantitative analysis.

The components of a sample that are to be determined are often referred to as **analytes.**

We shall be concerned principally with quantitative methods of analysis and methods of analytical separations. However, we refer occasionally to qualitative methods.

1A ROLE OF ANALYTICAL CHEMISTRY IN THE SCIENCES

Analytical chemistry has played a vital role in the development of science. For example, in 1894 Wilhelm Ostwald wrote,

> Analytical chemistry, or the art of recognizing different substances and determining their constituents, takes a prominent position among the applications of science, since the questions which it enables us to answer arise wherever chemical processes are employed for scientific or technical purposes. Its supreme importance has caused it to be assiduously cultivated from a very early period in the history of chemistry, and its records comprise a large part of the quantitative work which is spread over the whole domain of science.

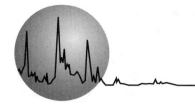

Since Ostwald's time, analytical chemistry has evolved from an art into a science with applications throughout industry, medicine, and all the sciences. To illustrate, consider a few examples. The amounts of hydrocarbons, nitrogen oxides, and carbon monoxide present in automobile exhaust gases must be measured in order to determine the effectiveness of smog-control devices. Quantitative measurements of ionized calcium in blood serum help diagnose parathyroid disease in human patients. The quantitative determination of nitrogen in foods establishes their protein content and thus their nutritional value. Analysis of steel during its production permits adjustment in the concentrations of such elements as carbon, nickel, and chromium to achieve a desired strength, hardness, corrosion resistance, and ductility. The mercaptan content of household gas supplies is monitored continually to assure that the gas has a sufficiently obnoxious odor to warn of dangerous leaks. Modern farmers tailor fertilization and irrigation schedules to meet changing plant needs during the growing season. They gauge these needs from quantitative analyses of the plants and the soil in which they grow.

Quantitative analytical measurements also play a vital role in many research areas in chemistry, biochemistry, biology, geology, and the other sciences. For example, chemists unravel the mechanisms of chemical reactions through reaction rate studies. The rates of consumption of reactants and formation of products in a chemical reaction can be calculated from quantitative measurements made at equal time intervals. Quantitative measurements of potassium, calcium, and sodium ions in the body fluids of animals permit physiologists to study the role these ions play in nerve-signal conduction and muscle contraction and relaxation. Materials scientists rely heavily upon quantitative analyses of crystalline germanium and silicon in their studies of the properties of semiconductor devices. Impurities in these devices are in the concentration range of 1×10^{-6} to 1×10^{-10} percent. Archaeologists identify the source of volcanic glasses (obsidian) by measuring concentrations of minor elements in samples taken from various locations. This knowledge in turn makes it possible to trace prehistoric trade routes for tools and weapons fashioned from obsidian.

Many chemists and biochemists devote a significant part of their time in the laboratory to gathering quantitative information about systems of interest to them. Analytical chemistry serves as an important tool for the scholarly efforts of such investigators.

1B CLASSIFICATION OF QUANTITATIVE METHODS OF ANALYSIS

We use a variety of measurements to complete analyses.

We compute the results of a typical quantitative analysis from two measurements. One is the mass or volume of sample to be analyzed. The second is the measurement of some quantity that is proportional to the amount of analyte in that sample and normally completes the analysis. Chemists classify analytical methods according to the nature of this final measurement. In a *gravimetric* method, the mass of the analyte or some compound chemically related to it is determined. In a *volumetric* method, the volume of a solution containing sufficient reagent to react completely with the analyte is measured. *Electroanalytical* methods involve the measurement of such electrical properties as potential, current, resistance, and quantity of electricity. *Spectroscopic* methods are based upon measurement of the interaction between electromagnetic radiation and analyte atoms or molecules or upon the production of such radiation by analytes. Finally, a group of miscellaneous methods should be mentioned. These include the measurement of such properties as mass-to-charge ratio *(mass spectrometry)*, rate of radioactive decay, heat of reaction, rate of reaction, thermal conductivity, optical activity, and refractive index.

Electromagnetic radiation includes X-ray, ultraviolet, visible, infrared, microwave, and radio-frequency radiation.

1C STEPS IN A TYPICAL QUANTITATIVE ANALYSIS

A typical quantitative analysis involves the sequence of steps shown in Figure 1-1. In some instances, we can leave out one or more of these steps. Ordinarily, though, all of them play important roles in the success of an analysis.

The first 27 chapters of this textbook focus on the last three steps in Figure 1-1. In the measurement step we determine one of the physical properties mentioned in Section 1B. In the calculation step we compute the relative amount of the analyte present in the samples. In the final step we evaluate the quality of the results and estimate their reliability.

We now describe each of these steps to give you an overview of a quantitative analysis.

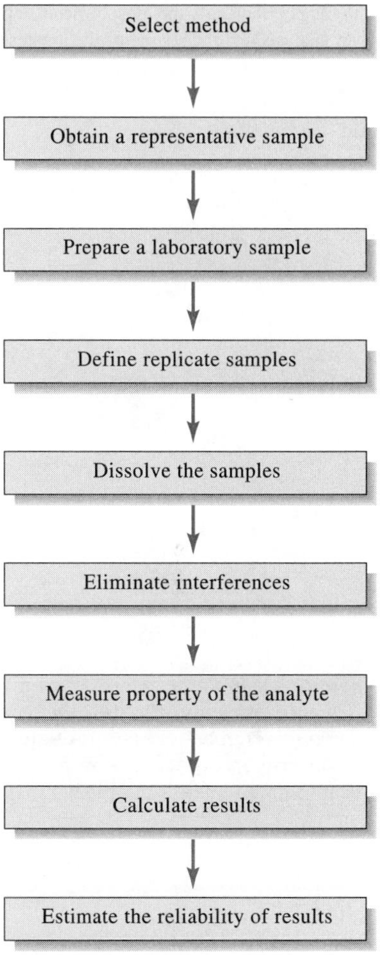

Figure 1-1
Steps in a quantitative analysis.

1C-1 Selecting a Method of Analysis

A vital first step in any quantitative analysis is the selection of a method. The choice can be difficult, requiring experience as well as intuition on the part of the chemist. An important consideration in selection is the level of accuracy required. Unfortunately, high reliability nearly always requires a large investment of time. The method that is selected ordinarily represents a compromise between accuracy and economics.

A second consideration related to economic factors is the number of samples to be analyzed. If there are many samples, we can afford to spend a good deal of time in such preliminary operations as assembling and calibrating instruments and equipment as well as preparing standard solutions. If we have only a single sample or a few samples at most, we may find it more expedient to select a procedure that avoids or minimizes such preliminary steps.

Finally, the choice of method is always governed by the complexity of the sample as well as the number of components in the sample. We discuss the choice of method in detail in Section 31A.

1C-2 Sampling

An **assay** is the process of determining how much of a given sample is the material indicated by its name. For example, a zinc alloy is assayed for its zinc content.

Sampling involves obtaining a small mass of a material whose composition accurately represents the bulk of the material being sampled.

To produce meaningful information, an analysis must be performed on a sample whose composition faithfully represents that of the bulk of material from which it was taken. Where the bulk is large and inhomogeneous, great effort is required to get a representative sample. Consider, for example, a railroad car containing 25 tons of silver ore. Buyer and seller must agree on a price based primarily upon the silver content of the shipment. The ore itself is inherently heterogeneous, consisting of many lumps that vary in size as well as in their silver content. The assay of this shipment will be performed upon a sample that weighs about one gram. For the analysis to have significance, this small sample must have a composition that is representative of the 25 tons (or approximately 22,700,000 g) of ore in the shipment. Isolation of 1 g of material that accurately reflects the average composition of the nearly 23 million g of bulk sample is a difficult undertaking that requires a careful, systematic manipulation of the entire shipment.

Sampling is frequently the most difficult step in an analysis and the source of the greatest error.

Many sampling problems are easier to solve than the one just described. Whether this problem is simple or complex, however, the chemist must have some assurance that the laboratory sample is representative of the whole before proceeding with an analysis. A detailed description of how various types of materials are sampled is presented in greater detail in Section 32A.

1C-3 Preparing a Laboratory Sample

A solid laboratory sample is ground to decrease particle size, mixed to assure homogeneity, and stored for various lengths of time before the analysis begins. Absorption or desorption of water may occur during each of these steps, depending upon the humidity of the environment. Because any loss or gain of water changes the chemical composition of solids, it is a good idea to dry samples just before starting an analysis. Alternatively, the moisture content of the sample can be determined at the time of the analysis in a separate analytical procedure. Further information about preparing samples for analysis and drying or determining their moisture content is given in Chapter 32.

1C-4 Defining Replicate Samples

Replicate samples are portions of a material of approximately the same size that are carried through an analytical procedure at the same time and in the same way.

Most chemical analyses are performed on *replicate samples* whose masses or volumes have been determined by careful measurements with an analytical balance or with a precise volumetric device. Replication improves the quality of the results and provides a measure of their reliability.

1C-5 Preparing Solutions of the Samples

Most analyses are performed on solutions of the sample. Ideally, the solvent should dissolve the entire sample (not just the analyte) rapidly and completely. The conditions of dissolution should be sufficiently mild that loss of the analyte cannot occur. Unfortunately, many materials that must be analyzed are insoluble in common solvents. Examples include silicate minerals, high-molecular-weight polymers, or specimens of animal tissue. Conversion of the analyte in such materials into a soluble form can be a difficult and time-consuming task that may involve heating the sample with aqueous solutions of strong acids, strong bases, oxidizing agents, reducing agents, or some combination of such reagents; ignition of the sample in air or oxygen; or high-temperature fusion of the sample in the presence of various fluxes. Methods for decomposing and dissolving samples are described in various parts of Chapter 33.

1C-6 Eliminating Interferences

An **interference** is a species that causes an error by enhancing or attenuating (making smaller) the quantity being measured in an analysis.

Few chemical or physical properties of importance in chemical analysis are unique to a single chemical species. Instead, the reactions used and the properties measured are characteristic of a group of elements or compounds. Species other than the analyte that affect the final measurement are called *interferences,* or *interferents.* A scheme must be devised to isolate the analyte from interferences before the final measurement is made. No hard and fast rules can be given for eliminating interferences; indeed, resolution of this problem can be the most demanding aspect of an analysis. Chapters 28 through 30 and Chapter 34 describe separation methods.

Techniques or reactions that work for only one analyte are said to be **specific.** Techniques or reactions that work for only a few analytes are **selective.**

The **matrix,** or **sample matrix,** is all of the components making up the sample containing an analyte.

1C-7 Calibration and Measurement

All analytical results depend on a final measurement X of a physical property of the analyte. This property must vary in a known and reproducible way with the concentration c_A of the analyte. Ideally, the measurement of the physical property is directly proportional to the concentration. That is,

$$c_A = kX$$

where k is a proportionality constant. With two exceptions, analytical methods require the empirical determination of k with chemical standards for which c_A is known.[1] The process of determining k is thus an important step in most analyses and is termed a *calibration.*

The empirical determination of the relationship between the quantity measured in an analysis and the analyte concentration is termed a **calibration.**

1C-8 Calculating Results

Computing analyte concentrations from experimental data is ordinarily a simple and straightforward task, particularly with modern calculators or computers. Such computations are based on the raw experimental data collected in the measurement step, the stoichiometry of the chemical reaction upon which the analysis is based, and instrumental factors. These calculations appear throughout this book.

1C-9 Evaluating Results from Estimating Their Reliability

Analytical results are incomplete without an estimate of their reliability. The experimenter must provide some measure of the uncertainties associated with computed results if the data are to have any value. Chapters 2, 3, and 4 present detailed methods for carrying out this important final step in the analytical process.

An analytical result without an estimate of reliability is of no value.

---

[1]The two exceptions are gravimetric methods, which are discussed in Chapter 5, and coulometric methods, which are considered in Chapter 20. In both of these methods, k can be computed from known physical constants.

1C-10 An Example Illustrating the Steps Outlined in Figure 1-1

FEATURE 1-1

Deer Kill: A Case Study Illustrating the Use of Analytical Chemistry in Solving Toxicological Problems

The tools of modern analytical chemistry are widely employed in environmental investigations. In this feature we describe a case study in which quantitative analysis was used to determine the agent that caused the deaths in a number of whitetail deer in a wildlife area of a Kentucky state park. We begin with a description of the problem and then show how the steps illustrated in Figure 1-1 were used to solve the analytical problem.

The Problem

The incident began when a park ranger found a dead whitetail deer near a pond in the Land Between the Lakes State Park in south central Kentucky. The park ranger enlisted the help of a chemist from the state veterinary diagnostic laboratory to help find the cause of death of the deer so that further deer kills might be prevented.

The ranger and the chemist examined the site where the badly decomposed carcass of the deer had been found. Because of the advanced state of decomposition, no fresh organ tissue samples could be gathered. A few days after the investigation, the ranger found two more dead deer at essentially the same location. The chemist came to the site of the kill, loaded the deer on a truck for transport to his laboratory, and he and the ranger conducted a careful examination of the surrounding area in hopes of finding some clue as to the cause of the deaths.

The search encompassed about two acres surrounding the pond. The investigators noticed that grass surrounding nearby power-line poles was wilted and discolored. They speculated that a herbicide had been used on the grass. A common ingredient in herbicides is arsenic in any one of a variety of forms such as arsenic trioxide, sodium arsenite, monosodium methanearsenate, or disodium methanearsenate. The last compound is the disodium salt of methanearsenic acid, $CH_3AsO(OH)_2$, which is very soluble in water and thus is used as the active ingredient in many herbicides. The herbicidal activity of disodium methanearsenate is due to its reactivity with the sulfydryl (S—H) groups in the amino acid cysteine. When cysteine in plant enzymes reacts with arsenical compounds, the enzyme function is inhibited and the plant eventually dies. Unfortunately, similar chemical effects occur in animals as well. The investigators therefore collected samples of the discolored dead grass so that they could be tested along with samples from the organs of the deer. They planned to analyze the samples to confirm the presence of arsenic and, if present, to determine its concentration in the samples.

Selecting a Method

A common method for the quantitative determination of arsenic in biological samples is found in the publication of the Association of Official Analytical Chemists (AOAC).[2] This method involves the distillation of arsenic as arsine, whose concentration is then determined by colorimetric measurements.

Obtaining Representative Samples

Back at the laboratory, the deer were dissected and the kidneys were removed for analysis. The kidneys were chosen because the suspected pathogen (arsenic) is rapidly eliminated from the animal body through the urinary tract, where the arsenic tends to be concentrated.

Preparing a Laboratory Sample

Each kidney was cut into pieces and macerated in a high-speed blender. This step served to reduce the size of the pieces of tissue and to homogenize the resulting laboratory sample.

Defining Replicate Samples

Three 10-g samples of the homogenized tissue from each deer were placed in porcelain crucibles.

Dissolving the Samples

In order to obtain an aqueous solution of the analyte for analysis, it was necessary to *dry ash* the sample in air to convert its organic matrix to carbon dioxide and water. This process involved heating each crucible and sample cautiously over an open flame until the sample stopped smoking. The crucible was then placed in a furnace and heated at 555°C for two hours. Dry ashing serves to free the analyte from organic material and convert it to arsenic pentoxide. The dry solid in each of the sample crucibles was then dissolved in dilute HCl, which converted the As_2O_5 to soluble H_3AsO_4.

Eliminating Interferences

Arsenic can be separated from other substances that might interfere in the analysis by converting it to arsine, AsH_3, a toxic, colorless gas that is evolved when a solution of H_3AsO_3 is treated with zinc. The solutions resulting from the deer and grass samples were combined with Sn^{2+}, and a small amount of iodide ion was added to catalyze the reduction of H_3AsO_4 to H_3AsO_3 according to the following reaction:

$$H_3AsO_4 + SnCl_2 + 2HCl \longrightarrow H_3AsO_3 + SnCl_4 + H_2O$$

[2] *Official Methods of Analysis,* 15th ed., p. 626. Washington, DC, 1990.

The H_3AsO_3 was then converted to AsH_3 by the addition of zinc metal as follows:

$$H_3AsO_3 + 3Zn + 6HCl \longrightarrow AsH_3(g) + 3ZnCl_2 + 3H_2O$$

The entire reaction was carried out in flasks equipped with a stopper and delivery tube so that the arsine gas could be collected in the absorber solution as shown in Figure 1-2. The arrangement ensures that interferences are left in the reaction flask and only the arsine is collected in the absorber in special transparent containers called *cuvettes*.

As the arsine bubbles into the solution in the cuvette, it reacts with silver diethyldithiocarbamate to form a red complex compound according to the following equation.

$$AsH_3 + 6\,Ag^+ + 3 \quad \begin{matrix} C_2H_5 \\ \diagdown \\ N-C \\ \diagup \\ C_2H_5 \end{matrix} \begin{matrix} S \\ \diagup\diagup \\ \\ S^- \end{matrix} \longrightarrow$$

$$As\left[\begin{matrix} C_2H_5 \\ \diagdown \\ N-C \\ \diagup \\ C_2H_5 \end{matrix} \begin{matrix} S \\ \diagup\diagup \\ \\ S \end{matrix}\right]_3 + 6\,Ag + 3H^+$$

red

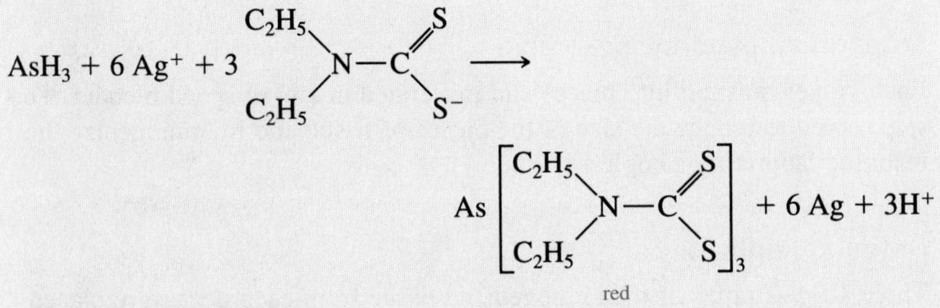

In addition to the flasks containing the deer and grass samples, several other flasks were prepared containing known concentrations of arsenic as well as a flask containing only the reagents and no arsenic. The latter is called a *blank*. The quantities of arsine collected from these flasks were used

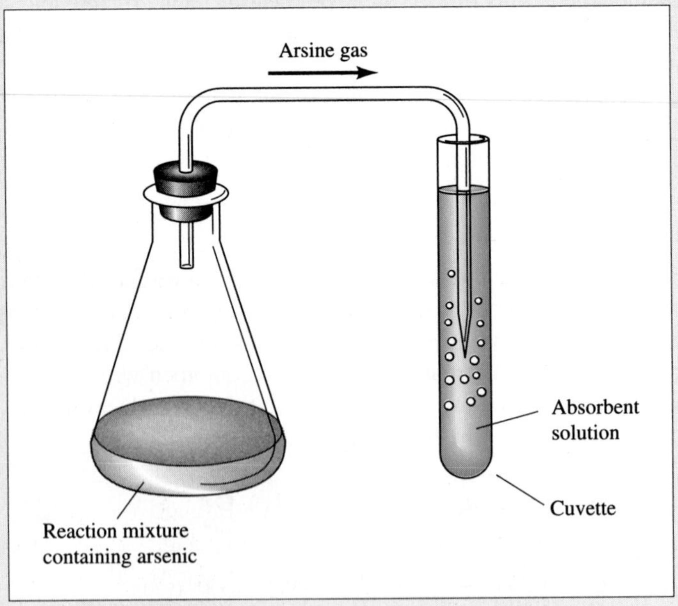

Figure 1-2 An arsine generator.

as standards for comparison with the unknown samples from the deer and grass.

Measuring the Amount of the Analyte

The amount of arsenic in each sample was determined by measuring the intensity of the red color formed in the solutions in the cuvettes with an instrument called a spectrophotometer. As will be shown in Chapters 23 and 24, a spectrophotometer provides a number called *absorbance* that is directly related to color intensity and is also proportional to the concentration of the species responsible for the color. In order to use absorbance for analytical purposes, it is necessary to prepare a calibration curve by measuring the absorbances of a series of standard analyte solutions having known concentrations of the arsenic complex. The standards for calibration are depicted in the upper part of Figure 1-3 where you can see that the color intensity increases as the arsenic content of the standards increase from 0 to 25 parts per million of arsenic.

Calculating the Concentration

The absorbances for the standard solutions containing known concentrations of arsenic are plotted to produce a *calibration curve* of absorbance versus concentration as shown in the lower part of Figure 1-3. The vertical lines between the upper and lower part of the figure show the correspondence between each solution and the corresponding point plotted on the graph. The intensity of the color of each solution is represented by its absorbance, which is plotted on the vertical axis of the calibration curve. Note that the absorbance increases from 0 to about 0.72 as the concentration of arsenic increases from 0 to 25 parts per million. The concentration of arsenic in parts per million in each standard solution corresponds to the vertical grid lines of the calibration curves as shown. The curve is then used to determine the concentration of the two unknown solutions shown on the right by finding the absorbances of the unknowns on the absorbance axis of the plot and reading the corresponding concentrations on the concentration axis. The lines leading from the cuvettes to the calibration curve show that the concentrations of arsenic in the two deer were 16 ppm and 22 ppm, respectively.

Arsenic in the kidney tissue of animals is toxic at levels above about 10 ppm, so it was probable that the deer were killed by ingesting an arsenic compound. The tests also showed that the samples of grass contained about 600 ppm arsenic. This very high level of arsenic suggested that the grass had been sprayed with an arsenical herbicide. The investigators concluded that the deer had probably died as a result of eating the poisoned grass.

Reliability of the Data

The data from these experiments were analyzed using the statistical methods described in Chapters 2, 3, and 4. For each of the standard arsenic solutions and the deer samples, the average of the three absorbance measurements was calculated. The average absorbance for the replicates is a more reliable measure of the concentration of arsenic than a single measurement. Least-squares analysis of the standard data (see Section 4E-2) was used to find the

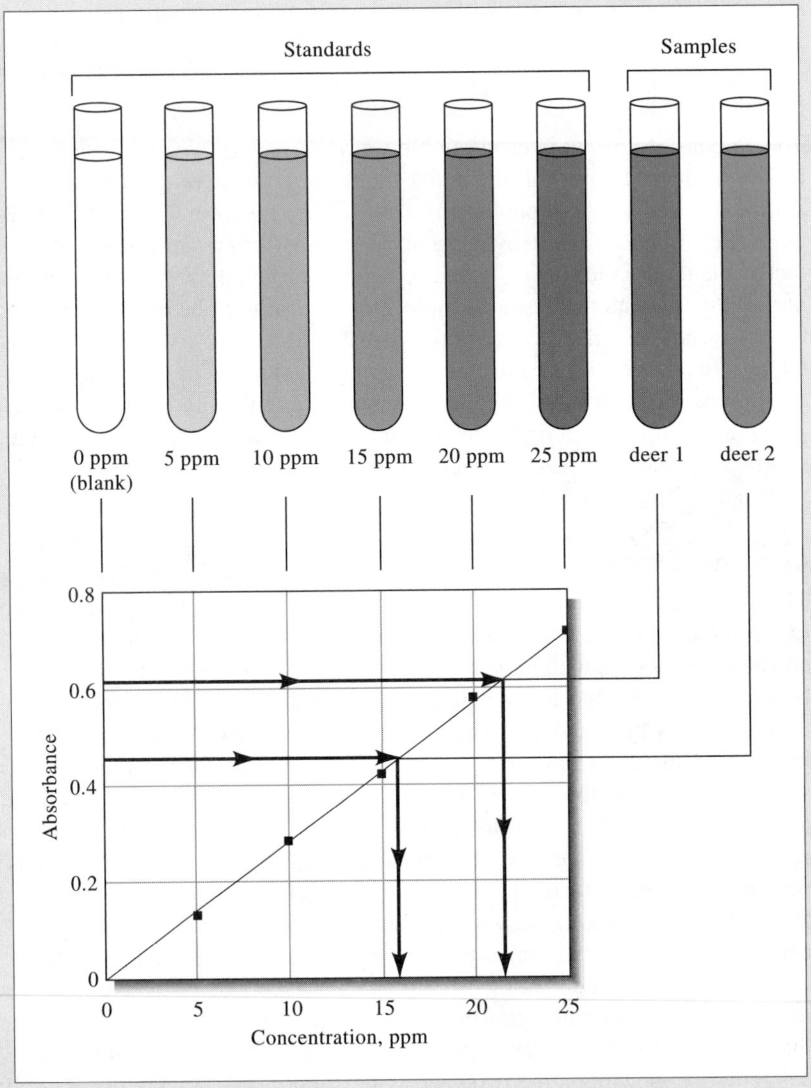

Figure 1-3 Obtaining a calibration curve for the determination of arsenic.

best straight line among the points and to calculate the concentrations of the unknown samples along with their uncertainties and confidence limits.

In this analysis, the formation of the highly colored product of the reaction served both to confirm the probable presence of arsenic and to provide a reliable estimate of its concentration in the deer and in the grass. Based on their results, the investigators recommended that the use of arsenical herbicides be suspended in the wildlife area to protect the deer and other animals that might eat plants in the area.

This case study illustrates how chemical analysis is used in the identification and determination of quantities of hazardous chemicals in the environment. Many of the methods and instruments of analytical chemistry are used routinely to provide vital information in environmental and toxicological studies of this type.

CHAPTER

2

Errors in Chemical Analyses

It is impossible to perform a chemical analysis that is totally free of errors, or uncertainties. All one can hope is to minimize these errors and to estimate their size with acceptable accuracy. In this and the next two chapters, we explore the nature of experimental errors and their effects on the results of chemical analyses.

The consequence of errors in analytical data is illustrated in Figure 2-1, which shows results for the quantitative determination of iron(III). Six equal portions of an aqueous solution that contained exactly 20.00 ppm of iron(III) were analyzed in exactly the same way. Note that the results range from a low of 19.4 ppm to a high of 20.3 ppm of iron(III). The average $\bar{x}$ of the data is 19.8 ppm.

Parts per million (ppm), that is, 20.00 parts of iron(III) per million parts of solution.

Every measurement is influenced by many uncertainties, which combine to produce a scatter of results like that in Figure 2-1. Measurement uncertainties can never be completely eliminated, so the true value for any quantity is always unknown.[1] The probable magnitude of the error in a measurement can often be evaluated, however. We can then define limits within which the true value of a measured quantity lies at a given level of probability.

The true value of a measurement is never known exactly.

It is seldom easy to estimate the reliability of experimental data. Nevertheless, we must make such estimates whenever we collect laboratory results *because data of unknown quality are worthless.* On the other hand, results that are not especially accurate may be of considerable value if the limits of uncertainty are known.

Unfortunately, there is no simple and widely applicable method for determining the reliability of data with absolute certainty. The evaluation of experimental data often requires an effort that is comparable to that involved in their acquisition. Reliability can be assessed in several ways. Experiments designed to reveal the presence of errors can be performed. Standards of known composition can be analyzed and the results compared with the known composition. A few minutes in the library to consult the literature of analytical chemistry can be profitable. Calibrating equipment enhances the quality of data. Finally, statistical tests can be applied to the data. None of these options is perfect, so in the end we have to make *judgments* as to the probable accuracy of our results. These judgments tend to become harsher and less optimistic with experience.

[1] Unfortunately, many people do not understand these truths. For example, when asked by a defense attorney in a celebrated homicide investigation what the rate of error in a blood test was, the assistant district attorney replied that their testing laboratories had no percentage of error because "they have not committed any errors" (*San Francisco Chronicle,* June 29, 1994, p. 4).

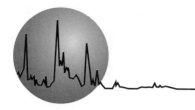

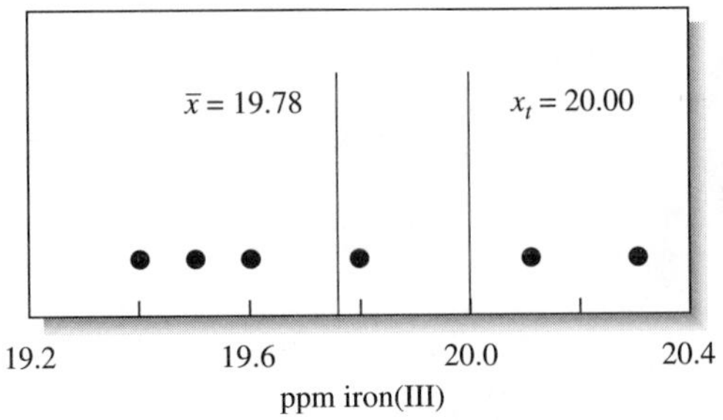

Figure 2-1
Results from six replicate determinations for iron in aqueous samples of a standard solution containing 20.00 ppm of iron(III).

One of the first questions to answer before beginning an analysis is, What is the maximum error that I can tolerate in the result? The answer to this question determines the time required to do the work. For example, a tenfold increase in accuracy may take hours, days, or even weeks of added labor. *No one can afford to waste time generating data that are more reliable than is needed.*

2A DEFINITION OF TERMS

Replicates are samples of about the same size that are carried through an analysis in *exactly* the same way.

Chemists usually carry two to five portions *(replicates)* of a sample through an entire analytical procedure. Individual results from a set of measurements are seldom the same (Figure 2-1), so a central or ''best'' value is used for the set. We justify the extra effort required to analyze several samples in two ways. First, the central value of a set should be more reliable than any of the individual results. Second, variation in the data should provide a measure of the uncertainty associated with the central result. Either the *mean* or the *median* may serve as the central value for a set of replicated measurements.

2A-1 The Mean and Median

The **mean** of two or more measurements is their average value.

Mean, arithmetic mean, and *average* $(\bar{x})$ are synonyms for the quantity obtained by dividing the sum of replicate measurements by the number of measurements in the set:

The symbol Σx_i means to add all of the values x_i for the replicates.

$$\bar{x} = \frac{\sum_{i=1}^{N} x_i}{N} \tag{2-1}$$

where x_i represents the individual values of x making up a set of N replicate measurements.

The *median* is the middle result when replicate data are arranged in order of size. There are equal numbers of data that are larger and smaller than the median. For an odd number of results, the median can be evaluated directly. For an even number, the mean of the middle pair is used.

The **median** is the middle value in a set of data that has been arranged in order of size. The median is used advantageously when a set of data contains an *outlier*—that is, a result that differs significantly from the rest of the data in the set. An outlier can have a significant effect on the mean of the set but has a lesser effect on the median.

EXAMPLE 2-1

Calculate the mean and median for the data shown in Figure 2-1.

$$\text{mean} = \bar{x} = \frac{19.4 + 19.5 + 19.6 + 19.8 + 20.1 + 20.3}{6}$$

$$= 19.78 \approx 19.8 \text{ ppm Fe}$$

Because the set contains an even number of measurements, the median is the average of the central pair:

$$\text{median} = \frac{19.6 + 19.8}{2} = 19.7 \text{ ppm Fe}$$

See *Mathcad Applications for Analytical Chemistry,* **pp. 15–18.**

Ideally, the mean and median are identical. Frequently they are not, however, particularly when the number of measurements in the set is small.

2A-2 Precision

Precision describes the reproducibility of measurements—that is, the closeness of results that have been obtained *in exactly the same way.* Generally, the precision of a measurement is readily determined by simply repeating the measurement.

Precision is the closeness of data to other data that have been obtained in exactly the same way.

Three terms are widely used to describe the precision of a set of replicate data: *standard deviation, variance,* and *coefficient of variation.* All of these terms are a function of the *deviation from the mean,* which is defined as

$$\text{deviation from the mean} = d_i = |x_i - \bar{x}| \tag{2-2}$$

The relationship between deviation from the mean and the three precision terms is given in Section 3B.

Note that deviations from the mean are calculated without regard to sign.

2A-3 Accuracy

Figure 2-2 illustrates the difference between accuracy and precision. *Accuracy* indicates the closeness of the measurement to its true or accepted value and is expressed by the *error.* Note the basic difference between accuracy and precision. Accuracy measures agreement between a result and its true value. Precision describes the agreement among several results that have been measured in the same way. Precision is determined by simply replicating a measurement. On the other hand, accuracy can never be determined exactly because the true value of a quantity can never be known exactly. An accepted value must be used instead.

Accuracy is the closeness of a result to its true or accepted value.

Accuracy is expressed in terms of either absolute or relative error.

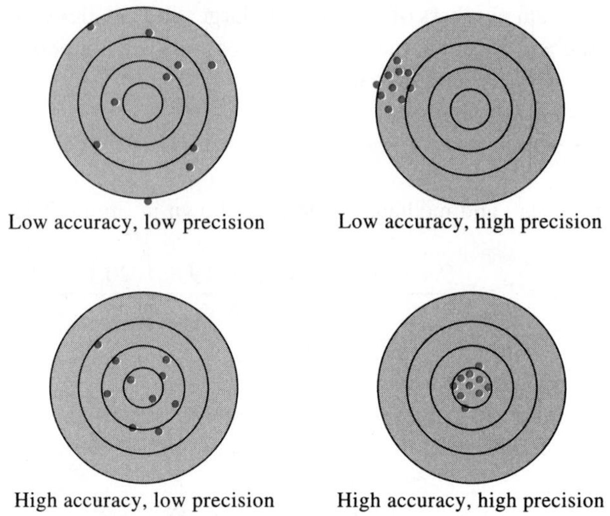

Figure 2-2
Accuracy and precision.

Absolute Error

The term "absolute" has a different meaning here than it does in mathematics. An absolute value in mathematics means the magnitude of a number *ignoring its sign.* As we shall use it, the absolute error is the difference between an *experimental result and the accepted value including its sign.*

The **absolute error** of a measurement is the difference between the measured value and the true value. It bears a sign.

The *absolute error E* in the measurement of a quantity x_i is given by the equation

$$E = x_i - x_t \tag{2-3}$$

where x_t is the true, or accepted, value of the quantity. Returning to the data displayed in Figure 2-1, the absolute error of the result immediately by the left of the true value of 20.00 ppm is -0.2 ppm Fe; the result at 20.10 ppm is in error by $+0.1$ ppm Fe. Note that we retain the sign in stating the error. Thus, the negative sign in the first case shows that the experimental result is smaller than the accepted value.

Relative Error

The **relative error** of a measurement is the absolute error divided by the true value.

Often, the *relative error* E_r is a more useful quantity than the absolute error. The percent relative error is given by the expression

$$E_r = \frac{x_i - x_t}{x_t} \times 100\% \tag{2-4}$$

Relative error is also expressed in parts per thousand (ppt). Thus, the relative error for the mean of the data in Figure 2-1 is

$$E_r = \frac{19.8 - 20.00}{20.00} \times 100\% = -1\% \text{ or } -10 \text{ ppt}$$

2A-4 Types of Errors in Experimental Data

The precision of a measurement is readily determined by comparing data from carefully replicated experiments. Unfortunately, an estimate of the accuracy is not so easy to obtain. To determine the accuracy, we have to know the true value, and ordinarily this is exactly what we are looking for.

It is tempting to assume that if we know the answer precisely, then we also know it accurately. The danger of this assumption is illustrated in Figure 2-3, which summarizes the results for the determination of nitrogen in the two pure compounds shown in the margin. The dots show the absolute errors of replicate results obtained by four analysts. Note that analyst 1 obtained relatively high precision and high accuracy. Analyst 2 had poor precision but good accuracy. The results of analyst 3 are surprisingly common. The precision is excellent, but there is significant error in the numerical average for the data. Both the precision and the accuracy are poor for the results of analyst 4.

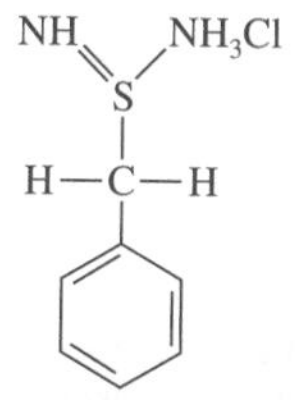

benzyl isothiourea hydrochloride

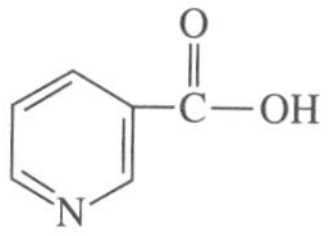

nicotinic acid

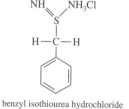

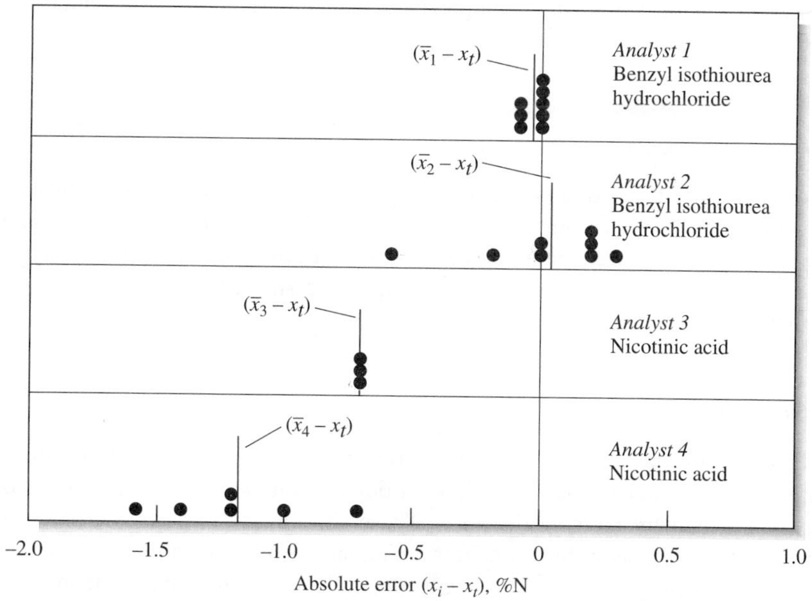

Figure 2-3

Absolute error in the micro-Kjeldahl determination of nitrogen. Each dot represents the error associated with a single determination. Each vertical line labeled $(\bar{x}_i - x_t)$ is the absolute average deviation of the set from the true value. (Data from C. O. Willits and C. L. Ogg, *J. Assoc. Offic. Anal. Chem.*, **1949,** *32*, 561. With permission.)

Random, or **indeterminate,** errors are errors that affect the precision of measurement.

Figures 2-1 and 2-3 suggest that chemical analyses are affected by at least two types of errors. One type, called *random* (or *indeterminate*) *error,* causes data to be scattered more or less symmetrically around a mean value. Refer again to Figure 2-3, and notice that the scatter in the data, and thus the random error, for analysts 1 and 3 is significantly less than that for analysts 2 and 4. In general, then, the random error in a measurement is reflected by its precision.

Systematic, or **determinate,** errors affect the accuracy of results.

A second type of error, called *systematic* (or *determinate*) *error,* causes the mean of a set of data to differ from the accepted value. For example, the results of analysts 1 and 2 in Figure 2-3 have little systematic error, but the data of analysts 3 and 4 show determinate errors of about -0.7 and -1.2% nitrogen. In general, a systematic error causes the results in a series of replicate measurements to be all high or all low.

An **outlier** is an occasional result in replicate measurements that obviously differs significantly from the rest of the results.

A third type of error is *gross error.* Gross errors differ from indeterminate and determinate errors. They usually occur only occasionally, are often large, and may cause a result to be either high or low. Gross errors lead to *outliers*—results that appear to differ markedly from all other data in a set of replicate measurements. There is no evidence of a gross error in Figures 2-1 and 2-3. Had one of the results shown in Figure 2-1 occurred at 22.5 ppm Fe, it might have been an outlier.

2B SYSTEMATIC ERRORS

Bias measures the systematic error associated with an analysis. It has a negative sign if it causes the results to be low and a positive sign otherwise.

Systematic errors have a definite value, an assignable cause, and are of the same magnitude for replicate measurements made in the same way. Systematic errors lead to *bias* in measurement technique. Note that bias affects all of the data in a set in the same way and that it bears a sign.

2B-1 Sources of Systematic Errors

There are three types of systematic errors: (1) *Instrument errors* are caused by imperfections in measuring devices and instabilities in their power supplies. (2) *Method errors* arise from nonideal chemical or physical behavior of analytical systems. (3) *Personal errors* result from the carelessness, inattention, or personal limitations of the experimenter.

Instrument Errors

All measuring devices are sources of systematic errors. For example, pipets, burets, and volumetric flasks may hold or deliver volumes slightly different from those indicated by their graduations. These differences may arise from using glassware at a temperature that differs significantly from the calibration temperature, from distortions in container walls due to heating while drying, from errors in the original calibration, or from contaminants on the inner surfaces of the containers. Calibration eliminates most systematic errors of this type.

Electronic instruments are subject to instrumental systematic errors. These uncertainties have many sources. For example, errors emerge as the voltage of a battery-operated power supply decreases with use. Errors result from increased resistance in circuits because of dirty electrical contacts. Temperature changes cause variations in resistors and standard potential sources. Currents induced from 110-V power lines affect electronic instruments. Errors from these and other sources are detectable and correctable.

Method Errors

The nonideal chemical or physical behavior of the reagents and reactions upon which an analysis is based often introduces systematic method errors. Such sources of nonideality include the slowness of some reactions, the incompleteness of others, the instability of some species, the nonspecificity of most reagents, and the possible occurrence of side reactions that interfere with the measurement process. For example, a common method error in volumetric methods results from the small excess of reagent required to cause an indicator to undergo the color change that signals completion of the reaction. The accuracy of such an analysis is thus limited by the very phenomenon that makes the titration possible.

Another example of method error is illustrated by the data in Figure 2-3, in which the results by analysts 3 and 4 show a negative bias that can be traced to the chemical nature of the sample, nicotinic acid. The analytical method used involves the decomposition of the organic samples in hot concentrated sulfuric acid, which converts the nitrogen in the samples to ammonium sulfate. The amount of ammonia in the ammonium sulfate is then determined in the measurement step. Experiments have shown that compounds containing a pyridine ring such as nicotinic acid (see page 15) are incompletely decomposed by the sulfuric acid unless special precautions are taken. Without these precautions, low results are obtained. It is highly likely that the negative errors, $(\bar{x}_3 - x_t)$ and $(\bar{x}_4 - x_t)$ in Figure 2-3 are systematic errors that can be blamed on incomplete decomposition of the samples.

Errors inherent in a method are often difficult to detect and are thus the most serious of the three types of systematic error.

Of the three types of systematic errors encountered in a chemical analysis, method errors are usually the most difficult to identify and correct.

Personal Errors

Many measurements require personal judgments. Examples include estimating the position of a pointer between two scale divisions, the color of a solution at the end point in a titration, or the level of a liquid with respect to a graduation in a pipet or buret (see Figure 3-5, page 41). Judgments of this type are often subject to systematic, unidirectional errors. For example, one person may read a pointer consistently high, another may be slightly slow in activating a timer, and a third may be less sensitive to color changes. An analyst who is insensitive to color changes tends to use excess reagent in a volumetric analysis. Physical handicaps are often sources of personal determinate errors.

Color blindness is a good example of a handicap that amplifies personal errors in a volumetric analysis. A famous color-blind analytical chemist enlisted his wife to come to the laboratory to help him detect color changes at end points of titrations.

A universal source of personal error is prejudice. Most of us, no matter how honest, have a natural tendency to estimate scale readings in a direction that improves the precision in a set of results, or we may have a preconceived notion of the true value for the measurement. We then subconsciously cause the results to fall close to this value. Number bias is another source of personal error that varies considerably from person to person. The most common number bias encountered in estimating the position of a needle on a scale involves a preference for the digits 0 and 5. Also prevalent is a prejudice favoring small digits over large and even numbers over odd.

Digital readouts on pH meters, laboratory balances, and other electronic instruments eliminate number bias because no judgment is involved in taking a reading.

Persons who make measurements must guard against personal bias to preserve the integrity of the collected data.

2B-2 The Effect of Systematic Errors upon Analytical Results

Constant errors are independent of the size of the sample being analyzed. **Proportional errors** decrease or increase in proportion to the size of the sample.

Systematic errors may be either *constant* or *proportional.* The magnitude of a constant error does not depend on the size of the quantity measured. Proportional errors increase or decrease in proportion to the size of the sample taken for analysis.

Constant Errors

Constant errors become more serious as the size of the quantity measured decreases. The effect of solubility losses on the results of a gravimetric analysis illustrates this behavior.

EXAMPLE 2-2

Suppose that 0.50 mg of precipitate is lost as a result of being washed with 200 mL of wash liquid. If the precipitate weighs 500 mg, the relative error due to solubility loss is $-(0.50/500) \times 100\% = -0.1\%$. Loss of the same quantity from 50 mg of precipitate results in a relative error of -1.0%.

The excess of reagent required to bring about a color change during a titration is another example of constant error. This volume, usually small, remains the same regardless of the total volume of reagent required for the titration. Again, the relative error from this source becomes more serious as the total volume decreases. One way of minimizing the effect of constant error is to use as large a sample as possible.

Proportional Errors

A common cause of proportional errors is the presence of interfering contaminants in the sample. For example, a widely used method for the determination of copper is based upon the reaction of copper(II) ion with potassium iodide to give iodine. The quantity of iodine is then measured and is proportional to the amount of copper in the sample. Iron(III), if present, also liberates iodine from potassium iodide. Unless steps are taken to prevent this interference, high results are observed for the percentage of copper because the iodine produced will be a measure of the copper(II) *and* iron(III) in the sample. The size of this error is fixed by the *fraction* of iron contamination, which is independent of the size of sample taken. If the sample size is doubled, for example, the amount of iodine liberated by both the copper and the iron contaminant is also doubled. Thus, the magnitude of the reported percentage of copper is independent of sample size.

2B-3 Detection of Systematic Instrument and Personal Errors

After entering a reading into the laboratory notebook, many scientists habitually make a second reading to assure the correctness of the entry.

Systematic instrument errors are usually found and corrected by calibration. Periodic calibration of equipment is always desirable because the response of most instruments changes with time as a result of wear, corrosion, or mistreatment.

Most personal errors can be minimized by care and self-discipline. It is a good habit to check instrument readings, notebook entries, and calculations systematically. Errors that result from a known physical handicap can usually be avoided by a careful choice of method.

2B-4 Detection of Systematic Method Errors

Bias in an analytical method is particularly difficult to detect. We may take one or more of the following steps to recognize and adjust for a systematic error in an analytical method.

Analysis of Standard Samples

The best way of estimating the bias of an analytical method is by the analysis of *standard reference materials*—materials that contain one or more analytes with exactly known concentration levels. Standard reference materials are obtained in several ways.

Standard materials can sometimes be prepared by synthesis. Here, carefully measured quantities of the pure components of a material are measured out and mixed so as to produce a homogeneous sample whose composition is known from the quantities taken. The overall composition of a synthetic standard material must approximate closely the composition of the samples to be analyzed. Great care must be taken to ensure that the concentration of analyte is known exactly. Unfortunately, the synthesis of such standard samples is often impossible or so difficult and time-consuming that this approach is not practical.

Standard reference materials can be purchased from a number of governmental and industrial sources. For example, the National Institute of Standards and Technology (NIST) (formerly the National Bureau of Standards) offers over 900 standard reference materials including rocks and minerals, gas mixtures, glasses, hydrocarbon mixtures, polymers, urban dusts, rainwaters, and river sediments.[2] The concentration of one or more of the components in these materials has been determined in one of three ways: (1) through analysis by a previously validated reference method, (2) through analysis by two or more independent, reliable measurement methods, or (3) through analysis by a network of cooperating laboratories, technically competent and thoroughly knowledgeable with the material being tested.

Several commercial supply houses also offer analyzed materials for method testing.[3]

One of the problems you will encounter in using standard reference materials to establish the presence or absence of bias is that the mean of your replicate analysis of the standard will ordinarily differ somewhat from the theoretical result. Then you are faced with the question whether this difference is due to random error of your measurements or to bias in the method. In Section 4B-1, we demonstrate a statistical test that can be applied to aid your judgment in answering this question.

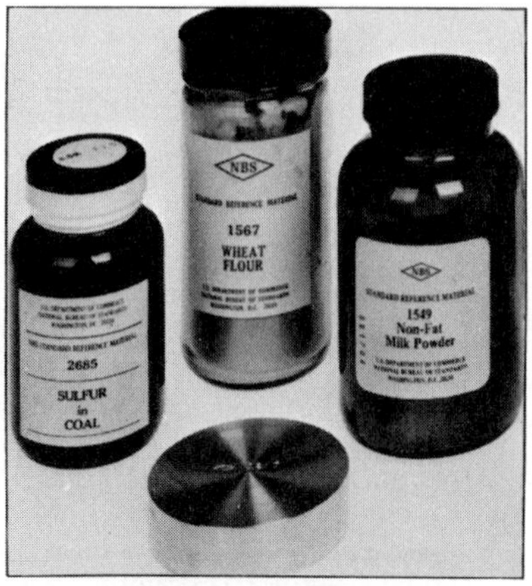

Figure 2-4
Standard reference materials from NIST. (Photo courtesy of the National Institute of Standards and Technology.)

A **standard reference material (SRM)** is a substance prepared and sold by the National Institute of Standards and Technology and certified to contain specified concentrations of one or more analytes.

In using SRMs it is often difficult to separate bias from ordinary random error.

---

[2]See U.S. Department of Commerce, *NIST Standard Reference Materials Catalog,* 1992–1993 ed., NIST Special Publication 260. Washington: Government Printing Office, 1992. For a description of the reference material programs of the NIST, see R. A. Alvarez, S. D. Rasberry, and G. A. Uriano, *Anal. Chem.,* **1982,** *54,* 1226A; G. A. Uriano, *ASTM Standardization News,* **1979**, *7*, 8.

[3]For sources of biological and environmental reference materials containing various elements, see C. Veillon, *Anal. Chem.,* **1986**, *58*, 851A.

Independent Analysis

If standard samples are not available, a second independent and reliable analytical method can be used in parallel with the method being evaluated. The independent method should differ as much as possible from the one under study. This minimizes the possibility that some common factor in the sample has the same effect on both methods. Here again, a statistical test must be used to determine whether any difference is a result of random errors in the two methods or to bias in the method under study (see Section 4B-2).

Blank Determinations

A **blank** solution contains the solvent and all of the reagents in an analysis, but none of the sample.

Blank determinations are useful for detecting certain types of constant errors. In a blank determination, or *blank*, all steps of the analysis are performed in the absence of a sample. The results from the blank are then applied as a correction to the sample measurements. Blank determinations reveal errors due to interfering contaminants from the reagents and vessels employed in analysis. Blanks also allow the analyst to correct titration data for the volume of reagent needed to cause an indicator to change color at the end point.

Variation in Sample Size

Example 2-2 demonstrates that as the size of a measurement increases, the effect of a constant error decreases. Thus, constant errors can often be detected by varying the sample size.

2C QUESTIONS AND PROBLEMS

2-1. Explain the difference between
- *(a) constant and proportional error.
- (b) random and systematic error.
- *(c) mean and median.
- (d) absolute and relative error.

***2-2.** Suggest some sources of random error in measuring the width of a 3-m table with a 1-m metal rule.

***2-3.** Name three types of systematic errors.

2-4. How are systematic method errors detected?

***2-5.** What kind of systematic errors are detected by varying the sample size?

2-6. A method of analysis yields weights for gold that are low by 0.3 mg. Calculate the percent relative error caused by this uncertainty if the weight of gold in the sample is
- *(a) 800 mg.
- (b) 500 mg.
- *(c) 100 mg.
- (d) 25 mg.

2-7. The method described in Problem 2-6 is to be used for the analysis of ores that assay about 1.2% gold. What minimum sample weight should be taken if the relative error resulting from a 0.3-mg loss is not to exceed
- *(a) −0.2%?
- (b) −0.5%?
- *(c) −0.8%?
- (d) −1.2%?

2-8. The color change of a chemical indicator requires an overtitration of 0.03 mL. Calculate the percent relative error if the total volume of titrant is
- *(a) 50.00 mL.
- (b) 10.0 mL.
- *(c) 25.0 mL.
- (d) 40.0 mL.

2-9. A loss of 0.4 mg of Zn occurs in the course of an analysis for that element. Calculate the percent relative error due to this loss if the weight of Zn in the sample is
- *(a) 40 mg.
- (b) 175 mg.
- *(c) 400 mg.
- (d) 600 mg.

CHAPTER

3

Random Errors in Analyses

In this chapter, we consider the sources of random errors, the determination of their magnitude, and their effects on computed results of chemical analyses.

3A THE NATURE OF RANDOM ERRORS

Random, or indeterminate, errors arise when a system of measurement is extended to its maximum sensitivity. This type of error is caused by the many uncontrollable variables that are an inevitable part of every physical or chemical measurement. There are many contributors to random error, but none can be positively identified or measured because most are so small that they cannot be detected individually. The accumulated effect of the individual indeterminate uncertainties, however, causes replicate measurements to fluctuate randomly around the mean of the set. For example, the scatter of data in Figures 2-1 and 2-3 is a direct result of the accumulation of small random uncertainties. Notice that in Figure 2-3 the random error in the results of analysts 2 and 4 is greater than in the results of analysts 1 and 3.

3A-1 Sources of Random Errors

We can get a qualitative idea of how small undetectable uncertainties produce a detectable random error in the following way. Imagine a situation in which just four small random errors combine to give an overall error. We will assume that each error has an equal probability of occurring and that each can cause the final result to be high or low by a fixed amount $\pm U$.

In our example, all the uncertainties have the same magnitude. This restriction is not necessary to derive the equation for a Gaussian curve.

Table 3-1 shows all the possible ways these four errors can combine to give the indicated deviations from the mean value. Note that only one combination leads to a deviation of $+4\ U$, four combinations give a deviation of $+2\ U$, and six give a deviation of $0\ U$. The negative errors have the same relationship. This ratio of 1 : 4 : 6 : 4 : 1 is a measure of the probability of a deviation of each magnitude. Therefore, if we make a sufficiently large number of measurements, we can expect a frequency distribution like that shown in Figure 3-1a. Note that the ordinate in the plot is the relative frequency of occurrence of the five possible combinations.

Figure 3-1b shows the theoretical distribution for ten equal-sized uncertainties. Again we see that the most frequent occurrence is zero deviation from the mean.

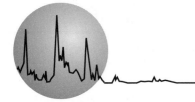

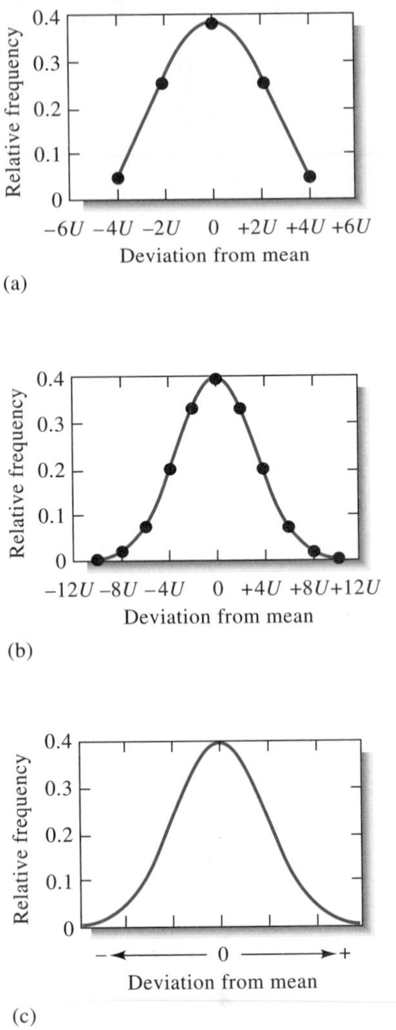

Figure 3-1
Frequency distribution for measurements containing (a) four random uncertainties; (b) ten random uncertainties; (c) a very large number of random uncertainties.

TABLE 3-1 Possible Combinations of Four Equal-Sized Uncertainties

Combinations of Uncertainties	Magnitude of Random Error	Number of Combinations	Relative Frequency
$+U_1 + U_2 + U_3 + U_4$	$+4U$	1	1/16 = 0.0625
$-U_1 + U_2 + U_3 + U_4$ $+U_1 - U_2 + U_3 + U_4$ $+U_1 + U_2 - U_3 + U_4$ $+U_1 + U_2 + U_3 - U_4$	$+2U$	4	4/16 = 0.250
$-U_1 - U_2 + U_3 + U_4$ $+U_1 + U_2 - U_3 - U_4$ $+U_1 - U_2 + U_3 - U_4$ $-U_1 + U_2 - U_3 + U_4$ $-U_1 + U_2 + U_3 - U_4$ $+U_1 - U_2 - U_3 + U_4$	0	6	6/16 = 0.375
$+U_1 - U_2 - U_3 - U_4$ $-U_1 + U_2 - U_3 - U_4$ $-U_1 - U_2 + U_3 - U_4$ $-U_1 - U_2 - U_3 + U_4$	$-2U$	4	4/16 = 0.250
$-U_1 - U_2 - U_3 - U_4$	$-4U$	1	1/16 = 0.0625

At the other extreme, a maximum deviation of 10 U occurs only about once in 500 measurements.

When the same procedure is applied to a very large number of individual errors, a bell-shaped curve like that shown in Figure 3-1c results. Such a plot is called a *Gaussian curve* or a *normal error curve.*

3A-2 Distribution of Experimental Data

We find empirically that the distribution of replicate data from most quantitative analytical experiments approaches that of the Gaussian curve shown in Figure 3-1c. As an example, consider the data in Table 3-2 for the calibration of a 10-mL pipet. In this experiment a small flask and stopper were weighed. Ten milliliters of water were transferred to the flask with the pipet, and the flask was stoppered. Finally, the flask, the stopper, and the water were weighed again. The temperature of the water was also measured to establish its density. The mass of the water was then calculated by taking the difference between the two masses; this difference was divided by the density of the water to find the volume delivered by the pipet. The experiment was repeated 50 times.

The data in Table 3-2 are typical of those obtained by an experienced worker weighing to the nearest milligram (which corresponds to 0.001 mL) on a top-loading balance and making every effort to avoid systematic error. Even so, the results vary from a low of 9.969 mL to a high of 9.994 mL. This 0.025 mL *spread* of data results directly from an accumulation of all of the random uncertainties in the experiment.

The **spread,** or **range,** of a set of replicate measurements is the difference between the highest and lowest result.

The information in Table 3-2 is easier to visualize when the data are rearranged into frequency distribution groups, as in Table 3-3. Here, we tabulate the number of data falling into a series of adjacent 0.003-mL *cells* and calculate the percentage of measurements falling into each cell. Note that 26% of the data reside in the cell containing the mean and median value of 9.982 mL and that more than half the data are within ± 0.004 mL of this mean.

The frequency distribution data in Table 3-3 are plotted as a bar graph, or *histogram* (labeled *A* in Figure 3-2). We can imagine that as the number of measurements increases, the histogram would approach the shape of the continuous curve shown as plot *B* in Figure 3-2. This curve is a *Gaussian curve,* or *normal error curve,* derived for an infinite set of data. These data have the same mean (9.982 mL), the same precision, and the same area under the curve as the histogram.

Variations in replicate results such as those in Table 3-2 result from numerous small and individually undetectable random errors that are attributable to uncontrollable variables in the experiment. Such small errors ordinarily tend to cancel one another and thus have a minimal effect. Occasionally, however, they occur in the same direction and produce a large positive or negative net error.

A **histogram** is a bar graph such as that shown by plot *A* in Figure 3-2.

A **Gaussian,** or **normal error, curve** is a curve that shows the symmetrical distribution of data around the mean of an infinite set of data such as the one in Figure 3-1c.

TABLE 3-2 Replicate Data on the Calibration of a 10-mL PIPET*

Trial	Volume, mL	Trial	Volume, mL	Trial	Volume, mL
1	9.988	18	9.975	35	9.976
2	9.973	19	9.980	36	9.990
3	9.986	20	9.994‡	37	9.988
4	9.980	21	9.992	38	9.971
5	9.975	22	9.984	39	9.986
6	9.982	23	9.981	40	9.978
7	9.986	24	9.987	41	9.986
8	9.982	25	9.978	42	9.982
9	9.981	26	9.983	43	9.977
10	9.990	27	9.982	44	9.977
11	9.980	28	9.991	45	9.986
12	9.989	29	9.981	46	9.978
13	9.978	30	9.969†	47	9.983
14	9.971	31	9.985	48	9.980
15	9.982	32	9.977	49	9.983
16	9.983	33	9.976	50	9.979
17	9.988	34	9.983		

Mean volume = 9.982 mL

Median volume = 9.982 mL

Spread = 0.025 mL

Standard deviation = 0.0056 mL

*Data listed in the order obtained.

†Minimum value.

‡Maximum value.

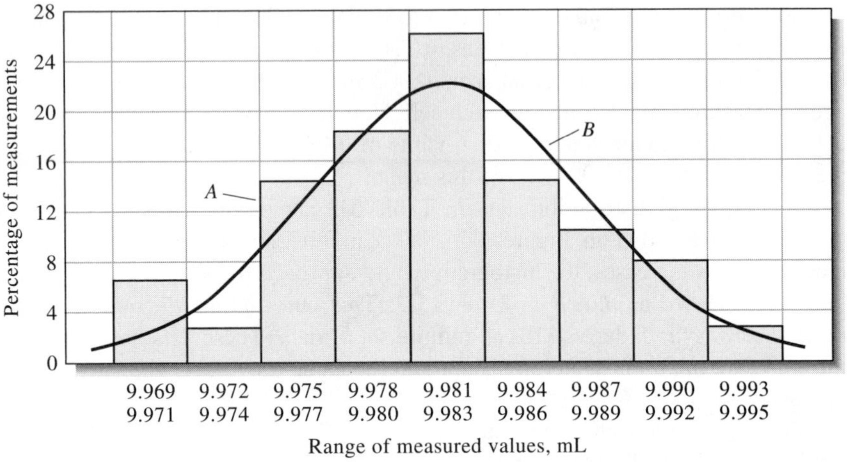

Figure 3-2
A histogram (*A*) showing distribution of the 50 results in Table 3-3 and a Gaussian curve (*B*) for data having the same mean and same standard deviation as the data in the histogram.

Sources of random uncertainties in the calibration of a pipet include (1) visual judgments, such as the level of the water with respect to the marking on the pipet and the mercury level in the thermometer; (2) variations in the drainage time and in the angle of the pipet as it drains; (3) temperature fluctuations, which affect the volume of the pipet, the viscosity of the liquid, and the performance of the balance; and (4) vibrations and drafts that cause small variations in the balance readings. Undoubtedly, numerous other sources of random uncertainty also operate in this calibration process. Thus, many small and uncontrollable variables affect even as simple a process as calibrating a pipet. We cannot identify the contribution of any one of these sources of random error to the error of the measurement, but their cumulative effect is responsible for the scatter of data around the mean.

TABLE 3-3
Frequency Distribution of Data from Table 3-2

Volume Range, mL	Number in Range	% in Range
9.969 to 9.971	3	6
9.972 to 9.974	1	2
9.975 to 9.977	7	14
9.978 to 9.980	9	18
9.981 to 9.983	13	26
9.984 to 9.986	7	14
9.987 to 9.989	5	10
9.990 to 9.992	4	8
9.993 to 9.995	1	2

FEATURE 3-1
Flipping Coins: A Student Activity to Illustrate a Normal Distribution

If you flip a coin ten times, how many heads will you get? Try it, and record your results. Repeat the experiment. Are your results the same? Ask friends or members of your class to perform the same experiment and tabulate the results. The table below contains the results obtained by several classes of analytical chemistry students over the period from 1980 to 1993.

No. Heads	0	1	2	3	4	5	6	7	8	9	10
Frequency	1	1	20	40	93	91	78	43	20	7	1

Add your results to those in the table, and plot a histogram similar to the one shown in Figure 3-3. Find the mean and the standard deviation (see Section 3B-3) for your results and compare them to the values shown in the plot. The smooth curve in Figure 3-3 is a normal error curve for an infinite num-

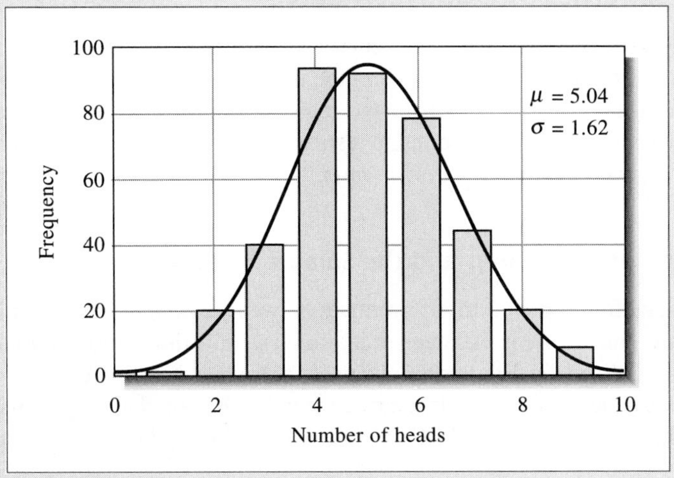

Figure 3-3 Results of a coin-flipping experiment by 395 students over a 13-year period.

ber of trials with the same mean and standard deviation as the data. Note that the mean of 5.04 is very close to the value of 5 that you would predict based on the laws of probability. As the number of trials increases, the histogram approaches the shape of the smooth curve and the mean approaches five.

3B THE STATISTICAL TREATMENT OF RANDOM ERROR

The random, or indeterminate, errors in the results of an analysis can be evaluated by the methods of statistics. Ordinarily, statistical analysis of analytical data is based upon the assumption that random errors in an analysis follow a Gaussian, or normal, distribution such as that illustrated in curve *B* in Figure 3-2 or in Figure 3-1c. Significant departures from Gaussian behavior can and do occur, but not often. Thus, we will base this discussion entirely upon normally distributed random errors.

You should understand that statistics only reveals information that is already present in a data set. That is, *no new information is created* by statistics. Statistics will allow us, however, to look at our data in different ways and make objective and intelligent decisions regarding their quality and use.

3B-1 The Sample and the Population

In statistics, a finite number of experimental observations is called a *sample* of data. The sample is treated as a tiny fraction of an infinite number of observations that could in principle be made given infinite time. Statisticians call the theoretical infinite number of data a *population,* or a *universe,* of data. Statistical laws have been derived assuming a population of data; often they must be modified substantially when applied to a small sample because a few data may not be representative of the population. In the discussion that follows, we shall first describe Gaussian statistics of populations. Then we will show how these relationships can be modified and applied to small samples of data.

Do not confuse the *statistical sample* with the *analytical sample.* Four analytical samples analyzed in the laboratory represent a single statistical sample. This is an unfortunate duplication of the term sample.

3B-2 Properties of a Gaussian Curve

An equation for a Gaussian curve can take the form

$$y = \frac{e^{-(x-\mu)^2/2\sigma^2}}{\sigma\sqrt{2\pi}}$$

Figure 3-4a shows two Gaussian curves in which the relative frequency y of occurrence of various deviations from the mean is plotted as a function of deviation from the mean. As shown in the margin, curves such as these can be described by an equation that contains just two parameters, the *population mean* μ and the *population standard deviation* σ.

The Population Mean μ and the Sample Mean $\bar{x}$

In the absence of systematic error, the population mean μ is the true value of a measured quantity.

Statisticians find it useful to differentiate between a *sample mean* and a *population mean.* The former is the mean of a limited sample drawn from a population of data. It is defined by Equation 2-1 (page 12) when N is a small number. The population mean, in contrast, is the true mean for the population. It is also defined by Equation 2-1 with the added provision that N is so large that it approaches

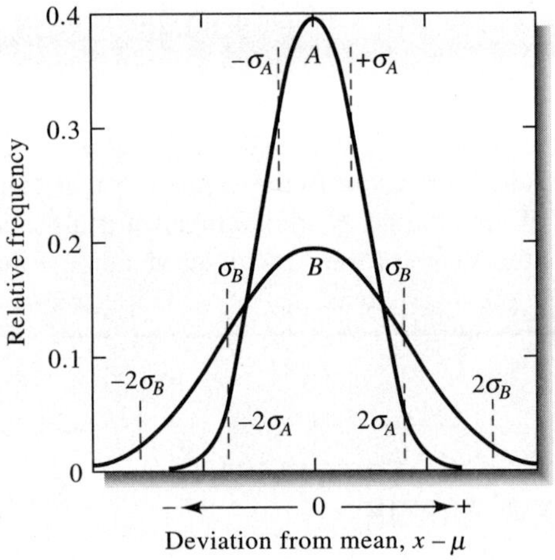

(a)

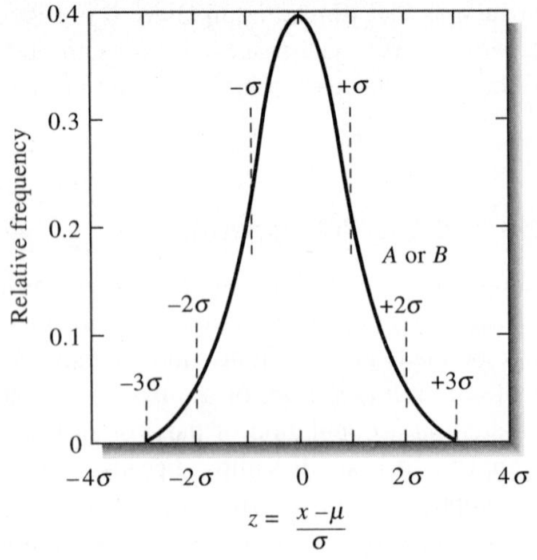

(b)

Figure 3-4
Normal error curves. The standard deviation for curve B is twice that for curve A, that is, $\sigma_B = 2\sigma_A$. (a) The abscissa is the deviation from the mean in the units of measurement. (b) The abscissa is the deviation from the mean in units of σ. Thus, the two curves A and B are identical here.

infinity. *In the absence of systematic error, the population mean is also the true value for the measured quantity.* To emphasize the difference between the two means, the sample mean is symbolized by $\bar{x}$ and the population mean by μ. More often than not, particularly when N is small, $\bar{x}$ differs from μ because a small sample of data does not exactly represent its population. The probable difference between $\bar{x}$ and μ decreases rapidly as the number of measurements making up the sample increases; ordinarily, by the time N reaches 20 to 30, this difference is negligible.

Sample mean = $\bar{x}$, where

$$\bar{x} = \frac{\sum_{i=1}^{N} x_i}{N} \text{ when } N \text{ is small}$$

Population mean = μ, where

$$\mu = \frac{\sum_{i=1}^{N} x_i}{N} \text{ when } N \longrightarrow \infty$$

The Population Standard Deviation (σ)

The *population standard deviation* σ, which is a measure of the *precision* of a population of data, is given by

See *Mathcad Applications for Analytical Chemistry,* **p. 19.**

$$\sigma = \sqrt{\frac{\sum_{i=1}^{N} (x_i - \mu)^2}{N}} \qquad (3\text{-}1)$$

where N is the number of replicate data making up the population.

The quantity $(x_i - \mu)$ in Equation 3-1 is the deviation of individual results from the population mean of data; compare with Equation 2-2, which is for a sample of data.

The two curves in Figure 3-4a are for two populations of data that differ only in their standard deviations. The standard deviation for the data set yielding the broader but lower curve B is twice that for the measurements yielding curve A. The breadth of these curves is a measure of the precision of the two sets of data. Thus, the precision of the data leading to curve A is twice as good as that of the data that are represented by curve B.

Figure 3-4b shows another type of normal error curve in which the abscissa is now a new variable z, which is defined as

$$z = \frac{(x - \mu)}{\sigma} \qquad (3\text{-}2)$$

Note that z is the deviation of the mean of a datum stated *in units of standard deviation.* That is, when $x - \mu = \sigma$, z is equal to one standard deviation; when $x - \mu = 2\sigma$, z is equal to two standard deviations; and so forth. Since z is the deviation of the mean in standard deviation units, a plot of relative frequency versus this parameter yields a single Gaussian curve that describes all populations of data regardless of standard deviation. Thus, Figure 3-4b is the normal error curve for both sets of data used to plot curves A and B in Figure 3-4a.

The quantity z represents the deviation of a result from the mean for a population of data in units of standard deviation.

A normal error curve has several general properties. (1) The mean occurs at the central point of maximum frequency. (2) There is a symmetrical distribution of positive and negative deviations about the maximum. (3) There is an exponential decrease in frequency as the magnitude of the deviations increases. Thus, small random uncertainties are observed much more often than very large ones.

Areas Under a Gaussian Curve

It can be shown that, regardless of its width, 68.3% of the area beneath a Gaussian curve for a population of data lies within one standard deviation ($\pm 1\sigma$) of the mean μ. Thus, 68.3% of the data making up the population lie within these bounds. Furthermore, approximately 95.5% of all data are within $\pm 2\sigma$ of the mean and 99.7% are within $\pm 3\sigma$. The vertical dashed lines show the areas bounded by $\pm 1\sigma$ and $\pm 2\sigma$ in Figure 3-4.

Because of area relationships such as these, the standard deviation of a population of data is a useful predictive tool. For example, we can say that the chances are 68.3 in 100 that the random uncertainty of any single measurement is no more than $\pm 1\sigma$. Similarly, the chances are 95.5 in 100 that the error is less than $\pm 2\sigma$, and so forth.

3B-3 The Sample Standard Deviation as a Measure of Precision

Equation 3-1 must be modified when it is applied to a small sample of data. Thus, the *sample standard deviation s* is given by the equation

$$s = \sqrt{\frac{\sum_{i=1}^{N} (x_i - \bar{x})^2}{N - 1}} \tag{3-3}$$

Equation 3-3 applies to small sets of data. It says, "Find the deviations from the mean d_i, square them, sum them, divide the sum by $N - 1$, and take the square root." The quantity $N - 1$ is called the **number of degrees of freedom.** Many scientific calculators have the standard deviation function built in.

Note that Equation 3-3 differs from Equation 3-1 in two ways. First, the sample mean, $\bar{x}$, appears in the numerator of Equation 3-3 in place of the population mean, μ. Second, N in Equation 3-1 is replaced by the *number of degrees of freedom* ($N - 1$). If this substitution is not used, the calculated s will on the average be less than the true standard deviation σ; that is, s will have a negative bias (see Feature 3-2).

FEATURE 3-2

The Significance of Number of Degrees of Freedom

The number of degrees of freedom indicates the number of *independent* data that go into the computation of a standard deviation. Thus, when μ is unknown, two quantities must be extracted from a set of replicate data: $\bar{x}$ and s. One degree of freedom is used to establish $\bar{x}$ because, with their signs retained, the sum of the individual deviations must add up to zero. Thus, when $N - 1$ deviations have been computed, the final one is known. Consequently, only $N - 1$ deviations provide an *independent* measure of the precision of the set. Failure to use $N - 1$ in calculating the standard deviation for small samples results in values of s, which are on the average smaller than the true standard deviation σ.

An Alternative Expression for Sample Standard Deviation

To calculate s with a calculator that does not have a standard deviation key the following rearrangement of Equation 3-3 is easier to use:

$$s = \sqrt{\frac{\sum_{i=1}^{N} x_i^2 - \frac{\left(\sum_{i=1}^{N} x_i\right)^2}{N}}{N - 1}} \qquad \text{(3-4)}$$

See *Mathcad Applications for Analytical Chemistry,* **pp. 19–21.**

EXAMPLE 3-1

The following results were obtained in the replicate analysis of a blood sample for its lead content: 0.752, 0.756, 0.752, 0.751, and 0.760 ppm Pb. Calculate the mean and the standard deviation of this set of data.

To apply Equation 3-4, we calculate Σx_i^2 and $(\Sigma x_i)^2/N$.

Sample	x_i	x_i^2
1	0.752	0.565504
2	0.756	0.571536
3	0.752	0.565504
4	0.751	0.564001
5	0.760	0.577600
	$\Sigma x_i = 3.771$	$\Sigma x_i^2 = 2.844145$

$$\bar{x} = 3.771/5 = 0.7542 \approx 0.754 \text{ ppm Pb}$$

$$\frac{(\Sigma x_i)^2}{N} = \frac{(3.771)^2}{5} = \frac{14.220441}{5} = 2.8440882$$

Substituting into Equation 3-4 leads to

$$s = \sqrt{\frac{2.844145 - 2.8440882}{5 - 1}} = \sqrt{\frac{0.0000568}{4}}$$

$$= 0.00377 \approx 0.004 \text{ ppm Pb}$$

Note in Example 3-1 that $(\Sigma x_i)^2$ and Σx_i^2 are quite different numerically (14.220441 and 2.844145).

Note that the difference between Σx_i^2 and $(\Sigma x_i)^2/N$ in Example 3-1 is very small. If we had rounded these numbers off before subtracting them, a serious error would have appeared in the computed value of s. To avoid this source of error, *never round a standard deviation calculation until the very end.* Furthermore, for this same reason, never use Equation 3-4 to calculate the standard deviation of numbers containing five or more digits. Use Equation 3-3 instead.[1]

Any time we subtract two large, approximately equal numbers, the difference will usually have a relatively large uncertainty.

[1] In most cases, the first two or three digits in a set of data are identical to each other. As an alternative, then, to using Equation 3-3, these identical digits can be dropped and the remaining digits used with Equation 3-4. For example, the standard deviation for the data in Example 3-1 could be based on 0.052, 0.056, 0.052, and so forth, or even 52, 56, 52, and so forth.

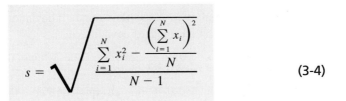

Note also that handheld calculators and small computers with a standard deviation function usually employ a version of Equation 3-4, so expect large errors in s when these devices are used to calculate the standard deviation of data that have five or more significant figures.

When $N \rightarrow \infty$, $\bar{x} \rightarrow \mu$ and $s \rightarrow \sigma$.

You should always keep in mind when you make statistical calculations that because of the uncertainty in $\bar{x}$, a sample standard deviation s may differ significantly from the population standard deviation μ.

Standard Error of a Mean

The **standard error** of a mean is the standard deviation of a set of data divided by the square root of the number of data in the set.

The figures on percentage distribution just quoted refer to the probable error for a *single* measurement. If a series of replicate samples, each containing N members, are taken randomly from a population of data, the mean of each set will show less and less scatter as N increases. The standard deviation of each mean is known as the *standard error* of the mean and is given the symbol s_m. It can be shown that the standard error is inversely proportional to the square root of the number of data N used to calculate the mean:

$$s_m = s/\sqrt{N} \tag{3-5}$$

3B-4 The Reliability of *s* as a Measure of Precision

CHALLENGE: Show that these statements are correct.

When $N > 20$, $s \approx \sigma$.

Most of the statistical tests we describe in Chapter 4 are based upon sample standard deviations, and the probability of correctness of the results of these tests improves as the reliability of s becomes greater. Uncertainty in the calculated value of s decreases as N in Equation 3-4 increases. When N is greater than about 20, s and σ can be assumed to be identical for all practical purposes. For example, if the 50 measurements in Table 3-2 are divided into 10 subgroups of 5 each, the value of s varies widely from one subgroup to another (0.0023 to 0.0079 mL) even though the average of the computed values of s is that for the entire set (0.0056 mL). In contrast, the computed values of s for two subsets of 25 each are nearly identical (0.0054 and 0.0058 mL).

The rapid improvement in the reliability of s with increases in N makes it feasible to obtain a good approximation of σ when the method of measurement is not excessively time-consuming and when an adequate supply of sample is available. For example, if the pH of numerous solutions is to be measured in the course of an investigation, it is useful to evaluate s in a series of preliminary experiments. This measurement is simple, requiring only that a pair of rinsed and dried electrodes be immersed in the test solution and the pH read from a scale or a display. To determine s, 20 to 30 portions of a well-buffered solution of fixed pH can be measured with all steps of the procedure being followed exactly. Normally, it is safe to assume that the random error in this test is the same as that in subsequent measurements. The value of s calculated from Equation 3-4 is a valid and accurate measure of the theoretical σ.

Pooling Data to Improve the Reliability of *s*

For analyses that are time-consuming, the foregoing procedure is seldom practical. In this situation, data from a series of similar samples accumulated over time can often be pooled to provide an estimate of s that is superior to the value for any individual subset. Again, we must assume the same sources of random error in all the measurements. This assumption is usually valid if the samples have similar compositions and have been analyzed in exactly the same way.

To obtain a pooled estimate of the standard deviation, s_{pooled}, deviations from the mean for each subset are squared; the squares of all subsets are then summed and divided by an appropriate number of degrees of freedom. The pooled s is obtained by extracting the square root of the quotient. One degree of freedom is lost for each subset. Thus, the number of degrees of freedom for the pooled s is equal to the total number of measurements minus the number of subsets. The equation for calculating a pooled estimate of s is given in Feature 3-3. Example 3-2 illustrates the application of this type of computation.

FEATURE 3-3
Equation for Calculating Pooled Standard Deviations

The equation for computing a pooled standard deviation from t sets of data takes the form

$$s_{\text{pooled}} = \sqrt{\frac{\sum_{i=1}^{N_1}(x_i - \bar{x}_1)^2 + \sum_{j=1}^{N_2}(x_j - \bar{x}_2)^2 + \sum_{k=1}^{N_3}(x_k - \bar{x}_3)^2 + \cdots}{N_1 + N_2 + N_3 + \cdots - N_t}}$$

where N_1 is the number of data in set 1, N_2 is the number in set 2, and so forth. The term N_t is the number of data sets that are pooled.

EXAMPLE 3-2

The mercury in samples of seven fish taken from Chesapeake Bay was determined by a method based upon the absorption of radiation by gaseous elemental mercury. Calculate a pooled estimate of the standard deviation for the method, based upon the first three columns of data:

See *Mathcad Applications for Analytical Chemistry,* **p. 32.**

Specimen	Number of Samples Measured	Hg Content, ppm	Mean, ppm Hg	Sum of Squares of Deviations from Mean
1	3	1.80, 1.58, 1.64	1.673	0.0258
2	4	0.96, 0.98, 1.02, 1.10	1.015	0.0115
3	2	3.13, 3.35	3.240	0.0242
4	6	2.06, 1.93, 2.12, 2.16, 1.89, 1.95	2.018	0.0611
5	4	0.57, 0.58, 0.64, 0.49	0.570	0.0114
6	5	2.35, 2.44, 2.70, 2.48, 2.44	2.482	0.0685
7	4	1.11, 1.15, 1.22, 1.04	1.130	0.0170
Number of measurements = 28			Sum of squares =	0.2196

The values in the last two columns for specimen 1 were computed as follows:

x_i	$\lvert(x_i - \bar{x})\rvert$	$(x_i - \bar{x})^2$
1.80	0.127	0.0161
1.58	0.093	0.0086
1.64	0.033	0.0011
5.02	Sum of squares =	0.0258

$$\bar{x} = \frac{5.02}{3} = 1.673$$

The other data in columns 4 and 5 were obtained similarly. Then

$$s_{\text{pooled}} = \sqrt{\frac{0.0258 + 0.0115 + 0.0242 + 0.0611 + 0.0114 + 0.0685 + 0.0170}{28 - 7}}$$

$$= 0.10 \text{ ppm Hg}$$

Note that one degree of freedom is lost for each of the seven samples. Because more than 20 degrees of freedom remain, however, the computed value of s can be considered a good approximation of σ; that is, $s \rightarrow \sigma = 0.10$ ppm Hg.

3B-5 Alternative Terms for Expressing the Precision of Samples of Data

Chemists ordinarily employ sample standard deviation to report the precision of their data. Three other terms are often encountered, however.

Variance (s^2)

The **variance** is equal to the square of the standard deviation.

The variance is simply the square of the standard deviation:

$$s^2 = \frac{\sum_{i=1}^{N} (x_i - \bar{x})^2}{N - 1} \tag{3-6}$$

Note that the standard deviation has the same units as the data, whereas the variance has the units of the data squared. People who do scientific work tend to use standard deviation rather than variance as a measure of precision. It is easier to relate the precision of a measurement to the measurement itself if they both have the same units. The advantage of using variance is that variances are additive, as we will see later in this chapter.

Relative Standard Deviation (RSD) and Coefficient of Variation (CV)

Chemists frequently quote standard deviations in relative rather than absolute terms. We calculate the relative standard deviation by dividing the standard deviation by the mean of the set of data. It is often expressed in parts per thousand (ppt) or in percent by multiplying this ratio by 1000 ppt or by 100%. For example,

$$\text{RSD} = (s/\bar{x}) \times 1000 \text{ ppt}$$

The relative standard deviation multiplied by 100% is called the *coefficient of variation* (CV).

The **coefficient of variation** is the percent relative standard deviation.

$$\text{CV} = (s/\bar{x}) \times 100\% \qquad (3\text{-}7)$$

Relative standard deviations often give a clearer picture of data quality than do absolute standard deviations. As an example, suppose that a sample contains about 50 mg of copper and that the standard deviation of a copper determination is 2 mg. The CV for this sample is 4%. For a sample containing only 10 mg, the CV is 20%.

Spread or Range (w). The *spread,* or *range,* is another term that is sometimes used to describe the precision of a set of replicate results. It is the difference between the largest value in the set and the smallest. Thus, the spread of the data in Example 3-1 is (0.760 − 0.751) = 0.009 ppm Pb.

EXAMPLE 3-3

For the set of data in Example 3-1, calculate (a) the variance, (b) the relative standard deviation in parts per thousand, (c) the coefficient of variation, (d) the spread.

In Example 3-1, we found

$$\bar{x} = 0.754 \text{ ppm Pb} \qquad \text{and} \qquad s = 0.0038 \text{ ppm Pb}$$

(a) $s^2 = (0.0038)^2 = 1.4 \times 10^{-5}$

(b) $\text{RSD} = \dfrac{0.0038}{0.754} \times 1000 \text{ ppt} = 5.0 \text{ ppt}$

(c) $\text{CV} = \dfrac{0.0038}{0.754} \times 100\% \times 0.50\%$

(d) $w = 0.760 - 0.751 = 0.009$

3C THE STANDARD DEVIATION OF COMPUTED RESULTS

We often need to estimate the standard deviation of a result that has been computed from two or more experimental data, each of which has a known sample standard deviation. As shown in Table 3-4, how such estimates are made depends upon the type of arithmetic that is involved. The relationships shown in this table are derived in Appendix 11.

TABLE 3-4 Error Propagation in Arithmetic Calculations

Type of Calculation	Example*	Standard Deviation of y†	
Addition or Subtraction	$y = a + b - c$	$s_y = \sqrt{s_a^2 + s_b^2 + s_c^2}$	(1)
Multiplication or Division	$y = a \cdot b/c$	$\frac{s_y}{y} = \sqrt{\left(\frac{s_a}{a}\right)^2 + \left(\frac{s_b}{b}\right)^2 + \left(\frac{s_c}{c}\right)^2}$	(2)
Exponentiation	$y = a^x$	$\frac{s_y}{y} = x\frac{s_a}{a}$	(3)
Logarithm	$y = \log_{10} a$	$s_y = 0.434\frac{s_a}{a}$	(4)
Antilogarithm	$y = \text{antilog}_{10}\, a$	$\frac{s_y}{y} = 2.303\, s_a$	(5)

*a, b, and c are experimental variables whose standard deviations are s_a, s_b, and s_c, respectively.

†These relationships are derived in Appendix 11.

3C-1 The Standard Deviation of Sums and Differences

Consider the summation

$$\begin{array}{rl} +0.50 & (\pm 0.02) \\ +4.10 & (\pm 0.03) \\ -1.97 & (\pm 0.05) \\ \hline 2.63 & \end{array}$$

In many of the examples in this chapter, we indicate the first uncertain digit in a number by printing it in color.

where the numbers in parentheses are absolute standard deviations. If the signs of the three individual standard deviations happen by chance to have the same sign, the standard deviation of the sum could be as large as $+0.02 + 0.03 + 0.05 = +0.10$ or $-0.02 - 0.03 - 0.05 = -0.10$. On the other hand, it is possible that the three standard deviations could combine to give an accumulated value of zero: $-0.02 - 0.03 + 0.05 = 0$ or $+0.02 + 0.03 - 0.05 = 0$. More likely, however, the standard deviation of the sum will lie between these two extremes. It can be shown from statistical theory that the variances of a sum or difference can be obtained by adding the individual variances. Thus, the most probable value for a standard deviation of a sum or difference can be found by taking the square root of the sum of the squares of the individual absolute standard deviations. So, for the computation

The variance of a sum or difference is equal to the *sum* of the variances of the numbers making up that sum or difference.

For a sum or a difference the *absolute standard deviation* of the answer is the square root of the sum of the squares of the *absolute standard deviations* of the numbers used to calculate the sum or difference.

$$y = a(\pm s_a) + b(\pm s_b) - c(\pm s_c)$$

the standard deviation of the result s_y is given by

$$s_y = \sqrt{s_a^2 + s_b^2 + s_c^2} \tag{3-8}$$

where s_a, s_b, and s_c are the standard deviations of the three terms making up the result.

Substituting the standard deviations from the example gives

$$s_y = \sqrt{(\pm 0.02)^2 + (\pm 0.03)^2 + (\pm 0.05)^2} = \pm 0.06$$

and the sum should be reported as 2.63 (± 0.06).

3C-2 The Standard Deviation of Products and Quotients

See *Mathcad Applications for Analytical Chemistry,* **pp. 114–116.**

Consider the following computation where the numbers in parentheses are again absolute standard deviations:

$$\frac{4.10(\pm 0.02) \times 0.0050(\pm 0.0001)}{1.97(\pm 0.04)} = 0.010406(\pm \,?)$$

In this situation, the standard deviations of two of the numbers in the calculation are larger than the result itself. Evidently, we need a different approach for multiplication and division. As shown in Table 3-4 the *relative standard deviation* of a product or quotient is determined by the *relative standard deviations* of the numbers forming the computed result. For example, for the calculation

$$y = \frac{a \times b}{c} \tag{3-9}$$

we obtain the relative standard deviation s_y/y of the result y by summing the squares of the relative standard deviations of a, b, and c and extracting the square root of the sum:

$$\frac{s_y}{y} = \sqrt{\left(\frac{s_a}{a}\right)^2 + \left(\frac{s_b}{b}\right)^2 + \left(\frac{s_c}{c}\right)^2} \tag{3-10}$$

For multiplication or division, the *relative standard deviation* of the answer is the square root of the sum of the squares of the *relative standard deviations* of the numbers that are multiplied or divided.

Applying this equation to the numerical example gives

$$\frac{s_y}{y} = \sqrt{\left(\frac{\pm 0.02}{4.10}\right)^2 + \left(\frac{\pm 0.0001}{0.005}\right)^2 + \left(\frac{\pm 0.04}{1.97}\right)^2}$$

$$= \sqrt{(0.0048)^2 + (0.0200)^2 + (0.0203)^2} = \pm 0.0289$$

In order to complete the calculation, we must find the absolute standard deviation of the result,

$$s_y = y \times (\pm 0.0289) = 0.0104 \times (\pm 0.0289) = \pm 0.000301$$

and we can write the answer and its uncertainty as 0.0104(± 0.0003).

To find the absolute standard deviation in a product or a quotient, first find the relative standard deviation in the result and then multiply it by the result.

The following example demonstrates the calculation of the standard deviation of the result for a more complex calculation.

EXAMPLE 3-4

Calculate the standard deviation of the result of

$$\frac{[14.3(\pm 0.2) - 11.6(\pm 0.2)] \times 0.050(\pm 0.001)}{[820(\pm 10) + 1030(\pm 5)] \times 42.3(\pm 0.4)} = 1.725(\pm ?) \times 10^{-6}$$

First, we must calculate the standard deviation of the sum and the difference. For the difference in the numerator,

$$s_a = \sqrt{(\pm 0.2)^2 + (\pm 0.2)^2} = \pm 0.283$$

and for the sum in the denominator,

$$s_b = \sqrt{(\pm 10)^2 + (\pm 5)^2} = 11.2$$

We may then rewrite the equation as

$$\frac{2.7(\pm 0.283) \times 0.050(\pm 0.001)}{1850(\pm 11.2) \times 42.3(\pm 0.4)} = 1.725 \times 10^{-6}$$

The equation now contains only products and quotients, and Equation 3-10 applies. Thus,

$$\frac{s_y}{y} = \sqrt{\left(\pm\frac{0.283}{2.7}\right)^2 + \left(\pm\frac{0.001}{0.050}\right)^2 + \left(\pm\frac{11.2}{1850}\right)^2 + \left(\pm\frac{0.4}{42.3}\right)^2} = 0.107$$

To obtain the absolute standard deviation, we write

$$s_y = y \times 0.107 = 1.725 \times 10^{-6} \times (\pm 0.107) = \pm 0.185 \times 10^{-6}$$

and round the answer to $1.7(\pm 0.2) \times 10^{-6}$.

3C-3 The Standard Deviation in Exponential Calculations

Consider the relationship

$$y = a^x$$

where the exponent x can be considered to be free of uncertainty. As shown in Table 3-4 and Appendix 11, the relative standard deviation in y resulting from the uncertainty in a is

$$\frac{s_y}{y} = x\left(\frac{s_a}{a}\right) \tag{3-11}$$

Thus, the relative standard deviation of the square of a number is twice the relative standard deviation of the number, the relative standard deviation of the cube root of a number is one third that of the number, and so forth.

EXAMPLE 3-5

The standard deviation in measuring the diameter d of a sphere is ± 0.02 cm. What is the standard deviation in the calculated volume V of the sphere if $d = 2.15$ cm?

From the equation for the volume of a sphere, we have

$$V = \frac{4}{3}\pi\left(\frac{d}{2}\right)^3 = \frac{4}{3}\pi\left(\frac{2.15}{2}\right)^3 = 5.20 \text{ cm}^3$$

Here we may write

$$\frac{s_V}{V} = 3 \times \frac{s_d}{d} = 3 \times \frac{0.02}{2.15} = 0.0279$$

The absolute standard deviation in V is then

$$s_V = 5.20 \times 0.0279 = 0.145$$

Thus,

$$V = 5.2(\pm 0.1)\text{cm}^3$$

EXAMPLE 3-6

The solubility product K_{sp} for the silver salt AgX is $4.0(\pm 0.4) \times 10^{-8}$. The solubility of AgX in water is

$$\text{solubility} = (K_{sp})^{1/2} = (4.0 \times 10^{-8})^{1/2} = 2.0 \times 10^{-4}$$

What is the uncertainty in the calculated solubility of AgX in water? Substituting y = solubility, $a = K_{sp}$, and $x = 1/2$ into Equation 3-11 gives

$$\frac{s_a}{a} = \frac{0.4 \times 10^{-8}}{4.0 \times 10^{-8}}$$

$$\frac{s_y}{y} = \frac{1}{2} \times \frac{0.4}{4.0} = 0.05$$

$$s_y = 2.0 \times 10^{-4} \times 0.05 = 0.1 \times 10^{-4}$$

$$\text{solubility} = 2.0(\pm 0.1) \times 10^{-4}\text{ M}$$

It is important to note that the error propagation in taking a number to a power is different from the error propagation in multiplication. For example, consider the uncertainty in the square of 4.0(±0.2). Here, the relative error in the result (16.0) is given by Equation 3-11:

$$\frac{s_y}{y} = 2\left(\frac{0.2}{4}\right) = 0.1 \qquad \text{or} \qquad 10\%$$

Consider now the situation where y is the product of *two independently measured* numbers that by chance happen to have identical values of $a_1 = 4.0(\pm 0.2)$ and $a_2 = 4.0(\pm 0.2)$. Here, the relative error of the product $a_1 a_2 = 16.0$ is given by Equation 3-10:

$$\frac{s_y}{y} = \sqrt{\left(\frac{0.2}{4}\right)^2 + \left(\frac{0.2}{4}\right)^2} = 0.07 \qquad \text{or} \qquad 7\%$$

The relative standard deviation of $y = a^3$ is *not* the same as the relative standard deviation of the product $y = abc$ where $a = b = c$.

The reason for this apparent anomaly is that, with measurements that are independent of one another, the sign associated with one error can be the same as or different from that of the other error. If they happen to be the same, the error is identical to that encountered in the first case, where the signs *must* be the same. On the other hand, if one sign is positive and the other negative, the relative errors tend to cancel. Thus, the probable error lies somewhere between the maximum (10%) and zero.

3C-4 The Standard Deviation of Logarithms and Antilogarithms

The last two entries in Table 3-4 show that for $y = \log a$

$$s_y = 0.434\,\frac{s_a}{a} \tag{3-12}$$

and for $y = \text{antilog } a$,

$$\frac{s_y}{y} = 2.303\,s_a \tag{3-13}$$

Thus, the *absolute* standard deviation of the logarithm of a number is determined by the *relative* standard deviation of the number; conversely, the *relative* standard deviation of the antilogarithm of a number is determined by the *absolute* standard deviation of the number.

EXAMPLE 3-7

Calculate the absolute standard deviations of the results of the following computations. The absolute standard deviation for each quantity is given in parentheses.

(a) $y = \log[2.00(\pm 0.02) \times 10^{-4}] = -3.6990 \pm ?$
(b) $y = \text{antilog}[1.200(\pm 0.003)] = 15.849 \pm ?$
(c) $y = \text{antilog}[45.4(\pm 0.3)] = 2.5119 \times 10^{45} \pm ?$

(a) Referring to Equation 3-12, we see that we must multiply the *relative* standard deviation by 0.434:

$$s_y = \pm 0.434 \times \frac{0.02 \times 10^{-4}}{2.00 \times 10^{-4}} = \pm 0.004$$

Thus,

$$\log[2.00(\pm 0.02) \times 10^{-4}] = -3.699(\pm 0.004)$$

(b) Applying Equation 3-13, we have

$$\frac{s_y}{y} = 2.303 \times (\pm 0.003) = \pm 0.0069$$

$$s_y = \pm 0.0069y = \pm 0.0069 \times 15.849 = \pm 0.11$$

Thus,

$$\text{antilog}[1.200(\pm 0.003)] = 15.8 \pm 0.1$$

(c) $\dfrac{s_y}{y} = 2.303 \times (\pm 0.3) = \pm 0.69$

$$s_y = +0.69y = +0.69 \times 2.5119 \times 10^{45} = +1.7 \times 10^{45}$$

Thus,

$$\text{antilog}[45.4(\pm 0.3)] = 2.5(\pm 1.7) \times 10^{45} = 3(\pm 2) \times 10^{45}$$

Example 3-7c demonstrates that a large absolute error is associated with the antilogarithm of a number with few digits beyond the decimal point. This large uncertainty is due to the fact that the numbers to the left of the decimal (the characteristic) serve only to locate the decimal point. The large error in the antilogarithm results from the relatively large uncertainty in the *mantissa* of the number (that is, 0.4 ± 0.3).

3D METHODS FOR REPORTING COMPUTED DATA

A numerical result is worthless to users of the data unless they know something about its accuracy. Therefore, it is always essential for you to indicate your best estimate of the reliability of your data. One of the best ways of indicating reliability is to give a confidence limit at the 90% or 95% confidence level, as we describe in Section 4A. Another method is to report the absolute standard deviation or the coefficient of variation of the data. Here, it is a good idea to indicate the number of data that were used to obtain the standard deviation so that the user of the data has some idea of the probable reliability of s. A less satisfactory but more common indicator of the quality of data is the *significant figure convention.*

3D-1 The Significant Figure Convention

The **significant figures** in a number are all of the certain digits plus the first uncertain digit.

See *Mathcad Applications for Analytical Chemistry,* **p. 4.**

A simple way of indicating the probable uncertainty associated with an experimental measurement is to round the result so that it contains only *significant figures.* By definition, the significant figures in a number are all of the certain digits *and the first uncertain digit.* For example, when you read the 50-mL buret section shown in Figure 3-5, you can easily tell that the liquid level is greater than 30.2 mL and less than 30.3 mL. You can also estimate the position of the liquid between the graduations to about ± 0.02 mL. So, using the significant figure convention, you should report the volume delivered as 30.24 mL, which is four significant figures. In this example the first three digits are certain, and the last digit (4) is uncertain.

Express data in scientific notation to avoid confusion in determining whether terminal zeros are significant.

Rules for determining the number of significant figures:

1. Disregard all initial zeros.
2. Disregard all final zeros *unless they follow a decimal point.*
3. All remaining digits including zeros between nonzero digits are significant.

A zero may or may not be significant depending upon its location in a number. A zero that is surrounded by other digits (such as in 30.24 mL) is always significant because it is read directly and with certainty from a scale or instrument readout. On the other hand, zeros that only locate the decimal point for us are not. If we write 30.24 mL as 0.03024 L, the number of significant figures is the same. The only function of the zero before the 3 is to locate the decimal point, so it is not significant. Terminal, or final, zeros may or may not be significant. For example, if the volume of a beaker is expressed as 2.0 L, the presence of the zero tells us that the volume is known to a few tenths of a liter. So, both the 2 and the zero are significant figures. If this same volume is reported as 2000 mL, the situation becomes confusing. The last two zeros are not significant because the uncertainty is still a few tenths of a liter or a few hundred milliliters. In order to follow the significant figure convention in a case such as this, use scientific notation and report the volume as 2.0×10^3 mL.

An obvious limitation of the significant figure convention as an indicator of reliability of data is its ambiguity. For example, when a result is reported as 61.6, the uncertainty could range from a high of ± 0.5 to a low of ± 0.05.

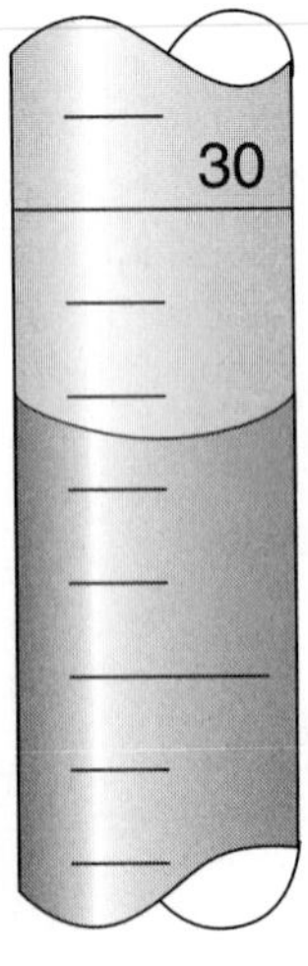

Figure 3-5
Buret section showing the liquid level and meniscus.

3D-2 Significant Figures in Numerical Computations

Care is required to determine the appropriate number of significant figures in the result of an arithmetic combination of two or more numbers.[2]

Sums and Differences

For addition and subtraction, the number of significant figures can be found by visual inspection. For example, in the expression

---

[2]For an extensive discussion of significant figures, see L. M. Schwartz, *J. Chem. Educ.*, **1985,** *62,* 693.

$$3.4 + 0.020 + 7.31 = 10.73 = 10.7$$

the second and third decimal places in the answer cannot be significant because 3.4 is uncertain in the first decimal place. Note that the result contains three significant digits even though two of the numbers involved have only two significant figures.

You have probably heard it said that a chain is only as strong as its weakest link. For addition and subtraction, the weak link is the number of decimal places in the number with the *smallest* number of decimal places.

Products and Quotients

A rule of thumb that is sometimes suggested for multiplication and division is that the answer should be rounded so that it contains the same number of significant digits as the original number with the smallest number of significant digits. Unfortunately, this procedure often leads to incorrect rounding. For example, consider the two calculations

$$\frac{24 \times 4.52}{100.0} = 1.08 \quad \text{and} \quad \frac{24 \times 4.02}{100.0} = 0.965$$

By this rule, the first answer would be rounded to 1.1 and the second to 0.96. If we assume a unit uncertainty in the last digit of each number in the first quotient, however, the relative uncertainties associated with each of these numbers are 1/24, 1/452, and 1/1000. Because the first relative uncertainty is much larger than the other two, the relative uncertainty in the result is also 1/24; the absolute uncertainty is then

$$1.08 \times 1/24 = 0.045 = 0.04$$

By the same argument the absolute uncertainty of the second answer is given by

$$0.965 \times 1/24 = 0.040 = 0.04$$

Therefore, the first result should be rounded to three significant figures or 1.08, but the second should be rounded to only two; that is 0.96.

When adding and subtracting numbers in scientific notation, express the numbers to the same power of ten. For example,

$$\begin{aligned} 2.432 \times 10^6 &= 2.432 \times 10^6 \\ +6.512 \times 10^4 &= +0.06512 \times 10^6 \\ -1.227 \times 10^5 &= \underline{-0.1227 \times 10^6} \\ & 2.37442 \times 10^6 \\ &= 2.374 \times 10^6 \end{aligned}$$

The weak link for multiplication and division is the number of *significant figures* in the number with the smallest number of significant figures. *Use this rule of thumb with caution.*

Logarithms and Antilogarithms

Be especially careful in rounding the results of calculations involving logarithms. The following rules apply to most situations:[3]

1. In a logarithm of a number, keep as many digits to the right of the decimal point as there are significant figures in the original number.
2. In an antilogarithm of a number, keep as many digits as there are digits to the right of the decimal point in the original number.

The number of significant figures in the *mantissa,* or the digits to the right of the decimal point of a logarithm, is the same as the number of significant figures in the original number.

$$\log(\mathbf{9.57} \times 10^4) = 4.\mathbf{981}$$

[3]D. E. Jones, *J. Chem. Educ.,* **1971,** *49,* 753.

EXAMPLE 3-8

Round the following answers so that only significant digits are retained. (a) log $6.000 \times 10^{-5} = -4.2218488$; (b) antilog $12.5 = 3.162277 \times 10^{12}$.

(a) Following rule 1, we retain 4 digits to the right of the decimal point:

$$\log \mathbf{6.000} \times 10^{-5} = -4.\mathbf{2218}$$

(b) Following rule 2, we may retain only 1 digit:

$$\text{antilog } 12.\mathbf{5} = \mathbf{3} \times 10^{12}$$

Here, the significant digits are in boldface type.

3D-3 Rounding Data

In rounding a number ending in 5, always round so that the result ends with an even number.

The computed results of a chemical analysis must be rounded in an appropriate way before being reported. For example, consider the replicate results: 61.60, 61.46, 61.55, and 61.61. The mean of these data is 61.555, and the standard deviation is 0.069. When you round the mean, do you take 61.55 or 61.56? A good guide to follow when rounding a 5 is always to round to the nearest even number. In this way, you eliminate any tendency to round in a set direction. In other words, there is an equal likelihood that the nearest even number will be the higher or the lower in any given situation. Accordingly, you might choose to report the result as 61.56 ± 0.07. If you had reason to doubt the reliability of the estimated standard deviation, you might report the result as 61.6 ± 0.1.

3D-4 Rounding the Results from Chemical Computations

Throughout this text and others, the reader is asked to perform calculations with data whose precision is indicated only by the significant figure convention. In these circumstances, common-sense assumptions must be made as to the uncertainty in each number. The uncertainty of the result is then estimated using the techniques presented in Section 3C. Finally, the result is rounded so that it contains only significant digits. *It is especially important to postpone rounding until the calculation is completed.* At least one extra digit beyond the significant digits should be carried through all of the computations in order to avoid a *rounding error.* This extra digit is sometimes called a "guard" digit. Modern calculators generally retain several extra digits that are not significant, and the user must be careful to round final results properly so that only significant figures are included. Example 3-9 illustrates this procedure.

EXAMPLE 3-9

A 3.4842-g sample of a solid mixture containing benzoic acid, C_6H_5COOH (122.123 g/mol) was dissolved and titrated with base to a phenolphthalein end point. The acid consumed 41.36 mL of 0.2328 M NaOH. Calculate the percent benzoic acid (HBz) in the sample.

As will be shown in Section 6C-5, the computation takes the following form:

$$\% \text{ HBz} = \frac{41.36 \cancel{\text{mL}} \times 0.2328 \frac{\cancel{\text{mmol NaOH}}}{\cancel{\text{mL NaOH}}} \times \frac{1 \cancel{\text{mmol HBz}}}{\cancel{\text{mmol NaOH}}} \times \frac{122.123 \text{ g HBz}}{1000 \cancel{\text{mmol HBz}}}}{3.4842 \text{ g sample}} \times 100\% = 33.749\%$$

Since all operations are either multiplication or division, the relative uncertainty of the answer is determined by the relative uncertainties of the experimental data. Let us estimate what these uncertainties are.

(a) The position of the liquid level in a buret can be estimated to ± 0.02 mL. Initial and final readings must be made, however, so that the standard deviation of the volume will be

$$\sqrt{(0.02)^2 + (0.02)^2} = \pm 0.028 \text{ mL} \qquad \text{(Equation 3-10)}$$

The relative uncertainty is then

$$\frac{\pm 0.028}{41.36} \times 1000 \text{ ppt} = \pm 0.68 \text{ ppt}$$

(b) Generally, the absolute uncertainty of a mass obtained with an analytical balance will be on the order of ± 0.0001 g. Thus, the relative uncertainty of the denominator is

$$\frac{0.0001}{3.4842} \times 1000 \text{ ppt} = 0.029 \text{ ppt}$$

(c) Usually we can assume that the absolute uncertainty in the molarity of a reagent solution is 0.0001, and so

$$\frac{0.0001}{0.2328} \times 1000 \text{ ppt} = 0.43 \text{ ppt}$$

(d) The relative uncertainty in the molar mass HBz is several orders of magnitude smaller than that of the three experimental data and thus is insignificant. Note, however, that we should retain enough digits in the calculation so that the molar mass is given to at least one more digit (the guard digit) than any of the experimental data. Thus, in the calculation, we use 122.123 for the molar mass (here we are carrying two extra digits).

(e) No uncertainty is associated with 100% and the 1000 mmol HBz, since these are exact numbers.

Substituting the three relative uncertainties into Equation 3-10, we obtain

$$\frac{s_y}{y} = \sqrt{\left(\frac{0.028}{41.36}\right)^2 + \left(\frac{0.0001}{3.4842}\right)^2 + \left(\frac{0.0001}{0.2328}\right)^2}$$
$$= 8.02 \times 10^{-4}$$

$$s_y = 8.02 \times 10^{-4} \times y = 8.02 \times 10^{-4} \times 33.749$$
$$= 0.027 \approx 0.03$$

Thus, the uncertainty in the calculated result is ±0.03% HBz, and we should report the result as 33.75% HBz or better 33.75 (±0.03)% HBz.

With a little practice, you can make rounding decisions such as those shown in Example 3-9 in your head.

There is no relationship between the number of digits displayed on a calculator and the true number of significant figures.

It is important to emphasize that rounding decisions are an important part of *every calculation* and that such decisions *cannot* be based on the number of digits displayed on the readout of a calculator.

3E QUESTIONS AND PROBLEMS

3-1. Define
- ***(a)** spread or range.
- **(b)** coefficient of variation.
- ***(c)** significant figures.
- **(d)** Gaussian distribution.

3-2. Differentiate between
- ***(a)** sample standard deviation and sample variance.
- **(b)** population mean and sample mean.
- ***(c)** accuracy and precision.
- **(d)** random and systematic error.

3-3. Distinguish between
- ***(a)** the meaning of the word "sample" as it is used in a chemical and in a statistical sense.
- **(b)** the sample standard deviation and the population standard deviation.

3-4. What is the standard error of a mean?

3-5. Consider the following sets of replicate measurements:

*A	B	*C	D	*E	F
2.4	69.94	0.0902	2.3	69.65	0.624
2.1	69.92	0.0884	2.6	69.63	0.613
2.1	69.80	0.0886	2.2	69.64	0.596
2.3		0.1000	2.4	69.21	0.607
1.5			2.9		0.582

For each set, calculate the (a) mean; (b) median; (c) spread, or range; (d) standard deviation; and (e) coefficient of variation.

3-6. The accepted values for the sets of data in Problem 3-5 are: *set A, 2.0; set B, 69.75; *set C, 0.0930; set D, 3.0; *set E, 69.05; set F, 0.635. For the mean of each set, calculate (a) the absolute error and (b) the relative error in parts per thousand.

3-7. Estimate the absolute deviation and the coefficient of variation for the results of the following calculations. Round each result so that it contains only significant digits. The numbers in parentheses are absolute standard deviations.

***(a)** $y = 6.75(+0.03) + 0.843(+0.001) - 7.021(+0.001) = 0.572$

(b) $y = 19.97(\pm 0.04) + 0.0030(\pm 0.0001) + 1.29(\pm 0.08) = 21.263$

***(c)** $y = 67.1(\pm 0.3) \times 1.03(\pm 0.02) \times 10^{-17} = 6.9113 \times 10^{-16}$

(d) $y = 243(\pm 1) \times \dfrac{760(\pm 2)}{1.006(\pm 0.006)} = 183578.5$

***(e)** $y = \dfrac{143(\pm 6) - 64(\pm 3)}{1249(\pm 1) + 77(\pm 8)} = 5.9578 \times 10^{-2}$

(f) $y = \dfrac{1.97(\pm 0.01)}{243(\pm 3)} = 8.106996 \times 10^{-3}$

3-8. Estimate the absolute standard deviation and the coefficient of variation for the results of the following calculations. Round each result to include only significant figures. The numbers in parentheses are absolute standard deviations.

*(a) $y = -1.02(\pm 0.02) \times 10^{-7} - 3.54(\pm 0.2) \times 10^{-8}$
$= -1.374 \times 10^{-7}$

(b) $y = 100.20(\pm 0.08) - 99.62(\pm 0.06) + 0.200(\pm 0.004)$
$= 0.780$

*(c) $y = 0.0010(\pm 0.0005) \times 18.10(\pm 0.02) \times 200(\pm 1)$
$= 3.62$

(d) $y = \dfrac{1.73(\pm 0.03) \times 10^{-14}}{1.63(\pm 0.04) \times 10^{-16}} = 106.1349693$

*(e) $y = \dfrac{100(\pm 1)}{2(\pm 1)} = 50$

(f) $y = \dfrac{1.43(\pm 0.02) \times 10^{-2} - 4.76(\pm 0.06) \times 10^{-3}}{24.3(\pm 0.7) + 8.06(\pm 0.08)}$
$= 2.948 \times 10^{-4}$

3-9. Calculate the absolute standard deviation and the coefficient of variation for the results of the following calculations. Round each result to include only significant figures. The numbers in parentheses are absolute standard deviations.

*(a) $y = \log[2.00(\pm 0.03) \times 10^{-4}] = -3.69897$
(b) $y = \log[4.42(\pm 0.01) \times 10^{37}] = 37.64542$
*(c) $y = \text{antilog}[1.200(\pm 0.003)] = 15.8489$
(d) $y = \text{antilog}[49.54(\pm 0.04)] = 3.4674 \times 10^{49}$

3-10. Calculate the absolute standard deviation and the coefficient of variation for the results of the following calculations. Round each result to include only significant figures. The numbers in parentheses are absolute standard deviations.

*(a) $y = [4.73(\pm 0.03) \times 10^{-4}]^3 = 105.8238$
(b) $y = [2.145(\pm 0.002)]^{1/4} = 1.210199$

***3-11.** The inside diameter of an open cylindrical tank was measured. The results of four replicate measurements were 5.4, 5.2, 5.5, and 5.2 m. Measurement of the height of the tank yielded 9.8, 9.9, and 9.6 m.

Calculate the volume in liters of the tank and the standard deviation of the result.

3-12. In Chapter 22, we show that quantitative molecular absorption spectrometry is based upon Beer's law, which can be written as

$$-\log T = \varepsilon b c_X$$

where T is the transmittance of a solution of an analyte X, b is the thickness of the absorbing solution, c_X is the molar concentration of X, and ε is an experimentally determined constant. By measuring a series of standard solutions of X, εb was found to have a value of 2505($\pm$ 12) where the number in parentheses is the absolute standard deviation.

An unknown solution of X was measured in a cell identical to the one used to determine εb. The replicate results were $T = 0.273, 0.276, 0.268$, and 0.274. Calculate (a) the molar concentration of the analyte c_X, (b) the absolute standard deviation of c_X, and (c) the coefficient of variation of c_X.

***3-13.** Analysis of several plant-food preparations for potassium ion yielded the following data:

Sample	Mean Percent K^+	Number of Observations	Deviation of Individual Results from Mean
1	4.80	5	0.13, 0.09, 0.07, 0.05, 0.06
2	8.04	3	0.09, 0.08, 0.12
3	3.77	4	0.02, 0.15, 0.07, 0.10
4	4.07	4	0.12, 0.06, 0.05, 0.11
5	6.84	5	0.06, 0.07, 0.13, 0.10, 0.09

(a) Evaluate the standard deviation s for each sample.
(b) Obtain a pooled estimate for s.

3-14. Six bottles of wine were analyzed for residual sugar, with the following results:

Bottle	Percent (w/v) Residual Sugar	Number of Observations	Deviation of Individual Results from Mean
1	0.94	3	0.050, 0.10, 0.08
2	1.08	4	0.060, 0.050, 0.090, 0.060
3	1.20	5	0.05, 0.12, 0.07, 0.00, 0.08
4	0.67	4	0.05, 0.10, 0.06, 0.09
5	0.83	3	0.07, 0.09, 0.10
6	0.76	4	0.06, 0.12, 0.04, 0.03

(a) Evaluate the standard deviation s for each set of data.
(b) Pool the data to establish an absolute standard deviation for the method.

***3-15.** Nine samples of illicit heroin preparations were analyzed in duplicate by a gas chromatographic technique. Pool the following data to establish an absolute standard deviation for the procedure:

Sample	Heroin, %	Sample	Heroin, %
1	2.24, 2.27	6	1.07, 1.02
2	8.4, 8.7	7	14.4, 14.8
3	7.6, 7.5	8	21.9, 21.1
4	11.9, 12.6	9	8.8, 8.4
5	4.3, 4.2		

3-16. Calculate a pooled estimate of s from the following spectrophotometric analysis for nitrilotriacetic acid (NTA) in water from the Ohio River:

Sample	NTA, ppb
1	13, 16, 14, 9
2	38, 37, 38
3	25, 29, 23, 29, 26

CHAPTER

4

Application of Statistics to Data Treatment and Evaluation

Experimentalists use statistical calculations to sharpen their judgments concerning the quality of experimental measurements. The most common applications of statistics to analytical chemistry include (1) establishing confidence limits for the mean of a set of replicate data, (2) determining the number of replications that are required to decrease the confidence limit for a mean to a given level, (3) determining at a given probability whether an experimental mean is different from the accepted value for the quantity being measured, (4) determining at a given probability level whether two experimental means are different, (5) determining at a given probability level whether the precision of two sets of measurements differs, (6) deciding whether an outlier is probably the result of a gross error and should be discarded in calculating a mean, (7) defining and estimating detection limits, (8) treating calibration data, and (9) in quality control of analytical data and of industrial products.

We will examine each of these applications in the sections that follow.

4A CONFIDENCE LIMITS

The exact value of the mean μ for a population of data can never be determined exactly because such a determination requires that an infinite number of measurements be made. Statistical theory, however, allows us to set limits around an experimentally determined mean $\bar{x}$ within which the population mean μ lies with a given degree of probability. These limits are called *confidence limits,* and the interval they define is known as the *confidence interval.*

Confidence limits define an interval around $\bar{x}$ that probably contains μ.

The confidence interval is the magnitude of the confidence limit.

The size of the confidence interval, which is derived from the sample standard deviation, depends upon how accurately we know s—that is, how closely we think our sample standard deviation approximates the population standard deviation σ. If we have reason to believe that s is a good approximation of σ, the confidence interval can be significantly narrower than if the estimate of s is based upon only two or three replicates.

The **confidence level** fixes the odds that the true mean will be within the defined limits.

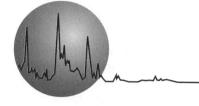

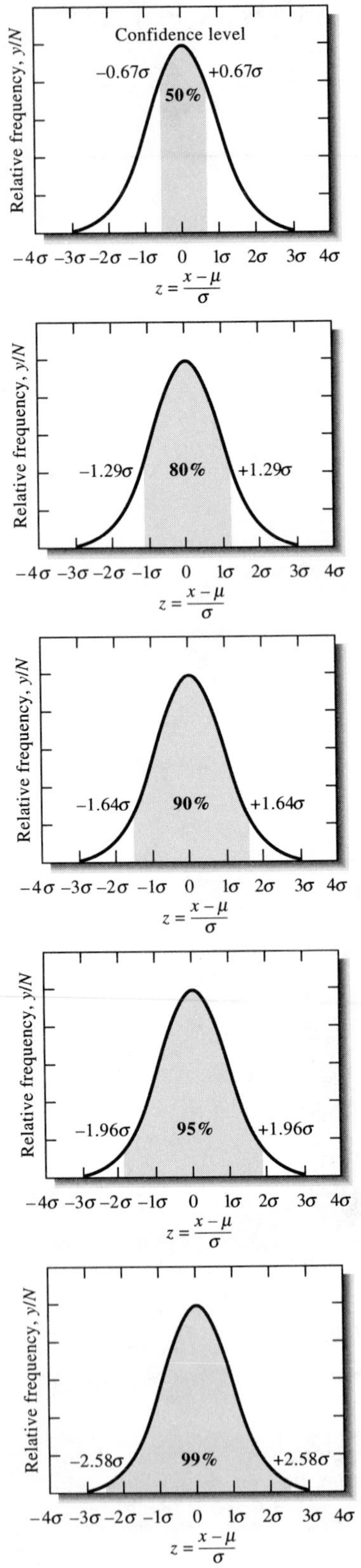

4A-1 The Confidence Interval when s Is a Good Approximation of σ

Figure 4-1 shows a series of five normal error curves. In each, the relative frequency is plotted as a function of the quantity z (Equation 3-2, page 27), which is the deviation from the mean *in units of the population standard deviation.* The shaded areas in each plot lie between the values of $-z$ and $+z$ that are indicated to the left and right of the curves. The numbers within the shaded areas are the percentage of the total area under the curve that is included within these values of z. For example, as shown in the top curve, 50% of the area under any Gaussian curve is located between -0.67σ and $+0.67\sigma$. Proceeding on down, we see that 80% of the total area lies between -1.29σ and $+1.29\sigma$, and 90% lies between -1.64σ and $+1.64\sigma$. Relationships such as these allow us to define a range of values around a measurement within which the true mean is likely to lie with a certain probability *provided we have a reasonable estimate of* σ. For example we may assume that 90 times out of 100, the true mean, μ, will be within $\pm 1.64\sigma$ of any measurement that we make. Here, the *confidence level* is 90% and the *confidence interval* is $\pm z\sigma = \pm 1.64\sigma$.

We find a general expression for the confidence limits (CL) of a single measurement by rearranging Equation 3-2 (remember that z can take positive or negative values). Thus,

$$\text{CL for } \mu = x \pm z\sigma \tag{4-1}$$

For the mean of N measurements, the standard error of the mean, $\sigma/\sqrt{N}$ (Equation 3-5), is employed in place of σ. That is,

$$\text{CL for } \mu = \bar{x} \pm \frac{z\sigma}{\sqrt{N}} \tag{4-2}$$

Values for z at various confidence levels are found in Table 4-1.

EXAMPLE 4-1

Calculate the 80% and 95% confidence limits for (a) the first entry (1.80 ppm Hg) in Example 3-2 (page 31) and (b) the mean value (1.67 ppm Hg) for specimen 1 in the example. Assume that in each part, $s \rightarrow \sigma = 0.1$.

(a) From Table 4-1, we see that $z = 1.29$ and 1.96 for the two confidence levels. Substituting into Equation 4-2,

$$80\%\ \text{CL} = 1.80 \pm \frac{1.29 \times 0.10}{\sqrt{1}} = 1.80 \pm 0.13$$

$$95\%\ \text{CL} = 1.80 \pm \frac{1.96 \times 0.10}{\sqrt{1}} = 1.80 \pm 0.20$$

Figure 4-1
Areas under a Gaussian curve for various values of $\pm z$.

See *Mathcad Applications for Analytical Chemistry,* **pp. 28–30.**

From these calculations, we conclude that the chances are 80 in 100 that μ, the population mean (and, *in the absence of systematic error,* the true value), lies between 1.67 and 1.93 ppm Hg. Furthermore, there is a 95% chance that it lies between 1.60 and 2.00 ppm Hg.

(b) For the three measurements,

$$80\%\ \text{CL} = 1.67 \pm \frac{1.29 \times 0.10}{\sqrt{3}} = 1.67 \pm 0.07$$

$$95\%\ \text{CL} = 1.67 \pm \frac{1.96 \times 0.10}{\sqrt{3}} = 1.67 \pm 0.11$$

Thus, the chances are 80 in 100 that the population mean is located within the limits of 1.60 to 1.74 ppm Hg and 95 in 100 that it lies between 1.56 and 1.78 ppm.

Number of Measurements Averaged, N	Relative Size of Confidence Interval
1	1.00
2	0.71
3	0.58
4	0.50
5	0.45
6	0.41
10	0.32

EXAMPLE 4-2

How many replicate measurements of specimen 1 in Example 3-2 are needed to decrease the 95% confidence interval to $\pm$ 0.07 ppm Hg?

The confidence interval is given by the second term on the right-hand side of Equation 4-2 ($\text{CI} = \pm z\sigma/\sqrt{N}$). Thus,

$$\text{CI} = 0.07 = \pm \frac{zs}{\sqrt{N}} = \pm \frac{1.96 \times 0.10}{\sqrt{N}}$$

$$\sqrt{N} = \pm \frac{1.96 \times 0.10}{0.07} = \pm 2.80$$

$$N = (\pm 2.8)^2 = 7.8$$

We conclude that eight measurements would provide a slightly better than 95% chance of the population mean lying within $\pm$ 0.07 ppm of the experimental mean.

Equation 4-2 tells us that the confidence interval for an analysis can be halved by carrying out four measurements. Sixteen measurements will narrow the interval by a factor of 4, and so on. We rapidly reach a point of diminishing returns in acquiring additional data. Ordinarily we take advantage of the relatively large gain attained by averaging two to four measurements, but we can seldom afford the time required for additional improvements in confidence.

It is essential to keep in mind at all times that confidence limits based on Equation 4-2 apply only *in the absence of bias and only if we can assume that* $s \approx \sigma$.

TABLE 4-1
Confidence Levels for Various Values of z

Confidence Levels, %	z
50	0.67
68	1.00
80	1.29
90	1.64
95	1.96
96	2.00
99	2.58
99.7	3.00
99.9	3.29

4A-2 The Confidence Interval when σ Is Not Known

Often we are faced with limitations in time or amount of available sample that prevent us from accurately estimating σ. In such a case, a single set of replicate measurements must provide not only a mean but also an estimate of precision. As indicated earlier, s calculated from a small set of data may be quite uncertain. Thus, confidence limits are necessarily broader when a good estimate of σ is not available.

The t statistic is often called *Student's t.* Student was the name used by W. S. Gossett when he wrote the classic paper on t that appeared in *Biometrika,* **1908**, *6*, 1. Gossett was employed by the Guinness Brewery to statistically analyze the results of determinations of the alcohol content in their products. As a result of this work, he discovered the now-famous statistical treatment of small sets of data. To avoid the disclosure of any trade secrets of his employer, Gossett published the paper under the name Student.

To account for the variability of s, we use the important statistical parameter t, which is defined in exactly the same way as z (Equation 3-2) except that s is substituted for σ:

$$t = \frac{x - \mu}{s} \tag{4-3}$$

Like z in Equation 3-2, t depends on the desired confidence level. In addition, however, t also depends on the number of degrees of freedom in the calculation of s. Table 4-2 provides values for t for a few degrees of freedom. More extensive tables are found in various mathematical and statistical handbooks. Note that $t \rightarrow z$ as the number of degrees of freedom becomes infinite.

Remember that the number of degrees of freedom is not equal to N but $N - 1$ instead.

The confidence limits for the mean $\bar{x}$ of N replicate measurements can be derived from t by an equation similar to Equation 4-2:

$$\text{CL for } \mu = \bar{x} \pm \frac{ts}{\sqrt{N}} \tag{4-4}$$

TABLE 4-2 Values of t for Various Levels of Probability

Degrees of Freedom	Factor for Confidence Interval				
	80%	**90%**	**95%**	**99%**	**99.9%**
1	3.08	6.31	12.7	63.7	637
2	1.89	2.92	4.30	9.92	31.6
3	1.64	2.35	3.18	5.84	12.9
4	1.53	2.13	2.78	4.60	8.60
5	1.48	2.02	2.57	4.03	6.86
6	1.44	1.94	2.45	3.71	5.96
7	1.42	1.90	2.36	3.50	5.40
8	1.40	1.86	2.31	3.36	5.04
9	1.38	1.83	2.26	3.25	4.78
10	1.37	1.81	2.23	3.17	4.59
11	1.36	1.80	2.20	3.11	4.44
12	1.36	1.78	2.18	3.06	4.32
13	1.35	1.77	2.16	3.01	4.22
14	1.34	1.76	2.14	2.98	4.14
∞	1.29	1.64	1.96	2.58	3.29

EXAMPLE 4-3

A chemist obtained the following data for the alcohol content of a sample of blood: % C_2H_5OH: 0.084, 0.089, and 0.079. Calculate the 95% confidence limits of the mean assuming (a) no additional knowledge about the precision of the method and (b) that on the basis of previous experience, it is known that $s \rightarrow \sigma = 0.005\%$ C_2H_5OH.

(a) $\Sigma x_i = 0.084 + 0.089 + 0.079 = 0.252$

$\Sigma x_i^2 = 0.007056 + 0.007921 + 0.006241 = 0.021218$

$$s = \sqrt{\frac{0.021218 - (0.252)^2/3}{3 - 1}} = 0.0050\%\ C_2H_5OH$$

Here, $\bar{x} = 0.252/3 = 0.084$. Table 4-2 indicates that $t = 4.30$ for two degrees of freedom and 95% confidence. Thus,

$$95\%\ \text{CL} = \bar{x} \pm \frac{ts}{\sqrt{N}} = 0.084 \pm \frac{4.30 \times 0.0050}{\sqrt{3}}$$
$$= 0.084 \pm 0.012\%\ C_2H_5OH$$

(b) Because a good value of σ is available,

$$95\%\ \text{CL} = \bar{x} \pm \frac{z\sigma}{\sqrt{N}} = 0.084 \pm \frac{1.96 \times 0.0050}{\sqrt{3}}$$
$$= 0.084 \pm 0.006\%\ C_2H_5OH$$

Note that a sure knowledge of σ decreases the confidence interval by a significant amount.

4B STATISTICAL AIDS TO HYPOTHESIS TESTING

Much of scientific and engineering endeavor is based upon hypothesis testing. Thus, in order to explain an observation, a hypothetical model is advanced and tested experimentally to determine its validity. If the results from these experiments do not support the model, we reject it and seek a new hypothesis. If agreement is found, the hypothetical model serves as the basis for further experiments. When the hypothesis is supported by sufficient experimental data, it becomes recognized as a useful theory until such time as data are obtained that refute it.

Experimental results seldom agree exactly with those predicted from a theoretical model. Consequently, scientists and engineers frequently must judge whether a numerical difference is simply a manifestation of the random errors inevitable in all measurements or of a systematic error in the measurement process. Certain statistical tests are useful in sharpening these judgments.

Tests of this kind make use of a *null hypothesis,* which assumes that the numerical quantities being compared are, in fact, the same. The probability of the observed differences appearing as a result of random error is then computed from

In statistics a **null hypothesis** postulates that two observed quantities are the same.

The notation to indicate the 5% probability level is $P > 0.05$.

statistical theory. Usually, if the observed difference is greater than or equal to the difference that would occur 5 times in 100 (the 5% probability level), the null hypothesis is considered questionable and the difference is judged to be significant. Other probability levels, such as 1 in 100 or 10 in 100, may also be adopted, depending upon the certainty desired in the judgment.

The kinds of testing that chemists use most often include the comparison of (1) the mean from an analysis $\bar{x}$ with what is believed to be the true value μ; (2) the means $\bar{x}_1$ and $\bar{x}_2$ from two sets of analyses; (3) the standard deviations s_1 and s_2 or σ_1 and σ_2 from two sets of measurements; and (4) the standard deviation s of a small set of data with the standard deviation σ of a larger set of measurements. The sections that follow consider some of the methods for making these comparisons.

4B-1 Comparison of an Experimental Mean with a True Value

A common way of testing for bias in an analytical method is to use the method to analyze a sample whose composition is accurately known. Bias in an analytical method is illustrated by the two curves shown in Figure 4-2, which show the frequency distribution of replicate results in the analysis of identical samples by two analytical methods having random errors of exactly the same size. Method A has no bias, so the population mean μ_A is the true value x_t. Method B has a systematic error, or bias that is given by

$$\text{bias} = \mu_B - x_t = \mu_B - \mu_A \tag{4-5}$$

Note that bias affects all of the data in the set in the same way and that it can be either positive or negative.

In testing for bias by analyzing a sample whose analyte concentration is known exactly, it is likely that the experimental mean $\bar{x}$ will differ from the accepted value x_t as shown in the figure; the judgment must then be made whether this difference is the consequence of random error or represents a systematic error.

In treating this type of problem statistically, the difference $\bar{x} - x_t$ is compared with the difference that could be caused by random error. If the observed difference is less than that computed for a chosen probability level, the null hypothesis that $\bar{x}$ and x_t are the same cannot be rejected; that is, no significant systematic error has been demonstrated. It is important to realize, however, that this statement does not say that there is no systematic error; it says only that whatever systematic error is present is so small that it cannot be detected. If $\bar{x} - x_t$ is significantly larger than either the expected or the critical value, we may assume that the difference is real and that the systematic error is significant.

The critical value for the rejection of the null hypothesis is calculated by rewriting Equation 4-4 (and remembering that x_t is postulated to be equal to μ) in the form

$$\bar{x} - x_t = \pm \frac{ts}{\sqrt{N}} \tag{4-6}$$

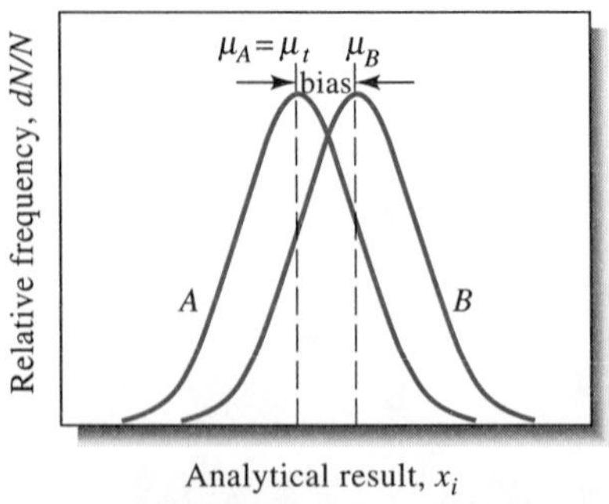

Figure 4-2
Illustration of bias:

$$\text{bias} = \mu_B - \mu_A = \mu_B - \mu_t$$

where N is the number of replicate measurements employed in the test. If a good estimate of σ is available, Equation 4-6 can be modified by replacing t with z and s with σ.

EXAMPLE 4-4

A new procedure for the rapid determination of sulfur in kerosenes was tested on a sample known from its method of preparation to contain 0.123% ($x_t = 0.123$) S. The results were % S = 0.112, 0.118, 0.115, and 0.119. Do the data indicate that there is a bias in the method?

$$\Sigma x_i = 0.112 + 0.118 + 0.115 + 0.119 = 0.464$$

$$\bar{x} = 0.464/4 = 0.116\%\text{S}$$

$$\bar{x} - x_t = 0.116 - 0.123 = -0.007\%\text{S}$$

$$\Sigma x_i^2 = 0.012544 + 0.013924 + 0.013225 + 0.014161 = 0.053854$$

$$s = \sqrt{\frac{0.053854 - (0.464)^2/4}{4 - 1}} = \sqrt{\frac{0.000030}{3}} = 0.0032$$

From Table 4-2, we find that at the 95% confidence level, t has a value of 3.18 for three degrees of freedom. Thus,

$$\frac{ts}{\sqrt{4}} = \frac{3.18 \times 0.0032}{\sqrt{4}} = \pm 0.0051$$

An experimental mean can be expected to deviate by $\pm$ 0.0051 or greater no more frequently than 5 times in 100. Thus, if we conclude that $\bar{x} - x_t = -0.007$ is a significant difference and that a systematic error is present, we will, on the average, be wrong fewer than 5 times in 100.

If we make a similar calculation employing the value for t at the 99% confidence level, $ts/\sqrt{N}$ assumes a value of 0.0093. Thus, if we insist upon being wrong no more often than 1 time in 100, we must conclude that no difference between the results has been *demonstrated.* Note that this statement is different from saying that there is no systematic error.

If it were confirmed by further experiments that the method always gave low results, we would say that the method had a *negative bias.*

Even if a mean value is shown to be equal to the true value at a given confidence level, we cannot conclude that there is *no* systematic error in the data.

4B-2 Comparison of Two Experimental Means

The results of chemical analyses are frequently used to determine whether two materials are identical. Here, the chemist must judge whether a difference in the means of two sets of identical analyses is real and constitutes evidence that the samples are different or whether the discrepancy is simply a consequence of random errors in the two sets. To illustrate, let us assume that N_1 replicate analyses of material 1 yielded a mean value of $\bar{x}_1$, and N_2 analyses of material 2 obtained by the same method gave a mean of $\bar{x}_2$. If the data were collected in an identical way, it is usually safe to assume that the standard deviations of the two sets of measurements are the same and modify Equation 4-6 to take into account that one set of results is being compared with a second rather than with the true mean of the data, x_t.

In this case, as with the previous one, we invoke the null hypothesis that the two samples are identical and that the observed difference in the results, $(\bar{x}_1 - \bar{x}_2)$, is the result of random errors. To test this hypothesis statistically, we modify Equation 4-6 in the following way. First we substitute $\bar{x}_2$ for x_t, thus making the left side of the equation the numerical difference between the two means $\bar{x}_1 - \bar{x}_2$. Since we know from Equation 3-5 that the standard deviation of the mean $\bar{x}_1$ is

$$s_{\text{m1}} = \frac{s_1}{\sqrt{N_1}} \qquad \text{and likewise for } \bar{x}_2 \qquad s_{\text{m2}} = \frac{s_2}{\sqrt{N_2}}$$

the variance s_d^2 of the difference ($d = x_1 - x_2$) between the means is given by

$$s_d^2 = s_{\text{m1}}^2 + s_{\text{m2}}^2$$

By substituting the values for s_d, s_{m1}, and s_{m2} into this equation, we have

$$\left(\frac{s_d}{\sqrt{N}}\right)^2 = \left(\frac{s_{\text{m1}}}{\sqrt{N_1}}\right)^2 + \left(\frac{s_{\text{m2}}}{\sqrt{N_2}}\right)^2$$

If we make the further assumption that the pooled standard deviation s_{pooled} is a good estimate of both s_{m1} and s_{m2}, then

$$\left(\frac{s_d}{\sqrt{N}}\right)^2 = \left(\frac{s_{\text{pooled}}}{\sqrt{N_1}}\right)^2 + \left(\frac{s_{\text{pooled}}}{\sqrt{N_2}}\right)^2 = s_{\text{pooled}}^2\left(\frac{N_1 + N_2}{N_1 N_2}\right)$$

and

$$\frac{s_d}{\sqrt{N}} = s_{\text{pooled}}\sqrt{\frac{N_1 + N_2}{N_1 N_2}}$$

Substituting this equation into Equation 4-6 (and also $\bar{x}_2$ for x_t), we find that

See *Mathcad Applications for Analytical Chemistry,* **p. 32.**

$$\bar{x}_1 - \bar{x}_2 = \pm t s_{\text{pooled}}\sqrt{\frac{N_1 + N_2}{N_1 N_2}} \qquad (4\text{-}7)$$

The numerical value for the term on the right is computed using t for the particular confidence level desired. The number of degrees of freedom for finding t in Table 4-2 is $N_1 + N_2 - 2$. If the experimental difference $\bar{x}_1 - \bar{x}_2$ is smaller than the computed value, the null hypothesis is not rejected and no significant difference between the means has been demonstrated. An experimental difference greater than the value computed from t indicates that there is a significant difference between the means.

If a good estimate of σ is available, Equation 4-7 can be modified by inserting z for t and σ for s.

EXAMPLE 4-5

Two barrels of wine were analyzed for their alcohol content in order to determine whether they were from different sources. On the basis of six analyses, the average content of the first barrel was established to be 12.61% ethanol. Four analyses of the second barrel gave a mean of 12.53% alcohol. The ten analyses yielded a pooled value of s of 0.070%. Do the data indicate a difference between the wines?

Here we employ Equation 4-7, using t for eight degrees of freedom (10 − 2). At the 95% confidence level,

$$\pm ts\sqrt{\frac{N_1 + N_2}{N_1N_2}} = \pm 2.31 \times 0.070\sqrt{\frac{6 + 4}{6 \times 4}} = \pm 0.10\%$$

The observed difference is

$$\bar{x}_1 - \bar{x}_2 = 12.61 - 12.53 = 0.08\%$$

As often as 5 times in 100, random error will be responsible for a difference as great as 0.10%. At the 95% confidence level, then, no difference in the alcohol content of the wine has been established.

In Example 4-5, no significant difference between the two wines was detected at the 95% probability level. Note that this statement is not equivalent to saying that $\bar{x}_1$ is equal to $\bar{x}_2$, nor do the tests prove that the wines come from the same source. Indeed, it is conceivable that one wine is a red and the other is a white. To establish with a reasonable probability that the two wines are from the same source would require extensive testing of other characteristics, such as taste, color, odor, and refractive index, as well as tartaric acid, sugar, and trace element content. If no significant differences are revealed by all of these tests and by others, then it might be possible to judge the two wines as having a common origin. In contrast, the finding of *one* significant difference in any test would clearly show that the two wines are different. Thus, the establishment of a significant difference by a single test is much more revealing than the establishment of an absence of difference.

4B-3 Comparison of the Precision of Measurements

The F test provides a simple method for comparing the precision of two sets of measurements. The sets do not necessarily have to be obtained from the same sample as long as the samples are sufficiently alike that the sources of random error can be assumed to be the same. The F test is also based upon the null hypothesis and thus assumes that the precisions are identical. The quantity F, which is defined as the ratio of the variances of the two measurements, is computed and compared with the maximum values of F expected (at a certain probability level) if there was no difference in precision between the two sets of measurements. If the experimental F exceeds the critical value found in probability tables, there is a statistical basis for questioning the null hypothesis that the two standard deviations are the same. Table 4-3 provides critical values for F at the

TABLE 4-3 Critical Values for F at the 5% Level

Degrees of Freedom (Denominator)	Degrees of Freedom (Numerator)							
	2	**3**	**4**	**5**	**6**	**12**	**20**	∞
2	19.00	19.16	19.25	19.30	19.33	19.41	19.45	19.50
3	9.55	9.28	9.12	9.01	8.94	8.74	8.66	8.53
4	6.94	6.59	6.39	6.26	6.16	5.91	5.80	5.63
5	5.79	5.41	5.19	5.05	4.95	4.68	4.56	4.36
6	5.14	4.76	4.53	4.39	4.28	4.00	3.87	3.67
12	3.89	3.49	3.26	3.11	3.00	2.69	2.54	2.30
20	3.49	3.10	2.87	2.71	2.60	2.28	2.12	1.84
∞	3.00	2.60	2.37	2.21	2.10	1.75	1.57	1.00

5% probability level. These values will be exceeded only 5 times in 100 if the standard deviations of the two measurements are the same. You will find more extensive tables of F values at various probability levels in most mathematics handbooks.

The F test may be used to provide insights into either of two questions: (1) whether method A is more precise than method B and (2) whether there is a difference in the precision of the two methods. For the first of these applications, the variance of the supposedly more precise procedure is always placed in the denominator and that of the less precise procedure is always placed in the numerator. For the second application, the larger variance always appears in the numerator. This arbitrary placement of the larger variance in the numerator makes the outcome of the test less certain; thus, the uncertainty level of the F values in Table 4-3 is doubled from 5% to 10%.

EXAMPLE 4-6

A standard method for the determination of carbon monoxide in gaseous mixtures is known from many hundreds of measurements to have a standard deviation of 0.21 ppm CO. A modification of the method yields a value for s of 0.15 ppm CO for a pooled data set with 12 degrees of freedom. A second modification, also based on 12 degrees of freedom, has a standard deviation of 0.12 ppm CO. Is either modification significantly more precise than the original?

Because an improvement is claimed, the variances of the modifications are placed in the denominator. For the first,

$$F_1 = \frac{s_{\text{std}}^2}{s_1^2} = \frac{(0.21)^2}{(0.15)^2} = 1.96$$

and for the second,

$$F_1 = \frac{(0.21)^2}{(0.12)^2} = 3.06$$

For the standard procedure, $s \rightarrow \sigma$ and the degrees of freedom for the numerator can be taken as infinite. The critical value of s is thus 2.30 (Table 4-3).

The value of F for the first modification is less than 2.30, so the null hypothesis has not been disproved at the 95% probability level. The second modification, however, does appear to have significantly greater precision.

It is interesting to note that if we ask whether the precision of the second modification is significantly better than that of the first, the F test indicates that such a difference has not been demonstrated. That is,

$$F = \frac{s_1^2}{s_2^2} = \frac{(0.15)^2}{(0.12)^2} = 1.56$$

In this instance, the critical value for F is 2.69.

4C DETECTION OF GROSS ERRORS

Outliers are the result of gross errors.

When a set of data contains an outlying result that appears to differ excessively from the average, the decision must be made whether to retain or reject it. The choice of criterion for the rejection of a suspected result has its perils. If we set a stringent standard that makes the rejection of a questionable measurement difficult, we run the risk of retaining results that are spurious and have an inordinate effect on the mean of the data. If we set lenient limits on precision and thereby make the rejection of a result easy, we are likely to discard measurements that rightfully belong in the set and thus introduce a bias to the data. It is an unfortunate fact that no universal rule can be invoked to settle the question of retention or rejection.[1]

4C-1 The Q Test

The Q test is a simple, widely used statistical test.[2] In this test the absolute value of the difference between the questionable result x_q and its nearest neighbor x_n is divided by the spread w of the entire set to give the quantity Q_{exp}:

$$Q_{exp} = |x_q - x_n|/w \tag{4-8}$$

This ratio is then compared with rejection values Q_{crit} found in Table 4-4. If Q_{exp} is greater than Q_{crit}, the questionable result can be rejected with the indicated degree of confidence (see Figure 4-3).

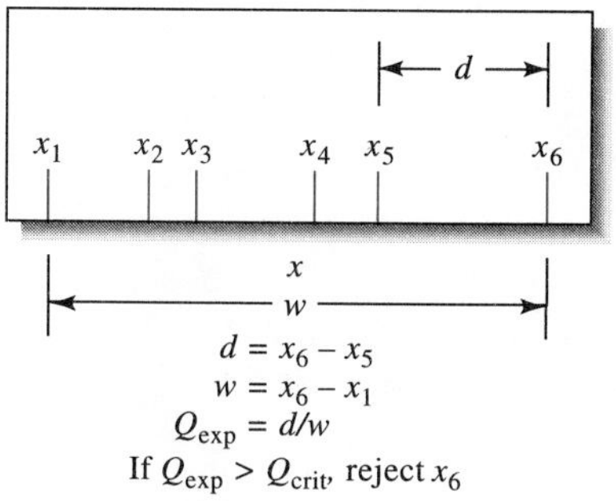

Figure 4-3
The Q test for outliers.

[1]J. Mandel, in *Treatise on Analytical Chemistry,* 2nd ed., I. M. Kolthoff and P. J. Elving, Eds., Part I, Vol. 1, pp. 282–289. New York: Wiley, 1978.

[2]R. B. Dean and W. J. Dixon, *Anal. Chem.,* **1951,** *23,* 626.

TABLE 4-4 Critical Values for the Rejection Quotient Q*

Number of Observations	Q_{crit} (Reject if $Q_{exp} > Q_{crit}$) 90% Confidence	95% Confidence	99% Confidence
3	0.941	0.970	0.994
4	0.765	0.829	0.926
5	0.642	0.710	0.821
6	0.560	0.625	0.740
7	0.507	0.568	0.680
8	0.468	0.526	0.634
9	0.437	0.493	0.598
10	0.412	0.466	0.568

*Reproduced from D. B. Rorabacher, *Anal. Chem.*, **1991,** *63,* 139. By courtesy of the American Chemical Society.

See *Mathcad Applications for Analytical Chemistry,* **pp. 30–31.**

EXAMPLE 4-7

The analysis of a calcite sample yielded CaO percentages of 55.95, 56.00, 56.04, 56.08, and 56.23. The last value appears anomalous; should it be retained or rejected?

The difference between 56.23 and 56.08 is 0.15%. The spread (56.23 − 55.95) is 0.28%. Thus,

$$Q_{exp} = \frac{0.15}{0.28} = 0.54$$

For five measurements, Q_{crit} at the 90% confidence level is 0.64. Because $0.54 < 0.64$, we must retain the outlier.

4C-2 Other Statistical Tests

Several other statistical tests have been developed to provide criteria for rejection or retention of outliers. Like the Q test, such tests assume that the distribution of the population data is normal, or Gaussian. Unfortunately, this condition cannot be proved or disproved for samples that have fewer than about 50 results. Consequently, statistical rules, which are perfectly reliable for normal distributions of data, should be *used with extreme caution* when applied to samples containing only a few data. J. Mandel, in discussing treatment of small sets of data, writes, "Those who believe that they can discard observations with statistical sanction by using statistical rules for the rejection of outliers are simply deluding themselves."[3] Thus, statistical tests for rejection should be used only as aids to common sense when small statistical samples are involved.

[3]J. Mandel, in *Treatise on Analytical Chemistry,* 2nd ed., I. M. Kolthoff and P. J. Elving, Eds., Part I, Vol. 1, p. 282. New York: Wiley, 1978.

The blind application of statistical tests to retain or reject a suspect measurement in a small set of data is not likely to be much more fruitful than an arbitrary decision. The application of good judgment based on broad experience with an analytical method is usually a sounder approach. In the end, the only valid reason for rejecting a result from a small set of data is the sure knowledge that a mistake was made in its acquisition. Without this knowledge, *a cautious approach to rejection of an outlier is wise.*

Use extreme caution when rejecting data for any reason.

4C-3 Recommendations for the Treatment of Outliers

Recommendations for the treatment of a small set of results that contains a suspect value follow:

1. Reexamine carefully all data relating to the outlying result to see if a gross error could have affected its value. This recommendation demands *a properly kept laboratory notebook containing careful notations of all observations* (see Section 35J).
2. If possible, estimate the precision that can be reasonably expected from the procedure to be sure that the outlying result actually is questionable.
3. Repeat the analysis if sufficient sample and time are available. Agreement between the newly acquired data and those of the original set that appear to be valid will lend weight to the notion that the outlying result should be rejected. Furthermore, if retention is still indicated, the questionable result will have a small effect on the mean of the larger set of data.
4. If more data cannot be secured, apply the Q test to the existing set to see if the doubtful result should be retained or rejected on statistical grounds.
5. If the Q test indicates retention, consider reporting the median of the set rather than the mean. The median has the great virtue of allowing inclusion of all data in a set without undue influence from an outlying value. In addition, the median of a normally distributed set containing three measurements provides a better estimate of the correct value than the mean of the set after the outlying value has been discarded.

4D ESTIMATION OF DETECTION LIMITS

Equation 4-7 is useful for estimating the detection limit for a measurement. Here, the standard deviation from several blank determinations is computed. The minimum detectable quantity $\Delta x_{\min}$ is

$$\Delta x_{\min} = \bar{x}_1 - \bar{x}_b > t\, s_b \sqrt{\frac{N_1 + N_2}{N_1 N_2}} \qquad (4\text{-}9)$$

where the subscript b refers to the blank determination.

EXAMPLE 4-8

A method for the analysis of DDT gave the following results when applied to pesticide-free foliage samples: μg DDT = 0.2, −0.5, −0.2, 1.0, 0.8, −0.6, 0.4, 1.2. Calculate the DDT detection limit (at the 99% confidence level) of the method for (a) a single analysis and (b) the mean of five analyses.

Here we find

$$\Sigma x_i = 0.2 - 0.5 - 0.2 + 1.0 + 0.8 - 0.6 + 0.4 + 1.2 = 2.3$$

$$\Sigma x_i^2 = 0.04 + 0.25 + 0.04 + 1.0 + 0.64 + 0.36 + 0.16 + 1.44 = 3.93$$

$$s_b = \sqrt{\frac{3.93 - (2.3)^2/8}{8 - 1}} = \sqrt{\frac{3.26875}{7}} = 0.68\ \mu\text{g}$$

(a) For a single analysis, $N_1 = 1$ and the number of degrees of freedom is $1 + 8 - 2 = 7$. From Table 4-2, we find $t = 3.50$, and so

$$\Delta x_{\text{min}} > 3.50 \times 0.68\sqrt{\frac{1 + 8}{1 \times 8}} > 2.5\ \mu\text{g DDT}$$

Thus, 99 times out of 100, a result greater than 2.5 μg DDT indicates the presence of the pesticide on the plant.

(b) Here, $N_1 = 5$, and the number of degrees of freedom is 11. Therefore, $t = 3.11$, and

$$\Delta x_{\text{min}} > 3.11 \times 0.68\sqrt{\frac{5 + 8}{5 \times 8}} > 1.2\ \mu\text{g DDT}$$

4E THE LEAST-SQUARES METHOD FOR DERIVING CALIBRATION PLOTS

Most analytical methods are based on an experimentally derived calibration curve in which a measured quantity y is plotted as a function of the known concentration x of a series of standards. The typical calibration curve shown in Figure 4-4 was obtained for the determination of isooctane in hydrocarbon samples. The ordinate (the dependent variable) is the area under the chromatographic peak for isooctane, and the abscissa (the independent variable) is the mole percent of isooctane. As is typical (and desirable), the plot approximates a straight line. Note, however, that because of the indeterminate errors in the measuring process, not all the data fall exactly on the line. Thus, the investigator must try to derive a "best" straight line from the points. A statistical technique called *regression analysis* provides the means for objectively obtaining such a line and also for specifying the uncertainties associated with its subsequent use. We consider here only the simplest regression procedure: the *method of least squares.*

4E-1 Assumptions

Linear least-squares analysis gives you the equation for the best straight line among a set of x, y data pairs.

When the method of least squares is used to generate a calibration curve, two assumptions are required. The first is that there is actually a linear relationship

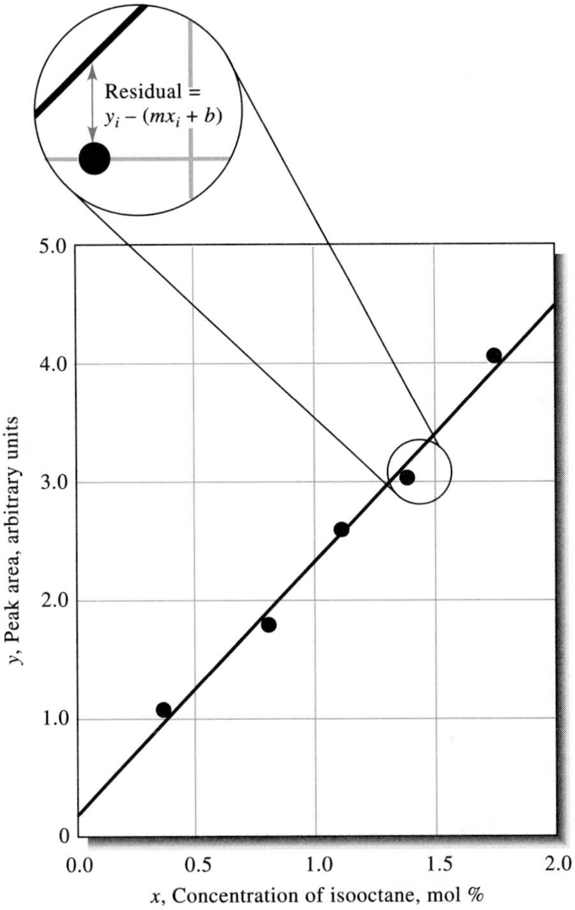

Figure 4-4
Calibration curve for the determination of isooctane in a hydrocarbon mixture.

between the measured variable (y) and the analyte concentration (x). This relationship is stated mathematically as

$$y = mx + b$$

where b is the y intercept (the value of y when x is zero) and m is the slope of the line (see Figure 4-5). We also assume that any deviation of individual points from the straight line results from error in the *measurement.* That is, we must assume that there is no error in the x values of the points. For the example of a calibration curve, we assume that exact concentrations of the standards are known from the way they were prepared. Both of these assumptions are appropriate for most analytical methods.

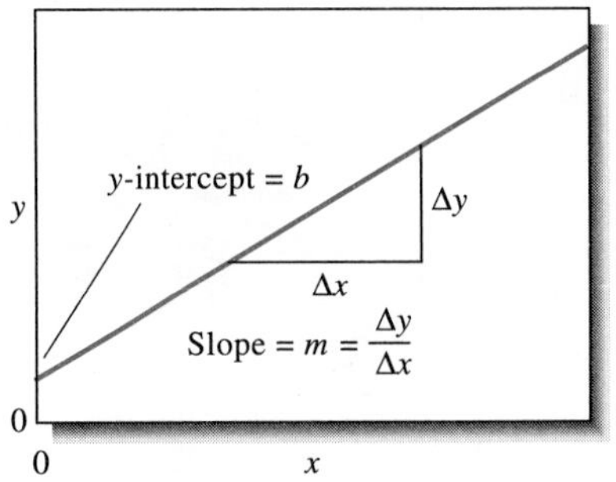

Figure 4-5
The slope-intercept form of a straight line.

4E-2 The Derivation of a Least-Squares Line

As illustrated in Figure 4-4, the vertical deviation of each point from the straight line is called a *residual.* The line generated by the least-squares method is the one that minimizes the sum of the squares of the residuals from all of the points. In addition to providing the best fit between the experimental points and the straight line, the method gives the standard deviations for m and b.[4]

The terms S_{xx} and S_{yy} are similar in form to the numerator in the equation for standard deviation.

For convenience, we define three quantities S_{xx}, S_{yy}, and S_{xy} as follows:

$$S_{xx} = \Sigma (x_i - \bar{x})^2 = \Sigma x_i^2 - \frac{(\Sigma x_i)^2}{N} \tag{4-10}$$

$$S_{yy} = \Sigma (y_i - \bar{y})^2 = \Sigma y_i^2 - \frac{(\Sigma y_i)^2}{N} \tag{4-11}$$

$$S_{xy} = \Sigma (x_i - \bar{x})(y_i - \bar{y}) = \Sigma x_i y_i - \frac{\Sigma x_i \Sigma y_i}{N} \tag{4-12}$$

where x_i and y_i are individual pairs of data for x and y, N is the number of pairs of data used in preparing the calibration curve, and $\bar{x}$ and $\bar{y}$ are the average values for the variables, that is,

$$\bar{x} = \frac{\Sigma x_i}{N} \quad \text{and} \quad \bar{y} = \frac{\Sigma y_i}{N}$$

Note that S_{xx} and S_{yy} are the sum of the squares of the deviations from the mean for the individual values of x and y. The equivalent expressions shown to the far right in Equations 4-10 to 4-12 are more convenient for use with a handheld calculator.

Six useful quantities can be derived from S_{xx}, S_{yy}, and S_{xy}:

1. The slope of the line m:

$$m = S_{xy}/S_{xx} \tag{4-13}$$

2. The intercept b:

$$b = \bar{y} - m\bar{x} \tag{4-14}$$

3. The standard deviation about regression s_r:

$$s_r = \sqrt{\frac{S_{yy} - m^2 S_{xx}}{N - 2}} \tag{4-15}$$

4. The standard deviation of the slope s_m:

$$s_m = \sqrt{s_r^2/S_{xx}} \tag{4-16}$$

5. The standard deviation of the intercept s_b:

$$s_b = s_r \sqrt{\frac{\Sigma x_i^2}{N \Sigma x_i^2 - (\Sigma x_i)^2}} = s_r \sqrt{\frac{1}{N - (\Sigma x_i)^2/\Sigma x_i^2}} \tag{4-17}$$

[4]R. L. Anderson, *Practical Statistics for Analytical Chemists,* pp. 89–121. New York: Van Nostrand–Reinhold, 1987.

6. The standard deviation for results obtained from the calibration curve s_c:

$$s_c = \frac{s_r}{m}\sqrt{\frac{1}{M} + \frac{1}{N} + \frac{(\bar{y}_c - \bar{y})^2}{m^2 S_{xx}}} \tag{4-18}$$

Equation 4-18 gives us a way to calculate the standard deviation from the mean $\bar{y}_c$ of a set of M replicate analyses of unknowns when a calibration curve that contains N points is used; recall that $\bar{y}$ is the mean value of y for the N calibration data.

The standard deviation about regression s_r (Equation 4-15) is the standard deviation for y when the deviations are measured not from the mean of y (as is usually the case) but from the derived straight line:

The standard deviation about regression is analogous to the standard deviation for one-dimensional data.

$$s_r = \sqrt{\frac{\sum_{i=1}^{N} [y_i - (b + m\,x_i)]^2}{N - 2}} \tag{4-19}$$

In this equation, the number of degrees of freedom is $N - 2$ since one degree is lost in calculating m and one in determining b.

EXAMPLE 4-9

Carry out a least-squares analysis of the experimental data provided in the first two columns in Table 4-5 and plotted in Figure 4-4.

See *Mathcaa Applications for Analytical Chemistry,* **pp. 111–114.**

Columns 3, 4, and 5 of the table contain computed values for x_i^2, y_i^2, and $x_i y_i$, with their sums appearing as the last entry in each column. Note that the number of digits carried in the computed values should be the *maximum allowed by the calculator or computer;* that is, *rounding off should not be performed until the calculation is complete.*

We now substitute into Equations 4-10, 4-11, and 4-12 and obtain

$$S_{xx} = \Sigma\, x_i^2 - (\Sigma\, x_i)^2/N = 6.90201 - (5.365)^2/5 = 1.14537$$

$$S_{yy} = \Sigma\, y_i^2 - (\Sigma\, y_i)^2/N = 36.3775 - (12.51)^2/5 = 5.07748$$

$$S_{xy} = \Sigma\, x_i y_i - \Sigma\, x_i \Sigma\, y_i/N = 15.81992 - 5.365 \times 12.51/5 = 2.39669$$

Substitution of these quantities into Equations 4-13 and 4-14 yields

TABLE 4-5 Calibration Data for a Chromatographic Method for the Determination of Isooctane in a Hydrocarbon Mixture

Mole Percent Isooctane, x_i	Peak Area y_i	x_i^2	y_i^2	$x_i y_i$
0.352	1.09	0.12390	1.1881	0.38368
0.803	1.78	0.64481	3.1684	1.42934
1.08	2.60	1.16640	6.7600	2.80800
1.38	3.03	1.90440	9.1809	4.18140
1.75	4.01	3.06250	16.0801	7.01750
5.365	12.51	6.90201	36.3775	15.81992

$$m = 2.39669/1.14537 = 2.0925 = 2.09$$

$$b = \frac{12.51}{5} - 2.0925 \times \frac{5.365}{5} = 0.2567 = 0.26$$

Thus, the equation for the least-squares line is

$$y = 2.09x + 0.26 \qquad \textbf{(4-20)}$$

Substitution into Equation 4-15 yields the standard deviation about regression:

$$s_r = \sqrt{\frac{S_{yy} - m^2 S_{xx}}{N - 2}} = \sqrt{\frac{5.07748 - (2.0925)^2 \times 1.14537}{5 - 2}}$$
$$= 0.144 = 0.14$$

and substitution into Equation 4-16 gives the standard deviation of the slope:

$$s_m = \sqrt{s_r^2/S_{xx}} = \sqrt{(0.144)^2/1.14537} = 0.13$$

Finally, we find the standard deviation of the intercept from Equation 4-17.

$$s_b = 0.144\sqrt{\frac{1}{5 - (5.365)^2/6.90201}} = 0.16$$

EXAMPLE 4-10

The calibration curve derived in Example 4-9 was used for the chromatographic determination of isooctane in a hydrocarbon mixture. A peak area of 2.65 was obtained. Calculate the mole percent of isooctane and the standard deviation for the result if the area was (a) the result of a single measurement and (b) the mean of four measurements.

In either case, substituting into Equation 4-20 and rearranging gives

$$x = \frac{y - 0.26}{2.09} = \frac{2.65 - 0.26}{2.09} = 1.14 \text{ mol } \%$$

(a) Substituting into Equation 4-18, we obtain

$$s_c = \frac{0.14}{2.09}\sqrt{\frac{1}{1} + \frac{1}{5} + \frac{(2.65 - 12.51/5)^2}{(2.09)^2 \times 1.145}} = 0.074 \text{ mol } \%$$

(b) For the mean of four measurements,

$$s_c = \frac{0.14}{2.09}\sqrt{\frac{1}{4} + \frac{1}{5} + \frac{(2.65 - 12.51/5)^2}{(2.09)^2 \times 1.145}} = 0.046 \text{ mol } \%$$

4F QUALITY ASSURANCE AND CONTROL CHARTS

When analytical methods are applied to real-world problems, the quality of results obtained by these methods as well as the performance quality of the tools and instruments used to accomplish the measurements must be evaluated constantly. We encounter three terms in discussions of measurement quality: *quality assurance, quality control,* and *quality assessment.* These terms are defined by Taylor in his authoritative text on measurement quality as follows:[5]

Quality assurance: A system of activities whose purpose is to provide to the producer or the user of a product or service the assurance that it meets the needs of users.

Quality control: The overall system of activities whose purpose is to control the quality of a product or service so that it meets the needs of users. The aim is to provide quality that is satisfactory, adequate, and dependable.

Quality assessment: The overall system of activities whose purpose is to provide assurance that the overall control job is being done effectively. It involves a continuing evaluation of the products produced and the performance of the production system.

4F-1 The Need for Quality Assurance

Chemical measurements are vital to a variety of human interests and endeavors including diagnosis and treatment of diseases, the protection of the environment, the production of safe and useful materials of commerce, and the performance of a wide range of scientific studies. The data from chemical measurements must be reliable, and there must be unequivocal evidence to prove it. Quality assurance studies are designed to provide such proof.

Two types of quality assessment measurements are encountered. The first involves the evaluation of the accuracy and precision of the methods of measurement. The second deals with the quality of a manufactured product that is being sold to the public. An example of each of these types of application follows.

An example of the need for quality assurance of measurement methods is found in the data produced by clinical instruments for the determination of blood gases. These devices must be calibrated frequently with standard samples to ensure that the results are sufficiently accurate and precise to be clinically useful. Inaccurate or imprecise measurements could lead to incorrect diagnosis, inappropriate patient care, or even death. Here, the end products of the laboratory are the measurements themselves.

As an example of quality assurance of manufactured products, consider the determination of fluoride in toothpaste (see Section 36I-5). There are at least three reasons why careful measurement and control of fluoride in toothpaste is important. First, fluoride is toxic at high concentration, and as a result, it is regulated by the Federal Government. In addition, product effectiveness often depends on the maintenance of a certain range of concentration of the active ingredient. If there is too little fluoride in toothpaste, then it may be ineffective in preventing tooth decay. If there is too much, fluoride is wasted, and in the extreme, high levels of

[5] J. K. Taylor, *Quality Assurance of Chemical Measurements,* p. 2. Chelsea, MI: Lewis, 1987.

fluoride could be dangerous. Finally, the effective formulation may be patented by the manufacturer or perhaps by one of its competitors. If the formulation is patented by a competitor, then the company may wish to maintain a fluoride concentration very close to, but not within the range covered by the patent. If the manufacturer holds the patent, it may wish to analyze samples of competitors' products to ensure that there is no patent infringement. Unless careful control is maintained over the amount of fluoride, patent infringement might occur which could lead to legal action by the offended company.

For applications such as these, quality assessment is often based upon the use of *control charts.* In the section that follows, we describe two typical applications of this useful statistical tool.

4F-2 The Application of Control Charts

A **control chart** is a sequential plot of some criterion of quality.

A control chart is a sequential plot of some quality characteristic that is important in quality assurance. The chart also shows the statistical limits of variation that are permissible for the characteristic being measured.

A Control Chart for Monitoring the Performance of an Analytical Instrument

As our first example of quality assessment let us consider how we might monitor the performance of one of the fundamental tools of the analyst, the modern analytical balance (see Section 35D). Both the accuracy and the precision of the balance can be monitored by periodically determining the mass of a standard. We then determine whether the measurements on consecutive days are within certain limits of the standard mass. These limits are called the upper control limit (UCL) and the lower control limit (LCL), and they are defined as

$$\text{UCL} = \mu + \frac{3\sigma}{\sqrt{N}}$$

$$\text{LCL} = \mu - \frac{3\sigma}{\sqrt{N}}$$

where μ is the population mean for a mass measurement, σ is the population standard deviation for the measurement, and N is the number of replicates that are to be obtained for each sample. The population mean and standard deviation for the standard mass must have been estimated from preliminary studies.

Note that the UCL and the LCL are three standard deviations on either side of the population mean and form a range within which a measured mass is expected to lie 99.7% of the time.

Figure 4-6 is a typical instrument control chart for an analytical balance. To construct this chart, mass data were collected on twenty consecutive days for a 20.000-g standard mass certified by the National Institute of Standards and Technology. To obtain the data, the balance was zeroed, the standard mass was placed on the balance pan, and the mass was determined. The entire process was repeated four times to provide a total of five replicate measurements for each day. From independent experiment, estimates of the population mean and population stan-

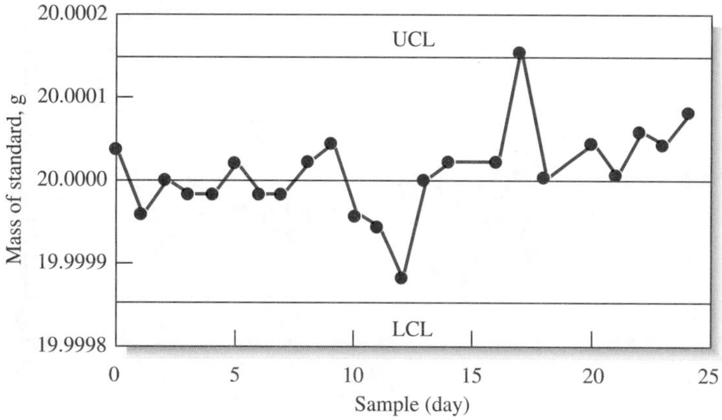

Figure 4-6
A control chart for a modern analytical balance. The data appear to fluctuate normally about the mean value except for day 17. Investigation led to the conclusion that the errant point resulted from a dirty balance pan.

dard deviation were found to be $\mu = 20.000$ g and $\sigma = 0.00012$ g, respectively. For the mean of five measurements, $3 \times 0.00012/\sqrt{5} = 0.000054$, and UCL = 20.00016 g and LCL = 19.99984 g. With these values and the mean masses for each day, we construct the control chart shown in Figure 4-6. Note that the mean mass for each day is plotted on the graph in the order in which it was determined. In this way, any systematic trends in the measurements can be easily spotted. As long as the sample mean mass remains between the LCL and the UCL, the balance is said to be in *statistical control.* When a sample mean falls outside the limits as occurred on day 17, the balance is said to be out of control. When an instrument or system goes out of control, we should then attempt to find a cause for the condition. For example, in the case of day 17, it was found that the balance had been improperly cleaned so that there was dust on the balance pan, and steps were taken to ensure that the problem did not recur. Systematic deviations from the mean should be relatively easy to spot on a control chart as we shall see in the next example.

An analytical measurement is in **statistical control** when essentially all of an extended sequence of measurements fall within the upper and lower control limits. Usually the control limits are set at $\mu \pm 3\sigma/\sqrt{N}$.

A Control Chart for Quality Assessment of a Pharmaceutical Product

As suggested previously, control charts in combination with analytical techniques may be used to monitor manufacturing processes. An interesting example of this use is found in the manufacture of both prescription and over-the-counter medications containing benzoyl peroxide, which are used for treating acne. Benzoyl peroxide is a bactericide which has been shown to be effective when applied to the skin in the form of a cream or gel containing 10% of the active ingredient. Because such substances are regulated by the Food and Drug Administration (FDA), the concentration of benzoyl peroxide must be monitored and maintained in statistical control. Benzoyl peroxide is an oxidizing agent that may be combined with an excess of iodide to produce iodine. The resulting iodine is then titrated with standard sodium thiosulfate (see Section 36G) to provide a measure of the benzoyl peroxide in the sample.

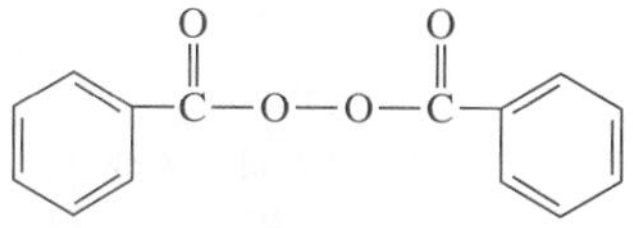

benzoyl peroxide

The control chart of Figure 4-7 shows the results of the analysis of 89 production runs measured on consecutive days of a cream containing a nominal 10% benzoyl peroxide. Each sample is represented by the mean percent benzoyl peroxide determined from the results of five titrations of different analytical samples of the cream obtained from the facility.

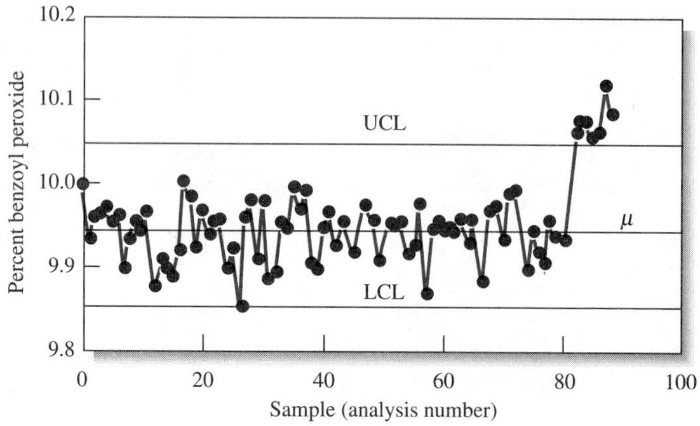

Figure 4-7
A control chart for monitoring the concentration of benzoyl peroxide in a commercial acne preparation. The manufacturing process became out of statistical control with sample 83 and exhibited a systematic change in the mean concentration.

The control chart shows that until day 83, the manufacturing process was in statistical control with normal random fluctuations in the amount of benzoyl peroxide that were within the range of the LCL and the UCL. Prior to day 83, only the results of day 26 strayed from the mean concentration of 9.94% and approached the lower control limit. On day 83, however, the system went out of control with a dramatic systematic increase above the UCL. This increase caused considerable concern at the manufacturing facility until its source was discovered and corrected. This example along with the example of the analytical balance shows how control charts provide an easy, effective means for presenting quality assurance data in a variety of situations.

These examples focus on the determination of the accuracy of devices and processes relative to accepted standards. Control charts can also be used to monitor precision. For further examples of quality assurance methodology, see the book by Taylor, and the references therein.[6]

[6]J. K. Taylor, *Quality Assurance of Chemical Measurements.* Chelsea, MI: Lewis, 1987.

4G QUESTIONS AND PROBLEMS

4-1. Consider the following sets of replicate measurements:

*A	B	*C	D	*E	F
2.4	69.94	0.0902	2.3	69.65	0.624
2.1	69.92	0.0884	2.6	69.63	0.613
2.1	69.80	0.0886	2.2	69.64	0.596
2.3		0.1000	2.4	69.21	0.607
1.5			2.9		0.582

Note that you were asked to calculate the mean and the standard deviation for each of these six data sets in Problem 3-5. Calculate the 95% confidence limit for each set of data. What do these confidence limits mean?

4-2. Calculate the 95% confidence limit for each set of data in Problem 4-1 if $s \rightarrow \sigma$ and has a value of: *set A, 0.20; set B, 0.050; *set C, 0.0070; set D, 0.50; *set E, 0.15; set F, 0.015.

4-3. The last result in each set of data in Problem 4-1 may be an outlier. Apply the Q test (95% confidence level) to determine whether or not there is a statistical basis for rejection.

***4-4.** An atomic absorption method for the determination of the amount of iron present in used jet engine oil was found, from pooling 30 triplicate analyses, to have a standard deviation $s \rightarrow \sigma = 2.4$ μg Fe/mL. Calculate the 80% and 95% confidence limits for the result 18.5 μg Fe/mL if it was based upon (a) a single analysis, (b) the mean of two analyses, (c) the mean of four analyses.

4-5. An atomic absorption method for determination of copper in fuels yielded a pooled standard deviation of $s \rightarrow \sigma =$

0.32 μg Cu/mL. The analysis of an oil from a reciprocating aircraft engine showed a copper content of 8.53 μg Cu/mL. Calculate the 90% and 99% confidence limits for the result if it was based upon (a) a single analysis, (b) the mean of 4 analyses, (c) the mean of 16 analyses.

*4-6. How many replicate measurements are needed to decrease the 95% and 99% confidence intervals for the analysis described in Problem 4-4 to ± 1.5 μg Fe/mL?

4-7. How many replicate measurements are necessary to decrease the 95% and 99% confidence intervals for the analysis described in Problem 4-5 to ± 0.2 μg Cu/mL?

*4-8. A volumetric calcium analysis on triplicate samples of the blood serum of a patient believed to be suffering from a hyperparathyroid condition produced the following data: meq Ca/L = 3.15, 3.25, 3.26. What is the 95% confidence limit for the mean of the data, assuming

(a) no prior information about the precision of the analysis?

(b) $s \rightarrow \sigma = 0.056$ meq Ca/L?

4-9. A chemist obtained the following data for percent lindane in the triplicate analysis of an insecticide preparation: 7.47, 6.98, 7.27. Calculate the 90% confidence limit for the mean of the three data, assuming that

(a) the only information about the precision of the method is the precision for the three data.

(b) on the basis of long experience with the method, it is believed that $s \rightarrow \sigma = 0.28\%$ lindane.

4-10. A standard method for the determination of tetraethyl lead (TEL) in gasoline is reported to have a standard deviation of 0.040 mL TEL per gallon. If $s \rightarrow \sigma = 0.040$, how many replicate analyses should be made in order for the mean for the analysis of a sample to be within

***(a)** ± 0.03 mL/gal of the true mean 99% of the time?

(b) ± 0.03 mL/gal of the true mean 95% of the time?

(c) ± 0.02 mL/gal of the true mean 90% of the time?

*4-11. A titrimetric method for the determination of calcium in limestone was tested by analysis of an NIST limestone containing 30.15% CaO. The mean result of four analyses was 30.26% CaO, with a standard deviation of 0.085%. By pooling data from several analyses, it was established that $s \rightarrow \sigma = 0.094\%$ CaO.

(a) Do the data indicate the presence of a systematic error at the 95% confidence level?

(b) Do the data indicate the presence of a systematic error at the 95% confidence level if no pooled value for σ was available?

4-12. In order to test the quality of the work of a commercial laboratory, duplicate analyses of a purified benzoic acid (68.8% C, 4.953% H) sample were requested. It is assumed that the relative standard deviation of the method is $s_r \rightarrow \sigma_r = 4$ ppt for carbon and 6 ppt for hydrogen. The means of the reported results are 68.5% C and 4.882% H. At the 95% confidence level, is there any indication of systematic error in either analysis?

*4-13. A prosecuting attorney in a criminal case presented as principal evidence small fragments of glass found imbedded in the coat of the accused. The attorney claimed that the fragments were identical in composition to a rare Belgian stained glass window broken during the crime. The average of triplicate analyses for five elements in the glass are shown below. On the basis of these data, does the defendant have grounds for claiming reasonable doubt as to guilt? Employ the 99% confidence level as a criterion for doubt.

Element	Concentration, ppm From clothes	From window	Standard Deviation $s \rightarrow \sigma$
As	129	119	9.5
Co	0.53	0.60	0.025
La	3.92	3.52	0.20
Sb	2.75	2.71	0.25
Th	0.61	0.73	0.043

4-14. The homogeneity of a standard chloride sample was tested by analyzing portions of the material from the top and the bottom of the container, with the following results:

% Chloride Top	Bottom
26.32	26.28
26.33	26.25
26.38	26.38
26.39	

(a) Is nonhomogeneity indicated at the 95% confidence level?

(b) Is nonhomogeneity indicated at the 95% confidence level if it is known that $s \rightarrow \sigma = 0.03\%$ Cl?

*4-15. A method for the analysis of codeine in prescription drugs yielded the following results when applied to a codeine-free blank: 0.1, − 0.2, 0.3, 0.2, 0.0, − 0.1 mg codeine. Calculate the detection limit (in terms of milligrams of codeine) at the 99% confidence level, based upon the mean of (a) two analyses, (b) four analyses, (c) six analyses.

4-16. A single alloy specimen was used to compare the results of two testing laboratories. The standard deviation s and degrees of freedom (DF) in pooled sets for four analyses are

	Element	Laboratory A s	DF	Laboratory B s	DF
*(a)	Fe	0.10	6	0.12	12
(b)	Ni	0.07	12	0.04	20
*(c)	Cr	0.05	20	0.07	6
(d)	Mn	0.020	20	0.035	6

Use the F test to determine whether the results from one laboratory are statistically more precise than those from the other.

*4-17. Apply the Q test to the following data sets to determine whether the outlying result should be retained or rejected at the 95% confidence level.
(a) 41.27, 41.61, 41.84, 41.70
(b) 7.295, 7.284, 7.388, 7.292

4-18. Apply the Q test to the following data sets to determine whether the outlying result should be retained or rejected at the 95% confidence level.
(a) 85.10, 84.62, 84.70
(b) 85.10, 84.62, 84.65, 84.70

4-19. The sulfate ion concentration in natural water can be determined by measuring the turbidity that results when an excess of $BaCl_2$ is added to a measured quantity of the sample. A turbidimeter, the instrument used for this analysis, was calibrated with a series of standard Na_2SO_4 solutions. The following data were obtained in the calibration:

mg SO_4^{2-}/L, C_x	Turbidimeter Reading, R
0.00	0.06
5.00	1.48
10.00	2.28
15.0	3.98
20.0	4.61

Assume that a linear relationship exists between the instrument reading and concentration.
(a) Plot the data and draw a straight line through the points by eye.
*(b) Derive a least-squares equation for the relationship between the variables.
(c) Compare the straight line from the relationship derived in (b) with that in (a).
*(d) Calculate the concentration of sulfate in a sample yielding a turbidimeter reading of 3.67. Calculate the absolute standard deviation of the result and the coefficient of variation.
*(e) Repeat the calculations in (d) assuming that the 3.67 was a mean of six turbidimeter readings.

4-20. The following data were obtained in calibrating a calcium ion electrode for the determination of pCa. A linear relationship between the potential E and pCa is known to exist.

pCa	E, mV
5.00	−53.8
4.00	−27.7
3.00	+2.7
2.00	+31.9
1.00	+65.1

(a) Plot the data and draw a line through the points by eye.
(b) Derive a least-squares expression for the best straight line through the points. Plot this line.
(c) Calculate the pCa of a serum solution in which the electrode potential was 20.3 mV. Calculate the absolute and relative standard deviations for pCa if the result was from a single voltage measurement.
(d) Calculate the absolute and relative standard deviations for pCa if the millivolt reading in (c) was the mean of two replicate measurements. Repeat the calculation based upon the mean of eight measurements.

4-21. The following are relative peak areas for chromatograms of standard solutions of methyl vinyl ketone (MVK).

Concn MVK, mmol/L	Relative Peak Area
0.500	3.76
1.50	9.16
2.50	15.03
3.50	20.42
4.50	25.33
5.50	31.97

(a) Derive a least-squares expression assuming the variables bear a linear relationship to one another.
(b) Plot the least-squares line as well as the experimental points.
(c) A sample containing MVK yielded a relative peak area of 6.3. Calculate the concentration of MVK in the solution.
(d) Assume that the result in (c) represents a single measurement as well as the mean of four measurements. Calculate the respective absolute and relative standard deviations.
(e) Repeat the calculations in (c) and (d) for a sample that gave a peak area of 27.5.

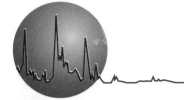

CHAPTER

5

Gravimetric Methods of Analysis

Gravimetric methods, which are based upon the measurement of mass, are of two major types.[1] In *precipitation methods,* the analyte is converted to a sparingly soluble precipitate. This precipitate is then filtered, washed free of impurities, and converted to a product of known composition by suitable heat treatment. This product is then weighed. For example, in a precipitation method for determining calcium in natural waters recommended by the Association of Official Analytical Chemists, an excess of oxalic acid, $H_2C_2O_4$, is added to a carefully measured volume of the sample. The addition of ammonia causes essentially all of the calcium in the sample to precipitate as calcium oxalate. The reaction is

$$Ca^{2+}(aq) + C_2O_4^{2-}(aq) \longrightarrow CaC_2O_4(s)$$

The precipitate is collected in a weighed filtering crucible, dried, and ignited at a red heat. This process converts the precipitate entirely to calcium oxide. The reaction is

$$CaC_2O_4(s) \longrightarrow CaO(s) + CO(g) + CO_2(g)$$

The crucible and precipitate are cooled, weighed, and the mass of calcium oxide determined by difference. The calcium content of the sample is then computed as shown in the various examples in Section 5A-3.

In *volatilization methods,* the analyte or its decomposition products are volatilized at a suitable temperature. The volatile product is then collected and weighed, or alternatively the mass of the product is determined indirectly from the loss in mass of the sample. An example of a gravimetric volatilization procedure is the determination of the sodium hydrogen carbonate content of antacid tablets. Here, a weighed sample of the finely ground tablets is treated with dilute sulfuric acid to convert the sodium hydrogen carbonate to carbon dioxide:

$$NaHCO_3(aq) + H_2SO_4(aq) \longrightarrow CO_2(g) + H_2O(l) + NaHSO_4(aq)$$

As shown in Figure 5-1 this reaction is carried out in a flask that is connected to a weighed tube that contains an absorbent (see Section 5D-5 for a description of the

Gravimetric methods are quantitative methods that are based upon determining the mass of a pure compound to which the analyte is chemically related.

Theodore W. Richards (1868–1928) and his graduate students at Harvard University developed or refined many of the techniques of gravimetric analysis involving silver and chlorine. These techniques were used to determine the atomic weights of 25 of the elements by preparing pure samples of the chlorides of the elements, decomposing known weights of these compounds, and determining their chloride content by gravimetric methods. For this work, Richards became the first American to receive the Nobel Prize in Chemistry in 1914.

Gravimetric methods of analysis are based upon mass measurements with an analytical balance, an instrument that yields highly accurate and precise data. In fact, if you perform a gravimetric chloride analysis in the laboratory, you may make some of the most accurate and precise measurements of your life.

[1]For an extensive treatment of gravimetric methods, see C. L. Rulfs, in *Treatise on Analytical Chemistry,* I. M. Kolthoff and P. J. Elving, Eds., Part I, Vol. 11, Chapter 13. New York: Wiley, 1975.

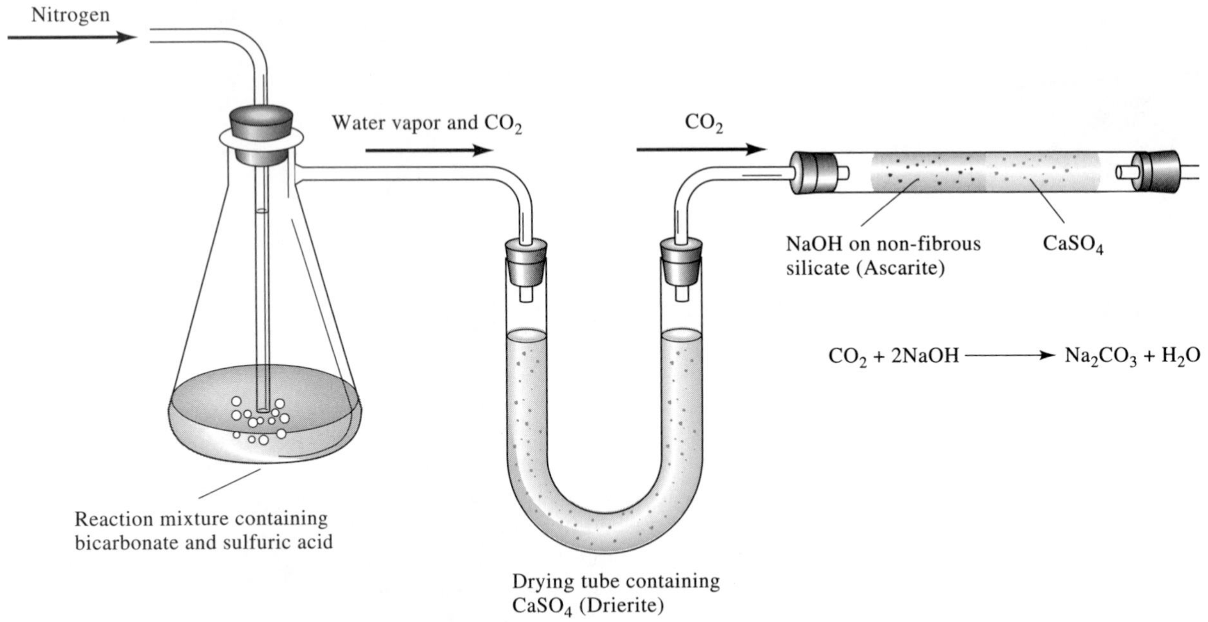

Figure 5-1
Apparatus for determining the sodium hydrogen carbonate content of antacid tablets by a gravimetric volatilization procedure.

absorbent) that retains the carbon dioxide selectively as it is removed from the solution by heating. As shown in Example 5-6, the difference in mass of the tube before and after absorption is used to calculate the amount of sodium hydrogen carbonate.

Mass is an invariant measure of the amount of matter. **Weight** is the force of gravitational attraction between that matter and the earth.

FEATURE 5-1
The Distinction Between Mass and Weight

It is important to understand the difference between mass and weight. *Mass* is an invariant measure of the amount of matter in an object. *Weight* is the force of attraction between an object and its surroundings, principally the earth. Because gravitational attraction varies with geographic location, the weight of an object depends upon where you weigh it. For example, a crucible *weighs* less in Denver than in Atlantic City (both cities are at approximately the same latitude) because the attractive force between crucible and earth is smaller at the higher altitude of Denver. Similarly, the crucible *weighs* more in Seattle than in Panama (both cities are at sea level) because the earth is somewhat flattened at the poles, and the force of attraction increases measurably with latitude. The *mass* of the crucible, however, remains constant regardless of where you measure it.

Weight and mass are related by the familiar expression

$$W = mg$$

where W is the weight of an object, m is its mass, and g is the acceleration due to gravity.

A chemical analysis is always based upon mass so that the results will not depend upon locality. A balance is used to compare the weight of an object with the weight of one or more standard masses. Because g affects both unknown and known equally, the mass of the object is identical to the standard masses with which it is compared.

The distinction between mass and weight is often lost in common usage, and the process of comparing masses is ordinarily called *weighing*. In addition, the objects of known mass as well as the results of weighing are frequently called *weights*. Always bear in mind, however, that analytical data are based upon mass rather than weight. Therefore, throughout this text we will use mass rather than weight to describe the amounts of substances or objects. On the other hand, for lack of a better word, we will use the verb "weigh" to describe the act of determining the mass of an object. Also, we will often use the term "weight" to describe the standard masses that are used in weighing.

Although as a matter of habit we speak of weighing and weights, we determine *mass* in the laboratory and use mass in chemical calculations.

5A GRAVIMETRIC CALCULATIONS

In this section, we describe how the raw laboratory data obtained in a gravimetric analysis are converted to concentrations of analytes in samples. By way of preparation, we will review briefly some elementary material on units of measurement and chemical stoichiometry.

5A-1 Some Important Units of Measurements

SI is the acronym for the French "Système International d'Unités."

SI Units

Scientists throughout the world have adopted a standardized system of units known as the *International System of Units* (SI). This system is based upon the seven fundamental base units shown in Table 5-1. Numerous other useful units, such as volts, hertz, joules, and coulombs, are derived from these base units.

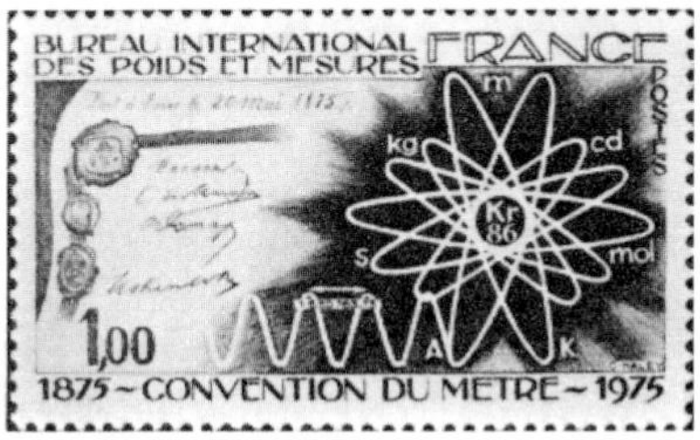

This French postage stamp commemorates the Meter Convention of 1875. The stamp shows some signatures from the Treaty of the Meter, the seven SI units, and the definition of the meter as 1,650,763.73 wavelengths of the orange light given off by Kr-86. Since the stamp was issued in 1975, the meter has been redefined as the distance that light travels in a vacuum during 1/299,792,458 of a second.

TABLE 5-1 SI Base Units

Physical Quantity	Name of Unit	Abbreviation
Mass	kilogram	kg
Length	meter	m
Time	second	s
Temperature	kelvin	K
Amount of substance	mole	mol
Electric current	ampere	A
Luminous intensity	candela	cd

The **ångström unit Å** is a non-SI unit of length that is widely used to express the wavelength of very short radiation such as X-rays (1 Å = 0.1 nm = 10^{-10} m). Thus, typical X-radiation lies in the range of 0.1 to 10 Å.

One megabyte in a computer memory is, in fact, 2^{20} bytes or 1.049×10^6 bytes, which is approximately 1 million bytes.

In order to express small or large measured quantities in terms of a few simple digits, a variety of prefixes are used with the base units in Table 5-1 as well as other derived units. As shown in Table 5-2, these prefixes multiply the unit by various powers of ten. For example, the wavelength of yellow radiation used for determining sodium by flame photometry is about 5.9×10^{-7} m, which can be expressed more compactly as 590 nm (nanometers); the volume of a liquid injected onto a chromatographic column is often roughly 50×10^{-6} L or 50 μL (microliters); and the amount of memory on some computer hard disks is about 200 million bytes or 200 Mbytes (megabytes).

In this course, we often measure the mass of chemical species with a balance. For such measurements, metric units of kilograms (kg), grams (g), milligrams (mg), or micrograms (μg) are used. Volumes of liquids are measured in units of liters (L), milliliters (mL), and sometimes microliters (μL). The liter is an SI unit that is *defined* as exactly 10^{-3} m^3. The milliliter is defined as 10^{-6} m^3 or 1 cm^3.

The Mole

A **mole** of a chemical species is 6.022×10^{23} atoms, molecules, ions, electrons, or ion pairs.

The **molar mass** $\mathcal{M}$ of a species is the mass of 1 mol or 6.022×10^{23} particles of the species.

The terms "formula weight" and "molecular weight" are often found in the literature as synonymous for "molar mass."

The mole (abbreviated mol) is the SI base unit for the amount of a chemical species. It is always associated with a chemical formula and refers to Avogadro's number (6.022×10^{23}) of particles represented by that formula. The *molar mass* ($\mathcal{M}$) of a substance is the mass in grams of 1 mol of that substance. Molar masses of compounds are derived by summing the masses of all the atoms appearing in a chemical formula. For example, the molar mass of formaldehyde, CH_2O, is

$$\mathcal{M}_{CH_2O} = \frac{1\ \cancel{\text{mol C}}}{\text{mol CH}_2\text{O}} \times \frac{12.0\ \text{g C}}{\cancel{\text{mol C}}} + \frac{2\ \cancel{\text{mol H}}}{\text{mol CH}_2\text{O}} \times \frac{1.0\ \text{g H}}{\cancel{\text{mol H}}} + \frac{1\ \cancel{\text{mol O}}}{\text{mol CH}_2\text{O}} \times \frac{16.0\ \text{g O}}{\cancel{\text{mol O}}} = 30.0\ \text{g/mol CH}_2\text{O}$$

and that of glucose ($C_6H_{12}O_6$) is

$$\mathcal{M}_{C_6H_{12}O_6} = \frac{6\ \cancel{\text{mol C}}}{\text{mol C}_6\text{H}_{12}\text{O}_6} \times \frac{12.0\ \text{g C}}{\cancel{\text{mol C}}} + \frac{12\ \cancel{\text{mol H}}}{\text{mol C}_6\text{H}_{12}\text{O}_6} \times \frac{1.0\ \text{g H}}{\cancel{\text{mol H}}} + \frac{6\ \cancel{\text{mol O}}}{\text{mol C}_6\text{H}_{12}\text{O}_6} \times \frac{16.0\ \text{g O}}{\cancel{\text{mol O}}} = 180.0\ \text{g/mol C}_6\text{H}_{12}\text{O}_6$$

Thus, 1 mol of formaldehyde has a mass of 30.0 g and 1 mol of glucose has a mass of 180.0 g.

TABLE 5-2
Prefixes for Units

Prefix	Abbreviation	Multiplier
giga-	G	10^9
mega-	M	10^6
kilo-	k	10^3
deci-	d	10^{-1}
centi-	c	10^{-2}
milli-	m	10^{-3}
micro-	μ	10^{-6}
nano-	n	10^{-9}
pico-	p	10^{-12}
femto-	f	10^{-15}
atto-	a	10^{-18}

FEATURE 5-2
Atomic Mass Units

You should keep in mind that the masses for the elements listed in the table inside the back cover of this text are *relative masses* in terms of *atomic mass units* (amu), or *daltons.* The atomic mass unit is based upon a relative scale in which the reference is the ^{12}C carbon isotope, which is *assigned* a mass of exactly 12 amu. Thus, the amu is by definition 1/12 of the mass of one neutral ^{12}C atom. The *molar mass* of ^{12}C is then defined as the mass in *grams* of 6.022×10^{23} atoms of the carbon-12 isotope, or exactly 12 g. Likewise, the molar mass of any other element is the mass in grams of 6.022×10^{23} atoms of that element and is numerically equal to the atomic mass of the element in amu units. Thus, the atomic mass of naturally occurring oxygen is 15.9994 amu; its molar mass is 15.9994 g.

The Millimole

Sometimes it is more convenient to make calculations with millimoles (mmol) rather than moles, where the millimole is 1/1000 of a mole. The mass in grams of a millimole, the millimolar mass $m\mathcal{M}$, is likewise 1/1000 of the molar mass.

1 mmol = 10^{-3} mol

Calculating the Number of Moles or Millimoles of a Substance

The two examples that follow illustrate how the number of moles or millimoles of a species can be derived from its mass in grams or from the mass of a chemically related species.

The number of moles of a species X is given by

$$\text{amount X} = \frac{\text{g X}}{\text{g X/mol X}} = \cancel{\text{g X}} \times \frac{\text{mol X}}{\cancel{\text{g X}}}$$

The number of millimoles is given by

$$\text{amount} = \frac{\text{g X}}{\text{g X/mmol X}} = \cancel{\text{g X}} \times \frac{\text{mmol X}}{\cancel{\text{g X}}}$$

EXAMPLE 5-1

How many moles and millimoles are contained in 2.00 g of pure benzoic acid (122.1 g/mol)?

If we use HBz to represent benzoic acid, we can write that 1 mol HBz has a mass of 122.1 g. Thus,

$$\text{amount of HBz} = 2.00\ \cancel{\text{g HBz}} \times \frac{1\ \text{mol HBz}}{122.1\ \cancel{\text{g HBz}}} = 0.0164\ \text{mol HBz}$$

To obtain the number of millimoles, we divide by the millimolar mass (0.1221 g/mol). That is,

$$\text{amount HBz} = 2.00\ \cancel{\text{g HBz}} \times \frac{1\ \text{mmol HBz}}{0.1221\ \cancel{\text{g HBz}}} = 16.4\ \text{mmol HBz}$$

EXAMPLE 5-2

How many grams of Na^+ (22.99 g/mol) are contained in 25.0 g of Na_2SO_4 (142.0 g/mol)?

The chemical formula tells us that 1 mol of Na_2SO_4 contains 2 mol of Na^+. That is,

$$\text{amount Na}^+ = \text{no.}\ \cancel{\text{mol Na}_2\text{SO}_4} \times \frac{2\ \text{mol Na}^+}{\cancel{\text{mol Na}_2\text{SO}_4}}$$

To obtain the number of moles Na_2SO_4 we proceed as in Example 5-1:

$$\text{amount Na}_2\text{SO}_4 = 25.0\ \cancel{\text{g Na}_2\text{SO}_4} \times \frac{1\ \text{mol Na}_2\text{SO}_4}{142.0\ \cancel{\text{g Na}_2\text{SO}_4}}$$

When you make calculations of this kind, include all units as we do throughout this chapter. This practice often reveals errors you have made in setting up equations.

Combining this equation with the first leads to

$$\text{amount Na}^+ = 25.0\ \text{g}\ \cancel{\text{Na}_2\text{SO}_4} \times \frac{1\ \cancel{\text{mol Na}_2\text{SO}_4}}{142.0\ \cancel{\text{g Na}_2\text{SO}_4}} \times \frac{2\ \text{mol Na}^+}{\cancel{\text{mol Na}_2\text{SO}_4}}$$

To obtain the mass of sodium in 25.0 g Na_2SO_4, we multiply the number of moles Na^+ by the molar mass of Na^+, or 22.99 g. That is,

$$\text{mass Na}^+ = \text{no.}\ \cancel{\text{mol Na}^+} \times \frac{22.99\ \text{g Na}^+}{\cancel{\text{mol Na}^+}}$$

Substituting the previous equation gives the number of grams of Na^+.

$$\text{mass Na}^+ = 25.0\ \text{g}\ \cancel{\text{Na}_2\text{SO}_4} \times \frac{1\ \cancel{\text{mol Na}_2\text{SO}_4}}{142.0\ \cancel{\text{g Na}_2\text{SO}_4}} \times \frac{2\ \cancel{\text{mol Na}^+}}{\cancel{\text{mol Na}_2\text{SO}_4}} \times \frac{22.99\ \text{g Na}^+}{\cancel{\text{mol Na}^+}} = 8.10\ \text{g Na}^+$$

5A-2 Chemical Stoichiometry

Stoichiometry describes the quantitative relationship among the amounts of reactants and products.

Stoichiometry is defined as the mass relationship among reacting chemical species. This section provides a review of stoichiometry.

Empirical Formulas and Molecular Formulas

An *empirical formula* gives the simplest whole number ratio of atoms in a chemical compound. In contrast, a *molecular formula* specifies the number of atoms in a molecule. Two or more substances may have the same empirical formula but different molecular formulas. For example, CH_2O is both the empirical and the molecular formula for formaldehyde; it is also the empirical formula for such diverse substances as acetic acid, $C_2H_4O_2$, glyceraldehyde, $C_3H_6O_3$, and glucose, $C_6H_{12}O_6$, as well as more than 50 other substances containing 6 or fewer carbon atoms. The empirical formula is obtained from the percent composition of a compound. The molecular formula requires, in addition, a knowledge of the molar mass of the species.

A *structural formula* provides additional information. For example, the chemically different ethanol and dimethyl ether share the same molecular formula, C_2H_6O. Their structural formulas, C_2H_5OH and CH_3OCH_3, reveal structural differences between these compounds that are not discernible in the molecular formula that they share.

Stoichiometric Calculations

A balanced chemical equation is a statement of the combining ratios, or stoichiometry, (in units of moles) between reacting substances and their products. Thus, the equation

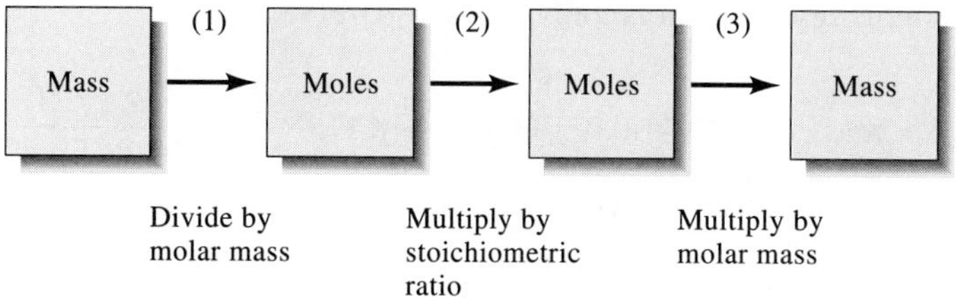

Figure 5-2 Flow diagram for making stoichiometric calculations.

$$2\text{NaI}(aq) + \text{Pb(NO}_3)_2(aq) = \text{PbI}_2(s) + 2\text{NaNO}_3(aq)$$

indicates that 2 mol of aqueous sodium iodide combines with 1 mol of aqueous lead nitrate to produce 1 mol of solid lead iodide and 2 mol of aqueous sodium nitrate.[2]

Often the physical state of substances appearing in equations is indicated by the letters (g), (l), (s), or (aq), which refer to gaseous, liquid, solid, and aqueous solution states, respectively.

Example 5-3 demonstrates how the masses in grams of reactants and products in a chemical reaction are related. As shown in Figure 5-2, a calculation of this type is a three-step process involving (1) transformation of the known mass of a substance in grams to a corresponding number of moles, (2) multiplication by a factor that accounts for the stoichiometry, and (3) reconversion of the data in moles back to the metric units called for in the answer.

EXAMPLE 5-3

(a) What mass of $AgNO_3$ (169.9 g/mol) is needed to convert 2.33 g of Na_2CO_3 (106.0 g/mol) to Ag_2CO_3? (b) What mass of Ag_2CO_3 (275.7 g/mol) will be formed?

(a) $Na_2CO_3(aq) + 2AgNO_3(aq) \rightarrow Ag_2CO_3(s) + 2NaNO_3(aq)$

Step 1. $$\text{no. mol Na}_2\text{CO}_3 = 2.33\ \cancel{\text{g Na}_2\text{CO}_3} \times \frac{1\text{ mol Na}_2\text{CO}_3}{106.0\ \cancel{\text{g Na}_2\text{CO}_3}}$$
$$= 0.02198\text{ mol Na}_2\text{CO}_3$$

Step 2. The balanced equation reveals that

$$\text{no. mol AgNO}_3 = 0.02198\ \cancel{\text{mol Na}_2\text{CO}_3} \times \frac{2\text{ mol AgNO}_3}{1\ \cancel{\text{mol Na}_2\text{CO}_3}}$$
$$= 0.04396\text{ mol AgNO}_3$$

Here the stoichiometric factor is (2 mol $AgNO_3$)/(1 mol Na_2CO_3).

Step 3. $$\text{mass AgNO}_3 = 0.04396\ \cancel{\text{mol AgNO}_3} \times \frac{169.9\text{ g AgNO}_3}{\cancel{\text{mol AgNO}_3}}$$
$$= 7.47\text{ g AgNO}_3$$

---

[2]Here it is advantageous to show the reaction in terms of chemical compounds. If we wish to focus on reacting species, the net ionic equation is preferable:

$$2\text{I}^-(aq) + \text{Pb}^{2+}(aq) \rightleftharpoons \text{PbI}_2(s)$$

(b) no. mol Ag_2CO_3 = no. mol Na_2CO_3 = 0.02198 mol

$$\text{mass } Ag_2CO_3 = 0.02198 \cancel{\text{mol } Ag_2CO_3} \times \frac{275.7 \text{ g } Ag_2CO_3}{\cancel{\text{mol } Ag_2CO_3}}$$
$$= 6.06 \text{ g } Ag_2CO_3$$

5A-3 Calculation of Results from Gravimetric Data

The results of a gravimetric analysis are generally computed from two experimental measurements: the mass of sample and the mass of a product of known composition. When the product is the analyte X, the concentration is given by the equation

$$\text{percent X} = \frac{\text{mass of X}}{\text{mass of sample}} \times 100\% \qquad (5\text{-}1)$$

where X represents the analyte. When the product is not the analyte, the numerator in Equation 5-1 is obtained by means of a calculation similar to that shown in Example 5-3.

EXAMPLE 5-4

A 0.3516-g sample of a commercial phosphate detergent was ignited at a red heat to destroy the organic matter. The residue was then taken up in hot HCl, which converted the P to H_3PO_4. The phosphate was precipitated as $MgNH_4PO_4 \cdot 6H_2O$ by addition of Mg^{2+} followed by aqueous NH_3. After being filtered and washed, the precipitate was converted to $Mg_2P_2O_7$ (222.57 g/mol) by ignition at 1000°C. This residue weighed 0.2161 g. Calculate the percent P (30.974 g/mol) in the sample.

The mass of analyte is

$$\text{mass P} = \underbrace{\cancel{\text{g } Mg_2P_2O_7}}_{\text{Metric quantity}} \times \underbrace{\frac{1 \cancel{\text{mol } Mg_2P_2O_7}}{222.57 \cancel{\text{g } Mg_2P_2O_7}}}_{\text{Conversion factor}} \times \underbrace{\frac{2 \cancel{\text{mol P}}}{1 \cancel{\text{mol } Mg_2P_2O_7}}}_{\text{Stoichiometric factor}} \times \underbrace{\frac{30.974 \text{ g P}}{\cancel{\text{mol P}}}}_{\text{Conversion factor}}$$

$$= 0.2161 \times \frac{2 \times 30.974}{222.57} = 0.060147 \approx 0.06015 \text{ g P}$$

Substituting into Equation 5-1 yields the percent analyte:

$$\text{percent P} = \frac{0.2161 \times \dfrac{2 \times 30.974}{222.57}}{0.3516} \times 100\% = 17.107 \approx 17.11\%$$

Note that we carry guard digits for the molar masses of P and $Mg_2P_2O_7$, and then round at the end.

EXAMPLE 5-5

Magnetite is a mineral having the formula Fe_3O_4 or $FeO \cdot Fe_2O_3$. A 1.1324-g sample of a magnetite ore was dissolved in concentrated HCl to give a solution that contained a mixture of Fe^{2+} and Fe^{3+}. Nitric acid was added and the solution was boiled for a few minutes, which converted all of the iron to Fe^{3+}. The Fe^{3+} was then precipitated as $Fe_2O_3 \cdot xH_2O$ by addition of NH_3. After filtration and washing, the residue was ignited at a high temperature to give 0.5394 g of pure Fe_2O_3 (159.69 g/mol). Calculate (a) the percent Fe (55.847 g/mol) and (b) the percent Fe_3O_4 (231.54 g/mol) in the sample.

See *Mathcad Applications for Analytical Chemistry,* **pp. 4–8.**

(a)
$$\text{mass Fe} = \cancel{\text{g Fe}_2\text{O}_3} \times \frac{1\ \cancel{\text{mol Fe}_2\text{O}_3}}{159.69\ \cancel{\text{g Fe}_2\text{O}_3}} \times \frac{2\ \cancel{\text{mol Fe}}}{1\ \cancel{\text{mol Fe}_2\text{O}_3}} \times \frac{55.847\ \text{g Fe}}{\cancel{\text{mol Fe}}}$$

$$= 0.5394 \times \frac{2 \times 55.847}{159.69}\ \text{g Fe}$$

$$\text{percent Fe} = \frac{0.5394 \times \dfrac{2 \times 55.847}{159.69}\ \text{g}}{1.1324\ \text{g}} \times 100\% = 33.317 \approx 33.32\%$$

(b) In this calculation, we assume that 3 mol of Fe_2O_3 are formed from 2 mol of Fe_3O_4:

$$3\ \text{Fe}_2\text{O}_3 \longrightarrow 2\ \text{Fe}_3\text{O}_4 + \tfrac{1}{2}\ \text{O}_2$$

Then,

$$\text{mass Fe}_3\text{O}_4 = \cancel{\text{g Fe}_2\text{O}_3} \times \frac{1\ \cancel{\text{mol Fe}_2\text{O}_3}}{159.69\ \cancel{\text{g Fe}_2\text{O}_3}} \times \frac{2\ \cancel{\text{mol Fe}_3\text{O}_4}}{3\ \cancel{\text{mol Fe}_2\text{O}_3}}$$
$$\times \frac{231.54\ \text{g Fe}_3\text{O}_4}{\cancel{\text{mol Fe}_3\text{O}_4}}$$

$$\text{percent Fe}_3\text{O}_4 = \frac{0.5394 \times \dfrac{2 \times 231.54}{3 \times 159.69}}{1.1324} \times 100\% = 46.043 \approx 46.04\%$$

Note again that we carried five figures throughout the calculation and then rounded the answer to four figures.

Note that the calculations in Examples 5-4 and 5-5 are similar in that the mass of analyte was obtained by multiplying the mass of the final product by a constant made up of the two conversion factors and the stoichiometric relationship between the analyte and the product weighed. This constant is sometimes called the *gravimetric factor* (GF). Thus, the gravimetric factor in Example 5-4 is

$$\text{GF} = \frac{2 \times \mathcal{M}_\text{P}}{\mathcal{M}_{\text{Mg}_2\text{P}_2\text{O}_7}} = \frac{2 \times 30.974}{222.57} = 0.27833$$

Similarly, the two gravimetric factors in Example 5-5 are

(a)
$$\text{GF} = \frac{2 \times \mathcal{M}_\text{Fe}}{\mathcal{M}_{\text{Fe}_2\text{O}_3}} = \frac{2 \times 55.847}{159.69} = 0.69944$$

$$\text{(b) GF} = \frac{2 \times \mathcal{M}_{Fe_3O_4}}{3 \times \mathcal{M}_{Fe_2O_3}} = \frac{2 \times 231.54}{3 \times 159.69} = 0.9662$$

A general definition for the gravimetric factor is

$$\text{GF} = \frac{a}{b} \times \frac{\mathcal{M}_{\text{sought}}}{\mathcal{M}_{\text{weighed}}}$$

where a and b are small whole numbers with values such that the number of moles in the numerator and the denominator are chemically equivalent.

A general equation for calculating the percent X from gravimetric data is

$$\text{percent X} = \frac{\text{mass product} \times \text{GF}}{\text{mass sample}} \times 100\% \qquad (5\text{-}2)$$

Jöns Jacob Berzelius (1779–1848), considered the leading chemist of his time, developed much of the apparatus and many of the techniques of nineteenth-century analytical chemistry. Examples include the use of ashless filter paper in gravimetry, the use of hydrofluoric acid to decompose silicates, and the use of the metric system in weight determinations. He performed thousands of analyses of pure compounds to determine the atomic weights of most of the elements known then. Berzelius also developed our present system of symbols for elements and compounds.

The symbol ≡ means equivalent to.

EXAMPLE 5-6

At elevated temperatures, $NaHCO_3$ is converted quantitatively to Na_2CO_3:

$$2NaHCO_3(s) \longrightarrow Na_2CO_3(s) + CO_2(g) + H_2O(g)$$

Ignition of a 0.3592-g sample of an antacid tablet containing $NaHCO_3$ and nonvolatile impurities yielded a residue weighing 0.2362 g. Calculate the percent purity of the sample.

The difference in mass before and after ignition represents the amount of CO_2 and H_2O evolved from the $NaHCO_3$ in the sample. The equation for the reaction indicates that

$$2 \text{ mol } NaHCO_3 \equiv 1 \text{ mol } CO_2 + 1 \text{ mol } H_2O$$

Thus,

$$\text{percent } NaHCO_3 = \frac{(0.3592 - 0.2362) \times \dfrac{2 \times \mathcal{M}_{NaHCO_3}}{\mathcal{M}_{CO_2} + \mathcal{M}_{H_2O}}}{0.3592} \times 100\%$$

$$= \frac{0.1230 \times \dfrac{2 \times 84.01}{44.01 + 18.015}}{0.3592} \times 100\% = 92.76 \approx 92.8\%$$

Note (1) that the denominator of the gravimetric factor in this example is the sum of the molar masses of the two volatile products and (2) that the combined mass of these products forms the basis for this analysis.

EXAMPLE 5-7

A 0.2795-g sample of an organic mixture containing only $C_6H_6Cl_6$ (290.83 g/mol) and $C_{14}H_9Cl_5$ (354.49 g/mol) was burned in a stream of oxygen in a quartz tube. The products (CO_2, H_2O, and HCl) were passed through a solution of $NaHCO_3$. After acidification, the chloride in this solution yielded 0.7161 g of AgCl (143.22 g/mol). Calculate the percentage of each halogen compound in the sample.

See *Mathcad Applications for Analytical Chemistry,* **pp. 8–12.**

There are two unknowns, and we must therefore develop two independent equations that can be solved simultaneously. One equation is

$$\text{mass } C_6H_6Cl_6 + \text{mass } C_{14}H_9Cl_5 = 0.2795 \text{ g}$$

A second equation is

$$\text{mass AgCl from } C_6H_6Cl_6 + \text{mass AgCl from } C_{14}H_9Cl_5 = 0.7161 \text{ g}$$

After we insert the appropriate gravimetric factors, the second equation becomes

$$\text{mass } C_6H_6Cl_6 \times \frac{6 \times \mathcal{M}_{AgCl}}{\mathcal{M}_{C_6H_6Cl_6}} + \text{mass } C_{14}H_9Cl_5 \times \frac{5 \times \mathcal{M}_{AgCl}}{\mathcal{M}_{C_{14}H_9Cl_5}}$$
$$= 0.7161 \text{ g}$$

Substituting numerical values for the molar masses leads to

$$\text{mass } C_6H_6Cl_6 \times \frac{6 \times 143.22}{290.83} + \text{mass } C_{14}H_9Cl_5 \times \frac{5 \times 143.22}{354.49} = 0.7161$$
$$\text{mass } C_6H_6Cl_6 \times 2.9568 + \text{mass } C_{14}H_9Cl_5 \times 2.0215 = 0.7161$$

The first equation can be rewritten as

$$\text{mass } C_{14}H_9Cl_6 = 0.1795 - \text{mass } C_6H_6Cl_6$$

Substitution of this relationship into the previous equation leads to

$$2.9568 \times \text{mass } C_6H_6Cl_6 + 2.02150(0.2795 - \text{mass } C_6H_6Cl_6) = 0.7161$$

Thus,

$$\text{mass } C_6H_6Cl_6 = 0.16154 \text{ g}$$
$$\text{percent } C_6H_6Cl_6 = \frac{0.16154}{0.2795} \times 100\% = 57.80\%$$
$$\text{percent } C_{14}H_9Cl_5 = 100\% - 57.80\% = 42.20\%$$

5B PROPERTIES OF PRECIPITATES AND PRECIPITATING AGENTS

An example of a selective reagent is $AgNO_3$. The only common ions that it precipitates from acidic solution are Cl^-, Br^-, I^-, and SCN^-. Dimethylglyoxime, which is discussed in Section 5D-3, is a specific reagent that precipitates only Ni^{2+} from alkaline solutions.

Ideally, a gravimetric precipitating agent should react *specifically*, or if not that, *selectively* with the analyte. Specific reagents, which are rare, react only with a single chemical species. Selective reagents, which are more common, react with only a limited number of species. In addition to specificity or selectivity, the ideal precipitating reagent would react with the analyte to give a product that is

1. readily filtered and washed free of contaminants;
2. of sufficiently low solubility so that no significant loss of the analyte occurs during filtration and washing;
3. unreactive with constituents of the atmosphere;
4. of known composition after it is dried or, if necessary, ignited.

Few if any reagents produce precipitates that possess all these desirable properties perfectly.

The variables that influence solubility (the second property in the above list) are discussed in Chapter 7. In this section we consider methods for obtaining pure and easily filtered solids of known composition.

5B-1 Particle Size and Filterability of Precipitates[3]

Precipitates made up of large particles are generally desirable in gravimetric work because large particles are easy to filter and wash free of impurities. In addition, such precipitates are usually purer than are finely divided precipitates.

Factors That Determine the Particle Size of Precipitates

A **colloid** is a solid made up of particles having diameters that are less than 10^{-4} cm.

In diffuse light, colloidal suspensions may be perfectly clear and appear to contain no solid. The presence of the second phase can be detected, however, by shining the beam of a flashlight into the solution. Because particles of colloidal dimensions scatter visible radiation, the path of the beam through the solution can be seen by the eye. This phenomenon is called the *Tyndall effect* (see color plate 6).

The particle size of solids formed by precipitation varies enormously. At one extreme are *colloidal suspensions,* whose tiny particles are invisible to the naked eye (10^{-7} to 10^{-4} cm in diameter). Colloidal particles show no tendency to settle from solution, nor are they easily filtered. At the other extreme are particles with dimensions on the order of tenths of a millimeter or greater. The temporary dispersion of such particles in the liquid phase is called a *crystalline suspension.* The particles of a crystalline suspension tend to settle spontaneously and are readily filtered.

Scientists have studied precipitate formation for many years, but the mechanism of the process is still not fully understood. It is certain, however, that the particle size of a precipitate is influenced by such experimental variables as precipitate solubility, temperature, reactant concentrations, and rate at which reactants are mixed. The net effect of these variables can be accounted for, at least qualitatively, by assuming that the particle size is related to a single property of the system called its *relative supersaturation,* where

[3]For a more detailed treatment of precipitates, see H. A. Laitinen and W. E. Harris, *Chemical Analysis,* 2nd ed., Chapters 8 and 9. New York: McGraw-Hill, 1975; A. E. Nielsen, in *Treatise on Analytical Chemistry,* 2nd ed., I. M. Kolthoff and P. J. Elving, Eds., Part I, Vol. 3, Chapter 27. New York: Wiley, 1983.

$$\text{relative supersaturation} = \frac{Q - S}{S} \qquad (5\text{-}3)$$

In this equation, Q is the concentration of a species at any instant and S is its equilibrium solubility.

Precipitation reactions are frequently so slow that a degree of supersaturation is likely even with dropwise addition of the reagent. Experimental evidence indicates that the particle size of a precipitate varies inversely with the average relative supersaturation during the time the reagent is being introduced. Thus, when $(Q - S)/S$ is large, the precipitate tends to be colloidal; when $(Q - S)$ is small, a crystalline solid is more likely.

Mechanism of Precipitate Formation

The effect of relative supersaturation on particle size can be explained if we assume that precipitates form by two different pathways, namely by *nucleation* and by *particle growth*. The particle size of a freshly formed precipitate is determined by which of these is the faster.

In nucleation, a few ions, atoms, or molecules (perhaps as few as four or five) come together to form a stable solid. Often, these nuclei form on the surface of suspended solid contaminants, such as dust particles. Further precipitation then involves a competition between additional nucleation and growth on existing nuclei (particle growth). If nucleation predominates, a precipitate containing a large number of small particles results; if growth predominates, a smaller number of larger particles is produced.

The rate of nucleation is believed to increase enormously with increasing relative supersaturation. In contrast, the rate of particle growth is only moderately enhanced under these circumstances. Thus, when a precipitate is formed at high relative supersaturation, nucleation is the major precipitation mechanism, and a large number of small particles are formed. At low relative supersaturations, on the other hand, the rate of particle growth tends to predominate, and deposition of solid on existing particles occurs to the exclusion of further nucleation; a crystalline suspension results.

Experimental Control of Particle Size

Experimental variables that minimize supersaturation and thus lead to crystalline precipitates include elevated temperatures to increase the solubility of the precipitate (S in Equation 5-3), dilute solutions (to minimize Q), and slow addition of the precipitating agent with good stirring. The last two measures also minimize the concentration of the solute (Q) at any given instant.

Larger particles can also be obtained by control of acidity, provided the solubility of the precipitate depends upon pH. For example, large, easily filtered crystals of calcium oxalate are obtained by forming the bulk of the precipitate in a mildly acidic environment in which the salt is moderately soluble. The precipitation is then completed by slowly adding aqueous ammonia until the acidity is sufficiently low that substantially all of the calcium oxalate is precipitated. The

The particles of colloidal suspensions are not easily filtered. To trap these particles, the pore size of the filtering medium must be so small that filtrations take too long. With suitable treatment, however, the individual colloidal particles can be made to stick together to give a filterable mass.

Equation 5-3 is known as the Von Weimarn equation in recognition of the scientist who proposed it in 1925.

> A **supersaturated solution** is an unstable solution that contains more solutes than a saturated solution. With time, supersaturation is relieved by precipitation of the excess solute.

To increase the particle size of a precipitate, minimize the relative supersaturation during precipitate formation.

> **Nucleation** is a process in which a minimum number of atoms, ions, or molecules join together to give a stable solid.

Precipitates form by nucleation and by particle growth. If nucleation predominates, a large number of very fine particles results; when particle growth predominates, a smaller number of larger particles are obtained.

additional precipitate produced during this step forms on the solid particles formed in the first step.

Unfortunately, many precipitates cannot be formed as crystals under practical laboratory conditions. A colloidal solid is generally encountered when a precipitate has such a low solubility that S in Equation 5-3 always remains negligible relative to Q. The relative supersaturation thus remains enormous throughout precipitate formation, and a colloidal suspension results. For example, under conditions feasible for an analysis, the hydrous oxides of iron(III), aluminum, and chromium(III) and the sulfides of most heavy-metal ions form only as colloids because of their very low solubilities.[4]

Precipitates having very low solubilities, such as many sulfides and hydrous oxides, generally form as colloids.

5B-2 Colloidal Precipitates

Colloidal suspensions are often stable for indefinite periods and are not directly usable for gravimetric analysis because their particles are too small to be readily filtered. Fortunately, the stability of most suspensions of this kind can be decreased by heating, by stirring, and by adding an electrolyte. These measures bind the individual colloidal particles together to give an amorphous mass that settles out of solution and is filterable. The process of converting a colloidal suspension into a filterable solid is called *coagulation* or *agglomeration.*

Coagulation of Colloids

Colloidal suspensions are stable because the particles are either all positively charged or all negatively charged and thus repel one another. This charge results from cations or anions that are bound to the surface of the particles. The process by which ions are retained *on the surface of a solid* is known as *adsorption.* We can readily demonstrate that colloidal particles are charged by observing their migration when placed in an electrical field.

Adsorption is a process in which a substance (gas, liquid, or solid) is held *on the surface* of a solid. In contrast, *absorption* involves retention of a substance *within* the pores of a solid.

The tendency of ions to be adsorbed on an ionic solid surface originates in the normal bonding forces that are responsible for crystal growth. For example, a silver ion at the surface of a silver chloride particle has a partially unsatisfied bonding capacity for anions because of its surface location. Negative ions are attracted to this site by the same forces that hold chloride ions in the silver chloride lattice. Chloride ions at the surface of the solid exert an analogous attraction for cations dissolved in the solvent.

The kind and number of ions retained on the surface of a colloidal particle depends, in a complex way, on several variables. For a suspension produced in the course of a gravimetric analysis, however, the species adsorbed, and hence the charge on the particles, is readily predicted because lattice ions are generally more strongly adsorbed than any other ions. For example, when sodium chloride is first added to a solution containing silver nitrate, the colloidal particles of silver chloride formed are positively charged, as shown in Figure 5-3. This charge is due to adsorption of some of the excess silver ions in the medium. The charge on the particles becomes negative, however, when enough sodium chloride has been added to provide an excess of chloride ions. Now, the adsorbed species is primar-

The charge on a colloidal particle formed in a gravimetric analysis is determined by the charge of the lattice ion that is in excess.

[4] Silver chloride illustrates that the relative supersaturation concept is imperfect. This compound ordinarily forms as a colloid, and yet its molar solubility is not significantly different from that of other compounds, such as $BaSO_4$, which generally form as crystals.

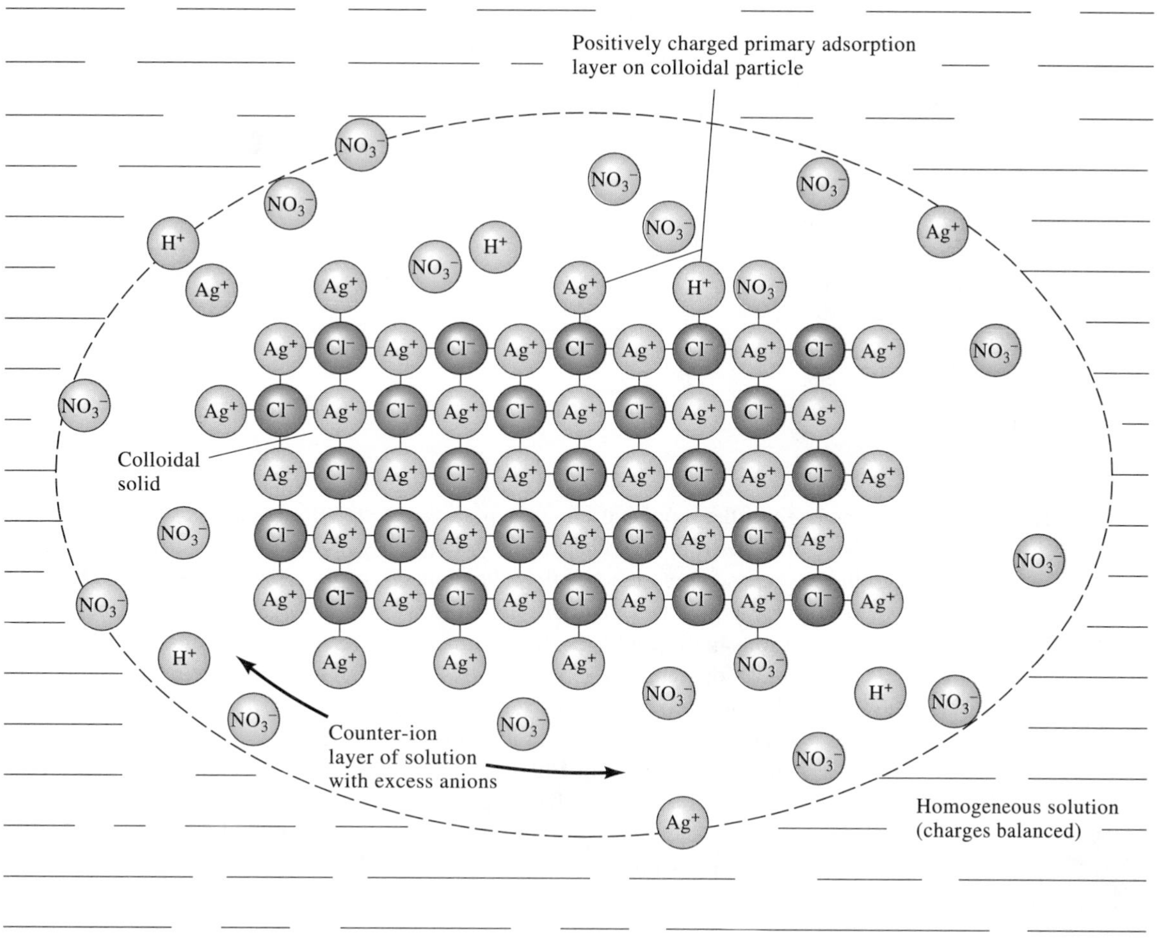

Figure 5-3
A colloidal silver chloride particle suspended in a solution of silver nitrate.

ily chloride ions. The surface charge is at a minimum when the supernatant liquid contains an excess of neither ion.

The extent of adsorption, and thus the charge on a given particle, increases rapidly as the concentration of the lattice ion becomes greater. Eventually, however, the surface of the particles becomes covered with the adsorbed ions, and the charge then becomes constant and independent of concentration.

Figure 5-3 shows a colloidal silver chloride particle in a solution that contains an excess of silver nitrate. Attached directly to the solid surface is the *primary adsorption layer,* which consists mainly of adsorbed silver ions. Surrounding the charged particle is a layer of solution, called *the counter-ion layer,* which contains an excess of negative ions (principally nitrate). The primarily adsorbed silver ions and the negative counter-ion layer constitute an *electric double layer* that stabilizes a colloidal suspension by preventing individual particles from coming close enough to one another to agglomerate.

The **electric double layer** of a colloid consists of a layer of charge adsorbed on the surface of the particles and a layer with a net opposite charge in the solution surrounding the particles.

Figure 5-4a shows the effective charge on two silver chloride particles. The upper curve is for a particle in a solution that contains a reasonably large excess of silver nitrate, whereas the lower curve depicts a particle in a solution that has a much lower silver nitrate content. The effective charge can be thought of as a

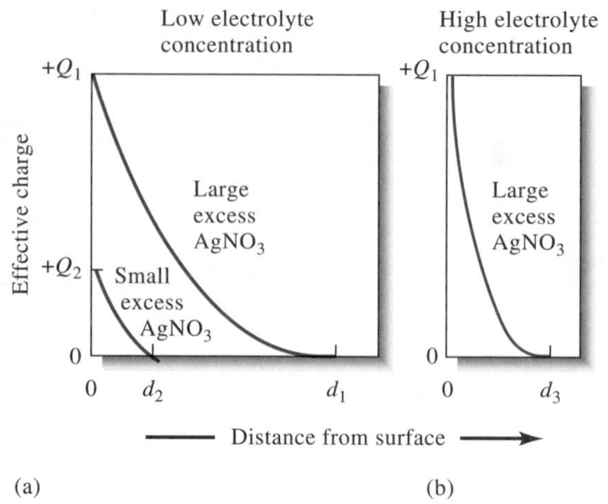

Figure 5-4
Effect of $AgNO_3$ and electrolyte concentration on the thickness of the double layer surrounding a colloidal AgCl particle in a solution containing excess of $AgNO_3$.

measure of the repulsive force the particle exerts on like particles in the solution. Note that the effective charge falls off rapidly as the distance from the surface increases, and approaches zero at the points d_1 or d_2. These decreases in effective charge (in both cases positive) are caused by the negative charge of the excess counter ions in the double layer surrounding each particle. At points d_1 and d_2, the number of counter ions in the layers is approximately equal to the number of primarily adsorbed ions on the surfaces of the particles, so the effective charge of the particles approaches zero at this point.

The upper schematic in Figure 5-5 depicts two silver chloride particles and their counter-ion layers as they approach one another in the concentrated silver nitrate just considered. Note that the effective charge on the particles prevents them from approaching one another more closely than about $2d_1$—a distance that is too great for coagulation to occur. As shown in the lower part of Figure 5-5 in the more dilute silver nitrate solution, the two particles can approach within $2d_2$ of one another. Ultimately, as the concentration of silver nitrate is further decreased, the distance between particles becomes small enough for the forces of agglomeration to take effect, and a coagulated precipitate begins to appear.

Colloidal suspensions can often be coagulated by heating, by stirring, and by adding an electrolyte.

Coagulation of a colloidal suspension can often be brought about by a short period of heating, particularly if accompanied by stirring. Heating decreases the number of adsorbed ions and thus the thickness d_i of the double layer. The particles may also gain enough kinetic energy at the higher temperature to overcome the barrier to close approach posed by the double layer.

An even more effective way to coagulate a colloid is to increase the electrolyte concentration of the solution. If we add a suitable ionic compound to a colloidal suspension, the concentration of counter ions increases in the vicinity of each particle. As a result, the volume of solution that contains sufficient counter ions to balance the charge of the primary adsorption layer decreases. The net effect of adding an electrolyte is thus a shrinkage of the counter-ion layer as shown in Figure 5-4b. The particles can then approach one another more closely and agglomerate.

Peptization of Colloids

Peptization is a process by which a coagulated colloid returns to its dispersed state.

Peptization refers to the process by which a coagulated colloid reverts to its original dispersed state. When a coagulated colloid is washed, some of the electrolyte responsible for its coagulation is leached from the internal liquid in contact with the solid particles. Removal of this electrolyte has the effect of increasing the volume of the counter-ion layer. The repulsive forces responsible for the original colloidal state are then reestablished, and particles detach themselves from the coagulated mass. The washings become cloudy as the freshly dispersed particles pass through the filter.

The chemist is thus faced with a dilemma in working with coagulated colloids. On the one hand, washing is needed to minimize contamination; on the other, there is the risk of losses resulting from peptization if pure water is used. The problem is commonly resolved by washing the precipitate with a solution containing an electrolyte that volatilizes during the subsequent drying or ignition step. For example, silver chloride is ordinarily washed with a dilute solution of nitric acid. While the precipitate undoubtedly becomes contaminated with the acid, no harm results because the nitric acid is volatilized during the ensuing drying step.

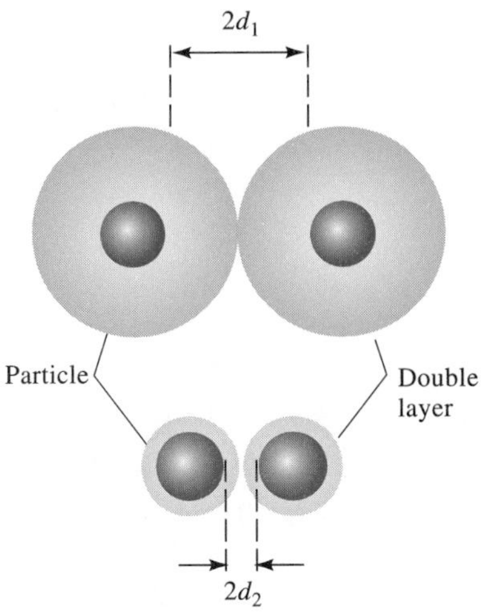

Figure 5-5
The electric double layer of a colloid consists of a layer of charge adsorbed on the surface of the particles and a layer of opposite charge in the solution surrounding the particles.

Practical Treatment of Colloidal Precipitates

Colloids are best precipitated from hot, stirred solutions containing sufficient electrolyte to ensure coagulation. The filterability of a coagulated colloid frequently improves if it is allowed to stand for an hour or more in contact with the hot solution from which it was formed. During this process, which is known as *digestion,* weakly bound water appears to be lost from the precipitate; the result is a denser mass that is easier to filter.

5B-3 Crystalline Precipitates

Digestion is a process in which a precipitate is heated for an hour or more in the solution from which it was formed (the *mother liquor*).

Mother liquor is the solution from which a precipitate forms.

Crystalline precipitates are generally more easily filtered and purified than are coagulated colloids. In addition, the size of individual crystalline particles, and thus their filterability, can be controlled to a degree.

Methods of Improving Particle Size and Filterability

The particle size of crystalline solids can often be improved significantly by minimizing Q and/or maximizing S in Equation 5-3. Minimization of Q is generally accomplished by using dilute solutions and adding the precipitating reagent slowly and with good mixing. Often S is increased by precipitating from hot solution or by adjusting the pH of the precipitation medium.

Digestion of crystalline precipitates (without stirring) for some time after formation frequently yields a purer, more filterable product. The improvement in filterability undoubtedly results from the dissolution and recrystallization that occur continuously and at an enhanced rate at elevated temperatures. Recrystallization apparently results in bridging between adjacent particles, a process that yields larger and more easily filtered crystalline aggregates. This view is supported by the observation that little improvement in filtering characteristics occurs if the mixture is stirred during digestion.

Digestion improves the purity and filterability of both colloidal and crystalline precipitates.

5B-4 Coprecipitation

Coprecipitation is a process in which *normally soluble* compounds are carried out of solution by a precipitate.

Coprecipitation is a phenomenon in which *otherwise soluble* compounds are removed from solution during precipitate formation. It is important to understand that the solution is *not* saturated with the coprecipitated species. Moreover, contamination of a precipitate by a second substance whose solubility product has been exceeded *does not constitute coprecipitation.*

There are four types of coprecipitation: *surface adsorption, mixed-crystal formation, occlusion,* and *mechanical entrapment.*[5] Surface adsorption and mixed-crystal formation are equilibrium processes, whereas occlusion and mechanical entrapment arise from the kinetics of crystal growth.

FEATURE 5-3

Specific Surface Area of Colloids

Specific surface area is defined as the surface area per unit mass of solid and is ordinarily expressed in terms of square centimeters per gram. For a given mass of solid, the specific surface area increases dramatically as particle size decreases, and it becomes enormous for colloids. For example, the solid cube shown in Figure 5-6, which has dimensions of 1 cm on a side, has a surface area of 6 cm^2. If this cube weighs 2 g, its specific surface area is 6 cm^2/2g = 3 cm^2/g. This cube could be divided into 1000 cubes each being 0.1 cm on a side. The surface area of each face of these cubes is now 0.1 cm $\times$ 0.1 cm = 0.01 cm^2 and the total area for the 6 faces of the cube is 0.06 cm^2. Because there are 1000 of these cubes, the total surface area for the 2 g of solid is now 60 cm^2; the specific surface area is then 30 cm^2/g. Continuing in this way we find that the specific surface area becomes 300 cm^2/g when we have 10^6 cubes that are 0.01 cm on a side. The particle size of a typical crystalline suspension lies in the region of 0.1 and 0.01 cm, so a typical crystalline precipitate has a specific surface area between 30 and 300 cm^2/g. Contrast these figures with that for 2 g of a colloid made up of 10^{18} particles having dimensions of 10^{-6} cm. Here, the specific area is 3×10^6 cm^2/g, which converts to something over 3000 ft^2/g. Thus, 1 g of a colloidal suspension has a surface area that is equivalent to the floor area of a good-sized home.

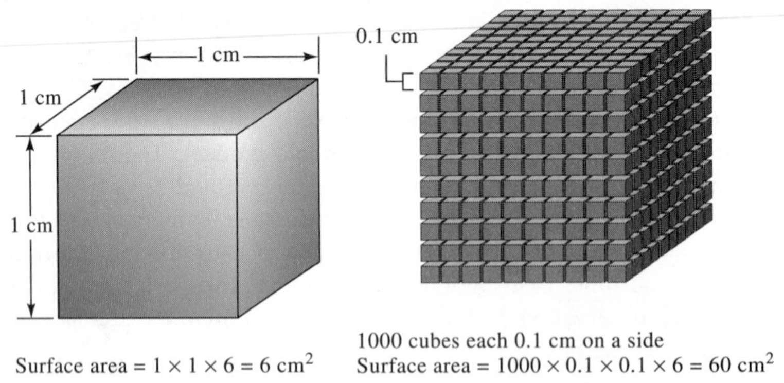

Figure 5-6
Increase in surface area per unit mass with decrease in particle size.

[5]Several systems of classification of coprecipitation phenomena have been suggested. We follow the simple system proposed by A. E. Nielsen, in *Treatise on Analytical Chemistry,* 2nd ed., I. M. Kolthoff and P. J. Elving, Eds., Part I, Vol. 3, p. 333. New York: Wiley, 1983.

Surface Adsorption

Adsorption is a common source of coprecipitation that is likely to cause significant contamination of precipitates with large specific surface areas—that is, coagulated colloids (see Feature 5-3 for the definition of "specific area"). Although adsorption does occur in crystalline solids, its effects on purity are usually undetectable because of the relatively small specific surface area of these solids.

Coagulation of a colloid does not significantly decrease the amount of adsorption because the coagulated solid still contains large internal surface areas that remain exposed to the solvent (Figure 5-7). The coprecipitated contaminant on the coagulated colloid consists of the lattice ion originally adsorbed on the surface before coagulation and the counter ion of opposite charge held in the film of solution immediately adjacent to the particle. *The net effect of surface adsorption is therefore the carrying down of an otherwise soluble compound as a surface contaminant.* For example, the coagulated silver chloride formed in the gravimetric determination of chloride ion is contaminated with primarily adsorbed silver ions along with nitrate or other anions in the counter-ion layer. As a consequence, silver nitrate, a normally soluble compound, is coprecipitated with the silver chloride.

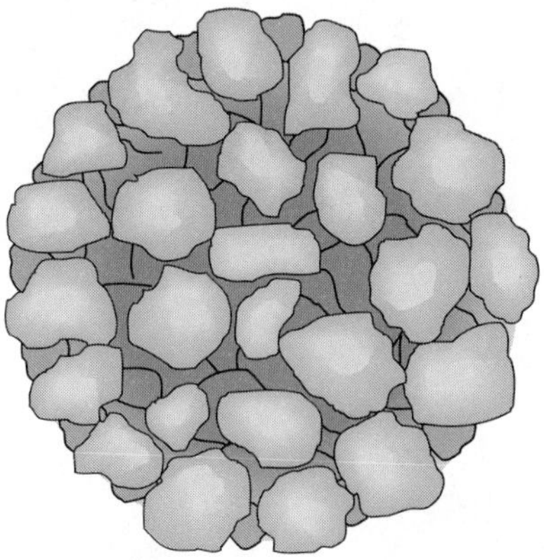

Figure 5-7
A coagulated colloid. Note the extensive *internal* surface area exposed to solvent.

Methods for Minimizing Adsorbed Impurities on Colloids The purity of many coagulated colloids is improved by digestion. During this process, water is expelled from the solid to give a denser mass that has a smaller specific surface area for adsorption.

Adsorption is often the major source of contamination in coagulated colloids but of no significance in crystalline precipitates.

Washing a coagulated colloid with a solution containing a volatile electrolyte may also be helpful because any nonvolatile electrolyte added earlier to cause coagulation is displaced by the volatile species. Washing generally does not remove many primarily adsorbed ions because the attraction between these ions and the surface of the solid is too strong. Exchange occurs, however, between existing *counter ions* and ions in the wash liquid. For example, in the determination of silver by precipitation with chloride ion, the primarily adsorbed species is chloride. Washing with an acidic solution converts the counter-ion layer largely to hydrogen ions, so both chloride and hydrogen ions are retained in the double layer. Volatile HCl is then given off when the precipitate is dried.

Adsorption is a process in which a normally soluble compound is carried out of solution on the surface of a coagulated colloid. This compound consists of the primarily adsorbed lattice ion and an ion of opposite charge from the counter-ion layer.

Regardless of the method of treatment, a coagulated colloid is always contaminated to some degree, even after extensive washing. The error introduced into the analysis from this source can be as low as 1 to 2 ppt, as in the coprecipitation of silver nitrate on silver chloride. On the other hand, coprecipitation of heavy-metal hydroxides on the hydrous oxides of trivalent iron or aluminum may introduce errors amounting to several percent, which is generally intolerable.

Surface adsorption does not usually affect the purity of a crystalline precipitate noticeably because the specific surface area of the solid is small.

Reprecipitation A drastic but effective way to minimize the effects of adsorption is *reprecipitation,* or *double precipitation.* Here, the filtered solid is redissolved and reprecipitated. The first precipitate ordinarily carries down only a fraction of the contaminant present in the original solvent. Thus, the solution containing the redissolved precipitate has a significantly lower contaminant concentration than the original, and even less adsorption occurs during the second precipitation. Reprecipitation adds substantially to the time required for an analysis but is often necessary for such precipitates as the hydrous oxides of iron(III) and aluminum, which have extraordinary tendencies to adsorb the hydroxides of heavy-metal cations, such as zinc, cadmium, and manganese.

Calcium oxalate is often freed from coprecipitated magnesium oxalate by reprecipitation. Here the precipitated calcium oxalate is dissolved in dilute sulfuric acid. Ammonium oxalate is added to the acid solution followed by ammonia, until a second precipitate of calcium oxalate forms. The new precipitate is largely free of magnesium because the concentration of this ion in the second solution is far less than it was in the first.

Mixed-Crystal Formation

Mixed-crystal formation is a type of coprecipitation in which a contaminant ion replaces an ion in the lattice of a crystal.

In mixed-crystal formation, one of the ions in the crystal lattice of a solid is replaced by an ion of another element. For this exchange to occur, it is necessary that the two ions have the same charge and that their sizes differ by no more than about 5%. Furthermore, the two salts must belong to the same crystal class. For example, barium sulfate formed by adding barium chloride to a solution containing sulfate, lead, and acetate ions is found to be severely contaminated by lead sulfate even though acetate ions normally prevent precipitation of lead sulfate by complexing the lead. Here, lead ions replace some of the barium ions in the barium sulfate crystals. Other examples of coprecipitation by mixed-crystal formation include $MgKPO_4$ in $MgNH_4PO_4$, $SrSO_4$ in $BaSO_4$, and MnS in CdS.

The extent of mixed-crystal contamination is governed by the law of mass action and increases as the ratio of contaminant to analyte concentration increases. Mixed-crystal formation is a particularly troublesome type of coprecipitation because little can be done about it when certain combinations of ions are present in a sample matrix. This problem is encountered with both colloidal suspensions and crystalline precipitates. When mixed-crystal formation occurs, separation of the interfering ion may have to be carried out before the final precipitation step. Alternatively, a different precipitating reagent that does not give mixed crystals with the ions in question may be used.

Occlusion and Mechanical Entrapment

Occlusion is a type of coprecipitation in which a compound is trapped within a pocket formed during rapid crystal growth.

When a crystal is growing rapidly during precipitate formation, foreign ions in the counter-ion layer may become trapped, or *occluded,* within the growing crystal. Because supersaturation, and thus growth rate, decrease as a precipitation progresses, the amount of occluded material is greatest in that part of a crystal that forms first.

Mechanical entrapment occurs when crystals lie close together during growth. Here, several crystals grow together and in so doing trap a portion of the solution in a tiny pocket.

Mixed-crystal formation may occur in both colloidal and crystalline precipitates, whereas occlusion and mechanical entrapment are confined to crystalline precipitates.

Both occlusion and mechanical entrapment are at a minimum when the rate of precipitate formation is low—that is, under conditions of low supersaturation. In addition, digestion is often remarkably helpful in reducing these types of coprecipitation. Undoubtedly, the rapid solution and reprecipitation that goes on at the elevated temperature of digestion opens up the pockets and allows the impurities to escape into the solution.

Coprecipitation Errors

Coprecipitation can cause either negative or positive errors.

Coprecipitated impurities may cause either negative or positive errors in an analysis. If the contaminant is not a compound of the ion being determined, positive errors always result. Thus, a positive error is observed whenever colloidal silver chloride adsorbs silver nitrate during a chloride analysis. In contrast, when the contaminant does contain the ion being determined, either positive or negative errors may be observed. For example, in the determination of barium by precipitation as barium sulfate, occlusion of other barium salts occurs. If the occluded contaminant is barium nitrate, a positive error is observed because this compound has a larger formula mass than the barium sulfate that would have formed had no coprecipitation occurred. If barium chloride is the contaminant, the error is negative because its molar mass is less than that of the sulfate salt.

5B-5 Precipitation from Homogeneous Solution

Precipitation from homogeneous solution is a technique in which a precipitating agent is generated in a solution of the analyte by a slow chemical reaction.[6] Local reagent excesses do not occur because the precipitating agent appears gradually and homogeneously throughout the solution and reacts immediately with the analyte. As a result, the relative supersaturation is kept low during the entire precipitation. In general, homogeneously formed precipitates, both colloidal and crystalline, are better suited to analysis than are solids formed by direct addition of a precipitating reagent.

Urea is often used for the homogeneous generation of hydroxide ion. The reaction can be expressed by the equation

$$(NH_2)_2CO + 3H_2O \longrightarrow CO_2 + 2NH_4^+ + 2OH^-$$

This reaction proceeds slowly just below 100°C, and a 1- to 2-hr heating period is needed to complete a typical precipitation. Urea is particularly valuable for the precipitation of hydrous oxides or basic salts. For example, hydrous oxides of iron(III) and aluminum formed by direct addition of base are bulky and gelatinous masses that are heavily contaminated and difficult to filter. In contrast, when these same products are produced by homogeneous generation of hydroxide ion, they are dense, readily filtered, and have considerably higher purity. Figure 5-8 shows

Figure 5-8
Hydrous aluminum oxide formed by homogeneous generation of hydroxide ion (left), and direct addition of base (right).

[6] For a general reference on this technique, see L. Gordon, M. L. Salutsky, and H. H. Willard, *Precipitation from Homogeneous Solution.* New York: Wiley, 1959.

TABLE 5-3 Methods for the Homogeneous Generation of Precipitating Agents

Precipitating Agent	Reagent	Generation Reaction	Elements Precipitated
OH^-	Urea	$(NH_2)_2CO + 3H_2O \rightarrow CO_2 + 2NH_4^+ + 2OH^-$	Al, Ga, Th, Bi, Fe, Sn
PO_4^{3-}	Trimethyl phosphate	$(CH_3O)_3PO + 3H_2O \rightarrow 3CH_3OH + H_3PO_4$	Zr, Hf
$C_2O_4^{2-}$	Ethyl oxalate	$(C_2H_5)_2C_2O_4 + 2H_2O \rightarrow 2C_2H_5OH + H_2C_2O_4$	Mg, Zn, Ca
SO_4^{2-}	Dimethyl sulfate	$(CH_3O)_2SO_2 + 4H_2O \rightarrow 2CH_3OH + SO_4^{2-} + 2H_3O^+$	Ba, Ca, Sr, Pb
CO_3^{2-}	Trichloroacetic acid	$Cl_3CCOOH + 2OH^- \rightarrow CHCl_3 + CO_3^{2-} + H_2O$	La, Ba, Ra
H_2S	Thioacetamide*	$CH_3CSNH_2 + H_2O \rightarrow CH_3CONH_2 + H_2S$	Sb, Mo, Cu, Cd
DMG†	Biacetyl + hydroxylamine	$CH_3COCOCH_3 + 2H_2NOH \rightarrow DMG + 2H_2O$	Ni
HOQ‡	8-Acetoxyquinoline§	$CH_3COOQ + H_2O \rightarrow CH_3COOH + HOQ$	Al, U, Mg, Zn

*
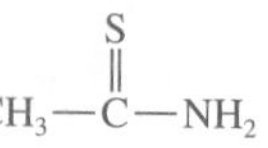

‡HOQ = 8-Hydroxyquinoline =
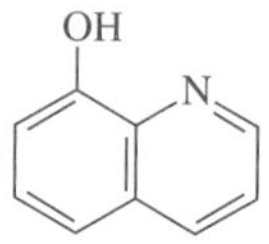

†DMG = Dimethylglyoxime =
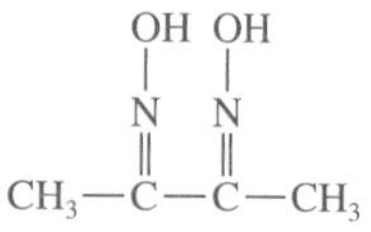

§
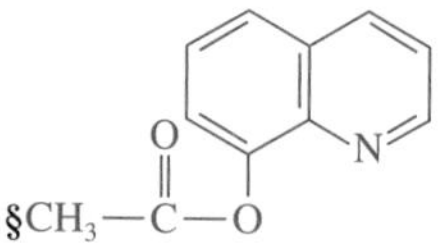

hydrous oxide precipitates of aluminum formed by direct addition of base and by homogeneous precipitation with urea. Similar pictures for hydrous oxides of iron(III) are shown in color plate 7. Homogeneous precipitation of crystalline precipitates also results in marked increases in crystal size as well as improvements in purity.

Homogeneous precipitation is a process in which a precipitate is formed by slow generation of a precipitating reagent homogeneously throughout a solution.

Representative methods based upon precipitation by homogeneously generated reagents are given in Table 5-3.

5C DRYING AND IGNITION OF PRECIPITATES

After filtration, a gravimetric precipitate is heated until its mass becomes constant. Heating removes the solvent and any volatile species carried down with the precipitate. Some precipitates are also ignited to decompose the solid and form a compound of known composition. This new compound is often called the *weighing form.*

The temperature required to dehydrate a precipitate completely may be as low as 100°C or as high as 1000°C.

The temperature required to produce a suitable product varies from precipitate to precipitate. Figure 5-9 shows mass loss as a function of temperature for several common analytical precipitates. These data were obtained with an automatic thermobalance,[7] an instrument that records the mass of a substance continuously as its temperature is increased at a constant rate in a furnace (see Figure 5-10). Heating three of the precipitates—silver chloride, barium sulfate, and aluminum oxide—simply causes removal of water and perhaps volatile contaminants. Note the vastly different temperatures required to produce an anhydrous precipitate of

[7]For descriptions of thermobalances, see W. W. Wendlandt, *Thermal Methods of Analysis,* 2nd ed. New York: Wiley, 1986.

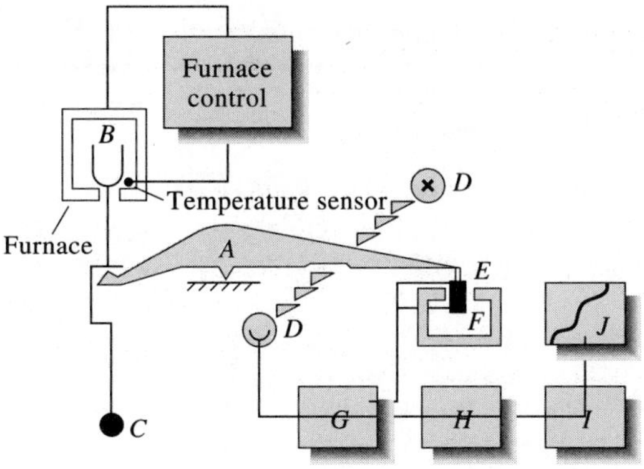

Figure 5-10
Schematic of a thermobalance: *A*: beam; *B*: sample cup and holder; *C*: counterweight; *D*: lamp and photodiodes; *E*: coil; *F*: magnet; *G*: control amplifier; *H*: tare calculator; *I*: amplifier; and *J*: recorder. (Courtesy of Mettler Instrument Corp., Hightstown, NJ.)

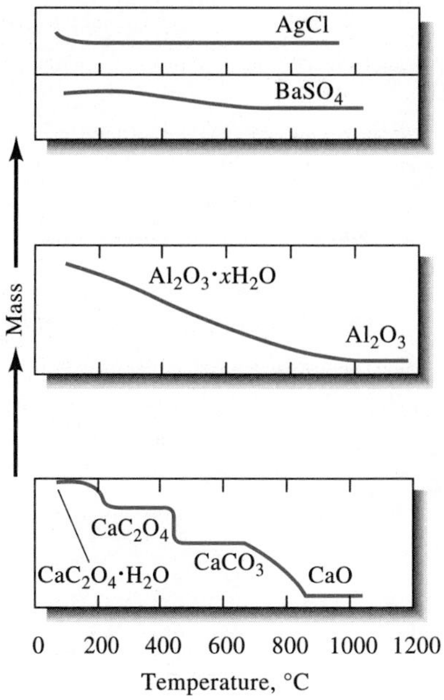

Figure 5-9
Effect of temperature on precipitate mass.

constant mass. Moisture is completely removed from silver chloride above 110°C, but dehydration of aluminum oxide is not complete until a temperature greater than 1000°C is achieved. It is of interest to note that aluminum oxide formed homogeneously with urea can be completely dehydrated at about 650°C.

The thermal curve for calcium oxalate is considerably more complex than the others in Figure 5-9. Below about 135°C, unbound water is eliminated to give the monohydrate $CaC_2O_4 \cdot H_2O$. This compound is then converted to the anhydrous oxalate CaC_2O_4 at 225°C. The abrupt change in mass at about 450°C signals the decomposition of calcium oxalate to calcium carbonate and carbon monoxide. The final step in the curve depicts the conversion of the carbonate to calcium oxide and carbon dioxide. It is evident that the compound finally weighed in a gravimetric calcium determination based upon precipitation as oxalate is highly dependent upon the ignition temperature.

5D APPLICATION OF GRAVIMETRIC METHODS

Gravimetric methods have been developed for most inorganic anions and cations, as well as for such neutral species as water, sulfur dioxide, carbon dioxide, and iodine. A variety of organic substances can also be readily determined gravimetrically. Examples include lactose in milk products, salicylates in drug preparations, phenolphthalein in laxatives, nicotine in pesticides, cholesterol in cereals, and benzaldehyde in almond extracts. Indeed, gravimetric methods are among the most widely applicable of all analytical procedures.

Gravimetric methods do not require a calibration or standardization step (as do all other analytical procedures except coulometry) because the results are calculated directly from the experimental data and molar masses. Thus, when only one or two samples are to be analyzed, a gravimetric procedure may be the method of choice because it involves less time and effort than a procedure that requires preparation of standards and calibration.

5D-1 Inorganic Precipitating Agents

Table 5-4 lists common inorganic precipitating agents. These reagents typically form slightly soluble salts or hydrous oxides with the analyte. As you can see from the many entries for each reagent, most inorganic reagents are not very selective.

TABLE 5-4 Some Inorganic Precipitating Agents*

Precipitating Agent	Element Precipitated†
$NH_3(aq)$	**Be** (BeO), **Al** (Al_2O_3), **Sc** (Sc_2O_3), Cr (Cr_2O_3),‡ **Fe** (Fe_2O_3), Ga (Ga_2O_3), Zr (ZrO_2), **In** (In_2O_3), Sn (SnO_2), U (U_3O_8)
H_2S	Cu (CuO),‡ **Zn** (ZnO, or $ZnSO_4$), **Ge** (GeO_2), As (As_2O_3, or As_2O_5), Mo (MoO_3), Sn (SnO_2),‡ Sb (Sb_2O_3, or Sb_2O_5), Bi (Bi_2S_3)
$(NH_4)_2S$	Hg (HgS), Co (Co_3O_4)
$(NH_4)_2HPO_4$	**Mg** ($Mg_2P_2O_7$), Al ($AlPO_4$), Mn ($Mn_2P_2O_7$), Zn ($Zn_2P_2O_7$), Zr ($Zr_2P_2O_7$), Cd ($Cd_2P_2O_7$), Bi ($BiPO_4$)
H_2SO_4	Li, Mn, **Sr, Cd, Pb, Ba** (all as sulfates)
H_2PtCl_6	K (K_2PtCl_6, or Pt), Rb (Rb_2PtCl_6), Cs (Cs_2PtCl_6)
$H_2C_2O_4$	Ca (CaO), Sr (SrO), **Th** (ThO_2)
$(NH_4)_2MoO_4$	Cd ($CdMoO_4$),‡ Pb ($PbMoO_4$)
HCl	**Ag** (AgCl), Hg (Hg_2Cl_2), Na (as NaCl from butyl alcohol), Si (SiO_2)
$AgNO_3$	**Cl** (AgCl), Br (AgBr), I(AgI)
$(NH_4)_2CO_3$	**Bi** (Bi_2O_3)
NH_4SCN	Cu [$Cu_2(SCN)_2$]
$NaHCO_3$	Ru, Os, Ir (precipitated as hydrous oxides; reduced with H_2 to metallic state)
HNO_3	Sn (SnO_2)
H_5IO_6	Hg [$Hg_5(IO_6)_2$]
NaCl, $Pb(NO_3)_2$	F (PbClF)
$BaCl_2$	$\mathbf{SO_4^{2-}}$ ($BaSO_4$)
$MgCl_2$, NH_4Cl	$\mathbf{PO_4^{3-}}$ ($Mg_2P_2O_7$)

*From W. F. Hillebrand, G. E. F. Lundell, H. A. Bright, and J. I. Hoffman, *Applied Inorganic Analysis.* New York: Wiley, 1953. By permission of John Wiley & Sons, Inc.

†Boldface type indicates that gravimetric analysis is the preferred method for the element or ion. The weighed form is indicated in parentheses.

‡A double dagger indicates that the gravimetric method is seldom used. An underscore indicates the most reliable gravimetric method.

5D-2 Reducing Agents

Table 5-5 lists several reagents that convert an analyte to its elemental form for weighing.

TABLE 5-5 Some Reducing Agents Employed in Gravimetric Methods

Reducing Agent	Analyte
SO_2	Se, Au
$SO_2 + H_2NOH$	Te
H_2NOH	Se
$H_2C_2O_4$	Au
H_2	Re, Ir
HCOOH	Pt
$NaNO_2$	Au
$SnCl_2$	Hg
Electrolytic reduction	Co, Ni, Cu, Zn, Ag, In, Sn, Sb, Cd, Re, Bi

5D-3 Organic Precipitating Agents

Numerous organic reagents have been developed for the gravimetric determination of inorganic species. Some of these reagents are significantly more selective in their reactions than are most of the inorganic reagents listed in Table 5-4.

We encounter two types of organic reagents. One forms slightly soluble nonionic products called *coordination compounds;* the other forms products in which the bonding between the inorganic species and the reagent is largely ionic.

Organic reagents that yield sparingly soluble coordination compounds typically contain at least two functional groups. Each of these groups is capable of bonding with a cation by donating a pair of electrons. The functional groups are located in the molecule such that a five- or six-membered ring results from the reaction. Reagents that form compounds of this type are called *chelating agents,* and their products are called *chelates.*

Metal chelates are relatively nonpolar and, as a consequence, have solubilities that are low in water but high in organic liquids. Usually these compounds possess low densities and are often intensely colored. Because they are not wetted by water, coordination compounds are readily freed of moisture at low temperatures. Two widely used chelating reagents are described in the paragraphs that follow.

Chelates are cyclical metal-organic compounds in which the metal is a part of one or more five- or six-membered rings.

The chelate pictured below is heme, which is a part of hemoglobin, the oxygen carrying molecule in human blood.

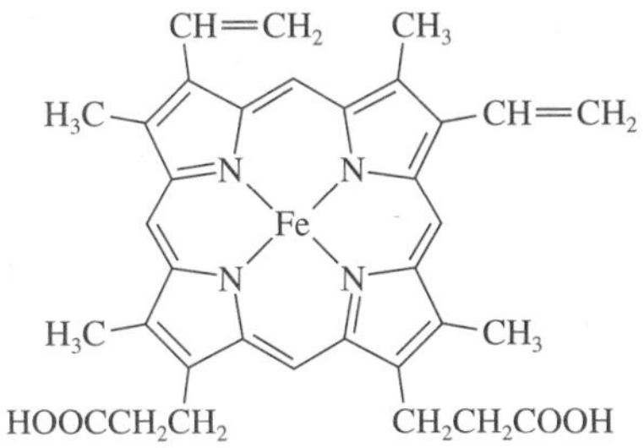

Notice the four six-membered rings that are formed with Fe(II) ions.

8-Hydroxyquinoline

Approximately two dozen cations form sparingly soluble chelates with 8-hydroxyquinoline. The structure of magnesium 8-hydroxyquinolate is typical of these chelates:

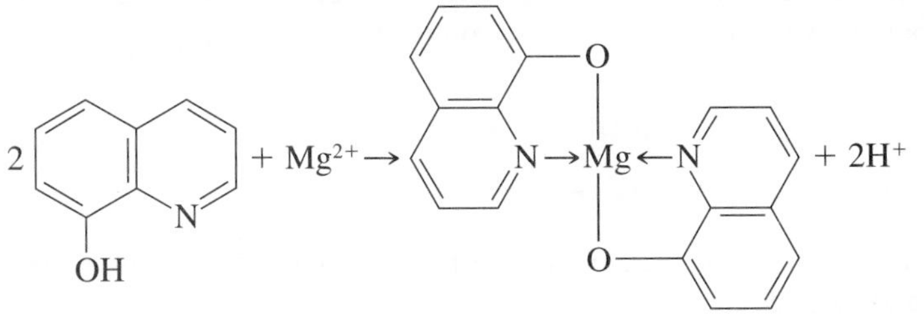

8-Hydroxyquinoline is sometimes called **oxine.**

The solubilities of metal 8-hydroxyquinolates vary widely from cation to cation and are pH-dependent because 8-hydroxyquinoline is always deprotonated during chelation reaction. Therefore, we can achieve a considerable degree of selectivity in the use of 8-hydroxyquinoline by controlling pH.

Dimethylglyoxime

Dimethylglyoxime is an organic precipitating agent of unparalleled specificity. Only nickel(II) is precipitated from a weakly alkaline solution. The reaction is

Nickel dimethylglyoxime is spectacular in appearance. It has a beautiful and vivid red color.

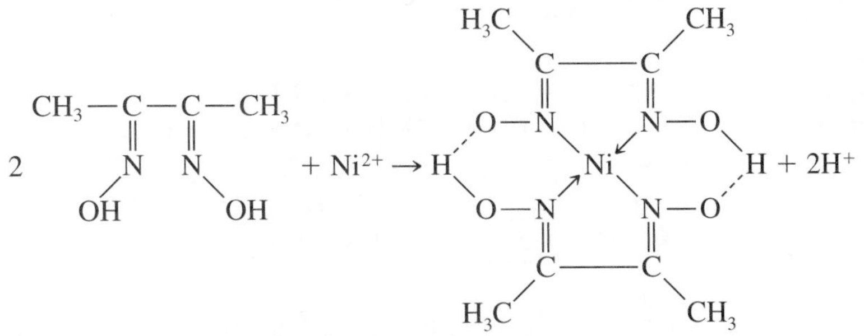

This precipitate is so bulky that only small amounts of nickel can be handled conveniently. It also has an exasperating tendency to creep up the sides of the container as it is filtered and washed. The solid is readily dried at 110°C and has the composition indicated by its formula.

Creeping is the process by which a precipitate (usually a metal-organic chelate) moves up the walls of a wetted surface of a glass container or a filter paper.

Sodium Tetraphenylboron

Sodium tetraphenylboron, $(C_6H_5)_4B^-Na^+$, is an important example of an organic precipitating reagent that forms salt-like precipitates. In cold mineral-acid solutions, it is a near-specific precipitating agent for potassium and ammonium ions. The composition of the precipitates is stoichiometric and contains one mole of potassium or ammonium ion for each mole of tetraphenylboron ion; these ionic compounds are readily filtered and can be brought to constant mass at 105° to 120°C. Only mercury(II), rubidium, and cesium interfere and must be removed by prior treatment.

Potassium tetraphenylboron is a salt rather than a chelate.

5D-4 Organic Functional Group Analysis

Several reagents react selectively with certain organic functional groups and thus can be used for the determination of most compounds containing these groups. A list of gravimetric functional-group reagents is given in Table 5-6. Many of the reactions shown can also be used for volumetric and spectrophotometric determinations.

5D-5 Volatilization Methods

The two most common gravimetric methods based on volatilization are those for water and carbon dioxide.

Water is quantitatively eliminated from many inorganic samples by ignition. In the direct determination, it is collected on any of several solid desiccants and its mass is determined from the mass gain of the desiccant.

TABLE 5-6 Gravimetric Methods for Organic Functional Groups

Functional Group	Basis for Method	Reaction and Product Weighed*
Carbonyl	Mass of precipitate with 2,4-dinitrophenylhydrazine	$RCHO + H_2NNHC_6H_3(NO_2)_2 \rightarrow$ $\underline{R{-}CH = NNHC_6H_3(NO_2)_2(s)} + H_2O$ ($RCOR'$ reacts similarly)
Aromatic carbonyl	Mass of CO_2 formed at 230°C in quinoline; CO_2 distilled, absorbed, and weighed	$ArCHO \xrightarrow[CuCO_3]{230°C} Ar + \underline{CO_2(g)}$
Methoxyl and ethoxyl	Mass of AgI formed after distillation and decomposition of CH_3I or C_2H_5I	$ROCH_3 + HI \rightarrow ROH + CH_3I$ $RCOOCH_3 + HI \rightarrow RCOOH + CH_3I$ $ROC_2H_5 + HI \rightarrow ROH + C_2H_5I$ } $CH_3I + Ag^+ + H_2O \rightarrow \underline{AgI(s)} + CH_3OH$
Aromatic nitro	Mass loss of Sn	$RNO_2 + \frac{3}{2}\underline{Sn(s)} + 6\,H^+ \rightarrow RNH_2 + \frac{3}{2}Sn^{4+} + 2\,H_2O$
Azo	Mass loss of Cu	$RN = NR' + \underline{2\,Cu(s)} + 4\,H^+ \rightarrow RNH_2 + R'NH_2 + 2\,Cu^{2+}$
Phosphate	Mass of Ba salt	$RO\overset{O}{\overset{\Vert}{P}}(OH)_2 + Ba^{2+} \rightarrow \underline{RO\overset{O}{\overset{\Vert}{P}}O_2Ba(s)} + 2\,H^+$
Sulfamic acid	Mass of $BaSO_4$ after oxidation with HNO_2	$RNHSO_3H + HNO_2 + Ba^{2+} \rightarrow ROH + \underline{BaSO_4(s)} + N_2 + 2\,H^+$
Sulfinic acid	Mass of Fe_2O_3 after ignition of Fe(III) sulfinate	$3\,ROSOH + Fe^{3+} \rightarrow (ROSO)_3Fe(s) + 3\,H^+$ $(ROSO)_3Fe \xrightarrow[O_2]{} CO_2 + H_2O + SO_2 + \underline{Fe_2O_3(s)}$

*The substance weighed is underlined.

The indirect method, in which the amount of water is determined by the loss of mass of the sample during heating, is less satisfactory because it must be assumed that water is the only component volatilized. This assumption is frequently unjustified, however, because heating of many substances results in their decomposition and a consequent change in mass, irrespective of the presence of water. Nevertheless, the indirect method has been widely used for the determination of water in items of commerce. For example, a semiautomated instrument for the determination of moisture in cereal grains can be purchased. It consists of a platform balance upon which a 10-g sample is heated with an infrared lamp. The percent residue is read directly.

Automatic instruments for the routine determination of water in various products of agriculture and commerce are marketed by several instrument manufacturers.

Carbonates are ordinarily decomposed by acids to give carbon dioxide, which is readily evolved from solution by heat. As in the direct analysis for water, the mass of carbon dioxide is established from the increase in the mass of a solid absorbent. Ascarite II,[8] which consists of sodium hydroxide on a nonfibrous silicate, retains carbon dioxide by the reaction

$$2NaOH + CO_2 \longrightarrow Na_2CO_3 + H_2O$$

The absorption tube must also contain a desiccant to prevent loss of the evolved water.

Sulfides and sulfites can also be determined by volatilization. The hydrogen sulfide or sulfur dioxide evolved from the sample after treatment with acid is collected in a suitable absorbent.

Finally, the classical method for the determination of carbon and hydrogen in organic compounds is a gravimetric procedure in which the combustion products (H_2O and CO_2) are collected selectively on weighed absorbents. The increase in mass serves as the analytical parameter.

---

[8]® Thomas Scientific, Swedesboro, NJ.

5E QUESTIONS AND PROBLEMS

5-1. Explain the difference between
- *(a) mass and weight.
- (b) an empirical formula and a molecular formula.
- *(c) a colloidal and a crystalline precipitate.
- (d) specific and selective precipitating reagents.
- *(e) precipitation and coprecipitation.
- (f) peptization and coagulation.
- *(g) occlusion and mixed-crystal formation.
- (h) nucleation and particle growth.

5-2. Define
- *(a) dalton.
- (b) mole.
- *(c) stoichiometry.
- (d) gravimetric method.
- *(e) digestion.
- (f) adsorption.
- *(g) reprecipitation.
- (h) precipitation from homogeneous solution.
- *(i) electric double layer.
- (j) specific surface area.
- *(k) relative supersaturation.

***5-3.** What are the structural characteristics of a chelating agent?

5-4. How can the relative supersaturation be varied during precipitate formation?

***5-5.** An aqueous solution contains $NaNO_3$ and KSCN. The thiocyanate ion is precipitated as AgSCN by addition of $AgNO_3$. After an excess of the precipitating reagent has been added,
- (a) what is the charge on the surface of the coagulated colloidal particles?
- (b) what is the source of the charge?
- (c) what ions predominate in the counter-ion layer?

5-6. Briefly explain why the AgCl produced in a gravimetric silver determination is inherently purer than the AgCl produced in a gravimetric chloride determination.

***5-7.** What is peptization and how is it avoided?

5-8. Suggest a precipitation method for the separation of K^+ from Na^+ and Li^+.

5-9. Propose a method for the homogeneous precipitation of
- *(a) Al^{3+} from a solution that also contains Na^+.
- (b) Ni^{2+} from a solution that also contains Cu^{2+}.

*(c) Pb^{2+} from a solution that also contains Ba^{2+}.
(d) Cu^{2+} from a solution that also contains Ca^{2+}.

*5-10. What is the atomic mass of Fe? What is the molar mass of Fe?

5-11. The rest mass of an electron is 9.109×10^{-31} kg. Calculate the molar mass of the electron.

*5-12. How many moles are contained in
(a) 6.84 g of B_2O_3?
(b) 296 mg of $Na_2B_4O_7 \cdot 10H_2O$?
(c) 8.75 g of Mn_3O_4?
(d) 67.4 mg of CaC_2O_4?

5-13. How many millimoles are contained in
(a) 79.8 mg of H_2?
(b) 8.43 g of SO_2?
(c) 64.4 g of Na_2CO_3?
(d) 411 mg of $KMnO_4$?

*5-14. How many milligrams are contained in
(a) 0.666 mol of HNO_3?
(b) 300 mmol of MgO?
(c) 19.0 mol of NH_4NO_3?
(d) 5.32 mol of $(NH_4)_2Ce(NO_3)_6$ (548.22 g/mol)?

5-15. How many grams are contained in
(a) 32.1 mmol of H_2O_2?
(b) 0.466 mol of NH_4VO_3 (117.0 g/mol)?
(c) 5.38 mol of $MgNH_4PO_4$?
(d) 26.7 mmol of $KH(IO_3)_2$?

5-16. Use chemical formulas to generate the stoichiometric factor needed to express the results of a gravimetric analysis in terms of the substance on the left if the weighing form is the substance on the right:

Sought	Weighed	Sought	Weighed
*(a) CO_2	$BaCO_3$	(f) UO_2	U_3O_8
(b) Mg	$Mg_2P_2O_7$	*(g) $C_8H_6O_3Cl_2$	AgCl
*(c) K_2O	$(C_6H_5)_4BK$	(h) $CoSiF_6 \cdot 6H_2O$	Co_3O_4
(d) Bi	Bi_2O_3	*(i) $CoSiF_6 \cdot 6H_2O$	H_2O
*(e) H_2S	$CdSO_4$	(j) $CoSiF_6 \cdot 6H_2O$	PbClF

5-17. Briefly explain why the numerically smallest stoichiometric factor is associated with the precipitate that yields the greatest mass for a given mass of analyte.

*5-18. How many grams of Ag_2CrO_4 can be produced from 1.200 g of
(a) $AgNO_3$? (b) K_2CrO_4?

5-19. How many grams of $Mg(OH)_2$ can be produced from 0.750 g of
(a) $Ba(OH)_2$? (b) $MgCl_2$?

5-20. What mass of AgCl can be produced from a 0.400-g sample that assays 41.3%
*(a) KCl? (b) $MgCl_2$? *(c) $FeCl_3$?

*5-21. A 50.0-mL portion of a solution containing 0.200 g of $BaCl_2 \cdot 2H_2O$ is mixed with 50.0 mL of a solution containing 0.300 g of $NaIO_3$. Assume that the solubility of $Ba(IO_3)_2$ in water is negligibly small and calculate
(a) the mass of the precipitated $Ba(IO_3)_2$.
(b) the mass of the unreacted compound that remains in solution.

5-22. How many grams of CO_2 are evolved from a 1.204-g sample that is 36.0% $MgCO_3$ and 44.0% K_2CO_3 by mass?

5-23. When a 100.0-mL portion of a solution containing 0.500 g of $AgNO_3$ is mixed with 100.0 mL of a solution containing 0.300 g of K_2CrO_4, a bright red precipitate of Ag_2CrO_4 forms.
(a) Assuming the solubility of Ag_2CrO_4 is negligible, calculate the mass of the precipitate.
(b) Calculate the mass of the unreacted component that remains in solution.

*5-24. Treatment of a 0.4000-g sample of impure potassium chloride with an excess of $AgNO_3$ resulted in the formation of 0.7332 g of AgCl. Calculate the percentage of KCl in the sample.

5-25. The aluminum in a 1.200-g sample of impure ammonium aluminum sulfate was precipitated with aqueous ammonia as the hydrous $Al_2O_3 \cdot xH_2O$. The precipitate was filtered and ignited at 1000°C to give anhydrous Al_2O_3, which weighed 0.1798 g. Express the result of this analysis in terms of
(a) % $NH_4Al(SO_4)_2$. (b) % Al_2O_3. (c) % Al.

*5-26. A 0.1799-g sample of an organic compound was burned in a stream of oxygen, and the CO_2 produced was collected in a solution of barium hydroxide. Calculate the percentage of carbon in the sample if 0.5613 g of $BaCO_3$ was formed.

5-27. A 0.7406-g sample of impure magnesite, $MgCO_3$, was decomposed with HCl; the liberated CO_2 was collected on absorbent and found to weigh 0.1881 g. Calculate the percentage of magnesium in the sample.

5-28. The hydrogen sulfide in a 50.0-g sample of crude petroleum was removed by distillation and collected in a solution of $CdCl_2$. The precipitated CdS was then filtered, washed, and ignited to $CdSO_4$. Calculate the percentage of H_2S in the sample if 0.108 g of $CdSO_4$ was recovered.

*5-29. Ammoniacal nitrogen can be determined by treatment of the sample with chloroplatinic acid; the product is slightly soluble ammonium chloroplatinate:

$$H_2PtCl_6 + 2NH_4^+ \longrightarrow (NH_4)_2PtCl_6 + 2H^+$$

The precipitate decomposes upon ignition, yielding metallic platinum and gaseous products:

$$(NH_4)_2PtCl_6 \longrightarrow Pt(s) + 2Cl_2(g) + 2NH_3(g) + 2HCl(g)$$

Calculate the percentage of ammonia in a sample if 0.2213 g gave rise to 0.5881 g of platinum.

*5-30. The sulfur in an 8-tablet sample of the hypnotic drug captodiamine, $C_{21}H_{29}NS_2$ (359.6 g/mol) was converted to sulfate and determined gravimetrically. Calculate the average mass of captodiamine per tablet if 0.3343 g of $BaSO_4$ was recovered.

*5-31. The phosphorus in a 0.2374-g sample was precipitated as the slightly soluble $(NH_4)_3PO_4 \cdot 12MoO_3$. This precipitate was filtered, washed, and then redissolved in acid. Treatment of the resulting solution with an excess of Pb^{2+} resulted in the formation of 0.2752 g of $PbMoO_4$. Express the results of this analysis in terms of percent P_2O_5.

5-32. A 0.6447-g portion of manganese dioxide was added to an acidic solution in which 1.1402 g of a chloride-containing sample was dissolved. Evolution of chlorine took place as a consequence of the following reaction:

$$MnO_2(s) + 2Cl^- + 4H^+ \longrightarrow Mn^{2+} + Cl_2(g) + 2H_2O$$

After the reaction was complete, the excess MnO_2 was collected by filtration, washed, and weighed, 0.3521 g being recovered. Express the results of this analysis in terms of percent aluminum chloride.

***5-33.** Nitrobenzene, $C_6H_5NO_2$ (123.11 g/mol), is quantitatively reduced to aniline $C_6H_5NH_2$ (93.12 g/mol) with metallic tin:

$$3Sn(s) + 2C_6H_5NO_2 + 12H^+ \longrightarrow 2C_6H_5NH_2 + 4H_2O + 3Sn^{4+}$$

A 0.5078-g sample of impure nitrobenzene was treated with 1.044 g of tin. When reaction was complete, the residual tin was found to weigh 0.338 g. Calculate the percentage of nitrobenzene in the sample.

5-34. A series of sulfate samples is to be analyzed by precipitation as $BaSO_4$. If it is known that the sulfate content in these samples ranges between 20% and 55%, what minimum sample mass should be taken to ensure that a precipitate mass no smaller than 0.300 g is produced? What is the maximum precipitate mass to be expected if this quantity of sample is taken?

***5-35.** The success of a particular catalyst is highly dependent upon its zirconium content. The starting material for this preparation is received in batches that assay between 68% and 84% $ZrCl_4$. Routine analysis based upon precipitation of AgCl is feasible, it having been established that there are no sources of chloride ion other than the $ZrCl_4$ in the sample.

(a) What sample mass should be taken to ensure a AgCl precipitate that weighs at least 0.400 g?

(b) If this sample mass is used, what is the maximum mass of AgCl that can be expected in this analysis?

(c) To simplify calculations, what sample mass should be taken in order to have the percentage of $ZrCl_4$ exceed the mass of AgCl produced by a factor of 100?

5-36. The addition of dimethylglyoxime, $H_2C_4H_6O_2N_2$, to a solution containing nickel(II) ion gives rise to a precipitate:

$$Ni^{2+} + 2H_2C_4H_6O_2N_2 \longrightarrow 2H^+ + Ni(HC_4H_6O_2N_2)_2$$

Nickel dimethylglyoxime is a bulky precipitate that is inconvenient to manipulate in amounts greater than 175 mg. The amount of nickel in a type of permanent-magnet alloy ranges between 24% and 35%. Calculate the sample size that should not be exceeded when analyzing these alloys for nickel.

***5-37.** A 0.6407-g sample containing chloride and iodide ions gave a silver halide precipitate weighing 0.4430 g. This precipitate was then strongly heated in a stream of Cl_2 gas to convert the AgI to AgCl; upon completion of this treatment, the precipitate weighed 0.3181 g. Calculate the percentage of chloride and iodide in the sample.

5-38. A 6.881-g sample containing magnesium chloride and sodium chloride was dissolved in sufficient water to give 500 mL of solution. Analysis for the chloride content of a 50.0-mL aliquot resulted in the formation of 0.5923 g of AgCl. The magnesium in a second 50.0-mL aliquot was precipitated as $MgNH_4PO_4$; upon ignition 0.1796 g of $Mg_2P_2O_7$ was found. Calculate the percentage of $MgCl_2 \cdot 6H_2O$ and of NaCl in the sample.

***5-39.** The iodide in a sample that also contained chloride was converted to iodate by treatment with an excess of bromine:

$$3H_2O + 3Br_2 + I^- \longrightarrow 6Br^- + IO_3^- + 6H^+$$

The unused bromine was removed by boiling; an excess of barium ion was then added to precipitate the iodate:

$$Ba^{2+} + 2IO_3^- \longrightarrow Ba(IO_3)_2$$

In the analysis of a 2.72-g sample, 0.0720 g of barium iodate was recovered. Express the results of this analysis as percent potassium iodide.

5-40. Several alloys that contained only Ag and Cu were analyzed by dissolving weighed quantities in HNO_3, introducing an excess of IO_3^-, and bringing the filtered mixture of $AgIO_3$ and $Cu(IO_3)_2$ to constant mass. Use the accompanying data to calculate the percentage composition of the alloys.

Mass Sample, g	Mass Precipitate, g
*(a) 0.2175	0.7391
(b) 0.1948	0.7225
*(c) 0.2473	0.7443
(d) 0.2386	0.9962
*(e) 0.1864	0.8506

CHAPTER

6

Titrimetric Methods of Analysis

> **Titrimetry** includes a group of analytical methods that are based upon determining the quantity of a reagent of known strength that is required to react completely with the analyte. The reagent may be a standard solution of a chemical or an electric current of known magnitude.

> **Volumetric titrimetry** is a type of titrimetry in which the standard reagent is measured volumetrically.

> **Gravimetric,** or **weight, titrimetry** is a type of titrimetry in which the mass of the standard reagent is measured.

> **Coulometric titrimetry** is a type of titrimetry in which the quantity of charge in coulombs required to complete a reaction with the analyte is measured.

> A **standard solution** is a reagent of exactly known concentration that is used in a titrimetric analysis.

> **Titration** is a process in which a standard reagent is added to a solution of an analyte until the reaction between the analyte and reagent is judged to be complete.

Titrimetric methods include a large and powerful group of quantitative procedures that are based upon measuring the amount of a reagent of known concentration that is consumed by the analyte. *Volumetric tritimetry* involves measuring the volume of a solution of known concentration that is needed to react essentially completely with the analyte. *Gravimetric titrimetry* differs only in that the mass of the reagent is measured instead of its volume. In *coulometric titrimetry,* the ''reagent'' is a constant direct electrical current of known magnitude that directly or indirectly reacts with the analyte; here, the time required to complete the electrochemical reaction is measured.

Titrimetric methods are widely used for routine analyses because they are rapid, convenient, accurate, and readily automated. Volumetric and gravimetric titrimetry are introduced in this chapter; additional information concerning theory and applications is provided in Chapters 10 through 17. Coulometric titrimetry is considered in Section 20D-5.

6A SOME GENERAL ASPECTS OF VOLUMETRIC TITRIMETRY[1]

6A-1 Definition of Some Terms

A *standard solution* (or a *standard titrant*) is a reagent of known concentration that is used to carry out a titrimetric analysis. A *titration* is performed by slowly adding a standard solution from a buret or other liquid dispensing device to a solution of the analyte until the reaction between the two is judged complete. The volume of reagent needed to complete the titration is determined from the difference between the initial and final volume readings.

The *equivalence point* in a titration is reached when the amount of added titrant is chemically equivalent to the amount of analyte in the sample. For example, the equivalence point in the titration of sodium chloride with silver nitrate occurs after exactly 1 mol of silver ion has been added for each mole of chloride ion in the sample. The equivalence point in the titration of sulfuric acid with sodium hydroxide is reached after introduction of 2 mol of base for each mole of acid.

---

[1]For a detailed discussion of volumetric methods, see J. I. Watters, in *Treatise on Analytical Chemistry,* I. M. Kolthoff and P. J. Elving, Eds., Part I, Vol. 11, Chapter 114. New York: Wiley, 1975.

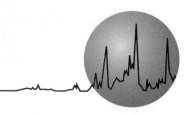

It is sometimes necessary to add an excess of the standard titrant and then determine the excess amount by *back-titration* with a second standard titrant. Here, the equivalence point corresponds to the point where the amount of initial titrant is chemically equivalent to the amount of analyte plus the amount of back-titrant.

Back-titration is a process in which the excess of a standard solution used to react with an analyte is determined by titration with a second standard solution. Back-titrations are often required when the rate of reaction between the analyte and reagent is slow or when the standard solution lacks stability.

6A-2 Equivalence Points and End Points

The equivalence point of a titration is a theoretical point that cannot be determined experimentally. Instead, we can only estimate its position by observing some physical change associated with the condition of equivalence. This change is called the *end point* for the titration. Every effort is made to ensure that any volume or mass difference between the equivalence point and the end point is small. Such differences do exist, however, as a result of inadequacies in the physical changes and in our ability to observe them. The difference in volume or mass between the equivalence point and the end point is the *titration error.*

The **equivalence point** is the point in a titration when the amount of added standard reagent is chemically equal to the amount of analyte.

Indicators are often added to the analyte solution in order to give an observable physical change (the end point) at or near the equivalence point. We shall see that large changes in the relative concentration of analyte and titrant occur in the equivalence-point region. These concentration changes cause the indicator to change in appearance. Typical indicator changes include the appearance or disappearance of a color, a change in color, or the appearance or disappearance of turbidity.

The **end point** is the point in a titration when a physical change that is associated with the condition of chemical equivalence occurs.

We often use instruments to detect end points. These instruments respond to certain properties of the solution that change in a characteristic way during the titration. Among such instruments are voltmeters, ammeters, and ohmmeters; colorimeters; temperature recorders; and refractometers.

In volumetric methods, the **titration error** E_t is given by

$$E_t = V_{eq} - V_{ep}$$

where V_{eq} is the theoretical volume of reagent required to reach the equivalence point and V_{ep} is the actual volume used to arrive at the end point.

6A-3 Primary Standards

A *primary standard* is a highly purified compound that serves as a reference material in all volumetric and mass titrimetric methods. The accuracy of a method is critically dependent on the properties of this compound. Important requirements for a primary standard are

A **primary standard** is an ultrapure compound that serves as the reference material for a titrimetric method of analysis.

1. High purity. Established methods for confirming purity should be available.
2. Stability toward air.
3. Absence of hydrate water so that the composition of the solid does not change with variations in relative humidity.
4. Ready availability at modest cost.
5. Reasonable solubility in the titration medium.
6. Reasonably large molar mass so that the relative error associated with weighing the standard is minimized.

Compounds that meet or even approach these criteria are very few, and only a limited number of primary standard substances are available to the chemist. As a consequence, compounds that are less pure must sometimes be employed in lieu of a primary standard. The purity of such a *secondary standard* must be established by careful analysis.

A **secondary standard** is a compound whose purity has been established by chemical analysis and serves as the reference material for a titrimetric method of analysis.

6B STANDARD SOLUTIONS

Standard solutions play a central role in all titrimetric methods of analysis. Therefore, we need to consider the desirable properties for such solutions, how they are prepared, and how their concentrations are expressed.

6B-1 Desirable Properties of Standard Solutions

The ideal standard solution for a titrimetric method will

1. be sufficiently stable so that it is only necessary to determine its concentration once;
2. react rapidly with the analyte so that the time required between additions of reagent is minimized;
3. react more or less completely with the analyte so that satisfactory end points are realized;
4. undergo a selective reaction with the analyte that can be described by a simple balanced equation.

Few reagents meet all these ideals perfectly.

6B-2 Methods for Establishing the Concentration of Standard Solutions

The accuracy of a titrimetric method can be no better than the accuracy of the concentration of the standard solution used in the titration. Two basic methods are used to establish the concentration of such solutions. The first is the *direct method,* in which a carefully weighed quantity of a primary standard is dissolved in a suitable solvent and diluted to an exactly known volume in a volumetric flask. The second is by standardization, in which the titrant to be standardized is used to titrate (1) a weighed quantity of a primary standard, (2) a weighed quantity of a secondary standard, or (3) a measured volume of another standard solution. A titrant that is standardized against a secondary standard or against another standard solution is sometimes referred to as a *secondary standard solution.* The concentration of a secondary standard solution is subject to a larger uncertainty than that for a primary standard solution. If there is a choice, then, solutions are best prepared by the direct method. On the other hand, many reagents lack the properties required for a primary standard and therefore require standardization.

Standardization is a process in which concentration of a volumetric solution is determined by using it to titrate a known mass of a primary or secondary standard or an exactly known volume of another standard solution.

6B-3 Methods for Expressing the Concentration of Standard Solutions

The concentrations of standard solutions are generally expressed in units of either *molarity* c or *normality* c_N. The first gives the number of moles of reagent contained in 1 L of solution, and the second gives the number of *equivalents* of reagent in the same volume.

Throughout this text we shall base volumetric calculations exclusively on molarity and molar masses. We have included in Appendix 9, however, a discussion of how volumetric calculations are carried out based upon normality and equivalent weights because the reader may encounter these terms and their uses in the industrial and health science literature.

6C VOLUMETRIC CALCULATIONS

6C-1 Solutions and Their Concentrations

Chemists express the concentration of solutes in solution in several ways. The most important of these are described in this section.

Molar Concentration

The molar concentration (c_X) of the solution of the chemical species X is the number of moles of that species that is contained in one liter of the solution *(not one liter of the solvent)*. The unit of molar concentration is *molarity,* M, which has the dimensions of mol L^{-1}. Molarity also expresses the number of millimoles of a solute per milliliter of solution.

$$c_X = \frac{\text{no. mol solute}}{\text{no. L solution}} = \frac{\text{no. mmol solute}}{\text{no. mL solution}} \tag{6-1}$$

EXAMPLE 6-1

Calculate the molar concentration of ethanol in an aqueous solution that contains 2.30 g of C_2H_5OH (46.07 g/mol) in 3.50 L of solution.

Because the unit of molarity is the number of moles of solute per liter of solution, both of these quantities will be needed. The number of liters is given as 3.50, so all we need to do is to convert the number of grams of ethanol to the corresponding number of moles.

$$\text{amount } C_2H_5OH = 2.30\ \cancel{\text{g } C_2H_5OH} \times \frac{1 \text{ mol } C_2H_5OH}{46.07\ \cancel{\text{g } C_2H_5OH}}$$
$$= 0.04992 \text{ mol } C_2H_5OH$$

To obtain the molar concentration, $c_{C_2H_5OH}$, we divide by the volume. Thus,

$$c_{C_2H_5OH} = \frac{2.30\ \cancel{\text{g } C_2H_5OH} \times \dfrac{1 \text{ mol } C_2H_5OH}{46.07\ \cancel{\text{g } C_2H_5OH}}}{3.50 \text{ L}}$$
$$= 0.0143 \text{ mol } C_2H_5OH/L = 0.0143 \text{ M}$$

Analytical molarity is the total number moles of a solute, regardless of its chemical state in one liter of solution. The analytical molarity describes how a solution of a given molarity can be prepared.

Analytical Molarity The *analytical molarity* of a solution gives the *total* number of moles of a solute in one liter of the solution (or the total number of millimoles in one milliliter). That is, the analytical molarity specifies a recipe by which the solution can be prepared. For example, a sulfuric acid solution that has an analytical concentration of 1.0 M can be prepared by dissolving 1.0 mol, or 98 g, of pure H_2SO_4 in water and diluting to exactly 1.0 L.

Equilibrium or **species, molarity** is the molar concentration of a particular species in a solution.

Some chemists prefer to distinguish between species and analytical concentrations in a different way. They use *molar concentration* for species concentration and *formal concentration* (F) for analytical concentration. Applying this convention to our example, we can say that the formal concentration of H_2SO_4 is 1.0 F, whereas its molar concentration is 0.0 M because no undissociated H_2SO_4 exists in the solution.

In this example the *analytical molarity* of H_2SO_4 is given by $c_{H_2SO_4} = [SO_4^{2-}] + [HSO_4^-]$ because these are the only two species that contain sulfate in the solution.

Equilibrium, or Species, Molarity The *equilibrium,* or *species, molarity* expresses the molar concentration of a particular species in a solution at equilibrium. In order to state the species molarity, it is necessary to know how the solute behaves when it is dissolved in a solvent. For example, the species molarity of H_2SO_4 in a solution with an analytical concentration of 1.0 M is 0.0 M because the sulfuric acid is entirely dissociated into a mixture of H_3O^+, HSO_4^-, and SO_4^{2-} ions. Essentially no H_2SO_4 molecules as such are present in this solution. The equilibrium concentrations and thus the species molarity of these three ions are 1.01, 0.99, and 0.01 M, respectively.

Equilibrium molar concentrations are often symbolized by placing square brackets around the chemical formula for the species, so for our solution of H_2SO_4 with an analytical concentration of 1.0 M, we can write

$$[H_2SO_4] = 0.00\ M \qquad [H_3O^+] = 1.01\ M$$
$$[HSO_4^-] = 0.99\ M \qquad [SO_4^{2-}] = 0.01\ M$$

EXAMPLE 6-2

Calculate the analytical and equilibrium molar concentrations of the solute species in an aqueous solution that contains 285 mg of trichloroacetic acid, Cl_3CCOOH (163.4 g/mol), in 10.0 mL (the acid is 73% ionized in water).

As in Example 6-1, we calculate the number of moles of Cl_3CCOOH, which we designate as HA, and divide by the volume of the solution, 10.0 mL or 0.01000 L. Thus,

$$\text{amount HA} = 285\ \cancel{\text{mg HA}} \times \frac{1\ \cancel{\text{g HA}}}{1000\ \cancel{\text{mg HA}}} \times \frac{1\ \text{mol HA}}{163.4\ \cancel{\text{g HA}}}$$
$$= 1.744 \times 10^{-3}\ \text{mol HA}$$

The analytical molar concentration, c_{HA}, is then

$$c_{HA} = \frac{1.744 \times 10^{-3}\ \text{mol HA}}{10.0\ \cancel{\text{mL}}} \times \frac{1000\ \cancel{\text{mL}}}{\text{L}} = 0.174\ \frac{\text{mol HA}}{\text{L}} = 0.174\ \text{M}$$

In this solution, 73% of the HA dissociates, giving H^+ and A^-:

$$HA \rightleftharpoons H^+ + A^-$$

The species molarity of HA is then 27% of c_{HA}. Thus,

$$[HA] = c_{HA} \times (100 - 73)/100 = 0.174 \times 0.27 = 0.047\ \text{mol/L}$$
$$= 0.047\ \text{M}$$

The species molarity of A^- is equal to 73% of the analytical concentration of HA. That is,

$$[A^-] = \frac{73\ \text{mol } A^-}{100\ \cancel{\text{mol HA}}} \times 0.174\ \frac{\cancel{\text{mol HA}}}{\text{L}} = 0.127\ \text{M}$$

Because one mole H^+ is formed for each mole A^-, we can also write

$$[H^+] = [A^-] = 0.127\ \text{M}$$

EXAMPLE 6-3

Describe the preparation of 2.00 L of 0.108 M $BaCl_2$ from $BaCl_2 \cdot 2H_2O$ (244.3 g/mol).

In order to determine the number of grams of solute to be dissolved and diluted to 2.00 L, we note that 1 mol of the dihydrate yields 1 mol of $BaCl_2$. Therefore, to produce this solution we will need

$$2.00\ \cancel{\text{L}} \times \frac{0.108\ \text{mol BaCl}_2 \cdot 2\text{H}_2\text{O}}{\cancel{\text{L}}} = 0.216\ \text{mol BaCl}_2 \cdot 2\text{H}_2\text{O}$$

The mass of $BaCl_2 \cdot 2H_2O$ is then

$$0.216\ \cancel{\text{mol BaCl}_2 \cdot 2\text{H}_2\text{O}} \times \frac{244.3\ \text{g BaCl}_2 \cdot 2\text{H}_2\text{O}}{\cancel{\text{mol BaCl}_2 \cdot 2\text{H}_2\text{O}}} = 52.8\ \text{g BaCl}_2 \cdot 2\text{H}_2\text{O}$$

Dissolve 52.8 g of $BaCl_2 \cdot 2H_2O$ in water and dilute to 2.00 L.

EXAMPLE 6-4

Describe the preparation of 500 mL of 0.0740 M Cl^- solution from solid $BaCl_2 \cdot 2H_2O$ (244.3 g/mol).

$$\text{mass BaCl}_2 \cdot 2\text{H}_2\text{O} = \frac{0.0740\ \cancel{\text{mol Cl}^-}}{\cancel{\text{L}}} \times 0.500\ \cancel{\text{L}} \times \frac{1\ \cancel{\text{mol BaCl}_2 \cdot 2\text{H}_2\text{O}}}{2\ \cancel{\text{mol Cl}^-}}$$
$$\times \frac{244.3\ \text{g BaCl}_2 \cdot 2\text{H}_2\text{O}}{\cancel{\text{mol BaCl}_2 \cdot 2\text{H}_2\text{O}}} = 4.52\ \text{g BaCl}_2 \cdot 2\text{H}_2\text{O}$$

Dissolve 4.52 g of $BaCl_2 \cdot 2H_2O$ in water and dilute to 0.500 L or 500 mL.

Percent Concentration

Chemists frequently express concentrations in terms of percent (parts per hundred). Unfortunately, this practice can be a source of ambiguity because percent composition of a solution can be expressed in several ways. Three common methods are

$$\text{weight percent (w/w)} = \frac{\text{mass solute}}{\text{mass solution}} \times 100\%$$

$$\text{volume percent (v/v)} = \frac{\text{volume solute}}{\text{volume solution}} \times 100\%$$

$$\text{weight/volume percent (w/v)} = \frac{\text{mass solute, g}}{\text{volume solution, mL}} \times 100\%$$

Note that the denominator in each of these expressions refers to the *solution* rather than to the solvent. Note also that the first two expressions do not depend on the units employed (provided, of course, that there is consistency between numerator and denominator). The units in the third expression do not cancel and therefore must be specified. Of the three expressions, only weight percent has the virtue of being temperature independent.

Weight percent would be more properly called mass percent and abbreviated m/m. The term ''weight percent'' is so widely used in the chemical literature, however, that we will use it throughout this text.

Weight percent is frequently employed to express the concentration of commercial aqueous reagents. For example, hydrochloric acid is sold as a 37% solution (see Figure 6-1), which means that the reagent contains 37 g of HCl per 100 g of solution.

Volume percent is commonly used to specify the concentration of a solution prepared by diluting a pure liquid with another liquid. For example, a 5% aqueous solution of methanol *usually* means a solution prepared by diluting 5.0 mL of pure methanol with enough water to give 100 mL.

Weight/volume percent is often employed to indicate the composition of dilute aqueous solutions of solid reagents. For example, 5% aqueous silver nitrate *often* refers to a solution prepared by dissolving 5 g of silver nitrate in sufficient water to give 100 mL of solution.

Always specify the type of percent when reporting concentrations in this way.

To avoid uncertainty, always specify explicitly the type of percent composition you are using. If this information is missing, the user must decide intuitively which of the several types is involved. The potential error resulting from a wrong choice is considerable. For example, commercial 50% (w/w) sodium hydroxide contains 763 g of the reagent per liter, which corresponds to 76.3% (w/v) sodium hydroxide.

Parts Per Million and Parts Per Billion

A handy rule in calculating parts per million is to remember that for dilute aqueous solutions whose densities are approximately 1.00 g/mL, 1 ppm = 1.00 mg/L. That is,

$$c_{\text{ppm}} = \frac{\text{mass solute (mg)}}{\text{volume solution (L)}} \qquad \text{(6-2)}$$

$$c_{\text{ppb}} = \frac{\text{mass solute g}}{\text{mass solution g}} \times 10^9 \text{ ppb}$$

For very dilute solutions, parts per million (ppm) is a convenient way to express concentration:

$$c_{\text{ppm}} = \frac{\text{mass of solute}}{\text{mass of solution}} \times 10^6 \text{ ppm}$$

where c_{ppm} is the concentration in parts per million. Obviously, the units of mass in the numerator and denominator must agree. For even more dilute solutions, 10^9 ppb rather than 10^6 ppm is employed in the foregoing equation to give the results in parts per billion (ppb). The term *parts per thousand* (ppt) is also encountered, especially in oceanography.

EXAMPLE 6-5

What is the molarity of K^+ in an aqueous solution that contains 63.3 ppm of $K_3Fe(CN)_6$ (329.2 g/mol)?

This solution contains 63.3 g of solute per 10^6 g of solution. The density of such a dilute aqueous solution will be that of pure water, or about 1.00 g/mL or 1000 g/L. Thus, we may write

$$[K^+] = \frac{63.3 \text{ g } K_3Fe(CN)_6}{10^6 \text{ g soln}} \times \frac{10^3 \text{ g soln}}{\text{L soln}} \times \frac{1 \text{ mol } K_3Fe(CN)_6}{329.2 \text{ g } K_3Fe(CN)_6}$$

$$\times \frac{3 \text{ mol } K^+}{\text{mol } K_3Fe(CN)_6} = 5.77 \times 10^{-4} \frac{\text{mol } K^+}{\text{L}}$$

Solution-Diluent Volume Ratios

The composition of a dilute solution is sometimes specified in terms of the volume of a more concentrated solution and the volume of solvent to be used in diluting it. The volume of the former is separated from that of the latter by a colon. Thus, a 1:4 HCl solution contains four volumes of water for each volume of concentrated hydrochloric acid. This method of notation is frequently ambiguous in that the concentration of the original solution is not always obvious to the reader. Unfortunately, some use the term 1:4 to mean dilute one volume with three volumes. In view of such possibilities for misunderstanding, you should be alert to uncertainties associated with solution-diluent ratios when you encounter them.

p-Functions

Scientists frequently express the concentration of a species in terms of its *p-function,* or *p-value.* The p-value is the negative logarithm (to the base 10) of the molar concentration of that species. So, for the species X,

$$pX = -\log [X]$$

A further discussion of the p-function will be found in Section 7B-3, page 130.

6C-2 Density and Specific Gravity of Solutions

Density and specific gravity are terms often encountered in the analytical literature. The *density* of a substance is its mass per unit volume, whereas its *specific gravity* is the ratio of its mass to the mass of an equal volume of water at a specified temperature (ordinarily 4°C). Density has units of kilograms per liter or grams per milliliter in the metric system. Specific gravity is dimensionless and so is not tied to any particular system of units. For this reason, specific gravity is widely used in describing items of commerce (see Figure 6-1). Since the density of water is approximately 1.00 g/mL and since we employ the metric system throughout this text, density and specific gravity are used interchangeably.

Density expresses the mass of a substance per unit volume. In SI units, density is expressed in units of kg/L.

Specific gravity is the ratio of the mass of a substance to the mass of an equal volume of water.

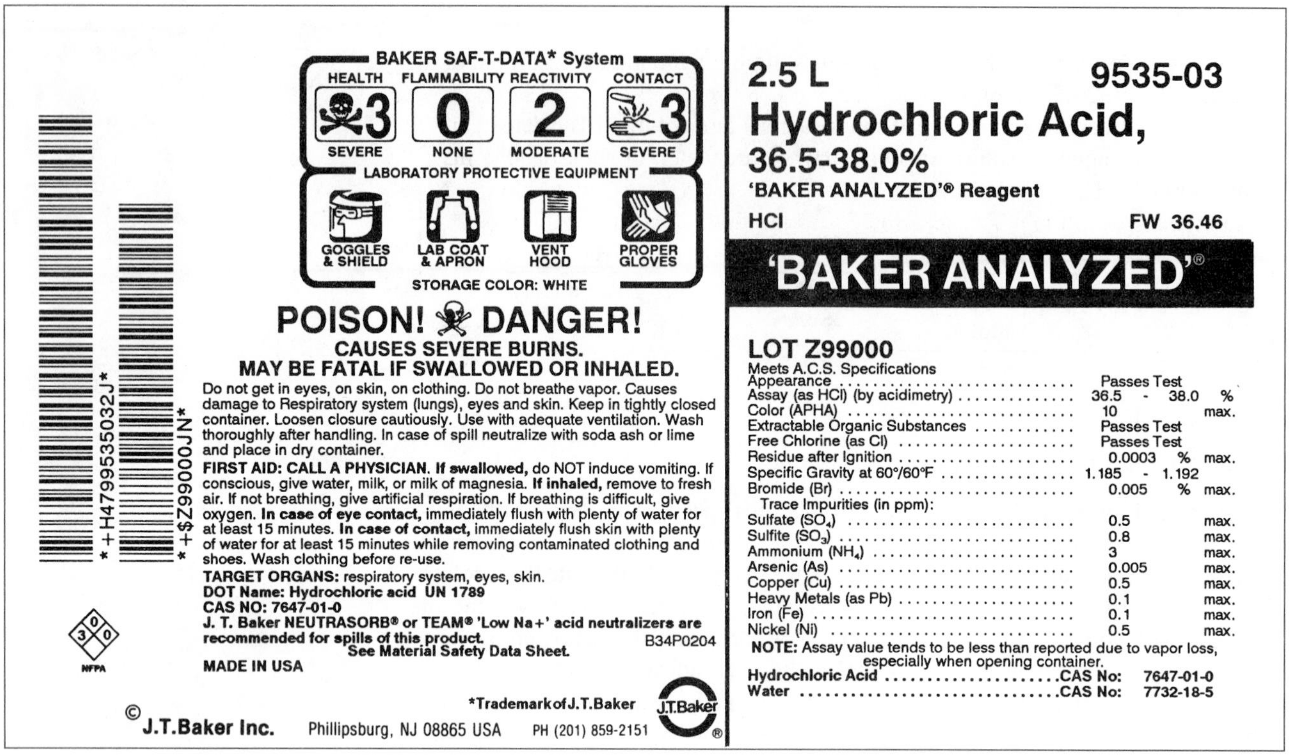

Figure 6-1
Label from a bottle of reagent-grade hydrochloric acid.

Specific Gravities of Commercially Available Concentrated Acids and Bases

Reagent	Typical Concentration, % (w/w)	Typical Specific Gravity
Acetic acid	99.7	1.05
Ammonia	29.0	0.90
Hydrochloric acid	37.2	1.19
Hydrofluoric acid	49.5	1.15
Nitric acid	70.5	1.42
Perchloric acid	71.0	1.67
Phosphoric acid	86.0	1.71
Sulfuric acid	96.5	1.84

The density of a gas is usually expressed in g/L. The density of a liquid or a solid is generally expressed in g/mL.

EXAMPLE 6-6

Calculate the molar concentration of HNO_3 (63.0 g/mol) in a solution that has a specific gravity of 1.42 and is 70% HNO_3 (w/w).

In order to obtain the molarity of the acid, we will first calculate the grams of acid in one liter of solution. We will then convert the grams of acid per liter to moles of acid per liter.

To calculate the grams of acid per liter of concentrated solution we write

$$\frac{\text{g HNO}_3}{\text{L reagent}} = \frac{1.42\ \cancel{\text{g reagent}}}{\cancel{\text{mL reagent}}} \times \frac{10^3\ \cancel{\text{mL reagent}}}{\text{L reagent}} \times \frac{70\ \text{g HNO}_3}{100\ \cancel{\text{g reagent}}}$$

$$= \frac{994\ \text{g HNO}_3}{\text{L reagent}}$$

To convert to moles per liter, we proceed as follows:

$$c_{\text{HNO}_3} = \frac{994\ \cancel{\text{g HNO}_3}}{\text{L reagent}} \times \frac{1\ \text{mol HNO}_3}{63.0\ \cancel{\text{g HNO}_3}} = \frac{15.8\ \text{mol HNO}_3}{\text{L reagent}} = 16\ \text{M}$$

EXAMPLE 6-7

Describe the preparation of 100 mL of 6.0 M HCl from a concentrated solution that has a specific gravity of 1.18 and is 37% (w/w) HCl (36.5 g/mol).

Proceeding as in Example 6-6, we first calculate the molarity of the concentrated reagent. We then calculate the number of moles of acid that we need for the diluted solution. Finally we divide the second figure by the first to obtain the volume of concentrated acid required. Thus, to obtain the molarity of the concentrated reagent, we write

$$c_{HCl} = \frac{1.18 \times 10^3 \ \cancel{\text{g reagent}}}{\text{L reagent}} \times \frac{37 \ \cancel{\text{g HCl}}}{100 \ \cancel{\text{g reagent}}} \times \frac{1 \text{ mol HCl}}{36.5 \ \cancel{\text{g HCl}}} = 12.0 \text{ M}$$

The number of moles HCl required is given by

$$\text{no. mol HCl} = 100 \ \cancel{\text{mL}} \times \frac{1 \ \cancel{\text{L}}}{1000 \ \cancel{\text{mL}}} \times \frac{6.0 \text{ mol HCl}}{\cancel{\text{L}}} = 0.600 \text{ mol}$$

Finally, to obtain the volume of concentrated reagent, we write

$$\text{vol concd reagent} = 0.600 \ \cancel{\text{mol HCl}} \times \frac{1 \text{ L reagent}}{12.0 \ \cancel{\text{mol HCl}}}$$
$$= 0.0500 \text{ L or } 50.0 \text{ mL}$$

Thus dilute 50 mL of the concentrated reagent to 100 mL.

The solution to Example 6-7 is based upon the following useful relationship, which we will be using countless times:

$$V_{concd} \times c_{concd} = V_{dil} \times c_{dil} \qquad \textbf{(6-3)}$$

where the two terms on the left are the volume and molar concentration of a concentrated solution that is being used to prepare a diluted solution with the volume and concentration given by the corresponding terms on the right. This equation is based upon the fact that the number of moles of solute in the diluted solution must equal the number of moles in the concentrated reagent. Note that the volumes can be in milliliters or liters as long as the same units are used for both solutions.

Equation 6-3 can be used with L and mol/L or mL and mmol/mL. Thus,

$$\text{L}_{concd} \times \frac{\text{mol}_{concd}}{\text{L}_{concd}} = \text{L}_{dil} \times \frac{\text{mol}_{dil}}{\text{L}_{dil}}$$

$$\text{mL}_{concd} \times \frac{\text{mmol}_{concd}}{\text{mL}_{concd}} = \text{mL}_{dil} \times \frac{\text{mmol}_{dil}}{\text{mL}_{dil}}$$

6C-3 Some Useful Algebraic Relationships

Most volumetric calculations are based on two pairs of simple equations that are derived from definitions of the millimole, the mole, and the molar concentration. For the chemical species A, we may write

$$\text{amount A (mmol)} = \frac{\text{mass A (g)}}{\text{millimolar mass A (g/mmol)}} \qquad \textbf{(6-4)}$$

$$\text{amount A (mol)} = \frac{\text{mass A (g)}}{\text{molar mass A (g/mol)}} \tag{6-5}$$

The second pair is derived from the definition of molar concentration. That is,

$$\text{amount A (mmol)} = V(\text{mL}) \times c_A(\text{mmol A/mL}) \tag{6-6}$$

$$\text{amount A (mol)} = V(\text{L}) \times c_A(\text{mol A/L}) \tag{6-7}$$

where V is the volume of the solution.

You should use Equations 6-4 and 6-6 when volumes are measured in milliliters and Equations 6-5 and 6-7 when the units are liters.

Units in Using Equations 6-4 Through 6-7 It is useful to know that any combination of grams, moles, and liters can be replaced with any analogous combination expressed in milligrams, millimoles, and milliliters. For example a 0.1 M solution contains 0.1 mol of a species per L or 0.1 mmol per mL. Similarly, the number of moles of a compound is equal to the mass in grams of that compound divided by its molar mass in grams or the mass in milligrams divided by its millimolar mass in milligrams.

6C-4 Calculation of the Molarity of Standard Solutions

The following four examples illustrate how volumetric reagents are prepared.

EXAMPLE 6-8

Describe the preparation of 5.000 L of 0.1000 M Na_2CO_3 (105.99 g/mol) from the primary standard solid.

Since the volume is in liters, we base our calculations on the mole rather than the millimole. Thus to obtain the amount of Na_2CO_3 needed, we write

$$\text{amount } Na_2CO_3 = V_{soln}(\text{L}) \times c_{Na_2CO_3}(\text{mol/L})$$

$$= 5.000\ \cancel{\text{L}} \times \frac{0.1000 \text{ mol } Na_2CO_3}{\cancel{\text{L}}} = 0.5000 \text{ mol } Na_2CO_3$$

To obtain the mass of Na_2CO_3, we rearrange Equation 6-5 to give

$$\text{mass } Na_2CO_3 = 0.5000\ \cancel{\text{mol } Na_2CO_3} \times \frac{105.99 \text{ g } Na_2CO_3}{\cancel{\text{mol } Na_2CO_3}}$$

$$= 53.00 \text{ g } Na_2CO_3$$

Therefore the solution is prepared by dissolving 53.00 g of Na_2CO_3 in water and diluting to exactly 5.000 L.

EXAMPLE 6-9

A standard 0.0100 M solution of Na^+ is required for calibrating a flame photometric method for determining the element. Describe how 500 mL of this solution can be prepared from primary standard Na_2CO_3.

We need to compute the mass of reagent required to give a species molarity of 0.0100. Here, we will use millimoles since the volume is in milliliters. Because Na_2CO_3 dissociates to give two Na^+ ions, we can write that the number of millimoles of Na_2CO_3 needed is

$$\text{amount } Na_2CO_3 = 500\ \cancel{\text{mL}} \times \frac{0.0100\ \cancel{\text{mmol } Na^+}}{\cancel{\text{mL}}} \times \frac{1 \text{ mmol } Na_2CO_3}{2\ \cancel{\text{mmol } Na^+}}$$

$$= 2.50 \text{ mmol}$$

From the definition of millimole, we write

$$\text{mass } Na_2CO_3 = 2.50 \text{ \sout{mmol Na$_2$CO$_3$}} \times 0.10599 \frac{\text{g } Na_2CO_3}{\text{\sout{mmol Na$_2$CO$_3$}}} = 0.265 \text{ g}$$

The solution is therefore prepared by dissolving 0.265 g of Na_2CO_3 in water and diluting to 500 mL.

EXAMPLE 6-10

How would you prepare 50.0-mL portions of standard solutions that are 0.00500 M, 0.00200 M, and 0.00100 M in Na^+ from the solution in Example 6-8?

The number of millimoles of Na^+ taken from the concentrated solution must equal the number in the diluted solutions. Thus,

$$\text{amount } Na^+ \text{ from concd soln} = \text{amount } Na^+ \text{ in dil soln}$$

Recall that the number of millimoles is equal to the number of millimoles per milliliter times the number of milliliters. That is,

$$V_{concd} \times c_{concd} = V_{dil} \times c_{dil}$$

where V_{concd} and V_{dil} are the volumes in milliliters of the concentrated and diluted solutions respectively and c_{concd} and c_{dil} are their Na^+ molar concentrations. This equation rearranges to

$$V_{concd} = \frac{V_{dil} \times c_{dil}}{c_{concd}} = \frac{50.0 \text{ mL} \times 0.00500 \text{ \sout{mmol Na$^+$/mL}}}{0.00100 \text{ \sout{mmol Na$^+$/mL}}} = 25.0 \text{ mL}$$

Thus, to produce 50.0 mL of 0.00500 M Na^+, 25.0 mL of the concentrated solution should be diluted to exactly 50.0 mL.

Repeat the calculation for the other two molarities to confirm that diluting 10.0 and 5.00 mL of the concentrated solution to 50.0 mL produces the desired solutions.

6C-5 Treatment of Titration Data

In this section, we describe two types of volumetric calculations. The first involves computing the molarity of solutions that have been standardized against either a primary standard or another standard solution. The second involves calculating the amount of analyte in a sample from titration data. Both types are based on three algebraic relationships. Two of these are Equations 6-4 and 6-6, both of which are based on millimoles and milliliters. The third relationship is the stoichiometric ratio of the number of millimoles of the analyte and the number of millimoles of titrant.

Calculation of Molarities from Standardization Data

Examples 6-11 and 6-12 illustrate how standardization data are treated.

EXAMPLE 6-11

Exactly 50.00 mL of an HCl solution required 29.71 mL of 0.01963 M $Ba(OH)_2$ to reach an end point with bromocresol green indicator. Calculate the molarity of the HCl.

In the titration, 1 mmol of $Ba(OH)_2$ reacts with 2 mmol of HCl, and thus the stoichiometric ratio is

$$\text{stoichiometric ratio} = \frac{2 \text{ mmol HCl}}{1 \text{ mmol Ba(OH)}_2}$$

The number of millimoles of the standard is obtained by substituting into Equation 6-6:

$$\text{amount Ba(OH)}_2 = 29.71 \cancel{\text{mL Ba(OH)}_2} \times 0.01963 \frac{\text{mmol Ba(OH)}_2}{\cancel{\text{mL Ba(OH)}_2}}$$

To obtain the number of millimoles of HCl, we multiply this result by the ratio derived initially:

$$\text{amount HCl} = (29.71 \times 0.01963) \cancel{\text{mmol Ba(OH)}_2} \times \frac{2 \text{ mmol HCl}}{1 \cancel{\text{mmol Ba(OH)}_2}}$$

To obtain the number of millimoles of HCl per mL, we divide by the volume of the acid. Thus,

$$c_{\text{HCl}} = \frac{(29.71 \times 0.01963 \times 2) \text{ mmol HCl}}{50.0 \text{ mL HCl}}$$

$$= 0.023328 \frac{\text{mmol HCl}}{\text{mL HCl}} = 0.02333 \text{ M}$$

EXAMPLE 6-12

Titration of 0.2121 g of pure $Na_2C_2O_4$ (134.00 g/mol) required 43.31 mL of $KMnO_4$. What is the molarity of the $KMnO_4$ solution? The chemical reaction is

$$2MnO_4^- + 5C_2O_4^{2-} + 16H^+ \longrightarrow 2Mn^{2+} + 10CO_2 + 8H_2O$$

From this equation, we see that the stoichiometric ratio is

$$\text{stoichiometric ratio} = \frac{2 \text{ mmol KMnO}_4}{5 \text{ mmol Na}_2\text{C}_2\text{O}_4}$$

The amount of primary standard $Na_2C_2O_4$ is given by Equation 6-4:

$$\text{amount Na}_2\text{C}_2\text{O}_4 = 0.2121 \cancel{\text{g Na}_2\text{C}_2\text{O}_4} \times \frac{1 \text{ mmol Na}_2\text{C}_2\text{O}_4}{0.13400 \cancel{\text{g Na}_2\text{C}_2\text{O}_4}}$$

To obtain the number of millimoles of $KMnO_4$, we multiply this result by the stoichiometric factor:

$$\text{amount } KMnO_4 = \frac{0.2121}{0.1340} \cancel{\text{mmol } Na_2C_2O_4} \times \frac{2 \text{ mmol } KMnO_4}{5 \cancel{\text{ mmol } Na_2C_2O_4}}$$

The molarity is then obtained by dividing by the volume of $KMnO_4$ consumed. Thus,

$$c_{KMnO_4} = \frac{\left(\frac{0.2121}{0.13400} \times \frac{2}{5}\right) \text{mmol } KMnO_4}{43.31 \text{ mL } KMnO_4} = 0.01462 \text{ M}$$

Note that units are carried through all calculations as a check on the correctness of the relationships used in Examples 6-11 and 6-12.

Calculation of Quantity of Analyte from Titration Data

As shown by the examples that follow, the same systematic approach just described is also used to compute analyte concentrations from titration data.

EXAMPLE 6-13

A 0.8040-g sample of an iron ore is dissolved in acid. The iron is then reduced to Fe^{2+} and titrated with 47.22 mL of 0.02242 M $KMnO_4$ solution. Calculate the results of this analysis in terms of (a) % Fe (55.847 g/mol) and (b) % Fe_3O_4 (231.54 g/mol). The reaction of the analyte with the reagent is described by the equation

$$MnO_4^- + 5Fe^{2+} + 8H^+ \longrightarrow Mn^{2+} + 5Fe^{3+} + 4H_2O$$

$$\text{(a) stoichiometric ratio} = \frac{5 \text{ mmol } Fe^{2+}}{1 \text{ mmol } KMnO_4}$$

$$\text{amount } KMnO_4 = 47.22 \cancel{\text{ mL } KMnO_4} \times \frac{0.02242 \text{ mmol } KMnO_4}{\cancel{\text{mL } KMnO_4}}$$

$$\text{amount } Fe^{2+} = (47.22 \times 0.02242) \cancel{\text{ mmol } KMnO_4} \times \frac{5 \text{ mmol } Fe^{2+}}{1 \cancel{\text{ mmol } KMnO_4}}$$

The mass Fe^{2+} is then given by

$$\text{mass } Fe^{2+} = (47.22 \times 0.02242 \times 5) \cancel{\text{ mmol } Fe^{2+}} \times 0.055847 \frac{\text{g } Fe^{2+}}{\cancel{\text{mmol } Fe^{2+}}}$$

The percent Fe^{2+} is

$$\% \text{ Fe}^{2+} = \frac{(47.22 \times 0.02242 \times 5 \times 0.055847) \text{ g Fe}^{2+}}{0.8040 \text{ g sample}} \times 100\%$$

$$= 36.77\%$$

(b) In order to derive a stoichiometric ratio, we note that

$$5\text{Fe}^{2+} \equiv 1\text{MnO}_4^-$$

Therefore,

$$5\text{Fe}_3\text{O}_4 \equiv 15\text{Fe}^{2+} \equiv 3\text{MnO}_4^-$$

and

$$\text{stoichiometric ratio} = \frac{5 \text{ mmol Fe}_3\text{O}_4}{3 \text{ mmol KMnO}_4}$$

As in part (a),

$$\text{amount KMnO}_4 = 47.22 \text{ }\cancel{\text{mL KMnO}_4} \times 0.02242 \text{ mmol KMnO}_4/\cancel{\text{mL KMnO}_4}$$

$$\text{amount Fe}_3\text{O}_4 = (47.22 \times 0.02242) \text{ }\cancel{\text{mmol KMnO}_4} \times \frac{5 \text{ mmol Fe}_3\text{O}_4}{3 \text{ }\cancel{\text{mmol KMnO}_4}}$$

$$\text{mass Fe}_3\text{O}_4 = (47.22 \times 0.02242 \times \tfrac{5}{3})\cancel{\text{mmol Fe}_3\text{O}_4} \times 0.23154 \frac{\text{g Fe}_3\text{O}_4}{\cancel{\text{mmol Fe}_3\text{O}_4}}$$

$$\% \text{ Fe}_3\text{O}_4 = \frac{(47.22 \times 0.02242 \times \tfrac{5}{3}) \times 0.23154 \text{ g Fe}_3\text{O}_4}{0.8040 \text{ g sample}} \times 100\%$$

$$= 50.81\%$$

FEATURE 6-1

Another Approach to Example 6-13(a)

Some people find it easier to write out the solution to a problem in such a way that the units in the denominator of each succeeding term eliminate the units in the numerator of the preceding one until the units of the answer are obtained. For example, the solution to part (a) of Example 6-13 can be written

$$47.22 \text{ }\cancel{\text{mL KMnO}_4} \times \frac{0.02242 \text{ }\cancel{\text{mmol KMnO}_4}}{\cancel{\text{mL KMnO}_4}} \times \frac{5 \text{ }\cancel{\text{mmol Fe}}}{1 \text{ }\cancel{\text{mmol KMnO}_4}} \times \frac{0.05585 \text{ g Fe}}{\cancel{\text{mmol Fe}}} \times \frac{1}{0.8040 \text{ g sample}} \times 100\% = 36.77\% \text{ Fe}$$

EXAMPLE 6-14

The organic matter in a 3.776-g sample of a mercuric ointment is decomposed with HNO_3. After dilution, the Hg^{2+} is titrated with 21.30 mL of a 0.1144 M solution of NH_4SCN. Calculate the percent Hg (200.59 g/mol) in the ointment.

This titration involves the formation of a stable neutral complex, $Hg(SCN)_2$:

$$Hg^{2+} + 2SCN^- \longrightarrow Hg(SCN)_2(aq)$$

At the equivalence point,

$$\text{stoichiometric ratio} = \frac{1 \text{ mmol } Hg^{2+}}{2 \text{ mmol } NH_4SCN}$$

$$\text{amount } NH_4SCN = 21.30 \cancel{\text{ mL } NH_4SCN} \times 0.1144 \frac{\text{mmol } NH_4SCN}{\cancel{\text{mL } NH_4SCN}}$$

$$\text{amount } Hg^{2+} = (21.30 \times 0.1144) \cancel{\text{ mmol } NH_4SCN} \times \frac{1 \text{ mmol } Hg^{2+}}{2 \cancel{\text{ mmol } NH_4SCN}}$$

$$\text{mass } Hg^{2+} = (21.30 \times 0.1144 \times \tfrac{1}{2}) \cancel{\text{mmol } Hg^{2+}} \times \frac{0.20059 \text{ g } Hg^{2+}}{\cancel{\text{mmol } Hg^{2+}}}$$

$$\% \text{ Hg} = \frac{(21.30 \times 0.1144 \times \frac{1}{2}) \times 0.20059 \text{ g } Hg^{2+}}{3.776 \text{ g sample}} \times 100\%$$

$$= 6.472\% = 6.47\%$$

Rounding the Answer to Example 6-14

You should note that the input data for Example 6-14 all contained four or more significant figures but the answer was rounded to three. Why is this?

Let us make the rounding decision by doing a couple of rough calculations in our heads. We will assume that the input data are uncertain to 1 part in the last significant figure. The largest *relative* error will then be associated with the molarity of the reagent. Here, the relative uncertainty is 0.0001/0.1144. But we really do not need to know the error this accurately, and we can simply estimate that the uncertainty is about 1 part in 1000 (compared with about 1 part in 2000 for the volume and 1 part in 3700 for the mass). We then assume the calculated result is uncertain to about the same amount as the least accurate measurement, or 1 part in 1000. The absolute uncertainty of the final result is then $6.472\% \times 1/1000 = 0.0065 = 0.01\%$, and we round to the second figure to the right of the decimal point. Thus, we report 6.47%.

You should practice making this rough type of rounding decision whenever you make a computation.

EXAMPLE 6-15

A 0.4755-g sample containing $(NH_4)_2C_2O_4$ and inert materials was dissolved in H_2O and made strongly alkaline with KOH, which converted NH_4^+ to NH_3. The liberated NH_3 was distilled into exactly 50.00 mL of 0.05035 M H_2SO_4. The excess H_2SO_4 was back-titrated with 11.13 mL of 0.1214 M NaOH. Calculate (a) the percent N (14.007 g/mol) and (b) the percent $(NH_4)_2C_2O_4$ (124.10 g/mol) in the sample.

(a) The H_2SO_4 reacts with both NH_3 and NaOH, and two stoichiometric ratios can be derived. These are

$$\frac{2 \text{ mmol } NH_3}{1 \text{ mmol } H_2SO_4} \quad \text{and} \quad \frac{1 \text{ mmol } H_2SO_4}{2 \text{ mmol NaOH}}$$

$$\text{total amount } H_2SO_4 = 50.00 \cancel{\text{ mL } H_2SO_4} \times 0.05035 \frac{\text{mmol } H_2SO_4}{\cancel{\text{mL } H_2SO_4}}$$

$$= 2.5175 \text{ mmol } H_2SO_4$$

The amount of H_2SO_4 consumed by the NaOH in the back-titration is

$$\text{amount } H_2SO_4 = (11.13 \times 0.1214) \cancel{\text{ mmol NaOH}} \times \frac{1 \text{ mmol } H_2SO_4}{2 \cancel{\text{ mmol NaOH}}}$$

$$= 0.6756 \text{ mmol } H_2SO_4$$

The amount of H_2SO_4 that reacted with NH_3 is then

$$\text{amount } H_2SO_4 = (2.5175 - 0.6756) \text{ mmol } H_2SO_4$$
$$= 1.8419 \text{ mmol } H_2SO_4$$

The amount of NH_3, which is equal to the number of millimoles of N, is

$$\text{amount N} = \text{no. mmol } NH_3 = 1.8419 \cancel{\text{mmol } H_2SO_4} \times \frac{2 \text{ mmol N}}{1 \cancel{\text{mmol } H_2SO_4}}$$
$$= 3.6838 \text{ mmol N}$$

$$\% \text{ N} = \frac{3.6838 \cancel{\text{mmol N}} \times 0.014007 \text{ g N}/\cancel{\text{mmol N}}}{0.4755 \text{ g sample}} \times 100\% = 10.85\%$$

(b) Since each millimole of $(NH_4)_2C_2O_4$ produces 2 mmol of NH_3, which reacts with 1 mmol of H_2SO_4,

$$\text{stoichiometric ratio} = \frac{1 \text{ mmol } (NH_4)_2C_2O_4}{1 \text{ mmol } H_2SO_4}$$

$$\text{amount } (NH_4)_2C_2O_4 = 1.8419 \cancel{\text{mmol } H_2SO_4} \times \frac{1 \text{ mmol } (NH_4)_2C_2O_4}{1 \cancel{\text{mmol } H_2SO_4}}$$

$$\text{mass } (NH_4)_2C_2O_4 = 1.81419 \cancel{\text{mmol } (NH_4)_2C_2O_4} \times \frac{0.12410 \text{ g } (NH_4)_2C_2O_4}{\cancel{\text{mmol } (NH_4)_2C_2O_4}}$$
$$= 0.22858 \text{ g}$$

$$\% \ (NH_4)_2C_2O_4 = \frac{0.22858 \text{ g } (NH_4)_2C_2O_4}{0.4755 \text{ g sample}} \times 100\% = 48.07\%$$

EXAMPLE 6-16

The CO in a 20.3-L sample of gas was converted to CO_2 by passing the gas over iodine pentoxide heated to 150°C:

$$I_2O_5(s) + 5CO(g) \longrightarrow 5CO_2(g) + I_2(g)$$

The iodine distilled at this temperature and was collected in an absorber containing 8.25 mL of 0.01101 M $Na_2S_2O_3$:

$$I_2(aq) + 2S_2O_3^{2-}(aq) \longrightarrow 2I^-(aq) + S_4O_6^{2-}(aq)$$

The excess $Na_2S_2O_3$ was back-titrated with 2.16 mL of 0.00947 M I_2 solution. Calculate the number of milligrams of CO (28.01 g/mol) per liter of sample.

Based on the two reactions, the stoichiometric ratios are

$$\frac{5 \text{ mmol CO}}{1 \text{ mmol } I_2} \quad \text{and} \quad \frac{2 \text{ mmol } Na_2S_2O_3}{1 \text{ mmol } I_2}$$

We divide the first ratio by the second to get a third useful ratio:

$$\frac{5 \text{ mmol CO}}{2 \text{ mmol Na}_2\text{S}_2\text{O}_3}$$

This relationship reveals that 5 mmol of CO are responsible for the consumption of 2 mmol $Na_2S_2O_3$. The total amount of $Na_2S_2O_3$ is

$$\text{amount Na}_2\text{S}_2\text{O}_3 = 8.25 \cancel{\text{mL Na}_2\text{S}_2\text{O}_3} \times 0.01101 \frac{\text{mmol Na}_2\text{S}_2\text{O}_3}{\cancel{\text{mL Na}_2\text{S}_2\text{O}_3}}$$
$$= 0.09083 \text{ mmol Na}_2\text{S}_2\text{O}_3$$

The amount of $Na_2S_2O_3$ used in the back-titration is

$$\text{amount Na}_2\text{S}_2\text{O}_3 = 2.16 \cancel{\text{mL I}_2} \times 0.00947 \frac{\cancel{\text{mmol I}_2}}{\cancel{\text{mL I}_2}} \times \frac{2 \text{ mmol Na}_2\text{S}_2\text{O}_3}{\cancel{\text{mmol I}_2}}$$
$$= 0.04091 \text{ mmol Na}_2\text{S}_2\text{O}_3$$

The number of millimoles of CO can then be obtained by employing the third stoichiometric ratio:

$$\text{amount CO} = (0.09083 - 0.04091) \cancel{\text{mmol Na}_2\text{S}_2\text{O}_3} \times \frac{5 \text{ mmol CO}}{2 \cancel{\text{mmol Na}_2\text{S}_2\text{O}_3}}$$
$$= 0.1248 \text{ mmol CO}$$

$$\text{mass CO} = 0.1248 \cancel{\text{mmol CO}} \times \frac{28.01 \text{ mg CO}}{\cancel{\text{mmol CO}}} = 3.4956 \text{ mg}$$

$$\frac{\text{mass CO}}{\text{vol sample}} = \frac{3.4956 \text{ mg CO}}{20.3 \text{ L sample}} = 0.172 \frac{\text{mg CO}}{\text{L}}$$

6D GRAVIMETRIC TITRIMETRY

Weight or *gravimetric titrimetry* differs from its volumetric counterpart in that the *mass* of titrant is measured rather than the volume. Thus, in a weight titration, a balance and a solution dispenser are substituted for a buret and its markings. Weight titrimetry actually predates volumetric titrimetry by more than 50 years.[2] With the advent of reliable burets, however, weight titrations were largely supplanted by volumetric methods because the former required relatively elaborate equipment and were tedious and time-consuming. The recent advent of sensitive top-loading balances and convenient plastic solution dispensers has changed this situation completely, and weight titrations can now be performed more easily and more rapidly than volumetric titrations.

---

[2]For a brief history of gravimetric and volumetric titrimetry, see B. Kratochvil and C. Maitra, *Amer. Lab.*, **1982** (1), 22.

6D-1 Calculations Associated with Weight Titrations

The most convenient unit of concentration for weight titrations is *weight molarity* M_w, which is the number of moles of a reagent in one kilogram of solution or the number of millimoles in one gram of solution. Thus aqueous 0.1 M_w NaCl contains 0.1 mol of the salt in 1 kg of solution or 0.1 mmol in 1 g of the solution.

The weight molarity $c_{w(A)}$ of a solution of a solute A is computed by means of either of two equations that are analogous to Equation 6-1:

$$c_{w(A)} = \frac{\text{no. mol A}}{\text{no. kg solution}} = \frac{\text{no. mmol A}}{\text{no. g solution}} \qquad (6\text{-}8)$$

Weight titration data can then be treated by the methods illustrated in Sections 6C-4 and 6C-5 after substituting weight molarity for molarity and grams and kilograms for milliliters and liters.

6D-2 Advantages of Weight Titrations

In addition to greater speed and convenience, weight titrations offer certain other advantages over their volumetric counterparts:

1. Calibration of glassware and tedious cleaning to ensure proper drainage is avoided.
2. Temperature corrections are unnecessary because weight molarity does not change with temperature in contrast to volume molarity. This advantage is particularly important in titrations that are performed in nonaqueous solutions because of the high coefficients of expansion of most organic liquids (about ten times that of water).
3. Weight measurements can be made with considerably greater precision and accuracy than can volume measurements. For example, 50 or 100 g of an aqueous solution can be readily measured to ± 1 mg, which corresponds to ± 0.001 mL. This greater sensitivity makes it possible to choose sample sizes that lead to significantly smaller consumption of standard reagents.
4. Weight titrations are more easily automated than are volumetric titrations.

6E QUESTIONS AND PROBLEMS

6-1. Write two equations that—along with the stoichiometric factor—form the basis for the calculations of volumetric titrimetry.

6-2. Define
- *(a) millimole.
- (b) titration.
- *(c) stoichiometric factor.
- (d) titration error.

6-3. Distinguish between
- *(a) the equivalence point and the end point of a titration.
- (b) the density and the specific gravity of a solution.
- *(c) a primary standard and a secondary standard.

***6-4.** Briefly explain why milligrams of solute per liter and parts per million can be used interchangeably to describe the concentration of a dilute aqueous solution.

6-5. Calculations of volumetric analysis ordinarily consist of transforming the quantity of titrant used (in chemical units) to a chemically equivalent quantity of analyte (also in chemical units) through use of a stoichiometric factor. Use chemical formulas (NO CALCULATIONS REQUIRED) to express this factor to calculate the percentage of
- *(a) hydrazine in rocket fuel by titration with standard iodine. Reaction:

$$H_2NNH_2 + 2I_2 \longrightarrow N_2(g) + 4I^- + 4H^+$$

(b) hydrogen peroxide in a cosmetic preparation by titration with standard permanganate. Reaction:

$$5H_2O_2 + 2MnO_4^- + 6H^+ \longrightarrow 2Mn^{2+} + 5O_2(g) + 8H_2O$$

*(c) boron in a sample of borax, $Na_2B_4O_7 \cdot 10\,H_2O$ by titration with standard acid. Reaction:

$$B_4O_7^{2-} + 2H^+ + 5H_2O \longrightarrow 4H_3BO_3$$

(d) sulfur in an agricultural spray that was converted to thiocyanate with an unmeasured excess of cyanide. Reaction:

$$S(s) + CN^- \longrightarrow SCN^-$$

After removal of the excess cyanide, the thiocyanate was titrated with a standard potassium iodate solution in strong HCl. Reaction:

$$2SCN^- + 3IO_3^- + 2H^+ + 6Cl^- \longrightarrow 2SO_4^{2-} + 2CN^- + 3ICl_2^- + H_2O$$

*6-6. How many millimoles of solute are contained in
(a) 2.00 L of 2.76×10^{-3} M $KMnO_4$?
(b) 750 mL of 0.0416 M KSCN?
(c) 250 mL of a solution that contains 4.20 ppm of $CuSO_4$?
(d) 3.50 L of 0.276 M KCl?

6-7. How many millimoles of solute are contained in
(a) 4.25 mL of 0.0917 M KH_2PO_4?
(b) 0.1020 L of 0.0643 M $HgCl_2$?
(c) 2.81 L of a 49.0 ppm solution of $Mg(NO_3)_2$?
(d) 79.8 mL of 0.1379 M NH_4VO_3?

6-8. How many milligrams of solute are contained in
*(a) 26.0 mL of 0.150 M sucrose (342 g/mol)?
*(b) 2.92 L of 5.23×10^{-3} M H_2O_2?
(c) 737 mL of a solution that contains 6.38 ppm of $Pb(NO_3)_2$?
(d) 6.75 mL of 0.0619 M KNO_3?

6-9. How many grams of solute are contained in
*(a) 450 mL of 0.164 M H_2O_2?
*(b) 27.0 mL of 8.75×10^{-4} M benzoic acid (122 g/mol)?
(c) 3.50 L of a solution that contains 21.7 ppm of $SnCl_2$?
(d) 21.7 mL of 0.0125 M $KBrO_3$?

*6-10. Calculate the molar concentration of a solution that is 50.0% in NaOH (w/w) and has a specific gravity of 1.52.

6-11. Calculate the molar concentration of a 20.0% solution (w/w) of KCl that has a specific gravity of 1.13.

6-12. Calculate the molar analytical concentration of solute in an aqueous solution that is
*(a) 11.00% (w/w) NH_3 and has a density of 0.9538.
(b) 18.00% (w/w) KBr and has a density of 1.149.
*(c) 28.00% (w/w) ethylene glycol (62.07 g/mol) and has a density of 1.0350.
(d) 15.00% (w/w) sucrose (342.5 g/mol) and has a density of 1.0592.

*6-13. How would you prepare
(a) 500 mL of 16.0% (w/v) aqueous ethanol (46.1 g/mol)?
(b) 500 mL of 16.0% (v/v) aqueous ethanol?
(c) 500 g of 16.0% (w/w) aqueous ethanol?

6-14. Describe the preparation of
(a) 250 mL of 20.0% (w/v) aqueous acetone (58.05 g/mol).
(b) 250 mL of 20.0% (v/v) aqueous acetone.
(c) 250 mL of 20.0% (w/w) aqueous acetone.

*6-15. Describe how you would prepare 2.00 L of 0.150 M perchloric acid from a concentrated solution that has a specific gravity of 1.66 and is 70% $HClO_4$ (w/w).

6-16. Describe how you would prepare 800 mL of 0.400 M aqueous ammonia from a concentrated solution that has a specific gravity of 0.90 and is 27% NH_3 (w/w).

*6-17. Describe the preparation of
(a) 500 mL of 0.0750 M $AgNO_3$ from the solid reagent.
(b) 1.00 L of 0.315 M HCl, starting with a 6.00 M solution of the reagent.
(c) 600 mL of a solution that is 0.0825 M in K^+, starting with solid $K_4Fe(CN)_6$.
(d) 400 mL of 3.00% (w/v) aqueous $BaCl_2$ from a 0.400 M $BaCl_2$ solution.
(e) 2.00 L of 0.120 M $HClO_4$ from the commercial reagent [60% $HClO_4$ (w/w), sp gr 1.60].
(f) 9.00 L of a solution that is 60.0 ppm in Na^+, starting with solid Na_2SO_4.

6-18. Describe the preparation of
(a) 5.00 L of 0.150 M $KMnO_4$ from the solid reagent.
(b) 4.00 L of 0.175 M $HClO_4$, starting with an 8.00 M solution of the reagent.
(c) 400 mL of a solution that is 0.0500 M in I^-, starting with MgI_2.
(d) 200 mL of 1.00% (w/v) aqueous $CuSO_4$, from a 0.218 M $CuSO_4$ solution.
(e) 1.50 L of 0.215 M NaOH from the concentrated commercial reagent [50% NaOH (w/w), sp gr 1.525].
(f) 1.50 L of a solution that is 12.0 ppm in K^+, starting with solid $K_4Fe(CN)_6$.

*6-19. A solution of $HClO_4$ was standardized by dissolving 0.3745 g of primary-standard-grade HgO in a solution of KBr:

$$HgO(s) + 4Br^- + H_2O \longrightarrow HgBr_4^{2-} + 2OH^-$$

The liberated OH^- required 37.79 mL of the acid to be neutralized. Calculate the molarity of the $HClO_4$.

6-20. A 0.3367-g sample of primary-standard-grade Na_2CO_3 required 28.66 mL of a H_2SO_4 solution to reach the end point in the reaction

$$CO_3^{2-} + 2H^+ \longrightarrow H_2O + CO_2(g)$$

What is the molarity of the H_2SO_4?

***6-21.** A 0.3396-g sample that assayed 96.4% Na_2SO_4 required 37.70 mL of a barium chloride solution. Reaction:

$$Ba^{2+} + SO_4^{2-} \longrightarrow BaSO_4$$

Calculate the analytical molarity of $BaCl_2$ in the solution.

***6-22.** A 0.4793-g sample of primary-standard Na_2CO_3 was treated with 40.00 mL of dilute perchloric acid. The solution was boiled to remove CO_2, following which the excess $HClO_4$ was back-titrated with 8.70 mL of dilute NaOH. In a separate experiment it was established that 27.43 mL of the $HClO_4$ neutralized the NaOH in a 25.00-mL portion. Calculate the molarities of the $HClO_4$ and NaOH.

6-23. Titration of 50.00 mL of 0.05251 M $Na_2C_2O_4$ required 38.71 mL of a potassium permanganate solution:

$$2MnO_4^- + 5H_2C_2O_4 + 6H^+ \longrightarrow 2Mn^{2+} + 10CO_2(g) + 8H_2O$$

Calculate the molarity of the $KMnO_4$ solution.

***6-24.** Titration of the I_2 produced from 0.1238 g of primary-standard KIO_3 required 41.27 mL of sodium thiosulfate:

$$IO_3^- + 5I^- + 6H^+ \longrightarrow 3I_2 + 3H_2O$$

$$I_2 + 2S_2O_3^{2-} \longrightarrow 2I^- + S_4O_6^{2-}$$

Calculate the concentration of the $Na_2S_2O_3$.

***6-25.** A 4.476-g sample of a petroleum product was burned in a tube furnace, and the SO_2 produced was collected in 3% H_2O_2. Reaction:

$$SO_2(g) + H_2O_2 \longrightarrow H_2SO_4$$

A 25.00-mL portion of 0.00923 M NaOH was introduced into the solution of H_2SO_4, following which the excess base was back-titrated with 13.33 mL of 0.01007 M HCl. Calculate the parts per million of sulfur in the sample.

6-26. A 100.0-mL sample of spring water was treated to convert any iron present to Fe^{2+}. Addition of 25.00 mL of 0.002107 M $K_2Cr_2O_7$ resulted in the reaction

$$6Fe^{2+} + Cr_2O_7^{2-} + 14H^+ \longrightarrow 6Fe^{3+} + 2Cr^{3+} + 7H_2O$$

The excess $K_2Cr_2O_7$ was back-titrated with 7.47 mL of a 0.00979 M Fe^{2+} solution. Calculate the parts per million of iron in the sample.

***6-27.** The arsenic in a 1.223-g sample of a pesticide was converted to H_3AsO_4 by suitable treatment. The acid was then neutralized, and exactly 40.00 mL of 0.07891 M $AgNO_3$ was added to precipitate the arsenic quantitatively as Ag_3AsO_4. The excess Ag^+ in the filtrate and washings from the precipitate was titrated with 11.27 mL of 0.1000 M KSCN; the reaction was

$$Ag^+ + SCN^- \longrightarrow AgSCN(s)$$

Calculate the percent As_2O_3 in the sample.

***6-28.** The thiourea in a 1.455-g sample of organic material was extracted into a dilute H_2SO_4 solution and titrated with 37.31 mL of 0.009372 M Hg^{2+} via the reaction

$$4(NH_2)_2CS + Hg^{2+} \longrightarrow [(NH_2)_2CS]_4Hg^{2+}$$

Calculate the percent $(NH_2)_2CS$ (76.12 g/mol) in the sample.

6-29. The ethyl acetate concentration in an alcoholic solution was determined by diluting a 10.00-mL sample to exactly 100 mL. A 20.00-mL portion of the diluted solution was refluxed with 40.00 mL of 0.04672 M KOH:

$$CH_3COOC_2H_5 + OH^- \longrightarrow CH_3COO^- + C_2H_5OH$$

After cooling, the excess OH^- was back-titrated with 3.41 mL of 0.05042 M H_2SO_4. Calculate the number of grams of ethyl acetate (88.11 g/mol) per 100 mL of the original sample.

***6-30.** A solution of $Ba(OH)_2$ was standardized against 0.1016 g of primary-standard-grade benzoic acid C_6H_5COOH (122.12 g/mol). An end point was observed after addition of 44.42 mL of base.

(a) Calculate the molarity of the base.

(b) Calculate the standard deviation of the molarity if the standard deviation for weighing was ±0.2 mg and that for the volume measurement was ±0.03 mL.

(c) Assuming an error of −0.3 mg in the weighing, calculate the absolute and relative systematic error in the molarity.

6-31. A 0.1475 M solution of $Ba(OH)_2$ was used to titrate the acetic acid (60.05 g/mol) in a dilute aqueous solution. The following results were obtained.

Sample	Sample Volume, mL	$Ba(OH)_2$ Volume, mL
1	50.00	43.17
2	49.50	42.68
3	25.00	21.47
4	50.00	43.33

(a) Calculate the mean w/v percentage of acetic acid in the sample.

(b) Calculate the standard deviation for the results.

(c) Calculate the 90% confidence interval for the mean.

(d) At the 90% confidence level, could any of the results be discarded?

(e) Assume that the buret used to measure out the acetic acid had a systematic error of −0.05 mL at all volumes delivered. Calculate the systematic error in the mean result.

*6-32. (a) A 0.3147-g sample of primary standard $Na_2C_2O_4$ was dissolved in dilute H_2SO_4 and titrated with 31.672 g of dilute $KMnO_4$:

$$2MnO_4^- + 5C_2O_4^{2-} + 16H^+ \longrightarrow 2Mn^{2+} + 10CO_2(g) + 8H_2O$$

Calculate the weight molarity of the $KMnO_4$ solution.

(b) The iron in a 0.6656-g ore sample was reduced quantitatively to the +2 state and then titrated with 26.753 g of the $KMnO_4$ solution. Calculate the percent Fe_2O_3 in the sample.

6-33. (a) A 0.1752-g sample of primary standard $AgNO_3$ was dissolved in 502.3 g of distilled water. Calculate the weight molarity of Ag^+ in this solution.

(b) The standard solution described in part (a) was used to titrate a 25.171-g sample of a KSCN solution. An end point was obtained after adding 23.765 g of the $AgNO_3$ solution. Calculate the weight molarity of the KSCN solution.

(c) The solutions described in parts (a) and (b) were used to determine the $BaCl_2 \cdot 2H_2O$ in a 0.7120-g sample. A 20.102-g sample of the $AgNO_3$ was added to a solution of the sample, and the excess $AgNO_3$ was back-titrated with 7.543 g of the KSCN solution. Calculate the percent $BaCl_2 \cdot 2H_2O$ in the sample.

*6-34. A solution was prepared by dissolving 10.12 g of $KCl \cdot MgCl_2 \cdot 6H_2O$ (277.85 g/mol) in sufficient water to give 2.000 L. Calculate

(a) the molar analytical concentration of $KCl \cdot MgCl_2$ in this solution.

(b) the molar concentration of Mg^{2+}.

(c) the molar concentration of Cl^-.

(d) the weight/volume percentage of $KCl \cdot MgCl_2 \cdot 6H_2O$.

(e) the millimoles of Cl^- in 25.0 mL of this solution.

(f) ppm K^+.

6-35. A solution was prepared by dissolving 367 mg of $K_3Fe(CN)_6$ (329.2 g/mol) in sufficient water to give 750 mL. Calculate

(a) the molar analytical concentration of $K_3Fe(CN)_6$.

(b) the molar concentration of K^+.

(c) the molar concentration of $Fe(CN)_6^{3-}$.

(d) the weight/volume percentage of $K_3Fe(CN)_6$.

(e) the millimoles of K^+ in 50.0 mL of this solution.

(f) ppm $Fe(CN)_6^{3-}$.

CHAPTER

7

Aqueous-Solution Chemistry

Chapter 7 provides a discussion of aqueous-solution chemistry, including chemical equilibrium and basic equilibrium-constant calculations. This material is treated in most general chemistry courses.

7A THE CHEMICAL COMPOSITION OF AQUEOUS SOLUTIONS

Water is the most plentiful solvent available on earth, is easily purified, and is not toxic. It therefore finds widespread use as a medium for carrying out chemical analyses.

7A-1 Solutions of Electrolytes

A **salt** is produced in the reaction of an acid with a base. Examples include NaCl, Na_2SO_4, and $NaOOCCH_3$ (sodium acetate).

An **acid** is a substance that donates protons; a **base** is a substance that accepts protons.

Most of the solutes we will discuss are *electrolytes,* which form ions when dissolved in water (or certain other solvents) to produce solutions that conduct electricity. *Strong electrolytes* ionize essentially completely in a solvent, whereas *weak electrolytes* ionize only partially. Therefore, weak electrolytes impart less conductivity to a solvent than do strong electrolytes. Table 7-1 is a compilation of solutes that act as strong and weak electrolytes in water. Among the strong electrolytes listed are acids, bases, and salts.

7A-2 Acids and Bases

An acid donates protons only in the presence of a proton acceptor (a base). Likewise, a base accepts protons only in the presence of a proton donor (an acid).

In 1923, two chemists, J. N. Brønsted in Denmark and J. M. Lowry in England, proposed independently a theory of acid/base behavior that is particularly useful in analytical chemistry. According to the Brønsted-Lowry theory, *an acid is a proton donor* and *a base is a proton acceptor.* In order for a species to behave as an acid, a proton acceptor (or base) must be present. The reverse is also true.

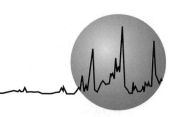

TABLE 7-1 Classification of Electrolytes

Strong	Weak
1. The inorganic acids HNO_3, $HClO_4$, H_2SO_4,* HCl, HI, HBr, $HClO_3$, $HBrO_3$ 2. Alkali and alkaline-earth hydroxides 3. Most salts	1. Many inorganic acids, including H_2CO_3, H_3BO_3, H_3PO_4, H_2S, H_2SO_3 2. Most organic acids 3. Ammonia and most organic bases 4. Halides, cyanides, and thiocyanates of Hg, Zn, and Cd

*H_2SO_4 is completely dissociated into HSO_4^- and H_3O^+ ions and for this reason is classified as a strong electrolyte. However, it should be noted that the HSO_4^- ion is a weak electrolyte, being only partially dissociated into SO_4^{2-} and H_3O^+.

Conjugate Acids and Bases

An important feature of the Brønsted-Lowry concept is the idea that the entity produced when an acid gives up a proton is itself a potential proton acceptor and is called a *conjugate base* of the parent acid. For example, when the species acid_1 gives up a proton, the species base_1 is formed, as shown by the reaction

$$\text{acid}_1 \rightleftharpoons \text{base}_1 + \text{proton}$$

Here, acid_1 and base_1 are a conjugate acid/base pair.

Similarly, every base produces a *conjugate acid* as a result of accepting a proton. That is,

$$\text{base}_2 + \text{proton} \rightleftharpoons \text{acid}_2$$

When these two processes are combined, the result is an acid/base, or neutralization, reaction:

$$\text{acid}_1 + \text{base}_2 \rightleftharpoons \text{base}_1 + \text{acid}_2$$

The extent to which this reaction proceeds depends upon the relative tendencies of the two bases to accept a proton (or the two acids to donate a proton).

Examples of conjugate acid/base relationships are shown in Equations 7-1 through 7-4.

Many solvents are proton donors or proton acceptors and can thus induce basic or acidic behavior in solutes dissolved in them. For example, in an aqueous solution of ammonia, water can donate a proton and thus acts as an acid with respect to the solute:

$$\underset{\text{base}_1}{NH_3} + \underset{\text{acid}_2}{H_2O} \rightleftharpoons \underset{\text{acid}_1}{NH_4^+} + \underset{\text{base}_2}{OH^-} \qquad (7\text{-}1)$$

In this reaction ammonia (base_1) reacts with water, which is labeled acid_2 to give the conjugate acid ammonium ion (acid_1) and hydroxide ion, which is the conju-

A **conjugate base** is the species formed when an acid loses a proton. For example, acetate ion is the conjugate base of acetic acid; similarly, ammonium ion is the conjugate acid of ammonia.

A **conjugate acid** is the species formed when a base accepts a proton.

A substance acts as an acid only in the presence of a base, and vice versa.

gate base (base$_2$) of the acid water. In contrast, water acts as a proton acceptor, or base in an aqueous solution of nitrous acid:

$$\underset{\text{base}_1}{H_2O} + \underset{\text{acid}_2}{HNO_2} \rightleftharpoons \underset{\text{acid}_1}{H_3O^+} + \underset{\text{base}_2}{NO_2^-} \qquad (7\text{-}2)$$

The conjugate base of the acid HNO_2 is nitrite ion. The conjugate acid of water is the hydrated proton written as H_3O^+. This species is called the *hydronium* ion and consists of a proton covalently bonded to one water molecule. Higher hydrates such as $H_5O_2^+$ and $H_9O_4^+$ also exist in aqueous solution of protons. For convenience, however, chemists generally use the notations H_3O^+ or, more simply, H^+ in writing chemical equations in which the proton is involved.

The species $H_9O_4^+$ appears to be the predominant form of the hydrated proton. Its structure is

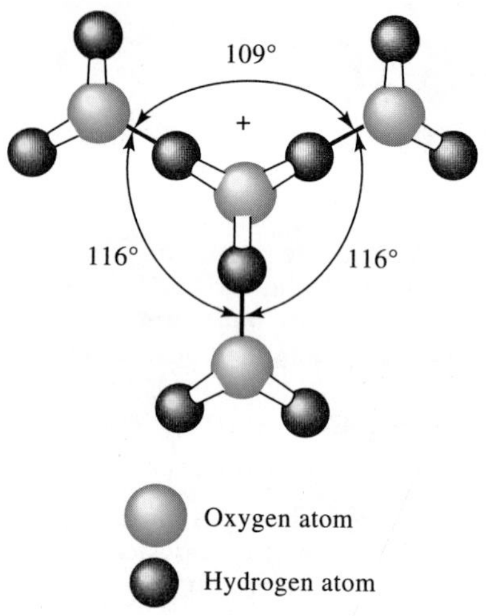

It is important to emphasize that an acid that has donated a proton becomes a conjugate base capable of accepting a proton to reform the original acid; the converse holds equally well. Thus, nitrite ion, the species produced by the loss of a proton from nitrous acid, is a potential acceptor of a proton from a suitable donor. It is this reaction that causes an aqueous solution of sodium nitrite to be slightly basic:

$$\underset{\text{base}_1}{NO_2^-} + \underset{\text{acid}_2}{H_2O} \rightleftharpoons \underset{\text{acid}_1}{HNO_2} + \underset{\text{base}_2}{OH^-}$$

FEATURE 7-1
Amphiprotic Species

Some solutes have both acidic and basic properties. An example is dihydrogen phosphate ion, $H_2PO_4^-$, which behaves as a base in the presence of a proton donor such as H_3O^+.

$$\underset{\text{base}_1}{H_2PO_4^-} + \underset{\text{acid}_2}{H_3O^+} \rightleftharpoons \underset{\text{acid}_1}{H_3PO_4} + \underset{\text{base}_2}{H_2O}$$

Here, H_3PO_4 is the conjugate acid of the original base. In the presence of a proton acceptor, such as hydroxide ion, however, $H_2PO_4^-$ behaves as an acid and forms the conjugate base HPO_4^{2-}.

$$\underset{\text{acid}_1}{H_2PO_4^-} + \underset{\text{base}_2}{OH^-} \rightleftharpoons \underset{\text{base}_1}{HPO_4^{2-}} + \underset{\text{acid}_2}{H_2O}$$

The simple amino acids are an important class of amphiprotic compounds that contain both a weak acid and a weak base functional group. When dissolved in water, an amino acid such as glycine undergoes a kind of internal acid/base reaction to produce a *zwitterion*—a species that bears both a positive and a negative charge. Thus,

A **zwitterion** is an ion that bears both a positive and a negative charge.

$$\underset{\text{glycine}}{NH_2CH_2COOH} \rightleftharpoons \underset{\text{zwitterion}}{NH_3^+CH_2COO^-}$$

This reaction is analogous to the acid/base reaction one observes between a carboxylic acid and an amine:

$$\underset{\text{acid}_1}{R'COOH} + \underset{\text{base}_2}{R''NH_2} \rightleftharpoons \underset{\text{base}_1}{R'COO^-} + \underset{\text{acid}_2}{R''NH_3^+}$$

Amphiprotic Solvents

Water is the classic example of an *amphiprotic* solvent—that is a solvent that can act either as an acid (Equation 7-1) or as a base (Equation 7-2) depending upon the solute. Other common amphiprotic solvents are methanol, ethanol, and anhydrous acetic acid. In methanol, for example, the equilibria analogous to those shown in Equations 7-1 and 7-2 are

Water can act as either an acid or a base.

Amphiprotic solvents behave as acids in the presence of basic solutes and bases in the presence of acidic solutes.

$$\underset{\text{base}_1}{NH_3} + \underset{\text{acid}_2}{CH_3OH} \rightleftharpoons \underset{\text{acid}_1}{NH_4^+} + \underset{\text{base}_2}{CH_3O^-} \qquad (7\text{-}3)$$

$$\underset{\text{base}_1}{CH_3OH} + \underset{\text{acid}_2}{HNO_2} \rightleftharpoons \underset{\text{acid}_1}{CH_3OH_2^+} + \underset{\text{base}_2}{NO_2^-} \qquad (7\text{-}4)$$

Autoprotolysis, also called autoionization, involves the spontaneous reaction of molecules of a substance to give a pair of ions.

7A-3 Autoprotolysis

Amphiprotic solvents undergo self-ionization, or *autoprotolysis,* to form a pair of ionic species. Autoprotolysis is yet another example of acid/base behavior, as illustrated by the following equations.

The **hydronium ion** is the hydrated proton formed when water reacts with an acid. It is usually formulated as H_3O^+ although several higher hydrates exist.

$$\text{base}_1 + \text{acid}_2 \rightleftharpoons \text{acid}_1 + \text{base}_2$$
$$H_2O + H_2O \rightleftharpoons H_3O^+ + OH^-$$
$$CH_3OH + CH_3OH \rightleftharpoons CH_3OH_2^+ + CH_3O^-$$
$$HCOOH + HCOOH \rightleftharpoons HCOOH_2^+ + HCOO^-$$
$$NH_3 + NH_3 \rightleftharpoons NH_4^+ + NH_2^-$$

In this text we shall use the symbol H_3O^+ in those chapters that deal with acid/base equilibria and acid/base equilibrium calculations. In the remaining chapters we simplify to the more convenient H^+, it being understood that this symbol represents the hydrated proton.

The extent to which water undergoes autoprotolysis is slight at room temperature. Thus, the hydronium and hydroxide ion concentrations in pure water are only about 10^{-7} M. Despite the small values of these concentrations, this dissociation reaction is of utmost importance in understanding the behavior of aqueous solutions.

Svante Arrhenius (1859–1927), Swedish chemist, formulated many ideas regarding ionic dissociation in solution. His ideas were not accepted at first; in fact, he was given the lowest possible passing grade for his Ph.D. examination. In 1903, Arrhenius was awarded the Nobel Prize in chemistry for these revolutionary ideas. He was one of the first scientists to suggest the relationship between the amount of carbon dioxide in the atmosphere and global temperature, a phenomenon that has come to be known as the greenhouse effect.

Strongest acid ↑ Weakest acid; Weakest base ↓ Strongest base

$$HClO_4 + H_2O \longrightarrow H_3O^+ + ClO_4^-$$
$$HCl + H_2O \longrightarrow H_3O^+ + Cl^-$$
$$H_3PO_4 + H_2O \rightleftharpoons H_3O^+ + H_2PO_4^-$$
$$Al(H_2O)_6^{3+} + H_2O \rightleftharpoons H_3O^+ + AlOH(H_2O)_5^{2+}$$
$$HC_2H_3O_2 + H_2O \rightleftharpoons H_3O^+ + C_2H_3O_2^-$$
$$H_2PO_4^- + H_2O \rightleftharpoons H_3O^+ + HPO_4^{2-}$$
$$NH_4^+ + H_2O \rightleftharpoons H_3O^+ + NH_3$$

Figure 7-1
Dissociation reactions and relative strengths of some common acids and their conjugate bases. Note that HCl and $HClO_4$ are completely dissociated in water.

7A-4 Strengths of Acids and Bases

The common strong bases include NaOH, KOH, $Ba(OH)_2$, and the quaternary ammonium hydroxides, R_4NOH, where R is an alkyl group such as CH_3 or C_2H_5.

The common strong acids include HCl, $HClO_4$, HNO_3, the first proton in H_2SO_4, HBr, HI, and the organic sulfonic acids, RSO_3H.

Figure 7-1 shows the dissociation reaction of a few common acids in water. The first two are *strong acids* because reaction with the solvent is sufficiently complete as to leave no undissociated solute molecules in aqueous solution. The remainder are *weak acids,* which react incompletely with water to give solutions that contain significant quantities of both the parent acid and its conjugate base. Note that acids can be cationic, anionic, or electrically neutral.

The acids in Figure 7-1 become progressively weaker from top to bottom. Perchloric acid and hydrochloric acid are completely dissociated in water. In contrast, only about 1% of acetic acid ($HC_2H_3O_2$) is dissociated. Ammonium ion is an even weaker acid. Only about 0.01% of this ion is dissociated into hydronium ions and ammonia molecules. Another generality illustrated in Figure 7-1 is that the weakest acid forms the strongest conjugate base; that is, ammonia has a much stronger affinity for protons than any base above it. Perchlorate and chloride ions have no affinity for protons.

The tendency of a solvent to accept or donate protons determines the strength of a solute acid or base dissolved in it. For example, perchloric and hydrochloric acids are strong acids in water. If anhydrous acetic acid, a weaker proton acceptor than water, is substituted *as the solvent,* neither of these acids undergoes complete dissociation; instead, equilibria such as the following are established:

Of all the acids listed in the previous marginal note and in Figure 7-1, only perchloric acid is a strong acid in methanol and ethanol. Thus, these two alcohols are also differentiating solvents.

$$\underset{\text{base}_1}{CH_3COOH} + \underset{\text{acid}_2}{HClO_4} \rightleftharpoons \underset{\text{acid}_1}{CH_3COOH_2^+} + \underset{\text{base}_2}{ClO_4^-}$$

In a **differentiating solvent,** various acids dissociate to different degrees and thus have different strengths. In a **leveling solvent,** several acids are completely dissociated and are thus of the same strength.

Perchloric acid is, however, considerably stronger than hydrochloric acid in this solvent, its dissociation being about 5000 times greater. Acetic acid thus acts as a *differentiating* solvent toward the two acids by revealing the inherent differences in their acidities. Water, on the other hand, is a *leveling* solvent for perchloric, hydrochloric, nitric, and sulfuric acids because all four are completely ionized in this solvent and thus exhibit no differences in strength. Differentiating and leveling solvents also exist for bases.

7B CHEMICAL EQUILIBRIUM

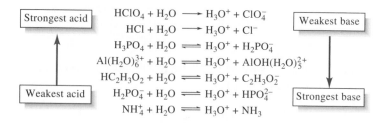

The reactions used in analytical chemistry never result in complete conversion of reactants to products. Instead, they proceed to a state of *chemical equilibrium* in which the ratio of concentrations of reactants and products is constant. *Equilibrium-constant expressions* are *algebraic* equations that describe the concentration relationships that exist among reactants and products at equilibrium. Among other things, equilibrium-constant expressions permit calculation of the error resulting from the quantity of unreacted analyte that remains when a steady state has been reached.

The discussion that follows deals with use of equilibrium-constant expressions to gain information about analytical systems in which no more than one or two equilibria are significant. Chapter 9 extends these methods to systems containing several simultaneous equilibria. Such complex systems are often encountered in analytical chemistry.

7B-1 The Equilibrium State

Consider the chemical equilibrium

$$H_3AsO_4 + 3I^- + 2H^+ \rightleftharpoons H_3AsO_3 + I_3^- + H_2O \qquad (7\text{-}5)$$

The rate of this reaction and the extent to which it proceeds to the right can be readily judged by observing the orange-red color of the triiodide ion I_3^- (the other participants in the reaction are colorless). If, for example, 1 mmol of arsenic acid H_3AsO_4 is added to 100 mL of a solution containing 3 mmol of potassium iodide, the red color of the triiodide ion appears almost immediately, and within a few seconds, the intensity of the color becomes constant, which shows that the triiodide concentration has become constant (see color plate 1b).

A solution of identical color intensity (and hence identical triiodide concentration) can also be produced by adding 1 mmol of arsenous acid H_3AsO_3 to 100 mL of a solution containing 1 mmol of triiodide ion (see color plate 1a). Here, the color intensity is initially greater than in the first solution but rapidly decreases as a result of the reaction

$$H_3AsO_3 + I_3^- + H_2O \rightleftharpoons H_3AsO_4 + 3I^- + 2H^+$$

Ultimately the color of the two solutions becomes identical. Many other combinations of the four reactants can be employed to yield solutions that are indistinguishable from the two just described.

The foregoing discussion illustrates that the concentration relationship at chemical equilibrium (that is, the *position of equilibrium*) is independent of the route by which the equilibrium state is achieved. However, this relationship can be altered by the application of stress to the system. Such stresses include changes in temperature, in pressure (if one of the reactants or products is a gas), or in total concentration of a reactant or a product. These effects can be predicted qualitatively from the *principle of Le Châtelier,* which states that the position of chemical equilibrium always shifts in a direction that tends to relieve the effect of an applied stress. Thus, an increase in temperature alters the concentration relationship in the direction that tends to absorb heat, and an increase in pressure favors those participants that occupy a smaller total volume.

The position of a chemical equilibrium is independent of the route by which equilibrium is reached.

The **Le Châtelier principle** states that the position of an equilibrium always shifts in such a direction as to relieve a stress that is applied to the system.

In an analysis, the effect of introducing an additional amount of a participating species to the reaction mixture is particularly important. Here, the resulting stress is relieved by a shift in equilibrium in the direction that partially uses up the added substance. Thus, for the equilibrium we have been considering (Equation 7-5), the addition of arsenic acid (H_3AsO_4) or hydrogen ions causes an increase in color as

> The **mass action effect** is a shift in the position of an equilibrium caused by changing the concentration of one of the reactants or products in a system.

Chemical reactions do not cease at equilibrium. Instead, the amounts of reactants and products are constant because the rates of the forward and reverse processes are identical.

> **Thermodynamics** is a branch of chemical science that deals with the flow of heat and energy in chemical reactions. The position of a chemical equilibrium can be related to these energy changes.

Equilibrium-constant expressions provide *no* information as to whether a chemical reaction is fast enough to be used for an analysis.

Cato Guldberg (1836–1902) and Peter Waage (1833–1900) were Norwegian chemists whose primary interests were in the field of thermodynamics. In 1864, these workers were the first to propose the law of mass action, which is expressed in Equation 7-7.

$[Z]^z$ in Equation 7-7 is replaced with p_Z^z in atmospheres if Z is a gas. If Z is a pure liquid or a pure solid present in excess, no term for this species is included in the equation.

The real reason that we omit pure liquids and solids from equilibrium-constant expressions is that the activity of a pure liquid or a pure solid is defined to be unity. For more details, see Chapter 8.

more triiodide ion and arsenous acid are formed; the addition of arsenous acid has the reverse effect. An equilibrium shift brought about by changing the amount of one of the participating species is called a *mass-action effect.*

If it were possible to examine the system under discussion at the molecular level, we would find that reactions among the participating species continue unabated even after equilibrium is achieved. The constant concentration ratio of reactants and products results from the equality in the rates of the forward and reverse reactions. In other words, chemical equilibrium is a dynamic state in which the rates of the forward and reverse reactions are identical.

7B-2 Equilibrium-Constant Expressions

The influence of concentration (or pressure if a gas is involved) on the position of a chemical equilibrium is conveniently described in quantitative terms by means of an equilibrium-constant expression. Such expressions are readily derived from thermodynamic theory. They are of great practical importance because they permit the chemist to predict the direction and completeness of a chemical reaction. We must emphasize, however, that an equilibrium-constant expression yields no information concerning the *rate* at which equilibrium is approached. In fact, we sometimes encounter reactions that have highly favorable equilibrium constants but are of little analytical use because their rates are low. This limitation can often be overcome by the use of a catalyst, which speeds the attainment of equilibrium without changing its position.

Let us consider a generalized equation for a chemical equilibrium

$$wW + xX \rightleftharpoons yY + zZ \qquad (7\text{-}6)$$

where the capital letters represent the formulas of participating chemical species and the lowercase italic letters are the small whole numbers required to balance the equation. Thus, the equation states that w mol of W reacts with x mol of X to form y mol of Y and z mol of Z. The equilibrium-constant expression for this reaction is

$$K = \frac{[Y]^y [Z]^z}{[W]^w [X]^x} \qquad (7\text{-}7)$$

where the square-bracketed terms have the following meanings:

1. molar concentration if the species is a dissolved solute.
2. partial pressure in atmospheres if the species is a gas. In fact, we will often replace the square-bracketed term (say [Z] in Equation 7-7) with the symbol p_Z^z, which stands for the partial pressure of the gas Z in atmospheres.

If one (or more) of the species in Equation 7-7 is a pure liquid, a pure solid, or the solvent present in excess, no term for this species appears in the equilibrium-constant expression. For example, if Z in Equation 7-6 is the solvent H_2O, the equilibrium-constant expression simplifies to

$$K = \frac{[Y]^y}{[W]^w[X]^x}$$

The reason for this simplification will become apparent in the sections that follow.

The constant K in Equation 7-7 is a temperature-dependent numerical quantity called the *equilibrium constant.* By convention, the concentrations of the products, *as the equation is written,* are always placed in the numerator and the concentrations of the reactants in the denominator.

Equation 7-7 is only an approximate form of a thermodynamic equilibrium-constant expression. The exact form is given by Equation 7-8 in the margin. Generally we will use the approximate form of this equation because it is less tedious and time-consuming to use. In Section 8B, we show when the use of Equation 7-7 is likely to lead to serious errors in equilibrium calculations and how Equation 7-8 is applied in these cases.

Remember, Equation 7-7 is only an approximate form of an equilibrium-constant expression. The exact expression takes the form

$$K = \frac{a_Y a_Z}{a_W a_X} \qquad (7\text{-}8)$$

where a_Y, a_Z, a_W, and a_X are the *activities* of species Y, Z, W, and X (see Section 8B).

7B-3 Types of Equilibrium Constants Encountered in Analytical Chemistry

Table 7-2 summarizes the types of chemical equilibria and equilibrium constants that are of importance in analytical chemistry. Simple applications of some of these constants are illustrated in the paragraphs that follow.

TABLE 7-2 Equilibria and Equilibrium Constants of Importance to Analytical Chemistry

Type of Equilibrium	Name and Symbol of Equilibrium Constant	Typical Example	Equilibrium-Constant Expression
Dissociation of water	Ion-product constant, K_w	$2\,H_2O \rightleftharpoons H_3O^+ + OH^-$	$K_w = [H_3O^+][OH^-]$
Heterogeneous equilibrium between a slightly soluble substance and its ions in a saturated solution	Solubility product, K_{sp}	$BaSO_4(s) \rightleftharpoons Ba^{2+} + SO_4^{2-}$	$K_{sp} = [Ba^{2+}][SO_4^{2-}]$
Dissociation of a weak acid or base	Dissociation constant, K_a or K_b	$CH_3COOH + H_2O \rightleftharpoons H_3O^+ + CH_3COO^-$	$K_a = \frac{[H_3O^+][CH_3COO^-]}{[CH_3COOH]}$
		$CH_3COO^- + H_2O \rightleftharpoons OH^- + CH_3COOH$	$K_b = \frac{[OH^-][CH_3COOH]}{[CH_3COO^-]}$
Formation of a complex ion	Formation constant, β_n	$Ni^{2+} + 4\,CN^- \rightleftharpoons Ni(CN)_4^{2-}$	$\beta_4 = \frac{[Ni(CN)_4^{2-}]}{[Ni^{2+}][CN^-]^4}$
Oxidation/reduction equilibrium	K_{redox}	$MnO_4^- + 5\,Fe^{2+} + 8\,H^+ \rightleftharpoons Mn^{2+} + 5\,Fe^{3+} + 4\,H_2O$	$K_{redox} = \frac{[Mn^{2+}][Fe^{3+}]^5}{[MnO_4^-][Fe^{2+}]^5[H^+]^8}$
Distribution equilibrium for a solute between immiscible solvents	K_d	$I_2(aq) \rightleftharpoons I_2(org)$	$K_d = \frac{[I_2]_{org}}{[I_2]_{aq}}$

FEATURE 7-2
Stepwise and Overall Formation Constants for Complex Ions

The formation of $Ni(CN)_4^{2-}$ (Table 7-2) is typical in that it occurs in steps as shown. Note that *stepwise formation* constants are symbolized by K_1, K_2, etc.

$$Ni^{2+} + CN^- \rightleftharpoons Ni(CN)^+ \qquad K_1 = \frac{[Ni(CN)^+]}{[Ni^{2+}][CN^-]}$$

$$Ni(CN)^+ + CN^- \rightleftharpoons Ni(CN)_2 \qquad K_2 = \frac{[Ni(CN)_2]}{[Ni(CN)^+][CN^-]}$$

$$Ni(CN)_2 + CN^- \rightleftharpoons Ni(CN)_3^- \qquad K_3 = \frac{[Ni(CN)_3^-]}{[Ni(CN)_2][CN^-]}$$

$$Ni(CN)_3^- + CN^- \rightleftharpoons Ni(CN)_4^{2-} \qquad K_4 = \frac{[Ni(CN)_4^{2-}]}{[Ni(CN)_3^-][CN^-]}$$

Overall constants are designated by the symbol β_n. Thus,

$$Ni^{2+} + 2CN^- \rightleftharpoons Ni(CN)_2 \qquad \beta_2 = K_1K_2 = \frac{[Ni(CN)_2]}{[Ni^{2+}][CN^-]^2}$$

$$Ni^{2+} + 3CN^- \rightleftharpoons Ni(CN)_3^- \qquad \beta_3 = K_1K_2K_3 = \frac{[Ni(CN)_3^-]}{[Ni^{2+}][CN^-]^3}$$

$$Ni^{2+} + 4CN^- \rightleftharpoons Ni(CN)_4^{2-} \qquad \beta_4 = K_1K_2K_3K_4 = \frac{[Ni(CN)_4^{2-}]}{[Ni^{2+}][CN^-]^4}$$

p-Functions

The **p-function** of a numerical datum is the negative logarithm to the base 10 of that datum.

In discussing chemical equilibria, we often are dealing with numbers that are very small and can vary over several orders of magnitude. For example, in the next section we talk about applications of the ion-product constant for water K_w, where $K_w = 1.00 \times 10^{-14}$ at room temperature. We use this constant to calculate hydronium ion or hydroxide ion concentrations that vary from greater than 0.1 M to smaller than 10^{-14} M. When dealing with a wide range of numbers, such as these, chemists find it helpful to express the numbers as p-functions where the *p-function,* or *p-value,* is the negative logarithm (to the base 10) of the number. Thus, the p-value of the number x is

$$px = -\log x$$

As shown by the following examples, p-values offer the advantage of allowing awkward numerical data to be expressed as small, usually positive, numbers.

EXAMPLE 7-1

Calculate the p-function of (a) the ion-product constant of water and (b) the concentrations of the various ions in a solution that is 2.00×10^{-3} M in NaCl and 5.4×10^{-4} M in HCl.

(a) For the ion product of water, pK_w is given by

$$pK_w = -\log (1.00 \times 10^{-14}) = 14.000$$

(b)
$$\begin{aligned} pH &= -\log [H^+] = -\log (5.4 \times 10^{-4}) \\ &= -\log 5.4 - \log 10^{-4} = -0.73 - (-4) = 3.27 \end{aligned}$$

$$\begin{aligned} pNa &= -\log (2.00 \times 10^{-3}) = -\log 2.00 - \log 10^{-3} \\ &= -0.301 - (-3.00) = 2.699 \end{aligned}$$

The total Cl^- concentration is given by the sum of the concentrations of the two solutes:

$$[Cl^-] = 2.00 \times 10^{-3}\ M + 5.4 \times 10^{-4}\ M = 2.54 \times 10^{-3}\ M$$

$$pCl = -(\log 2.54 \times 10^{-3}) = 2.595$$

Note that in Example 7-1, and in the one that follows, the results are rounded according to the rules listed on page 41.

Logarithmic functions such as p-functions are often used for measurements that have a wide **dynamic range,** that is, those that range over many orders of magnitude.

EXAMPLE 7-2

Calculate the molar concentration of Ag^+ in a solution that has pAg of 6.372.

$$pAg = -\log [Ag^+] = 6.372$$

$$\log [Ag^+] = -6.372 = 0.628 - 7.000$$

$$\begin{aligned} [Ag^+] &= \text{antilog}(0.628) \times \text{antilog}(-7.000) = 4.246 \times 10^{-7} \\ &= 4.25 \times 10^{-7} \end{aligned}$$

The antilog key on handheld calculators is usually labeled 10^x.

Ion-Product Constant for Water

Aqueous solutions contain small amounts of hydronium and hydroxide ions as a consequence of the dissociation reaction

$$2H_2O \rightleftharpoons H_3O^+ + OH^- \qquad (7\text{-}9)$$

An equilibrium constant for this reaction can be formulated as shown in Equation 7-7:

$$K = \frac{[H_3O^+][OH^-]}{[H_2O]^2} \qquad (7\text{-}10)$$

The concentration of water in dilute aqueous solutions is enormous, however, when compared with the concentration of hydrogen and hydroxide ions. As a consequence, $[H_2O]$ in Equation 7-10 can be taken as constant, and we write

We obtain a useful relationship by taking the negative logarithm of Equation 7-11

$$-\log K_w = -\log [H_3O^+] - \log [OH^-]$$

By the definition of p-functions,

$$pK_w = pH + pOH$$

At 25°C, $pK_w = 14.00$.

$$K[H_2O]^2 = K_w = [H_3O^+][OH^-] \tag{7-11}$$

where the new constant K_w is given a special name, the *ion-product constant for water.*

FEATURE 7-3
Why $[H_2O]$ Does Not Appear in Equilibrium-Constant Expressions for Aqueous Solutions

In a dilute aqueous solution the molar concentration of water is

$$[H_2O] = \frac{1000 \text{ g } H_2O}{\text{L } H_2O} \times \frac{1 \text{ mol } H_2O}{18.0 \text{ g } H_2O} = 55.6 \text{ M}$$

Let us suppose we have 0.1 mol of HCl in 1 L of water. The presence of this acid will shift the equilibrium shown in Equation 7-9 to the left. Originally, however, there were only 10^{-7} mol/L OH^- to consume the added protons. Thus, even if all the OH^- ions are converted to H_2O, the water concentration will increase only to

$$[H_2O] = 55.6 \frac{\text{mol } H_2O}{\text{L } H_2O} + 1 \times 10^{-7} \frac{\text{mol } OH^-}{\text{L } H_2O} \times \frac{1 \text{ mol } H_2O}{\text{mol } OH^-} \approx 55.6 \text{ M}$$

The percent change in water concentration is

$$\frac{10^{-7} \text{ M}}{55.6 \text{ M}} \times 100\% = 2 \times 10^{-7}\%$$

which is certainly inconsequential. Thus, $K[H_2O]^2$ in Equation 7-10 is, for all practical purposes, a constant. That is,

$$K(55.6)^2 = K_w = 1.00 \times 10^{-14}$$

TABLE 7-3
Variation of K_w with Temperature

Temperature, °C	K_w
0	0.114×10^{-14}
25	1.01×10^{-14}
50	5.47×10^{-14}
100	49×10^{-14}

At 25°C, the ion-product constant for water is 1.008×10^{-14}. For convenience, we shall use the approximation that at room temperature $K_w \approx 1.00 \times 10^{-14}$. Table 7-3 shows the dependence of this constant upon temperature.

The ion-product constant for water permits the ready calculation of the hydronium and hydroxide ion concentrations of aqueous solutions.

EXAMPLE 7-3

Calculate the hydronium and hydroxide ion concentrations of pure water at 25°C and 100°C.

Because OH^- and H_3O^+ are formed only from the dissociation of water, their concentrations must be equal:

$$[H_3O^+] = [OH^-]$$

Substitution into Equation 7-11 gives

$$[H_3O^+]^2 = [OH^-]^2 = K_w$$
$$[H_3O^+] = [OH^-] = \sqrt{K_w}$$

At 25°C,

$$[H_3O^+] = [OH^-] = \sqrt{1.00 \times 10^{-14}} = 1.00 \times 10^{-7}$$

At 100°C, from Table 7-3,

$$[H_3O^+] = [OH^-] = \sqrt{49 \times 10^{-14}} = 7.0 \times 10^{-7}$$

EXAMPLE 7-4

Calculate the hydronium and hydroxide ion concentrations in 0.200 M aqueous NaOH.

Sodium hydroxide is a strong electrolyte, and its contribution to the hydroxide ion concentration in this solution is 0.200 mol/L. As in Example 7-3, hydroxide ions and hydronium ions are formed *in equal amounts* from dissociation of water. Therefore, we write

$$[OH^-] = 0.200 + [H_3O^+]$$

where $[H_3O^+]$ accounts for the hydroxide ions contributed by the solvent. The concentration of OH^- from the water is insignificant, however, when compared with 0.200, so we can write

$$[OH^-] \approx 0.200$$

Equation 7-11 is then used to calculate the hydronium ion concentration:

$$[H_3O^+] = \frac{K_w}{[OH^-]} = \frac{1.00 \times 10^{-14}}{0.200} = 5.00 \times 10^{-14}$$

Note that the approximation

$$[OH^-] = 0.200 + 5.00 \times 10^{-14} \approx 0.200$$

causes no significant error.

Solubility-Product Constants

When we say that a sparingly soluble salt is completely dissociated, *we do not imply* that all of the salt dissolves. Instead, the very small amount that *does* go into solution dissociates completely.

Most sparingly soluble salts are essentially completely dissociated in saturated aqueous solution. For example, when an excess of barium iodate is equilibrated with water, the dissociation process is adequately described by the equation

$$Ba(IO_3)_2(s) \rightleftharpoons Ba^{2+}(aq) + 2IO_3^-(aq)$$

Application of Equation 7-7 leads to

$$K = \frac{[Ba^{2+}][IO_3^-]^2}{[Ba(IO_3)_2(s)]}$$

The denominator represents the molar concentration of $Ba(IO_3)_2$ *in the solid,* a phase that is separate from but in contact with the saturated solution. The concentration of a compound in its solid state is, however, constant. In other words, the number of moles of $Ba(IO_3)_2$ divided by the *volume* of the solid $Ba(IO_3)_2$ is constant no matter how much excess solid is present. Therefore, the foregoing equation can be rewritten in the form

$$K[Ba(IO_3)_2(s)] = K_{sp} = [Ba^{2+}][IO_3^-]^2 \quad (7\text{-}12)$$

For Equation 7-12 to apply, it is necessary only that *some solid be present. You should always keep in mind that in the absence of $Ba(IO_3)_2(s)$, Equation 7-12 is not valid.*

where the new constant is called the *solubility-product constant* or the *solubility product.* It is important to appreciate that Equation 7-12 shows that the position of this equilibrium is independent of the *amount* of $Ba(IO_3)_2$ so long as some solid is present; that is, it does not matter whether the amount is a few milligrams or several grams.

A table of solubility-product constants for numerous inorganic salts is found in Appendix 2. The examples that follow demonstrate some typical uses of solubility-product expressions. Further applications are considered in Chapter 9.

The Solubility of a Precipitate in Pure Water

The solubility-product expression permits the ready calculation of the solubility of a sparingly soluble substance that ionizes in water.

EXAMPLE 7-5

How many grams of $Ba(IO_3)_2$ (487 g/mol) can be dissolved in 500 mL of water at 25°C?

The solubility-product constant for $Ba(IO_3)_2$ is 1.57×10^{-9} (Appendix 2). The equilibrium between the solid and its ions in solution is described by the equation

$$Ba(IO_3)_2(s) \rightleftharpoons Ba^{2+} + 2IO_3^-$$

and so

$$K_{sp} = [Ba^{2+}][IO_3^-]^2 = 1.57 \times 10^{-9}$$

The equation describing the equilibrium reveals that 1 mol of Ba^{2+} is formed for each mole of $Ba(IO_3)_2$ that dissolves. Therefore,

$$\text{molar solubility of } Ba(IO_3)_2 = [Ba^{2+}]$$

Note that the molar solubility is equal to $[Ba^{2+}]$ or to $\frac{1}{2}[IO_3^-]$.

The iodate concentration is clearly twice that for barium ion:

$$[IO_3^-] = 2[Ba^{2+}]$$

Substituting this last equation into the equilibrium-constant expression gives

$$[Ba^{2+}](2[Ba^{2+}])^2 = 4[Ba^{2+}]^3 = 1.57 \times 10^{-9}$$

$$[Ba^{2+}] = \left(\frac{1.57 \times 10^{-9}}{4}\right)^{1/3} = 7.32 \times 10^{-4}\text{ M}$$

Since 1 mol Ba^{2+} is produced for every mole of $Ba(IO_3)_2$

$$\text{solubility} = 7.32 \times 10^{-4}\text{ M}$$

To compute the number of millimoles of $Ba(IO_3)_2$ dissolved in 500 mL of solution, we write

$$\text{no. mmol } Ba(IO_3)_2 = 7.32 \times 10^{-4}\,\frac{\text{mmol } Ba(IO_3)_2}{\cancel{\text{mL}}} \times 500\,\cancel{\text{mL}}$$

The mass of $Ba(IO_3)_2$ in 500 mL is given by

$$\text{mass } Ba(IO_3)_2 = (7.32 \times 10^{-4} \times 500)\,\cancel{\text{mmol } Ba(IO_3)_2}$$

$$\times\, 0.487\,\frac{\text{g } Ba(IO_3)_2}{\cancel{\text{mmol } Ba(IO_3)_2}}$$

$$= 0.178\text{ g}$$

The Effect of a Common Ion on the Solubility of a Precipitate The *common-ion effect* is a mass-action effect predicted from the Le Châtelier principle and is demonstrated by the following examples.

The **common-ion effect** for solids is the reduction in solubility of an ionic precipitate when a soluble compound containing one of the ions of the precipitate is added to a saturated solution of the solid.

EXAMPLE 7-6

Calculate the molar solubility of $Ba(IO_3)_2$ in a solution that is 0.0200 M in $Ba(NO_3)_2$.

The solubility is no longer equal to $[Ba^{2+}]$ because $Ba(NO_3)_2$ is also a source of barium ions. We know, however, that solubility is related to $[IO_3^-]$:

$$\text{molar solubility of } Ba(IO_3)_2 = \tfrac{1}{2}[IO_3^-]$$

There are two sources of barium ions: $Ba(NO_3)_2$ and $Ba(IO_3)_2$. The contribution from the former is 0.0200 M, and that from the latter is equal to the molar solubility, or $\frac{1}{2}[IO_3^-]$. Thus,

$$[Ba^{2+}] = 0.0200 + \tfrac{1}{2}[IO_3^-]$$

Substitution of these quantities into the solubility-product expression yields

$$(0.0200 + \tfrac{1}{2}[IO_3^-])[IO_3^-]^2 = 1.57 \times 10^{-9}$$

See *Mathcad Applications for Analytical Chemistry,* **pp. 35–37.**

Since the exact solution for $[IO_3^-]$ requires solving a cubic equation, we seek an approximation that simplifies the algebra. The small numerical value of K_{sp} suggests that the solubility of $Ba(IO_3)_2$ is not large, and this is confirmed by the result obtained in Example 7-5. Moreover, barium ion from $Ba(NO_3)_2$ will further repress the already limited solubility of $Ba(IO_3)_2$. Thus, it is reasonable to seek a provisional answer to the problem by assuming that 0.0200 is large with respect to $\frac{1}{2}[IO_3^-]$. That is, $\frac{1}{2}[IO_3^-] \ll 0.0200$, and

$$[Ba^{2+}] = 0.0200 + \tfrac{1}{2}[IO_3^-] \approx 0.0200$$

The original equation then simplifies to

$$0.0200[IO_3^-]^2 = 1.57 \times 10^{-9}$$

$$[IO_3^-] = \sqrt{1.57 \times 10^{-9}/0.0200} = \sqrt{7.85 \times 10^{-8}} = 2.80 \times 10^{-4}\ \text{M}$$

The assumption that $(0.0200 + \frac{1}{2} \times 2.80 \times 10^{-4}) \approx 0.0200$ does not appear to cause serious error because the second term, representing the amount of Ba^{2+} arising from the dissociation of $Ba(IO_3)_2$, is only about 0.7% of 0.0200. Ordinarily, we consider an assumption of this type to be satisfactory if the discrepancy is less than 10%. Finally, then,

$$\text{solubility of } Ba(IO_3)_2 = \tfrac{1}{2}[IO_3^-] = \tfrac{1}{2} \times 2.80 \times 10^{-4} = 1.40 \times 10^{-4}\ \text{M}$$

If we compare this result with the solubility of barium iodate in pure water (Example 7-5), we see that the presence of a small concentration of the common ion has lowered the molar solubility of $Ba(IO_3)_2$ by a factor of about five.

EXAMPLE 7-7

Calculate the solubility of $Ba(IO_3)_2$ in a solution prepared by mixing 200 mL of 0.0100 M $Ba(NO_3)_2$ with 100 mL of 0.100 M $NaIO_3$.

We must first establish whether either reactant is present in excess at equilibrium. The amounts taken are

$$\text{no. mmol } Ba^{2+} = 200\ \text{mL} \times 0.0100\ \text{mmol/mL} = 2.00$$

$$\text{no. mmol } IO_3^- = 100\ \text{mL} \times 0.100\ \text{mmol/mL} = 10.0$$

If formation of $Ba(IO_3)_2$ is complete,

$$\text{no. mmol excess NaIO}_3 = 10.0 - 2(2.00) = 6.00$$

Thus,

$$[IO_3^-] = \frac{6.00 \text{ mmol}}{200 \text{ mL} + 100 \text{ mL}} = \frac{6.00 \text{ mmol}}{300 \text{ mL}} = 0.0200 \text{ M}$$

The uncertainty in $[IO_3^-]$ is 0.1 part in 6.0 or 1 part in 60. Thus $0.0200 \times (1/60) = 0.0003$ and we round to 0.0200 M.

As in Example 7-5,

$$\text{molar solubility of Ba(IO}_3)_2 = [Ba^{2+}]$$

Here, however,

$$[IO_3^-] = 0.0200 + 2[Ba^{2+}]$$

where $2[Ba^{2+}]$ represents the iodate contributed by the sparingly soluble $Ba(IO_3)_2$. We can obtain a provisional answer after making the assumption that $[IO_3^-] \approx 0.0200$; thus,

$$\text{solubility of Ba(IO}_3)_2 = [Ba^{2+}] = \frac{K_{sp}}{[IO_3^-]^2} = \frac{1.57 \times 10^{-9}}{(0.0200)^2}$$
$$= 3.93 \times 10^{-6} \text{ mol/L}$$

The approximation appears to be reasonable.

The last two examples demonstrate that an excess of iodate ions is more effective in decreasing the solubility of $Ba(IO_3)_2$ than is the same excess of barium ions.

A 0.02 M excess of Ba^{2+} decreases the solubility of $Ba(IO_3)_2$ by a factor of about 5; this same excess of IO_3^- lowers the solubility by about 200.

Acid and Base Dissociation Constants

When a weak acid or a weak base is dissolved in water, partial dissociation occurs. Thus, for nitrous acid, we can write

$$HNO_2 + H_2O \rightleftharpoons H_3O^+ + NO_2^- \qquad K_a = \frac{[H_3O^+][NO_2^-]}{[HNO_2]}$$

where K_a is the *acid dissociation constant* for nitrous acid. In an analogous way, the *base dissociation constant* for ammonia is

$$NH_3 + H_2O \rightleftharpoons NH_4^+ + OH^- \qquad K_b = \frac{[NH_4^+][OH^-]}{[NH_3]}$$

Note that $[H_2O]$ does not appear in the denominator in either dissociation-constant expression. The reason for this omission is discussed in Feature 7-3. Thus in common with the ion-product constant for water, $[H_2O]$ is incorporated in the equilibrium constants K_a and K_b.

Dissociation Constants for Conjugate Acid/Base Pairs

Consider the base dissociation-constant expression for ammonia and the acid dissociation-constant expression for its conjugate acid, ammonium ion:

$$NH_3 + H_2O \rightleftharpoons NH_4^+ + OH^- \qquad K_b = \frac{[NH_4^+][OH^-]}{[NH_3]}$$

$$NH_4^+ + H_2O \rightleftharpoons NH_3 + H_3O^+ \qquad K_a = \frac{[NH_3][H_3O^+]}{[NH_4^+]}$$

Multiplication of one equilibrium-constant expression by the other gives

$$K_aK_b = \frac{\cancel{[NH_3]}[H_3O^+]}{\cancel{[NH_4^+]}} \times \frac{\cancel{[NH_4^+]}[OH^-]}{\cancel{[NH_3]}} = [H_3O^+][OH^-]$$

but

$$K_w = [H_3O^+][OH^-]$$

and therefore

$$K_w = K_aK_b \qquad (7\text{-}13)$$

This relationship is general for all conjugate acid/base pairs. Many compilations of equilibrium-constant data list only acid dissociation constants because it is so easy to calculate basic dissociation constants with Equation 7-13. For example, in Appendix 3 we find no data on the basic dissociation of ammonia (nor for any other bases). Instead, we find the acid dissociation constant for the conjugate acid, ammonium ion. That is,

$$NH_4^+ + H_2O \rightleftharpoons H_3O^+ + NH_3 \qquad K_a = \frac{[H_3O^+][NH_3]}{[NH_4^+]} = 5.70 \times 10^{-10}$$

and we can write

In order to obtain a dissociation constant for a base, we look in Appendix 3 for the dissociation constant for its conjugate acid and then divide this number into 1.00×10^{-14}.

$$NH_3 + H_2O \rightleftharpoons NH_4^+ + OH^-$$

$$K_b = \frac{[NH_4^+][OH^-]}{[NH_3]} = \frac{1.00 \times 10^{-14}}{5.70 \times 10^{-10}} = 1.75 \times 10^{-5}$$

FEATURE 7-4
Relative Strengths of Conjugate Acid/Base Pairs

Equation 7-13 confirms the observation in Figure 7-1 that as the acid of a conjugate acid/base pair becomes weaker, its conjugate base becomes stronger and vice versa. Thus the conjugate base of an acid with a dissociation constant of 10^{-2} will have a basic dissociation constant of 10^{-12}, whereas an acid with a dissociation constant of 10^{-9} has a conjugate base with a dissociation constant of 10^{-5}.

EXAMPLE 7-8

What is K_b for the equilibrium

$$CN^- + H_2O \rightleftharpoons HCN + OH^-$$

Appendix 3 lists a K_a value of 6.2×10^{-10} for HCN. Thus,

$$K_b = \frac{K_w}{K_{HCN}} = \frac{[HCN][OH^-]}{[CN^-]}$$

$$K_b = \frac{1.00 \times 10^{-14}}{6.2 \times 10^{-10}} = 1.62 \times 10^{-5}$$

Hydronium Ion Concentration in Solutions of Weak Acids

When the weak acid HA is dissolved in water, two equilibria are established that yield hydronium ions:

$$HA + H_2O \rightleftharpoons H_3O^+ + A^- \qquad K_a = \frac{[H_3O^+][A^-]}{[HA]}$$

$$2H_2O \rightleftharpoons H_3O^+ + OH^- \qquad K_w = [H_3O^+][OH^-]$$

Ordinarily, the hydronium ions produced from the first equilibrium suppress the dissociation of water to such an extent that the contribution of hydronium ions from the second equilibrium is negligible. Under these circumstances, one H_3O^+ ion is formed for each A^- ion and we write

$$[A^-] \approx [H_3O^+] \tag{7-14}$$

Furthermore, the sum of the molar concentrations of the weak acid and its conjugate base must equal the analytical concentration of the acid c_{HA} because the solution contains no other source of A^- ions. Thus,

$$c_{HA} = [A^-] + [HA] \tag{7-15}$$

Substituting $[H_3O^+]$ for $[A^-]$ (Equation 7-14) into Equation 7-15 yields

$$c_{HA} = [H_3O^+] + [HA]$$

which rearranges to

$$[HA] = c_{HA} - [H_3O^+] \quad (7\text{-}16)$$

When $[A^-]$ and [HA] are replaced by their equivalent terms from Equations 7-14 and 7-16, the equilibrium-constant expression becomes

$$K_a = \frac{[H_3O^+]^2}{c_{HA} - [H_3O^+]} \quad (7\text{-}17)$$

which rearranges to

$$[H_3O^+]^2 + K_a[H_3O^+] - K_a c_{HA} = 0 \quad (7\text{-}18)$$

The positive solution to this quadratic equation is

$$[H_3O^+] = \frac{-K_a + \sqrt{K_a^2 + 4K_a c_{HA}}}{2} \quad (7\text{-}19)$$

As an alternative to using Equation 7-19, Equation 7-18 may be solved by successive approximations as shown in Feature 7-5.

TABLE 7-4 Error Introduced by Assuming H_3O^+ Concentration Is Small Relative to c_{HA} in Equation 7-16

K_a	c_{HA}	$[H_3O^+]$ Using Assumption	$\frac{c_{HA}}{K_a}$	$[H_3O^+]$ Using More Exact Equation	Percent Error
1.00×10^{-2}	1.00×10^{-3}	3.16×10^{-3}	10^{-1}	0.92×10^{-3}	244
	1.00×10^{-2}	1.00×10^{-2}	10^{0}	0.62×10^{-2}	61
	1.00×10^{-1}	3.16×10^{-2}	10^{1}	2.70×10^{-2}	17
1.00×10^{-4}	1.00×10^{-4}	1.00×10^{-4}	10^{0}	0.62×10^{-4}	61
	1.00×10^{-3}	3.16×10^{-4}	10^{1}	2.70×10^{-4}	17
	1.00×10^{-2}	1.00×10^{-3}	10^{2}	0.95×10^{-3}	5.3
	1.00×10^{-1}	3.16×10^{-3}	10^{3}	3.11×10^{-3}	1.6
1.00×10^{-6}	1.00×10^{-5}	3.16×10^{-6}	10^{1}	2.70×10^{-6}	17
	1.00×10^{-4}	1.00×10^{-5}	10^{2}	0.95×10^{-5}	5.3
	1.00×10^{-3}	3.16×10^{-5}	10^{3}	3.11×10^{-5}	1.6
	1.00×10^{-2}	1.00×10^{-4}	10^{4}	9.95×10^{-5}	0.5
	1.00×10^{-1}	3.16×10^{-4}	10^{5}	3.16×10^{-4}	0.0

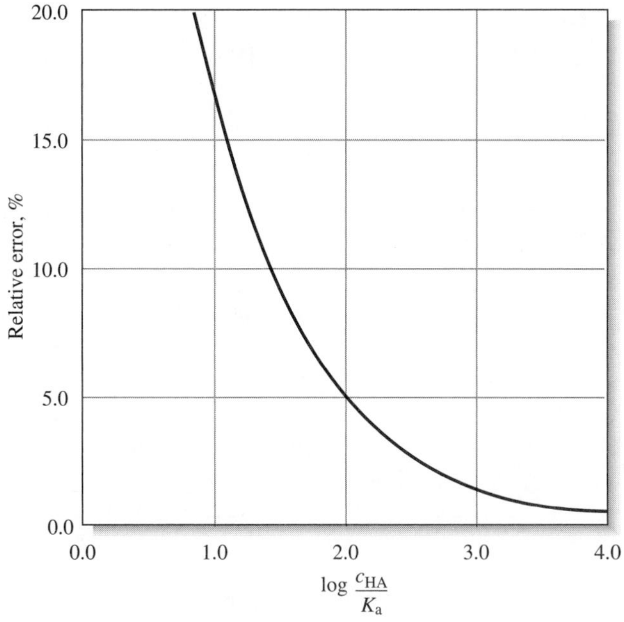

Figure 7-2
Relative error resulting from the assumption that $[H_3O^+] \ll c_{HA}$ in Equation 7-15.

Equation 7-16 can frequently be simplified by making the assumption that dissociation does not appreciably decrease the molar concentration of HA. Thus, provided $[H_3O^+] \ll c_{HA}$, $c_{HA} - [H_3O^+] \approx c_{HA}$ and Equation 7-17 reduces to

$$\frac{[H_3O^+]^2}{c_{HA}} = K_a \tag{7-20}$$

and

$$[H_3O^+] = \sqrt{K_a c_{HA}} \tag{7-21}$$

The magnitude of the error introduced by the assumption that $[H_3O^+] \ll c_{HA}$ increases as the molar concentration of acid becomes smaller and as the dissociation constant for the acid becomes larger. This statement is supported by the data in Table 7-4. Note that the error introduced by the assumption is about 0.5% when the ratio c_{HA}/K_a is 10^4. The error increases to about 1.6% when the ratio is 10^3, to about 5% when it is 10^2, and to about 17% when it is 10. Figure 7-2 illustrates the effect graphically. It is noteworthy that the hydronium ion concentration computed with the approximation becomes equal to or greater than the molar concentration of the acid when the ratio is unity or smaller, which is clearly a meaningless result.

In general, it is good practice to make the simplifying assumption and obtain a trial value for $[H_3O^+]$ that can be compared with c_{HA} in Equation 7-16. If the trial value alters [HA] by an amount smaller than the allowable error in the calculation, the solution may be considered satisfactory. Otherwise, the quadratic equation must be solved to obtain a better value for $[H_3O^+]$. Alternatively, the method of successive approximations (Feature 7-5) may be employed.

FEATURE 7-5
The Method of Successive Approximations

For convenience let us write the quadratic equation in Example 7-10 in the form

$$x^2 + 2.51 \times 10^{-5}\,x - 5.02 \times 10^{-9} = 0$$

where $x = [H_3O^+]$.

As a first step, let us rearrange the equation to the form

See *Mathcad Applications for Analytical Chemistry,* **pp. 50–51.**

$$x = \sqrt{5.02 \times 10^{-9} - 2.51 \times 10^{-5}\,x}$$

We then assume that x on the right-hand side of the equation is zero and calculate a first value, x_1.

$$x_1 = \sqrt{5.02 \times 10^{-9} - 2.51 \times 10^{-5} \times 0} = 7.09 \times 10^{-5}$$

We then substitute this value into the original equation and derive a second value x_2. That is,

$$x_2 = \sqrt{5.02 \times 10^{-9} - 2.51 \times 10^{-5} \times 7.09 \times 10^{-5}} = 5.69 \times 10^{-5}$$

Repeating this calculation gives

$$x_3 = \sqrt{5.02 \times 10^{-9} - 2.51 \times 10^{-5} \times 5.69 \times 10^{-5}} = 5.99 \times 10^{-5}$$

Continuing in the same way we obtain

$$x_4 = 5.93 \times 10^{-5}$$
$$x_5 = 5.94 \times 10^{-5}$$
$$x_6 = 5.94 \times 10^{-5}$$

Note that after three iterations x_3 is 5.99×10^{-5}, which is within about 0.8% of the final value of 5.94×10^{-5} M.

Occasionally, the final result will oscillate between a high and a low value. In this case you may have to solve the quadratic equation.

The method of successive approximations is particularly useful when you need to solve cubic or higher power equations.

EXAMPLE 7-9

Calculate the hydronium ion concentration in 0.120 M nitrous acid. The principal equilibrium is

$$HNO_2 + H_2O \rightleftharpoons H_3O^+ + NO_2^-$$

for which (Appendix 3)

$$K_a = 7.1 \times 10^{-4} = \frac{[H_3O^+][NO_2^-]}{[HNO_2]}$$

Substitution into Equations 7-14 and 7-16 gives

$$[NO_2^-] = [H_3O^+]$$
$$[HNO_2] = 0.120 - [H_3O^+]$$

When these relationships are introduced into the expression for K_a, we obtain

$$K_a = \frac{[H_3O^+]^2}{0.120 - [H_3O^+]} = 7.1 \times 10^{-4}$$

If we now assume $[H_3O^+] \ll 0.120$, we find

$$\frac{[H_3O^+]^2}{0.120} = 7.1 \times 10^{-4}$$

$$[H_3O^+] = \sqrt{0.120 \times 7.1 \times 10^{-4}} = 9.2 \times 10^{-3}\ M$$

We now examine the assumption that $0.120 - 0.0092 \approx 0.120$ and see that the error is about 8%. The relative error in $[H_3O^+]$ is actually smaller than this figure, however, as we can see by calculating $\log (c_{HA}/K_a) = 2.2$, which, from Figure 7-2, suggests an error of about 4%. If a more accurate figure is needed, solution of the quadratic equation yields 8.9×10^{-3} M for the hydronium ion concentration.

EXAMPLE 7-10

Calculate the hydronium ion concentration in a solution that is 2.0×10^{-4} M in aniline hydrochloride, $C_6H_5NH_3Cl$.

In aqueous solution, dissociation of the salt to Cl^- and $C_6H_5NH_3^+$ is complete. The weak acid $C_6H_5NH_3^+$ dissociates as follows:

$$C_6H_5NH_3^+ + H_2O \rightleftharpoons C_6H_5NH_2 + H_3O^+ \qquad K_a = \frac{[H_3O^+][C_6H_5NH_2]}{[C_6H_5NH_3^+]}$$

Inspection of Appendix 3 gives K_a for $C_6H_5NH_3^+$ of 2.51×10^{-5}. Proceeding as in Example 7-9, we have

$$[H_3O^+] = [C_6H_5NH_2]$$
$$[C_6H_5NH_3^+] = 2.0 \times 10^{-4} - [H_3O^+]$$

Let us assume that $[H_3O^+] \ll 2.0 \times 10^{-4}$ and substitute the simplified value for $[C_6H_5NH_3^+]$ into the dissociation-constant expression to obtain (see Equation 7-20)

$$\frac{[H_3O^+]^2}{2.0 \times 10^{-4}} = 2.51 \times 10^{-5}$$

$$[H_3O^+] = \sqrt{5.02 \times 10^{-9}} = 7.09 \times 10^{-5}\text{ M}$$

Comparison of 7.09×10^{-5} with 2.0×10^{-4} suggests that a significant error has been introduced by the assumption that $[H_3O^+] \ll c_{C_6H_5NH_3^+}$ (Figure 7-2 indicates that this error is about 20%). Thus, unless only an approximate value for $[H_3O^+]$ is needed, it is necessary to use the more nearly exact expression (Equation 7-17)

$$\frac{[H_3O^+]^2}{2.0 \times 10^{-4} - [H_3O^+]} = 2.51 \times 10^{-5}$$

See *Mathcad Applications for Analytical Chemistry*, **pp. 35–36.**

which rearranges to

$$[H_3O^+]^2 + 2.51 \times 10^{-5}[H_3O^+] - 5.02 \times 10^{-9} = 0$$

$$[H_3O^+] = \frac{-2.51 \times 10^{-5} + \sqrt{(2.54 \times 10^{-5})^2 + 4 \times 5.02 \times 10^{-9}}}{2}$$

$$= 5.94 \times 10^{-5}$$

The quadratic equation can also be solved by the iterative method shown in Feature 7-5.

Hydronium Ion Concentration in Solutions of Weak Bases

The techniques discussed in previous sections are readily adapted to the calculation of the hydroxide or hydronium ion concentration in solutions of weak bases.

Aqueous ammonia is basic by virtue of the reaction

$$NH_3 + H_2O \rightleftharpoons NH_4^+ + OH^-$$

The predominant solute species in such solutions has been clearly demonstrated to be NH_3. Nevertheless, solutions of ammonia are still called ammonium hydroxide occasionally, because at one time chemists thought that NH_4OH rather than NH_3 was the undissociated form of the base. Application of the mass law to the equilibrium as written yields

$$K_b = \frac{[NH_4^+][OH^-]}{[NH_3]}$$

EXAMPLE 7-11

Calculate the hydroxide ion concentration of a 0.0750 M NH_3 solution. The predominant equilibrium is

$$NH_3 + H_2O \rightleftharpoons NH_4^+ + OH^-$$

As shown on page 138,

$$K_b = \frac{[NH_4^+][OH^-]}{[NH_3]} = \frac{1.00 \times 10^{-14}}{5.70 \times 10^{-10}} = 1.75 \times 10^{-5}$$

The chemical equation shows that

$$[NH_4^+] = [OH^-]$$

Both NH_4^+ and NH_3 come from the 0.0750 M solution. Thus,

$$[NH_4^+] + [NH_3] = c_{NH_3} = 0.0750 \text{ M}$$

If we substitute $[OH^-]$ for $[NH_4^+]$ in the second of these equations and rearrange, we find that

$$[NH_3] = 0.0750 - [OH^-]$$

Substituting these quantities into the expression for K_b yields

$$\frac{[OH^-]^2}{7.50 \times 10^{-2} - [OH^-]} = 1.75 \times 10^{-5}$$

which is analogous to Equation 7-17 for weak acids. Provided that $[OH^-] \ll 7.50 \times 10^{-2}$, this equation simplifies to

$$[OH^-]^2 \approx 7.50 \times 10^{-2} \times 1.75 \times 10^{-5}$$

$$[OH^-] = 1.15 \times 10^{-3} \text{ M}$$

Upon comparing the calculated value for $[OH^-]$ with 7.50×10^{-2}, we see that the error in $[OH^-]$ is less than 2%. If needed, a better value for $[OH^-]$ can be obtained by solving the quadratic equation.

EXAMPLE 7-12

Calculate the hydroxide ion concentration in a 0.0100 M sodium hypochlorite solution.

The equilibrium between OCl^- and water is

$$OCl^- + H_2O \rightleftharpoons HOCl + OH^-$$

for which

$$K_b = \frac{[HOCl][OH^-]}{[OCl^-]}$$

Appendix 3 reveals that the acid dissociation constant for HOCl is 3.0×10^{-8}. Therefore, we rearrange Equation 7-13 and write

$$K_b = \frac{K_w}{K_a} = \frac{1.00 \times 10^{-14}}{3.0 \times 10^{-8}} = 3.33 \times 10^{-7}$$

Proceeding as in Example 7-11, we have

$$[OH^-] = [HOCl]$$
$$[OCl^-] + [HOCl] = 0.0100$$
$$[OCl^-] = 0.0100 - [OH^-] \approx 0.0100$$

Here we have assumed that $[OH^-] \ll 0.0100$. Substitution into the equilibrium-constant expression gives

$$\frac{[OH^-]^2}{0.0100} = 3.33 \times 10^{-7}$$
$$[OH^-] = 5.8 \times 10^{-5}\ M$$

Note that the error resulting from the approximation is small.

7C QUESTIONS AND PROBLEMS

7-1. Briefly describe or define
*(a) a weak electrolyte.
(b) a Brønsted-Lowry acid.
*(c) the conjugate base of a Brønsted-Lowry acid.
(d) neutralization, in terms of the Brønsted-Lowry concepts.
*(e) an amphiprotic solute.
(f) a zwitterion.
*(g) autoprotolysis.
(h) a strong acid.
*(i) the Le Châtelier principle.
(j) the common-ion effect.

7-2. Briefly describe or define
*(a) an amphiprotic solvent.
(b) a differentiating solvent.
*(c) a leveling solvent.
(d) a mass-action effect.

*7-3. Briefly explain why there is no term in an equilibrium-constant expression for water or for a pure solid, even though one (or both) appear in the balanced net ionic equation for the equilibrium.

7-4. Identify the acid on the left and its conjugate base on the right in the following equations:
*(a) $HCN + H_2O \rightleftharpoons H_3O^+ + CN^-$
(b) $HONH_2 + H_2O \rightleftharpoons HONH_3^+ + OH^-$
*(c) $NH_4^+ + H_2O \rightleftharpoons NH_3 + H_3O^+$
(d) $2HCO_3^- \rightleftharpoons H_2CO_3 + CO_3^{2-}$
*(e) $PO_4^{3-} + H_2PO_4^- \rightleftharpoons 2HPO_4^{2-}$

7-5. Identify the base on the left and its conjugate acid on the right in the equations for Problem 7-4.

7-6. Write expressions for the autoprotolysis of
*(a) H_2O.
(b) CH_3COOH.
*(c) CH_3NH_2.
(d) CH_3CH_2OH.

7-7. Write equilibrium-constant expressions for
*(a) solid AgI and its ions in solution.
(b) solid $PbCl_2$ and its ions in solution.
*(c) solid Ag_2CrO_4 and its ions in solution.
(d) solid $BaCrO_4$ and its ions in solution.

7-8. Write the equilibrium-constant expressions and obtain numerical values for each constant in
*(a) the basic dissociation of ethylamine, $C_2H_5NH_2$.
(b) the acidic dissociation of hydrogen cyanide, HCN.
*(c) the acidic dissociation of pyridine hydrochloride, C_5H_5NHCl.
(d) the basic dissociation of NaCN.
*(e) the dissociation of H_3AsO_4 to H_3O^+ and AsO_4^{3-}.
(f) the reaction of CO_3^{2-} with H_2O to give H_2CO_3 and OH^-.

7-9. Generate the solubility-product expression for
*(a) $AgIO_3$.
*(b) Ag_2SO_3.
*(c) Ag_3AsO_4.
*(d) PbClF.
(e) CuI.
(f) PbI_2.
(g) BiI_3.
(h) $MgHN_4PO_4$.

*7-10. Express the solubility-product constant for each substance in Problem 7-9 in terms of its molar solubility *S*.

7-11. Calculate the solubility-product constant for each of the following substances, given that the molar concentrations of their saturated solutions are as indicated:
*(a) AgSeCN (2.0×10^{-8} mol/L; products are Ag^+ and $SeCN^-$).
(b) $RaSO_4$ (6.6×10^{-6} mol/L).
*(c) $Ba(BrO_3)_2$ (9.2×10^{-3} mol/L).
(d) PbF_2 (1.9×10^{-3} mol/L).
*(e) $Ce(IO_3)_3$ (1.9×10^{-3} mol/L).
(f) BiI_3 (1.3×10^{-5} mol/L).

*7-12. Calculate the solubility of the solutes in Problem 7-11 for solutions in which the cation concentration is 0.050 M.

***7-13.** Calculate the solubility of the solutes in Problem 7-11 for solutions in which the anion concentration is 0.050 M.

***7-14.** The solubility product for Tl_2CrO_4 is 9.8×10^{-13}. What CrO_4^{2-} concentration is required to
(a) initiate precipitation of Tl_2CrO_4 from a solution that is 2.12×10^{-3} M in Tl^+?
(b) lower the concentration of Tl^+ in a solution to 1.00×10^{-6} M?

7-15. What hydroxide concentration is required to
(a) initiate precipitation of Fe^{3+} from a 1.00×10^{-3} M solution of $Fe_2(SO_4)_3$?
(b) lower the Fe^{3+} concentration in the foregoing solution to 1.00×10^{-9} M?

***7-16.** The solubility-product constant for $Ce(IO_3)_3$ is 3.2×10^{-10}. What is the Ce^{3+} concentration in a solution prepared by mixing 50.0 mL of 0.0500 M Ce^{3+} with 50.00 mL of
(a) water?
(b) 0.050 M IO_3^-?
(c) 0.150 M IO_3^-?
(d) 0.300 M IO_3^-?

7-17. The solubility-product constant for K_2PtCl_6 is 1.1×10^{-5} ($K_2PtCl_6 \rightleftharpoons 2K^+ + PtCl_6^{2-}$). What is the K^+ concentration of a solution prepared by mixing 50.0 mL of 0.400 M KCl with 50.0 mL of
(a) 0.100 M $PtCl_6^{2-}$?
(b) 0.200 M $PtCl_6^{2-}$?
(c) 0.400 M $PtCl_6^{2-}$?

***7-18.** The solubility products for a series of iodides are

TlI	$K_{sp} = 6.5 \times 10^{-8}$
AgI	$K_{sp} = 8.3 \times 10^{-17}$
PbI_2	$K_{sp} = 7.1 \times 10^{-9}$
BiI_3	$K_{sp} = 8.1 \times 10^{-19}$

List these four compounds in order of decreasing molar solubility in
(a) water.
(b) 0.10 M NaI.
(c) a 0.010 M solution of the solute cation.

7-19. The solubility products for a series of hydroxides are

BiOOH	$K_{sp} = 4.0 \times 10^{-10} = [BiO^+][OH^-]$
$Be(OH)_2$	$K_{sp} = 7.0 \times 10^{-22}$
$Tm(OH)_3$	$K_{sp} = 3.0 \times 10^{-24}$
$Hf(OH)_4$	$K_{sp} = 4.0 \times 10^{-26}$

Which hydroxide has
(a) the lowest molar solubility in H_2O?
(b) the lowest molar solubility in a solution that is 0.10 M in NaOH?

7-20. Calculate the hydronium ion concentration of water at 100°C.

7-21. At 25°C, what are the molar H_3O^+ and OH^- concentrations in
***(a)** 0.0200 M HOCl?
(b) 0.0800 M propanoic acid?
***(c)** 0.200 M methylamine?
(d) 0.100 M trimethylamine?
***(e)** 0.120 M NaOCl?
(f) 0.0860 M CH_3COONa?
***(g)** 0.100 M hydroxylamine hydrochloride?
(h) 0.0500 M ethanolamine hydrochloride?

7-22. At 25°C, what is the hydronium ion concentration in
***(a)** 0.100 M chloroacetic acid?
***(b)** 0.100 M sodium chloroacetate?
(c) 0.0100 M methylamine?
(d) 0.0100 M methylamine hydrochloride?
***(e)** 1.00×10^{-3} M aniline hydrochloride?
(f) 0.200 M HIO_3?

CHAPTER

8

Effect of Electrolytes on Ionic Equilibria

We have noted that Equation 7-7 (page 128) is an approximate form of the thermodynamic equilibrium-constant expression derived from theory. In this chapter, we describe when the use of this approximate form is likely to lead to error. We also show how the equation is modified to give a more exact description of a system at chemical equilibrium.

8A THERMODYNAMIC VERSUS CONCENTRATION-BASED EQUILIBRIUM CONSTANTS

Experimentally we find that the position of most ionic solution equilibria depends upon the electrolyte concentration of the medium. This happens even when the added electrolyte contains no ion in common with those involved in the equilibrium. For example, consider again the oxidation of iodide ion by arsenic acid:

$$H_3AsO_4 + 3I^- + 2H^+ \rightleftharpoons H_3AsO_3 + I_3^- + H_2O$$

If an electrolyte, such as barium nitrate, potassium sulfate, or sodium perchlorate, is added to this solution, the color becomes less intense. This decrease in color intensity indicates that the concentration of I_3^- has decreased and that the equilibrium has been shifted to the left by the added electrolyte.

Figure 8-1 further illustrates the effect of electrolytes. Curve *A* is a plot of the product of the molar hydronium and hydroxide ion *concentrations* ($\times 10^{14}$) as a function of the concentration of sodium chloride. This *concentration*-based ion product is designated K'_w. At low sodium chloride concentrations, K'_w becomes independent of the electrolyte concentration and is equal to 1.00×10^{-14}, which is the *thermodynamic* ion-product constant for water, K_w. A relationship that approaches a constant value as some parameter (here, the electrolyte concentration) approaches zero is called a *limiting law*; the constant numerical value observed at this limit is referred to as a *limiting value.*

Concentration-based equilibrium constants are commonly indicated by adding a prime mark. For example: K'_w, K'_{sp}, K'_a.

The vertical axis for curve *B* in Figure 8-1 is the product of the molar concentrations of barium and sulfate ions ($\times 10^{10}$) in saturated solutions of barium sulfate. This concentration-based solubility product is designated as K'_{sp}. At low

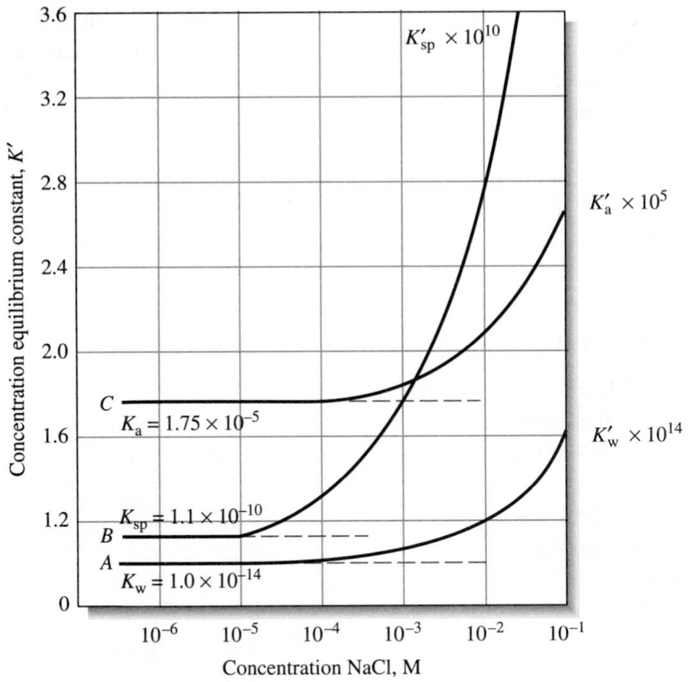

Figure 8-1
Effect of electrolyte concentration on concentration-based equilibrium constants.

electrolyte concentrations, K'_{sp} has a limiting value of 1.1×10^{-10}, which is the accepted thermodynamic value of K_{sp} for barium sulfate.

Curve C is a plot of K'_a ($\times 10^5$), the concentration quotient for the equilibrium involving the dissociation of acetic acid, as a function of electrolyte concentration. Here again, the ordinate function approaches a limiting value K_a, which is the thermodynamic acid dissociation constant for acetic acid.

As the electrolyte concentration becomes very small, concentration-based equilibrium constants approach their thermodynamic values: K_w, K_{sp}, K_a.

The dashed lines in Figure 8-1 represent ideal behavior of the solutes. Note that departures from ideality can be significant. For example, the product of the molar concentrations of hydrogen and hydroxide ion increases from 1.0×10^{-14} in pure water to about 1.7×10^{-14} in a solution that is 0.1 M in sodium chloride. The effect is even more pronounced with barium sulfate; here, K'_{sp} in 0.1 M sodium chloride is more than double that of its limiting value.

The electrolyte effect shown in Figure 8-1 is not peculiar to sodium chloride. Indeed, we would see identical curves if potassium nitrate or sodium perchlorate were substituted for sodium chloride. In each case, the effect has as its origin the electrostatic attraction between the ions of the electrolyte and the ions involved in the equilibrium. Since the electrostatic forces associated with all singly charged ions are approximately the same, the three salts exhibit essentially identical effects on equilibria.

We must now consider how we can take the electrolyte effect into account when we wish to make more accurate equilibrium calculations.

8A-1 Effect of the Charge of Reactants and Products

Extensive studies have revealed that the magnitude of the electrolyte effect is highly dependent upon the charges of the participants in an equilibrium. When only neutral species are involved, the position of equilibrium is essentially inde-

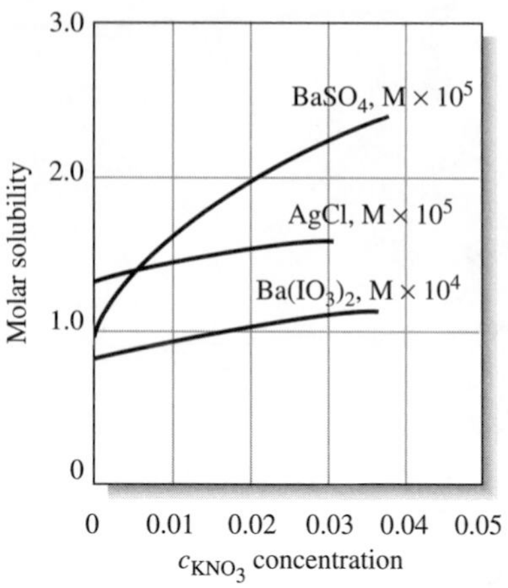

Figure 8-2
Effect of electrolyte concentration on the solubility of some salts.

pendent of electrolyte concentration. With ionic participants, the magnitude of the electrolyte effect increases with charge. This generality is demonstrated by the three solubility curves in Figure 8-2. Note, for example, that in a 0.02 M solution of potassium nitrate, the solubility of barium sulfate, with its pair of doubly charged ions, is larger than it is in pure water by a factor of 2. This same change in electrolyte concentration increases the solubility of barium iodate by a factor of only 1.25 and that of silver chloride by 1.2. The enhanced effect due to doubly charged ions is also reflected in the greater slope of curve *B* in Figure 8-1.

8A-2 Effect of Ionic Strength

Systematic studies have shown that the effect of added electrolyte on equilibria is *independent* of the chemical nature of the electrolyte but depends upon a property of the solution called the *ionic strength.* This quantity is defined as

$$\text{ionic strength} = \mu = \tfrac{1}{2}([A]Z_A^2 + [B]Z_B^2 + [C]Z_C^2 + \cdots) \qquad (8\text{-}1)$$

where [A], [B], [C], . . . represent the species molar concentrations of ions A, B, C, . . . and Z_A, Z_B, Z_C, . . . are their charges.

EXAMPLE 8-1

Calculate the ionic strength of (a) a 0.1 M solution of KNO_3 and (b) a 0.1 M solution of Na_2SO_4.

(a) For the KNO_3 solution, $[K^+]$ and $[NO_3^-]$ are 0.1 M and

$$\mu = \tfrac{1}{2}(0.1 \times 1^2 + 0.1 \times 1^2) = 0.1$$

(b) For the Na_2SO_4 solution, $[Na^+] = 0.2$ and $[SO_4^{2-}] = 0.1$. Therefore,

$$\mu = \tfrac{1}{2}(0.2 \times 1^2 + 0.1 \times 2^2) = 0.3$$

EXAMPLE 8-2

What is the ionic strength of a solution that is 0.05 M in KNO_3 and 0.1 M in Na_2SO_4?

$$\mu = \tfrac{1}{2}(0.05 \times 1^2 + 0.05 \times 1^2 + 0.2 \times 1^2 + 0.1 \times 2^2) = 0.35$$

It is apparent from these examples that the ionic strength of a solution of a strong electrolyte consisting solely of singly charged ions is identical with its total molar salt concentration. The ionic strength is, however, greater than the molar concentration if the solution contains ions with multiple charges.

For solutions with ionic strengths of 0.1 M or less, the electrolyte is *independent* of the *kind* of ions and dependent *only upon the ionic strength.* Thus, the solubility of barium sulfate is the same in aqueous sodium iodide, potassium nitrate, or aluminum chloride, provided the molar concentrations of these species are such that the ionic strengths are identical. Note that this independence with respect to electrolyte species disappears at high ionic strengths.

Effect of Charge on Ionic Strength

Type of Electrolyte	Example	Ionic Strength*
1:1	NaCl	*c*
1:2	$Ba(NO_3)_2$, Na_2SO_4	3*c*
1:3	$Al(NO_3)_3$, Na_3PO_4	6*c*
2:2	$MgSO_4$	4*c*

**c* = molarity of the salt

8A-3 Source of the Salt Effect

The electrolyte effect, which we have just described, results from the electrostatic attractive and repulsive forces that exist between the ions of an electrolyte and the ions involved in an equilibrium. These forces cause each ion from the dissociated reactant to be surrounded by a sheath of solution that contains a slight excess of electrolyte ions of opposite charge. For example, when a barium sulfate precipitate is equilibrated with a sodium chloride solution, each dissolved barium ion is surrounded by an ionic atmosphere that, because of electrostatic attraction and repulsion, carries a small net negative charge on the average due to repulsion of sodium ions and an attraction of chloride ions. Similarly, each sulfate ion is surrounded by ionic atmosphere that tends to be slightly positive. These charged layers make the barium ions somewhat less positive and the sulfate ions somewhat less negative than they would be in the absence of electrolyte. The consequence of this effect is a decrease in overall attraction between barium and sulfate ions and an increase in solubility, which becomes greater as the number of electrolyte ions in the solution becomes larger. That is, the *effective concentration* of barium ions and of sulfate ions becomes less as the ionic strength of the medium becomes greater.

8B ACTIVITY COEFFICIENTS

In order to describe quantitatively the effective concentration of participants in an equilibrium at any given ionic strength, chemists use a term called *activity, a,* which for the species X is defined as

The activity of a species is a measure of its effective concentration as determined by the lowering of the freezing point of water, by electrical conductivity, and by the mass action effect.

$$a_X = \gamma_X[X] \tag{8-2}$$

where a_X is the activity of the species X, [X] is its molar concentration, and γ_X is a dimensionless quantity called the *activity coefficient.* The activity coefficient and thus the activity of X vary with ionic strength such that substitution of a_X for [X] in any equilibrium-constant expression frees the numerical value of the constant from dependence on the ionic strength. To illustrate, if X_mY_n is a precipitate, the thermodynamic solubility product expression is defined by the equation

$$K_{sp} = a_X^m \cdot a_Y^n \tag{8-3}$$

Applying Equation 8-2 gives

$$K_{sp} = \gamma_X^m \gamma_Y^n \cdot [X]^m[Y]^n = \gamma_X^m \gamma_Y^n \cdot K'_{sp} \tag{8-4}$$

Here K'_{sp} is the *concentration solubility product constant* and K_{sp} is the thermodynamic equilibrium constant.[1] The activity coefficients γ_X and γ_Y vary with ionic strength in such a way as to keep K_{sp} numerically constant and independent of ionic strength (in contrast to the concentration constant K'_{sp}).

[1] In the chapters that follow we use the prime notation only when it is necessary to distinguish between thermodynamic and concentration equilibrium constant.

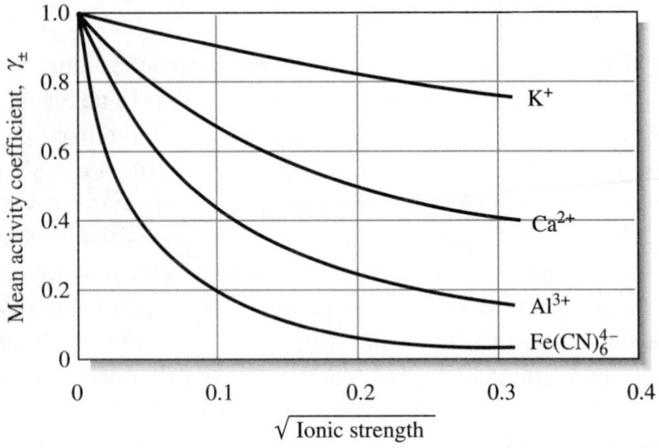

Figure 8-3
Effect of ionic strength on activity coefficients.

8B-1 Properties of Activity Coefficients

Activity coefficients have the following properties:

As $\mu \rightarrow 0$, $\gamma_X \rightarrow 1$, $a_X \rightarrow [X]$, and $K'_{sp} \rightarrow K_{sp}$.

1. The activity coefficient of a species is a measure of the effectiveness with which that species influences an equilibrium in which it is a participant. In very dilute solutions, where the ionic strength is minimal, this effectiveness becomes constant, and the activity coefficient is unity. Under such circumstances, the activity and the molar concentration of the species are identical (as are thermodynamic and concentration equilibrium constants). As the ionic strength increases, however, an ion loses some of its effectiveness, and its activity coefficient decreases. We may summarize this behavior in terms of Equations 8-2 and 8-3. At moderate ionic strengths, $\gamma_X < 1$; as the solution approaches infinite dilution, however, $\gamma_X \rightarrow 1$ and thus $a_X \rightarrow [X]$ and $K'_{sp} \rightarrow K_{sp}$. At high ionic strengths ($\mu > 0.1$ M), activity coefficients often increase and may even become greater than unity. Because interpretation of the behavior of such solutions is difficult, we shall confine our discussion to regions of low or moderate ionic strength (that is, where $\mu \leq 0.1$ M).
   The variation of typical activity coefficients as a function of ionic strength is shown in Figure 8-3.
2. In solutions that are not too concentrated, the activity coefficient for a given species is independent of the nature of the electrolyte and dependent only upon the ionic strength.
3. For a given ionic strength, the activity coefficient of an ion departs farther from unity as the charge carried by the species increases. This effect is shown in Figure 8-2. The activity coefficient of an uncharged molecule is approximately unity, regardless of ionic strength.
4. At any given ionic strength, the activity coefficients of ions of the same charge are approximately equal. The small variations that do exist can be correlated with the effective diameter of the hydrated ions.
5. The activity coefficient of a given ion describes its effective behavior in all equilibria in which it participates. For example, at a given ionic strength, a single activity coefficient for cyanide ion describes the influence of that species upon any of the following equilibria:

$$HCN + H_2O \rightleftharpoons H_3O^+ + CN^-$$
$$Ag^+ + CN^- \rightleftharpoons AgCN(s)$$
$$Ni^{2+} + 4CN^- \rightleftharpoons Ni(CN)_4^{2-}$$

8B-2 The Debye-Hückel Equation

In 1923, P. Debye and E. Hückel used the ionic atmosphere model, described in Section 8A-3, to derive a theoretical expression that permits the calculation of activity coefficients of ions from their charge and their average size.[2] This equation, which has become known as the *Debye-Hückel equation,* takes the form

$$-\log \gamma_X = \frac{0.51\, Z_X^2 \sqrt{\mu}}{1 + 3.3\, \alpha_X \sqrt{\mu}} \qquad (8\text{-}5)$$

where

γ_X = activity coefficient of the species X
Z_X = charge on the species X
μ = ionic strength of the solution
α_X = effective diameter of the hydrated ion X in nanometers (10^{-9} m)

The constants 0.51 and 3.3 are applicable to aqueous solutions at 25°C; other values must be employed at other temperatures.

Unfortunately, there is considerable uncertainty regarding the magnitude of α_X in Equation 8-5. Its value appears to be approximately 0.3 nm for most singly charged ions; for these species, then, the denominator of the Debye-Hückel equation simplifies to approximately $1 + \sqrt{\mu}$. For ions with higher charge, α_X may be as large as 1.0 nm. This increase in size with increase in charge makes good chemical sense. The larger the charge on an ion, the larger the number of polar water molecules that will be held in the solvation shell about the ion. It should be noted that the second term of the denominator is small with respect to the first when the ionic strength is less than 0.01, so at these ionic strengths, uncertainties in α_A are of little significance in calculating activity coefficients.

Kielland[3] has derived values of α_X for numerous ions from a variety of experimental data. His best values for effective diameters are given in Table 8-1. Also presented are activity coefficients calculated from Equation 8-5 using these values for the size parameter.

Experimental determination of single-ion activity coefficients such as those shown in Table 8-1 is unfortunately impossible because all experimental methods give only a mean activity coefficient for the positively and negatively charged ions in a solution. In other words, it is impossible to measure the properties of individual ions in the presence of counter ions of opposite charge and solvent

Peter Debye (1884–1966) was born and educated in Europe but became Professor of Chemistry at Cornell University in 1940. He was noted for his work in several different areas of chemistry including electrolyte solutions, X-ray diffraction, and the properties of polar molecules. He received the 1936 Nobel Prize in Chemistry.

When μ is less than 0.01, $1 + \sqrt{\mu} \approx 1$, and Equation 8-5 becomes

$$-\log \gamma_X = 0.51\, Z_X^2 \sqrt{\mu}$$

This equation is referred to as the Debye-Hückel Limiting Law (DHLL). Thus, in solutions of very low ionic strength, the DHLL can be used to calculate approximate activity coefficients.

[2] P. Debye and E. Hückel, *Physik. Z.,* **1923,** *24,* 185.

[3] J. Kielland, *J. Amer. Chem. Soc.,* **1937,** *59,* 1675.

TABLE 8-1 Activity Coefficients for Ions at 25°C*

Ion	α_X, nm	Activity Coefficient at Indicated Ionic Strength 0.001	0.005	0.01	0.05	0.1
H_3O^+	0.9	0.967	0.933	0.914	0.86	0.83
Li^+, $C_6H_5COO^-$	0.6	0.965	0.929	0.907	0.84	0.80
Na^+, IO_3^-, HSO_3^-, HCO_3^-, $H_2PO_4^-$, $H_2AsO_4^-$, OAc^-	0.4–0.45	0.964	0.928	0.902	0.82	0.78
OH^-, F^-, SCN^-, HS^-, ClO_3^-, ClO_4^-, BrO_3^-, IO_4^-, MnO_4^-	0.35	0.964	0.926	0.900	0.81	0.76
K^+, Cl^-, Br^-, I^-, CN^-, NO_2^-, NO_3^-, $HCOO^-$	0.3	0.964	0.925	0.899	0.80	0.76
Rb^+, Cs^+, Tl^+, Ag^+, NH_4^+	0.25	0.964	0.924	0.898	0.80	0.75
Mg^{2+}, Be^{2+}	0.8	0.872	0.755	0.69	0.52	0.45
Ca^{2+}, Cu^{2+}, Zn^{2+}, Sn^{2+}, Mn^{2+}, Fe^{2+}, Ni^{2+}, Co^{2+}, Phthalate^{2-}	0.6	0.870	0.749	0.675	0.48	0.40
Sr^{2+}, Ba^{2+}, Cd^{2+}, Hg^{2+}, S^{2-}	0.5	0.868	0.744	0.67	0.46	0.38
Pb^{2+}, CO_3^{2-}, SO_3^{2-}, $C_2O_4^{2-}$	0.45	0.868	0.742	0.665	0.46	0.37
Hg_2^{2+}, SO_4^{2-}, $S_2O_3^{2-}$, CrO_4^{2-}, HPO_4^{2-}	0.40	0.867	0.740	0.660	0.44	0.36
Al^{3+}, Fe^{3+}, Cr^{3+}, La^{3+}, Ce^{3+}	0.9	0.738	0.54	0.44	0.24	0.18
PO_4^{3-}, $Fe(CN)_6^{3-}$	0.4	0.725	0.50	0.40	0.16	0.095
Th^{4+}, Zr^{4+}, Ce^{4+}, Sn^{4+}	1.1	0.588	0.35	0.255	0.10	0.065
$Fe(CN)_6^{4-}$	0.5	0.57	0.31	0.20	0.048	0.021

*From J. Kielland, *J. Am. Chem. Soc.*, **1937,** *59,* 1675. By courtesy of the American Chemical Society.

See *Mathcad Applications for Analytical Chemistry,* **pp. 38, 40.**

molecules. It should be pointed out, however, that mean activity coefficients calculated from the data in Table 8-1 agree satisfactorily with the experimental values.

EXAMPLE 8-3

Use Equation 8-5 to calculate the activity coefficient for Hg^{2+} in a solution that has an ionic strength of 0.085. Use 0.5 nm for the effective diameter of the ion. Compare the calculated value with $\gamma_{Hg^{2+}}$ obtained by interpolation of data from Table 8-1.

$$-\log \gamma_{Hg^{2+}} = \frac{(0.51)\,(2)^2\,\sqrt{0.085}}{1 + (3.3)\,(0.5)\,\sqrt{0.085}} = 0.4016$$

$$\frac{1}{\gamma_{Hg^{2+}}} = 2.52$$

$$\gamma_{Hg^{2+}} = 0.397 = 0.40$$

Table 8-1 indicates that $\gamma_{Hg^{2+}} = 0.46$ when $\mu = 0.100$ and 0.38 when $\mu =$ 0.050. Thus, for $\mu = 0.85$

$$\gamma_{Hg^{2+}} = 0.38 + \frac{0.015}{0.050}(0.46 - 0.38) = 0.404 = 0.40$$

The Debye-Hückel relationship and the data in Table 8-1 give satisfactory activity coefficients for ionic strengths up to about 0.1. Beyond this value, the equation fails, and experimentally determined mean activity coefficients must be used.

Values for activity coefficients at ionic strengths not listed in Table 8-1 can be approximated by interpolation as shown here.

FEATURE 8-1
Mean Activity Coefficients

The mean activity coefficient of the electrolyte A_mB_n is defined as

$$\gamma_{\pm} = \text{mean activity coefficient} = (\gamma_A^m \gamma_B^n)^{1/(m+n)}$$

The mean activity coefficient can be measured in any of several ways, but it is impossible experimentally to resolve this term into the individual activity coefficients for γ_A and γ_B. For example, if

$$K_{sp} = [A]^m [B]^n \cdot \gamma_A^m \gamma_B^n = [A]^m [B]^n \cdot \gamma_{\pm}^{m+n}$$

we can obtain K_{sp} by measuring the solubility of A_mB_n in a solution in which the electrolyte concentration approaches zero (that is, where both γ_A and $\gamma_B \rightarrow 1$). A second solubility measurement at some ionic strength μ_1 gives values for [A] and [B]. These data then permit the calculation of $\gamma_A^m \gamma_B^n = \gamma_{\pm}^{m+n}$ for ionic strength μ_1.

It is important to understand that this procedure does not provide enough experimental data to permit the calculation of the *individual* quantities γ_A and γ_B and that there appears to be no additional experimental information that would permit evaluation of these quantities. This situation is general, and the *experimental* determination of individual activity coefficients is impossible.

8B-3 Equilibrium Calculations Employing Activity Coefficients

Equilibrium calculations with activities yield values that are in better agreement with experimental results than those obtained with molar concentrations. Unless otherwise specified, equilibrium constants found in tables are generally based upon activities and are thus thermodynamic equilibrium constants. The examples that follow illustrate how activity coefficients from Table 8-1 are applied to such data.

EXAMPLE 8-4

Find the relative error introduced by neglecting activities in calculating the solubility of $Ba(IO_3)_2$ in a 0.033 M solution of $Mg(IO_3)_2$. The thermodynamic solubility product for $Ba(IO_3)_2$ is 1.57×10^{-9} (Appendix 2).

At the outset, we write the solubility-product expression in terms of activities:

$$a_{Ba^{2+}} \cdot a^2_{IO_3^-} = K_{sp} = 1.57 \times 10^{-9}$$

where $a_{Ba^{2+}}$ and $a^2_{IO_3^-}$ are the activities of barium and iodate ions. Replacing activities in this equation by activity coefficients and concentrations from Equation 8-2 yields

$$\gamma_{Ba^{2+}}\,[Ba^{2+}] \cdot \gamma^2_{IO_3^-}\,[IO_3^-]^2 = K_{sp}$$

where $\gamma_{Ba^{2+}}$ and $\gamma_{IO_3^-}$ are the activity coefficients for the two ions. Rearranging this expression gives

$$K'_{sp} = \frac{K_{sp}}{\gamma_{Ba^{2+}}\gamma^2_{IO_3^-}} = [Ba^{2+}][IO_3^-]^2 \qquad \textbf{(8-6)}$$

where K'_{sp} is the *concentration-based solubility product.*

The ionic strength of the solution is obtained by substituting into Equation 8-1:

$$\mu = \tfrac{1}{2}([Mg^{2+}] \times 2^2 + [IO_3^-] \times 1^2)$$
$$= \tfrac{1}{2}(0.033 \times 4 + 0.066 \times 1) = 0.099 \approx 0.1$$

In calculating μ, we have assumed that the Ba^{2+} and IO_3^- ions from the precipitate do not significantly affect the ionic strength of the solution. This simplification seems justified, considering the low solubility of barium iodate and the relatively high concentration of $Mg(IO_3)_2$. In situations where it is not possible to make such an assumption, the concentrations of the two ions can be approximated by solubility calculation in which activities and concentrations are assumed to be identical (as in Examples 7-5, 7-6, and 7-7). These concentrations can then be introduced to give a better value for μ.

Turning now to Table 8-1, we find that at an ionic strength of 0.1,

$$\gamma_{Ba^{2+}} = 0.38 \qquad \gamma_{IO_3^-} = 0.78$$

If the calculated ionic strength did not match that of one of the columns in the table, $\gamma_{Ba^{2+}}$ and $\gamma_{IO_3^-}$ could be calculated from Equation 8-5.

Substituting into the thermodynamic solubility-product expression gives

$$K'_{sp} = \frac{1.57 \times 10^{-9}}{(0.38)\,(0.78)^2} = 6.8 \times 10^{-9}$$
$$[Ba^{2+}][IO_3^-]^2 = 6.8 \times 10^{-9}$$

Proceeding now as in earlier solubility calculations,

$$\text{solubility} = [Ba^{2+}]$$
$$[IO_3^-] = 2 \times 0.033 + 2[Ba^{2+}] \approx 0.066$$

$$[Ba^{2+}](0.066)^2 = 6.8 \times 10^{-9}$$

$$[Ba^{2+}] = \text{solubility} = 1.56 \times 10^{-6}\ M$$

If we neglect activities, the solubility is obtained as shown earlier. That is,

$$[Ba^{2+}](0.066)^2 = 1.57 \times 10^{-9}$$

$$[Ba^{2+}] = \text{solubility} = 3.60 \times 10^{-7}\ M$$

$$\text{relative error} = \frac{3.60 \times 10^{-7} - 1.56 \times 10^{-6}}{1.56 \times 10^{-6}} \times 100\% = -77\%$$

EXAMPLE 8-5

Use activities to calculate the hydronium ion concentration in a 0.120 M solution of HNO_2 that is also 0.050 M in NaCl.

The ionic strength of this solution is

$$\mu = \tfrac{1}{2}(0.0500 \times 1^2 + 0.0500 \times 1^2) = 0.0500$$

In Table 8-1, at ionic strength 0.050, we find

$$\gamma_{H_3O^+} = 0.86 \qquad \gamma_{NO_2^-} = 0.80$$

Also, from rule 3 (page 152), we can write

$$\gamma_{HNO_2} = 1.0$$

These three values for γ permit the calculation of a concentration-based dissociation constant from the thermodynamic constant of 5.1×10^{-4} (Appendix 3):

$$K_a' = \frac{[H_3O^+][NO_2^-]}{[HNO_2]} = \frac{K_a \cdot \gamma_{HNO_2}}{\gamma_{H_3O^+}\gamma_{NO_2^-}} = \frac{5.1 \times 10^{-4} \times 1.0}{0.86 \times 0.81} = 7.3 \times 10^{-4}$$

Proceeding as in Example 7-9, we write

$$[H_3O^+] = \sqrt{0.120 \times 7.3 \times 10^{-4}} = 9.4 \times 10^{-3}$$

Note that neglecting the activity coefficient gives $[H_3O^+] = 7.8 \times 10^{-3}$.

8B-4 Omission of Activity Coefficients in Equilibrium Calculations

We shall ordinarily neglect activity coefficients and simply use molar concentrations in applications of the equilibrium law. This recourse simplifies the calculations and greatly decreases the amount of data needed. For most purposes, the error introduced by the assumption of unity for the activity coefficient is not large enough to lead to false conclusions. It should be apparent from the preceding examples, however, that disregard of activity coefficients may introduce a significant numerical error in calculations of this kind. Note, for example, that neglect of activities in Example 8-4 resulted in an error of about −77% relative. The reader should be alert to the conditions under which the substitution of concentration for activity is likely to lead to the largest error. Significant discrepancies occur when the ionic strength is large (0.01 or larger) or when the ions involved have multiple charges (Table 8-1). With dilute solutions (ionic strength < 0.01) of nonelectrolytes or of singly charged ions, the use of concentrations in a mass-law calculation often provides reasonably accurate results.

It is also important to note that the decrease in solubility resulting from the presence of an ion common to the precipitate is in part counteracted by the larger electrolyte concentration associated with the presence of the salt containing the common ion.

8C QUESTIONS AND PROBLEMS

***8-1.** Make a distinction between
(a) activity and activity coefficient.
(b) thermodynamic and concentration equilibrium constants.

8-2. List general properties of activity coefficients.

***8-3.** Neglecting any effects caused by volume changes, would you expect the ionic strength to (1) increase, (2) decrease, or (3) remain essentially unchanged by the addition of NaOH to a dilute solution of
(a) magnesium chloride [$Mg(OH)_2(s)$ forms]?
(b) hydrochloric acid?
(c) acetic acid?

8-4. Neglecting any effects caused by volume changes, would you expect the ionic strength to (1) increase, (2) decrease, or (3) remain essentially unchanged by the addition of iron(III) chloride to
(a) HCl?
(b) NaOH?
(c) $AgNO_3$?

***8-5.** Why is the slope of curve *B* in Figure 8-3 greater than that of curve *A*?

8-6. What is the numerical value of the activity coefficient of aqueous ammonia (NH_3) at an ionic strength of 0.1?

8-7. Calculate the ionic strength of a solution that is
*(a) 0.040 M in $FeSO_4$.
(b) 0.20 M in $(NH_4)_2CrO_4$.
*(c) 0.10 M in $FeCl_3$ and 0.20 M in $FeCl_2$.
(d) 0.060 M in $La(NO_3)_3$ and 0.030 M in $Fe(NO_3)_2$.

8-8. Use Equation 8-5 to calculate the activity coefficient of
*(a) Fe^{3+} at $\mu = 0.075$.
(b) Pb^{2+} at $\mu = 0.012$.
*(c) Ce^{4+} at $\mu = 0.080$.
(d) Sn^{4+} at $\mu = 0.060$.

8-9. Calculate activity coefficients for the species in Problem 8-8 by linear interpolation of the data in Table 8-1.

8-10. For a solution in which μ is 5.0×10^{-2}, calculate K'_{sp} for
*(a) AgSCN.
(b) PbI_2.
*(c) $La(IO_3)_3$.
(d) $MgNH_4PO_4$.

***8-11.** Use activities to calculate the molar solubility of $Zn(OH)_2$ in
(a) 0.0100 M KCl.
(b) 0.0167 M K_2SO_4.
(c) the solution that results when you mix 20.0 mL of 0.250 M KOH with 80.0 mL of 0.0250 M $ZnCl_2$.
(d) the solution that results when you mix 20.0 mL of 0.100 M KOH with 80.0 mL of 0.0250 M $ZnCl_2$.

***8-12.** Calculate the solubilities of the following compounds in a 0.0333 M solution of $Mg(ClO_4)_2$ employing (1) activities and (2) molar concentrations:
(a) AgSCN.
(b) PbI_2.
(c) $BaSO_4$.
(d) $Cd_2Fe(CN)_6$.

$$Cd_2Fe(CN)_6(s) \rightleftharpoons 2Cd^{2+} + Fe(CN)_6^{4-}$$
$$K_{sp} = 3.2 \times 10^{-17}$$

8-13. Calculate the solubilities of the following compounds in a 0.0167 M solution of $Ba(NO_3)_2$ employing (1) activities and (2) molar concentrations:
(a) $AgIO_3$.
(b) $Mg(OH)_2$.
(c) $BaSO_4$.
(d) $La(IO_3)_3$.

CHAPTER

9

Application of Equilibrium Calculations to Complex Systems

Aqueous solutions encountered in the laboratory often contain several species that interact with one another and water to yield two or more simultaneous equilibria. For example, when water is saturated with the sparingly soluble barium sulfate, three equilibria develop:

The introduction of a new equilibrium system into a solution does not change the equilibrium constants for any existing equilibria.

$$BaSO_4(s) \rightleftharpoons Ba^{2+} + SO_4^{2-} \quad (9\text{-}1)$$

$$SO_4^{2-} + H_3O^+ \rightleftharpoons HSO_4^- + H_2O \quad (9\text{-}2)$$

$$2H_2O \rightleftharpoons H_3O^+ + OH^- \quad (9\text{-}3)$$

If hydronium ions are added to this system, the second equilibrium is shifted to the right by the common-ion effect. The resulting decrease in sulfate concentration causes the first equilibrium to shift to the right as well, thus increasing the solubility of the barium sulfate.

The solubility of barium sulfate is also increased when acetate ions are added to an aqueous suspension of the solid because acetate ions tend to form a soluble complex with barium ions as shown by the reaction

$$Ba^{2+} + OAc^- \rightleftharpoons BaOAc^+ \quad (9\text{-}4)$$

Again, the common-ion effect causes both this equilibrium and the solubility equilibrium to shift to the right, thus causing a solubility increase.

If we wish to calculate the solubility of barium sulfate in a system containing hydronium and acetate ions, we must take into account not only the solubility equilibrium, but the other three equilibria as well. It immediately becomes clear, however, that a solubility calculation involving four equilibrium-constant expressions is much more difficult and complex than the rather simple procedure that we illustrated in Examples 7-5, 7-6, and 7-7. In order to solve this type of problem, we will find it helpful to use a systematic approach that is described in the section that follows. We then use this approach to illustrate the effect of pH and complex formation on the solubility of typical analytical precipitates. In subsequent chapters, we will use this same systematic method to solve problems involving multiple equilibria of several types.

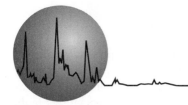

9A A SYSTEMATIC METHOD FOR SOLVING MULTIPLE-EQUILIBRIUM PROBLEMS

Solving a multiple-equilibrium problem requires us to develop as many independent equations as there are participants in the system being studied. For example, if we wish to compute the solubility of barium sulfate in a solution of acid we need to be able to calculate the concentration of all of the species present in the solution. Their number is five: Ba^{2+}, SO_4^{2-}, HSO_4^-, H_3O^+, and OH^-. In order to calculate the solubility of barium sulfate in this solution rigorously, it would then be necessary to develop five independent algebraic equations containing these five unknowns and solve them simultaneously.

Three types of algebraic equations are used in solving multiple-equilibrium problems: (1) equilibrium-constant expressions, (2) *mass-balance* equations, and (3) a single *charge-balance* equation. We have already shown in Section 7B how equilibrium-constant expressions are written; we must now turn our attention to the development of the other two types of equations.

9A-1 Mass-Balance Equations

The term "mass-balance equation," although widely used, is slightly misleading because such equations are really based upon balancing *concentrations* rather than *masses.*

Mass-balance equations relate the equilibrium concentrations of various species in a solution to one another and to the analytical concentrations of the various solutes. They are derived from information about how the solution was prepared and from a knowledge of the kinds of equilibria established in the solution.

EXAMPLE 9-1

Write mass-balance expressions for a 0.0100 M solution of HCl that is in equilibrium with an excess of solid $BaSO_4$.

From our general knowledge of the behavior of aqueous solutions, we can write equations for three equilibria that must be present in this solution. These are given as Equations 9-1, 9-2, and 9-3.

Because the only source for the two sulfate species is the dissolved $BaSO_4$, we may write the mass-balance equation

$$[Ba^{2+}] = [SO_4^{2-}] + [HSO_4^-]$$

We may write a second mass-balance expression based upon the fact that the hydronium ions in the solution have two sources, namely, from the HCl and from the dissociation of the solvent water. Thus,

$$[H_3O^+] = c_{HCl} + [OH^-] = 0.0100 + [OH^-]$$

Here, the second term on the right side of the equation accounts for the concentration of hydronium ions that result from the dissociation of water.

EXAMPLE 9-2

Write mass-balance expressions for the system formed when a 0.010 M NH_3 solution is saturated with AgBr.

Here, equations for the pertinent equilibria in the solution are

$$AgBr(s) \rightleftharpoons Ag^+ + Br^-$$
$$Ag^+ + 2NH_3 \rightleftharpoons Ag(NH_3)_2^+$$
$$NH_3 + H_2O \rightleftharpoons NH_4^+ + OH^-$$
$$2H_2O \rightleftharpoons H_3O^+ + OH^-$$

Because the only source of Br^-, Ag^+, and $Ag(NH_3)_2^+$ is AgBr and because silver and bromide ions exist in a 1 : 1 ratio in the starting material, it follows that one mass-balance equation is

$$[Ag^+] + [Ag(NH_3)_2^+] = [Br^-]$$

where the bracketed terms are molar species concentrations. Also, we know that the only source of ammonia containing species is the 0.010 M NH_3. Therefore,

$$c_{NH_3} = [NH_3] + [NH_4^+] + 2[Ag(NH_3)_2^+] = 0.010$$

From the last two equilibria, we see that one hydroxide ion is formed for each NH_4^+ and each hydronium ion. Therefore,

$$[OH^-] = [NH_4^+] + [H_3O^+]$$

For slightly soluble salts with stoichiometries other than 1 : 1, the mass-balance expression is obtained by multiplying the concentration of one of the ions by the stoichiometric ratio. For example, the mass-balance expression for a saturated solution of Ag_2CrO_4 is

$$2[CrO_4^{2-}] = [Ag^+]$$

9A-2 Charge-Balance Equation

We know that electrolyte solutions are electrically neutral even though they may contain many millions of charged ions. Solutions are neutral because the *molar concentration of positive charge* in an electrolyte solution always equals *the molar concentration of negative charge.* That is, for any solution containing electrolytes, we may write

$$\text{no. mol/L positive charge} = \text{no. mol/L negative charge}$$

This equation represents the condition of charge balance and is called the charge-balance equation. To be useful for equilibrium calculations, this equality must be expressed in terms of the molar concentrations of the species that carry the charge in the solution.

How much charge is contributed to a solution by 1 mol of Na^+? Or, how about 1 mol of Mg^{2+} or 1 mol of PO_4^{3-}? The concentration of charge contributed to a solution by an ion is equal to the molar concentration of that ion multiplied by its charge. Thus, the molar concentration of positive charge in a solution due to the presence of sodium ions is

Always remember that a charge-balance equation is based upon the equality in *molar charge concentrations* and that to obtain the charge concentration of an ion, you must multiply the molar concentration of the ion by its charge.

$$\frac{\text{no. mol}}{\text{L}} = \frac{\cancel{\text{mol Na}^+}}{\text{L}} \times \frac{1 \text{ mol positive charge}}{\cancel{\text{mol Na}^+}}$$
$$= 1 \times [Na^+]$$

The concentration of positive charge due to magnesium ions is

$$\frac{\text{no. mol}}{\text{L}} = \frac{\cancel{\text{mol Mg}^{2+}}}{\text{L}} \times \frac{\text{2 mol positive charge}}{\cancel{\text{mol Mg}^{2+}}}$$
$$= [\text{Mg}^{2+}] \times 2$$

since each mole of magnesium ion contributes 2 mol of positive charge to the solution. Similarly, we may write for phosphate ion

$$\frac{\text{no. mol negative charge}}{\text{L}} = \frac{\cancel{\text{mol PO}_4^{3-}}}{\text{L}} \times \frac{\text{3 mol negative charge}}{\cancel{\text{mol PO}_4^{3-}}}$$
$$= [\text{PO}_4^{3-}] \times 3$$

In some systems, a charge-balance equation cannot be written for want of information, is of no use, or is identical to one of the mass-balance equations.

Now, consider how we would write a charge-balance equation for a 0.100 M solution of sodium chloride. Positive charges in this solution are supplied by Na^+ and H_3O^+ (from dissociation of water). Negative charges come from Cl^- and OH^-. The molarity of positive and negative charges are

$$\text{mol/L positive charge} = [\text{Na}^+] + [\text{H}_3\text{O}^+] = 0.100 + 1 \times 10^{-7}$$
$$\text{mol/L negative charge} = [\text{Cl}^-] + [\text{OH}^-] = 0.100 + 1 \times 10^{-7}$$

We write the charge-balance equation by equating the concentrations of positive and negative charges. That is,

$$[\text{Na}^+] + [\text{H}_3\text{O}^+] = [\text{Cl}^-] + [\text{OH}^-] = 0.100 + 1 \times 10^{-7}$$

Let us now consider a solution that has an analytical concentration of magnesium chloride of 0.100 M. Here, the molarities of positive and negative charge are given by

$$\text{mol/L positive charge} = 2[\text{Mg}^{2+}] + [\text{H}_3\text{O}^+] = 2 \times 0.100 + 1 \times 10^{-7}$$
$$\text{mol/L negative charge} = [\text{Cl}^-] + [\text{OH}^-] = 2 \times 0.100 + 1 \times 10^{-7}$$

In the first equation, the molar concentration of magnesium ion is multiplied by two (2×0.100) because 1 mol of that ion contributes 2 mol of positive charge to the solution. In the second equation the molar chloride ion concentration is twice that of the magnesium chloride concentration or 2×0.100. To obtain the charge-balance equation, we equate the concentration of positive charge with the concentration of negative charge to obtain

$$2[\text{Mg}^{2+}] + [\text{H}_3\text{O}^+] = [\text{Cl}^-] + [\text{OH}^-] = 0.200 + 1 \times 10^{-7}$$

For a neutral solution, $[H_3O^+]$ and $[OH^-]$ are very small and equal, so we can ordinarily simplify the charge-balance equation to

$$2[\text{Mg}^{2+}] \approx [\text{Cl}^-] \approx 0.200$$

EXAMPLE 9-3

Write a charge-balance equation for the system in Example 9-2.

$$[Ag^+] + [Ag(NH_3)_2^+] + [H_3O^+] + [NH_4^+] = [OH^-] + [Br^-]$$

EXAMPLE 9-4

Neglecting the dissociation of water, write a charge-balance equation for a solution that contains NaCl, $Ba(ClO_4)_2$, and $Al_2(SO_4)_3$.

$$[Na^+] + 2[Ba^+] + 3[Al^{3+}] = [ClO_4^-] + [NO_3^-] + 2[SO_4^{2-}]$$

9A-3 Steps for Solving Problems Involving Several Equilibria

Step 1. Write a set of balanced chemical equations for all pertinent equilibria.
Step 2. State in terms of equilibrium concentrations what quantity is being sought.
Step 3. Write equilibrium-constant expressions for all equilibria developed in Step 1, and find numerical values for the constants in tables of equilibrium constants.
Step 4. Write mass-balance expressions for the system.
Step 5. If possible, write a charge-balance expression for the system.
Step 6. Count the number of unknown concentrations in the equations developed in Steps 3, 4, and 5, and compare this number with the number of independent equations. If the number of equations is equal to the number of unknowns, proceed to Step 7. If the number is not, seek additional equations. If enough equations cannot be developed, try to eliminate unknowns by suitable approximations regarding the concentration of one or more of the unknowns. If such approximations cannot be found, the problem cannot be solved.
Step 7. Make suitable approximations to simplify the algebra.
Step 8. Solve the algebraic equations for the equilibrium concentrations needed to give a provisional answer as defined in Step 2.
Step 9. Check the validity of the approximations made in Step 7 using the provisional concentrations computed in Step 8.

Step 6 is critical because it shows whether or not an exact solution to the problem is possible. If the number of unknowns is identical to the number of equations, the problem has been reduced to one of *algebra* alone. That is, answers can be obtained with sufficient perseverance. On the other hand, if there are not enough equations even after approximations are made, the problem should be abandoned.

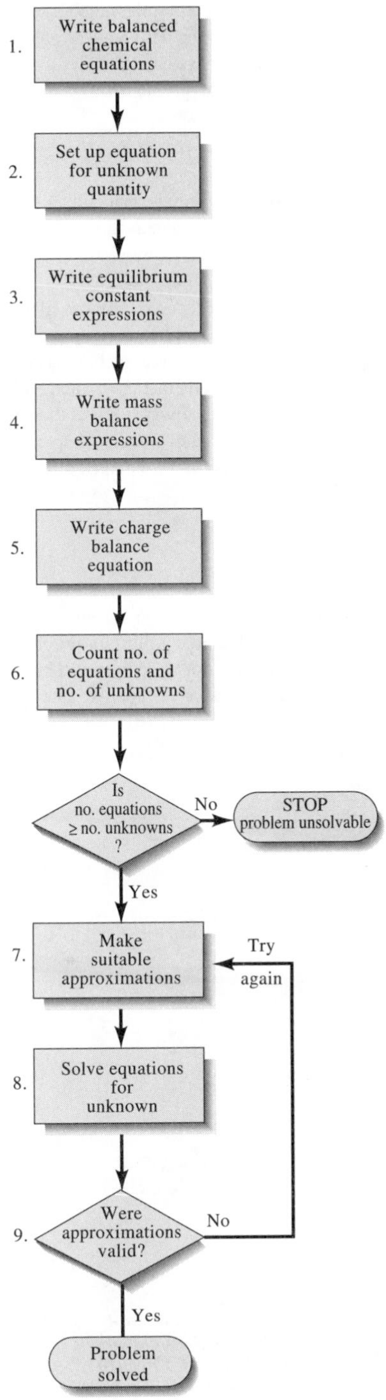

Figure 9-1
A systematic method for solving multiple-equation problems.

9A-4 The Use of Approximations in Solving Equilibrium Calculations

When Step 6 of the systematic approach is complete, we have the *mathematical* problem of solving several nonlinear simultaneous equations. This job is often formidable, tedious, and time-consuming unless a suitable computer program is

Do not waste time starting the algebra in an equilibrium calculation until you are absolutely sure that you have enough independent equations to solve the problem.

available, or unless approximations can be found that decrease the number of unknowns and equations. In this section, we consider in general terms how equations describing equilibrium relationships can be simplified by suitable approximations.

Approximations can only be made in charge-balance and mass-balance equations—never in equilibrium-constant expressions.

Bear in mind that *only* the mass-balance and charge-balance equations can be simplified because only in these equations do the concentration terms appear as sums or differences rather than as products or quotients. It is always possible to assume that one (or more) of the terms in a sum or difference is so much smaller than the others that it can be ignored without significantly affecting the equality. The assumption that a concentration term in an equilibrium-constant expression is zero, makes the expression meaningless.

Several software packages are now available for solving multiple nonlinear simultaneous equations rigorously. Three such programs are Mathcad®, Mathematica®, and TK Solver Plus®. When software such as these is available, Steps 7, 8, and 9 can be dispensed with, and the results can be obtained directly.

Many students find Step 7 to be the most troublesome because they fear that making invalid approximations will lead to serious errors in their computed results. *Such fears are groundless.* Experienced scientists are often as puzzled as beginners when making an approximation that simplifies an equilibrium calculation. Nonetheless, they make such approximations without fear because they know that the effects of an invalid assumption will become obvious by the time a computation is completed. Generally, questionable assumptions should be tried at the outset and provisional answers computed. If the assumption leads to an intolerable error (which is easily recognized), a recalculation without the faulty approximation is then performed. Usually, it is more efficient to try a questionable assumption at the outset than to make a more time-consuming and tedious calculation without the assumption.

Never be afraid to make an assumption in attempting to solve an equilibrium problem. If the assumption is not valid, you will know it as soon as you have an approximate answer.

9B THE CALCULATION OF SOLUBILITY BY THE SYSTEMATIC METHOD

The use of the systematic method is illustrated in this section with examples involving the solubility of precipitates under various conditions. In later chapters we shall apply this method to other types of equilibria as well.

9B-1 Metal Hydroxides

Examples 9-5 and 9-6 involve calculating the solubilities of two metal hydroxides. One is relatively soluble compared with the second. These examples illustrate the importance of making approximations and provisional calculations in the systematic procedure for solving mass-law problems.

See *Mathcad Applications for Analytical Chemistry,* **pp. 44–46.**

EXAMPLE 9-5

Calculate the molar solubility of $Mg(OH)_2$ in water.

Step 1. Pertinent Equilibria
Two equilibria that need to be considered are

$$Mg(OH)_2(s) \rightleftharpoons Mg^{2+} + 2OH^-$$

$$2H_2O \rightleftharpoons H_3O^+ + OH^-$$

Step 2. Definition of Unknown
Since 1 mol of Mg^{2+} is formed for each mole of $Mg(OH)_2$ dissolved,

$$\text{solubility } Mg(OH)_2 = [Mg^{2+}]$$

Step 3. Equilibrium-Constant Expressions

$$[Mg^{2+}][OH^-]^2 = 7.1 \times 10^{-12} \qquad (9\text{-}5)$$

$$[H_3O^+][OH^-] = 1.00 \times 10^{-14} \qquad (9\text{-}6)$$

Step 4. Mass-Balance Expression

$$[OH^-] = 2[Mg^{2+}] + [H_3O^+] \qquad (9\text{-}7)$$

The first term on the right-hand side of Equation 9-7 represents the hydroxide ion concentration resulting from dissolved $Mg(OH)_2$ in the solution, and the second term is the hydroxide ion concentration resulting from the dissociation of water.

We arrived at Equation 9-7 with the following reasoning. If $[OH^-]_{H_2O}$ and $[OH^-]_{Mg(OH)_2}$ are the concentrations of OH^- produced from H_2O and $Mg(OH)_2$ respectively, then

$$[OH^-]_{H_2O} = [H_3O^+]$$

$$[OH^-]_{Mg(OH)_2} = 2[Mg^{2+}]$$

$$[OH^-]_{total} = [OH^-]_{H_2O} + [OH^-]_{Mg(OH)_2} = [H_3O^+] + 2[Mg^{2+}]$$

Step 5. Charge-Balance Expression

$$[OH^-] = 2[Mg^{2+}] + [H_3O^+]$$

Note that this equation is identical to Equation 9-7. Often a mass-balance and a charge-balance equation are the same.

Step 6. Number of Independent Equations and Unknowns
We have developed three independent algebraic equations (Equations 9-5, 9-6, and 9-7) and have three unknowns ($[Mg^{2+}]$, $[OH^-]$, and $[H_3O^+]$). Therefore, the problem can be solved rigorously.

Step 7. Approximations
We can make approximations only in Equation 9-7. Since the solubility-product constant for $Mg(OH)_2$ is relatively large, the solution will be somewhat basic. Therefore, it is reasonable to assume that $[H_3O^+] \ll [Mg^{2+}]$. Equation 9-7 then simplifies to

$$2[Mg^{2+}] \approx [OH^-] \qquad (9\text{-}8)$$

Step 8. Solution to Equations
Substitution of Equation 9-8 into Equation 9-5 gives

$$[Mg^{2+}](2[Mg^{2+}])^2 = 7.1 \times 10^{-12}$$

$$[Mg^{2+}]^3 = \frac{7.1 \times 10^{-12}}{4} = 1.78 \times 10^{-12}$$

$$[Mg^{2+}] = \text{solubility} = 1.21 \times 10^{-4} = 1.2 \times 10^{-4} \text{ mol/L}$$

Step 9. Check of Approximations
Substitution of this provisional value for $[Mg^{2+}]$ into Equation 9-8 yields

$$[OH^-] = 2 \times 1.21 \times 10^{-4} = 2.42 \times 10^{-4}$$

$$[H_3O^+] = \frac{1.00 \times 10^{-14}}{2.42 \times 10^{-4}} = 4.1 \times 10^{-11}$$

Thus, our assumption that $4.1 \times 10^{-11} \ll 1.6 \times 10^{-4}$ is certainly valid.

EXAMPLE 9-6

Calculate the solubility of $Fe(OH_3)$ in water.

Proceeding by the systematic approach used in Example 9-5, we write

Step 1. Pertinent Equilibrium

$$Fe(OH)_3(s) \rightleftharpoons Fe^{3+} + 3OH^-$$

$$2H_2O \rightleftharpoons H_3O^+ + OH^-$$

Step 2. Definition of Unknown

$$\text{solubility} = [Fe^{3+}]$$

Step 3. Equilibrium-Constant Expressions

$$[Fe^{3+}][OH^-]^3 = 2 \times 10^{-39}$$

$$[H_3O^+][OH^-] = 1.00 \times 10^{-14}$$

Steps 4 and 5.

As in Example 9-5, the mass-balance equation and the charge-balance equation are identical. That is,

$$[OH^-] = 3[Fe^{3+}] + [H_3O^+]$$

Step 6.

We see that we have enough equations to calculate the three unknowns.

Step 7. Approximations

As in Example 9-5, let us assume $[H_3O^+] \ll 3[Fe^{3+}]$, so that

$$3[Fe^{3+}] \approx [OH^-]$$

Step 8. Solution to Equations

Substituting this equation into the solubility-product expression gives

$$[Fe^{3+}](3[Fe^{3+}])^3 = 2 \times 10^{-39}$$

$$[Fe^{3+}] = \left(\frac{2 \times 10^{-39}}{27}\right)^{1/4} = 9 \times 10^{-11}$$

$$\text{solubility} = [Fe^{3+}] = 9 \times 10^{-11} \text{ mol/L}$$

Step 9. Check of Assumptions
From the assumption made in Step 7, we can calculate a provisional value of $[OH^-]$. That is,

$$[OH^-] \approx 3[Fe^{3+}] = 3 \times 9 \times 10^{-11} = 3 \times 10^{-10}$$

Let us use this value of $[OH^-]$ to compute a *provisional* value for $[H_3O^+]$:

$$[H_3O^+] = \frac{1.00 \times 10^{-14}}{3 \times 10^{-10}} = 3 \times 10^{-5}$$

But 3×10^{-5} is not much smaller than three times our provisional value of $[Fe^{3+}]$. This discrepancy means that our assumption was invalid and the provisional values for $[Fe^{3+}]$, $[OH^-]$, and $[H_3O^+]$ are all significantly in error. Therefore, let us go back to Step 7 and assume that

$$3[Fe^{3+}] \ll [H_3O^+]$$

Now the mass-balance expression becomes

$$[H_3O^+] = [OH^-]$$

Substituting this equality into the expression for K_w gives

$$[H_3O^+] = [OH^-] = 1.00 \times 10^{-7}$$

Substituting this number into the solubility-product expression developed in Step 3 gives

$$[Fe^{3+}] = \frac{2 \times 10^{-39}}{(1.00 \times 10^{-7})^3} = 2 \times 10^{-18} \text{ mol/L}$$

In this case we have assumed that $3[Fe^{3+}] \ll [OH^-]$ or $3 \times 2 \times 10^{-18} \ll 10^{-7}$. Clearly, the assumption is valid and we may write

$$\text{solubility} = 2 \times 10^{-18} \text{ mol/L}$$

Note the very large error introduced by the invalid assumption.

Example 9-6 illustrates how readily the effects of an invalid assumption are detected. It may bother the reader that we used a faulty value for $[OH^-]$ to obtain $[H_3O^+]$, which we then compared with $3[Fe^{3+}]$, but the point is that the faulty value of $[OH^-]$ also arose because of the invalid assumption. Had the assumption been valid we would have had an internally consistent set of calculated concentrations. The presence of even one internal inconsistency is a clear indication of an invalid assumption. A valid assumption leads to concentrations that satisfy all of the algebraic equalities developed in Steps 3, 4, and 5.

The faulty assumption in Example 9-6 was readily recognized.

9B-2 The Effect of pH on Solubility

All precipitates that contain an anion that is the conjugate base of a weak acid are more soluble at low than at high pH.

The solubility of precipitates containing an anion with basic properties, a cation with acidic properties, or both will be dependent upon pH. The example that follows illustrates how the effect of pH on solubility can be treated in quantitative terms.

Solubility Calculations when the pH Is Fixed and Known

A **buffer** keeps the pH of a solution relatively constant (see Section 10D-2).

Analytical precipitations are frequently performed in buffered solutions in which the pH is fixed at some predetermined and known value. The calculation of solubility under this circumstance is illustrated by the following example.

See *Mathcad Applications for Analytical Chemistry,* **pp. 46–50.**

EXAMPLE 9-7

Calculate the molar solubility of calcium oxalate in a solution that has been buffered so that its pH is constant and equal to 4.00.

Step 1. Pertinent Equilibria

$$CaC_2O_4 \rightleftharpoons Ca^{2+} + C_2O_4^{2-} \tag{9-9}$$

Oxalate ions react with water to form $HC_2O_4^-$ and $H_2C_2O_4$. Thus, two other equilibria that will be present in this solution are (see Feature 9-1)

$$H_2C_2O_4 + H_2O \rightleftharpoons H_3O^+ + HC_2O_4^- \tag{9-10}$$

$$HC_2O_4^- + H_2O \rightleftharpoons H_3O^+ + C_2O_4^{2-} \tag{9-11}$$

Step 2. Definition of the Unknown
Calcium oxalate is a strong electrolyte so that its molar analytical concentration is equal to the equilibrium calcium ion concentration. That is,

$$\text{solubility} = [Ca^{2+}] \tag{9-12}$$

Step 3. Equilibrium-Constant Expressions

$$[Ca^{2+}][C_2O_4^{2-}] = K_{sp} = 1.7 \times 10^{-9} \tag{9-13}$$

$$\frac{[H_3O^+][HC_2O_4^-]}{[H_2C_2O_4]} = K_1 = 5.60 \times 10^{-2} \tag{9-14}$$

$$\frac{[H_3O^+][C_2O_4^{2-}]}{[HC_2O_4^-]} = K_2 = 5.42 \times 10^{-5} \tag{9-15}$$

Step 4. Mass-Balance Expressions
Because CaC_2O_4 is the only source of Ca^{2+} and the three oxalate species,

$$[Ca^{2+}] = [C_2O_4^{2-}] + [HC_2O_4^-] + [H_2C_2O_4] \tag{9-16}$$

Moreover, the problem states that the pH is 4.00. Thus,

$$[H_3O^+] = 1.00 \times 10^{-4}$$

Step 5. Charge-Balance Expressions
A buffer is required to maintain the pH at 4.00. The buffer most likely consists of some weak acid HA and its conjugate base, A^- (Section 10D-1). The nature of the three species and their concentrations have not been specified, however, so we do not have enough information to write a charge-balance equation.

Step 6. Number of Independent Equations and Unknowns
We have four unknowns ($[Ca^{2+}]$, $[C_2O_4^{2-}]$, $[HC_2O_4^-]$, and $[H_2C_2O_4]$) as well as four independent algebraic relationships (Equations 9-13, 9-14, 9-15, and 9-16). Therefore, an exact solution can be obtained, and the problem becomes one of algebra.

Step 7. Approximations
An exact solution is so readily obtained in this case that we will not bother with approximations.

Step 8. Solution of the Equations
A convenient way to solve the problem is to substitute Equations 9-14 and 9-15 into 9-16 in such a way as to develop a relationship between $[Ca^{2+}]$, $[C_2O_4^{2-}]$, and $[H_3O^+]$. Thus, we rearrange Equation 9-15 to give

$$[HC_2O_4^-] = \frac{[H_3O^+][C_2O_4^{2-}]}{K_2}$$

Substituting numerical values for $[H_3O^+]$ and K_2 gives

$$[HC_2O_4^-] = \frac{1.00 \times 10^{-4}[C_2O_4^{2-}]}{5.42 \times 10^{-5}} = 1.85[C_2O_4^{2-}]$$

Substituting this relationship into Equation 9-14 gives upon rearranging

$$[H_2C_2O_4] = \frac{[H_3O^+][C_2O_4^{2-}] \times 1.85}{K_1}$$

Substituting numerical values for $[H_3O^+]$ and K_1 yields

$$[H_2C_2O_4] = \frac{1.85 \times 10^{-4}[C_2O_4^{2-}]}{5.60 \times 10^{-2}} = 3.30 \times 10^{-3}[C_2O_4^{2-}]$$

Substituting these expressions for $[HC_2O_4^-]$ and $[H_2C_2O_4]$ into Equation 9-16 gives

$$[Ca^{2+}] = [C_2O_4^{2-}] + 1.85[C_2O_4^{2-}] + 3.30 \times 10^{-3}[C_2O_4^{2-}]$$
$$= 2.85[C_2O_4^{2-}]$$

or

$$[C_2O_4^{2-}] = [Ca^{2+}]/2.85$$

Substituting into Equation 9-13 gives

$$\frac{[Ca^{2+}][Ca^{2+}]}{2.85} = 1.7 \times 10^{-9}$$

$$[Ca^{2+}] = \text{solubility} = \sqrt{2.85 \times 1.7 \times 10^{-9}} = 7.0 \times 10^{-5} \text{ mol/L}$$

FEATURE 9-1
An Alternate Set of Equations for Solving Example 9-7

You may find it more rational to write the equilibria describing the solubility of CaC_2O_4 at a fixed and known hydronium ion concentration as follows:

Step 1. $CaC_2O_4(s) \rightleftharpoons Ca^{2+} + C_2O_4^{2-}$
The $C_2O_4^{2-}$ formed by this reaction then reacts with water as follows:

$$C_2O_4^{2-} + H_2O \rightleftharpoons HC_2O_4^- + OH^-$$
$$HC_2O_4^- + H_2O \rightleftharpoons H_2C_2O_4 + OH^-$$

Since we were given data on $[H_3O^+]$ but not $[OH^-]$ we must also write

$$2H_2O \rightleftharpoons H_3O^+ + OH^-$$

Step 2. solubility = $[Ca^{2+}]$

Step 3. $[Ca^{2+}][C_2O_4^{2-}] = K_{sp} = 1.7 \times 10^{-9}$ (1)

$$\frac{[HC_2O_4^-][OH^-]}{[C_2O_4^-]} = \frac{K_w}{K_2} = \frac{1.00 \times 10^{-14}}{5.42 \times 10^{-5}} = 1.85 \times 10^{-10} \quad (2)$$

$$\frac{[H_2C_2O_4][OH^-]}{[HC_2O_4^-]} = \frac{K_w}{K_1} = \frac{1.00 \times 10^{-14}}{5.60 \times 10^{-2}} = 1.79 \times 10^{-13} \quad (3)$$

$$[H_3O^+][OH^-] = K_w = 1.00 \times 10^{-14} \quad (4)$$

Step 4. $[Ca^+] = [C_2O_4^{2-}] + [HC_2O_4^-] + [H_2C_2O_4] = 1.00 \times 10^{-14}$ (5)
$[H_3O^+] = 1.00 \times 10^{-4}$

In this case we have five unknowns rather than four, the extra one being $[OH^-]$. But we also have five equations.

If you go ahead and solve the five equations you will obtain an identical result to that shown in Example 9-7. The need for the added equilibrium (5), however, complicates the calculation somewhat.

Solubility Calculations when the pH Is Variable

Saturating an unbuffered solution with a sparingly soluble salt containing a basic anion or an acidic cation causes the pH of the solution to change. For example, pure water saturated with barium carbonate is basic as a consequence of the reactions

$$BaCO_3(s) \rightleftharpoons Ba^{2+} + CO_3^{2-}$$
$$CO_3^{2-} + H_2O \rightleftharpoons HCO_3^- + OH^-$$
$$HCO_3^- + H_2O \rightleftharpoons H_2CO_3 + OH^-$$

In contrast to Example 9-9, the hydroxide ion concentration now becomes an unknown, and an additional algebraic equation must therefore be developed if the solubility of barium carbonate is to be calculated.

In many instances, the reaction of a precipitate with water cannot be neglected without introducing an appreciable error in the calculation. As shown by the data in Table 9-1, the magnitude of the error depends upon the solubility of the precipitate as well as on the base dissociation constant of the anion. The solubilities of the hypothetical precipitate MA, shown in column 4, were obtained by taking into account the reaction of A^- with water. Column 5 gives the calculated results when the basic properties of A^- are neglected; here, the solubility is simply the square root of the solubility product. Two solubility products, 1.0×10^{-10} and 1.0×10^{-20}, have been assumed for these calculations as well as several values for K_b (column 3). It is apparent that neglecting the reaction of the anions with water leads to a negative error that becomes more pronounced both as the solubility of the precipitate decreases (smaller K_{sp}) and as the conjugate base becomes stronger. Note that the error becomes insignificant for anions derived from acids with dissociation constants greater than about 10^{-6}.

It is not difficult to write the algebraic relationships needed to calculate the solubility of such a precipitate. Solving the equations, however, is tedious. Fortunately, it is ordinarily possible to invoke one of two simplifying assumptions to decrease the algebraic labor:

1. The first simplification is applicable to moderately soluble compounds containing an anion that reacts extensively with water. Here it is assumed that the hydroxide ion concentration is large enough to render the hydronium ion concentration negligible in the calculations. Another way of stating this assumption is to say that the hydroxide ion concentration of the solution is determined exclusively by the reaction of the anion with water and that the contribution of hydroxide ions from the dissociation of water is negligible by comparison. This type of calculation is illustrated by Example 9-8.
2. The second simplification is applicable to precipitates of very low solubility, particularly those containing an anion that does not react extensively with water. In such a system, it can be assumed that dissolution of the precipitate does not significantly change the hydronium or hydroxide ion concentrations and that, at room temperature, these concentrations remain essentially 10^{-7} mol/L. The solubility calculation then follows the course shown in Example 9-7. This type of calculation is shown in Example 9-9.

TABLE 9-1 Calculated Solubility of MA from Various Assumed Values of K_{sp} and K_b

Assumed K_{sp} for MA	Assumed K_{HA}	Dissociation Constant for A^- $K_b = K_w/K_{HA}$	Calculated Solubility of MA, M	Calculated Solubility of MA Neglecting Reaction of A^- with Water, M
1.0×10^{-10}	1.0×10^{-6}	1.0×10^{-8}	1.02×10^{-5}	1.0×10^{-5}
	1.0×10^{-8}	1.0×10^{-6}	1.2×10^{-5}	1.0×10^{-5}
	1.0×10^{-10}	1.0×10^{-4}	2.4×10^{-5}	1.0×10^{-5}
	1.0×10^{-12}	1.0×10^{-2}	10×10^{-5}	1.0×10^{-5}
1.0×10^{-20}	1.0×10^{-6}	1.0×10^{-8}	1.05×10^{-10}	1.0×10^{-10}
	1.0×10^{-8}	1.0×10^{-6}	3.3×10^{-10}	1.0×10^{-10}
	1.0×10^{-10}	1.0×10^{-4}	32×10^{-10}	1.0×10^{-10}
	1.0×10^{-12}	1.0×10^{-2}	290×10^{-10}	1.0×10^{-10}

See *Mathcad Applications for Analytical Chemistry,* **pp. 50–51.**

EXAMPLE 9-8

Calculate the solubility of $BaCO_3$ in water.

Step 1. Pertinent Equilibria

$$BaCO_3(s) \rightleftharpoons Ba^{2+} + CO_3^{2-} \tag{9-17}$$

$$CO_3^{2-} + H_2O \rightleftharpoons HCO_3^- + OH^- \tag{9-18}$$

$$HCO_3^- + H_2O \rightleftharpoons H_2CO_3 + OH^- \tag{9-19}$$

$$2H_2O \rightleftharpoons H_3O^+ + OH^- \tag{9-20}$$

Step 2. Definition of the Unknown

$$\text{solubility} = [Ba^{2+}] = [CO_3^{2-}] + [HCO_3^-] + [H_2CO_3]$$

Step 3. Equilibrium-Constant Expressions

$$[Ba^{2+}][CO_3^{2-}] = K_{sp} = 5.0 \times 10^{-9} \tag{9-21}$$

$$\frac{[HCO_3^-][OH^-]}{[CO_3^{2-}]} = \frac{K_w}{K_2} = \frac{1.00 \times 10^{-14}}{4.69 \times 10^{-11}} = 2.13 \times 10^{-4} \tag{9-22}$$

$$\frac{[H_2CO_3][OH^-]}{[HCO_3^-]} = \frac{K_w}{K_1} = \frac{1.00 \times 10^{-14}}{4.45 \times 10^{-7}} = 2.25 \times 10^{-8} \tag{9-23}$$

and

$$[H_3O^+][OH^-] = 1.00 \times 10^{-14} \tag{9-24}$$

Step 4. Mass-Balance Expression

$$[Ba^{2+}] = [CO_3^{2-}] + [HCO_3^-] + [H_2CO_3] \tag{9-25}$$

Step 5. Charge-Balance Expression

$$2[Ba^{2+}] + [H_3O^+] = 2[CO_3^{2-}] + [HCO_3^-] + [OH^-] \tag{9-26}$$

Step 6. Number of Equations and Unknowns
We have developed six equations (Equations 9-21 through 9-26), which are sufficient to solve for the six unknowns ($[Ba^{2+}]$, $[CO_3^{2-}]$, $[HCO_3^-]$, $[H_2CO_3]$, $[OH^-]$, and $[H_3O^+]$).

Step 7. Approximations
We examine Equations 9-25 and 9-26 with the goal of eliminating one or more terms on the ground that their elimination does not create a significant error. One such candidate is $[H_3O^+]$ in Equation 9-26. Because the solution is basic as a consequence of reactions 9-18 and 9-19, $[H_3O^+]$ must be smaller than 10^{-7}. Also, $[Ba^{2+}]$ must be considerably larger than 10^{-7} because, if no reaction occurs between CO_3^{2-} and H_2O, $[Ba^{2+}]$ is simply the square root of K_{sp}, or 7×10^{-5} M. The fact that reaction does occur means that $[CO_3^{2-}]$ is smaller than 7×10^{-5} and therefore $[Ba^{2+}] > 7 \times 10^{-5}$ M. Thus, the assumption that $[H_3O^+] \ll 2[Ba^{2+}]$ in Equation 9-26 appears entirely reasonable. With this assumption, we no longer need Equation 9-24.

A second possible assumption is that $[H_2CO_3]$ is so much smaller than $[HCO_3^-]$ that the former can be deleted from Equation 9-25. We base this assumption on our knowledge that the solution is basic and therefore $[OH^-] > 10^{-7}$. If $[OH^-]$ is 10^{-6} M, substitution into Equation 9-23 reveals that

$$\frac{[H_2CO_3]}{[HCO_3^-]} = \frac{2.25 \times 10^{-8}}{10^{-6}} = 0.0225$$

If $[OH^-]$ is 10^{-5}, this ratio is about 0.002. Although these calculations suggest that $[H_2CO_3] \ll [HCO_3^-]$, we cannot be sure that this assumption is valid. It is surely worth trying, however.

As a result of the first assumption, Equation 9-26 becomes

$$2[Ba^{2+}] = 2[CO_3^{2-}] + [HCO_3^-] + [OH^-] \qquad \text{(9-27)}$$

Further, because $[HCO_3^-] \gg [H_2CO_3]$, the mass-balance expression simplifies to

$$[Ba^{2+}] = [CO_3^{2-}] + [HCO_3^-] \qquad \text{(9-28)}$$

Equations 9-23 and 9-24 are now no longer needed. Thus, we have reduced both the number of equations and the number of unknowns to four.

Step 8. Solution of Equations
If we multiply Equation 9-28 by 2 and subtract the product from Equation 9-27, we obtain, upon rearrangement,

$$[OH^-] = [HCO_3^-] \qquad \text{(9-29)}$$

Substitution of $[HCO_3^-]$ for $[OH^-]$ in Equation 9-22 gives

$$\frac{[HCO_3^-]^2}{[CO_3^{2-}]} = \frac{K_w}{K_2}$$

$$[HCO_3^-] = \sqrt{\frac{K_w}{K_2}[CO_3^{2-}]}$$

This expression permits elimination of $[HCO_3^-]$ from Equation 9-28:

$$[Ba^{2+}] = [CO_3^{2-}] + \sqrt{\frac{K_w}{K_2}[CO_3^{2-}]} \quad (9\text{-}30)$$

From Equation 9-21, we have

$$[CO_3^{2-}] = \frac{K_{sp}}{[Ba^{2+}]}$$

Substituting for $[CO_3^{2-}]$ in Equation 9-30 yields

$$[Ba^{2+}] = \frac{K_{sp}}{[Ba^{2+}]} + \sqrt{\frac{K_w K_{sp}}{K_2[Ba^{2+}]}}$$

It is convenient to multiply through by $[Ba^{2+}]$ and rearrange:

$$[Ba^{2+}]^2 - \sqrt{\frac{K_w}{K_2}K_{sp}[Ba^{2+}]} - K_{sp} = 0 \quad (9\text{-}31)$$

Finally, after numerical values are supplied for the constants, we obtain

$$[Ba^{2+}]^2 - (1.03 \times 10^{-6})[Ba^{2+}]^{1/2} - 5.0 \times 10^{-9} = 0$$

In order to solve this equation by successive approximations (see Feature 7-5, page 142), we let $x = [Ba^{2+}]$ and rewrite the equation in the form

$$x_n = (1.03 \times 10^{-6}\sqrt{x_{n-1}} + 5.0 \times 10^{-9})^{1/2}$$

If we let $x_1 = 0$ and solve for x_2, we obtain

$$x_2 = (0 + 5.0 \times 10^{-9})^{1/2} = 7.07 \times 10^{-5}$$

Substituting this value for x in the original equation gives

$$x_3 = (1.03 \times 10^{-6} \times \sqrt{7.07 \times 10^{-5}} + 5.0 \times 10^{-9})^{1/2} = 1.17 \times 10^{-4}$$

Further iteration leads to $x_4 = 1.27 \times 10^{-4}$, $x_5 = 1.29 \times 10^{-4}$, and $x_6 = 1.29 \times 10^{-4}$. Thus,

$$\text{solubility} = [Ba^{2+}] = 1.29 \times 10^{-4} \approx 1.3 \times 10^{-4}\ M$$

Step 9. Check of Approximations

To check the two approximations that were made, we must calculate the concentrations of most of the other ions in the solution. We can evaluate $[CO_3^{2-}]$ from Equation 9-21:

$$[CO_3^{2-}] = \frac{5.0 \times 10^{-9}}{1.3 \times 10^{-4}} = 3.9 \times 10^{-5}$$

From Equation 9-28,

$$[HCO_3^-] = 12.9 \times 10^{-5} - 3.9 \times 10^{-5} = 9.0 \times 10^{-5}$$

From Equation 9-29,

$$[OH^-] = [HCO_3^-] = 9.0 \times 10^{-5}$$

From Equation 9-23,

$$\frac{[H_2CO_3](9.0 \times 10^{-5})}{9.0 \times 10^{-5}} = 2.25 \times 10^{-8}$$

$$[H_2CO_3] = 2.2 \times 10^{-8}$$

Finally, from Equation 9-24,

$$[H_3O^+] = \frac{1.00 \times 10^{-14}}{9.0 \times 10^{-5}} = 1.1 \times 10^{-10}$$

We see that the two approximations do not lead to large errors: $[HCO_3^-]$ is about 4000 times greater than $[H_2CO_3]$, and $[H_3O^+]$ is clearly much smaller than any of the species in Equation 9-26.

Failure to take into account the basic reaction of CO_3^{2-} would have yielded a solubility of 7.1×10^{-5}, which is only about one half the value yielded by the more rigorous method.

An example of the second type of calculation referred to on page 171 follows. Here, it is assumed that the hydronium and hydroxide ion concentrations are 10^{-7} M after the solid has dissolved.

EXAMPLE 9-9

Calculate the solubility of silver sulfide in pure water.

Step 1. Pertinent Equilibria

$$Ag_2S(s) \rightleftharpoons 2Ag^+ + S^{2-} \quad (9\text{-}32)$$

$$S^{2-} + H_2O \rightleftharpoons HS^- + OH^- \quad (9\text{-}33)$$

$$HS^- + H_2O \rightleftharpoons H_2S + OH^- \quad (9\text{-}34)$$

$$2H_2O \rightleftharpoons H_3O^+ + OH^- \quad (9\text{-}35)$$

Step 2. Definition of Unknown

$$\text{solubility} = \tfrac{1}{2}[Ag^+] = [S^{2-}] + [HS^-] + [H_2S]$$

Step 3. Equilibrium-Constant Expressions

$$[Ag^+]^2[S^{2-}] = 8 \times 10^{-51} \quad (9\text{-}36)$$

$$\frac{[HS^-][OH^-]}{[S^{2-}]} = \frac{K_w}{K_2} = \frac{1.0 \times 10^{-14}}{1.3 \times 10^{-14}} = 0.769 \quad (9\text{-}37)$$

$$\frac{[H_2S][OH^-]}{[HS^-]} = \frac{K_w}{K_1} = \frac{1.0 \times 10^{-14}}{9.6 \times 10^{-8}} = 1.04 \times 10^{-7} \quad (9\text{-}38)$$

$$[H_3O^+][OH^-] = 1.00 \times 10^{-14}$$

Step 4. Mass-Balance Expression

$$\tfrac{1}{2}[Ag^+] = [S^{2-}] + [HS^-] + [H_2S] \quad (9\text{-}39)$$

Step 5. Charge-Balance Expression

$$[Ag^+] + [H_3O^+] = 2[S^{2-}] + [HS^-] + [OH^-] \quad (9\text{-}40)$$

Step 6. Comparison of Equations and Unknowns
We have six unknowns and six equations. Thus, an exact solution is feasible.

Step 7. Approximations
The solubility product for Ag_2S is very small; therefore, it is probable that there is little change in hydroxide ion concentration as the precipitate dissolves. As a consequence, we can assume tentatively that

$$[OH^-] \approx [H_3O^+] = 1.0 \times 10^{-7}$$

This assumption will be correct if, in Equation 9-40,

$$[Ag^+] \ll [H_3O^+] \qquad \text{and} \qquad 2[S^{2-}] + [HS^-] \ll [OH^-]$$

Step 8. Solution of Equations
We now proceed exactly as we did in Feature 9-1. Substitution of 1.0×10^{-7} for $[OH^-]$ in Equations 9-37 and 9-38 gives

$$\frac{[HS^-]}{[S^{2-}]} = \frac{0.769}{[OH^-]} = \frac{0.769}{1.0 \times 10^{-7}} = 7.69 \times 10^6 \quad (9\text{-}41)$$

$$[H_2S] = \frac{(1.04 \times 10^{-7})[HS^-]}{1.0 \times 10^{-7}} = 1.04[HS^-] \quad (9\text{-}42)$$

Substituting Equation 9-41 into 9-42 yields

$$[H_2S] = 1.04 \times 7.69 \times 10^6[S^{2-}] = 8.00 \times 10^6[S^{2-}]$$

When these relationships are substituted into Equation 9-39, we obtain

$$\tfrac{1}{2}[Ag^+] = [S^{2-}] + (7.69 \times 10^6)[S^{2-}] + (8.00 \times 10^6)[S^{2-}]$$
$$= 1.57 \times 10^7[S^{2-}]$$

$$[S^{2-}] = \frac{[Ag^+]}{2 \times 1.57 \times 10^7} = 3.19 \times 10^{-8}[Ag^+]$$

Substituting this relationship into the solubility-product expression (Equation 9-36) gives

$$[Ag^+]^2(3.19 \times 10^{-8}[Ag^+]) = 8 \times 10^{-51}$$

$$[Ag^+] = (8 \times 10^{-51}/3.19 \times 10^{-8})^{1/3} = (2.51 \times 10^{-43})^{1/3} = 6.3 \times 10^{-15}$$

$$\text{solubility} = \tfrac{1}{2}[Ag^+] = \frac{6.3 \times 10^{-15}}{2} = 3 \times 10^{-15} \text{ mol/L}$$

Step 9. Check of Assumptions

The assumption that $[Ag^+]$ is much smaller than $[H_3O^+]$ is clearly valid. We can readily calculate a value for $2[S^{2-}] + [HS^-]$ and confirm that this sum is likewise much smaller than $[OH^-]$. Therefore, we conclude that the assumptions made are reasonable and that the approximate solution is satisfactory.

9B-3 The Solubility of Precipitates in the Presence of Complexing Agents

The solubility of a precipitate may increase dramatically in the presence of reagents that form complexes with the anion or the cation of the precipitate. For example, fluoride ions prevent the quantitative precipitation of aluminum hydroxide even though the solubility product of this precipitate is remarkably small (3×10^{-34}). The cause of the increase in solubility is shown by the equations

The solubility of a precipitate always increases in the presence of a complexing agent that reacts with the cation of the precipitate.

$$\begin{array}{ccc} Al(OH)_3(s) \rightleftharpoons & Al^{3+} & + \; 3OH^- \\ & + & \\ & 6F^- & \\ & \upharpoons\downharpoonright & \\ & AlF_6^{3-} & \end{array}$$

The fluoride complex is sufficiently stable to permit fluoride ions to compete successfully with hydroxide ions for aluminum ions.

EXAMPLE 9-10

The solubility product of CuI is 1.0×10^{-12}. The formation constant K_2 for the reaction of CuI with I^- to give CuI_2^- is 7.9×10^{-4}. Calculate the molar solubility of CuI in a 1.0×10^{-4} M solution of KI.

Step 1. Pertinent Equilibria

From the input data, we assume that the following equilibria exist in a solution of KI that is saturated with CuI:

$$CuI(s) \rightleftharpoons Cu^+ + I^-$$

$$CuI(s) + I^- \rightleftharpoons CuI_2^-$$

Since neither H_3O^+ nor OH^- reacts to any significant extent with any of the participants in these equilibria, we need not include the dissociation of water in our list.

Step 2. Definition of Unknown
From inspection of the two equations, we see that the dissolved copper(I) is present as either Cu^+ or CuI_2^-. Therefore,

$$\text{solubility} = [Cu^+] + [CuI_2^-]$$

Step 3. Equilibrium-Constant Expression

$$[Cu^+][I^-] = K_{sp} = 1.0 \times 10^{-12} \qquad (9\text{-}43)$$

$$\frac{[CuI_2^-]}{[I^-]} = K_2 = 7.9 \times 10^{-4} \qquad (9\text{-}44)$$

Step 4. Mass-Balance Expression
Here, we may write

$$[I^-] = c_{KI} + [Cu^+] - [CuI_2^-] \qquad (9\text{-}45)$$

The first term on the right-hand side of this equation represents the concentration of iodide from the KI, and the second term corresponds to the contribution of iodide from the dissolution of CuI. The third term gives the concentration of iodide needed to form the complex from CuI.

Step 5. Charge-Balance Equation

$$[Cu^+] + [K^+] = [I^-] + [CuI_2^-] \qquad (9\text{-}46)$$

The K^+ concentration is known to be 1.00×10^{-4} M. Substituting this figure into Equation 9-46 gives, after rearranging,

$$[I^-] = 1.0 \times 10^{-4} + [Cu^+] - [CuI_2^-] \qquad (9\text{-}47)$$

Note, however, that as in Example 9-5, the charge-balance and mass-balance equations are identical. Therefore, we have only three independent equations (Equations 9-43, 9-44, and 9-47).

Step 6. Number of Independent Equations and Unknowns
We have three unknowns, $[Cu^+]$, $[I^-]$, and $[CuI_2^-]$, and as noted in Step 5, three algebraic equations. Therefore, an exact solution to the equations is possible.

Step 7. Approximations
Two approximations are possible in Equation 9-47. The first is that $[Cu^+]$ is so much smaller than 1.0×10^{-4} that the former can be eliminated from the equation. The second is that $[CuI_2^-]$ is also small enough to be ignored. The first assumption is surely valid in light of the very small numerical value for the solubility-product constant for CuI. The validity of the second assump-

tion is less obvious. As we have noted, however, it is usually wise to make the assumption, calculate a provisional value for the concentration, and see whether the provisional value is indeed much smaller than the number with which it is being compared. If it is not, a recalculation must be made with the questionable concentration term retained in the equation.

With the assumption that $([Cu^+] - [CuI_2^-]) \ll 1.0 \times 10^{-4}$, Equation 9-47 simplifies to

$$[I^-] = 1.0 \times 10^{-4}$$

Step 8. Solution to Equations
Substituting for $[I^-]$ in Equation 9-43 gives, after rearranging,

$$[Cu^+] = (1.0 \times 10^{-12})/(1.0 \times 10^{-4}) = 1.0 \times 10^{-8}$$

Substituting the value for $[I^-]$ into Equation 9-44 and rearranging yield

$$[CuI_2^-] = 1.0 \times 10^{-4} \times 7.9 \times 10^{-4} = 7.9 \times 10^{-8}$$

When these concentrations are substituted into the relationship developed in Step 2, we obtain

$$\text{solubility} = 1.0 \times 10^{-8} + 7.9 \times 10^{-8} = 8.9 \times 10^{-8} \text{ mol/L}$$

Step 9. Check of Assumptions
We see that 1.0×10^{-4} is much larger than either of the calculated values for $[Cu^+]$ and $[CuI_2^-]$. The provisional values for these species are therefore acceptable.

EXAMPLE 9-11

Calculate the solubility of AgBr in 0.0200 M NH_3, given that the formation constant for $Ag(NH_3)_2^+$ is 1.7×10^7, the solubility product for AgBr is 5.0×10^{-13}, and the basic dissociation constant for NH_3 is 1.75×10^{-5}.

See *Mathcad Applications for Analytical Chemistry,* **pp. 51–56.**

Step 1. Pertinent Equilibria

$$AgBr \rightleftharpoons Ag^+ + Br^-$$
$$Ag^+ + 2NH_3 \rightleftharpoons Ag(NH_3)_2^+$$
$$NH_3 + H_2O \rightleftharpoons NH_4^+ + OH^-$$

Step 2. Definition of Unknown
It is apparent for these equations that 1 mol of AgBr produces 1 mol of Br^- and 1 mol of silver containing species. Therefore,

$$\text{solubility} = [Br^-] = [Ag^+] + [Ag(NH_3)_2^+]$$

Step 3. Equilibrium-Constant Expressions

$$[Ag^+][Br^-] = K_{sp} = 5.0 \times 10^{-13} \qquad (9\text{-}48)$$

$$\frac{[Ag(NH_3)_2^+]}{[Ag^+][NH_3]^2} = \beta_2 = 1.7 \times 10^7 \tag{9-49}$$

$$\frac{[NH_4^+][OH^-]}{[NH_3]} = K_b = 1.75 \times 10^{-5} \tag{9-50}$$

Step 4. Mass-Balance Expressions
As shown in Example 9-2 and in Step 2,

$$[Br^-] = [Ag^+] + [Ag(NH_3)_2^+] \tag{9-51}$$

Since the analytical concentration of NH_3 is 0.0200,

$$0.0200 = [NH_3] + [NH_4^+] + 2[Ag(NH_3)_2^+] \tag{9-52}$$

Furthermore, reaction of NH_3 with water produces one NH_4^+ for each OH^-. Therefore,

$$[OH^-] = [NH_4^+] \tag{9-53}$$

We do not include $[H_3O^+]$ in the charge-balance equation because its concentration is negligible in a solution made basic with NH_3.

Step 5. Charge-Balance Expression

$$[Ag^+] + [NH_4^+] + [Ag(NH_3)_2^+] = [Br^-] + [OH^-] \tag{9-54}$$

Step 6. Number of Equations and Unknowns
We count seven equations (9-48 through 9-54) but only six unknowns ($[OH^-]$, $[Br^-]$, $[Ag^+]$, $[Ag(NH_3)_2^+]$, $[NH_4^+]$, and $[NH_3]$). This discrepancy suggests that the algebraic equations are not all independent. Close examination shows that Equation 9-54 is in fact the sum of Equations 9-51 and 9-53 and therefore not independent. Thus, we remove Equation 9-54 from consideration and now have six independent equations and an equal number of unknowns.

Step 7. Approximations
To locate possible approximations, we again turn to those equations in which concentrations appear as sums or differences—that is, Equations 9-51, 9-52, and 9-54.

1. Examining Equation 9-51 first, we speculate that $[Ag^+]$ may be considerably smaller than $[Ag(NH_3)_2^+]$ because the formation constant for the complex is so very large. Therefore, we assume provisionally that

$$[Ag^+] \ll [Ag(NH_3)_2^+]$$

2. The value for K_b for ammonia is small, which suggests that the concentration of NH_3 is significantly greater than the concentration of NH_4^+. We therefore assume that in Equation 9-52

$$[NH_4^+] \ll [NH_3] + 2[Ag(NH_3)_2^+]$$

3. A second possible assumption regarding Equation 9-52 is that $[Ag(NH_3)_2^+]$ is also significantly smaller than $[NH_3]$. The reason for this assumption is not immediately obvious from the equations at hand. If Equations 9-48 and 9-49 are multiplied together, however, we obtain

$$\frac{[Ag(NH_3)_2^+][Br^-]}{[NH_3]^2} = \beta_2 K_{sp} = 1.7 \times 10^7 \times 5.0 \times 10^{-13} \quad (9\text{-}55)$$

$$= 8.5 \times 10^{-6}$$

This equation suggests that the numerator is significantly smaller than $[NH_3]^2$. Therefore, we shall try the additional assumption that

$$2[Ag(NH_3)_2^+] \ll [NH_3]$$

Each of these three assumptions is open to some doubt; however, no harm is done in making them because a lack of validity in any one will become obvious by the time the calculation is completed. With the assumptions, Equations 9-49 and 9-50 are no longer needed and Equations 9-51 and 9-52 simplify to

$$[Br^-] = [Ag(NH_3)_2^+] \quad (9\text{-}56)$$

$$0.0200 = [NH_3]$$

We now have just three equations (9-55, 9-48, and 9-49) and three unknowns ($[Ag^+]$, $[Ag(NH_3)_2^+]$, and $[Br^-]$).

Step 8. Solution to Equations

Substituting $[NH_3] = 0.0200$ into Equation 9-49 and rearranging give

$$[Ag(NH_3)_2^+] = (1.7 \times 10^7)(0.0200)^2[Ag^+] = (6.8 \times 10^3)[Ag^+]$$

Equation 9-56 then becomes

$$[Br^-] = (6.8 \times 10^3)[Ag^+]$$

Replacing $[Ag^+]$ with $K_{sp}/[Br^-]$ (Equation 9-48) gives

$$[Br^-] = \frac{6.8 \times 10^3 \times 5.0 \times 10^{-13}}{[Br^-]}$$

$$[Br^-] = \sqrt{6.8 \times 10^3 \times 5.0 \times 10^{-13}} = 5.9 \times 10^{-5}$$

Referring to Step 2, we see

$$\text{solubility} = [Br^-] = 5.9 \times 10^{-5} \text{ mol/L}$$

Step 9. Check of Assumptions

To check assumption 1, we turn to Equation 9-56 and find

$$[Ag(NH_3)_2^+] = [Br^-] = 5.9 \times 10^{-5}$$

We assumed that this concentration is much smaller than $[NH_3]$, that is, $5.9 \times 10^{-5} \ll 0.0200$. Essentially no error is introduced by this assumption.

To check assumption 2, we substitute $[NH_3] = 0.0200$ and Equation 9-53 into Equation 9-50, which gives

$$\frac{[NH_4^+]^2}{0.0200} = 1.75 \times 10^{-5}$$

$$[NH_4^+] = \sqrt{1.75 \times 10^{-5} \times 0.0200} = 5.9 \times 10^{-4}$$

We have assumed that

$$5.9 \times 10^{-4} = 0.00059 \ll 0.0200$$

The resulting error is then less than $(0.00059/0.0200) \times 100\% = 3.0\%$, which is acceptable in most cases.

To check assumption 3, we must calculate $[Ag^+]$ by substituting $[Br^-] = 5.9 \times 10^{-5}$ into the expression for K_{sp}:

$$[Ag^+] = (5.0 \times 10^{-13})/(5.9 \times 10^{-5}) = 8.5 \times 10^{-9}$$

We know from Equation 9-56 that

$$[Ag(NH_3)_2^+] = [Br^-] = 5.9 \times 10^{-5}$$

and we assumed that $8.5 \times 10^{-9} \ll 5.9 \times 10^{-5}$. The percent error introduced by this assumption is less than 0.1%.

Complex Formation with an Ion That Is Common to the Precipitate

In gravimetric procedures, a small excess of precipitating agent minimizes solubility losses but a large excess often causes increased losses due to complex formation.

Many precipitates react with the precipitating reagent to form soluble complexes. In a gravimetric analysis, this tendency may have the unfortunate effect of decreasing the recovery of analytes if too large an excess of reagent is used. For example, in solutions containing high concentrations of chloride, silver chloride forms chloro complexes such as $AgCl(aq)$, $AgCl_2^-(aq)$, and $AgCl_3^{2-}(aq)$. The effect of these complexes is illustrated in Figure 9-2, in which the experimentally determined solubility of silver chloride is plotted against the logarithm of the potassium chloride concentration. For low anion concentrations, the experimental solubilities do not differ greatly from those calculated with the solubility-product constant for silver chloride; beyond a chloride ion concentration of about 10^{-3} M, however, the calculated solubilities approach zero while the measured values rise steeply. Note that the solubility of silver chloride is about the same in 0.3 M KCl as in pure water, and is about eight times that figure in a 1 M solution. We can describe these effects quantitatively if the compositions of the complexes and their formation constants are known.

Increases in solubility caused by large excesses of a common ion are not unusual. Of particular interest are amphoteric hydroxides, which are sparingly soluble in dilute base but are redissolved by excess hydroxide ion. The hydroxides of

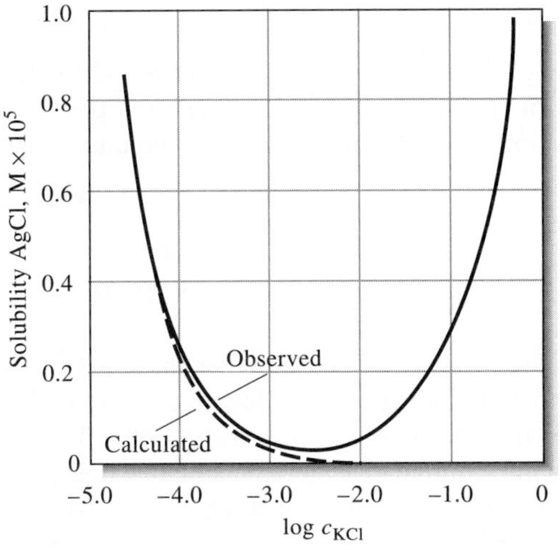

Figure 9-2
Solubility of silver chloride in potassium chloride solutions. The dashed curve is calculated from K_{sp}; the solid curve is plotted from experimental data of A. Pinkus and A. M. Timmermans, *Bull. Soc. Belges,* **1937,** *46,* 46–73.

zinc and aluminum, for example, are converted to the soluble zincate and aluminate ions upon treatment with excess base. For zinc, the equilibria can be represented as

$$Zn^{2+} + 2OH^- \rightleftharpoons Zn(OH)_2(s)$$
$$Zn(OH)_2(s) + 2OH^- \rightleftharpoons Zn(OH)_4^{2-}$$

As with silver chloride, the solubilities of amphoteric hydroxides pass through minima and then increase rapidly with increasing concentrations of base. The hydroxide ion concentration at which the solubility is a minimum can be calculated, provided equilibrium constants for the reactions are available.

9C SEPARATION OF IONS BY CONTROL OF THE CONCENTRATION OF THE PRECIPITATING AGENT

Several precipitating agents permit separation of ions based upon solubility differences. Such separations require close control of the active reagent concentration at a suitable and predetermined level. Most often, such control is achieved by controlling the pH of the solution with suitable buffers. This technique is applicable to anionic reagents in which the anion is the conjugate base of a weak acid. Examples include sulfide ion (the conjugate base of hydrogen sulfide), hydroxide ion (the conjugate base of water), and the anions of several organic weak acids.

9C-1 Calculation of the Feasibility of Separations

The following examples illustrate how solubility-product calculations are used to determine the feasibility of separations based upon solubility differences.

Note that the units for the two solubility-product constants in Example 9-12 are not strictly comparable because they have different units [mol^4/L^4 for $Fe(OH)_3$ and mol^3/L^3 for $Mg(OH)_2$]. It is only the enormous difference in the numerical values for these constants that permits us to state with confidence that the solubility of the one solute (in units of mol/L) is much smaller than that of the other.

EXAMPLE 9-12

Can Fe^{3+} and Mg^{2+} be separated quantitatively as hydroxides from a solution that is 0.10 M in each cation? If the separation is possible, what range of OH^- concentrations is permissible? Solubility-product constants for the two precipitates are

$$[Fe^{3+}][OH^-]^3 = 2 \times 10^{-39}$$
$$[Mg^{2+}][OH^-]^2 = 7.1 \times 10^{-12}$$

The K_{sp} for $Fe(OH)_3$ is so much smaller than that for $Mg(OH)_2$ that it appears likely that the former will precipitate at a lower OH^- concentration. We can answer the questions posed in this problem by (1) calculating the OH^- concentration required to achieve quantitative precipitation of Fe^{3+} and (2) computing the OH^- concentration at which $Mg(OH)_2$ just begins to precipitate. If (1) is smaller than (2), a separation is feasible in principle, and the range of permissible OH^- concentrations is defined by the two values.

To determine (1), we must first specify what constitutes a quantitative removal of Fe^{3+} from the solution. The decision here is arbitrary and depends upon the purpose of the separation. In this example and the next, we shall consider a precipitation to be quantitative when all but 1 part in 1000 of the ion has been removed from the solution—that is, when $[Fe^{3+}] \ll 1 \times 10^{-4}$.

We can readily calculate the OH^- concentration in equilibrium with 1×10^{-1} M Fe^{3+} by substituting directly into the solubility-product expression:

$$(1.0 \times 10^{-4})[OH^-]^3 = 2 \times 10^{-39}$$
$$[OH^-] = [(2 \times 10^{-39})/(1.0 \times 10^{-4})]^{1/3} = 3 \times 10^{-12}\ M$$

Thus, if we maintain the OH^- concentration at about 3×10^{-12} mol/L, the Fe^{3+} concentration will be lowered to 1×10^{-4} mol/L. Note that quantitative precipitation of $Fe(OH)_3$ is achieved in a distinctly acidic medium.

To determine the maximum OH^- concentration that can exist in the solution without causing formation of $Mg(OH)_2$, we note that precipitation cannot occur until the product $[Mg^{2+}][OH^-]^2$ exceeds the solubility product, 7.1×10^{-12}. Substitution of 0.1 (the molar Mg^{2+} concentration of the solution) into the solubility-product expression permits the calculation of the *maximum* OH^- concentration that can be tolerated:

$$0.10[OH^-]^2 = 7.1 \times 10^{-12}$$
$$[OH^-] = 8.4 \times 10^{-6}$$

When the OH^- concentration exceeds this level, the solution will be supersaturated with respect to $Mg(OH)_2$, and precipitation may begin.

From these calculations, we conclude that quantitative separation of $Fe(OH)_3$ can be achieved if the OH^- concentration is greater than 3×10^{-12} mol/L and that $Mg(OH)_2$ will not precipitate until a OH^- concentration of 8.4×10^{-6} mol/L is reached. Therefore, it is possible, in principle, to separate Fe^{3+} from Mg^{2+} by maintaining the OH^- concentration between these levels. In practice, the concentration of OH^- is kept as low as practical—often about 10^{-10} M.

9C-2 Sulfide Separations

Sulfide ion forms precipitates with heavy metal cations that have solubility products that vary from 10^{-10} to 10^{-90} or smaller. In addition the concentration of S^{2-} can be varied over a range of about 0.1 M to 10^{-22} M by controlling the pH of a saturated solution of hydrogen sulfide. These two properties make possible a number of useful cation separations. To illustrate the use of hydrogen sulfide to separate cations based upon pH control, let us consider the precipitation of the divalent cation M^{2+} from a solution that is kept saturated with hydrogen sulfide by bubbling the gas continuously through the solution. The important equilibria in this solution are

$$MS(s) \rightleftharpoons M^{2+} + S^{2-} \qquad [M^{2+}][S^{2-}] = K_{sp}$$

$$H_2S + H_2O \rightleftharpoons H_3O^+ + HS^- \qquad \frac{[H_3O^+][HS^-]}{[H_2S]} = K_1 = 9.6 \times 10^{-8}$$

$$HS^- + H_2O \rightleftharpoons H_3O^+ + S^{2-} \qquad \frac{[H_3O^+][S^{2-}]}{[HS^-]} = K_2 = 1.3 \times 10^{-14}$$

We may also write

$$\text{solubility} = [M^{2+}]$$

The concentration of hydrogen sulfide in a saturated solution of the gas is approximately 0.1 M. Thus, we may write as a mass-balance expression

$$0.1 = [S^{2-}] + [HS^-] + [H_2S]$$

Because we know the hydronium ion concentration, we have four unknowns: the concentration of the metal ion and the three sulfide species.

We can simplify the calculation greatly by assuming that $([S^{2-}] + [HS^-]) \ll [H_2S]$ so that

$$[H_2S] \approx 0.10 \text{ mol/L}$$

An expression for the overall dissociation of hydrogen sulfide to sulfide ion is given by

$$H_2S + 2H_2O \rightleftharpoons 2H_3O^+ + S^{2-} \qquad \frac{[H_3O^+]^2[S^{2-}]}{[H_2S]} = K_1K_2 = 1.2 \times 10^{-21}$$

The constant for this overall reaction is simply the product of K_1 and K_2.

Substituting the numerical value for $[H_2S]$ into this equation gives

$$\frac{[H_3O^+]^2[S^{2-}]}{0.10} = 1.2 \times 10^{-21}$$

Upon rearranging this equation, we obtain

$$[S^{2-}] = \frac{1.2 \times 10^{-22}}{[H_3O^+]^2} \qquad (9\text{-}57)$$

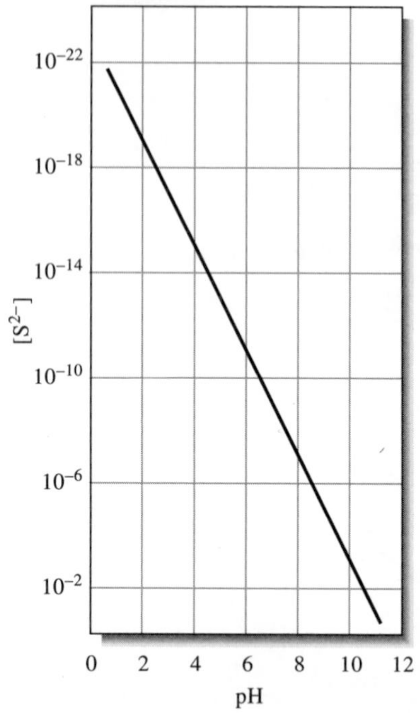

Figure 9-3
Sulfide ion concentration as a function of pH in a saturated H_2S solution.

Thus, we see that the sulfide ion concentration of a saturated hydrogen sulfide solution varies inversely as the square of the hydrogen ion concentration. Figure 9-3, which was obtained with this equation, reveals that the sulfide ion concentration of an aqueous solution can be varied by over 20 orders of magnitude by varying the pH from 1 to 11.

Substituting Equation 9-57 into the solubility-product expression gives

$$\frac{[M^{2+}] \times 1.2 \times 10^{-22}}{[H_3O^+]^2} = K_{sp}$$

$$[M^{2+}] = \text{solubility} = \frac{[H_3O^+]^2 K_{sp}}{1.2 \times 10^{-22}}$$

Thus, the solubility of a divalent metal sulfide increases as the square of the hydronium ion concentration.

EXAMPLE 9-13

Cadmium sulfide is less soluble than thallium(I) sulfide. Find the conditions under which Cd^{2+} and Tl^+ can, in theory, be separated quantitatively with H_2S from a solution that is 0.1 M in each cation.

The constants for the two solubility equilibria are:

$$CdS(s) \rightleftharpoons Cd^{2+} + S^{2-} \qquad [Cd^{2+}][S^{2-}] = 1 \times 10^{-27}$$

$$Tl_2S(s) \rightleftharpoons 2Tl^+ + S^{2-} \qquad [Tl^+]^2[S^{2-}] = 6 \times 10^{-22}$$

Since CdS precipitates at a lower $[S^{2-}]$ than does Tl_2S, we first compute the sulfide ion concentration necessary for quantitative removal of Cd^{2+} from solution. In order to make such a calculation, we must first specify what constitutes a quantitative removal. As in Example 9-12, we shall arbitrarily consider a separation to be quantitative when all but 1 part in 1000 of the Cd^{2+} has been removed; that is, the concentration of the cation has been lowered to 1.00×10^{-4} M. Substituting this value into the solubility-product expression gives

$$10^{-4}[S^{2-}] = 1 \times 10^{-27}$$

$$[S^{2-}] = 2 \times 10^{-23}$$

Thus, if we maintain the sulfide concentration at this level or greater, we may assume that quantitative removal of the cadmium will take place. Next, we compute the $[S^{2-}]$ needed to initiate precipitation of Tl_2S from a 0.1 M solution. Precipitation will begin when the solubility product is just exceeded. Since the solution is 0.1 M in Tl^+,

$$(0.1)^2[S^{2-}] = 6 \times 10^{-22}$$

$$[S^{2-}] = 6 \times 10^{-20}$$

These two calculations show that quantitative precipitation of Cd^{2+} takes place if $[S^{2-}]$ is made greater than 1×10^{-23}. No precipitation of Tl^+ occurs, however, until $[S^{2-}]$ becomes greater than 6×10^{-20} M.

Substituting these two values for $[S^{2-}]$ into Equation 9-57 permits calculation of the $[H_3O^+]$ range required for the separation.

$$[H_3O^+]^2 = \frac{1.2 \times 10^{-22}}{1 \times 10^{-23}} = 12$$

$$[H_3O^+] = 3.5$$

and

$$[H_3O^+]^2 = \frac{1.2 \times 10^{-22}}{6 \times 10^{-20}} = 2.0 \times 10^{-3}$$

$$[H_3O^+] = 0.045$$

By maintaining $[H_3O^+]$ between approximately 0.045 and 3.5 M, we can in theory separate CdS quantitatively from Tl_2S.

9D QUESTIONS AND PROBLEMS

NOTE: Use molar concentrations rather than activities for these problems unless you are directed otherwise.

***9-1.** How does calculation of the molar solubility for $Fe(OH)_2$ differ from that for $Fe(OH)_3$?

9-2. Demonstrate how the sulfide ion concentration is related to the hydronium ion concentration of a solution that is kept saturated with hydrogen sulfide.

9-3. Why are simplifying assumptions restricted to relationships that are sums or differences?

***9-4.** Why do molar concentrations of some species appear as multiples in charge-balance equations?

9-5. Write the mass-balance expressions for a solution that is
- ***(a)** 0.10 M in H_3PO_4.
- **(b)** 0.10 M in Na_2HPO_4.
- ***(c)** 0.100 M in HNO_2 and 0.0500 M in $NaNO_2$.
- **(d)** 0.25 M in NaF and saturated with CaF_2.
- ***(e)** 0.100 M in NaOH and saturated with $Zn(OH)_2$ (which undergoes the reaction $Zn(OH)_2 + 2OH^- \rightleftharpoons Zn(OH)_4^{2-}$).
- **(f)** saturated with $MgCO_3$.
- ***(g)** saturated with CaF_2.

9-6. Write the charge-balance equations for the solutions in Problem 9-5. c.

9-7. Calculate the molar solubility of Ag_2CO_3 in a solution that has a fixed H_3O^+ concentration of
- ***(a)** 1.0×10^{-6} M.
- **(b)** 1.0×10^{-7} M.
- ***(c)** 1.0×10^{-9} M.
- **(d)** 1.0×10^{-11} M.

9-8. Calculate the molar solubility of $BaSO_4$ in a solution in which $[H_3O^+]$ is
***(a)** 2.0 M. **(b)** 1.0 M. ***(c)** 0.50 M. **(d)** 0.10 M.

***9-9.** Calculate the molar solubility of CuS in a solution in which the $[H_3O^+]$ concentration is held constant at (a) 1.0×10^{-1} M and (b) 1.0×10^{-4} M.

9-10. Calculate the concentration of CdS in a solution in which the $[H_3O^+]$ is held constant at (a) 1.0×10^{-1} M and (b) 1.0×10^{-4} M.

9-11. Calculate the molar solubility of MnS (green) in a solution with a constant $[H_3O^+]$ of *(a) 1.00×10^{-5} and (b) 1.00×10^{-7}.

***9-12.** Calculate the molar solubility of $PbCO_3$
- **(a)** in a solution buffered to a pH of 7.00.
- **(b)** in water.

9-13. Calculate the molar solubility of Ag_2SO_3 ($K_{sp} = 1.5 \times 10^{-14}$)
- **(a)** in a solution buffered to a pH of 7.00.
- **(b)** in water.

***9-14.** Dilute NaOH is introduced into a solution that is 0.050 M in Cu^{2+} and 0.040 M in Mn^{2+}.
- **(a)** Which hydroxide precipitates first?
- **(b)** What OH^- concentration is needed to initiate precipitation of the first hydroxide?
- **(c)** What is the concentration of the cation forming the less soluble hydroxide when the more soluble hydroxide begins to form?

9-15. A solution is 0.040 M in Na_2SO_4 and 0.050 M in $NaIO_3$. To this is added a solution containing Ba^{2+}. Assuming that no HSO_4^- is present in the original solution,
- **(a)** which barium salt will precipitate first?
- **(b)** what is the Ba^{2+} concentration as the first precipitate forms?
- **(c)** what is the concentration of the anion that forms the less soluble barium salt when the more soluble precipitate begins to form?

***9-16.** Silver ion is being considered as a reagent for separating I^- from SCN^- in a solution that is 0.060 M in KI and 0.070 M in NaSCN.
- **(a)** What Ag^+ concentration is needed to lower the I^- concentration to 1.0×10^{-6} M?
- **(b)** What is the Ag^+ concentration of the solution when AgSCN begins to precipitate?
- **(c)** What is the ratio of SCN^- to I^- when AgSCN begins to precipitate?

(d) What is the ratio of SCN^- to I^- when the Ag^+ concentration is 1.0×10^{-3} M?

9-17. Using 1.0×10^{-6} M as the criterion for quantitative removal, determine whether it is feasible to use

***(a)** SO_4^{2-} to separate Ba^{2+} and Sr^{2+} in a solution that is initially 0.10 M in Sr^{2+} and 0.25 M in Ba^{2+}.

***(b)** SO_4^{2-} to separate Ba^{2+} and Ag^+ in a solution that is initially 0.040 M in each cation. For Ag_2SO_4, $K_{sp} = 1.6 \times 10^{-5}$.

(c) OH^- to separate Be^{3+} and Hf^{4+} in a solution that is initially 0.020 M in Be^{2+} and 0.010 M in Hf^{4+}. For $Be(OH)_3$, $K_{sp} = 7.0 \times 10^{-22}$ and for $Hf(OH)_4$, $K_{sp} = 4.0 \times 10^{-26}$.

(d) IO_3^- to separate In^{3+} and Tl^+ in a solution that is initially 0.110 M in In^{3+} and 0.060 M in Tl^+. For $In(IO_3)_3$, $K_{sp} = 3.3 \times 10^{-11}$; for $TlIO_3$, $K_{sp} = 3.1 \times 10^{-6}$.

***9-18.** What weight of AgBr dissolves in 200 mL of 0.100 M NaCN?

$$Ag^+ + 2CN^- \rightleftharpoons Ag(CN)_2^- \qquad \beta_2 = 1.3 \times 10^{21}$$

9-19. The equilibrium constant for formation of $CuCl_2^-$ is given by

$$Cu^+ + 2Cl^- \rightleftharpoons CuCl_2^-$$

$$\beta_2 = \frac{[CuCl_2^-]}{[Cu^+][Cl^-]^2} = 3.2 \times 10^5$$

What is the solubility of CuCl in solutions having the following analytical NaCl concentrations:

(a) 1.0 M?

(b) 1.0×10^{-1} M?

(c) 1.0×10^{-2} M?

(d) 1.0×10^{-3} M?

(e) 1.0×10^{-4} M?

***9-20.** In contrast to many salts, calcium sulfate is only partially dissociated in aqueous solution:

$$CaSO_4(aq) \rightleftharpoons Ca^{2+} + SO_4^{2-} \qquad K_d = 5.2 \times 10^{-3}$$

The solubility-product constant for $CaSO_4$ is 2.4×10^{-5}. Calculate the solubility of $CaSO_4$ in (a) water and (b) 0.0100 M Na_2SO_4. In addition, calculate the percent of undissociated $CaSO_4$ in each solution.

CHAPTER

10

Theory of Neutralization Titrations

Standard solutions of strong acids and strong bases are used extensively for determining analytes that are themselves acids or bases or analytes that can be converted to such species by chemical treatment. This chapter deals with theoretical aspects of titrations with such reagents.

10A SOLUTIONS AND INDICATORS FOR ACID/BASE TITRATIONS

We begin our discussion by describing the types of standard reagents that are used for neutralization titrations and the behavior of indicators during such titrations.

10A-1 Standard Solutions

The standard solutions employed in neutralization titrations are strong acids or strong bases because these substances react more completely with an analyte than do their weaker counterparts and thus yield sharper end points. Standard solutions of acids are prepared by diluting concentrated hydrochloric, perchloric, or sulfuric acid. Nitric acid is seldom used because its oxidizing properties offer the potential for undesirable side reactions. It should be pointed out that hot concentrated perchloric and sulfuric acids are potent oxidizing agents and thus hazardous. Fortunately, cold dilute solutions of these reagents are relatively benign and can be used in the analytical laboratory without any special precautions other than eye protection.

The standard reagents used in acid/base titrations are always strong acids or strong bases, most commonly HCl, $HClO_4$, H_2SO_4, NaOH, and KOH. Weak acids and bases are never used as standard reagents because they react incompletely with analytes.

Standard solutions of bases are ordinarily prepared from solid sodium, potassium, and occasionally barium hydroxides. Again, eye protection should always be used when handling dilute solutions of these reagents.

10A-2 Acid/Base Indicators

Many substances, both naturally occurring and synthetic, display colors that depend upon the pH of the solutions in which they are dissolved. Some of these substances, which have been used for centuries to indicate the acidity or alkalinity of water, still find current application as acid/base indicators.

For a list of common acid/base indicators and their colors, look inside the front cover of this book.

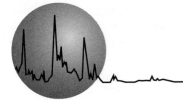

An acid/base indicator is a weak organic acid or a weak organic base whose undissociated form differs in color from its conjugate base or its conjugate acid form. For example, the behavior of an acid-type indicator, HIn, is described by the equilibrium

$$\underset{\text{acid color}}{\mathrm{HIn}} + \mathrm{H_2O} \rightleftharpoons \underset{\text{base color}}{\mathrm{In^-}} + \mathrm{H_3O^+}$$

Here, internal structural changes accompany dissociation and cause the color change. The equilibrium for a base-type indicator, In, is

$$\underset{\text{base color}}{\mathrm{In}} + \mathrm{H_2O} \rightleftharpoons \underset{\text{acid color}}{\mathrm{InH^+}} + \mathrm{OH^-}$$

In the paragraphs that follow, we focus on the behavior of acid-type indicators. The discussion, however, can be readily extended to base-type indicators as well.

The equilibrium-constant expression for the dissociation of an acid-type indicator takes the form

$$K_a = \frac{[\mathrm{H_3O^+}][\mathrm{In^-}]}{[\mathrm{HIn}]} \tag{10-1}$$

Rearranging leads to

$$[\mathrm{H_3O^+}] = K_a \frac{[\mathrm{HIn}]}{[\mathrm{In^-}]} \tag{10-2}$$

We see then that the hydronium ion concentration determines the ratio between the acid and conjugate base forms of the indicator.

The human eye is not very sensitive to color differences in a solution containing a mixture of $\mathrm{In^-}$ and HIn, particularly when the ratio $[\mathrm{In^-}]/[\mathrm{HIn}]$ is greater than about 10 or smaller than about 0.1. Consequently, the color imparted to a solution by a typical indicator appears to the average observer to change rapidly only within the limited concentration ratio of approximately 10 to 0.1. At greater or smaller ratios, the color becomes essentially constant to the human eye and is independent of the ratio. Therefore, we can write that the average indicator HIn exhibits its pure acid color when

$$\frac{[\mathrm{HIn}]}{[\mathrm{In^-}]} \geq \frac{10}{1}$$

and its base color when

$$\frac{[\mathrm{HIn}]}{[\mathrm{In^-}]} \leq \frac{1}{10}$$

The color appears to be intermediate for ratios between these two values. These ratios vary considerably from indicator to indicator, of course. Furthermore, people differ significantly in their ability to distinguish between colors, with a color-blind person representing one extreme.

If these two concentration ratios are substituted into Equation 10-1, the range of hydronium ion concentrations needed to effect the complete indicator color change can be evaluated. Thus, for the full acid color,

$$[H_3O^+] \geq 10K_a$$

and similarly for the full base color,

$$[H_3O^+] \leq \frac{1}{10} K_a$$

To obtain the indicator pH range, we take the negative logarithms of the two expressions:

$$\text{pH (acid color)} = -\log(10K_a) = pK_a + 1$$
$$\text{pH (basic color)} = -\log(\tfrac{1}{10} K_a) = pK_a - 1$$

$$\text{indicator range} = pK_a \pm 1 \qquad (10\text{-}3)$$

Thus, an indicator with an acid dissociation constant of 1×10^{-5} ($pK_a = 5$) typically shows a complete color change when the pH of the solution in which it is dissolved changes from 4 to 6 (see Figure 10-1). A similar relationship is easily derived for a basic-type indicator.

The list of acid/base indicators is large and includes a number of organic structures. Indicators are available for any desired pH range. A few common indicators and their properties are listed in Table 10-1. Note that the transition ranges vary from 1.1 to 2.2, with the average being about 1.6 units.

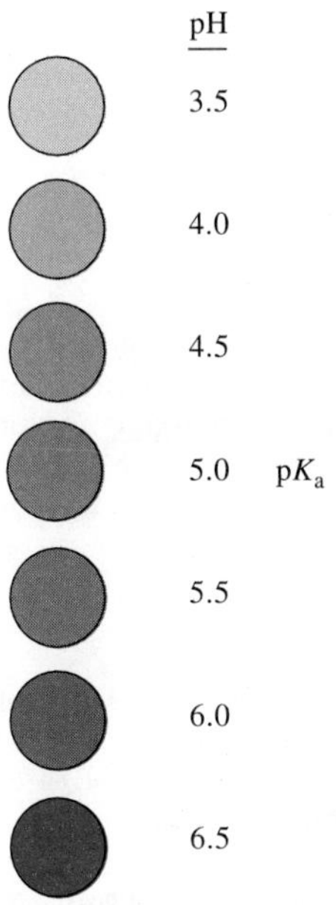

Figure 10-1
Indicator color as a function of pH ($pK_a = 5.0$).

The approximate pH transition range of most acid type indicators is roughly $pK_a \pm 1$.

TABLE 10-1 Some Important Acid/Base Indicators

Common Name	Transition Range, pH	pK_a*	Color Change†	Indicator Type‡
Thymol blue	1.2–2.8	1.65§	R–Y	1
	8.0–9.6	8.90§	Y–B	
Methyl yellow	2.9–4.0		R–Y	2
Methyl orange	3.1–4.4	3.46§	R–O	2
Bromocresol green	3.8–5.4	4.66§	Y–B	1
Methyl red	4.2–6.3	5.00§	R–Y	2
Bromocresol purple	5.2–6.8	6.12§	Y–P	1
Bromothymol blue	6.2–7.6	7.10§	Y–B	1
Phenol red	6.8–8.4	7.81§	Y–R	1
Cresol purple	7.6–9.2		Y–P	1
Phenolphthalein	8.3–10.0		C–R	1
Thymolphthalein	9.3–10.5		C–B	1
Alizarin yellow GG	10–12		C–Y	2

*At ionic strength of 0.1.

†B = blue; C = colorless; O = orange; P = purple; R = red; Y = yellow.

‡(1) Acid type: $HIn + H_2O \rightleftharpoons H_3O^+ + In^-$
(2) Base type: $In + H_2O \rightleftharpoons InH^+ + OH^-$

§For the reaction $InH^+ + H_2O \rightleftharpoons H_3O^+ + In$.

10B TITRATION CURVES

An end point is an observable physical change that occurs near the equivalence point of a titration. The two most widely used end points involve (1) changes in color due to the reagent, the analyte, or an indicator and (2) a change in potential of an electrode that responds to the concentration of the reagent or the analyte.

Titration curves are plots of a concentration-related variable as a function of reagent volume.

To help us understand the theoretical basis of end points and the sources of titration errors we will derive *titration curves* for the system under consideration. Titration curves consist of a plot of reagent volume as the horizontal axis and some function of the analyte or reagent concentration as the vertical.

10B-1 Types of Titration Curves

The vertical axis in a sigmoidal titration curve is either the p-function of the analyte or reagent or the potential of an analyte- or reagent-sensitive electrode.

The vertical axis for a linear-segment titration curve is an instrumental signal that is proportional to the concentration of the analyte or reagent.

Two general types of titration curves (and thus two general types of end points) are encountered in titrimetric methods. In the first type, called a *sigmoidal curve,* important observations are confined to a small region (typically ± 0.1 to ± 0.5 mL) surrounding the equivalence point. A sigmoidal curve, in which the p-function of the analyte (or sometimes the reagent) is plotted as a function of reagent volume, is shown in Figure 10-2a. In the second type of curve, called a *linear-segment curve,* measurements are made on both sides of, but well away from, the equivalence point. Measurements near equivalence are avoided. In this type of curve, the vertical axis is an instrument reading that is directly proportional to the concentration of the analyte or the reagent. A typical linear segment curve is found in Figure 10-2b. The sigmoidal type offers the advantages of speed and convenience. The linear segment type is advantageous for reactions that are complete only in the presence of a goodly excess of the reagent or analyte.

In this, and the several chapters that follow, we will be dealing exclusively with sigmoidal titration curves. Linear-segment curves are considered in Sections 21B-4 and 24A-5.

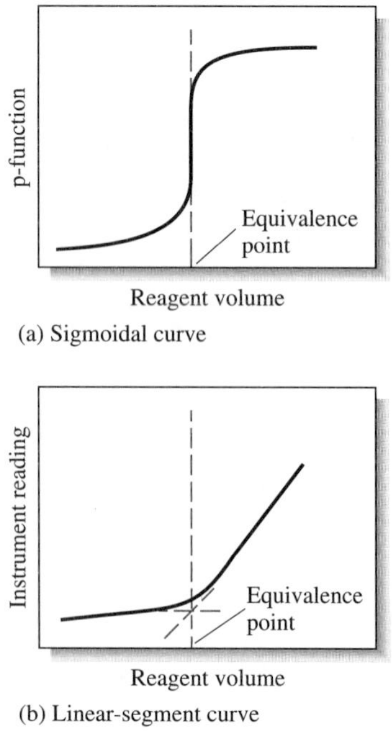

Figure 10-2
Two types of titration curves.

TABLE 10-2 Concentration Changes During a Titration of 50.00 mL of 0.1000 M HCl

Volume of 0.100 M NaOH, mL	$[H_3O^+]$ mol/L	Volume of NaOH to Cause a Tenfold Decrease in $[H_3O^+]$, mL	pH	pOH
0.00	1.000×10^{-1}		1.00	13.00
40.91	1.000×10^{-2}	40.91	2.00	12.00
49.01	1.000×10^{-3}	8.11	3.00	11.00
49.90	1.000×10^{-4}	0.89	4.00	10.00
49.99	1.000×10^{-5}	0.09	5.00	9.00
49.999	1.000×10^{-6}	0.009	6.00	8.00
50.00	1.000×10^{-7}	0.001	7.00	7.00
50.001	1.000×10^{-8}	0.001	8.00	6.00
50.01	1.000×10^{-9}	0.009	9.00	5.00
50.10	1.000×10^{-10}	0.09	10.00	4.00
51.01	1.000×10^{-11}	0.91	11.00	3.00
61.11	1.000×10^{-12}	10.10	12.00	2.00

10B-2 Concentration Changes During Titrations

The equivalence point in a titration is characterized by major changes in the *relative* concentrations of reagent and analyte. Table 10-2 illustrates this phenomenon. The data in the second column of the table show the changes in concentration of hydronium ion as a 50.00-mL aliquot of a 0.1000 M solution of hydrochloric acid is titrated with a 0.1000 M solution of sodium hydroxide. The neutralization reaction is described by the equation

$$H_3O^+ + OH^- \rightleftharpoons 2H_2O$$

In order to emphasize the changes in *relative* concentration that occur in the equivalence region, the volume increments computed are those required to cause tenfold decreases in the concentration of H_3O^+ (or tenfold increase in hydroxide ion concentration). Thus, we see in the third column that an addition of 40.91 mL of base is needed to decrease the concentration of the hydronium ion by one order of magnitude from 0.100 M to 0.0100 M. An addition of only 8.11 mL is required to lower the concentration by another factor of 10 to 0.00100 M; 0.89 mL causes yet another tenfold decrease. Corresponding increases in hydroxide concentration occur at the same time. End-point detection, then, is based upon this large change in *relative* concentration of the analyte (or the reagent) that occurs at the equivalence point for every type of titration.

At the beginning of the titration described in Table 10-2, about 41 mL of reagent brings about a tenfold decrease in the concentration of H_3O^+; only 0.001 mL is required to cause this same change at the equivalence point.

The large relative concentration changes that occur in the region of chemical equivalence are shown by plotting the negative logarithm of the analyte or the reagent concentration (the p-function) against reagent volume as has been done in Figure 10-3. The data for this plot are found in the fourth column of Table 10-2. Titration curves for reactions involving complex formation, precipitation, and oxidation/reduction all exhibit the same sharp increase or decrease in p-function as those shown in Figure 10-3 in the equivalence-point region. Titration curves define the properties required of an indicator and allow us to estimate the error associated with titration methods.

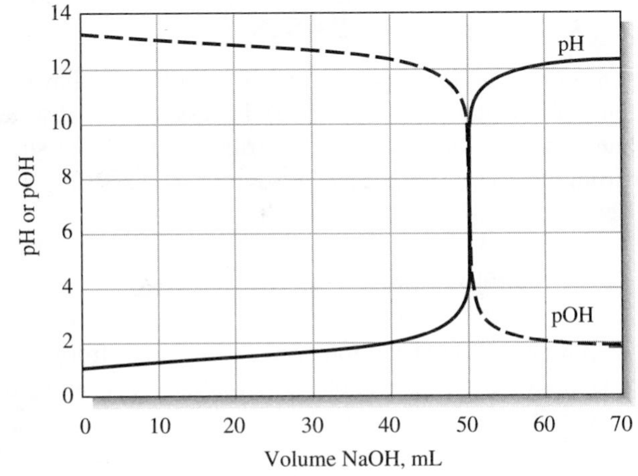

Figure 10-3
Titration curve from data in Table 10-2.

FEATURE 10-1

How Did We Calculate the Volumes of NaOH Shown in the First Column of Table 10-2?

Up until the equivalence point, $[H_3O^+]$ will equal the concentration of unreacted HCl (c_{HCl}). The concentration of HCl is equal to the original number of millimoles HCl (50.00×0.1000) minus the number of millimoles NaOH added ($V_{NaOH} \times 0.1000$) divided by the total volume of the solution. That is,

$$c_{HCl} = [H_3O^+] = \frac{50.00 \times 0.1000 - V_{NaOH} \times 0.1000}{50.00 + V_{NaOH}}$$

where V_{NaOH} is the volume of 0.1000 M NaOH added. This equation reduces to

$$50.00[H_3O^+] + V_{NaOH}[H_3O^+] = 5.000 - 0.1000\ V_{NaOH}$$

Collecting the terms containing V_{NaOH} gives

$$V_{NaOH}(0.1000 + [H_3O^+]) = 5.000 - [H_3O^+]$$

or

$$V_{NaOH} = \frac{5.000 - 50.00[H_3O^+]}{0.1000 + [H_3O^+]}$$

Thus, to obtain $[H_3O^+] = 1.000 \times 10^{-2}$, we find

$$V_{NaOH} = \frac{5.000 - 50.00 \times 1.000 \times 10^{-2}}{0.1000 + 1.000 \times 10^{-2}} = 40.91\ \text{mL}$$

10C TITRATION CURVES FOR STRONG ACIDS AND STRONG BASES

In solutions of a strong acid that are more concentrated than about 1×10^{-6} M, it is proper to assume that the equilibrium concentration of H_3O^+ is equal to the analytical concentration of the acid. The same is true for the hydroxide concentration in solutions of strong bases.

The hydronium ions in an aqueous solution of a strong acid have two sources: (1) the reaction of the acid with water and (2) the dissociation of water. In all but the most dilute solutions, however, the contribution from the strong acid far exceeds that from the solvent. Thus, for a solution of HCl with a concentration greater than about 1×10^{-6} M, we can write

$$[H_3O^+] = c_{HCl} + [OH^-] \approx c_{HCl}$$

where $[OH^-]$ represents the contribution of hydronium ions from the dissociation of water. An analogous relationship applies for a solution of a strong base, such as sodium hydroxide. That is,

$$[OH^-] = c_{NaOH} + [H_3O^+] \approx c_{NaOH}$$

10C-1 The Titration of a Strong Acid with a Strong Base

To derive a titration curve for a solution of a strong acid with a strong base, three types of calculations are required, each of which corresponds to a distinct stage in the titration: (1) preequivalence, (2) equivalence, and (3) postequivalence. In the preequivalence stage, we compute the concentration of the acid from its starting concentration and the amount of base added. At the equivalence point, the hydronium and hydroxide ions are present in equal concentrations, and the hydronium ion concentration is derived directly from the ion-product constant for water. In the postequivalence stage, the analytical concentration of the excess base is computed, and the hydroxide ion concentration will be equal to (or a multiple of) the analytical concentration. A convenient way of converting hydroxide concentration to pH can be developed by taking the negative logarithm of each side of the ion-product constant for water. Thus,

Before the equivalence point we calculate the pH from the molar concentration of unreacted acid.

At the equivalence point, the solution is neutral and pH = 7.00.

Beyond the equivalence point, we first calculate pOH and then pH.

Remember: $\text{pH} = \text{p}K_w - \text{pOH} = 14.00 - \text{pOH}$

$$K_w = [H_3O^+][OH^-]$$

$$-\log K_w = -\log [H_3O^+][OH^-] = -\log [H_3O^+] - \log [OH^-]$$

$$\text{p}K_w = \text{pH} + \text{pOH}$$

$$-\log 10^{-14} = 14.00 = \text{pH} + \text{pOH}$$

$$\text{pH} + \text{pOH} = 14.00 = \text{p}K_w$$

EXAMPLE 10-1

Derive a curve for the titration of 50.00 mL of 0.0500 M HCl with 0.1000 M NaOH.

Initial Point

At the outset, the solution is 0.0500 M in H_3O^+, and

$$\text{pH} = -\log [H_3O^+] = -\log 0.0500 = 1.30$$

After Addition of 10.00 mL of Reagent

The hydronium ion concentration decreases as a result of both reaction with the base and dilution. Thus, the analytical concentration of HCl is

$$c_{HCl} = \frac{\text{no. mmol HCl remaining after addition of NaOH}}{\text{total volume solution}}$$

$$= \frac{\text{original no. mmol HCl} - \text{no. mmol NaOH added}}{\text{total volume solution}}$$

$$= \frac{(50.00\text{ mL} \times 0.0500\text{ M}) - (10.00\text{ mL} \times 0.1000\text{ M})}{50.00\text{ mL} + 10.00\text{ mL}}$$

$$= \frac{(2.500\text{ mmol} - 1.000\text{ mmol})}{60.00\text{ mL}} = 2.500 \times 10^{-2}\text{ M}$$

$$[H_3O^+] = 2.50 \times 10^{-2}$$

OH⁻ $-\log\left[\frac{5}{100}\right] = 1.3 \rightarrow 12.699$

$-\log .04 = 12.60206$

and

$$pH = -\log [H_3O^+] = -\log (2.500 \times 10^{-2}) = 1.602$$

Additional points defining the curve in the region before the equivalence point are obtained in the same way. The results of such calculations are shown in the second column of Table 10-3.

Equivalence Point

At the equivalence point, neither HCl nor NaOH is in excess, and so the concentrations of hydronium and hydroxide ions must be equal. Substituting this equality into the ion-product constant for water yields

$$[H_3O^+] = \sqrt{K_w} = \sqrt{1.00 \times 10^{-14}} = 1.00 \times 10^{-7}$$

$$pH = -\log (1.00 \times 10^{-7}) = 7.00$$

After Addition of 25.10 mL of Reagent

The solution now contains an excess of NaOH, and we can write

$$c_{NaOH} = \frac{25.10 \times 0.1000 - 50.00 \times 0.0500}{75.10} = 1.33 \times 10^{-4}\ M$$

and the equilibrium concentration of hydroxide ion is

$$[OH^-] = c_{NaOH} = 1.33 \times 10^{-4}\ M$$

$$pOH = -\log (1.33 \times 10^{-4}) = 3.88$$

and

$$pH = 14.00 - 3.88 = 10.12$$

Additional data defining the curve beyond the equivalence point are computed in the same way. The results of such computations are shown in Table 10-3.

TABLE 10-3 Changes in pH During the Titration of a Strong Acid with a Strong Base

	pH	
Volume of NaOH, mL	**50.00 mL of 0.0500 M HCl with 0.1000 M NaOH**	**50.00 mL of 0.000500 M HCl with 0.001000 M NaOH**
0.00	1.30	3.30
10.00	1.60	3.60
20.00	2.15	4.15
24.00	2.87	4.87
24.90	3.87	5.87
25.00	7.00	7.00
25.10	10.12	8.12
26.00	11.12	9.12
30.00	11.80	9.80

FEATURE 10-2
Deriving Titration Curves from the Charge-Balance Equation

In Example 10-1 we generated an acid/base titration curve from the reaction stoichiometry. We can show that all points on the curve can be derived from the charge-balance equation also.

For the system treated in Example 10-1, the charge-balance equation is given by

$$[H_3O^+] + [Na^+] = [OH^-] + [Cl^-]$$

where the sodium and chloride ion concentrations are given by

$$[Na^+] = \frac{V_{NaOH}\, c_{NaOH}}{V_{NaOH} + V_{HCl}}$$

$$[Cl^-] = \frac{V_{HCl}\, c_{HCl}}{V_{NaOH} + V_{HCl}}$$

For volumes of NaOH short of the equivalence point $[OH^-] \ll [Cl^-]$, and we can rewrite the first equation in the form

$$[H_3O^+] = [Cl^-] - [Na^+]$$

and

$$[H_3O^+] = \frac{V_{HCl}\, c_{HCl}}{V_{HCl} + V_{NaOH}} - \frac{V_{NaOH}\, c_{NaOH}}{V_{HCl} + V_{NaOH}} = \frac{V_{HCl}\, c_{HCl} - V_{NaOH}\, c_{NaOH}}{V_{HCl} + V_{NaOH}}$$

At the equivalence point, $[Na^+] = [Cl^-]$ and

$$[H_3O^+] = [OH^-]$$

$$[H_3O^+] = \sqrt{K_w}$$

Beyond the equivalence point, $[H_3O^+] \ll [Na^+]$, and the original equation rearranges to

$$[OH^-] = [Na^+] - [Cl^-]$$

$$= \frac{V_{NaOH}\, c_{NaOH}}{V_{NaOH} + V_{HCl}} - \frac{V_{HCl}\, c_{HCl}}{V_{NaOH} + V_{HCl}} = \frac{V_{NaOH}\, c_{NaOH} - V_{HCl}\, c_{HCl}}{V_{NaOH} + V_{HCl}}$$

FEATURE 10-3
Significant Figures in Titration Curve Calculations

Concentrations calculated in the equivalence-point region of titration curves are generally of low precision because they are based upon small differences between large numbers. For example, in the calculation of c_{NaOH} after introduction of 25.10 mL of NaOH in Example 10-1, the numerator (2.510 − 2.500 = 0.010) is known to only two significant figures. To minimize rounding error, however, three digits were retained in c_{NaOH} (1.33×10^{-4}), and rounding was postponed until pOH and pH were computed.

In rounding the calculated values for p-functions, you should remember (page 41) that it is the *mantissa of a logarithm (that is, the digits to the right of the decimal point) that should be rounded to include only significant figures* because the characteristic (the digits to the left of the decimal point) serves merely to locate the decimal point. Fortunately, the large changes in p-functions characteristic of most equivalence points are not obscured by the limited precision of the calculated data. Generally, in deriving data for titration curves, we will round p-functions to two places to the right of the decimal point whether or not such rounding is called for.

The Effect of Concentration

The effects of reagent and analyte concentration on the neutralization titration curves for strong acids are shown by the two sets of data in Table 10-3 and the plots in Figure 10-4. With 0.1 M NaOH as the titrant (curve *A*), the change in pH in the equivalence-point region is large. With 0.001 M NaOH, the change is markedly less but still pronounced.

Indicator Choice

Figure 10-4 shows that the selection of an indicator is not critical when the reagent concentration is approximately 0.1 M. Here, the volume differences in titrations with the three indicators shown are of the same magnitude as the uncertainties associated with reading the buret and are thus negligible. Note, however, that bromocresol green is clearly unsuited for a titration involving the 0.001 M reagent because the color change occurs over a 5-mL range well before the equivalence point. The use of phenolphthalein is subject to similar objections. Of the three indicators, then, only bromothymol blue provides a satisfactory end

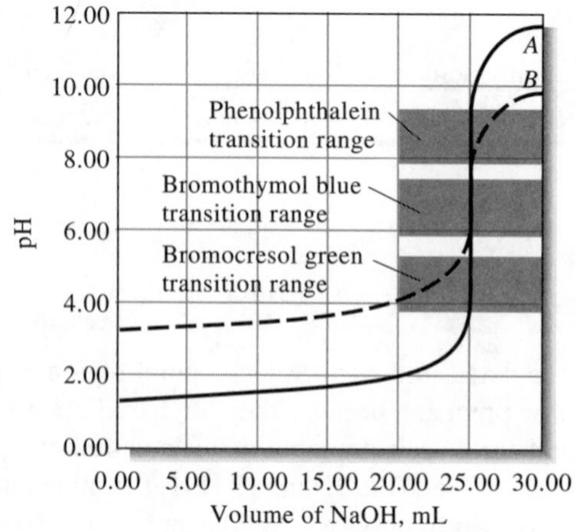

Figure 10-4
Titration curves for HCl with NaOH. *A*: 50.00 mL of 0.0500 M HCl with 0.1000 M NaOH. *B*: 50.00 mL of 0.000500 M HCl with 0.001000 M NaOH.

point with a minimal systematic error in the titration of the more dilute solution.

10C-2 The Titration of a Strong Base with a Strong Acid

Titration curves for strong bases are derived in an analogous way to those for strong acids. Short of the equivalence point, the solution is highly basic, the hydroxide ion concentration being numerically equal to the analytical molarity of the base. The solution is neutral at the equivalence point and becomes acidic in the region beyond the equivalence point; here, the hydronium ion concentration is equal to the analytical concentration of the excess strong acid.

EXAMPLE 10-2

Calculate the pH during the titration of 50.00 mL of 0.0500 M NaOH with 0.1000 M HCl after the addition of the following volumes of reagent: (a) 24.50 mL, (b) 25.00 mL, (c) 25.50 mL.

(a) At 24.50 mL added, $[H_3O^+]$ is very small and cannot be computed from stoichiometric considerations but $[OH^-]$ can be obtained readily

$$[OH^-] = c_{NaOH} = \frac{\text{original no. mmol NaOH} - \text{no. mmol HCl added}}{\text{total volume of solution}}$$

$$= \frac{(50.00 \times 0.0500 - 24.50 \times 0.1000)}{50.00 + 24.50} = 6.71 \times 10^{-4}$$

$$[H_3O^+] = K_w/(6.71 \times 10^{-4}) = (1.00 \times 10^{-14})/(6.71 \times 10^{-4})$$

$$= 1.49 \times 10^{-11}$$

$$pH = -\log(1.49 \times 10^{-11}) = 10.83$$

(b) This is the equivalence point where $[H_3O^+] = [OH^-]$

$$[H_3O^+] = \sqrt{K_w} = \sqrt{1.00 \times 10^{-14}} = 1.00 \times 10^{-7}$$

$$pH = -\log(1.00 \times 10^{-7}) = 7.00$$

(c) $$[H_3O^+] = c_{HCl} = \frac{(25.50 \times 0.1000 - 50.00 \times 0.0500)}{75.50}$$

$$= 6.62 \times 10^{-4}$$

$$pH = -\log(6.62 \times 10^{-4}) = 3.18$$

$[OH^-] = \frac{0.05 \text{ mmol}}{1 \text{ mL}}$
pOH = 12.69897
pH = 9.3299
$K_w = [OH]^-[H_3O^+]$
1.00×10^{-14}
$[H_3O^+] = 4.678 \times 10^{-10}$
$[OH^-] = \frac{1.00 \times 10^{-14}}{4.678 \times 10^{-10}}$
$[OH^-] = 2.1374 \times 10^{-5}$
= 4.6701 ?

Curves for the titration of 0.0500 M and 0.00500 M NaOH with 0.1000 M and 0.0100 M HCl are shown in Figure 10-5. Indicator selection is based upon the same considerations described for the titration of a strong acid with a strong base.

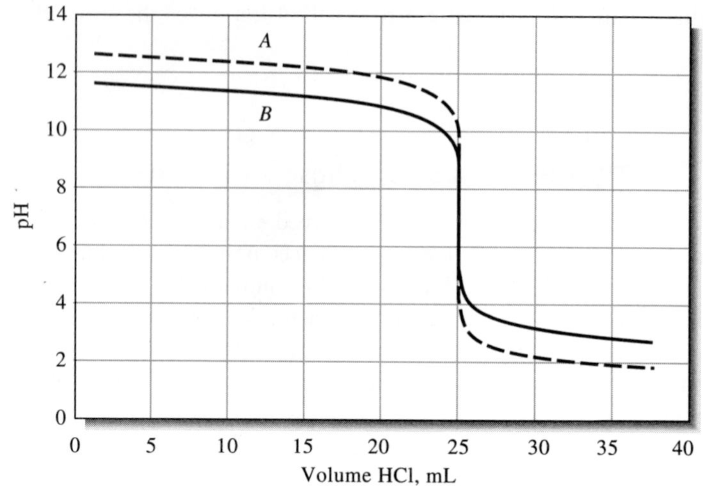

Figure 10-5
Titration curves for NaOH with HCl. *A*: 50.00 mL of 0.0500 M NaOH with 0.1000 M HCl. *B*: 50.00 mL of 0.00500 M NaOH with 0.01000 M NaOH.

10D BUFFER SOLUTIONS

A **buffer** is a mixture of a weak acid and its conjugate base or a weak base and its conjugate acid that resists changes in pH of a solution.

Buffers are used in all types of chemistry whenever it is desirable to maintain the pH of a solution at a constant and predetermined level.

Buffered aspirin contains buffers to help prevent stomach irritation from the acidity of the carboxylic acid group in aspirin.

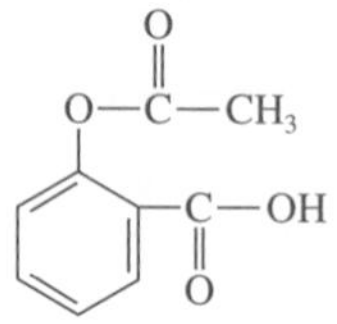

aspirin

Whenever a weak acid is titrated with a strong base or a weak base with a strong acid, a *buffer solution* consisting of a conjugate acid/base pair is formed. Thus, before we can show how titration curves for weak acids and weak bases are derived we must investigate in some detail the properties and behavior of buffer solutions. By definition, a *buffer solution* is a solution of a conjugate acid/base pair that resists changes in pH. Chemists employ buffers whenever they need to maintain the pH of a solution at a constant and predetermined level. You will find many references to the use of buffers throughout this text.

10D-1 Calculation of the pH of Buffer Solutions

Weak Acid/Conjugate Base Buffers

A solution containing a weak acid, HA, and its conjugate base, A^-, may be acidic, neutral, or basic, depending upon the position of two competitive equilibria:

$$HA + H_2O \rightleftharpoons H_3O^+ + A^- \qquad K_a = \frac{[H_3O^+][A^-]}{[HA]} \tag{10-4}$$

$$A^- + H_2O \rightleftharpoons OH^- + HA \qquad K_b = \frac{K_w}{K_a} = \frac{[OH^-][HA]}{[A^-]} \tag{10-5}$$

If the first equilibrium lies farther to the right than the second, the solution is acidic. If the second equilibrium is more favorable, the solution is basic. These two equilibrium-constant expressions show that the relative concentrations of the hydronium and hydroxide ions depend not only upon the magnitudes of K_a and K_b but also upon the ratio of the concentrations of the acid and its conjugate base.

In order to compute the pH of a solution containing both an acid, HA, and its salt, NaA, we need to express the equilibrium concentrations of HA and NaA in terms of their analytical concentrations, c_{HA} and c_{NaA}. An examination of the two equilibria reveals that the first reaction decreases the concentration of HA by an amount equal to $[H_3O^+]$, whereas the second increases the HA concentration by an amount equal to $[OH^-]$. Thus, the species concentration of HA is related to its analytical concentration by the mass-balance equation

$$[HA] = c_{HA} - [H_3O^+] + [OH^-] \qquad (10\text{-}6)$$

Similarly, the first equilibrium will increase the concentration of A^- by an amount equal to $[H_3O^+]$, and the second will decrease this concentration by the amount $[OH^-]$. Thus, the equilibrium concentration is given by a second mass-balance equation

$$[A^-] = c_{NaA} + [H_3O^+] - [OH^-] \qquad (10\text{-}7)$$

FEATURE 10-4

Application of the Systematic Method to Buffer Calculations

Equations 10-6 and 10-7 can also be derived from mass- and charge-balance expressions. Thus, mass-balance considerations require that

$$c_{HA} + c_{NaA} = [HA] + [A^-]$$

Electrical neutrality considerations require that

$$[Na^+] + [H_3O^+] = [A^-] + [OH^-]$$

but

$$[Na^+] = c_{NaA}$$

Therefore, the charge-balance equation is

$$c_{NaA} + [H_3O^+] = [A^-] + [OH^-]$$

which rearranges to Equation 10-7:

$$[A^-] = c_{NaA} + [H_3O^+] - [OH^-]$$

Subtraction of the first equation from the fourth gives, upon rearrangement,

$$[HA] = c_{HA} - [H_3O^+] + [OH^-]$$

which is identical to Equation 10-6.

Because of the inverse relationship between $[H_3O^+]$ and $[OH^-]$, it is *always* possible to eliminate one or the other from Equations 10-6 and 10-7. Moreover, the *difference* in concentration between these two species is often so small relative to the molar concentrations of acid and conjugate base that Equations 10-6 and 10-7 simplify to

$$[HA] = c_{HA} \qquad (10\text{-}8)$$

$$[A^-] = c_{NaA} \qquad (10\text{-}9)$$

Substitution of Equations 10-8 and 10-9 into the dissociation-constant expression and rearrangement yields

$$[H_3O^+] = K_a \frac{c_{HA}}{c_{NaA}} \qquad (10\text{-}10)$$

The assumption leading to Equations 10-8 and 10-9 sometimes breaks down with acids or bases that have dissociation constants greater than about 10^{-3} or when the molar concentration of either the acid or its conjugate base (or both) is very small. In these circumstances, either $[OH^-]$ or $[H_3O^+]$ must be retained in Equations 10-6 and 10-7, depending upon whether the solution is acidic or basic. In any case, Equations 10-8 and 10-9 should always be used initially. The provisional values for $[H_3O^+]$ and $[OH^-]$ can then be employed to test the assumptions.

Within the limits imposed by the assumptions made in its derivation, Equation 10-10 says that the hydronium ion concentration of a solution containing a weak acid and its conjugate base is dependent only upon the *ratio* of the molar concentrations of these two solutes. Furthermore, this ratio is *independent of dilution* because the concentration of each component changes proportionately when the volume changes.

FEATURE 10-5

The Henderson-Hasselbalch Equation

The Henderson-Hasselbalch equation is an alternative form of Equation 10-10 that is frequently encountered in biological and biochemical literature. It is obtained by expressing each term in the equation in the form of its negative logarithm and inverting the concentration ratio to keep all signs positive:

$$-\log [H_3O^+] = -\log K_a - \log \frac{c_{HA}}{c_{NaA}}$$

Therefore,

$$pH = pK_a + \log \frac{c_{NaA}}{c_{HA}} \qquad (10\text{-}11)$$

EXAMPLE 10-3

What is the pH of a solution that is 0.400 M in formic acid and 1.00 M in sodium formate?

The equilibrium governing the hydronium ion concentration in this solution is

$$H_2O + HCOOH \rightleftharpoons H_3O^+ + HCOO^-$$

for which (Appendix 3)

$$K_a = \frac{[H_3O^+][HCOO^-]}{[HCOOH]} = 1.80 \times 10^{-4}$$

$$[HCOO^-] \approx c_{HCOO^-} = 1.00$$

$$[HCOOH] \approx c_{HCOOH} = 0.400$$

Substitution into Equation 10-10 gives with rearrangement

$$[H_3O^+] = 1.80 \times 10^{-4} \times \frac{0.400}{1.00} = 7.20 \times 10^{-5}$$

Note that the assumption that $[H_3O^+] \ll c_{HCOOH}$ and $[H_3O^+] \ll c_{HCOO^-}$ is valid. Thus,

$$pH = -\log(7.20 \times 10^{-5}) = 4.14$$

Weak Base/Conjugate Acid Buffers

As shown in Example 10-4, Equations 10-6 and 10-7 also apply to buffer systems consisting of a weak base and its conjugate acid. Furthermore, in most cases it is possible to simplify these equations so that Equation 10-10 can be used.

EXAMPLE 10-4

Calculate the pH of a solution that is 0.200 M in NH_3 and 0.300 M in NH_4Cl. In Appendix 3, we find that the acid dissociation constant K_a for NH_4^+ is 5.70×10^{-10}.

The equilibria we must consider are

$$NH_4^+ + H_2O \rightleftharpoons NH_3 + H_3O^+ \qquad K_a = 5.70 \times 10^{-10}$$

$$NH_3 + H_2O \rightleftharpoons NH_4^+ + OH^- \qquad K_b = \frac{K_w}{K_a} = \frac{1.00 \times 10^{-14}}{5.70 \times 10^{-10}} = 1.75 \times 10^{-5}$$

Using the arguments that led to Equations 10-6 and 10-7, we obtain

$$[NH_4^+] = c_{NH_4Cl} + [OH^-] - [H_3O^+] \approx c_{NH_4Cl} + [OH^-]$$

$$[NH_3] = c_{NH_3} + [H_3O^+] - [OH^-] \approx c_{NH_3} - [OH^-]$$

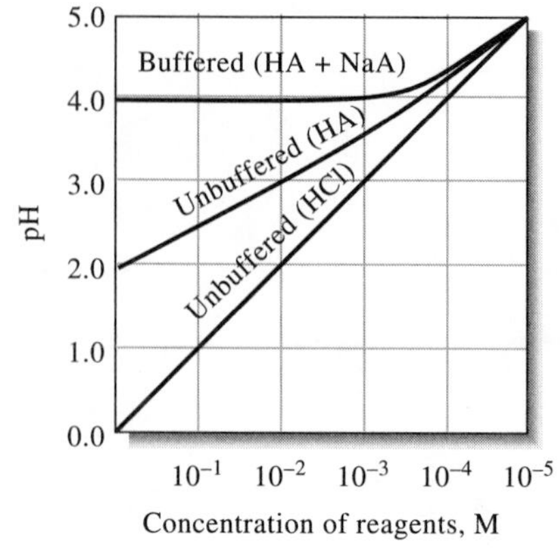

Figure 10-6
The effect of dilution on the pH of buffered and unbuffered solutions. The dissociation constant for HA is 1.00×10^{-4}. Initial solute concentrations are 1.00 M.

Here, we have assumed on the basis of the relative sizes of the two equilibrium constants that $[H_3O^+]$ is negligibly small and can be neglected.

Let us also assume that $[OH^-]$ is much smaller than c_{NH_4Cl} and c_{NH_3} so that

$$[NH_4^+] = c_{NH_4Cl} = 0.300$$

$$[NH_3] = c_{NH_3} = 0.200$$

Substituting into the acid-dissociation constant for NH_4^+, we obtain a relationship similar to Equation 10-10. That is,

$$[H_3O^+] = \frac{K_a \times [NH_4^+]}{[NH_3]} = \frac{5.70 \times 10^{-10} \times c_{NH_4^+}}{c_{NH_3}}$$

$$= \frac{5.70 \times 10^{-10} \times 0.300}{0.200} = 8.55 \times 10^{-10}$$

To check the validity of our approximations, we calculate $[OH^-]$. Thus,

$$[OH^-] = 1.00 \times 10^{-14}/(8.55 \times 10^{-10}) = 1.17 \times 10^{-5}$$

which is certainly much smaller than either $c_{NH_4^+}$ or c_{NH_3}. Thus, we may write

$$pH = -\log(8.55 \times 10^{-10}) = 9.07$$

10D-2 Properties of Buffer Solutions

In this section we illustrate the resistance of buffers to changes of pH brought about by dilution or addition of strong acids or bases.

The Effect of Dilution

The pH of a buffer solution remains essentially independent of dilution until the concentrations of the species it contains are decreased to the point where the approximations used to develop Equations 10-8 and 10-9 become invalid. Figure 10-6 contrasts the behavior of buffered and unbuffered solutions with dilution. For each, the initial solute concentration is 1.00 M. The resistance of the buffered solution to changes in pH during dilution is clear.

The Effect of Added Acids and Bases

Example 10-5 illustrates a second property of buffer solutions, their resistance to pH change after addition of small amounts of strong acids or bases.

EXAMPLE 10-5

Calculate the pH change that takes place when a 100-mL portion of (a) 0.0500 M NaOH and (b) 0.0500 M HCl is added to 400 mL of the buffer solution that was described in Example 10-4.

(a) Addition of NaOH converts part of the NH_4^+ in the buffer to NH_3:

$$NH_4^+ + OH^- \longrightarrow NH_3 + H_2O$$

The analytical concentrations of NH_3 and NH_4Cl then become

$$c_{NH_3} = \frac{400 \times 0.200 + 100 \times 0.0500}{500} = \frac{85.0}{500} = 0.170 \text{ M}$$

$$c_{NH_4Cl} = \frac{400 \times 0.300 - 100 \times 0.0500}{500} = \frac{115}{500} = 0.230 \text{ M}$$

When substituted into the acid-dissociation constant expression for NH_4^+, these values yield

$$[H_3O^+] = 5.70 \times 10^{-10} \times \frac{0.230}{0.170} = 7.71 \times 10^{-10}$$

$$pH = -\log (7.71 \times 10^{-10}) = 9.11$$

and the change in pH is

$$\Delta pH = 9.11 - 9.07 = 0.04$$

Buffers do not maintain pH at an absolutely constant value, but changes in pH are relatively small when small amounts of acid or base are added.

(b) Addition of HCl converts part of the NH_3 to NH_4^+; thus,

$$NH_3 + H_3O^+ \longrightarrow NH_4^+ + H_2O$$

$$c_{NH_3} = \frac{400 \times 0.200 - 100 \times 0.0500}{500} = \frac{75}{500} = 0.150 \text{ M}$$

$$c_{NH_4^+} = \frac{400 \times 0.300 + 100 \times 0.0500}{500} = \frac{125}{500} = 0.250 \text{ M}$$

$$[H_3O^+] = 5.70 \times 10^{-10} \times \frac{0.250}{0.150} = 9.50 \times 10^{-10}$$

$$pH = -\log (9.50 \times 10^{-10}) = 9.02$$

$$\Delta pH = 9.02 - 9.07 = -0.05$$

It is of interest to contrast the behavior of an unbuffered solution with a pH of 9.07 to that of the buffer in Example 10-5. It is readily shown that adding the same quantity of base to the unbuffered solution would increase the pH to 12.00—a pH change of 2.93 units. Adding the acid would decrease the pH by slightly more than 7 units.

Buffer Capacity

Figure 10-6 and Example 10-5 demonstrate that a solution containing a conjugate acid/base pair possesses remarkable resistance to changes in pH. The ability of a buffer to prevent a significant change in pH is directly related to the total concentration of the buffering species as well as to their concentration ratio. For example, the pH of a 400-mL portion of a buffer formed by diluting the solution

See *Mathcad Applications for Analytical Chemistry,* **p. 68.**

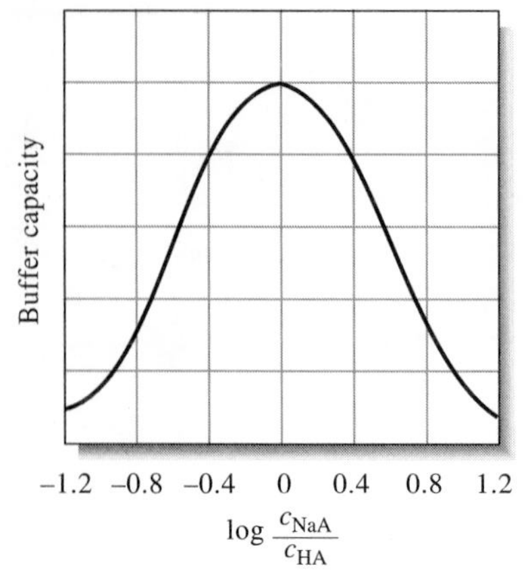

Figure 10-7
Buffer capacity as a function of the ratio c_{NaA}/c_{HA}.

The **buffer capacity** of a buffer is the number of moles of strong acid or strong base that 1 L of the buffer can absorb without changing pH by more than 1.

Table 19-2 gives the components of buffers that are recommended by the National Institute of Standards and Technology.

described in Example 10-4 by 10 would change by about 0.4 to 0.5 unit when treated with 100 mL of 0.0500 M sodium hydroxide or 0.0500 M hydrochloric acid. We have shown in Example 10-5 that the change is only about 0.04 to 0.05 unit for the more concentrated buffer.

The *buffer capacity* of a solution is defined as the number of moles of a strong acid or a strong base that causes 1.00 L of the buffer to undergo a 1.00-unit change in pH. The capacity of a buffer depends not only on the total concentration of its components but also on their concentration ratio. Buffer capacity falls off moderately rapidly as the concentration ratio of acid to conjugate base becomes larger or smaller than unity (see Figure 10-7). For this reason, the pK_a of the acid chosen for a given application should lie within ± 1 unit of the desired pH in order for the buffer to have a reasonable capacity.

Preparation of Buffers

In principle, a buffer solution of any desired pH can be prepared by combining calculated quantities of a suitable conjugate acid/base pair. In practice, however, the pH values of buffers prepared from theoretically generated recipes differ from the predicted values because of uncertainties in the numerical values of many dissociation constants and because of the simplifications used in calculations. Of more importance is the fact that the ionic strength of a buffer is usually so high that good values for the activity coefficients of the ions in the solution cannot be obtained from the Debye-Hückel relationship. Because of these uncertainties, we prepare buffers by making up a solution of approximately the desired pH and then adjust by adding acid or conjugate base until the required pH is indicated by a pH meter.

Alternatively, empirically derived recipes for preparing buffer solutions of known pH are available in chemical handbooks and reference works.[1]

Buffers are of tremendous importance in biological and biochemical studies where a low but constant concentration of hydronium ions (10^{-6} to 10^{-10} M) must be maintained throughout experiments. Several biological supply houses offer a variety of such buffers.

FEATURE 10-6
Acid Rain and the Buffer Capacity of Lakes

Acid rain has been the subject of considerable controversy over the past two decades. Acid rain forms when the gaseous oxides of nitrogen and sulfur dissolve in water droplets in the air. These gases form at high temperatures in power plants, automobiles, and other combustion sources. The combustion products pass into the atmosphere, where they react with water to form nitric acid and sulfuric acid.

$$4NO_2(g) + 2H_2O(l) + O_2(g) \longrightarrow 4HNO_2(aq)$$

$$SO_3(g) + H_2O(l) \longrightarrow H_2SO_4(aq)$$

[1]See, for example, L. Meites, Ed., *Handbook of Analytical Chemistry,* pp. **5**-112 and **11**-3 to **11**-8. New York: McGraw-Hill, 1963.

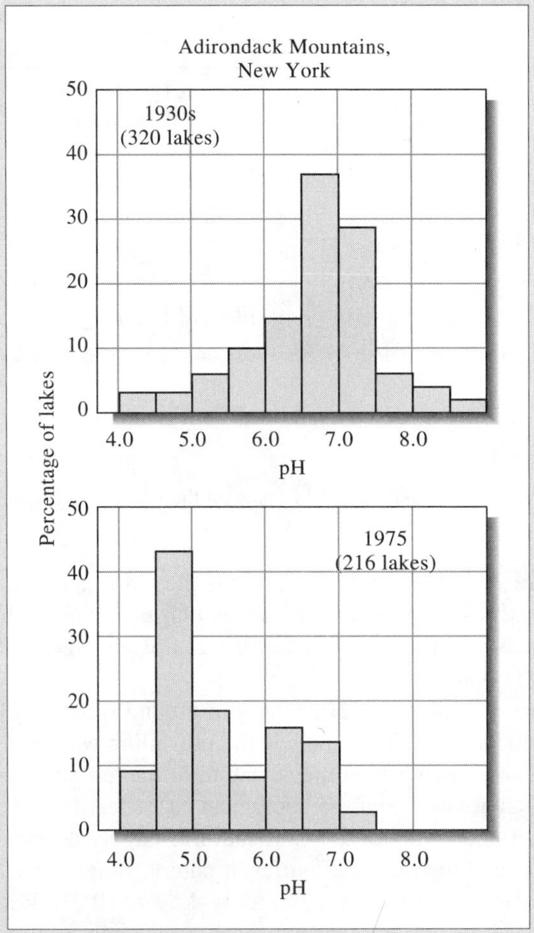

Figure 10-8 Changes in pH of lakes between 1930 and 1975.

Eventually the droplets coalesce with other droplets of acid to form acid rain. The profound effects of acid rain have been highly publicized. Stone buildings and monuments literally dissolve away as acid rain flows over their surfaces. Forests are slowly being killed off in some locations. To illustrate the effects on aquatic life, let us consider the changes in pH that have occurred in the lakes of the Adirondack Mountains area of New York illustrated in the bar graphs of Figure 10-8.

The graphs show the distribution of pH in these lakes, which were studied first in the 1930s and then again in 1975.[2] The shift in pH of the lakes over a 40-year period is dramatic. The average pH of the lakes changed from 6.4 to about 5.1, which represents a 20-fold change in the hydronium ion concentration. Such changes in pH have a profound effect upon aquatic life as shown by a study of the fish population in lakes in the same area.[3] In the

[2] R. F. Wright and E. T. Gjessing, *Ambio,* **1976,** *5,* 219.

[3] C. L. Schofield, *Ambio,* **1976,** *5,* 228.

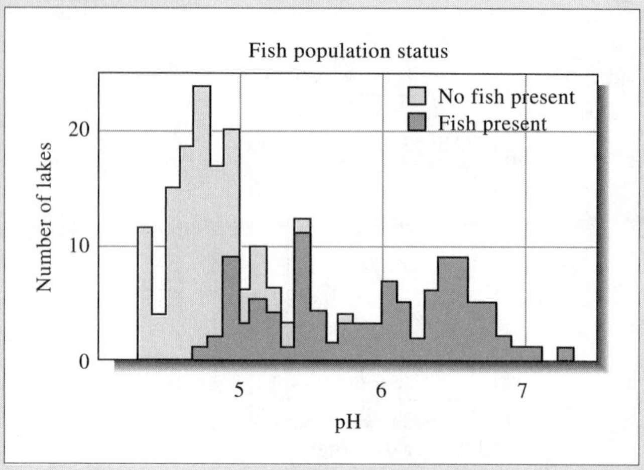

Figure 10-9 Effect of pH of lakes on their fish population.

graph of Figure 10-9, the number of lakes is plotted as a function of pH. The dark bars represent lakes containing fish, and lakes having no fish are light in color. There is a distinct correlation between pH changes in the lakes and diminished fish population.

Many factors contribute to pH changes in ground water and lakes in a given geographical area. These include the prevailing wind patterns and weather, types of soils, water sources, nature of the terrain, characteristics of plant life, human activity, and the geological characteristics of the area. The susceptibility of natural water to acidification is largely determined by its buffer capacity, and the principal buffer of natural water is bicarbonate ion (for carbonic acid, $K_1 = 4.45 \times 10^{-7}$, $K_2 = 4.69 \times 10^{-11}$). Recall that the buffer capacity of a solution is proportional to the concentration of the buffering agent. So, the higher the concentration of dissolved bicarbonate, the greater is the capacity of the water to neutralize acid from acid rain. The most important source of bicarbonate ion in natural water is limestone, or calcium carbonate, which reacts with hydronium ion as shown in the following equation.

$$CaCO_3(s) + H_3O^+(aq) \rightleftharpoons HCO_3^-(aq) + Ca^{2+}(aq)$$

Limestone-rich areas have lakes with relatively high concentrations of dissolved bicarbonate, and thus low susceptibility to acidification. Granite, sandstone, shale, and other rock containing little or no calcium carbonate are associated with lakes having high susceptibility to acidification. The map of the United States shown in Figure 10-10 vividly illustrates the correlation between the absence of limestone-bearing rocks and the acidification of ground waters.[4]

[4]J. Root, et al., cited in *The Effects of Air Pollution and Acid Rain on Fish, Wildlife, and Their Habitats—Introduction.* U.S. Fish and Wildlife Service, Biological Services Program, Eastern Energy and Land Use Team, M. A. Peterson, Ed., p. 63. U.S. Government Publication FWS/OBS-80/40.3.

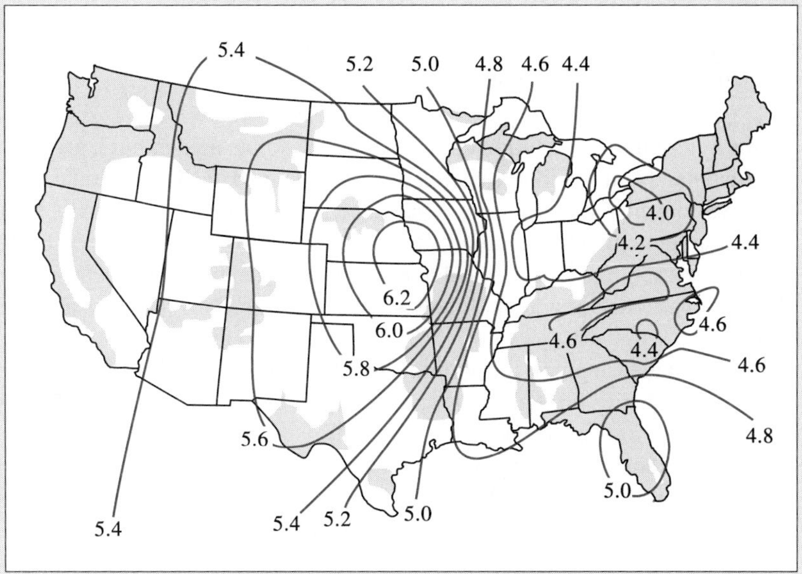

Figure 10-10 Effect of presence of limestone on pH of lakes in the United States. Shaded areas contain little limestone.

Areas containing little limestone are shaded in color; areas rich in limestone are white. Contour lines of equal pH for ground water during the period 1978–1979 are superimposed on the map. The Adirondack Mountains area, located in northeastern New York, contains little limestone and exhibits pH in the range of 4.2 to 4.4. The low buffer capacity of the lakes in this region combined with the low pH of precipitation appears to have caused the decline of fish populations. Similar correlations among acid rain, buffer capacity of lakes, and wildlife decline occur throughout the industrialized world.

Although natural sources such as volcanoes produce sulfur trioxide, and lightning discharges in the atmosphere generate nitrogen dioxide, large quantities of these compounds come from the burning of high-sulfur coal and from automobile emissions. To minimize emissions of these pollutants, some states have enacted legislation imposing strict standards on automobiles sold and operated within their borders. Some states have required the installation of scrubbers to remove oxides of sulfur from the emissions of coal-fired power plants. To minimize the effects of acid rain on lakes, powdered limestone is dumped on the surface to increase the buffer capacity of the water. Solutions to these problems require the expenditure of much time, energy, and money. We must sometimes make difficult economic decisions in order to preserve the quality of our environment and to reverse trends that have operated for many decades. For a comprehensive discussion of the problem of acid rain, refer to the book by Park.[5]

---

[5] C. C. Park, *Acid Rain.* New York: Methuen, 1987.

10E TITRATION CURVES FOR WEAK ACIDS

Four distinctly different types of calculations are needed to derive a titration curve for a weak acid (or a weak base):

1. At the beginning, the solution contains only the solute acid or base, and the pH is calculated from the concentration of that solute and its dissociation constant.
2. After various increments of titrant have been added (in quantities up to, but not including, an equivalent amount), the solution consists of a series of buffers. The pH of each buffer can be calculated from the analytical concentrations of the conjugate base or acid and the residual concentrations of the weak acid or base.
3. At the equivalence point, the solution contains only the conjugate of the weak acid or base being titrated (that is, a salt), and the pH is calculated from the concentration of this product.
4. Beyond the equivalence point, the excess of strong acid or base titrant represses the acidic or basic character of the reaction product to such an extent that the pH is governed largely by the concentration of the excess titrant.

EXAMPLE 10-6

Derive a curve for the titration of 50.00 mL of 0.1000 M acetic acid ($K_a = 1.75 \times 10^{-5}$) with 0.1000 M sodium hydroxide.

Initial pH

Initially, we must calculate the pH of a 0.1000 M solution of HOAc using Equation 7-21 (page 141):

$$[H_3O^+] = \sqrt{K_a\, c_{HOAc}} = \sqrt{1.75 \times 10^{-5} \times 0.1000} = 1.32 \times 10^{-3}$$

$$pH = -\log(1.32 \times 10^{-3}) = 2.88$$

pH after Addition of 10.00 mL of Reagent

A buffer solution consisting of NaOAc and HOAc has now been produced. The analytical concentrations of the two constituents are

$$c_{HOAc} = \frac{50.00\text{ mL} \times 0.1000\text{ M} - 10.00\text{ mL} \times 0.1000\text{ M}}{60.00\text{ mL}} = \frac{4.000}{60.00}\text{ M}$$

$$c_{NaOAc} = \frac{10.00\text{ mL} \times 0.1000\text{ M}}{60.00\text{ mL}} = \frac{1.000}{60.00}\text{ M}$$

We substitute these concentrations into the dissociation-constant expression for acetic acid and obtain

$$\frac{[H_3O^+](1.000/60.00)}{4.000/60.00} = K_a = 1.75 \times 10^{-5}$$

$$[H_3O^+] = 7.00 \times 10^{-5}$$

$$pH = 4.16$$

Calculations similar to this provide points on the curve throughout the buffer region. Data from such calculations are given in column 2 of Table 10-4.

Equivalence-Point pH

At the equivalence point, all of the acetic acid has been converted to sodium acetate. The solution is therefore similar to one formed by dissolving that base in water, and the pH calculation is identical to that shown in Example 7-12 (page 145) for a weak base. In the present example, the NaOAc concentration is 0.0500 M. Thus,

$$OAc^- + H_2O \rightleftharpoons HOAc + OH^-$$

$$[OH^-] = [HOAc]$$

$$[OAc^-] = 0.0500 - [OH^-] \approx 0.0500$$

Substituting in the base dissociation-constant expression for OAc^- gives

$$\frac{[OH^-]^2}{0.0500} = \frac{K_w}{K_a} = \frac{1.00 \times 10^{-14}}{1.75 \times 10^{-5}} = 5.71 \times 10^{-10}$$

$$[OH^-] = \sqrt{0.0500 \times 5.71 \times 10^{-10}} = 5.34 \times 10^{-6}$$

$$pH = 14.00 - (-\log 5.34 \times 10^{-6}) = 8.73$$

Note that the pH at the equivalence point of this titration is greater than 7. The solution is alkaline.

Titration curves for strong and weak acids are identical just slightly beyond the equivalence point. The same is true for strong and weak bases.

pH after Addition of 50.10 mL of Base

After the addition of 50.10 mL of NaOH, both the excess base and the acetate ion are sources of the hydroxide ion. The contribution of the latter is small, however, because the excess of strong base represses the reaction. This fact becomes evident when we consider that the hydroxide ion concentration is only 5.34×10^{-6} M at the equivalence point; once an excess of strong base is added, the contribution from the reaction of the acetate is even smaller. Thus,

$$[OH^-] \approx c_{NaOH} = \frac{50.10 \text{ mL} \times 0.1000 \text{ M} - 50.00 \text{ mL} \times 0.1000 \text{ M}}{100.0 \text{ mL}}$$

$$= 1.00 \times 10^{-4} \text{ M}$$

$$pH = 14.00 - (-\log 1.00 \times 10^{-4}) = 10.00$$

Note that the titration curve for a weak acid with a strong base is identical to that for a strong acid with a strong base in the region slightly beyond the equivalence point.

Note that the analytical concentrations of acid and conjugate base are identical when an acid has been half neutralized (in Example 10-6, after the addition of exactly 25.00 mL of base). Thus, these terms cancel in the equilibrium-constant expression, and the hydronium ion concentration is numerically equal to the dissociation constant. Likewise, in the titration of a weak base, the hydroxide ion concentration is numerically equal to the dissociation constant of the base at the midpoint in the titration curve. In addition, the buffer capacities of each of the solutions are at a maximum at this point.

At the half-neutralization point in the titration of a weak acid $[H_3O^+] = K_a$ or $pH = pK_a$.

At the half-neutralization point in the titration of a weak base $[OH^-] = K_b$ or $pOH = pK_b$.

FEATURE 10-7
Determination of Dissociation Constants for Weak Acids and Bases

The dissociation constants of weak acids or weak bases are often determined by monitoring the pH of the solution while the acid or base is being titrated. A pH meter and a glass electrode are used for the measurements. For an acid, the measured pH when the acid is exactly half neutralized is numerically equal to pK_a. For a weak base, the pH at half neutralization must be converted to pOH, which is then equal to pK_b.

10E-1 The Effect of Concentration

CHALLENGE: Compute the data in the third column of Table 10-4.

The second and third columns of Table 10-4 contain pH data for the titration of 0.1000 M and of 0.001000 M acetic acid with sodium hydroxide solutions with the same two concentrations. In deriving the data for the more dilute acid, none of the approximations shown in Example 10-5 was valid, and solution of a quadratic equation was necessary throughout.

Figure 10-11 is a plot of the data in Table 10-4. Note that the initial pH values are higher and the equivalence-point pH is lower for the more dilute solution. At intermediate titrant volumes, however, the pH values differ only slightly because of the buffering action of the acetic acid/sodium acetate system that exists in this region. Figure 10-11 is graphical confirmation of the fact that the pH of buffers is largely independent of dilution.

10E-2 The Effect of Reaction Completeness

Titration curves for 0.1000 M solutions of acids with different dissociation constants are shown in Figure 10-12. Note that the pH change in the equivalence-point region becomes smaller as the acid becomes weaker—that is, as the reac-

TABLE 10-4 Changes in pH During the Titration of a Weak Acid with a Strong Base

	pH	
Volume of NaOH, mL	**50.00 mL of 0.1000 M HOAc Titrated with 0.1000 M NaOH**	**50.00 mL of 0.001000 M HOAc Titrated with 0.001000 M NaOH**
0.00	2.88	3.91
10.00	4.16	4.30
25.00	4.76	4.80
40.00	5.36	5.38
49.00	6.45	6.46
49.90	7.46	7.47
50.00	8.73	7.73
50.10	10.00	8.09
51.00	11.00	9.00
60.00	11.96	9.96
75.00	12.30	10.30

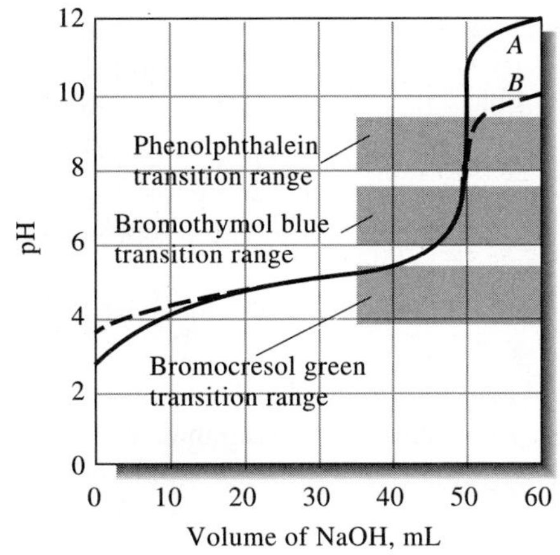

Figure 10-11
Curve for the titration of acetic acid with sodium hydroxide. *A*: 0.1000 M acid with 0.1000 M base. *B*: 0.001000 M acid with 0.001000 M base.

tion between the acid and the base becomes less complete. The effects of the reactant concentrations and the completeness of a reaction illustrated by Figures 10-11 and 10-12 are analogous to the effects on the strong acid titrations shown in Figure 10-4.

10E-3 Indicator Choice; The Feasibility of Titration

Figures 10-11 and 10-12 show clearly that the choice of indicator for the titration of a weak acid is more limited than that for a strong acid. For example, from Figure 10-11, it is obvious that bromocresol green is totally unsuited for titration of 0.1000 M acetic acid. Bromothymol blue is also unsatisfactory because its full color change occurs over a range from about 47 mL to 50 mL of 0.1000 M base. On the other hand, an indicator exhibiting a color change in the basic region, such as phenolphthalein, should provide a sharp end point with a minimal titration error.

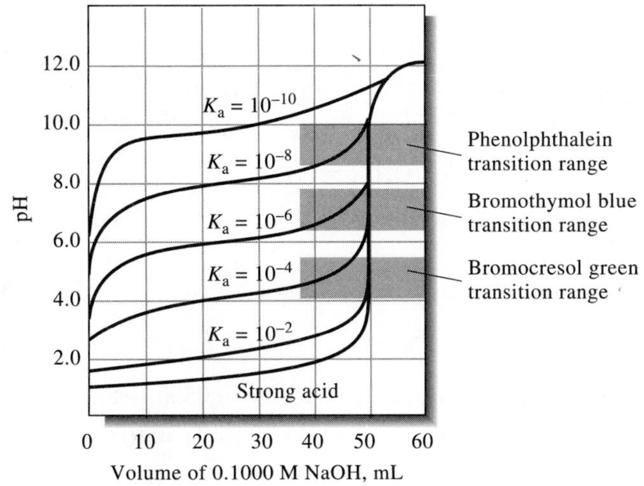

Figure 10-12
The effect of acid strength on titration curves. Each curve represents the titration of 50.0 mL of 0.1000 M acid with 0.1000 M NaOH.

The end-point pH change associated with the titration of 0.001000 M acetic acid (curve *B*, Figure 10-11) is so small that a significant titration error is likely to be introduced regardless of indicator. However, use of an indicator with a transition range between that of phenolphthalein and that of bromothymol blue in conjunction with a suitable color comparison standard makes it possible to establish the end point in this titration with a reproducibility of a few percent relative.

Figure 10-12 illustrates that similar problems exist as the strength of the acid being titrated decreases. Precision on the order of ±2 ppt can be achieved in the titration of a 0.1000 M acid solution with a dissociation constant of 10^{-8} provided a suitable color comparison standard is available. With more concentrated solutions, somewhat weaker acids can be titrated with reasonable precision.

10F TITRATION CURVES FOR WEAK BASES

The derivation of a curve for the titration of a weak base is analogous to that of a weak acid.

EXAMPLE 10-7

A 50.00-mL aliquot of 0.0500 M NaCN is titrated with 0.1000 M HCl. The reaction is

$$CN^- + H_3O^+ \rightleftharpoons HCN + H_2O$$

Calculate the pH after the addition of (a) 0.00, (b) 10.00, (c) 25.00, and (d) 26.00 mL of acid.

(a) **0.00 mL of Reagent**
The pH of a solution of NaCN can be derived by the method shown in Example 7-12, page 145:

$$CN^- + H_2O \rightleftharpoons HCN + OH^-$$

$$\frac{[OH^-][HCN]}{[CN^-]} = K_b = \frac{K_w}{K_a} = \frac{1.00 \times 10^{-14}}{6.2 \times 10^{-10}} = 1.61 \times 10^{-5}$$

$$[OH^-] = [HCN]$$

$$[CN^-] = c_{NaCN} - [OH^-] \approx c_{NaCN} = 0.0500$$

Substitution into the dissociation-constant expression gives, after rearrangement,

$$[OH^-] = \sqrt{K_b\, c_{NaCN}} = \sqrt{1.61 \times 10^{-5} \times 0.0500} = 8.97 \times 10^{-4}$$

$$pH = 14.00 - (-\log 8.97 \times 10^{-4}) = 10.95$$

(b) **10.00 mL of Reagent**
Addition of acid produces a buffer with a composition given by

$$c_{\text{NaCN}} = \frac{50.00 \times 0.0500 - 10.00 \times 0.1000}{60.00} = \frac{1.500}{60.00}\text{ M}$$

$$c_{\text{HCN}} = \frac{10.00 \times 0.1000}{60.00} = \frac{1.000}{60.00}\text{ M}$$

These values are then substituted into the expression for the acid dissociation constant of HCN to give $[H_3O^+]$ directly:

$$[H_3O^+] = \frac{6.2 \times 10^{-10} \times (1.000/60.00)}{1.500/60.00} = 4.13 \times 10^{-10}$$

$$\text{pH} = -\log(4.13 \times 10^{-10}) = 9.38$$

CHALLENGE: Show that the pH of the buffer can be calculated with K_a for HCN, as was done here, or equally well with K_b. We used K_a because it gives $[H_3O^+]$ directly; K_b gives $[OH^-]$.

(c) **25.00 mL of Reagent**
This volume corresponds to the equivalence point, where the principal solute species is the weak acid HCN. Thus,

$$c_{\text{HCN}} = \frac{25.00 \times 0.1000}{75.00} = 0.03333\text{ M}$$

Applying Equation 7-21 gives

$$[H_3O^+] = \sqrt{K_a\, c_{\text{HCN}}} = \sqrt{6.2 \times 10^{-10} \times 0.03333} = 4.55 \times 10^{-6}$$

$$\text{pH} = -\log(4.55 \times 10^{-6}) = 5.34$$

Since the principal solute species at the equivalence point is HCN, the pH is acidic.

(d) **26.00 mL of Reagent**
The excess of strong acid now present represses the dissociation of the HCN to the point where its contribution to the pH is negligible. Thus,

$$[H_3O^+] = c_{\text{HCl}} = \frac{26.00 \times 0.1000 - 50.00 \times 0.0500}{76.00} = 1.32 \times 10^{-3}$$

$$\text{pH} = -\log(1.32 \times 10^{-3}) = 2.88$$

Figure 10-13 shows theoretical curves for a series of weak bases of different strengths. Clearly, indicators with *acidic* transition ranges must be employed for weak bases.

When titrating a weak base, use an indicator with an acidic transition range. When titrating a weak acid, use an indicator with a basic transition range.

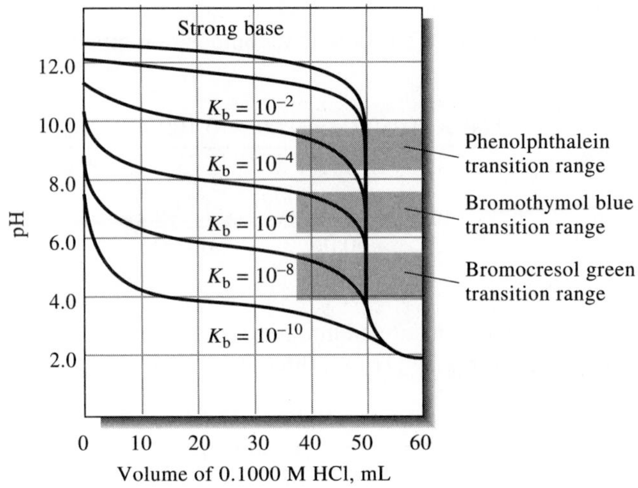

Figure 10-13

The effect of base strength on titration curves. Each curve represents the titration of 50.0 mL of 0.1000 M base with 0.1000 M HCl.

10G THE COMPOSITION OF BUFFER SOLUTIONS AS A FUNCTION OF pH

The changes in composition that occur while a solution of a weak acid or a weak base is being titrated are sometimes of interest and can be visualized by plotting the *relative* concentration of the weak acid as well as the relative concentration of the conjugate base as a function of the pH of the solution. These relative concentrations are called *alpha values.* For example, if we let c_T be the sum of the analytical concentrations of acetic acid and sodium acetate at any point in the titration curve derived in Example 10-6, we may write

$$c_T = c_{HOAc} + c_{NaOAc} \tag{10-12}$$

We then define α_0 as

$$\alpha_0 = \frac{[HOAc]}{c_T} \tag{10-13}$$

and α_1 as

$$\alpha_1 = \frac{[OAc^-]}{c_T} \tag{10-14}$$

Alpha values are unitless ratios whose sum must equal unity. That is,

$$\alpha_0 + \alpha_1 = 1$$

Alpha values are determined by $[H_3O^+]$ and K_a alone. To obtain expressions for α_0, we rearrange the dissociation-constant expression to

$$[OAc^-] = \frac{K_a[HOAc]}{[H_3O^+]}$$

From mass-balance consideration, we write

$$c_T = [HOAc] + [OAc^-] \tag{10-15}$$

Substituting the previous equation into Equation 10-15 gives

$$c_T = [HOAc] + \frac{K_a[HOAc]}{[H_3O^+]} = [HOAc]\left(\frac{[H_2O^+] + K_a}{[H_3O^+]}\right)$$

Upon rearrangement we obtain

$$\frac{[HOAc]}{c_T} = \frac{[H_3O^+]}{[H_3O^+] + K_a}$$

But by definition $[HOAc]/c_T = \alpha_0$ (Equation 10-13), or

$$\alpha_0 = \frac{[HOAc]}{c_T} = \frac{[H_3O^+]}{[H_3O^+] + K_a} \tag{10-16}$$

To obtain an expression for α_1, we rearrange the dissociation-constant expression to

$$[HOAc] = \frac{[H_3O^+][OAc^-]}{K_a}$$

and substitute into Equation 10-15

$$c_T = \frac{[H_3O^+][OAc^-]}{K_a} + [OAc^-] = [OAc^-]\left(\frac{[H_3O^+] + K_a}{K_a}\right)$$

Rearranging this gives α_1 as defined by Equation 10-14

$$\alpha_1 = \frac{[OAc^-]}{c_T} = \frac{K_a}{[H_3O^+] + K_a} \tag{10-17}$$

See *Mathcad Applications for Analytical Chemistry,* **p. 64.**

Note that the denominator is the same in Equations 10-16 and 10-17.

The solid straight lines labeled α_0 and α_1 in Figure 10-14 were derived with Equations 10-16 and 10-17 using values for $[H_3O^+]$ shown in column 2 of Table 10-4. The actual titration curve is shown as the curved line in Figure 10-14. Note that at the outset of the titration α_0 is nearly 1 (0.987), meaning that 98.7% of the acetate containing species is present as HOAc and only 1.3% is present as OAc^-. At the equivalence point, α_0 has decreased to 1.1×10^{-4} and α_1 is approaching 1. Thus, only about 0.011% of the acetate containing species is HOAc. Note that when the acid is half neutralized (25.00 mL), α_0 and α_1 are each 0.5.

Alternatively,

$$\begin{aligned}\alpha_1 &= 1 - \alpha_0 \\ &= 1 - \frac{[H_3O^+]}{[H_3O^+] + K_a} \\ &= \frac{[H_3O^+] + K_a - [H_3O^+]}{[H_3O^+] + K_a} \\ &= \frac{K_a}{[H_3O^+] + K_a}\end{aligned}$$

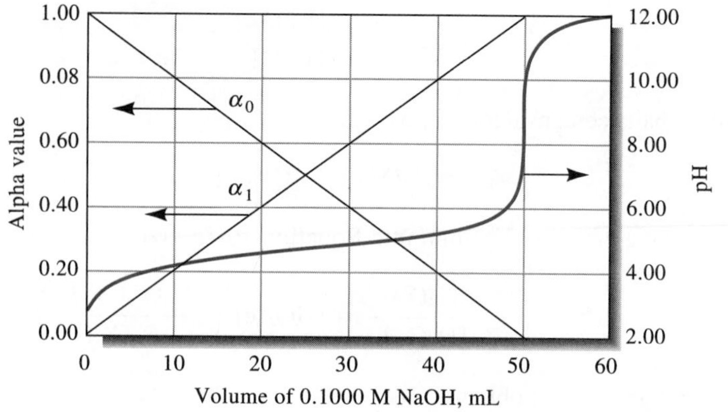

Figure 10-14
The straight lines show the change in relative amounts of HOAc (α_0) and OAc^- (α_1) during the titration of 50.00 mL of 0.1000 M acetic acid. The curved line is the titration curve for the system.

10H COMMON TYPES OF ACID/BASE INDICATORS

Numerous organic compounds serve as indicators for neutralization titrations.

10H-1 Common Indicator Structures

The majority of acid/base indicators possess structural properties that permit classification into perhaps half a dozen categories.[6] Three of these classes are described in the following paragraphs.

Phthalein Indicators

Most phthalein indicators are colorless in moderately acidic solutions and exhibit a variety of colors in alkaline media. These colors tend to fade slowly in strongly alkaline solutions, which is an inconvenience in some applications. As a group, the phthaleins are sparingly soluble in water but readily dissolve in ethanol to give dilute solutions of the indicator.

The best-known phthalein indicator is *phenolphthalein,* whose structures can be represented as

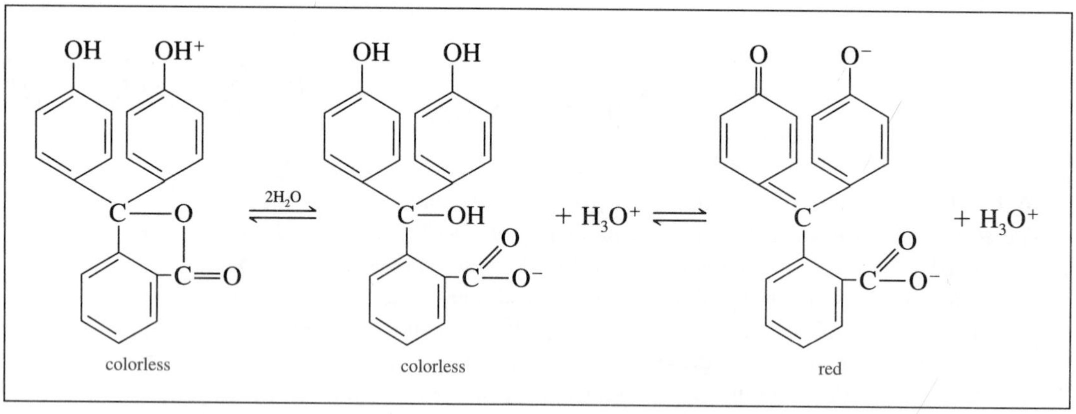

[6]See E. Banyai, in *Indicators,* E. Bishop, Ed., Chapter 3. New York: Pergamon, 1972.

Note that a quinoid ring, which imparts color to most organic compounds, is formed in the second reaction. The pH at which the pink color of this quinoid structure first becomes detectable depends on the concentration of the indicator and on the visual acuity of the observer. For most people, however, the pink appears in the pH range of 8.0 to 8.2.

The other phthalein indicators have various functional groups substituted on the phenolic rings. For example, *thymolphthalein* contains two alkyl groups on each ring. The structural alterations associated with the color change of this indicator are similar to those of phenolphthalein.

Sulfonphthalein Indicators

Many of the sulfonphthaleins exhibit two useful color-change ranges—one in somewhat acidic solutions and the other in neutral or moderately basic media. In contrast to the phthaleins, the basic color shows good stability toward strong alkali.

The sodium salts of the sulfonphthaleins are ordinarily used for the preparation of indicator solutions, owing to the appreciable acidity of the parent molecule. Solutions can be prepared directly from the sodium salt or indirectly by dissolving the sulfonphthalein in its acidic form in an appropriate volume of dilute aqueous sodium hydroxide.

The simplest sulfonphthalein indicator is *phenolsulfonphthalein,* known also as *phenol red.* The principal equilibria for a solution of the sodium salt of this compound are

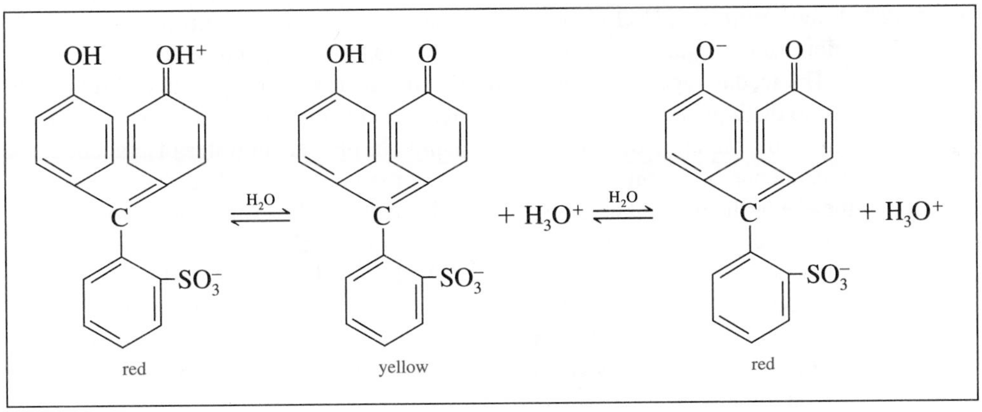

Only the second color change, which occurs in the pH range between 6.4 and 8.0, is useful.

Substitution of halogens or alkyl groups for the hydrogens in the phenolic rings of the parent compound yields sulfonphthaleins that differ in color and pH range.

Azo Indicators

Most azo indicators exhibit a color change from red to yellow with increasing basicity; their transition ranges are generally on the acidic side of neutrality. The most commonly encountered examples are *methyl orange* and *methyl red.* The behavior of the former is described by the equations

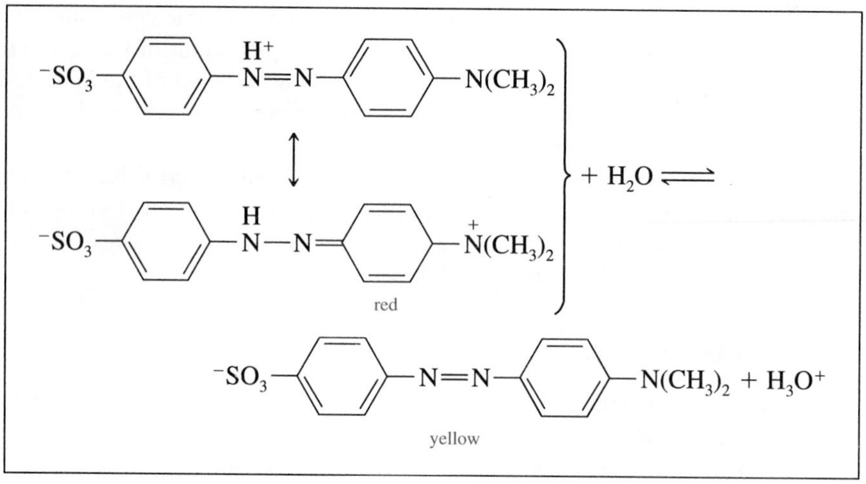

Methyl red contains a carboxylic acid group in place of the sulfonic acid group. Variations in the substituents on the amino nitrogen and in the rings give rise to a series of indicators with slightly different properties.

10H-2 Titration Errors with Acid/Base Indicators

Two types of titration errors are encountered in neutralization titrations. The first is a systematic error that occurs when the pH at which the indicator changes color differs from the pH at chemical equivalence. This type of error can usually be minimized by judicious indicator selection or by a blank correction.

The second type is a random error that originates from the limited ability of the eye to distinguish reproducibly the intermediate color of the indicator. The magnitude of this error depends on the change in pH per milliliter of reagent at the equivalence point, on the concentration of the indicator, and on the sensitivity of the eye to the two indicator colors. On the average, the visual uncertainty with an acid/base indicator is in the range of ± 0.5 to ± 1 pH unit. This uncertainty can often be decreased to as little as ± 0.1 pH unit by matching the color of the solution being titrated with that of a reference standard containing a similar amount of indicator at the appropriate pH. These uncertainties are of course approximations that vary considerably from indicator to indicator as well as from person to person.

10H-3 Variables That Influence the Behavior of Indicators

The pH interval over which a given indicator exhibits a color change is influenced by temperature, by the ionic strength of the medium, and by the presence of organic solvents and colloidal particles. Some of these effects, particularly the last two, can cause the transition range to shift by one or more pH units.[7]

---

[7]For a discussion of these effects, see H. A. Laitinen and W. E. Harris, *Chemical Analysis,* 2nd ed., pp. 48–51. New York: McGraw-Hill, 1975.

10I QUESTIONS AND PROBLEMS

In this chapter, round all calculated values for pH and pOH to two figures to the right of the decimal point unless otherwise instructed.

***10-1.** Consider curves for the titration of 0.10 M NaOH and 0.010 M NH_3 with 0.10 M HCl.
(a) Briefly account for the differences between curves for the two titrations.
(b) In what respect will the two curves be indistinguishable?

10-2. What factors affect end-point sharpness in an acid/base titration?

***10-3.** Why does the typical acid/base indicator exhibit its color change over a range of about 2 pH units?

10-4. What variables can cause the pH range of an indicator to shift?

***10-5.** Why are the standard reagents used in neutralization titrations generally strong acids and bases rather than weak acids and bases?

10-6. What is a buffer solution and what are its properties?

***10-7.** Define buffer capacity.

10-8. Which has the greater buffer capacity, (a) a mixture containing 0.100 mol of NH_3 and 0.200 mol of NH_4Cl or (b) a mixture containing 0.0500 mol of NH_3 and 0.100 mol of NH_4Cl?

***10-9.** Consider solutions prepared by
(a) dissolving 8.00 mmol of NaOAc in 200 mL of 0.100 M HOAc.
(b) adding 100 mL of 0.0500 M NaOH to 100 mL of 0.175 M HOAc.
(c) adding 40.0 mL of 0.1200 M HCl to 160.0 mL of 0.0420 M NaOAc.
In what respects do these solutions resemble one another? How do they differ?

10-10. Consult Appendix 3 and pick out a suitable acid/base pair to prepare a buffer with a pH of
*(a) 3.5. (b) 7.6. *(c) 9.3. (d) 5.1.

10-11. Which solute would provide the sharper end point in a titration with 0.10 M HCl,
***(a)** 0.10 M NaOCl or 0.10 M hydroxylamine?
(b) 0.10 M NH_3 or 0.10 M sodium phenolate?
***(c)** 0.10 M methylamine or 0.10 M hydroxylamine?
(d) 0.10 M hydrazine or 0.10 M NaCN?

10-12. Which solute would provide the sharper end point in a titration with 0.10 M NaOH,
***(a)** 0.10 M nitrous acid or 0.10 M iodic acid?
(b) 0.10 M anilinium hydrochloride ($C_6H_5NH_3Cl$) or 0.10 M benzoic acid?
***(c)** 0.10 M hypochlorous acid or 0.10 M pyruvic acid?
(d) 0.10 M salicylic acid or 0.10 M acetic acid?

10-13. Before glass electrodes and pH meters became so widely used, pH was often determined by measuring the concentration of the acid and base forms of the indicator colorimetrically. If bromothymol blue is introduced into a solution and the concentration ratio of acid to base form is found to be 1.43, what is the pH of the solution?

***10-14.** The procedure described in Question 10-13 was used to determine pH with methyl orange as the indicator. The concentration ratio of the acid to base form of the indicator was 1.64. Calculate the pH of the solution.

10-15. Values of K_w at 0°, 50°, and 100°C are 1.14×10^{-15}, 5.47×10^{-14}, and 4.9×10^{-13}, respectively. Calculate the pH for a neutral solution at each of these temperatures.

10-16. Calculate pK_w at
*(a) 0°C. (b) 50°C. (c) 100°C.

10-17. Calculate the pH of a 1.00×10^{-2} M NaOH solution at
*(a) 0°C. (b) 50°C. (c) 100°C.

***10-18.** What is the pH of an aqueous solution that is 14.0% HCl by weight and has a density of 1.054 g/mL?

10-19. Calculate the pH of a solution that contains 9.00% (w/w) NaOH and has a density of 1.098 g/mL.

***10-20.** What is the pH of a solution that is 2.0×10^{-8} M in NaOH? (*Hint:* In such a dilute solution you must take into account the contribution of H_2O to the hydroxide ion concentration.)

10-21. What is the pH of a 2.0×10^{-8} M HCl solution?

***10-22.** Calculate the pH of the solution that results upon mixing 20.0 mL of 0.2000 M HCl with 25.0 mL of
(a) distilled water.
(b) 0.132 M $AgNO_3$.
(c) 0.132 M NaOH.
(d) 0.132 M NH_3.
(e) 0.232 M NaOH.

10-23. What is the pH of the solution that results when 0.102 g of $Mg(OH)_2$ is mixed with
(a) 75.0 mL of 0.0600 M HCl?
(b) 15.0 mL of 0.0600 M HCl?
(c) 30.0 mL of 0.0600 M HCl?
(d) 30.0 mL of 0.0600 M $MgCl_2$?

***10-24.** Calculate the hydronium ion concentration and pH of a solution that is 0.0500 M in HCl
(a) neglecting activities.
(b) using activities.

10-25. Calculate the hydroxide ion concentration and the pH of a 0.0167 M $Ba(OH)_2$ solution
(a) neglecting activities.
(b) using activities.

***10-26.** Calculate the pH of a HOCl solution that is (a) 1.00×10^{-1} M, (b) 1.00×10^{-2} M, (c) 1.00×10^{-4} M.

10-27. Calculate the pH of a NaOCl solution that is (a) 1.00×10^{-1} M, (b) 1.00×10^{-2}, (c) 1.00×10^{-4} M.

***10-28.** Calculate the pH of an ammonia solution that is (a) 1.00×10^{-1}, (b) 1.00×10^{-2}, (c) 1.00×10^{-4} M.

10-29. Calculate the pH of an NH_4Cl solution that is (a) 1.00×10^{-1} M, (b) 1.00×10^{-2} M, (c) 1.00×10^{-4} M.

***10-30.** Calculate the pH of a solution in which the concentration of piperidine is (a) 1.00×10^{-1}, (b) 1.00×10^{-2}, (c) 1.00×10^{-4} M.

10-31. Calculate the pH of an iodic acid solution that is (a) 1.00×10^{-1} M, (b) 1.00×10^{-2}, (c) 1.00×10^{-4} M.

***10-32.** Calculate the pH of a solution prepared by
(a) dissolving 43.0 g of lactic acid in water and diluting to 500 mL.
(b) diluting 25.0 mL of the solution in (a) to 250 mL.
(c) diluting 10.0 mL of the solution in (b) to 1.00 L.

10-33. Calculate the pH of a solution prepared by
(a) dissolving 1.05 g of picric acid, $(NO_2)_3C_6H_2OH$ (229.11 g/mol), in 100 mL of water.
(b) diluting 10.0 mL of the solution in (a) to 100 mL.
(c) diluting 10.0 mL of the solution in (b) to 1.00 L.

***10-34.** Calculate the pH of the solution that results when 20.0 mL of 0.200 M formic acid are
(a) diluted to 45.0 mL with distilled water.
(b) mixed with 25.0 mL of 0.160 M NaOH solution.
(c) mixed with 25.0 mL of 0.200 M NaOH solution.
(d) mixed with 25.0 mL of 0.200 sodium formate solution.

10-35. Calculate the pH of the solution that results when 40.0 mL of 0.100 M NH_3 are
(a) diluted to 20.0 mL with distilled water.
(b) mixed with 20.0 mL of 0.200 M HCl solution.
(c) mixed with 20.0 mL of 0.250 M HCl solution.
(d) mixed with 20.0 mL of 0.200 M NH_4Cl solution.
(e) mixed with 20.0 mL of 0.100 M HCl solution.

10-36. A solution is 0.0500 M in NH_4Cl and 0.0300 M in NH_3. Calculate its OH^- concentration and its pH
(a) neglecting activities.
(b) taking activities into account.

***10-37.** What is the pH of a solution that is
(a) prepared by dissolving 9.20 g of lactic acid (90.08 g/mol) and 11.15 g of sodium lactate (112.06 g/mol) in water and diluting to 1.00 L?
(b) 0.0550 M in acetic acid and 0.0110 M in sodium acetate?
(c) prepared by dissolving 3.00 g of salicylic acid, $C_6H_4(OH)COOH$ (138.12 g/mol), in 50.0 mL of 0.1130 M NaOH and diluting to 500.0 mL?
(d) 0.0100 M in picric acid and 0.100 M in sodium picrate?

10-38. What is the pH of a solution that is
(a) prepared by dissolving 3.30 g of $(NH_4)_2SO_4$ in water, adding 125.0 mL of 0.1011 M NaOH, and diluting to 500.0 mL?
(b) 0.120 M in piperidine and 0.080 M in its chloride salt?
(c) 0.050 M in ethylamine and 0.167 M in its chloride salt?
(d) prepared by dissolving 2.32 g of aniline (93.13 g/mol) in 100 mL of 0.0200 M HCl and diluting to 250.0 mL?

10-39. Calculate the change in pH that occurs in each of the solutions listed below as a result of a tenfold dilution with water. Round calculated values for pH to three figures to the right of the decimal point.
*(a) H_2O.
(b) 0.0500 M HCl.
*(c) 0.0500 M NaOH.
(d) 0.0500 M CH_3COOH.
*(e) 0.0500 M CH_3COONa.
(f) 0.0500 M CH_3COOH + 0.0500 M CH_3COONa.
*(g) 0.500 M CH_3COOH + 0.500 M CH_3COONa.

10-40. Calculate the change in pH that occurs when 1.00 mmol of a strong acid is added to 100 mL of the solutions listed in Problem 10-39.

10-41. Calculate the change in pH that occurs when 1.00 mmol of a strong base is added to 100 mL of the solutions listed in Problem 10-39. Calculate values to three decimal places.

10-42. Calculate the change in pH to three decimal places that occurs when 0.50 mmol of a strong acid is added to 100 mL of
(a) 0.0200 M lactic acid + 0.0800 M sodium lactate.
*(b) 0.0800 M lactic acid + 0.0200 M sodium lactate.
(c) 0.0500 M lactic acid + 0.0500 M sodium lactate.

***10-43.** What weight of sodium formate must be added to 400 mL of 1.00 M formic acid to produce a buffer solution that has a pH of 3.50?

10-44. What weight of sodium glycolate should be added to 300 mL of 1.00 M glycolic acid to produce a buffer solution with a pH of 4.00?

***10-45.** What volume of 0.200 M HCl must be added to 250 mL of 0.300 M sodium mandelate to produce a buffer solution with a pH of 3.37?

10-46. What volume of 2.00 M NaOH must be added to 300 mL of 1.00 M glycolic acid to produce a buffer solution having a pH of 4.00?

***10-47.** A 50.00-mL aliquot of 0.1000 M NaOH is titrated with 0.1000 M HCl. Calculate the pH of the solution after the addition of 0.00, 10.00, 25.00, 40.00, 45.00, 49.00, 50.00, 51.00, 55.00, and 60.00 mL of acid and prepare a titration curve from the data.

***10-48.** In a titration of 50.00 mL of 0.05000 M formic acid with 0.1000 M KOH, the titration error must be smaller than ±0.05 mL. What indicator can be chosen to realize this goal?

10-49. In a titration of 50.00 mL of 0.1000 M ethylamine with 0.1000 M $HClO_4$, the titration error must be no more than ±0.05 mL. What indicator can be chosen to realize this goal?

10-50. Calculate the pH after addition of 0.00, 5.00, 15.00, 25.00, 40.00, 45.00, 49.00, 50.00, 51,00, 55.00, and 60.00 mL of 0.1000 M NaOH in the titration of 50.00 mL of
*(a) 0.1000 M HNO_2.
(b) 0.1000 M lactic acid.
*(c) 0.1000 M pyridinium chloride.

10-51. Calculate the pH after addition of 0.00, 5.00, 15.00, 25.00, 40.00, 45.00, 49.00, 50.00, 51.00, 55.00, and 60.00 mL of 0.1000 M HCl in the titration of 50.00 mL of
*(a) 0.1000 M ammonia.
(b) 0.1000 M hydrazine.
(c) 0.1000 M sodium cyanide.

10-52. Calculate the pH after addition of 0.00, 5.00, 15.00, 25.00, 40.00, 49.00, 50.00, 51.00, 55.00, and 60.00 mL of reagent in the titration of 50.0 mL of
*(a) 0.1000 M anilinium chloride with 0.1000 M NaOH.
(b) 0.01000 M chloroacetic acid with 0.01000 M NaOH.
*(c) 0.1000 M hypochlorous acid with 0.1000 M NaOH.
(d) 0.1000 M hydroxylamine with 0.1000 M HCl.
Construct titration curves from the data.

10-53. Calculate α_0 and α_1 for
*(a) acetic acid species in a solution with a pH of 5.320.
(b) picric acid species in a solution with a pH of 1.250.

*(c) hypochlorous acid species in a solution with a pH of 7.000.

(d) hydroxylamine acid species in a solution with a pH of 5.120.

*(e) piperidine species in a solution with a pH of 10.080.

***10-54.** Calculate the equilibrium concentration of undissociated HCOOH in a formic acid solution with an analytical formic acid concentration of 0.0850 and a pH of 3.200.

10-55. Calculate the equilibrium concentration of methylamine in a solution that has a molar analytical CH_3NH_2 concentration of 0.120 and a pH of 11.471.

10-56. Supply the missing data.

Acid	Molar Analytical Concentration, c_T ($c_T = c_{HA} + c_{A^-}$)	pH	[HA]	[A^-]	α_0	α_1
*Lactic	0.120				0.640	
Iodic	0.200					0.765
Butanoic		5.00	0.0644			
Hypochlorous	0.280	7.00				
Nitrous				0.105	0.413	0.587
Hydrogen cyanide			0.145	0.221		
*Sulfamic	0.250	1.20				

CHAPTER

11

Titration Curves for Complex Acid/Base Systems

In this chapter, we describe methods for deriving titration curves for complex acid/base systems. For the purpose of this discussion, complex systems are defined as solutions made up of (1) two acids or two bases of different strengths, (2) an acid or a base that has two or more acidic or basic functional groups, or (3) an amphiprotic substance, which is capable of acting both as an acid and a base. Equations for more than one equilibrium are required to describe the characteristics of any of these systems.

11A MIXTURES OF STRONG AND WEAK ACIDS OR STRONG AND WEAK BASES

It is possible to determine each of the components in a mixture containing a strong acid and a weak acid (or a strong base and a weak base) provided that the concentrations of the two are of the same order of magnitude and provided also that the dissociation constant for the weak acid or base is somewhat less than about 10^{-4}. In order to demonstrate that this statement is true, let us show how a titration curve can be derived for a solution containing roughly equal concentrations of HCl and HA, where HA is a weak acid that has a dissociation constant of 10^{-4}.

EXAMPLE 11-1

Calculate the pH of a 25.00-mL mixture that is 0.1200 M in hydrochloric acid and 0.0800 M in the weak acid HA ($K_a = 1.00 \times 10^{-4}$) during its titration with 0.1000 M KOH. Derive data for additions of the following mL of base: (a) 0.00 and (b) 5.00.

(a) **0.00 mL KOH Added**

The molar hydronium concentration in this mixture is equal to the concentration of HCl plus the concentration of hydronium ions that results from dissociation of HA and H_2O. In the presence of the two acids, however, the concentration of hydronium ions from the dissociation of water is surely negligibly small, so we need to only take into account the other two sources of protons. Thus, we may write

$$[H_3O^+] = c_{HCl} + [A^-] = 0.1200 + [A^-]$$

Note that $[A^-]$ is equal to the concentration of hydronium ions from the dissociation of HA.

Let us now assume that the presence of the strong acid so represses the dissociation of HA that $[A^-] \ll 0.1200$; then $[H_3O^+] \approx 0.1200$ and the pH is 0.92. To check this assumption, the provisional value for $[H_3O^+]$ is substituted into the dissociation-constant expression for HA, which upon rearrangement gives

$$\frac{[A^-]}{[HA]} = \frac{K_a}{[H_3O^+]} = \frac{1.00 \times 10^{-4}}{0.1200} = 8.33 \times 10^{-4}$$

which can be rearranged to

$$[HA] = [A^-]/(8.33 \times 10^{-4})$$

From mass-balance considerations, we can write

$$0.0800 = c_{HA} = [HA] + [A^-]$$

Substituting the value of [HA] from the previous equation gives

$$0.0800 = [A^-]/(8.33 \times 10^{-4}) + [A^-] = (1.20 \times 10^3)[A^-]$$

$$[A^-] = 6.7 \times 10^{-5}$$

We see that $[A^-]$ is indeed much smaller than 0.1200 M, as assumed.

(b) **Upon Addition of 5.00 mL of Base**

$$c_{HCl} = \frac{25.00 \times 0.1200 - 5.00 \times 0.1000}{25.00 + 5.00} = 0.0833$$

and we may write

$$[H_3O^+] = 0.0833 + [A^-] \approx 0.0833$$

$$\text{pH} = 1.08$$

To determine whether our assumption is still valid, we compute $[A^-]$ as we did in part (a), knowing that the concentration of HA is now $0.0800 \times 25.00/30.00 = 0.0667$, and find

$$[A^-] = 8.0 \times 10^{-5}$$

which is still much smaller than 0.0833.

Example 11-1 demonstrates that hydrochloric acid represses the dissociation of the weak acid in the early stages of the titration to such an extent that we can assume that $[A^-] \ll c_{HCl}$ and $[H_3O^+] = c_{HCl}$. The hydronium ion concentration is then simply the molar concentration of the strong acid.

The approximation employed in Example 11-1 can be shown to apply until most of the hydrochloric acid has been neutralized by the titrant. Therefore, the curve in this region *is identical to the titration curve for a 0.1200 M solution of a strong acid by itself.*

As shown by Example 11-2, the presence of HA must be taken into account as the first end point in the titration is approached.

EXAMPLE 11-2

Calculate the pH of the solution that results when 29.00 mL of 0.1000 M KOH are added to 25.00 mL of the solution described in Example 11-1.

Here,

$$c_{HCl} = \frac{25.00 \times 0.1200 - 29.00 \times 0.1000}{25.00 + 29.00} = 1.85 \times 10^{-3} \text{ M}$$

$$c_{HA} = \frac{25.00 \times 0.0800}{54.00} = 3.70 \times 10^{-2} \text{ M}$$

A provisional result based (as in the previous example) on the assumption that $[H_3O^+] = 1.85 \times 10^{-3}$ yields a value of 1.90×10^{-3} for $[A^-]$. Clearly, $[A^-]$ is no longer much smaller than $[H_3O^+]$, and we must write

$$[H_3O^+] = c_{HCl} + [A^-] = 1.85 \times 10^{-3} + [A^-] \qquad \textbf{(11-1)}$$

In addition, from mass-balance considerations, we know that

$$[HA] + [A^-] = c_{HA} = 3.70 \times 10^{-2} \qquad \textbf{(11-2)}$$

We rearrange the acid dissociation-constant expression for HA and obtain

$$[HA] = \frac{[H_3O^+][A^-]}{1.00 \times 10^{-4}}$$

Substitution of this expression into Equation 11-2 yields

$$\frac{[H_3O^+][A^-]}{1.00 \times 10^{-4}} + [A^-] = 3.70 \times 10^{-2}$$

$$[A^-] = \frac{3.70 \times 10^{-6}}{[H_3O^+] + 1.00 \times 10^{-4}}$$

Substitution for $[A^-]$ and c_{HCl} in Equation 11-1 yields

$$[H_3O^+] = 1.85 \times 10^{-3} + \frac{3.70 \times 10^{-6}}{[H_3O^+] + 1.00 \times 10^{-4}}$$

$$[H_3O^+]^2 + (1.00 \times 10^{-4})[H_3O^+]$$
$$= (1.85 \times 10^{-3})[H_3O^+] + 1.85 \times 10^{-7} + 3.7 \times 10^{-6}$$

Collecting terms gives

$$[H_3O^+]^2 - (1.75 \times 10^{-3})[H_3O^+] - 3.885 \times 10^{-6} = 0$$
$$[H_3O^+] = 3.03 \times 10^{-3}$$
$$pH = 2.52$$

Note that the contributions to the hydronium ion concentration from HCl (1.85×10^{-3} M) and HA (3.03×10^{-3} M $- 1.85 \times 10^{-3}$ M) are of comparable magnitude.

When the amount of base added is equivalent to the amount of hydrochloric acid originally present, the solution is identical in all respects to one prepared by dissolving appropriate quantities of the weak acid and potassium chloride in a suitable volume of water. The potassium chloride, however, has no effect on the pH (neglecting the influence of increased ionic strength); thus, the remainder of the titration curve is identical to that for a dilute solution of HA.

The shape of the curve for a mixture of weak and strong acids, and hence the information obtainable from it, depends in large measure upon the strength of the weak acid. Figure 11-1 depicts the pH changes that occur during the titration of mixtures containing hydrochloric acid and several weak acids. Note that the rise in pH at the first equivalence point is small or essentially nonexistent when the weak acid has a relatively large dissociation constant (curves *A* and *B*). For titrations such as these, only the total number of millimoles of weak and strong acid can be ascertained accurately. Conversely, when the weak acid has a very small dissociation constant, only the strong acid content can be determined. For weak acids of intermediate strength (K_a somewhat less than 10^{-4} but greater than 10^{-8}), there are usually two useful end points.

The composition of a mixture of a strong acid and a weak acid can be determined by titration with suitable indicators provided that the weak acid has a dissociation constant that lies between 10^{-4} and 10^{-8} and provided that the concentrations of the two acids are of the same order of magnitude.

Determination of the amount of each component in a mixture that contains a strong base and a weak base is also possible, subject to the constraints just described for the strong acid/weak acid system. The derivation of a curve for such a titration is analogous to that for a mixture of acids.

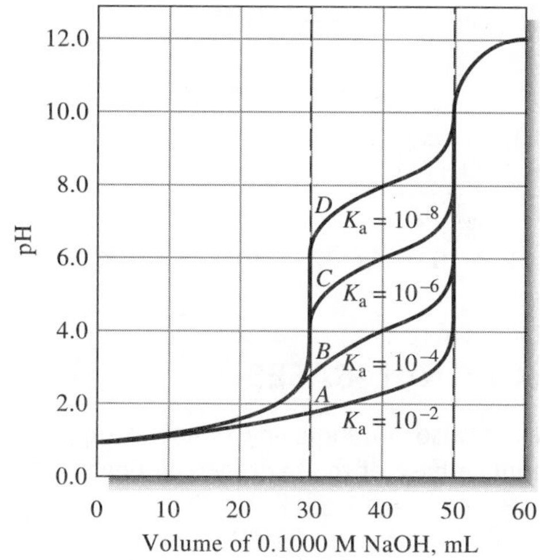

Figure 11-1
Curves for the titration of strong acid/weak acid mixtures with 0.1000 M NaOH. Each titration is on 25.00 mL of a solution that is 0.1200 M in HCl and 0.0800 M in HA.

11B POLYFUNCTIONAL ACIDS

Phosphoric acid is a typical polyfunctional acid. In aqueous solution it undergoes the following three dissociation reactions

Throughout the remainder of this chapter we will find it useful to use K_{a1}, K_{a2}, etc. to represent the first and second dissociation constants of acids and K_{b1}, K_{b2}, etc. for the stepwise constants for bases.

For most polyfunctional weak acids, $K_{a1} > K_{a2}$, often by a factor of 10^4 to 10^5 because of electrostatic forces. That is, the first dissociation involves separating a single positively charged hydronium ion from a singly charged anion. In the second step, a hydronium ion is separated from a doubly charged anion, a process that requires considerably more energy.

A second reason that $K_{a1} > K_{a2}$ is a statistical one. In the first step, a proton can be removed from two locations; in the second step, only one can be removed. Thus, the first dissociation is twice as probable as the second.

$$H_3PO_4 + H_2O \rightleftharpoons H_2PO_4^- + H_3O^+ \qquad K_{a1} = \frac{[H_3O^+][H_2PO_4^-]}{[H_3PO_4]} = 7.11 \times 10^{-3}$$

$$H_2PO_4^- + H_2O \rightleftharpoons HPO_4^{2-} + H_3O^+ \qquad K_{a2} = \frac{[H_3O^+][HPO_4^{2-}]}{[H_2PO_4^-]} = 6.32 \times 10^{-8}$$

$$HPO_4^{2-} + H_2O \rightleftharpoons PO_4^{3-} + H_3O^+ \qquad K_{a3} = \frac{[H_3O^+][PO_4^{3-}]}{[HPO_4^{2-}]} = 4.5 \times 10^{-13}$$

With this acid, as with other polyprotic acids, $K_{a1} > K_{a2} > K_{a3}$.

When two adjacent stepwise equilibria are *added,* the equilibrium constant for the resulting overall reaction is the *product* of the two constants. Thus, for the first two dissociation equilibria for H_3PO_4, we may write

$$\begin{array}{l} H_3PO_4 + H_2O \rightleftharpoons H_2PO_4^- + H_3O^+ \\ H_2PO_4^- + H_2O \rightleftharpoons HPO_4^{2-} + H_3O^+ \\ \hline H_3PO_4 + 2H_2O \rightleftharpoons HPO_4^{2-} + 2H_3O^+ \end{array}$$

and

$$K_{a1}K_{a2} = \frac{[H_3O^+]^2[HPO_4^{2-}]}{[H_3PO_4]} = 7.11 \times 10^{-3} \times 6.32 \times 10^{-8} = 4.49 \times 10^{-10}$$

Similarly, for the reaction

$$H_3PO_4 \rightleftharpoons 3H_3O^+ + PO_4^{3-}$$

we may write

$$K_{a1}K_{a2}K_{a3} = \frac{[H_3O^+]^3[PO_4^{3-}]}{H_3PO_4} = 7.11 \times 10^{-3} \times 6.32 \times 10^{-8} \times 4.5 \times 10^{-13} = 2.0 \times 10^{-22}$$

11C POLYFUNCTIONAL BASES

Polyfunctional bases are also common, an example being sodium carbonate. Carbonate ion, the conjugate base of the hydrogen carbonate ion, is involved in the stepwise equilibria:

$$CO_3^{2-} + H_2O \rightleftharpoons HCO_3^- + OH^- \qquad K_{b1} = \frac{[HCO_3^-][OH^-]}{[CO_3^{2-}]} = \frac{K_w}{K_{a2}}$$

$$= \frac{1.00 \times 10^{-14}}{4.69 \times 10^{-11}} = 2.13 \times 10^{-4}$$

$$HCO_3^- + H_2O \rightleftharpoons H_2CO_3 + OH^- \qquad K_{b2} = \frac{[H_2CO_3][OH^-]}{[HCO_3^-]} = \frac{K_w}{K_{a1}}$$

$$= \frac{1.00 \times 10^{-14}}{4.45 \times 10^{-7}} = 2.25 \times 10^{-8}$$

where K_{a1} and K_{a2} are the first and second dissociation constants for carbonic acid, and K_{b1} and K_{b2} are the first and second dissociation constants of the base CO_3^{2-}.

The overall basic dissociation reaction of sodium carbonate is described by the equations

$$CO_3^{2-} + 2H_2O \rightleftharpoons H_2CO_3 + 2OH^- \qquad K_{b1}K_{b2} = \frac{[H_2CO_3][OH^-]^2}{[CO_3^{2-}]}$$

$$= 2.13 \times 10^{-4} \times 2.25 \times 10^{-8}$$

$$= 4.79 \times 10^{-12}$$

The pH of polyfunctional systems, such as phosphoric acid or sodium carbonate, can be computed rigorously through use of the systematic approach to multiple-equilibrium problems described in Chapter 9. Solution of the several simultaneous equations that are involved is difficult and time-consuming, however. Fortunately, simplifying assumptions can be invoked when the successive equilibrium constants for the acid (or base) differ by a factor of about 10^3 (or more). With one exception, these assumptions make it possible to derive pH data for titration curves by the techniques we have discussed in earlier chapters.

CHALLENGE: Write a sufficient number of equations to permit the calculation of all of the species present in a solution containing known molar analytical concentrations of Na_2CO_3 and $NaHCO_3$.

11D BUFFER SOLUTIONS INVOLVING POLYPROTIC ACIDS

Two buffer systems can be prepared from a weak dibasic acid and its salts. The first consists of free acid H_2A and its conjugate base NaHA, and the second makes use of the acid NaHA and its conjugate base Na_2A. The pH of the latter system is higher than that of the former because the acid dissociation constant for HA^- is usually less than that for H_2A.

Sufficient independent equations are readily written to permit a rigorous evaluation of the hydronium ion concentration for either of these systems. Ordinarily, however, it is permissible to introduce the simplifying assumption that only one of the equilibria is important in determining the hydronium ion concentration of the solution. Thus, for a buffer prepared from H_2A and NaHA, the dissociation of HA^- to yield A^{2-} is neglected, and the calculation is based on the first dissociation only. With this simplification, the hydronium ion concentration is calculated by the method described in Section 10C-1 for a simple buffer solution. As before, it is an easy matter to check the validity of the assumption by calculating an approximate concentration of A^{2-} and comparing this value with the concentrations of H_2A and HA^-.

EXAMPLE 11-3

Calculate the hydronium ion concentration for a buffer solution that is 2.00 M in phosphoric acid and 1.50 M in potassium dihydrogen phosphate.

The principal equilibrium in this solution is the dissociation of H_3PO_4.

$$H_3PO_4 + H_2O \rightleftharpoons H_3O^+ + H_2PO_4^- \qquad K_{a1} = 7.11 \times 10^{-3} = \frac{[H_3O^+][H_2PO_4^-]}{[H_3PO_4]}$$

The dissociation of $H_2PO_4^-$ is assumed to be negligible; that is, $[HPO_4^{2-}]$ and $[PO_4^{3-}] \ll [H_2PO_4^-]$ or $[H_3PO_4]$. Then,

$$[H_3PO_4] \approx c_{H_3PO_4} = 2.00$$

$$[H_2PO_4^-] \approx c_{KH_2PO_4} = 1.50$$

$$[H_3O^+] = \frac{7.11 \times 10^{-3} \times 2.00}{1.50} = 9.48 \times 10^{-3}$$

We now use the equilibrium-constant expression for K_{a2} to show that $[HPO_4^{2-}]$ can be neglected:

$$K_{a2} = 6.34 \times 10^{-8} = \frac{[H_3O^+][HPO_4^{2-}]}{[H_2PO_4^-]} = \frac{9.48 \times 10^{-3}[HPO_4^{2-}]}{1.50}$$

$$[HPO_4^{2-}] = 1.00 \times 10^{-5}$$

and our assumption is valid. Note that $[PO_4^{3-}]$ is even smaller than $[HPO_4^{2-}]$.

For a buffer prepared from NaHA and Na_2A, the second dissociation will ordinarily predominate, and the reaction

$$HA^- + H_2O \rightleftharpoons H_2A + OH^-$$

is disregarded. The concentration of H_2A is negligible compared with that of HA^- or A^{2-}; the hydronium ion concentration can then be calculated from the second dissociation constant, again employing the techniques for a simple buffer solution. To test the assumption, an estimate of the H_2A concentration is compared with the concentrations of HA^- and A^{2-}.

EXAMPLE 11-4

Calculate the hydronium ion concentration of a buffer that is 0.0500 M in potassium hydrogen phthalate (KHP) and 0.150 M in potassium phthalate (K_2P).

$$HP^- + H_2O \rightleftharpoons H_3O^+ + P^{2-} \qquad K_{a2} = 3.91 \times 10^{-6} = \frac{[H_3O^+][P^{2-}]}{[HP^-]}$$

Provided the concentration of H_2P in this solution is negligible,

$$[HP^-] \approx c_{KHP} = 0.0500$$

$$[P^{2-}] \approx c_{K_2P} = 0.150$$

$$[H_3O^+] = \frac{3.91 \times 10^{-6} \times 0.0500}{0.150} = 1.30 \times 10^{-6}$$

To check the first assumption, an approximate value for $[H_2P]$ is calculated by substituting numerical values for $[H_3O^+]$ and $[HP^-]$ into the expression for K_{a1}:

$$K_{a1} = 1.12 \times 10^{-3} = \frac{(1.30 \times 10^{-6})(0.0500)}{[H_2P]}$$

$$[H_2P] = 6 \times 10^{-5}$$

This result justifies the assumption that $[H_2P] \ll [HP^-]$ and $[P^{2-}]$, that is, that the reaction of HP^- as a base can be neglected.

In all but a few situations, the assumption of a single principal equilibrium, as invoked in Examples 11-3 and 11-4, provides a satisfactory estimate of the pH of buffer mixtures derived from polybasic acids. Appreciable errors occur, however, when the concentration of the acid or the salt is very low or when the two dissociation constants are numerically close to one another. A more laborious and rigorous calculation is then required.

11E CALCULATION OF THE pH OF SOLUTIONS OF AMPHIPROTIC SALTS

Thus far, we have not considered how to calculate the pH of solutions of salts that have both acidic and basic properties—that is, salts that are *amphiprotic.* Such salts are formed during neutralization titration of polyfunctional acids and bases. For example, when 1 mol of NaOH is added to a solution containing 1 mol of the acid H_2A, 1 mol of NaHA is formed. The pH of this solution is determined by two equilibria established between HA^- and water:

$$HA^- + H_2O \rightleftharpoons A^{2-} + H_3O^+$$

and

$$HA^- + H_2O \rightleftharpoons H_2A + OH^-$$

One of these reactions produces hydronium ions and the other hydroxide ions. A solution of NaHA will be acidic or basic depending upon the relative magnitude of the equilibrium constants for these processes:

$$K_{a2} = \frac{[H_3O^+][A^{2-}]}{[HA^-]} \tag{11-3}$$

$$K_{b2} = \frac{K_w}{K_{a1}} = \frac{[H_2A][OH^-]}{[HA^-]} \tag{11-4}$$

where K_{a1} and K_{a2} are the acid dissociation constants for H_2A and K_{b2} is the *basic* dissociation constant for HA^-. If K_{b2} is greater than K_{a2}, the solution is basic; otherwise, it is acidic.

In order to derive an expression for the hydronium ion concentration of a solution of HA^-, we employ the systematic approach described in Section 9A. We first write a mass-balance expression. That is,

$$c_{NaHA} = [HA^-] + [H_2A] + [A^{2-}] \qquad (11\text{-}5)$$

The charge-balance equation takes the form:

$$[Na^+] + [H_3O^+] = [HA^-] + 2[A^{2-}] + [OH^-]$$

Since the sodium ion concentration is equal to the molar analytical concentration of the salt, the last equation can be rewritten as

$$c_{NaHA} + [H_3O^+] = [HA^-] + 2[A^{2-}] + [OH^-] \qquad (11\text{-}6)$$

We now have four algebraic equations (Equations 11-5 and 11-6 and the two dissociation constant expressions for H_2A) and need one additional to solve for the five unknowns. The ion-product constant for water serves this purpose:

$$K_w = [H_3O^+][OH^-]$$

The rigorous computation of the hydronium ion concentration from these five equations is difficult. However, a reasonable approximation, applicable to solutions of most acid salts, can be obtained as follows.

We first subtract the mass-balance equation from the charge-balance equation.

$$\begin{array}{lll} c_{NaHA} + [H_3O^+] = [HA^-] + 2[A^{2-}] + [OH^-] & & \text{charge balance} \\ c_{NaHA} \qquad\qquad\;\; = [H_2A] + [HA^-] + [A^{2-}] & & \text{mass balance} \\ \hline [H_3O^+] = [A^{2-}] + [OH^-] - [H_2A] & & (11\text{-}7) \end{array}$$

We then rearrange the acid dissociation-constant expressions for H_2A to obtain

$$[H_2A] = \frac{[H_3O^+][HA^-]}{K_{a1}}$$

and for HA^- to give

$$[A^{2-}] = \frac{K_{a2}[HA^-]}{[H_3O^+]}$$

Substituting these expressions and the expression for K_w into Equation 11-7 yields

$$[H_3O^+] = \frac{K_{a2}[HA^-]}{[H_3O^+]} + \frac{K_w}{[H_3O^+]} - \frac{[H_3O^+][HA^-]}{K_{a1}}$$

Multiplication through by $[H_3O^+]$ gives

$$[H_3O^+]^2 = K_{a2}[HA^-] + K_w - \frac{[H_3O^+]^2[HA^-]}{K_{a1}}$$

We collect terms to obtain

$$[H_3O^+]^2\left(\frac{[HA^-]}{K_{a1}} + 1\right) = K_{a2}[HA^-] + K_w$$

Finally, this equation rearranges to

$$[H_3O^+] = \sqrt{\frac{K_{a2}[HA^-] + K_w}{1 + [HA^-]/K_{a1}}} \qquad (11\text{-}8)$$

Under most circumstances, we can make the approximation that

$$[HA^-] \approx c_{NaHA} \qquad (11\text{-}9)$$

Introduction of this relationship into Equation 11-8 gives

$$[H_3O^+] = \sqrt{\frac{K_{a2}c_{NaHA} + K_w}{1 + c_{NaHA}/K_{a1}}} \qquad (11\text{-}10)$$

It is important to understand that the approximation shown as Equation 11-9 requires that $[HA^-]$ be much larger than any of the other equilibrium concentrations in Equations 11-5 and 11-6. This assumption is not valid for very dilute solutions of NaHA or when K_{a2} or K_w/K_{a1} is relatively large.

Frequently, the ratio c_{NaHA}/K_{a1} is much larger than unity in the denominator of Equation 11-10 and $K_{a2}c_{NaHA}$ is considerably greater than K_w in the numerator. With these assumptions, the equation simplifies to

$$[H_3O^+] \approx \sqrt{K_{a1}K_{a2}} \qquad (11\text{-}11)$$

Note that Equation 11-11 does not contain c_{NaHA}, which implies that the pH of solutions of this type remains constant over a considerable range of solute concentrations.

Make sure that you always check the assumptions that are inherent in Equation 11-11.

EXAMPLE 11-5

Calculate the hydronium ion concentration of a 0.100 M $NaHCO_3$ solution.

We first examine the assumptions leading to Equation 11-11. The dissociation constants for H_2CO_3 are $K_{a1} = 4.45 \times 10^{-7}$ and $K_{a2} = 4.69 \times 10^{-11}$. Clearly, c_{NaHA}/K_{a1} in the denominator is much larger than unity; in addition, $K_{a2}c_{NaHA}$ has a value of 4.69×10^{-12}, which is substantially greater than K_w. Thus, Equation 11-11 applies and

$$[H_3O^+] = \sqrt{4.45 \times 10^{-7} \times 4.69 \times 10^{-11}} = 4.6 \times 10^{-9}$$

EXAMPLE 11-6

Calculate the hydronium ion concentration of a 1.00×10^{-3} M Na_2HPO_4 solution.

The pertinent dissociation constants are K_{a2} and K_{a3}, which both contain $[HPO_4^{2-}]$. Their values are $K_{a2} = 6.32 \times 10^{-8}$ and $K_{a3} = 4.5 \times 10^{-13}$. Considering again the assumptions that led to Equation 11-11, we find that $(1.0 \times 10^{-3})/(6.32 \times 10^{-8})$ is much larger than 1, so the denominator can be simplified. The product $K_{a2}c_{Na_2HPO_4}$ is by no means much larger than K_w, however. We therefore use a partially simplified version of Equation 11-10:

$$[H_3O^+] = \sqrt{\frac{4.5 \times 10^{-13} \times 1.00 \times 10^{-3} + 1.00 \times 10^{-14}}{(1.00 \times 10^{-3})/(6.32 \times 10^{-8})}} = 8.1 \times 10^{-10}$$

Use of Equation 11-11 yields a value of 1.6×10^{-10} M.

EXAMPLE 11-7

Find the hydronium ion concentration of a 0.0100 M NaH_2PO_4 solution.

The two dissociation constants of importance (those containing $[H_2PO_4^-]$) are $K_{a1} = 7.11 \times 10^{-3}$ and $K_{a2} = 6.32 \times 10^{-8}$. We see that the denominator of Equation 11-10 cannot be simplified, but the numerator reduces to $K_{a2}c_{NaH_2PO_4}$. Thus, Equation 11-10 becomes

$$[H_3O^+] = \sqrt{\frac{6.32 \times 10^{-8} \times 1.00 \times 10^{-2}}{1.00 + (1.00 \times 10^{-2})/(7.11 \times 10^{-3})}} = 1.62 \times 10^{-5}$$

11F TITRATION CURVES FOR POLYFUNCTIONAL ACIDS

For a different approach, see *Mathcad Applications for Analytical Chemistry*, **pp. 79–85.**

Compounds with two or more acid functional groups yield multiple end points in a titration provided the functional groups differ sufficiently in strengths as acids. The computational techniques described in Chapter 10 permit derivation of reasonably accurate theoretical titration curves for polyprotic acids if the ratio K_{a1}/K_{a2} is somewhat greater than 10^3. If this ratio is smaller, the error, particularly in the region of the first equivalence point, becomes excessive, and a more rigorous treatment of the equilibrium relationships is required.

Figure 11-2 shows the titration curve for a diprotic acid H_2A with dissociation constants of $K_{a1} = 1.00 \times 10^{-3}$ and $K_{a2} = 1.00 \times 10^{-7}$. Because K_{a1}/K_{a2} is significantly greater than 10^3, we can derive this curve (except for the first equivalence point) using the techniques developed in Chapter 10 for simple monoprotic weak acids. Thus, to obtain the initial pH (point *A*), we treat the system as if it contained a single monoprotic acid with a dissociation constant of $K_{a1} = 1.00 \times 10^{-3}$. In region *B* we have the equivalent of a simple buffer solution consisting of the weak acid H_2A and its conjugate base NaHA. That is, we assume that the concentration of A^{2-} is negligible with respect to the other two A-containing species and employ Equation 10-10 (page 202) to obtain $[H_3O^+]$. At the first equivalence point (point *C*), we have a solution of an acid salt and use Equation 11-10 or one of its simplifications to compute the hydronium ion concentration. In the region labeled *D*, we have a second buffer consisting of a weak acid HA^- and its conjugate base Na_2A, and we calculate the pH employing the second dissocia-

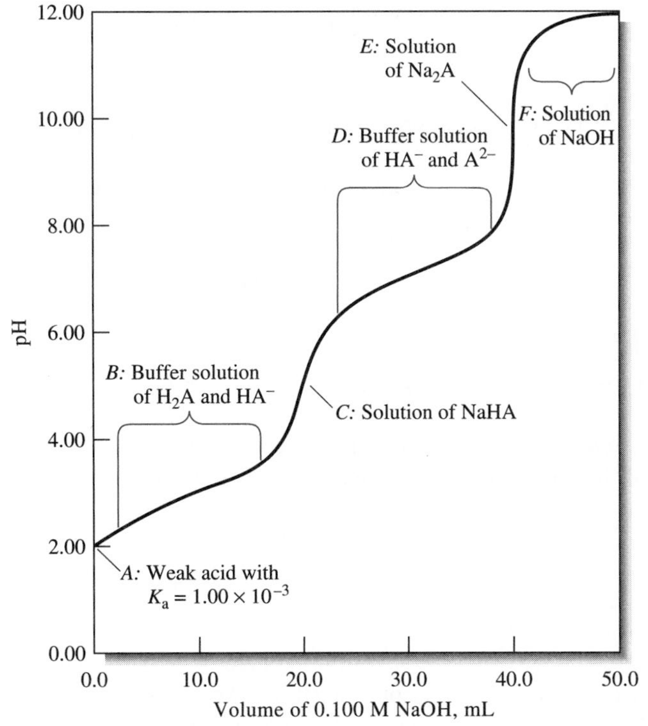

Figure 11-2
Titration of 20.0 mL of 0.100 M H_2A with 0.100 M NaOH. For H_2A, $K_{a1} = 1.00 \times 10^{-3}$ and $K_{a2} = 1.00 \times 10^{-7}$. Method of pH calculation is shown for several points and regions on the titration curve.

tion constant, $K_{a2} = 1.00 \times 10^{-7}$. At point E, the solution contains the conjugate base of a weak acid with a dissociation constant of 1.00×10^{-7}. That is, we assume that the hydroxide concentration of the solution is determined solely by the reaction of A^{2-} with water to form HA^- and OH^-. Finally, in the region labeled F, we compute the hydroxide concentration from the molarity of the excess NaOH and derive the pH from this quantity.

EXAMPLE 11-8

Derive a curve for the titration of 25.00 mL of 0.1000 M maleic acid, HOOC—CH═CH—COOH, with 0.1000 M NaOH.

Symbolizing the acid as H_2M, we can write the two dissociation equilibria as

$$H_2M + H_2O \rightleftharpoons H_3O^+ + HM^- \qquad K_{a1} = 1.3 \times 10^{-2}$$

$$HM^- + H_2O \rightleftharpoons H_3O^+ + M^{2-} \qquad K_{a2} = 5.9 \times 10^{-7}$$

Because the ratio K_{a1}/K_{a2} is large (2×10^4), we proceed as just described.

Initial pH

Only the first dissociation makes an appreciable contribution to $[H_3O^+]$; thus,

$$[H_3O^+] \approx [HM^-]$$

Mass balance requires that

$$[H_2M] + [HM^-] \approx 0.1000$$

or

$$[H_2M] = 0.1000 - [HM^-] = 0.1000 - [H_3O^+]$$

Substituting these relationships into the expression for K_{a1} gives

$$K_{a1} = 1.3 \times 10^{-2} = \frac{[H_3O^+]^2}{0.1000 - [H_3O^+]}$$

Rearranging yields

$$[H_3O^+]^2 + 1.3 \times 10^{-2}[H_3O^+] - 1.3 \times 10^{-3} = 0$$

Because K_{a1} for maleic acid is large, we must solve the quadratic equation exactly or by successive approximations. When we do so, we obtain

$$[H_3O^+] = 3.01 \times 10^{-2}$$
$$pH = 2 - \log 3.01 = 1.52$$

First Buffer Region

The addition of 5.00 mL of base results in the formation of a buffer consisting of the weak acid H_2M and its conjugate base HM^-. To the extent that dissociation of HM^- to give M^{2-} is negligible, the solution can be treated as a simple buffer system. Thus, applying Equations 10-8 and 10-9 (page 202) gives

$$c_{NaHM} \approx [HM^-] = \frac{5.00 \times 0.1000}{30.00} = 1.67 \times 10^{-2}\ M$$

$$c_{H_2M} \approx [H_2M] = \frac{25.00 \times 0.1000 - 5.00 \times 0.1000}{30.00} = 6.67 \times 10^{-2}\ M$$

Substitution of these values into the equilibrium-constant expression for K_{a1} yields a tentative value of 5.2×10^{-2} M for $[H_3O^+]$. It is clear, however, that the approximation $[H_3O^+] \ll c_{H_2M}$ or c_{HM^-} is not valid; therefore Equations 10-6 and 10-7 must be used, and

$$[HM^-] = 1.67 \times 10^{-2} + [H_3O^+] - \cancel{[OH^-]}$$
$$[H_2M] = 6.67 \times 10^{-2} - [H_3O^+] + \cancel{[OH^-]}$$

Because the solution is quite acidic, the approximation that $[OH^-]$ is very small is surely justified. Substitution of these expressions into the dissociation-constant relationship gives

$$\frac{[H_3O^+](1.67 \times 10^{-2} + [H_3O^+])}{6.67 \times 10^{-2} - [H_3O^+]} = 1.3 \times 10^{-2} = K_{a1}$$
$$[H_3O^+]^2 + (2.97 \times 10^{-2})[H_3O^+] - 8.67 \times 10^{-4} = 0$$
$$[H_3O^+] = 1.81 \times 10^{-2}$$
$$pH = -\log(1.81 \times 10^{-2}) = 1.74$$

Additional points in the first buffer region can be computed in a similar way.

First Equivalence Point

At the first equivalence point,

$$[HM^-] \approx c_{NaHM} = \frac{25.00 \times 0.1000}{50.00} = 5.00 \times 10^{-2}$$

Simplification of the numerator in Equation 11-10 is clearly justified. On the other hand, the second term in the denominator is not $\ll 1$. Hence,

$$[H_3O^+] \approx \sqrt{\frac{K_{a2}c_{HM^-}}{1 + c_{HM^-}/K_{a1}}} = \sqrt{\frac{5.9 \times 10^{-7} \times 5.00 \times 10^{-2}}{1 + (5.00 \times 10^{-2})/(1.3 \times 10^{-2})}}$$
$$= 7.80 \times 10^{-5}$$
$$pH = -\log(7.80 \times 10^{-5}) = 4.11$$

Second Buffer Region

Further additions of base to the solution create a new buffer system consisting of HM^- and M^{2-}. When enough base has been added so that the reaction of HM^- with water to give OH^- can be neglected (a few tenths of a milliliter beyond the first equivalence point), the pH of the mixture is readily obtained from K_{a2}. With the introduction of 25.50 mL of NaOH, for example,

$$[M^{2-}] \approx c_{Na_2M} \approx \frac{(25.50 - 25.00)(0.1000)}{50.50} = \frac{0.050}{50.50}\,M$$

and the molar concentration of NaHM is

$$[HM^-] \approx c_{NaHM} \approx \frac{(25.00 \times 0.1000) - (25.50 - 25.00)(0.1000)}{50.50}$$
$$= \frac{2.45}{50.50}\,M$$

Substituting these values into the expression for K_{a2} gives

$$\frac{[H_3O^+](0.050/\cancel{50.50})}{2.45/\cancel{50.50}} = 5.9 \times 10^{-7}$$
$$[H_3O^+] = 2.89 \times 10^{-5}$$

The assumption that $[H_3O^+]$ is small relative to c_{HM^-} and $c_{M^{2-}}$ is valid and pH = 4.54.

Second Equivalence Point

After the addition of 50.00 mL of 0.1000 M sodium hydroxide, the solution is 0.0333 M in Na_2M. Reaction of the base M^{2-} with water is the predominant equilibrium in the system and the only one that we need to take into account. Thus,

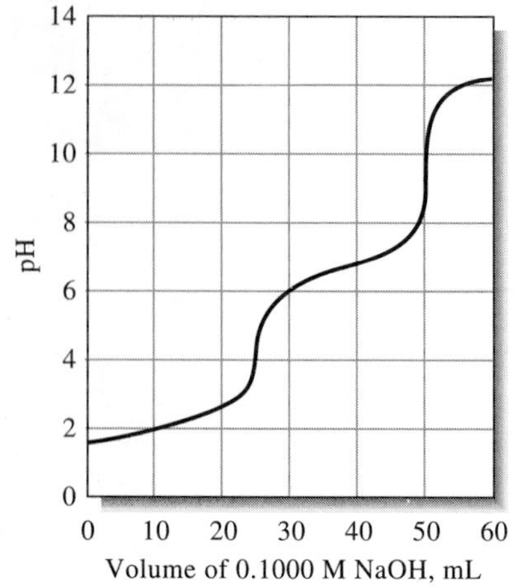

Figure 11-3
Titration curve for 25.00 mL of 0.1000 M maleic acid, H_2M, with 0.1000 M NaOH.

$$M^{2-} + H_2O \rightleftharpoons OH^- + HM^-$$

$$K_{b1} = \frac{K_w}{K_{a2}} = \frac{1.00 \times 10^{-14}}{5.9 \times 10^{-7}} = 1.69 \times 10^{-8} = \frac{[OH^-][HM^-]}{[M^{2-}]}$$

$$[OH^-] \approx [HM^-]$$

$$[M^{2-}] = 0.0333 - [OH^-] \approx 0.0333$$

$$\frac{[OH^-]^2}{0.0333} = \frac{1.00 \times 10^{-14}}{5.9 \times 10^{-7}}$$

$$[OH^-] = 2.38 \times 10^{-5} \quad \text{and} \quad pOH = -\log(2.38 \times 10^{-5}) = 4.62$$

$$pH = 14.00 - 4.62 = 9.38$$

pH Beyond the Second Equivalence Point

Further additions of sodium hydroxide repress the basic dissociation of M^{2-}. The pH is calculated from the concentration of NaOH added in excess of that required for the complete neutralization of H_2M. Thus, when 51.00 mL of NaOH have been added, we have a 1.00 mL excess of 0.1000 M NaOH and

$$[OH^-] = \frac{1.00 \times 0.1000}{76.00} = 1.32 \times 10^{-3} \quad \text{and} \quad pOH = 2.88$$

$$pH = 14.00 - 2.88 = 11.12$$

Figure 11-3 is the titration curve for 0.1000 M maleic acid derived as shown in Example 11-8. Two end points are apparent, either of which could in principle be used as a measure of the concentration of the acid. The second end point is clearly more satisfactory, however, inasmuch as the pH change is more pronounced.

In titrating a polyprotic acid or base, two usable end points are obtained if the ratio of dissociation constants is greater than 10^4 and if the weaker acid or base has a dissociation constant greater than 10^{-8}.

Figure 11-4 shows titration curves for three other polyprotic acids. These curves illustrate that a well-defined end point corresponding to the first equivalence point is observed only when the degree of dissociation of the two acids is sufficiently different. The ratio of K_{a1} to K_{a2} for oxalic acid (curve *B*) is approximately 1000. Although the curve for this titration shows an inflection corre-

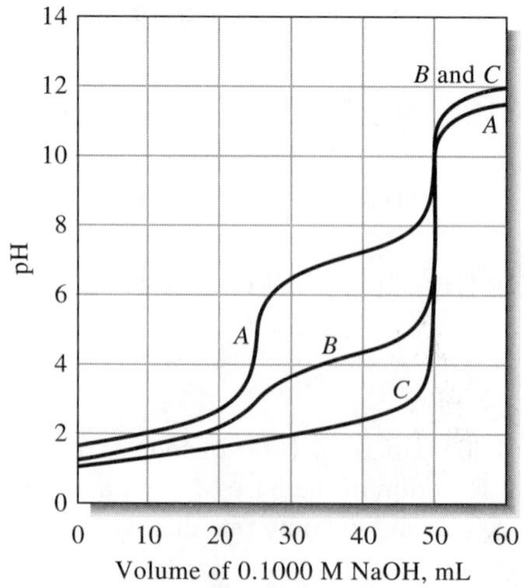

Figure 11-4
Curves for the titration of polyprotic acids. A 0.1000 M NaOH solution is used to titrate 25.00 mL of 0.1000 M H_3PO_4 (*A*), 0.1000 M oxalic acid (*B*), and 0.1000 M H_2SO_4 (*C*).

sponding to the first equivalence point, the magnitude of the pH change is too small to permit precise location of equivalence with an indicator. The second end point, however, provides a means for the accurate determination of oxalic acid.

Curve *A* in Figure 11-4 is the theoretical titration curve for triprotic phosphoric acid. Here, the ratio K_{a1}/K_{a2} is approximately 10^5, as is K_{a2}/K_{a3}. This results in two well-defined end points, either of which is satisfactory for analytical purposes. An acid-range indicator will provide a color change when 1 mol of base has been introduced for each mole of acid; a base-range indicator will require 2 mol of base per mole of acid. The third hydrogen of phosphoric acid is so slightly dissociated ($K_{a3} = 4.5 \times 10^{-13}$) that no practical end point is associated with its neutralization. The buffering effect of the third dissociation is noticeable, however, and causes the pH for curve *A* to be lower than the pH for the other two curves in the region beyond the second equivalence point.

Curve *C* is the titration curve for sulfuric acid, a substance that has one fully dissociated proton and one that is dissociated to a relatively large extent ($K_{a2} = 1.02 \times 10^{-2}$). Because of the similarity in strengths of the two acids, only a single end point, corresponding to the titration of both protons, is observed.

FEATURE 11-1
The Dissociation of Sulfuric Acid

Sulfuric acid is unusual in that one of its protons behaves as a strong acid and the other as a weak acid ($K_{a2} = 1.02 \times 10^{-2}$). Let us consider how the hydronium ion concentration of sulfuric acid solutions is computed using a 0.0400 M solution as an example.

We will first assume that the dissociation of HSO_4^- is negligible because of the large excess of H_3O^+ resulting from the complete dissociation of H_2SO_4. Therefore,

$$[H_3O^+] \approx [HSO_4^-] \approx 0.0400$$

However, an estimate of $[SO_4^{2-}]$ based upon this approximation and the expression for K_{a2} reveals that

$$\frac{\cancel{0.0400}[SO_4^{2-}]}{\cancel{0.0400}} = 1.02 \times 10^{-2}$$

Clearly, $[SO_4^{2-}]$ is *not* small relative to $[HSO_4^-]$, and a more rigorous solution is required.

From stoichiometric considerations, it is necessary that

$$[H_3O^+] = 0.0400 + [SO_4^{2-}]$$

The first term on the right is the concentration of H_3O^+ from dissociation of the H_2SO_4 to HSO_4^-. The second term is the contribution of the dissociation of HSO_4^-. Rearrangement yields

$$[SO_4^{2-}] = [H_3O^+] - 0.0400$$

CHALLENGE: Derive a titration curve for 50.0 mL of 0.0500 M H_2SO_4 with 0.100 M NaOH.

Mass-balance considerations require that

$$c_{H_2SO_4} = 0.0400 = [HSO_4^-] + [SO_4^{2-}]$$

Combining the last two equations and rearranging yield

$$[HSO_4^-] = 0.0800 - [H_3O^+]$$

Introduction of these equations for $[SO_4^{2-}]$ and $[HSO_4^-]$ into the expression for K_{a2} yields

$$\frac{[H_3O^+]([H_3O^+] - 0.0400)}{0.0800 - [H_3O^+]} = 1.02 \times 10^{-2}$$

$$[H_3O^+]^2 - (0.0298)[H_3O^+] - 8.16 \times 10^{-4} = 0$$

$$[H_3O^+] = 0.0471$$

11G TITRATION CURVES FOR POLYFUNCTIONAL BASES

See *Mathcad Applications for Analytical Chemistry,* **pp. 85–88.**

The derivation of a titration curve for a polyfunctional base involves no new principles. To illustrate, consider the titration of a sodium carbonate solution with standard hydrochloric acid. The important equilibrium constants are

$$CO_3^{2-} + H_2O \rightleftharpoons OH^- + HCO_3^- \qquad K_{b1} = \frac{K_w}{K_{a2}} = \frac{1.00 \times 10^{-14}}{4.69 \times 10^{-11}} = 2.13 \times 10^{-4}$$

$$HCO_3^- + H_2O \rightleftharpoons OH^- + H_2CO_3 \qquad K_{b2} = \frac{K_w}{K_{a1}} = \frac{1.00 \times 10^{-14}}{4.45 \times 10^{-7}} = 2.25 \times 10^{-8}$$

The Composition of a Solution of CO_2
It has been shown that the equilibrium

$$H_2CO_3(aq) \rightleftharpoons H_2O(l) + CO_2(aq)$$

lies far to the right so that the solution formed by the addition of 2 mol of strong acid to 1 mol of Na_2CO_3 is largely made up of dissolved CO_2. Such a solution is acidic because of the reaction

$$CO_2(aq) + 2H_2O(l) \rightleftharpoons H_3O^+(aq) + HCO_3^-(aq)$$

Thus, a more correct formulation of the acidic nature of this type of solution (but one that is seldom encountered) would be

$$K_{a1} = \frac{[H_3O^+][HCO_3^-]}{[CO_2]} = 4.45 \times 10^{-7}$$

CHALLENGE: Show that either K_{b2} or K_{a1} can be used to calculate the pH of a buffer that is 0.100 M in Na_2CO_3 and 0.100 M in $NaHCO_3$.

The reaction of carbonate ion with water governs the initial pH of the solution, which can be computed by the method shown for the second equivalence point in Example 11-8. With the first additions of acid, a carbonate/hydrogen carbonate buffer is established. In this region, the pH can be derived from *either* the hydroxide ion concentration calculated from K_{b1} *or* the hydronium ion concentration calculated from K_{a2}. Because we are usually interested in calculating $[H_3O^+]$ and pH, the expression for K_{a2} gives the answer more directly.

Sodium hydrogen carbonate is the principal solute species at the first equivalence point, and Equation 11-11 is used to compute the hydronium ion concentration (see Example 11-5). With the addition of more acid, a new buffer consisting of sodium hydrogen carbonate and carbonic acid is formed. The pH of this buffer is readily obtained from either K_{b2} or K_{a1}.

At the second equivalence point, the solution consists of carbonic acid and sodium chloride. The carbonic acid can be treated as a simple weak acid having a dissociation constant K_{a1}. Finally, after excess hydrochloric acid has been introduced, the dissociation of the weak acid is repressed to a point where the hydronium ion concentration is essentially that of the molar concentration of the strong acid.

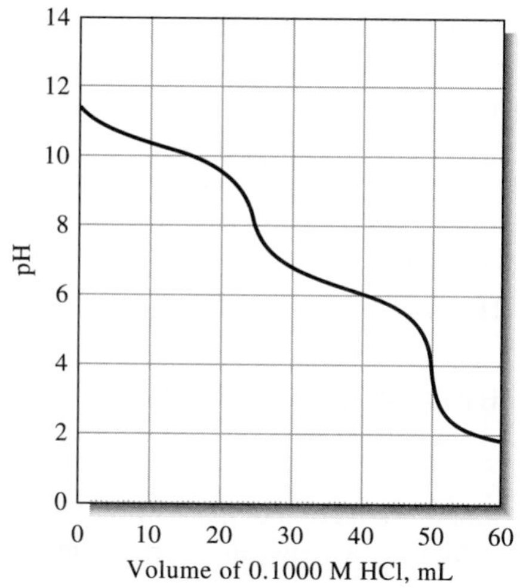

Figure 11-5
Curve for the titration of 25.00 mL of 0.1000 M Na_2CO_3 with 0.1000 M HCl.

Figure 11-5 illustrates that two end points are observed in the titration of sodium carbonate, the second being appreciably sharper than the first. It is apparent that the individual components in mixtures of sodium carbonate and sodium hydrogen carbonate can be determined by neutralization methods.

In general, the titration of acids or bases that have two reactive groups yields individual end points that are of practical value only when the ratio between the two dissociation constants is at least 10^4. If the ratio is much smaller than this, the pH change at the first equivalence point will prove less satisfactory for an analysis.

11H TITRATION CURVES FOR AMPHIPROTIC SPECIES

See *Mathcad Applications for Analytical Chemistry,* **pp. 88–90.**

As noted earlier, an amphiprotic substance, when dissolved in a suitable solvent, behaves both as a weak acid and as a weak base. If either its acidic or basic character predominates sufficiently, titration of the species with a strong base or a strong acid may be feasible. For example, in sodium dihydrogen phosphate solution, the principal equilibria are

$$H_2PO_4^- + H_2O \rightleftharpoons H_3O^+ + HPO_4^{2-} \qquad K_{a2} = 6.32 \times 10^{-8}$$

$$H_2PO_4^- + H_2O \rightleftharpoons OH^- + H_3PO_4 \qquad K_{b3} = \frac{K_w}{K_{a1}} = \frac{1.00 \times 10^{-14}}{7.11 \times 10^{-3}} = 1.41 \times 10^{-12}$$

Note that K_{b3} is much too small to permit titration of $H_2PO_4^-$ with an acid, but K_{a2} is large enough for a successful titration of the ion with a standard base solution.

A different situation prevails in solutions containing disodium hydrogen phosphate, for which the analogous equilibria are

$$HPO_4^{2-} + H_2O \rightleftharpoons H_3O^+ + PO_4^{3-} \qquad K_{a3} = 4.5 \times 10^{-13}$$

$$HPO_4^{2-} + H_2O \rightleftharpoons OH^- + H_2PO_4^- \qquad K_{b2} = \frac{K_w}{K_{a2}} = \frac{1.00 \times 10^{-14}}{6.32 \times 10^{-8}} = 1.58 \times 10^{-7}$$

The magnitude of the constants indicates that HPO_4^{2-} can be titrated with a standard acid but not with a standard base.

Amino acids are amphiprotic.

The simple amino acids are an important class of amphiprotic compounds that contain both a weak acid and a weak base functional group. In an aqueous solution of a typical amino acid, such as glycine, three important equilibria operate:

$$NH_2CH_2COOH \rightleftharpoons NH_3^+CH_2COO^- \qquad (11\text{-}12)$$

$$NH_3^+CH_2COO^- + H_2O \rightleftharpoons NH_2CH_2COO^- + H_3O^+ \qquad K_a = 2 \times 10^{-10} \qquad (11\text{-}13)$$

$$NH_3^+CH_2COO^- + H_2O \rightleftharpoons NH_3^+CH_2COOH + OH^- \qquad K_b = 2 \times 10^{-12} \qquad (11\text{-}14)$$

The first reaction constitutes a kind of internal acid/base reaction and is analogous to the reaction one would observe between a carboxylic acid and an amine:

$$R_1NH_2 + R_2COOH \rightleftharpoons R_1NH_3^+ + R_2COO^- \qquad (11\text{-}15)$$

The typical aliphatic amine has a base dissociation constant of 10^{-4} to 10^{-5}, and many carboxylic acids have acid dissociation constants of about the same magnitude. The consequence is that both reaction 11-12 and reaction 11-15 proceed far to the right, with the product or products being the predominant species in the solution.

A zwitterion is an ionic species that has both a positive and a negative charge.

The amino acid species in Equation 11-12, bearing both a positive and a negative charge, is called a *zwitterion.* As shown by Equations 11-13 and 11-14, the zwitterion of glycine is stronger as an acid than as a base. Thus, an aqueous solution of glycine is somewhat acidic.

The zwitterion of an amino acid, containing as it does a positive and a negative charge, has no tendency to migrate to an electric field, whereas the singly charged anionic and cationic species are attracted to electrodes of opposite charge. No *net* migration of the amino acid occurs in an electric field when the pH of the solvent is such that the concentrations of the anionic and cationic forms are identical. The pH at which no net migration occurs is called the *isoelectric point* and is an important physical constant for characterizing amino acids. The isoelectric point is readily related to the ionization constants for the species. Thus, for glycine

The isoelectric point is the pH at which no net migration of amino acids occurs when they are placed in an electric field.

$$K_a = \frac{[H_3O^+][NH_2CH_2COO^-]}{[NH_3^+CH_3COO^-]}$$

$$K_b = \frac{[OH^-][NH_3^+CH_2COOH]}{[NH_3^+CH_2COO^-]}$$

At the isoelectric point,

$$[NH_2CH_2COO^-] = [NH_3^+CH_2COOH]$$

Thus, if we divide K_a by K_b and substitute this relationship, we obtain for the isoelectric point

$$\frac{K_a}{K_b} = \frac{[H_3O^+]\cancel{[NH_2CH_2COO^-]}}{[OH^-]\cancel{[NH_3^+CH_2COOH]}} = \frac{[H_3O^+]}{[OH^-]}$$

Substitution of $K_w/[H_3O^+]$ for $[OH^-]$ and rearrangement yield

$$[H_3O^+] = \sqrt{\frac{K_a K_w}{K_b}}$$

The isoelectric point for glycine occurs at a pH of 6.0 as shown by the following:

$$[H_3O^+] = \left(\frac{2 \times 10^{-10}}{2 \times 10^{-12}} \times 1 \times 10^{-14}\right)^{1/2} = 1 \times 10^{-6}$$

For simple amino acids, K_a and K_b are generally so small that their determination by direct neutralization is impossible. Addition of formaldehyde removes the amine functional group, however, and leaves the carboxylic acid available for titration with a standard base. For example, with glycine,

$$NH_3^+CH_2COO^- + CH_2O \longrightarrow CH_2{=}NCH_2COOH + H_2O$$

The titration curve for the product is that of a typical carboxylic acid.

11I THE COMPOSITION OF SOLUTIONS OF A POLYPROTIC ACID AS A FUNCTION OF pH

In Section 10G, we showed how alpha values are useful in visualizing the changes in the concentration of various species that occur in the titration of a simple weak acid. Alpha values can also be derived for polyfunctional acids and bases. For example, if we let c_T be the sum of the molar concentrations of the maleate containing species in the solution throughout the titration described in Example 11-8, the alpha value for the free acid α_0 is defined as

See *Mathcad Applications for Analytical Chemistry,* **pp. 77–79.**

$$\alpha_0 = \frac{[H_2M]}{c_T}$$

where

$$c_T = [H_2M] + [HM^-] + [M^{2-}] \qquad (11\text{-}16)$$

The alpha values for HM^- and M^{2-} are given by similar equations:

$$\alpha_1 = \frac{[HM^-]}{c_T}$$

$$\alpha_2 = \frac{[M^{2-}]}{c_T}$$

As noted earlier, the sum of the alpha values for a system must equal unity:

$$\alpha_0 + \alpha_1 + \alpha_2 = 1$$

The alpha values for the maleic acid system are readily expressed in terms of $[H_3O^+]$, K_{a1}, and K_{a2}. To obtain such expressions, we follow the method used to derive Equations 10-16 and 10-17 in Section 10G and obtain

CHALLENGE: Derive Equations 11-17, 11-18, and 11-19.

$$\alpha_0 = \frac{[H_3O^+]^2}{[H_3O^+]^2 + K_{a1}[H_3O^+] + K_{a1}K_{a2}} \tag{11-17}$$

$$\alpha_1 = \frac{K_{a1}[H_3O^+]}{[H_3O^+]^2 + K_{a1}[H_3O^+] + K_{a1}K_{a2}} \tag{11-18}$$

$$\alpha_2 = \frac{K_{a1}K_{a2}}{[H_3O^+]^2 + K_{a1}[H_3O^+] + K_{a1}K_{a2}} \tag{11-19}$$

Note that the denominator is the same for each expression. Note also that the fractional amount of each species is fixed at any pH and is *independent* of the total concentration, c_T.

FEATURE 11-2
A General Expression for Alpha Values

For the weak acid H_nA, the denominator in all alpha-value expressions takes the form:

$$[H_3O^+]^n + K_{a1}[H_3O^+]^{(n-1)} + K_{a1}K_{a2}[H_3O^+]^{(n-2)} + \cdots + K_{a1}K_{a2}\cdots K_{an}$$

The numerator for α_0 is the first term in the denominator; for α_1, it is the second term, and so forth. Thus, if we let D be the denominator, $\alpha_0 = [H_3O^+]^n/D$ and $\alpha_1 = K_{a1}[H_3O^+]^{(n-1)}/D$.

Alpha values for polyfunctional bases are generated in an analogous way, with the equations being written in terms of base dissociation constants and $[OH^-]$.

The three curves plotted in Figure 11-6 show the alpha value for each maleate containing species as a function of pH. The solid curves in Figure 11-7 depict the

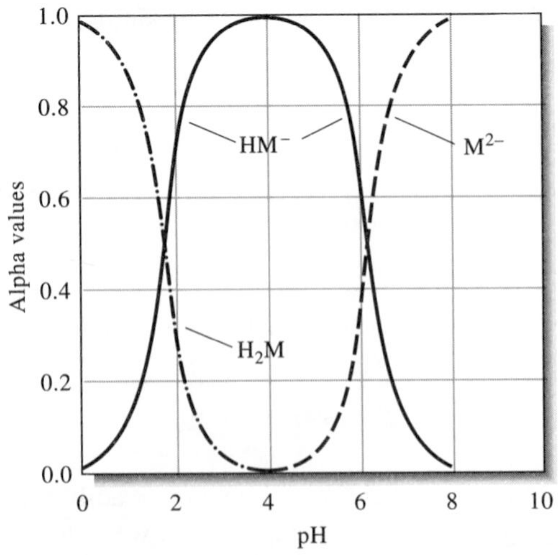

Figure 11-6
Composition of H_2M solutions as a function of pH.

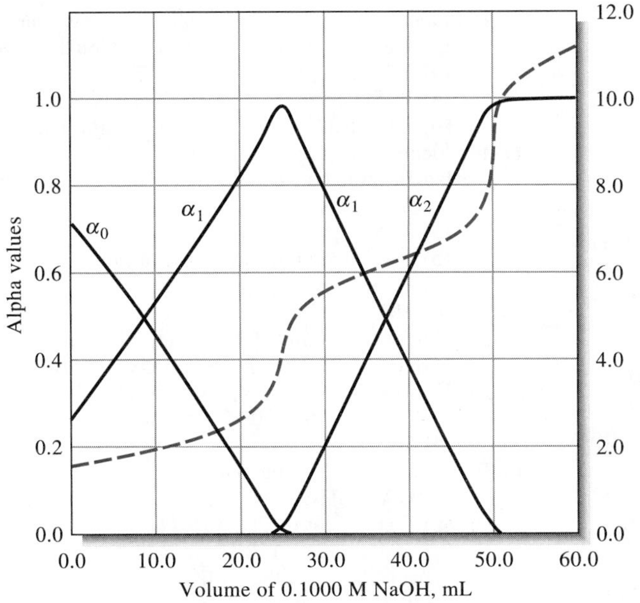

Figure 11-7
Titration of 25.00 mL of 0.1000 M maleic acid with 0.1000 M NaOH. The solid curves are plots of alpha values as a function of volume. The broken curve is a plot of pH as a function of volume.

same alpha values but now plotted as a function of volume of sodium hydroxide as the acid is titrated. The titration curve is also shown by the dashed line in Figure 11-7. Consideration of these curves gives a clear picture of all concentration changes that occur during the titration. For example, Figure 11-7 reveals that before the addition of any base, α_0 for H_2M is roughly 0.7 and α_1 for HM^- is approximately 0.3. For all practical purposes, α_2 is zero. Thus, approximately 70% of the maleic acid exists as H_2M and 30% as HM^-. With addition of base, the pH rises, as does the fraction of HM^-. At the first equivalence point (pH = 4.11), essentially all of the maleate is present as HM^- ($\alpha_1 \rightarrow 1$). Beyond the first equivalence point, HM^- decreases and M^{2-} increases. At the second equivalence point (pH = 9.38) and beyond, essentially all of the maleate exists as M^{2-}.

11J QUESTIONS AND PROBLEMS

***11-1.** As its name implies, NaHA is an ''acid salt'' because it has a proton available to donate to a base. Briefly explain why a pH calculation for a solution of NaHA differs from that for a weak acid of the type HA.

11-2. Solutions of acid salts such as NaHA may be acidic or basic. From the data in Appendix 3, choose an acid salt that will yield a solution that is **(a)** acidic and **(b)** basic. Explain your choice.

***11-3.** Why is it impossible to titrate all three protons of phosphoric acid in aqueous solution?

11-4. Indicate whether an aqueous solution of the following compounds is acidic, neutral, or basic. Explain your answer.

- *(a) NH_4OAc
- (b) $NaNO_2$
- *(c) $NaNO_3$
- (d) $NaHC_2O_4$
- *(e) $Na_2C_2O_4$
- (f) Na_2HPO_4
- *(g) NaH_2PO_4
- (h) Na_3PO_4

11-5. Suggest an indicator that could be used to provide an end point for the titration of the first proton in H_3AsO_4.

***11-6.** Suggest an indicator that would give an end point for the titration of the first two protons in H_3AsO_4.

11-7. Suggest a method for the determination of the amounts of H_3PO_4 and NaH_2PO_4 in an aqueous solution.

11-8. Suggest a suitable indicator for a titration based upon the following reactions; use 0.05 M if an equivalence-point concentration is needed.

- *(a) $H_2CO_3 + NaOH \rightarrow NaHCO_3 + H_2O$
- (b) $H_2P + 2NaOH \rightarrow Na_2P + 2H_2O$ (H_2P = *o*-phthalic acid)
- *(c) $H_2T + 2NaOH \rightarrow Na_2T + 2H_2O$ (H_2T = tartaric acid)
- (d) $NH_2C_2H_4NH_2 + HCl \rightarrow NH_2C_2H_4NH_3Cl$
- *(e) $NH_2C_2H_4NH_2 + 2HCl \rightarrow ClNH_3C_2H_4NH_3Cl$
- (f) $H_2SO_3 + NaOH \rightarrow NaHSO_3 + H_2O$
- *(g) $H_2SO_3 + 2NaOH \rightarrow Na_2SO_3 + 2H_2O$

11-9. Calculate the pH of a solution that is 0.0400 M in

- *(a) H_3PO_4.
- (b) $H_2C_2O_4$.
- *(c) H_3PO_3.
- (d) H_2SO_3.
- *(e) H_2S.
- (f) $H_2NC_2H_4NH_2$.

11-10. Calculate the pH of a solution that is 0.0400 M in
*(a) NaH_2PO_4. (b) $NaHC_2O_4$. *(c) NaH_2PO_3. (d) $NaHSO_3$. *(e) NaHS. (f) $H_2NC_2H_4NH_3^+Cl^-$.

11-11. Calculate the pH of a solution that is 0.0400 M in
*(a) Na_3PO_4. (b) $Na_2C_2O_4$. *(c) Na_2HPO_3. (d) Na_2SO_3. *(e) Na_2S. (g) $C_2H_4(NH_3^+Cl^-)_2$.

***11-12.** Calculate the pH of a solution that is made up to contain the following analytical concentrations:
(a) 0.0500 M in H_3AsO_4 and 0.0200 M in NaH_2AsO_4.
(b) 0.0300 M in NaH_2AsO_4 and 0.0500 M in Na_2HAsO_4.
(c) 0.0600 M in Na_2CO_3 and 0.0300 M in $NaHCO_3$.
(d) 0.0400 M in H_3PO_4 and 0.0200 M in Na_2HPO_4.
(e) 0.0500 M in $NaHSO_4$ and 0.0400 M in Na_2SO_4.

11-13. Calculate the pH of a solution made up to contain the following analytical concentrations:
(a) 0.240 M in H_3PO_4 and 0.480 M in NaH_2PO_3.
(b) 0.0670 M in Na_2SO_3 and 0.0315 M in $NaHSO_3$.
(c) 0.640 M in $HOC_2H_4NH_2$ and 0.750 M in $HOC_2H_4NH_3Cl$.
(d) 0.0240 in $H_2C_2O_4$ (oxalic acid) and 0.0360 M in $Na_2C_2O_4$.
(e) 0.0100 M in $Na_2C_2O_4$ and 0.0400 M in $NaHC_2O_4$.

***11-14.** Calculate the pH of a solution that is
(a) 0.0100 M in HCl and 0.0200 M in picric acid.
(b) 0.0100 M in HCl and 0.0200 M in benzoic acid.
(c) 0.0100 M in NaOH and 0.100 M in Na_2CO_3.
(d) 0.0100 M in NaOH and 0.100 M in NH_3.

11-15. Calculate the pH of a solution that is
(a) 0.0100 M in $HClO_4$ and 0.0300 M in monochloroacetic acid.
(b) 0.0100 M in HCl and 0.0150 M in H_2SO_4.
(c) 0.0100 M in NaOH and 0.0300 M in Na_2S.
(d) 0.0100 M in NaOH and 0.0300 M in sodium acetate.

***11-16.** Identify the principal conjugate acid/base pair and calculate the ratio between them in a solution that is buffered to pH 6.00 and contains
(a) H_2SO_3. (b) citric acid. (c) malonic acid. (d) tartaric acid.

11-17. Identify the principal conjugate acid/base pair and calculate the ratio between them in a solution that is buffered to pH 9.00 and contains
(a) H_2S.
(b) ethylenediamine dihydrochloride.
(c) H_3AsO_4.
(d) H_2CO_3.

***11-18.** How many grams of $Na_2HPO_4 \cdot 2H_2O$ must be added to 400 mL of 0.200 M H_3PO_4 to give a buffer of pH 7.30?

11-19. How many grams of dipotassium phthalate must be added to 750 mL of 0.0500 M phthalic acid to give a buffer of pH 5.75?

***11-20.** What is the pH of the buffer formed by mixing 50.0 mL of 0.200 M NaH_2PO_4 with
(a) 50.0 mL of 0.120 M HCl?
(b) 50.0 mL of 0.120 M NaOH?

11-21. What is the pH of the buffer formed by adding 100 mL of 0.150 M potassium hydrogen phthalate to
(a) 100 mL of 0.0800 M NaOH?
(b) 100 mL of 0.0800 M HCl?

***11-22.** How would you prepare 1.00 L of a buffer with a pH of 9.60 from 0.300 M Na_2CO_3 and 0.200 M HCl?

11-23. How would you prepare 1.00 L of a buffer with a pH of 7.00 from 0.200 M H_3PO_4 and 0.160 M NaOH?

***11-24.** How would you prepare 1.00 L of a buffer with a pH of 6.00 from 0.500 M Na_3AsO_4 and 0.400 M HCl?

11-25. Identify by letter the curve you would expect in the titration of a solution containing
(a) disodium maleate, Na_2M, with standard acid.
(b) pyruvic acid, HP, with standard base.
(c) sodium carbonate, Na_2CO_3, with standard acid.

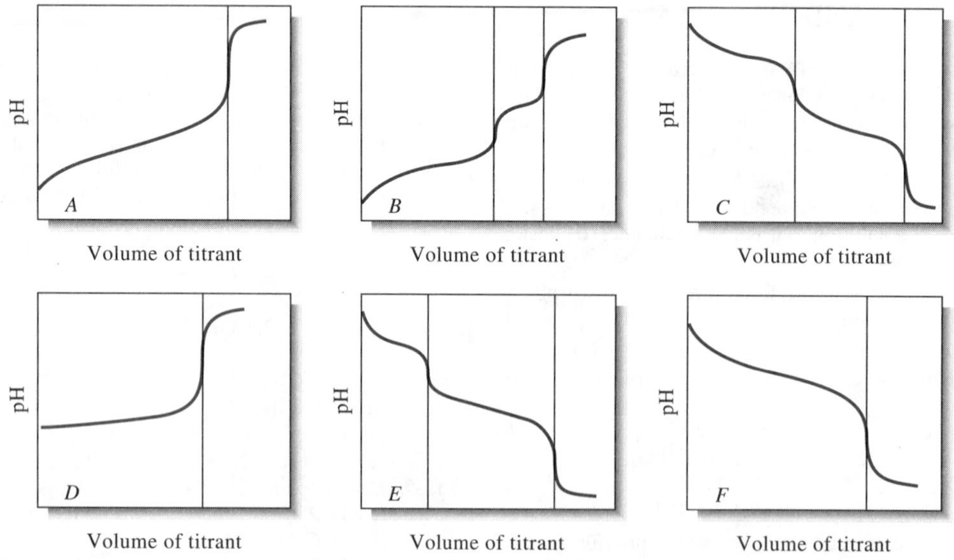

11-26. Describe the composition of a solution that would be expected to yield a curve resembling (see Problem 11-25)
(a) curve *B*. **(b)** curve *A*. **(c)** curve *E*.

11-27. Briefly explain why curve *B cannot* describe the titration of a mixture consisting of H_3PO_4 and NaH_2PO_4.

11-28. Derive a curve for the titration of 50.00 mL of a 0.1000 M solution of compound A with a 0.2000 M solution of compound B in the following list. For each titration, calculate the pH after the addition of 0.00, 12.50, 20.00, 24.00, 25.00, 26.00, 37.50, 45.00, 49.00, 50.00, 51.00, and 60.00 mL of compound B:

A	B
*(a) Na_2CO_3	HCl
(b) ethylenediamine	HCl
*(c) H_2SO_4	NaOH
(d) H_2SO_3	NaOH

***11-29.** Generate a curve for the titration of 50.00 mL of a solution in which the analytical concentration of NaOH is 0.1000 M and that for hydrazine is 0.0800 M. Calculate the pH after addition of 0.00, 10.00, 20.00, 24.00, 25.00, 26.00, 35.00, 44.00, 45.00, 46.00, and 50.00 mL of 0.2000 M $HClO_4$.

11-30. Generate a curve for the titration of 50.00 mL of a solution in which the analytical concentration of $HClO_4$ is 0.1000 M and that for formic acid is 0.0800 M. Calculate the pH after addition of 0.00, 10.00, 20.00, 24.00, 25.00, 26.00, 35.00, 44.00, 45.00, 46.00, and 50.00 mL of 0.2000 M KOH.

11-31. Formulate equilibrium constants for the following equilibria, giving numerical values for the constants:
*(a) $H_2AsO_4^- + H_2AsO_4^- \rightleftharpoons H_3AsO_4 + HAsO_4^{2-}$
(b) $HAsO_4^{2-} + HAsO_4^{2-} \rightleftharpoons AsO_4^{3-} + H_2AsO_4^-$

***11-32.** Derive a numerical value for the equilibrium constant for the reaction:

$$NH_4^+ + OAc^- \rightleftharpoons NH_3 + HOAc$$

11-33. For pH values of 2.00, 6.00, and 10.00, calculate the alpha value for each species in an aqueous solution of
*(a) phthalic acid. (d) arsenic acid.
(b) phosphoric acid. *(e) phosphorous acid.
*(c) citric acid. (f) oxalic acid.

11-34. Derive equations that define α_0, α_1, α_2, and α_3 for the acid H_3AsO_4.

CHAPTER

12

Applications of Neutralization Titrations

Neutralization titrations are widely used to determine the concentration of analytes that are themselves acids or bases or are convertible to such species by suitable treatment.[1] Water is the usual solvent for neutralization titrations because it is readily available, inexpensive, and nontoxic. Its low temperature coefficient of expansion is an added virtue. Some analytes, however, are not titratable in aqueous media because their solubilities are too low or because their strengths as acids or bases are not sufficiently great to provide satisfactory end points. Such substances can often be titrated in a solvent other than water.[2] We shall restrict our discussions to aqueous systems.

Nonaqueous solvents, such as methyl and ethyl alcohol, glacial acetic acid, and methyl isobutyl ketone, often make it possible to titrate acids or bases that are too weak to titrate in aqueous solution.

12A REAGENTS FOR NEUTRALIZATION TITRATIONS

In Chapter 11, we noted that strong acids and strong bases cause the most pronounced change in pH at the equivalence point. For this reason, standard solutions for neutralization titrations are always prepared from these reagents.

12A-1 Preparation of Standard Acid Solutions

Hydrochloric acid is widely used for titration of bases. Dilute solutions of the reagent are stable indefinitely and do not cause troublesome precipitation reactions with most cations. It is reported that 0.1 M solutions of HCl can be boiled for as long as an hour without loss of acid, provided that the water lost by evaporation is periodically replaced; 0.5 M solutions can be boiled for at least ten minutes without significant loss.

Solutions of HCl, $HClO_4$, and H_2SO_4 are stable indefinitely. Restandardization is never required.

Solutions of perchloric acid and sulfuric acid are also stable and are useful for titrations where chloride ion interferes by forming precipitates. Standard solutions of nitric acid are seldom encountered because of their oxidizing properties.

---

[1]For a review of applications of neutralization titrations, see D. Rosenthal and P. Zuman, in *Treatise on Analytical Chemistry,* 2nd ed., I. M. Kolthoff and P. J. Elving, Eds., Part I, Vol. 2, Chapter 18. New York: Wiley, 1979.

[2]For a review of nonaqueous acid/base titrimetry, see I. M. Kolthoff and M. K. Chantoon Jr., in *Treatise on Analytical Chemistry,* 2nd ed., I. M. Kolthoff and P. J. Elving, Eds., Part I, Vol. 2, Chapters 19A–19E. New York: Wiley, 1979.

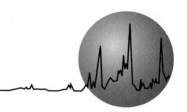

Standard acid solutions are ordinarily prepared by diluting an approximate volume of the concentrated reagent and subsequently standardizing the diluted solution against a primary-standard base. Less frequently, the composition of the concentrated acid is established through careful density measurement; a weighed quantity is then diluted to an exact volume. (Tables relating reagent density to composition are found in most chemistry and chemical engineering handbooks.) A stock solution with an exactly known hydrochloric acid concentration can also be prepared by dilution of a quantity of the concentrated reagent with an equal volume of water followed by distillation. Under controlled conditions, the final quarter of the distillate, which is known as *constant-boiling* HCl, has a fixed and known composition, its acid content being dependent only upon atmospheric pressure. For a pressure P between 670 and 780 torr, the mass in air of the distillate that contains exactly one mole of H_3O^+ is[3]

$$\frac{\text{mass constant-boiling HCl in g}}{\text{mol } H_3O^+} = 164.673 + 0.02039\,P \qquad (12\text{-}1)$$

Standard solutions are prepared by diluting weighed quantities of this acid to accurately known volumes.

12A-2 The Standardization of Acids

Sodium Carbonate

Acids are frequently standardized against weighed quantities of sodium carbonate. Primary-standard-grade sodium carbonate is available commercially or can be prepared by heating purified sodium hydrogen carbonate between 270°C to 300°C for 1 hr:

$$2NaHCO_3(s) \longrightarrow Na_2CO_3(s) + H_2O(g) + CO_2(g)$$

Sodium carbonate occurs naturally in large deposits as *washing soda,* $Na_2CO_3 \cdot 10H_2O$, and as *trona,* $Na_2CO_3 \cdot NaHCO_3 \cdot 2H_2O$. These minerals find wide use in the glass industry and many others as well. Primary standard sodium carbonate is manufactured by extensive purification of these minerals.

As shown in Figure 11-5, two end points are observed in the titration of sodium carbonate. The first, corresponding to the conversion of carbonate to hydrogen carbonate, occurs at about pH 8.3; the second, involving the formation of carbonic acid and carbon dioxide, is observed at about pH 3.8. The second end point is always used for standardization because the change in pH is greater than that of the first. An even sharper end point can be achieved by boiling the solution briefly to eliminate the reaction product, carbonic acid and carbon dioxide. The sample is titrated to the first appearance of the acid color of the indicator (such as bromocresol green or methyl orange). At this point, the solution contains a large amount of dissolved carbon dioxide and small amounts of carbonic acid and unreacted hydrogen carbonate. Boiling effectively destroys this buffer by eliminating the carbonic acid:

$$H_2CO_3(aq) \longrightarrow CO_2(g) + H_2O(l)$$

[3] *Official Methods of Analysis of the AOAC,* 15th ed., p. 692. Washington, DC: Association of Official Analytical Chemists, 1990.

The solution then becomes alkaline again due to the residual hydrogen carbonate ion. The titration is completed after the solution has cooled. Now, however, a substantially larger decrease in pH occurs during the final additions of acid, thus giving a more abrupt color change (see Figure 12-1).

As an alternative, the acid can be introduced in an amount slightly in excess of that needed to convert the sodium carbonate to carbonic acid. The solution is boiled as before to remove carbon dioxide and cooled; the excess acid is then back-titrated with a dilute solution of base. Any indicator suitable for a strong acid/strong base titration is satisfactory. The volume ratio of acid to base must of course be established by an independent titration.

Other Primary Standards for Acids

Tris-(hydroxymethyl)aminomethane, $(HOCH_2)_3CNH_2$, known also as TRIS or THAM, is available in primary-standard purity from commercial sources. It possesses the advantage of a substantially greater mass per mole of protons consumed (121.1) than sodium carbonate (53.0) (see Example 12-2).

Sodium tetraborate decahydrate and mercury(II) oxide have also been recommended as primary standards. The reaction of an acid with the tetraborate is

$$B_4O_7^{2-} + 2H_3O^+ + 3H_2O \longrightarrow 4H_3BO_3$$

A high mass per proton consumed is desirable in a primary standard because a larger mass of reagent must be used, thus decreasing the relative weighing error.

$$
\begin{array}{c}
CH_2OH \\
| \\
H_2N-C-CH_2OH \\
| \\
CH_2OH
\end{array}
$$

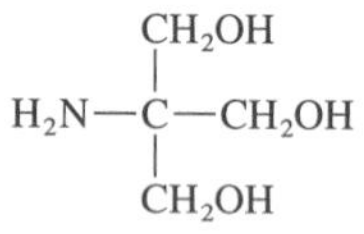

TRIS

TRIS (THAM) reacts in a 1:1 molar ratio with hydronium ions.

Borax, $Na_2B_4O_7 \cdot 10H_2O$, is a mineral that is mined in desert country and is widely used in cleaning preparations. A highly purified form of borax is used as a primary standard for bases.

EXAMPLE 12-1

Compare the masses of (a) TRIS (121 g/mol), (b) Na_2CO_3 (106 g/mol), and (c) $Na_2B_4O_7 \cdot 10H_2O$ (381 g/mol) that should be taken to standardize an approximately 0.020 molar solution of HCl if it is desired that at least 30 mL of the acid be used for the titration.

In each case,

$$\text{amount of HCl} = 30.0 \text{ mL HCl} \times 0.020 \frac{\text{mmol HCl}}{\text{mL HCl}} = 0.60 \text{ mmol HCl}$$

(a) $\text{mass TRIS} = 0.60 \text{ mmol HCl} \times \dfrac{1 \text{ mmol TRIS}}{\text{mmol HCl}} \times \dfrac{0.121 \text{ g TRIS}}{\text{mmol TRIS}}$

$$= 0.073 \text{ g}$$

(b) $\text{mass Na}_2\text{CO}_3 = 0.60 \text{ mmol HCl} \times \dfrac{1 \text{ mmol Na}_2\text{CO}_3}{2 \text{ mmol HCl}}$

$$\times \dfrac{0.106 \text{ g Na}_2\text{CO}_3}{\text{mmol Na}_2\text{CO}_3} = 0.032 \text{ g}$$

(c) $\text{mass Na}_2\text{B}_4\text{O}_7 \cdot 10\text{H}_2\text{O} = 0.60 \text{ mmol HCl} \times \dfrac{1 \text{ mmol Na}_2\text{B}_4\text{O}_7 \cdot 10\text{H}_2\text{O}}{2 \text{ mmol HCl}}$

$$\times \dfrac{0.381 \text{ g Na}_2\text{B}_4\text{O}_7 \cdot 10\text{H}_2\text{O}}{\text{mmol Na}_2\text{B}_4\text{O}_7 \cdot 10\text{H}_2\text{O}} = 0.11 \text{ g}$$

EXAMPLE 12-2

Assume that the standard deviation associated with weighing out the primary standards in Example 12-1 was 0.1 mg. Calculate the relative standard deviation that this uncertainty would introduce into each of the calculated molarities.

Here, the relative uncertainty in the molarity due to weighing would be equal to the relative uncertainty in the weighing process.
(a) For TRIS,

$$\text{rel std dev} = \frac{0.0001\text{ g}}{0.073\text{ g}} \times 100\% = 0.14\%$$

(b) Similarly, for Na_2CO_3,

$$\text{rel std dev} = \frac{0.0001\text{ g}}{0.032\text{ g}} \times 100\% = 0.31\%$$

(c) For $Na_2B_4O_7 \cdot 10H_2O$,

$$\text{rel std dev} = \frac{0.0001\text{ g}}{0.11\text{ g}} \times 100\% = 0.09\%$$

12A-3 Preparation of Standard Solutions of Bases

Sodium hydroxide is the most common base for preparing standard solutions although potassium hydroxide and barium hydroxide are also encountered. None of these is obtainable in primary-standard purity, so standardization is required after preparation.

The Effect of Carbon Dioxide upon Standard Base Solutions

In solution as well as in the solid state, the hydroxides of sodium, potassium, and barium react rapidly with atmospheric carbon dioxide to produce the corresponding carbonate:

$$CO_2(g) + 2OH^- \longrightarrow CO_3^{2-} + H_2O$$

Although production of each carbonate ion uses up two hydroxide ions, the uptake of carbon dioxide by a solution of base does not necessarily alter its combining capacity for hydronium ions. Thus, at the end point of a titration that requires an acid-range indicator (such as bromocresol green), each carbonate ion produced from sodium or potassium hydroxide will have reacted with two hydronium ions of the acid (Figure 12-1):

$$CO_3^{2-} + 2H_3O^+ \longrightarrow H_2CO_3 + 2H_2O$$

Because the amount of hydronium ion consumed by this reaction is identical to the amount of hydroxide lost during formation of the carbonate ion, no error is incurred.

Absorption of carbon dioxide by a standardized solution of sodium or potassium hydroxide leads to a negative systematic error in analyses that require an indicator with a basic range; no systematic error is incurred when an indicator with an acidic range is used.

Unfortunately, most applications of standard base require an indicator with a basic transition range (phenolphthalein, for example). Here, each carbonate ion has reacted with only one hydronium ion when the color change of the indicator is observed:

$$CO_3^{2-} + H_3O^+ \longrightarrow HCO_3^- + H_2O$$

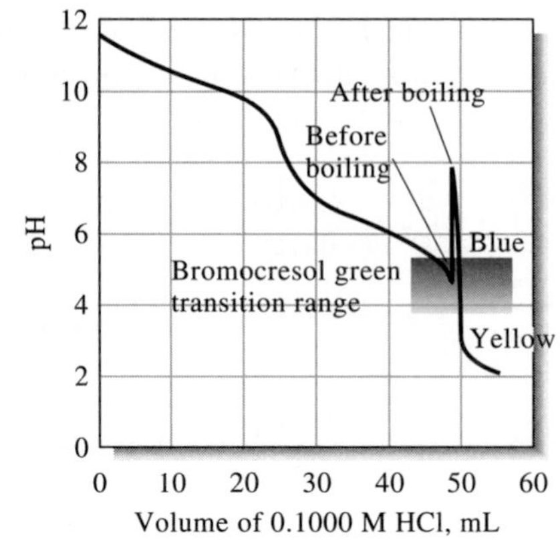

Figure 12-1
Titration of 25.00 mL of 0.1000 M Na_2CO_3 with 0.1000 M HCl. After about 49 mL of HCl have been added, the solution is boiled, causing the increase in pH shown. The change in pH on further addition of HCl is much larger.

The effective concentration of the base is thus diminished by absorption of carbon dioxide, and a systematic error (called a *carbonate error*) results.

EXAMPLE 12-3

A carbonate-free NaOH solution was found to be 0.05118 M immediately after preparation. If exactly 1.000 L of this solution was exposed to air for some time and absorbed 0.1962 g of CO_2, calculate the relative carbonate error that would arise in the determination of acetic acid with the contaminated solution if phenolphthalein were used as an indicator.

$$2NaOH + CO_2 \longrightarrow Na_2CO_3 + H_2O$$

$$c_{Na_2CO_3} = 0.1962\ \cancel{g\ CO_2} \times \frac{1\ \cancel{mol\ CO_2}}{44.01\ \cancel{g\ CO_2}} \times \frac{1\ mol\ Na_2CO_3}{\cancel{mol\ CO_2}} \times \frac{1}{1.000\ L\ soln}$$

$$= 4.458 \times 10^{-3}\ M$$

Effective concentration c_{NaOH} of NaOH for acetic acid is

$$c_{NaOH} = 0.05118\ \frac{mol\ NaOH}{L} - \frac{4.458 \times 10^{-3}\ \cancel{mol\ Na_2CO_3}}{L}$$

$$\times \frac{1\ \cancel{mol\ HCl}}{\cancel{mol\ Na_2CO_3}} \times \frac{1\ mol\ NaOH}{\cancel{mol\ HCl}} = 0.04672\ M$$

$$\text{rel error} = \frac{0.04672 - 0.05118}{0.05118} \times 100\% = -8.7\%$$

Generally, carbonate ion in standard solutions of bases is undesirable because it decreases the sharpness of end points.

The solid reagents used to prepare standard solutions of base are always contaminated by significant amounts of carbonate ion. The presence of this contaminant does not cause a carbonate error provided the same indicator is used for both standardization and analysis. It does, however, lead to less sharp end points. Consequently steps are usually taken to remove carbonate ion before a solution of a base is standardized.

The best method for preparing carbonate-free sodium hydroxide solutions takes advantage of the very low solubility of sodium carbonate in concentrated solutions of the base. An approximately 50% aqueous solution of sodium hydroxide is prepared (or purchased from commercial sources). The solid sodium carbonate is allowed to settle to give a clear liquid that is decanted and diluted to give the desired concentration (alternatively the solid is removed by vacuum filtration).

Water that is used to prepare carbonate-free solutions of base must also be free of carbon dioxide. Distilled water, which is sometimes supersaturated with carbon dioxide, should be boiled briefly to eliminate the gas. The water is then allowed to cool to room temperature before the introduction of base because hot alkali solutions rapidly absorb carbon dioxide. Deionized water ordinarily does not contain significant amounts of carbon dioxide.

A tightly capped polyethylene bottle usually provides sufficient short-term protection against the uptake of atmospheric carbon dioxide. Before capping, the bottle is squeezed to minimize the interior air space. Care should also be taken to keep the bottle closed except during the brief periods when the contents are being transferred to a buret. Sodium hydroxide solutions will ultimately cause a polyethylene bottle to become brittle.

The concentration of a sodium hydroxide solution will decrease slowly (0.1% to 0.3% per week) if the base is stored in glass bottles. The loss in strength is caused by the reaction of the base with the glass to form sodium silicates. For this reason, standard solutions of base should not be stored for extended periods (longer than 1 or 2 weeks) in glass containers. In addition, bases should never be kept in glass-stoppered containers because the reaction between the base and the stopper may cause the latter to "freeze" after a brief period. Finally, to avoid the same type of freezing, burets with glass stopcocks should be promptly drained and thoroughly rinsed with water after use with standard base solutions. This problem is avoided with burets equipped with Teflon stopcocks.

WARNING: Concentrated solutions of NaOH (and KOH) are extremely corrosive to the skin. In making up standard solutions of NaOH, a face shield, rubber gloves, and protective clothing **must be worn at all times.**

Water that is in equilibrium with atmospheric constituents contains only about 1.5×10^{-5} mol CO_2/L, an amount that has a negligible effect on the strength of most standard bases. As an alternative to boiling to remove CO_2 from supersaturated solutions of CO_2, the excess gas can be removed by bubbling air through the water for several hours. This process is called *sparging* and produces a solution that contains the equilibrium concentration of CO_2.

Sparging is the process of removing a gas from a solution by bubbling an inert gas through the solution.

Solutions of bases are preferably stored in polyethylene bottles rather than glass because of the reaction between bases and glass. Such solutions should never be stored in glass-stoppered bottles; after the solution has stood for a period, removal of the stopper often becomes impossible.

12A-4 The Standardization of Bases

Several excellent primary standards are available for the standardization of bases. Most are weak organic acids that require the use of an indicator with a basic transition range.

Standard solutions of strong bases cannot be prepared directly by mass and must always be standardized against a primary standard with acidic properties.

Potassium Hydrogen Phthalate, $KHC_8H_4O_4$

Potassium hydrogen phthalate is an ideal primary standard. It is a nonhygroscopic crystalline solid with a high molar mass (204.2 g/mol). For most purposes, the commercial analytical-grade salt can be used without further purification. For the most exacting work, potassium hydrogen phthalate of certified purity is available from the National Institute of Standards and Technology.

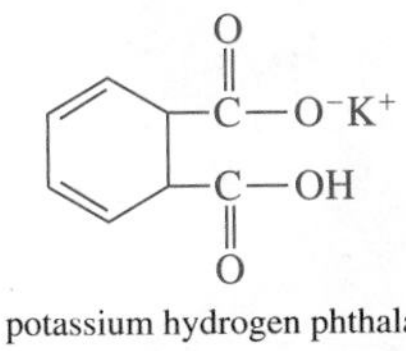

potassium hydrogen phthalate

Other Primary Standards for Bases

Benzoic acid is obtainable in primary-standard purity and can be used for the standardization of bases. Because its solubility in water is limited, this reagent is ordinarily dissolved in ethanol prior to dilution with water and titration. A blank should always be carried through this standardization because commercial alcohol is sometimes slightly acidic.

Potassium hydrogen iodate, $KH(IO_3)_2$, is an excellent primary standard with a high molecular mass per mole of protons (389.9 g/mol). It is also a strong acid that can be titrated using virtually any indicator with a transition range between pH 4 and 10.

$KH(IO_3)_2$, in contrast to all other primary standards for bases, has the advantage of being a strong acid, thus making the choice of indicator less critical.

12B TYPICAL APPLICATIONS OF NEUTRALIZATION TITRATIONS

Neutralization titrations are still among the most widely used analytical methods.

Neutralization titrations are used to determine the innumerable inorganic, organic, and biological species that possess inherent acidic or basic properties. Equally important, however, are the many applications that involve conversion of an analyte to an acid or base by suitable chemical treatment followed by titration with a standard strong base or acid.

Two major types of end points find widespread use in neutralization titrations. The first is a visual end point based on indicators such as those described in Section 10A. The second is a *potentiometric* end point in which the potential of a glass/reference electrode system is determined with a voltage-measuring device. The measured potential is directly proportional to pH. Potentiometric end points are described in Section 19C.

12B-1 Elemental Analysis

Kjeldahl is pronounced Kyell′ dahl. Hundreds of thousands of Kjeldahl nitrogen determinations are performed each year, primarily to provide a measure of the protein content of meats, grains, and animal feeds.

Several important elements that occur in organic and biological systems are conveniently determined by methods that involve an acid/base titration as the final step. Generally, the elements susceptible to this type of analysis are nonmetallic and include carbon, nitrogen, chlorine, bromine, fluorine, as well as a few other less common species. Pretreatment converts the element to an inorganic acid or base that is then titrated. A few examples follow.

Nitrogen

Nitrogen is found in a wide variety of substances of interest in research, industry, and agriculture. Examples include amino acids, proteins, synthetic drugs, fertilizers, explosives, soils, potable water supplies, and dyes. Thus, analytical methods for the determination of nitrogen, particularly in organic substrates, are of singular importance.

The most common method for determining organic nitrogen is the *Kjeldahl method,* which is based on a neutralization titration. The procedure is straightforward, requires no special equipment, and is readily adapted to the routine analysis of large numbers of samples. It (or one of its modifications) is the standard means for determining the protein content of grains, meats, and other biological materials. Since most proteins contain approximately the same percentage of nitrogen, multiplication of this percentage by a suitable factor (6.25 for meats, 6.38 for dairy products, and 5.70 for cereals) gives the percentage of protein in a sample.

The Kjeldahl method was developed by a Danish chemist who first described it in 1883: J. Kjeldahl, *Z. Anal. Chem.* **1883,** *22,* 366.

FEATURE 12-1
Other Methods for Determining Organic Nitrogen

Several other methods are used to determine the nitrogen content of organic materials. In the *Dumas method,* the sample is mixed with powdered copper(II) oxide and ignited in a combustion tube to give carbon dioxide, water, nitrogen, and small amounts of nitrogen oxides. A stream of carbon dioxide carries these products through a packing of hot copper, which reduces any oxides of nitrogen to elemental nitrogen. The mixture then is passed into a gas buret filled with concentrated potassium hydroxide. The only component not absorbed by the base is nitrogen, and its volume is measured directly.

The newest method for determining organic nitrogen involves combusting the sample at 1100°C for a few minutes to convert the nitrogen to nitric oxide, NO. Ozone is then introduced into the gaseous mixture, which oxidizes the nitric oxide to nitrogen dioxide. This reaction gives off visible radiation *(chemiluminescence)*, the intensity of which is measured and is proportional to the nitrogen content of the sample. An instrument for this procedure is available from commercial sources.

In the Kjeldahl method, the sample is decomposed in hot, concentrated sulfuric acid to convert the bound nitrogen to ammonium ion. The resulting solution is then cooled, diluted, and made basic. The liberated ammonia is distilled, collected in an acidic solution, and determined by a neutralization titration.

The critical step in the Kjeldahl method is the decomposition with sulfuric acid, which oxidizes the carbon and hydrogen in the sample to carbon dioxide and water. The fate of the nitrogen, however, depends upon its state of combination in the original sample. Amine and amide nitrogens are quantitatively converted to ammonium ion. In contrast, nitro, azo, and azoxy groups are likely to yield elemental nitrogen or various oxides of nitrogen, all of which are lost from the hot acidic medium. This loss can be avoided by pretreating the sample with a reducing agent to form products that behave as amide or amine nitrogen. In one such prereduction scheme, salicylic acid and sodium thiosulfate are added to the concentrated sulfuric acid solution containing the sample. After a brief period, the digestion is performed in the usual way.

$-NO_2$ nitro group

$-N{=}N-$ azo group

$-N^{+}{=}N-$ (with O^- on N$^+$) azoxy group

Certain aromatic heterocyclic compounds, such as pyridine and its derivatives, are particularly resistant to complete decomposition by sulfuric acid. Such compounds yield low results as a consequence (Figure 2-3) unless special precautions are taken.

The decomposition step is frequently the most time-consuming aspect of a Kjeldahl determination. Some samples may require heating periods in excess of 1 hr. Numerous modifications of the original procedure have been proposed with the aim of shortening the digestion time. In the most widely used modification, a neutral salt, such as potassium sulfate, is added to increase the boiling point of the sulfuric acid solution and thus the temperature at which the decomposition occurs. In another modification, a solution of hydrogen peroxide is added to the mixture after the digestion has decomposed most of the organic matrix.

Many substances catalyze the decomposition of organic compounds by sulfuric acid. Mercury, copper, and selenium, either combined or in the elemental state, are effective. Mercury(II), if present, must be precipitated with hydrogen sulfide prior to distillation to prevent retention of ammonia as a mercury(II) ammine complex.

EXAMPLE 12-4

A 0.7121-g sample of a wheat flour was analyzed by the Kjeldahl method. The ammonia formed by addition of concentrated base after digestion with

H_2SO_4 was distilled into 25.00 mL of 0.04977 M HCl. The excess HCl was then back-titrated with 3.97 mL of 0.04012 M NaOH. Calculate the percent protein in the flour.

$$\text{amount HCl} = 25.00 \cancel{\text{mL HCl}} \times 0.04977 \frac{\text{mmol HCl}}{\cancel{\text{mL HCl}}} = 1.2443 \text{ mmol}$$

$$\text{amount NaOH} = 3.97 \cancel{\text{mL NaOH}} \times 0.04012 \frac{\text{mmol NaOH}}{\cancel{\text{mL NaOH}}} = \underline{0.1593} \text{ mmol}$$

$$\text{amount N} = 1.0850 \text{ mmol}$$

$$\% \text{ N} = \frac{1.0850 \cancel{\text{mmol N}} \times \frac{0.014007 \text{ g N}}{\cancel{\text{mmol N}}}}{0.7121 \text{ g sample}} \times 100\% = 2.1341$$

$$\% \text{ protein} = 2.1341 \cancel{\% \text{N}} \times \frac{5.70\% \text{ protein}}{\cancel{\% \text{N}}} = 12.16$$

Sulfur

Sulfur dioxide in the atmosphere is often determined by drawing a sample through a hydrogen peroxide solution and then titrating the sulfuric acid.

Sulfur in organic and biological materials is conveniently determined by burning the sample in a stream of oxygen. The sulfur dioxide (as well as the sulfur trioxide) formed during the oxidation is collected by distillation into a dilute solution of hydrogen peroxide:

$$SO_2(g) + H_2O_2 \longrightarrow H_2SO_4$$

The sulfuric acid is then titrated with standard base.

Other Elements

Table 12-1 lists other elements that can be determined by neutralization methods.

12B-2 The Determination of Inorganic Substances

Numerous inorganic species can be determined by titration with strong acids or bases. A few examples follow.

TABLE 12-1 Elemental Analyses Based on Neutralization Titrations

Element	Converted to	Absorption or Precipitation Products	Titration
N	NH_3	$NH_3(g) + H_3O^+ \rightarrow NH_4^+ + H_2O$	Excess HCl with NaOH
S	SO_2	$SO_2(g) + H_2O_2 \rightarrow H_2SO_4$	NaOH
C	CO_2	$CO_2(g) + Ba(OH)_2 \rightarrow BaCO_3(s) + H_2O$	Excess $Ba(OH)_2$ with HCl
Cl (Br)	HCl	$HCl(g) + H_2O \rightarrow Cl^- + H_3O^+$	NaOH
F	SiF_4	$SiF_4(g) + H_2O \rightarrow H_2SiF_6$	NaOH
P	H_3PO_4	$12H_2MoO_4 + 3NH_4^+ + H_3PO_4 \rightarrow (NH_4)_3PO_4 \cdot 12MoO_3(s) + 12H_2O + 3H^+$	
		$(NH_4)_3PO_4 \cdot 12MoO_3(s) + 26OH^- \rightarrow HPO_4^{2-} + 12MoO_4^{2-} + 14H_2O + 3NH_3(g)$	Excess NaOH with HCl

Ammonium Salts

Ammonium salts are conveniently determined by conversion to ammonia with strong base followed by distillation. The ammonia is collected and titrated as in the Kjeldahl method.

Nitrates and Nitrites

The method just described for ammonium salts can be extended to the determination of inorganic nitrate or nitrite. These ions are first reduced to ammonium ion by Devarda's alloy (50% Cu, 45% Al, 5% Zn). Granules of the alloy are introduced into a strongly alkaline solution of the sample in a Kjeldahl flask. The ammonia is distilled after reaction is complete. Arnd's alloy (60% Cu, 40% Mg) has also been used as the reducing agent.

Carbonates and Carbonate Mixtures

The qualitative and quantitative determination of the constituents in a solution containing sodium carbonate, sodium hydrogen carbonate, and sodium hydroxide, either alone or admixed, provides interesting examples of how neutralization titrations can be employed to analyze mixtures. No more than two of these three constituents can exist in an appreciable amount in any solution because reaction will eliminate the third. Thus, mixing sodium hydroxide with sodium hydrogen carbonate results in the formation of sodium carbonate until one or the other (or both) of the original reactants is exhausted. If the sodium hydroxide is used up, the solution will contain sodium carbonate and sodium hydrogen carbonate; if sodium hydrogen carbonate is depleted, sodium carbonate and sodium hydroxide will remain; if equimolar amounts of sodium hydrogen carbonate and sodium hydroxide are mixed, the principal solute species will be sodium carbonate.

The analysis of such mixtures requires two titrations: one with an alkaline-range indicator, such as phenolphthalein, and the other with an acid-range indicator, such as bromocresol green. The composition of the solution can then be deduced from the relative volumes of acid needed to titrate equal volumes of the sample (Table 12-2 and Figure 12-2). Once the composition of the solution has been established, the volume data can be used to determine the concentration of each component in the sample.

TABLE 12-2 Volume of Acid Relationships in the Analysis of Mixtures Containing Hydroxide, Carbonate, and Hydrogen Carbonate Ions

Constituent(s) in Sample	Relationship between V_{phth} and V_{bcg} in the Titration of an Equal Volume of Sample*
NaOH	$V_{phth} = V_{bcg}$
Na_2CO_3	$V_{phth} = \frac{1}{2}V_{bcg}$
$NaHCO_3$	$V_{phth} = 0$; $V_{bcg} > 0$
NaOH, Na_2CO_3	$V_{phth} > \frac{1}{2}V_{bcg}$
Na_2CO_3, $NaHCO_3$	$V_{phth} < \frac{1}{2}V_{bcg}$

*V_{phth} = volume of acid needed for a phenolphthalein end point; V_{bcg} = volume of acid needed for a bromocresol green end point.

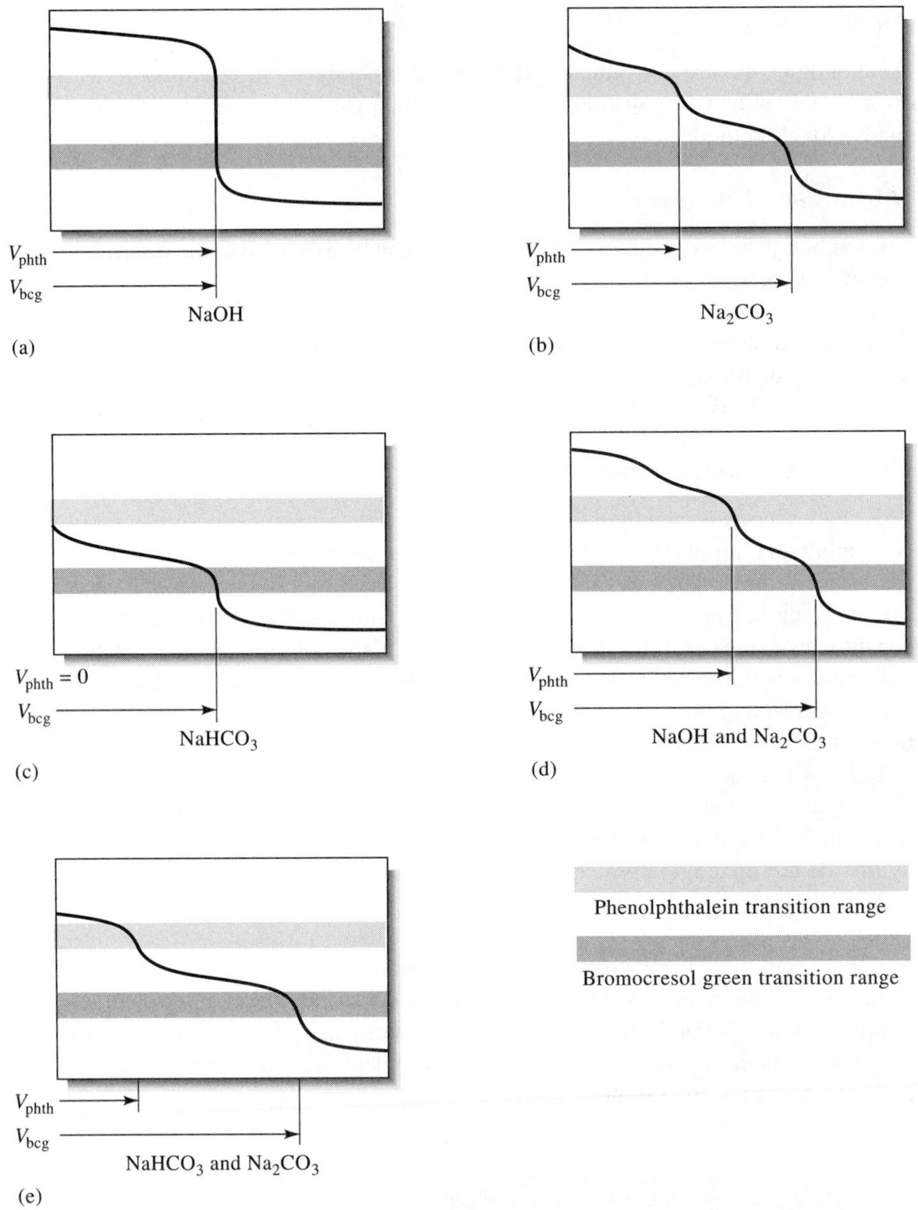

Figure 12-2
Titration curves and indicator transition ranges for the analysis of mixtures containing hydroxide, carbonate, and hydrogen carbonate ions.

EXAMPLE 12-5

A solution contains $NaHCO_3$, Na_2CO_3, and NaOH, either alone or in a permissible combination. Titration of a 50.0-mL portion to a phenolphthalein end point requires 22.1 mL of 0.100 M HCl. A second 50.0-mL aliquot requires 48.4 mL of the HCl when titrated to a bromocresol green end point. Deduce the composition, and calculate the molar solute concentrations of the original solution.

If the solution contained only NaOH, the volume of acid required would be the same regardless of indicator (see Figure 12-2a). Similarly, we can rule out the presence of Na_2CO_3 alone because titration of this compound to a bromocresol green end point would consume just twice the volume of acid required to reach the phenolphthalein end point (see Figure 12-2b). In fact, however, the second titration requires 48.4 mL. Because less than half of this amount is involved in the first titration, the solution must contain some $NaHCO_3$ in addition to Na_2CO_3 (see Figure 12-2e). We can now calculate the concentration of the two constituents.

When the phenolphthalein end point is reached, the CO_3^{2-} originally present is converted to HCO_3^-. Thus,

$$\text{no. mmol Na}_2\text{CO}_3 = 22.1 \text{ mL} \times 0.100 \text{ mmol/mL} = 2.21$$

The titration from the phenolphthalein to the bromocresol green end point (48.4 − 22.1 = 26.3 mL) involves both the hydrogen carbonate originally present and that formed by titration of the carbonate. Thus,

$$\text{no. mmol NaHCO}_3 + \text{no. mmol Na}_2\text{CO}_3 = 26.3 \times 0.100 = 2.63$$

Hence,

$$\text{no. mmol NaHCO}_3 = 2.63 - 2.21 = 0.42$$

The molar concentrations are readily calculated from these data:

$$c_{\text{Na}_2\text{CO}_3} = \frac{2.21 \text{ mmol}}{50.0 \text{ mL}} = 0.0442 \text{ M}$$

$$c_{\text{NaHCO}_3} = \frac{0.42 \text{ mmol}}{50.0 \text{ mL}} = 0.084 \text{ M}$$

Compatible mixtures containing two of the following can also be analyzed in a similar way: HCl, H_3PO_4, NaH_2PO_4, Na_2HPO_4, Na_3PO_4, and NaOH.

The method described in Example 12-5 is not entirely satisfactory because the pH change corresponding to the hydrogen carbonate equivalence point is not sufficient to give a sharp color change with a chemical indicator (Figure 11-5). Relative errors of 1% or more must be expected as a consequence.

How could you analyze a mixture of HCl and H_3PO_4? A mixture of Na_3PO_4 and Na_2HPO_4? See Figure 11-4, curve *A*.

The accuracy of methods for analyzing solutions containing mixtures of carbonate and hydrogen carbonate ions or carbonate and hydroxide ions can be greatly improved by taking advantage of the limited solubility of barium carbonate in neutral and basic solutions. For example, in the *Winkler method* for the analysis of carbonate/hydroxide mixtures, both components are titrated with a standard acid to the end point with an acid-range indicator, such as bromocresol green (the end point is established after the solution is boiled to remove carbon dioxide). An unmeasured excess of neutral barium chloride is then added to a second aliquot of the sample solution to precipitate barium carbonate, following which the hydroxide ion is titrated to a phenolphthalein end point. The presence of the sparingly soluble barium carbonate does not interfere as long as the concentration of barium ion is greater than 0.1 M.

Carbonate and hydrogen carbonate ions can be accurately determined in mixtures by first titrating both ions with standard acid to an end point with an acid-range indicator (with boiling to eliminate carbon dioxide). The hydrogen carbonate in a second aliquot is converted to carbonate by the addition of a known excess of standard base. After a large excess of barium chloride has been introduced, the excess base is titrated with standard acid to a phenolphthalein end point.

The presence of solid barium carbonate does not hamper end-point detection in either of these methods.

12B-3 The Determination of Organic Functional Groups

Neutralization titrations provide convenient methods for the direct or indirect determination of several organic functional groups. Brief descriptions of methods for the more common groups follow.

Carboxylic and Sulfonic Acid Groups

Carboxylic and sulfonic acid groups are the two most common structures that impart acidity to organic compounds. Most carboxylic acids have dissociation constants that range between 10^{-4} and 10^{-6}, and are thus readily titrated. A base-range indicator, such as phenolphthalein, is required.

Many carboxylic acids are not sufficiently soluble in water to permit direct titration in this medium. Where this problem exists, the acid can be dissolved in ethanol and titrated with aqueous base. Alternatively, the acid can be dissolved in an excess of standard base followed by back-titration with standard acid.

Sulfonic acids are generally strong acids and readily dissolve in water. Their titration with a base is therefore straightforward.

Neutralization titrations are often employed to determine the equivalent weight of purified organic acids (see Feature 12-2). Equivalent weights serve as an aid in qualitative identification of organic acids.

The *equivalent weight* of an acid or base is the mass of the acid or base in grams that reacts with or contains one mole of protons. Thus, the equivalent weight of KOH (56.11 g/mol) is its molar mass; for $Ba(OH)_2$ it is the molar mass divided by 2 ($\frac{1}{2} \times 171.3$ g/mol).

FEATURE 12-2
Equivalent Weights of Acids and Bases

The equivalent weight of a participant in a neutralization reaction is the weight that reacts with or supplies one mole of protons *in a particular reaction.* For example, the equivalent weight of H_2SO_4 is one half of its formula weight. The equivalent weight of Na_2CO_3 is usually one half of its formula weight because in most applications its reaction is

$$Na_2CO_3 + 2H_3O^+ \longrightarrow H_2O + H_2CO_3 + 2Na^+$$

When titrated with some indicators, however, it consumes but a single proton:

$$Na_2CO_3 + H_3O^+ \longrightarrow NaHCO_3 + Na^+$$

Here, the equivalent weight and the formula weight of Na_2CO_3 are identical. These observations demonstrate that the equivalent weight of a compound cannot be defined without having a particular reaction in mind (Appendix 9).

Amine Groups

Aliphatic amines generally have base dissociation constants on the order of 10^{-5} and can thus be titrated directly with a solution of a strong acid. In contrast, aromatic amines, such as aniline and its derivatives, are usually too weak for titration in aqueous medium ($K_b \approx 10^{-10}$). The same is true for cyclic amines with aromatic character, such as pyridine and its derivatives. Many saturated cyclic amines, such as piperidine, tend to resemble aliphatic amines in their acid/base behavior and thus can be titrated in aqueous media.

Many amines that are too weak to be titrated as bases in water are readily titrated in nonaqueous solvents, such as anhydrous acetic acid, which enhance their basicity.

Ester Groups

Esters are commonly determined by *saponification* with a measured quantity of standard base:

$$R_1COOR_2 + OH^- \longrightarrow R_1COO^- + HOR_2$$

The excess base is then titrated with standard acid.

Esters vary widely in their rate of saponification. Some require several hours of heating with a base to complete the process. A few react rapidly enough to permit direct titration with the base. Typically, the ester is refluxed with standard 0.5 M KOH for 1 to 2 hr. After cooling, the excess base is determined with standard acid.

> **Saponification** is the process by which an ester is hydrolyzed in alkaline solution to give an alcohol and a conjugate base. For example,
>
> $$CH_3\overset{\overset{O}{\|}}{C}OCH_3 + OH^- \longrightarrow$$
>
> $$CH_3\overset{\overset{O}{\|}}{C}-O^- + CH_3OH$$

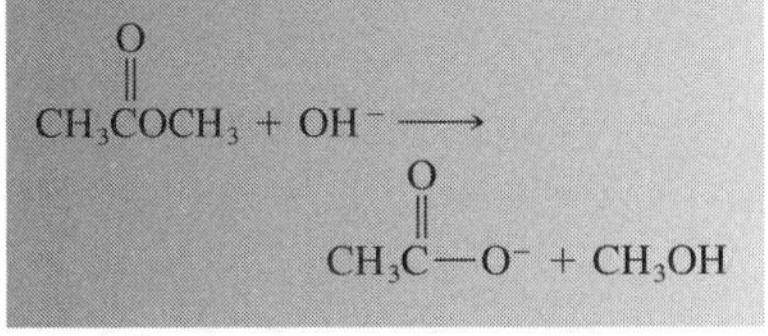

Hydroxyl Groups

Hydroxyl groups in organic compounds can be determined by esterification with various carboxylic acid anhydrides or chlorides; the two most common reagents are acetic anhydride and phthalic anhydride. With acetic anhydride, the reaction is

$$(CH_3CO)_2O + ROH \longrightarrow CH_3COOR + CH_3COOH$$

The acetylation is ordinarily carried out by mixing the sample with a carefully measured volume of acetic anhydride in pyridine. After this mixture is heated, water is added to hydrolyze the unreacted anhydride:

$$(CH_3CO)_2O + H_2O \longrightarrow 2CH_3COOH$$

The acetic acid is then titrated with a standard solution of alcoholic sodium or potassium hydroxide. A blank is carried through the analysis to establish the original amount of anhydride.

Amines, if present, are converted quantitatively to amides by acetic anhydride; a correction for this source of interference is frequently possible by a direct titration of another portion of the sample with standard acid.

Carbonyl Groups

Many aldehydes and ketones can be determined with a solution of hydroxylamine hydrochloride. The reaction, which produces an oxime, is

$$\underset{R_2}{\overset{R_1}{>}}C{=}O + NH_2OH \cdot HCl \longrightarrow \underset{R_2}{\overset{R_1}{>}}C{=}NOH + HCl + H_2O$$

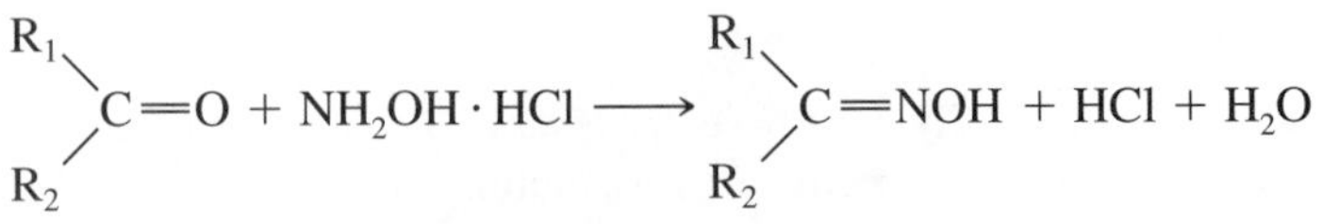

where R_2 may be an atom of hydrogen. The liberated hydrochloric acid is titrated with base. Here again, the conditions necessary for quantitative reaction vary. Typically, 30 min suffices for aldehydes. Many ketones require refluxing with the reagent for an hour or more.

12B-4 The Determination of Salts

The total salt content of a solution can be accurately and readily determined by an acid/base titration. The salt is converted to an equivalent amount of an acid or a base by passage through a column packed with an ion-exchange resin. (This application is considered in more detail in Section 34E-3).

Standard acid or base solutions can also be prepared with ion-exchange resins. Here, a solution containing a known mass of a pure compound, such as sodium chloride, is washed through the resin column and diluted to a known volume. The salt liberates an equivalent amount of acid or base from the resin, permitting calculation of the molarity of the reagent in a straightforward way.

12C QUESTIONS AND PROBLEMS

*12-1. The boiling points of HCl and CO_2 are nearly the same $(-85°$ and $-78°C)$. Explain why CO_2 can be removed from an aqueous solution by boiling briefly while essentially no HCl is lost even after boiling for 1 hr or more.

12-2. Why is HNO_3 seldom used to prepare standard acid solutions?

*12-3. Explain how Na_2CO_3 of primary-standard grade can be prepared from primary-standard $NaHCO_3$.

12-4. Why is it common practice to boil the solution near the equivalence point in the standardization of Na_2CO_3 with acid?

*12-5. Give two reasons why $KH(IO_3)_2$ would be preferred over benzoic acid as a primary standard for a 0.010 M NaOH solution.

12-6. Briefly describe the circumstance where the molarity of a sodium hydroxide solution will apparently be unaffected by the absorption of carbon dioxide.

12-7. What types of compounds containing organic nitrogen tend to yield low results with the Kjeldahl method unless special precautions are taken?

*12-8. How would you prepare 2.00 L of
(a) 0.15 M KOH from the solid?
(b) 0.015 M $Ba(OH)_2 \cdot 8H_2O$ from the solid?
(c) 0.200 M HCl from a reagent that has a density of 1.0579 g/mL and is 11.50% HCl (w/w)?

12-9. How would you prepare 500 mL of
(a) 0.250 M H_2SO_4 from a reagent that has a density of 1.1539 g/mL and is 21.8% H_2SO_4 (w/w)?
(b) 0.30 M NaOH from the solid?
(c) 0.08000 M Na_2CO_3 from the pure solid?

*12-10. Standardization of a sodium hydroxide solution against potassium hydrogen phthalate (KHP) yielded the accompanying results.

mass KHP, g	0.7987	0.8365	0.8104	0.8039
volume NaOH, mL	38.29	39.96	38.51	38.29

Calculate
(a) the average molarity of the base.
(b) the standard deviation and the coefficient of variation for the data.

12-11. The molarity of a perchloric acid solution was established by titration against primary standard sodium carbonate (product: CO_2); the following data were obtained.

mass Na_2CO_3, g	0.2068	0.1997	0.2245	0.2137
volume $HClO_4$, mL	36.31	35.11	39.00	37.54

(a) Calculate the average molarity of the acid.
(b) Calculate the standard deviation for the data and the coefficient of variation for the data.
(c) Does statistical justification exist for disregarding the outlying result?

*12-12. If 1.000 L of 0.1500 M NaOH was unprotected from the air after standardization and absorbed 11.2 mmol of CO_2, what is its new molarity when it is standardized against a standard solution of HCl using
 (a) phenolphthalein?
 (b) bromocresol green?

12-13. A NaOH solution was 0.1019 M immediately after standardization. Exactly 500.0 mL of the reagent was left exposed to air for several days and absorbed 0.652 g of CO_2. Calculate the relative carbonate error in the determination of acetic acid with this solution if the titrations were performed with phenolphthalein.

*12-14. Calculate the molar concentration of a dilute HCl solution if
 (a) a 50.00-mL aliquot yielded 0.6010 g of AgCl.
 (b) titration of 25.00 mL of 0.04010 M $Ba(OH)_2$ required 19.92 mL of the acid.
 (c) titration of 0.2694 g of primary standard Na_2CO_3 required 38.77 mL of the acid (products: CO_2 and H_2O).

12-15. Calculate the molarity of a dilute $Ba(OH)_2$ solution if
 (a) 50.00 mL yielded 0.1684 g of $BaSO_4$.
 (b) titration of 0.4815 g of primary standard potassium hydrogen phthalate (KHP) required 29.41 mL of the base.
 (c) addition of 50.00 mL of the base to 0.3614 g of benzoic acid required a 4.13-mL back-titration with 0.05317 M HCl.

12-16. Suggest a range of sample masses for the indicated primary standard if it is desired to use between 35 and 45 mL of titrant:
 *(a) 0.150 M $HClO_4$ titrated against Na_2CO_3 (CO_2 product).
 (b) 0.075 M HCl titrated against $Na_2C_2O_4$:

$$Na_2C_2O_4 \longrightarrow Na_2CO_3 + CO$$
$$CO_3^{2-} + 2H^+ \longrightarrow H_2O + CO_2$$

 *(c) 0.20 M NaOH titrated against benzoic acid.
 (d) 0.030 M $Ba(OH)_2$ titrated against $KH(IO_3)_2$.
 *(e) 0.040 M $HClO_4$ titrated against TRIS.
 (f) 0.080 M H_2SO_4 titrated against $Na_2B_4O_7 \cdot 10H_2O$. Reaction:

$$B_4O_7^{2-} + 2H_3O^+ + 3H_2O \longrightarrow 4H_3BO_3$$

*12-17. Calculate the absolute standard deviation in the computed molarity of 0.0200 M HCl if this acid was standardized against the masses calculated in Example 12-1 for (a) TRIS, (b) Na_2CO_3, and (c) $Na_2B_4O_7 \cdot 10H_2O$. Assume that the absolute standard deviation in the mass measurement is 0.0002 g and that this measurement limits the precision of the computed molarity.

12-18. (a) Compare the masses of potassium hydrogen phthalate (204.22 g/mol); potassium hydrogen iodate (389.91 g/mol); and benzoic acid (122.12 g/mol) needed for a 30.00-mL standardization of 0.0400 M NaOH.
 (b) What would be the relative standard deviation in the molarity of the base if the standard deviation in the

measurement of mass in (a) is 0.002 g and this uncertainty limits the precision of the calculation?

*12-19. A 50.00-mL sample of a white dinner wine required 21.48 mL of 0.03776 M NaOH to achieve a phenolphthalein end point. Express the acidity of the wine in terms of grams of tartaric acid ($H_2C_4H_4O_6$, 150.09 g/mol) per 100 mL. (Assume that two hydrogens of the acid are titrated.)

12-20. A 25.0-mL aliquot of vinegar was diluted to 250 mL in a volumetric flask. Titration of 50.0-mL aliquots of the diluted solution required an average of 34.88 mL of 0.09600 M NaOH. Express the acidity of the vinegar in terms of the percentage (w/v) of acetic acid.

*12-21. Titration of a 0.7439-g sample of impure $Na_2B_4O_7$ required 31.64 mL of 0.1081 M HCl [see Problem 12-16(f) for reaction]. Express the results of this analysis in terms of percent
 (a) $Na_2B_4O_7$.
 (b) $Na_2B_4O_7 \cdot 10H_2O$.
 (c) B_2O_3.
 (d) B.

12-22. A 0.6334-g sample of impure mercury(II) oxide was dissolved in an unmeasured excess of potassium iodide. Reaction:

$$HgO(s) + 4I^- + H_2O \longrightarrow HgI_4^{2-} + 2OH^-$$

Calculate the percentage of HgO in the sample if titration of the liberated hydroxide required 42.59 mL of 0.1178 M HCl.

*12-23. The formaldehyde content of a pesticide preparation was determined by weighing 0.3124 g of the liquid sample into a flask containing 50.0 mL of 0.0996 M NaOH and 50 mL of 3% H_2O_2. Upon heating, the following reaction took place:

$$OH^- + HCHO + H_2O_2 \longrightarrow HCOO^- + 2H_2O$$

After cooling, the excess base was titrated with 23.3 mL of 0.05250 M H_2SO_4. Calculate the percentage of HCHO (30.026 g/mol) in the sample.

12-24. The benzoic acid extracted from 106.3 g of catsup required a 14.76-mL titration with 0.0514 M NaOH. Express the results of this analysis in terms of percent sodium benzoate (144.10 g/mol).

*12-25. The active ingredient in disulfiram, a drug used for the treatment of chronic alcoholism, is tetraethylthiuram disulfide,

$$\underset{(C_2H_5)_2NCSSCN(C_2H_5)_2}{\overset{\textstyle S \quad\; S}{\overset{\textstyle \| \quad\; \|}{}}}$$

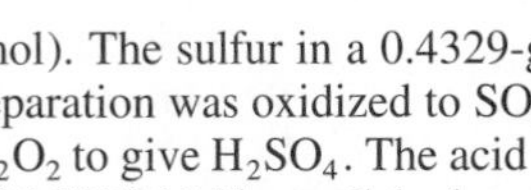

(296.54 g/mol). The sulfur in a 0.4329-g sample of a disulfiram preparation was oxidized to SO_2, which was absorbed in H_2O_2 to give H_2SO_4. The acid was titrated with 22.13 mL of 0.03736 M base. Calculate the percentage of active ingredient in the preparation.

12-26. A 25.00-mL sample of a household cleaning solution was diluted to 250.0 mL in a volumetric flask. A 50.00-mL aliquot of this solution required 40.38 mL of 0.2506 M

HCl to reach a bromocresol green end point. Calculate the weight/volume percentage of NH_3 in the sample. (Assume that all the alkalinity results from the ammonia.)

***12-27.** A 0.1401-g sample of a purified carbonate was dissolved in 50.00 mL of 0.1140 M HCl and boiled to eliminate CO_2. Back-titration of the excess HCl required 24.21 mL of 0.09802 M NaOH. Identify the carbonate.

12-28. A dilute solution of an unknown weak acid required a 28.62-mL titration with 0.1084 M NaOH to reach a phenolphthalein end point. The titrated solution was evaporated to dryness. Calculate the equivalent weight (see margin note, page 260) of the acid if the sodium salt was found to weigh 0.2110 g.

***12-29.** A 3.00-L sample of urban air was bubbled through a solution containing 50.0 mL of 0.0116 M $Ba(OH)_2$, which caused the CO_2 in the sample to precipitate as $BaCO_3$. The excess base was back-titrated to a phenolphthalein end point with 23.6 mL of 0.0108 M HCl. Calculate the parts per million of CO_2 in the air (that is, mL CO_2/10^6 mL air); use 1.98 g/L for the density of CO_2.

12-30. Air was bubbled at a rate of 30.0 L/min through a trap containing 75 mL of 1% H_2O_2 ($H_2O_2 + SO_2 \rightarrow H_2SO_4$). After 10.0 min, the H_2SO_4 was titrated with 11.1 mL of 0.00204 M NaOH. Calculate the parts per million of SO_2 (that is, mL SO_2/10^6 mL air) if the density of SO_2 is 0.00285 g/mL.

***12-31.** The digestion of a 0.1417-g sample of a compound containing phosphorus in a mixture of HNO_3 and H_2SO_4 resulted in the formation of CO_2, H_2O, and H_3PO_4. Addition of ammonium molybdate yielded a solid having the composition $(NH_4)_3PO_4 \cdot 12MoO_3$ (1876.3 g/mol). This precipitate was filtered, washed, and dissolved in 50.00 mL of 0.2000 M NaOH:

$$(NH_4)_3 \cdot 12MoO_3(s) + 26OH^- \longrightarrow HPO_4^{2-} + 12MoO_4^{2-} + 14H_2O + 3NH_3(g)$$

After the solution was boiled to remove the NH_3, the excess NaOH was titrated with 14.17 mL of 0.1741 M HCl to a phenolphthalein end point. Calculate the percentage of phosphorus in the sample.

***12-32.** A 0.8160-g sample containing dimethylphthalate, $C_6H_4(COOCH_3)_2$ (194.19 g/mol), and unreactive species was refluxed with 50.00 mL of 0.1031 M NaOH to hydrolyze the ester groups (this process is called *saponification*).

$$C_6H_4(COOCH_3)_2 + 2OH^- \longrightarrow C_6H_4(COO)_2^{2-} + 2CH_3OH$$

After the reaction was complete, the excess NaOH was back-titrated with 24.27 mL of 0.1644 M HCl. Calculate the percentage of dimethylphthalate in the sample.

***12-33.** Neohetramine, $C_{16}H_{21}ON_4$ (285.37 g/mol), is a common antihistamine. A 0.1247-g sample containing this compound was analyzed by the Kjeldahl method. The ammonia produced was collected in H_3BO_3; the resulting $H_2BO_3^-$ was titrated with 26.13 mL of 0.01477 M HCl. Calculate the percentage of neohetramine in the sample.

12-34. The *Merck Index* indicates that 10 mg of guanidine, CH_5N_3, may be administered for each kilogram of body weight in the treatment of myasthenia gravis. The nitrogen in a 4-tablet sample that weighed a total of 7.50 g was converted to ammonia by a Kjeldahl digestion, followed by distillation into 100.0 mL of 0.1750 M HCl. The analysis was completed by titrating the excess acid with 11.37 mL of 0.1080 M NaOH. How many of these tablets represent a proper dose for a 48-kg patient?

***12-35.** A 1.047-g sample of canned tuna was analyzed by the Kjeldahl method; 24.61 mL of 0.1180 M HCl were required to titrate the liberated ammonia. Calculate the percentage of nitrogen in the sample.

12-36. Calculate the grams of protein in a 6.50-oz can of tuna in Problem 12-35 (1 oz = 28.3 g).

***12-37.** A 0.5843-g sample of a plant food preparation was analyzed for its N content by the Kjeldahl method, the liberated NH_3 being collected in 50.00 mL of 0.1062 M HCl. The excess acid required an 11.89 mL back titration with 0.0925 M NaOH. Express the results of this analysis in terms of percent

*(a) N.
(b) urea, H_2NCONH_2.
*(c) $(NH_4)_2SO_4$.
(d) $(NH_4)_3PO_4$.

12-38. A 0.9092-g sample of a wheat flour was analyzed by the Kjeldahl procedure. The ammonia formed was distilled into 50.00 mL of 0.05063 M HCl; a 7.46-mL back-titration with 0.04917 M NaOH was required. Calculate the percentage of protein in the flour.

***12-39.** A 1.219-g sample containing $(NH_4)_2SO_4$, NH_4NO_3, and nonreactive substances was diluted to 200 mL in a volumetric flask. A 50.00-mL aliquot was made basic with strong alkali, and the liberated NH_3 was distilled into 30.00 mL of 0.08421 M HCl. The excess HCl required 10.17 mL of 0.08802 M NaOH. A 25.00-mL aliquot of the sample was made alkaline after the addition of Devarda's alloy, and the NO_3^- was reduced to NH_3. The NH_3 from both NH_4^+ and NO_3^- was then distilled into 30.00 mL of the standard acid and back-titrated with 14.16 mL of the base. Calculate the percentage of $(NH_4)_2SO_4$ and NH_4NO_3 in the sample.

***12-40.** A 1.217-g sample of commercial KOH contaminated by K_2CO_3 was dissolved in water, and the resulting solution was diluted to 500.0 mL. A 50.00-mL aliquot of this solution was treated with 40.00 mL of 0.05304 M HCl and boiled to remove CO_2. The excess acid consumed 4.74 mL of 0.04983 M NaOH (phenolphthalein indicator). An excess of neutral $BaCl_2$ was added to another 50.00-mL aliquot to precipitate the carbonate as $BaCO_3$. The solution was then titrated with 28.56 mL of the acid to a phenolphthalein end point. Calculate the percentage of KOH, K_2CO_3, and H_2O in the sample, assuming that these are the only compounds present.

12-41. A 0.5000-g sample containing $NaHCO_3$, Na_2CO_3, and H_2O was dissolved and diluted to 250.0 mL. A 25.00-mL aliquot was then boiled with 50.00 mL of 0.01255 M HCl. After cooling, the excess acid in the solution required 2.34 mL of 0.01063 M NaOH when titrated to a phenolphthalein end point. A second 25.00-mL aliquot was then treated with an excess of $BaCl_2$ and 25.00 mL of the base;

precipitation of all the carbonate resulted, and 7.63 mL of the HCl were required to titrate the excess base. Calculate the composition of the mixture.

***12-42.** Calculate the volume of 0.06122 M HCl needed to titrate
(a) 20.00 mL of 0.05555 M Na_3PO_4 to a thymolphthalein end point.
(b) 25.00 mL of 0.05555 M Na_3PO_4 to a bromocresol green end point.
(c) 40.00 mL of a solution that is 0.02102 M in Na_3PO_4 and 0.01655 M in Na_2HPO_4 to a bromocresol green end point.
(d) 20.00 mL of a solution that is 0.02102 M in Na_3PO_4 and 0.01655 M in NaOH to a thymolphthalein end point.

12-43. Calculate the volume of 0.07731 M NaOH needed to titrate
(a) 25.00 mL of a solution that is 0.03000 M in HCl and 0.01000 M in H_3PO_4 to a bromocresol green end point.
(b) the solution in (a) to a thymolphthalein end point.
(c) 30.00 mL of 0.06407 M NaH_2PO_4 to a thymolphthalein end point.
(d) 25.00 mL of a solution that is 0.02000 M in H_3PO_4 and 0.03000 M in NaH_2PO_4 to a thymolphthalein end point.

***12-44.** A series of solutions containing NaOH, Na_2CO_3, and $NaHCO_3$, alone or in compatible combination, was titrated with 0.1202 M HCl. The volumes of acid needed to titrate 25.00-mL portions of each solution to a (1) phenolphthalein and (2) bromocresol green end point are given in the table that follows. Use this information to deduce the composition of the solutions. In addition, calculate the number of milligrams of each solute per milliliter of solution.

	(1)	(2)
(a)	22.42	22.44
(b)	15.67	42.13
(c)	29.64	36.42
(d)	16.12	32.23
(e)	0.00	33.333

12-45. A series of solutions containing NaOH, Na_3AsO_4, and Na_2HAsO_4, alone or in compatible combination, was titrated with 0.08601 M HCl. The volumes of acid needed to titrate 25.00-mL portions of each solution to a (1) phenolphthalein and (2) bromocresol green end point are given in the table that follows. Use this information to deduce the composition of the solutions. In addition, calculate the number of milligrams of each solute per milliliter of solution.

	(1)	(2)
(a)	0.00	18.15
(b)	21.00	28.15
(c)	19.80	39.61
(d)	18.04	18.03
(e)	16.00	37.37

***12-46.** Define the equivalent weight of (a) an acid and (b) a base.

12-47. Calculate the equivalent weight of oxalic acid dihydrate ($H_2C_2O_4 \cdot 2H_2O$, 126.1 g/mol) when it is titrated to (a) a bromocresol green end point and (b) a phenolphthalein end point.

CHAPTER

13

Precipitation Titrimetry

Precipitation titrimetry, which is based upon reactions that yield ionic compounds of limited solubility, is one of the oldest analytical techniques, dating back to the mid-1800s. The slow rate of formation of most precipitates, however, limits the number of precipitating agents that can be used in titrimetry to a handful. By far the most important precipitating reagent is silver nitrate, which is used for the determination of the halides, the halide-like anions (SCN^-, CN^-, CNO^-), mercaptans, fatty acids, and several divalent inorganic anions. Titrimetric methods based upon silver nitrate are sometimes termed *argentometric methods.* In this text, we will limit our discussion of precipitation titrimetry to argentometric methods.

Argentometric is derived from the Latin noun *argentum,* which means silver.

13A TITRATION CURVES

In this section, we first show how titration curves are derived for anions that react with silver ion to form a precipitate of limited solubility. We then show how titration curves can be derived for mixtures of anions that react with silver ion to produce precipitates that differ in their solubilities. Most indicators for argentometric iterations respond to changes in the silver ion concentration in the equivalence-point region. For this reason, argentometric titration curves usually involve a plot of pAg as a function of volume of silver nitrate.

13A-1 Titration Curves for a Single Anion

Titration curves for a single anion are derived in a way completely analogous to the methods described in Section 10C-1 for titrations involving strong acids and strong bases. The only difference is that the solubility product of the precipitate is substituted for the ion-product constant for water. The example that follows illustrates how p-functions are derived for the preequivalence-point region, the postequivalence-point region, and the equivalence point for a typical precipitation titration. Compare these calculations with those for the strong acid/base system shown in Example 10-2.

See *Mathcad Applications for Analytical Chemistry,* **pp. 61–64.** Note that this example involves NaCl rather than NaBr and that the concentrations are different.

EXAMPLE 13-1

Calculate the pAg of the solution during the titration of 50.00 mL of 0.0500 M NaCl with 0.1000 M $AgNO_3$ after the addition of the following volumes of reagent: (a) 0.00 mL, (b) 24.50 mL, (c) 25.00 mL, (d) 25.50 mL.

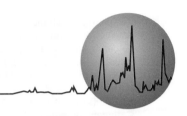

(a) Because no $AgNO_3$ has been added, $[Ag^+] = 0$ and pAg is indeterminate.

(b) At 24.50 mL added, $[Ag^+]$ is very small and cannot be computed from stoichiometric considerations alone but $[Cl^-]$ can still be obtained directly.

$$[Cl^-] \approx c_{NaCl} = \frac{\text{original no. mmol } Cl^- - \text{no. mmol } AgNO_3 \text{ added}}{\text{total volume of solution}}$$

$$= \frac{(50.00 \times 0.0500 - 24.50 \times 0.1000)}{50.00 + 24.50} = 6.71 \times 10^{-4}$$

$$[Ag^+] = K_{sp}/(6.71 \times 10^{-4}) = 1.82 \times 10^{-10}/(6.71 \times 10^{-4})$$

$$= 2.71 \times 10^{-7}$$

$$pAg = -\log (2.71 \times 10^{-7}) = 6.57$$

(c) This is the equivalence point where $[Ag^+] = [Cl^-]$ and

$$[Ag^+] = \sqrt{K_{sp}} = \sqrt{1.82 \times 10^{-10}} = 1.35 \times 10^{-5}$$

$$pAg = -\log (1.35 \times 10^{-5}) = 4.87$$

(d) The solution now has an excess of Ag^+. Therefore,

$$[Ag^+] = c_{AgNO_3} = \frac{(25.50 \times 0.1000 - 50.00 \times 0.0500)}{75.50} = 6.62 \times 10^{-4}$$

$$pAg = -\log (6.62 \times 10^{-4}) = 3.18$$

Most indicators for argentometric titrations respond to change in concentrations of silver ions. As a consequence, titration curves for precipitation reactions usually consist of a plot of pAg versus volume of the silver reagent (usually $AgNO_3$).

A useful relationship can be derived by taking the negative logarithm of both sides of a solubility-product expression. Thus, for silver chloride

$$-\log K_{sp} = -\log ([Ag^+][Cl^-])$$
$$= -\log [Ag^+] - \log [Cl^-]$$
$$pK_{sp} = pAg + pCl$$

Compare this expression with the one in the margin note on page 132, Chapter 7.

Figure 13-1 shows titration curves for chloride ion that have been derived employing the techniques shown in Example 13-1. Note the effect of analyte and reagent concentrations on the magnitude of the change in pAg in the equivalence-point region. The effect here is analogous to that illustrated for acid/base titrations in Figure 10-4. As shown by the shaded region in Figure 13-1, an indicator with a pAg range of 4 to 6 should provide a sharp end point in the titration of the 0.05 M chloride solution. For the more dilute solution, in contrast, the end point

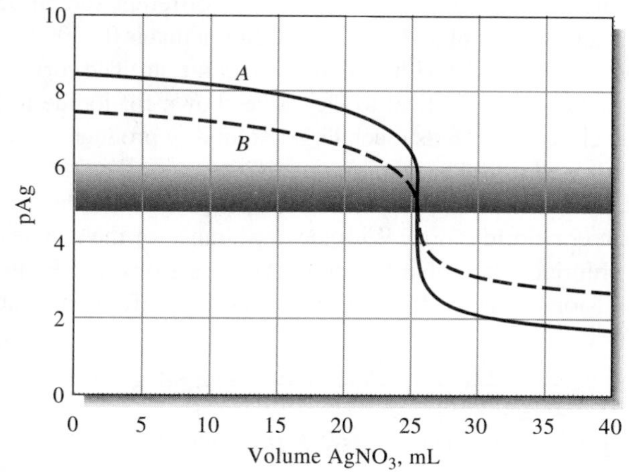

Figure 13-1
Effect of titrant concentration on titration curves: *A*: 50.00 mL of 0.0500 M NaCl with 0.1000 M $AgNO_3$; *B*: 50.00 mL of 0.00500 M NaCl with 0.01000 M $AgNO_3$.

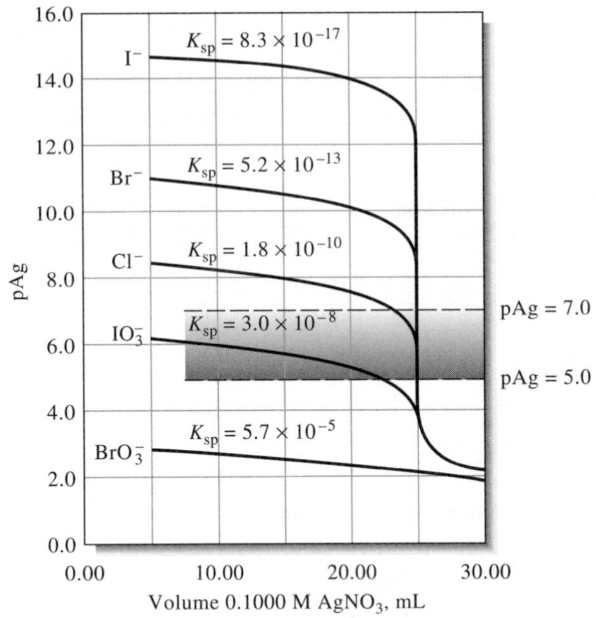

Figure 13-2
Effect of reaction completeness on titration curves. For each curve, 50.00 mL of a 0.0500 M solution of the anion was titrated with 0.1000 M $AgNO_3$.

would be drawn out over a large enough volume of reagent to make accurate establishment of the end point impossible. That is, the color change would begin at about 24 mL and be complete at about 26 mL.

Figure 13-2 illustrates the effect of solubility product on the sharpness of the end point on titrations with 0.1 M silver nitrate. Clearly, the change in pAg at the equivalence point becomes greater as the solubility products become smaller—that is, as the reaction between the analyte and silver nitrate becomes more complete. By careful choice of indicator—one that changes color in the region of pAg from 4 to 6—titration of chloride ions should be possible with a minimal titration error. Note that ions forming precipitates with solubility products much larger than about 10^{-10} do not yield satisfactory end points.

13A-2 Titration Curves for Mixtures of Anions

The methods developed in the previous section for deriving titration curves can be extended to mixtures that form precipitates with different solubilities. To illustrate, consider the titration of 50.00 mL of a solution that is 0.0500 M in iodide ion and 0.0800 M in chloride ion with 0.1000 M silver nitrate. The curve for the initial stages of this titration is identical to the curve shown for iodide in Figure 13-3 because silver chloride, with its much larger solubility product, does not begin to precipitate until well into the titration.

It is of interest to determine how much iodide is precipitated before appreciable amounts of silver chloride form. With the appearance of the smallest amount of solid silver chloride, the solubility-product expressions for both precipitates apply, and division of one by the other provides the useful relationship

$$\frac{[Ag^+][I^-]}{[Ag^+][Cl^-]} = \frac{8.3 \times 10^{-17}}{1.82 \times 10^{-10}} = 4.56 \times 10^{-7}$$

$$[I^-] = (4.56 \times 10^{-7})[Cl^-]$$

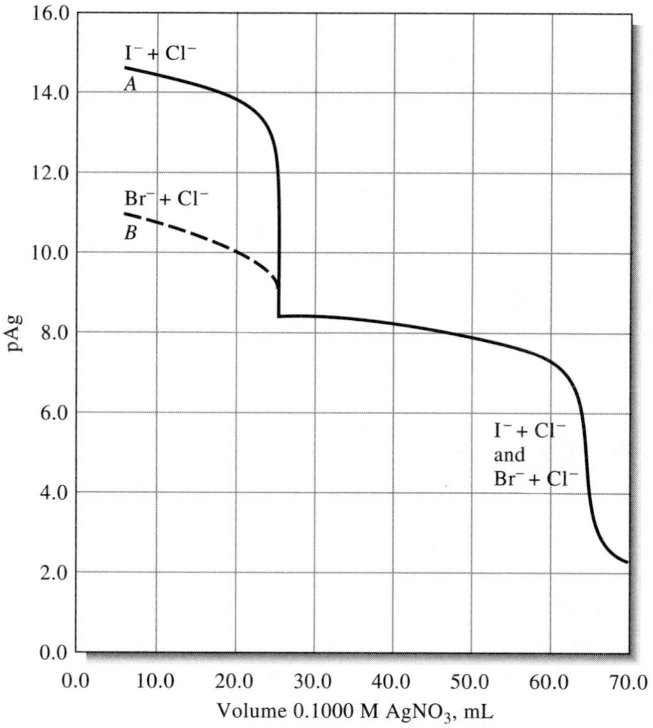

Figure 13-3
Titration curves for 50.00 mL of a solution 0.0800 M in Cl^- and 0.0500 M in I^- or Br^-.

It is apparent from this relationship that the iodide concentration is decreased to a tiny fraction of the chloride ion concentration prior to the onset of silver chloride precipitation. Thus, for all practical purposes, formation of silver chloride will occur only after 25.00 mL of titrant have been added in this titration. At this point because of dilution, the chloride ion concentration is approximately

$$c_{Cl} = [Cl^-] = \frac{50.00 \times 0.0800}{50.00 + 25.00} = 0.0533 \text{ M}$$

Substituting into the previous equation yields

$$[I^-] = 4.56 \times 10^{-7} \times 0.0533 = 2.43 \times 10^{-8}$$

The percentage of iodide unprecipitated at this point can be calculated as follows:

$$\text{amount } I^- = (75.00 \cancel{\text{mL}})(2.43 \times 10^{-8} \text{ mmol } I^-/\cancel{\text{mL}}) = 1.82 \times 10^{-6} \text{ mmol}$$

$$\text{original amount } I^- = (50.00 \cancel{\text{mL}})(0.0500 \text{ mmol}/\cancel{\text{mL}}) = 2.50 \text{ mmol}$$

$$I^- \text{ unprecipitated} = \frac{1.82 \times 10^{-6}}{2.50} \times 100\% = 7.3 \times 10^{-5}\%$$

Thus, to within about 7.3×10^{-5} percent of the equivalence point for iodide, no silver chloride forms, and up to this point, the titration curve is indistinguishable from that for iodide alone (Figure 13-2). The first part of the titration curve shown by the solid line in Figure 13-3 was derived on this basis.

As chloride ion begins to precipitate, the rapid decrease in pAg is terminated abruptly at a level that can be calculated from the solubility-product constant for silver chloride and the computed chloride concentration:

$$[Ag^+] = \frac{1.82 \times 10^{-10}}{0.0533} = 3.41 \times 10^{-9}$$

$$pAg = -\log(3.41 \times 10^{-9}) = 8.47$$

Further additions of silver nitrate decrease the chloride ion concentration, and the curve then becomes that for the titration of chloride by itself. For example, after 30.00 mL of titrant have been added,

$$c_{Cl} = [Cl^-] = \frac{50.00 \times 0.0800 + 50.00 \times 0.0500 - 30.00 \times 0.100}{50.00 + 30.00}$$

Here, the first two terms in the numerator give the number of millimoles of chloride and iodide, respectively, and the third is the number of millimoles of titrant added. Thus,

$$[Cl^-] = 0.0438$$

$$[Ag^+] = \frac{1.82 \times 10^{-10}}{0.0438} = 4.16 \times 10^{-9}$$

$$pAg = 8.38$$

The remainder of the curve can be derived in the same way as a curve for chloride by itself.

Curve *A* in Figure 13-3, which is the titration curve for the chloride/iodide mixture just considered, is a composite of the individual curves for the two anionic species. Two equivalence points are evident. Curve *B* is the titration curve for a mixture of bromide and chloride ions. Clearly, the change associated with the first equivalence point becomes less distinct as the solubilities of the two precipitates approach one another. In the bromide/chloride titration, the initial pAg values are lower than they are in the iodide/chloride titration because the solubility of silver bromide exceeds that of silver iodide. Beyond the first equivalence point, however, where chloride ion is being titrated, the two titration curves are identical.

13B END POINTS FOR ARGENTOMETRIC TITRATIONS

Three types of end points are encountered in titrations with silver nitrate: (1) chemical, (2) potentiometric, and (3) amperometric. Some chemical indicators are described in the section that follows. Potentiometric end points are obtained by measuring the potential between a silver electrode and a reference electrode whose potential is constant and independent of the added reagent. Titration curves similar to those shown in Figures 13-2 and 13-3 are obtained. Potentiometric end points are discussed in Section 19C-2. To obtain an amperometric end point, the current generated between a pair of silver microelectrodes in the solution of the analyte is measured and plotted as a function of reagent volume. Amperometric methods are considered at the end of Section 21B-4.

13B-1 Chemical Indicators for Precipitation Titrations

The end point produced by a chemical indicator usually consists of a color change or, occasionally, the appearance or disappearance of turbidity in the solution being titrated. The requirements for an indicator for a precipitation titration are analogous to those for an indicator for a neutralization titration: (1) the color change should occur over a limited range in p-function of the reagent or the analyte and (2) the color change should take place within the steep portion of the titration curve for the analyte. For example, turning to Figure 13-2 we see that the indicator shown should provide a satisfactory end point for the titration of iodide and bromide ions but not chloride. An indicator with a pAg range of 6.0 to 4.0, however, should be satisfactory for chloride ions (Figure 13-1). Three indicators that have found extensive use for argentometric titrations are discussed in the sections that follow.

Chromate Ion; The Mohr Method

Sodium chromate can serve as an indicator for the argentometric determination of chloride, bromide, and cyanide ions by reacting with silver ion to form a brick-red silver chromate (Ag_2CrO_4) precipitate in the equivalence-point region. The silver ion concentration at chemical equivalence in the titration of chloride with silver ions is given by

$$[Ag^+] = \sqrt{K_{sp}} = \sqrt{1.82 \times 10^{-10}} = 1.35 \times 10^{-5}\ M$$

The Mohr method was first described in 1865 by K. F. Mohr, a German pharmaceutical chemist who did much pioneering work in the development of titrimetry.

Mohr Method for Chloride

$$\underset{\text{white}}{Ag^+ + Cl^- \rightleftharpoons AgCl(s)}$$

$$\underset{\text{red}}{2Ag^+ + CrO_4^{2-} \rightleftharpoons Ag_2CrO_4(s)}$$

The chromate ion concentration required to initiate formation of silver chromate under this condition can be computed from the solubility product constant for silver chromate,

$$[CrO_4^{2-}] = \frac{K_{sp}}{[Ag^+]^2} = \frac{1.2 \times 10^{-12}}{(1.35 \times 10^{-5})^2} = 6.6 \times 10^{-3}$$

In principle, then, an amount of chromate ion to give this concentration should be added for appearance of the red precipitate just after the equivalence point. In fact, however, a chromate ion concentration of 6.6×10^{-3} M imparts such an intense yellow color to the solution that formation of the red silver chromate is not readily detected; for this reason, lower concentrations of chromate ion are generally used. An excess of silver nitrate is therefore required before precipitation begins. An additional excess of the reagent must also be added to produce enough silver chromate to be seen. These two factors create a positive systematic error in the Mohr method that becomes significant in magnitude at reagent concentrations lower than about 0.1 M. A correction for this error can readily be made by a blank titration of a chloride-free suspension of calcium carbonate. Alternatively, the silver nitrate solution can be standardized against primary-standard-grade sodium chloride using the same conditions as are to be used in the analysis. This latter technique not only compensates for the overconsumption of reagent but also for the acuity of the analyst in detecting the appearance of the color.

The Mohr titration must be carried out at a pH between 7 and 10 because chromate ion is the conjugate base of the weak chromic acid. Consequently, in more acidic solutions the chromate ion concentration is too low to produce the precipitate at the equivalence point. Normally, a suitable pH is achieved by saturating the analyte solution with sodium hydrogen carbonate.

Adsorption Indicators; The Fajans Method

Adsorption indicators were first described by K. Fajans, a Polish chemist, in 1926. His name is pronounced Fay′ yahns.

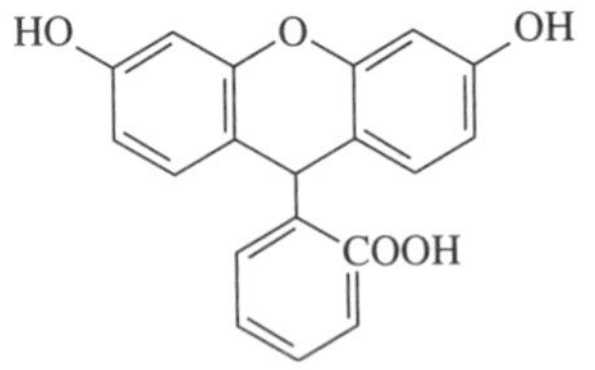

fluorescein

An *adsorption indicator* is an organic compound that tends to be adsorbed onto the surface of the solid in a precipitation titration. Ideally, the adsorption (or desorption) occurs near the equivalence point and results not only in a color change but also in a transfer of color from the solution to the solid (or the reverse).

Fluorescein is a typical adsorption indicator that is useful for the titration of chloride ion with silver nitrate. In aqueous solution, fluorescein partially dissociates into hydronium ions and negatively charged fluoresceinate ions that are yellow-green. The fluoresceinate ion forms an intensely red silver salt. Whenever this dye is used as an indicator, however, *its concentration is never large enough to precipitate as silver fluoresceinate.*

In the early stages of the titration of chloride ion with silver nitrate, the colloidal silver chloride particles are negatively charged because of surface adsorption of excess chloride ions (Section 5B-2). The dye anions are repelled from this surface by electrostatic repulsion and impart a yellow-green color to the solution. Beyond the equivalence point, however, the silver chloride particles strongly adsorb silver ions and thereby acquire a positive charge. Fluoresceinate anions are now attracted *into the counter-ion layer* that surrounds each colloidal silver chloride particle. The net result is the appearance of the red color of silver fluoresceinate *in the surface layer of the solution surrounding the solid.* It is important to emphasize that the color change is an *adsorption* (not a precipitation) process inasmuch as the solubility product of the silver fluoresceinate is never exceeded. The adsorption is reversible, the dye being desorbed upon back-titration with chloride ion.

See color plates 10 and 11 for the end point color change with an adsorption indicator.

Titrations involving adsorption indicators are rapid, accurate, and reliable, but their application is limited to the relatively few precipitation reactions in which a colloidal precipitate is formed rapidly.

Iron(III) Ion; The Volhard Method

The Volhard method was first described by Jacob Volhard, a German chemist, in 1874.

Volhard Method for Chloride

$$\underset{\text{excess}}{Ag^+} + Cl^- \rightleftharpoons \underset{\text{white}}{AgCl(s)}$$

$$SCN^- + Ag^+ \rightleftharpoons \underset{\text{white}}{AgSCN(s)}$$

$$Fe^{3+} + SCN^- \rightleftharpoons \underset{\text{red}}{Fe(SCN)^{2+}}$$

In the Volhard method, silver ions are titrated with a standard solution of thiocyanate ion:

$$Ag^+ + SCN^- \rightleftharpoons AgSCN(s)$$

Iron(III) serves as the indicator. The solution turns red with the first slight excess of thiocyanate ion:

$$Fe^{3+} + SCN^- \rightleftharpoons \underset{\text{red}}{Fe(SCN)^{2+}} \qquad K_f = 1.05 \times 10^3 = \frac{[Fe(SCN)^{2+}]}{[Fe^{3+}][SCN^-]}$$

The titration must be carried out in acidic solution to prevent precipitation of iron(III) as the hydrated oxide.

FEATURE 13-1
Calculation of $[Fe^{3+}]$ to Give an End Point in the Volhard Titration

Experiment shows that the average observer can just detect the red color of $Fe(SCN)^{2+}$ when its concentration is 6.4×10^{-6} M. In the titration of 50.0 mL of 0.050 M Ag^+ with 0.100 M KSCN, what concentration of Fe^{3+} should be used to lower the titration error to zero?

For a zero titration error, the $Fe(SCN)^{2+}$ color should appear when the concentration of Ag^+ remaining in the solution is identical to the sum of the two thiocyanate species. That is, at the equivalence point

$$[Ag^+] = [SCN^-] + [Fe(SCN)^{2+}]$$

Substituting the detectable concentration of $FeSCN^{2+}$ gives

$$[Ag^+] = [SCN^-] + 6.4 \times 10^{-6}$$

or

$$[Ag^+] = \frac{K_{sp}}{[SCN^-]} = \frac{1.1 \times 10^{-12}}{[SCN^-]} = [SCN^-] + 6.4 \times 10^{-6}$$

which rearranges to

$$[SCN^-]^2 + 6.4 \times 10^{-6}\,[SCN^-] - 1.1 \times 10^{-12} = 0$$

$$[SCN^-] = 1.7 \times 10^{-7}$$

The formation constant for $Fe(SCN)^{2+}$ is

$$K_f = 1.05 \times 10^3 = \frac{[Fe(SCN)^{2+}]}{[Fe^{3+}][SCN^-]}$$

If we now substitute the $[SCN^-]$ necessary to give a detectable concentration of $Fe(SCN)^{2+}$ at the equivalence point, we obtain

$$1.05 \times 10^3 = \frac{6.4 \times 10^{-6}}{[Fe^{3+}]1.7 \times 10^{-7}}$$

$$[Fe^{3+}] = 0.036$$

The indicator concentration is not critical in the Volhard titration. In fact, calculations similar to those shown in Feature 13-1 demonstrate that a titration error of one part in a thousand or less is possible if the iron(III) concentration is held between 0.002 and 1.6 M. In practice, it is found that an indicator concentration greater than 0.2 M imparts sufficient color to the solution to make detection of the thiocyanate complex difficult because of the yellow color of Fe^{3+}. Therefore, lower concentrations (usually about 0.01 M) of iron(III) are employed.

The most important application of the Volhard method is for the indirect determination of halide ions. A measured excess of standard silver nitrate solution is added to the sample, and the excess silver ion is determined by back-titration with a standard thiocyanate solution. The strong acid environment required for the Volhard procedure represents a distinct advantage over other titrimetric methods of halide analysis because such ions as carbonate, oxalate, and arsenate (which form slightly soluble silver salts in neutral media but not in acidic media) do not interfere.

The Volhard procedure requires that the analyte solution be distinctly acidic.

Silver chloride is more soluble than silver thiocyanate. As a consequence, in chloride determinations by the Volhard method, the reaction

$$AgCl(s) + SCN^- \rightleftharpoons AgSCN(s) + Cl^-$$

occurs to a significant extent near the end of the back-titration of the excess silver ion. This reaction causes the end point to fade and results in an overconsumption of thiocyanate ion, which in turn leads to low values for the chloride analysis. This error can be circumvented by filtering the silver chloride before undertaking the back-titration. Filtration is not required in the determination of other halides because they all form silver salts that are less soluble than silver thiocyanate.

EXAMPLE 13-2

The arsenic in a 9.13-g sample of pesticide was converted to AsO_4^{3-} and precipitated as Ag_3AsO_4 with 50.00 mL of 0.02105 M $AgNO_3$. The excess Ag^+ was then titrated with 4.75 mL of 0.04321 M KSCN. Calculate the percentage of As_2O_3 in the sample.

$$\text{no. mmol } AgNO_3 = 50.00 \cancel{\text{mL } AgNO_3} \times 0.02105 \frac{\text{mmol } AgNO_3}{\cancel{\text{mL } AgNO_3}} = 1.0075$$

$$\text{no. mmol KSCN} = 4.75 \cancel{\text{mL KSCN}} \times 0.04321 \frac{\text{mmol KSCN}}{\cancel{\text{mL KSCN}}} = \underline{0.3052}$$

$$\text{no. mmol } AgNO_3 \text{ consumed by } AsO_4^{3-} = 0.8023$$

$$As_2O_3 \equiv 2AsO_4^{3-} \equiv 6AgNO_3$$

$$\% \, As_2O_3 = \frac{0.8023 \cancel{\text{mmol } AgNO_3} \times \dfrac{1 \cancel{\text{mmol } As_2O_3}}{6 \cancel{\text{mmol } AgNO_3}} \times 0.1984 \dfrac{\text{g } As_2O_3}{\cancel{\text{mmol } As_2O_3}}}{9.13 \text{ g sample}} \times 100\%$$

$$= 0.2906$$

13B-2 Potentiometric End Points

A useful method for determining end points in argentometric titrations is to measure the potential of a silver electrode immersed in the analyte solution. As we shall see (Chapter 18), the measured potential is directly proportional to pAg.

TABLE 13-1 Typical Argentometric Precipitation Methods

Substance Being Determined	End Point	Remarks
AsO_4^{3-}, Br^-, I^-, CNO^-, SCN^-	Volhard	Removal of silver salt not required
CO_3^{2-}, CrO_4^{2-}, CN^-, Cl^-, $C_2O_4^{2-}$, PO_4^{3-}, S^{2-}, NCN^{2-}	Volhard	Removal of silver salt required before back-titration of excess Ag^+
BH_4^-	Modified Volhard	Titration of excess Ag^+ following $BH_4^- + 8Ag^+ + 8OH^- \rightarrow 8Ag(s) + H_2BO_3^- + 5H_2O$
Epoxide	Volhard	Titration of excess Cl^- following hydrohalogenation
K^+	Modified Volhard	Precipitation of K^+ with known excess of $B(C_6H_5)_4^-$, addition of excess Ag^+ giving $AgB(C_6H_5)_4(s)$, and back-titration of the excess
Br^-, Cl^-	$2Ag^+ + CrO_4^{2-} \rightarrow \underset{\text{red}}{Ag_2CrO_4(s)}$	In neutral solution
Br^-, Cl^-, I^-, SeO_3^{2-}	Adsorption indicator	
$V(OH)_4^+$, fatty acids, mercaptans	Electroanalytical	Direct titration with Ag^+
Zn^{2+}	Modified Volhard	Precipitation as $ZnHg(SCN)_4$, filtration, dissolution in acid, addition of excess Ag^+, back-titration of excess Ag^+
F^-	Modified Volhard	Precipitation as PbClF, filtration, dissolution in acid, addition of excess Ag^+, back-titration of excess Ag^+

Thus, curves such as those shown in Figures 13-1, 13-2, and 13-3 can be derived experimentally. The end point is then the volume corresponding to the inflection point in the curve.

13C APPLICATIONS OF STANDARD SILVER NITRATE SOLUTIONS

Table 13-1 lists some typical applications of precipitation titrations in which silver nitrate is the standard solution. In most of these methods, the analyte is precipitated with a measured excess of silver nitrate and the excess determined by a Volhard titration with standard potassium thiocyanate.

Both silver nitrate and potassium thiocyanate are obtainable in primary-standard quality. The latter is somewhat hygroscopic, however, and thiocyanate solutions are ordinarily standardized against silver nitrate. Both silver nitrate and potassium thiocyanate solutions are stable indefinitely.

13D QUESTIONS AND PROBLEMS

***13-1.** In what respect is the Fajans method superior to the Volhard method for the titration of chloride ion?

13-2. Briefly explain why the sparingly soluble product must be removed by filtration before you back-titrate the excess silver ion in the Volhard determination of
(a) chloride ion.
(b) cyanide ion.
(c) carbonate ion.

***13-3.** Why does a Volhard determination of iodide ion require fewer steps than a Volhard determination of

(a) carbonate ion?
(b) cyanide ion?

13-4. Why does the charge on the surface of precipitate particles change sign at the equivalence point in a titration?

*13-5. Outline a method for the determination of K^+ based on argentometry. Write balanced equations for the reactions.

13-6. Suggest an argentometric method for the determination of F^-. Write balanced equations for the reactions.

13-7. A silver nitrate solution contains 14.77 g of primary-standard $AgNO_3$ in 1.00 L. What volume of this solution will be needed to react with
*(a) 0.2631 g of NaCl?
(b) 0.1799 g of Na_2CrO_4?
*(c) 64.13 mg of Na_3AsO_4?
(d) 381.1 mg of $BaCl_2 \cdot 2H_2O$?
*(e) 25.00 mL of 0.05361 M Na_3PO_4?
(f) 50.00 mL of 0.01808 M H_2S?

13-8. What is the molar analytical concentration of a silver nitrate solution if a 25.00-mL aliquot reacts with the amount of solutes listed in Problem 13-7?

13-9. What minimum volume of 0.09621 M $AgNO_3$ will be needed to assure an excess of silver ion in the titration of
*(a) an impure NaCl sample that weighs 0.2513 g?
(b) a 0.3462-g sample that is 74.52% (w/w) $ZnCl_2$?
*(c) 25.00 mL of 0.01907 M $AlCl_3$?

13-10. A Fajans titration of a 0.7908-g sample required 45.32 mL of 0.1046 M $AgNO_3$. Express the results of this analysis in terms of the percentage of
(a) Cl^-.
(b) $BaCl_2 \cdot 2H_2O$.
(c) $ZnCl_2 \cdot 2NH_4Cl$ (243.28 g/mol).

*13-11. Titration of a 0.485-g sample by the Mohr method required 36.8 mL of standard 0.1060 M $AgNO_3$ solution. Calculate the percentage of chloride in the sample.

13-12. A 0.1064-g sample of a pesticide was decomposed by the action of sodium biphenyl in toluene. The liberated Cl^- was extracted with water and titrated with 23.28 mL of 0.03337 M $AgNO_3$ using an adsorption indicator. Express the results of this analysis in terms of percent aldrin, $C_{12}H_8Cl_6$ (364.92 g/mol).

*13-13. A 100-mL sample of brackish water was made ammoniacal and the sulfide it contained titrated with 8.47 mL of 0.01310 M $AgNO_3$. The net reaction is

$$2Ag^+ + S^{2-} \longrightarrow Ag_2S(s)$$

Calculate the parts per million of H_2S in the water.

13-14. A 2.000-L water sample was evaporated to a small volume and treated with an excess of sodium tetraphenylboron, $NaB(C_6H_5)_4$. The precipitated $KB(C_6H_5)_4$ was filtered and then redissolved in acetone. The analysis was completed by a Mohr titration, with 37.90 mL of 0.03981 M $AgNO_3$ being used. The net reaction is

$$KB(C_6H_5)_4 + Ag^+ \longrightarrow AgB(C_6H_5)_4(s) + K^+$$

Express the results of this analysis in terms of parts per million of K (that is mg K/L).

*13-15. The phosphate in a 4.258-g sample of plant food was precipitated as Ag_3PO_4 through the addition of 50.00 mL of 0.0820 M $AgNO_3$:

$$3Ag^+ + HPO_4^{2-} \longrightarrow Ag_3PO_4(s) + H^+$$

The solid was filtered and washed, following which the filtrate and washings were diluted to exactly 250.0 mL. Titration of a 50.00-mL aliquot of this solution required a 4.64-mL back-titration with 0.0625 M KSCN. Express the results of this analysis in terms of the percentage of P_2O_5.

*13-16. The monochloroacetic acid ($ClCH_2COOH$) preservative in 100.0 mL of a carbonated beverage was extracted into diethyl ether and then returned to aqueous solution as $ClCH_3COO^-$ by extraction with 1 M NaOH. This aqueous extract was acidified and treated with 50.00 mL of 0.04521 M $AgNO_3$. The reaction is

$$ClCH_2COOH + Ag^+ + H_2O \longrightarrow HOCH_2COOH + H^+ + AgCl(s)$$

After filtration of the AgCl, titration of the filtrate and washings required 10.43 mL of an NH_4SCN solution. Titration of a blank taken through the entire process used 22.98 mL of the NH_4SCN. Calculate the weight (in milligrams) of $ClCH_2COOH$ in the sample.

13-17. An analysis for borohydride ion is based upon its reaction with Ag^+:

$$BH_4^- + 8Ag^+ + 8OH^- \longrightarrow H_2BO_3^- + 8Ag(s) + 5H_2O$$

The purity of a quantity of KBH_4 for use in an organic synthesis was established by diluting 3.213 g of the material to exactly 500.0 mL, treating a 100.0-mL aliquot with 50.00 mL of 0.2221 M $AgNO_3$, and titrating the excess silver ion with 3.36 mL of 0.0397 M KSCN. Calculate the percent purity of the KBH_4 (53.941 g/mol).

13-18. What volume of 0.04642 M KSCN would be needed if the analysis in Problem 13-17 were completed by filtering off the metallic Ag, dissolving it in acid, diluting to 250.0 mL, and titrating a 50.00-mL aliquot?

*13-19. An adsorption indicator is to be used in the routine analysis of samples that contain chloride. It is desired that the volume of standard $AgNO_3$ used in these titrations be numerically equal to the percent Cl^- when 0.2500-g samples are used. What should be the molarity of the $AgNO_3$ solution?

*13-20. The Association of Official Analytical Chemists recommends a Volhard titration for analysis of the insecticide heptachlor, $C_{10}H_5Cl_7$. The percentage of heptachlor is given by

$$\% \text{ heptachlor} = \frac{(\text{mL}_{\text{Ag}} \times c_{\text{Ag}} - \text{mL}_{\text{SCN}} \times c_{\text{SCN}}) \times 37.33}{\text{mass sample}}$$

What does this calculation reveal concerning the stoichiometry of this titration?

13-21. A carbonate fusion was needed to free the Bi from a 0.6423-g sample containing the mineral eulytite ($2Bi_2O_3 \cdot 3SiO_2$). The fused mass was dissolved in dilute acid, following which the Bi^{3+} was titrated with 27.36 mL of 0.03369 M NaH_2PO_4. The reaction is

$$Bi^{3+} + H_2PO_4^- \longrightarrow BiPO_4(s) + 2H^+$$

Calculate the percentage purity of eulytite (1112 g/mol) in the sample.

13-22. The theobromine ($C_7H_8N_4O_2$) in a 2.95-g sample of ground cocoa beans was converted to the sparingly soluble silver salt $C_7H_7N_4O_2Ag$ by warming in an ammoniacal solution containing 25.0 mL of 0.0100 M $AgNO_3$. After reaction was complete, all solids were removed by filtration. Calculate the percentage of theobromine (180.1 g/mol) in the sample if the combined filtrate and washings required a back-titration with 7.69 mL of 0.0108 M KSCN.

***13-23.** A 20-tablet sample of soluble saccharin was treated with 20.00 mL of 0.08181 M $AgNO_3$. The reaction is

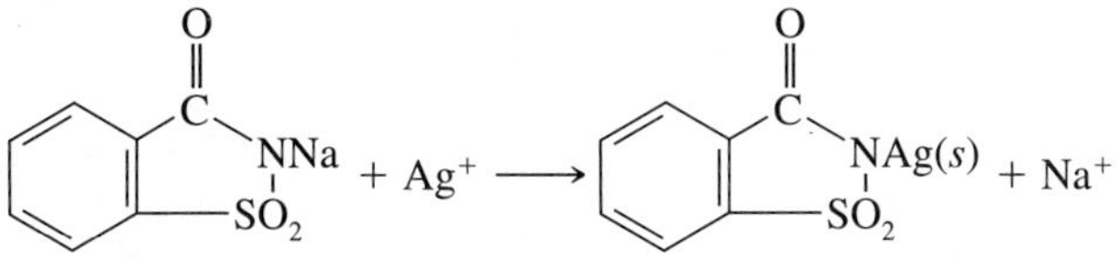

After removal of the solid, titration of the filtrate and washings required 2.81 mL of 0.04124 M KSCN. Calculate the average number of milligrams of saccharin (205.17 g/mol) in each tablet.

***13-24.** The formaldehyde in a 5.00-g sample of a seed disinfectant was steam distilled, and the aqueous distillate was collected in a 500-mL volumetric flask. After dilution to volume, a 25.0-mL aliquot was treated with 30.0 mL of 0.121 M KCN solution to convert the formaldehyde to potassium cyanohydrin.

$$K^+ + CH_2O + CN^- \longrightarrow KOCH_2CN$$

The excess KCN was then removed by addition of 40.0 mL of 0.100 M $AgNO_3$.

$$2CN^- + 2Ag^+ \longrightarrow Ag_2(CN)_2(s)$$

The excess Ag^+ in the filtrate and washings required a 16.1-mL titration with 0.134 M NH_4SCN. Calculate the percent CH_2O in the sample.

***13-25.** The action of an alkaline I_2 solution upon the rodenticide warfarin, $C_{19}H_{16}O_4$ (308.34 g/mol), results in the formation of 1 mol of iodoform, CHI_3 (393.73 g/mol), for each mole of the parent compound reacted. Analysis for warfarin can then be based upon the reaction between CHI_3 and Ag^+:

$$CHI_3 + 3Ag^+ + H_2O \longrightarrow 3AgI(s) + 3H^+ + CO(g)$$

The CHI_3 produced from a 13.96-g sample was treated with 25.00 mL of 0.02979 M $AgNO_3$, and the excess Ag^+ was then titrated with 2.85 mL of 0.05411 M KSCN. Calculate the percentage of warfarin in the sample.

13-26. A 5.00-mL aqueous suspension of elemental selenium was treated with 25.00 mL of ammoniacal 0.0360 M $AgNO_3$. The reaction is

$$6Ag(NH_3)_2^+ + 3Se(s) + 3H_2O \longrightarrow 2Ag_2Se(s) + Ag_2SeO_3(s) + 6NH_4^+$$

After this reaction was complete, nitric acid was added to dissolve the Ag_2SeO_3 but not the Ag_2Se. The Ag^+ from the dissolved Ag_2SeO_3 and the excess reagent required 16.74 mL of 0.01370 M KSCN in a Volhard titration. How many milligrams of Se were contained in each milliliter of sample?

13-27. A 2.4414-g sample containing KCl, K_2SO_4, and inert materials was dissolved in sufficient water to give 250.0 mL of solution. A Mohr titration of a 50.00-mL aliquot required 41.36 mL of 0.05818 M $AgNO_3$. A second 50.00-mL aliquot was treated with 40.00 mL of 0.1083 M $NaB(C_6H_5)_4$. The reaction is

$$NaB(C_6H_5)_4 + K^+ \longrightarrow KB(C_6H_5)_4(s) + Na^+$$

The solid was filtered, redissolved in acetone, and titrated with 49.98 mL of the $AgNO_3$ solution (see Problem 13-14). Calculate the percentages of KCl and K_2SO_4 in the sample.

***13-28.** A 1.998-g sample containing Cl^- and ClO_4^- was dissolved in sufficient water to give 250.0 mL of solution. A 50.00-mL aliquot required 13.97 mL of 0.08551 M $AgNO_3$ to titrate the Cl^-. A second 50.00-mL aliquot was treated with $V_2(SO_4)_3$ to reduce the ClO_4^- to Cl^-:

$$ClO_4^- + 4V_2(SO_4)_3 + 4H_2O \longrightarrow Cl^- + 12SO_4^{2-} + 8VO^{2+} + 8H^+$$

Titration of the reduced sample required 40.12 mL of the $AgNO_3$ solution. Calculate the percentages of Cl^- and ClO_4^- in the sample.

13-29. For each of the following precipitation titrations, calculate the cation and anion concentrations at equivalence as well as at reagent volumes corresponding to ± 20.00 mL, ± 10.00 mL, and ± 1.00 mL of equivalence. Construct a titration curve from the data, plotting the p-function of the cation versus reagent volume.

***(a)** 25.00 mL of 0.05000 M $AgNO_3$ with 0.02500 M NH_4SCN
(b) 20.00 mL of 0.06000 M $AgNO_3$ with 0.03000 M KI
***(c)** 30.00 mL of 0.07500 M $AgNO_3$ with 0.07500 M NaCl
(d) 35.00 mL of 0.4000 M Na_2SO_4 with 0.2000 M $Pb(NO_3)_2$
***(e)** 40.00 mL of 0.02500 M $BaCl_2$ with 0.05000 M Na_2SO_4
(f) 50.00 mL of 0.2000 M NaI with 0.4000 M $TlNO_3$ (K_{sp} for TlI = 6.5×10^{-8})

13-30. Calculate the silver ion concentration after the addition of 5.00*, 15.00, 25.00, 30.00, 35.00, 39.00, 40.00*, 41.00, 45.00*, and 50.00 mL of 0.05000 M $AgNO_3$ to 50.0 mL of 0.0400 M KBr. Construct a titration curve from these data plotting pAg as a function of titrant volume.

CHAPTER

14

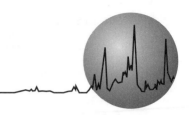

Complex-Formation Titrations

Complex-formation reagents are widely used for titrating cations. The most versatile of these reagents are organic compounds that contain several electron-donor groups that form multiple covalent bonds with metal ions.

14A COMPLEX-FORMATION REACTIONS

A **ligand** is an ion or a molecule that forms a covalent bond with a cation or a neutral metal atom by donating a pair of electrons, which are then shared by the two.

The **coordination number** of a cation is the number of covalent bonds it tends to form with electron-donor species.

A **chelate** is a cyclic complex formed when a cation is bonded by two or more donor groups contained in a single ligand.

Chelate is pronounced keé late.

Dentate (from Latin) means having tooth-like projections.

Most metal ions react with electron-pair donors to form coordination compounds or complexes. The donor species, or *ligand,* must have at least one pair of unshared electrons available for bond formation. Water, ammonia, and halide ions are common inorganic ligands.

The number of covalent bonds that a cation tends to form with electron donors is its *coordination number.* Typical values for coordination numbers are two, four, and six. The species formed as a result of coordination can be electrically positive, neutral, or negative. For example, copper(II), which has a coordination number of four, forms a cationic ammine complex, $Cu(NH_3)_4^{2+}$; a neutral complex with glycine, $Cu(NH_2CH_2COO)_2$; and an anionic complex with chloride ion, $CuCl_4^{2-}$.

Titrimetric methods based on complex formation (sometimes called *complexometric methods*) have been used for more than a century. The truly remarkable growth in their analytical application began in the 1940s, however, and is based upon a particular class of coordination compounds called *chelates.* A chelate is produced when a metal ion coordinates with two (or more) donor groups of a single ligand to form a five- or six-membered heterocyclic ring. The copper complex of glycine, mentioned in the previous paragraph, is an example. Here, copper bonds to both the oxygen of the carboxyl group and the nitrogen of the amine group:

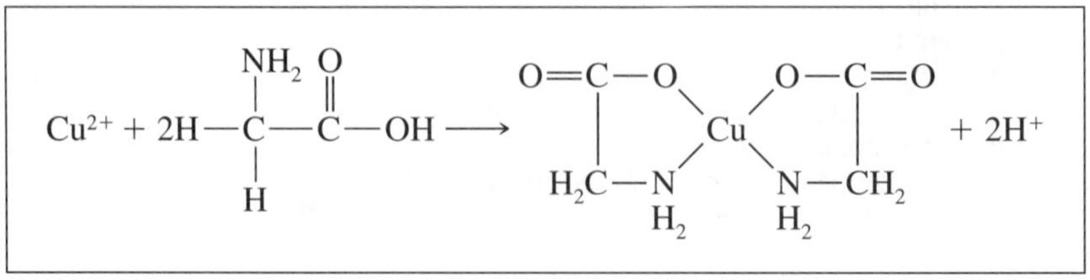

A ligand that has a single donor group, such as ammonia, is called *unidentate* (single-toothed), whereas one such as glycine, which has two groups available for

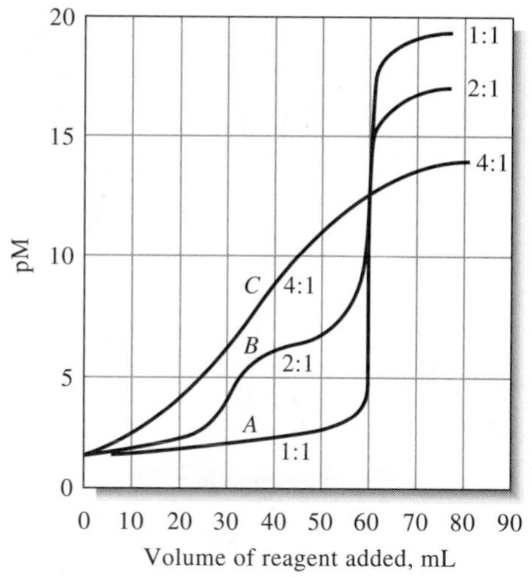

Figure 14-1
Curves for complex-formation titrations. Titration of 60.0 mL of a solution that is 0.020 M in M with (*A*) a 0.020 M solution of the tetradentate ligand D to give MD as product; (*B*) a 0.040 M solution of the bidentate ligand B to give MB_2; and (*C*) a 0.080 M solution of the unidentate ligand A to give MA_4. The overall formation constant for each product is 1.0×10^{20}.

covalent bonding, is called *bidentate. Tridentate, tetradentate, pentadentate,* and *hexadentate* chelating agents are also known.

As titrants, multidentate ligands, particularly those having four or six donor groups, have two advantages over their unidentate counterparts. First, they generally react more completely with cations and thus provide sharper end points. Second, they ordinarily react with metal ions in a single-step process, whereas complex formation with unidentate ligands usually involves two or more intermediate species.

Tetradentate or hexadentate ligands are more satisfactory as titrants than ligands with a lesser number of donor groups because their reactions with cations are more complete and because they tend to form 1 : 1 complexes.

The advantage of a single-step reaction is illustrated by the titration curves shown in Figure 14-1. Each of the titrations involves a reaction that has an overall equilibrium constant of 10^{20}. Curve *A* is derived for a reaction in which a metal ion M having a coordination number of four reacts with a tetradentate ligand D to form the complex of MD (here we have omitted the charges on the two reactants for convenience). Curve *B* is for the reaction of M with a hypothetical bidentate ligand B to give MB_2 in two steps. The formation constant for the first step is 10^{12} and for the second 10^8. Curve *C* involves a unidentate ligand, A, that forms MA_4 in four steps with successive formation constants of 10^8, 10^6, 10^4, and 10^2. These curves demonstrate that a much sharper end point is obtained with a reaction that takes place in a single step. For this reason, multidentate ligands are ordinarily preferred for complexometric titrations.

14B TITRATIONS WITH AMINOCARBOXYLIC ACIDS

Tertiary amines that also contain carboxylic acid groups form remarkably stable chelates with many metal ions.[1] Schwarzenbach first recognized their potential as analytical reagents in 1945. Since this original work, investigators throughout the world have described applications of these compounds to the volumetric determination of most of the metals in the periodic table.

[1]These reagents are the subject of several excellent monographs. See, for example, R. Pribil, *Applied Complexometry.* New York: Pergamon, 1982; A. Ringbom and E. Wanninen, in *Treatise on Analytical Chemistry,* 2nd ed., I. M. Kolthoff and P. J. Elving, Eds., Part I, Vol. 2, Chapter 11. New York: Wiley, 1979.

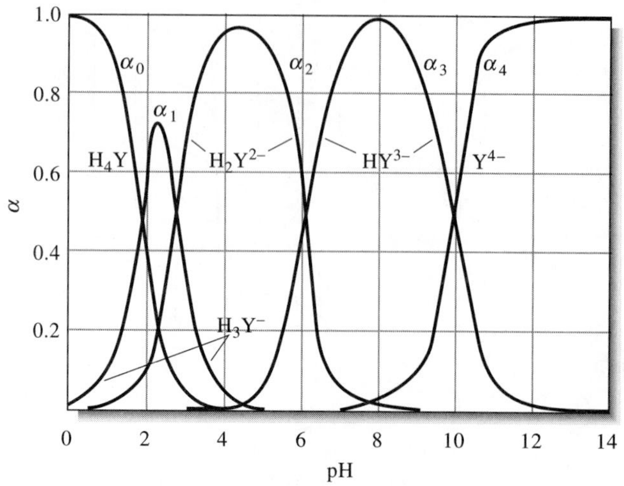

Figure 14-2

Composition of EDTA solutions as a function of pH.

14B-1 Ethylenediaminetetraacetic Acid (EDTA)

See *Mathcad Applications for Analytical Chemistry,* **pp. 96–98.**

Ethylenediaminetetraacetic acid [also called (ethylenedinitrilo)tetraacetic acid], which is commonly shortened to EDTA, is the most widely used complexometric titrant. EDTA has the structure

$$\text{HOOC}-\text{CH}_2 \quad \quad \quad \text{CH}_2-\text{COOH}$$
$$\ddot{\text{N}}-\text{CH}_2-\text{CH}_2-\text{N}\ddot{:}$$
$$\text{HOOC}-\text{CH}_2 \quad \quad \quad \text{CH}_2-\text{COOH}$$

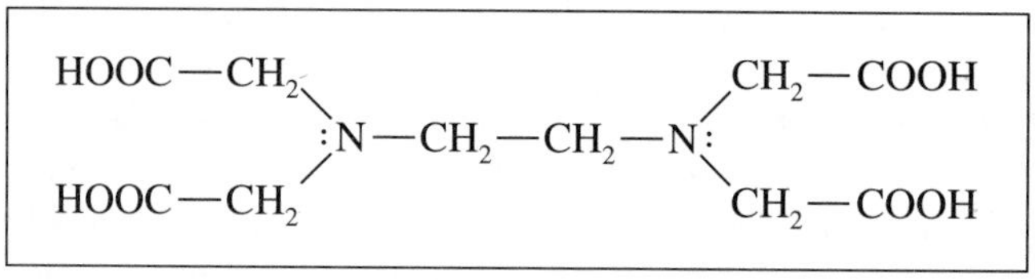

> **EDTA** stands for ethylenediaminetetraacetic acid.

The EDTA molecule has six potential sites for bonding a metal ion: the four carboxyl groups and the two amino groups, each of the latter with an unshared pair of electrons. Thus, EDTA is a hexadentate ligand.

Acidic Properties of EDTA

EDTA, a hexadentate ligand, is among the most important and widely used reagents in titrimetry.

The dissociation constants for the acidic groups in EDTA are $K_1 = 1.02 \times 10^{-2}$, $K_2 = 2.14 \times 10^{-3}$, $K_3 = 6.92 \times 10^{-7}$, and $K_4 = 5.50 \times 10^{-11}$. It is of interest that the first two constants are of the same order of magnitude, which suggests that the two protons involved dissociate from opposite ends of the rather long molecule. As a consequence of their physical separation, the negative charge created by the first dissociation does not greatly affect the removal of the second proton. The same cannot be said for the dissociation of the other two protons, however, which are much closer to the negatively charged carboxylate ions created by the initial dissociations.

The various EDTA species are often abbreviated H_4Y, H_3Y^-, H_2Y^{2-}, HY^{3-}, and Y^{4-}. Figure 14-2 illustrates how the relative amounts of these five species vary as a function of pH. It is apparent that H_2Y^{2-} predominates in moderately acidic media (pH 3 to 6). Only at pH values greater than 10 does Y^{4-} become a major component of solutions.

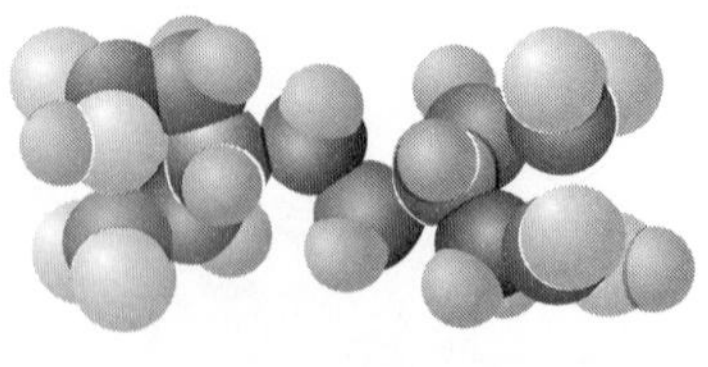

Space-filling model of H_4Y zwitterion.

FEATURE 14-1

Species Present in a Solution of EDTA

When it is dissolved in water, EDTA behaves like an amino acid, such as glycine (see Section 11H). Here, however, a double zwitterion forms, which has the structure shown in Figure 14-3a. Note that the net charge on this species is zero and that it contains four dissociatable protons, two associated with two of the carboxyl groups and the other two with the two amine groups. For simplicity, we generally formulate the double zwitterion as H_4Y, where Y^{4-} is the fully deprotonated ion shown in Figure 14-3e. The first and second steps in the dissociation process involve successive loss of protons from the two carboxylic acid groups; the third and fourth steps involve dissociation of the protonated amine groups. The structures of H_3Y^-, H_2Y^{2-}, and HY^{3-} are shown in Figure 14-3b, c, and d.

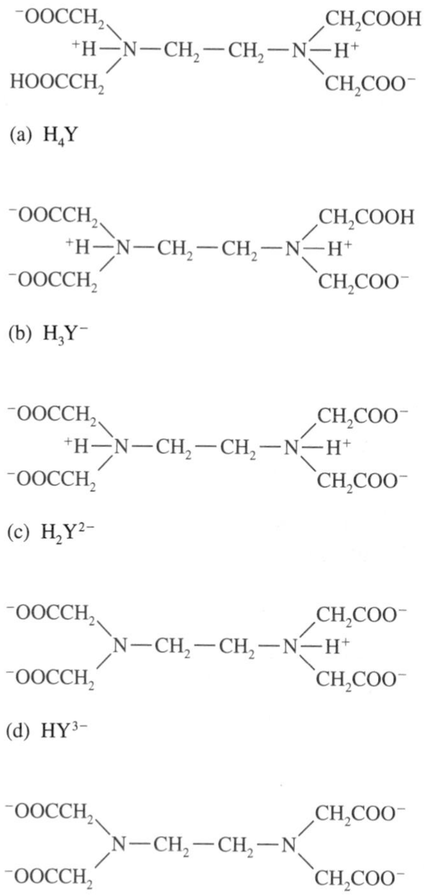

Figure 14-3
Structure of H_4Y and its dissociation products.

Reagents

The free acid, H_4Y, and the dihydrate of the sodium salt, $Na_2H_2Y \cdot 2H_2O$, are commercially available in reagent quality. The former can serve as a primary standard after it has been dried for several hours at 130°C to 145°C. It is then dissolved in the minimum amount of base required for complete solution.

Under normal atmospheric conditions, the dihydrate $Na_2H_2Y \cdot 2H_2O$ contains 0.3% moisture in excess of the stoichiometric amount. For all but the most exacting work, this excess is sufficiently reproducible to permit use of a corrected mass of the salt in the direct preparation of a standard solution. If necessary, the pure dihydrate can be prepared by drying at 80°C for several days in an atmosphere of 50% relative humidity.

A number of compounds that are chemically similar to EDTA have also been investigated but do not appear to offer advantages over the original species. We shall thus confine our discussion to the properties and applications of EDTA.

Standard EDTA solutions are ordinarily prepared by dissolving weighed quantities of $Na_2H_2Y \cdot 2H_2O$ and diluting to the mark in a volumetric flask.

Nitrilotriacetic acid (NTA) is the second most common aminopolycarboxylic acid used for titrimetry. It is a tetradentate ligand that has the structure

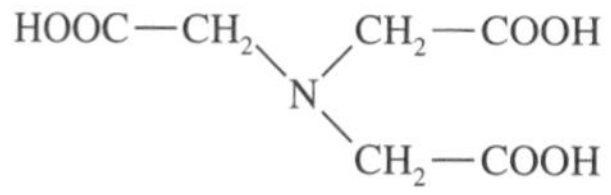

14B-2 Complexes of EDTA and Metal Ions

Solutions of EDTA are particularly valuable as titrants because the reagent combines with metal ions in a 1 : 1 ratio regardless of the charge on the cation. For example, formation of the silver and aluminum complexes is described by the equations:

$$Ag^+ + Y^{4-} \rightleftharpoons AgY^{3-}$$
$$Al^{3+} + Y^{4-} \rightleftharpoons AlY^-$$

The reaction of the EDTA anion with a metal ion M^{n+} is described by the equation

$$M^{n+} + Y^{4-} \rightleftharpoons MY^{(n-4)+}$$

EDTA is a remarkable reagent not only because it forms chelates with all cations but also because most of these chelates are sufficiently stable to form the basis for a titrimetric method. This great stability undoubtedly results from the several complexing sites within the molecule, which give rise to a cagelike structure in which the cation is effectively surrounded and isolated from solvent mole-

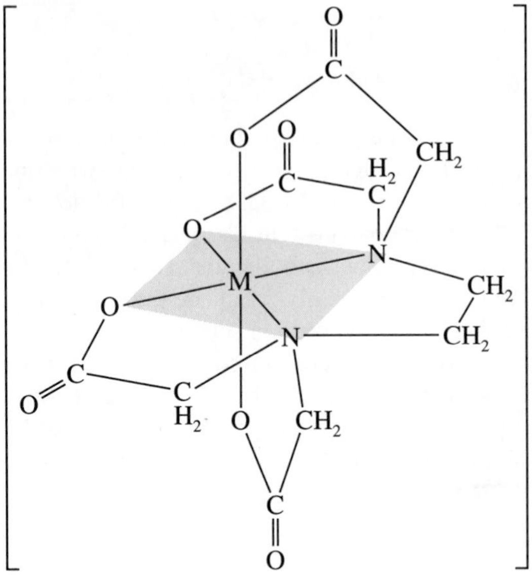

Figure 14-4
Structure of a metal/EDTA chelate.

cules. One form of the complex is depicted in Figure 14-4. Note that six donor atoms are involved in bonding the divalent metal ion.

Table 14-1 lists formation constants K_{MY} for common EDTA complexes. Note that the constant refers to the equilibrium involving the deprotonated species Y^{4-} with the metal ion:

$$M^{n+} + Y^{4-} \rightleftharpoons MY^{(n-4)+} \qquad K_{MY} = \frac{[MY^{(n-4)+}]}{[M^{n+}][Y^{4-}]} \tag{14-1}$$

14B-3 Equilibrium Calculations Involving EDTA

A titration curve for the reaction of a cation M^{n+} with EDTA consists of a plot of pM versus reagent volume. Values for pM are readily computed in the early stage

TABLE 14-1 Formation Constants for EDTA Complexes*

Cation	K_{MY}	log K_{MY}	Cation	K_{MY}	log K_{MY}
Ag^+	2.1×10^7	7.32	Cu^{2+}	6.3×10^{18}	18.80
Mg^{2+}	4.9×10^8	8.69	Zn^{2+}	3.2×10^{16}	16.50
Ca^{2+}	5.0×10^{10}	10.70	Cd^{2+}	2.9×10^{16}	16.46
Sr^{2+}	4.3×10^8	8.63	Hg^{2+}	6.3×10^{21}	21.80
Ba^{2+}	5.8×10^7	7.76	Pb^{2+}	1.1×10^{18}	18.04
Mn^{2+}	6.2×10^{13}	13.79	Al^{3+}	1.3×10^{16}	16.13
Fe^{2+}	2.1×10^{14}	14.33	Fe^{3+}	1.3×10^{25}	25.1
Co^{2+}	2.0×10^{16}	16.31	V^{3+}	7.9×10^{25}	25.9
Ni^{2+}	4.2×10^{18}	18.62	Th^{4+}	1.6×10^{23}	23.2

*Data from G. Schwarzenbach, *Complexometric Titrations,* p. 8. London: Chapman and Hall, 1957. With permission. Constants are valid at 20°C and an ionic strength of 0.1.

of a titration by assuming that the equilibrium concentration of M^{n+} is equal to its analytical concentration, which in turn is readily derived from stoichiometric data.

Calculation of M^{n+} at and beyond the equivalent point requires the use of Equation 14-1. The computation in this region is troublesome and time-consuming if the pH is unknown and variable because both $[MY^{(n-4)+}]$ and $[M^{n+}]$ are pH dependent. Fortunately, EDTA titrations are always performed in solutions that are buffered to a known pH to avoid interference by other cations and to ensure satisfactory indicator behavior. Calculating $[M^{n+}]$ in a buffered solution containing EDTA is a relatively straightforward procedure provided the pH is known. In this computation, use is made of an alpha value for H_4Y. Recall (Section 11I) that α_4 for H_4Y would be defined as

$$\alpha_4 = \frac{[Y^{4-}]}{c_T} \tag{14-2}$$

where c_T is the total molar concentration of *uncomplexed* EDTA:

$$c_T = [Y^{4-}] + [HY^{3-}] + [H_2Y^{2-}] + [H_3Y^-] + [H_4Y]$$

Conditional Formation Constants

Conditional, or *effective, formation constants* are pH-dependent equilibrium constants that apply *at a single pH only.* To obtain the conditional constant for the equilibrium shown in Equation 14-1, we substitute $\alpha_4 c_T$ from Equation 14-2 for $[Y^{4-}]$ in the formation-constant expression (Equation 14-1):

Conditional formation constants are pH dependent.

$$M^{n+} + Y^{4-} \rightleftharpoons MY^{(n-4)+} \qquad K_{MY} = \frac{[MY^{(n-4)+}]}{[M^{n+}]\,\alpha_4 c_T} \tag{14-3}$$

Combining the two constants yields a new constant K'_{MY}

$$K'_{MY} = \frac{[MY^{(n-4)+}]}{[M^{n+}]\,c_T} = \alpha_4 K_{MY} \tag{14-4}$$

where K'_{MY}, the conditional formation constant, describes equilibrium relationships *only at the pH for which α_4 is applicable.*

Conditional constants are readily computed and provide a simple means by which the equilibrium concentrations of the metal ion and the complex can be calculated at the equivalence point and where there is an excess of chelating reagent. Note that replacement of $[Y^{4-}]$ with c_T in the equilibrium-constant expression greatly simplifies calculations because c_T is readily determined from the reaction stoichiometry, whereas $[Y^{4-}]$ is not.

Alpha values for the other EDTA species are readily obtained in a similar way and are found to be

$\alpha_0 = [H^+]^4/D$ $\qquad \alpha_2 = K_1K_2[H^+]^2/D$

$\alpha_1 = K_1[H^+]^3/D$ $\qquad \alpha_3 = K_1K_2K_3[H^+]/D$

To derive titration curves, however, only α_4 is needed.

The Computation of α_4 Values for EDTA Solutions

An expression for calculating α_4 at a given hydrogen ion concentration is readily derived by the method demonstrated in Section 11I (see also Feature 11-2). Thus, α_4 for EDTA is

TABLE 14-2
Values for α_4 for EDTA at Selected pH Values

pH	α_4	pH	α_4
2.0	3.7×10^{-14}	7.0	4.8×10^{-4}
3.0	2.5×10^{-11}	8.0	5.4×10^{-3}
4.0	3.6×10^{-9}	9.0	5.2×10^{-2}
5.0	3.5×10^{-7}	10.0	3.5×10^{-1}
6.0	2.2×10^{-5}	11.0	8.5×10^{-1}
		12.0	9.8×10^{-1}

In this chapter and those that follow, we shall revert to the use of H^+ as a convenient shorthand notation for H_3O^+; from time to time, we shall refer to the species represented by H^+ as the hydrogen ion.

$$\alpha_4 = \frac{K_1K_2K_3K_4}{[H^+]^4 + K_1[H^+]^3 + K_1K_2[H^+]^2 + K_1K_2K_3[H^+] + K_1K_2K_3K_4} \quad (14\text{-}5)$$

$$\alpha_4 = \frac{K_1K_2K_3K_4}{D} \quad (14\text{-}6)$$

where K_1, K_2, K_3, and K_4 are the four dissociation constants for H_4Y and D is the denominator of Equation 14-5.

Table 14-2 lists α_4 at selected pH values. Note that only about 4×10^{-12} percent of the EDTA exists as Y^{4-} at pH 2.00. Example 14-1 illustrates how $[Y^{4-}]$ is calculated for a solution of known pH.

EXAMPLE 14-1

Calculate α_4 and the mole percent of Y^{4-} in a solution of EDTA that is buffered to pH 10.20.

$$[H^+] = \text{antilog}\,(-10.20) = 6.31 \times 10^{-11} \approx 6.3 \times 10^{-11}$$

From the values for the dissociation constants for H_4Y (page 280), we obtain

$$K_1 = 1.02 \times 10^{-2} \qquad K_1K_2K_3 = 1.51 \times 10^{-11}$$
$$K_1K_2 = 2.18 \times 10^{-5} \qquad K_1K_2K_3K_4 = 8.31 \times 10^{-22}$$

Numerical values for the several terms in the denominator in Equation 14-5 are

$$\begin{aligned}
[H^+]^4 &= (6.31 \times 10^{-11})^4 &&= 1.58 \times 10^{-41}\\
K_1[H^+]^3 &= (1.02 \times 10^{-2})\,(6.31 \times 10^{-11})^3 &&= 2.56 \times 10^{-33}\\
K_1K_2[H^+]^2 &= (2.18 \times 10^{-5})(6.31 \times 10^{-11})^2 &&= 8.68 \times 10^{-26}\\
K_1K_2K_3[H^+] &= (1.51 \times 10^{-11})\,(6.31 \times 10^{-11}) &&= 9.53 \times 10^{-22}\\
K_1K_2K_3K_4 & &&= \underline{8.31 \times 10^{-22}}\\
& & D &= 1.78 \times 10^{-21}
\end{aligned}$$

and Equation 14-6 becomes

$$\alpha_4 = \frac{K_1K_2K_3K_4}{D} = \frac{8.31 \times 10^{-22}}{1.78 \times 10^{-21}} = 0.466 \approx 0.47$$

$$\text{mol \% } Y^{4-} = 0.47 \times 100\% = 47\%$$

Note that only the last two terms in the denominator contribute significantly to the sum D at pH 10.20. At low pH values, in contrast, only the first two or three terms are important.

Calculation of the Cation Concentration in EDTA Solutions

Example 14-2 demonstrates how the cation concentration in a solution of an EDTA complex is computed. Example 14-3 illustrates this calculation for a solution that contains an excess of EDTA.

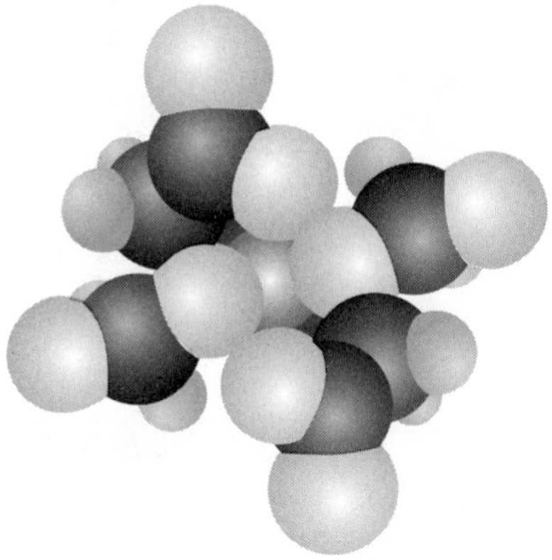

Space-filling model of NiY^{2-}.

EXAMPLE 14-2

Calculate the equilibrium concentration of Ni^{2+} in a solution with an analytical NiY^{2-} concentration of 0.0150 M at pH (a) 3.0 and (b) 8.0.

From Table 14-1,

$$Ni^{2+} + Y^{4-} \rightleftharpoons NiY^{2-} \qquad K_{MY} = \frac{[NiY^{2-}]}{[Ni^{2+}][Y^{4-}]} = 4.2 \times 10^{18}$$

The equilibrium concentration of NiY^{2-} is equal to the analytical concentration of the complex minus the concentration lost by dissociation. The latter is given by the equilibrium nickel ion concentration. Thus,

$$[NiY^{2-}] = 0.0150 - [Ni^{2+}]$$

If we assume $[Ni^{2+}] \ll 0.0150$, an assumption that is almost certainly valid in light of the large formation constant of the complex, the foregoing equation simplifies to

$$[NiY^{2-}] \approx 0.0150$$

Since the complex is the only source of both Ni^{2+} and the EDTA species,

$$[Ni^{2+}] = c_T = [Y^{4-}] + [HY^{3-}] + [H_2Y^{2-}] + [H_3Y^-] + [H_4Y]$$

Substitution of this equality into Equation 14-4 gives

$$K'_{MY} = \alpha_4 K_{MY} = \frac{[NiY^{2-}]}{[Ni^{2+}]c_T} = \frac{[NiY^{2-}]}{[Ni^{2+}]^2}$$

(a) Table 14-2 indicates that α_4 is 2.5×10^{-11} at pH 3.0. Substitution of this value and the concentration of NiY^{2-} into the equation for K'_{MY} gives

$$\frac{0.0150}{[Ni^{2+}]^2} = \alpha_4 K'_{MY} = 2.5 \times 10^{-11} \times 4.2 \times 10^{18} = 1.05 \times 10^8$$

$$[Ni^{2+}] = \sqrt{\frac{0.0150}{1.05 \times 10^8}} = \sqrt{1.43 \times 10^{-10}} = 1.2 \times 10^{-5}\ M$$

Note that $[Ni^{2+}] \ll 0.0150$, as assumed.

Our assumption is valid.

(b) At pH 8.0, the conditional constant is much larger. Thus,

$$K'_{MY} = 5.4 \times 10^{-3} \times 4.2 \times 10^{18} = 2.27 \times 10^{16}$$

and substitution into the equation for K'_{MY} followed by rearrangement gives

$$[Ni^{2+}] = \sqrt{0.0150/(2.27 \times 10^{16})} = 8.1 \times 10^{-10}\ M$$

EXAMPLE 14-3

Calculate the concentration of Ni^{2+} in a solution that was prepared by mixing 50.0 mL of 0.0300 M Ni^{2+} with 50.0 mL of 0.0500 M EDTA. The mixture was buffered to a pH of 3.00.

Here, the solution has an excess of EDTA, and the analytical concentration of the complex is determined by the amount of Ni^{2+} originally present. Thus,

$$c_{NiY^{2-}} = 50.0 \times \frac{0.03000}{100} = 0.0150\ M$$

$$c_{EDTA} = \frac{50.0 \times 0.0500 - 50.0 \times 0.0300}{100} = 0.0100\ M$$

Our assumption that $[Ni^{2+}] \ll [NiY^{2-}]$ is seen to be valid.

Again let us assume that $[Ni^{2+}] \ll [NiY^{2-}]$ so that

$$[NiY^{2-}] = 0.0150 - [Ni^{2+}] \approx 0.0150$$

At this point, the total concentration of uncomplexed EDTA is given by its analytical molarity:

$$c_T = 0.0100\ M$$

Substitution into Equation 14-4 gives

$$K'_{MY} = \frac{0.0150}{[Ni^{2+}]\,0.0100} = \alpha_4 K_{MY}$$

$$= 2.5 \times 10^{-11} \times 4.2 \times 10^{18} = 1.05 \times 10^{8}$$

$$[Ni^{2+}] = \frac{0.0150}{0.0100 \times 1.05 \times 10^{8}} = 1.4 \times 10^{-8}\ M$$

14B-4 EDTA Titration Curves

Example 14-4 demonstrates how data for an EDTA titration curve for a metal ion are computed for a solution of fixed pH.

See *Mathcad Applications for Analytical Chemistry,* **pp. 98–100.**

EXAMPLE 14-4

Derive a curve (pCa as a function of volume of EDTA) for the titration of 50.0 mL of 0.00500 M Ca^{2+} with 0.0100 M EDTA in a solution buffered to a constant pH of 10.0.

Calculation of a Conditional Constant

The conditional formation constant for the calcium/EDTA complex at pH 10.0 is obtained from the formation constant of the complex (Table 14-1) and the α_4 value for EDTA at pH 10.0 (Table 14-2). Thus, substitution into Equation 14-4 gives

$$K'_{CaY} = \frac{[CaY^{2-}]}{[Ca^{2+}]\, c_T} = \alpha_4 K_{CaY}$$
$$= 0.35 \times 5.0 \times 10^{10} = 1.75 \times 10^{10}$$

Preequivalence-Point Values for pCa

Before the equivalence point is reached, the equilibrium concentration of Ca^{2+} is equal to the sum of the contributions from the untitrated excess of the cation and from the dissociation of the complex, the latter being numerically equal to c_T. It is ordinarily reasonable to assume that c_T is small relative to the analytical concentration of the uncomplexed calcium ion. Thus, for example, after the addition of 10.0 mL of reagent,

$$[Ca^{2+}] = \frac{50.0 \times 0.00500 - 10.0 \times 0.0100}{60.0} + c_T \approx 2.50 \times 10^{-3}\ M$$
$$pCa = -\log (2.50 \times 10^{-3}) = 2.60$$

Other preequivalence-point data are derived in this same way.

The Equivalence-Point pCa

Following the method shown in Example 14-2, we first compute the analytical concentration of CaY^{2-}:

$$c_{CaY^{2-}} = \frac{50.0 \times 0.00500}{50.0 + 25.0} = 3.33 \times 10^{-3}\ M$$

The only source of Ca^{2+} ions is the dissociation of this complex. It also follows that the calcium ion concentration must be identical to the sum of the concentrations of the uncomplexed EDTA ions, c_T. Thus,

$$[Ca^{2+}] = c_T$$
$$[CaY^{2-}] = 0.00333 - [Ca^{2+}] \approx 0.00333\ M$$

Substituting into the conditional formation-constant expression gives

$$\frac{[CaY^{2-}]}{[Ca^{2+}]\, c_T} = \frac{0.00333}{[Ca^{2+}]^2} = 1.75 \times 10^{10}$$
$$[Ca^{2+}] = \sqrt{\frac{0.00333}{1.75 \times 10^{10}}} = 4.36 \times 10^{-7}\ M$$
$$pCa = -\log (4.36 \times 10^{-7}) = 6.36$$

Postequivalence-Point pCa

Beyond the equivalence point, analytical concentrations of CaY^{2-} and EDTA are obtained directly from the stoichiometric data. A calculation similar to that in Example 14-3 is then performed. Thus, after the addition of 35.0 mL of reagent,

$$c_{CaY^{2-}} = \frac{50.0 \times 0.00500}{50.0 + 35.0} = 2.94 \times 10^{-3}\ M$$

$$c_{EDTA} = \frac{35.0 \times 0.0100 - 50.0 \times 0.00500}{85.0} = 1.18 \times 10^{-3}\ M$$

As an approximation, we can write

$$[CaY^{2-}] = 2.94 \times 10^{-3} - [Ca^{2+}] \approx 2.94 \times 10^{-3}$$
$$c_T = 1.18 \times 10^{-3} + [Ca^{2+}] \approx 1.18 \times 10^{-3}\ M$$

and substitution into the conditional formation-constant expression gives

$$K'_{CaY} = \frac{2.94 \times 10^{-3}}{[Ca^{2+}] \times 1.18 \times 10^{-3}} = 1.75 \times 10^{10}$$

$$[Ca^{2+}] = \frac{2.94 \times 10^{-3}}{1.18 \times 10^{-3} \times 1.75 \times 10^{10}} = 1.42 \times 10^{-10}$$

$$pCa = -\log(1.42 \times 10^{-10}) = 9.85$$

The approximation was clearly valid.

Other postequivalence-point data are computed in this same way.

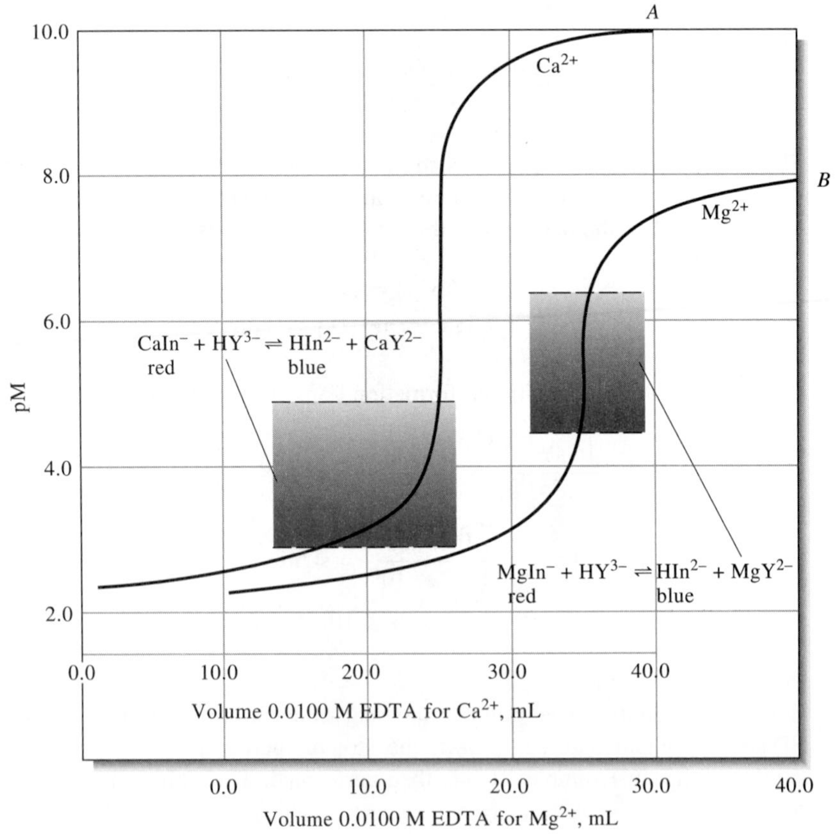

Figure 14-5
EDTA titration curves for 50.0 mL of 0.00500 M Ca^{2+} (K' for $CaY^{2-} = 1.75 \times 10^{10}$) and Mg^{2+} (K' for $MgY^{2-} = 1.72 \times 10^{8}$) at pH 10.0. The shaded areas show the transition range for Eriochrome Black T.

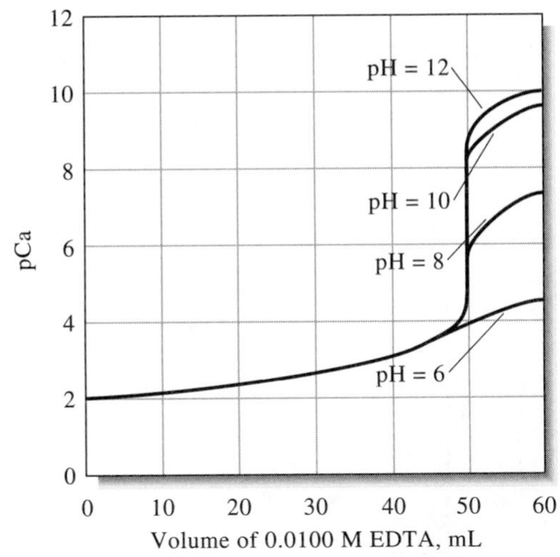

See *Mathcad Applications for Analytical Chemistry,* **p. 100.**

Figure 14-6
Influence of pH on the titration of 0.0100 M Ca^{2+} with 0.0100 M EDTA.

Curve *A* in Figure 14-5 is a plot of data for the titration in Example 14-4. Curve *B* is the titration curve for a solution of magnesium ion under identical conditions. The formation constant for the EDTA complex of magnesium is smaller than that of the calcium complex. Consequently, the reaction of calcium ion with the EDTA is more complete, and a larger change in p-function is observed in the equivalence region. The effect is analogous to what we have seen earlier for precipitation and neutralization titrations.

Figure 14-6 provides titration curves for calcium ion in solutions buffered to various pH levels. Recall that α_4, and hence K'_{CaY}, becomes smaller as the pH decreases. The less favorable equilibrium constant leads to a smaller change in pCa in the equivalence-point region. It is apparent from Figure 14-6 that an adequate end point in the titration of calcium requires a pH of about 8 or greater. As shown in Figure 14-7, however, cations with larger formation constants provide good end points even in acidic media. Figure 14-8 shows the minimum permissible pH for a satisfactory end point in the titration of various metal ions in the absence of competing complexing agents. Note that a moderately acidic environment is satisfactory for many divalent heavy-metal cations and that a strongly acidic medium can be tolerated in the titration of such ions as iron(III) and indium(III).

End points for EDTA titrations become less sharp with pH decreases because the complex formation reaction is less complete under these circumstances.

Figure 14-8 shows that most cations having a +3 or +4 charge can be titrated in a distinctly acidic solution.

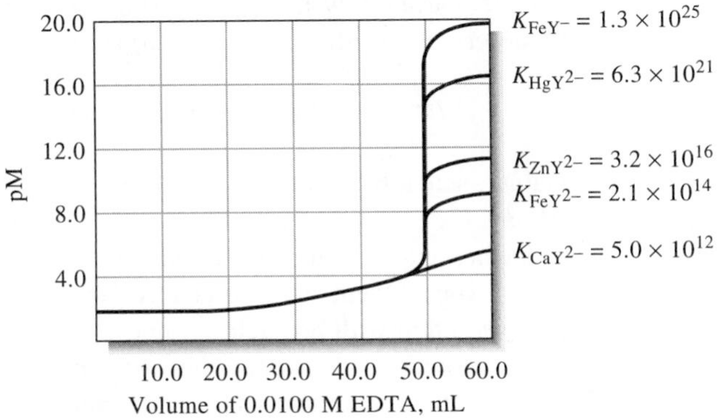

Figure 14-7
Titration curves for 50.0 mL of 0.0100 M cation solutions at pH 6.0.

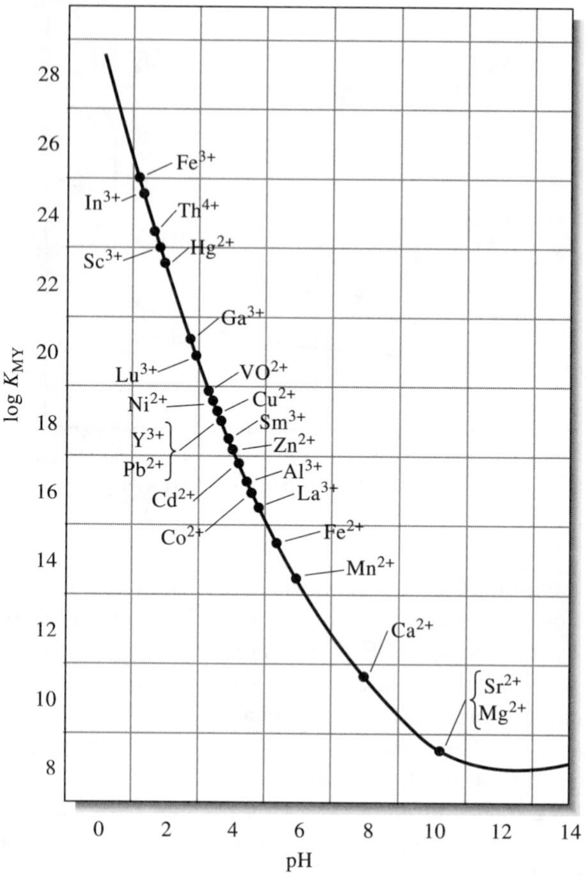

Figure 14-8
Minimum pH needed for satisfactory titration of various cations with EDTA. (Reprinted with permission, C. N. Reilley and R. W. Schmid, *Anal. Chem.*, **1958,** *30,* 947. Copyright 1958 American Chemical Society.)

14B-5 The Effect of Other Complexing Agents on EDTA Titration Curves

Many cations form hydrous oxide precipitates when the pH is raised to the level required for their successful titration with EDTA. When this problem is encountered, an auxiliary complexing agent is needed to keep the cation in solution. For example, zinc(II) is ordinarily titrated in a medium that has fairly high concentrations of ammonia and ammonium chloride. These species buffer the solution to a pH that ensures complete reaction between cation and titrant; in addition, ammonia forms ammine complexes with zinc(II) and prevents formation of the sparingly soluble zinc hydroxide, particularly in the early stages of the titration. A somewhat more realistic description of the reaction, then, is

$$Zn(NH_3)_4^{2+} + HY^{3-} \longrightarrow ZnY^{2-} + 3NH_3 + NH_4^+$$

Often, auxiliary complexing agents must be used in EDTA titrations to prevent precipitation of the analyte as a hydrous oxide. Such reagents decrease the sharpness of end points.

The solution also contains such other zinc/ammonia species as $Zn(NH_3)_3^{2+}$, $Zn(NH_3)_2^{2+}$, and $Zn(NH_3)^{2+}$. Calculation of pZn in a solution that contains ammonia must take these species into account. Qualitatively, complexation of a cation by an auxiliary complexing reagent causes preequivalence pM values to be larger than in a comparable solution with no such reagent.

Figure 14-9 shows two theoretical curves for the titration of zinc(II) with EDTA at pH 9.00. The equilibrium concentration of ammonia was 0.100 M for

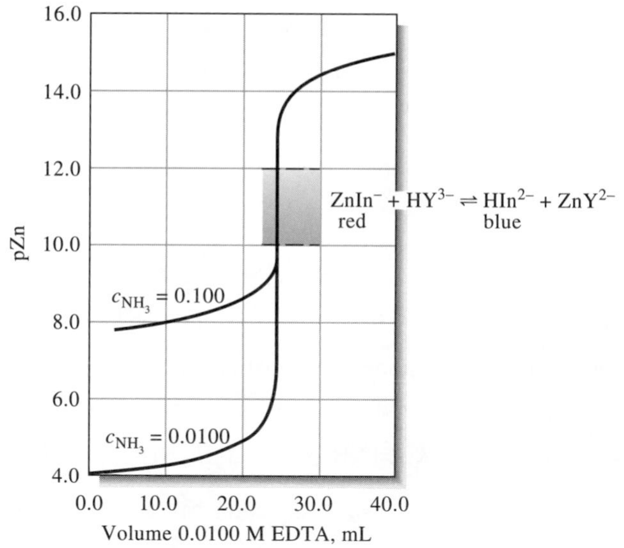

Figure 14-9
Influence of ammonia concentration on the end point for the titration of 50.0 mL of 0.00500 M Zn^{2+}. Solutions are buffered to pH 9.00. The shaded area shows the transition range for Eriochrome Black T.

one titration and 0.0100 M for the other. Note that the presence of ammonia decreases the change in pZn near the equivalence point. For this reason, the concentration of auxiliary complexing reagents should always be kept to the minimum required to prevent precipitation of the analyte. Note that the auxiliary complexing agent does not affect pZn beyond the equivalence point. On the other hand, keep in mind that α_4, and thus pH, plays an important role in defining this part of the titration curve (Figure 14-6).

FEATURE 14-2
EDTA Titration Curves When a Complexing Agent Is Present

A quantitative description of the effects of an auxilliary complexing reagent can be derived by a procedure similar to that used to determine the influence of pH on EDTA titration curves. Here, a quantity α_M is defined that is analogous to α_4:

$$\alpha_M = \frac{[M^{n+}]}{c_M} \qquad (14\text{-}7)$$

where c_M is the sum of the concentrations of species containing the metal ion *exclusive* of that combined with EDTA. For solutions containing zinc(II) and ammonia,

$$c_M = [Zn^{2+}] + [Zn(NH_3)^{2+}] + [Zn(NH_3)_2^{2+}] + [Zn(NH_3)_3^{2+}] + [Zn(NH_3)_4^{2+}] \qquad (14\text{-}8)$$

The value of α_M can be expressed readily in terms of the ammonia concentration and the formation constants for the various ammine complexes. To arrive at such an expression, we write

$$K_1 = \frac{[Zn(NH_3)^{2+}]}{[Zn^{2+}][NH_3]}$$

$$[Zn(NH_3)^{2+}] = K_1[Zn^{2+}][NH_3]$$

Similarly, it is readily shown that

$$[Zn(NH_3)_2^{2+}] = K_1K_2[Zn^{2+}][NH_3]^2$$
$$[Zn(NH_3)_3^{2+}] = K_1K_2K_3[Zn^{2+}][NH_3]^3$$
$$[Zn(NH_3)_4^{2+}] = K_1K_2K_3K_4[Zn^{2+}][NH_3]^4$$

Substitution of these expressions into Equation 14-8 gives

$$c_M = [Zn^{2+}](1 + K_1[NH_3] + K_1K_2[NH_3]^2 + K_1K_2K_3[NH_3]^3 + K_1K_2K_3K_4[NH_3]^4)$$

Substituting this expression for c_M in Equation 14-7 (here, $[M^{n+}] = [Zn^{2+}]$), leads to

$$\alpha_M = \frac{1}{1 + K_1[NH_3] + K_1K_2[NH_3]^2 + K_1K_2K_3[NH_3]^3 + K_1K_2K_3K_4[NH_3]^4} \quad (14\text{-}9)$$

Finally, a conditional constant for the equilibrium between EDTA and zinc(II) in an ammonia/ammonium chloride buffer is obtained by substituting Equation 14-7 into Equation 14-4 and rearranging

$$K''_{ZnY} = \frac{[ZnY^{2-}]}{c_M c_T} = \alpha_4 \alpha_M K_{ZnY} \quad (14\text{-}10)$$

where K''_{ZnY} is a new conditional constant that applies at a single pH as well as a single concentration of ammonia.

To show how Equations 14-7 to 14-10 can be used to obtain a titration curve, let us calculate the pZn for solutions prepared by adding 20.0, 25.0, and 30.0 mL of 0.0100 M EDTA to 50.0 mL of 0.00500 M Zn^{2+}. Assume that both the Zn^{2+} and EDTA solutions are 0.100 M in NH_3 and 0.175 M in NH_4Cl to provide a constant pH of 9.0.

In Appendix 4, we find that the logarithms of the stepwise formation constants for the four zinc complexes with ammonia are 2.21, 2.29, 2.36, and 2.03. Thus,

$$K_1 = \text{antilog } 2.21 = 1.62 \times 10^2$$
$$K_1K_2 = \text{antilog } (2.21 + 2.29) = 3.16 \times 10^4$$
$$K_1K_2K_3 = \text{antilog } (2.21 + 2.29 + 2.36) = 7.24 \times 10^6$$
$$K_1K_2K_3K_4 = \text{antilog } (2.21 + 2.29 + 2.36 + 2.03) = 7.76 \times 10^8$$

Calculation of a Conditional Constant

A value of α_M can be obtained from Equation 14-9 by assuming that the equilibrium molar and the analytical concentrations of ammonia are essentially the same; thus, for $[NH_3] = 0.100$,

$$\alpha_M = \frac{1}{1 + 16 + 316 + 7.24 \times 10^3 + 7.76 \times 10^4} = 1.17 \times 10^{-5}$$

A value of K_{ZnY} is found in Table 14-1, and α_4 for pH 9.0 is given in Table 14-2. Substituting into Equation 14-10, we find

$$K''_{ZnY} = 5.2 \times 10^{-2} \times 1.17 \times 10^{-5} \times 3.2 \times 10^{16}$$
$$= 1.9 \times 10^{10}$$

Calculation of pZn after Addition of 20.0 mL of EDTA

At this point, only part of the zinc has been complexed by EDTA. The remainder is present as Zn^{2+} and the four ammine complexes. By definition, the sum of the concentrations of these five species is c_M. Therefore,

$$c_M = \frac{50.0 \times 0.00500 - 20.0 \times 0.0100}{70.0} = 7.14 \times 10^{-4}\ M$$

Substitution of this value into Equation 14-7 gives

$$[Zn^{2+}] = c_M\alpha_M = (7.14 \times 10^{-4})(1.17 \times 10^{-5}) = 8.35 \times 10^{-9}\ M$$
$$pZn = 8.08$$

Calculation of pZn after Addition of 25.0 mL of EDTA

At the equivalence point, the analytical concentration of ZnY^{2-} is

$$c_{ZnY^{2-}} = \frac{50.0 \times 0.00500}{50.0 + 25.0} = 3.33 \times 10^{-3}\ M$$

The sum of the concentrations of the various zinc species not combined with EDTA equals the sum of the concentrations of the uncomplexed EDTA species:

$$c_M = c_T$$

and

$$[ZnY^{2-}] = 3.33 \times 10^{-3} - c_M \approx 3.33 \times 10^{-3}\ M$$

Substituting into Equation 14-10, we have

$$\frac{3.33 \times 10^{-3}}{c_M^2} = 1.9 \times 10^{10} \times K''_{ZnY}$$
$$c_M = 4.18 \times 10^{-7}\ M$$

Employing Equation 14-7, we obtain

$$[Zn^{2+}] = c_M\alpha_M = (4.18 \times 10^{-7})(1.17 \times 10^{-5}) = 4.90 \times 10^{-12}$$

$$pZn = 11.31$$

Calculation of pZn after Addition of 30.0 mL of EDTA

The solution now contains an excess of EDTA; thus,

$$c_{EDTA} = c_T = \frac{30.0 \times 0.0100 - 50.0 \times 0.00500}{80.0} = 6.25 \times 10^{-4}\ M$$

and since essentially all of the original Zn^{2+} is now complexed,

$$c_{ZnY^{2-}} = [ZnY^{2-}] = \frac{50.0 \times 0.00500}{80.0} = 3.12 \times 10^{-3}\ M$$

Rearranging Equation 14-10 gives

$$c_M = \frac{[ZnY^{2-}]}{c_T K''_{ZnY}} = \frac{3.12 \times 10^{-3}}{(6.25 \times 10^{-4})(1.9 \times 10^{10})} = 2.63 \times 10^{-10}\ M$$

and, from Equation 14-7,

$$[Zn^{2+}] = c_M\alpha_M = (2.63 \times 10^{-10})(1.17 \times 10^{-5}) = 3.07 \times 10^{-15}$$

$$pZn = 14.51$$

14B-6 Indicators for EDTA Titrations

Reilley and Barnard[2] have listed nearly 200 organic compounds that have been investigated as indicators for metal ions in EDTA titrations. In general, these indicators are organic dyes that form colored chelates with metal ions in a pM range that is characteristic of the particular cation and dye. The complexes are often intensely colored and are discernible to the eye at concentrations in the range of 10^{-6} to 10^{-7} M.

Eriochrome Black T® is a typical metal-ion indicator that is used in the titration of several common cations. As shown in Figure 14-10, this compound contains a sulfonic acid group, which is completely dissociated in water, and two phenolic groups that only partially dissociate. Its behavior as a weak acid is described by the equations

$$\underset{\text{red}}{H_2O + H_2In^-} \rightleftharpoons \underset{\text{blue}}{HIn^{2-}} + H_3O^+ \qquad K_1 = 5 \times 10^{-7}$$

$$\underset{\text{blue}}{H_2O + HIn^{2-}} \rightleftharpoons \underset{\text{orange}}{In^{3-}} + H_3O^+ \qquad K_2 = 2.8 \times 10^{-12}$$

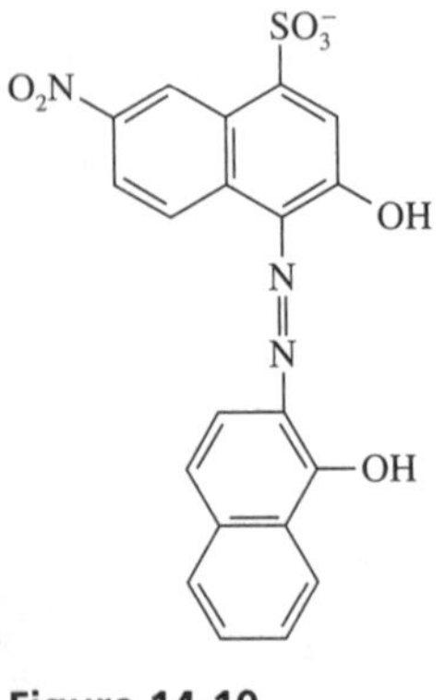

Figure 14-10
Eriochrome Black T.

[2]C. N. Reilley and A. J. Barnard Jr., in *Handbook of Analytical Chemistry,* L. Meites, Ed., p. **3**-77. New York: McGraw-Hill, 1963.

Note that the acids and their conjugate bases have different colors. Thus, Eriochrome Black T behaves as an acid/base indicator as well as a metal ion indicator.

The metal complexes of Eriochrome Black T are generally red as is H_2In^-. Thus, for metal ion detection, it is necessary to adjust the pH to 7 or above so that the blue form of the species, HIn^{2-}, predominates in the absence of a metal ion. Until the equivalence point in a titration, the indicator complexes the excess metal ion, so the solution is red. When EDTA becomes present in slight excess, the solution turns blue as a consequence of the reaction

$$\underset{\text{red}}{MIn^-} + HY^{3-} \rightleftharpoons \underset{\text{blue}}{HIn^{2-}} + MY^{2-}$$

Eriochrome Black T forms red complexes with more than two dozen metal ions, but the formation constants of only a few are appropriate for end-point detection. Example 14-5 demonstrates how the formation constant for a metal/indicator complex can be used to calculate the pM range over which a color change for the indicator can be expected.[3]

EXAMPLE 14-5

Determine the transition ranges for Eriochrome Black T in titrations of Mg^{2+} and Ca^{2+} at pH 10.0, given (a) that the second acid dissociation constant for the indicator is

$$HIn^{2-} + H_2O \rightleftharpoons In^{3-} + H_3O^+ \qquad K_2 = 2.8 \times 10^{-12} = \frac{[H_3O^+][In^{3-}]}{[HIn^{2-}]}$$

(b) that the formation constant for $MgIn^-$ is

$$Mg^{2+} + In^{3-} \rightleftharpoons MgIn^- \qquad K_f = 1.0 \times 10^7 = \frac{[MgIn^-]}{[Mg^{2+}][In^{3-}]}$$

and (c) that the analogous constant for Ca^{2+} is 2.5×10^5.

We shall assume, as we did earlier (Section 10A-2), that a detectable color change requires a tenfold excess of one or the other of the colored species; that is, a detectable color change is observed when $[MgIn^-]/[HIn^{2-}]$ changes from 10 to 0.10.

Multiplication of K_2 for the indicator by K_f for $MgIn^-$ gives an expression that contains the foregoing ratio:

$$\frac{[MgIn^-][H_3O^+]}{[HIn^{2-}][Mg^{2+}]} = 2.8 \times 10^{-12} \times 1.0 \times 10^7 = 2.8 \times 10^{-5}$$

which rearranges to

[3]For a complete discussion of the principles of indicator choice in complex-formation titrations, see C. N. Reilley and R. W. Schmid, *Anal. Chem.*, **1959**, *31*, 887.

$$[Mg^{2+}] = \frac{[MgIn^-]}{[HIn^{2-}]} \times \frac{[H_3O^+]}{2.8 \times 10^{-5}}$$

Substitution of 1.0×10^{-10} for $[H_3O^+]$ and 10 and 0.10 for the ratio yields the range of $[Mg^{2+}]$ over which the color change occurs:

$$[Mg^{2+}] = 3.6 \times 10^{-5} \quad \text{to} \quad 3.6 \times 10^{-7}$$

$$pMg = 5.4 \pm 1.0$$

Proceeding in the same way, we find the range for pCa to be 3.8 ± 1.0.

The ranges for magnesium and calcium are indicated on the titration curves in Figure 14-5. Eriochrome Black T is clearly an ideal indicator for magnesium but a totally unsatisfactory one for calcium. Note that the formation constant for $CaIn^-$ is only about one fortieth that for magnesium. As a consequence, significant conversion of $CaIn^-$ to HIn^{2-} occurs well before equivalence.

A similar calculation reveals that Eriochrome Black T is also well suited for the titration of zinc with EDTA (Figure 14-9).

A limitation of Eriochrome Black T is that its solutions decompose slowly with standing. It is claimed that solutions of Calmagite, an indicator that for all practical purposes is identical in behavior to Eriochrome Black T, do not suffer this disadvantage. The structure of Calmagite is similar to that of Eriochrome Black T. Many other metal indicators have been developed for EDTA titrations.[4]

14B-7 Titration Methods Employing EDTA

The paragraphs that follow describe several methods for the determination of cations with EDTA.

Direct Titration

Direct-titration procedures with a metal ion indicator are the easiest and the most convenient to use.

Methods Based on Indicators for the Analyte Ion Reilley and Barnard[5] list 40 elements that can be determined by direct titration with EDTA using metal ion indicators. The direct method is not applicable to all cations, however, either because no good indicators have been developed or because the reaction between the metal ion and EDTA is so slow as to make titration impractical. Such cations can usually be determined by one of the several methods described in the paragraphs that follow.

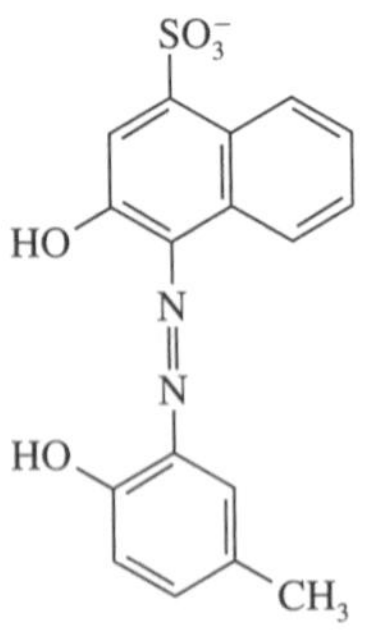

Figure 14-11
Calmagite.

Methods Based on Indicators for an Added Metal Ion It is often convenient to introduce a small amount of a cation that forms an EDTA complex that is less stable than the analyte complex and for which a good indicator exists. For example, indicators for calcium ion are generally less satisfactory than those we have

[4] See, for example, L. Meites, *Handbook of Analytical Chemistry*, pp. **3**-101 to **3**-165. New York: McGraw-Hill, 1963.

[5] C. N. Reilley and A. J. Barnard Jr., in *Handbook of Analytical Chemistry*, pp. **3**-166 to **3**-200. New York: McGraw-Hill, 1963.

described for magnesium ion. Consequently, a small amount of magnesium chloride is often added to an EDTA solution that is to be used for the determination of calcium with Eriochrome Black T. In the initial stages in the titration, magnesium ions are displaced from the EDTA complex by calcium ions and are free to combine with the Eriochrome Black T, thus imparting a red color to the solution. When all of the calcium ions have been complexed, however, the liberated magnesium ions again combine with the EDTA and the end point is observed. This procedure requires standardization of the EDTA solution against primary-standard calcium carbonate.

Potentiometric Methods Potential measurements can be used for end-point detection in the EDTA titration of those metal ions for which specific ion electrodes are available. Electrodes of this type are described in Section 18D. In addition, a mercury electrode can be made sensitive to EDTA ions and used i titrations with this reagent (Equation 18-5).

Back-Titration Methods

Back-titration is useful for the determination of cations that form stable EDTA complexes and for which a satisfactory indicator is not available. The method is also useful for cations that react only slowly with EDTA. A measured excess of standard EDTA solution is added to the analyte solution. After the reaction is judged complete, the excess EDTA is back-titrated with a standard magnesium or zinc ion solution to an Eriochrome Black T or Calmagite end point.[6] For this procedure to be successful, it is necessary that the magnesium or zinc ions form an EDTA complex that is less stable than the corresponding analyte complex.

Back-titration is also useful for analyzing samples that contain anions that would otherwise form sparingly soluble precipitates with the analyte under the analytical conditions. Here, the excess EDTA prevents precipitate formation.

Back-titration procedures are used when no suitable indicator is available, when the reaction between analyte and EDTA is slow, or when the analyte forms a precipitate at the pH required for its titration.

Displacement Methods

In displacement titrations, an unmeasured excess of a solution containing the magnesium or zinc complex of EDTA is introduced into the analyte solution. If the analyte forms a more stable complex than that of magnesium or zinc, the following displacement reaction occurs:

$$MgY^{2-} + M^{2+} \longrightarrow MY^{2-} + Mg^{2+}$$

The liberated cation is then titrated with the standard EDTA.

Displacement titrations are used when no indicator for an analyte is available.

14B-8 The Scope of EDTA Titrations

Complexometric titrations with EDTA have been applied to the determination of virtually every metal cation with the exception of the alkali metal ions. Because EDTA forms complexes with most cations, the reagent might appear at first

[6]For a discussion of the back-titration procedure, see C. Macca and M. Fiorana, *J. Chem. Educ.,* **1986**, *63*, 121.

glance to be totally lacking in selectivity. In fact, however, considerable control over interferences can be realized by pH regulation. For example, trivalent cations can usually be titrated without interference from divalent species by maintaining the solution at a pH of about 1 (Figure 14-8). At this pH, the less stable divalent chelates do not form to any significant extent, but the trivalent ions are quantitatively complexed.

Similarly, ions such as cadmium and zinc, which form more stable EDTA chelates than does magnesium, can be determined in the presence of the latter ion by buffering the mixture to pH 7 before titration. Eriochrome Black T serves as an indicator for the cadmium or zinc end points without interference from magnesium because the indicator chelate with magnesium is not formed at this pH.

A **masking agent** is a complexing agent that reacts selectively with a component in a solution and in so doing prevents that component from interfering in an analysis.

Finally, interference from a particular cation can sometimes be eliminated by adding a suitable *masking agent,* an auxiliary ligand that preferentially forms highly stable complexes with the potential interference.[7] For example, cyanide ion is often employed as a masking agent to permit the titration of magnesium and calcium ions in the presence of ions such as cadmium, cobalt, copper, nickel, zinc, and palladium. All of the latter form sufficiently stable cyanide complexes to prevent reaction with EDTA. Feature 14-3 illustrates how masking and demasking reagents are used to improve the selectivity of EDTA reactions.

FEATURE 14-3

Illustration Showing How Masking and Demasking Agents Can Be Used to Enhance the Selectivity of EDTA Titrations

Lead, magnesium, and zinc can be determined on a single sample by two titrations with standard EDTA and one titration with standard Mg^{2+}. The sample is first treated with an excess of NaCN, which masks Zn^{2+} and prevents it from reacting with EDTA.

$$Zn^{2+} + 4CN^- \rightleftharpoons Zn(CN)_4^{2-}$$

BAL is an abbreviation for British Anti Lewisite, a complexing agent that was developed in Great Britain to combat the effects of Lewisite. Lewisite is an arsenic containing gas that was used in World War I for chemical warfare.

The Pb^{2+} and Mg^{2+} are then titrated with standard EDTA. After the equivalence point has been reached, a solution of the complexing agent BAL (2-3-dimercapto-1-propanol, $CH_2SHCHSHCH_2OH$), which we will formulate $R(SH)_2$, is added to the solution. This bidentate ligand reacts selectively to form a complex with Pb^{2+} that is much more stable than PbY^{2-}:

$$PbY^{2-} + 2R(SH)_2 \longrightarrow Pb(RS)_2^{2-} + 2H^+ + Y^{4-}$$

The liberated Y^{4-} is then titrated with a standard solution of Mg^{2+}. Finally, the zinc is demasked by adding formaldehyde

$$Zn(CN)_4^{2-} + 4HCHO + 4H_2O \longrightarrow Zn^{2+} + 4HOCH_2CN + 4OH^-$$

[7]For further information, see D. D. Perrin, *Masking and Demasking of Chemical Reactions.* New York: Wiley-Interscience, 1970: C. N. Reilley and A. J. Barnard Jr., in *Handbook of Analytical Chemistry,* L. Meites, Ed., pp. **3**-208 to **3**-225. New York: McGraw-Hill, 1963.

The liberated Zn^{2+} is then titrated with the standard EDTA solution.

Suppose the initial titration of Mg^{2+} and Pb^{2+} required 42.22 mL of 0.02064 M EDTA. Titration of the Y^{4-} liberated by the BAL consumed 19.35 mL of 0.007657 M Mg^{2+}. Finally, after addition of formaldehyde, the liberated Zn^{2+} was titrated with 28.63 mL of the EDTA. Calculate the percentages of the three elements in a 0.4085 g sample.

The initial titration reveals the number of millimoles of Pb^{2+} and Mg^{2+} present. That is,

$$\text{no. mmol } (Pb^{2+} + Mg^{2+}) = 42.22 \times 0.02064 = 0.87142$$

The second titration gives the number of millimoles of Pb^{2+}. Thus,

$$\text{no. mmol } Pb^{2+} = 19.35 \times 0.007657 = 0.14816$$

$$\text{no. mmol } Mg^{2+} = 0.87142 - 0.14816 = 0.72326$$

Finally, from the third titration we obtain

$$\text{no. mmol } Zn^{2+} = 28.63 \times 0.02064 = 0.59092$$

To obtain the percentages, we write

$$\frac{0.14816 \cancel{\text{mmol Pb}} \times 0.2072 \cancel{\text{g}} \text{ Pb}/\cancel{\text{mmol}}}{0.4085 \cancel{\text{g}} \text{ sample}} \times 100\% = 7.515\% \text{ Pb}$$

$$\frac{0.72326 \cancel{\text{mmol Mg}} \times 0.024305 \cancel{\text{g}} \text{ Mg}/\cancel{\text{mmol}}}{0.4085 \cancel{\text{g}} \text{ sample}} \times 100\% = 4.303\% \text{ Mg}$$

$$\frac{0.59092 \cancel{\text{mmol Zn}} \times 0.06539 \cancel{\text{g}} \text{ Zn}/\cancel{\text{mmol}}}{0.4085 \cancel{\text{g}} \text{ sample}} \times 100\% = 9.459\% \text{ Zn}$$

14B-9 The Determination of Water Hardness

Historically, water "hardness" was defined in terms of the capacity of cations in the water to replace the sodium or potassium ions in soaps and form sparingly soluble products. Most multiply charged cations share this undesirable property. In natural waters, however, the concentrations of calcium and magnesium ions generally far exceed that of any other metal ion. Consequently, hardness is now expressed in terms of the concentration of calcium carbonate that is equivalent to the total concentration of all the multivalent cations in the sample.

Hard water contains calcium, magnesium, and heavy-metal ions that form precipitates with soap (but not detergents).

The determination of hardness is a useful analytical test that provides a measure of water quality for household and industrial uses. The test is important to industry because, upon being heated, hard water precipitates calcium carbonate, which then clogs boilers and pipes.

Water hardness is ordinarily determined by an EDTA titration after the sample has been buffered to pH 10. Magnesium, which forms the least stable EDTA complex of all of the common multivalent cations in typical water samples, is not titrated until enough reagent has been added to complex all of the other cations in the sample. Therefore, a magnesium ion indicator, such as Calmagite or Eriochrome Black T, can serve as indicator in water-hardness titrations. Often, a small concentration of the magnesium-EDTA chelate is incorporated in the buffer or in the titrant to ensure the presence of sufficient magnesium ions for satisfactory indicator action.

FEATURE 14-4
Test Kits for Water Hardness

Test kits for determining the hardness of household water are available at stores that sell water softeners and plumbing supplies. They usually consist of a vessel calibrated to contain a known volume of water, a measuring scoop to deliver an appropriate amount of a solid buffer mixture, an indicator solution, and a bottle of standard EDTA, which is equipped with a medicine dropper. The drops of standard reagent needed to cause a color change are counted. The concentration of the EDTA solution is ordinarily such that one drop corresponds to one grain (about 0.065 g) of calcium carbonate per gallon of water.

14C TITRATIONS WITH INORGANIC COMPLEXING AGENTS

Complexometric titrations with inorganic reagents are among the oldest volumetric methods.[8] For example, the titration of iodide ion with mercury(II) ions,

$$Hg^{2+} + 4I^- \rightleftharpoons HgI_4^{2-}$$

was first described in 1833. Table 14-3 lists the common inorganic complexing agents as well as some of their applications.

TABLE 14-3 Typical Inorganic Complex-Formation Titrations*

Titrant	Analyte	Remarks
$Hg(NO_3)_2$	Br^-, Cl^-, SCN^-, CN^-, thiourea	Products are neutral mercury(II) complexes; various indicators used
$AgNO_3$	CN^-	Product is $Ag(CN)_2^-$; indicator is I^-; titrate to first turbidity of AgI
$NiSO_4$	CN^-	Product is $Ni(CN)_4^{2-}$; indicator is AgI; titrate to first turbidity of AgI
KCN	Cu^{2+}, Hg^{2+}, Ni^{2+}	Products are $Cu(CN)_4^{2-}$, $Hg(CN)_2$, $Ni(CN)_4^{2-}$; various indicators used

*For further applications and selected references, see L. Meites, *Handbook of Analytical Chemistry,* p. **3**-226. New York: McGraw-Hill, 1963.

[8]For further information, see I. M. Kolthoff and V. A. Stenger, *Volumetric Analysis,* Vol. 2, pp. 282–331. New York: Interscience, 1947.

14D QUESTIONS AND PROBLEMS

14-1. Define
*(a) chelate.
(b) tetradentate chelating agent.
*(c) ligand.
(d) coordination number.
*(e) conditional formation constant.
(f) NTA.
*(g) water hardness.
(h) EDTA displacement titration.

***14-2.** Describe three general methods for performing EDTA titrations. What are the advantages of each?

14-3. Why are multidentate ligands preferable to unidentate ligands for complexometric titrations?

14-4. Write chemical equations and equilibrium-constant expressions for the stepwise formation of
*(a) $Ag(S_2O_3)_2^{3-}$. (b) $Ni(SCN)_3^-$.

***14-5.** Explain how stepwise and overall formation constants are related.

14-6. Propose a complexometric method for the determination of the individual components in a solution containing In^{3+}, Zn^{2+}, and Mg^{2+}.

14-7. Why is a small amount of MgY^{2-} often added to a water specimen that is to be titrated for hardness?

***14-8.** An EDTA solution was prepared by dissolving 3.853 g of purified and dried $Na_2H_2Y \cdot 2H_2O$ in sufficient water to give 1.000 L. Calculate the molar concentration, given that the solute contained 0.3% excess moisture (page 281).

14-9. A solution was prepared by dissolving about 3.0 g of $Na_2H_2Y \cdot 2H_2O$ in approximately 1 L of water and standardizing against 50.00-mL aliquots of 0.004517 M Mg^{2+}. An average titration of 32.22 mL was required. Calculate the molar concentration of the EDTA.

14-10. Calculate the volume of 0.0500 M EDTA needed to titrate
*(a) 26.37 mL of 0.0741 M $Mg(NO_3)_2$.
(b) the Ca in 0.2145 g of $CaCO_3$.
*(c) the Ca in a 0.4397-g mineral specimen that is 81.4% brushite, $CaHPO_4 \cdot 2H_2O$ (172.09 g/mol).
(d) the Mg in a 0.2080-g sample of the mineral hydromagnesite, $3MgCO_3 \cdot Mg(OH)_2 \cdot 3H_2O$ (365.3 g/mol).
*(e) the Ca and Mg in a 0.1557-g sample that is 92.5% dolomite, $CaCO_3 \cdot MgCO_3$ (184.4 g/mol).

14-11. A solution contains 1.694 mg of $CoSO_4$ (155.0 g/mol) per milliliter. Calculate
(a) the volume of 0.08640 M EDTA needed to titrate a 25.00-mL aliquot of this solution.
(b) the volume of 0.009450 M Zn^{2+} needed to titrate the excess reagent after addition of 50.00 mL of 0.008640 M EDTA to a 25.00-mL aliquot of this solution.
(c) the volume of 0.008640 M EDTA needed to titrate the Zn^{2+} displaced by Co^{2+} following addition of an unmeasured excess of ZnY^{2-} to a 25.00-mL aliquot of the $CoSO_4$ solution. The reaction is

$$Co^{2+} + ZnY^{2-} \longrightarrow CoY^{2-} + Zn^{2+}$$

***14-12.** The Zn in a 0.7556-g sample of foot powder was titrated with 21.27 mL of 0.01645 M EDTA. Calculate the percent Zn in this sample.

14-13. The Cr plating on a surface that measured 3.00 × 4.00 cm was dissolved in HCl. The pH was suitably adjusted, following which 15.00 mL of 0.01768 M EDTA were introduced. The excess reagent required a 4.30-mL back-titration with 0.008120 M Cu^{2+}. Calculate the average mass of Cr on each square centimeter of surface.

***14-14.** The Tl in a 9.76-g sample of rodenticide was oxidized to the trivalent state and treated with an unmeasured excess of Mg/EDTA solution. The reaction is

$$Tl^{3+} + MgY^{2-} \longrightarrow TlY^- + Mg^{2+}$$

Titration of the liberated Mg^{2+} required 13.34 mL of 0.03560 M EDTA. Calculate the percentage of Tl_2SO_4 (504.8 g/mol) in the sample.

14-15. An EDTA solution was prepared by dissolving approximately 4 g of the disodium salt in approximately 1 L of water. An average of 42.35 mL of this solution was required to titrate 50.00-mL aliquots of a standard that contained 0.7682 g of $MgCO_3$ per liter. Titration of a 25.00-mL sample of mineral water at pH 10 required 18.81 mL of the EDTA solution. A 50.00-mL aliquot of the mineral water was rendered strongly alkaline to precipitate the magnesium as $Mg(OH)_2$. Titration with a calcium-specific indicator required 31.54 mL of the EDTA solution. Calculate
(a) the molarity of the EDTA solution.
(b) the ppm of $CaCO_3$ in the mineral water.
(c) the ppm of $MgCO_3$ in the mineral water.

***14-16.** A 50.00-mL aliquot of a solution containing iron(II) and iron(III) required 13.73 mL of 0.01200 M EDTA when titrated at pH 2.0 and 29.62 mL when titrated at pH 6.0. Express the concentration of the solution in terms of the parts per million of each solute.

14-17. A 24-hour urine specimen was diluted to 2.000 L. After the solution was buffered to pH 10, a 10.00-mL aliquot was titrated with 26.81 mL of 0.003474 M EDTA. The calcium in a second 10.00-mL aliquot was isolated as $CaC_2O_4(s)$, redissolved in acid, and titrated with 11.63 mL of the EDTA solution. Assuming that 15 to 300 mg of magnesium and 50 to 400 mg of calcium per day are normal, did this specimen fall within these ranges?

***14-18.** A 1.509-g sample of a Pb/Cd alloy was dissolved in acid and diluted to exactly 250.0 mL in a volumetric flask. A 50.00-mL aliquot of the diluted solution was brought to a pH of 10.0 with an NH_4^+/NH_3 buffer. The subsequent titration involved both cations and required 28.89 mL of 0.06950 M EDTA. A second 50.00-mL aliquot was brought to a pH of 10.0 with an HCN/NaCN buffer, which also served to mask the Cd^{2+}; 11.56 mL of the EDTA solution were needed to titrate the Pb^{2+}. Calculate the percent Pb and Cd in the sample.

14-19. A 0.6004-g sample of Ni/Cu condenser tubing was dissolved in acid and diluted to 100.0 mL in a volumetric flask. Titration of both cations in a 25.00-mL aliquot of this solution required 45.81 mL of 0.05285 M EDTA. Mercaptoacetic acid and NH_3 were then introduced; production of the Cu complex with the former resulted in the release of an equivalent amount of EDTA, which required a 22.85-mL titration of 0.07238 M Mg^{2+}. Calculate the percent Cu and Ni in the alloy.

***14-20.** Calamine, which is used for relief of skin irritations, is a mixture of zinc and iron oxides. A 1.022-g sample of dried calamine was dissolved in acid and diluted to 250.0 mL. Potassium fluoride was added to a 10.00-mL aliquot of the diluted solution to mask the iron; after suitable adjustment of the pH, Zn^{2+} consumed 38.71 mL of 0.01294 M EDTA. A second 50.00-mL aliquot was suitably buffered and titrated with 2.40 mL of 0.002727 M ZnY^{2-} solution:

$$Fe^{3+} + ZnY^{2-} \longrightarrow FeY^{-} + Zn^{2+}$$

Calculate the percentages of ZnO and Fe_2O_3 in the sample.

14-21. A 3.650-g sample containing bromate and bromide was dissolved in sufficient water to give 250.0 mL. After acidification, silver nitrate was introduced to a 25.00-mL aliquot to precipitate AgBr, which was filtered, washed, and then redissolved in an ammoniacal solution of potassium tetracyanonickelate(II):

$$Ni(CN)_4^{2-} + 2AgBr(s) \longrightarrow 2Ag(CN)_2^{-} + Ni^{2+} + 2Br^{-}$$

The liberated nickel ion required 26.73 mL of 0.02089 M EDTA. The bromate in a 10.00-mL aliquot was reduced to bromide with arsenic(III) prior to the addition of silver nitrate. The same procedure was followed, and the released nickel ion was titrated with 21.94 mL of the EDTA solution. Calculate the percentages of NaBr and $NaBrO_3$ in the sample.

***14-22.** The potassium ion in a 250.0-mL sample of mineral water was precipitated with sodium tetraphenylboron:

$$K^{+} + B(C_6H_4)_4^{-} \longrightarrow KB(C_6H_5)_4(s)$$

The precipitate was filtered, washed, and redissolved in an organic solvent. An excess of the mercury(II)/EDTA chelate was added:

$$4HgY^{2-} + B(C_6H_4)_4^{-} + 4H_2O \longrightarrow H_3BO_3 + 4C_6H_5Hg^{+} + 4HY^{3-} + OH^{-}$$

The liberated EDTA was titrated with 29.64 mL of 0.05581 M Mg^{2+}. Express the potassium ion concentration in parts per million.

14-23. Chromel is an alloy composed of nickel, iron, and chromium. A 0.6472-g sample was dissolved and diluted to 250.0 mL. When a 50.00-mL aliquot of 0.05182 M EDTA was mixed with an equal volume of the diluted sample, all three ions were chelated, and a 5.11-mL back-titration with 0.06241 M copper(II) was required. The chromium in a second 50.0-mL aliquot was masked through the addition of hexamethylenetetramine; titration of the Fe and Ni required 36.28 mL of 0.05182 M EDTA. Iron and chromium were masked with pyrophosphate in a third 50.0-mL aliquot, and the nickel was titrated with 25.91 mL of the EDTA solution. Calculate the percentages of nickel, chromium, and iron in the alloy.

***14-24.** A 0.3284-g sample of brass (containing lead, zinc, copper, and tin) was dissolved in nitric acid. The sparingly soluble $SnO_2 \cdot 4H_2O$ was removed by filtration, and the combined filtrate and washings were then diluted to 500.0 mL. A 10.00-mL aliquot was suitably buffered; titration of the lead, zinc, and copper in this aliquot required 37.56 mL of 0.002500 M EDTA. The copper in a 25.00-mL aliquot was masked with thiosulfate; the lead and zinc were then titrated with 27.67 mL of the EDTA solution. Cyanide ion was used to mask the copper and zinc in a 100-mL aliquot; 10.80 mL of the EDTA solution were needed to titrate the lead ion. Determine the composition of the brass sample; evaluate the percentage of tin by difference.

***14-25.** Calculate conditional constants for the formation of the EDTA complex of Fe^{2+} at a pH of (a) 6.0, (b) 8.0, (c) 10.0.

14-26. Calculate conditional constants for the formation of the EDTA complex of Ba^{2+} at a pH of (a) 7.0, (b) 9.0, (c) 11.0.

***14-27.** Derive a titration curve for 50.00 mL of 0.01000 M Sr^{2+} with 0.02000 M EDTA in a solution buffered to pH 11.0. Calculate pSr values after the addition of 0.00, 10.00, 24.00, 24.90, 25.00, 25.10, 26.00, and 30.00 mL of titrant.

14-28. Derive a titration curve for 50.00 mL of 0.0150 M Fe^{2+} with 0.0300 M EDTA in a solution buffered to pH 7.0. Calculate pFe values after the addition of 0.00, 10.00, 24.00, 24.90, 25.00, 25.10, 26.00, and 30.00 mL of titrant.

CHAPTER

15

An Introduction to Electrochemistry

We now turn our attention to several analytical methods that are based upon oxidation/reduction reactions. These methods, which are described in Chapters 17 through 21, include oxidation/reduction titrimetry, potentiometry, coulometry, electrogravimetry, and voltammetry. Fundamentals of electrochemistry that are necessary for the treatment of the theory of these topics are presented in this chapter and the next.

15A OXIDATION/REDUCTION REACTIONS

An *oxidation/reduction reaction* is one in which electrons are transferred from one reactant to another. An example is the oxidation of iron(II) ions by cerium(IV) ions. The reaction is described by the equation

Oxidation/reduction reactions are sometimes called **redox** reactions.

$$Ce^{4+} + Fe^{2+} \rightleftharpoons Ce^{3+} + Fe^{3+} \qquad (15\text{-}1)$$

Here, Ce^{4+} ion extracts an electron from Fe^{2+} to form Ce^{3+} and Fe^{3+} ions. A substance, such as Ce^{4+}, that has a strong affinity for electrons, and thus tends to remove them from other species, is called an *oxidizing agent,* or an *oxidant.* A *reducing agent,* or *reductant,* is a species, such as Fe^{2+}, that readily donates electrons to another species. With regard to Equation 15-1, we say that Fe^{2+} is *oxidized* by Ce^{4+}; similarly, Ce^{4+} is *reduced* by Fe^{2+}.

A **reducing agent** is an electron donor. An **oxidizing agent** is an electron acceptor.

We can split any oxidation/reduction equation into two *half-reactions* that show clearly which species gains electrons and which loses them. For example, Equation 15-1 is the sum of the two half-reactions

$$Ce^{4+} + e^- \rightleftharpoons Ce^{3+} \qquad \text{(reduction of } Ce^{4+}\text{)}$$
$$Fe^{2+} \rightleftharpoons Fe^{3+} + e^- \qquad \text{(oxidation of } Fe^{2+}\text{)}$$

It is important to understand that while we can readily write an equation for a half-reaction in which electrons are consumed or generated, we cannot observe an isolated half-reaction experimentally because there must always be a second half-reaction that serves as a source of electrons or a recipient of electrons—that is, an individual half-reaction is a theoretical concept.

The rules for balancing half-reactions are the same as those for other reaction types; that is, the number of atoms of each element as well as the net charge on each side of the equation must be the same. Thus, for the oxidation of Fe^{2+} by MnO_4^- ion, the half-reactions are

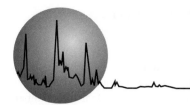

$$MnO_4^- + 5e^- + 8H^+ \rightleftharpoons Mn^{2+} + 4H_2O$$

$$5Fe^{2+} \rightleftharpoons 5Fe^{3+} + 5e^-$$

In the first half-reaction, the net charge on the left side is $(-1 - 5 + 8) = +2$, which is the same as that on the right. Note also that we had to multiply the second half-reaction by 5 so that the number of electrons lost by Fe^{2+} equals the number gained by MnO_4^-. A balanced net-ionic equation for the overall reaction is then obtained by adding the two half-reactions

$$MnO_4^- + 5Fe^{2+} + 8H^+ \rightleftharpoons Mn^{2+} + 5Fe^{3+} + 4H_2O$$

FEATURE 15-1

Balancing Redox Equations

A knowledge of how to balance oxidation/reduction reactions is essential to understanding all the concepts covered in this chapter. Although you probably remember this technique from your general chemistry course, we present a quick review here to remind you how the process works. For practice, let us complete and balance the following equation, adding H^+, OH^-, or H_2O as needed.

$$MnO_4^- + NO_2^- \rightleftharpoons Mn^{2+} + NO_3^-$$

First we write and balance the two half-reactions involved. For MnO_4^- we first write

$$MnO_4^- \rightleftharpoons Mn^{2+}$$

To account for the 4 oxygen atoms on the left-hand side of the equation we add 4 H_2O on the right-hand side of the equation, which means we must provide $8H^+$ on the left:

$$MnO_4^- + 8H^+ \rightleftharpoons Mn^{2+} + 4H_2O$$

To balance the charge we need to add 5 electrons to the left side of the equation. Thus,

$$MnO_4^- + 8H^+ + 5e^- \rightleftharpoons Mn^{2+} + 4H_2O$$

For the other half-reaction

$$NO_2^- \rightleftharpoons NO_3^-$$

we add one H_2O to the left side of the equation to supply the needed oxygen and $2H^+$ on the right.

$$NO_2^- + H_2O \rightleftharpoons NO_3^- + 2H^+$$

Then we add two electrons to the right-hand side to balance the charge.

$$NO_2^- + H_2O \rightleftharpoons NO_3^- + 2H^+ + 2e^-$$

Before combining the two equations, we must multiply the first by 2 and the second by 5 so that the electrons will cancel. We then add the two equations to obtain

$$2MnO_4^- + 16H^+ + \cancel{10e^-} + 5NO_2^- + 5H_2O \rightleftharpoons 2Mn^{2+} + 8H_2O + 5NO_3^- + 10H^+ + \cancel{10e^-}$$

which then rearranges to the balanced equation

$$2MnO_4^- + 6H^+ + 5NO_2^- \rightleftharpoons 2Mn^{2+} + 5NO_3^- + 3H_2O$$

15A-1 Comparison of Oxidation/Reduction Reactions with Acid/Base Reactions

Oxidation/reduction reactions can be viewed in a way that is analogous to the Brønsted-Lowry concept of acid/base reactions (Section 7A-2). Both involve the transfer of one or more charged particles from a donor to an acceptor—the particles being electrons in oxidation/reduction and protons in neutralization. When an acid donates a proton, it becomes a conjugate base that is capable of accepting a proton. By analogy, when a reducing agent donates an electron, it becomes an oxidizing agent that can accept an electron. This product could be called a conjugate oxidant (but seldom, if ever, is). We may then write a generalized equation for a redox reaction as

$$A_{red} + B_{ox} \rightleftharpoons A_{ox} + B_{red} \qquad (15\text{-}2)$$

Here, B_{ox}, the oxidized form of species B, accepts electrons from A_{red} to form the new reductant, B_{red}. At the same time, reductant A_{red}, having given up electrons, becomes an oxidizing agent, A_{ox}. If we know from chemical evidence that the equilibrium in Equation 15-2 lies to the right, we can state that B_{ox} is a better electron acceptor (stronger oxidant) than A_{ox} and, furthermore, that A_{red} is a more effective electron donor (better reductant) than B_{red}.

Recall that in the Brønsted/Lowry concept an acid/base reaction is described by the equation

$$\text{acid}_1 + \text{base}_2 \rightleftharpoons \text{base}_1 + \text{acid}_2$$

EXAMPLE 15-1

The following reactions proceed to the right, as written.

$$2H^+ + Cd(s) \rightleftharpoons H_2(g) + Cd^{2+}$$
$$2Ag^+ + H_2(g) \rightleftharpoons 2Ag(s) + 2H^+$$
$$Cd^{2+} + Zn(s) \rightleftharpoons Cd(s) + Zn^{2+}$$

What can we deduce regarding the strengths of H^+, Ag^+, Cd^{2+}, and Zn^{2+} as electron acceptors (or oxidizing agents)?

The second reaction establishes that Ag^+ is a more effective electron acceptor than H^+; the first reaction demonstrates that H^+ is more effective than Cd^{2+}. Finally, the third equation shows that Cd^{2+} is more effective than Zn^{2+}. Thus, the order of oxidizing strength is $Ag^+ > H^+ > Cd^{2+} > Zn^{2+}$.

Figure 15-1
Photograph of a "silver tree."

To demonstrate this reaction, immerse a piece of copper in a solution of silver nitrate. The result is the deposition of silver on the copper in the form of a "silver tree." See Figure 15-1 and color plate 13.

15A-2 Oxidation/Reduction Reactions in Electrochemical Cells

Many oxidation/reduction reactions can be carried out in either of two ways that are physically quite different. In one, the reaction is performed by direct contact between the oxidant and the reductant in a suitable container. In the second, the reaction is carried out in an electrochemical cell in which the reactants do not come in direct contact with one another. An example of the first process involves immersing a strip of copper in a solution containing silver nitrate. Here, silver ions migrate to the metal and are reduced:

$$Ag^+ + e^- \longrightarrow Ag(s)$$

At the same time, an equivalent quantity of copper is oxidized:

$$Cu(s) \rightleftharpoons Cu^{2+} + 2e^-$$

Multiplication of the silver half-reaction by two and addition yields a net-ionic equation for the overall process.

$$2Ag^+ + Cu(s) \rightleftharpoons 2Ag(s) + Cu^{2+} \qquad (15\text{-}3)$$

A unique aspect of oxidation/reduction reactions is that the transfer of electrons—and thus an identical net reaction—can often be brought about in an *electrochemical cell* in which the oxidizing agent and the reducing agent are physically separated from one another. Figure 15-2a shows such an arrangement. Note that a *salt bridge* isolates the reactants but maintains electrical contact between the two halves of the cell. The bridge is necessary to prevent Ag^+ from reacting directly with the copper metal. An external metallic conductor connects the two metals. In this cell, metallic copper is oxidized, silver ions are reduced, and electrons flow through the external circuit to the silver electrode. The voltmeter measures the potential difference between the two metals at any instant and is a measure of the tendency of the cell reaction to proceed toward equilibrium. With time, this tendency, and thus the potential, decreases continuously and approaches zero as the state of equilibrium for the overall reaction is approached.

When the $CuSO_4$ and $AgNO_3$ solutions are 0.0200 M, the cell develops a potential of 0.412 V as shown in Figure 15-2a.

When zero voltage is reached, the concentrations of Cu(II) and Ag(I) ions will have values that satisfy the equilibrium-constant expression for the reaction shown in Equation 15-3. At this point, no further net flow of electrons will occur. *It is important to realize that the overall reaction and its position of equilibrium are totally independent of the way the reaction is carried out,* whether it is by direct reaction in a solution or by indirect action in an electrochemical cell.

The equilibrium-constant expression for the reaction shown in Equation 15-3 is

$$K_{eq} = \frac{[Cu^{2+}]}{[Ag^+]^2} = 4.1 \times 10^{15} \qquad (15\text{-}4)$$

This expression applies regardless of whether the reaction occurs directly between reactants or indirectly in an electrochemical cell.

15B ELECTROCHEMICAL CELLS

Oxidation/reduction equilibria are conveniently studied by measuring the potentials of electrochemical cells in which the two half-reactions making up the equilibrium are participants. For this reason, we need to consider some of the characteristics of cells.

At equilibrium, the two half-reactions in a cell continue, but their rates become equal.

Salt bridges are widely used in electrochemistry to prevent mixing of the contents of the two electrolyte solutions making up electrochemical cells. Ordinarily, the two ends of the bridge are equipped with sintered glass disks or other devices to prevent liquid from siphoning from one part of the cell to the other.

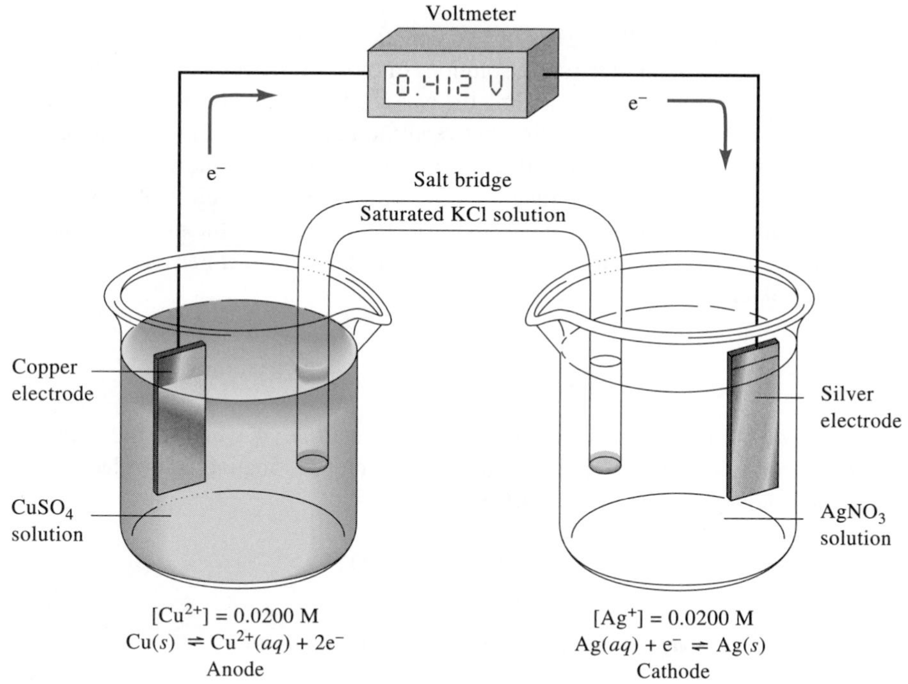

(a)

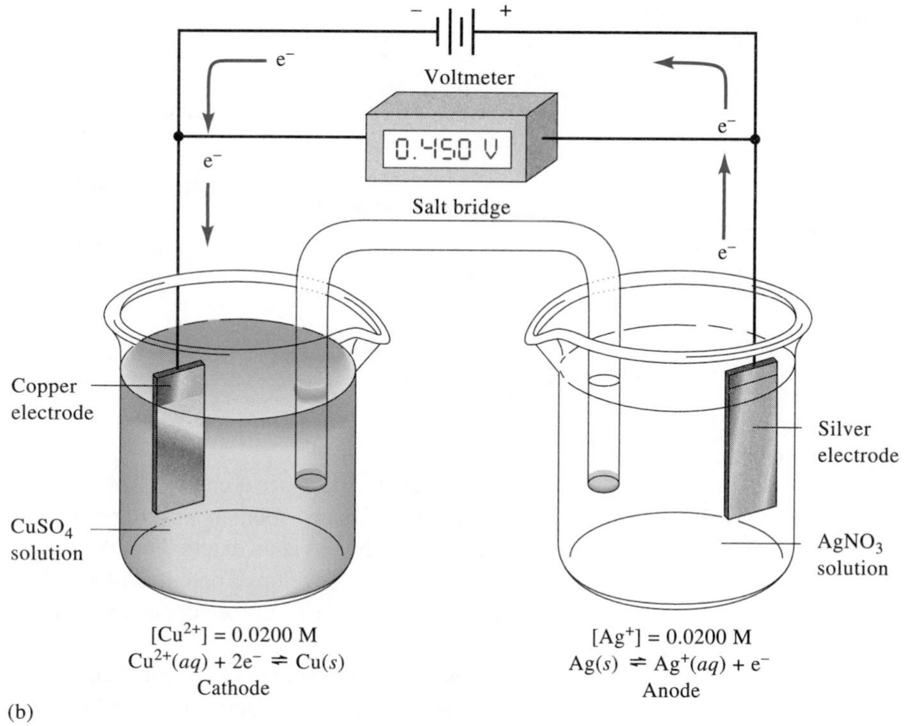

(b)

Figure 15-2
(a) A galvanic cell. (b) An electrolytic cell.

An electrochemical cell consists of two conductors called *electrodes,* each of which is immersed in an electrolyte solution. In most of the cells that will be of interest to us, the solutions surrounding the two electrodes are different and must be separated to avoid direct reaction between the reactants. The most common way of avoiding mixing is to insert a *salt bridge,* such as that shown in Figure 15-2, between the solutions. Conduction of electricity from one electrolyte solution to the other then occurs by migration of potassium ions in the bridge in one direction and chloride ions in the other. However, direct contact between copper metal and silver ions is prevented.

The electrodes in some cells share a common electrolyte; these are known as *cells without liquid junction.* For an example of such a cell, see Figure 16-2 and Example 16-4.

15B-1 Cathodes and Anodes

A cathode is the electrode where reduction occurs. An **anode** is the electrode where oxidation occurs.

The *cathode* in an electrochemical cell is the electrode at which a reduction reaction occurs. The *anode* is the electrode at which an oxidation takes place.

Examples of typical cathodic reactions include

$$Ag^+ + e^- \rightleftharpoons Ag(s)$$

$$Fe^{3+} + e^- \rightleftharpoons Fe^{2+}$$

$$NO_3^- + 10H^+ + 8e^- \rightleftharpoons NH_4^+ + 3H_2O$$

The reaction $2H^+ + 2e^- \rightleftharpoons H_2(g)$ occurs at a cathode when an aqueous solution contains no easily reduced species.

We can make these reactions occur by applying a suitable potential to an inert electrode such as platinum. Note that the third reaction reveals that *anions* can migrate to a cathode and be reduced.

Typical anodic reactions include

$$Cu(s) \rightleftharpoons Cu^{2+} + 2e^-$$

$$2Cl^- \rightleftharpoons Cl_2(g) + 2e^-$$

$$Fe^{2+} \rightleftharpoons Fe^{3+} + e^-$$

The Fe^{2+} half-reaction may seem somewhat unusual because a cation rather than an anion migrates to the electrode and gives up an electron. As we shall see, oxidation of a cation at an anode or reduction of an anion at a cathode is not uncommon.

The first reaction requires a copper anode, but the other two can be carried out at the surface of an inert platinum electrode.

The reaction $2H_2O \rightleftharpoons O_2(g) + 4H^+ + 4e^-$ occurs at an anode when an aqueous solution contains no other easily oxidized species.

15B-2 Types of Electrochemical Cells

Galvanic cells store electrical energy; **electrolytic cells** consume electricity.

For both galvanic and electrolytic cells, remember that (1) reduction always takes place at the cathode and (2) oxidation always takes place at the anode.

Electrochemical cells are either galvanic or electrolytic. They can also be classified as reversible or irreversible.

Galvanic, or *voltaic,* cells are batteries that store electrical energy. The reactions at the two electrodes in such cells tend to proceed spontaneously and produce a flow of electrons from the anode to the cathode via an external conductor. The cell shown in Figure 15-2a is a galvanic cell that develops a potential of about 0.412 V as electrons move through the external circuit from the copper anode to the silver cathode.

An *electrolytic* cell, in contrast, requires an external source of electrical energy for operation. The cell just considered can be operated electrolytically by connecting the positive terminal of a battery having a potential of somewhat greater than 0.412 V to the silver electrode and the negative terminal to the copper electrode. Here, the direction of the current is reversed, and the reactions at the electrodes are reversed as well (see Figure 15-2b). Oxidation then occurs at the silver anode and reduction takes place at the copper cathode. That is,

$$2Ag(s) + Cu^{2+} \rightleftharpoons 2Ag^{+} + Cu(s) \quad (15\text{-}5)$$

The cell in Figure 15-2 is an example of a *reversible cell* in which the direction of the electrochemical reaction is reversed when the direction of electron flow is changed. In an *irreversible* cell, changing the direction of current causes an entirely different half-reaction to occur at one or both electrodes.

Alessandro Volta (1745–1827), Italian physicist, was the inventor of the first battery, the so-called voltaic pile. It consisted of alternating disks of copper and zinc separated by disks of cardboard soaked with salt solution. In honor of his many contributions to electrical science, the unit of potential difference, the volt, is named for Volta.

In a **reversible cell,** reversing the current reverses the cell reaction. In an **irreversible cell,** reversing the current causes a different half-reaction to occur at one or both of the electrodes.

FEATURE 15-2
The Daniell Gravity Cell

The Daniell gravity cell was one of the earliest galvanic cells to find widespread practical application. It was used in the mid-1800s as a battery to power telegraphic communication systems. As shown in Figure 15-3, the cathode was a piece of copper immersed in a saturated solution of copper sulfate. A much less dense solution of dilute zinc sulfate was layered on top of the copper sulfate, and a massive zinc electrode was located in this solution. The electrode reactions were

$$Zn(s) \rightleftharpoons Zn^{2+} + e^{-} \quad \text{(anode)}$$

$$Cu^{2+} + e^{-} \rightleftharpoons Cu(s) \quad \text{(cathode)}$$

This cell develops an initial voltage of 1.18 V, which gradually decreases with use.

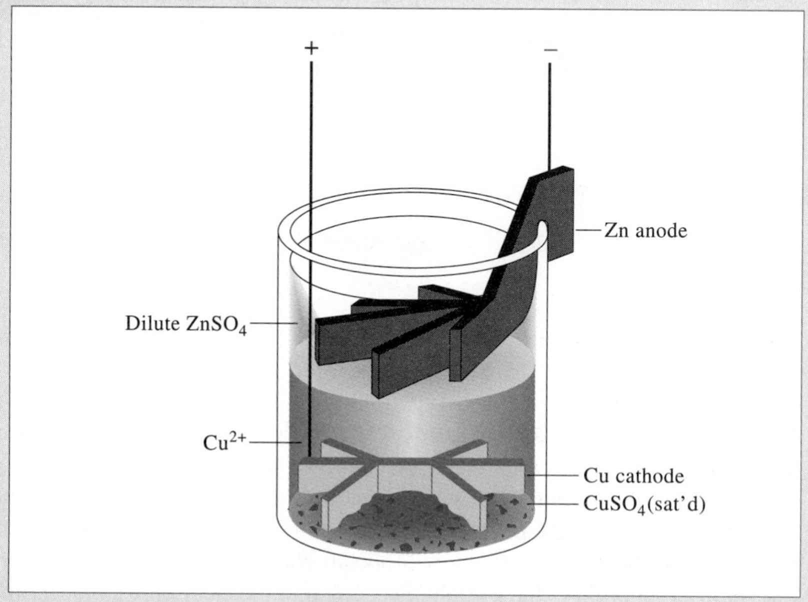

Figure 15-3
A Daniell gravity cell.

15B-3 Schematic Representation of Cells

In a schematic cell, the anode constituents always appear on the left and the cathode on the right.

Chemists frequently use a shorthand notation to describe electrochemical cells. The cell in Figure 15-2a, for example, is described by

$$Cu|Cu^{2+}(0.0200\text{ M})\|Ag^{+}(0.0200\text{ M})|Ag \quad (15\text{-}6)$$

By convention, the anode is *always* displayed on the left in these representations. A single vertical line indicates a phase boundary, or interface, at which a potential develops. For example, the first vertical line in this schematic indicates that a potential develops at the phase boundary between the copper anode and the copper sulfate solution. The double vertical line represents two phase boundaries, one at each end of the salt bridge. A *liquid-junction potential* develops at each of these interfaces. This potential results from differences in rates with which the ions in the cell compartments and the salt bridge migrate across the interfaces. A liquid-junction potential can amount to as much as several hundredths of a volt, but the junction potentials at the two ends of a salt bridge tend to cancel each other. The net effect on the overall potential of a cell is thus a few millivolts or less. For our purposes, we will neglect the contribution of liquid-junction potentials to the total potential of the cell.

A **liquid-junction potential** is a potential that develops across the interface between two solutions that differ in their electrolyte composition.

An alternative way of writing the cell shown in Figure 15-2a is

$$Cu|CuSO_4(0.0200\text{ M})\|AgNO_3(0.0200\text{ M})|Ag$$

Here, the compounds used to prepare the cell are indicated rather than the active participants in the cell half-reactions.

The cell in Figure 15-2b can be formulated as

$$Ag|Ag^{+}(0.0200\text{ M})\|Cu^{2+}(0.0200\text{ M})|Cu$$

Placing the silver half-cell on the left indicates that the silver couple is acting as the anode.

15B-4 Currents in Electrochemical Cells

As shown in Figure 15-4, electricity is transported through an electrochemical cell by three mechanisms:

1. Electrons carry electricity within the electrodes as well as the external conductor.
2. Anions and cations carry electricity within the cell. As shown in Figure 15-4, copper ions, silver ions, and other positively charged species move away from the copper anode and toward the silver cathode, whereas anions, such as sulfate and hydrogen sulfate ions, are attracted toward the copper anode. Within the salt bridge, chloride ions migrate toward and into the copper compartment, whereas potassium ions move in the opposite direction.
3. The ionic conduction of the solution is coupled to the electronic conduction in the electrodes by the reduction reaction at the cathode and the oxidation reaction at the anode.

In a cell, electricity is carried by movement of anions toward the anode and cations toward the cathode.

The phase boundary between an electrode and its solution is called an **interface.**

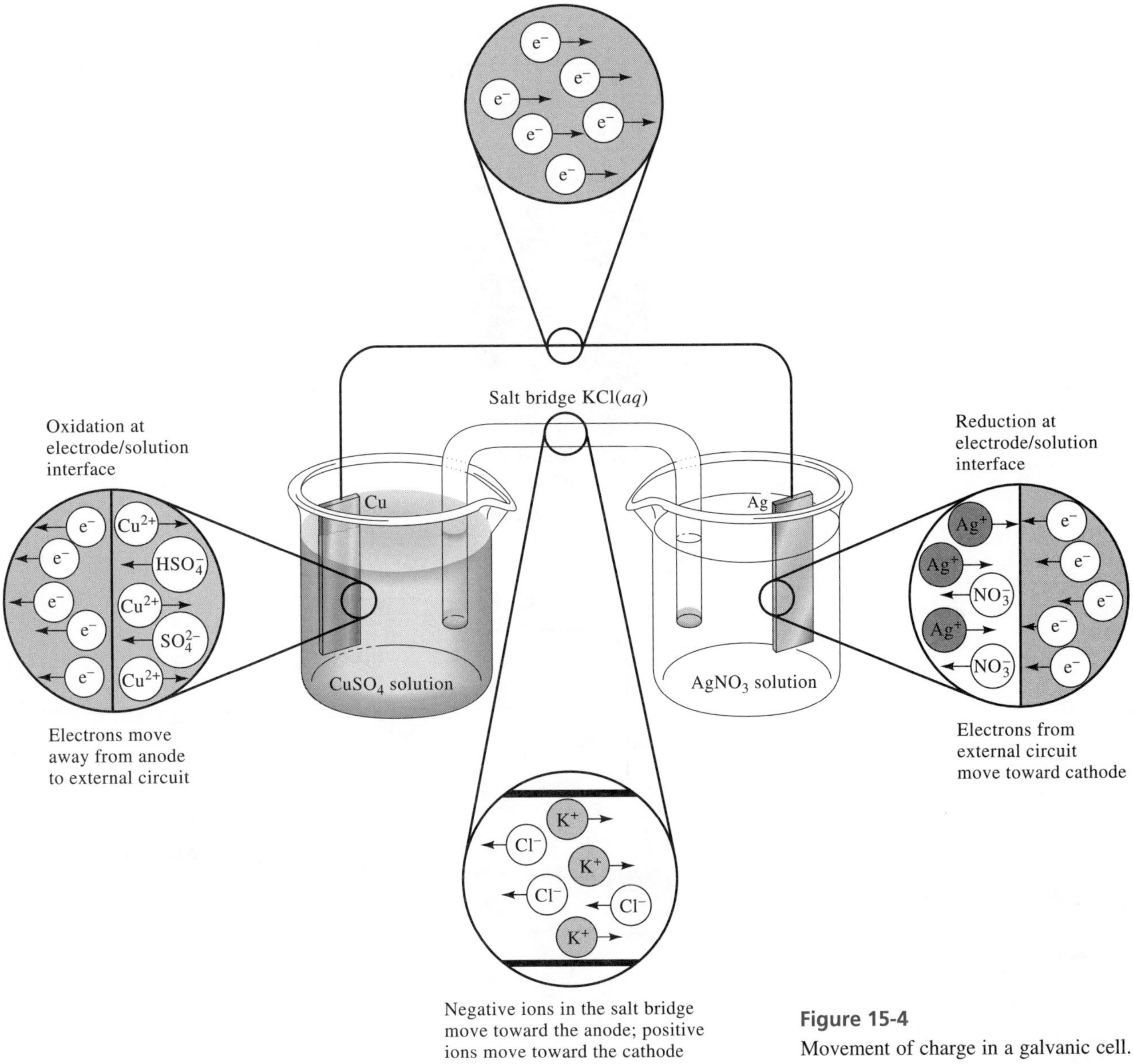

Figure 15-4
Movement of charge in a galvanic cell.

15C ELECTRODE POTENTIALS

The potential difference that develops between the cathode and the anode of the cell in Figure 15-5a is a measure of the tendency for the reaction

$$2Ag^+ + Cu(s) \rightleftharpoons 2Ag(s) + Cu^{2+}$$

to proceed from a nonequilibrium state to the condition of equilibrium. Thus, as shown in the upper figure, when the copper and silver ion concentrations in the cell are both 0.0200 M, a potential of 0.412 V develops, which shows that this reaction is far from equilibrium. As the reaction proceeds, this potential becomes smaller and smaller until at equilibrium the meter reads 0.000 V. As shown in

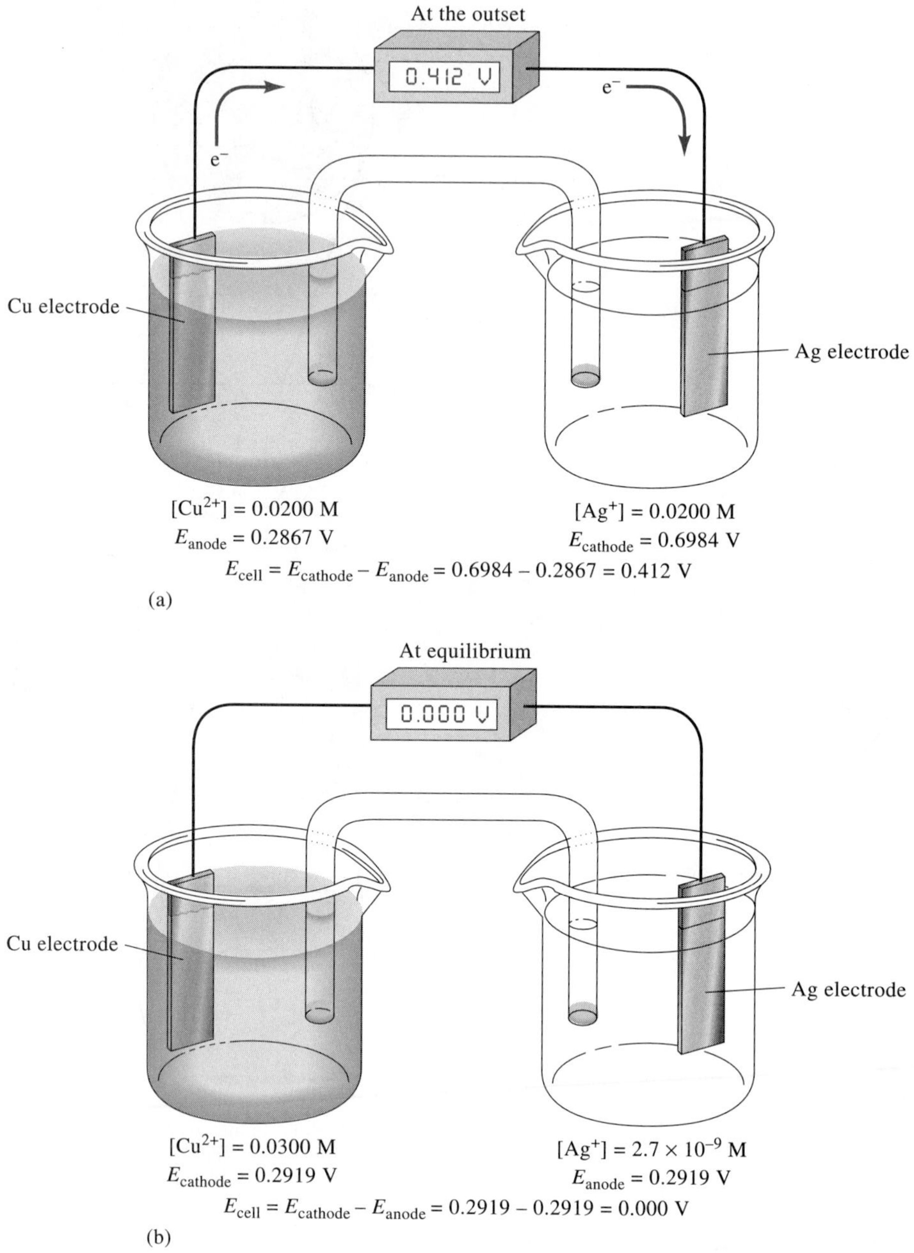

Figure 15-5
Change in cell potential after passage of current until equilibrium is reached.

Figure 15-5b, the copper ion equilibrium concentration is then just slightly less than 0.0300 M and the silver ion concentration is 2.7×10^{-9} M.

The potential of a cell such as that shown in Figure 15-5a is the difference between two *half-cell* or *single-electrode potentials,* one associated with the half-reaction at the silver cathode ($E_{cathode}$), the other with the half-reaction at the copper anode (E_{anode}).

Although we cannot determine absolute potentials of electrodes such as these (see Feature 15-3), we can readily determine *relative* electrode potentials. For example, if we replace the silver cathode in the cell in Figure 15-2 with a palladium electrode immersed in a palladium sulfate solution, the voltmeter reads about 0.19 V greater than the original cell. Since the anode compartment remains unaltered, we conclude that the half-cell potential for the reduction of palladium is about 0.19 V greater than that for silver (that is, palladium is a stronger oxidant than is silver). The tendency for other ions to take on electrons can also be compared by substituting other cathode half-cells while keeping the anode half-cell unchanged.

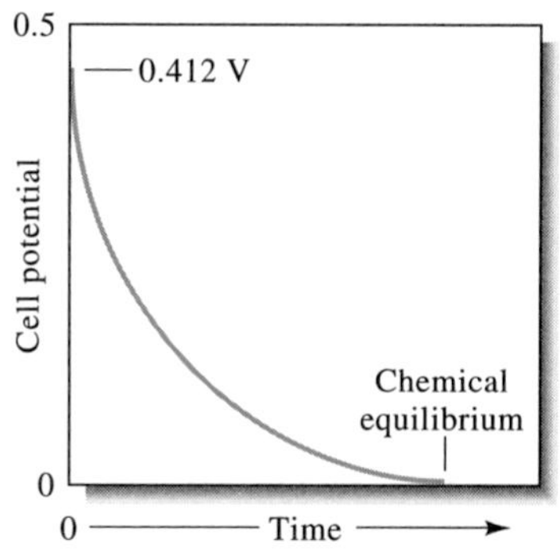

Figure 15-6
Potential of cell shown in Figure 15-5a as a function of time.

FEATURE 15-3
Why Absolute Electrode Potentials Cannot Be Measured

Although it is not difficult to measure *relative* half-cell potentials, it is impossible to determine absolute half-cell potentials because all voltage-measuring devices measure only *differences* in potential. To measure the potential of an electrode, one contact of a voltmeter is connected to the electrode in question. The other contact from the meter must then be brought into electrical contact with the solution in the electrode compartment via another conductor. This second contact, however, inevitably involves a solid/solution interface that acts as a second half-cell at which chemical change *must occur* if charge is to flow and the potential measured. A potential is associated with this second reaction. Thus, an absolute half-cell potential is not obtained but rather the difference between the half-cell potential of interest and a half-cell consisting of the second contact and the solution.

Our inability to measure absolute half-cell potentials presents no real obstacle because relative half-cell potentials are just as useful, provided they are all measured against the same half-cell. Relative potentials can be combined to give cell potentials. We can also use them to calculate equilibrium constants and generate titration curves.

15C-1 The Standard Hydrogen Reference Electrode

For relative electrode potential data to be widely applicable and useful, we must have a generally agreed-upon reference half-cell against which all others are compared. Such an electrode must be easy to construct, be reversible, and be highly reproducible in its behavior. The *standard hydrogen electrode* (SHE) meets these specifications and has been accepted throughout the world for many years as a universal reference electrode. It is a typical *gas electrode.*

The **standard hydrogen electrode** is sometimes called the **normal hydrogen electrode.**

SHE is the abbreviation for standard hydrogen electrode.

Figure 15-7 shows how a hydrogen electrode is constructed. The metal conductor is a piece of platinum that has been coated, or *platinized,* with finely divided platinum *(platinum black)* to increase its specific surface area. This electrode is immersed in an aqueous acid solution of known, constant hydrogen ion activity. The solution is kept saturated with hydrogen by bubbling the gas at constant pressure over the surface of the electrode. The platinum does not take part in the electrochemical reaction and serves only as the site where electrons are transferred. The half-reaction responsible for the potential that develops at this electrode is

Platinum black is a layer of finely divided platinum that is formed on the surface of a smooth platinum electrode by electrolytic deposition of the metal from a solution of chloroplatinic acid H_2PtCl_6. The platinum black provides a large specific surface area of platinum at which reaction can occur.

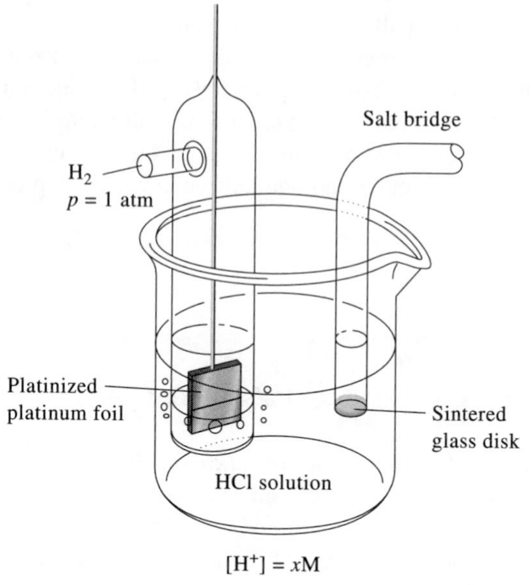

Figure 15-7
A hydrogen gas electrode.

$$2H^+(aq) + 2e^- \rightleftharpoons H_2(g) \qquad (15\text{-}7)$$

Platinum black catalyzes the reaction shown in Equation 15-7. Remember that catalysts do not change the position of equilibrium but simply shorten the time it takes to reach equilibrium.

The reaction shown as Equation 15-7 involves two equilibria:

$$2H^+ + 2e^- \rightleftharpoons H_2(aq)$$

$$H_2(aq) \rightleftharpoons H_2(g)$$

The continuous stream of gas at constant pressure provides the solution with a constant molecular hydrogen concentration.

At $p_{H_2} = 1.00$ and $a_{H^+} = 1.00$, the potential of the hydrogen electrode is assigned a value of exactly 0.000 V at all temperatures.

The standard hydrogen electrode is reversible and can act either as an anode or a cathode depending upon the electrode to which it is coupled.

The hydrogen electrode shown in Figure 15-7 can be represented schematically as

$$\mathrm{Pt,H_2}(p = 1.00 \text{ atm})|([\mathrm{H}^+] = x \text{ M})||$$

Here, the hydrogen is specified as having a partial pressure of one atmosphere and the hydrogen ion concentration in the solution is x M.

The hydrogen electrode is reversible and acts either as an anode or as a cathode, depending upon the half-cell with which it is coupled. Hydrogen is oxidized to hydrogen ion when the electrode is the anode. Hydrogen ion is reduced to molecular hydrogen when it acts as a cathode.

The potential of a hydrogen electrode depends upon temperature and the activities of hydrogen ion and molecular hydrogen in the solution. The latter, in turn, is proportional to the pressure of the gas that is used to keep the solution saturated in hydrogen. For the SHE, the activity of hydrogen ions is specified as unity and the partial pressure of the gas is specified as one atmosphere. *By convention, the potential of the standard hydrogen electrode is assigned a value of 0.000 V at all temperatures.* As a consequence of this definition, any potential developed in a galvanic cell consisting of a standard hydrogen electrode and some other electrode is attributed entirely to the other electrode.

Several other reference electrodes that are more convenient for routine measurements have been developed. Some of these are described in Section 18B.

15C-2 Definition of Electrode Potential and Standard Electrode Potential

An *electrode potential* is defined as the potential of a cell consisting of the electrode in question *acting as a cathode* and the standard hydrogen electrode

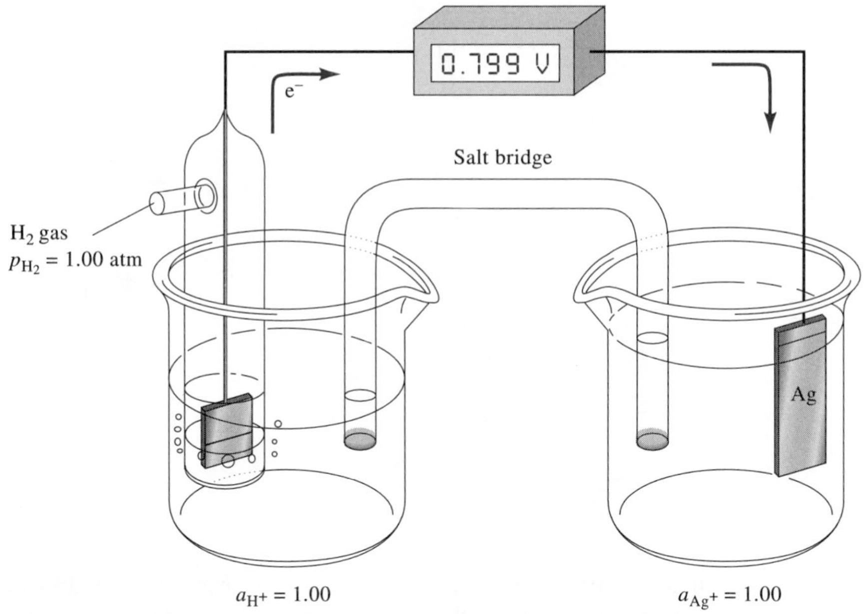

Figure 15-8
Definition of the standard electrode potential for $Ag^+ + e^- \rightleftharpoons Ag(s)$.

acting as an anode. It should be emphasized that, despite its name, *an electrode potential is in fact the potential of an electrochemical cell involving a carefully defined reference electrode.* It could be more properly called a ''relative electrode potential'' (but seldom is).

An electrode potential is the potential of a cell that has a standard hydrogen electrode *as the anode.*

The *standard electrode potential, E^0,* of a half-reaction is defined as its electrode potential when the activities of all reactants and products are unity. The cell in Figure 15-8 illustrates the definition of the standard electrode potential for the half-reaction

$$Ag^+ + e^- \rightleftharpoons Ag(s)$$

Here, the half-cell on the right consists of a strip of pure silver in contact with a solution that has a silver ion activity of 1.00; the electrode on the left is the standard hydrogen electrode.

The cell shown in Figure 15-8 can be represented schematically as

$$Pt, H_2(p = 1.00 \text{ atm})|H^+(a_{H^+} = 1.00)\|Ag^+(a_{Ag^+} = 1.00)|Ag$$

or alternatively as

$$SHE\|Ag^+(a_{Ag^+} = 1.00)|Ag$$

where SHE is a shorthand notation for the standard hydrogen electrode. This galvanic cell develops a potential of 0.799 V with the silver electrode functioning *as the cathode;* that is, the spontaneous cell reaction is

$$2Ag^+ + H_2(g) \rightleftharpoons 2Ag(s) + 2H^+$$

Because the silver electrode serves as the cathode, the measured potential is, *by definition,* the standard electrode potential for the silver half-reaction (or the silver

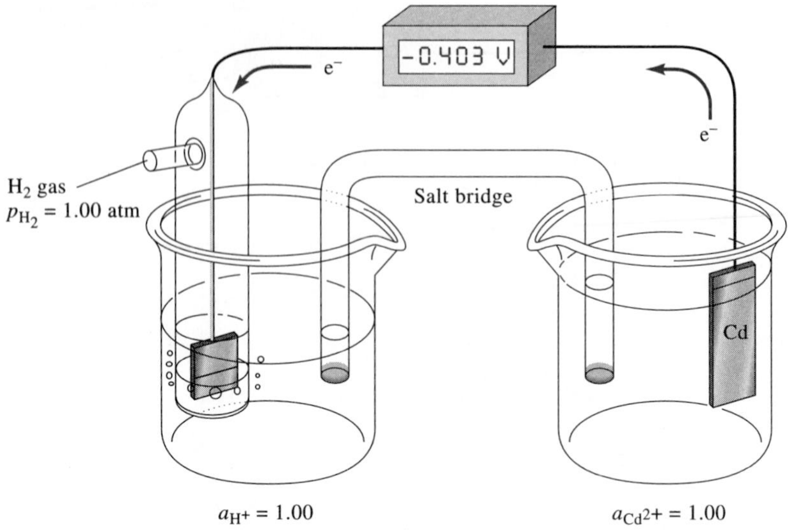

Figure 15-9
Definition of the standard electrode potential for $Cd^{2+} + 2e^- \rightleftharpoons Cd(s)$.

A half-cell is sometimes called a **couple.**

couple). Note that the silver electrode is positive with respect to the hydrogen anode (that is, electrons flow from the negative hydrogen anode to the silver cathode). Therefore, the electrode potential is given a positive sign, and we write

$$Ag^+ + e^- \rightleftharpoons Ag(s) \qquad E^0_{Ag^+} = +0.799 \text{ V}$$

Figure 15-9 illustrates the definition of the standard electrode potential for the half-reaction

$$Cd^{2+} + 2e^- \rightleftharpoons Cd(s)$$

In contrast to the silver electrode, the cadmium electrode acts as the *anode* of the galvanic cell. That is, the spontaneous cell reaction is

$$Cd(s) + 2H^+ \rightleftharpoons H_2(g) + Cd^{2+}$$

Here, the cadmium electrode bears a negative charge with respect to the standard hydrogen electrode. In order to reverse this reaction so that the cadmium electrode acts as a cathode, a potential more negative than −0.403 V must be applied to the cell. Consequently, the standard electrode potential of the Cd/Cd^{2+} couple is *by convention* given a negative sign and is equal to $E^0_{Cd^{2+}} = -0.403$ V.

A zinc electrode immersed in a solution with a zinc ion activity of unity develops a potential of −0.763 V when paired with a standard hydrogen electrode. Because the zinc electrode also behaves as an anode in the galvanic cell, its electrode potential is also negative.

The standard electrode potentials for the four half-cells just described can be arranged in the order

Half-Reaction	Standard Electrode Potential, V
$Ag^+ + e^- \rightleftharpoons Ag(s)$	+0.799
$2H^+ + 2e^- \rightleftharpoons H_2(g)$	0.000
$Cd^{2+} + 2e^- \rightleftharpoons Cd(s)$	−0.403
$Zn^{2+} + 2e^- \rightleftharpoons Zn(s)$	−0.763

The magnitudes of these electrode potentials indicate the relative strength of the four ionic species as electron acceptors (oxidizing agents); that is, in decreasing strength, $Ag^+ > H^+ > Cd^{2+} > Zn^{2+}$.

15C-3 Sign Convention for Electrode Potentials

Historically, electrochemists have not always used the sign convention just described. Indeed, disagreements regarding the conventions to be used in specifying signs for half-cell processes caused much controversy and confusion in the development of electrochemistry. The International Union of Pure and Applied Chemistry (IUPAC) addressed itself to this problem at its 1953 meeting in Stockholm. The usages adopted at that meeting are collectively referred to as the *Stockholm Convention* or the *IUPAC Convention* and are now generally accepted. The sign convention described in the previous section and in the paragraphs that follow is based upon the IUPAC recommendations.

Any sign convention must be based upon expressing half-cell processes in a single way—that is, either as oxidations or as reductions. According to the IUPAC convention, the term *electrode potential* (or, more exactly, *relative electrode potential*) *is reserved exclusively to describe half-reactions written as reductions.* There is no objection to the use of the term *oxidation potential* to indicate a process written in the opposite sense, but it is not proper to refer to such a potential as an electrode potential.

An electrode potential is, by definition, a reduction potential. An oxidation potential is the potential for the half-reaction written in the opposite way. The sign of an oxidation potential is therefore opposite that for a reduction potential, but the magnitude is the same.

The sign of an electrode potential is determined by the sign of the half-cell in question when it is coupled to a standard hydrogen electrode. When the half-cell of interest behaves spontaneously as a cathode, it is the positive electrode of the galvanic cell (Figure 15-8). Thus, its electrode potential is positive. When the half-cell of interest behaves as an anode, the electrode is negative, and so is its electrode potential (Figure 15-9).

It is important to emphasize that the electrode potential refers to a half-cell process written *as a reduction.* For the zinc and cadmium couples we have been considering, the spontaneous reactions are oxidations. *It is evident, then, that the sign of an electrode potential indicates whether the reduction is spontaneous with respect to the standard hydrogen electrode.* The positive sign associated with the electrode potential for silver indicates that the process

The IUPAC sign convention is based upon the actual sign of the half-cell of interest when it is coupled with the standard hydrogen electrode.

$$2Ag^+ + H_2(g) \rightleftharpoons 2Ag(s) + 2H^+$$

favors the products under ordinary conditions. Similarly, the negative sign of the electrode potential for zinc means that the analogous reaction

$$Zn^{2+} + H_2(g) \rightleftharpoons Zn(s) + 2H^+$$

is not spontaneous. In other words the reaction tends to go in the other direction. That is,

$$Zn(s) + 2H^+ \rightleftharpoons Zn^{2+} + H_2(g)$$

15C-4 Effect of Concentration on Electrode Potentials: The Nernst Equation

An electrode potential is a measure of the extent to which the existing concentrations in a half-cell differ from their equilibrium values. Thus, for example, there is a greater tendency for the process

$$Ag^+ + e^- \rightleftharpoons Ag(s)$$

to occur in a concentrated solution of silver(I) than in a dilute solution of that ion. It follows that the magnitude of the electrode potential for this process must also become larger (more positive) as the silver ion concentration of a solution is increased. We now examine the quantitative relationship between concentration and electrode potential.

Consider the reversible half-reaction

$$aA + bB + \cdots + ne^- \rightleftharpoons cC + dD + \cdots \tag{15-8}$$

where the capital letters represent formulas for the participating species (atoms, molecules, or ions), e^- represents electrons, and the lower-case italic letters indicate the number of moles of each species appearing in the half-reaction as it has been written. The electrode potential E for this process is described by the equation

$$E = E^0 - \frac{RT}{nF} \ln \frac{[C]^c [D]^d \cdots}{[A]^a [B]^b \cdots} \tag{15-9}$$

where

E^0 = the *standard electrode potential,* which is a characteristic constant for each half-reaction

R = the gas constant 8.314 J K^{-1} mol^{-1}

T = temperature in kelvins

n = number of moles of electrons that appear in the half-reaction for the electrode process as it has been written

F = the faraday = 96,485 C (coulombs)

ln = the natural logarithm = 2.303 log

Substituting numerical values for the constants, converting to base 10 logarithms, and specifying 25°C for the temperature give

The meanings of the bracketed terms in Equations 15-9 and 15-10 are,

for a solute A, [A] = molar concentration

for a gas B, [B] = p_B = partial pressure in atmospheres

If one or more of the species appearing in Equation 15-9 is a pure liquid, a pure solid, or the solvent present in excess, then no bracketed term for this species appears in the quotient.

$$E = E^0 - \frac{0.0592}{n} \log \frac{[C]^c [D]^d \cdots}{[A]^a [B]^b \cdots} \tag{15-10}$$

The letters in brackets strictly represent activities, but we shall ordinarily follow our practice of substituting molar concentrations for activities in most calculations. Thus, if some participating species A is a solute, [A] is the concentration of A in moles per liter. If A is a gas, [A] in Equation 15-10 is replaced by p_A, the partial pressure of A in atmospheres. If A is a pure liquid, a pure solid, or the solvent, no term for A is included in the equation because its concentration is constant. The rationale for these assumptions is the same as that described in Section 7B, which deals with equilibrium-constant expressions.

Equation 15-10 is known as the *Nernst equation* in honor of the German chemist who was responsible for its development.

EXAMPLE 15-2

Typical half-cell reactions and their corresponding Nernst expressions follow.

(1) $Zn^{2+} + 2e^- \rightleftharpoons Zn(s) \qquad E = E^0 - \dfrac{0.0592}{2}\log\dfrac{1}{[Zn^{2+}]}$

No term for elemental zinc is included in the logarithmic term because it is a pure second phase. Thus, the electrode potential varies linearly with the logarithm of the reciprocal of the zinc ion concentration.

(2) $Fe^{3+} + e^- \rightleftharpoons Fe^{2+} \qquad E = E^0 - \dfrac{0.0592}{1}\log\dfrac{[Fe^{2+}]}{[Fe^{3+}]}$

The potential for this couple can be measured with an inert metallic electrode immersed in a solution containing both iron species. The potential depends upon the logarithm of the ratio between the molar concentrations of these ions.

(3) $2H^+ + 2e^- \rightleftharpoons H_2(g) \qquad E = E^0 - \dfrac{0.0592}{2}\log\dfrac{p_{H_2}}{[H^+]^2}$

In this example, p_{H_2} is the partial pressure of hydrogen (in atmospheres) at the surface of the electrode. Ordinarily, its value will be the same as the atmospheric pressure.

(4) $MnO_4^- + 5e^- + 8H^+ \rightleftharpoons Mn^{2+} + H_2O$

$$E = E^0 - \frac{0.0592}{5}\log\frac{[Mn^{2+}]}{[MnO_4^-][H^+]^8}$$

Here, the potential depends not only upon the concentrations of the manganese species but also on the pH of the solution.

(5) $AgCl(s) + e^- \rightleftharpoons Ag(s) + Cl^- \qquad E = E^0 - \dfrac{0.0592}{1}\log\,[Cl^-]$

This half-reaction describes the behavior of a silver electrode immersed in a chloride solution that is *saturated* with AgCl. To ensure this condition, an excess of the solid must always be present. Note that this electrode reaction is the sum of two reactions, namely,

$$AgCl(s) \rightleftharpoons Ag^+ + Cl^-$$
$$Ag^+ + e^- \rightleftharpoons Ag(s)$$

The Nernst expression in part 5 of Example 15-2 requires an excess of solid AgCl *so that the solution is saturated* in that compound at all times.

Walther Hermann Nernst (1864–1941) was a German physical chemist who made many contributions to our understanding of electrochemistry. He is probably most famous for the equation that bears his name (Equation 15-9), but he was also known for discoveries and inventions in other fields. He invented the Nernst glower, a source of infrared radiation that is shown on the stamp. Nernst received the Nobel Prize in Chemistry in 1920 for his numerous contributions to the field of chemical thermodynamics.

Note also that the electrode potential is independent of the amount of AgCl as long as there is some present to keep the solution saturated.

15C-5 The Standard Electrode Potential, E^0

The **standard electrode potential** for a half-reaction, E^0, is defined as the electrode potential when all reactants and products of a half-reaction are at unit activity.

Examination of Equations 15-9 and 15-10 reveals that the constant E^0 is the electrode potential whenever the activity quotient has a value of one. This constant is by definition *the standard electrode potential* for the half-reaction. Note that the quotient is always unity when the activities of the reactants and products of a half-reaction are unity.

The standard electrode potential is an important physical constant that provides quantitative information regarding the driving force for a half-cell reaction.[1] The major characteristics of these constants are the following:

1. The standard electrode potential is a relative quantity in the sense that it is the potential of an electrochemical cell in which the anode is the standard hydrogen electrode, whose potential has been arbitrarily set at 0 V.
2. The standard electrode potential for a half-reaction refers exclusively to a reduction reaction; that is, it is a relative reduction potential.
3. The standard electrode potential measures the relative force tending to drive the half-reaction from a state in which the reactants and products are at unit activity to a state in which the reactants and products are at their equilibrium activities relative to the standard hydrogen electrode.
4. The standard electrode is independent of the number of moles of reactant and product shown in the balanced half-reaction. Thus, the standard electrode for the half-reaction

$$Fe^{3+} + e^- \rightleftharpoons Fe^{2+} \qquad E^0 = +0.771$$

does not change if we choose to write the reaction as

$$5Fe^{3+} + 5e^- \rightleftharpoons 5Fe^{2+} \qquad E^0 = +0.771$$

Note, however, that the Nernst equation must be consistent with the half-reaction as written. For the first case, it will be

$$E = 0.771 - \frac{0.0592}{1}\log\frac{[Fe^{2+}]}{[Fe^{3+}]}$$

and for the second

$$E = 0.771 - \frac{0.0592}{5}\log\frac{[Fe^{2+}]^5}{[Fe^{3+}]^5} = 0.771 - \frac{0.0592}{5}\log\left(\frac{[Fe^{2+}]}{[Fe^{3+}]}\right)^5$$

$$= 0.771 - \frac{\cancel{5} \times 0.0592}{\cancel{5}}\log\frac{[Fe^{2+}]}{[Fe^{3+}]}$$

Note that the two log terms have identical values. That is,

$$\frac{0.0592}{1}\log\frac{[Fe^{2+}]}{[Fe^{3+}]} = \frac{0.0592}{\cancel{5}}\log\frac{[Fe^{2+}]^{\cancel{5}}}{[Fe^{3+}]^{\cancel{5}}}$$

[1]For further reading on standard electrode potentials, see R. G. Bates, in *Treatise on Analytical Chemistry,* 2nd ed., I. M. Kolthoff and P. J. Elving, Eds., Part I, Vol. 1, Chapter 13. New York: Wiley, 1978.

TABLE 15-1 Standard Electrode Potentials*

Reaction	E^0 at 25°C, V
$Cl_2(g) + 2e^- \rightleftharpoons 2Cl^-$	+1.359
$O_2(g) + 4H^+ + 4e^- \rightleftharpoons 2H_2O$	+1.229
$Br_2(aq) + 2e^- \rightleftharpoons 2Br^-$	+1.087
$Br_2(l) + 2e^- \rightleftharpoons 2Br^-$	+1.065
$Ag^+ + e^- \rightleftharpoons Ag(s)$	+0.799
$Fe^{3+} + e^- \rightleftharpoons Fe^{2+}$	+0.771
$I_3^- + 2e^- \rightleftharpoons 3I^-$	+0.536
$Cu^{2+} + 2e^- \rightleftharpoons Cu(s)$	+0.337
$UO_2^{2+} + 4H^+ + 2e^- + \rightleftharpoons U^{4+} + 2H_2O$	+0.334
$Hg_2Cl_2(s) + 2e^- \rightleftharpoons 2Hg(l) + 2Cl^-$	+0.268
$AgCl(s) + e^- \rightleftharpoons Ag(s) + Cl^-$	+0.222
$Ag(S_2O_3)_2^{3-} + e^- \rightleftharpoons Ag(s) + 2S_2O_3^{2-}$	+0.017
$2H^+ + 2e^- \rightleftharpoons H_2(g)$	0.000
$AgI(s) + e^- \rightleftharpoons Ag(s) + I^-$	−0.151
$PbSO_4(s) + 2e^- \rightleftharpoons Pb(s) + SO_4^{2-}$	−0.350
$Cd^{2+} + 2e^- \rightleftharpoons Cd(s)$	−0.403
$Zn^{2+} + 2e^- \rightleftharpoons Zn(s)$	−0.763

*See Appendix 5 for a more extensive list.

5. A positive electrode potential indicates that the half-reaction in question is spontaneous with respect to the standard hydrogen electrode half-reaction. That is, the oxidant in the half-reaction is a stronger oxidant or electron acceptor than is hydrogen ion. A negative sign indicates just the opposite.
6. The standard electrode potential for a half-reaction is temperature dependent.

Standard electrode potential data are available for an enormous number of half-reactions. Many have been determined directly from electrochemical measurements. Others have been computed from equilibrium studies of oxidation/reduction systems and from thermochemical data associated with such reactions. Table 15-1 contains standard electrode data for several half-reactions that we will be considering in the pages that follow. A more extensive listing is found in Appendix 5.[2]

FEATURE 15-4

Sign Convention in the Older Literature

Reference works, particularly those published before 1953, often contain tabulations of electrode potentials that are not in accord with the IUPAC

[2]Comprehensive sources for standard electrode potentials include *Standard Electrode Potentials in Aqueous Solutions,* A. J. Bard, R. Parsons, and J. Jordan, Eds. New York: Marcel Dekker, 1985; G. Milazzo and S. Caroli, *Tables of Standard Electrode Potentials.* New York: Wiley-Interscience, 1977; M. S. Antelman and F. J. Harris, *Chemical Electrode Potentials.* New York: Plenum Press, 1982. Some compilations are arranged alphabetically by element; others are tabulated according to the numerical value of E^0.

recommendations. For example, in a classic source of standard-potential data complied by Latimer,[3] one finds

$$Zn(s) \rightleftharpoons Zn^{2+} + 2e^- \qquad E = +0.76\ V$$

$$Cu(s) \rightleftharpoons Cu^{2+} + 2e^- \qquad E = -0.34\ V$$

In converting these oxidation potentials to electrode potentials as defined by the IUPAC convention, one must mentally (1) express the half-reactions as reductions and (2) change the signs of the potentials.

The sign convention used in a tabulation of electrode potentials may not be explicitly stated. This information can be readily deduced, however, by noting the direction and sign of the potential for a half-reaction with which you are familiar. If the sign agrees with the IUPAC convention, the table can be used as is; if not, the signs of all of the data must be reversed. For example, the reaction

$$O_2(g) + 4H^+ + 4e^- \rightleftharpoons 2H_2O \qquad E = +1.229\ V$$

occurs spontaneously with respect to the standard hydrogen electrode and thus carries a positive sign. If the potential for this half-reaction is negative in a tabulation, it and all the other potentials should be multiplied by -1.

[3]W. M. Latimer, *The Oxidation States of the Elements and Their Potentials in Aqueous Solutions,* 2nd ed. Englewood Cliffs, NJ: Prentice-Hall, 1952.

Table 15-1 and Appendix 5 illustrate the two common ways for tabulating standard potential data. In Table 15-1, potentials are listed in decreasing numerical order. As a consequence, the species in the upper left part are the most effective electron acceptors, as evidenced by their large positive E^0 values. They are therefore the strongest *oxidizing agents.* As we proceed down the left side of such a table, each succeeding species is less effective as an electron acceptor than the one above it. The half-cell reactions at the bottom of the table have little or no tendency to take place as they are written. On the other hand, they do tend to occur in the opposite sense. The most effective *reducing agents,* then, are those species that appear in the lower right portion of the table.

Based upon the E^0 values in Table 15-1 for Fe^{3+} and I_3^-, which species would you expect to predominate in a solution produced by mixing iron(III) and iodide ions? (See color plate 15.)

Compilations of electrode-potential data, such as that shown in Table 15-1, provide the chemist with qualitative insights into the extent and direction of electron-transfer reactions. For example, the standard potential for silver(I) (+0.799 V) is more positive than that for copper(II) (+0.337 V). We therefore conclude that a piece of copper immersed in a silver(I) solution will cause the reduction of that ion and the oxidation of the copper. On the other hand, we would expect no reaction if we place a piece of silver in a copper(II) solution.

In contrast to the data in Table 15-1, standard potentials in Appendix 5 are arranged alphabetically by element to make it easier to locate data for a given electrode reaction.

Standard Potentials for Systems Involving Precipitates or Complex Ions

In Table 15-1, we find several entries involving Ag(I) including:

$$Ag^+ + e^- \rightleftharpoons Ag(s) \qquad E^0_{Ag^+} = +0.799\ V$$

$$AgCl(s) + e^- \rightleftharpoons Ag(s) + Cl^- \qquad E^0_{AgCl} = +0.222\ V$$

$$Ag(S_2O_3)_2^{3-} + e^- \rightleftharpoons Ag(s) + 2S_2O_3^{2-} \qquad E^0_{Ag(S_2O_3)_2^{3-}} = +0.17\ V$$

Each gives the potential of a silver electrode in a different environment. Let us see how the three potentials are related.

The Nernst expression for the first half-reaction is

$$E = E^0_{Ag^+} - \frac{0.0592}{1}\log\frac{1}{[Ag^+]}$$

If we replace $[Ag^+]$ with $K_{sp}/[Cl^-]$, we obtain

$$E = E^0_{Ag^+} - \frac{0.0592}{1}\log\frac{[Cl^-]}{K_{sp}} = E^0_{Ag^+} + 0.0592\log K_{sp} - 0.0592\log[Cl^-]$$

By definition, standard potential for the second half-reaction is the potential where $[Cl^-] = 1.00$. That is, when $[Cl^-] = 1.00$, $E = E^0_{AgCl}$. Substituting these values gives

$$E^0_{AgCl} = E^0_{Ag^+} + 0.0592\log 1.82 \times 10^{-10} - 0.0592\log(1.00)$$
$$= 0.799 + (-0.577) - 0.000 = 0.222\ V$$

Figure 15-10 illustrates the definition of the standard electrode potential for the Ag/AgCl electrode.

If we proceed in the same way, we can obtain an expression for the standard electrode potential for the reduction of the thiosulfate complex of silver ion shown by the third equilibrium at the start of this section. Here the standard potential is given by

$$E^0_{Ag(S_2O_3)_2^{3-}} = E^0_{Ag^+} - 0.0592\log\beta_2 \qquad (15\text{-}11)$$

CHALLENGE: Derive Equation 15-11.

where β_2 is the formation constant for the complex. That is,

$$\beta_2 = \frac{[Ag(S_2O_3)_2^{3-}]}{[Ag^+][S_2O_3^{2-}]^2}$$

EXAMPLE 15-3

Calculate the electrode potential of a silver electrode immersed in a 0.0500 M solution of NaCl using (a) $E^0_{Ag^+} = 0.799$ V and (b) $E^0_{AgCl} = 0.222$ V.

(a) $Ag^+ + e^- \rightleftharpoons Ag(s) \qquad E^0 = 0.799\ V$

The Ag^+ concentration of this solution is given by

$$[Ag^+] = K_{sp}/[Cl^-] = 1.82 \times 10^{-10}/0.0500 = 3.64 \times 10^{-9}\ M$$

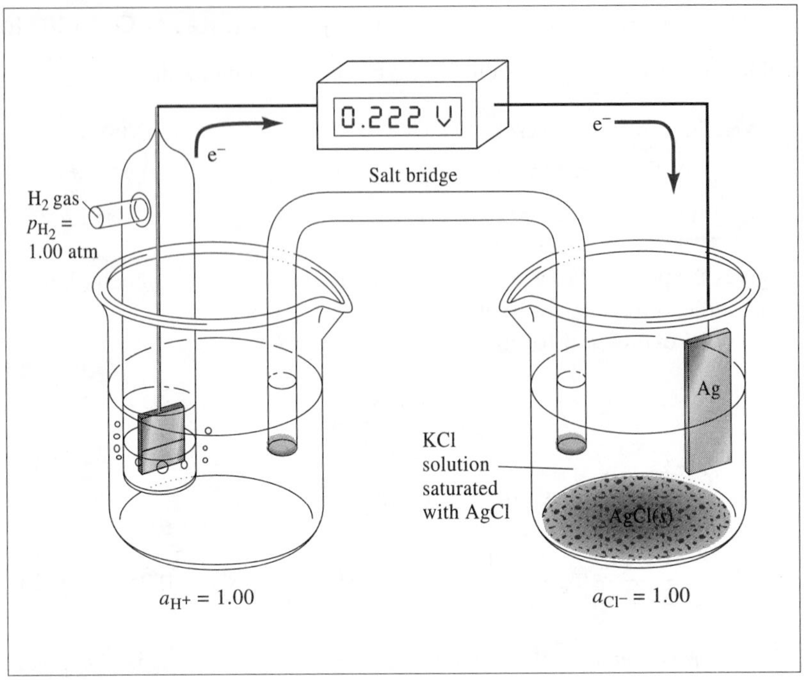

Figure 15-10
Definition of the standard electrode potential for an Ag/AgCl electrode.

Substituting into the Nernst expression gives

$$E = 0.799 - 0.0592 \log \frac{1}{3.64 \times 10^{-9}} = 0.299 \text{ V}$$

(b) Here we may write

$$E = 0.222 - 0.0592 \log [\text{Cl}^-] = 0.222 - 0.0592 \log 0.0500$$
$$= 0.299 \text{ V}$$

FEATURE 15-5
Why Are There Two Electrode Potentials for Br_2 in Table 15-1?

In Table 15-1, we find the following data for Br_2:

$$\text{Br}_2(aq) + 2e^- \rightleftharpoons 2\text{Br}^- \qquad E^0 = +1.087 \text{ V}$$
$$\text{Br}_2(l) + 2e^- \rightleftharpoons 2\text{Br}^- \qquad E^0 = +1.065 \text{ V}$$

The second standard potential applies only to a solution that is saturated with Br_2 and not to undersaturated solutions of the element. You should use 1.065 V to calculate the electrode potential of a 0.0100 M solution of KBr that is saturated with Br_2 and in contact with an excess of the liquid element. In such a case,

$$E = 1.065 - \frac{0.0592}{2}\log [Br^-]^2 = 1.065 - \frac{0.0592}{2}\log (0.0100)^2$$

$$= 1.065 - \frac{0.0592}{2}(-4.00) = 1.183\ V$$

In this calculation, no term for Br_2 appears in the logarithmic term because it is a pure liquid present in excess.

The standard electrode potential shown in the first entry for $Br_2(aq)$ is a *hypothetical* standard potential because the solubility of Br_2 at 25°C is only about 0.18 M. Thus, the recorded value of 1.087 V is based upon a system that—in terms of our definition of E^0—cannot be realized experimentally. Nevertheless, the hypothetical potential does permit us to calculate electrode potentials for solutions that are *undersaturated* in Br_2. For example, if we wish to calculate the electrode potential for a solution that was 0.0100 M in KBr and 0.00100 M in Br_2, we would write

$$E = 1.087 - \frac{0.0592}{2}\log \frac{[Br^-]^2}{[Br_2(aq)]} = 1.087 - \frac{0.0592}{2}\log \frac{(0.0100)^2}{0.00100}$$

$$= 1.087 - \frac{0.0592}{2}\log 0.100 = 1.117\ V$$

15C-6 Limitations to the Use of Standard Electrode Potentials

We shall be using standard electrode potentials throughout the rest of this text to calculate cell potentials and equilibrium constants for redox reactions as well as to derive data for redox titration curves. You should be aware that such calculations sometimes lead to results that are significantly different from those you would obtain experimentally in the laboratory. The main sources of these differences are two: (1) the necessity of using concentrations in place of activities in the Nernst equation and (2) failure to take into account other equilibria such as dissociation, association, complex formation, and solvolysis.

Use of Concentrations Instead of Activities

Most analytical oxidation/reduction reactions are carried out in solutions that have such high ionic strengths that activity coefficients cannot be obtained via the Debye-Hückel equation (Section 8B-2). Significant errors may result, however, if concentrations are used in the Nernst equation rather than activities. For example, the standard potential for the half-reaction

$$Fe^{3+} + e^- \rightleftharpoons Fe^{2+} \qquad E^0 = +0.771\ V$$

is $+0.771$ V. When the potential of a platinum electrode, immersed in a solution that is 10^{-4} M in iron(III) ion, iron(II) ion, and perchloric acid, is measured against a standard hydrogen electrode, a reading of close to $+0.77$ V is obtained as predicted by theory. If perchloric acid is added to this mixture until the acid concentration is 0.1 M, however, the potential is found to decrease to about

+ 0.75 V. This difference is attributable to the fact that the activity coefficient of iron(III) is considerably smaller than that of iron(II) (0.4 vs. 0.18) at the high ionic strength of the 0.1 M perchloric acid medium (see Table 8-1, page 154). As a consequence, the ratio of activities of the two species ($[Fe^{2+}]/[Fe^{3+}]$) in the Nernst equation is greater than unity, a condition that leads to a decrease in the electrode potential. In 1 M $HClO_4$, the electrode potential is still smaller (~0.73 V).

Effect of Other Equilibria

The application of standard electrode potential data to many systems of interest in analytical chemistry is further complicated by association, dissociation, complex formation, and solvolysis equilibria involving the species that appear in the Nernst equation. These phenomena can be taken into account only if their existence is known and appropriate equilibrium constants are available. More often than not, neither of these requirements is met and significant discrepancies arise as a consequence. For example, substitution of 1 M hydrochloric acid in the iron(II)/iron(III) mixture we have just discussed leads to a measured potential of + 0.70 V; in 1 M sulfuric acid, a potential of + 0.68 V is observed; and in a 2 M phosphoric acid, the potential is + 0.46 V. In each of these cases, the iron(II)/iron(III) activity ratio is larger because the complexes of iron(III) with chloride, sulfate, and phosphate ions are more stable than those of iron(II); thus, the ratio of the *species* concentrations, $[Fe^{2+}]/[Fe^{3+}]$, in the Nernst equation is greater than unity and the measured potential is less than the standard potential. If formation constants for these complexes were available, it would be possible to make appropriate corrections. Unfortunately, such data are often not available or, if they are, they are not very reliable.

Formal Potentials

A **formal potential** is the electrode potential when the ratio of *analytical concentrations* of reactants and products of a half-reaction are exactly 1.00 and the molar concentrations of any other solutes are specified.

Formal potentials are empirically derived potentials that compensate for the types of activity and competing equilibria effects that we have just described. The formal potential $E^{0\prime}$ of a system is the potential of the half-cell with respect to the standard hydrogen electrode measured under conditions such that the ratio of *analytical concentrations* of reactants and products as they appear in the Nernst equation is exactly unity and the concentrations of other species in the system are all carefully specified. For example, the formal potential for the half-reaction

$$Ag^+ + e^- \rightleftharpoons Ag(s) \qquad E^{0\prime} = 0.792 \text{ V}$$

in 1.00 M $HClO_4$ could be obtained by measuring the potential of the cell shown in Figure 15-11. Here, the cathode is a silver electrode immersed in a solution that is 1.00 M in $AgNO_3$ and 1.00 M in $HClO_4$; the anode is a standard hydrogen electrode. This cell develops a potential of 0.792 V, which is the formal potential of the Ag/Ag^+ couple in 1.00 M $HClO_4$. Note that the standard potential for this couple is 0.799 V.

Formal potentials for many half-reactions are listed in Appendix 5. Note that large differences exist between the formal and standard potentials for some half-reactions. For example, the formal potential for

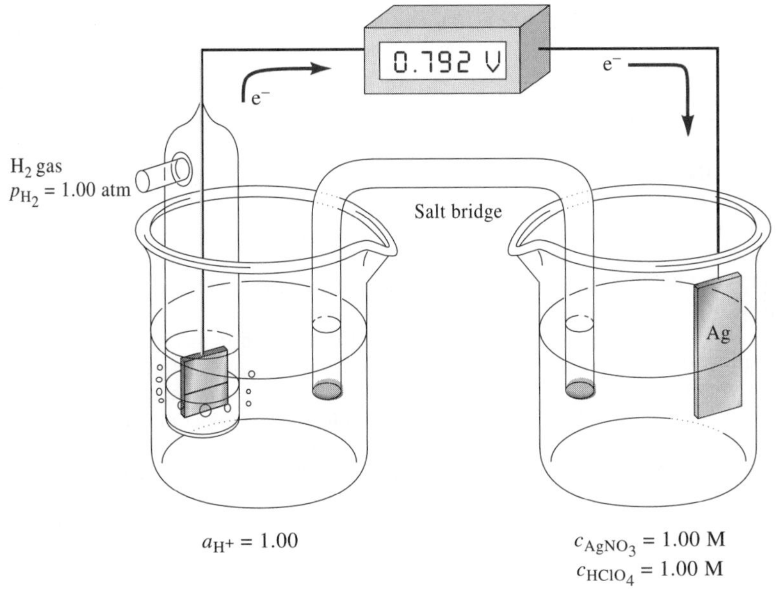

Figure 15-11
Definition of the formal potential for Ag^+/Ag couple in 1 M $HClO_4$.

$$Fe(CN)_6^{3-} + e^- \rightleftharpoons Fe(CN)_6^{4-} \qquad E^0 = +0.36 \text{ V}$$

is 0.72 V in 1 M perchloric or sulfuric acids, which is 0.36 V greater than the standard electrode potential for the half-reaction. The reason for this difference is that in the presence of high concentrations of hydrogen ion, ferrocyanide ions ($Fe(CN)_6^{4-}$) and ferricyanide ions ($Fe(CN)_6^{3-}$) combine with one or more protons to form hydroferrocyanic and hydroferricyanic acid species. Because the former are weaker, the ratio of the *species* concentrations, $[Fe(CN)_6^{4-}]/[Fe(CN)_6^{3-}]$, in the Nernst equation is less than unity, thus making the observed potentials greater.

Substitution of formal potentials for standard electrode potentials in the Nernst equation yields better agreement between calculated and experimental results—provided that the electrolyte concentration of the solution approximates that for which the formal potential is applicable. Attempts to apply formal potentials to systems that differ substantially in type and in concentration of electrolyte can result in errors that are larger than those associated with the use of standard electrode potentials. In this text, we use whichever is the more appropriate.

15D QUESTIONS AND PROBLEMS

Note: Numerical data are molar analytical concentrations where the full formula of a species is provided. Molar equilibrium concentrations are supplied for species displayed as ions.

15-1. Briefly describe or define
- *(a) oxidation.
- (b) oxidizing agent.
- *(c) salt bridge.
- (d) liquid junction.
- *(e) Nernst equation.

15-2. Briefly describe or define
- *(a) electrode potential.
- (b) formal potential.
- *(c) standard electrode potential.
- (d) liquid-junction potential.
- *(e) oxidation potential.

***15-3.** Make a clear distinction between
- *(a) reduction and reducing agent.
- (b) a galvanic cell and an electrolytic cell.
- *(c) the anode and the cathode of an electrochemical cell.
- (d) a reversible electrochemical cell and an irreversible electrochemical cell.
- *(e) the standard electrode potential and formal potential.

***15-4.** The following entries are found in a table of standard electrode potentials:

$$I_2(s) + 2e^- \rightleftharpoons 2I^- \qquad E^0 = 0.5355\ V$$
$$I_2(aq) + 2e^- \rightleftharpoons 2I^- \qquad E^0 = 0.615\ V$$

What is the significance of the difference between these two standard potentials?

***15-5.** Why is it necessary to bubble hydrogen through the electrolyte in a hydrogen electrode?

15-6. The standard electrode potential for the reduction of Ni^{2+} to Ni is −0.25 V. Would the potential of a nickel electrode immersed in a 1.00 M NaOH solution saturated with $Ni(OH)_2$ be more negative than $E^0_{Ni^{2+}}$ or less? Explain.

15-7. Write balanced net-ionic equations for the following reactions. Supply H^+ and/or H_2O, as needed, to obtain balance.

*(a) $Fe^{3+} + Sn^{2+} \rightarrow Fe^{2+} + Sn^{4+}$
(b) $Cr(s) + Ag^+ \rightarrow Cr^{3+} + Ag(s)$
*(c) $NO_3^- + Cu(s) \rightarrow NO_2(g) + Cu^{2+}$
(d) $MnO_4^- + H_2SO_3 \rightarrow Mn^{2+} + SO_4^{2-}$
*(e) $Ti^{3+} + Fe(CN)_6^{3-} \rightarrow TiO^{2+} + Fe(CN)_6^{4-}$
(f) $H_2O_2 + Ce^{4+} \rightarrow O_2(g) + Ce^{3+}$
*(g) $Ag(s) + I^- + Sn^{4+} \rightarrow AgI(s) + Sn^{2+}$
(h) $UO_2^{2+} + Zn(s) \rightarrow U^{4+} + Zn^{2+}$
*(i) $HNO_2 + MnO_4^- \rightarrow NO_3^- + Mn^{2+}$
(j) $H_2NNH_2 + IO_3^- + Cl^- \rightarrow N_2(g) + ICl_2^-$

***15-8.** Identify the oxidizing agent and the reducing agent on the left side of each equation in Problem 15-7; write a balanced equation for each half-reaction.

***15-9.** Write balanced net-ionic equations for the following reactions. Supply H^+ and/or H_2O, as needed, to obtain balance.

*(a) $MnO_4^- + VO^{2+} \rightarrow Mn^{2+} + V(OH)_4^+$
(b) $I_2 + H_2S(g) \rightarrow I^- + S(s)$
*(c) $Cr_2O_7^{2-} + U^{4+} \rightarrow Cr^{3+} + UO_2^{2+}$
(d) $Cl^- + MnO_2(s) \rightarrow Cl_2(g) + Mn^{2+}$
*(e) $IO_3^- + I^- \rightarrow I_2(aq)$
(f) $IO_3^- + I^- + Cl^- \rightarrow ICl_2^-$
*(g) $HPO_3^{2-} + MnO_4^- + OH^- \rightarrow PO_4^{3-} + MnO_4^{2-}$
(h) $SCN^- + BrO_3^- \rightarrow Br^- + SO_4^{2-} + HCN$
*(i) $V^{2+} + V(OH)_4^+ \rightarrow VO^{2+}$
(j) $MnO_4^- + Mn^{2+} + OH^- \rightarrow MnO_2(s)$

***15-10.** Identify the oxidizing agent and the reducing agent on the left side of each equation in Problem 15-9; write a balanced equation for each half-reaction.

***15-11.** Consider the following oxidation/reduction reactions:

$$AgBr(s) + V^{2+} \longrightarrow Ag(s) + V^{3+} + Br^-$$
$$Tl^{3+} + 2Fe(CN)_6^{4-} \longrightarrow Tl^+ + 2Fe(CN)_6^{3-}$$
$$2V^{3+} + Zn(s) \longrightarrow 2V^{2+} + Zn^{2+}$$
$$Fe(CN)_6^{3-} + Ag(s) + Br^- \longrightarrow Fe(CN)_6^{4-} + AgBr(s)$$
$$S_2O_8^{2-} + Tl^+ \longrightarrow 2SO_4^{2-} + Tl^{3+}$$

(a) Write each net process in terms of two balanced half-reactions.
(b) Express each half-reaction as a reduction.
(c) Arrange the half-reactions in (b) in order of decreasing effectiveness as electron acceptors.

15-12. Consider the following oxidation/reduction reactions:

$$2H^+ + Sn(s) \longrightarrow H_2(g) + Sn^{2+}$$
$$Ag^+ + Fe^{2+} \longrightarrow Ag(s) + Fe^{3+}$$
$$Sn^{4+} + H_2(g) \longrightarrow Sn^{2+} + 2H^+$$
$$2Fe^{3+} + Sn^{2+} \longrightarrow 2Fe^{2+} + Sn^{4+}$$
$$Sn^{2+} + Co(s) \longrightarrow Sn(s) + Co^{2+}$$

(a) Write each net process in terms of two balanced half-reactions.
(b) Express each half-reaction as a reduction.
(c) Arrange the half-reactions in (b) in order of decreasing effectiveness as electron acceptors.

***15-13.** Calculate the potential of a copper electrode immersed in
(a) 0.0440 M $Cu(NO_3)_2$.
(b) 0.0750 M in NaCl and saturated with CuCl.
(c) 0.0400 M in NaOH and saturated with $Cu(OH)_2$.
(d) 0.0250 M in $Cu(NH_3)_4^{2+}$ and 0.128 M in NH_3 (β_4 for $Cu(NH_3)_4^{2+} = 5.62 \times 10^{11}$).
(e) a solution in which the molar analytical concentration of $Cu(NO_3)_2$ is 4.00×10^{-3} M, that for H_2Y^{2-} is 2.90×10^{-2} M (Y = EDTA), and the pH is fixed at 4.00.

15-14. Calculate the potential of a zinc electrode immersed in
(a) 0.0600 M $Zn(NO_3)_2$.
(b) 0.01000 M in NaOH and saturated with $Zn(OH)_2$.
(c) 0.0100 M in $Zn(NH_3)_4^{2+}$ and 0.250 M in NH_3 (β_4 for $Zn(NH_3)_4^{2+} = 7.76 \times 10^8$).
(d) a solution in which the molar analytical concentration of $Zn(NO_3)_2$ is 5.00×10^{-3} M, that for H_2Y^{2-} is 0.0445 M, and the pH is fixed at 9.00.

15-15. Use activities to calculate the electrode potential of a hydrogen electrode in which the electrolyte is 0.0100 M HCl and the activity of H_2 is 1.00 atm.

***15-16.** Calculate the potential of a platinum electrode immersed in a solution that is
(a) 0.0263 M in K_2PtCl_4 and 0.1492 M in KCl.
(b) 0.0750 M in $Sn(SO_4)_2$ and 2.50×10^{-3} M in $SnSO_4$.
(c) buffered to a pH of 6.00 and saturated with $H_2(g)$ at 1.00 atm.
(d) 0.0353 M in $VOSO_4$, 0.0586 M in $V_2(SO_4)_3$, and 0.100 M in $HClO_4$.
(e) prepared by mixing 25.00 mL of 0.0918 M $SnCl_2$ with an equal volume of 0.1568 M $FeCl_3$.
(f) prepared by mixing 25.00 mL of 0.0832 M $V(OH)_4^+$ with 50.00 mL of 0.01087 M $V_2(SO)_4)_3$ and has a pH of 1.00.

15-17. Calculate the potential of a platinum electrode immersed in a solution that is
(a) 0.0813 M in $K_4Fe(CN)_6$ and 0.00566 M in $K_3Fe(CN)_6$.
(b) 0.0400 M in $FeSO_4$ and 0.00845 M in $Fe_2(SO_4)_3$.
(c) buffered to a pH of 5.55 and saturated with H_2 at 1.00 atm.
(d) 0.1996 M in $V(OH)_4^+$, 0.0789 M in VO^{2+}, and 0.0800 M in $HClO_4$.
(e) prepared by mixing 50.00 mL of 0.0607 M $Ce(SO_4)_2$ with an equal volume of 0.100 M $FeCl_2$. Assume so-

lutions were 1.00 M in H_2SO_4 and use formal potentials.

(f) prepared by mixing 25.00 mL of 0.0832 M $V_2(SO_4)_3$ with 50.00 mL of 0.00628 M $V(OH)_4^+$ and has a pH of 1.00.

*15-18. Indicate whether the following half-cells would act as an anode or a cathode when coupled with a standard hydrogen electrode in a galvanic cell. Calculate the cell potential.

(a) $Ni|Ni^{2+}(0.0943\ M)$
(b) $Ag|AgI(sat'd),KI(0.0922\ M)$
(c) $Pt,O_2(780\ torr)|HCl(1.50 \times 10^{-4}\ M)$
(d) $Pt|Sn^{2+}(0.0944),Sn^{4+}(0.350\ M)$
(e) $Ag|Ag(S_2O_3)_2^{3-}(0.00753\ M),Na_2S_2O_3(0.1439\ M)$

15-19. Indicate whether the following half-cells behave as an anode or a cathode when coupled with a standard hydrogen electrode in a galvanic cell. Calculate the cell potential.

(a) $Cu|Cu^{2+}(0.0897\ M)$
(b) $Pt|CuI(sat'd),KI(0.1214\ M)$
(c) $Pt,H_2(0.984\ atm)|HCl(1.00 \times 10^{-4}\ M)$
(d) $Pt|Fe^{3+}(0.0906\ M),Fe^{2+}(0.1628\ M)$
(e) $Ag|Ag(CN)_2^-(0.0827\ M),KCN(0.0699\ M)$

*15-20. The solubility-product constant for Ag_2SO_3 is 1.5×10^{-14}. Calculate E^0 for the process

$$Ag_2SO_3(s) + 2e^- \rightleftharpoons 2Ag(s) + SO_3^{2-}$$

15-21. The solubility-product constant for $Ni_2P_2O_7$ is 1.7×10^{-13}. Calculate E^0 for the process

$$Ni_2P_2O_7(s) + 4e^- \rightleftharpoons 2Ni(s) + P_2O_7^{4-}$$

*15-22. The solubility-product constant for Tl_2S is 6×10^{-22}. Calculate E^0 for the reaction

$$Tl_2S(s) + 2e^- \rightleftharpoons 2Tl(s) + S^{2-}$$

15-23. The solubility product for $Pb_3(AsO_4)_2$ is 4.1×10^{-36}. Calculate E^0 for the reaction

$$Pb_3(AsO_4)_2(s) + 6e^- \rightleftharpoons 3Pb(s) + 2AsO_4^{3-}$$

*15-24. Compute E^0 for the process

$$ZnY^{2-} + 2e^- \rightleftharpoons Zn(s) + Y^{4-}$$

where Y^{4-} is the completely deprotonated anion of EDTA. The formation constant for ZnY^{2-} is 3.2×10^{16}.

*15-25. Given the formation constants

$$Fe^{3+} + Y^{4-} \rightleftharpoons FeY^- \qquad K_f = 1.3 \times 10^{25}$$

$$Fe^{2+} + Y^{4-} \rightleftharpoons FeY^{2-} \qquad K_f = 2.1 \times 10^{14}$$

calculate E^0 for the process

$$FeY^- + e^- \rightleftharpoons FeY^{2-}$$

15-26. Calculate E^0 for the process

$$Cu(NH_3)_4^{2+} + e^- \rightleftharpoons Cu(NH_3)_2^+ + 2NH_3$$

given that

$$Cu^+ + 2NH_3 \rightleftharpoons Cu(NH_3)_2^+ \qquad \beta_2 = 7.2 \times 10^{10}$$

$$Cu^{2+} + 4NH_3 \rightleftharpoons Cu(NH_3)_4^{2+} \qquad \beta_4 = 5.62 \times 10^{11}$$

CHAPTER

16

Applications of Standard Electrode Potentials

In this chapter, we show how standard electrode potentials can be used for (1) calculating thermodynamic cell potentials, (2) calculating equilibrium constants for redox reactions, and (3) deriving redox titration curves.

16A THE THERMODYNAMIC POTENTIAL OF ELECTROCHEMICAL CELLS

We can use standard electrode potentials and the Nernst equation to calculate the potential obtainable from a galvanic cell or the potential required to operate an electrolytic cell. The calculated potentials (sometimes called *thermodynamic potentials*) are theoretical in the sense that they refer to cells in which there is no current. As we show in Chapter 20, additional factors must be taken into account if a current is involved.

The **thermodynamic potential** of a cell is defined as the electrode potential of the cathode minus the electrode potential of the anode.

The thermodynamic potential of an electrochemical cell E_{cell} is the difference between the *electrode potential* of the cathode and the *electrode potential* of the anode. That is,

$$E_{cell} = E_{cathode} - E_{anode} \quad (16\text{-}1)$$

It is important to note that $E_{cathode}$ and E_{anode} in Equation 16-1 are both *electrode potentials* as defined at the beginning of Section 15C-2.

where $E_{cathode}$ and E_{anode} are the electrode potentials of the cathode and anode, respectively.

EXAMPLE 16-1

Calculate the thermodynamic potential of the following cell

$$Cu|Cu^{2+}(0.0200\text{ M})||Ag^{+}(0.0200\text{ M})|Ag$$

Note that this cell is the galvanic cell shown in Figure 15-2a.

The two half-reactions and standard potentials are

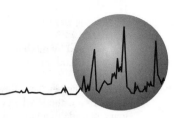

$$Ag^+ + e^- \rightleftharpoons Ag(s) \qquad E^0 = 0.799\ \text{V} \qquad (16\text{-}2)$$

$$Cu^{2+} + 2e^- \rightleftharpoons Cu(s) \qquad E^0 = 0.337\ \text{V} \qquad (16\text{-}3)$$

The electrode potentials are

$$E_{Ag^+} = 0.799 - 0.0592 \log \frac{1}{0.0200} = 0.6984\ \text{V}$$

$$E_{Cu^{2+}} = 0.337 - \frac{0.0592}{2} \log \frac{1}{0.0200} = 0.2867\ \text{V}$$

We see from the cell diagram that the silver electrode is the cathode and the copper electrode is the anode. Therefore, application of Equation 16-1 gives

$$E_{cell} = E_{Ag^+} - E_{Cu^{2+}} = 0.6984 - 0.2867 = +0.412\ \text{V}$$

Gustav Robert Kirchhoff (1824–1877) was a German physicist who made many important contributions to physics and chemistry. In addition to his work in spectroscopy, he is known for Kirchhoff's laws of electrical circuits. These laws are shown in the form of equations on this stamp.

EXAMPLE 16-2

Calculate the thermodynamic potential for the cell

$$Ag|Ag^+(0.0200\ \text{M})\|Cu^{2+}(0.0200\ \text{M})|Cu$$

The electrode potentials for the two half-reactions are identical to the electrode potentials calculated in Example 16-1. That is,

$$E_{Ag^+} = 0.6984\ \text{V} \qquad \text{and} \qquad E_{Cu^{2+}} = 0.2867\ \text{V}$$

In contrast to the previous example, the silver electrode is now the anode and the copper electrode the cathode. Substituting these electrode potentials into Equation 16-1 gives

$$E_{cell} = E_{Cu^{2+}} - E_{Ag^+} = 0.2867 - 0.6984 = -0.412\ \text{V}$$

In order to develop a current in the electrolytic cell in Example 16-2, it would be necessary to apply a potential that was more negative than -0.412 V. See Figure 15-2b.

Examples 16-1 and 16-2 illustrate an important fact. The cell potential computed from Equation 16-1 is positive for a galvanic cell and negative for an electrolytic cell.

If you calculate a positive cell potential, you know that the cell reaction proceeds just as it is written. If you calculate a negative cell potential, the cell reaction proceeds in the opposite direction. That is, the cell reaction must be forced to take place as it is written by applying an external potential opposite in sense and greater in magnitude than the calculated cell potential. Cells with positive calculated potentials are galvanic cells. Those with negative potentials are called electrolytic cells because they must be driven by an external source of energy to carry out the electrolysis.

EXAMPLE 16-3

Calculate the thermodynamic potential of the following cell and indicate whether it is galvanic or electrolytic (see Figure 16-1).

$$Pt|UO_2^{2+}(0.0150\ \text{M}),U^{4+}(0.200\ \text{M}),H^+(0.0300\ \text{M})\|$$
$$Fe^{2+}(0.0100\ \text{M}),Fe^{3+}(0.0250\ \text{M})|Pt$$

The two half-reactions are

$$Fe^{3+} + e^- \rightleftharpoons Fe^{2+} \qquad E^0 = +0.771\ \text{V}$$

$$UO_2^{2+} + 4H^+ + 2e^- \rightleftharpoons U^{4+} + 2H_2O \qquad E^0 = +0.334\ \text{V}$$

The *electrode potential* for the cathode is

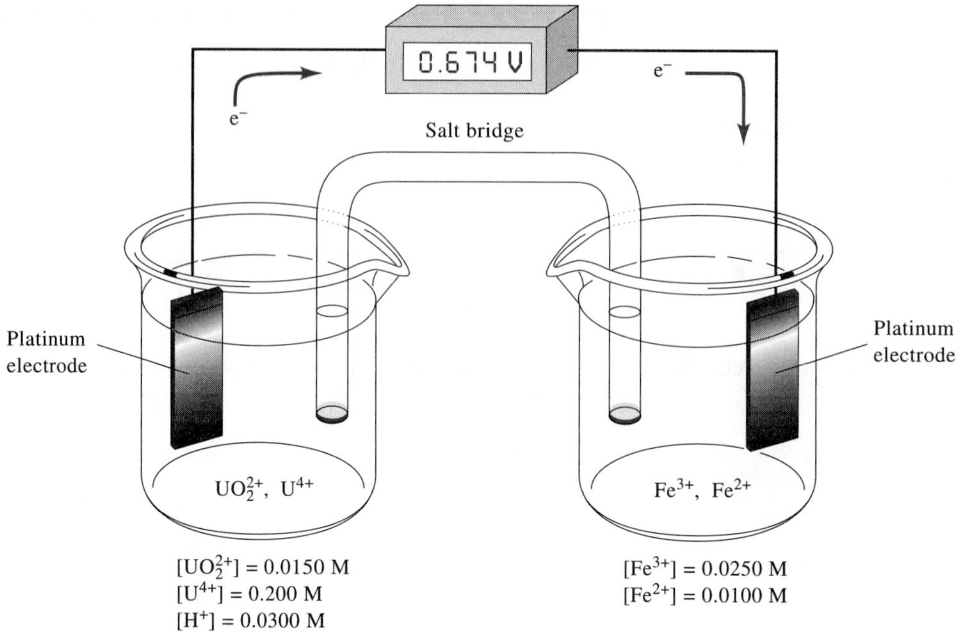

Figure 16-1
Cell for Example 16-3.

$$E_{cathode} = 0.771 - 0.0592 \log \frac{[Fe^{2+}]}{[Fe^{3+}]}$$

$$= 0.771 - 0.0592 \log \frac{0.0100}{0.0250} = 0.799 - (-0.0236)$$

$$= 0.7946 \text{ V}$$

The *electrode potential* for the anode is

$$E_{anode} = 0.334 - \frac{0.0592}{2} \log \frac{[U^{4+}]}{[UO_2^{2+}][H^+]^4}$$

$$= 0.334 - \frac{0.0592}{2} \log \frac{0.200}{(0.0150)(0.0300)^4}$$

$$= 0.334 - 0.2136 = 0.1204 \text{ V}$$

and

$$E_{cell} = E_{cathode} - E_{anode} = 0.7946 - 0.1204 = 0.674 \text{ V}$$

The positive sign means that the cell is galvanic.

EXAMPLE 16-4

Calculate the theoretical potential for the cell

$$Ag|AgCl(sat'd),HCl(0.0200\text{ M})|H_2(0.800\text{ atm}),Pt$$

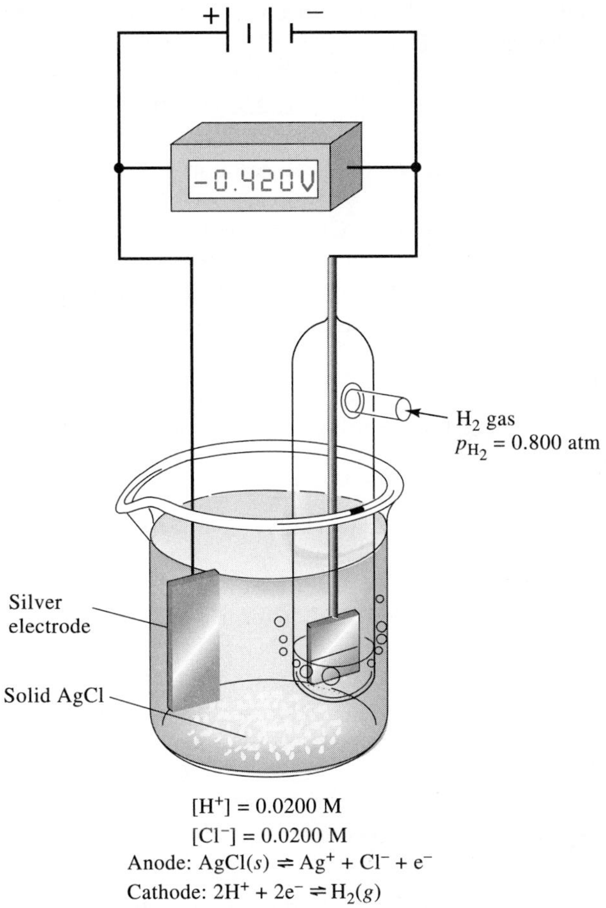

Figure 16-2
Cell without liquid junction for Example 16-4.

Note that this cell does not require two compartments (nor a salt bridge) because molecular H_2 has little tendency to react directly with the low concentration of Ag^+ in the electrolyte solution. This is an example of a cell *without liquid junction* (see Figure 16-2).

The two half-reactions and their corresponding standard electrode potentials are (Table 15-1)

$$2H^+ + 2e^- \rightleftharpoons H_2(g) \qquad E^0_{H^+} = 0.000 \text{ V}$$

$$AgCl(s) + e^- \rightleftharpoons Ag(s) + Cl^- \qquad E^0_{AgCl} = 0.222 \text{ V}$$

The two electrode potentials are

$$E_{cathode} = 0.000 - \frac{0.0592}{2} \log \frac{p_{H_2}}{[H^+]^2} = \frac{0.0592}{2} \log \frac{0.800}{(0.0200)^2}$$

$$= -0.0977 \text{ V}$$

$$E_{anode} = E^0_{AgCl} - \frac{0.0592}{1} \log [Cl^-] = 0.222 - \frac{0.0592}{1} \log 0.0200$$

$$= 0.3226 \text{ V}$$

The cell diagram specifies the silver electrode as the anode and the hydrogen electrode as the cathode. Thus,

$$E_{cell} = -0.0977 - 0.3226 = -0.420 \text{ V}$$

The negative sign indicates that the cell reaction

$$2H^+ + 2Ag(s) + 2Cl^- \rightleftharpoons H_2(g) + 2AgCl(s)$$

is nonspontaneous, and thus the cell is electrolytic. It would therefore require an external power source for operation in this way.

EXAMPLE 16-5

Calculate the potential for the following cell employing (a) concentrations and (b) activities:

$$Zn|ZnSO_4(x \text{ M}),PbSO_4(\text{sat'd})|Pb$$

where $x = 5.00 \times 10^{-4}$, 2.00×10^{-3}, 1.00×10^{-2}, 5.00×10^{-2}.

(a) In a neutral solution, little HSO_4^- is formed and we can assume that

$$[SO_4^{2-}] = c_{ZnSO_4} = x = 5.00 \times 10^{-4}$$

The half-reactions and standard potentials are

$$PbSO_4(s) + 2e^- \rightleftharpoons Pb(s) + SO_4^{2-} \qquad E^0 = -0.350 \text{ V}$$

$$Zn^{2+} + 2e^- \rightleftharpoons Zn \qquad E^0 = -0.763 \text{ V}$$

The electrode potential for the lead electrode half-reaction is

$$E_{PbSO_4} = E^0_{PbSO_4} = \frac{0.0592}{2} \log [SO_4^{2-}]$$

$$= -0.350 - \frac{0.0592}{2} \log (5.00 \times 10^{-4}) = -0.252 \text{ V}$$

For the zinc half-reaction,

$$[Zn^{2+}] = 5.00 \times 10^{-4}$$

$$E_{Zn^{2+}} = -0.763 - \frac{0.0592}{2} \log \frac{1}{5.00 \times 10^{-4}} = -0.860 \text{ V}$$

Since the lead electrode is specified as the cathode,

$$E_{cell} = -0.252 - (-0.860) = 0.608 \text{ V}$$

Cell potentials at the other concentrations can be derived in the same way. Their values are given in Table 16-1.

(b) To obtain activity coefficients for Zn^{2+} and SO_4^{2-}, we must first calculate the ionic strength with the aid of Equation 8-1:

$$\mu = [5.00 \times 10^{-4} \times (2)^2 + 5.00 \times 10^{-4} \times (2)^2]/2 = 2.00 \times 10^{-3}$$

In Table 8-1, we find $\alpha_X = 0.4$ nm for SO_4^{2-} and $\alpha_X = 0.6$ nm for Zn^{2+}. Substituting these values into Equation 8-5 gives for sulfate ion

$$-\log \gamma_{SO_4^{2-}} = \frac{0.51 \times (2)^2 \times \sqrt{2.00 \times 10^{-3}}}{1 + 3.3 \times 0.4\sqrt{2.00 \times 10^{-3}}} = 8.61 \times 10^{-2}$$

$$\gamma_{SO_4^{2-}} = 0.820$$

Repeating the calculations for Zn^{2+}, for which $\alpha_X = 0.60$, yields

$$\gamma_{Zn^{2+}} = 0.825$$

The Nernst equation for the lead electrode now becomes

$$E_{PbSO_4} = E^0_{PbSO_4} - \frac{0.0592}{2} \log (c_{SO_4^{2-}} \times \gamma_{SO_4^{2-}})$$

$$= -0.350 - \frac{0.0592}{2} \log (5.00 \times 10^{-4} \times 0.820) = -0.250$$

Similarly, for the zinc electrode

$$E_{Zn^{2+}} = E^0_{Zn^{2+}} - \frac{0.0592}{2} \log \frac{1}{(c_{Zn^{2+}} \times \gamma_{Zn^{2+}})}$$

$$= -0.763 - \frac{0.0592}{2} \log \frac{1}{5.00 \times 10^{-4} \times 0.825} = -0.863$$

Thus,

$$E_{cell} = -0.250 - (-0.863) = 0.613 \text{ V}$$

Values for other concentrations and experimentally determined potentials for the cell are found in Table 16-1.

It is evident from the data in Table 16-1 that neglect of activity coefficients causes significant errors in calculated cell potentials. It is also evident from the data in the fifth column of this table that potentials computed with activities agree reasonably well with experiment.

EXAMPLE 16-6

Calculate the potential required to initiate deposition of copper from a solution that is 0.010 M in $CuSO_4$ and contains sufficient H_2SO_4 to give a pH of 4.00.

The deposition of copper necessarily occurs at the cathode. Since there is no more easily oxidizable species than water in the system, O_2 will be evolved at the anode. The required standard electrode potentials are (Table 15-1)

TABLE 16-1 Effect of Ionic Strength on the Potential of a Galvanic Cell*

Concentration, $ZnSO_4$	Ionic Strength μ	(a) E, Based on Concentrations	(b) E, Based on Activities	E, Experimental Values†
5.00×10^{-4}	2.00×10^{-3}	0.608	0.613	0.611
2.00×10^{-3}	8.00×10^{-3}	0.573	0.582	0.583
1.00×10^{-2}	4.00×10^{-2}	0.531	0.550	0.553
2.00×10^{-2}	8.00×10^{-2}	0.513	0.537	0.542
5.00×10^{-2}	2.00×10^{-1}	0.490	0.521	0.529

*Cell described in Example 16-5.

†Experimental data from I. A. Cowperthwaite and V. K. LaMer, *J. Amer. Chem. Soc.*, **1931,** *53,* 4333.

$$O_2(g) + 4H^+ + 4e^- \rightleftharpoons 2H_2O \qquad E^0 = +1.229\ V$$

$$Cu^{2+} + 2e^- \rightleftharpoons Cu(s) \qquad E^0 = +0.337\ V$$

The electrode potential for the Cu electrode is given by

$$E = +0.337 - \frac{0.0592}{2}\log\frac{1}{0.010} = +0.278\ V$$

If O_2 is evolved at 1.00 atm, the electrode potential for the oxygen electrode is

$$E = +1.229 - \frac{0.0592}{4}\log\frac{1}{p_{O_2} \times [H^+]^4}$$

$$= +1.229 - \frac{0.0592}{4}\log\frac{1}{(1.00)(1.00 \times 10^{-4})^4} = +0.992\ V$$

and the cell potential is

$$E_{cell} = +0.278 - 0.992 = -0.714\ V$$

Thus, initiation of the reaction

$$2Cu^{2+} + 2H_2O \rightleftharpoons O_2(g) + 4H^+ + 2Cu(s)$$

requires the application of a potential greater than -0.714 V.

16A-1 The Experimental Determination of Standard Potentials

Although standard electrode potentials for hundreds of half-reactions are found in compilations of electrochemical data, it is noteworthy that neither the standard hydrogen electrode nor any other electrode in its standard state can be realized in the laboratory. The standard hydrogen electrode is in fact a *hypothetical elec-*

trode, as is any other electrode system in which the reactants and products are at unit activity or pressure. The reason such electrode systems cannot be prepared experimentally is that chemists lack the knowledge to produce solutions that have ionic activities of exactly unity. That is, no adequate theory exists that permits calculation of the *concentration* of a compound needed to give a solution of unit ionic activity. At such high ionic strengths, the Debye-Hückel relationship (Section 8B-2) is not valid, and no independent experimental method exists for the determination of activity coefficients in such solutions. Thus, for example, the *concentration* of HCl or other acids required to give the unit hydrogen ion activity specified in the standard hydrogen electrode *cannot be calculated or determined experimentally.* Nonetheless, data taken in solutions of low ionic strengths can be extrapolated to give valid measures of standard electrode potentials as theoretically defined. An example of how hypothetical standard electrode potentials might be obtained from experimental data follows.

EXAMPLE 16-7

D. A. MacInnes[1] found that a cell similar to that shown in Figure 16-2 had a potential of 0.52053 V. The cell is described by

$$\text{Pt,H}_2(1.00\text{ atm})|\text{HCl}(3.215\times10^{-3}\text{ M}),\text{AgCl(sat'd)}|\text{Ag}$$

Calculate the standard electrode potential for the half-reaction

$$\text{AgCl}(s) + e^- \rightleftharpoons \text{Ag}(s) + \text{Cl}^-$$

Here, the electrode potential for the cathode is

$$E_{\text{cathode}} = E^0_{\text{AgCl}} - 0.0592 \log c_{\text{HCl}}\gamma_{\text{Cl}^-}$$

where γ_{Cl^-} is the activity coefficient of Cl^-. The second half-cell reaction is

$$\text{H}^+ + e^- \rightleftharpoons \tfrac{1}{2}\text{H}_2(g)$$

and

$$E_{\text{anode}} = E^0_{\text{H}^+} = \frac{0.0592}{1}\log\frac{p_{\text{H}_2}^{1/2}}{c_{\text{HCl}}\gamma_{\text{H}^+}}$$

The measured potential is the difference between these potentials (Equation 16-1):

$$E_{\text{cell}} = E_{\text{cathode}} - E_{\text{anode}}$$

$$E_{\text{cell}} = (E^0_{\text{AgCl}} - 0.0592\log c_{\text{HCl}}\gamma_{\text{Cl}^-}) - \left(E^0_{\text{H}^+} - 0.0592\log\frac{p_{\text{H}_2}^{1/2}}{c_{\text{HCl}}\gamma_{\text{H}^+}}\right)$$

$$= E^0_{\text{AgCl}} - 0.0592\log c_{\text{HCl}}\gamma_{\text{Cl}^-} - 0.000 - 0.0592\log\frac{c_{\text{HCl}}\gamma_{\text{H}^+}}{p_{\text{H}_2}^{1/2}}$$

[1]D. A. MacInnes, *The Principles of Electrochemistry,* p. 187. New York: Reinhold, 1939.

Note that we have inverted the terms in the second logarithmic relationship. Combining the two logarithmic terms and substituting the experimental value for E_{cell} gives

$$E_{cell} = 0.52053 = E^0_{AgCl} - 0.0592 \log \frac{c^2_{HCl}\gamma_{H^+}\gamma_{Cl^-}}{p^{1/2}_{H_2}}$$

The activity coefficients for H^+ and Cl^- can be calculated from Equation 8-5 using 3.215×10^{-3} M for the ionic strength μ. These values are 0.945 and 0.939, respectively. Substitution of these activity coefficients and the experimental data into the foregoing equation gives, upon rearrangement,

$$E^0_{AgCl} = 0.52053 + 0.0592 \log \frac{(3.215 \times 10^{-3})^2(0.945)(0.939)}{1.00^{1/2}}$$

$$= 0.2223 \approx 0.222 \text{ V}$$

(The mean for this and similar measurements at other concentrations was 0.222 V.)

16B CALCULATION OF REDOX EQUILIBRIUM CONSTANTS

Let us again consider the equilibrium that is established when a piece of copper is immersed in a dilute solution of silver nitrate:

$$Cu(s) + 2Ag^+ \rightleftharpoons Cu^{2+} + 2Ag(s) \qquad \textbf{(16-4)}$$

The equilibrium constant for this reaction is

$$K_{eq} = \frac{[Cu^{2+}]}{[Ag^+]^2} \qquad \textbf{(16-5)}$$

As we showed in Example 16-1, this reaction can be carried out in the cell

$$Cu|Cu^{2+}(x\text{M})\|Ag^+(y\text{M})|Ag$$

A sketch of a cell similar to this is shown in Figure 15-2a. Its cell potential at any instant is given by Equation 16-1:

$$E_{cell} = E_{cathode} - E_{anode} = E_{Ag^+} - E_{Cu^{2+}}$$

As the reaction proceeds, the concentration of Cu(II) ions increases and the concentration of Ag(I) ions decreases. These changes make the potential of the copper electrode more positive and that of the silver electrode less positive. As shown in Figure 15-5, the net effect of these changes is a continuous decrease in the potential of the cell as it discharges. Ultimately, the concentrations of Cu(II) and Ag(I) attain their equilibrium values as determined by Equation 16-5, and the current ceases. Under these conditions, *the potential of the cell becomes zero. Thus, at chemical equilibrium,* we may write

$$E_{cell} = 0 = E_{cathode} - E_{anode} = E_{Ag^+} - E_{Cu^{2+}}$$

or

$$E_{cathode} = E_{anode} = E_{Ag^+} = E_{Cu^{2+}} \tag{16-6}$$

We can generalize Equation 16-6 by stating that *at equilibrium, the electrode potentials for all half-reactions in an oxidation/reduction system are equal.* This generalization applies regardless of the number of half-reactions present in the system because interactions among *all* must take place until the electrode potentials are identical. For example, if we have four oxidation/reduction systems in a solution, interaction among all four takes place until

Remember that *when redox systems are at equilibrium, the electrode potentials of all systems that are present are identical.* This generality applies whether the reactions take place directly in solution or indirectly in a galvanic cell.

$$E_{Ox_1} = E_{Ox_2} = E_{Ox_3} = E_{Ox_4} \tag{16-7}$$

where E_{Ox_1}, E_{Ox_2}, E_{Ox_3}, and E_{Ox_4} are the electrode potentials for the four half-reactions.

Returning to the reaction shown in Equation 16-4, let us substitute Nernst expressions for the two electrode potentials in Equation 16-6, which gives

$$E^0_{Ag^+} - \frac{0.0592}{2}\log\frac{1}{[Ag^+]^2} = E^0_{Cu^{2+}} - \frac{0.0592}{2}\log\frac{1}{[Cu^{2+}]} \tag{16-8}$$

It is important to note that the Nernst equation is applied to the silver half-reaction as it appears in the balanced equation (Equation 16-2):

$$2Ag^+ + 2e^- \rightleftharpoons 2Ag(s) \qquad E^0 = 0.799\text{ V}$$

Rearrangement of Equation 16-8 gives

$$E^0_{Ag^+} - E^0_{Cu^{2+}} = \frac{0.0592}{2}\log\frac{1}{[Ag^+]^2} - \frac{0.0592}{2}\log\frac{1}{[Cu^{2+}]}$$

Inverting the second ratio and changing the sign of the log gives

$$E^0_{Ag^+} - E^0_{Cu^{2+}} = \frac{0.0592}{2}\log\frac{1}{[Ag^+]^2} + \frac{0.0592}{2}\log\frac{[Cu^{2+}]}{1}$$

Finally, combining the log terms and rearranging gives

$$\frac{2(E^0_{Ag^+} - E^0_{Cu^{2+}})}{0.0592} = \log\frac{[Cu^{2+}]}{[Ag^+]^2} = \log K_{eq} \tag{16-9}$$

The concentration terms in Equation 16-9 are *equilibrium concentrations; the ratio $[Cu^{2+}]/[Ag^+]^2$ in the logarithmic term is therefore the equilibrium constant for the reaction.*

In making calculations of the sort shown in Example 16-8, you should follow the rounding rule for antilogs that is given on page 41.

EXAMPLE 16-8

Calculate the equilibrium constant for the reaction shown in Equation 16-4.

Substituting numerical values into Equation 16-9 yields

$$\log K_{eq} = \log \frac{[Cu^{2+}]}{[Ag^+]^2} = \frac{2(0.799 - 0.337)}{0.0592}$$

$$= 15.61$$

$$K_{eq} = \text{antilog } 15.61 = 4.1 \times 10^{15}$$

EXAMPLE 16-9

Calculate the equilibrium constant for the reaction

$$2Fe^{3+} + 3I^- \rightleftharpoons 2Fe^{2+} + I_3^-$$

In Appendix 5, we find

$$2Fe^{3+} + 2e^- \rightleftharpoons 2Fe^{3+} \qquad 0.771 \text{ V}$$

$$I_3^- + 2e^- \rightleftharpoons 3I^- \qquad 0.536 \text{ V}$$

We have multiplied the first half-reaction by 2 so that the number of moles of Fe^{3+} and Fe^{2+} are the same as in the balanced overall equation. We write the Nernst equation for Fe^{3+} based upon the half-reaction for a 2-electron transfer. That is,

$$E_{Fe^{3+}} = E^0_{Fe^{3+}} - \frac{0.0592}{2} \log \frac{[Fe^{2+}]^2}{[Fe^{3+}]^2}$$

and

$$E_{I_3^-} = E^0_{I_3^-} - \frac{0.0592}{2} \log \frac{[I^-]^3}{[I_3^-]}$$

At equilibrium, the electrode potentials are equal and

$$E_{Fe^{3+}} = E_{I_3^-}$$

$$E^0_{Fe^{3+}} - \frac{0.0592}{2} \log \frac{[Fe^{2+}]^2}{[Fe^{3+}]^2} = E^0_{I_3^-} - \frac{0.0592}{2} \log \frac{[I^-]^3}{[I_3^-]}$$

This equation rearranges to

$$\frac{2(E^0_{Fe^{3+}} - E^0_{I_3^-})}{0.0592} = \log \frac{[Fe^{2+}]^2}{[Fe^{3+}]^2} - \log \frac{[I^-]^3}{[I_3^-]} = \log \frac{[Fe^{2+}]^2}{[Fe^{3+}]^2} + \log \frac{[I_3^-]}{[I^-]^3}$$

$$= \log \frac{[Fe^{2+}]^2[I_3^-]}{[Fe^{3+}]^2[I^-]^3}$$

Notice that we have changed the sign of the second logarithmic term by inverting the fraction. Further arrangement gives

$$\log \frac{[Fe^{2+}]^2[I_3^-]}{[Fe^{3+}]^2[I^-]^3} = \frac{2(E^0_{Fe^{3+}} - E^0_{I_3^-})}{0.0592}$$

Recall, however, that here the concentration terms are *equilibrium concentrations* and

$$\log K_{eq} = \frac{2(E^0_{Fe^{3+}} - E^0_{I_3^-})}{0.0592} = \frac{2(0.771 - 0.536)}{0.0592} = 7.939$$

$$K_{eq} = \text{antilog } 7.94 = 8.7 \times 10^7$$

We round the answer to two figures because log K_{eq} contains only 2 significant figures (the two to the right of the decimal point).

EXAMPLE 16-10

Calculate the equilibrium constant for the reaction

$$2MnO_4^- + 3Mn^{2+} + 2H_2O \rightleftharpoons 5MnO_2(s) + 4H^+$$

In Appendix 5 we find

$$2MnO_4^- + 8H^+ + 6e^- \rightleftharpoons 2MnO_2(s) + 4H_2O \qquad E^0 = +1.695 \text{ V}$$

$$3MnO_2(s) + 12H^+ + 6e^- \rightleftharpoons 3Mn^{2+} + 6H_2O \qquad E^0 = +1.23 \text{ V}$$

Again, we have multiplied both equations by integers so that the number of electrons are equal. When this system is at equilibrium,

$$E_{MnO_4^-} = E_{MnO_2}$$

$$1.695 - \frac{0.0592}{6} \log \frac{1}{[MnO_4^-]^2[H^+]^8} = 1.23 - \frac{0.0592}{6} \log \frac{[Mn^{2+}]^3}{[H^+]^{12}}$$

Inverting the log term on the right and rearranging lead to

$$\frac{6(1.695 - 1.23)}{0.0592} = \log \frac{1}{[MnO_4^-]^2[H^+]^8} + \log \frac{[H^+]^{12}}{[Mn^{2+}]}$$

Adding the two log terms gives

$$\frac{6(1.695 - 1.23)}{0.0592} = \log \frac{[H^+]^{12}}{[MnO_4^-]^2[Mn^{2+}]^3[H^+]^8}$$

$$47.1 = \log \frac{[H^+]^4}{[MnO_4^-]^2[Mn^{2+}]^3} = \log K_{eq}$$

$$K_{eq} = \text{antilog } 47.1 = 1 \times 10^{47}$$

Note that the final result has only one significant figure.

FEATURE 16-1
A General Expression for Calculating Equilibrium Constants from Standard Potentials

In order to derive a general relationship for computing equilibrium constants from standard-potential data, let us consider a reaction in which a species A_{red} reacts with a species B_{ox} to yield A_{ox} and B_{red}. The two electrode reactions are

$$A_{ox} + ae^- \rightleftharpoons A_{red}$$
$$B_{ox} + be^- \rightleftharpoons B_{red}$$

Note that the product ab is the total number of electrons gained in the reduction (and lost in the oxidation) for the balanced redox equation. Thus, if $a = b$, it is not necessary to multiply the half-reactions by a and b. If $a = b = n$, the equilibrium constant is determined from

$$\log K_{eq} = \frac{n\,(E^0_B - E^0_A)}{0.0592}$$

In order to obtain a balanced equation for the desired reaction, we must multiply the first equation by b and the second by a to give

$$bA_{ox} + bae^- \rightleftharpoons bA_{red}$$
$$aB_{ox} + bae^- \rightleftharpoons aB_{red}$$

Subtraction of the first equation from the second yields a balanced equation for the redox reaction

$$bA_{red} + aB_{ox} \rightleftharpoons bA_{ox} + aB_{red}$$

When this system is in equilibrium, the two electrode potentials E_A and E_B are numerically identical, that is,

$$E_A = E_B$$

Substitution of the Nernst expression into this equation reveals that, *at equilibrium,*

$$E^0_A - \frac{0.0592}{ab}\log\frac{[A_{red}]^b}{[A_{ox}]^b} = E^0_B - \frac{0.0592}{ab}\log\frac{[B_{red}]^a}{[B_{ox}]^a}$$

which rearranges to

$$E^0_B - E^0_A = \frac{0.0592}{ab}\log\frac{[A_{ox}]^b[B_{red}]^a}{[A_{red}]^b[B_{ox}]^a} = \frac{0.0592}{ab}\log K_{eq}$$

Finally, then,

$$\log K_{eq} = \frac{ab(E^0_B - E^0_A)}{0.0592} \qquad (16\text{-}10)$$

16C REDOX TITRATION CURVES

Because most redox indicators respond to changes in electrode potential, the vertical axis in oxidation/reduction titration curves is generally an electrode potential instead of the logarithmic p-functions that were used for precipitation, complex-formation, and neutralization titration curves. We have seen in Chapter 15 that there is a logarithmic relationship between electrode potential and concentration of the analyte or titrant; as a result, the redox titration curves are similar in appearance to those for other types of titrations.

See *Mathcad Applications for Analytical Chemistry,* **pp. 107–109** for a different approach to redox titration curves.

16C-1 Electrode Potentials During Redox Titrations

Let us now consider the redox titration of iron(II) with a standard solution of cerium(IV). This reaction is widely used to determine iron in various kinds of samples. The titration is described by the equation

$$Fe^{2+} + Ce^{4+} \rightleftharpoons Fe^{3+} + Ce^{3+}$$

This reaction is rapid and reversible, so the system is at equilibrium at all times throughout the titration. Consequently, the electrode potentials for the two half-reactions are always identical (Equation 16-7); that is,

Remember that *at equilibrium, the electrode potentials of all systems are identical.* This generality applies whether the reactions take place directly in solution or indirectly in a galvanic cell.

$$E_{Ce^{4+}} = E_{Fe^{3+}} = E_{system}$$

where we have termed E_{system} as *the potential of the system.* If a redox indicator has been added to this solution, the ratio of the concentrations of its oxidized and reduced forms must adjust so that the electrode potential for the indicator, E_{In}, is also equal to the system potential; thus, employing Equation 16-7, we may write

$$E_{In} = E_{Ce^{4+}} = E_{Fe^{3+}} = E_{system}$$

The electrode potential of a system is readily derived from standard potential data. Thus, for the reaction under consideration, the titration mixture is treated as if it were part of the hypothetical cell

$$\text{SHE}\|Ce^{4+},Ce^{3+},Fe^{3+},Fe^{2+}|\text{Pt}$$

where SHE symbolizes the standard hydrogen electrode. The potential of the platinum electrode with respect to the standard hydrogen electrode is determined by the tendencies of iron(III) and cerium(IV) to accept electrons—that is, by the tendencies of the following half-reactions to occur:

$$Fe^{3+} + e^- \rightleftharpoons Fe^{2+}$$
$$Ce^{4+} + e^- \rightleftharpoons Ce^{3+}$$

At equilibrium, the concentration ratios of the oxidized and reduced forms of the two species are such that their attraction for electrons (and thus their electrode potentials) are identical. Note that these concentration ratios vary continuously

Most end points in oxidation/reduction titrations are based upon the rapid changes in E_{system} that occur at or near chemical equivalence.

Before the equivalence point, E_{system} calculations are easiest to make using the Nernst equation for the analyte. After the equivalence point, the Nernst equation for the reagent is more convenient.

throughout the titration, as must E_{system}. End points are determined from the characteristic variation in E_{system} that occurs during the titration.

Because $E_{system} = E_{Fe^{3+}} = E_{Ce^{4+}}$, data for a titration curve can be obtained by applying the Nernst equation for *either* the cerium(IV) half-reaction or the iron(III) half-reaction. It turns out, however, that one or the other will be more convenient, depending upon the stage of the titration. For example, the iron(III) potential is easier to compute in the region short of the equivalence point because here the species concentrations of iron(II) and iron(III) are appreciable and are approximately equal to their analytical concentrations. In contrast, the concentration of cerium(IV), which is negligible prior to the equivalence point because of the large excess of iron(II), can be obtained at this stage only by calculations based upon the equilibrium constant for the reaction. Beyond the equivalence point, the analytical concentrations of cerium(IV) and cerium(III) are readily computed directly from the volumetric data, whereas that for iron(II) is not. In this region, then, the cerium(IV) electrode potential is the easier to use. Equivalence-point potentials are derived by the method shown in the next section.

Equivalence-Point Potentials

At the equivalence point, the concentrations of cerium(IV) and iron(II) are minute and cannot be obtained from the stoichiometry of the reaction. Fortunately, equivalence-point potentials are readily obtained by taking advantage of the fact that the two reactant species and the two product species have known concentration ratios at chemical equivalence.

At the equivalence point in the titration of iron(II) with cerium(IV), the potential of the system E_{eq} is given by both

$$E_{eq} = E^0_{Ce^{4+}} - \frac{0.0592}{1} \log \frac{[Ce^{3+}]}{[Ce^{4+}]}$$

and

$$E_{eq} = E^0_{Fe^{3+}} - \frac{0.0592}{1} \log \frac{[Fe^{2+}]}{[Fe^{3+}]}$$

Adding these two expressions gives

The concentration quotient in Equation 16-11 is *not* the usual ratio of product concentrations and reactant concentrations that appear in equilibrium-constant expressions.

$$2E_{eq} = E^0_{Ce^{4+}} + E^0_{Fe^{3+}} - \frac{0.0592}{1} \log \frac{[Ce^{3+}][Fe^{2+}]}{[Ce^{4+}][Fe^{3+}]} \tag{16-11}$$

The definition of equivalence point requires that

$$[Fe^{3+}] = [Ce^{3+}]$$
$$[Fe^{2+}] = [Ce^{4+}]$$

Substitution of these equalities into Equation 16-11 results in the concentration quotient becoming unity and the logarithmic term becoming zero:

$$2E_{eq} = E^0_{Ce^{4+}} + E^0_{Fe^{3+}} - \frac{0.0592}{1} \log \frac{\cancel{[Ce^{3+}]}\cancel{[Ce^{4+}]}}{\cancel{[Ce^{4+}]}\cancel{[Ce^{3+}]}} = E^0_{Ce^{4+}} + E^0_{Fe^{3+}}$$

Thus,

$$E_{eq} = \frac{E^0_{Ce^{4+}} + E^0_{Fe^{3+}}}{2} \qquad (16\text{-}12)$$

Example 16-11 illustrates how the equivalence-point potential is derived for a more complex reaction.

EXAMPLE 16-11

Derive an expression for the equivalence-point potential in the titration of 0.0500 M U^{4+} with 0.1000 M Ce^{4+}. Assume both solutions are 1.0 M in H_2SO_4.

$$U^{4+} + 2Ce^{4+} + 2H_2O \rightleftharpoons UO_2^{2+} + 2Ce^{3+} + 4H^+$$

In Appendix 5, we find

$$UO_2^{2+} + 4H^+ + 2e^- \longrightarrow U^{4+} + 2H_2O \qquad E^0 = 0.334\ \text{V}$$

$$Ce^{4+} + e^- \rightleftharpoons Ce^{3+} \qquad E^{0\prime} = 1.44\ \text{V}$$

Here we use the formal potential for Ce^{4+} in 1.0 M H_2SO_4.

Proceeding as in the cerium(IV)/iron(II) equivalence-point calculation we write

$$E_{eq} = E^0_{UO_2^{2+}} - \frac{0.0592}{2} \log \frac{[U^{4+}]}{[UO_2^{2+}][H^+]^4}$$

$$E_{eq} = E^{0\prime}_{Ce^{4+}} - 0.0592 \log \frac{[Ce^{3+}]}{[Ce^{4+}]}$$

In order to combine the log terms we must multiply the first equation by 2 to give

$$2E_{eq} = 2E^0_{UO_2^{2+}} - 0.0592 \log \frac{[U^{4+}]}{[UO_2^{2+}][H^+]^4}$$

Adding this to the previous equation leads to

$$3E_{eq} = 2E^0_{UO_2^{2+}} + E^{0\prime}_{Ce^{4+}} - 0.0592 \log \frac{[U^{4+}][Ce^{3+}]}{[UO_2^{2+}][Ce^{4+}][H^+]^4}$$

But at equivalence

$$[U^{4+}] = [Ce^{4+}]/2$$

and

$$[UO_2^{2+}] = [Ce^{3+}]/2$$

Substituting these equations gives upon rearranging

$$E_{eq} = \frac{2E^0_{UO_2^{2+}} + E^{0'}_{Ce^{4+}}}{3} - \frac{0.0592}{3}\log\frac{\cancel{2}\cancel{[Ce^{4+}]}\cancel{[Ce^{3+}]}}{\cancel{2}\cancel{[Ce^{3+}]}\cancel{[Ce^{4+}]}[H^+]^4}$$

$$= \frac{2E^0_{UO_2^{2+}} + E^{0'}_{Ce^{4+}}}{3} - \frac{0.0592}{3}\log\frac{1}{[H^+]^4}$$

We see that the equivalence-point potential is pH dependent in this titration.

16C-2 Derivation of Titration Curves

See *Mathcad Applications for Analytical Chemistry,* **pp. 109–110.**

Let us first consider the titration of 50.00 mL of 0.0500 M Fe^{2+} with 0.1000 M Ce^{4+} in a medium that is 1.0 M in H_2SO_4 at all times. Formal potential data for both half-cell processes are available in Appendix 5 and are used for these calculations. That is,

$$Ce^{4+} + e^- \rightleftharpoons Ce^{3+} \qquad E^{0'} = 1.44\ \text{V}\ (1\ \text{M}\ H_2SO_4)$$

$$Fe^{3+} + e^- \rightleftharpoons Fe^{2+} \qquad E^{0'} = 0.68\ \text{V}\ (1\ \text{M}\ H_2SO_4)$$

Initial Potential

The solution contains no cerium species at the outset. In all likelihood, there is a small but unknown amount of Fe^{3+} present due to air oxidation of Fe^{2+}. In any event, we lack sufficient information to calculate an initial potential.

Potential after the Addition of 5.00 mL of Cerium(IV)

Remember, the equation for this reaction is $Fe^{2+} + Ce^{4+} \rightleftharpoons Fe^{3+} + Ce^{3+}$

With the first addition of titrant, the solutions will contain readily calculated amounts of the three species, Fe^{2+}, Fe^{3+}, and Ce^{3+}; that of the fourth, Ce^{4+}, is vanishingly small. Therefore, it is more convenient to use the concentrations of the two iron species to calculate the electrode potential of the system.

The equilibrium concentration of Fe(III) is equal to its analytical concentration less the equilibrium concentration of the unreacted Ce(IV):

$$[Fe^{3+}] = \frac{5.00 \times 0.100}{50.00 + 5.00} - [Ce^{4+}] = \frac{0.500}{55.00} - [Ce^{4+}]$$

Similarly, the Fe^{2+} concentration is given by its molarity plus the equilibrium concentration of unreacted $[Ce^{4+}]$:

$$[Fe^{2+}] = \frac{50.00 \times 0.0500 - 5.00 \times 0.1000}{55.00} + [Ce^{4+}] = \frac{2.00}{55.00} + [Ce^{4+}]$$

Generally, the reactions used in redox titrimetry are sufficiently complete that the equilibrium concentration of one of the species (here $[Ce^{4+}]$) is minuscule with respect to the other species present in the solution. Thus, the foregoing two equations can be simplified to

$$[Fe^{3+}] \approx \frac{0.500}{55.00} \qquad \text{and} \qquad [Fe^{2+}] \approx \frac{2.00}{55.00}$$

Substitution for $[Fe^{2+}]$ and $[Fe^{3+}]$ in the Nernst equation gives

$$E_{system} = +0.68 - \frac{0.0592}{1} \log \frac{2.00/\cancel{55.00}}{0.500/\cancel{55.00}} = 0.64 \text{ V}$$

Note that the volumes in the numerator and denominator cancel, which indicates that the potential is independent of dilution. This independence persists until the solution becomes so dilute that the two assumptions made in the calculation become invalid.

The quantity $[Ce^{4+}]$ is equal to the concentration of $[Fe^{2+}]$ that *does not react* with Ce^{4+}. Thus, $[Fe^{3+}]$ must be decreased by this amount and $[Fe^{2+}]$ must be increased by this amount. $[Ce^{4+}]$ is ordinarily so small that we can neglect it in both cases.

It is worth emphasizing again that the use of the Nernst equation for the Ce(IV)/Ce(III) system would yield the same value for E_{system}, but to do so would require the extra step of computing $[Ce^{4+}]$ by means of the equilibrium constant for the reaction.

Additional potentials needed to define the titration curve short of the equivalence point can be obtained similarly. Such data are given in Table 16-2. You may want to confirm one or two of these values.

Equivalence-Point Potential

Substitution of the two formal potentials into Equation 16-11 yields

$$E_{eq} = \frac{E^{0'}_{Ce^{4+}} + E^{0'}_{Fe^{3+}}}{2} = \frac{1.44 + 0.68}{2} = 1.06 \text{ V}$$

Potential after the Addition of 25.10 mL of Cerium(IV)

The molar concentrations of Ce(III), Ce(IV), and Fe(III) are readily computed at this point but that for Fe(II) is not. Therefore, E_{system} computations based on the

TABLE 16-2 Electrode Potential versus SHE in Titrations with 0.1000 M Ce^{4+}

Reagent Volume, mL	Potential, V vs. SHE: **50.00 mL of 0.05000 M Fe^{2+}**		Potential, V vs. SHE: **50.00 mL of 0.02500 M U^{4+}***
5.00	0.64		0.316
15.00	0.69		0.339
20.00	0.72		0.352
24.00	0.76		0.375
24.90	0.82	Equivalence	0.405
25.00	1.06 ⟵	point	⟶ 0.703
25.10	1.30		1.30
26.00	1.36		1.36
30.00	1.40		1.40

*H_2SO_4 concentration is such that $[H^+] = 1.0$ throughout.

cerium half-reaction are more convenient. The concentrations of the two cerium ion species are

$$[Ce^{3+}] = \frac{25.00 \times 0.1000}{75.10} - [Fe^{2+}] \approx \frac{2.500}{75.10}$$

$$[Ce^{4+}] = \frac{25.10 \times 0.1000 - 50.00 \times 0.0500}{75.10} + [Fe^{2+}] \approx \frac{0.010}{75.10}$$

Here, the iron(II) concentration is negligible with respect to the analytical concentrations of the two cerium species. Substitution into the Nernst equation for the cerium couple gives

$$E = +1.44 - \frac{0.0592}{1} \log \frac{[Ce^{3+}]}{[Ce^{4+}]} = +1.44 - \frac{0.0592}{1} \log \frac{2.500/\cancel{75.10}}{0.010/\cancel{75.10}}$$

$$= +1.30 \text{ V}$$

The additional postequivalence potentials in Table 16-2 were derived in a similar fashion.

In contrast to the titration curves we have seen so far, oxidation/reduction curves are *independent* of the concentration of the reactants except when the solution becomes very dilute.

Why is it impossible to calculate the potential of the system before titrant is added?

The titration curve of iron(II) with cerium(IV) appears as *A* in Figure 16-3. This plot resembles closely the curves encountered in neutralization, precipitation, and complex-formation titrations, with the equivalence point being signaled by a rapid change in the ordinate function. A titration involving 0.00500 M iron(II) and 0.01000 M cerium(IV) yields a curve that, for all practical purposes, is identical to the one we have derived, since the electrode potential of the system is independent of dilution.

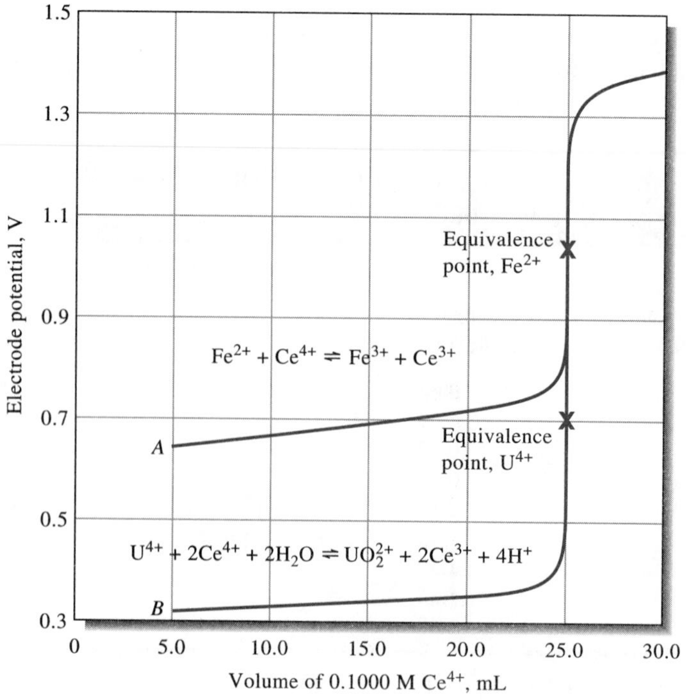

Figure 16-3
Titration curves for 0.1000 M Ce^{4+} titration. *A*: Titration of 50.00 mL of 0.05000 M Fe^{2+}. *B*: Titration of 50.00 mL of 0.02500 M U^{4+}.

EXAMPLE 16-12

Derive a curve for the titration of 50.00 mL of 0.02500 M U^{4+} with 0.1000 M Ce^{4+}. The solution is 1.0 M in H_2SO_4 throughout the titration (for the sake of simplicity, assume that $[H^+]$ for this solution is also about 1.0 M).

The analytical reaction is

$$U^{4+} + 2H_2O + 2Ce^{4+} \rightleftharpoons UO_2^{2+} + 2Ce^{3+} + 4H^+$$

and in Appendix 5 we find

$$Ce^{4+} + e^- \rightleftharpoons Ce^{3+} \qquad E^{0'} = +1.44 \text{ V}$$

$$UO_2^{2+} + 4H^+ + 2e^- \rightleftharpoons U^{4+} + 2H_2O \qquad E^0 = +0.334 \text{ V}$$

Potential after Adding 5.00 mL of Ce^{4+}

$$\text{original amount } U^{4+} = 50.00 \cancel{\text{mL } U^{4+}} \times 0.02500 \frac{\text{mmol } U^{4+}}{\cancel{\text{mL } U^{4+}}}$$

$$= 1.250 \text{ mmol } U^{4+}$$

$$\text{amount } Ce^{4+} \text{ added} = 5.00 \cancel{\text{mL } Ce^{4+}} \times 0.1000 \frac{\text{mmol } Ce^{4+}}{\cancel{\text{mL } Ce^{4+}}}$$

$$= 0.5000 \text{ mmol } Ce^{4+}$$

$$\text{amount } UO_2^{2+} \text{ formed} = 0.5000 \cancel{\text{mmol } Ce^{4+}} \times \frac{1 \text{ mmol } UO_2^{2+}}{2 \cancel{\text{mmol } Ce^{4+}}}$$

$$= 0.2500 \text{ mmol } UO_2^{2+}$$

$$\text{amount } U^{4+} \text{ remaining} = 1.250 \text{ mmol } U^{4+} - 0.2500 \cancel{\text{mmol } UO_2^{2+}}$$

$$\times \frac{1 \text{ mmol } U^{4+}}{\cancel{\text{mmol } UO_2^{2+}}} = 1.000 \text{ mmol } U^{4+}$$

$$\text{Total volume of solution} = (50.00 + 5.00) \text{ mL} = 55.00 \text{ mL}$$

Applying the Nernst equation for UO_2^{2+}, we obtain

$$E = 0.334 - \frac{0.0592}{2} \log \frac{[U^{4+}]}{[UO_2^{2+}][H^+]^4}$$

$$= 0.334 - \frac{0.0592}{2} \log \frac{[U^{4+}]}{[UO_2^{2+}](1.00)^4}$$

Substituting concentrations of the two uranium species gives

$$E = 0.334 - \frac{0.0592}{2} \log \frac{1.000 \text{ mmol } U^{4+}/\cancel{55.00 \text{ mL}}}{0.2500 \text{ mmol } UO_2^{2+}/\cancel{55.00 \text{ mL}}}$$

$$= 0.316 \text{ V}$$

Other preequivalence-point data, calculated in the same way, are given in the third column in Table 16-2.

Equivalence-Point Potential

Following the procedure shown in Example 16-11, we obtain

$$E_{eq} = \frac{(2E^0_{UO_2^{2+}} + E^{0'}_{Ce^{4+}})}{3} - \frac{0.0592}{3}\log\frac{1}{[H^+]^4}$$

Substituting gives

$$E_{eq} = \frac{2 \times 0.334 + 1.44}{3} - \frac{0.0592}{3}\log\frac{1}{(1.00)^4} = 0.782\ V$$

Potential after Adding 25.10 mL of Ce^{4+}

$$\text{original amount } U^{4+} = 50.00\ \cancel{\text{mL } U^{4+}} \times 0.02500\ \frac{\text{mmol } U^{4+}}{\cancel{\text{mL } U^{4+}}}$$

$$= 1.250\ \text{mmol } U^{4+}$$

$$\text{amount } Ce^{4+} \text{ added} = 25.10\ \cancel{\text{mL } Ce^{4+}} \times 0.1000\ \frac{\text{mmol } Ce^{4+}}{\cancel{\text{mL } Ce^{4+}}}$$

$$= 2.510\ \text{mmol } Ce^{4+}$$

$$\text{amount } Ce^{3+} \text{ formed} = 1.250\ \cancel{\text{mmol } U^{4+}} \times \frac{2\ \text{mmol } Ce^{3+}}{\cancel{\text{mmol } U^{4+}}}$$

$$= 2.500\ \text{mmol } Ce^{3+}$$

$$\text{amount } Ce^{4+} \text{ remaining} = 2.510\ \text{mmol } Ce^{4+} - 2.500\ \cancel{\text{mmol } Ce^{3+}} \times \frac{1\ \text{mmol } Ce^{4+}}{\cancel{\text{mmol } Ce^{3+}}} = 0.010\ \text{mmol } Ce^{4+}$$

$$\text{volume of solution} = 75.10\ \text{mL}$$

$$[Ce^{3+}] = 2.500\ \text{mmol } Ce^{3+}/75.10\ \text{mL}$$

$$[Ce^{4+}] = 0.010\ \text{mmol } Ce^{4+}/75.10\ \text{mL}$$

Substituting into the expression for the formal potential gives

$$E = 1.44 - 0.0592 \log \frac{2.500/\cancel{75.10}}{0.010/\cancel{75.10}} = 1.30\ V$$

Table 16-2 contains other postequivalence-point data obtained in this same way.

Redox titration curves are symmetric when the reactants combine in a 1 : 1 ratio. Otherwise, they are asymmetric.

The data in the third column of Table 16-2 are plotted as curve *B* in Figure 16-3. The two curves in the figure are seen to be identical for volumes greater than 25.10 mL because the concentrations of the two cerium species are identical in this region. It is also interesting that the curve for iron(II) is symmetric about the equivalence point, while the curve for uranium(IV) is not symmetric. In general, symmetric curves are obtained when the analyte and titrant react in a 1 : 1 molar ratio.

16C-3 The Effect of System Variables on Redox Titration Curves

In earlier chapters, we have considered the effects of reactant concentrations and completeness of the reaction on titration curves. Here, we describe the effects of these variables on oxidation/reduction titration curves.

Reactant Concentration

As we have just seen, E_{system} for an oxidation/reduction titration is ordinarily independent of dilution. Consequently, titration curves for oxidation/reduction reactions are usually independent of analyte and reagent concentrations. This independence is in distinct contrast to that observed in the other types of titration curves we have encountered.

Completeness of the Reaction

The change in E_{system} in the equivalence-point region of an oxidation/reduction titration becomes larger as the reaction becomes more complete. This effect is demonstrated by the two curves in Figure 16-3. The equilibrium constant for the reaction of cerium(IV) with iron(II) is 7×10^{12} whereas that for U(IV) is 2×10^{37}. The effect of reaction completeness is further demonstrated in Figure 16-4, which shows curves for the titration of a hypothetical reductant having a standard electrode potential of 0.20 V with several hypothetical oxidants with standard potentials ranging from 0.40 to 1.20 V; the corresponding equilibrium constants lie between about 2×10^3 and 8×10^{16}. Clearly, the greatest change in potential of the system is associated with the reaction that is most complete. In this respect, then, oxidation/reduction titration curves are similar to those involving other types of reactions.

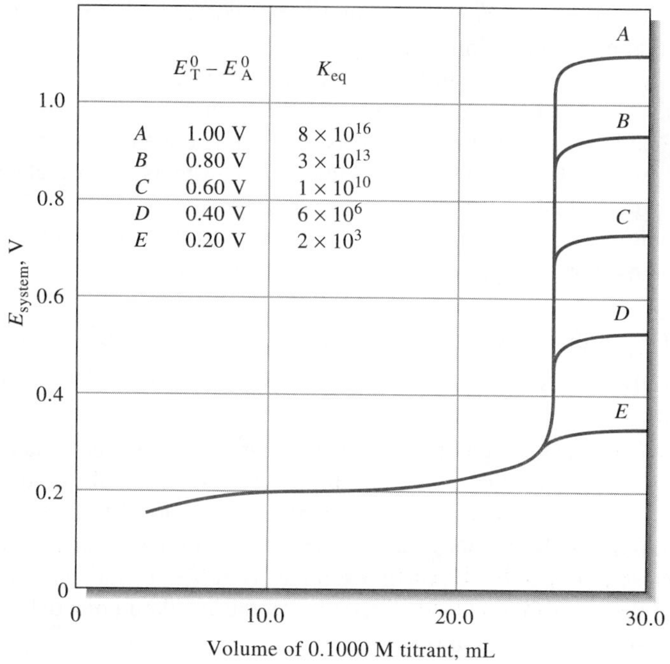

Figure 16-4
Effect of titrant electrode potential upon reaction completeness. The standard electrode potential for the analyte (E^0_A) is 0.200 V; starting with curve *A*, standard electrode potentials for the titrant (E^0_T) are 1.20, 1.00, 0.80, 0.60, and 0.40 V, respectively. Both analyte and titrant undergo a one-electron change.

FEATURE 16-2
Reaction Rates and Electrode Potentials

Standard potentials reveal whether or not a reaction proceeds far enough toward completion to be useful in a particular analytical problem, but they provide no information about the rate at which the equilibrium state is approached. Consequently, a reaction that appears extremely favorable from equilibrium considerations may be totally unacceptable from a kinetic standpoint. The oxidation of arsenic(III) with cerium(IV) in dilute sulfuric acid is a typical example. The reaction is

$$H_3AsO_3 + 2Ce^{4+} + H_2O \rightleftharpoons H_3AsO_4 + 2Ce^{3+} + 2H^+$$

The formal potentials $E^{0\prime}$ for these two systems are

$$Ce^{4+} + e^- \rightleftharpoons Ce^{3+} \qquad E^{0\prime} = +1.3 \text{ V}$$

$$H_3AsO_4 + 2H^+ + 2e^- \rightleftharpoons H_3AsO_3 + H_2O \qquad E^{0\prime} = +0.56 \text{ V}$$

and an equilibrium constant of about 10^{28} can be deduced from these data. Even though this reaction is highly favorable from an equilibrium standpoint, titration of arsenic(III) with cerium(IV) is impossible without a catalyst because several hours are needed for the attainment of equilibrium. Fortunately, several substances catalyze the reaction and thus make the titration feasible.

16C-4 The Titration of Mixtures

Solutions containing two oxidizing agents or two reducing agents yield titration curves that contain two inflection points, provided the standard potentials of the two analytes involved are sufficiently different from each other. If this difference is greater than about 0.2 V, the end points are usually distinct enough to permit determination of each component. This situation is quite comparable to the titration of two acids with different dissociation constants or to that of two ions forming precipitates that have different solubilities.

In addition, the behavior of a few redox systems is analogous to that of polyprotic acids. For example, consider the two half-reactions

$$VO^{2+} + 2H^+ + e^- \rightleftharpoons V^{3+} + H_2O \qquad E^0 = +0.359 \text{ V}$$

$$V(OH)_4^+ + 2H^+ + e^- \rightleftharpoons VO^{2+} + 3H_2O \qquad E^0 = +1.00 \text{ V}$$

The curve for the titration of V^{3+} with a strong oxidizing agent, such as permanganate, has two inflection points, the first corresponding to the oxidation of V^{3+} to VO^{2+} and the second to the oxidation of VO^{2+} to $V(OH)_4^+$. The stepwise oxidation of molybdenum(III), first to the +5 oxidation state and subsequently to the +6 state, is another common example. Here again, satisfactory inflections occur in the curves because the difference in the standard potentials of the pertinent half-reactions is about 0.4 V.

The derivation of titration curves for either of these types of reactions is not difficult if the difference in standard potentials is sufficiently great. An example is the titration of a solution containing iron(II) and titanium(III) ions with potassium permanganate. The standard potentials are

$$TiO^{2+} + 2H^+ + e^- \rightleftharpoons Ti^{3+} + H_2O \qquad E^0 = +0.099\ V$$

$$Fe^{3+} + e^- \rightleftharpoons Fe^{2+} \qquad E^0 = +0.771\ V$$

The first additions of permanganate are used up by the more readily oxidized titanium(III) ion. As long as an appreciable concentration of this species remains in solution, the potential of the system cannot become high enough to change the concentration of iron(II) ions appreciably. Thus, points defining the first part of the titration curve can be obtained by substituting the stoichiometric concentrations of titanium(III) and titanium(IV) ion into the equation

$$E = +0.099 - 0.0592 \log \frac{[Ti^{3+}]}{[TiO^{2+}][H^+]^2}$$

For all practical purposes then, the first part of this curve is identical to the titration curve for titanium(III) ion by itself. Beyond the first equivalence point, the solution contains both iron(II) and iron(III) ions in significant concentrations, so points on the curve can be most conveniently obtained from the relationship

$$E = E^0_{Fe^{3+}} - 0.0592 \log \frac{[Fe^{2+}]}{[Fe^{3+}]}$$

Throughout this region and beyond the second equivalence point, the curve is essentially identical to that for the titration of iron(II) ion alone. Such calculations leave undefined only the potential at the first equivalence point. A convenient way of estimating its value is to add the Nernst equations for the iron(II) and titanium(III) potentials. Since the electrode potentials for the two systems are identical at equilibrium, we can write

$$2E = +0.099 + 0.771 - 0.0592 \log \frac{[Ti^{3+}][Fe^{2+}]}{[TiO^{2+}][Fe^{3+}][H^+]^2}$$

Iron(III) and titanium(III) ions exist in small and equal amounts as a result of the equilibrium

$$2H^+ + TiO^{2+} + Fe^{2+} \rightleftharpoons Fe^{3+} + Ti^{3+} + H_2O$$

Thus, we can write

$$[Fe^{3+}] = [Ti^{3+}]$$

Substitution of this equality into the previous equation for the potential yields

$$E = \frac{+0.87}{2} - \frac{0.0592}{2} \log \frac{[Fe^{2+}]}{[TiO^{2+}][H^+]^2}$$

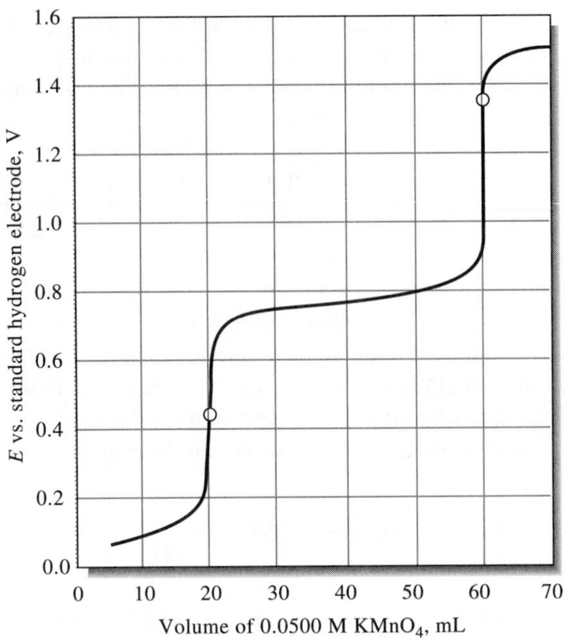

Figure 16-5
Curve for the titration of 50.0 mL of a solution that is 0.0500 M in Ti^{3+} and 0.200 M in Fe^{2+} with 0.0500 M $KMnO_4$. The concentration of H^+ is 1.0 M throughout.

Finally, if $[TiO^{2+}]$ and $[Fe^{2+}]$ are assumed to be essentially identical to their analytical concentrations, we can compute the equivalence potential.

A curve for a mixture of iron(II) and titanium(III) ions titrated with a standard permanganate solution is shown in Figure 16-5.

16D OXIDATION/REDUCTION INDICATORS

Two types of chemical indicators are used to detect end points for oxidation/reduction titrations: *general redox indicators* and *specific indicators.*

16D-1 General Redox Indicators

Color changes for general redox indicators depend only upon the potential of the system.

General oxidation/reduction indicators are substances that change color upon being oxidized or reduced. In contrast to specific indicators, the color changes of true redox indicators are largely independent of the chemical nature of the analyte and titrant and depend instead upon the changes in the electrode potential of the system that occur as the titration progresses.

The half-reaction responsible for color change in a typical general oxidation/reduction indicator can be written as

$$\mathrm{In_{ox}} + ne \rightleftharpoons \mathrm{In_{red}}$$

If the indicator reaction is reversible, we can write

$$E = E^0_{\mathrm{In}} - \frac{0.0592}{n}\log\frac{[\mathrm{In_{red}}]}{[\mathrm{In_{ox}}]} \qquad (16\text{-}13)$$

Typically, a change from the color of the oxidized form of the indicator to the color of the reduced form requires a change of about 100 in the ratio of reactant concentrations; that is, a color change is seen when

$$\frac{[\mathrm{In_{red}}]}{[\mathrm{In_{ox}}]} \leq \frac{1}{10}$$

changes to

$$\frac{[\mathrm{In_{red}}]}{[\mathrm{In_{ox}}]} \geq 10$$

The potential change required to produce the full color change of a typical general indicator can be found by substituting these two values into Equation 16-13, which gives

$$E = E^0_{\mathrm{In}} \pm \frac{0.0592}{n}$$

This equation shows that a typical general indicator exhibits a detectable color change when a titrant causes the system potential to shift from $E^0_{\mathrm{In}} + 0.0592/n$ to $E^0_{\mathrm{In}} - 0.0592/n$, or about $(0.118/n)$ V. For many indicators, $n = 2$ and a change of 0.059 V is thus sufficient.

Protons are involved in the reduction of many indicators, so the range of potentials over which a color change occurs (the *transition potential*) is often pH dependent.

Table 16-3 lists transition potentials for several redox indicators. Note that indicators functioning in any desired potential range up to about + 1.25 V are available. Structures for and reactions of a few of the indicators listed in the table are considered in the paragraphs that follow.

Iron(II) Complexes of Orthophenanthrolines

A class of organic compounds known as 1,10-phenanthrolines (or orthophenanthrolines) form stable complexes with iron(II) and certain other ions. The parent compound has a pair of nitrogen atoms located in such positions that each can form a covalent bond with the iron(II) ion. Three orthophenanthroline molecules combine with each iron ion to yield a complex with the structure.

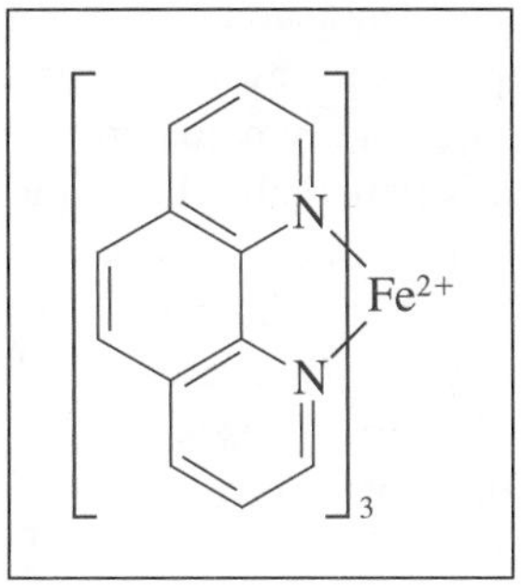

This complex, which is sometimes called ''ferroin,'' is conveniently formulated as $(\mathrm{phen})_3\mathrm{Fe}^{2+}$.

TABLE 16-3 Selected Oxidation/Reduction Indicators*

Indicator	Color: Oxidized	Color: Reduced	Transition Potential, V	Conditions
5-Nitro-1, 10-phenanthroline iron(II) complex	Pale blue	Red-violet	+ 1.25	1 M H_2SO_4
2,3′-Diphenylamine dicarboxylic acid	Blue-violet	Colorless	+ 1.12	7–10 M H_2SO_4
1,10-Phenanthroline iron(II) complex	Pale blue	Red	+ 1.11	1 M H_2SO_4
5-Methyl 1, 10-phenanthroline iron(II) complex	Pale blue	Red	+ 1.02	1 M H_2SO_4
Erioglaucin A	Blue-red	Yellow-green	+ 0.98	0.5 M H_2SO_4
Diphenylamine sulfonic acid	Red-violet	Colorless	+ 0.85	Dilute acid
Diphenylamine	Violet	Colorless	+ 0.76	Dilute acid
p-Ethoxychrysoidine	Yellow	Red	+ 0.76	Dilute acid
Methylene blue	Blue	Colorless	+ 0.53	1 M acid
Indigo tetrasulfonate	Blue	Colorless	+ 0.36	1 M acid
Phenosafranine	Red	Colorless	+ 0.28	1 M acid

*Data in part from I. M. Kolthoff and V. A. Stenger, *Volumetric Analysis,* 2nd ed., Vol. 1, p. 140. New York: Interscience, 1942.

The complexed iron in the ferroin undergoes a reversible oxidation/reduction reaction that can be written

$$\underset{\text{pale blue}}{(\text{phen})_3\text{Fe}^{3+}} + e^- \rightleftharpoons \underset{\text{red}}{(\text{phen})_3\text{Fe}^{2+}} \qquad E^0 = +1.06 \text{ V}$$

In practice, the color of the oxidized form is so slight as to go undetected, and the color change associated with this reduction is thus from nearly colorless to red. Because of the difference in color intensity, the end point is usually taken when only about 10% of the indicator is in the iron(II) form. The transition potential is thus approximately + 1.11 V in 1 M sulfuric acid.

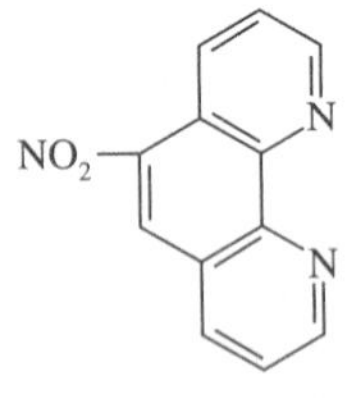

5-nitro-1,10-phenanthroline

Of all the oxidation/reduction indicators, ferroin approaches most closely the ideal substance. It reacts rapidly and reversibly, its color change is pronounced, and its solutions are stable and readily prepared. In contrast to many indicators, the oxidized form of ferroin is remarkably inert toward strong oxidizing agents. At temperatures above 60°C, ferroin decomposes.

Several substituted phenanthrolines have been investigated for their indicator properties, and some have proved to be as useful as the parent compound. Among these, the 5-nitro and 5-methyl derivatives are noteworthy; their transition potentials are + 1.25 V and + 1.02 V, respectively.

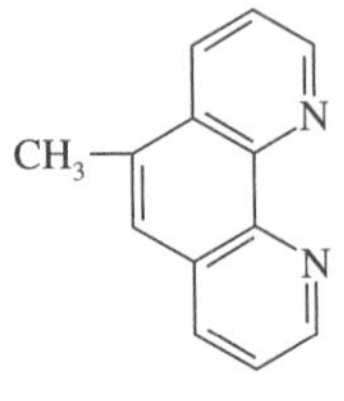

5-methyl-1,10-phenanthroline

Diphenylamine and Its Derivatives

Diphenylamine, $C_{12}H_{11}N$, which was one of the first redox indicators discovered, was used by Knop in 1924 for the titration of iron(II) with potassium dichromate.

In the presence of a strong oxidizing agent, diphenylamine is believed to undergo the reactions

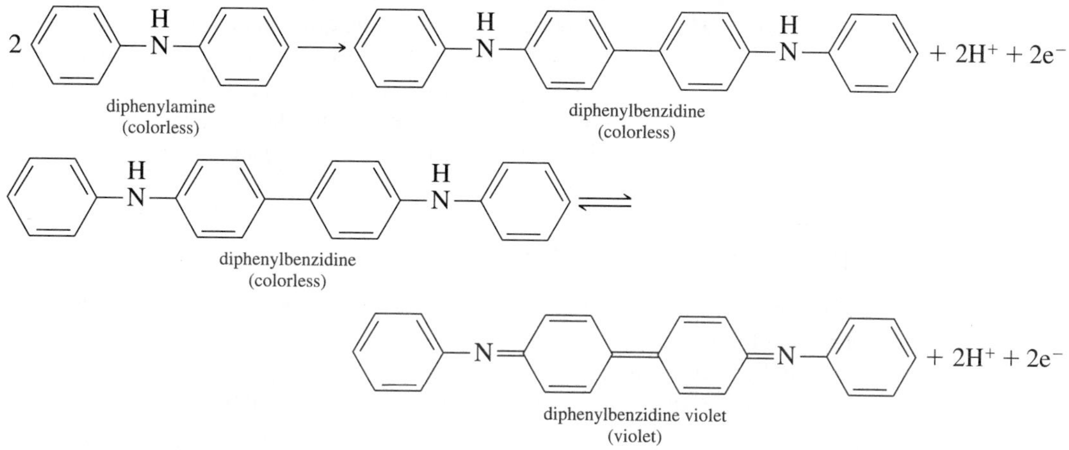

The first reaction is irreversible, but the second can be reversed and constitutes the actual indicator reaction.

The reduction potential for the second reaction is about + 0.76 V. Despite the fact that hydrogen ions appear in the equation, variations in acidity have little effect upon the magnitude of this potential, perhaps because the colored product is a protonated species.

Diphenylamine is not very soluble in water, and indicator solutions must be prepared in rather concentrated sulfuric acid. In addition, the indicator is not applicable to solutions containing tungstate ion, which precipitates the violet reaction product. Mercury(II) ions also interfere by inhibiting the indicator reaction.

The sulfonic acid derivative of diphenylamine, which has the structure

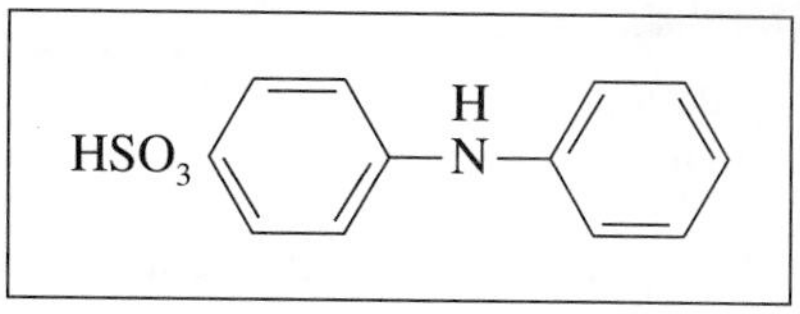

does not suffer from these disadvantages. The barium or sodium salt of this acid is used to prepare aqueous indicator solutions, which behave in essentially the same manner as the parent substance. The color change is somewhat sharper, however, passing from colorless through green to deep violet. The transition potential is about + 0.8 V and again is independent of the acid concentration. The sulfonic acid derivative is now widely used in redox titrations.

Diphenylbenzidine, the intermediate in the oxidation of diphenylamine, should behave in redox reactions just as diphenylamine does but should consume less oxidizing agent. Unfortunately, the low solubility of diphenylbenzidine in water and in sulfuric acid has precluded its widespread use. As might be expected, the sulfonic acid derivative of diphenylbenzidine is a satisfactory indicator.

Starch/Iodine Solutions

Starch, which forms a blue complex with triiodide ion, is a widely used specific indicator in oxidation/reduction reactions involving iodine as an oxidant or iodide ion as a reductant. A starch solution containing a little triiodide or iodide ion can also function as a true redox indicator, however. In the presence of excess oxidizing agent, the concentration ratio of iodine to iodide is high, giving a blue color to the solution. With excess reducing agent, on the other hand, iodide ion predominates, and the blue color is absent. Thus, the indicator system changes from colorless to blue in the titration of many reducing agents with various oxidizing agents. This color change is quite independent of the chemical composition of the reactants, depending only upon the potential of the system at the equivalence point.

The Choice of Redox Indicator

It is apparent from Figure 16-3 that all the indicators in Table 16-3 except for the first and the last could be employed with reagent A. In contrast, with reagent D, only indigo tetrasulfonate could be employed. The change in potential with reagent E is too small to be satisfactorily detected by an indicator.

16D-2 Specific Indicators

Perhaps the best-known specific indicator is starch, which forms a dark blue complex with triiodide ion. This complex signals the end point in titrations in which iodine is either produced or consumed.

Another specific indicator is potassium thiocyanate, which may be employed, for example, in the titration of iron(III) with solutions of titanium(III) sulfate. The end point involves the disappearance of the red color of the iron(III)/thiocyanate complex as a result of the marked decrease in the iron(III) concentration at the equivalence point.

16E POTENTIOMETRIC END POINTS

End points for many oxidation/reduction titrations are readily observed by making the solution of the analyte part of the cell:

reference electrode‖analyte solution|Pt

By measuring the potential of this cell during a titration, data for curves analogous to those shown in Figures 16-3 and 16-4 can be generated. End points are readily estimated from such curves. Potentiometric end points are considered in detail in Chapter 18.

16F QUESTIONS AND PROBLEMS

***16-1.** Briefly define the electrode potential of the system that contains two or more redox couples.

16-2. For an oxidation/reduction reaction, briefly distinguish between

- ***(a)** equilibrium and equivalence.
- **(b)** a true oxidation/reduction indicator and a specific indicator.

16-3. What is unique about the condition of equilibrium in an oxidation/reduction reaction?

***16-4.** How is an oxidation/reduction titration curve generated through the standard electrode potential for the analyte species and the volumetric titrant?

16-5. How does calculation of the electrode potential of the system at the equivalence point differ from that for any other point of an oxidation/reduction titration?

***16-6.** Under what circumstance is the curve for an oxidation/reduction titration asymmetric about the equivalence point?

16-7. Calculate the theoretical potential of the following cells. Indicate whether the cell, as written, is galvanic or electrolytic.

(a) $Pb|Pb^{2+}(0.1393\ M)\|Cd^{2+}(0.0511)|Cd$

(b) $Zn|Zn^{2+}(0.0364\ M)\|Tl^{3+}(9.06 \times 10^{-3}\ M),$ $Tl^{+}(0.0620\ M)|Pt$

(c) $Pt,H_2(765\ torr)|HCl(1.00 \times 10^{-4}\ M)\|$ $Ni^{2+}(0.0214\ M)|Ni$

(d) $Pb|PbI_2(sat'd),I^{-}(0.0120\ M)\|$ $Hg^{2+}(4.59 \times 10^{-3}\ M)|\ Hg$

(e) $Pt,H_2(1.00\ atm)|NH_3(0.438\ M),NH_4^{+}(0.379)\ M)\|$ SHE

(f) $Pt|TiO^{2+}(0.0790\ M),Ti^{3+}(0.00918\ M),$ $H^{+}(1.47 \times 10^{-2}\ M)\|VO^{2+}(0.1340\ M),$ $V^{3+}(0.0784\ M),\ H^{+}(0.0538\ M)|Pt$

***16-8.** Calculate the theoretical cell potential of the following cells. Indicate whether each cell, as written, is galvanic or electrolytic.

(a) $Zn|Zn^{2+}(0.0955\ M)\|Co^{2+}(6.78 \times 10^{-3}\ M)|Co$

(b) $Pt|Fe^{3+}(0.1310\ M),Fe^{2+}(0.0681\ M)\|$ $Hg^{2+}(0.0671\ M)|Hg$

(c) $Ag|Ag^{+}(0.1544\ M)|H^{+}(0.0794\ M)|O_2(1.12\ atm),Pt$

(d) $Cu|Cu^{2+}(0.0601\ M)\|I^{-}(0.1350\ M),AgI(sat'd)|Ag$

(e) $SHE\|HCOOH(0.1302\ M),HCOO^{-}(0.0764\ M)|$ $H_2(1.00\ atm),Pt$

(f) $Pt|UO_2^{2+}(7.93 \times 10^{-3}\ M),U^{4+}(6.37 \times 10^{-2}\ M),$ $H^{+}(1.16 \times 10^{-3}\ M)\|Fe^{3+}(0.003876\ M),$ $Fe^{2+}(0.1134\ M)|Pt$

16-9. Calculate the potential of

*(a) a galvanic cell consisting of a lead electrode immersed in 0.0848 M Pb^{2+} and a zinc electrode in contact with 0.1364 M Zn^{2+}.

(b) a galvanic cell with two platinum electrodes, the one immersed in a solution that is 0.0301 M in Fe^{3+} and 0.0760 M in Fe^{2+}, the other in a solution that is 0.00309 M in $Fe(CN)_6^{4-}$ and 0.1564 M in $Fe(CN)_6^{3-}$.

*(c) a galvanic cell consisting of a standard hydrogen electrode and a platinum electrode immersed in a solution that is 1.46×10^{-3} M in TiO^{2+}, 0.02723 M in Ti^{3+}, and buffered to a pH of 3.00.

(d) an electrolytic cell consisting of a cobalt electrode immersed in 0.0767 M Co^{2+} and a platinum electrode in contact with 0.200 M HCl; oxygen is evolved from the latter electrode at a pressure of 1.00 atm.

*(e) an electrolytic cell consisting of two silver electrodes, the one in contact with 0.2058 M Ag^{+}, the other immersed in 0.0791 M KBr and saturated with AgBr.

(f) an electrolytic cell with two platinum electrodes, the one immersed in a solution that is 0.1523 M in I^{-} and 0.0364 M in I_3^{-}, the other in a solution that is 0.02105 M in Fe^{3+} and 0.1037 M in Fe^{2+}.

16-10. Use the shorthand notation (page 310) to describe the cells in Problem 16-9. Each cell is supplied with a salt bridge to provide electrical contact between the solutions in the two cell compartments.

16-11. Generate equilibrium constant expressions for the following reactions. Calculate numerical values for $K_{eq.}$.

*(a) $Fe^{3+} + V^{2+} \rightleftharpoons Fe^{2+} + V^{3+}$

(b) $Fe(CN)_6^{3-} + Cr^{2+} \rightleftharpoons Fe(CN)_6^{4-} + Cr^{3+}$

*(c) $2V(OH)_4^{+} + U^{4+} \rightleftharpoons 2VO^{2+} + UO_2^{2+} + 4H_2O$

(d) $Tl^{3+} + 2Fe^{2+} \rightleftharpoons Tl^{+} + 2Fe^{3+}$

*(e) $2Ce^{4+} + H_3AsO_3 + H_2O \rightleftharpoons 2Ce^{3+} + H_3AsO_4 + 2H^{+}$ (1 M $HClO_4$)

(f) $2V(OH)_4^{+} + H_2SO_3 \rightleftharpoons SO_4^{2-} + 2VO^{2+} + 5H_2O$

*(g) $VO^{2+} + V^{2+} + 2H^{+} \rightleftharpoons 2V^{3+} + H_2O$

(h) $TiO^{2+} + Ti^{2+} + 2H^{+} \rightleftharpoons 2Ti^{3+} + H_2O$

16-12. Calculate the electrode potential of the system at the equivalence point for each of the reactions in Problem 16-11. Use 0.100 M where a value for $[H^{+}]$ is needed and is not otherwise specified.

***16-13.** Select an indicator from Table 16-2 that might be suitable for each of the titrations in Problem 16-11. Write NONE if no indicator is suitable.

16-14. Construct curves for the following titrations. Calculate potentials after the addition of 10.00, 25.00, 49.00, 49.90, 50.00, 50.10, 51.00, and 60.00 mL of the reagent. Where necessary, assume that $[H^{+}] = 1.00$ M throughout.

*(a) 50.00 mL of 0.1000 M V^{2+} with 0.05000 M Sn^{4+}

(b) 50.00 mL of 0.1000 M $Fe(CN)_6^{3-}$ with 0.1000 M Cr^{2+}

*(c) 50.00 mL of 0.1000 M $Fe(CN)_6^{4-}$ with 0.05000 M Tl^{3+}

(d) 50.00 mL of 0.1000 M Fe^{3+} with 0.05000 M Sn^{2+}

*(e) 50.00 mL of 0.05000 M U^{4+} with 0.02000 M MnO_4^{-}

CHAPTER

17

Applications of Oxidation/ Reduction Titrations

This chapter is concerned with the preparation of standard solutions of oxidants and reductants and with their applications in analytical chemistry. In addition, auxiliary reagents that convert an analyte to a single oxidation state are described.[1]

17A AUXILIARY OXIDIZING AND REDUCING REAGENTS

The analyte in an oxidation/reduction titration must be in a single oxidation state at the outset. The steps that precede the titration (dissolution of the sample and separation of interferences), however, frequently convert the analyte to a mixture of oxidation states. For example, the solution formed when an iron-containing sample is dissolved usually contains a mixture of iron(II) and iron(III) ions. If we choose to use a standard oxidant for the determination of iron, we must first treat the sample solution with an auxiliary reducing agent; on the other hand, if we plan to titrate with a standard reductant, pretreatment with an auxiliary oxidizing reagent will be needed.[2]

To be useful as a preoxidant or a prereductant, a reagent must react quantitatively with the analyte. In addition, excesses of the reagent must be readily removable because such excesses will inevitably interfere by consuming standard solution.

17A-1 Auxiliary Reducing Reagents

A number of metals are good reducing agents and have been used for the prereduction of analytes. Included among these are zinc, aluminum, cadmium, lead, nickel, copper, and silver (in the presence of chloride ion). Sticks or coils of the

---

[1]For further reading on redox titrimetry, see J. A. Goldman and V. A. Stenger, in *Treatise on Analytical Chemistry,* I. M. Kolthoff and P. J. Elving, Eds., Part I, Vol. 11, Chapter 119. New York: Wiley, 1975; I. M. Kolthoff and R. Belcher, *Volumetric Analysis,* Vol. 3. New York: Interscience, 1957.

[2]For a brief summary of auxiliary reagents, see J. A. Goldman and V. A. Stenger, in *Treatise on Analytical Chemistry,* I. M. Kolthoff and P. J. Elving, Eds., Part I, Vol. 11, pp. 7204–7206. New York: Wiley, 1975.

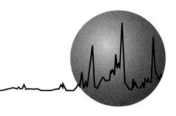

TABLE 17-1 Uses of the Walden Reductor and the Jones Reductor*

Walden	Jones
$Ag(s) + Cl^- \rightarrow AgCl(s) + e^-$	**$Zn(Hg)(s) \rightarrow Zn^{2+} + Hg + 2e^-$**
$Fe^{3+} + e^- \rightarrow Fe^{2+}$	$Fe^{3+} + e^- \rightleftharpoons Fe^{2+}$
$Cu^{2+} + e^- \rightarrow Cu^+$	$Cu^{2+} + 2e^- \rightleftharpoons Cu(s)$
$H_2MoO_4 + 2H^+ + e^- \rightarrow MoO_2^+ + 2H_2O$	$H_2MoO_4 + 6H^+ + 3e^- \rightleftharpoons Mo^{3+} + 4H_2O$
$UO_2^{2+} + 4H^+ + 2e^- \rightarrow U^{4+} + 2H_2O$	$UO_2^{2+} + 4H^+ + 2e^- \rightleftharpoons U^{4+} + 2H_2O$
	$UO_2^{2+} + 4H^+ + 3e^- \rightleftharpoons U^{3+} + 2H_2O$†
$V(OH)_4^+ + 2H^+ + e^- \rightarrow VO^{2+} + 3H_2O$	$V(OH)_4^+ + 4H^+ + 3e^- \rightleftharpoons V^{2+} + 4H_2O$
TiO^{2+} not reduced	$TiO^{2+} + 2H^+ + e^- \rightleftharpoons Ti^{3+} + H_2O$
Cr^{3+} not reduced	$Cr^{3+} + e^- \rightleftharpoons Cr^{2+}$

*From I. M. Kolthoff and R. Belcher, *Volumetric Analysis,* Vol. 3, p. 12. New York: Interscience, 1957. With permission.

†A mixture of oxidation states is obtained. The reductor is still usable, however, because any U^{3+} formed can be converted to U^{4+} by shaking the solution with air for a few minutes.

metal can be immersed directly in the analyte solution. After reduction is judged complete, the solid is removed manually and rinsed with water. It is necessary to filter the analyte solution to remove granular or powdered forms of the metal. An alternative to filtration is the use of a *reductor,* such as that shown in Figure 17-1.[3] Here, the finely divided metal is held in a vertical glass tube through which the analyte solution is drawn under a mild vacuum. The metal in a reductor is ordinarily sufficient for hundreds of reductions.

A typical *Jones reductor* has a diameter of about 2 cm and holds a 40- to 50-cm column of amalgamated zinc. Amalgamation is accomplished by allowing zinc granules to stand briefly in a solution of mercury(II) chloride:

$$2Zn(s) + Hg^{2+} \longrightarrow Zn^{2+} + Zn(Hg)(s)$$

Zinc amalgam is nearly as effective a reducing agent as the pure metal and has the important virtue of inhibiting the reduction of hydrogen ions by zinc, a parasitic reaction that not only needlessly uses up the reducing agent but also contaminates the sample solution with zinc(II) ions. Solutions that are quite acidic can be passed through a Jones reductor without significant hydrogen formation.

Table 17-1 lists the principal applications of the Jones reductor. Also listed in this table are reductions that can be accomplished with a *Walden reductor,* in which granular metallic silver held in a narrow glass column is the reductant. Silver is not a good reducing agent unless chloride or some other ion that forms a silver salt of low solubility is present. For this reason, prereductions with a Walden reductor are generally carried out from hydrochloric acid solutions of the analyte. The coating of silver chloride produced on the metal is removed periodically by dipping a zinc rod into the solution that covers the packing.

Table 17-1 suggests that the Walden reductor is somewhat more selective in its action than is the Jones reductor.

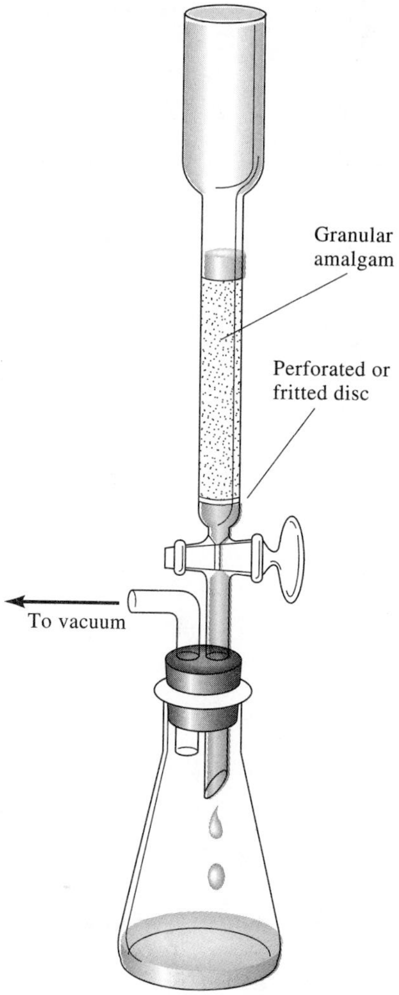

Figure 17-1
A Jones reductor.

[3]For a discussion of reductors, see F. Hecht, in *Treatise on Analytical Chemistry,* I. M. Kolthoff and P. J. Elving, Eds., Part I, Vol. 11, pp. 6703–6707. New York: Wiley, 1975.

17A-2 Auxiliary Oxidizing Reagents

Sodium Bismuthate

Sodium bismuthate is a powerful oxidizing agent—capable, for example, of converting manganese(II) quantitatively to permanganate ion. This bismuth salt is a sparingly soluble solid with a formula that is usually written as $NaBiO_3$, although its exact composition is somewhat uncertain. Oxidations are performed by suspending the bismuthate in the analyte solution and boiling for a brief period. The unused reagent is then removed by filtration. The half-reaction for the reduction of sodium bismuthate can be written as

$$NaBiO_3(s) + 4H^+ + 2e^- \rightleftharpoons BiO^+ + Na^+ + 2H_2O$$

Ammonium Peroxydisulfate

Ammonium peroxydisulfate, $(NH_4)_2S_2O_8$, is also a powerful oxidizing agent. In acidic solution, it converts chromium(III) to dichromate, cerium(III) to cerium(IV), and manganese(II) to permanganate. The half-reaction is

$$S_2O_8^{2-} + 2e^- \rightleftharpoons 2SO_4^{2-} \qquad E^0 = 2.01 \text{ V}$$

The oxidations are catalyzed by traces of silver ion. The excess reagent is readily decomposed by a brief period of boiling:

$$2S_2O_8^{2-} + 2H_2O \longrightarrow 4SO_4^{2-} + O_2(g) + 4H^+$$

Sodium Peroxide and Hydrogen Peroxide

Peroxide is a convenient oxidizing agent either as the solid sodium salt or as a dilute solution of the acid. The half-reaction for hydrogen peroxide in acidic solution is

$$H_2O_2 + 2H^+ + 2e^- \rightleftharpoons 2H_2O \qquad E^0 = 1.78 \text{ V}$$

After oxidation is complete, the solution is freed of excess reagent by boiling:

$$2H_2O_2 \longrightarrow 2H_2O + O_2(g)$$

17B APPLICATION OF STANDARD REDUCTANTS

Standard solutions of most reducing agents tend to react with atmospheric oxygen. For this reason, reductants are seldom used for the direct titration of oxidizing analytes; indirect methods are used instead. The two most common indirect methods, which are discussed in the paragraphs that follow, are based upon standard solutions of iron(II) ions and standard solutions of sodium thiosulfate.

17B-1 Iron(II) Solutions

Solutions of iron(II) are readily prepared from iron(II) ammonium sulfate, $Fe(NH_4)_2(SO_4)_2 \cdot 6H_2O$ (Mohr's salt), or from the closely related iron(II) ethylenediamine sulfate, $FeC_2H_4(NH_3)_2(SO_4)_2 \cdot 4H_2O$ (Oesper's salt). Air-oxidation of iron(II) takes place rapidly in neutral solutions but is inhibited in the presence of acids, with the most stable preparations being about 0.5 M in H_2SO_4. Such solutions are stable for no longer than one day, if that long.

Numerous oxidizing agents are conveniently determined by treatment of the analyte solution with a measured excess of standard iron(II) followed by immediate titration of the excess with a standard solution of potassium dichromate or cerium(IV) (Sections 17C-1 and 17C-2). Just before or after the analyte is titrated, the volumetric ratio between the standard oxidant and the iron(II) solution is established by titrating two or three aliquots of the latter with the former.

This procedure has been applied to the determination of organic peroxides; hydroxylamine; chromium(VI); cerium(IV); molybdenum(VI); nitrate, chlorate, and perchlorate ions; and numerous other oxidants (see for example, Problems 17-37 and 17-39).

17B-2 Sodium Thiosulfate

Thiosulfate ion is a moderately strong reducing agent that has been widely used to determine oxidizing agents by an indirect procedure that involves iodine as an intermediate. With iodine, thiosulfate ion is oxidized quantitatively to tetrathionate ion, the half-reaction being

$$2S_2O_3^{2-} \rightleftharpoons S_4O_6^{2-} + 2e^-$$

In its reaction with iodine, each thiosulfate ion gains one electron.

The quantitative aspect of this reaction with iodine is unique. Other oxidants oxidize the tetrathionate ion, wholly or in part, to sulfate ion.

The scheme used to determine oxidizing agents involves adding an unmeasured excess of potassium iodide to a slightly acidic solution of the analyte. Reduction of the analyte produces a stoichiometrically equivalent amount of iodine. The liberated iodine is then titrated with a standard solution of sodium thiosulfate, $Na_2S_2O_3$, one of the few reducing agents that is stable toward air-oxidation. An example of this procedure is the determination of sodium hypochlorite in bleaches. The reactions are

Sodium thiosulfate is one of the few reducing agents that is not oxidized by air.

$$\begin{aligned} OCl^- + 2I^- + 2H^+ &\longrightarrow Cl^- + I_2 + H_2O \qquad \text{(excess KI)} \\ I_2 + 2S_2O_3^{2-} &\longrightarrow 2I^- + S_4O_6^{2-} \end{aligned} \tag{17-1}$$

The quantitative conversion of thiosulfate ion to tetrathionate ion shown in Equation 17-1 requires a pH smaller than 7. If strongly acidic solutions must be titrated, air-oxidation of the excess iodide must be prevented by blanketing the solution with an inert gas, such as carbon dioxide or nitrogen.

End Points in Iodine/Thiosulfate Titrations

A solution that is about 5×10^{-6} M in I_2 has a discernible color, which corresponds to less than one drop of a 0.05 M iodine solution in 100 mL. Thus, provided the analyte solution is colorless, the disappearance of the iodine color can serve as the indicator in titrations with sodium thiosulfate.

More commonly, titrations involving iodine are performed with a suspension of starch as an indicator. The deep blue color that develops in the presence of iodine is believed to arise from the absorption of iodine into the helical chain of β-amylose, a macromolecular component of most starches. The closely related α-amylose forms a red adduct with iodine. This reaction is not readily reversible and is thus undesirable. So-called *soluble starch,* which is available from commercial sources, consists principally of β-amylose, the alpha fraction having been removed; indicator solutions are readily prepared from this product.

Aqueous starch suspensions decompose within a few days, primarily because of bacterial action. The decomposition products tend to interfere with the indicator properties of the preparation and may also be oxidized by iodine. The rate of decomposition can be inhibited by preparing and storing the indicator under sterile conditions and by adding mercury(II) iodide or chloroform as a bacteriostat. Perhaps the simplest alternative is to prepare a fresh suspension of the indicator, which requires only a few minutes, on each day it is to be used.

Starch undergoes decomposition in solutions with high I_2 concentrations. In titrations of excess I_2 with $Na_2S_2O_3$, addition of the indicator must be deferred until most of the I_2 has been reduced.

Starch decomposes irreversibly in solutions containing large concentrations of iodine. Therefore, in titrating solutions of iodine with sodium thiosulfate ion, as in the indirect determination of oxidants, addition of the indicator is delayed until the titration is nearly complete, as shown by a color change from deep red-brown to faint yellow. The indicator can be introduced at the outset when thiosulfate solutions are being titrated directly with iodine.

The Stability of Sodium Thiosulfate Solutions

Although sodium thiosulfate solutions are resistant to air-oxidation, they do tend to decompose to give sulfur and hydrogen sulfite ion:

$$S_2O_3^{2-} + H^+ \rightleftharpoons HSO_3^- + S(s)$$

Variables that influence the rate of this reaction include pH, the presence of microorganisms, the concentration of the solution, the presence of copper(II) ions, and exposure to sunlight. These variables may cause the concentration of a thiosulfate solution to change by several percent over a period of a few weeks. On the other hand, proper attention to detail will yield solutions that need only occasional restandardization.

When sodium thiosulfate is added to a strongly acidic medium, a cloudiness develops almost immediately as a consequence of the precipitation of elemental sulfur. Even in neutral solution, this reaction proceeds at such a rate that standard sodium thiosulfate must be restandardized periodically.

The rate of the decomposition reaction increases markedly as the solution becomes acidic.

The most important single cause for the instability of neutral or slightly basic thiosulfate solutions is bacteria that metabolize thiosulfate ion to sulfite and sulfate ions as well as elemental sulfur. To minimize this problem, standard solutions of the reagent are prepared under reasonably sterile conditions. Bacterial activity appears to be at a minimum at a pH between 9 and 10, which accounts, at least in part, for the reagent's greater stability in slightly basic solutions. The presence of a bactericide, such as chloroform, sodium benzoate, or mercury(II) iodide, also slows decomposition.

The Standardization of Thiosulfate Solutions

Potassium iodate is an excellent primary standard for thiosulfate solutions. In this application, weighed amounts of primary-standard-grade reagent are dissolved in water containing an excess of potassium iodide. When this mixture is acidified with a strong acid, the reaction

$$IO_3^- + 5I^- + 6H^+ \rightleftharpoons 3I_2 + 2H_2O$$

occurs instantaneously. The liberated iodine is then titrated with the thiosulfate solution. The stoichiometry of the reactions is

$$1 \text{ mol } IO_3^- \equiv 3 \text{ mol } I_2 \equiv 6 \text{ mol } S_2O_3^{2-}$$

EXAMPLE 17-1

A solution of sodium thiosulfate was standardized by dissolving 0.1210 g KIO_3 (214.00 g/mol) in water, adding a large excess of KI, and acidifying with HCl. The liberated iodine required 41.64 mL of the thiosulfate solution to decolorize the blue starch/iodine complex. Calculate the molarity of the $Na_2S_2O_3$.

$$\text{amount } Na_2S_2O_3 = 0.1210 \cancel{\text{ g } KIO_3} \times \frac{1 \cancel{\text{ mmol } KIO_3}}{0.21400 \cancel{\text{ g } KIO_3}} \times \frac{6 \text{ mmol } Na_2S_2O_3}{\cancel{\text{mmol } KIO_3}}$$

$$= 3.3925 \text{ mmol } Na_2S_2O_3$$

$$c_{Na_2S_2O_3} = \frac{3.3925 \text{ mmol } Na_2S_2O_3}{41.64 \text{ mL } Na_2S_2O_3} = 0.08147 \text{ M}$$

Other primary standards for sodium thiosulfate are potassium dichromate, potassium bromate, potassium hydrogen iodate, potassium ferricyanide, and metallic copper. Solutions of all these compounds liberate stoichiometric amounts of iodine when treated with excess potassium iodide.

Applications of Sodium Thiosulfate Solutions

Numerous substances can be determined by the indirect method involving titration with sodium thiosulfate; typical applications are summarized in Table 17-2.

TABLE 17-2 Some Applications of Sodium Thiosulfate as a Reductant

Analyte	Half-Reaction	Special Conditions
IO_4^-	$IO_4^- + 8H^+ + 7e^- \rightleftharpoons \frac{1}{2}I_2 + 4H_2O$	Acidic solution
	$IO_4^- + 2H^+ + 2e^- \rightleftharpoons IO_3^- + H_2O$	Neutral solution
IO_3^-	$IO_3^- + 6H^+ + 5e^- \rightleftharpoons \frac{1}{2}I_2 + 3H_2O$	Strong acid
BrO_3^-, ClO_3^-	$XO_3^- + 6H^+ + 6e^- \rightleftharpoons X^- + 3H_2O$	Strong acid
Br_2, Cl_2	$X_2 + 2I^- \rightleftharpoons I_2 + 2X^-$	
NO_2^-	$HNO_2 + H^+ + e^- \rightleftharpoons NO(g) + H_2O$	
Cu^{2+}	$Cu^{2+} + I^- + e^- \rightleftharpoons CuI(s)$	
O_2	$O_2 + 4Mn(OH)_2(s) + 2H_2O \rightleftharpoons 4Mn(OH)_3(s)$	Basic solution
	$Mn(OH)_3(s) + 3H^+ + e^- \rightleftharpoons Mn^{2+} + 3H_2O$	Acidic solution
O_3	$O_3(g) + 2H^+ + 2e^- \rightleftharpoons O_2(g) + H_2O$	
Organic peroxide	$ROOH + 2H^+ + 2e^- \rightleftharpoons ROH + H_2O$	

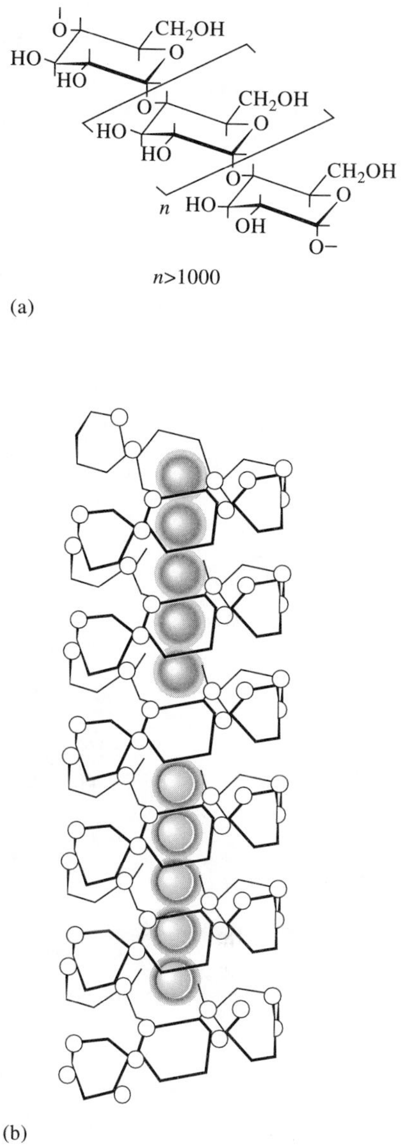

Figure 17-2
Thousands of glucose molecules polymerize to form huge molecules of β-amylose as shown schematically in (a) above. Molecules of β-amylose tend to assume a helical structure. The iodine species I_5^- as shown in (b) is incorporated into the amylose helix. For further details see R. C. Teitelbaum, S. L. Ruby, and T. J. Marks, *J. Amer. Chem. Soc.*, **1980,** *102*, 3322.

17C APPLICATIONS OF STANDARD OXIDANTS

Table 17-3 summarizes the properties of five of the most widely used volumetric oxidizing reagents. Note that the standard potentials for these reagents vary from 0.5 to 1.5 V. The choice among them depends upon the strength of the analyte as a reducing agent, the rate of reaction between oxidant and analyte, the stability of the standard oxidant solutions, the cost, and the availability of a satisfactory indicator.

17C-1 The Strong Oxidants—Potassium Permanganate and Cerium(IV)

Solutions of permanganate ion and cerium(IV) ion are strong oxidizing reagents whose applications closely parallel one another. Half-reactions for the two are

$$MnO_4^- + 8H^+ + 5e^- \rightleftharpoons Mn^{2+} + 4H_2O \qquad E^0 = 1.51\text{ V}$$

$$Ce^{4+} + e^- \rightleftharpoons Ce^{3+} \qquad E^{0\prime} = 1.44\text{ V} \qquad (1\text{ M } H_2SO_4)$$

The formal potential shown for the reduction of cerium(IV) is for solutions that are 1 M in sulfuric acid. In 1 M perchloric acid and 1 M nitric acid, the potentials are 1.70 and 1.61 V, respectively. Solutions of cerium(IV) in the latter two acids are not very stable and thus find limited application.

The half-reaction shown for permanganate ion occurs only in solutions that are 0.1 M or greater in strong acid. In less acidic media, the product may be Mn(III), Mn(IV), or Mn(VI), depending upon conditions.

TABLE 17-3 Some Common Oxidants Used as Standard Solutions

Reagent Formula	Reduction Product	Standard Potential, V	Standardized with	Indicator*	Stability†
Potassium permanganate, $KMnO_4$	Mn^{2+}	1.51‡	$Na_2C_2O_4$, Fe, As_2O_3	MnO_4^-	(b)
Potassium bromate, $KBrO_3$	Br^-	1.44‡	$KBrO_3$	(1)	(a)
Cerium(IV), Ce^{4+}	Ce^{3+}	1.44‡	$Na_2C_2O_4$, Fe, As_2O_3	(2)	(a)
Potassium dichromate, $K_2Cr_2O_7$	Cr^{3+}	1.33‡	$K_2Cr_2O_7$, Fe	(3)	(a)
Iodine, I_2	I^-	0.536‡	$BaS_2O_3 \cdot H_2O$, $Na_2S_2O_3$	starch	(c)

*(1) α-Naphthoflavone; (2) 1,10-phenanthroline iron(II) complex (ferroin); (3) diphenylamine sulfonic acid.

†(a) Indefinitely stable; (b) moderately stable, requires periodic standarization; (c) somewhat unstable, requires frequent standardization.

‡$E^{0\prime}$ in 1 M H_2SO_4.

Comparison of the Two Reagents

For all practical purposes, the oxidizing strengths of permanganate and cerium(IV) solutions are comparable. Solutions of cerium(IV) in sulfuric acid, however, are stable indefinitely, whereas permanganate solutions decompose slowly and thus require occasional restandardization. Furthermore, cerium(IV) solutions in sulfuric acid do not oxidize chloride ion and thus can be used to titrate hydrochloric acid solutions of analytes; in contrast, permanganate ion cannot be used with hydrochloric acid solutions unless special precautions are taken to prevent the slow oxidation of chloride ion that leads to overconsumption of the standard reagent. A further advantage of cerium(IV) is that a primary-standard-grade salt of the reagent is available, thus making possible the direct preparation of standard solutions.

Despite these advantages of cerium solutions over permanganate solutions, the latter are more widely used. One reason is the color of permanganate solutions, which is intense enough to serve as an indicator in titrations. A second reason for the popularity of permanganate solutions is their modest cost. The cost of a liter of 0.02 M solution is about \$0.08, whereas a liter of cerium(IV) of comparable strength costs about \$2.20 (\$4.40 if reagent of primary-standard grade is employed). Another disadvantage of cerium(IV) solutions is their tendency to form sparingly soluble basic salts in solutions that are less than 0.1 M in strong acid.

End Points

A useful property of a potassium permanganate solution is its intense purple color, which is sufficient to serve as an indicator for most titrations. As little as 0.01 to 0.02 mL of a 0.02 M solution imparts a perceptible color to 100 mL of water. If the permanganate solution is very dilute, diphenylamine sulfonic acid or the 1,10-phenanthroline complex of iron(II) (Table 16-3) provides a sharper end point.

The permanganate end point is not permanent because excess permanganate ions react slowly with the relatively large concentration of manganese(II) ions present at the end point:

$$2MnO_4^- + 3Mn^{2+} + 2H_2O \longrightarrow 5MnO_2(s) + 4H^+$$

The equilibrium constant for this reaction is about 10^{47}, which indicates that the *equilibrium* concentration of permanganate ion is vanishingly small even in highly acidic media. Fortunately, the rate at which this equilibrium is approached is so slow that the end point fades only gradually over a period of perhaps 30 seconds.

Solutions of cerium(IV) are yellow-orange, but the color is not intense enough to act as an indicator in titrations. Several oxidation/reduction indicators are available for titrations with standard solutions of cerium(IV). The most widely used of these is the iron(II) complex of 1,10-phenanthroline or one of its substituted derivatives (Table 16-3).

The Preparation and Stability of Standard Solutions

Aqueous solutions of permanganate are not entirely stable because the ion tends to oxidize water:

$$4MnO_4^- + 2H_2O \longrightarrow 4MnO_2(s) + 3O_2(g) + 4OH^-$$

Although the equilibrium constant for this reaction indicates that the products are favored, permanganate solutions, when properly prepared, are reasonably stable because the decomposition reaction is slow. It is catalyzed by light, heat, acids, bases, manganese(II), and manganese dioxide.

Permanganate solutions are moderately stable provided they are free of manganese dioxide and stored in a dark container.

Moderately stable solutions of permanganate ion can be prepared if the effects of these catalysts, particularly manganese dioxide, are minimized. Manganese dioxide is a contaminant in even the best grade of solid potassium permanganate. Furthermore, this compound forms in freshly prepared solutions of the reagent as a consequence of the reaction of permanganate ion with organic matter and dust present in the water used to prepare the solution. Removal of manganese dioxide by filtration before standardization markedly improves the stability of standard permanganate solutions. Before filtration, the reagent solution is allowed to stand for about 24 hours or is heated for a brief period to hasten oxidation of the organic species generally present in small amounts in distilled and deionized water. Paper cannot be used for filtering because permanganate ion reacts with it to form additional manganese dioxide.

Standardized permanganate solutions should be stored in the dark. Filtration and restandardization are required if any solid is detected in the solution or on the walls of the storage bottle. In any event, restandardization every 1 or 2 weeks is a good precautionary measure.

Solutions containing excess standard permanganate should never be heated because they decompose by oxidizing water. This decomposition cannot be compensated for with a blank. On the other hand, it is possible to titrate hot, acidic solutions of reductants with permanganate without error provided the reagent is added slowly enough so that large excesses do not accumulate.

EXAMPLE 17-2

Describe how you would prepare 2.0 L of an approximately 0.010 M solution of $KMnO_4$ (158.03 g/mol).

$$\text{mass } KMnO_4 \text{ needed} = 2.0\ \cancel{L} \times 0.010\ \frac{\cancel{\text{mol } KMnO_4}}{\cancel{L}} \times \frac{158.03\text{ g } KMnO_4}{\cancel{\text{mol } KMnO_4}}$$

$$= 3.16\text{ g}$$

Dissolve about 3.2 g of $KMnO_4$ in a little water. After solution is complete, add water to bring the volume to about 2.0 L. Heat the solution to boiling for a brief period, and let stand until it is cool. Filter through a glass filtering crucible and store in a clean, dark bottle.

The most widely used compounds for the preparation of solutions of cerium(IV) are listed in Table 17-4. Primary-standard-grade cerium ammonium nitrate is available commercially and can be used to prepare standard solutions of the cation directly by weight. More commonly, less expensive reagent-grade cerium(IV) ammonium nitrate or ceric hydroxide is used to prepare solutions that are subsequently standardized. In either case, the reagent is dissolved in a solution that is at least 0.1 M in sulfuric acid to prevent the precipitation of basic salts.

TABLE 17-4 Analytically Useful Cerium(IV) Compounds

Name	Formula	Molar Mass
Cerium(IV) ammonium nitrate	$Ce(NO_3)_4 \cdot 2NH_4NO_3$	548.2
Cerium(IV) ammonium sulfate	$Ce(SO_4)_2 \cdot 2(NH_4)_2SO_4 \cdot 2H_2O$	632.6
Cerium(IV) hydroxide	$Ce(OH)_4$	208.1
Cerium(IV) hydrogen sulfate	$Ce(HSO_4)_4$	528.4

Sulfuric acid solutions of cerium(IV) are remarkably stable and can be stored for months or heated at 100°C for prolonged periods without loss of oxidizing capacity.

Primary Standards

Several excellent primary standards are available for the standardization of solutions of potassium permanganate and cerium(IV).

Sodium Oxalate Sodium oxalate is widely used to standardize permanganate and cerium(IV) solutions. In acidic solutions, the oxalate ion is converted to the undissociated acid. Thus, its reaction with the permanganate ion can be depicted as

$$2MnO_4^- + 5H_2C_2O_4 + 6H^+ \rightleftharpoons 2Mn^{2+} + 10CO_2(g) + 8H_2O$$

The same oxidation products are formed with cerium(IV).

The reaction between permanganate ion and oxalic acid is complex and proceeds slowly even at elevated temperature unless manganese(II) is present as a catalyst. Thus, when the first few milliliters of standard permanganate are added to a hot solution of oxalic acid, several seconds are required before the color of the permanganate ion disappears. As the concentration of manganese(II) builds up, however, the reaction proceeds more and more rapidly as a result of autocatalysis.

The reaction between Ce^{4+} and $H_2C_2O_4$ is

$$2Ce^{4+} + H_2C_2O_4 \rightleftharpoons 2Ce^{3+} + 2H^+ + 2CO_2$$

Autocatalysis is a type of catalysis in which the product of a reaction catalyses the reaction. This phenomenon causes the rate of the reaction to increase with time.

It has been found that when solutions of sodium oxalate are titrated at 60° to 90°C, the consumption of permanganate is from 0.1% to 0.4% less than theoretical, probably due to the air-oxidation of a fraction of the oxalic acid. This small error can be avoided by adding 90% to 95% of the required permanganate to a cool solution of the oxalate. After the added permanganate is completely consumed (as indicated by the disappearance of color), the solution is heated to about 60°C and titrated to a pink color that persists for about 30 seconds. The disadvantage of this procedure is that prior knowledge of the approximate concentration of the permanganate solution is needed so that a proper initial volume of it can be added. For most purposes, direct titration of the hot oxalic acid solution provides adequate data (usually 0.2% to 0.3% high). If greater accuracy is required, a direct titration of the hot solution of one portion of the primary standard can be followed by titration of two or three portions in which the solution is not heated until the end.

Cerium(IV) standardizations against sodium oxalate are usually performed at 50°C in a hydrochloric acid solution containing iodine monochloride as a catalyst.

Solutions of $KMnO_4$ and Ce^{4+} can also be standardized with electrolytic iron wire or with potassium iodide.

EXAMPLE 17-3

You wish to standardize the solution in Example 17-2 against primary $Na_2C_2O_4$ (134.00 g/mol). If you want to use between 30 and 45 mL of the reagent for the standardization, what range of masses of the primary standard should you weigh out?

For a 30-mL titration:

$$\text{amount KMnO}_4 = 30 \text{ mL } \cancel{\text{KMnO}_4} \times 0.010 \frac{\text{mmol KMnO}_4}{\cancel{\text{mL KMnO}_4}}$$

$$= 0.30 \text{ mmol KMnO}_4$$

$$\text{mass Na}_2\text{C}_2\text{O}_4 = 0.30 \cancel{\text{mmol KMnO}_4} \times \frac{5 \cancel{\text{mmol Na}_2\text{C}_2\text{O}_4}}{2 \cancel{\text{mmol KMnO}_4}}$$

$$\times\ 0.134 \frac{\text{g Na}_2\text{C}_2\text{O}_4}{\cancel{\text{mmol Na}_2\text{C}_2\text{O}_4}}$$

$$= 0.101 \text{ g Na}_2\text{C}_2\text{O}_4$$

Proceeding in the same way, we find for a 45-mL titration:

$$\text{mass Na}_2\text{C}_2\text{O}_4 = 45 \times 0.010 \times \tfrac{5}{2} \times 0.134 = 0.151 \text{ g Na}_2\text{C}_2\text{O}_4$$

Thus, you should weigh between 0.10 and 0.15 g samples of the primary standard.

EXAMPLE 17-4

Exactly 33.31 mL of the solution in Example 17-2 were required to titrate a 0.1278-g sample of primary-standard $Na_2C_2O_4$. What was the molarity of the $KMnO_4$ reagent?

$$\text{amount Na}_2\text{C}_2\text{O}_4 = 0.1278 \cancel{\text{g Na}_2\text{C}_2\text{O}_4} \times \frac{1 \text{ mmol Na}_2\text{C}_2\text{O}_4}{0.13400 \cancel{\text{g Na}_2\text{C}_2\text{O}_4}}$$

$$= 0.95373 \text{ mmol Na}_2\text{C}_2\text{O}_4$$

$$c_{\text{KMnO}_4} = 0.95373 \cancel{\text{mmol Na}_2\text{C}_2\text{O}_4} \times \frac{2 \text{ mmol KMnO}_4}{5 \cancel{\text{mmol Na}_2\text{C}_2\text{O}_4}}$$

$$\times \frac{1}{33.31 \text{ mL KMnO}_4} = 0.01145 \text{ M}$$

Applications of Potassium Permanganate and Cerium(IV) Solutions

Table 17-5 lists some of the many applications of permanganate and cerium(IV) solutions to the volumetric determination of inorganic species. Both reagents have also been applied to the determination of organic compounds with oxidizable functional groups.

EXAMPLE 17-5

Aqueous solutions containing approximately 3% (w/w) H_2O_2 are sold in drug stores as a disinfectant. Propose a method for determining the peroxide con-

TABLE 17-5 Some Applications of Potassium Permanganate and Cerium(IV) in Acid Solution

Analyte	Half-Reaction	Special Conditions
Sn	$Sn^{2+} \rightleftharpoons Sn^{4+} + 2e^-$	Prereduction with Zn
H_2O_2	$H_2O_2 \rightleftharpoons O_2(g) + 2H^+ + 2e^-$	
Fe	$Fe^{2+} \rightleftharpoons Fe^{3+} + e^-$	Prereduction with $SnCl_2$ or with Jones or Walden reductor
$Fe(CN)_6^{4-}$	$Fe(CN)_6^{4-} \rightleftharpoons Fe(CN)_6^{3-} + e^-$	
V	$VO^{2+} + 3H_2O \rightleftharpoons V(OH)_4^+ + 2H^+ + e^-$	Prereduction with Bi amalgam or SO_2
Mo	$Mo^{3+} + 4H_2O \rightleftharpoons MoO_4^{2-} + 8H^+ + 3e^-$	Prereduction with Jones reductor
W	$W^{3+} + 4H_2O \rightleftharpoons WO_4^{2-} + 8H^+ + 3e^-$	Prereduction with Zn or Cd
U	$U^{4+} + 2H_2O \rightleftharpoons UO_2^{2+} + 4H^+ + 2e^-$	Prereduction with Jones reductor
Ti	$Ti^{3+} + H_2O \rightleftharpoons TiO^{2+} + 2H^+ + e^-$	Prereduction with Jones reductor
$H_2C_2O_4$	$H_2C_2O_4 \rightleftharpoons 2CO_2 + 2H^+ + 2e^-$	
Mg, Ca, Zn, Co, Pb, Ag	$H_2C_2O_4 \rightleftharpoons 2CO_2 + 2H^+ + 2e^-$	Sparingly soluble metal oxalates filtered, washed, and dissolved in acid; liberated oxalic acid titrated
HNO_2	$HNO_2 + H_2O \rightleftharpoons NO_3^- + 3H^+ + 2e^-$	15-min reaction time; excess $KMnO_4$ back-titrated
K	$K_2NaCo(NO_2)_6 + 6H_2O \rightleftharpoons Co^{2+} + 6NO_3^- + 12H^+ + 2K^+ + Na^+ + 11e^-$	Precipitated as $K_2NaCo(NO_2)_6$; filtered and dissolved in $KMnO_4$; excess $KMnO_4$ back-titrated
Na	$U^{4+} + 2H_2O \rightleftharpoons UO_2^{2+} + 4H^+ + 2e^-$	Precipitated as $NaZn(UO_2)_3(OAc)_9$; filtered, washed, dissolved; U determined as above

tent of such a preparation using the standard solution described in Examples 17-3 and 17-4. Assume that you wish to use between 30 and 45 mL of the reagent for a titration. The reaction is

$$5H_2O_2 + 2MnO_4^- + 6H^+ \longrightarrow 5O_2 + 2Mn^{2+} + 8H_2O$$

The amount of $KMnO_4$ in 35 to 45 mL of the reagent is between

$$\text{amount } KMnO_4 = 35 \text{ mL KMnO}_4 \times 0.01145 \frac{\text{mmol KMnO}_4}{\text{mL KMnO}_4}$$
$$= 0.401 \text{ mmol KMnO}_4$$

and

$$\text{amount } KMnO_4 = 45 \times 0.1145 = 0.515 \text{ mmol KMnO}_4$$

The amount of H_2O_2 consumed by 0.401 mmol of $KMnO_4$ is

$$\text{amount } H_2O_2 = 0.401 \text{ mmol KMnO}_4 \times \frac{5 \text{ mmol } H_2O_2}{2 \text{ mmol KMnO}_4}$$
$$= 1.00 \text{ mmol } H_2O_2$$

and by 0.515 mmol of $KMnO_4$ is

$$\text{amount } H_2O_2 = 0.515 \times 5/2 = 1.29 \text{ mmol } H_2O_2$$

We, therefore, need to take samples that contain from 1.00 to 1.29 mmol H_2O_2

$$\text{mass sample} = 1.00 \cancel{\text{mmol } H_2O_2} \times 0.03401 \frac{\cancel{\text{g } H_2O_2}}{\cancel{\text{mmol } H_2O_2}} \times \frac{100 \text{ g sample}}{3 \cancel{\text{g } H_2O_2}}$$

$$= 1.1 \text{ g}$$

to

$$\text{mass sample} = 1.29 \times 0.03401 \times \frac{100}{3} = 1.5 \text{ g}$$

Thus, we could weigh out from 1.1- to 1.5-g samples. These should be diluted to perhaps 75 to 100 mL with water and made slightly acidic with dilute H_2SO_4 before titration.

17C-2 Potassium Dichromate

In its analytical applications, dichromate ion is reduced to green chromium(III) ion:

$$Cr_2O_7^{2-} + 14H^+ + 6e^- \rightleftharpoons 2Cr^{3+} + 7H_2O \qquad E^0 = 1.33 \text{ V}$$

Dichromate titrations are generally carried out in solutions that are about 1 M in hydrochloric or sulfuric acid. In these media, the formal potential for the half-reaction is 1.0 to 1.1 V.

Potassium dichromate solutions are indefinitely stable, can be boiled without decomposition, and do not react with hydrochloric acid. Moreover, the primary-standard reagent is available commercially and at a modest cost. The disadvantages of potassium dichromate compared with cerium(IV) and permanganate ion are its lower electrode potential and the slowness of its reaction with certain reducing agents.

The Preparation, Properties, and Uses of Dichromate Solutions

For most purposes, reagent-grade potassium dichromate is sufficiently pure to permit the direct preparation of standard solutions; the solid simply is dried at 150° to 200°C before weighing.

The orange color of a dichromate solution is not intense enough for use in end-point detection. However, diphenylamine sulfonic acid (Table 16-3) is an excellent indicator for titrations with this reagent. The oxidized form of the indicator is violet, and its reduced form is essentially colorless; thus, the color change observed at the end point in a direct titration is from the green of chromium(III) to violet.

Applications of Potassium Dichromate Solutions

The principal use of dichromate is for the volumetric titration of iron(II) based upon the reaction

$$Cr_2O_7^{2-} + 6Fe^{2+} + 14H^+ \longrightarrow 2Cr^{3+} + 6Fe^{3+} + 7H_2O$$

This titration can be performed in the presence of moderate concentrations of hydrochloric acid.

The reaction of dichromate with iron(II) has been widely used for the indirect determination of a variety of oxidizing agents. In these applications, a measured excess of an iron(II) solution is added to an acidic solution of the analyte. The excess iron(II) is then back-titrated with standard potassium dichromate (Section 17B-1). The iron(II) solution is titrated concurrently with the analyte to account for any air-oxidation that may have occurred. This method has been applied to the determination of nitrate, chlorate, permanganate, and dichromate ions as well as organic peroxides and several other oxidizing agents.

Standard solutions of $K_2Cr_2O_7$ have the great advantage that they are indefinitely stable and do not oxidize HCl. Furthermore, primary-standard grade is inexpensive and readily available commercially.

EXAMPLE 17-6

A 5.00-mL sample of brandy was diluted to 1.000 L in a volumetric flask. The ethanol (C_2H_5OH) in a 25.00-mL aliquot of the diluted solution was distilled into 50.00 mL of 0.02000 M $K_2Cr_2O_7$ and oxidized to acetic acid with heating. The reaction is

$$3C_2H_5OH + 2Cr_2O_7^{2-} + 16H^+ \longrightarrow 4Cr^{3+} + 3CH_3COOH + 11H_2O$$

After cooling, 20.00 mL of 0.1253 M Fe^{2+} were pipetted into the flask. The excess Fe^{2+} was then titrated with 7.46 mL of the standard $K_2Cr_2O_7$ to a diphenylamine sulfonic acid end point. Calculate the percent (w/v) C_2H_5OH (46.07 g/mol) in the brandy.

$$\text{total amount } K_2Cr_2O_7 = (50.00 + 7.46)\ \cancel{\text{mL } K_2Cr_2O_7} \times 0.02000\ \frac{\text{mmol } K_2Cr_2O_7}{\cancel{\text{mL } K_2Cr_2O_7}} = 1.1492 \text{ mmol } K_2Cr_2O_7$$

$$\text{amount } K_2Cr_2O_7 \text{ consumed by } Fe^{2+} = 220.00\ \cancel{\text{mL } Fe^{2+}} \times 0.1253\ \frac{\cancel{\text{mmol } Fe^{2+}}}{\cancel{\text{mL } Fe^{2+}}} \times \frac{1 \text{ mmol } K_2Cr_2O_7}{6\ \cancel{\text{mmol } Fe^{2+}}} = 0.41767 \text{ mmol } K_2Cr_2O_7$$

$$\text{amount } K_2Cr_2O_7 \text{ consumed by } C_2H_5OH = (1.1492 - 0.41767) \text{ mmol } K_2Cr_2O_7 = 0.73153 \text{ mmol } K_2Cr_2O_7$$

$$\text{mass } C_2H_5OH = 0.73153\ \cancel{\text{mmol } K_2Cr_2O_7} \times \frac{3\ \cancel{\text{mmol } C_2H_5OH}}{2\ \cancel{\text{mmol } K_2Cr_2O_7}} \times 0.04607\ \frac{\text{g } C_2H_5OH}{\cancel{\text{mmol } C_2H_5OH}} = 0.050552 \text{ g } C_2H_5OH$$

$$\%\ C_2H_5OH\ (\text{w/v}) = \frac{0.050552 \text{ g } C_2H_5OH}{5.00 \text{ mL sample} \times 25.00\ \cancel{\text{mL}}/1000\ \cancel{\text{mL}}} \times 100\% = 40.44\% = 40.4\%\ C_2H_5OH$$

17C-3 Iodine

Solutions of iodine are weak oxidizing agents that are used for the determination of strong reductants. The most accurate description of the half-reaction for iodine in these applications is

$$I_3^- + 2e^- \rightleftharpoons 3I^- \qquad E^0 = 0.536 \text{ V}$$

where I_3^- is the triiodide ion.

Standard iodine solutions have relatively limited application compared with the other oxidants we have described because of their significantly smaller electrode potential. Occasionally, however, this low potential is advantageous because it imparts a degree of selectivity that makes possible the determination of strong reducing agents in the presence of weak ones. An important advantage of iodine is the availability of a sensitive and reversible indicator for the titrations. On the other hand, iodine solutions lack stability and must be restandardized regularly.

The Preparation and Properties of Iodine Solutions

Solutions prepared by dissolving iodine in a concentrated solution of potassium iodide are properly called *triiodide solutions.* In practice, however, they are often termed *iodine solutions* because this terminology accounts for the stoichiometric behavior of these solutions ($I_2 + 2e^- \rightarrow 2I^-$).

Iodine is not very soluble in water (~0.001 M). To obtain solutions having analytically useful concentrations of the element, iodine is ordinarily dissolved in moderately concentrated solutions of potassium iodide. In this medium, iodine is reasonably soluble as a consequence of the reaction

$$I_2(s) + I^- \rightleftharpoons I_3^- \qquad K = 7.1 \times 10^2$$

Iodine dissolves only slowly in solutions of potassium iodide, particularly if the iodide concentration is low. To ensure complete solution, the iodine is always dissolved in a small volume of concentrated potassium iodide, care being taken to avoid dilution of the concentrated solution until the last trace of solid iodine has disappeared. Otherwise, the molarity of the diluted solution gradually increases with time. This problem can be avoided by filtering the solution through a sintered glass crucible before standardization.

Iodine solutions lack stability for several reasons, one being the volatility of the solute. Losses of iodine from an open vessel will occur in a relatively short time even in the presence of an excess of iodide ion. In addition, iodine slowly attacks most organic materials. Consequently, cork or rubber stoppers are never used to close containers of the reagent, and precautions must be taken to protect standard solutions from contact with organic dusts and fumes.

Air-oxidation of iodide ion also causes changes in the molarity of an iodine solution:

$$4I^- + O_2(g) + 4H^+ \longrightarrow 2I_2 + 2H_2O$$

In contrast to the other effects, this reaction causes the molarity of an iodine solution to increase. Air-oxidation is promoted by acids, heat, and light.

The Standardization and Application of Iodine Solutions

Iodine solutions can be standardized against anhydrous sodium thiosulfate or barium thiosulfate monohydrate, both of which are available commercially. The

TABLE 17-6 Some Applications of Iodine Solutions

Substance Analyzed	Half-Reaction
As	$H_3AsO_3 + H_2O \rightleftharpoons H_3AsO_4 + 2H^+ + 2e^-$
Sb	$H_3SbO_3 + H_2O \rightleftharpoons H_3SbO_4 + 2H^+ + 2e^-$
Sn	$Sn^{2+} \rightleftharpoons Sn^{4+} + 2e^-$
H_2S	$H_2S \rightleftharpoons S(s) + 2H^+ + 2e^-$
SO_2	$SO_3^{2-} + H_2O \rightleftharpoons SO_4^{2-} + 2H^+ + 2e^-$
$S_2O_3^{2-}$	$2\ S_2O_3^{2-} \rightleftharpoons S_4O_6^{2-} + 2e^-$
N_2H_4	$N_2H_4 \rightleftharpoons N_2(g) + 4H^+ + 4e^-$
Ascorbic acid*	$C_6H_8O_6 \rightleftharpoons C_6H_6O_6 + 2H^+ + 2e^-$

*For the structure of ascorbic acid, see Section 36H-3.

reaction between iodine and sodium thiosulfate is discussed in detail in Section 17B-2. Often, solutions of iodine are standardized against solutions of sodium thiosulfate that have in turn been standardized against potassium iodate or potassium dichromate (Section 17B-2).

Table 17-6 summarizes methods that use iodine as an oxidizing agent.

17D SOME SPECIALIZED OXIDANTS

In this section, we describe three oxidizing agents that are used primarily for determining certain special groups of compounds. Potassium bromate is used for the determination of organic compounds that contain olefinic and certain types of aromatic functional groups; periodic acid reacts selectively with organic compounds having hydroxyl, carbonyl, or amine groups on adjacent carbon atoms; and Karl Fischer reagent is widely employed for the determination of water in a variety of organic and inorganic samples.

17D-1 Potassium Bromate as a Source of Bromine

Primary-standard potassium bromate is available from commercial sources and can be used directly to prepare standard solutions that are stable indefinitely. Direct titrations with potassium bromate are relatively few. Instead, the reagent is a convenient and widely used stable source of bromine.[4] In this application, an unmeasured excess of potassium bromide is added to an acidic solution of the analyte. Upon introduction of a measured volume of standard potassium bromate, a stoichiometric quantity of bromine is produced.

$$\underset{\text{standard solution}}{BrO_3^-} + \underset{\text{excess}}{5Br^-} + 6H^+ \longrightarrow 3Br_2 + 3H_2O$$

1 mol $KBrO_3^- \equiv 3$ mol Br_2

This indirect generation circumvents the problems associated with the use of standard bromine solutions, which lack stability.

[4]For a discussion of bromate solutions and their applications, see M. R. F. Ashworth, *Titrimetric Organic Analysis,* Part I, pp. 118–130. New York: Interscience, 1964.

The primary use of standard potassium bromate is for the determination of organic compounds that react with bromine. Few of these reactions are rapid enough to permit direct titration. Instead, a measured excess of standard bromate is added to the solution that contains the sample plus an excess of potassium bromide. After acidification, the mixture is allowed to stand in a glass-stoppered vessel until the bromine/analyte reaction is judged complete. To determine the excess bromine, an excess of potassium iodide is introduced so that the following reaction occurs:

1 mol $BrO_3^- \equiv 6$ mol e^-

$$2I^- + Br_2 \longrightarrow I_2 + 2Br^-$$

The liberated iodine is then titrated with standard sodium thiosulfate (Equation 17-1).

Bromine is incorporated into an organic molecule either by substitution or by addition.

Substitution Reactions

Halogen substitution involves the replacement of hydrogen in an aromatic ring by a halogen. Substitution methods have been successfully applied to the determination of aromatic compounds that contain strong ortho-para-directing groups, particularly amines and phenols.

EXAMPLE 17-7

A 0.2981-g sample of an antibiotic powder containing sulfanilamide was dissolved in HCl and the solution diluted to 100.0 mL. A 20.00-mL aliquot was transferred to a flask, and followed by 25.00 mL of 0.01767 M $KBrO_3$. An excess of KBr was added to form Br_2, and the flask was stoppered. After 10 min, during which time the Br_2 brominated the sulfanilamide, an excess of KI was added. The liberated iodine was titrated with 12.92 mL of 0.1215 M sodium thiosulfate. The reactions are

$$BrO_3^- + 5Br^- + 6H^+ \longrightarrow 3Br_2 + 3H_2O$$

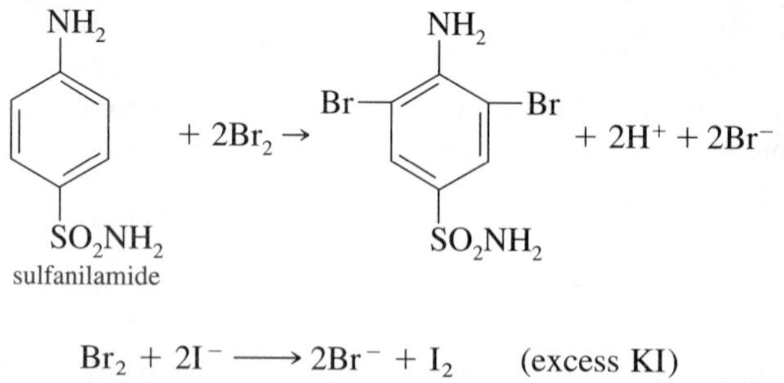

sulfanilamide

$$Br_2 + 2I^- \longrightarrow 2Br^- + I_2 \quad \text{(excess KI)}$$

$$I_2 + 2S_2O_3^{2-} \longrightarrow S_4O_6^{2-} + 2I^-$$

Calculate the percent $NH_2C_6H_4SO_2NH_2$ (172.21 g/mol) in the powder.

First we determine the amount of Br_2:

$$\text{total amount Br}_2 = 25.00\ \cancel{\text{mL KBrO}_3} \times 0.01767\ \frac{\cancel{\text{mmol KBrO}_3}}{\cancel{\text{mL KBrO}_3}} \times \frac{3\ \text{mmol Br}_2}{\cancel{\text{mmol KBrO}_3}} = 1.32525\ \text{mmol Br}_2$$

We next calculate how much Br_2 was in excess over that required to brominate the analyte:

$$\text{amount excess Br}_2 = \text{amount I}_2 = 12.92\ \cancel{\text{mL Na}_2\text{S}_2\text{O}_3} \times 0.1215\ \frac{\cancel{\text{mmol Na}_2\text{S}_2\text{O}_3}}{\cancel{\text{mL Na}_2\text{S}_2\text{O}_3}} \times \frac{1\ \text{mmol I}_2}{2\ \cancel{\text{mmol Na}_2\text{S}_2\text{O}_3}} = 0.78489\ \text{mmol I}_2 = 0.78489\ \text{mmol Br}_2$$

The amount Br_2 consumed by the sample is given by

$$\text{amount Br}_2 = 1.32525 - 0.78489 = 0.54036\ \text{mmol Br}_2$$

$$\text{mass analyte} = 0.54036\ \cancel{\text{mmol Br}_2} \times \frac{1\ \cancel{\text{mmol analyte}}}{2\ \cancel{\text{mmol Br}_2}} \times 0.17221\ \frac{\text{g analyte}}{\cancel{\text{mmol analyte}}} = 0.046528\ \text{g analyte}$$

$$\%\ \text{analyte} = \frac{0.046528\ \text{g analyte}}{0.2891\ \text{g sample} \times 20.00\ \cancel{\text{mL}}/100\ \cancel{\text{mL}}} \times 100\% = 80.47\%\ \text{sulfanilamide}$$

An important application of a bromine substitution reaction is the determination of 8-hydroxyquinoline:

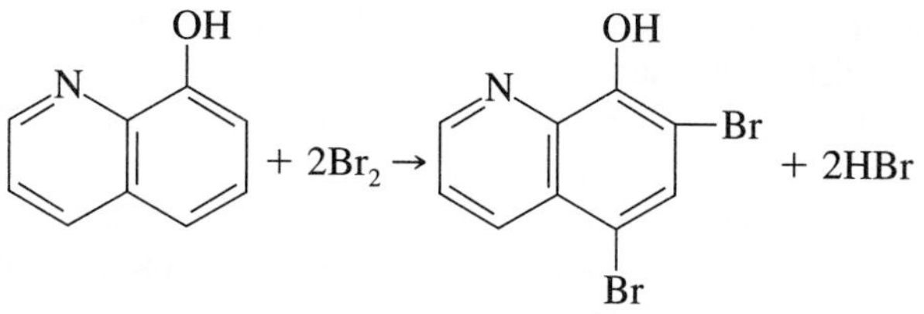

In contrast to most bromine substitutions, this reaction takes place rapidly enough in hydrochloric acid solution to make direct titration feasible. The titration of 8-hydroxyquinoline with bromine is particularly significant because the former is an excellent precipitating reagent for cations (Section 5D-3). For example, aluminum can be determined according to the sequence

$$Al^{3+} + 3HOC_9H_6N \xrightarrow{\text{pH 4–9}} Al(OC_9H_6N)_3\,(s) + 3H^+$$

$$Al(OC_9H_6N)_3\,(s) \xrightarrow{\text{hot 4 M HCl}} 3HOC_9H_6N + Al^{3+}$$

$$3HOC_9H_6N + 6Br_2 \longrightarrow 3HOC_9H_4NBr_2 + 6HBr$$

The stoichiometric relationships in this case are

$$1 \text{ mol } Al^{3+} \equiv 3 \text{ mol } HOC_9H_6N \equiv 6 \text{ mol } Br_2 \equiv 2 \text{ mol } KBrO_3$$

Addition Reactions

Addition reactions involve the opening of an olefinic double bond. For example, 1 mol of ethylene reacts with 1 mol of bromine in the reaction

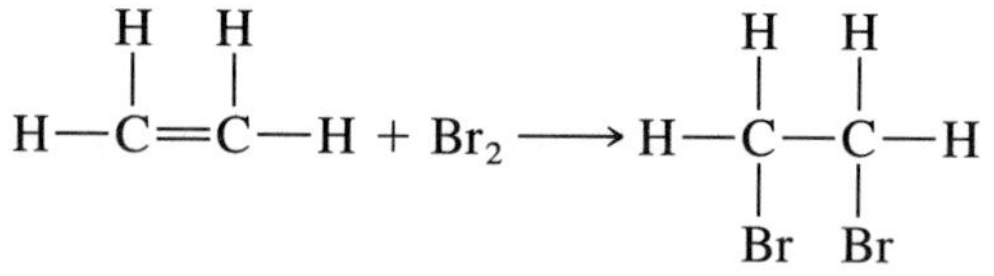

The literature contains numerous references to the use of bromine for the estimation of olefinic unsaturation in fats, oils, and petroleum products.

17D-2 Periodic Acid

Aqueous solutions of iodine in the +7 oxidation state are highly complex. In the presence of strong acid, paraperiodic acid, H_5IO_6, and its conjugate base predominate, although the metaperiodates HIO_4 and IO_4^- are undoubtedly present as well. The reduction of periodic acid to iodate ion is best described by the half-reaction

$$H_5IO_6 + H^+ + 2e^- \rightleftharpoons IO_3^- + 3H_2O \qquad E^0 = 1.6 \text{ V}$$

The Preparation and Properties of Periodic Acid Solutions

Several periodates are available for the preparation of standard solutions. Among these is paraperiodic acid itself, a crystalline, readily soluble, hygroscopic solid. An even more useful compound is sodium metaperiodate, $NaIO_4$, which is soluble in water to the extent of 0.06 M at 25°C. Sodium paraperiodate, Na_5IO_6, is not sufficiently soluble for the preparation of standard solutions; it is, however, readily converted to the more soluble metaperiodate by recrystallization from hot concentrated nitric acid. Potassium metaperiodate can be used as a primary standard for the preparation of periodate solutions. Although its solubility is only about 5 g/L at room temperature, it readily dissolves at elevated temperatures and can be subsequently converted to a more soluble form by the addition of base.

Periodate solutions vary considerably in stability, depending on their mode of preparation and storage. A solution prepared by dissolving sodium metaperiodate in water decomposes at the rate of several percent per week. In contrast, a solution of potassium metaperiodate in excess alkali changes no more than 0.3% to 0.4%

in 100 days. The most stable periodate solutions appear to be those containing an excess of sulfuric acid; such solutions decrease in molarity by less than 0.1% in 4 months.

The Standardization of Periodate Solutions

Periodate solutions are most conveniently standardized by buffering aliquots of the reagent with solid borax or sodium hydrogen carbonate to ensure that they remain slightly alkaline. An excess of potassium iodide is then introduced, which results in the formation of 1 mol of iodine for each mole of periodate:

$$H_4IO_6^- + 2I^- \longrightarrow IO_3^- + I_2 + OH^- + H_2O$$

As long as the solution is kept slightly alkaline, further reduction of iodate does not occur, and the liberated iodine can be titrated directly with a standard solution of sodium thiosulfate or sodium arsenite.

Applications of Periodic Acid

The reason periodic acid is so widely used is that it reacts remarkably selectively with organic compounds containing certain combinations of functional groups.[5] Ordinarily, these oxidations are performed at room temperature in the presence of a measured excess of the periodate; most reactions are complete in 0.5 to 1 hr. After oxidation, the excess periodate is determined by the method described for standardization. Alternatively, a reaction product such as ammonia, formaldehyde, or a carboxylic acid must be determined. In this application, the exact quantity of periodate used need not be known.

Periodate oxidations are usually carried out in aqueous solution, although solvents such as methanol, ethanol, or dioxane may be added to enhance the solubility of the sample.

Compounds Attacked by Periodate

At room temperature, organic compounds containing aldehyde, ketone, or alcohol groups *on adjacent carbon atoms* are rapidly oxidized by periodic acid. Primary and secondary α-hydroxylamines are also readily attacked; α-diamines are not. With few exceptions, other organic compounds do not react at a significant rate. Thus, compounds containing isolated aldehydes, ketone, alcohol, or amine groups are not affected by periodic acid, nor are compounds with a carboxylic acid group either isolated from or adjacent to any of the reactive groups. At elevated temperatures, the extraordinary selectivity of periodic acid tends to disappear.

Periodate oxidations of organic compounds follow a regular and predictable set of rules:

1. Attack of adjacent functional groups always results in rupture of the carbon-to-carbon bond between the groups.

---

[5]The use of periodic acid for this purpose was first reported by L. Malaprade, *Compt. Rend.*, **1928,** *186,* 382. See also M. R. F. Ashworth, *Titrimetric Organic Analysis,* Part II, pp. 724–744. New York: Interscience, 1965.

2. A carbon atom bonded to a hydroxyl group is oxidized to an aldehyde or ketone.
3. A carbonyl group is converted to a carboxylic acid group.
4. A carbon atom bonded to an amine group loses ammonia (or a substituted amine) and is itself converted to an aldehyde.

The following half-reactions illustrate these rules:

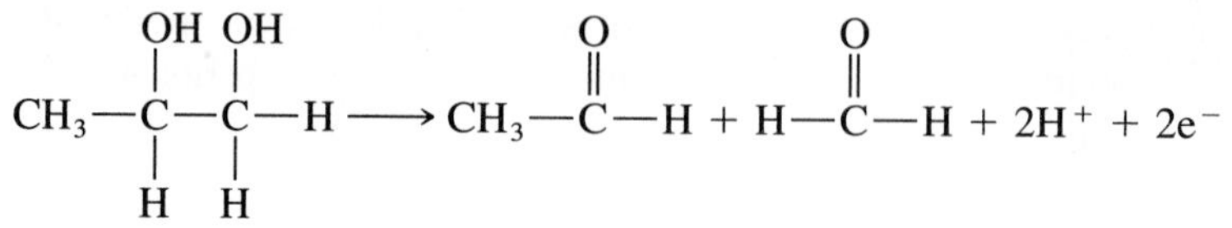

propylene glycol

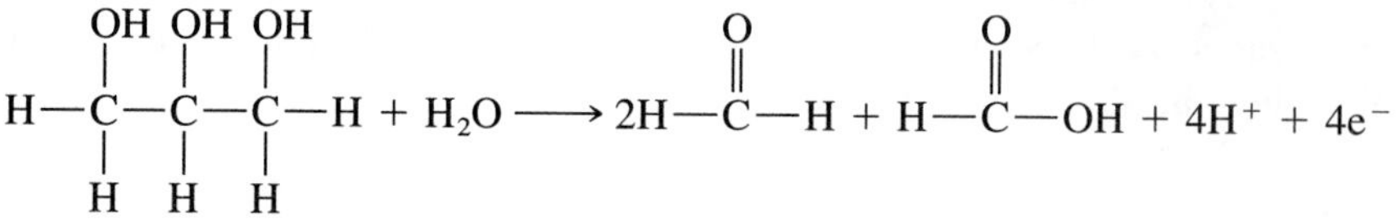

glycerol

For predicting the reaction products of the oxidation of glycerol, the first step can be thought of as producing 1 mol of formaldehyde and 1 mol of an α-hydroxyaldehyde (glycolic aldehyde). The latter is then further oxidized and according to the second rule produces a second mole of formaldehyde plus 1 mol of formic acid.

Additional examples of the four rules include

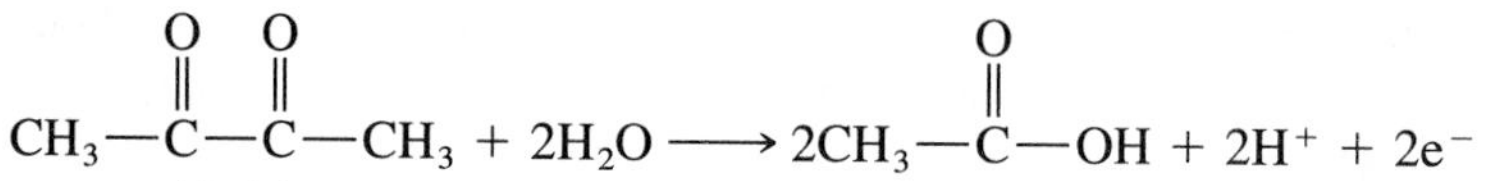

biacetyl

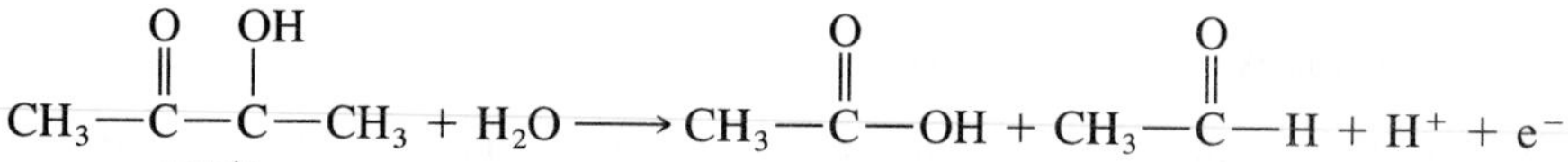

acetoin

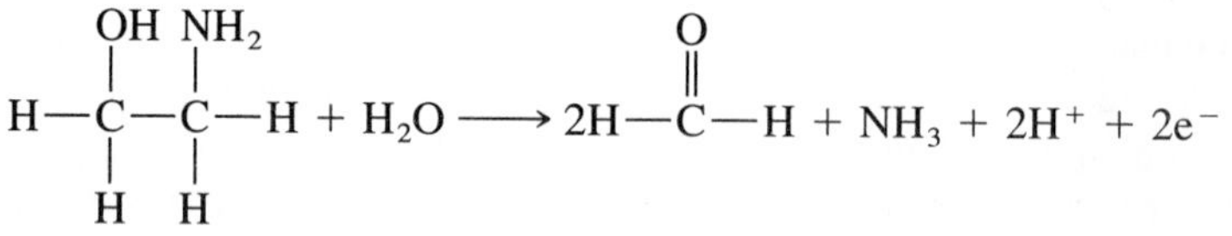

ethanolamine

The Determination of α-Hydroxylamines As mentioned earlier, the periodate oxidation of compounds with hydroxyl and amino groups on adjacent carbon atoms results in the formation of aldehydes and the liberation of ammonia (rule 4). The latter is readily distilled from the alkaline oxidation mixture and determined by a neutralization titration. This procedure is particularly useful in the analysis of mixtures of the various amino acids that occur in proteins. Because only serine, threonine, β-hydroxyglutamic acid, and hydroxylysine have the requisite structure for the liberation of ammonia, the method is selective for these species.

17D-3 Karl Fischer Reagent for Water Determination

A number of chemical methods for the determination of water in solids and organic solvents have been devised (Section 32C). Unquestionably, the most important of these involves the use of Karl Fischer reagent, which is relatively specific for water.[6]

The Reaction and Stoichiometry

Karl Fischer reagent is composed of iodine, sulfur dioxide, pyridine, and methanol. This mixture reacts with water according to the equation

$$C_5H_5N \cdot I_2 + C_5H_5N \cdot SO_2 + C_5H_5N + H_2O \longrightarrow 2C_5H_5N \cdot HI + C_5H_5N \cdot SO_3 \quad (17\text{-}2)$$

$$C_5H_5 \cdot SO_3 + CH_3OH \longrightarrow C_5H_5N(H)SO_4CH_3 \quad (17\text{-}3)$$

Note that only the first step, which involves the oxidation of sulfur dioxide by iodine to give sulfur trioxide and hydrogen iodide, consumes water. In the presence of a large amount of pyridine, C_5H_5N, all reactants and products exist as complexes, as indicated in the equations. The second step, which occurs when an excess of methanol is present, is important to the success of the titration because the pyridine/sulfur trioxide complex is also capable of consuming water:

$$C_5H_5N \cdot SO_3 + H_2O \longrightarrow C_5H_5NHSO_4H \quad (17\text{-}4)$$

This last reaction is undesirable because it is not as specific for water as the reaction shown in Equation 17-2; it can be prevented completely by having a large excess of methanol present.

Properties of the Reagent

Equation 17-3 indicates that the stoichiometry of the Karl Fischer titration involves the consumption of 1 mol of iodine, 1 mol of sulfur dioxide, and 3 mol of pyridine for each mole of water. In practice, excesses of both sulfur dioxide and pyridine are employed so that the reagent's combining capacity for water is determined by its iodine content. For typical applications, between 2 and 5 mg of water react with each milliliter of reagent; a twofold excess of sulfur dioxide and a threefold to fourfold excess of pyridine are provided.

Karl Fischer reagent decomposes on standing. Because decomposition is particularly rapid immediately after preparation, it is common practice to prepare the reagent a day or two before it is to be used. Ordinarily, its strength must be established at least daily against a standard solution of water in methanol. A proprietary commercial Karl Fischer reagent that is reported to require only occasional restandardization is now available.

It is obvious that great care must be exercised to keep atmospheric moisture from contaminating the Karl Fischer reagent and the sample. All glassware must be carefully dried before use, and the standard solution must be stored out of

---

[6]For a monograph on this subject, see J. Mitchell and D. M. Smith, *Aquametry,* 2nd ed. New York: Interscience, 1977.

contact with air. It is also necessary to minimize contact between the atmosphere and the solution during the titration.

End-Point Detection

The end point in a Karl Fischer titration is signaled by the appearance of the first excess of the pyridine/iodine complex when all water has been consumed. The color of the reagent is intense enough for a visual end point; the change is from the yellow of the reaction products to the brown of the excess reagent. With some practice, and in the absence of other colored materials, the end point can be established with a reasonable degree of certainty (that is, to perhaps ± 0.2 mL).

Various electrometric end points are also employed for Karl Fischer titrations, the most widely used being the amperometric technique with twin microelectrodes (see end of Section 21B-4). Several instrument manufacturers offer fully automated instruments for performing such titrations.

Applications

Karl Fischer reagent has been applied to the determination of water in numerous types of samples.[7] There are several variations of the basic technique, depending upon the solubility of the material, the state in which the water is retained, and the physical state of the sample. If the sample can be dissolved completely in methanol, a direct and rapid titration is usually feasible. This method has been applied to the determination of water in many organic acids, alcohols, esters, ethers, anhydrides, and halides. The hydrated salts of most organic acids, as well as the hydrates of a number of inorganic salts that are soluble in methanol, can also be determined by direct titration.

Direct titration of samples that are only partially dissolved in the reagent usually leads to incomplete recovery of the water. Satisfactory results with this type of sample are often obtained, however, by the addition of an excess of reagent and back-titration with a standard solution of water in methanol after a suitable reaction time. An effective alternative is to extract the water from the sample by refluxing with anhydrous methanol or other organic solvents. The resulting solution is then titrated directly with the Karl Fischer reagent.

Difficulty is also encountered in the analysis of sorbed moisture and tightly bound hydrate water. For these, the extraction techniques just described are frequently effective.

Certain substances interfere with the Karl Fischer method. Among these are compounds that react with one of the components of the reagent to produce water. For example, carbonyl compounds combine with methanol to give acetals:

$$\mathrm{RCHO} + 2\mathrm{CH_3OH} \longrightarrow \mathrm{R{-}CH}\begin{matrix}\diagup \mathrm{OCH_3} \\ \diagdown \mathrm{OCH_3}\end{matrix} + \mathrm{H_2O}$$

---

[7]For a complete discussion of the applications of the reagent, see J. Mitchell and D. M. Smith, *Aquametry,* 2nd ed. New York: Interscience, 1977.

The result is a fading end point. Many metal oxides react with the hydrogen iodide formed in the titration to give water:

$$MO + 2HI \rightleftharpoons MI_2 + H_2O$$

Again, erroneous data result. Preliminary treatment of the sample can sometimes prevent these interferences.

Oxidizing or reducing substances frequently interfere with the Karl Fischer titration by reoxidizing the iodide produced or reducing the iodine in the reagent.

17E QUESTIONS AND PROBLEMS

***17-1.** Write balanced net-ionic equations to describe
(a) the oxidation of Mn^{2+} to MnO_4^- by ammonium peroxydisulfate.
(b) the oxidation of Ce^{3+} to Ce^{4+} by sodium bismuthate.
(c) the oxidation of U^{4+} to UO_2^{2+} by H_2O_2.
(d) the reaction of $V(OH)_4^+$ in a Walden reductor.
(e) the titration of H_2O_2 with $KMnO_4$.
(f) the reaction between KI and ClO_3^- in acidic solution.

17-2. Write balanced net-ionic equations to describe
(a) the reduction of Fe^{3+} to Fe^{2+} by SO_2.
(b) the reaction of H_2MoO_4 in a Jones reductor.
(c) the oxidation of HNO_2 by a solution of MnO_4^-.
(d) the reaction of aniline ($C_6H_4NH_2$) with a mixture of $KBrO_3$ and KBr in acidic solution.
(e) the air-oxidation of $HAsO_3^{2-}$ to $HAsO_4^{2-}$.
(f) the reaction of KI with HNO_2 in acidic solution.

***17-3.** Why is a Walden reductor always used with solutions that contain appreciable concentrations of HCl?

17-4. Why is zinc amalgam preferable to pure zinc in a Jones reductor?

***17-5.** Write a balanced net-ionic equation for the reduction of UO_2^{2+} in a Walden reductor.

17-6. Write a balanced net-ionic equation for the reduction of TiO^{2+} in a Jones reductor.

***17-7.** Why are standard solutions of reductants less often used for titrations than standard solutions of oxidants?

***17-8.** Why are standard $KMnO_4$ solutions seldom used for the titration of solutions containing HCl?

17-9. Why are Ce^{4+} solutions never used for the titration of reductants in basic solutions?

***17-10.** Write a balanced net-ionic equation showing why $KMnO_4$ end points fade.

17-11. Why are $KMnO_4$ solutions filtered before they are standardized?

17-12. Why are solutions of $KMnO_4$ and $Na_2S_2O_3$ generally stored in dark reagent bottles?

***17-13.** When a solution of $KMnO_4$ was left standing in a buret for 3 hr, a brownish ring formed at the surface of the liquid. Write a balanced net-ionic equation to account for this observation.

17-14. What is the primary use of standard $K_2Cr_2O_7$ solutions?

***17-15.** Why are iodine solutions prepared by dissolving I_2 in concentrated KI?

17-16. A standard solution of I_2 increased in molarity with standing. Write a balanced net-ionic equation that accounts for the increase.

***17-17.** When a solution of $Na_2S_2O_3$ is introduced into a solution of HCl, a cloudiness develops almost immediately. Write a balanced net-ionic equation that accounts for this phenomenon.

17-18. Suggest a way in which a solution of KIO_3 could be used as a source of known quantities of I_2.

***17-19.** Write balanced equations showing how $KBrO_3$ could be used as a primary standard for solutions of $Na_2S_2O_3$.

17-20. Write balanced equations showing how $K_2Cr_2O_7$ could be used as a primary standard for solutions of $Na_2S_2O_3$.

***17-21.** Write a balanced net-ionic equation describing the titration of hydrazine (N_2H_4) with standard iodine.

17-22. In the titration of I_2 solutions with $Na_2S_2O_3$, starch indicator is never added until just before chemical equivalence. Why?

17-23. A solution prepared by dissolving a 0.2464-g sample of electrolytic iron wire in acid was passed through a Jones reductor. The iron(II) in the resulting solution required a 39.31-mL titration. Calculate the molar oxidant concentration if the titrant used was
*(a) Ce^{4+} (product: Ce^{3+}).
(b) $Cr_2O_7^{2-}$ (product: Cr^{3+}).
*(c) MnO_4^- (product: Mn^{2+}).
(d) $V(OH)_4^+$ (product: VO^{2+}).
*(e) IO_3^- (product: ICl_2^-).

***17-24.** How would you prepare 250.0 mL of 0.03500 M $K_2Cr_2O_7$?

17-25. How would you prepare 2.000 L of 0.02500 M $KBrO_3$?

***17-26.** How would you prepare 1.5 L of approximately 0.1 M $KMnO_4$?

17-27. How would you prepare 2.0 L of approximately 0.05 M I_3^-?

***17-28.** Titration of 0.1467 g of primary-standard $Na_2C_2O_4$ required 28.85 mL of a potassium permanganate solution. Calculate the molar concentration of $KMnO_4$ in this solution.

17-29. A 0.1809-g sample of pure iron wire was dissolved in acid, reduced to the +2 state, and titrated with 31.33 mL of cerium(IV). Calculate the molar concentration of the Ce^{4+} solution.

***17-30.** The iodine produced when an excess of KI was added to a solution containing 0.1518 g of $K_2Cr_2O_7$ required a 46.13-mL titration with $Na_2S_2O_3$. Calculate the molar concentration of the thiosulfate solution.

17-31. A 0.1017-g sample of $KBrO_3$ was dissolved in dilute HCl and treated with an unmeasured excess of KI. The liberated iodine required 39.75 mL of a sodium thiosulfate solution. Calculate the molar concentration of the $Na_2S_2O_3$.

***17-32.** The Sb(III) in a 1.080-g ore sample required a 41.67-mL titration with 0.03134 M I_2 [reaction product: Sb(V)]. Express the results of this analysis in terms of (a) percentage of Sb and (b) percentage of stibnite (Sb_2S_3).

17-33. Calculate the percentage of MnO_2 in a mineral specimen if the I_2 liberated by a 0.1344-g sample in the net reaction

$$MnO_2(s) + 4H^+ + 2I^- \longrightarrow Mn^{2+} + I_2 + 2H_2O$$

required 32.30 mL of 0.07220 M $Na_2S_2O_3$.

***17-34.** Under suitable conditions, thiourea is oxidized to sulfate by solutions of bromate

$$3CS(NH_2)_2 + 4BrO_3^- + 3H_2O \rightleftharpoons 3CO(NH_2)_2 + 3SO_4^{2-} + 4Br^- + 6H^+$$

A 0.0715-g sample of a material was found to consume 14.10 mL of 0.00833 M $KBrO_3$. What was the percent purity of the thiourea sample?

***17-35.** A 0.7120-g specimen of iron ore was brought into solution and passed through a Jones reductor. Titration of the Fe(II) produced required 39.21 mL of 0.02086 M $KMnO_4$. Express the results of this analysis in terms of percent (a) Fe and (b) Fe_2O_3.

17-36. The Sn in a 0.4352-g mineral specimen was reduced to the +2 state with Pb and titrated with 29.77 mL of 0.01735 M $K_2Cr_2O_7$. Calculate the results of this analysis in terms of percent (a) Sn and (b) SnO_2.

***17-37.** Treatment of hydroxylamine (H_2NOH) with an excess of Fe(II) results in the formation of N_2O and an equivalent amount of Fe(II):

$$2H_2NOH + 4Fe^{3+} \longrightarrow N_2O(g) + 4Fe^{2+} + 4H^+ + H_2O$$

Calculate the molar concentration of an H_2NOH solution if the Fe(II) produced by treatment of a 50.00-mL aliquot required 23.61 mL of 0.02170 M $K_2Cr_2O_7$.

17-38. The organic matter in a 0.9280-g sample of burn ointment was eliminated by ashing, following which the solid residue of ZnO was dissolved in acid. Treatment with $(NH_4)_2C_2O_4$ resulted in the formation of the sparingly soluble ZnC_2O_4. The solid was filtered, washed, and then redissolved in dilute acid. The liberated $H_2C_2O_4$ required 37.81 mL of 0.01508 M $KMnO_4$. Calculate the percentage of ZnO in the medication.

***17-39.** The $KClO_3$ in a 0.1342-g sample of an explosive was determined by reaction with 50.00 mL of 0.09601 M Fe^{2+}:

$$ClO_3^- + 6Fe^{2+} + 6H^+ \longrightarrow Cl^- + 3H_2O + 6Fe^{3+}$$

When the reaction was complete, the excess Fe^{2+} was back-titrated with 12.99 mL of 0.08362 M Ce^{4+}. Calculate the percentage of $KClO_3$ in the sample.

17-40. The tetraethyl lead [$Pb(C_2H_5)_4$] in a 25.00-mL sample of aviation gasoline was shaken with 15.00 mL of 0.02095 M I_2. The reaction is

$$Pb(C_2H_5)_4 + I_2 \longrightarrow Pb(C_2H_5)_3I + C_2H_5I$$

After the reaction was complete, the unused I_2 was titrated with 6.09 mL of 0.03465 M $Na_2S_2O_3$. Calculate the weight (in milligrams) of $Pb(C_2H_5)_4$ (323.4 g/mol) in each liter of the gasoline.

***17-41.** An 8.13-g sample of an ant-control preparation was decomposed by wet-ashing with H_2SO_4 and HNO_3. The As in the residue was reduced to the trivalent state with hydrazine. After removal of the excess reducing agent, the As(III) required a 23.77-mL titration with 0.02425 M I_2 in a faintly alkaline medium. Express the results of this analysis in terms of percentage of As_2O_3 in the original sample.

17-42. A sample of alkali metal chlorides was analyzed for sodium by dissolving a 0.800-g sample in water and diluting to exactly 500 mL. A 25.0-mL aliquot of the diluted solution was treated to precipitate the sodium as $NaZn(UO_2)_3(OAc)_9 \cdot 6H_2O$. The precipitate was filtered, dissolved in acid, and passed through a lead reductor, which converted the uranium to U^{4+}. Oxidation of this to UO_2^{2+} required 19.9 mL of 0.0100 M $K_2Cr_2O_7$. Calculate the percentage of NaCl in the sample.

***17-43.** The ethyl mercaptan concentration in a mixture was determined by shaking a 1.657-g sample with 50.0 mL of 0.01194 M I_2 in a tightly stoppered flask:

$$2C_2H_5SH + I_2 \longrightarrow C_2H_5SSC_2H_5 + 2I^- + 2H^+$$

The excess I_2 was back-titrated with 16.77 mL of 0.01325 M $Na_2S_2O_3$. Calculate the percentage of C_2H_5SH (62.13 g/mol).

17-44. A 4.971-g sample containing the mineral tellurite was dissolved and then treated with 50.00 mL of 0.03114 M $K_2Cr_2O_7$:

$$3TeO_2 + Cr_2O_7^{2-} + 8H^+ \longrightarrow 3H_2TeO_4 + 2Cr^{3+} + H_2O$$

Upon completion of the reaction, the excess $Cr_2O_7^{2-}$ required a 10.05-mL back-titration with 0.1135 M Fe^{2+}. Calculate the percentage of TeO_2 in the sample.

***17-45.** A sensitive method for I^- in the presence of Cl^- and Br^- entails oxidation of the I^- to IO_3^- with Br_2. The excess Br_2 is then removed by boiling or by reduction with formate ion. The IO_3^- produced is determined by addition of excess I^- and titration of the resulting I_2. A 1.204-g sample of mixed halides was dissolved and analyzed by the foregoing procedure; 20.66 mL of 0.05551 M thiosulfate were required. Calculate the percentage of KI in the sample.

***17-46.** A 1.065-g sample of stainless steel was dissolved in HCl (this treatment converts the Cr present to Cr^{3+}) and diluted to 500.0 mL in a volumetric flask. One 50.00-mL aliquot was passed through a Walden reductor and then titrated with 13.72 mL of 0.01920 M $KMnO_4$. A 100.0-mL aliquot was passed through a Jones reductor into 50 mL of 0.1000 M Fe^{3+}. Titration of the resulting solution required 36.43 mL of the $KMnO_4$ solution. Calculate the percentages of Fe and Cr in the alloy.

17-47. A 2.559-g sample containing both Fe and V was dissolved under conditions that converted the elements to Fe(III) and V(V). The solution was diluted to 500.0 mL, and a 50.00-mL aliquot was passed through a Walden reductor and titrated with 17.74 mL of 0.1000 M Ce^{4+}. A second 50.00-mL aliquot was passed through a Jones reductor and required 44.67 mL of the same Ce^{4+} solution to reach an end point. Calculate the percentage of Fe_2O_3 and V_2O_5 in the sample.

***17-48.** A 25.00-mL aliquot of a solution containing Tl(I) ion was treated with K_2CrO_4. The Tl_2CrO_4 was filtered, washed free of excess precipitating agent, and dissolved in dilute H_2SO_4. The $Cr_2O_7^{2-}$ produced was titrated with 40.60 mL of 0.1004 M Fe^{2+} solution. What was the mass of Tl in the sample? The reactions are

$$2Tl^+ + CrO_4^{2-} \longrightarrow Tl_2CrO_4(s)$$

$$2Tl_2CrO_4(s) + 2H^+ \longrightarrow 4Tl^+ + Cr_2O_7^{2-} + H_2O$$

$$Cr_2O_7^{2-} + 6Fe^{2+} + 14H^+ \longrightarrow 6Fe^{3+} + 2Cr^{3+} + 7H_2O$$

***17-49.** A gas mixture was passed at the rate of 2.50 L/min through a solution of sodium hydroxide for a total of 64.00 min. The SO_2 in the mixture was retained as sulfite ion

$$SO_2(g) + 2OH^- \longrightarrow SO_3^{2-}$$

After acidification with HCl, the sulfite was titrated with 4.98 mL of 0.003125 M KIO_3

$$IO_3^- + 2H_2SO_3 + 2Cl^- \longrightarrow ICl_2^- + 2SO_4^{2-} + 2H^+$$

Use 1.20 g/L for the density of the mixture and calculate the concentration of SO_2 in ppm.

17-50. A 24.7-L sample of air drawn from the vicinity of a coking oven was passed over iodine pentoxide at 150°C, where CO was converted to CO_2 and a chemically equivalent quantity of I_2 was produced:

$$I_2O_5(s) + 5CO(g) \longrightarrow 5CO_2(g) + I_2(g)$$

The I_2 distilled at this temperature and was collected in a solution of KI. The I_3^- produced was titrated with 7.76 mL of 0.00221 M $Na_2S_2O_3$. Does the air in this space comply with federal regulations that mandate a maximum CO level no greater than 50 ppm?

***17-51.** A 30.00-L air sample was passed through an absorption tower containing a solution of Cd^{2+}, where H_2S was retained as CdS. The mixture was acidified and treated with 10.00 mL of 0.01070 M I_2. After the reaction

$$S^{2-} + I_2 \longrightarrow S(s) + 2I^-$$

was complete, the excess iodine was titrated with 12.85 mL of 0.01344 M thiosulfate. Calculate the concentration of H_2S in ppm; use 1.20 g/L for the density of the gas stream.

17-52. A square of photographic film 2.0 cm on an edge was suspended in a 5% solution of $Na_2S_2O_3$ to dissolve the silver halides. After removal and washing of the film, the solution was treated with an excess of Br_2 to oxidize the iodide present to IO_3^- and destroy the excess thiosulfate ion. The solution was boiled to remove the bromine, and an excess of iodide was added. The liberated iodine was titrated with 13.7 mL of 0.0352 M thiosulfate solution.

(a) Write balanced equations for the reactions involved in the method.

(b) Calculate the milligrams of AgI per square centimeter of film.

CHAPTER

18

Theory of Potentiometry

In Chapter 15, we showed that the potential of an electrode, relative to the standard hydrogen electrode, is determined by the activity of one or more of the species in the solution in which the electrode is immersed. This chapter describes how electrode potential data are used to determine the concentration of analytes.[1] Analytical methods that are based upon potential measurements are termed *potentiometric methods,* or *potentiometry.*

18A GENERAL PRINCIPLES

In Feature 15-3, we showed that absolute values for individual half-cell potentials cannot be determined in the laboratory. That is, only cell potentials can be obtained experimentally. Figure 18-1 shows a typical cell for potentiometric analysis. This cell can be depicted as

$$\underbrace{\text{reference electrode}}_{E_{\text{ref}}}\,|\,\underbrace{\text{salt bridge}}_{E_{\text{j}}}\,|\,\text{analyte solution}\,|\,\underbrace{\text{indicator electrode}}_{E_{\text{ind}}}$$

A **reference electrode** is a half-cell having a known electrode potential that remains constant and is independent of the composition of the analyte solution.

Reference electrodes are *always* treated as anodes in this text.

An **indicator electrode** is an electrode system having a potential that varies in a known way with variations in the concentration of an analyte.

The *reference electrode* E_{ref} in this diagram is a half-cell with an accurately known electrode potential that is independent of the concentration of the analyte or any other ions in the solution under study. It can be a standard hydrogen electrode but seldom is because a standard hydrogen electrode is somewhat troublesome to maintain and use. By convention, the reference electrode is always treated as the anode in potentiometric measurements. The *indicator electrode,* which is immersed in a solution of the analyte, develops a potential, E_{ind}, that depends upon the activity of the analyte. Most indicator electrodes used in potentiometry are highly selective in their responses. The third component of a potentiometric cell is a salt bridge that prevents the components of the analyte solution from mixing with those of the reference electrode. As noted in Chapter 15, a potential develops across the liquid junctions at each end of the salt bridge. These two potentials tend to cancel one another if the mobilities of the cation and the anion in the bridge solution are about alike. The net junction across the bridge E_{j}

[1]For further reading on potentiometric methods, see E. P. Serjeant, *Potentiometry and Potentiometric Titrations.* New York: Wiley, 1984.

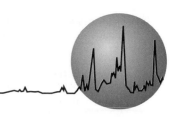

is therefore usually less than a few millivolts. For most electroanalytical methods, this net junction potential is small enough to be neglected. Here, however, we shall see that uncertainty in the junction potential places a limit on the accuracy of potentiometric analyses.

The potential of the cell we have just considered is given by the equation

$$E_{cell} = E_{ind} - E_{ref} + E_j \qquad (18\text{-}1)$$

The first term in this equation, E_{ind}, contains the information we seek about the concentration of the analyte. A potentiometric analysis, then, involves measuring a cell potential, correcting this potential for the reference and junction potentials, and computing the analyte concentration from the indicator electrode potential.

The sections that follow deal with the sources of the three potentials shown on the right side of Equation 18-1.

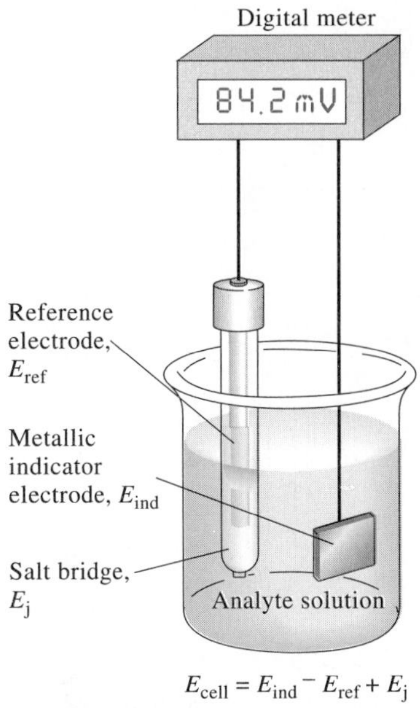

Figure 18-1
A cell for potentiometric analysis.

18B REFERENCE ELECTRODES

The ideal reference electrode has a potential (versus the standard hydrogen electrode) that is accurately known, constant, and completely insensitive to the composition of the analyte solution. In addition, this electrode should be rugged and easy to assemble and should maintain a constant potential while passing small currents.

A hydrogen electrode is seldom used as a reference electrode for day-to-day potentiometric measurements because it is somewhat inconvenient and is also a fire hazard.

18B-1 Calomel Reference Electrodes

A calomel electrode can be represented schematically as

$$Hg \,|\, Hg_2Cl_2(\text{sat'd}), KCl(x\text{M}) \,\|$$

where x represents the molar concentration of potassium chloride in the solution. Three concentrations of potassium chloride are common, 0.1 M, 1 M, and saturated (about 4.6 M). The saturated calomel electrode (SCE) is the most widely used because it is easily prepared. Its main disadvantage is its somewhat larger temperature coefficient than the other two. This disadvantage is important only in those rare circumstances where substantial temperature changes occur during a measurement. The *electrode* potential of the saturated calomel electrode is 0.2444 V at 25°C.

Calomel is Hg_2Cl_2, which has a limited solubility in water.

The ''saturated'' in a saturated calomel electrode refers to the KCl concentration and not the calomel concentration. All calomel electrodes are saturated with Hg_2Cl_2 (calomel).

The electrode reaction in calomel half-cells is

$$Hg_2Cl_2(s) + 2e^- \rightleftharpoons 2Hg(l) + 2Cl^-(aq)$$

Table 18-1 lists the compositions and electrode potentials for the three most common calomel electrodes. Note that the electrodes differ only in their potassium chloride concentrations; all are saturated with calomel (Hg_2Cl_2). Figure 18-2 illustrates a typical commercial saturated calomel electrode. It consists of a 5- to 15-cm long tube that is 0.5 to 1.0 cm in diameter. A mercury/mercury(I) chloride paste in saturated potassium chloride is contained in an inner tube and connected to the saturated potassium chloride solution in the outer tube through a

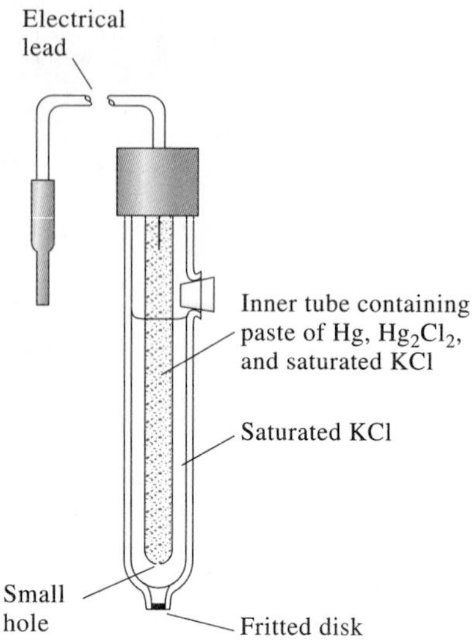

Figure 18-2
Diagram of a typical commercial saturated calomel electrode.

TABLE 18-1 Electrode Potentials for Reference Electrodes as a Function of Composition and Temperature

Temperature, °C	Potential (vs. SHE), V				
	0.1 M Calomel*	**3.5 M Calomel†**	**Sat'd Calomel***	**3.5 M Ag/AgCl†**	**Sat'd Ag/AgCl†**
12	0.3362		0.2528		
15	0.3362	0.254	0.2511	0.212	0.209
20	0.3359	0.252	0.2479	0.208	0.204
25	0.3356	0.250	0.2444	0.205	0.199
30	0.3351	0.248	0.2411	0.201	0.194
35	0.3344	0.246	0.2376	0.197	0.189

*From R. G. Bates, in *Treatise on Analytical Chemistry*, 2nd ed., I. M. Kolthoff and P. J. Elving, Eds., Part I, Vol. 1, p. 793. New York: Wiley, 1978.

†From D. T. Sawyer and J. L. Roberts Jr., *Experimental Electrochemistry for Chemists*, p. 42. New York: Wiley, 1974.

small opening. An inert metal electrode is immersed in the paste. Contact with the analyte solution is made through a fritted disk, a porous fiber, or a piece of porous Vycor® (''thirsty glass'') sealed in the end of the outer tubing.

Figure 18-3 shows a saturated calomel electrode that you can easily construct from materials available in your laboratory. A salt bridge (Section 15A-2) provides electrical contact with the analyte solution.

Agar, which is available as translucent flakes, is a heteropolysaccharide that is extracted from certain East Indian seaweed. Solutions of agar in hot water set to a gel when they cool.

A salt bridge is readily constructed by filling a U-tube with a conducting gel prepared by heating about 5 g of agar in 100 mL of an aqueous solution containing about 35 g of potassium chloride. When the liquid cools, it sets up into a gel that is a good conductor but prevents the two solutions at either end of the tube from mixing.

18B-2 Silver/Silver Chloride Reference Electrodes

A system analogous to the saturated calomel electrode consists of a silver electrode immersed in a solution that is saturated in both potassium chloride and silver chloride:

$$Ag|AgCl(sat'd),KCl(sat'd)||$$

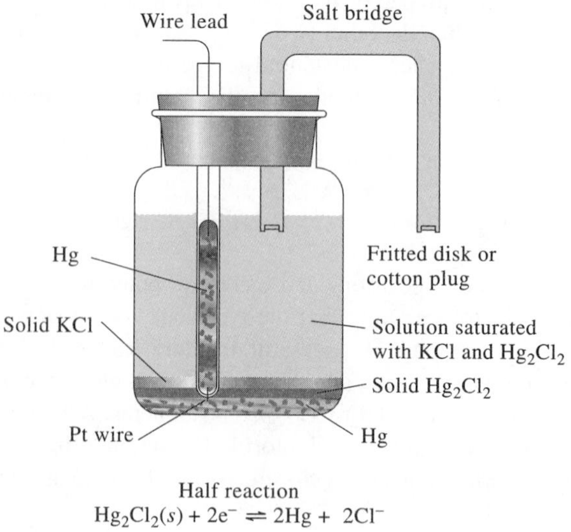

Figure 18-3
A saturated calomel electrode made from materials readily available in any laboratory.

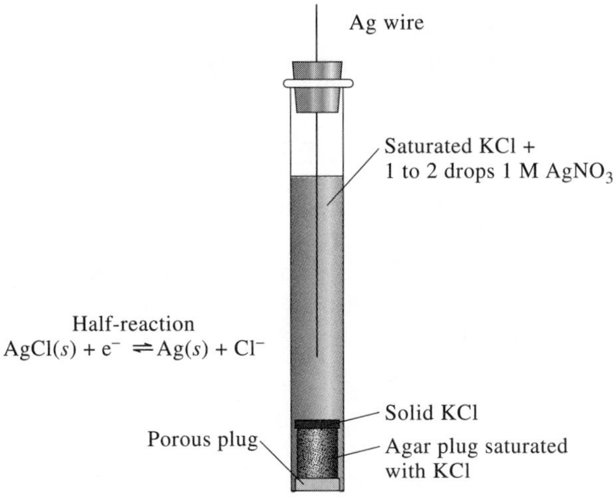

Figure 18-4
Diagram of a silver/silver chloride electrode.

The half-reaction is

$$AgCl(s) + e^- \rightleftharpoons Ag(s) + Cl^-(aq)$$

The potential of this electrode is 0.199 V at 25°C.

At 25°C, the potential of the saturated calomel electrode versus the standard hydrogen electrode is 0.244 V; for the saturated silver/silver chloride electrode it is 0.199 V.

Silver/silver chloride electrodes of various sizes and shapes are on the market (Table 18-1). A simple and easily constructed electrode of this type is shown in Figure 18-4.

18C LIQUID-JUNCTION POTENTIALS

A liquid-junction potential develops across the boundary between two electrolyte solutions that have different compositions. Figure 18-5 shows a very simple liquid junction consisting of a 1 M hydrochloric acid solution that is in contact with a solution that is 0.01 M in that acid. An inert porous barrier, such as a fritted glass plate, prevents the two solutions from mixing. Both hydrogen ions and chloride ions tend to diffuse across this boundary from the more concentrated to the more dilute solution. The driving force for each ion is proportional to the activity difference between the two solutions. In the present example, hydrogen ions are substantially more mobile than chloride ions. Thus, hydrogen ions diffuse more rapidly than chloride ions, and, as shown in the figure, a separation of charge results. The more dilute side of the boundary becomes positively charged because of the more rapid diffusion of hydrogen ions. The concentrated side therefore acquires a negative charge from the excess of slower moving chloride ions. The charge developed tends to counteract the differences in diffusion rates of the two ions so that a condition of equilibrium is attained rapidly. The potential difference resulting from this charge separation may be several hundredths of a volt.

The net-junction potential across a typical salt bridge is a few millivolts.

The magnitude of the liquid-junction potential can be minimized by placing a *salt bridge* between the two solutions. The salt bridge is most effective if the mobilities of the negative and positive ions in the bridge are nearly equal and if their concentrations are large. A saturated solution of potassium chloride is good from both standpoints. The net-junction potential with such a bridge is typically a few millivolts.

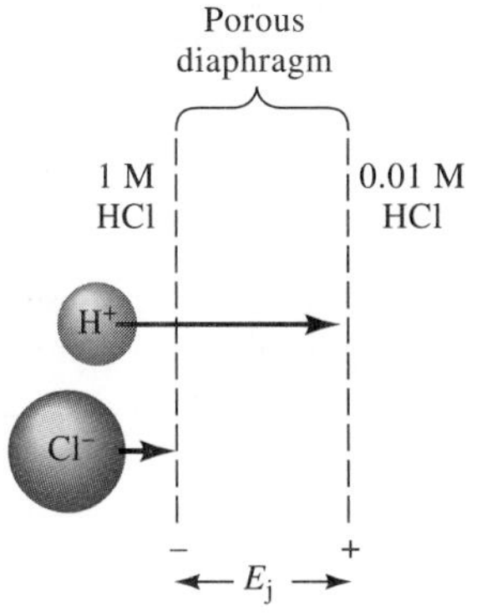

Figure 18-5
Schematic representation of a liquid junction, showing the source of the junction potential E_j. The length of the arrows corresponds to the relative mobility of the two ions.

All cells used for potentiometric analyses contain a salt bridge that connects the reference electrode to the analyte solution. As we have noted, uncertainties in the

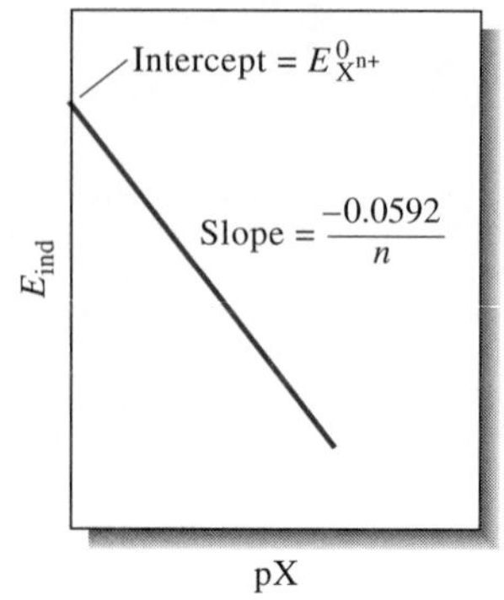

Figure 18-6
A plot of Equation 18-3 for an electrode of the first kind.

magnitude of the junction potential across this bridge place a fundamental limit on the accuracy of potentiometric methods of analysis.

18D INDICATOR ELECTRODES

An ideal indicator electrode responds rapidly and reproducibly to changes in the concentration of an analyte ion (or group of analyte ions). Although no indicator electrode is absolutely specific in its response, a few are now available that are remarkably selective. Indicator electrodes are of three types: metallic, membrane, and ion-selective field effect transistors.

18D-1 Metallic Indicator Electrodes

It is convenient to classify metallic indicator electrodes as *electrodes of the first kind, electrodes of the second kind,* and inert *redox electrodes.*

Electrodes of the First Kind

An electrode of the first kind is a pure metal electrode that is in direct equilibrium with its cation in the solution. A single reaction is involved. For example, the equilibrium between a metal X and its cation X^{n+} is

$$X^{n+}(aq) + ne^- \rightleftharpoons X(s)$$

for which

$$E_{ind} = E^0_{X^{n+}} - \frac{0.0592}{n}\log\frac{1}{a_{X^{n+}}} = E^0_{X^{n+}} + \frac{0.0592}{n}\log a_{X^{n+}} \qquad (18\text{-}2)$$

where E_{ind} is the electrode potential of the metal electrode and $a_{X^{n+}}$ is the activity of the ion (or approximately its molar concentration, $[X^{n+}]$).

We often express the electrode potential of the indicator electrode in terms of the p-function of the cation. Thus, substituting the definition of pX into Equation 18-2 gives

$$E_{ind} = E^0_{X^{n+}} + \frac{0.0592}{n}\log a_{X^{n+}} = E^0_{X^{n+}} - \frac{0.0592}{n}\text{pX} \qquad (18\text{-}3)$$

Potentiometric analyses provide data in terms of *activities* of analytes in contrast to the other methods we consider in this text, which give the *concentrations* of analytes. Recall that the activity of a species a_X is related to the molar concentration of X by Equation 8-2

$$a_X = \gamma_X[X]$$

where γ_X is the activity coefficient of X, a parameter that varies with the ionic strength of the solution. Because potentiometric data are dependent on activities, we will not in most cases make the usual approximation that $a_X \approx [X]$ in this chapter.

Electrode systems of the first kind are not widely used for potentiometric analyses for several reasons. For one, they are not very selective and respond not only to their own cations but also to other more easily reduced cations. For example, a copper electrode cannot be used for the determination of copper(II) ions in the presence of silver(I) ions, which are also reduced at the copper surface. In addition, many metal electrodes, such as zinc and cadmium, can only be used in neutral or basic solutions because they dissolve in the presence of acids. Third, some metals are so easily oxidized that their use is restricted to solutions that have been deaerated. Finally, certain harder metals—such as iron, chromium, cobalt, and nickel—do not provide reproducible potentials. Moreover, for these electrodes, plots of pX versus activity yield slopes that differ significantly and irregularly from the theoretical ($-0.0592/n$). For these reasons, the only electrode systems of the first kind that have been used are Ag/Ag^+ and Hg/Hg_2^{2+} in neutral solutions and Cu/Cu^{2+}, Zn/Zn^{2+}, Cd/Cd^{2+}, Bi/Bi^{3+}, Tl/Tl^+, and Pb/Pb^{2+} in deaerated solutions.

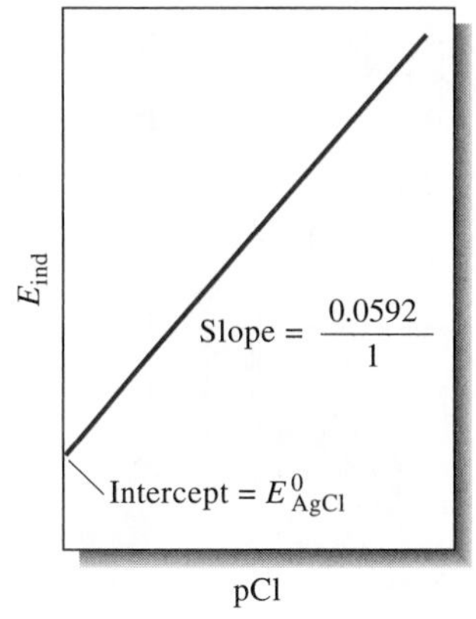

Figure 18-7
A plot of Equation 18-4 for an electrode of the second kind for Cl^-.

Electrodes of the Second Kind

Metals not only serve as indicator electrodes for their own cations but also respond to the activities of anions that form sparingly soluble precipitates or stable complexes with such cations. The potential of a silver electrode, for example, correlates reproducibly with the activity of chloride ion in a solution saturated with silver chloride. Here, the electrode reaction can be written as

$$AgCl(s) + e^- \rightleftharpoons Ag(s) + Cl^-(aq) \qquad E^0_{AgCl} = 0.222 \text{ V}$$

The Nernst expression for this process is

$$E_{ind} = E^0_{AgCl} - 0.0592 \log a_{Cl^-} = E^0_{AgCl} + 0.0592 \text{ pCl} \qquad (18\text{-}4)$$

Equation 18-4 shows that the potential of a silver electrode is proportional to pCl, the negative logarithm of the chloride ion activity. Thus, in a solution saturated with silver chloride, a silver electrode can serve as an indicator electrode of the second kind for chloride ion. Note that the sign of the log term for an electrode of this type is opposite that for an electrode of the first kind (see Equation 18-3).

Mercury serves as an indicator electrode of the second kind for the EDTA anion Y^{4-}. For example, when a small amount of HgY^{2-} is added to a solution containing Y^{4-}, the half-reaction at a mercury cathode is

$$HgY^{2-} + 2e^- \rightleftharpoons Hg(l) + Y^{4-} \qquad E^0 = 0.21 \text{ V}$$

for which

$$E_{ind} = 0.21 - \frac{0.0592}{2} \log \frac{a_{Y^{4-}}}{a_{HgY^{2-}}}$$

The formation constant for HgY^{2-} is very large (6.3×10^{21}), so the concentration of the complex remains essentially constant over a large range of Y^{4-} concentrations. The Nernst equation for the process can therefore be written as

$$E = K - \frac{0.0592}{2} \log a_{Y^{4-}} = K + \frac{0.0592}{2} \text{ pY} \qquad (18\text{-}5)$$

where

$$K = 0.21 - \frac{0.0592}{2} \log \frac{1}{a_{HgY^{2-}}}$$

The mercury electrode is thus a valuable electrode of the second kind for EDTA titrations.

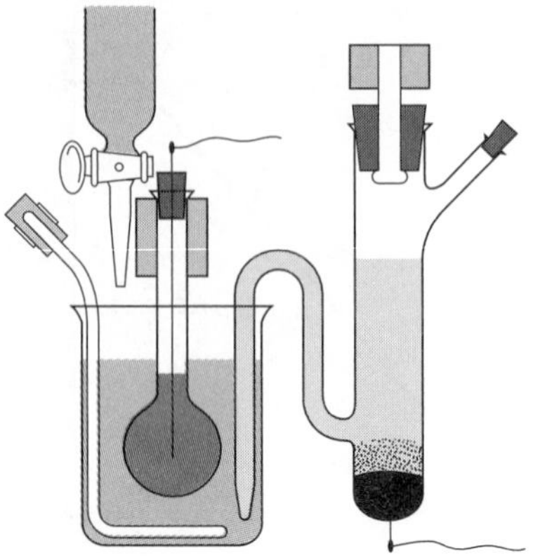

Figure 18-8
The first practical glass electrode. (From Haber and Klemensiewicz, *Z. Phys. Chem.*, **1909,** *65*, 385.)

Inert Metallic Electrodes for Redox Systems

As noted in Chapter 15, an inert conductor—such as platinum, gold, palladium, or carbon—responds to the potential of a redox system with which it is in contact. For example, the potential of a platinum electrode immersed in a solution containing cerium(III) and cerium(IV) is

$$E_{\text{ind}} = E^0_{\text{Ce(IV)}} - 0.0592 \log \frac{a_{\text{Ce}^{3+}}}{a_{\text{Ce}^{4+}}}$$

A platinum electrode is a convenient indicator electrode for titrations involving standard cerium(IV) solutions.

18D-2 Membrane Electrodes[2]

For many years, the most convenient method for determining pH has involved measurement of the potential that develops across a thin glass membrane that separates two solutions with different hydrogen ion concentrations. The phenomenon upon which the measurement is based was first reported in 1906 and by now has been extensively studied by many investigators. As a result, the sensitivity and selectivity of glass membranes toward hydrogen ions are reasonably well understood. Furthermore, this understanding has led to the development of other types of membranes that respond selectively to more than two dozen other ions.

Membrane electrodes are sometimes called *pIon electrodes* because the data obtained from them are usually presented as p-functions, such as pH, pCa, or pNO_3. In this section, we consider several types of pIon membranes.

It is important to note at the outset of this discussion that membrane electrodes are *fundamentally different* from metal electrodes both in design and in principle. We shall use the glass electrode for pH measurements to illustrate these differences.

18D-3 The Glass Electrode for pH Measurements

The membrane of a typical glass electrode (with a thickness of 0.03 to 0.1 mm) has an electrical resistance of 50 to 500 MΩ.

Figure 18-9 shows a typical *cell* for measuring pH. The cell consists of a glass indicator electrode and a saturated calomel reference electrode immersed in the solution whose pH is sought. The indicator electrode consists of a thin, pH-sensitive glass membrane sealed onto one end of a heavy-walled glass or plastic tube. A small volume of dilute hydrochloric acid saturated with silver chloride is contained in the tube (the inner solution in some electrodes is a buffer containing chloride ion). A silver wire in this solution forms a silver/silver chloride reference electrode, which is connected to one of the terminals of a potential-measuring device. The calomel electrode is connected to the other terminal.

The external reference electrode can be a Ag/AgCl electrode rather than the SCE shown in Figures 18-9 and 18-10.

Figure 18-9 and the schematic representation of this cell in Figure 18-10 show that a glass-electrode system contains *two* reference electrodes: (1) the *external*

[2]Some suggested sources for additional information on this topic are A. Evans, *Potentiometry and Ion-Selective Electrodes,* New York: Wiley, 1987; *Ion-Selective Methodology,* A. K. Covington, Ed., Boca Raton, FL: CRC Press, 1979; R. P. Buck, in *Comprehensive Treatise of Electrochemistry,* J. D. M. Bockris, B. C. Conway, and H. E. Yeager, Eds., Vol. 8, Chapter 3. New York: Plenum Press, 1984; J. Koryta, *Ions, Electrodes, and Membranes,* 2nd ed. New York: Wiley, 1991.

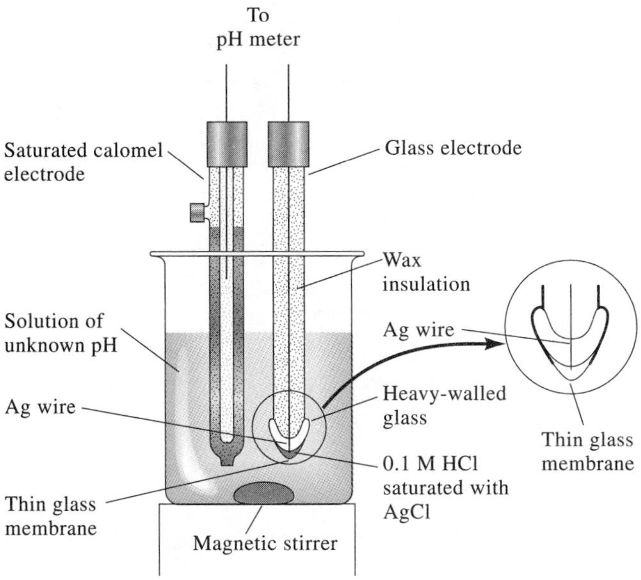

Figure 18-9
Typical electrode system for measuring pH.

calomel electrode and (2) the *internal* silver/silver chloride electrode. While the internal reference electrode is a part of the glass electrode, *it is not the pH-sensing element.* Instead, *it is the thin glass membrane at the tip of the electrode that responds to pH.*

The Composition and Structure of Glass Membranes

Much systematic investigation has been devoted to the effects of glass composition on the sensitivity of membranes to protons and other cations, and a number of formulations are now used for the manufacture of electrodes. Corning 015 glass, which has been widely used for membranes, consists of approximately 22% Na_2O, 6% CaO, and 72% SiO_2. This membrane shows excellent specificity toward hydrogen ions up to a pH of about 9. At higher pH values, however, the glass becomes somewhat responsive to sodium as well as to other singly charged cations. Other glass formulations are now in use in which sodium and calcium ions are replaced to various degrees by barium and lithium ions. These membranes have superior selectivity and lifetime.

As shown in Figure 18-11, a silicate glass used for membranes consists of an infinite three-dimensional network of SiO_4^{4-} groups in which each silicon is bonded to four oxygens and each oxygen is shared by two silicons. Within the interstices of this structure are sufficient cations to balance the negative charge of

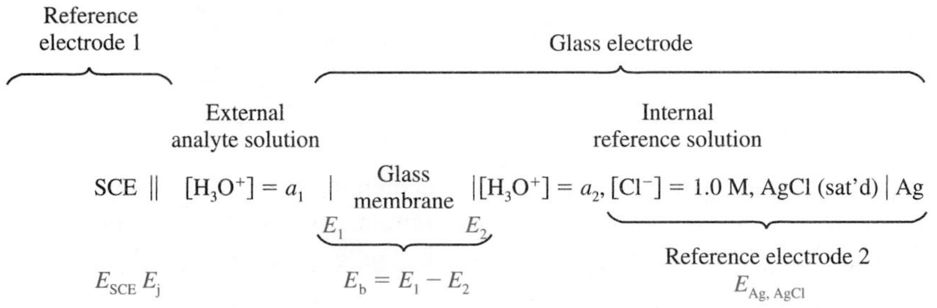

Figure 18-10
Diagram of a glass/calomel cell for measurement of pH.

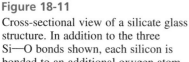

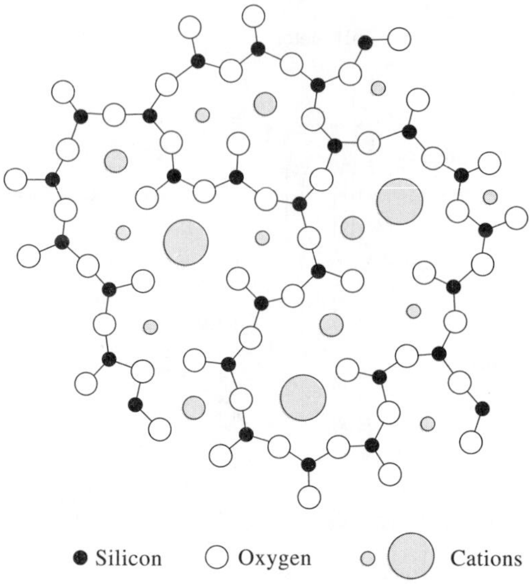

Figure 18-11
Cross-sectional view of a silicate glass structure. In addition to the three Si—O bonds shown, each silicon is bonded to an additional oxygen atom, either above or below the plane of the paper. (Adapted with permission from G. A. Perley, *Anal. Chem.*, **1949,** *21*, 395. Copyright 1949 American Chemical Society.)

the silicate groups. Singly charged cations, such as sodium and lithium, are mobile in the lattice and are responsible for electrical conduction within the membrane.

Glasses that absorb water are said to be **hygroscopic.**

Both membrane surfaces must be hydrated before a glass electrode will function as a pH electrode. Nonhygroscopic glasses show no pH function. Even hygroscopic glasses lose their pH sensitivity after dehydration by storage over a desiccant. The effect is reversible, however, and response is restored by soaking a glass electrode in water.

The hydration of a pH-sensitive glass membrane involves an ion-exchange reaction between singly charged cations in the interstices of the glass lattice and protons from the solution. The process involves univalent cations exclusively because di- and trivalent cations are too strongly held within the silicate structure to exchange with ions in the solution. Typically, then, the ion-exchange reaction can be written as

$$\underset{\text{soln}}{H^+} + \underset{\text{glass}}{Na^+Gl^-} \rightleftharpoons \underset{\text{soln}}{Na^+} + \underset{\text{glass}}{H^+Gl^-} \quad (18\text{-}6)$$

Oxygen atoms attached to only one silicon atom are the negatively charged Gl^- sites shown in this equation. The exchange process is depicted in Figure 18-12. The equilibrium constant for this process is so large that the surfaces of a hydrated glass membrane ordinarily consist entirely of silicic acid (H^+Gl^-). An exception to this situation exists in highly alkaline media, where the hydrogen ion concentration is extremely small and the sodium ion concentration is large; here, a significant fraction of the sites are occupied by sodium ions.

Electrical Conduction Across Membranes

To serve as an indicator for cations, a glass membrane must conduct electricity. Conduction within the hydrated glass membrane involves the movement of sodium and hydrogen ions. Sodium ions are the charge carriers in the dry interior of

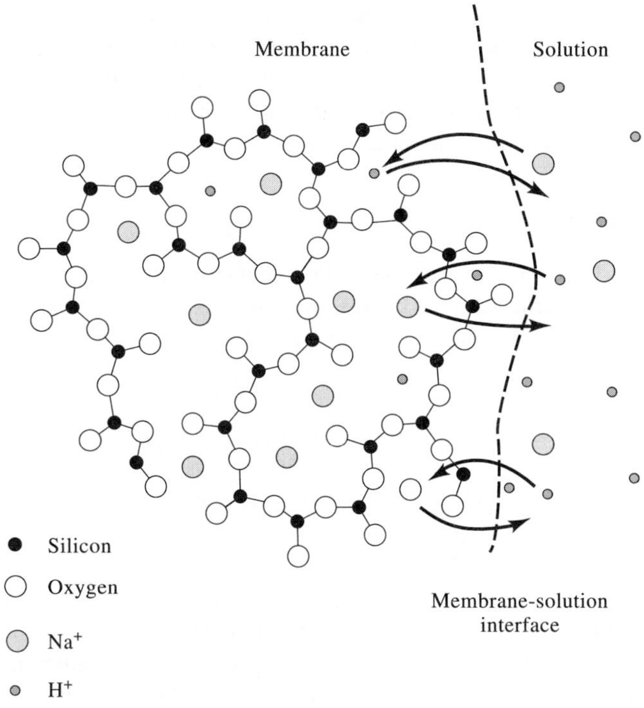

Figure 18-12
Ion exchange at the membrane/solution interface. Protons exchange with sodium ions in the silicate structure of the glass membrane.

the membrane, and the protons are mobile in the gel layer. Conduction across the solution/gel interfaces occurs by the reactions

$$\underset{\text{soln}_1}{H^+} + \underset{\text{glass}_1}{Gl^-} \rightleftharpoons \underset{\text{glass}_1}{H^+Gl^-} \tag{18-7}$$

$$\underset{\text{glass}_2}{H^+Gl^-} \rightleftharpoons \underset{\text{soln}_2}{H^+} + \underset{\text{glass}_2}{Gl^-} \tag{18-8}$$

where the subscript 1 refers to the interface between glass and analyte solution and the subscript 2 refers to the interface between internal solution and glass. The positions of these two equilibria are determined by the hydrogen ion concentrations in the solutions on the two sides of the membrane. Where these positions differ from each other, the surface at which the greater dissociation has occurred is negative with respect to the other surface. A boundary potential E_b thus develops across the membrane. The magnitude of the boundary potential depends upon the ratio of the hydrogen ion concentrations of the two solutions. It is this potential difference that serves as the analytical parameter in a potentiometric pH measurement.

Membrane Potentials

The lower part of Figure 18-10 shows four potentials that develop in a cell for the measurement of pH with a glass electrode. Two of these, $E_{Ag,AgCl}$ and E_{SCE}, are reference electrode potentials that are constant. A third potential is the net potential across the salt bridge that separates the calomel electrode from the analyte solution. This junction and its associated *junction potential,* E_j, are found in all cells used for the potentiometric measurement of ion concentrations. The fourth, and most important, potential shown in Figure 18-10 is the *boundary potential, E_b, which varies with the pH of the analyte solution.* The two reference electrodes simply provide electrical contacts with the solutions so that changes in the boundary potential can be measured.

The Boundary Potential

In Figure 18-10, the boundary potential is shown as being made up of two potentials, E_1 and E_2, which develop at the two *surfaces* of the glass membrane. The source of these two potentials is the charge that develops as a consequence of the reactions

$$\underset{\text{glass}_1}{H^+Gl^-(s)} \rightleftharpoons \underset{\text{soln}_1}{H^+(aq)} + \underset{\text{glass}_1}{Gl^-(s)} \qquad (18\text{-}9)$$

$$\underset{\text{glass}_2}{H^+Gl^-(s)} \rightleftharpoons \underset{\text{soln}_2}{H^+(aq)} + \underset{\text{glass}_2}{Gl^-(s)} \qquad (18\text{-}10)$$

where the subscript 1 refers to the interface between the exterior of the glass and the analyte solution and the subscript 2 refers to the interface between the internal solution and the interior of the glass. These two reactions cause the two glass surfaces to be negatively charged with respect to the solutions in which they are immersed, thus giving rise to the two potentials E_1 and E_2 shown in Figure 18-10. The positions of the two equilibria that cause the two potentials to develop are determined by the hydrogen ion concentrations in the solutions on the two sides of the membrane. Where these positions differ, the surface at which the greater dissociation has occurred is negative with respect to the other surface. This difference in charge is the boundary potential E_b, which is related to the activities of hydrogen ion in each of the solutions by the Nernst-like equation

$$E_b = E_1 - E_2 = 0.0592 \log \frac{a_1}{a_2} \qquad (18\text{-}11)$$

where a_1 is the activity of the analyte solution and a_2 is that of the internal solution. For a glass pH electrode, the hydrogen ion activity of the internal solution is held constant so that Equation 18-11 simplifies to

$$E_b = L' + 0.0592 \log a_1 = L' - 0.0592 \text{ pH} \qquad (18\text{-}12)$$

where

$$L' = -0.0592 \log a_2$$

The boundary potential is then a measure of the hydrogen ion activity of the external solution.

The significance of the potentials and the potential differences shown in Equation 18-11 is illustrated by the potential profiles shown in Figure 18-13. The profiles are plotted across the membrane from the analyte solution on the left through the membrane to the internal solution on the right.

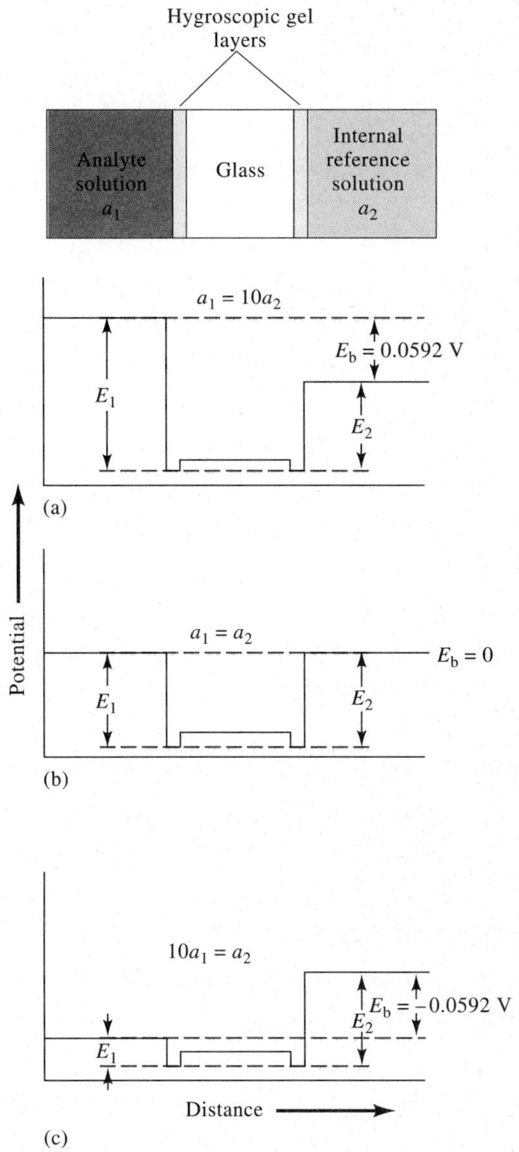

Figure 18-13
Potential profile across a glass membrane from the analyte solution to the internal reference solution. The reference electrode potentials are not shown.

The Asymmetry Potential

When identical solutions and reference electrodes are placed on the two sides of a glass membrane, the boundary potential should in principle be zero. In fact, however, a small *asymmetry potential* that changes gradually with time is frequently encountered.

The sources of the asymmetry potential are obscure but undoubtedly include such causes as differences in strain on the two surfaces of the membrane imparted during manufacture, mechanical abrasion on the outer surface during use, and chemical etching of the outer surface. In order to eliminate the bias caused by the asymmetry potential, all membrane electrodes must be calibrated against one or more standard analyte solutions. Such calibrations should be carried out at least daily, and more often when the electrode receives heavy use.

The Potential of the Glass Electrode

The potential of a glass indicator electrode E_{ind} has three components: (1) the boundary potential, given by Equation 18-11, (2) the potential of the internal Ag/AgCl reference electrode, and (3) a small asymmetry potential, E_{asy}, which changes slowly with time. In equation form, we may write

$$E_{ind} = E_b + E_{Ag/AgCl} + E_{asy}$$

Substitution of Equation 18-12 for E_b gives

$$E_{ind} = L' + 0.0592 \log a_1 + E_{Ag/AgCl} + E_{asy}$$

or

$$E_{ind} = L + 0.0592 \log a_1 = L - 0.0592 \text{ pH} \tag{18-13}$$

where L is a combination of the three constant terms. Compare Equations 18-13 and 18-3. Although these two equations are similar in form, remember that the sources of the potential of the electrodes that they describe *are totally different.*

The Alkaline Error

In basic solutions, glass electrodes respond to the concentration of both hydrogen ion and alkali metal ions. The magnitude of the resulting *alkaline error* for four different glass membranes is shown in Figure 18-14 (curves C to F). These curves refer to solutions in which the sodium ion concentration was held constant at 1 M while the pH was varied. Note that the error is negative (that is, the measured pH values are lower than the true values), which suggests that the electrode is re-

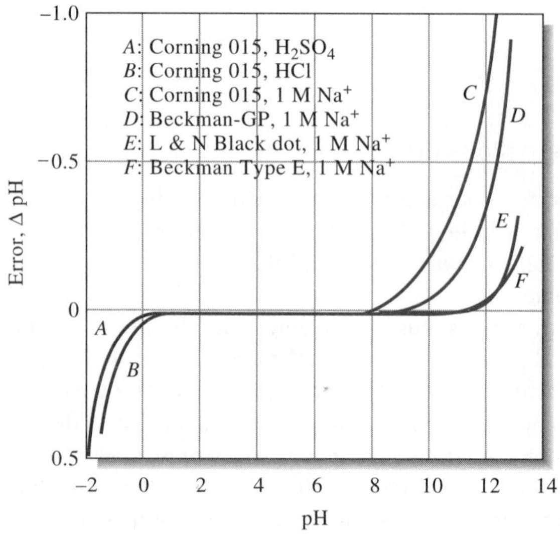

Figure 18-14
Acid and alkaline error for selected glass electrodes at 25°C. (From R. G. Bates, *Determination of pH,* 2nd ed., p. 365. New York: Wiley, 1973. With permission.)

sponding to sodium ions as well as to protons. This observation is confirmed by data obtained for solutions containing different sodium ion concentrations. Thus at pH 12, the electrode with a Corning 015 membrane (curve *C* in Figure 18-14) registered a pH of 11.3 when immersed in a solution having a sodium ion concentration of 1 M but 11.7 in a solution that was 0.1 M in this ion. All singly charged cations induce an alkaline error whose magnitude depends upon both the cation in question and the composition of the glass membrane.

The alkaline error can be satisfactorily explained by assuming an exchange equilibrium between the hydrogen ions on the glass surface and the cations in solution. This process is simply the reverse of that shown in Equation 18-6:

$$\underset{\text{glass}}{H^+Gl^-} + \underset{\text{soln}}{B^+} \rightleftharpoons \underset{\text{glass}}{B^+Gl^-} + \underset{\text{soln}}{H^+}$$

Where B^+ represents some singly charged cation, such as sodium ion.

The equilibrium constant for this reaction is

$$K_{ex} = \frac{a_1 b_1'}{a_1' b_1} \qquad \text{(18-14)}$$

In Equation 18-14, b_1 represents the activity of some singly charged cation such as Na^+ or K^+.

where a_1 and b_1 represent the activities of H^+ and B^+ in solution and a_1' and b_1' are the activities of these ions on the gel surface. Equation 18-14 can be rearranged to give a ratio of the activities B^+ to H^+ on the glass surface:

$$\frac{b_1'}{a_1'} = \frac{b_1}{a_1} K_{ex}$$

For the glasses used for pH electrodes, K_{ex} is so small that the activity ratio b_1'/a_1' is ordinarily minuscule. The situation differs in strongly alkaline media, however. For example, b_1'/a_1' for an electrode immersed in a pH 11 solution that is 1 M in sodium ions (Figure 18-14) is 10^{11} K_{ex}. Here, the activity of sodium ions relative to that of hydrogen ions becomes so large that the electrode responds to both species.

Selectivity Coefficients

The effect of an alkali metal ion on the potential across a membrane can be accounted for by inserting an additional term in Equation 18-12 to give

$$E_b = L' + 0.0592 \log (a_1 + k_{H,B} b_1) \qquad \text{(18-15)}$$

where $k_{H,B}$ is the *selectivity coefficient* for the electrode. Equation 18-15 applies not only to glass indicator electrodes for hydrogen ion but also to all other types of membrane electrodes. Selectivity coefficients range from zero (no interference) to values greater than unity. Thus, if an electrode for ion A responds 20 times more strongly to ion B than to ion A, $k_{A,B}$ has a value of 20. If the response of the

The **selectivity coefficient** is a measure of the response of an ion selective electrode to other ions.

electrode to ion C is 0.001 of its response to A (a much more desirable situation), $k_{A,C}$ is 0.001.[3]

The product $k_{H,B}b_1$ for a glass pH electrode is ordinarily small relative to a_1 provided the pH is less than 9; under these conditions, Equation 18-15 simplifies to Equation 18-12. At high pH values and at high concentrations of a singly charged ion, however, the second term in Equation 18-15 assumes a more important role in determining E_b, and an alkaline error is encountered. For electrodes specifically designed for work in highly alkaline media (curve *E* in Figure 18-14), the magnitude of $k_{H,B}b_1$ is appreciably smaller than for ordinary glass electrodes.

The Acid Error

As shown in Figure 18-14, the typical glass electrode exhibits an error, opposite in sign to the alkaline error, in solutions of pH less than about 0.5; pH readings tend to be too high in this region. The magnitude of the error depends upon a variety of factors and is generally not very reproducible. The causes of the acid error are not well understood.

18D-4 Glass Electrodes for Cations Other than Protons

The alkaline error in early glass electrodes led to investigations concerning the effect of glass composition upon the magnitude of this error. One consequence has been the development of glasses for which the alkaline error is negligible below about pH 12 (see curves *E* and *F*, Figure 18-14). Other studies have discovered glass compositions that permit the determination of cations other than hydrogen. Incorporation of Al_2O_3 or B_2O_3 in the glass has the desired effect. Glass electrodes that permit the direct potentiometric measurement of such singly charged species as Na^+, K^+, NH_4^+, Rb^+, Li^+, and Ag^+ have been developed. Some of these glasses are reasonably selective toward particular singly charged cations. Glass electrodes for Na^+, Li^+, NH_4^+, and total concentration of univalent cations are now available from commercial sources.

18D-5 Liquid-Membrane Electrodes

A liquid-membrane electrode develops a potential across the interface between the solution containing the analyte and a liquid-ion exchanger that selectively bonds with the analyte ion. These electrodes have been developed for the direct potentiometric measurement of numerous polyvalent cations as well as certain anions.

Figure 18-15 is a schematic of a liquid-membrane electrode for calcium. It consists of a conducting membrane that selectively bonds calcium ions, an internal solution containing a fixed concentration of calcium chloride, and a silver electrode that is coated with silver chloride to form an internal reference electrode. Notice the similarities between the liquid membrane electrode and the glass electrode as shown in Figure 18-16. The active membrane ingredient is an ion

---

[3]For tables of selectivity coefficients for a variety of membranes and ionic species, see Y. Umezawa, *CRC Handbook of Ion Selective Electrodes: Selectivity Coefficients.* Boca Raton, FL: CRC Press, 1990.

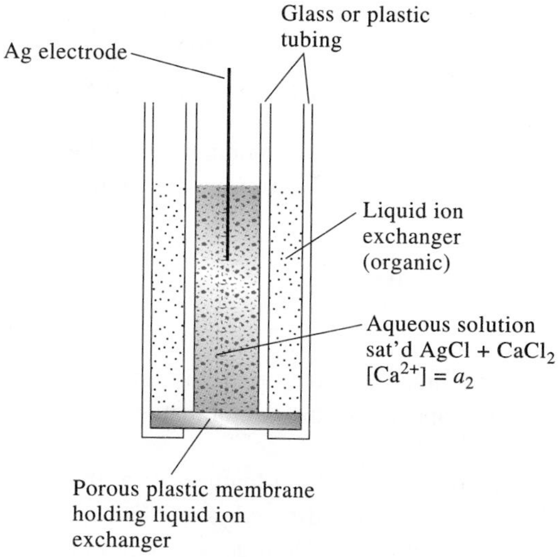

Figure 18-15
Diagram of a liquid-membrane electrode for Ca^{2+}.

exchanger that consists of a calcium dialkyl phosphate that is nearly insoluble in water. In the electrode shown in Figures 18-15 and 18-16, the ion exchanger is dissolved in an immiscible organic liquid that is forced by gravity into the pores of a hydrophobic porous disk. This disk then serves as the membrane that separates the internal solution from the analyte solution. In a more recent design, the ion exchanger is immobilized in a tough polyvinyl chloride gel cemented to the end of a tube that holds the internal solution and reference electrode. In either design, a dissociation equilibrium develops at each membrane interface that is analogous to Equations 18-7 and 18-8:

Hydrophobic means *water hating.* The hydrophobic disk is porous toward organic liquids but repels water.

$$\underset{\text{organic}}{[(RO)_2POO]_2Ca} \rightleftharpoons \underset{\text{organic}}{2(RO)_2POO^-} + \underset{\text{aqueous}}{Ca^{2+}}$$

where R is a high-molecular-weight aliphatic group. As with the glass electrode, a potential develops across the membrane when dissociation of the ion exchanger at one surface differs from that at the other surface. This potential is a result of differences in the calcium ion activity of the internal and external solutions. The relationship between the membrane potential and the calcium ion activities is given by an equation that is similar to Equation 18-11:

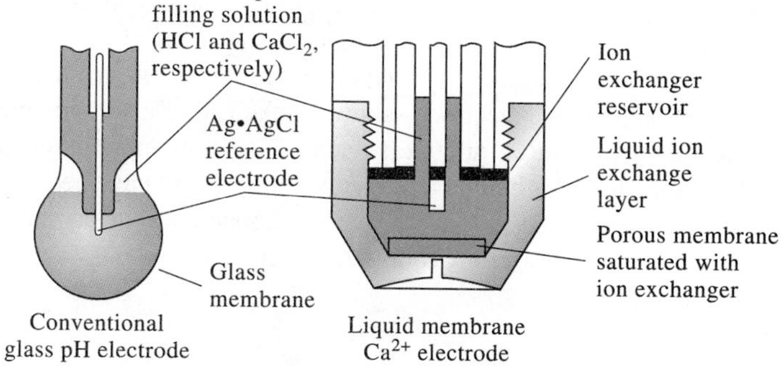

Figure 18-16
Comparison of a liquid-membrane calcium ion electrode with a glass electrode. (Courtesy of Orion Research, Boston, MA.)

$$E_b = E_1 - E_2 = \frac{0.0592}{2} \log \frac{a_1}{a_2} \tag{18-16}$$

where a_1 and a_2 are the activities of calcium ion in the external and internal solutions, respectively. Since the calcium ion activity of the internal solution is constant,

$$E_b = N + \frac{0.0592}{2} \log a_1 = N - \frac{0.0592}{2} \text{pCa} \tag{18-17}$$

where N is a constant (compare Equations 18-17 and 18-12). Note that because calcium is divalent, a 2 appears in the denominator of the coefficient of the log term.

Ion-selective microelectrodes can be used to make measurements of ion activities within a living organism.

Figure 18-16 compares the structural features of a glass-membrane electrode and a commercially available liquid-membrane electrode for calcium ion. The sensitivity of the liquid-membrane electrode for calcium ion is reported to be 50 times greater than that for magnesium ion and 1000 times greater than that for sodium or potassium ions. Calcium ion activities as low as 5×10^{-7} M can be measured. Performance of the electrode is independent of pH in the range between 5.5 and 11. At lower pH levels, hydrogen ions undoubtedly replace some of the calcium ions on the exchanger; the electrode then becomes sensitive to pH as well as to pCa.

The calcium ion liquid-membrane electrode is a valuable tool for physiological investigations because this ion plays important roles in such processes as nerve conduction, bone formation, muscle contraction, cardiac expansion and contraction, renal tubular function, and perhaps hypertension. Most of these processes are more influenced by the *activity* than the concentration of the calcium ion; activity, of course, is the parameter measured by the membrane electrode. Thus the calcium ion electrode (as well as the potassium ion electrode and others) is an important tool in studying physiological processes.

A liquid-membrane electrode specific for potassium ion is also of great value for physiologists because the transport of neural signals appears to involve movement of this ion across nerve membranes. Investigation of this process requires an electrode that can detect small concentrations of potassium ion in media that contain much larger concentrations of sodium ion. Several liquid-membrane electrodes show promise in meeting this requirement. One is based upon the antibiotic valinomycin, a cyclic ether that has a strong affinity for potassium ion. Of equal importance is the observation that a liquid membrane consisting of valinomycin in diphenyl ether is about 10^4 times as responsive to potassium ion as to sodium ion.[4] Figure 18-17 is a photomicrograph of a tiny electrode used for determining the potassium content of a single cell.

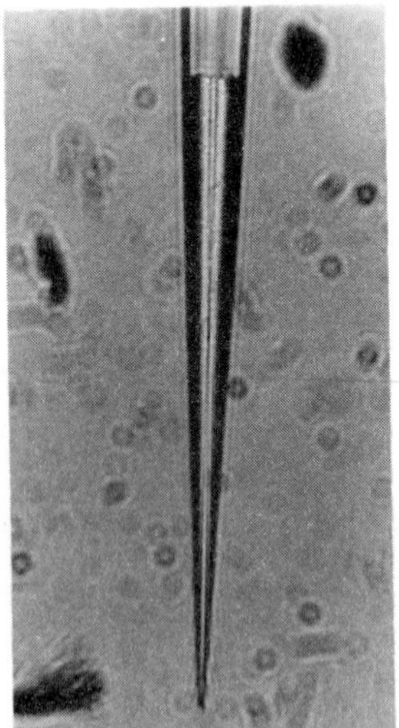

Figure 18-17
Photograph of a potassium liquid ion exchanger microelectrode with 125 μm of ion exchanger inside the tip. The magnification of the original photo was 400×. (From J. L. Walker, *Anal. Chem.*, **1971,** *43(3)N*, 91A. Reproduced by permission of the American Chemical Society.)

Table 18-2 lists some liquid-membrane electrodes available from commercial sources. The anion-sensitive electrodes shown make use of a solution containing an anion-exchange resin in an organic solvent. Liquid-membrane electrodes in which the exchange liquid is held in a polyvinyl chloride gel have been developed for Ca^{2+}, K^+, NO_3^-, and BF_4^-. These have the appearance of crystalline electrodes, which are considered in the following section.

[4]M. S. Frant and J. W. Ross Jr., *Science*, **1970,** *167*, 987.

TABLE 18-2 Characteristics of Liquid-Membrane Electrodes*

Analyte Ion	Concentration Range, M	Major Interferences
Ca^{2+}	10^0 to 5×10^{-7}	Pb^{2+}, Fe^{2+}, Ni^{2+}, Hg^{2+}, Sr^{2+}
Cl^-	10^0 to 5×10^{-6}	I^-, OH^-, SO_4^{2-}
NO_3^-	10^0 to 7×10^{-6}	ClO_4^-, I^-, ClO_3^-, CN^-, Br^-
ClO_4^-	10^0 to 7×10^{-6}	I^-, ClO_3^-, CN^-, Br^-
K^+	10^0 to 1×10^{-6}	Cs^+, NH_4^+, Tl^+
Water hardness ($Ca^{2+} + Mg^{2+}$)	10^0 to 6×10^{-6}	Cu^{2+}, Zn^{2+}, Ni^{2+}, Sr^{2+}, Fe^{2+}, Ba^{2+}

*From *Orion Guide to Ion Analysis*. Boston, MA: Orion Research, 1992. With permission.

FEATURE 18-1

An Easily Constructed Liquid-Membrane Ion-Selective Electrode

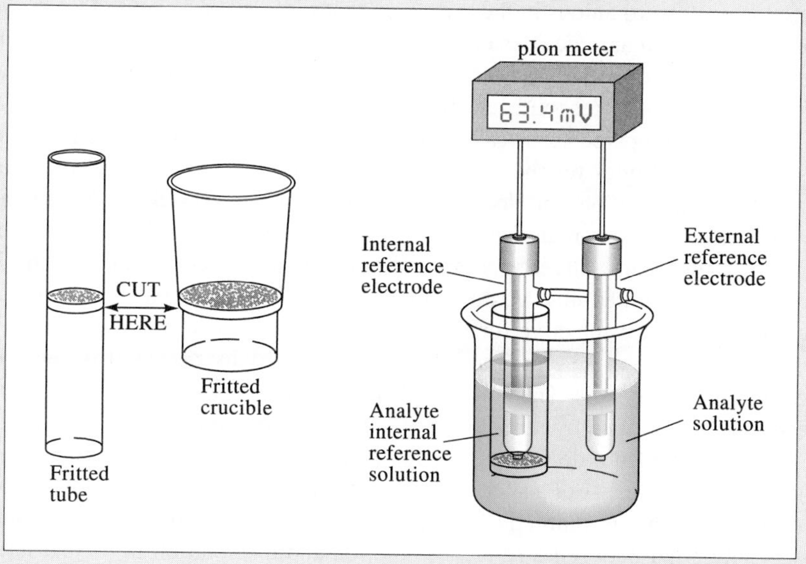

Figure 18-18 A homemade liquid-membrane electrode.

You can make a liquid-membrane ion-selective electrode with glassware and chemicals available in most laboratories.[5] All you need are a pH meter, a pair of reference electrodes, a fritted-glass filter crucible or tube, trimethylchlorosilane, and a liquid ion exchanger.

First, cut the filter crucible (or alternatively, a fritted tube) as shown in Figure 18-18, above. Carefully clean and dry the crucible, and then draw a small amount of trimethylchlorosilane into the frit. This coating makes the glass in the frit hydrophobic. Rinse the frit with water, dry, and apply a commercial liquid ion exchanger to it. After a minute, remove the excess exchanger. Add a few milliliters of a 10^{-2} M solution of the ion of interest to the crucible, insert a reference electrode into the solution, and voilà, you have a very nice ion-selective electrode. The exact details of washing, drying, and preparing the electrode are provided in the original article.

Connect the ion-selective electrode and the second reference electrode to the pH meter as shown in Figure 18-18. Prepare a series of standard solutions of the ion of interest, measure the cell potential for each concentration, plot a working curve of E_{cell} versus log c, and perform a least-squares analysis on the data. Compare the slope of the line with the theoretical slope of (0.0592 V)/n. Measure the potential for an unknown solution of the ion and calculate the concentration from the least-squares parameters.

[5]T. K. Christopoulos and E. P. Diamandis, *J. Chem. Educ.*, **1988,** *65*, 648.

18D-6 Crystalline-Membrane Electrodes

Considerable work has been devoted to the development of solid membranes that are selective toward anions in the same way that some glasses respond to cations. We have seen that anionic sites on a glass surface account for the selectivity of a membrane toward certain cations. By analogy, a membrane with cationic sites might be expected to respond selectively toward anions.

Membranes prepared from cast pellets of silver halides have been used successfully in electrodes for the selective determination of chloride, bromide, and iodide ions. In addition, an electrode based upon a polycrystalline Ag_2S membrane is offered by one manufacturer for the determination of sulfide ion. In both types of membranes, silver ions are sufficiently mobile to conduct electricity through the solid medium. Mixtures of PbS, CdS, and CuS with Ag_2S provide membranes that are selective for Pb^{2+}, Cd^{2+}, and Cu^{2+}, respectively. Silver ion must be present in these membranes to conduct electricity because divalent ions are immobile in crystals. The potential that develops across crystalline solid state electrodes is described by a relationship similar to Equation 18-12.

A crystalline electrode for fluoride ion is available from commercial sources. The membrane consists of a slice of a single crystal of lanthanum fluoride that has been doped with europium(II) fluoride to improve its conductivity. The membrane, supported between a reference solution and the solution to be measured, shows a theoretical response to changes in fluoride ion activity from 10^0 to 10^{-6} M. The electrode is selective for fluoride ion over other common anions by several orders of magnitude; only hydroxide ion appears to offer serious interference.

Some solid state electrodes available from commercial sources are listed in Table 18-3. Several types of crystalline electrodes are shown in Figure 18-19.

18D-7 Ion-Selective Field Effect Transistors (ISFETs)

The field effect transistor, or the *metal oxide field effect transistor (MOSFET)* is a tiny solid state semiconductor device that is widely used in computers and other electronic circuits as a switch to control current flow in circuits. One of the problems in employing this type of device in electronic circuits has been its

TABLE 18-3 Characteristics of Solid State Crystalline Electrodes*

Analyte Ion	Concentration Range, M	Major Interferences
Br^-	10^0 to 5×10^{-6}	CN^-, I^-, S^{2-}
Cd^{2+}	10^{-1} to 1×10^{-7}	Fe^{2+}, Pb^{2+}, Hg^{2+}, Ag^+, Cu^{2+}
Cl^-	10^0 to 5×10^{-5}	CN^-, I^-, Br^-, S^{2-}, OH^-, NH_3
Cu^{2+}	10^{-1} to 1×10^{-8}	Hg^{2+}, Ag^+, Cd^{2+}
CN^-	10^{-2} to 1×10^{-6}	S^{2-}, I^-
F^-	Sat'd to 1×10^{-6}	OH^-
I^-	10^0 to 5×10^{-8}	CN^-
Pb^{2+}	10^{-1} to 1×10^{-6}	Hg^{2+}, Ag^+, Cu^{2+}
Ag^+/S^{2-}	Ag^+: 10^0 to 1×10^{-7} S^{2-}: 10^0 to 1×10^{-7}	Hg^{2+}
SCN^-	10^0 to 5×10^{-6}	I^-, Br^-, CN^-, S^{2-}

*From *Orion Guide to Ion Analysis*. Cambridge, MA: Orion Research, 1983. With permission.

pronounced sensitivity to ionic surface impurities, and a great deal of money and effort has been expended by the electronic industry in minimizing or eliminating this sensitivity in order to produce stable transistors.

Beginning in 1970, a number of scientists have attempted to exploit the sensitivities of MOSFETs to surface ionic impurities for the selective potentiometric determination of various ions. These studies have led to the development of a number of different ion-selective field effect transistors termed ISFETs. The theory of their selective ion sensitivity is well understood and is described in Feature 18-2.

ISFETs stands for **ion-selective field effect transistors.**

ISFETs offer a number of significant advantages over membrane electrodes including ruggedness, small size, inertness toward harsh environments, rapid response, and low electrical impedance. In contrast to membrane electrodes,

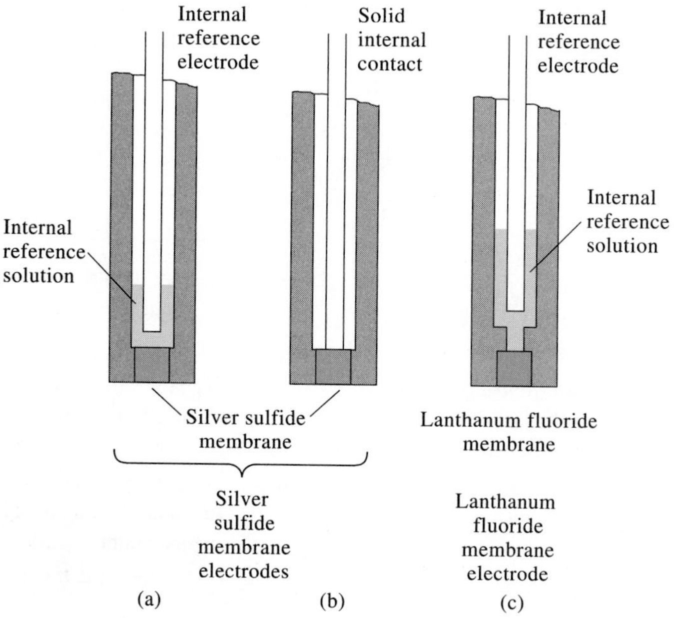

Figure 18-19
Types of crystalline ion-selective electrodes. (a) Solid-membrane electrode sensitive to Ag^+ and S^{2-}; (b) solid state solid-membrane electrode sensitive to Ag^+ and S^{2-}; (c) membrane-configuration ion-sensing electrode for fluoride ion.

ISFETs do not require hydration before use and can be stored indefinitely in the dry state. Despite these many advantages, no ISFET specific ion electrode appeared upon the market until the early 1990s, over 20 years after their invention. The reason for this delay is that manufacturers were unable to develop the technology of encapsulating the devices to give a product that did not exhibit drift and instability.

In the early 1990s several companies came out with an ISFET for the determination of pH. At the time of writing, it has not been possible to learn details of the surface layer on these electrodes that is responsible for their selective behavior toward protons, nor is information comparing the performance characteristics of these electrodes with the classical glass electrode. Certainly, they offer the several advantages listed in the previous paragraph. Just what their disadvantages are, if any, is not clear at this time.

FEATURE 18-2
The Structure and Performance Characteristics of Ion-Selective Field Effect Transistors

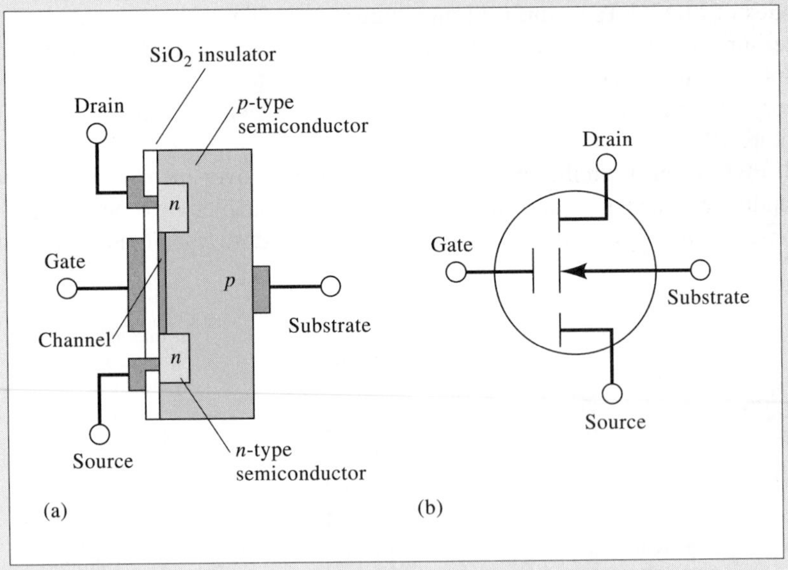

Figure 18-20 A metal oxide field effect transistor (MOSFET). (a) Cross-sectional diagram; (b) circuit symbol.

The *metal oxide field effect transistor (MOSFET)* is a solid state semiconductor device that is used widely for switching signals in computers and many types of electronic circuits. Figure 18-20 shows a cross-sectional diagram and a circuit symbol for an *n*-channel enhancement mode MOSFET. Modern semiconductor fabrication techniques are used to construct the MOSFET on the surface of a piece of *p*-type semiconductor called the substrate. For a discussion of the characteristics of *p*-type and *n*-type semicon-

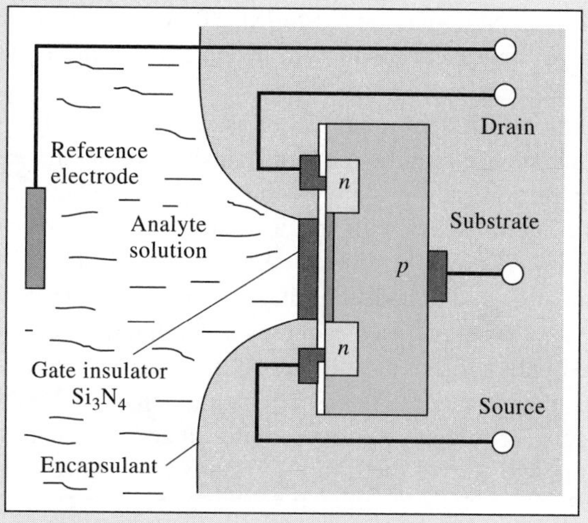

Figure 18-21 An ion-selective field effect transistor (ISFET) for measuring pH.

ductors, refer to the paragraphs on silicon photodiodes in Section 23A-4. As shown in the figure, two islands of *n*-type semiconductors are formed on the surface of the *p*-type substrate, and the surface is then covered by insulating SiO_2. The last step in the fabrication process is the deposition of metallic conductors that are used to connect the MOSFET to external circuits. There are a total of four such connections to the drain, the gate, the source, and the substrate as shown in Figure 18-20.

The area on the surface of the *p*-type substrate between the drain and the source is called the channel (see figure). Note that the channel is separated from the gate connection by an insulating layer of SiO_2. When an electrical potential is applied between the gate and the source, the electrical conductivity of the channel is enhanced by a factor that is related to the magnitude of the potential.

The *ion-selective field effect transistor*, or ISFET, is very similar in construction and function to an *n*-channel enhancement mode MOSFET. The ISFET differs only in that variation in the concentration of the ions of interest provides the variable gate voltage to control the conductivity of the channel. As shown in Figure 18-21, instead of the usual metallic contact, the gate of the ISFET is covered with an insulating layer of silicon nitride. The analyte solution, containing hydronium ions in this example, is in contact with this insulating layer and with a reference electrode. The surface of the gate insulator functions very much like the surface of a glass electrode. Protons from the hydronium ions in the test solution are adsorbed by available microscopic sites on the silicon nitride. Any change in the hydronium ion concentration of the solution results in a change in the concentration of adsorbed protons. The change in concentration of adsorbed protons then gives rise to a changing electrochemical potential between the gate and source, which in turn changes the conductivity of the channel of the ISFET. The conductivity of the channel can be monitored electronically to provide a signal that is proportional to the logarithm of the concentration of hydronium ion in the solution. Note that the entire ISFET except the gate insulator is coated with a polymeric encapsulant to insulate all electrical connections from the analyte solution.

The ion-sensitive surface of the ISFET is naturally sensitive to pH changes, but the device may be rendered sensitive to other species by coating the silicon nitride gate insulator with a polymer containing molecules that tend to form complexes with species other than hydronium ion. Furthermore, several ISFETs may be fabricated on the same substrate so that multiple measurements may be made simultaneously. All of the ISFETs may detect the same species to enhance accuracy and reliability, or each ISFET may be coated with a different polymer so that measurements of several different species may be made. Their small size (about 1 to 2 mm^2), rapid response time relative to glass electrodes, and ruggedness suggest that ISFETs may be the ion detectors of the future for many applications.

18D-8 Gas-Sensing Probes

A **gas-sensing probe** is a galvanic *cell* whose potential is related to the concentration of a gas in a solution. In instrument brochures, these devices are often called gas-sensing electrodes, which is a misnomer.

Figure 18-22 illustrates the essential features of a potentiometric gas-sensing probe, which consists of a tube containing a reference electrode, a specific ion electrode, and an electrolyte solution. A thin, replaceable, gas-permeable membrane attached to one end of the tube serves as a barrier between the internal and analyte solutions. As can be seen from Figure 18-22, this device is a complete electrochemical cell and is more properly referred to as a probe rather than an electrode, a term that is frequently encountered in advertisements by instrument manufacturers. Gas-sensing probes have found widespread use for determining gas dissolved in water and other solvents.

Membrane Composition

A *microporous membrane* is fabricated from a hydrophobic polymer. As the name implies, the membrane is highly porous (the average pore size is less than

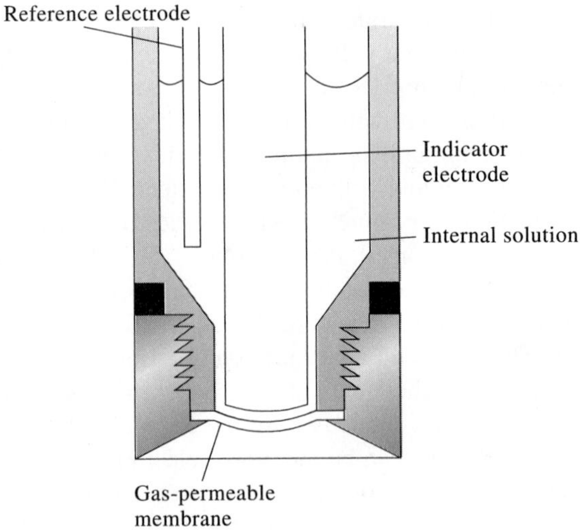

Figure 18-22
Diagram of a gas-sensing probe.

1 μm) and allows the free passage of gases; at the same time, the water-repellent polymer prevents water and solute ions from entering the pores. The thickness of the membrane is about 0.1 mm.

The Mechanism of Response

Using carbon dioxide as an example, we can represent the transfer of gas to the internal solution in Figure 18-22 by the following set of equations:

$$\underset{\text{analyte solution}}{CO_2(aq)} \rightleftharpoons \underset{\text{membrane pores}}{CO_2(g)}$$

$$\underset{\text{membrane pores}}{CO_2(g)} \rightleftharpoons \underset{\text{internal solution}}{CO_2(aq)}$$

$$\underset{\text{internal solution}}{CO_2(aq)} + 2H_2O \rightleftharpoons \underset{\text{internal solution}}{HCO_3^- + H_3O^+}$$

The last equilibrium causes the pH of the internal surface film to change. This change is then detected by the internal glass/calomel electrode system. A description of the overall process is obtained by adding the equations for the three equilibria to give

$$\underset{\text{analyte solution}}{CO_2(aq)} + 2H_2O \rightleftharpoons \underset{\text{internal solution}}{H_3O^+ + HCO_3^-}$$

The thermodynamic equilibrium constant K for this overall reaction is

$$K = \frac{(a_{H_3O^+})_{int} \times (a_{HCO_3^-})_{int}}{(a_{CO_2})_{ext}}$$

For a neutral species such as CO_2, $a_{CO_2} = [CO_2(aq)]$

$$K = \frac{(a_{H_3O^+})_{int} \times (a_{HCO_3^-})_{int}}{[CO_2(aq)]_{ext}}$$

where $[CO_2(aq)]_{ext}$ is the molar concentration of the gas in the analyte solution. In order for the measured cell potential to vary linearly with the logarithm of the carbon dioxide concentration of the external solution, the hydrogen carbonate activity of the internal solution must be sufficiently large that it is not altered significantly by the carbon dioxide entering from the external solution. Assuming then that $(a_{HCO_3^-})_{int}$ is constant, we can rearrange the previous equation to

$$\frac{(a_{H_3O^+})_{int}}{[CO_2(aq)]_{ext}} = \frac{K}{(a_{HCO_3^-})_{int}} = K_g$$

Letting a_1 be the hydrogen ion activity of the internal solution, we rearrange this equation to give

$$(a_{H_3O^+})_{int} = a_1 = K_g[CO_2(aq)]_{ext} \tag{18-18}$$

Substitution of Equation 18-18 for a_1 in Equation 18-13 yields

$$E_{ind} = L + 0.0592 \log a_1 = L + 0.0592 \log K_g[CO_2(aq)]_{ext}$$
$$= L + 0.0592 \log K_g + 0.0592 \log [CO_2(aq)]_{ext}$$

Combining the two constant terms to give a new constant L' leads to

$$E_{ind} = L' + 0.0592 \log [CO_2(aq)]_{ext} \quad (18\text{-}19)$$

Finally, since

$$E_{cell} = E_{ind} - E_{ref}$$

then

$$E_{cell} = L' + 0.0592 \log [CO_2(aq)]_{ext} - E_{ref} \quad (18\text{-}20)$$

or

$$E_{cell} = L'' + 0.0592 \log [CO_2(aq)]_{ext}$$

where

$$L'' = L' + 0.0592 \log K_g - E_{ref}$$

Although sold as gas-sensing electrodes, these devices are complete electrochemical cells and should be called gas-sensing probes.

Thus, the potential between the glass electrode and the reference electrode in the internal solution is determined by the CO_2 concentration in the external solution. *Note that no electrode comes in direct contact with the analyte solution.* Therefore, these devices are gas-sensing *cells*, or *probes*, rather than gas-sensing electrodes. Nevertheless, they continue to be called electrodes in some literature and many advertising brochures.

The only species that interfere are other dissolved gases that permeate the membrane and then affect the pH of the internal solution. The specificity of gas probes depends only upon the permeability of the gas membrane. Gas-sensing probes for CO_2, NO_2, H_2S, SO_2, HF, HCN, and NH_3 are now available from commercial sources.

18E QUESTIONS AND PROBLEMS

18-1. Briefly describe or define
*(a) indicator electrode.
(b) reference electrode.
*(c) electrode of the first kind.
(d) electrode of the second kind.

18-2. Briefly describe or define
*(a) liquid-junction potential.
(b) boundary potential.
(c) asymmetry potential.

***18-3.** Describe how a mercury electrode could function as
(a) an electrode of the first kind for Hg(II).
(b) an electrode of the second kind for EDTA.

18-4. What is meant by Nernstian behavior in an indicator electrode?

***18-5.** Describe the source of pH-dependence in a glass membrane electrode.

18-6. Why is it necessary for the glass in the membrane of a pH-sensitive electrode to be appreciably hygroscopic?

*18-7. List several sources of uncertainty in pH measurements with a glass/calomel electrode system.

18-8. What experimental factor places a limit on the number of significant figures in the response of a membrane electrode?

*18-9. Describe the alkaline error in the measurement of pH. Under what circumstances is this error appreciable? How are pH data affected by alkaline error?

18-10. How does a gas-sensing probe differ from other membrane electrodes?

18-11. What is the source of
(a) the asymmetry potential in a membrane electrode?
*__(b)__ the boundary potential in a membrane electrode?
(c) a junction potential in a glass/calomel electrode system?
*__(d)__ the potential of a crystalline membrane electrode used to determine the concentration of F^-?

CHAPTER

19

Applications of Potentiometry

Potentiometry finds widespread use by chemists (1) for direct and selective measurement of analyte concentrations, (2) for establishing end points in various types of titrations, and (3) for the determination of several types of equilibrium constants. We consider each of these applications in this chapter.

19A INSTRUMENTS FOR MEASURING CELL POTENTIALS

Most cells containing a membrane electrode have very high electrical resistance (as much as 10^8 ohms or more). In order to measure potentials of such high-resistance circuits accurately, it is necessary that the voltmeter have an electrical resistance that is several orders of magnitude greater than the resistance of the cell being measured. If the meter resistance is too low, current is drawn from the cell, which has the effect of lowering its output potential, thus creating a negative *loading error* (see Feature 19-1). This effect is shown in Figure 19-1, which is a plot of the relative error in potential reading as a function of the ratio of the resistance of a meter to the resistance of a cell. When the meter and the cell have the same resistance, a relative error of −50% results. When this ratio is 10, the error is about −9%. When it is 1000, the error is less than 0.1% relative.

A **loading error** in potentiometry is an error that arises if the electrical resistance of the meter used for potential measurements is not significantly greater than the resistance of the cell containing the analyte solution.

FEATURE 19-1
The Loading Error in Potential Measurements

In measuring potential differences, the presence of the meter tends to perturb the circuit and cause a *loading error*. This situation is not peculiar to potential measurements. In fact, it is a simple example of a general limitation to any physical measurement. That is, the process of measurement inevitably disturbs the system of interest so that the quantity actually measured differs from its value prior to the measurement. This type of error can never be completely eliminated, but it can often be reduced to insignificant proportions.

The magnitude of the loading error in potential measurements depends on the ratio of the internal resistance of the meter to the resistance of the circuit under consideration. The percent relative loading error E_r associated with the measured potential V_M in Figure 19-2 is given by

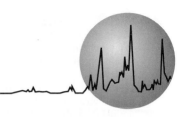

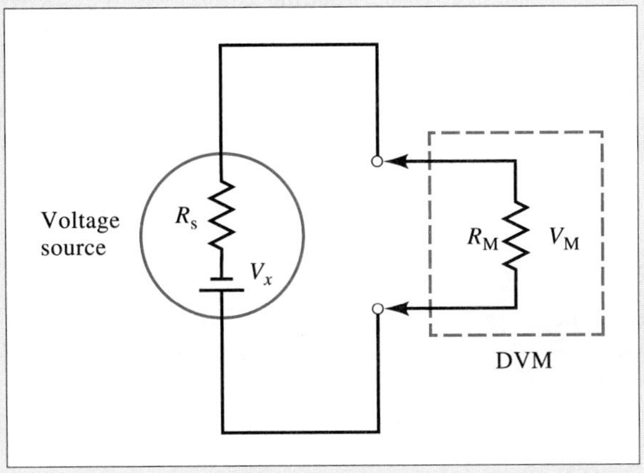

Figure 19-2 Measurement of output V_x from a potential source with a digital voltmeter.

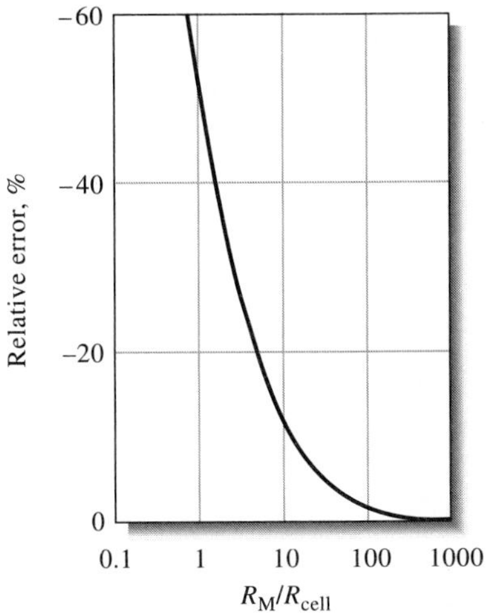

Figure 19-1
Relative error in a cell-potential measurement as a function of the ratio of the electrical resistance of the meter R_M to the resistance of the cell R_{cell}.

$$E_r = \frac{V_M - V_x}{V_x} \times 100\%$$

where V_x is the true voltage of the power source. The voltage drop across the resistance of the meter is given by

$$V_M = V_x \frac{R_M}{R_M + R_s}$$

Substituting this equation into the previous one and rearranging gives

$$E_r = \frac{R_s}{R_M + R_s} \times 100\% \tag{19-1}$$

Note in this equation that the relative loading error becomes smaller as the meter resistance R_M becomes larger relative to the source resistance R_s. Table 19-1 illustrates this effect. Digital voltmeters offer the great advantage of having enormous internal resistances (10^{11} to 10^{12} ohms), thus avoiding loading errors except in circuits having load resistances of greater than about 10^9 ohms.

TABLE 19-1 Effect of Meter Resistance on the Accuracy of Potential Measurements*

Meter Resistance R_M, Ω	Resistance of Source R_s, Ω	R_M/R_s	Relative Error, %
10	20	0.50	−67
50	20	2.5	−29
500	20	25	−3.8
1.0×10^3	20	50	−2.0
1.0×10^4	20	500	−0.20

*See Figure 19-2.

Numerous high-resistance, direct-reading digital voltmeters with internal resistances of 10^{11} to 10^{12} ohms are now on the market. These meters are commonly called *pH meters* but could more properly be referred to as *pIon meters* or *ion meters* since they are frequently used for the measurement of concentrations of other ions as well. Feature 19-2 describes the design of a typical modern ion meter.

FEATURE 19-2
Operational-Amplifier Voltage Measurements

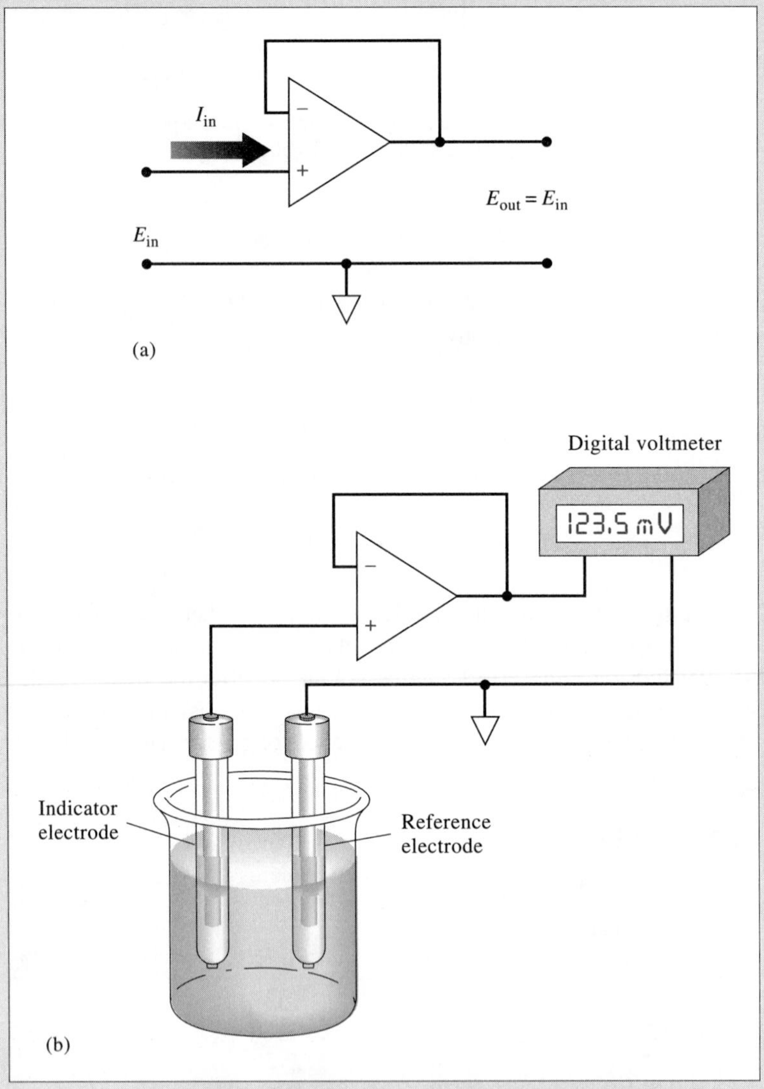

Figure 19-3 (a) A voltage-follower operational amplifier. (b) Typical arrangement for potentiometric measurements with a membrane electrode.

One of the most important developments in chemical instrumentation over the last several years has been the advent of compact, inexpensive, versatile integrated-circuit amplifiers (op amps).[1] These devices allow us to make potential measurements on high-resistance cells, such as those that contain a glass electrode, without drawing appreciable current. Even a small current (10^{-7} to 10^{-10} A) in a glass electrode results in a large error in the measured voltage. One of the most important uses for operational amplifiers is to isolate voltage sources from their measurement circuits. The basic *voltage follower,* which allows this type of measurement, is shown in Figure 19-3a. This circuit has two important characteristics. The output voltage E_{out} is equal to the input voltage E_{in}, and the input current I_{in} is essentially zero (10^{-9} to 10^{-11} A).

A practical application of this circuit is in measuring cell potentials. The cell is connected to the op amp input as shown in Figure 19-3b, and the output of the op amp is connected to a digital voltmeter to measure the voltage. Modern op amps are nearly ideal voltage-measurement devices and are incorporated in most ion meters and pH meters to monitor high-resistance indicator electrodes with little error.

[1]For a detailed description of op amp circuits, see H. V. Malmstadt, C. G. Enke, and S. R. Crouch, *Micro Computers and Electronic Instrumentation: Making the Right Connections,* Chapter 5. Washington, DC: American Chemical Society, 1994.

The readout of ion meters is either digital or analog (in the latter, a needle sweeps a range on a scale from 0 to 14 pH units). Some meters are capable of precision on the order of 0.001 to 0.005 pH unit. Seldom is it possible to measure pH with a comparable degree of *accuracy.* Inaccuracies of ± 0.02 to ± 0.03 pH unit are typical.

19B DIRECT POTENTIOMETRIC MEASUREMENTS

Direct potentiometric measurements provide a rapid and convenient method for determining the activity of a variety of cations and anions. The technique requires only a comparison of the potential developed in a cell containing the indicator electrode in the analyte solution with its potential when immersed in one or more standard solutions of known analyte concentration. If the response of the electrode is specific for the analyte, as it often is, no preliminary separation steps are required. Direct potentiometric measurements are also readily adapted to applications requiring continuous and automatic recording of analytical data.

19B-1 Equations for Direct Potentiometry

The sign convention for potentiometry is consistent with the convention described in Section 15C-3 for standard electrode potentials.[2] In this convention, the indica-

[2]According to Bates, the convention being described here has been endorsed by standardizing groups in the United States and Great Britain as well as IUPAC. See R. G. Bates, in *Treatise on Analytical Chemistry,* 2nd ed., I. M. Kolthoff and P. J. Elving, Eds., Part I, Vol. 1, pp. 831–832. New York: Wiley, 1978.

In potentiometry, the indicator electrode is *always* treated as the cathode and the reference electrode as the anode.

tor electrode is *always* treated as the *cathode* and the reference electrode as the *anode.*[3] For direct potentiometric measurements, the potential of a cell can then be expressed in terms of the potentials developed by the indicator electrode, the reference electrode, and a junction potential:

$$E_{cell} = E_{ind} - E_{ref} + E_j \tag{19-2}$$

In Section 18D we have described the response of various types of indicator electrodes to analyte activities. For the cation X^{n+} at 25°C, the electrode response takes the general *Nernstian* form

$$E_{ind} = L - \frac{0.0592}{n}\text{pX} = L + \frac{0.0592}{n}\log a_X \tag{19-3}$$

where L is a constant and a_X is the activity of the cation. For metallic indicator electrodes, L is ordinarily the standard electrode potential; for membrane electrodes, L is the summation of several constants, including the time-dependent asymmetry potential of uncertain magnitude.

Substitution of Equation 19-3 into Equation 19-2 yields with rearrangement

$$\text{pX} = -\log a_X = -\frac{E_{cell} - (E_j - E_{ref} + L)}{0.0592/n} \tag{19-4}$$

The constant terms in parentheses can be combined to give a new constant K.

$$\text{pX} = -\log a_X = -\frac{E_{cell} - K}{0.0592/n} \tag{19-5}$$

For an anion A^{n-}, the sign of Equation 19-5 is reversed:

$$\text{pA} = \frac{E_{cell} - K}{0.0592/n} = \frac{n\,(E_{cell} - K)}{0.0592} \tag{19-6}$$

All direct potentiometric methods are based upon Equation 19-5 or 19-6. The difference in sign in the two equations has a subtle but important consequence in the way that ion-selective electrodes are connected to pH meters and pIon meters. When the two equations are solved for E_{cell}, we find that for cations

$$E_{cell} = K - \frac{0.0592}{n}\text{pX} \tag{19-7}$$

and for anions

[3]In effect, the sign convention for electrode potentials described in Section 15C-3 also designates the indicator electrode as the cathode by stipulating that half-reactions always be written as reductions; the standard hydrogen electrode, which is the reference electrode in this case, is then the anode.

$$E_{cell} = K + \frac{0.0592}{n}\text{pA} \tag{19-7}$$

Equation 19-7 shows that for a cation-selective electrode, an increase in pX results in a *decrease* in E_{cell} (Figure 19-4). Thus, when a high-resistance voltmeter is connected to the cell in the usual way, with the indicator electrode attached to the positive terminal, the meter reading decreases as pX increases. To eliminate this problem, instrument manufacturers generally reverse the leads so that cation-sensitive electrodes are connected to the *negative* terminal of the voltage measuring device. Meter readings then increase with increases of pX. Anion-selective electrodes, on the other hand, are connected to the *positive* terminal of the meter so that increases in pA also yield larger readings.

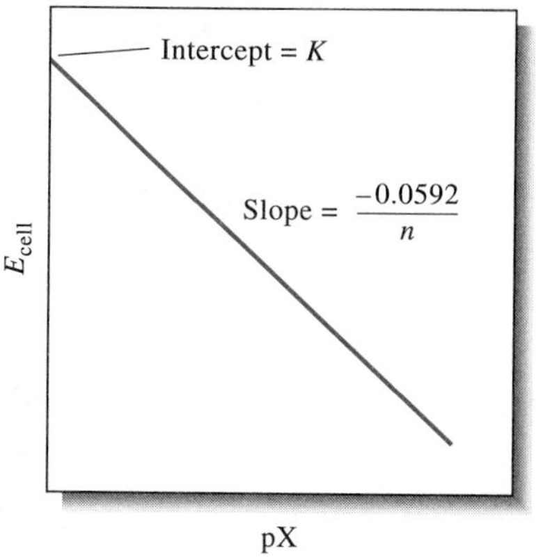

Figure 19-4
A plot of Equation 19-7 for cationic electrodes.

19B-2 The Electrode-Calibration Method

As we have seen from our discussions in Section 18D, the constant K in Equations 19-5 and 19-6 is made up of several constants at least one of which, the junction potential, cannot be computed from theory nor measured directly. Thus, before these equations can be used for the determination of pX or pA, K must be evaluated *experimentally* with a standard solution of the analyte.

In the electrode-calibration method, K in Equations 19-5 and 19-6 is determined by measuring E_{cell} for one or more standard solutions of known pX or pA. The assumption is then made that K is unchanged when the standard is replaced by the analyte solution. The calibration is ordinarily performed at the time pX or pA for the unknown is determined. With membrane electrodes, recalibration may be required if measurements extend over several hours because of slow changes in the asymmetry potential.

The electrode-calibration method offers the advantages of simplicity, speed, and applicability to the continuous monitoring of pX or pA, but suffers from a somewhat limited accuracy because of uncertainties in junction potentials.

Inherent Error in the Electrode-Calibration Procedure

A serious disadvantage of the electrode-calibration method is the inherent error that results from the assumption that K in Equations 19-5 and 19-6 remains constant after calibration. This assumption can seldom, if ever, be exactly true because the electrolyte composition of the unknown almost inevitably differs from that of the solution employed for calibration. The junction-potential term contained in K varies slightly as a consequence, even when a salt bridge is used. This error is frequently on the order of 1 mV or more. Unfortunately, because of the nature of the potential/activity relationship, such an uncertainty has an amplified effect on the inherent accuracy of the analysis.

The magnitude of the error in analyte concentration can be estimated by differentiating Equation 19-5 while holding E_{cell} constant

$$-\log_{10} e\frac{da_1}{a_1} = -0.434\frac{da_1}{a_1} = -\frac{dK}{0.0592/n}$$

$$\frac{da_1}{a_1} = \frac{ndK}{0.0257} = 38.9\,ndK$$

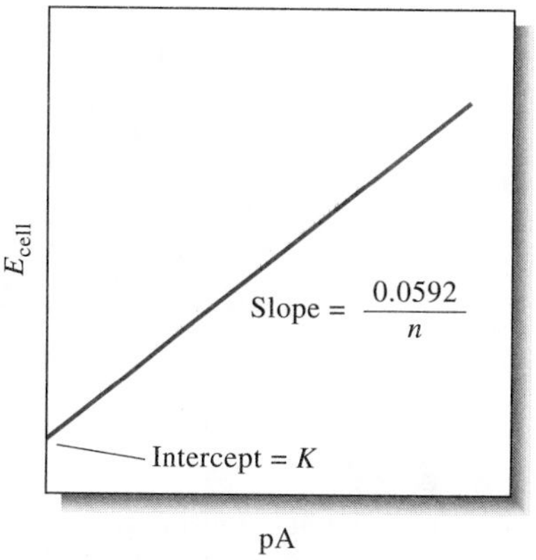

Figure 19-5
A plot of Equation 19-8 for anionic electrodes.

Upon replacing da_1 and dK with finite increments and multiplying both sides of the equation by 100%, we obtain

$$\text{percent relative error} = \frac{\Delta a_1}{a_1} \times 100\% = 38.9\, n\Delta K \times 100\%$$
$$= 3.89 \times 10^3 n\Delta K\% \approx 4000\, n\Delta K\%$$

The quantity $\Delta a_1/a_1$ is the relative error in a_1 associated with an absolute uncertainty ΔK in K. If, for example, ΔK is ± 0.001 V, a relative error in activity of about $\pm 4n\%$ can be expected. *It is important to appreciate that this error is characteristic of all measurements involving cells that contain a salt bridge and that this error cannot be eliminated by even the most careful measurements of cell potentials or the most sensitive and precise measuring devices.*

Activity Versus Concentration

Electrode response is related to analyte activity rather than analyte concentration. We are usually interested in concentration, however, and the determination of this quantity from a potentiometric measurement requires activity coefficient data. Activity coefficients are seldom available because the ionic strength of the solution is either unknown or else is so large that the Debye-Hückel equation is not applicable.

The difference between activity and concentration is illustrated by Figure 19-6, in which the response of a calcium ion electrode is plotted against a logarithmic function of calcium chloride *concentration.* The nonlinearity is due to the increase in ionic strength—and the consequent decrease in the activity of calcium ion—with increasing electrolyte concentration. The upper curve is obtained when these concentrations are converted to activities. This straight line has the theoretical slope of 0.0296 (0.0592/2).

Activity coefficients for singly charged species are less affected by changes in ionic strength than are the coefficients for ions with multiple charges. Thus, the effect shown in Figure 19-6 is less pronounced for electrodes that respond to H^+, Na^+, and other univalent ions.

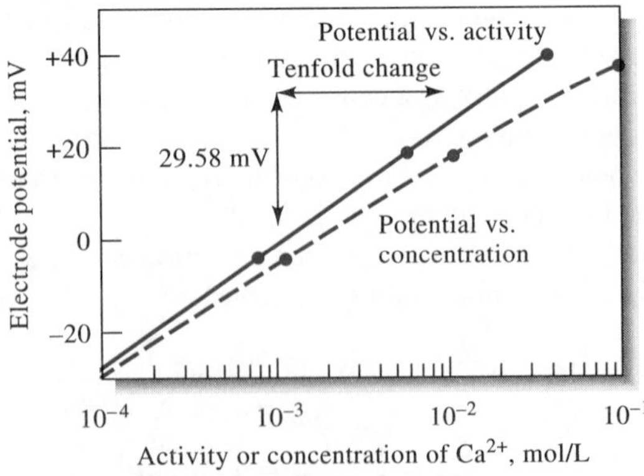

Figure 19-6
Response of a liquid-membrane electrode to variations in the concentration and activity of calcium ion. (Courtesy of Orion Research, Boston, MA.)

In potentiometric pH measurements, the pH of the standard buffer used for calibration is generally based on the activity of hydrogen ions. Thus, the results are also on an activity scale. If the unknown sample has a high ionic strength, the hydrogen ion *concentration* will differ appreciably from the activity measured.

Many chemical reactions of physiological importance depend upon the activity of metal ions rather than their concentration.

An obvious way to convert potentiometric measurements from activity to concentration is to make use of an empirical calibration curve, such as the lower plot in Figure 19-6. For this approach to be successful, it is necessary to make the ionic composition of the standards essentially the same as that of the analyte solution. Matching the ionic strength of standards to that of samples is often difficult, particularly for samples that are chemically complex.

Where electrolyte concentrations are not too great, it is often useful to swamp both samples and standards with a measured excess of an inert electrolyte. The added effect of the electrolyte from the sample matrix becomes negligible under these circumstances, and the empirical calibration curve yields results in terms of concentration. This approach has been used, for example, in the potentiometric determination of fluoride ion in drinking water. Both samples and standards are diluted with a solution that contains sodium chloride, an acetate buffer, and a citrate buffer; the diluent is sufficiently concentrated so that the samples and standards have essentially identical ionic strengths. This method provides a rapid means for measuring fluoride concentrations in the parts-per-million range with an accuracy of about 5% relative.

A **total ionic strength adjustment buffer (TISAB)** is used to control the ionic strength and the pH of samples and standards for measurement with ion-selective electrodes.

19B-3 The Standard-Addition Method

The standard-addition method involves determining the potential of the electrode system before and after a measured volume of a standard has been added to a known volume of the analyte solution. Often an excess of an electrolyte is incorporated into the analyte solution at the outset to prevent any major shift in ionic strength that might accompany the addition of standard. It is also necessary to assume that the junction potential remains constant during the two measurements.

EXAMPLE 19-1

A cell consisting of a saturated calomel electrode and a lead ion electrode developed a potential of −0.4706 V when immersed in 50.00 mL of a sample. A 5.00-mL addition of standard 0.02000 M lead solution caused the potential to shift to −0.4490 V. Calculate the molar concentration of lead in the sample.

We shall assume that the activity of Pb^{2+} is approximately equal to $[Pb^{2+}]$ and apply Equation 19-5. Thus,

$$\text{pPb} = -\log[\text{Pb}^{2+}] = -\frac{E'_{\text{cell}} - K}{0.0592/2}$$

where E'_{cell} is the initial measured potential (−0.4706 V).

After the standard solution is made, the potential becomes E''_{cell} (−0.4490 V), and

$$-\log \frac{50.00 \times [Pb^{2+}] + 5.00 \times 0.0200}{50.00 + 5.00} = -\frac{E''_{cell} - K}{0.0592/2}$$

$$-\log (0.9091\,[Pb^{2+}] + 1.818 \times 10^{-3}) = -\frac{E''_{cell} - K}{0.0592/2}$$

Subtracting this equation from the first leads to

$$-\log \frac{[Pb^{2+}]}{0.9091\,[Pb^{2+}] + 1.818 \times 10^{-3}} = \frac{2(E''_{cell} - E'_{cell})}{0.0592}$$

$$= \frac{2[-0.4490 - (-0.4706)]}{0.0592}$$

$$= 0.7297$$

$$\frac{[Pb^{2+}]}{0.9091\,[Pb^{2+}] + 1.818 \times 10^{-3}} = \text{antilog}\,(-0.7297) = 0.1863$$

$$[Pb^{2+}] = 4.08 \times 10^{-4}\ M$$

19B-4 Potentiometric pH Measurements with a Glass Electrode[4]

The glass electrode is unquestionably the most important electrode for hydrogen ion. It is convenient to use and subject to few of the interferences that affect other pH-sensing electrodes.

The glass/calomel electrode system is a remarkably versatile tool for the measurement of pH under many conditions. It can be used without interference in solutions containing strong oxidants, strong reductants, proteins, and gases; the pH of viscous or even semisolid fluids can be determined. Electrodes for special applications are available. Included among these are small electrodes for pH measurements in one drop (or less) of solution, in a tooth cavity, or in the sweat on the skin; microelectrodes that permit the measurement of pH inside a living cell; rugged electrodes for insertion in a flowing liquid stream to provide a continuous monitoring of pH; and small electrodes that can be swallowed to measure the acidity of the stomach contents (the calomel electrode is kept in the mouth).

Errors That Affect pH Measurements with the Glass Electrode

The ubiquity of the pH meter and the general applicability of the glass electrode tend to lull the chemist into the attitude that any measurement obtained with such equipment is surely correct. The reader must be alert to the fact that there are distinct limitations to the electrode, some of which have been discussed in earlier sections:

1. *The alkaline error.* The ordinary glass electrode becomes somewhat sensitive to alkali metal ions and gives low readings at pH values greater than 9.
2. *The acid error.* Values registered by the glass electrode tend to be somewhat high when the pH is less than about 0.5.

[4]For a detailed discussion of potentiometric pH measurements, see R. G. Bates, *Determination of pH,* 2nd ed. New York: Wiley, 1973.

3. *Dehydration.* Dehydration may cause erratic electrode performance.
4. *Errors in low ionic strength solutions.* It has been found that significant errors (as much as 1 or 2 pH units) may occur when the pH of samples of low ionic strength, such as lake or stream water, is measured with a glass/calomel electrode system.[5] The prime source of such errors has been shown to be nonreproducible junction potentials, which apparently result from partial clogging of the fritted plug or porous fiber that is used to restrict the flow of liquid from the salt bridge into the analyte solution. In order to overcome this problem, free diffusion junctions (FDJs) of various types have been designed, and one is produced commercially.
5. *Variation in junction potential.* A fundamental source of uncertainty for which a correction cannot be applied is the junction-potential variation resulting from differences in the composition of the standard and the unknown solution.
6. *Error in the pH of the standard buffer.* Any inaccuracies in the preparation of the buffer used for calibration or any changes in its composition during storage cause an error in subsequent pH measurements. The action of bacteria on organic buffer components is a common cause for deterioration.

Particular care must be taken in measuring the pH of approximately neutral unbuffered solutions, such as samples from lakes and streams.

The Operational Definition of pH

The utility of pH as a measure of the acidity and alkalinity of aqueous media, the wide availability of commercial glass electrodes, and the relatively recent proliferation of inexpensive solid state pH meters have made the potentiometric measurement of pH perhaps the most common analytical technique in all of science. It is thus extremely important that pH be defined in a manner that is easily duplicated at various times and in various laboratories throughout the world. To meet this requirement, it is necessary to define pH in operational terms—that is, by the way the measurement is made. Only then will the pH measured by one worker be the same as that by another.

Perhaps the most common analytical instrumental technique is the measurement of pH.

The operational definition of pH, which is endorsed by the National Institute of Standards and Technology (NIST), similar organizations in other countries, and the IUPAC, is based upon the direct calibration of the meter with carefully prescribed standard buffers followed by potentiometric determination of the pH of unknown solutions.

Consider, for example, the glass/calomel system in Figure 18-9. When these electrodes are immersed in a standard buffer, Equation 19-5 applies and we can write

$$pH_S = -\frac{E_S - K}{0.0592}$$

where E_S is the cell potential when the electrodes are immersed in the buffer. Similarly, if the cell potential is E_U when the electrodes are immersed in a solution of unknown pH, we have

$$pH_U = -\frac{E_U - K}{0.0592}$$

[5]See W. Davison and C. Woof, *Anal. Chem.,* **1985,** *57,* 2567; T. R. Harbinson and W. Davison, *Anal. Chem.,* **1987,** *59,* 2450; A. Kopelove, S. Franklin, and G. M. Miller, *Amer. Lab.,* **1989** (6), 40.

TABLE 19-2 Values of NIST Primary-Standard pH Solutions from 0° to 60°C*

Temperature, °C	Sat'd (25°C) KH Tartrate	0.05 *m* KH_2 Citrate†	0.05 *m* KH Phthalate	0.025 *m* KH_2PO_4/ 0.025 *m* Na_2HPO_4	0.008695 *m* KH_2PO_4/ 0.03043 *m* Na_2HPO_4	0.01 *m* $Na_2B_4O_7$	0.025 *m* $NaHCO_3$/ 0.025 *m* Na_2CO_3
0	—	3.863	4.003	6.984	7.534	9.464	10.317
5	—	3.840	3.999	6.951	7.500	9.395	10.245
10	—	3.820	3.998	6.923	7.472	9.332	10.179
15	—	3.802	3.999	6.900	7.448	9.276	10.118
20	—	3.788	4.002	6.881	7.429	9.225	10.062
25	3.557	3.776	4.008	6.865	7.413	9.180	10.012
30	3.552	3.766	4.015	6.853	7.400	9.139	9.966
35	3.549	3.759	4.024	6.844	7.389	9.102	9.925
40	3.547	3.753	4.035	6.838	7.380	9.068	9.889
45	3.547	3.750	4.047	6.834	7.373	9.038	9.856
50	3.549	3.749	4.060	6.833	7.367	9.011	9.828
55	3.554	—	4.075	6.834	—	8.985	—
60	3.560	—	4.091	6.836	—	8.962	—
Buffer capacity‡ (mol/pH unit)	0.027	0.034	0.016	0.029	0.016	0.020	0.029
$\Delta pH_{1/2}$ for 1 : 1 dilutions§	0.049	0.024	0.052	0.080	0.07	0.01	0.079

*Adapted from R. G. Bates, *Determination of pH*, 2nd ed., p. 73. New York: Wiley, 1973.

†*m* = molality (mol solute/kg H_2O).

‡See page 205.

§Change in pH that occurs when one volume of buffer is diluted with one volume of H_2O.

By definition, pH is what you measure with a glass electrode and a pH meter. It is approximately equal to the theoretical definition of pH = log a_{H^+}.

An operational definition of a quantity defines the quantity in terms of how it is measured.

By subtracting the first equation from the second and solving for pH_U, we find

$$pH_U = pH_S - \frac{(E_U - E_S)}{0.0592} \tag{19-9}$$

Equation 19-9 has been adopted throughout the world as the *operational definition of pH.*

Workers at the National Institute of Standards and Technology and elsewhere have used cells without liquid junctions to study primary-standard buffers extensively. Some of the properties of these buffers are presented in Table 19-2 and are discussed in detail elsewhere.[6] Note that the NIST buffers are described by their *molal* concentrations (mol solute/kg solvent) for accuracy and precision of preparation. For general use, the buffers can be prepared from relatively inexpensive laboratory reagents; for careful work, however, certified buffers can be purchased from the NIST. As Table 19-2 shows, the buffers form an internally consistent series over the pH range from 3.5 to 10.3, and their pH values are presented as a function of temperature. The buffer capacity listed for each solution indicates its pH stability with the addition of acid or base.

[6] R. G. Bates, *Determination of pH,* 2nd ed., Chapter 4. New York: Wiley, 1973.

It should be emphasized that the strength of the operational definition of pH is that it provides a coherent scale for the determination of acidity or alkalinity that can be reproduced throughout the world. However, measured pH values cannot be expected to yield a detailed picture of solution composition that is entirely consistent with solution theory. This uncertainty stems from our fundamental inability to measure single ion activities. That is, the operational definition of pH does not yield the exact pH as defined by the equation

$$pH = -\log \gamma_{H^+} [H^+]$$

19C POTENTIOMETRIC TITRATIONS

A *potentiometric titration* involves measurement of the potential of a suitable indicator electrode as a function of titrant volume. The information provided by a potentiometric titration is not the same as that obtained from a direct potentiometric measurement. For example, the direct measurement of 0.100 M solutions of hydrochloric and acetic acids would yield two substantially different hydrogen ion concentrations because the latter is only partially dissociated. In contrast, the potentiometric titration of equal volumes of the two acids would require the same amount of standard base because both solutes have the same number of titratable protons.

Potentiometric titrations provide data that are more reliable than data from titrations that use chemical indicators and are particularly useful with colored or turbid solutions and for detecting the presence of unsuspected species. Such titrations are also readily automated. Manual potentiometric titrations of this kind, on the other hand, suffer from the disadvantage of being more time-consuming than those involving indicators.

Automatic *titrators* for carrying out potentiometric titrations are available from several manufacturers.

Figure 19-7 illustrates a typical apparatus for performing a manual potentiometric titration. Its use involves measuring and recording the cell potential (in

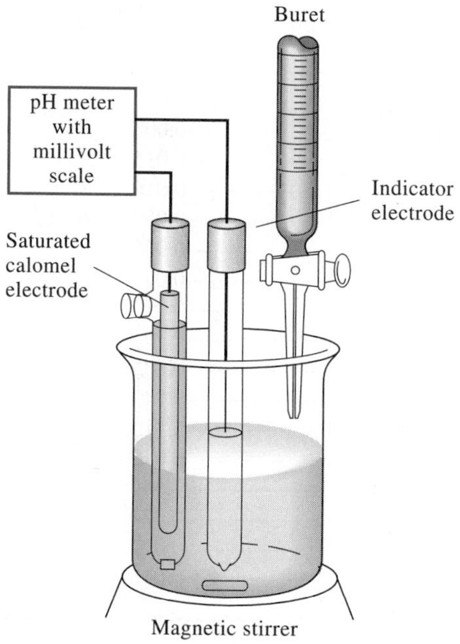

Figure 19-7
Apparatus for a potentiometric titration.

units of millivolts or pH, as appropriate) after each addition of reagent. The titrant is added in large increments at the outset and in smaller and smaller increments as the end point is approached (as indicated by larger changes in response per unit volume).

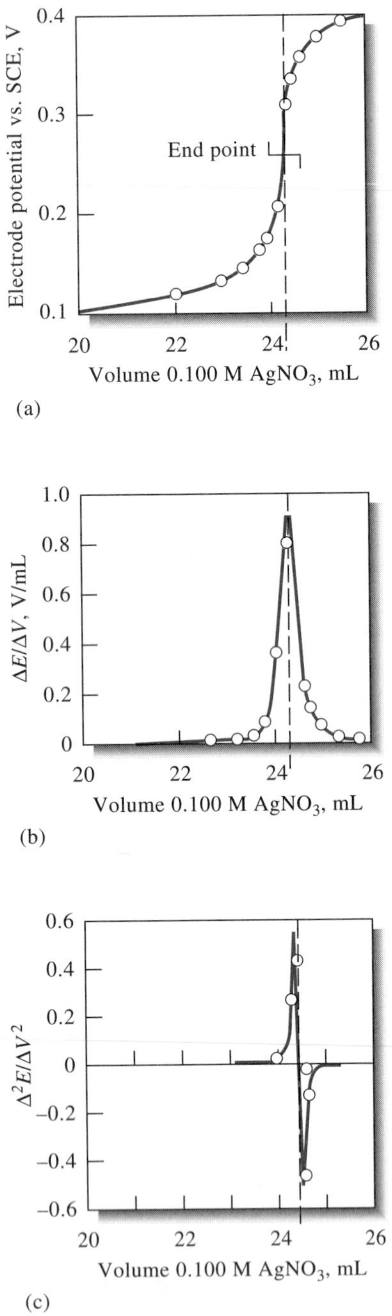

Figure 19-8
Titration of 2.433 mmol of chloride ion with 0.1000 M silver nitrate.
(a) Titration curve. (b) First-derivative curve. (c) Second-derivative curve.

19C-1 End-Point Detection

Several methods can be used to determine the end point of a potentiometric titration. The most straightforward involves a direct plot of potential as a function of reagent volume, as in Figure 19-8a; the midpoint in the steeply rising portion of the curve is estimated visually and taken as the end point. Various graphical methods have been proposed to aid in the establishment of the midpoint, but it is doubtful that these procedures significantly improve its determination.

A second approach to end-point detection is to calculate the change in potential per unit volume of titrant (that is, $\Delta E/\Delta V$), as in column 3 of Table 19-3. A plot of these data as a function of the average volume V produces a curve with a maximum that corresponds to the point of inflection (Figure 19-8b). Alternatively, this ratio can be evaluated during the titration and recorded in lieu of the potential. Inspection of column 3 of Table 19-3 reveals that the maximum is located between 24.30 and 24.40 mL; 24.35 mL would be adequate for most purposes.

Column 4 of Table 19-3 and Figure 19-8c show that the second derivative for the data changes sign at the point of inflection. This change is used as the analytical signal in some automatic titrators.

TABLE 19-3 Potentiometric Titration Data for 2.433 mmol of Chloride with 0.1000 M Silver Nitrate

Volume $AgNO_3$, mL	E vs. SCE, V	$\Delta E/\Delta V$, V/mL	$\Delta^2E/\Delta V^2$, V^2/mL^2
5.0	0.062		
		0.002	
15.0	0.085		
		0.004	
20.0	0.107		
		0.008	
22.0	0.123		
		0.015	
23.0	0.138		
		0.016	
23.50	0.146		
		0.050	
23.80	0.161		
		0.065	
24.00	0.174		
		0.09	
24.10	0.183		
		0.11	
24.20	0.194		2.8
		0.39	
24.30	0.233		4.4
		0.83	
24.40	0.316		−5.9
		0.24	
24.50	0.340		−1.3
		0.11	
24.60	0.351		−0.4
		0.07	
24.70	0.358		
		0.050	
25.00	0.373		
		0.024	
25.5	0.385		
		0.022	
26.0	0.396		
		0.015	
28.0	0.426		

All the foregoing methods of end-point evaluation are predicated on the assumption that the titration curve is symmetric about the equivalence point and that the inflection in the curve corresponds to this point. This assumption is perfectly valid, provided the participants in the titration react with one another in an equimolar ratio and also provided the electrode reaction is perfectly reversible. The former condition is lacking in many oxidation/reduction titrations; the titration of iron(II) with permanganate is an example. The curve for such titrations is ordinarily so steep, however, that failure to account for asymmetry results in a vanishingly small titration error.

19C-2 Potentiometric Precipitation Titrations

Electrode Systems

The indicator electrode for a precipitation titration is often the metal from which the reacting cation is derived. A membrane electrode responsive to this cation or to the anion in the titration can also be used.

Silver nitrate is without question the most versatile reagent for precipitation titrations. A silver wire serves as the indicator electrode. For reagent and analyte concentrations of 0.1 M or greater, a calomel reference electrode can be located directly in the titration vessel without serious error from the slight leakage of chloride ions from the salt bridge. This leakage can be a source of significant error in titrations that involve very dilute solutions or require high precision, however. The difficulty is eliminated by immersing the calomel electrode in a potassium nitrate solution that is connected to the analyte solution by a salt bridge containing potassium nitrate. Reference electrodes with bridges of this type can be purchased from laboratory supply houses.

Titration Curves

A theoretical curve for a potentiometric titration is readily derived. For example, the potential of a silver electrode in the argentometric titration of chloride can be described by

$$E_{Ag} = E^0_{AgCl} - 0.0592 \log [Cl^-] = 0.222 - 0.0592 \log [Cl^-]$$

where E^0_{AgCl} is the standard potential for the reduction of AgCl to Ag(s). Alternatively, the standard potential for the reduction of silver ion can be used:

$$E_{Ag} = E^0_{Ag^+} - 0.0592 \log \frac{1}{[Ag^+]} = 0.799 - 0.0592 \log \frac{1}{[Ag^+]}$$

The former potential is more convenient for calculating the potential of the silver electrode when an excess of chloride exists, whereas the latter is preferable for solutions containing an excess of silver ion.

Potentiometric measurements are particularly useful for titrations of mixtures of anions with standard silver nitrate. For example Figure 13-3 shows a theoretical curve for the titration of a chloride/iodide mixture. An experimental curve has the same general appearance, although the ordinate units are different.

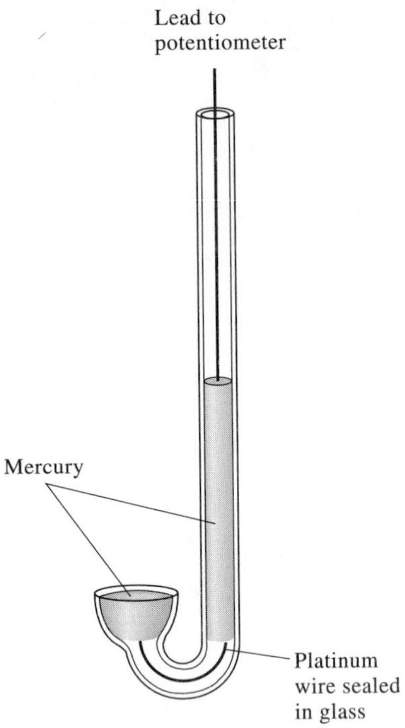

Figure 19-9
A typical mercury electrode.

19C-3 Complex-Formation Titrations

Both metallic and membrane electrodes have been used to detect end points in potentiometric titrations involving complex formation. The mercury electrode (Figure 19-9) is particularly useful for EDTA titrations of cations that form complexes that are less stable than HgY^{2-} (see page 391 for the half-reactions involved).

19C-4 Neutralization Titrations

Experimental neutralization curves closely approximate the theoretical curves described in Chapters 10 and 11. Ordinarily, however, the experimental curves are somewhat displaced from the theoretical along the pH axis because concentrations rather than activities are used in their derivation; this displacement is of no consequence in locating end points. Potentiometric neutralization titrations are particularly valuable for the analysis of mixtures of acids or polyprotic acids. The same considerations apply to bases.

The Determination of Dissociation Constants

An approximate numerical value for the dissociation constant of a weak acid or base can be estimated from potentiometric titration curves. This quantity can be computed from the pH at any point along the curve, but as a practical matter, the value at half-neutralization is most convenient. Here,

$$[HA] \approx [A^-]$$

Therefore,

$$K_a = \frac{[H_3O^+]\cancel{[A^-]}}{\cancel{[HA]}} = [H_3O^+]$$

$$pK_a = pH$$

It is important to note that the use of concentrations instead of activities may cause the value for K_a to differ from its published value by a factor of 2 or more. A more correct form of the dissociation constant for HA is

$$K_a = \frac{a_{H_3O^+} a_{A^-}}{a_{HA}} = \frac{a_{H_3O^+} \gamma_{A^-} \cancel{[A^-]}}{\gamma_{HA} \cancel{[HA]}} \qquad (19\text{-}10)$$

$$K_a = \frac{a_{H_3O^+} \gamma_{A^-}}{\gamma_{HA}}$$

Since the glass electrode provides a good approximation of $a_{H_3O^+}$, the value of K_a measured differs from the thermodynamic value by the ratio of the two activity coefficients. The activity coefficient in the denominator of Equation 19-10 does not change significantly as ionic strength increases because HA is a neutral species. The activity coefficient for A^-, on the other hand, decreases as the electrolyte concentration increases. Therefore, the observed hydrogen ion activity is numerically larger than the thermodynamic dissociation constant.

EXAMPLE 19-2

In order to determine K_1 and K_2 for H_3PO_4 from titration data, careful pH measurements are made after 0.5 and 1.5 mol of base are added for each mole of acid. It is then assumed that the hydrogen ion activities computed from these data are identical to the desired dissociation constants. Calculate the relative error incurred by this assumption if the ionic strength is 0.1 at the time of each measurement. (From Appendix 3, K_1 and K_2 for H_3PO_4 are 7.11×10^{-3} and 6.34×10^{-8}.)

Rearrangement of Equation 19-10 gives

$$K_a(\text{exptl}) = a_{H_3O^+} = K_a(\gamma_{HA}/\gamma_{A^-})$$

The activity coefficient for H_3PO_4 is approximately unity since this species is uncharged. In Table 8-1, we find that the activity coefficient for $H_2PO_4^-$ is 0.78 and that for HPO_4^{2-} is 0.36. Substituting these values into the equations for K_1 and K_2 gives

$$K_1(\text{exptl}) = 7.11 \times 10^{-3} \times 1.00/0.78 = 9.1 \times 10^{-3}$$

$$\text{error} = \frac{9.1 \times 10^{-3} - 7.11 \times 10^{-3}}{7.11 \times 10^{-3}} \times 100\% = 28\%$$

$$K_2(\text{exptl}) = 6.34 \times 10^{-8} \times 0.78/0.36 = 1.37 \times 10^{-7}$$

$$\text{error} = \frac{1.37 \times 10^{-7} - 6.34 \times 10^{-8}}{6.34 \times 10^{-8}} \times 100\% = 116\%$$

A single titration of a pure acid can provide sufficient data (equivalent weight and dissociation constant) to make identification of the acid possible.

19C-5 Oxidation/Reduction Titrations

An inert indicator electrode constructed of platinum is ordinarily used to detect end points in oxidation/reduction titrations. Occasionally, other inert metals, such as silver, palladium, gold, and mercury, are used instead. Titration curves similar to those derived in Chapter 16 are ordinarily obtained, although they may be displaced along the ordinate axis as a consequence of the effects of high ionic strengths. End points are determined by the methods described earlier in this chapter.

19D THE DETERMINATION OF EQUILIBRIUM CONSTANTS FROM ELECTRODE POTENTIAL MEASUREMENTS

Numerical values for solubility-product constants, dissociation constants, and formation constants are conveniently evaluated through the measurement of cell potentials. One important virtue of this technique is that the measurement can be made without appreciably affecting any equilibria that may exist in the solution. For example, the potential of a silver electrode in a solution containing silver ion, cyanide ion, and the complex formed between them depends upon the activities of the three species. It is possible to measure this potential with negligible current. Since the activities of the participants are not sensibly altered during the measurement, the position of the equilibrium

$$Ag^+ + 2CN^- \rightleftharpoons Ag(CN)_2^-$$

is likewise undisturbed.

EXAMPLE 19-3

Calculate the formation constant K_f for $Ag(CN)_2^-$:

$$Ag^+ + 2CN^- \rightleftharpoons Ag(CN)_2^-$$

if the cell

$$SCE \| Ag(CN)_2^-(7.50 \times 10^{-3}\ M),\ CN^-(0.0250\ M) | Ag$$

develops a potential of -0.625 V.

Proceeding as in the earlier examples, we have

$$Ag^+ + e^- \rightleftharpoons Ag(s) \qquad E^0 = +0.799\ V$$

$$-0.625 = E_{cathode} - E_{anode} = E_{Ag^+} - 0.244$$

$$E_{Ag^+} = -0.625 + 0.244 = -0.381\ V$$

Application of the Nernst equation for the silver electrode gives

$$-0.381 = 0.799 - \frac{0.0592}{1} \log \frac{1}{[Ag^+]}$$

$$\log [Ag^+] = \frac{-0.381 - 0.799}{0.0592} = -19.93$$

$$[Ag^+] = 1.2 \times 10^{-20}$$

$$K_f = \frac{[Ag(CN)_2^-]}{[Ag^+][CN^-]^2} = \frac{7.50 \times 10^{-3}}{(1.2 \times 10^{-20})(2.50 \times 10^{-2})^2}$$

$$= 1.0 \times 10^{21} \approx 1 \times 10^{21}$$

In theory, any electrode system in which hydrogen ions are participants can be used to evaluate dissociation constants for acids and bases.

EXAMPLE 19-4

Calculate the dissociation constant for the weak acid HP if the cell

$$SCE \| HP(0.010\ M), NaP(0.040\ M) | Pt, H_2(1.00\ atm)$$

develops a potential of -0.591 V.

The diagram for this cell indicates that the saturated calomel electrode is the anode. Thus,

$$E_{cell} = E_{cathode} - 0.244$$

$$E_{cathode} = -0.591 + 0.244 = -0.347 \text{ V}$$

Application of the Nernst equation for the hydrogen electrode gives

$$-0.347 = 0.000 - \frac{0.0592}{2}\log\frac{1.00}{[H_3O^+]^2}$$

$$= 0.000 + \frac{2 \times 0.0592}{2}\log[H_3O^+]$$

$$\log[H_3O^+] = \frac{-0.347 - 0.000}{0.0592} = -5.86$$

$$[H_3O^+] = 1.38 \times 10^{-6}$$

Substitution of this value for $[H^+]$ as well as the concentrations of the weak acid and its conjugate base into the dissociation-constant expression gives

$$K_{HP} = \frac{[H_3O^+][P^-]}{[HP]} = \frac{(1.38 \times 10^{-6})(0.040)}{0.010} = 5.5 \times 10^{-6}$$

CHALLENGE: Calculate the thermodynamic dissociation constant for HP in Example 19-4 if $\gamma_{P^-} = 0.81$ and $\gamma_{H_3O^+} = 0.86$.

19E QUESTIONS AND PROBLEMS

Note: Numerical data are molar analytical concentrations where the full formula of a species is provided. Molar equilibrium concentrations are supplied for species displayed as ions. Unless otherwise noted, neglect the effects of liquid-junction potentials in your calculations.

***19-1.** How does information supplied by a direct potentiometric measurement of pH differ from that obtained from a potentiometric acid/base titration?

***19-2.** **(a)** Calculate E^0 for the process

$$AgIO_3(s) + e^- \rightleftharpoons Ag(s) + IO_3^-$$

(b) Use the shorthand notation to describe a cell consisting of a saturated calomel electrode as anode and a silver cathode that could be used to measure pIO_3.

(c) Develop an equation that relates the potential of the cell in (b) to pIO_3.

(d) Calculate pIO_3 if the cell in (b) has a potential of 0.294 V.

19-3. **(a)** Calculate E^0 for the process

$$PbI_2(s) + e^- \rightleftharpoons Pb(s) + 2I^-$$

(b) Use the shorthand notation to describe a cell consisting of a saturated calomel electrode as anode and a lead cathode that could be used for the measurement of pI.

(c) Generate an equation that relates the potential of this cell to pI.

(d) Calculate pI if this cell has a potential of −0.348 V.

19-4. Use the shorthand notation to describe a cell consisting of a saturated calomel electrode as anode and a silver cathode for the measurement of

*(a) pSCN. (b) pI. *(c) pSO_3. (d) pPO_4.

***19-5.** Generate an equation that relates pAnion to E_{cell} for each of the cells in Problem 19-4. (For Ag_2SO_3, $K_{sp} = 1.5 \times 10^{-14}$; for Ag_3PO_4, $K_{sp} = 1.3 \times 10^{-20}$.)

19-6. Calculate

*(a) pSCN if the cell in Problem 19-4(a) has a potential of 0.122 V.

(b) pI if the cell in Problem 19-4(b) has a potential of −211 mV.

*(c) pSO_3 if the cell in Problem 19-4(c) has a potential of 300 mV.

(d) pPO_4 if the cell in Problem 19-4(d) has a potential of 0.244 V.

***19-7.** The cell

$$SCE \| Ag_2CrO_4(\text{sat'd}), CrO_4^{2-}(x\text{ M}) | Ag$$

is employed for the determination of $pCrO_4$. Calculate $pCrO_4$ when the cell potential is 0.402 V.

19-8. Calculate the potential of the cell

$$SCE \| \text{aqueous solution} | Hg$$

when the aqueous solution is

(a) 7.40×10^{-3} M Hg^{2+}.

(b) 7.40×10^{-3} M Hg_2^{2+}.
(c) Hg_2SO_4(sat'd),SO_4^{2-}(0.0250 M).
(d) $Hg^{2+}(2.00 \times 10^{-3})$,$OAc^-$(0.100 M).

$$Hg^{2+} + 2OAc^- \rightleftharpoons Hg(OAc)_2(aq) \qquad K_f = 2.7 \times 10^8$$

***19-9.** The standard potential for the reduction of the EDTA complex of Hg(II) is

$$HgY^{2-} + 2e^- \rightleftharpoons Hg(l) + Y^{4-} \qquad E^0 = 0.210 \text{ V}$$

Calculate the potential of the cell

$$SCE \| HgY^{4-}(2.00 \times 10^{-4}),Z | Hg$$

when Z is
(a) $H^+(1.00 \times 10^{-4}$ M),EDTA(0.0200 M).
(b) $H^+(1.00 \times 10^{-8}$ M),EDTA(0.0200 M).

19-10. Calculate the potential of the cell described in Problem 19-9 when Z is
(a) $H^+(1.00 \times 10^{-6}$ M),EDTA(0.0100 M).
(b) $H^+(1.00 \times 10^{-10}$ M),EDTA(0.0100 M).

***19-11.** The cell

$$SCE \| H^+(a = x) | \text{glass electrode}$$

has a potential of 0.2094 V when the solution in the right-hand compartment is a buffer of pH 4.006. The following potentials are obtained when the buffer is replaced with unknowns: (a) −0.3011 V and (b) +0.1163 V. Calculate the pH and the hydrogen ion activity of each unknown. (c) Assuming an uncertainty of ±0.002 V in the junction potential, what is the range of hydrogen ion activities within which the true value might be expected to lie?

19-12. The following cell

$$SCE \| MgA_2(a_{Mg^{2+}} = 9.62 \times 10^{-3}) | \text{membrane electrode for } Mg^{2+}$$

has a potential of 0.367 V.
(a) When the solution of known magnesium activity is replaced with an unknown solution, the potential is +0.244 V. What is the pMg of this unknown solution?
(b) Assuming an uncertainty of ±0.002 V in the junction potential, what is the range of Mg^{2+} activities within which the true value might be expected?

***19-13.** The cell

$$SCE \| CdA_2(\text{sat'd}),A^-(0.0250 \text{ M}) | Cd$$

has a potential of −0.721 V. Calculate the solubility product of CdA_2, neglecting the junction potential.

19-14. The cell

$$SCE \| HA(0.250 \text{ M}),NaA(0.170 \text{ M}) | H_2(1.00 \text{ atm}),Pt$$

has a potential of −0.797 V. Calculate the dissociation constant of HA, neglecting the junction potential.

***19-15.** A 0.3798-g sample of a purified organic acid was dissolved in water and titrated potentiometrically. A plot of the data revealed a single end point after 27.22 mL of 0.1025 M NaOH had been introduced. Calculate the molar mass of the acid assuming it contained just one reactive proton.

19-16. How could you use the data obtained in Problem 19-15 to evaluate K_{HA} for the unknown acid?

19-17. A 40.00-mL aliquot of 0.05000 M HNO_2 is diluted to 75.00 mL and titrated with 0.0800 M Ce^{4+}. The pH of the solution is maintained at 1.00 throughout the titration; the formal potential of the cerium system is 1.44 V.
***(a)** Calculate the potential of the indicator electrode with respect to a saturated calomel reference electrode after the addition of 5.00, 10.00, 15.00, 25.00, 40.00, 49.00, 50.00, 51.00, and 60.00 mL of cerium(IV).
(b) Draw a titration curve for these data.

19-18. Calculate the potential of a silver cathode versus the standard calomel electrode after the addition of 5.00, 15.00, 25.00, 30.00, 35.00, 39.00, 40.00, 41.00, 45.00, and 50.00 mL of 0.1000 M $AgNO_3$ to 50.00 mL of 0.0800 M KSeCN. Construct a titration curve from these data. (K_{sp} for AgSeCN = 4.20×10^{-16}.)

***19-19.** What is the ''operational definition of pH''? Why is it used?

19-20. A membrane electrode has an electrical resistance of 450 MΩ. What would be the relative loading error of the potential if this electrode were measured with a meter having an internal resistance of
***(a)** 50 MΩ?
(b) 500 MΩ?
***(c)** 1000 MΩ?
(d) 5000 MΩ?

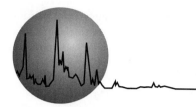

CHAPTER

20

Electrogravimetric and Coulometric Methods

In this chapter we describe three related electroanalytical methods: *electrogravimetry, potentiostatic coulometry,* and *amperostatic coulometry,* or *coulometric titrimetry.*[1] These methods differ from potentiometry in several regards. First, all three require that a current be developed in a cell containing the analyte, whereas in potentiometry currents must be kept at a vanishingly small level. Second, in these procedures, the analyte is quantitatively converted to a new oxidation state, whereas in potentiometry the effect of the measurement on the analyte concentration is usually undetectable. Third, in most cases, the cells in the methods discussed in this chapter are operated as electrolytic cells, whereas the cells in potentiometry are galvanic. Fourth, the final measurement in these procedures is either the mass of a product formed electrolytically from the analyte or the quantity of charge required to convert the analyte to a new oxidation state, whereas, as we have seen, the final measurement in potentiometry is that of a cell potential.

Electrogravimetry and coulometry often have relative errors of a few parts per thousand.

Electrogravimetry and the two coulometric methods are moderately sensitive and among the most accurate and precise techniques available to the chemist. Like gravimetry, these methods require no preliminary calibration against standards because the functional relationship between the quantity measured and the analyte concentration can be derived from theory and atomic-mass data.

André Marie Ampère (1775–1836), French mathematician and physicist, was the first to apply mathematics to the study of electrical current. Consistent with Benjamin Franklin's definitions of positive and negative charge, Ampère defined a positive current to be the direction of flow of positive charge. Although we now know that negative electrons carry current in metals, Ampère's definition has survived to the present. The unit of current, the ampere, is named in his honor.

20A THE EFFECT OF CURRENT ON THE POTENTIAL OF ELECTROCHEMICAL CELLS

When a current develops in an electrochemical cell, the measured potential across the two electrodes is no longer simply the difference between the electrode potentials of the cathode and the anode (the thermodynamic cell potential). Two additional phenomena, *IR drop* and *polarization,* require application of potentials greater than the thermodynamic potential to operate an electrolytic cell and result in the development of potentials smaller than theoretical in a galvanic cell. Before proceeding, we must examine these two phenomena in detail. To do so, we will

[1]For further information concerning the methods in this chapter, see J. A. Plambeck, *Electroanalytical Chemistry.* New York: Wiley, 1982; *Laboratory Techniques in Electroanalytical Chemistry,* P. T. Kissinger and W. R. Heineman, Eds., New York: Marcel Dekker, 1984.

employ as an example an electrolytic cell that has found application in determining such metal ions as copper(II), cadmium(II), and zinc(II) in hydrochloric acid solutions of the metals by electrogravimetry and coulometry. A typical cell for the determination of cadmium ion takes the following form:

$$\text{Ag}|\text{AgCl}(s),\text{Cl}^-(0.200\text{ M}),\text{Cd}^{2+}(0.00500\text{ M})|\text{Cd}$$

Here, the working electrode is a metal electrode that has been coated with a layer of cadmium. The auxiliary electrode is a silver/silver chloride electrode whose electrode potential remains more or less constant during the analysis. Note that this is an example of a cell without liquid junction. As shown in Example 20-1, this cell, as written, has a thermodynamic potential of -0.734 V. Here the negative sign implies that the cell shown in the diagram is electrolytic and would require the application of a potential somewhat larger than -0.734 V to behave in the way indicated by Equation 20-1.

$$\text{Cd}^{2+} + 2\text{Ag}(s) + 2\text{Cl}^- \rightleftharpoons \text{Cd}(s) + \text{AgCl}(s) \tag{20-1}$$

Note that the cell being considered is reversible, so in the absence of the external potential shown in the figure, the cell reaction would be the reverse of Equation 20-1, and the cell would behave as a galvanic cell with an output potential of $+0.734$ V.

20A-1 Ohmic Potential; *IR* Drop

Ohm's law: $E = IR$ or $I = \left(\frac{1}{R}\right)E$

Electrochemical cells, like metallic conductors, resist the flow of charge. Ohm's law describes the effect of this resistance on the magnitude of the current in the cell. The product of the resistance R of a cell in ohms (Ω) and the current I in amperes (A) is called the *ohmic potential* or the *IR drop* of the cell. In Figure 20-1b we have used a resistor R to represent the cell resistance of the cell in Figure 20-1a. In order to generate a current of I amperes in this cell, we must apply a potential E_{applied} that is $-IR$ volts more negative than the thermodynamic potential. That is,

$$E_{\text{applied}} = E_{\text{cathode}} - E_{\text{anode}} - IR \tag{20-2}$$

where E_{cathode} and E_{anode} are electrode potentials computed with the Nernst equation.

EXAMPLE 20-1

The following cell has been used for the determination of cadmium in the presence of chloride ions by both electrogravimetry and coulometry.

$$\text{Ag}|\text{AgCl}(s),\text{Cl}^-(0.200\text{ M}),\text{Cd}^{2+}(0.00500\text{ M})|\text{Cd}$$

Calculate the potential that (a) must be applied to prevent a current from developing in the cell when the two electrodes are connected and (b) must be

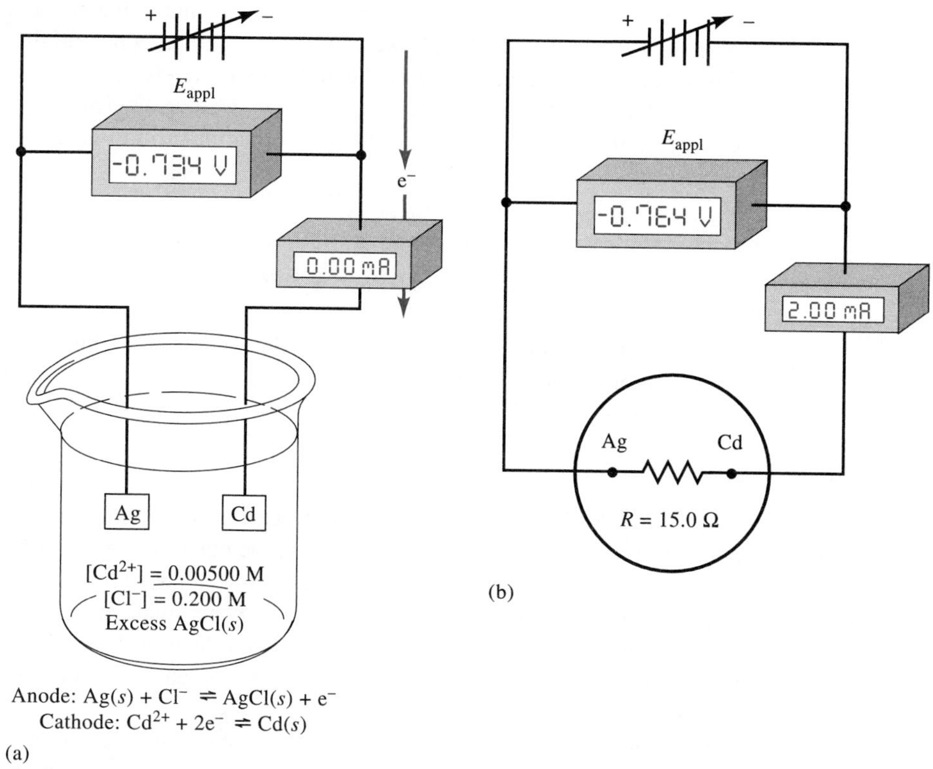

Figure 20-1
An electrolytic cell for determining Cd^{2+}. (a) Current = 0.00 mA. (b) Schematic of cell in (a) with internal resistance of cell represented by a 15.0 Ω resistor and $E_{applied}$ increased to give a current of 2.00 mA.

applied to cause an electrolytic current of 2.00 mA to develop. Assume that the internal resistance of the cell is 15.0 Ω.

(a) In Appendix 5, we find

$$Cd^{2+} + 2e^- \rightleftharpoons Cd(s) \qquad E^0 = -0.403 \text{ V}$$

$$AgCl(s) + e^- \rightleftharpoons Ag(s) + Cl^- \qquad E^0 = 0.222 \text{ V}$$

The electrode potential of the cadmium cathode is

$$E_{cathode} = -0.403 - \frac{0.0592}{2} \log \frac{1}{0.00500} = -0.471 \text{ V}$$

and of the silver anode is

$$E_{anode} = 0.222 - 0.0592 \log (0.200) = 0.263 \text{ V}$$

$$E_{cell} = E_{applied} = E_{cathode} - E_{anode}$$

$$= -0.471 - 0.263 = -0.734 \text{ V}$$

Thus, to prevent this cell from behaving as a galvanic cell with a cadmium anode and a silver cathode we would need to apply a potential of 0.734 V from an external source as shown in Figure 20-1a.

(b) In order to calculate the applied potential needed to develop a current of 2.00 mA, or 2.00×10^{-3} A, we substitute into Equation 20-2 to give

$$\begin{aligned} E_{\text{applied}} &= E_{\text{cathode}} - E_{\text{anode}} - IR \\ &= -0.734 - 2.00 \times 10^{-3}\ \text{A} \times 15\ \Omega \\ &= -0.734 - 0.030 = -0.764\ \text{V} \end{aligned}$$

Thus, to obtain a 2.00 mA current as shown in Figure 20-1b, an applied potential of 0.764 V would be required.

20A-2 Polarization Effects

According to Equation 20-2, a plot of current in an electrolytic cell as a function of applied potential should be a straight line with a slope equal to the negative reciprocal of the resistance. As shown in Figure 20-2, the plot is indeed linear with small currents. (In this experiment, the measurements were made in a brief enough time so that neither E_{cathode} nor E_{anode} changed significantly as a consequence of the electrolytic reaction.) As the applied voltage increases, the current ultimately begins to deviate from linearity.

Overvoltage is the potential difference between the theoretical cell potential from Equation 20-2 and the actual cell potential at a given level of current.

Cells that exhibit nonlinear behavior at higher currents are said to be *polarized,* and the degree of polarization is given by an *overvoltage,* or *overpotential,* which is symbolized by Π in the figure. Note that polarization requires the application of a potential greater than the theoretical value to give a current of the expected magnitude. Thus, the overpotential required to achieve a current of 7.00 mA in the electrolytic cell in Figure 20-2 is about −0.23 V. For an electrolytic cell affected by overvoltage, Equation 20-2 then becomes

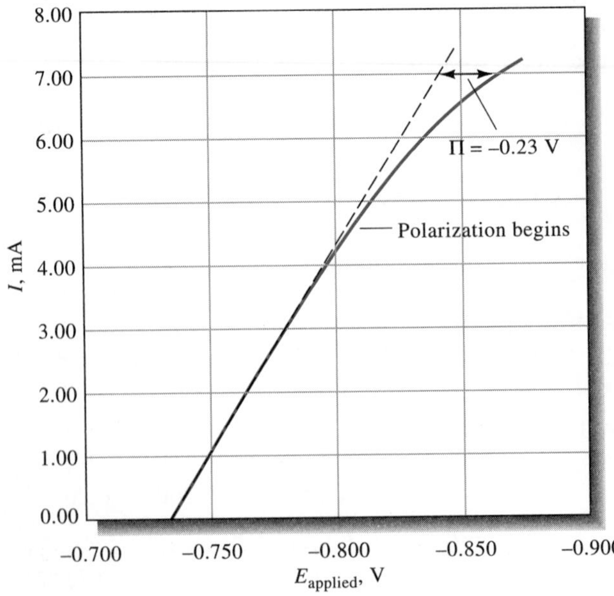

Figure 20-2
Experimental current/voltage curve for operation of the cell shown in Figure 20-1. Dashed line is the theoretical curve assuming no polarization. Overvoltage Π is the potential difference between the theoretical curve and the experimental.

$$E_{\text{applied}} = E_{\text{cathode}} - E_{\text{anode}} - IR - \Pi \qquad (20\text{-}3)$$

A similar equation can be written for the output E_{output} of a galvanic cell. That is,

$$E_{\text{output}} = E_{\text{cathode}} - E_{\text{anode}} - IR - \Pi \qquad (20\text{-}4)$$

Polarization is an electrode phenomenon that may affect either or both of the electrodes in a cell. The degree of polarization of an electrode varies widely. In some instances it approaches zero, but in others it can be so large that the current in the cell becomes independent of potential. Under this circumstance, polarization is said to be complete. Polarization phenomena are conveniently divided into two categories: *concentration polarization* and *kinetic polarization.*

Factors that influence polarization: (1) electrode size, shape, and composition; (2) composition of the electrolyte solution; (3) temperature and stirring rate; (4) current level; (5) physical state of species involved in the cell reaction.

Concentration Polarization

Electron transfer between a reactive species in a solution and an electrode can take place only from a thin film of solution located immediately adjacent to the surface of the electrode; this film is only a fraction of a nanometer in thickness and contains a limited number of reactive ions or molecules. In order for there to be a steady current in a cell, this film must be continuously replenished with reactant from the bulk of the solution. That is, as reactant ions or molecules are consumed by the electrochemical reaction, more must be transported into the surface film at a rate that is sufficient to maintain the current. For example, in order to have a current of 2.0 mA in the cell described in Figure 20-1b, it is necessary to transport cadmium ions to the cathode surface at a rate of about 1×10^{-8} mol/s or 6×10^{15} cadmium ions per second. (Similarly, silver ions must be removed from the surface film of the anode at a rate of 2×10^{-8} mol/s.)

Concentration polarization occurs when reactant species do not arrive at the surface of the electrode or product species do not leave the surface of the electrode fast enough to maintain the desired current. When this happens the current is limited to values less than that predicted by Equation 20-2.

Reactants are transported to the surface of an electrode by three mechanisms: (1) *diffusion,* (2) *migration,* and (3) *convection.* Products are removed from electrode surfaces in the same ways. We will focus on mass-transport processes to the cathode, but our discussion applies equally to anodes.

Reactants are transported to an electrode by (1) diffusion, (2) migration, and (3) convection.

Diffusion When there is a concentration difference between two regions of a solution, ions or molecules move from the more concentrated region to the more dilute. This process is called *diffusion* and ultimately leads to a disappearance of the concentration gradient. The rate of diffusion is directly proportional to the concentration difference. For example, when cadmium ions are deposited at a cathode by a current as illustrated in Figure 20-3, the concentration of Cd^{2+} at the electrode surface $[Cd^{2+}]_0$ is very small. The difference between the concentration at the surface and the concentration in the bulk solution $[Cd^{2+}]$ creates a concentration gradient that causes cadmium ions to diffuse from the bulk of the solution to the surface film. The rate of diffusion is given by

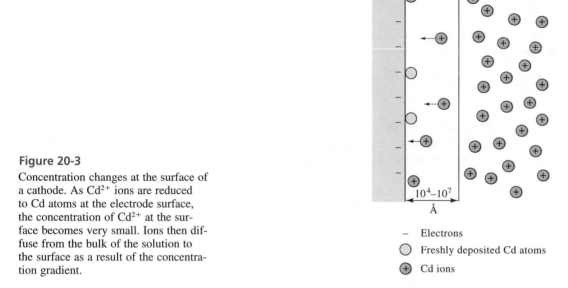

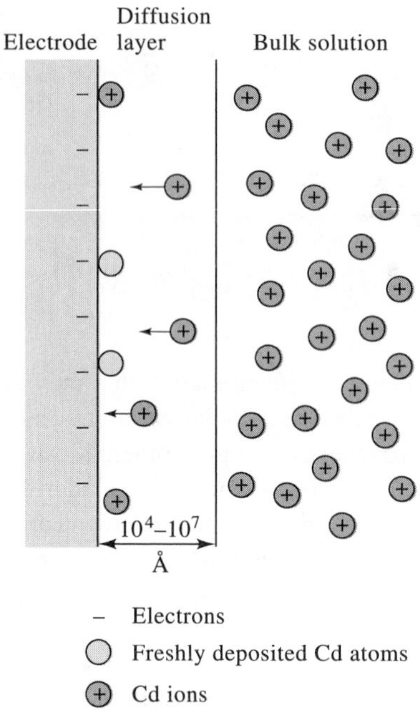

Figure 20-3
Concentration changes at the surface of a cathode. As Cd^{2+} ions are reduced to Cd atoms at the electrode surface, the concentration of Cd^{2+} at the surface becomes very small. Ions then diffuse from the bulk of the solution to the surface as a result of the concentration gradient.

$$\text{rate of diffusion to cathode surface} = k'([Cd^{2+}] - [Cd^{2+}]_0) \quad \textbf{(20-5)}$$

where $[Cd^{2+}]$ is the reactant concentration in the bulk of the solution, $[Cd^{2+}]_0$ is its equilibrium concentration at the surface of the cathode, and k' is a proportionality constant. The value of $[Cd^{2+}]_0$ at any instant is fixed by *the potential of the electrode* and can be calculated from the Nernst equation. In the present example, we find the surface cadmium ion concentration from the relationship

$$E_{\text{cathode}} = E^0_{Cd^{2+}} - \frac{0.0592}{2} \log \frac{1}{[Cd^{2+}]_0}$$

Diffusion is a process in which ions or molecules move from a more concentrated part of a solution to a more dilute.

where E_{cathode} is the potential applied to the cathode. As the applied potential becomes more and more negative, $[Cd^{2+}]_0$ becomes smaller and smaller. The result is that the rate of diffusion and the current become correspondingly larger.

Migration involves the movement of ions through a solution as a result of electrostatic attraction between the ions and the electrodes.

Migration The process by which ions move under the influence of an electric field is called *migration.* This process, shown schematically in Figure 20-4, is the primary cause of mass transfer in the bulk of the solution in a cell. The rate at which ions migrate to or away from an electrode surface generally increases as the electrode potential increases. This charge movement constitutes a current, which also increases with potential.

Convection is the mechanical transport of ions or molecules through a solution as a result of stirring, vibration, or temperature gradients.

Convection Reactants can also be transferred to or from an electrode by mechanical means. Forced *convection,* such as stirring or agitation, will tend to decrease the thickness of the diffusion layer at the surface of an electrode and thus decrease concentration polarization. Natural convection processes caused by temperature or density differences also contribute to the transport of molecules to and from an electrode.

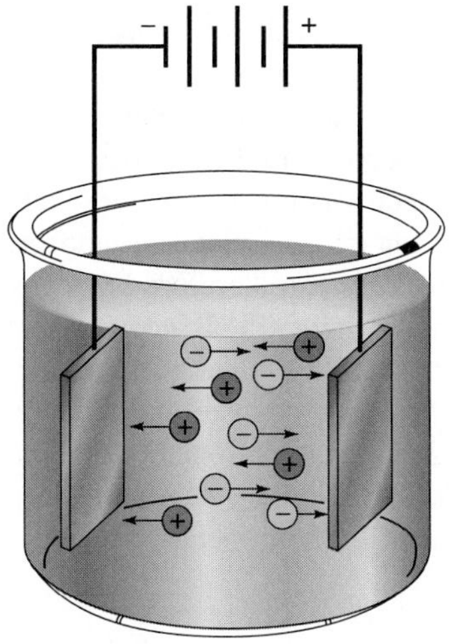

Figure 20-4
Migration is the movement of ions through a solution as a result of electrostatic attraction between the ions and the electrodes.

The Importance of Concentration Polarization As noted earlier, concentration polarization sets in when the effects of diffusion, migration, and convection are insufficient to transport a reactant to or from an electrode surface at a rate that produces a current of the magnitude given by Equation 20-2. Concentration polarization requires applied potentials that are more negative than theoretical to maintain a given current in an electrolytic cell (Figure 20-2). Similarly, the phenomenon causes a galvanic cell potential to be smaller than the value predicted on the basis of the theoretical potential and the *IR* drop.

Concentration polarization is important in several electroanalytical methods. In some applications, its effects are undesirable, and steps are taken to eliminate it. In others, it is essential to the analytical method, and every effort is made to promote its occurrence.

The experimental variables that influence the degree of concentration polarization are (1) reactant concentration, (2) total electrolyte concentration, (3) mechanical agitation, and (4) electrode size.

Kinetic Polarization

In kinetic polarization, the magnitude of the current is limited by the rate of one or both of the electrode reactions—that is, the rate of electron transfer between the reactants and the electrodes. In order to offset kinetic polarization, an additional potential, or overvoltage, is required to overcome the energy barrier to the half-reaction.

The *current in a kinetically polarized cell is governed by the rate of electron transfer* rather than the rate of mass transfer.

Kinetic polarization is most pronounced for electrode processes that yield gaseous products and is often negligible for reactions that involve the deposition or solution of a metal. Kinetic effects usually decrease with increasing temperature and decreasing current density. These effects also depend upon the composition of the electrode and are most pronounced with softer metals, particularly mercury. The magnitude of overvoltage effects cannot be predicted from present theory and can only be estimated from empirical information in the literature.[2] In common with *IR* drop, overvoltage effects cause the potential of a galvanic cell to be

Current density is defined as amperes per square centimeter (A/cm^2) of electrode surface.

[2]Overvoltage data for various gaseous species on different electrode surfaces have been compiled by J. A. Page, in *Handbook of Analytical Chemistry,* L. Meites, Ed., p. **5**-184 to **5**-186. New York: McGraw-Hill, 1963.

smaller than theoretical and to require potentials greater than theoretical to operate an electrolytic cell at a desired current.

Kinetic polarization is most commonly encountered when the reactant or product in an electrochemical cell is gas.

The overvoltages associated with the formation of hydrogen and oxygen are often 1 V or more and are of considerable importance because these molecules are frequently produced by electrochemical reactions. Of particular interest is the high overvoltage of hydrogen on such metals as copper, zinc, lead, and mercury. These metals and several others can therefore be deposited without interference from hydrogen evolution. In theory, it is not possible to deposit zinc from a neutral aqueous solution because hydrogen forms at a potential that is considerably less than that required for zinc deposition. In fact, zinc can be deposited on a copper electrode with no significant hydrogen formation because the rate at which the gas forms on both zinc and copper is negligible, as shown by the high hydrogen overvoltage associated with these metals.

FEATURE 20-1
Overvoltage and the Lead/Acid Battery

If it were not for the high overvoltage of hydrogen on lead and lead oxide electrodes, the lead/acid storage batteries found in automobiles and trucks would not operate because of hydrogen formation at the cathode during both charging and use. Certain trace metals in the system lower this overvoltage and eventually lead to gassing, or hydrogen formation, which limits the lifetime of the battery. The basic difference between a battery with a 48-month warranty and a 72-month warranty is the concentration of these trace metals in the system.

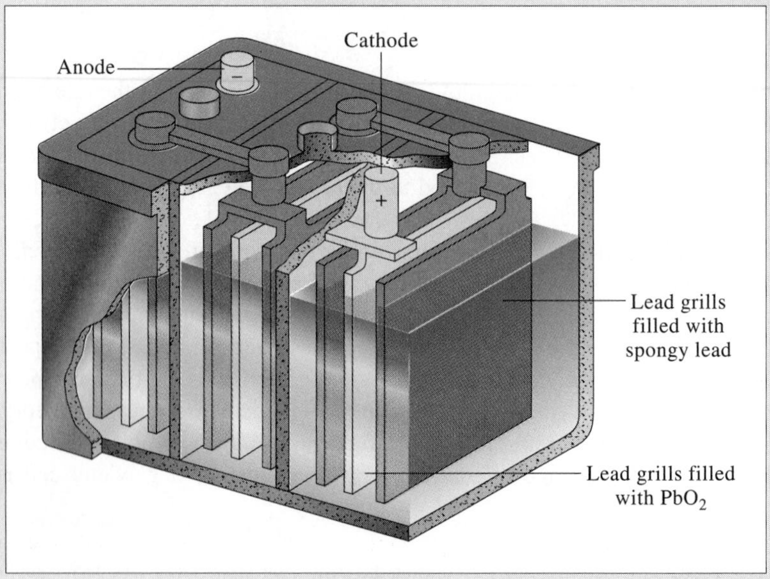

20B THE POTENTIAL SELECTIVITY OF ELECTROLYTIC METHODS

In principle, electrolytic methods offer a reasonably selective means for separating and determining a number of ions. The feasibility of and theoretical conditions for accomplishing a given separation can be readily derived from the standard electrode potentials of the species of interest.

EXAMPLE 20-2

Is a quantitative separation of Cu^{2+} and Pb^{2+} by electrolytic deposition feasible in principle? If so, what range of cathode potentials (vs. SCE) can be used? Assume that the sample solution is initially 0.1000 M in each ion and that quantitative removal of an ion is realized when only 1 part in 10,000 remains undeposited.

In Appendix 5, we find

$$Cu^{2+} + 2e^- \rightleftharpoons Cu(s) \qquad E^0 = 0.337\ \text{V}$$

$$Pb^{2+} + 2e^- \rightleftharpoons Pb(s) \qquad E^0 = -0.126\ \text{V}$$

It is apparent that copper will begin to deposit before lead because the electrode potential for Cu^{2+} is larger than that for Pb^{2+}. Let us first calculate the cathode potential required to reduce the Cu^{2+} concentration to 10^{-4} of its original concentration (that is, to 1.00×10^{-5} M). Substituting into the Nernst equation, we obtain

$$E = 0.337 - \frac{0.0592}{2} \log \frac{1}{1.00 \times 10^{-5}} = 0.189\ \text{V}$$

Similarly, we can derive the cathode potential at which lead begins to deposit:

$$E = -0.126 - \frac{0.0592}{2} \log \frac{1}{0.100} = -0.156\ \text{V}$$

Therefore, if the cathode potential is maintained between 0.189 V and -0.156 V (vs. SHE), a quantitative separation should in theory occur. To convert these potentials to potentials relative to a saturated calomel electrode, we treat the reference electrode as the anode and write

$$E_{\text{cell}} = E_{\text{cathode}} - E_{\text{SCE}} = 0.189 - 0.244 = -0.055\ \text{V}$$

and

$$E_{\text{cell}} = -0.156 - 0.244 = -0.400\ \text{V}$$

Therefore, the cathode potential should be kept between -0.055 V and -0.400 V versus the SCE.

Calculations such as the foregoing make it possible to compute the differences in standard electrode potentials theoretically needed to determine one ion without interference from another; these differences range from about 0.04 V for triply charged ions to approximately 0.24 V for singly charged species.

These theoretical separation limits can be approached only by maintaining the potential of the *working electrode* (usually the cathode, at which a metal deposits) at a level required by theory. The potential of this electrode can *be controlled only by variation of the potential applied to the cell.* It is evident from Equation 20-3, however, that variations in $E_{applied}$ affect not only the cathode potential but also the anode potential, the *IR* drop, and the overpotential. As a consequence, the only *practical* way of achieving separation of species whose electrode potentials differ by a few tenths of a volt is to *measure* the cathode potential continuously against a reference electrode whose potential is known; the applied cell potential can then be adjusted to maintain the cathode potential at the desired level. An analysis performed in this way is called a *controlled-cathode-potential electrolysis,* or a *potentiostatic electrolysis.* Controlled-potential methods are discussed in Sections 20C-2 and 20D-4.

20C ELECTROGRAVIMETRIC METHODS OF ANALYSIS

Electrolytic precipitation has been used for over a century for the gravimetric determination of metals. In most applications, the metal is deposited on a weighed platinum cathode, and the increase in mass is determined. Important exceptions to this procedure include the anodic deposition of lead as lead dioxide on platinum and of chloride as silver chloride on silver.

A **working electrode** is the electrode at which the analytical reaction occurs.

A **potentiostatic method** is an electrolytic procedure in which the potential of the working electrode is maintained at a constant level versus a reference electrode, such as a SCE.

There are two general types of electrogravimetric methods. In one, no control over the potential of the working electrode is exercised, and the applied cell potential is held at a more or less constant level that provides current that is adequate to complete the electrolysis in a reasonable length of time. The second type is the *potentiostatic method.* This procedure is also called the *controlled-cathode-potential* or the *controlled-anode-potential method,* depending upon whether the working electrode is a cathode or an anode.

20C-1 Electrogravimetry Without Potential Control of the Working Electrode

Electrolytic procedures in which no effort is made to control the potential of the working electrode make use of simple and inexpensive equipment and require little operator attention. In these procedures, the potential applied across the cell is maintained at a more or less constant level throughout the electrolysis.

Instrumentation

As shown in Figure 20-5, the apparatus for an analytical electrodeposition without cathode potential control consists of a suitable cell and a direct-current power supply. The dc power supply usually consists of an ac rectifier, although a 6-V storage battery can be used instead. The voltage applied to the cell is controlled by the rheostat, *R*. An ammeter and a voltmeter indicate the approximate current and applied voltage. In performing an analytical electrolysis with this apparatus, the applied voltage is adjusted with the rheostat to give a current of several tenths of an ampere. The voltage is then maintained at about the initial level until the deposition is judged to be complete.

Electrolysis Cells

Figure 20-5 shows a typical cell for the deposition of a metal on a solid electrode. Ordinarily, the working electrode is a platinum gauze cylinder 2 or 3 cm in diameter and perhaps 6 cm in length. Often, as shown, the anode takes the form of a solid platinum stirring paddle that is located inside and connected to the cathode through the external circuit.

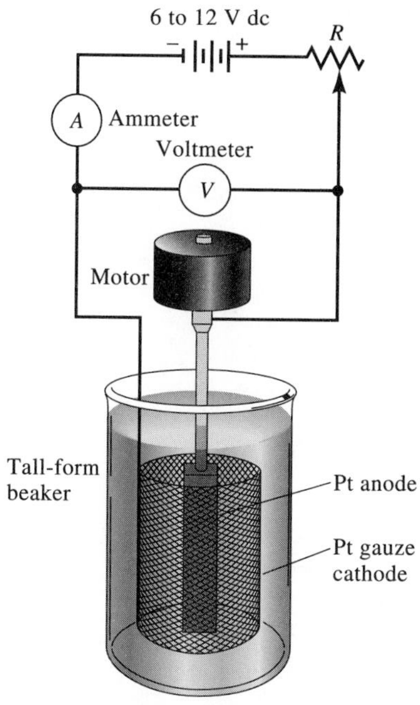

Figure 20-5
Apparatus for the electrodeposition of metals without cathode-potential control.

The Physical Properties of Electrolytic Precipitates

Ideally, an electrolytically deposited metal should be strongly adherent, dense, and smooth so that it can be washed, dried, and weighed without mechanical loss or reaction with the atmosphere. Good metallic deposits are fine-grained and have a metallic luster; spongy, powdery, or flaky precipitates are likely to be less pure and less adherent.

The principal factors that influence the physical characteristics of deposits are current density, temperature, and the presence of complexing agents. Ordinarily, the best deposits are formed at current densities that are less than 0.1 A/cm^2. Stirring generally improves the quality of a deposit. The effects of temperature are unpredictable and must be determined empirically.

Many metals form smoother and more adherent films when deposited from solutions in which their ions exist primarily as complexes. Cyanide and ammonia complexes often provide the best deposits. The reasons for this effect are not obvious.

Applications

In practice, electrolysis at a constant cell potential is limited to the separation of easily reduced cations from those that are more difficult to reduce than hydrogen ion or nitrate ion. The reason for this limitation is illustrated in Figure 20-6, which shows the changes of current, IR drop, and cathode potential during an electrolysis in the cell in Figure 20-5. The analyte here is copper(II) ions in a solution containing an excess of acid. Initially, R is adjusted so that the potential applied to the cell is about -2.5 V, which, as shown in Figure 20-6a, provides a current of about 1.5 A. The electrolytic deposition of copper is then completed at this applied potential.

As shown in Figure 20-6b, the IR drop decreases continually as the reaction proceeds. The reason for this decrease is primarily concentration polarization at the cathode, which limits the rate at which copper ions are brought to the electrode surface and thus the current. From Equation 20-3, it is apparent that the decrease in IR must be offset by an increase in the cathode potential since the applied cell potential is constant.

Ultimately, the decrease in current and the increase in cathode potential is slowed at point B by the reduction of hydrogen ions. Because the solution contains a large excess of acid, the current is now no longer limited by concentration polarization, and codeposition of copper and hydrogen goes on simultaneously

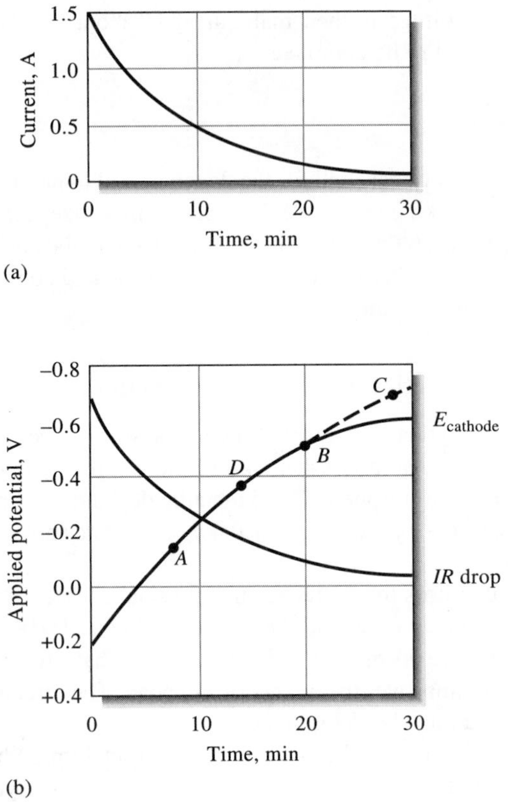

Figure 20-6
(a) Current; (b) *IR* drop and cathode-potential change during the electrolytic deposition of copper at a constant-applied-cell potential.

until the remainder of the copper ions are deposited. Under these conditions, the cathode is said to be *depolarized* by hydrogen ions.

Consider now the fate of some metal ion, such as lead(II), which begins to deposit at point *A* on the cathode potential curve. Clearly, it would codeposit well before copper deposition was complete, and an interference would result. In contrast, a metal ion, such as cobalt(II), which reacts at a cathode potential corresponding to point *C* on the curve would not interfere because depolarization by hydrogen gas formation prevents the cathode from reaching this potential.

A **depolarizer** is a chemical that is easily reduced (or oxidized) and stabilizes the potential of a working electrode by minimizing concentration polarization.

Codeposition of hydrogen during electrolysis often leads to formation of nonadherent deposits, which are unsatisfactory for analytical purposes. This problem can be resolved by introducing another species that is reduced at a less negative potential than hydrogen ion and does not adversely affect the physical properties of the deposit. One such cathode depolarizer is nitrate ion, its reaction being

$$NO_3^- + 10H^+ + 8e^- \rightleftharpoons NH_4^+ + 3H_2O \qquad (20\text{-}6)$$

Electrolytic methods performed without electrode-potential control, while limited by their lack of selectivity, do have several applications of practical importance. Table 20-1 lists the common elements that are often determined by this procedure.

TABLE 20-1 Typical Applications of Electrogravimetric Methods Without Potential Control

Analyte	Weighed As	Cathode	Anode	Conditions
Ag^+	Ag	Pt	Pt	Alkaline CN^- solution
Br^-	AgBr (on anode)	Pt	Ag	
Cd^{2+}	Cd	Cu on Pt	Pt	Alkaline CN^- solution
Cu^{2+}	Cu	Pt	Pt	H_2SO_4/HNO_3 solution
Mn^{2+}	MnO_2 (on anode)	Pt	Pt dish	HCOOH/HCOONa solution
Ni^{2+}	Ni	Cu on Pt	Pt	Ammoniacal solution
Pb^{2+}	PbO_2 (on anode)	Pt	Pt	Strong HNO_3 solution
Zn^{2+}	Zn	Cu on Pt	Pt	Acidic citrate solution

20C-2 Potentiostatic Gravimetry

In the discussion that follows, we shall assume that the working electrode is a cathode at which an analyte is deposited as a metal. The discussion is, however, readily extended to an anodic working electrode where nonmetallic deposits are formed.

Instrumentation

In order to separate species with electrode potentials that differ by only a few tenths of a volt, it is necessary to use a more elaborate approach than the one just described. Such techniques are required because concentration polarization at the cathode, if unchecked, causes the potential of that electrode to become so negative that codeposition of the other species present begins before the analyte is completely deposited (Figure 20-6b). A large negative drift in the cathode potential can be avoided by employing a three-electrode system, such as that shown in Figure 20-7.

The controlled-potential apparatus shown in Figure 20-7 is made up of two independent electrical circuits that share a common electrode, the *working elec-*

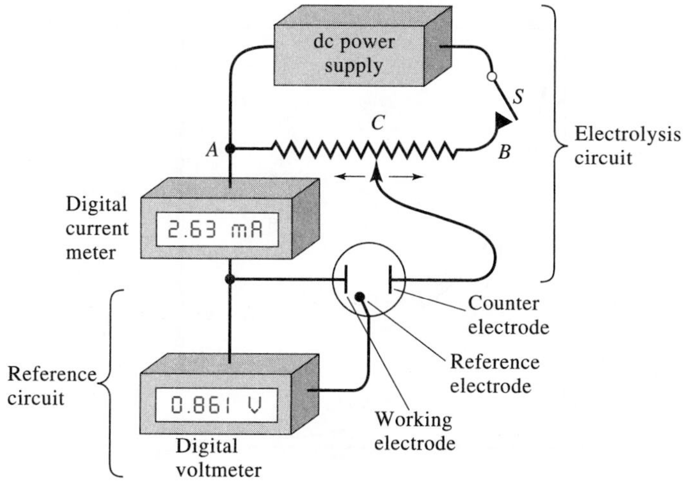

Figure 20-7 Apparatus for controlled-potential, or potentiostatic, electrolysis. Contact C is adjusted as necessary to maintain the working electrode (cathode in this example) at a constant potential. The current in the reference-electrode circuit is essentially zero at all times.

A **counter electrode** has no effect on the reaction at the working electrode. It simply serves to feed electrons to the working cathode.

trode at which the analyte is deposited. The *electrolysis circuit* consists of a dc source, a potential divider *(ACB)* that permits continuous variation in the potential applied across the working electrode, a *counter electrode,* and a current meter. The *control circuit* is made up of a reference electrode (often a SCE), a high-resistance digital voltmeter, and the working electrode. The electrical resistance of the control circuit is so large that the electrolysis circuit supplies essentially all of the current for the electrolysis.

The purpose of the control circuit is to monitor continuously the potential between the working electrode and the reference electrode. When this potential reaches a level at which codeposition of an interfering species is about to begin, the potential across the working and counter electrodes is decreased by moving contact *C* to the left. Since the potential of the counter electrode remains constant during this change, the cathode potential becomes smaller, thus preventing interference from codeposition.

The current and the applied, or cell, voltage changes that occur in a typical constant-cathode-potential electrolysis are depicted in Figure 20-8. Note that the applied cell potential has to be decreased continuously throughout the electrolysis. Manual adjustment of the potential is tedious (particularly at the outset) and, above all, time-consuming. To avoid such waste of operator time, controlled-cathode-potential electrolyses are generally performed with automated instruments called *potentiostats,* which maintain a constant cathode potential electronically.

A **potentiostat** is an instrument that maintains the working electrode potential at a constant value.

Cells

Figure 20-5 shows a typical cell for the deposition of a metal on a solid electrode. Tall-form beakers are ordinarily employed, and mechanical stirring is provided to minimize concentration polarization; frequently, the anode serves as a mechanical stirrer.

Electrodes

Ordinarily, the working electrode in electrogravimetry is a metallic gauze cylinder, as shown in Figure 20-5. Electrodes are usually constructed of platinum,

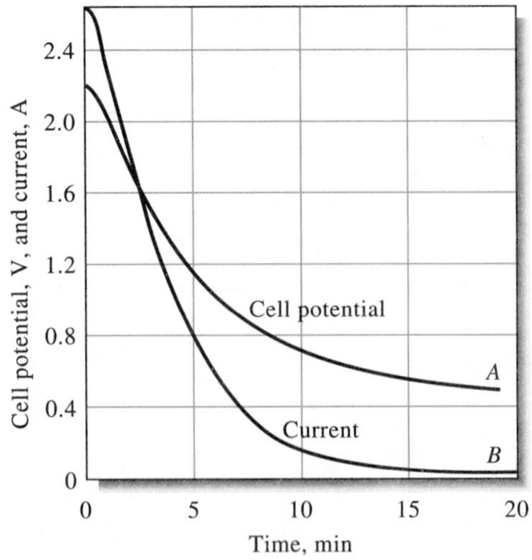

Figure 20-8
Changes in cell potential (*A*) and current (*B*) during a controlled-cathode-potential deposition of copper. The cathode is maintained at −0.36 V (vs. SCE) throughout the experiment. (Data from J. J. Lingane, *Anal. Chem. Acta,* **1948,** *2*, 590. With permission.)

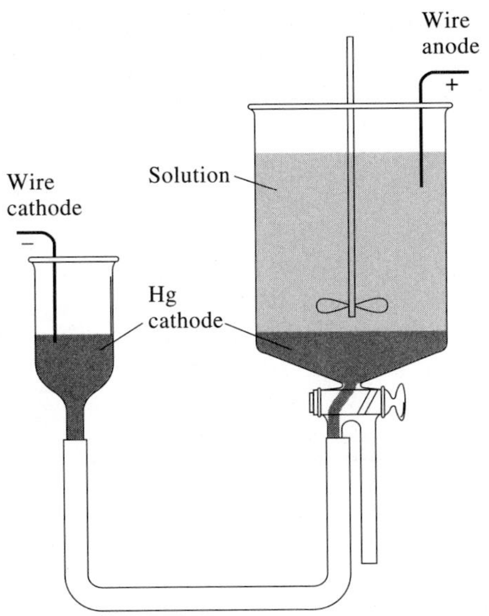

Figure 20-9
A mercury cathode for the electrolytic removal of metal ions from solution.

although copper, brass, and other metals find occasional use. Platinum electrodes have the advantage of being relatively unreactive and can be ignited to remove any grease, organic matter, or gases that could have a deleterious effect on the physical properties of the deposit.

Certain metals (notably bismuth, zinc, and gallium) cannot be deposited directly onto platinum without causing permanent damage to the electrode. Consequently, a protective coating of copper is always deposited on a platinum electrode before the electrolysis of these metals is undertaken.

The Mercury Cathode

A mercury cathode, such as that shown in Figure 20-9, is particularly useful for removing easily reduced elements as a preliminary step in an analysis. For example, copper, nickel, cobalt, silver, and cadmium are readily separated at this electrode from such ions as aluminum, titanium, the alkali metals, sulfates, and phosphates. The precipitated elements dissolve in the mercury with little hydrogen evolution because even at high applied potentials, formation of the gas is prevented by the high overvoltage associated with its evolution on mercury. Ordinarily, the deposited metals are not determined after electrolysis, the goal being simply the removal of interfering ions from a solution of an analyte.

Applications

The controlled-cathode-potential method is a potent tool for separating and determining metallic species having standard potentials that differ by only a few tenths of a volt. An example illustrating the power of the method involves determining copper, bismuth, lead, cadmium, zinc, and tin in mixtures by successive deposition of the metals on a weighed platinum cathode. The first three elements are deposited from a nearly neutral solution containing tartrate ion to complex the tin(IV) and prevent its deposition. Copper is first reduced quantitatively by maintaining the cathode potential at -0.2 V with respect to a saturated calomel electrode. After being weighed, the copper-plated cathode is returned to the solution,

TABLE 20-2 Some Applications of Controlled-Cathode-Potential Electrolysis

Element Determined	Other Elements That Can Be Present
Ag	Cu and heavy metals
Cu	Bi, Sb, Pb, Sn, Ni, Cd, Zn
Bi	Cu, Pb, Zn, Sb, Cd, Sn
Sb	Pb, Sn
Sn	Cd, Zn, Mn, Fe
Pb	Cd, Sn, Ni, Zn, Mn, Al, Fe
Cd	Zn
Ni	Zn, Al, Fe

and bismuth is removed at a potential of −0.4 V. Lead is then deposited quantitatively by increasing the cathode potential to −0.6 V. When lead deposition is complete, the solution is made strongly ammoniacal, and cadmium and zinc are deposited successively at −1.2 and −1.5 V. Finally, the solution is acidified in order to decompose the tin/tartrate complex by the formation of undissociated tartaric acid. Tin is then deposited at a cathode potential of −0.65 V. A fresh cathode must be used here because the zinc redissolves under these conditions. A procedure such as this is particularly attractive for use with a potentiostat because little operator time is required for the complete analysis.

Table 20-2 lists some other separations performed by the controlled-cathode-potential method.

20D COULOMETRIC METHODS OF ANALYSIS

Coulometric methods are performed by measuring the quantity of electrical charge (electrons) required to convert a sample of an analyte quantitatively to a different oxidation state. Coulometric and gravimetric methods share the common advantage that the proportionality constant between the quantity measured and the analyte mass is derived from accurately known physical constants, thus eliminating the need for calibration standards. In contrast to gravimetric methods, coulometric procedures are usually rapid and do not require that the product of the electrochemical reaction be a weighable solid. Coulometric methods are as accurate as conventional gravimetric and volumetric procedures and in addition are readily automated.[3]

20D-1 The Quantity of Electrical Charge

1 coulomb = 1 ampere · second = 1 A · s

Units for the quantity of charge include the coulomb (C) and the faraday (F). *The coulomb is the quantity of electrical charge transported by a constant current of*

[3]For additional information about coulometric methods, see E. Bishop, in *Comprehensive Analytical Chemistry,* C. L. Wilson and D. W. Wilson, Eds., Vol. 11D. New York: Elsevier Scientific, 1975; J. A. Plambeck, *Electroanalytical Chemistry,* Chapter 12. New York: Wiley, 1982; D. J. Curran, in *Laboratory Techniques in Electroanalytical Chemistry,* P. T. Kissinger and W. R. Heineman, Eds., pp. 539–568. New York: Marcel Dekker, 1984.

one ampere in one second. Thus, the number of coulombs (Q) resulting from a constant current of I amperes operated for t seconds is

$$Q = It \tag{20-7}$$

For a variable current i, the number of coulombs is given by the integral

$$Q = \int_0^t i\,dt \tag{20-8}$$

The faraday is the quantity of charge that corresponds to one mole or 6.022×10^{23} electrons. The faraday also equals 96,485 C. As shown in Example 20-3, we can use these definitions to calculate the mass of a chemical species that is formed at an electrode by a current of known magnitude.

A **faraday** of charge is equivalent to one mole of electrons, or 6.022×10^{23} electrons.

EXAMPLE 20-3

A constant current of 0.800 A is used to deposit copper at the cathode and oxygen at the anode of an electrolytic cell. Calculate the number of grams of each product formed in 15.2 min, assuming no other redox reaction.

The two half-reactions are

$$Cu^{2+} + 2e^- \longrightarrow Cu(s)$$
$$2H_2O \longrightarrow 4e^- + O_2(g) + 4H^+$$

Thus, 1 mol of copper is equivalent to 2 mol of electrons and 1 mol of oxygen corresponds to 4 mol of electrons.

Substituting into Equation 20-7 yields

$$Q = 0.800\text{ A} \times 15.2\text{ min} \times 60\text{ s/min} = 729.6\text{ A}\cdot\text{s} = 729.6\text{ C}$$

$$\text{no. } F = \frac{729.6\text{ C}}{96{,}485\text{ C}/F} = 7.56 \times 10^{-3}\,F \equiv 7.56 \times 10^{-3}\text{ mol of electrons}$$

The masses of Cu and O_2 are given by

$$\text{mass Cu} = 7.56 \times 10^{-3}\,\cancel{\text{mol e}^-} \times \frac{1\,\cancel{\text{mol Cu}}}{2\,\cancel{\text{mol e}^-}} \times \frac{63.55\text{ g Cu}}{1\,\cancel{\text{mol Cu}}} = 0.240\text{ g Cu}$$

$$\text{mass O}_2 = 7.56 \times 10^{-3}\,\cancel{\text{mol e}^-} \times \frac{1\,\cancel{\text{mol O}_2}}{4\,\cancel{\text{mol e}^-}} \times \frac{32.00\text{ g O}_2}{1\,\cancel{\text{mol O}_2}} = 0.0605\text{ g O}_2$$

Michael Faraday (1791–1867) was one of the foremost chemists and physicists of his time. Among his most important discoveries were Faraday's laws of electrolysis. Faraday, a simple man who lacked mathematical sophistication, was a superb experimentalist and an inspiring teacher and lecturer. The quantity of charge equal to a mole of electrons is named in his honor.

20D-2 Types of Coulometric Methods

Two methods have been developed that are based on measuring the quantity of charge: *potentiostatic coulometry* and *amperostatic coulometry,* or *coulometric titrimetry.* Potentiostatic methods are performed in much the same way as con-

Amperostatic coulometry is also called **coulometric titrimetry.**

trolled-potential gravimetric methods, with the potential of the working electrode being maintained at a constant level relative to a reference electrode throughout the electrolysis. Here, however, the electrolysis current is recorded as a function of time to give a curve similar to curve *B* in Figure 20-8. The analysis is then completed by integrating the current–time curve to obtain the number of coulombs and thus the number of faradays of charge consumed or produced by the analyte.

Electrons are the reagent in a coulometric titration.

Coulometric titrations are similar to other titrimetric methods in that analyses are based on measuring the combining capacity of the analyte with a standard reagent. In the coulometric procedure, the reagent is electrons and the standard solution is a constant current of known magnitude. Electrons are added to the analyte (via the direct current) or to some species that immediately reacts with the analyte until an end point is reached. At that point, the electrolysis is discontinued. The amount of analyte is determined from the magnitude of the current and the time required to complete the titration. The magnitude of the current in amperes is analogous to the molarity of a standard solution, and the time measurement is analogous to the volume measurement in conventional titrimetry.

20D-3 Current-Efficiency Requirements

One equivalent of chemical change is the change that corresponds to 1 mol of electrons. Thus, for the two half-reactions in Example 20-3, one equivalent of chemical change involves production of $\frac{1}{2}$ mol of Cu or $\frac{1}{4}$ mol of O_2.

A fundamental requirement for all coulometric methods is 100% current efficiency; that is, each faraday of electricity must bring about one equivalent of chemical change in the analyte. Note that 100% current efficiency can be achieved without direct participation of the analyte in electron transfer at an electrode. For example, chloride ions are readily determined by potentiostatic coulometry or by coulometric titrations by generating silver ions at a silver anode. These ions then react with the analyte to form a precipitate or deposit of silver chloride. The quantity of electricity required to complete the silver chloride formation serves as the analytical parameter. In this instance, 100% current efficiency is realized because the number of moles of electrons is exactly equal to the number of moles of chloride ion in the sample despite the fact that these ions do not react directly at the electrode surface.

20D-4 Controlled-Potential Coulometry

In controlled-potential coulometry, the potential of the working electrode is maintained at a constant level such that only the analyte is responsible for the conduction of charge across the electrode/solution interface. The number of coulombs required to convert the analyte to its reaction product is then determined by recording and integrating the current-versus-time curve during the electrolysis.

Instrumentation

The instrumentation for potentiostatic coulometry consists of an electrolysis cell, a potentiostat, and a device for determining the number of coulombs consumed by the analyte.

Cells Figure 20-10 illustrates two types of cells that are used for potentiostatic coulometry. The first consists of a platinum-gauze working electrode, a plati-

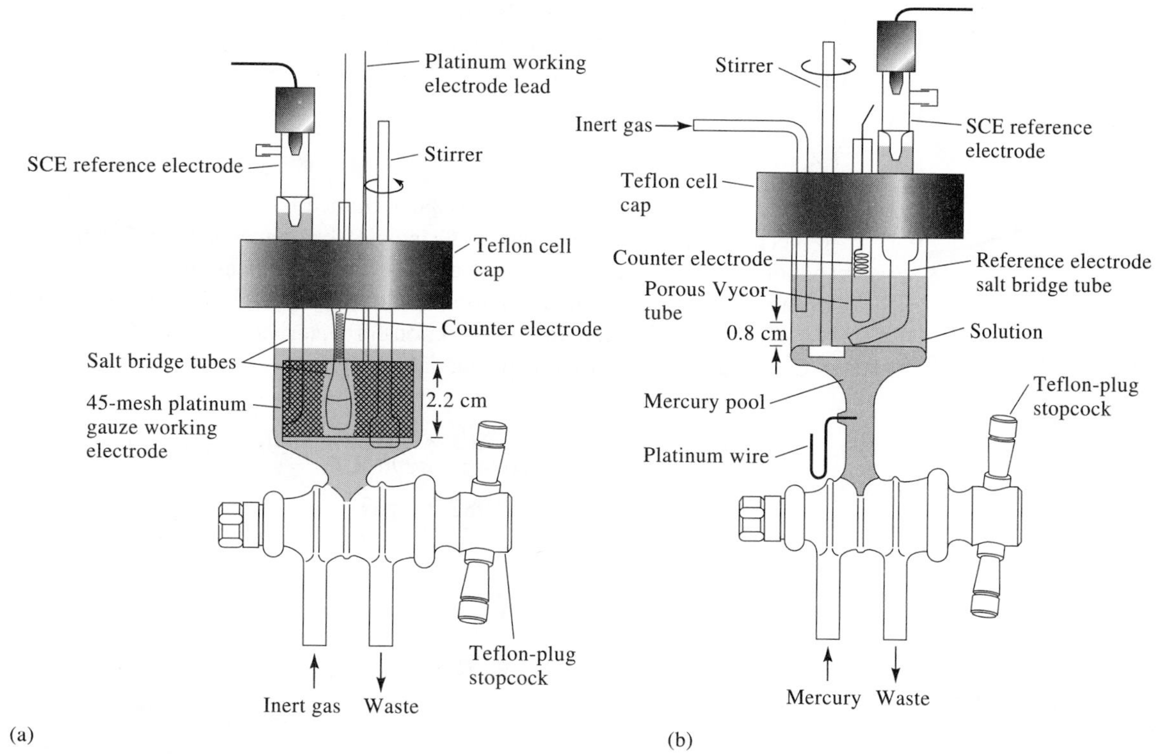

Figure 20-10
Electrolysis cells for potentiostatic coulometry. Working electrode: (a) platinum-gauze, (b) mercury-pool. (Reprinted with permission from J. E. Harrar and C. L. Pomernacki, *Anal. Chem.,* **1973,** *45,* 57. Copyright 1973 American Chemical Society.)

num-wire counter electrode, and a saturated calomel reference electrode. The counter electrode is separated from the analyte solution by a salt bridge that usually contains the same electrolyte as the solution being analyzed. This bridge is needed to prevent the reaction products formed at the counter electrode from diffusing into the analyte solution and interfering. For example, hydrogen gas is a common product at a cathodic counter electrode. Unless this gas is physically isolated from the analyte solution by the bridge, it will react directly with many of the analytes that are determined by oxidation at the working anode.

The second type of cell, shown in Figure 20-10b, is a mercury-pool type. A mercury cathode is particularly useful for separating easily reduced elements as a preliminary step in an analysis. In addition, however, it has found considerable use for the coulometric determination of several cations that form metals that are soluble in mercury. In these applications, little or no hydrogen evolution occurs even at high applied potentials because of the large overvoltage effects. A coulometric cell such as that shown in Figure 20-10b is also useful for the coulometric determination of certain types of organic compounds.

Potentiostats and Coulometers For controlled-potential coulometry, a potentiostat similar in design to that described in Figure 20-7 is required. Generally, however, the potentiostat is automated and is equipped with a recorder that pro-

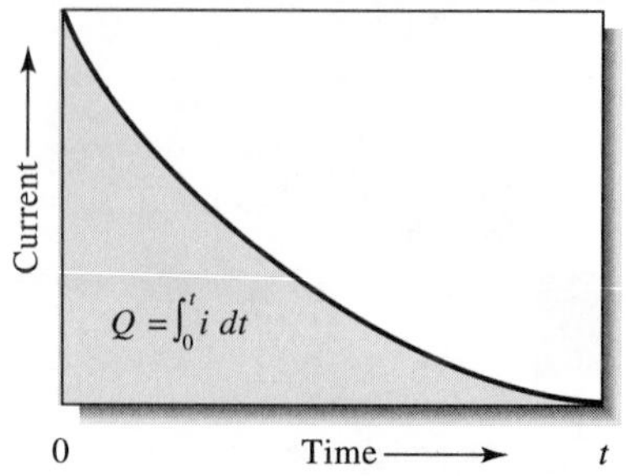

Figure 20-11
For a current that varies with time, the quantity of charge Q in a time t is the shaded area under the curve.

vides a plot of current as a function of time as shown in Figure 20-11. An electronic current integrator then gives the number of coulombs necessary to complete the reaction.

EXAMPLE 20-4

The Fe(III) in a 0.8202-g sample was determined by coulometric reduction to Fe(II) at a platinum cathode. Calculate the percentage of $Fe_2(SO_4)_3$ (g/mol = 399.88) in the sample if a hydrogen/oxygen coulometer arranged in series with the cell containing the sample evolved 19.37 mL of gas (H_2 + O_2) at 23°C and 765 torr (after correction for water vapor).

Converting the gas volume to standard conditions gives

$$V = 19.37 \text{ mL} \times \frac{765 \text{ torr}}{760 \text{ torr}} \times \frac{273 \text{ K}}{296 \text{ K}} = 17.982 \text{ mL (STP)}$$

$$\text{no. mol gas} = \frac{17.982 \text{ mL gas}}{22{,}400 \text{ mL gas/mol gas}} = 8.028 \times 10^{-4}$$

The reactions in the coulometer are

$$4H^+ + 4e^- \longrightarrow 2H_2(g)$$
$$2H_2O \longrightarrow O_2(g) + 4H^+ + 4e^-$$

Thus, 3 mol of gas is produced by the passage of 4 mol of electrons, which is equivalent to 0.750 mol of gas per faraday, and the number of faradays is

$$8.028 \times 10^{-4} \text{ mol gas} \times 1\ F/0.750 \text{ mol gas} = 1.0704 \times 10^{-3}\ F$$

Since 1 mol of $Fe_2(SO_4)_3$ consumes 2 mol of electrons, we may write

$$\text{mass } Fe_2(SO_4)_3 = 1.0704 \times 10^{-3}\ F \times \frac{1 \text{ mol e}^-}{F} \times \frac{1 \text{ mol } Fe_2(SO_4)_3}{2 \text{ mol e}^-} \times \frac{399.88 \text{ g } Fe_2(SO_4)_3}{\text{mol } Fe_2(SO_4)_3}$$
$$= 0.21401 \text{ g}$$

$$\%\ Fe_2(SO_4)_3 = \frac{0.21401 \text{ g } Fe_2(SO_4)_3}{0.8202 \text{ g sample}} \times 100\% = 26.09\%$$

Applications of Controlled-Potential Coulometry

Controlled-potential coulometric methods have been applied to the determination of some 55 elements in inorganic compounds.[4] Mercury appears to be favored as the cathode, and methods for the deposition at this electrode of more than two

[4]For a summary of the applications, see J. E. Harrar, in *Electroanalytical Chemistry,* A. J. Bard, Ed., Vol. 8. New York: Marcel Dekker, 1975; E. Bishop, in *Comprehensive Analytical Chemistry,* C. L. Wilson and D. W. Wilson, Eds., Vol. 11D, Chapter XV. New York: Elsevier, 1975.

dozen metals have been described. The method has found widespread use in the nuclear-energy field for the relatively interference-free determination of uranium and plutonium.

Controlled-potential coulometric procedures also offer possibilities for the electrolytic determination (and synthesis) of organic compounds. For example, trichloroacetic acid and picric acid are quantitatively reduced at a mercury cathode whose potential is suitably controlled:

$$Cl_3CCOO^- + H^+ + 2e^- \longrightarrow Cl_2HCCOO^- + Cl^-$$

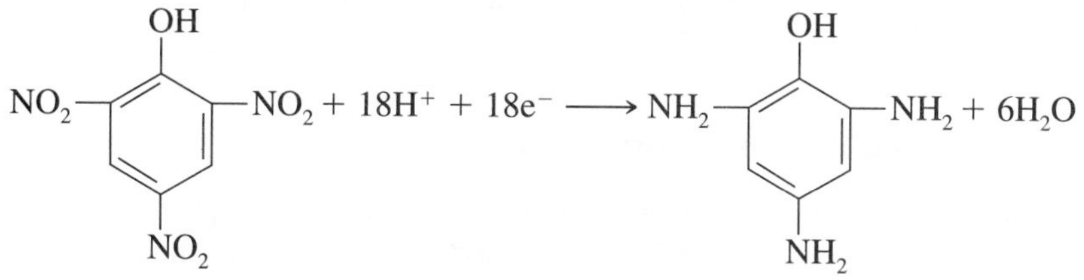

Coulometric measurements permit the determination of these compounds with a relative error of a few tenths of a percent.

20D-5 Coulometric Titrations[5]

Coulometric titrations are carried out with a constant-current source called an *amperostat,* which senses decreases in current in a cell and responds by increasing the potential applied to the cell until the current is restored to its original level. Because of the effects of concentration polarization, 100% current efficiency with respect to the analyte can be maintained only by having present in large excess an auxiliary reagent that is oxidized or reduced at the electrode to give a product that reacts with the analyte. As an example, consider the coulometric titration of iron(II) at a platinum anode. At the beginning of the titration, the primary anodic reaction is

Constant current generators are called **amperostats** or **galvanostats.**

Auxiliary reagents are essential in coulometric titrations.

$$Fe^{2+} \longrightarrow Fe^{3+} + e^-$$

As the concentration of iron(II) decreases, however, the requirement of a constant current results in an increase in the applied cell potential. Because of concentration polarization, this increase in potential causes the anode potential to increase to the point where the decomposition of water becomes a competing process:

$$2H_2O \longrightarrow O_2(g) + 4H^+ + 4e^-$$

The quantity of electricity required to complete the oxidation of iron(II) then exceeds that demanded by theory, and the current efficiency is less than 100%. The lowered current efficiency is prevented, however, by introducing at the outset an unmeasured excess of cerium(III), which is oxidized at a lower potential than is water:

[5]For further details on this technique, see D. J. Curran, in *Laboratory Techniques in Electroanalytical Chemistry,* P. T. Kissinger and W. R. Heineman, Eds., Chapter 20. New York: Marcel Dekker, 1984.

$$Ce^{3+} \longrightarrow Ce^{4+} + e^-$$

With stirring, the cerium(IV) produced is rapidly transported from the surface of the electrode to the bulk of the solution, where it oxidizes an equivalent amount of iron(II):

$$Ce^{4+} + Fe^{2+} \longrightarrow Ce^{3+} + Fe^{3+}$$

The net effect is an electrochemical oxidation of iron(II) with 100% current efficiency, even though only a fraction of that species is directly oxidized at the electrode surface.

End Points in Coulometric Titrations

Coulometric titrations, like their volumetric counterparts, require a means for determining when the reaction between analyte and reagent is complete. Generally, the end points described in the chapters on volumetric methods are applicable to coulometric titrations as well. Thus, for the titration of iron(II) just described, an oxidation/reduction indicator, such as 1,10-phenanthroline, can be used; as an alternative, the end point can be established potentiometrically. Similarly, an adsorption indicator or a potentiometric end point can be employed in the coulometric titration of chloride ion by the silver ions generated at a silver anode.

Instrumentation

As shown in Figure 20-12, the equipment required for a coulometric titration includes a source of constant current, a titration vessel, a switch, an electric timer, and a device for monitoring current. Movement of the switch to position 1 simultaneously starts the timer and initiates a current in the titration cell. When the

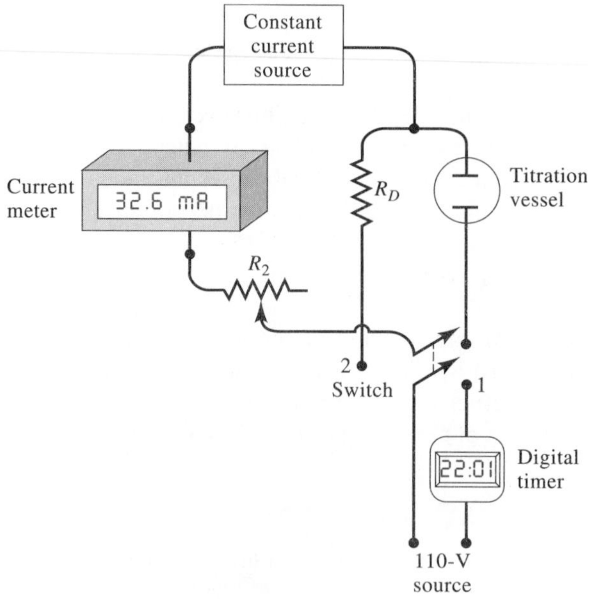

Figure 20-12
Diagram of a coulometric titration apparatus.

switch is moved to position 2, the electrolysis and the timing are discontinued. With the switch in this position, however, electricity continues to be drawn from the source and passes through a dummy resistor R_D that has about the same electrical resistance as the cell. This arrangement ensures continuous operation of the source, which aids in maintaining the current at a constant level.

The constant-current source for a coulometric titration is often an *amperostat,* an electronic device capable of maintaining a current of 200 mA or more that is constant to a few hundredths of a percent. Amperostats are available from several instrument manufacturers. A much less expensive source that produces reasonably constant currents can be constructed from several heavy-duty batteries connected in series to give an output potential of 100 to 300 V. This output is applied to the cell that is in series with a resistor whose resistance is large relative to the resistance of the cell. Small changes in the cell conductance then have a negligible effect on the current in the resistor and cell.

An ordinary motor-driven electric clock is unsatisfactory for the measurement of the electrolysis time because the rotor of such a device tends to coast when stopped and lag when started. Digital electronic timers eliminate this problem.

Cells for Coulometric Titrations Figure 20-13 shows a typical coulometric titration cell consisting of a generator electrode at which the reagent is produced and a counter electrode to complete the circuit. The generator electrode—ordinarily a platinum rectangle, a coil of wire, or a gauze cylinder—has a relatively large surface area to minimize polarization effects. Ordinarily, the counter electrode is isolated from the reaction medium by a sintered disk or some other porous medium in order to prevent interference by the reaction products from this electrode. For example, hydrogen is often evolved at the cathode as an oxidizing agent is being generated at an anode. Hydrogen reacts rapidly with most oxidizing agents which leads to a positive systematic error unless the gas is generated in a separate compartment.

An alternative to isolation of the auxiliary electrode is a device in which the reagent is generated externally, such as that shown in Figure 20-14. The apparatus is so arranged that electrolyte flow continues briefly after the current is discontin-

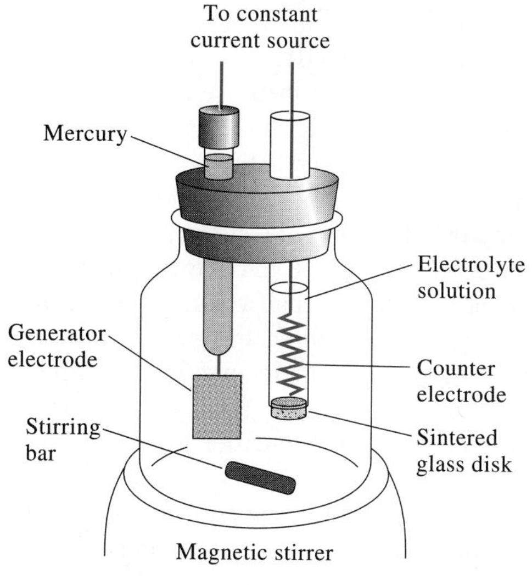

Figure 20-13
A typical coulometric titration cell.

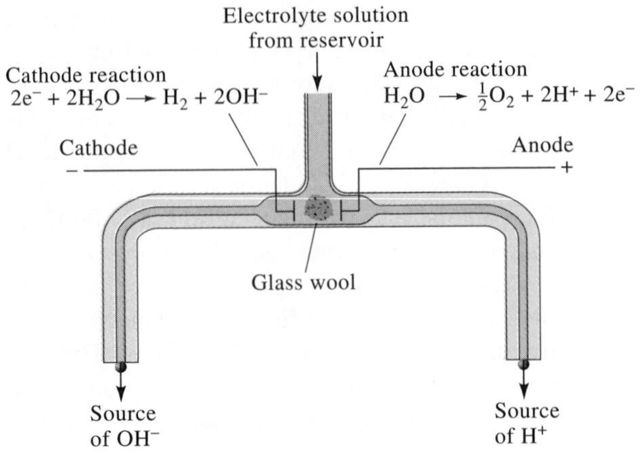

Figure 20-14
A cell for the external coulometric generation of acid and base.

ued, thus flushing the residual reagent into the titration vessel. Note that the apparatus shown in Figure 20-14 provides either hydrogen or hydroxide ions, depending upon which arm is used. The apparatus has also been used for the generation of other reagents. An example is the iodine produced by oxidation of iodide at the anode.

A Comparison of Coulometric and Conventional Titrations

The various components of the titrator in Figure 20-12 have their counterparts in the reagents and apparatus required for a volumetric titration. The constant-current source of known magnitude serves the same function as the standard solution in a volumetric method. The electronic timer and switch correspond to the buret and stopcock, respectively. Electricity is passed through the cell for relatively long periods of time at the outset of a coulometric titration, but the time intervals are made smaller and smaller as chemical equivalence is approached. Note that these steps are analogous to the operation of a buret in a conventional titration.

A coulometric titration offers several significant advantages over a conventional volumetric procedure. Principal among these is the elimination of the problems associated with the preparation, standardization, and storage of standard solutions. This advantage is particularly significant with labile reagents such as chlorine, bromine, and titanium(III) ion, which are sufficiently unstable in aqueous solution to seriously limit their use as volumetric reagents. In contrast, their utilization in a coulometric determination is straightforward because they are consumed as soon as they are generated.

Coulometric methods are as accurate and precise as comparable volumetric methods.

Coulometric methods also excel when small amounts of analyte have to be titrated because tiny quantities of reagent are generated with ease and accuracy through the proper choice of current. In contrast, the use of very dilute solutions and the accurate measurement of small volumes are inconvenient at best.

A further advantage of the coulometric procedure is that a single constant-current source provides reagents for precipitation, complex formation, neutralization, or oxidation/reduction titrations. Finally, coulometric titrations are more readily automated since currents are easier to control than is control of liquid flow.

The current–time measurements required for a coulometric titration are inherently as accurate as or more accurate than the comparable volume/molarity measurements of a conventional volumetric method, particularly where small quantities of reagent are involved. When the accuracy of a titration is limited by the sensitivity of the end point, the two titration methods have comparable accuracies.

Applications of Coulometric Titrations

Coulometric titrations have been developed for all types of volumetric reactions.[6] Selected applications are described in this section.

Neutralization Titrations Hydroxide ion can be generated at the surface of a platinum cathode immersed in a solution containing the analyte acid:

$$2H_2O + 2e^- \longrightarrow 2OH^- + H_2(g)$$

The platinum anode must be isolated by a diaphragm to eliminate potential interference from the hydrogen ions produced by anodic oxidation of water. As a convenient alternative, a silver wire can be substituted for the platinum anode, provided chloride or bromide ions are added to the analyte solution. The anode reaction then becomes

$$Ag(s) + Br^- \longrightarrow AgBr(s) + e^-$$

Silver bromide does not interfere with the neutralization reaction.

Coulometric titrations of acids are much less susceptible to the carbonate error encountered in volumetric methods (Section 12A-3). The only measure required to avoid this error is to remove the carbon dioxide from the solvent by boiling or by bubbling an inert gas, such as nitrogen, through the solution for a brief period.

Hydrogen ions generated at the surface of a platinum anode can be used for the coulometric titration of strong as well as weak bases:

$$2H_2O \longrightarrow O_2 + 4H^+ + 4e^-$$

Here, the cathode must be isolated from the analyte solution to prevent interference from hydroxide ion.

Precipitation and Complex-Formation Reactions Coulometric titrations with EDTA are carried out by reduction of the ammine mercury(II) EDTA chelate at a mercury cathode:

$$HgNH_3Y^{2-} + NH_4^+ + 2e^- \longrightarrow Hg(l) + 2NH_3 + HY^{3-} \qquad (20\text{-}9)$$

Because the mercury chelate is more stable than the corresponding complexes of such cations as calcium, zinc, lead, or copper, complexation of these ions occurs only after the ligand has been freed by the electrode process.

As shown in Table 20-3, several precipitating reagents can be generated coulometrically. The most widely used of these is silver ion, which is generated at a silver anode.

---

[6]For additional information, see E. Bishop, in *Comprehensive Analytical Chemistry,* C. L. Wilson and D. W. Wilson, Eds., Vol. 11D, Chapters XVIII to XXIV. New York: Elsevier, 1975; J. T. Stock, *Anal. Chem.,* **1984,** *56,* 1R and **1980,** *52,* 1R.

TABLE 20-3 Summary of Coulometric Titrations Involving Neutralization, Precipitation, and Complex-Formation Reactions

Species Determined	Generator Electrode Reaction	Secondary Analytical Reaction
Acids	$2H_2O + 2e^- \rightleftharpoons 2OH^- + H_2$	$OH^- + H^+ \rightleftharpoons H_2O$
Bases	$H_2O \rightleftharpoons 2H^+ + \frac{1}{2}O_2 + 2e^-$	$H^+ + OH^- \rightleftharpoons H_2O$
Cl^-, Br^-, I^-	$Ag \rightleftharpoons Ag^+ + e^-$	$Ag^+ + Cl^- \rightleftharpoons AgCl(s)$, etc.
Mercaptans	$Ag \rightleftharpoons Ag^+ + e^-$	$Ag^+ + RSH \rightleftharpoons AgSR(s) + H^+$
Cl^-, Br^-, I^-	$2Hg \rightleftharpoons Hg_2^{2+} + 2e^-$	$Hg_2^{2+} + 2Cl^- \rightleftharpoons Hg_2Cl_2(s)$, etc.
Zn^{2+}	$Fe(CN)_6^{3-} + e^- \rightleftharpoons Fe(CN)_6^{4-}$	$3Zn^{2+} + 2K^+ + 2Fe(CN)_6^{4-} \rightleftharpoons K_2Zn_3[Fe(CN)_6]_2(s)$
Ca^{2+}, Cu^{2+}, Zn^{2+}, Pb^{2+}	See Equation 20-9.	$HY^{3-} + Ca^{2+} \rightleftharpoons CaY^{2-} + H^+$, etc.

Oxidation/Reduction Titrations Table 20-4 reveals that a variety of redox reagents can be generated coulometrically. Of particular interest is bromine, the coulometric generation of which forms the basis for a large number of coulometric methods. Of interest as well are reagents not ordinarily encountered in conventional volumetric analysis owing to the instability of their solutions; silver(II), manganese(III), and the chloride complex of copper(I) are examples.

Automatic Coulometric Titrators

A number of instrument manufacturers offer automatic coulometric titrators, most of which employ a potentiometric end point. Some of these instruments are multipurpose and can be used for the determination of a variety of species. Others are designed for a single type of analysis. Examples of the latter are chloride titrators, in which silver ion is generated coulometrically; sulfur dioxide monitors, where anodically generated bromine oxidizes the analyte to sulfate ions; carbon dioxide monitors, in which the gas, absorbed in monoethanolamine, is titrated with coulometrically generated base; and water titrators, in which Karl Fischer reagent (see Section 17D-3, page 381) is generated electrolytically.

TABLE 20-4 Summary of Coulometric Titrations Involving Oxidation/Reduction Reactions

Reagent	Generator Electrode Reaction	Substance Determined
Br_2	$2Br^- \rightleftharpoons Br_2 + 2e^-$	As(III), Sb(III), U(IV), Tl(I), I^-, SCN^-, NH_3, N_2H_4, NH_2OH, phenol, aniline, mustard gas, mercaptans, 8-hydroxyquinoline, olefins
Cl_2	$2Cl^- \rightleftharpoons Cl_2 + 2e^-$	As(III), I^-, styrene, fatty acids
I_2	$2I^- \rightleftharpoons I_2 + 2e^-$	As(III), Sb(III), $S_2O_3^{2-}$, H_2S, ascorbic acid
Ce^{4+}	$Ce^{3+} \rightleftharpoons Ce^{4+} + e^-$	Fe(II), Ti(III), U(IV), As(III), I^-, $Fe(CN)_6^{4-}$
Mn^{3+}	$Mn^{2+} \rightleftharpoons Mn^{3+} + e^-$	$H_2C_2O_4$, Fe(II), As(III)
Ag^{2+}	$Ag^+ \rightleftharpoons Ag^{2+} + e^-$	Ce(III), V(IV), $H_2C_2O_4$, As(III)
Fe^{2+}	$Fe^{3+} + e^- \rightleftharpoons Fe^{2+}$	Cr(VI), Mn(VII), V(V), Ce(IV)
Ti^{3+}	$TiO^{2+} + 2H^+ + e^- \rightleftharpoons Ti^{3+} + H_2O$	Fe(III), V(V), Ce(IV), U(VI)
$CuCl_3^{2-}$	$Cu^{2+} + 3Cl^- + e^- \rightleftharpoons CuCl_3^{2-}$	V(V), Cr(VI), IO_3^-
U^{4+}	$UO_2^{2+} + 4H^+ + 2e^- \rightleftharpoons U^{4+} + 2H_2O$	Cr(VI), Ce(IV)

20E QUESTIONS AND PROBLEMS

Note: Numerical data are molar analytical concentrations where the full formula of a species is provided. Molar equilibrium concentrations are supplied for species displayed as ions.

20-1. Briefly distinguish between
- *(a) concentration polarization and kinetic polarization.
- (b) an amperostat and a potentiostat.
- *(c) a coulomb and a faraday.
- (d) a working electrode and a counter electrode.
- *(e) the electrolysis circuit and the control circuit for controlled-potential methods.

20-2. Briefly define
- *(a) current density.
- (b) ohmic potential.
- *(c) coulometric titration.
- (d) controlled-potential electrolysis.
- *(e) current efficiency.
- (f) an electrochemical equivalent.

***20-3.** Describe three mechanisms responsible for the transport of dissolved species to and from an electrode surface.

20-4. How does the existence of a current affect the potential of an electrochemical cell?

***20-5.** How do concentration polarization and kinetic polarization resemble one another? How do they differ?

20-6. What experimental variables affect concentration polarization in an electrochemical cell?

***20-7.** Describe conditions that favor kinetic polarization in an electrochemical cell.

20-8. How do electrogravimetric and coulometric methods differ from potentiometric methods?

***20-9.** Identify three factors that influence the physical characteristics of an electrolytic deposit.

20-10. What is the purpose of a cathode depolarizer?

***20-11.** What is the function of (a) an amperostat and (b) a potentiostat?

***20-12.** Differentiate between amperostatic coulometry and potentiostatic coulometry.

***20-13.** Why is it ordinarily necessary to isolate the working electrode from the counter electrode in a controlled-potential coulometric analysis?

20-14. Why is an auxiliary reagent usually required in a coulometric titration?

20-15. Calculate the number of ions involved at the surface of an electrode during each second that an electrochemical cell is operated at 0.020 A and the ions involved are
- (a) univalent.
- *(b) divalent.
- (c) trivalent.

20-16. Calculate the theoretical potential needed to initiate the deposition of
- *(a) copper from a solution that is 0.150 M in Cu^{2+} and buffered to a pH of 3.00. Oxygen is evolved at the anode at 1.00 atm.
- (b) tin from a solution that is 0.120 M in Sn^{2+} and buffered to a pH of 4.00. Oxygen is evolved at the anode at 770 torr.
- *(c) silver bromide on a silver anode from a solution that is 0.0864 M in Br^- and buffered to a pH of 3.40. Hydrogen is evolved at the cathode at 765 torr.
- (d) Tl_2O_3 from a solution that is 4.00×10^{-3} M in Tl^+ and buffered to a pH of 8.00. The solution is also made 0.010 M in Cu^{2+}, which acts as a cathode depolarizer. For the process

$$Tl_2O_3(s) + 3H_2O + 4e^- \rightleftharpoons 2Tl^+ + 6OH^- \qquad E^0 = 0.020 \text{ V}$$

***20-17.** Calculate the initial potential needed for a current of 0.078 A in the cell

$$Co|Co^{2+}(6.40 \times 10^{-2} \text{ M})\|Zn^{2+}(3.75 \times 10^{-3} \text{ M})|Zn$$

if this cell has a resistance of 5.00 Ω.

***20-18.** The cell

$$Sn|Sn^{2+}(8.22 \times 10^{-4} \text{ M})\|Cd^{2+}(7.50 \times 10^{-2} \text{ M})\|Cd$$

has a resistance of 3.95 Ω. Calculate the initial potential that will be needed for a current of 0.072 A in this cell.

20-19. Copper is to be deposited from a solution that is 0.200 M in Cu(II) and is buffered to a pH of 4.00. Oxygen is evolved from the anode at a partial pressure of 740 torr. The cell has a resistance of 3.60 Ω; the temperature is 25°C. Calculate
- (a) the theoretical potential needed to initiate deposition of copper from this solution.
- (b) the *IR* drop associated with a current of 0.10 A in this cell.
- (c) the initial potential, given that the overvoltage of oxygen is 0.50 V under these conditions.
- (d) the potential of the cell when $[Cu^{2+}]$ is 8.00×10^{-6}, assuming that *IR* drop and O_2 overvoltage remain unchanged.

20-20. Nickel is to be deposited on a platinum cathode (area = 120 cm²) from a solution that is 0.200 M in Ni^{2+} and buffered to a pH of 2.00. Oxygen is evolved at a partial pressure of 1.00 atm at a platinum anode with an area of 80 cm². The cell has a resistance of 3.15 Ω; the temperature is 25°C. Calculate
- (a) the thermodynamic potential needed to initiate the deposition of nickel.
- (b) the *IR* drop for a current of 1.10 A.
- (c) the current density at the anode and the cathode.
- (d) the initial applied potential, given that the overvoltage of oxygen on platinum is approximately 0.52 V under these conditions.
- (e) the applied potential when the nickel concentration has decreased to 2.00×10^{-4} M (assume that all variables other than $[Ni^{2+}]$ remain constant).

***20-21.** Silver is to be deposited from a solution that is 0.150 M in $Ag(CN)_2^-$, 0.320 M in KCN, and buffered to a pH of 10.00. Oxygen is evolved at the anode at a partial pressure of 1.00 atm. The cell has a resistance of 2.90 Ω; the temperature is 25°C. Calculate
- (a) the theoretical potential needed to initiate deposition of silver from this solution.
- (b) the *IR* drop associated with a current of 0.12 A.

(c) the initial applied potential, given that the O_2 overvoltage is 0.80 V.
(d) the applied potential when $[Ag(CN)_2^-]$ is 1.00×10^{-5} M, assuming no changes in *IR* drop and O_2 overvoltage.

***20-22.** A solution is 0.150 M in Co^{2+} and 0.0750 M in Cd^{2+}. Calculate
(a) the Co^{2+} concentration in the solution as the first cadmium starts to deposit.
(b) the cathode potential needed to lower the Co^{2+} concentration to 1×10^{-5} M.

20-23. A solution is 0.0500 M in BiO^+ and 0.0400 M in Co^{2+} and has a pH of 2.50.
(a) What is the concentration of the more readily reduced cation at the onset of deposition of the less reducible one?
(b) What is the potential of the cathode when the concentration of the more easily reduced species is 1.00×10^{-6} M?

20-24. Electrogravimetric analysis involving control of the cathode potential is proposed as a means for separating BiO^+ and Sn^{2+} in a solution that is 0.200 M in each ion and buffered to pH 1.50.
(a) Calculate the theoretical cathode potential at the start of deposition of the more readily reduced ion.
(b) Calculate the residual concentration of the more readily reduced species at the outset of the deposition of the less readily reduced species.
(c) Propose a range (vs. SCE), if such exists, within which the cathode potential should be maintained; consider a residual concentration less than 10^{-6} M as constituting quantitative removal.

***20-25.** Halide ions can be deposited on a silver anode via the reaction

$$Ag(s) + X^- \longrightarrow AgX(s) + e^-$$

(a) If 1.00×10^{-5} M is used as the criterion for quantitative removal, is it theoretically feasible to separate Br^- from I^- through control of the anode potential in a solution that is initially 0.250 M in each ion?
(b) Is a separation of Cl^- and I^- theoretically feasible in a solution that is initially 0.250 M in each ion?
(c) If a separation is feasible in either (a) or (b), what range of anode potentials (vs. SCE) should be used?

20-26. A solution is 0.100 M in each of two reducible cations, A and B. Removal of the more reducible species A is deemed complete when [A] has been decreased to 1.00×10^{-5} M. What minimum difference in standard electrode potentials will permit the isolation of A without interference from B when

	A is	B is
*(a)	univalent	univalent
(b)	divalent	univalent
*(c)	trivalent	univalent
(d)	univalent	divalent
*(e)	divalent	divalent
(f)	trivalent	divalent
*(g)	univalent	trivalent
(h)	divalent	trivalent
*(i)	trivalent	trivalent

***20-27.** Calculate the time needed for a constant current of 0.961 A to deposit 0.500 g of Co(II) as
(a) elemental cobalt on the surface of a cathode.
(b) Co_3O_4 on an anode.

20-28. Calculate the time needed for a constant current of 1.20 A to deposit 0.500 g of
(a) Tl(III) as the element on a cathode.
(b) Tl(I) as Tl_2O_3 on an anode.
(c) Tl(I) as the element on a cathode.

***20-29.** A 0.1516-g sample of a purified organic acid was neutralized by the hydroxide ion produced in 5 min and 24 s by a constant current of 0.401 A. Calculate the equivalent weight of the acid (the mass of acid that contains 1 mol of protons).

20-30. The CN^- concentration of 10.0 mL of a plating solution was determined by titration with electrogenerated hydrogen ion to a methyl orange end point. A color change occurred after 3 min and 22 s with a current of 43.4 mA. Calculate the number of grams of NaCN per liter of solution.

***20-31.** An excess of $HgNH_3Y^{2-}$ was introduced to 25.00 mL of well water. Express the hardness of the water in terms of ppm $CaCO_3$ if the EDTA needed for the titration was generated at a mercury cathode (Equation 20-9) in 2.02 min by a constant current of 31.6 mA.

20-32. Electrolytically generated I_2 was used to determine the amount of H_2S in 100.0 mL of brackish water. Following addition of excess KI, a titration required a constant current of 36.32 mA for 10.12 min. The reaction was

$$H_2S + I_2 \longrightarrow S(s) + 2H^+ + 2I^-$$

Express the results of the analysis in terms of ppm H_2S.

***20-33.** The nitrobenzene in 210 mg of an organic mixture was reduced to phenylhydroxylamine at a constant potential of −0.96 V (vs. SCE) applied to a mercury cathode:

$$C_6H_5NO_2 + 4H^+ + 4e^- \longrightarrow C_6H_5NHOH + H_2O$$

The sample was dissolved in 100 mL of methanol; after electrolysis for 30 min, the reaction was judged complete. An electronic coulometer in series with the cell indicated that the reduction required 26.74 C. Calculate the percentage of $C_6H_5NO_2$ in the sample.

20-34. The phenol content of water downstream from a coking furnace was determined by coulometric analysis. A 100-mL sample was rendered slightly acidic, and an excess of KBr was introduced. To produce Br_2 for the reaction

$$C_6H_5OH + 3Br_2 \longrightarrow Br_3C_6H_2OH(s) + 2HBr$$

a steady current of 0.0313 A for 7 min and 33 s was required. Express the results of this analysis in terms of parts of C_6H_5OH per million parts of water. (Assume that the density of water is 1.00 g/mL.)

***20-35.** At a potential of −1.0 V (vs. SCE), CCl_4 in methanol is reduced to $CHCl_3$ at a mercury cathode:

$$2CCl_4 + 2H^+ + 2e^- + 2Hg(l) \longrightarrow 2CHCl_3 + Hg_2Cl_2(s)$$

At −1.80 V, the $CHCl_3$ further reacts to give CH_4:

$$2CHCl_3 + 6H^+ + 6e^- + 6Hg(l) \longrightarrow$$
$$2CH_4 + 3Hg_2Cl_2(s)$$

A 0.750-g sample containing CCl_4, $CHCl_3$, and inert organic species was dissolved in methanol and electrolyzed at -1.0 V until the current approached zero. A coulometer indicated that 11.63 C were required to complete the reaction. The potential of the cathode was adjusted to -1.8 V. Completion of the titration at this potential required an additional 68.6 C. Calculate the percentage of CCl_4 and $CHCl_3$ in the mixture.

20-36. A 0.1309-g sample containing only $CHCl_3$ and CH_2Cl_2 was dissolved in methanol and electrolyzed in a cell containing a mercury cathode; the potential of the cathode was held constant at -1.80 V (vs. SCE). Both compounds were reduced to CH_4 (see Problem 20-35 for the reaction). Calculate the percentage of $CHCl_3$ and CH_2Cl_2 if 306.7 C were required to complete the reduction.

***20-37.** Traces of $C_6H_5NH_2$ can be determined by reaction with an excess of electrolytically generated Br_2:

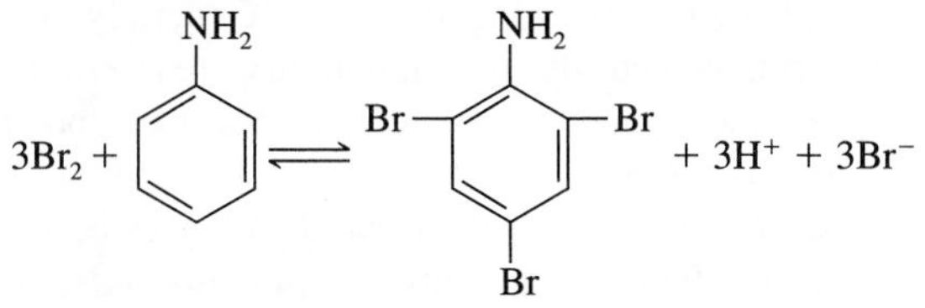

$$3Br_2 + \text{[aniline]} \rightleftharpoons \text{[tribromoaniline]} + 3H^+ + 3Br^-$$

The polarity of the working electrode is then reversed, and the excess Br_2 is determined by a coulometric titration involving the generation of Cu(I):

$$Br_2 + 2Cu^+ \longrightarrow 2Br^- + 2Cu^{2+}$$

Suitable quantities of KBr and $CuSO_4$ were added to a 25.0-mL sample containing aniline. Calculate the number of micrograms of $C_6H_5NH_2$ in the sample from the data:

Working Electrode Functioning As	Generation Time with a Constant Current of 1.51 mA, min
anode	3.76
cathode	0.270

20-38. Quinone can be reduced by hydroquinone with an excess of electrolytically generated Sn(II):

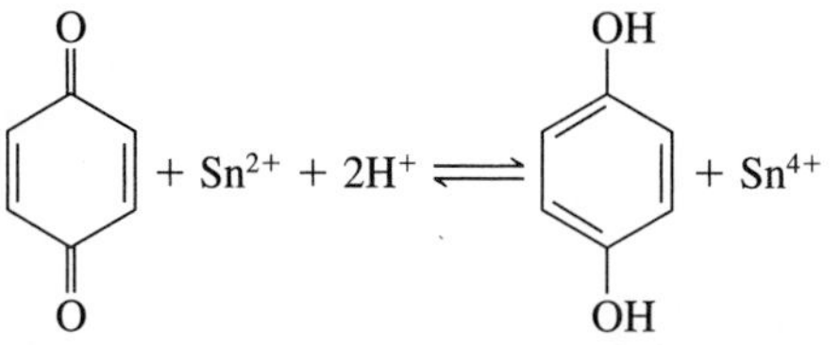

$$\text{[quinone]} + Sn^{2+} + 2H^+ \rightleftharpoons \text{[hydroquinone]} + Sn^{4+}$$

The polarity of the working electrode is then reversed, and the excess Sn(II) is oxidized with Br_2 generated in a coulometric titration:

$$Sn^{2+} + Br_2 \rightleftharpoons Sn^{4+} + 2Br^-$$

Appropriate quantities of $SnCl_4$ and KBr were added to a 50.0-mL sample. Calculate the weight of $C_6H_4O_2$ in the sample from the data:

Working Electrode Functioning As	Generation Time with a Constant Current of 1.062 mA, min
cathode	8.34
anode	0.691

CHAPTER

21

Voltammetry

Voltammetry comprises a group of electroanalytical methods in which information about the analyte is derived from the measurement of current as a function of applied potential, the measurements being obtained under conditions that encourage polarization of an indicator, or working, electrode. Generally, in order to enhance polarization, the working electrodes in voltammetry are *microelectrodes* that have surface areas of a few square millimeters at the most and in some applications, a few square micrometers.

Working electrodes with surface areas smaller than a few square millimeters are called **microelectrodes.** Electrodes with areas of a few square micrometers or less are sometimes called **ultramicroelectrodes.**

Voltammetry is an electrochemical method that is based upon measurement of current as a function of the potential applied to a tiny microelectrode.

Polarography is voltammetry performed with a dropping mercury electrode.

Heyrovsky was awarded the 1959 Nobel Prize in chemistry for his discovery and development of polarography.

Voltammetry is used as a detection technique for liquid chromatography.

At the outset, it is worthwhile pointing out the basic differences between voltammetry and the types of electrochemical methods that were discussed in earlier chapters. Voltammetry is based upon the measurement of a current that develops in an electrochemical cell under conditions of complete concentration polarization. In contrast, potentiometric measurements are made at currents that approach zero and where polarization is absent. Voltammetry differs from electrogravimetry and coulometry in the respect that with the latter pair, measures are taken to minimize or compensate for the effects of concentration polarization. Furthermore, in voltammetry a minimal consumption of analyte takes place, whereas in electrogravimetry and coulometry essentially all of the analyte is converted to another state.

Historically, the field of voltammetry developed from *polarography,* a particular type of voltammetry that was discovered by the Czechoslovakian chemist Jaroslav Heyrovsky in the early 1920s.[1] Polarography, which is still an important branch of voltammetry, differs from other types of voltammetry in the respect that the working microelectrode takes the form of a *dropping mercury electrode* (DME). The construction and unique properties of this electrode are discussed in Section 21B-5.

Voltammetry is widely used by inorganic, physical, and biological chemists for nonanalytical purposes including fundamental studies of oxidation and reduction processes in various media, adsorption processes on surfaces, and electron transfer mechanisms at chemically modified electrode surfaces. At one time, voltammetry (particularly classical polarography) was an important tool used by chemists for the determination of inorganic ions and certain organic species in aqueous solutions. In the late 1950s and the early 1960s, however, these analytical applications were largely supplanted by various atomic spectroscopic

[1]J. Heyrovsky, *Chem. Listy,* **1922**, *16,* 256.

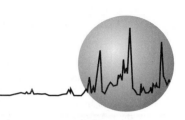

methods and voltammetry ceased to be important in analysis except for certain special applications, such as the determination of molecular oxygen in solutions.

In the mid-1960s, several major modifications of classical voltammetric techniques were developed that enhance significantly the sensitivity and selectivity of the method. At about this same time, the advent of low-cost operational amplifiers made possible the commercial development of relatively inexpensive instruments that incorporated many of these modifications and made them available to all chemists. The result has been a recent resurgence of interest in applying voltammetric methods to the determination of a host of species, particularly those of pharmaceutical interest.[2] Furthermore, voltammetry coupled with liquid chromatography has become a powerful tool for the analysis of complex mixtures of different kinds. Modern voltammetry also continues to be a potent tool used by various kinds of chemists interested in studying oxidation and reduction processes as well as adsorption processes.[3]

21A EXCITATION SIGNALS IN VOLTAMMETRY

In voltammetry, a variable potential *excitation signal* is impressed upon an electrochemical cell containing a microelectrode. This excitation signal elicits a characteristic current response upon which the method is based. The waveforms of four of the most common excitation signals used in voltammetry are shown in Figure 21-1. The classical voltammetric excitation signal is the linear scan shown in Figure 21-1a, in which the dc potential applied to the cell increases linearly (usually over a 2 to 3 V range) as a function of time. The current that develops in the cell is then recorded as a function of time (and thus as a function of the applied potential).

Two pulse-type excitation signals are shown in Figures 21-1b and c. Currents are measured at various times during the lifetime of these pulses. With the triangular waveform shown in Figure 21-1d, the potential is cycled between two values, first increasing linearly to a maximum and then decreasing linearly, with the same absolute numerical slope, to its original value. This process may be repeated numerous times, with the current being recorded as a function of time. A complete cycle may take 100 or more seconds.

The last column in Figure 21-1 lists the types of voltammetry that employ the various excitation signals. The first three of these techniques are discussed in the sections that follow. The fourth, cyclic voltammetry, has found considerable application as a diagnostic tool that provides information about the mechanism of oxidation/reduction reactions under various conditions. It has limited applications for quantitative analysis, however, and for this reason, will not be discussed further in this book.

---

[2]For a brief summary of several of these modified voltammetric techniques, see J. B. Flato, *Anal. Chem.*, **1972**, *44*, 75A.

[3]Some general references that deal with voltammetry include *Analytical Voltammetry,* M. R. Smyth and F. G. Vos, Eds. New York: Elsevier, 1992; A. J. Bard and L. R. Faulkner, *Electrochemical Methods.* New York: Wiley, 1980; *Laboratory Techniques in Electrochemical Methods,* P. T. Kissinger and W. R. Heineman, Eds. New York: Marcel Dekker, 1984.

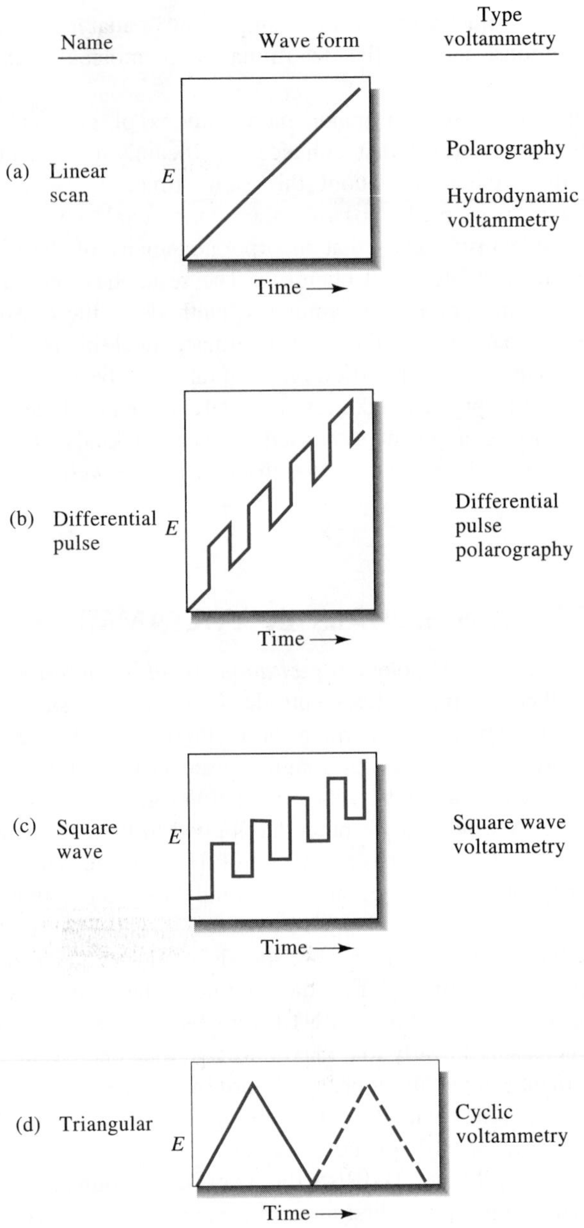

Figure 21-1
Potential excitation signals used in voltammetry.

21B LINEAR-SCAN VOLTAMMETRY

A **voltammogram** is a plot of current versus applied voltage.

The earliest and simplest voltammetric methods were of the linear-scan type, in which the potential of the working electrode is increased or decreased at a typical rate of 2 to 5 mV/s. The current, usually in microamperes, is then recorded to give a *voltammogram,* which is a plot of current as a function of potential applied to the working electrode. Linear-scan voltammetry is of two types: *hydrodynamic voltammetry* and *polarography.*

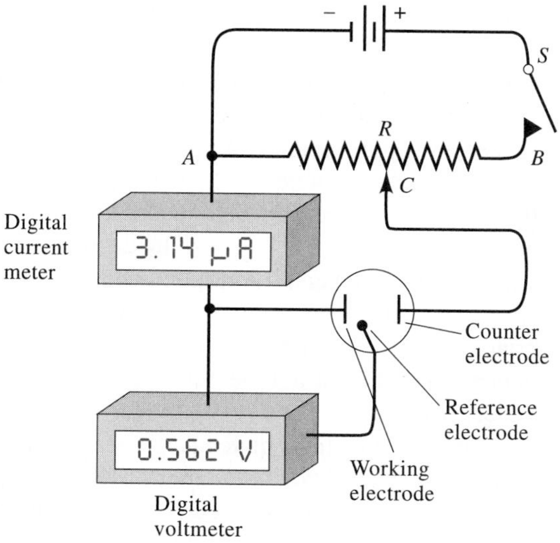

Figure 21-2
A manual potentiostat for voltammetry.

21B-1 Voltammetric Instruments

Figure 21-2 is a schematic showing the components of a simple apparatus for carrying out linear-scan voltammetric measurements. The cell is made up of three electrodes immersed in a solution containing the analyte and also an excess of a nonreactive electrolyte called a *supporting electrolyte.* (Note the similarity of this cell to the one for controlled potential electrolysis shown in Figure 20-7.) One of the three electrodes is the microelectrode, or *working electrode,* whose potential is varied linearly with time. Its dimensions are kept small in order to enhance its tendency to become polarized. The second electrode is a reference electrode whose potential remains constant throughout the experiment. The third electrode is a *counter electrode,* which is often a coil of platinum wire or a pool of mercury that simply serves to conduct electricity from the source through the solution to the microelectrode.[4] The signal source is a variable dc power supply consisting of a battery in series with a variable resistor *R*. The desired potential is selected by moving the contact *C* to the proper position on the resistor. The electrical resistance of the circuit containing the reference electrode is so large ($> 10^{11}\ \Omega$) that essentially no current is present in it. Thus, the entire current from the source is carried from the counter electrode to the microelectrode. A voltammogram is recorded by moving the contact *C* in Figure 21-2 and recording the resulting current as a function of the potential between the working electrode and the reference electrode.

A **supporting electrolyte** in voltammetry is a salt added in excess to the analyte solution. Most commonly, it is an alkali metal salt that does not react at the microelectrode at the potentials being used to determine the analyte.

The **working electrode** in electroanalytical chemistry is the electrode at which the analyte is oxidized or reduced.

A **counter electrode** in electroanalytical chemistry is the electrode that is coupled to the working electrode but plays no part in determining the magnitude of the potential being measured.

[4]Early voltammetry was performed with a two-electrode system rather than the three-electrode system shown in Figure 21-2. With a two-electrode system, the second electrode is either a large metal electrode, such as a pool of mercury, or a reference electrode large enough to prevent its polarization during an experiment. This second electrode combines the functions of the reference electrode and the counter electrode in Figure 21-2. Here, it is assumed that the potential of this second electrode remains constant throughout a scan so that the microelectrode potential is simply the difference between the applied potential and the potential of the second electrode. With solutions of high electrical resistance, however, this assumption is not valid because the *IR* drop becomes significant and increases as the current increases. Distorted voltammograms are the consequence. Almost all voltammetry is now performed with three-electrode systems.

In principle, the manual potentiostat of Figure 21-2 could be used to generate a linear-scan voltammogram. In such an experiment, contact *C* would be moved at a constant rate from *A* to *B* to produce the excitation signal shown in Figure 21-1a. The current and voltage would then be recorded at consecutive equal time intervals during the voltage (or time) scan. Unfortunately, it is difficult to maintain a constant rate of motion of the sliding contact and equally difficult to record the current manually. Thus, in practice we would record a manual voltammogram by moving contact *C* in small increments indicated by the digital voltmeter in Figure 21-2. For each increment of voltage, the resulting current would be recorded to produce a voltammogram point by point. It is important to emphasize that the independent variable in this experiment is the potential of the *microelectrode* versus the reference electrode and not the potential between the microelectrode and the counter electrode.

The circuit of Figure 21-2 functions quite nicely for recording current at a fixed potential or for generating manual voltammograms. However, complex excitation signals such as those shown in Figures 21-1b and 21-1c must be generated electronically. Modern voltammetric instruments automatically vary the potential in a prescribed way with respect to the reference electrode and record the resulting current. A potentiostat based on operational amplifiers that is designed to carry out this task is described in Feature 21-1.

FEATURE 21-1
Voltammetric Instruments Based on Operational Amplifiers

In Feature 19-2 we describe the use of operational amplifiers to measure the potential of electrochemical cells. Op amps also can be used to measure currents and carry out a variety of other control and measurement tasks. Let us consider the measurement of current as illustrated in Figure 21-3.

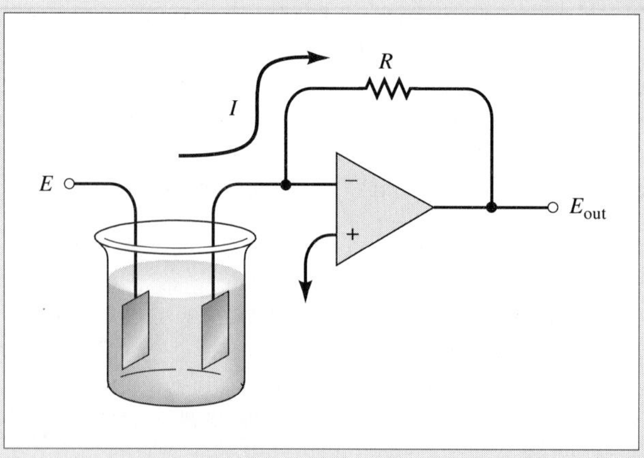

Figure 21-3 An op amp circuit for measuring voltammetric current.

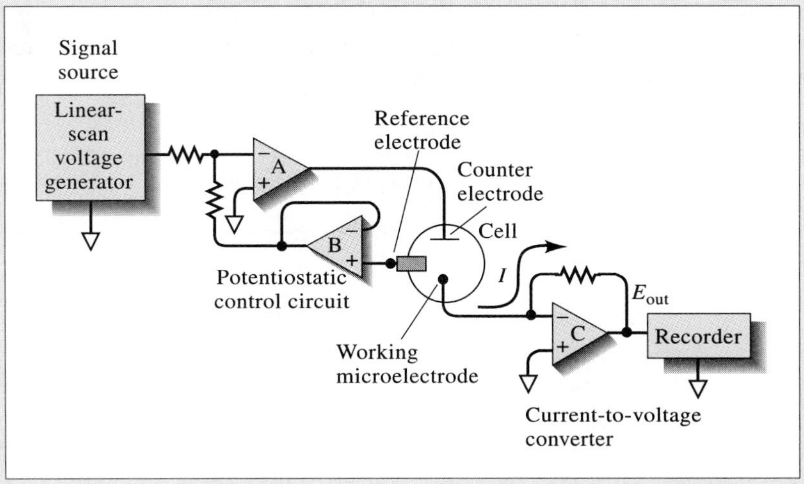

Figure 21-4 An op amp potentiostat.

In this circuit a voltage source is attached to one electrode of an electrochemical cell, which produces a current I in the cell. Because of the very high input resistance of the op amp, essentially all of the current passes through the resistor R to the output of the op amp. The voltage at the output of the op amp is given by Ohm's law, $E_{out} = -IR$. By solving this equation for I, we have

$$I = -\frac{E_{out}}{R}$$

In other words, the current in the electrochemical cell is proportional to the voltage output of the op amp. The value of the current then can be calculated from the measured values of E_{out} and the resistance R. The circuit is called a *current-to-voltage converter.*

Op amps can be used to construct an automatic three-electrode potentiostat as illustrated in Figure 21-4. Notice that the current-measuring circuit of Figure 21-3 is connected to the working electrode of the cell (op amp C). The reference electrode is attached to a voltage follower (op amp B). As discussed in Feature 19-2, the voltage follower monitors the potential of the reference electrode without drawing any current from the cell. The output of op amp B, which is the reference electrode potential, is fed back to the input of op amp A to complete the circuit. The functions of op amp A are (1) to provide the current in the electrochemical cell between the counter electrode and the working electrode and (2) to maintain the potential difference between the reference electrode and the working electrode at a value provided by the linear-scan voltage generator.

In operation, the linear-scan voltage generator sweeps the potential between the reference and working electrodes, and the current in the cell is monitored by op amp C. The output voltage E_{out} of op amp B, which is proportional to the current I in the cell, is recorded on a stripchart or by a computer for data analysis and presentation.[5]

[5]For a complete discussion of op amp three-electrode potentiostats, see *Laboratory Techniques in Electroanalytical Chemistry,* P. T. Kissinger and W. R. Heineman, Eds., pp. 171–186. New York: Marcel Dekker, 1984.

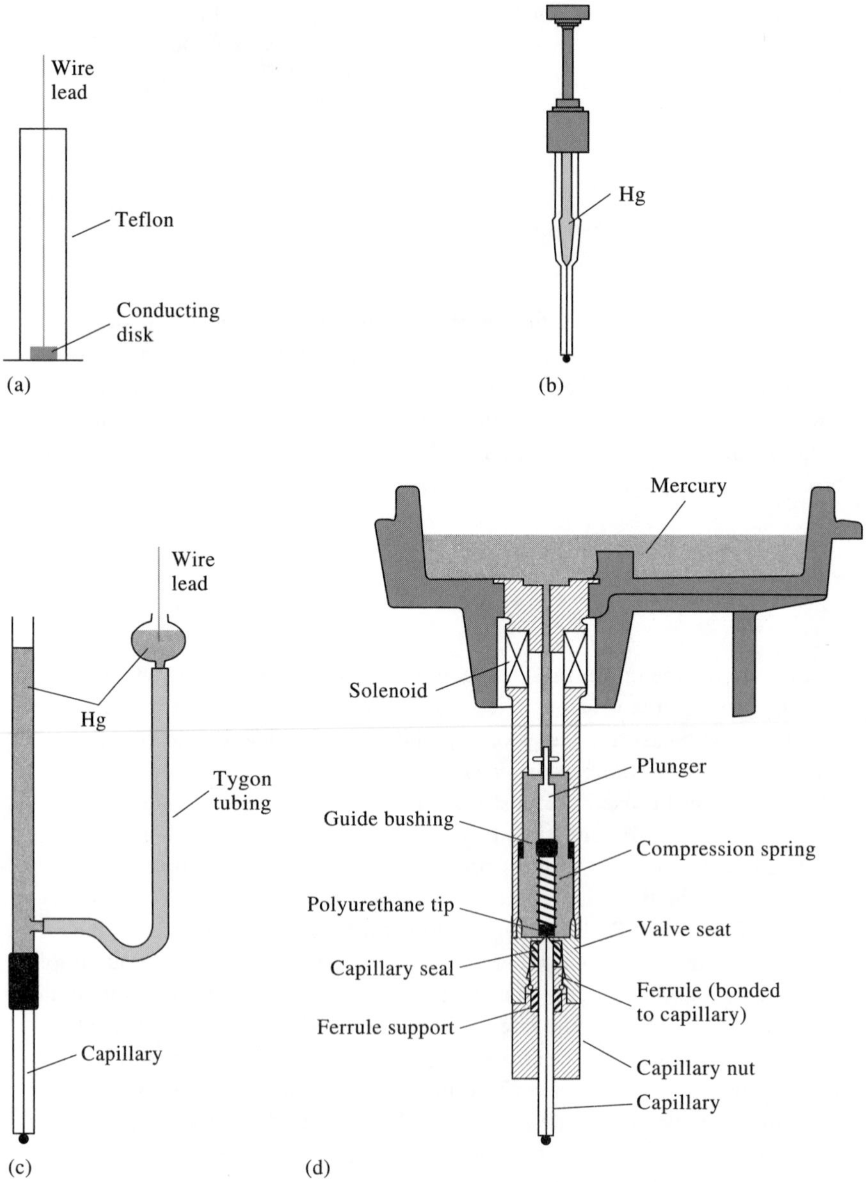

Figure 21-5

Some common types of microelectrodes: (a) a disk electrode; (b) a mercury hanging drop electrode; (c) a dropping mercury electrode; (d) a static mercury dropping electrode.

21B-2 Microelectrodes

The microelectrodes employed in voltammetry take a variety of shapes and forms. Often, they are small flat disks of a conductor that are press fitted into a rod of an inert material, such as Teflon or Kel-F, that has a wire contact imbedded in it (see Figure 21-5a). The conductor may be an inert metal, such as platinum or gold; pyrolytic graphite or glassy carbon; a semiconductor, such as tin or indium oxide; or a metal coated with a film of mercury. As shown in Figure 21-6, the range of potentials that can be used with these electrodes in aqueous solutions varies and depends not only upon electrode material but also upon the composition of the solution in which it is immersed. Generally, the positive potential limitations are caused by the large currents that develop due to oxidation of the water to give molecular oxygen. Negative limits arise from the reduction of water, giving hydrogen. Note that relatively large negative potentials can be tolerated with mercury electrodes owing to the high overvoltage of hydrogen on this metal.

Mercury microelectrodes have been widely employed in voltammetry for several reasons. One is the relatively large negative potential range just described. Furthermore, a fresh metallic surface is readily formed by simply producing a new drop. In addition, many metal ions are reversibly reduced to amalgams at the surface of a mercury electrode, which simplifies the chemistry. Mercury microelectrodes take several forms. The simplest of these is a mercury film electrode formed by electrodeposition of the metal onto a disk electrode, such as that shown in Figure 21-5a. Figure 21-5b illustrates a *hanging mercury drop electrode* (HMDE). The electrode, which is available from commercial sources, consists of a very fine capillary tube connected to a reservoir containing mercury. The metal is forced out of the capillary by a piston arrangement driven by a micrometer screw. The micrometer permits formation of drops having surface areas that are reproducible to five percent or better.

Large negative potentials can be used with mercury electrodes.

Figure 21-5c shows a typical *dropping mercury electrode* (DME), which was used in nearly all early polarographic experiments. It consists of roughly 10 cm of a fine capillary tubing (inside diameter ~0.05 mm) through which mercury is forced by a mercury head of perhaps 50 cm. The diameter of the capillary is such that a new drop forms and breaks every 2 to 6 s. The diameter of the drop is 0.5 to 1 mm and is highly reproducible. In some applications, the drop time is controlled by a mechanical knocker that dislodges the drop at a fixed time after it begins to form.

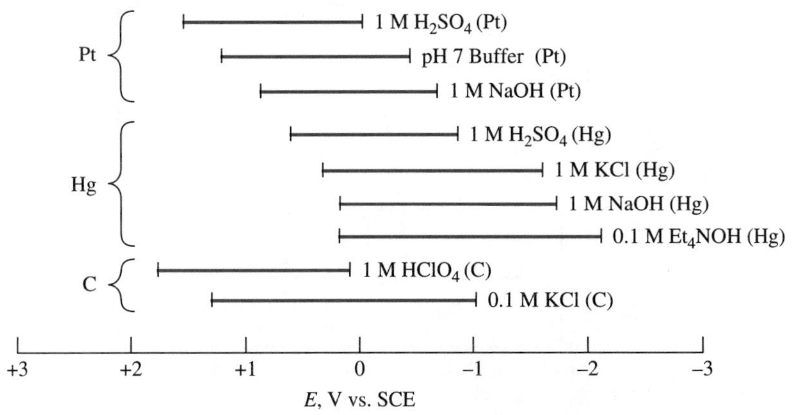

Figure 21-6
Potential ranges for three types of electrodes in various supporting electrolytes. (Adapted from A. J. Bard and L. R. Faulkner, *Electrochemical Methods,* back cover. New York: Wiley, 1980. With permission.)

Figure 21-5d shows a commercially available mercury electrode, which can be operated as a dropping mercury electrode or a hanging drop electrode. The mercury is contained in a plastic-lined reservoir about 15 cm above the upper end of the capillary. A compression spring forces the polyurethane-tipped plunger against the head of the capillary, thus preventing a flow of mercury. This plunger is lifted upon activation of the solenoid by a signal from the control system. The capillary is much larger in diameter (0.15 mm) than the typical one. As a result, the formation of the drop is extremely rapid. After 50, 100, or 200 ms, the valve is closed, leaving a full-sized drop in place until it is dislodged by a mechanical knocker that is built into the electrode support block. This system has the advantage that the full-sized drop forms quickly and current measurements can be delayed until the surface area is stable and constant. This procedure largely eliminates the large current fluctuations that are encountered with the classical dropping electrode.

21B-3 Voltammograms

Figure 21-7 illustrates the appearance of a typical linear-scan voltammogram for an electrolysis involving the reduction of an analyte species A to give a product P at a mercury film microelectrode. Here, the microelectrode is assumed to be connected to the negative terminal of the linear-scan generator so that the applied potentials are given a negative sign as shown. By convention, cathodic currents are always treated as being positive, whereas anodic currents are given a negative sign. In this hypothetical experiment, the solution is assumed to be about 10^{-4} M in A, 0.0 M in P, and 0.1 M in KCl, which serves as the supporting electrolyte. The half-reaction at the microelectrode is the reversible reaction

$$A + ne^- \rightleftharpoons P \qquad E^0 = -0.26 \text{ V} \tag{21-1}$$

For convenience, we have neglected the charges on A and P and also assumed that the standard potential for the half-reaction is -0.26 V.

Linear-scan voltammograms generally take the shape of a sigmoidal (S-shaped) curve called a *voltammetric wave*. The constant current beyond the steep rise is

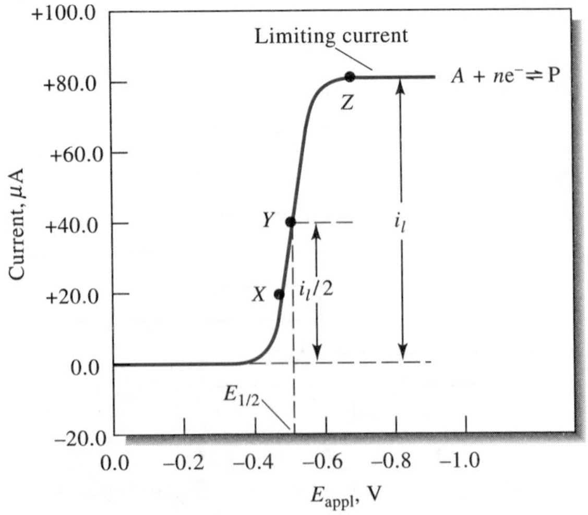

Figure 21-7
Linear-scan voltammogram for the reduction of a hypothetical species A to give a product P.

called the *limiting current* i_l because it is limited by the rate at which the reactant can be brought to the surface of the electrode by mass-transport processes. Limiting currents are generally directly proportional to reactant concentration. Thus, we may write

A **voltammetric wave** is an S-shaped wave obtained in current-voltage plots in voltammetry.

$$i_l = kc_A$$

where c_A is the analyte concentration and k is a constant. Quantitative linear-scan voltammetry is based upon this relationship.

The potential at which the current is equal to one half the limiting current is called the *half-wave potential* and given the symbol $E_{1/2}$. The half-wave potential is closely related to the standard potential for the half-reaction but is usually not identical to that constant. Half-wave potentials are sometimes useful for identification of the components of a solution.

The **half-wave potential** occurs when the current is equal to one half of the limiting value.

In order to obtain reproducible limiting currents rapidly, it is necessary either (1) that the solution or the microelectrode be in continuous and reproducible motion or (2) that a dropping mercury electrode be employed. Linear-scan voltammetry in which the solution or the electrode is kept in motion is called *hydrodynamic voltammetry*. Voltammetry that employs a dropping mercury electrode is called *polarography*. We shall consider both.

A **limiting current** in voltammetry is the current plateau that is observed at the top of a voltammetric wave.

21B-4 Hydrodynamic Voltammetry

Hydrodynamic voltammetry is performed in several ways. One method involves stirring the solution vigorously while it is in contact with a fixed microelectrode. Alternatively, the microelectrode is rotated at a constant high speed in the solution, thus providing the stirring action. Still another way of carrying out hydrodynamic voltammetry involves causing the analyte solution to flow through a tube in which the microelectrode is mounted. The last technique is becoming widely used for detecting oxidizable or reducible analytes as they exit from a liquid chromatographic column.

Hydrodynamic voltammetry is a type of voltammetry in which the analyte solution is kept in continuous motion.

As described in Section 20A-2, during an electrolysis, reactant is carried to the surface of an electrode by three mechanisms: (1) *migration* under the influence of an electric field, (2) *convection* resulting from stirring or vibration, and (3) *diffusion* due to concentration differences between the film of liquid at the electrode surface and the bulk of the solution. In voltammetry, every effort is made to minimize the effect of migration by introducing an excess of an inactive supporting electrolyte. When the concentration of supporting electrolyte exceeds that of the analyte by 50- to 100-fold, the fraction of the total current carried by the analyte approaches zero. As a result, the rate of migration of the analyte toward the electrode of opposite charge becomes essentially independent of applied potential.

Mass-transport processes include diffusion, migration, and convection.

Concentration Profiles at Microelectrode Surfaces During Electrolysis

Throughout this discussion we will consider that the electrode reaction shown in Equation 21-1 takes place at a mercury-coated microelectrode in a solution of A that also contains an excess of a supporting electrolyte. We will further assume that the initial concentration of A is c_A while that of P is zero and that P is not soluble in the mercury. Finally, we assume that the reduction reaction is rapid and reversible so that the concentrations of A and P in the film of solution immediately adjacent to the electrode is given at any instant by the Nernst equation:

$$E_{appl} = E_A^0 - \frac{0.0592}{n} \log \frac{c_P^0}{c_A^0} - E_{ref} \tag{21-2}$$

where E_{appl} is the potential between the microelectrode and the reference electrode and c_P^0 and c_A^0 are the molar concentrations of P and A *in a thin layer of solution at the electrode surface only.* We also assume that because the electrode is so very small, the electrolysis, over short periods of time, does not alter the bulk concentration of the solution appreciably. As a consequence, the concentration of A in the bulk of the solution c_A is substantially unchanged by the electrolysis and the concentration of P in the bulk of the solution c_P continues to be, for all practical purposes, zero ($c_P \approx 0$).

Electrolysis at a microelectrode does not alter significantly the bulk concentration of the analyte solution during the course of a voltammetry experiment.

Let us now consider concentration/distance profiles when the reduction described in the previous section is performed at a planar microelectrode immersed in a solution that is stirred vigorously. In order to understand the effect of stirring, it is necessary to develop a picture of liquid flow patterns in a stirred solution containing a small planar electrode. Three types of flow can be identified as shown in Figure 21-8. (1) *Turbulent flow,* in which liquid motion has no regular pattern, occurs in the bulk of the solution away from the electrode. (2) As the surface is approached, a transition to *laminar flow* takes place. In laminar flow, layers of liquid slide by one another in a direction parallel to the electrode surface. (3) At δ cm from the surface of the electrode, the rate of laminar flow approaches zero as a result of friction between the liquid and the electrode and gives a thin layer of stagnant solution, which is called the *Nernst diffusion layer.* It is only

In **laminar flow,** layers of liquid slide by one another in a direction that is parallel to the surface containing the liquid.

The **Nernst diffusion layer** is a layer of liquid in a flowing system that is immediately adjacent to a surface. The velocity of flow in this layer approaches zero.

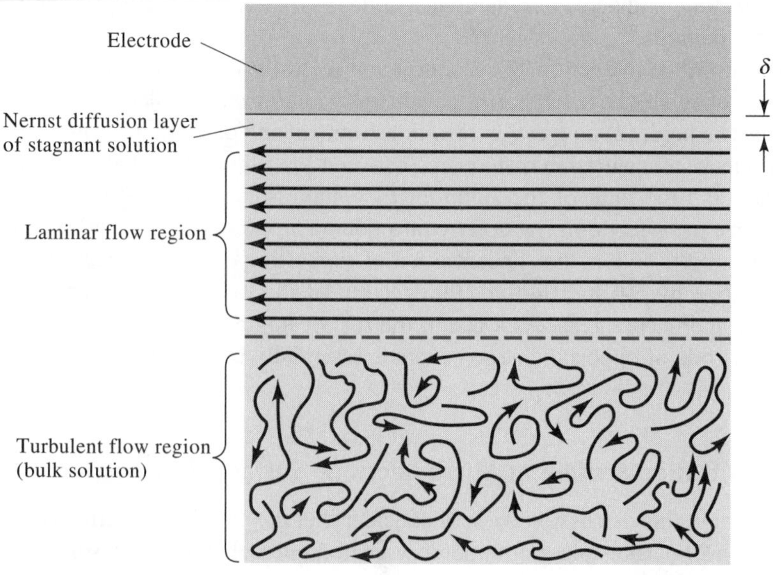

Figure 21-8
Flow patterns and regions of interest near the working electrode in hydrodynamic voltammetry.

within the Nernst diffusion layer that the concentrations of reactant and product vary as a function of distance from the electrode surface. That is, throughout the laminar flow and turbulent flow regions, convection maintains the concentration of A at its original value and the concentration of P at a level that is extremely small.

Figure 21-9a shows the concentration profiles for A at the three potentials labeled *X*, *Y*, and *Z* in Figure 21-7. The solution is divided into two regions. One makes up the bulk of the solution, where mass transport takes place by mechanical convection brought about by the stirrer. The concentration of A throughout this region is c_A. The second region is the Nernst diffusion layer, which is immediately adjacent to the electrode surface and has a thickness of δ cm. Typically, δ ranges from 0.01 to 0.001 cm, depending upon the efficiency of the stirring and the viscosity of the liquid. Within the diffusion layer, mass transport takes place by diffusion alone, as would be the case with the unstirred solution. With the stirred solution, however, diffusion is limited to a narrow layer of liquid and

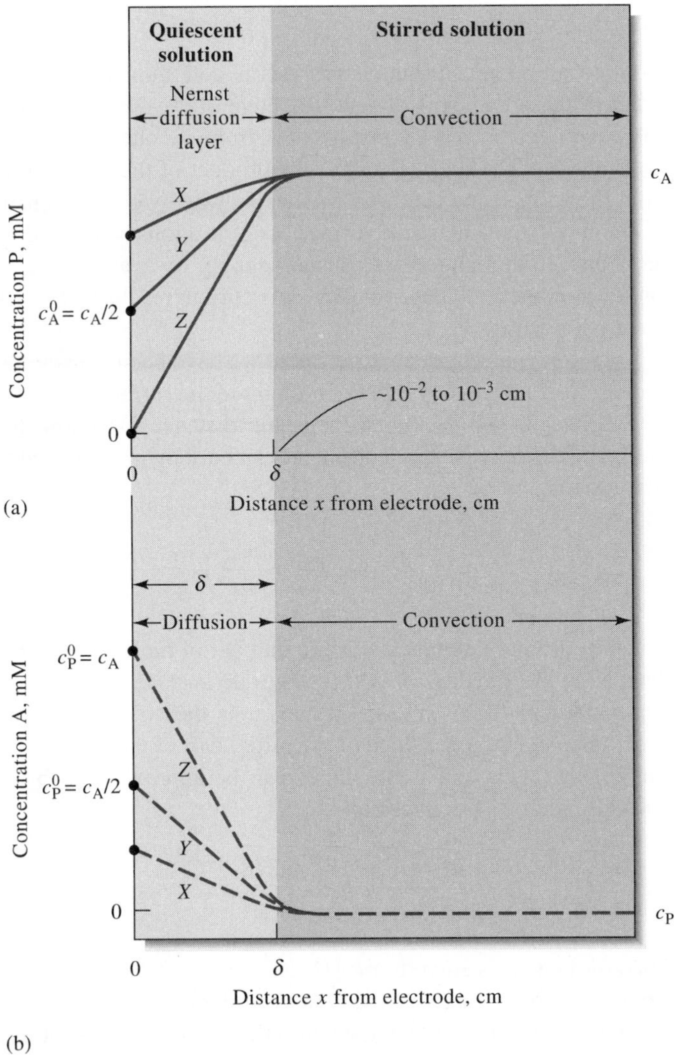

Figure 21-9
Concentration profiles at an electrode/solution interface during the electrolysis $A + ne^- \rightarrow P$ from a stirred solution of A. See Figure 21-7 for potentials corresponding to curves *X*, *Y*, and *Z*.

cannot extend out indefinitely into the solution. As a consequence, steady, diffusion-controlled currents are realized shortly after application of a voltage.

Figure 21-9b gives concentration profiles for P at the three potentials X, Y, and Z. In the Nernst diffusion region, the concentration of P decreases linearly with distance from the electrode surface and approaches zero at δ.

Note in the figures that at potential X, the equilibrium concentration of A at the electrode surface has been reduced to about 80% of its original value while the equilibrium concentration P has increased by an equivalent amount. (That is, $c_P^0 = c_A - c_A^0$.) At potential Y, which is the half-wave potential, the equilibrium concentrations of the two species at the surface are approximately the same and equal to $c_A/2$. Finally, at potential Z and beyond, the surface concentration of A approaches zero, whereas that of P approaches the original concentration of A, c_A. At potentials more negative than Z, essentially all A ions approaching the electrode surface are instantaneously reduced to P. The P ions formed in this way rapidly diffuse into the bulk of the solution, so the concentration of P in the surface layer remains constant at c_A.

Voltammetric Currents

The current at any point in this electrolysis is determined by a combination of (1) the rate of mass transport of A to the edge of the Nernst diffusion layer by convection and (2) the rate of transport of A from the outer edge of the diffusion layer to the electrode surface. Because the product of the electrolysis P diffuses away from the surface and is ultimately swept away by convection, a continuous current is required to maintain the surface concentrations demanded by the Nernst equation. Convection maintains a constant supply of A at the outer edge of the diffusion layer, however. Thus, a steady state current results that is determined by the applied potential.

The current in the electrolysis experiment we have been considering is a quantitative measure of how fast A is being brought to the surface of the electrode, and this rate is given by $\partial c_A/\partial x$ where x is the distance in centimeters from the electrode surface. For a planar electrode, it can be shown that the current is given by the expression

$$i = nFAD_A\left(\frac{\partial c_A}{\partial x}\right) \tag{21-3}$$

where i is the current in amperes, n is the number of moles of electrons per mole of analyte, F is the faraday, A is the electrode surface area in cm^2, D_A is the diffusion coefficient for A in cm^2s^{-1}, and c_A is the concentration of A in mol cm^{-3}. Note that $\partial c_A/\partial x$ is the slope of the initial part of the concentration profiles shown in Figure 21-9a and these slopes can be approximated by $(c_A - c_A^0)/\delta$. Therefore, Equation 21-3 reduces to

$$i = \frac{nFAD_A}{\delta}(c_A - c_A^0) = k_A(c_A - c_A^0) \tag{21-4}$$

where the constant k_A is equal to $nFAD_A/\delta$.

Equation 21-4 shows that as c_A^0 becomes smaller as a result of a larger negative applied potential, the current increases until the surface concentration approaches zero at which point the current becomes constant and independent of the applied potential. Thus, when $c_A^0 \rightarrow 0$, the current becomes the limiting current i_l, and

$$i_l = \frac{nFAD_A}{\delta} c_A = k_A c_A \qquad (21\text{-}5)$$

CHALLENGE: Show that the units of Equation 21-5 are amperes if the units of the quantities in the equation are as follows:

Quantity	Units
n	mol electrons/mol analyte
F	coulomb/faraday = coulomb/mol electrons
A	centimeters^2
D_A	centimeters^2/second
c_A	mol analyte/centimeter^3
δ	cm

This derivation is based upon an oversimplified picture of the diffusion layer in that the interface between the moving and stationary layers is viewed as a sharply defined edge where transport by convection ceases and transport by diffusion begins. Nevertheless, this simplified model does provide a reasonable approximation of the relationship between current and the variables that affect the current.

Although our model is somewhat oversimplified, it provides a reasonably accurate picture of the processes that occur at the electrode-solution interface.

Current/Voltage Relationships for Reversible Reactions

In order to develop an equation for the sigmoid curve shown in Figure 21-7, we subtract Equation 21-4 from Equation 21-5 and rearrange, which gives

$$c_A^0 = \frac{i_l - i}{k_A} \qquad (21\text{-}6)$$

The surface concentration of P can also be expressed in terms of the current by employing a relationship similar to Equation 21-4. That is,

$$i = -\frac{nFAD_P}{\delta}(c_P - c_P^0) \qquad (21\text{-}7)$$

where the minus sign results from the negative slope of the concentration profile for P. Note that D_P is now the diffusion coefficient of P. But we have said earlier that throughout the electrolysis the concentration of P approaches zero in the bulk of the solution and, therefore, when $c_P \approx 0.0$,

$$i = -\frac{nFAD_P c_P^0}{\delta} = k_P c_P^0$$

where $k_P = -nAD_P/\delta$. Rearranging gives

$$c_P^0 = i/k_P \qquad (21\text{-}8)$$

Substituting Equations 21-6 and 21-8 into Equation 21-2 yields after rearrangement

$$E_{appl} = E_A^0 - \frac{0.0592}{n}\log\frac{k_A}{k_P} - \frac{0.0592}{n}\log\frac{i}{i_l - i} - E_{ref} \qquad (21\text{-}9)$$

When $i = i_l/2$, the third term on the right side of this equation becomes equal to zero, and, by definition, E_{appl} is the half-wave potential. That is,

$$E_{appl} = E_{1/2} = E_A^0 - \frac{0.0592}{n} \log \frac{k_A}{k_P} - E_{ref} \tag{21-10}$$

Substituting this expression into Equation 21-9 gives an equation for the voltammogram in Figure 21-7. That is,

$$E_{appl} = E_{1/2} - \frac{0.0592}{n} \log \frac{i}{i_l - i} \tag{21-11}$$

Often, the ratio k_A/k_P in Equation 21-10 is nearly unity, so we may write for the species A

$$E_{1/2} \approx E_A^0 - E_{ref} \tag{21-12}$$

Current/Voltage Relationships for Irreversible Reactions

Many voltammetric electrode processes, particularly those associated with organic systems, are irreversible, which leads to drawn-out and less-well-defined waves. The quantitative description of such waves requires an additional term (involving the activation energy of the reaction) in Equation 21-11 to account for the kinetics of the electrode process. Although half-wave potentials for irreversible reactions ordinarily show some dependence upon concentration, diffusion currents remain linearly related to concentration; such processes are, therefore, readily adapted to quantitative analysis.

Voltammograms for Mixtures of Reactants

Ordinarily, the reactants of a mixture will behave independently of one another at a microelectrode; a voltammogram for a mixture is thus simply the summation of the waves for the individual components. Figure 21-10 shows the voltammograms for a pair of two-component mixtures. The half-wave potentials of the two reactants differ by about 0.1 V in curve *A* and by about 0.2 V in curve *B*. Note that a single polarogram may permit the quantitative determination of two or more

Voltammetric half-wave potentials must differ by 0.1 to 0.3 V in order to obtain resolvable waves for mixtures of reactants.

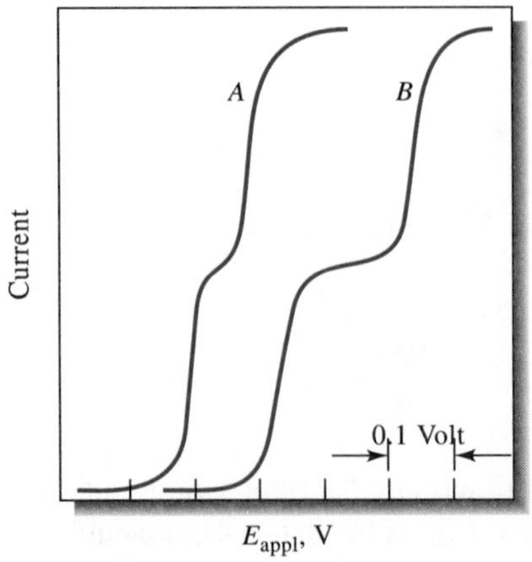

Figure 21-10
Voltammograms for two-component mixtures. Half-wave potentials differ by 0.1 V in curve *A* and by 0.2 V in curve *B*.

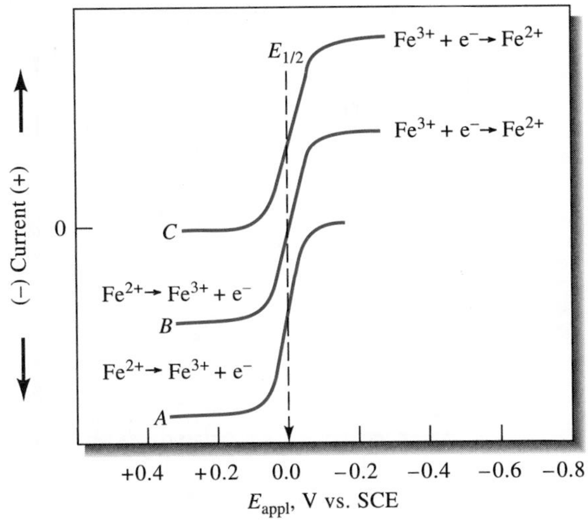

Figure 21-11
Voltammetric behavior of iron(II) and iron(III) in a citrate medium. Curve *A*: anodic wave for a solution in which $[Fe^{2+}] = 1 \times 10^{-4}$. Curve *B*: anodic/cathodic wave for a solution in which $[Fe^{2+}] = [Fe^{3+}] = 0.5 \times 10^{-4}$. Curve *C*: cathodic wave for a solution in which $[Fe^{3+}] = 1 \times 10^{-4}$.

species provided there is sufficient difference between succeeding half-wave potentials to permit evaluation of individual diffusion currents. Generally, 0.1 to 0.2 V is required if the more reducible species undergoes a two-electron reduction; a minimum of about 0.3 V is needed if the first reduction is a one-electron process.

Anodic and Mixed Anodic/Cathodic Voltammograms

Anodic waves as well as cathodic waves are encountered in voltammetry. An example of an anodic wave is illustrated in curve *A* of Figure 21-11, where the electrode reaction involves the oxidation of iron(II) to iron(III) in the presence of citrate ion. A limiting current is obtained at about +0.1 V, which is due to the half-reaction

$$Fe^{2+} \rightleftharpoons Fe^{3+} + e^-$$

As the potential is made more negative, a decrease in the anodic current occurs; at about −0.02 V, the current becomes zero because the oxidation of iron(II) ion has ceased.

Curve *C* represents the voltammogram for a solution of iron(III) in the same medium. Here, a cathodic wave results from reduction of the iron(III) to the divalent state. The half-wave potential is identical with that for the anodic wave, indicating that the oxidation and reduction of the two iron species are perfectly reversible at the microelectrode.

Curve *B* is the voltammogram of an equimolar mixture of iron(II) and iron(III). The portion of the curve below the zero-current line corresponds to the oxidation of the iron(II); this reaction ceases at an applied potential equal to the half-wave potential. The upper portion of the curve is due to the reduction of iron(III).

Oxygen Waves

Dissolved oxygen is readily reduced at various microelectrodes; an aqueous solution saturated with air exhibits two distinct waves attributable to this element (see Figure 21-12). The first results from the reduction of oxygen to peroxide

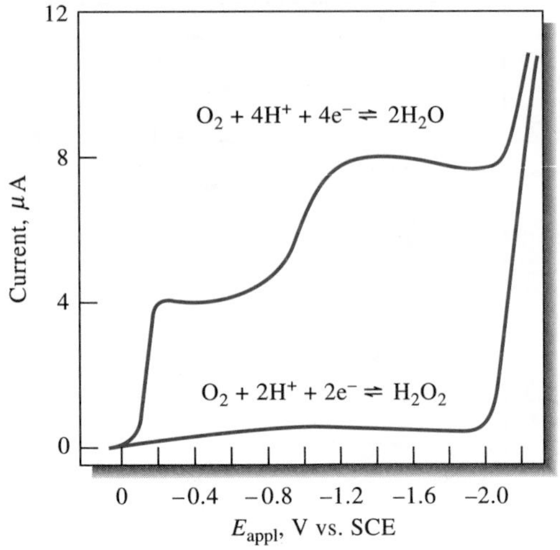

Figure 21-12
Voltammogram for the reduction of oxygen in an air-saturated 0.1 M KCl solution. The lower curve is for oxygen-free 0.1 M KCl.

$$O_2(g) + 2H^+ + 2e^- \rightleftharpoons H_2O_2$$

The second corresponds to the further reduction of the hydrogen peroxide

$$H_2O_2 + 2H^+ + 2e^- \rightleftharpoons 2H_2O$$

As would be expected from stoichiometric considerations, the two waves are of equal height.

Voltammetric measurements offer a convenient and widely used method for determining dissolved oxygen in solutions. However, the presence of oxygen often interferes with the accurate determination of other species. Thus, oxygen removal is ordinarily the first step in voltammetric procedures. Deaeration of the solution for several minutes with an inert gas *(sparging)* accomplishes this end; a stream of the same gas, usually nitrogen, is passed over the surface during analysis to prevent reabsorption of oxygen.

Sparging is a process in which dissolved gases are swept out of a solvent by bubbling an inert gas, such as nitrogen or helium, through the solution.

Remove oxygen from analyte solutions by sparging with nitrogen.

Applications of Hydrodynamic Voltammetry

Currently, the most important uses of hydrodynamic voltammetry include (1) detection and determination of chemical species as they exit from chromatographic columns or flow-injection apparatus; (2) routine determination of oxygen and certain species of biochemical interest, such as glucose, lactose, and sucrose; (3) detection of end points in coulometric and volumetric titrations; and (4) fundamental studies of electrochemical processes.

Flow-injection analysis is discussed in Section 24B-2.

Voltammetric Detectors in Chromatography and Flow-Injection Analysis Hydrodynamic voltammetry is becoming widely used for detection and determination of oxidizable or reducible compounds or ions that have been separated by high-performance liquid chromatography or by flow-injection methods. In these applications a thin layer cell such as that shown in Figure 21-13 is used. In these cells the working electrode is typically imbedded in the wall of an insulating block that is separated from a counter electrode by a thin spacer as shown. The

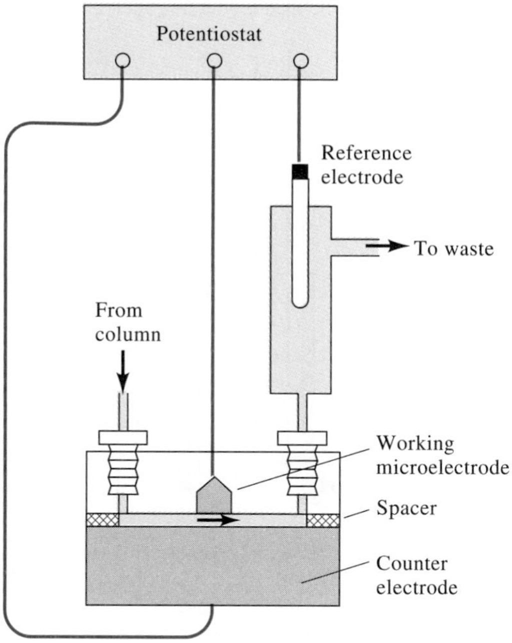

Figure 21-13
A voltammetric system for detecting electroactive species as they exit from a column.

volume of such a cell is typically 0.1 to 1 μL. A potential corresponding to the limiting current region for analytes is applied between the metal or glassy carbon working electrode and a silver/silver chloride reference electrode that is located downstream from the detector. In this type of application, detection limits as low as 10^{-9} to 10^{-10} M of analyte are obtained.

Voltammetric Sensors A number of voltammetric systems are produced commercially for the determination of specific species that are of interest in industry and research. These devices are sometimes called electrodes but are, in fact, complete voltammetric cells and are better referred to as sensors. Two of these devices are described here.

The determination of dissolved oxygen in a variety of aqueous environments, such as seawater, blood, sewage, effluents from chemical plants, and soils, is of tremendous importance to industry, biomedical and environmental research, and clinical medicine. One of the most common and convenient methods for making such measurements is with the *Clark oxygen sensor,* which was patented by L. C. Clark, Jr., in 1956.[6] A schematic of the Clark oxygen sensor is shown in Figure 21-14. The cell consists of a cathodic platinum disk working electrode imbedded in a centrally located cylindrical insulator. Surrounding the lower end of this insulator is a ring-shaped silver anode. The tubular insulator and electrodes are mounted inside a second cylinder that contains a buffered solution of potassium chloride. A thin (~ 20 μm) replaceable, oxygen-permeable membrane of Teflon or polyethylene is held in place at the bottom end of the tube by an O-ring. The thickness of the electrolyte solution between the cathode and the membrane is approximately 10 μm.

The Clark oxygen sensor is widely used in clinical laboratories for the determination of O_2 in blood and other bodily fluids.

[6]For a detailed discussion of the Clark oxygen sensor, see M. L. Hitchman, *Measurement of Dissolved Oxygen,* Chapters 3–5. New York: Wiley, 1978.

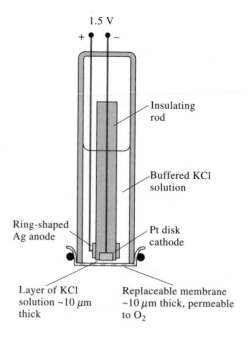

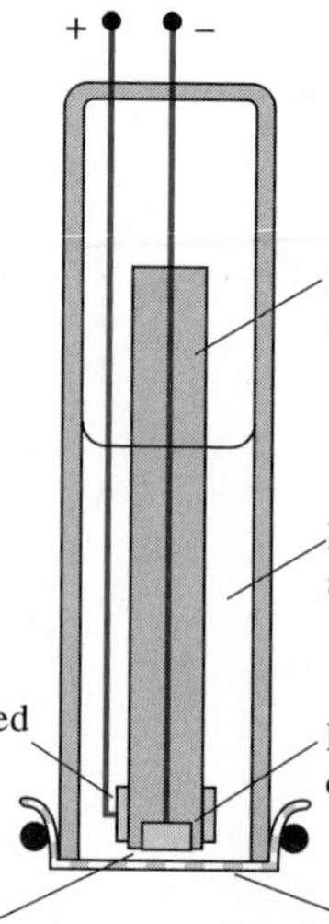

Figure 21-14
The Clark voltammetric oxygen sensor. Cathode reaction:

$$O_2 + 4H^+ + 4e^- \rightleftharpoons H_2O$$

Anodic reaction:

$$Ag + Cl^- \longrightarrow AgCl(s) + e^-$$

When the oxygen sensor is immersed in a flowing or stirred solution of the analyte, oxygen diffuses through the membrane into the thin layer of electrolyte immediately adjacent to the disk cathode, where it diffuses to the electrode and is immediately reduced to water. In contrast to a normal hydrodynamic electrode, two diffusion processes are involved—one through the membrane and the other through the solution between the membrane and the electrode surface. In order for a steady state condition to be reached in a reasonable period (10 to 20 s), the thickness of the membrane and the electrolyte film must be 20 μm or less. Under these conditions, it is the rate of equilibration of oxygen transfer across the membrane that determines the rate at which steady state currents are achieved.

A number of enzyme-based voltammetric sensors are offered commercially. An example is a glucose sensor that is widely used in clinical laboratories for the routine determination of glucose in blood sera. This device is similar in construction to the oxygen sensor shown in Figure 21-14. The membrane in this case is more complex and consists of three layers. The outer layer is a polycarbonate film that is permeable to glucose but impermeable to proteins and other constituents of blood. The middle layer is an immobilized enzyme—here, glucose oxidase. The inner layer is a cellulose acetate membrane, which is permeable to small molecules, such as hydrogen peroxide. When this device is immersed in a glucose containing solution, glucose diffuses through the outer membrane into the immobilized enzyme, where the following catalytic reaction occurs:

$$\text{glucose} + O_2 \longrightarrow H_2O_2 + \text{gluconic acid}$$

The hydrogen peroxide then diffuses through the inner layer of membrane and to the electrode surface, where it is oxidized to give oxygen. That is,

$$H_2O_2 + 2OH^- \rightleftharpoons O_2 + H_2O + 2e^-$$

The resulting current is directly proportional to the glucose concentration of the analyte solution.

Several other sensors are available that are based upon the voltammetric measurement of hydrogen peroxide produced by enzymatic oxidations of other species of clinical interest. These analytes include sucrose, lactose, ethanol, and L-lactate. A different enzyme is, of course, required for each species.

Amperometric Titrations Hydrodynamic voltammetry can be employed to estimate the equivalence point of titrations, provided at least one of the participants or products of the reaction involved is oxidized or reduced at a microelectrode. Here, the current at some fixed potential in the limiting current region is measured as a function of the reagent volume (or of time if the reagent is generated by a constant-current coulometric process). Plots of the data on either side of the equivalence point are straight lines with differing slopes; the end point is established by extrapolation to their intersection.

Amperometric titration curves typically take one of the forms shown in Figure 21-15. Curve (a) represents a titration in which the analyte reacts at the electrode-while the reagent does not. Figure 21-15b is typical of a titration in which the reagent reacts at the microelectrode and the analyte does not. Figure 21-15c corresponds to a titration in which both the analyte and the titrant react at the microelectrode.

Two types of amperometric electrode systems are encountered. One employs a single polarizable microelectrode coupled to a reference; the other uses a pair of identical solid state microelectrodes immersed in a stirred solution. For the first, the microelectrode is often a rotating platinum electrode constructed by sealing a platinum wire into the side of a glass tube that is connected to a stirring motor (see Figure 21-16). A dropping mercury electrode may also be used, in which case the solution is not stirred.

Amperometric titrations with one indicator electrode have, with one notable exception, been confined to titrations in which a precipitate or a stable complex is

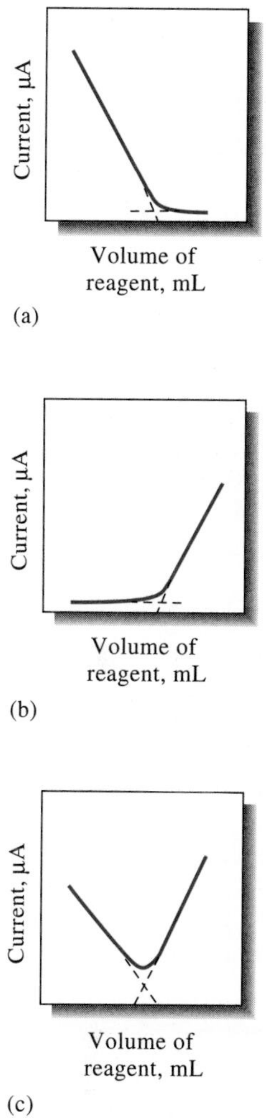

Figure 21-15

Typical amperometric titration curves: (a) analyte is reduced, reagent is not; (b) reagent is reduced, analyte is not; (c) both reagent and analyte are reduced.

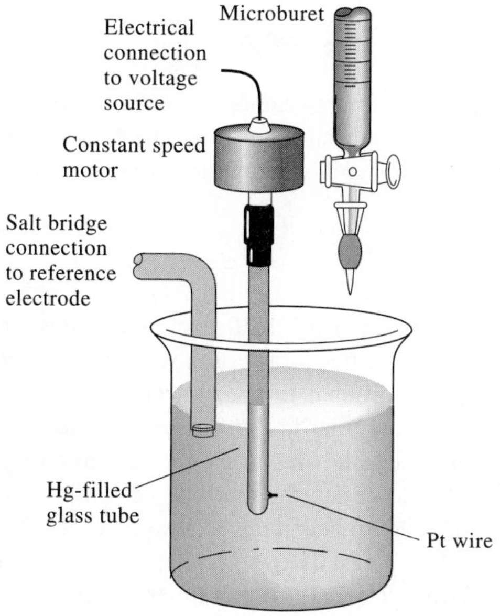

Figure 21-16

Typical cell arrangement for amperometric titrations with a rotating platinum electrode.

the product. Precipitating reagents include silver nitrate for halide ions, lead nitrate for sulfate ion, and several organic reagents, such as 8-hydroxyquinoline, dimethylglyoxime, and cupferron, for various metallic ions that are reducible at microelectrodes. Several metal ions have also been determined by titration with standard solutions of EDTA. The exception just noted involves titrations of organic compounds, such as certain phenols, aromatic amines, and olefins; hydrazine; and arsenic(III) and antimony(III) with bromine. The bromine is often generated coulometrically; it has also been formed by adding a standard solution of potassium bromate to an acidic solution of the analyte that also contains an excess of potassium bromide. Bromine is formed in the acidic medium by the reaction

$$BrO_3^- + 5Br^- + 6H^+ \longrightarrow 3Br_2 + 3H_2O$$

This type of titration has been carried out with a rotating platinum electrode or twin platinum microelectrodes. No current is observed prior to the equivalence point; after chemical equivalence, a rapid increase in current takes place due to electrochemical reduction of the excess bromine.

The use of a pair of identical metallic microelectrodes to establish the equivalence point in amperometric titrations offers the advantages of simplicity of equipment and not having to prepare and maintain a reference electrode. This type of system has been incorporated into equipment designed for the routine automatic determination of a single species, usually with a coulometrically generated reagent. An example of this type of system is an instrument for the automatic determination of chloride in samples of serum, sweat, tissue extracts, pesticides, and food products. Here, the reagent is silver ion coulometrically generated from a silver anode. The indicator system consists of a pair of twin silver microelectrodes that are maintained at a potential of perhaps 0.1 V. Short of the equivalence point in the titration of chloride ion, there is essentially no current because no easily reduced species is present in the solution. Consequently, electron transfer at the cathode is precluded and that electrode is completely polarized. Note that the anode is not polarized because the reaction

$$Ag \rightleftharpoons Ag^+ + e^-$$

occurs in the presence of a suitable cathodic reactant or depolarizer.

After the equivalence point has been passed, the cathode becomes depolarized owing to the presence of a significant amount of silver ions, which can react to give silver. That is,

$$Ag^+ + e^- \rightleftharpoons Ag$$

A current develops as a result of this half-reaction and the corresponding oxidation of silver at the anode. The magnitude of the current is, as in other amperometric methods, directly proportional to the concentration of the excess reagent. Thus, the titration curve is similar to that shown in Figure 21-15b. In the automatic titrator just mentioned, the amperometric current signal causes the coulometric generator current to cease; the chloride concentration is then computed from the magnitude of the current and the generation time. The instrument is said to have a range of 1 to 999.9 mM Cl^- per liter, a precision of 0.1% relative, and an accuracy of 0.5% relative. Typical titration times are 20 s.

21B-5 Polarography

Polarographic currents are controlled by diffusion alone, not by convection.

Linear-scan polarography was the first type of voltammetry to be discovered and used. It differs from hydrodynamic voltammetry in two regards. First, convection is avoided and second, a dropping mercury electrode, such as that shown in Figure 21-5c is used as the working electrode. A consequence of the first difference is that polarographic limiting currents are controlled by diffusion alone rather than by both diffusion and convection. Because convection is absent, polarographic limiting currents are generally one or more orders of magnitude smaller than hydrodynamic limiting currents.[7]

Polarographic Currents

The current in a cell containing a dropping electrode undergoes periodic fluctuations corresponding in frequency to the drop rate. As a drop dislodges from the capillary, the current falls to zero. As the surface area of a new drop increases, so also does the current. The *average current* is the hypothetical *constant* current that, in the drop time t, would produce the same quantity of charge as the fluctuating current does during this same period. In order to determine the average current, it is necessary to reduce the large fluctuations in the current by an electronic filter or by sampling the current near the end of each drop, where the change in current with time is relatively small. As shown in Figure 21-17, curve *A*, electronic filtering limits the oscillations to a reasonable magnitude; the average current (or, alternatively, the maximum current) is then readily determined, provided the drop rate t is reproducible. Note the effect of irregular drops in the upper part of curve *A*, probably caused by vibration of the apparatus.

Polarograms

Figure 21-17 shows two polarograms: one for a solution that is 1.0 M in hydrochloric acid and 5×10^{-4} M in cadmium ion (curve *A*), the second for the 0.1 M acid alone (curve *B*). The polarographic wave in curve *A* arises from the reaction

$$Cd^{2+} + 2e^- + Hg \rightleftharpoons Cd(Hg) \qquad (21\text{-}13)$$

where Cd(Hg) represents elemental cadmium dissolved in mercury to form an amalgam. The sharp increase in current at about -1 V in both polarograms is caused by the reduction of hydrogen ions to give hydrogen. Examination of the polarogram for the supporting electrolyte alone reveals that a small current, called the *residual current,* is present in the cell even in the absence of cadmium ions.

Residual current in polarography is the small charging current observed in the absence of a reactive species.

As in hydrodynamic voltammetry, limiting currents are observed when the magnitude of the current is limited by the rate at which analyte can be brought to the electrode surface. In polarography, however, the only mechanism of mass transport is diffusion, and for this reason, polarographic limiting currents are

---

[7]References dealing with polarography include R. C. Kapoor and B. S. Aggarwal, *Principles of Polarography.* New York: Wiley, 1991; A. M. Bond, *Modern Polarographic Methods in Analytical Chemistry.* New York: Marcel Dekker, 1980; J. Heyrovsky and J. Kuta, *Principles of Polarography.* New York: Academic Press, 1966; I. M. Kolthoff and J. J. Lingane, *Polarography,* 2nd ed. New York: Interscience, 1952; T. Riley and A. Watson, *Polarography and Other Voltammetric Methods.* New York: Wiley, 1987.

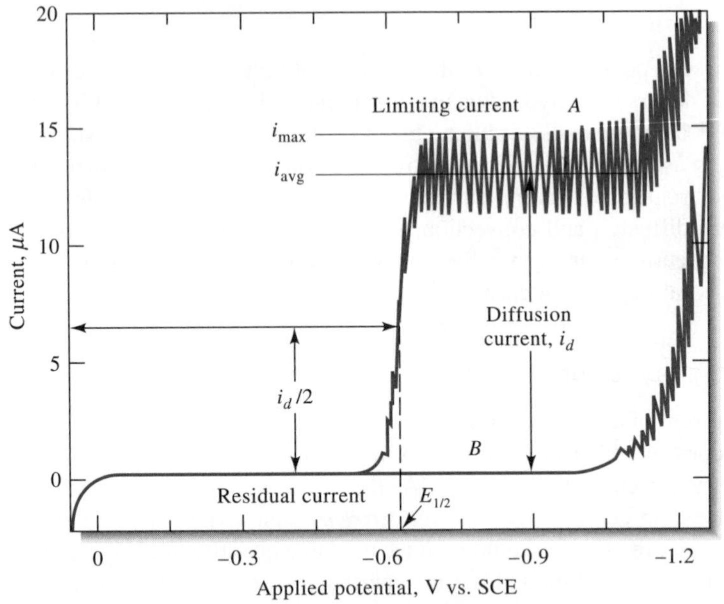

Figure 21-17
Polarograms for (*A*) a 1 M solution of HCl that is 5×10^{-4} M in Cd^{2+} and (*B*) a 1 M solution of HCl. (From D. T. Sawyer and J. L. Roberts, Jr., *Experimental Electrochemistry for Chemists.* New York: Wiley, 1974. Reprinted by permission of John Wiley & Sons, Inc.)

Diffusion current is the limiting current observed in polarography when the magnitude of the current is limited only by the rate of diffusion of the reactant to the dropping mercury electrode surface.

Diffusion current is proportional to concentration of analyte.

usually termed *diffusion currents* and given the symbol i_d. As shown in Figure 21-17, the diffusion current is the difference between the limiting and the residual current. The diffusion current is directly proportional to analyte concentration.

Diffusion Current at Dropping Electrodes

In deriving an equation for polarographic diffusion currents, it is necessary to take into account the rate of growth of the spherical electrode, which is related to the drop time *t* in seconds, the rate of flow of mercury through the capillary *m* in mg/s, and the diffusion coefficient of the analyte *D* in cm^2/s. These variables are taken into account in the *Ilkovic equation:*

$$(i_d)_{avg} = 607\, nD^{1/2}m^{2/3}t^{1/6}c \qquad (21\text{-}14)$$

In polarography, currents are often recorded in microamperes. Concentrations calculated when milliamperes are used in Equation 21-14 bear units of millimoles per liter.

where $(i_d)_{avg}$ is the average current in microamperes, and *c* is the analyte concentration in millimoles per liter. Note that either the average or the maximum current can be used in quantitative polarography.

Residual Currents

Figure 21-18 shows a residual current curve (obtained at high sensitivity) for a 0.1 M solution of hydrogen chloride. This current has two sources. The first is the reduction of trace impurities that are almost inevitably present in the blank solution; contributors here include small amounts of dissolved oxygen, heavy-metal ions from the distilled water, and impurities present in the salt used as the supporting electrolyte.

A **faradaic current** in an electrochemical cell is the current that is carried across electrode solution interfaces by an oxidation/reduction process.

A second component of the residual current is the *charging* or *condenser current* resulting from a flow of electrons that charge the mercury droplets with

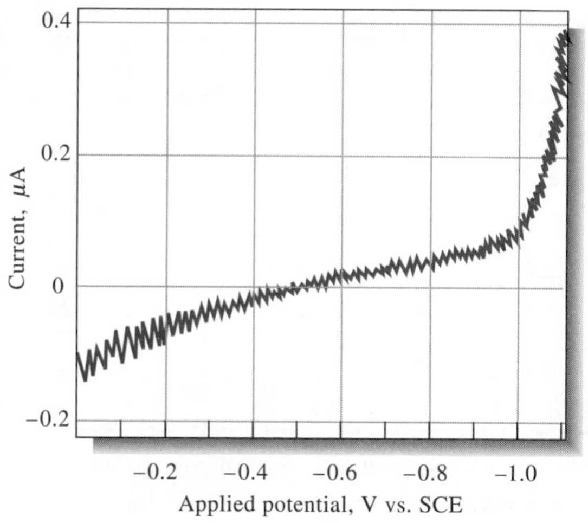

Figure 21-18
Residual current curve for a 0.1 M solution of HCl.

respect to the solution; this current may be either negative or positive. At potentials more negative than about − 0.4 V, an excess of electrons from the dc source provides the surface of each droplet with a negative charge. These excess electrons are carried down with the drop as it breaks; since each new drop is charged as it forms, a small but continuous current results. At applied potentials less negative than about − 0.4 V, the mercury tends to be positive with respect to the solution; thus, as each drop is formed, electrons are repelled from the surface toward the bulk of mercury, and a negative current is the result. At about − 0.4 V, the mercury surface is uncharged, and the charging current is zero. The charging current is a type of *nonfaradaic current* in the sense that charge is carried across an electrode/solution interface without an accompanying oxidation/reduction process.

A **nonfaradaic current** is a charging current that produces a charged double layer across electrode/solution interfaces. An example of a nonfaradaic current is the current that charges a condenser.

Ultimately, the accuracy and sensitivity of the polarographic method depend upon the magnitude of the nonfaradaic residual current and the accuracy with which a correction for its effect can be determined.

Most of the current making up the residual current in polarography is a nonfaradaic current that charges each mercury drop as it forms.

Comparison of Currents from Dropping and Stationary Planar Electrodes

Constant currents are not obtained in reasonable periods of time with a planar electrode in an unstirred solution because concentration gradients out from the electrode surface are constantly changing with time. In contrast, the dropping electrode exhibits constant reproducible currents essentially instantaneously after an applied voltage adjustment. This behavior represents an advantage of the dropping mercury electrode that accounts for its widespread use in the early years of voltammetry.

The rapid achievement of constant currents arises from the highly reproducible nature of the drop formation process and, equally important, the fact that the solution in the electrode area becomes homogenized each time a drop breaks from the capillary. Thus, a concentration gradient is developed only during the brief lifetime of the drop. As we have noted, current changes due to an increase in surface area occur during each lifetime. Changes in the diffusion gradient dc/dx also occur during this period, but these changes are entirely reproducible, leading to currents that are also highly reproducible.

TABLE 21-1 Effect of Complexing Agents on Polarographic Half-Wave Potentials at the Dropping Mercury Electrode

Ion	Noncomplexing Media	1 M KCN	1 M KCl	1 M NH_3, 1 M NH_4Cl
Cd^{2+}	−0.59	−1.18	−0.64	−0.81
Zn^{2+}	−1.00	NR*	−1.00	−1.35
Pb^{2+}	−0.40	−0.72	−0.44	−0.67
Ni^{2+}	−1.01	−1.36	−1.20	−1.10
Co^{2+}	—	−1.45	−1.20	−1.29
Cu^{2+}	+0.02	NR*	+0.04 and −0.22†	−0.24 and −0.51†

*No reduction occurs before involvement of the supporting electrolyte.

†Reduction occurs in two steps having different electrode potentials:

$$Cu^{2+} + 2Cl^- + e^- \rightleftharpoons CuCl_2^-$$

$$CuCl_2^- + Hg + e^- \rightleftharpoons Cu(Hg) + 2\,Cl^-$$

Effect of Complex Formation on Polarographic Waves

We have already seen (Section 15C-6) that the potential for the oxidation or reduction of a metallic ion is greatly affected by the presence of species that form complexes with that ion. It is not surprising, therefore, that similar effects are observed with polarographic half-wave potentials. The data in Table 21-1 show clearly that the half-wave potential for the reduction of a metal complex is generally more negative than that for reduction of the corresponding simple metal ion. In fact, this negative shift in potential permits the elucidation of the composition of the complex ion and the determination of its formation constant *provided that the electrode reaction is reversible.* Thus, for the reactions

$$M^{n+} + Hg + ne^- \rightleftharpoons M(Hg)$$

and

$$M^{n+} + xA^- \rightleftharpoons MA_x^{(n-x)+}$$

You can evaluate the formula of a complex by this method only if the electrode reaction is reversible.

Lingane[8] derived the following relationship between the molar concentrations of the ligand c_L and the shift in half-wave potential brought about by its presence:

$$(E_{1/2})_c - E_{1/2} = -\frac{0.0592}{n}\log K_f - \frac{0.0592x}{n}\log c_L \qquad \textbf{(21-15)}$$

where $(E_{1/2})_c$ and $E_{1/2}$ are the half-wave potentials for the complexed and uncomplexed cations, respectively, K_f is the formation constant for the complex, and x is the molar combining ratio of complexing agent to cation.

Equation 21-15 makes it possible to evaluate the formula for the complex. Thus, a plot of the half-wave potential against log c_L for several ligand concentra-

[8]J. J. Lingane, *Chem. Rev.*, **1941**, *29*, 1.

tions gives a straight line, the slope of which is $0.0592x/n$. If n is known, the combining ratio of ligand to metal ion x is readily calculated. Equation 21-15 can then be employed to calculate K_f.

Effect of pH on Polarograms

Most organic electrode processes and a few inorganic ones involve hydrogen ions, the typical reaction being represented as

$$R + nH^+ + ne^- \rightleftharpoons RH_n$$

where R and RH_n are the oxidized and reduced forms of the reactant species. Half-wave potentials for compounds of this type are therefore markedly pH-dependent. Furthermore, alteration of the pH may result in a change in the reaction product.

It should be emphasized that an electrode process that consumes or produces hydrogen ions will alter the pH of the solution *at the electrode surface,* often drastically, unless the solution is well-buffered. These changes affect the reduction potential of the reaction and cause drawn-out, poorly defined waves. Moreover, where the electrode process is altered by pH, nonlinearity in the diffusion current/concentration relationship will also be encountered. Thus, good buffering is generally vital for the generation of reproducible half-wave potentials and diffusion currents for organic polarography.

Advantages and Disadvantages of the Dropping Mercury Electrode

In the past, the dropping mercury electrode was the most widely used microelectrode for voltammetry because of its several unusual features. The first is the unusually high overvoltage associated with the reduction of hydrogen ions. As a consequence, metal ions such as zinc and cadmium can be deposited from acidic solution even though their thermodynamic potentials suggest that deposition of these metals without hydrogen formation is impossible. A second advantage is that a new metal surface is generated continuously; thus, the behavior of the electrode is independent of its past history. In contrast, solid metal electrodes are notorious for their irregular behavior, which is related to adsorbed or deposited impurities. A third unusual feature of the dropping electrode, which has already been described, is that reproducible average currents are *immediately* realized at any given potential whether this potential is approached from lower or higher settings.

One serious limitation of the dropping electrode is the ease with which mercury is oxidized; this property severely limits the use of the electrode as an anode. At potentials greater than about +0.4 V, formation of mercury(I) gives a wave that masks the curves of other oxidizable species. In the presence of ions that form precipitates or complexes with mercury(I), this behavior occurs at even lower potentials. For example, in Figure 21-18, the beginning of an anodic wave can be seen at 0 V due to the reaction

$$2Hg + 2Cl^- \longrightarrow Hg_2Cl_2(s) + 2e^-$$

Incidentally, this anodic wave can be used for the determination of chloride ion.

Another important disadvantage of the dropping mercury electrode is the non-faradaic residual or charging current, which limits the sensitivity of the classical method to concentrations of about 10^{-5} M. At lower concentrations, the residual current is likely to be greater than the diffusion current, a situation that prohibits accurate measurement of the latter. As will be shown later, methods are now available for enhancing detection limits by one to two orders of magnitude.

The detection limit for classical polarography is about 10^{-5} M.

Finally, the dropping mercury electrode is cumbersome to use and tends to malfunction as a result of clogging.

21C PULSE POLAROGRAPHIC AND VOLTAMMETRIC METHODS

By the 1960s, linear-scan polarography ceased to be an important analytical tool in most laboratories. The reason for the decline in use of this once popular technique was not only the appearance of several more convenient spectroscopic methods but also the inherent disadvantages of the method, including slowness, inconvenient apparatus, and, particularly, poor detection limits. These limitations were largely overcome by pulse methods and the development of electrodes such as those shown in Figure 21-5d. We shall discuss the two most important pulse techniques, *differential pulse polarography* and *square-wave polarography*. Both methods have also been applied with electrodes other than the dropping mercury electrode, in which case the procedures are termed differential pulse and square-wave voltammetry.[9]

21C-1 Differential Pulse Polarography

Figure 21-19 shows the two most common excitation signals that are employed in commercial instruments for differential pulse polarography. The first (19a), which is used in analog instruments, is obtained by superimposing a periodic pulse on a linear scan. The second (19b), which is ordinarily used in digital instruments, involves combination of a pulse output with a staircase signal. In

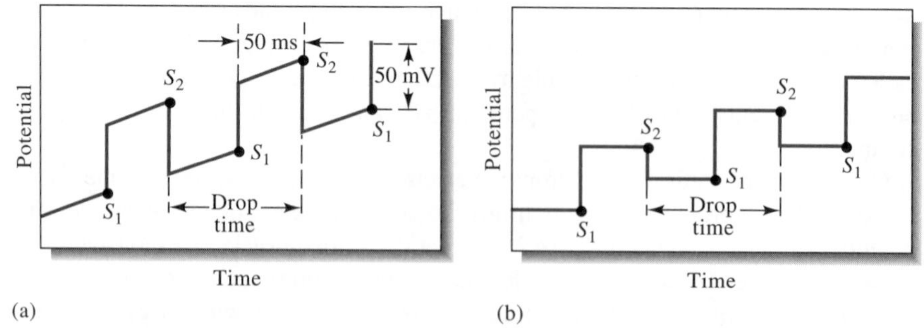

Figure 21-19
Excitation signals for differential pulse polarography.

[9]For review on pulse and square-wave voltammetry, see G. N. Eccles, *Crit. Rev. Anal. Chem.*, **1991**, *22*, 345; J. Osteryoung, *Acc. Chem. Res.*, **1993**, *26*, 77.

either case, a 50 mV pulse is applied during the last 50 ms of the lifetime of the mercury drop. Here again, to synchronize the pulse with the drop, the latter is detached at an appropriate time by a mechanical means.

As shown in Figure 21-19b, two current measurements are made alternately—one (at S_1) that is 16.7 ms prior to the dc pulse and one for 16.7 ms (at S_2) at the end of the pulse. The *difference in current per pulse* (Δi) is recorded as a function of the linearly increasing voltage. A differential curve results consisting of a peak (see Figure 21-20) the height of which is directly proportional to concentration. For a reversible reaction, the peak potential is approximately equal to the standard potential for the half-reaction.

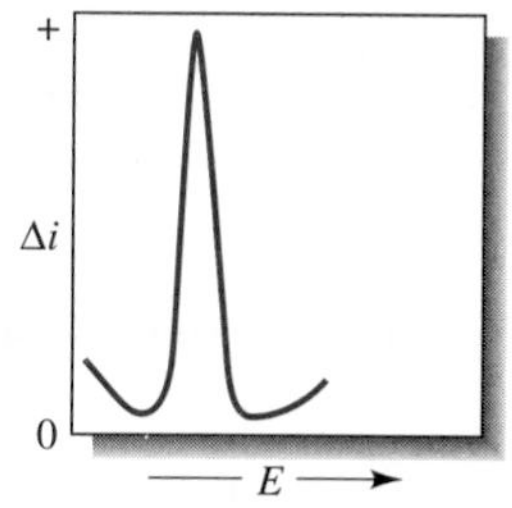

Figure 21-20
Voltammogram for a differential pulse polarography experiment. Here $\Delta i = E_{S_2} - E_{S_1}$. (See Figure 21-19.)

Derivative polarograms yield peaks that are convenient for qualitative identification of analytes.

Detection limits for differential pulse polarography are two to three orders of magnitude lower than for classical polarography.

One advantage of the derivative-type polarogram is that individual peak maxima can be observed for substances with half-wave potentials differing by as little as 0.04 to 0.05 V; in contrast, classical and normal pulse polarography requires a potential difference of about 0.2 V for resolution of waves. More important, however, differential pulse polarography increases the sensitivity of the polarographic method. This enhancement is illustrated in Figure 21-21. Note that a classical polarogram for a solution containing 180 ppm of the antibiotic tetracycline gives two barely discernible waves; differential pulse polarography, in contrast, provides well-defined peaks at a concentration level that is 2×10^{-3} that for the classic wave, or 0.36 ppm. Note also that the current scale for Δi is in nanoamperes. Generally, detection limits with differential pulse polarography are two to three orders of magnitude lower than those for classical polarography and lie in the range of 10^{-7} to 10^{-8} M.

The greater sensitivity of differential pulse polarography can be attributed to two sources. The first is an enhancement of the faradaic current; the second is a decrease in the nonfaradaic charging current. To account for the former, let us consider the events that must occur in the surface layer around an electrode as the potential is suddenly increased by 50 mV. If a reactive species is present in

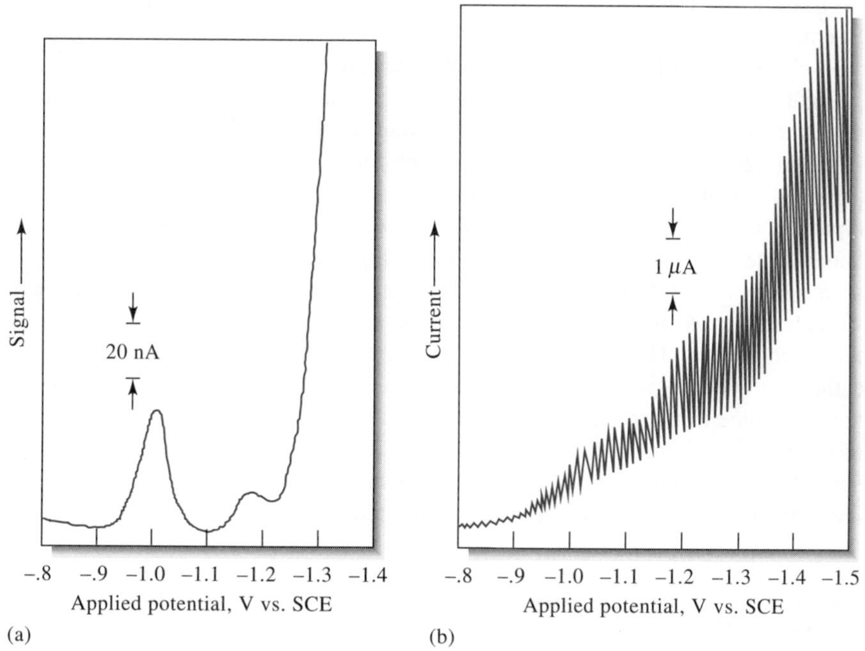

Figure 21-21
(a) Differential pulse polarogram: 0.36 ppm tetracycline · HCl in 0.1 M acetate buffer, pH 4, PAR Model 174 polarographic analyzer, dropping mercury electrode, 50-mV pulse amplitude, 1-s drop. (b) DC polarogram: 180 ppm tetracycline · HCl in 0.1 M acetate buffer, pH 4, similar conditions. (Reprinted with permission from J. B. Flato, *Anal. Chem.*, **1972,** *44* (11), 75A. Copyright 1972 American Chemical Society.)

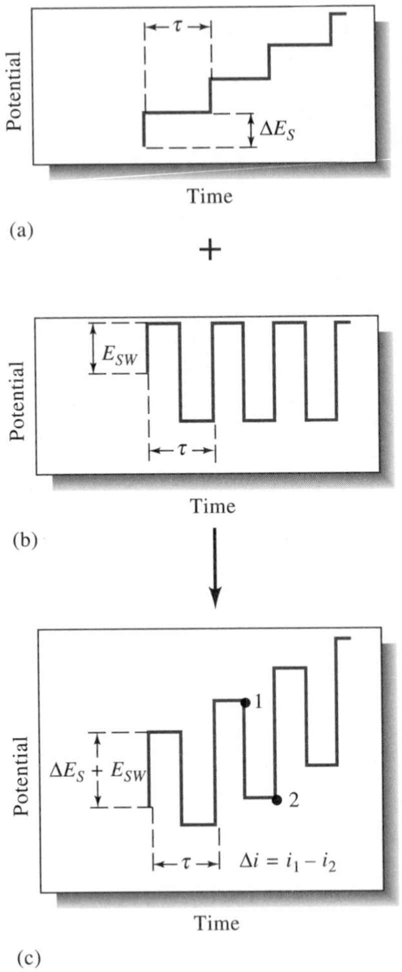

Figure 21-22
Generation of a square-wave voltammetry excitation signal. The staircase signal in (a) is added to the pulse train in (b) to give the square-wave excitation signal in (c). The current response Δi is equal to the current at potential 1 minus the current at potential 2.

Multiple scans from multiple drops can be summed to improve the signal-to-noise ratio of a square-wave voltammogram.

this layer, there will be a surge of current that lowers the reactant concentration to that demanded by the new potential. As the equilibrium concentration for that potential is approached, however, the current decays to a level just sufficient to counteract diffusion, that is, to the diffusion-controlled current. In classical polarography, the initial surge of current is not observed because the time scale of the measurement is long relative to the lifetime of the momentary current. On the other hand, in pulse polarography, the current measurement is made before the surge has completely decayed. Thus, the current measured contains both a diffusion-controlled component and a component that has to do with reducing the surface layer to the concentration demanded by the Nernst expression; the total current is typically several times larger than the diffusion current. It should be noted that when the drop is detached, the solution again becomes homogeneous with respect to the analyte. Thus, at any given voltage, an identical current surge accompanies each voltage pulse.

When the potential pulse is first applied to the electrode, a surge in the nonfaradaic current also occurs as the charge on the drop increases. This current, however, decays exponentially with time and approaches zero near the end of the life of a drop when its surface area is changing only slightly (see Figure 21-17). By measuring currents at this time only, the nonfaradaic residual current is greatly reduced, and the signal-to-noise ratio is larger. Enhanced sensitivity results.

Reliable instruments for differential pulse polarography are now available commercially at reasonable cost. The method has thus become the most widely used analytical polarographic procedure.

21C-2 Square-Wave Polarography and Voltammetry[10]

Square-wave polarography is a type of pulse polarography that offers the advantage of great speed and high sensitivity. An entire voltammogram is obtained in less than 10 ms. With a dropping mercury electrode, the scan is performed during the last few milliseconds of the life of a single drop, when the charging current is essentially constant. Square-wave voltammetry has also been used with hanging drop electrodes and with chromatographic detectors.

Figure 21-22c shows the excitation signal in square-wave voltammetry, which is obtained by superimposing the pulse train shown in 22b onto the staircase signal in 22a. The length of each step of the staircase and the period of the pulses (τ) are identical and usually about 5 ms. The potential step of the staircase ΔE_S is typically 10 mV. The magnitude of the pulse $2E_{SW}$ is often 50 mV. Operating under these conditions, which correspond to a pulse frequency of 200 Hz, a 1 V scan requires 0.5 s. For a reversible reduction reaction, the size of a pulse is great enough that oxidation of the product formed on the forward pulse occurs during the reverse pulse. Thus, as shown in Figure 21-23, the forward pulse produces a cathodic current i_1, whereas the reverse pulse gives an anodic current i_2. Usually the difference in these currents, Δi, is plotted to give voltammograms. This difference is directly proportional to concentration; the potential of the peak corresponds to the polarographic half-wave potential. Because of the speed of the measurement, it is possible and practical to increase the precision of analyses by

[10]For further information on square-wave voltammetry, see J. G. Osteryoung and R. A. Osteryoung, *Anal. Chem.*, **1985**, *57*, 101A.

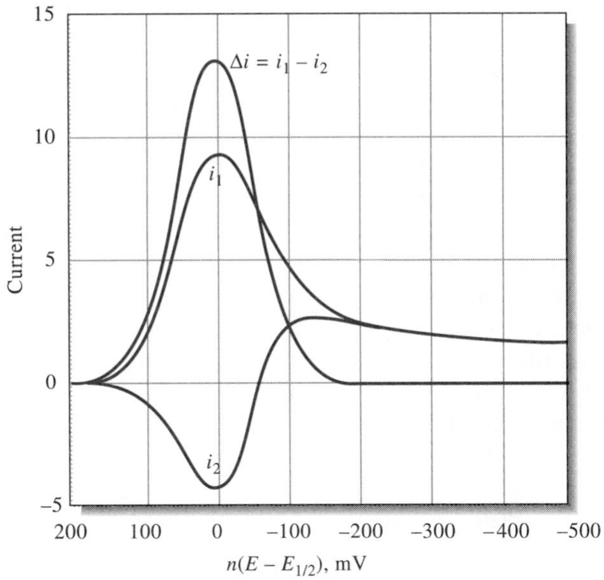

Figure 21-23
Current response for a reversible reaction to excitation signal in Figure 21-21c. i_1, forward current; i_2, reverse current; $i_1 - i_2$, current difference. (From J. J. O'Dea, J. Osteryoung, and R. A. Osteryoung, *Anal. Chem.*, **1981,** *53*, 695. With permission. Copyright 1981 American Chemical Society.)

signal averaging data from several voltammetric scans. Detection limits for square-wave voltammetry are reported to be 10^{-7} to 10^{-8} M.

Detection limits for both differential pulse polarography and square-wave voltammetry are 10^{-7} to 10^{-8} M.

Commercial instruments for square-wave voltammetry have recently become available from several manufacturers, and as a consequence, it seems likely that this technique will gain considerable use for analysis of inorganic and organic species. It has also been suggested that square-wave voltammetry can be used in detectors for HPLC.

21C-3 Applications of Pulse Polarography

In the past, linear-scan polarography was used for the quantitative determination of a wide variety of inorganic and organic species including molecules of biological and biochemical interest. Currently, pulse methods have supplanted the classical method almost completely because of their greater sensitivity, convenience, and selectivity. Generally, quantitative applications are based upon calibration curves in which peak heights are plotted as a function of analyte concentration. In some instances, the standard addition method is employed in lieu of calibration curves. In either case, it is essential that the composition of standards resemble as closely as possible the composition of the sample as to both electrolyte concentrations and pH. When this is done, relative precisions and accuracies in the 1% to 3% range can often be realized.

Inorganic Applications

The polarographic method is widely applicable to the analysis of inorganic substances. Most metallic cations, for example, are reduced at the dropping electrode. Even the alkali and alkaline-earth metals are reducible, provided the supporting electrolyte does not react at the high potentials required; here, the tetraalkyl ammonium halides are useful electrolytes because of their high reduction potentials.

The successful polarographic determination of cations frequently depends upon the supporting electrolyte that is used. To aid in this selection, tabular compilations of half-wave potential data are available.[11] The judicious choice of anion often enhances the selectivity of the method. For example, with potassium chloride as a supporting electrolyte, the waves for iron(III) and copper(II) interfere with one another; in a fluoride medium, however, the half-wave potential of the former is shifted by about − 0.5 V, while that for the latter is altered by only a few hundredths of a volt. The presence of fluoride thus results in the appearance of well-separated waves for the two ions.

The polarographic method is also applicable to the analysis of such inorganic anions as bromate, iodate, dichromate, vanadate, selenite, and nitrite. In general, polarograms for these substances are affected by the pH of the solution because the hydrogen ion is a participant in their reduction. As a consequence, strong buffering to some fixed pH is necessary to obtain reproducible data.

21C-4 Organic Polarographic Analysis

The following organic functional groups produce one or more polarographic waves.

1. Carbonyl groups
2. Certain carboxylic acids
3. Most peroxides and epoxides
4. Nitro, nitroso, amine oxide, and azo groups
5. Most organic halogen groups
6. Carbon/carbon double bonds
7. Hydroquinones and mercaptans

Almost from its inception, the polarographic method has been used for the study and analysis of organic compounds, with many papers being devoted to this subject. Several common functional groups are reduced at the dropping electrode, thus making possible the determination of a wide variety of organic compounds.[12]

In general, the reactions of organic compounds at a microelectrode are slower and more complex than those for inorganic species. Consequently, theoretical interpretation of the data is more difficult and often impossible; moreover, a much stricter adherence to detail is required for quantitative work. Despite these handicaps, organic polarography has proved fruitful for the determination of structure, the quantitative analysis of mixtures, and occasionally the qualitative identification of compounds.

21D STRIPPING METHODS

Stripping methods encompass a variety of electrochemical procedures having a common, characteristic initial step.[13] In all of these procedures, the analyte is first deposited on a microelectrode, usually from a stirred solution. After an accurately measured period, the electrolysis is discontinued, the stirring is stopped, and the deposited analyte is determined by one of the voltammetric procedures that have been described in the previous section. During this second step in the analysis, the analyte is redissolved or stripped from the microelectrode; hence the name at-

---

[11]For example, see *Handbook of Analytical Chemistry,* L. Meites, Ed. New York: McGraw-Hill, 1963; D. T. Sawyer and J. L. Roberts, *Experimental Electrochemistry for Chemists.* New York: Wiley, 1974.

[12]For a detailed discussion of organic polarographic analysis, see P. Zuman, *Organic Polarographic Analysis.* Oxford: Pergamon Press, 1964 and *Topics in Organic Polarography,* P. Zuman, Ed. New York: Plenum Press, 1970; W. F. Smyth, *Polarography of Molecules of Biological Significance.* New York: Academic Press, 1979.

[13]For detailed discussions of stripping methods, see J. Wang, *Stripping Analysis.* Deerfield Beach, FL: VCH Publishers, 1985; A. M. Bond, *Modern Polarographic Methods in Analytical Chemistry,* Chapter 9. New York: Marcel Dekker, 1980.

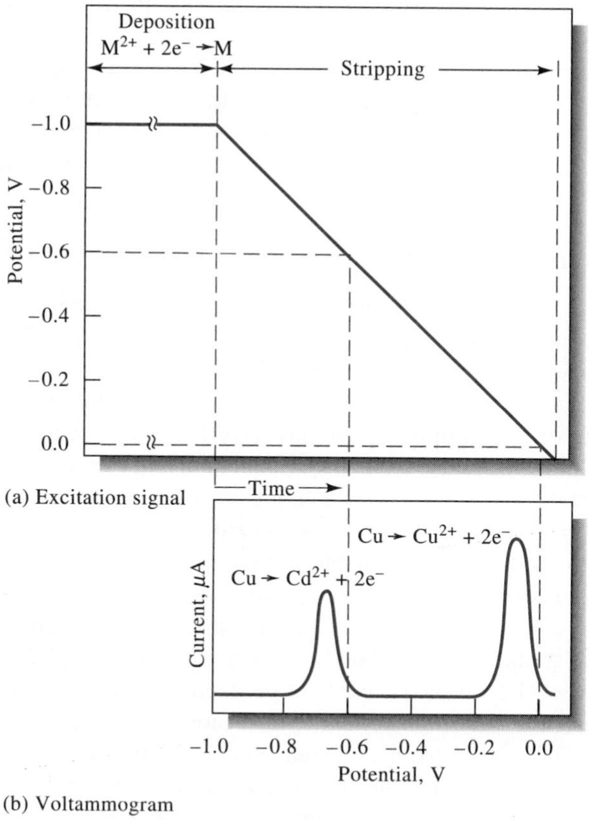

Figure 21-24
(a) Excitation signal for stripping determination of Cd^{2+} and Cu^{2+}.
(b) Voltammogram.

tached to these methods. In *anodic stripping methods,* the microelectrode behaves as a cathode during the deposition step and an anode during the stripping step, with the analyte being oxidized back to its original form. In a *cathodic stripping method,* the microelectrode behaves as an anode during the deposition step and a cathode during stripping. The deposition step amounts to an electrochemical preconcentration of the analyte; that is, the concentration of the analyte in the surface of the microelectrode is far greater than it is in the bulk solution.

Figure 21-24a illustrates the voltage excitation program that is followed in an anodic stripping method for determining cadmium and copper in an aqueous solution of these ions. A linear-scan voltammetric method is used to complete the analysis. Initially, a constant cathodic potential of about − 1 V is applied to the microelectrode, which causes both cadmium and copper ions to be reduced and deposited as metals. The electrode is maintained at this potential for several minutes until a significant amount of the two metals has accumulated at the electrode. The stirring is then stopped for perhaps 30 s while the electrode is maintained at − 1 V. The potential of the electrode is then decreased linearly to less negative values while the current in the cell is recorded as a function of time, or potential. Figure 21-24b shows the resulting voltammogram. At a potential somewhat more negative than − 0.6 V, cadmium starts to be oxidized, causing a sharp increase in the current. As the deposited cadmium is consumed, the current peaks and then decreases to its original level. A second peak for oxidation of the copper is then observed when the potential has decreased to approximately

− 0.1 V. The heights of the two peaks are proportional to the weights of deposited metal.

Stripping methods are of prime importance in trace work because the concentrating aspects of the electrolysis permit the determination of minute amounts of an analyte with reasonable accuracy. Thus, the analysis of solutions in the 10^{-6} to 10^{-9} M range becomes feasible by methods that are both simple and rapid.

21D-1 Electrodeposition Step

A major advantage of stripping analysis is its capability for electrochemically preconcentrating the analyte prior to the measurement step.

Ordinarily, only a fraction of the analyte is deposited during the electrodeposition step; hence, quantitative results depend not only upon control of electrode potential but also upon such factors as electrode size, length of deposition, and stirring rate for both the sample and standard solutions employed for calibration.

Microelectrodes for stripping methods have been formed from a variety of materials including mercury, gold, silver, platinum, and carbon in various forms. The most popular electrode is the *hanging mercury drop electrode* (HMDE), which consists of a single drop of mercury in contact with a platinum wire. Hanging drop electrodes are available from several commercial sources. These electrodes often consist of a microsyringe with a micrometer for exact control of drop size. The drop is then formed at the tip of a capillary by displacement of the mercury in the syringe-controlled delivery system (see Figure 21-5b). The system shown in Figure 21-5d is also capable of producing a hanging drop electrode.

To carry out the determination of a metal ion by anodic stripping, a fresh hanging drop is formed, stirring is begun, and a potential is applied that is a few tenths of a volt more negative than the half-wave potential for the ion of interest. Deposition is allowed to occur for a carefully measured period that can vary from a minute or less for 10^{-7} M solutions to 30 min or longer for 10^{-9} M solutions. It should be emphasized that these times seldom result in complete removal of the analyte. The electrolysis period is determined by the sensitivity of the method ultimately employed for completion of the analysis.

21D-2 Voltammetric Completion of the Analysis

The analyte collected in the hanging drop electrode can be determined by any of several voltammetric procedures. For example, in the linear anodic scan procedure, described at the beginning of this section, stirring is discontinued for perhaps 30 s after termination of the deposition. The voltage is then decreased at a linear fixed rate from its original cathodic value, and the resulting anodic current is recorded as a function of the applied voltage. This linear scan produces a curve of the type shown in Figure 21-24b. Analyses of this type are generally based on calibration with standard solutions of the cations of interest. With reasonable care, analytical precisions of about 2% relative can be obtained.

Most of the other voltammetric procedures described in the previous section have also been applied to the stripping step. The most widely used of these appears to be an anodic differential pulse technique. Often, narrower peaks are produced by this procedure, which is desirable when mixtures are to be analyzed. Another method of obtaining narrower peaks is to use a mercury film electrode. Here, a thin mercury film is electrodeposited on an inert microelectrode such as glassy carbon. Usually, the mercury deposition is carried out simultaneously with

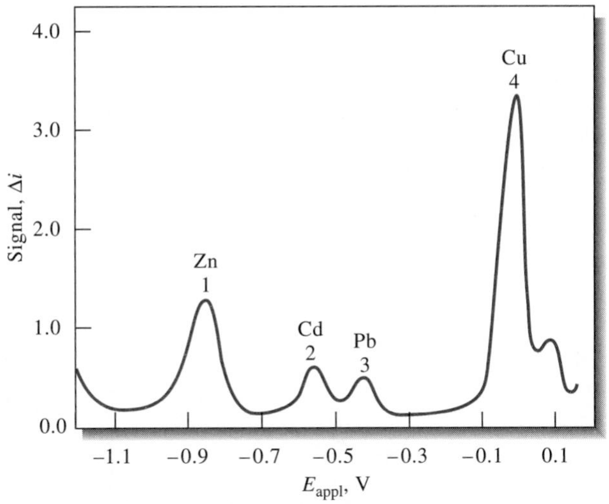

Figure 21-25
Differential-pulse anodic stripping voltammogram of 25 ppm zinc, cadmium, lead, and copper. (Reprinted with permission from W. M. Peterson and R. V. Wong, *Amer. Lab.*, **1981,** *13*(11), 116. Copyright 1981 by International Scientific Communications, Inc.)

the analyte deposition. Because the average diffusion path length from the film to the solution interface is much shorter than that in a drop of mercury, escape of the analyte is hastened; the consequence is narrower and larger voltammetric peaks, which leads to greater sensitivity and better resolution of mixtures. On the other hand, the hanging drop electrode appears to give more reproducible results, especially at higher analyte concentrations. Thus, for most applications the hanging drop electrode is employed. Figure 21-25 is a differential-pulse anodic stripping polarogram for a mixture of cations present at concentrations of 25 ppb, showing good resolution and adequate sensitivity for many purposes.

Many other variations of the stripping technique have been developed. For example, a number of cations have been determined by electrodeposition on a platinum cathode. The quantity of electricity required to remove the deposit is then measured coulometrically. Here again, the method is particularly advantageous for trace analyses. Cathodic stripping methods for the halides have also been developed. Here, the halide ions are first deposited as mercury(I) salts on a mercury anode. Stripping is then performed by a cathodic current.

21D-3 Adsorption Stripping Methods

Adsorption stripping methods are quite similar to the anodic and cathodic stripping methods we have just considered. Here, a microelectrode, most commonly a hanging mercury drop electrode, is immersed in a stirred solution of the analyte for several minutes. Deposition of the analyte then occurs by physical adsorption on the electrode surface rather than by electrolytic deposition. After sufficient analyte has accumulated, the stirring is discontinued and the deposited material determined by linear-scan or pulsed voltammetric measurements. Quantitative information is based upon calibration with standard solutions that are treated in the same way as samples.

Many organic molecules of clinical and pharmaceutical interest have a strong tendency to be adsorbed from aqueous solutions onto a mercury surface, particularly if the surface is maintained at about − 0.4 V (vs. SCE) where the charge on

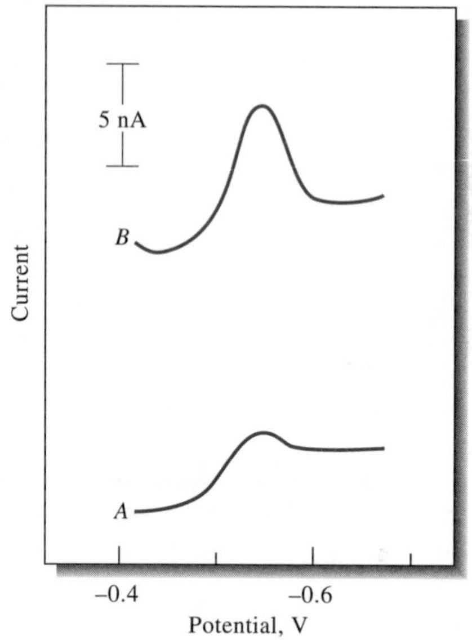

Figure 21-26

Differential pulse voltammogram for 5×10^{-10} M riboflavin. Adsorptive preconcentrations for 5 (*A*) and 30 (*B*) min at -0.2 V. (From J. Wang, *Amer. Lab.*, **1985** (5), 43. Copyright 1985 by International Scientific Communications, Inc.)

the mercury is zero (see page 483). With good stirring, adsorption is rapid, and only 1 to 5 min are required to accumulate sufficient analyte for analysis from 10^{-7} M solutions and 10 to 20 min for 10^{-9} M solutions. Figure 21-26 illustrates the sensitivity of differential pulse adsorptive stripping voltammetry when it is applied to the determination of riboflavin in a 5×10^{-10} M solution. Many other examples of this type can be found in the recent literature.

Adsorptive stripping voltammetry has also been applied to the determination of a variety of inorganic cations at very low concentrations. In these applications the cations are generally complexed with surface-active complexing agents, such as dimethylglyoxime, catechol, and bipyridine. Detection limits in the 10^{-10} to 10^{-11} M range have been reported.

21E VOLTAMMETRY WITH ULTRAMICROELECTRODES

During the last decade, a number of voltammetric studies have been carried out with microelectrodes that have dimensions that are smaller by an order of magnitude or more than the microelectrodes we have described so far. The electrochemical behavior of these tiny electrodes is significantly different from classical microelectrodes and appears to offer advantages in certain analytical applications.[14] Such electrodes have sometimes been called *microscopic electrodes,* or *ultramicroelectrodes,* to distinguish them from classical microelectrodes. The dimensions of such electrodes are generally smaller than about 20 μm and may be as small as a few tenths of a micrometer. These miniature microelectrodes take several forms. The most common is a planar electrode formed by sealing a carbon fiber with a radius of 5 μm or a gold or platinum wire having dimensions from 0.3 to 20 μm into a fine capillary tube; the fiber or wires are then cut flush with the

[14]See R. M. Wightman, *Science*, **1988**, *240*, 415; S. Pons and M. Fleischmann, *Anal. Chem.*, **1987**, *59*, 1391A.

ends of the tubes. Cylindrical electrodes are also used in which a small portion of the wire extends from the end of the tube. Several other forms of these electrodes have also been used.

Generally, the instrumentation used with ultramicroelectrodes is simpler than that shown in Figure 21-2 because there is no need to employ a three-electrode system. The reason that the reference electrode can be dispensed with is that the currents are so very small (in the picoampere to nanoampere range) that the *IR* drop does not distort the voltammetric waves the way microampere currents do.

One of the reasons for the early interest in microscopic microelectrodes was the desire to study the chemical processes going on inside organs of living species, such as in mammalian brains. An approach to this problem was to use electrodes that are small enough to cause no significant alteration in the function of the organ. One of the outcomes from these studies was the realization that ultramicroelectrodes have certain advantages that justify their application to other kinds of analytical problems. These advantages include very small *IR* losses, which make them applicable to solvents having low dielectric constants, such as toluene. Second, capacitive charging currents, which often limit detection with ordinary microelectrodes, are reduced to insignificant proportions as the electrode size is diminished. Third, the rate of mass transport to and from an electrode increases as the size of an electrode diminishes; as a consequence, steady state currents are established in unstirred solutions in less than a microsecond rather than in a millisecond or more as is required with classical microelectrodes. Such high-speed measurements permit the study of intermediates in rapid electrochemical reactions. Undoubtedly, the future will see many more applications of these ultramicroelectrodes.

21F QUESTIONS AND PROBLEMS

21-1. Distinguish between
 *(a) voltammetry and polarography.
 (b) linear-scan polarography and pulse polarography.
 *(c) differential-pulse polarography and square-wave polarography.
 (d) a hanging drop mercury electrode and a dropping mercury electrode.
 *(e) a limiting current and a residual current.
 (f) a limiting current and a diffusion current.
 *(g) laminar flow and turbulent flow.
 (h) the standard electrode potential and the half-wave potential for a reversible reaction at a microelectrode.

21-2. Define
 *(a) voltammogram.
 (b) hydrodynamic voltammetry.
 *(c) Nernst diffusion layer.
 (d) mercury film electrode.
 *(e) half-wave potential.

***21-3.** Why is it necessary to buffer solutions in organic voltammetry?

21-4. List the advantages and disadvantages of the dropping mercury electrode compared with platinum or carbon microelectrodes.

***21-5.** Suggest how Equation 21-11 could be employed to determine the number of electrons *n* involved in a reversible reaction at a microelectrode.

21-6. Quinone undergoes a reversible reduction at a dropping mercury electrode. The reaction is

$$Q + 2H^+ + 2e^- \rightleftharpoons H_2Q \qquad E^0 = 0.599\ V$$

 *(a) Assume that the diffusion coefficient for quinone and hydroquinone are approximately the same and calculate the approximate half-wave potential (vs. SCE) for the reduction of hydroquinone at a planar electrode from a stirred solution buffered to a pH of 7.0.
 (b) Repeat the calculation in (a) for a solution buffered to a pH of 5.0.

21-7. What are the sources of the residual current in linear-scan polarography? Why are residual currents smaller with current sampled polarography?

***21-8.** The polarogram for 20.0 mL of solution that was 3.65×10^{-3} M in Cd^{2+} gave a wave for that ion with a diffusion current of 31.3 μA. Calculate the percentage change in concentration of the solution if the current in the limiting current region were allowed to continue for (a) 5 min; (b) 10 min; (c) 30 min.

21-9. Calculate the milligrams of cadmium in each milliliter of sample, based upon the following data (corrected for residual current):

Solution	Sample	0.400 M KCl	2.00×10^{-3} M Cd^{2+}	H_2O	Current, μA
	Volumes Used, mL				
*(a)	15.0	20.0	0.00	15.0	79.7
	15.0	20.0	5.00	10.0	95.9
(b)	10.0	20.0	0.00	20.0	49.9
	10.0	20.0	10.0	10.0	82.3
*(c)	20.0	20.0	0.00	10.0	41.4
	20.0	20.0	5.00	5.00	57.6
(d)	15.0	20.0	0.00	15.0	67.9
	15.0	20.0	10.0	5.00	100.3

21-10. The polarogram that follows is for a solution that was 1.0×10^{-4} M in KBr and 0.1 M in KNO_3. Offer an explanation of the wave that occurs at +0.12 V and the rapid change in current that starts at about +0.48 V. Would the wave at 0.12 V have any analytical applications? Explain.

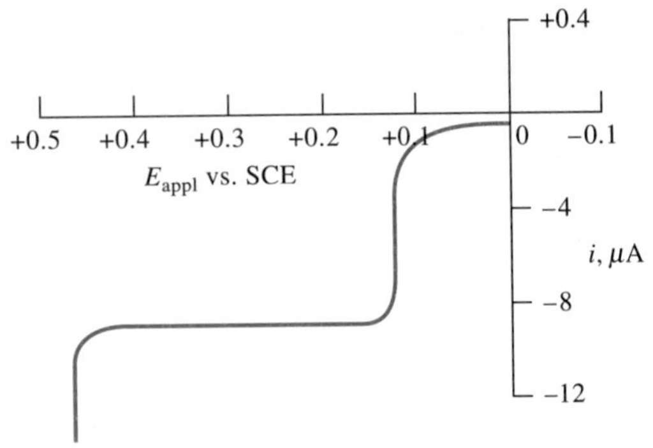

21-11. The following reaction is reversible and has a half-wave potential of −0.349 V when carried out at a dropping mercury electrode from a solution buffered to pH 2.5.

$$Ox + 4H^+ + 4e^- \rightleftharpoons R$$

Predict the half-wave potential at pH *(a) 1.0; (b) 3.5; (c) 7.0.

***21-12.** Why are stripping methods more sensitive than other voltammetric procedures?

21-13. What is the purpose of the electrodeposition step in stripping analysis?

***21-14.** What are the advantages of performing voltammetry with ultramicroelectrodes?

CHAPTER

22

An Introduction to Spectrochemical Methods

Historically, the term *spectroscopy* referred to a branch of science in which light (that is, visible radiation) was resolved into its component wavelengths to produce *spectra,* which were plots of some function of radiant intensity versus wavelength or frequency. In modern times, the meaning of spectroscopy has been broadened to include studies not only with visible radiation but also with other types of electromagnetic radiation, such as X-ray, ultraviolet, infrared, microwave, and radio-frequency radiation. Indeed, current usage extends the meaning of spectroscopy yet further to include techniques that do not even involve electromagnetic radiation. Examples of the last include acoustic, mass, and electron spectroscopy.

Spectroscopy has played a vital role in the development of modern atomic theory. In addition, *spectrochemical methods* have provided perhaps the most widely used tools for the elucidation of the structure of molecular species as well as the quantitative and qualitative determination of both inorganic and organic compounds.

This chapter contains introductory material dealing with the properties of electromagnetic radiation and its various interactions with matter. The next four chapters are then devoted to spectrochemical methods that are based upon emission and absorption of ultraviolet, visible, and infrared radiation by atomic and molecular species.[1]

22A GENERAL PROPERTIES OF ELECTROMAGNETIC RADIATION

Electromagnetic radiation is a type of energy that takes numerous forms, the most easily recognizable being visible light and radiant heat. Less obvious manifestations include gamma-ray, X-ray, ultraviolet, microwave, and radio-frequency radiation. Many of the properties of electromagnetic radiation are conveniently

---

[1] For further study, see E. J. Meehan, in *Treatise on Analytical Chemistry,* P. J. Elving, E. J. Meehan, and I. M. Kolthoff, Eds., 2nd ed., Part I, Vol. 7, Chapters 1–3. New York: Wiley, 1981; J. D. Ingle Jr. and S. R. Crouch, *Analytical Spectroscopy.* Englewood Cliffs, NJ: Prentice-Hall, 1988; J. E. Crooks, *The Spectrum in Chemistry.* New York: Academic Press, 1978.

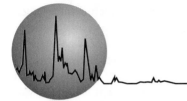

described by means of a classical sinusoidal wave model, which employs such parameters as wavelength, frequency, velocity, and amplitude. In contrast to other wave phenomena, such as sound, electromagnetic radiation requires no supporting medium for its transmission and thus passes readily through a vacuum.

The wave model fails to account for phenomena associated with the absorption and emission of radiant energy. To understand these processes, it is necessary to invoke a particle model in which electromagnetic radiation is viewed as a stream of discrete particles or wave packets of energy called *photons* with the energy of a photon being proportional to the frequency of the radiation. These dual views of radiation as particles and as waves are not mutually exclusive but, rather, complementary. Indeed, the duality is found to apply to the behavior of streams of electrons and other elementary particles such as protons and is completely rationalized by wave mechanics.

22A-1 Wave Properties of Electromagnetic Radiation

A wave in which the direction of displacement is perpendicular to the direction of propagation, such as the wave in Figure 22-1, is called a **transverse wave.**

For many purposes, electromagnetic radiation is conveniently represented as electric and magnetic fields that undergo in-phase, sinusoidal oscillations at right angles to each other and to the direction of propagation. Figure 22-1a is such a representation of a single ray of plane-polarized electromagnetic radiation. Plane-polarized implies that all oscillations of either the electric or the magnetic fields lie within a single plane. Figure 22-1b is a two-dimensional representation of the electric component of the ray in Figure 22-1a. The electric force in this figure is represented as a vector whose length is proportional to the field strength. The abscissa of this plot is either time, as the radiation passes a fixed point in space, or distance, when time is held constant. Throughout this chapter and most of the remaining text, only the electric component of radiation will be considered because the electric field is responsible for most of the phenomena that are of interest to us including transmission, reflection, refraction, and absorption. It is noteworthy, however, that the magnetic component of electromagnetic radiation is responsible for the absorption of radio-frequency waves in nuclear magnetic resonance.

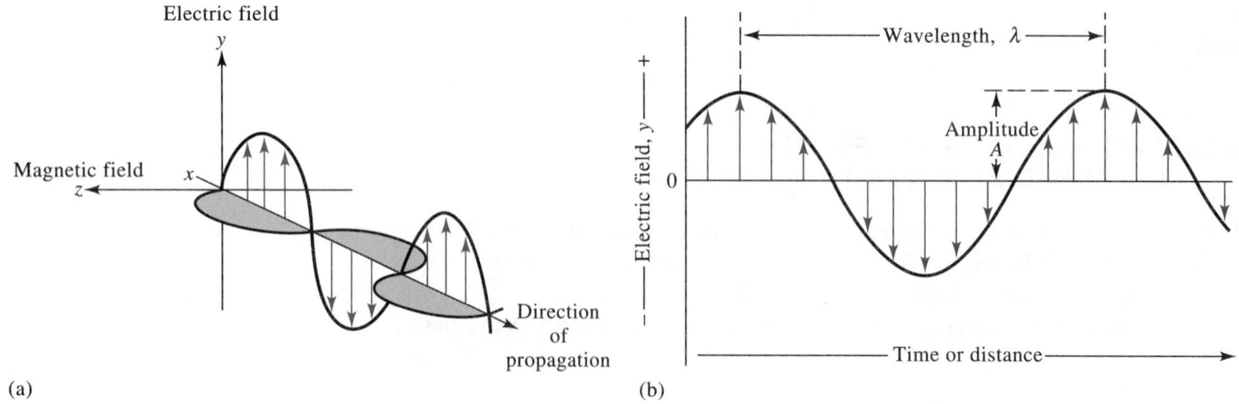

Figure 22-1

Representation of a beam of monochromatic, plane-polarized radiation. (a) Electrical and magnetic fields at right angles to one another and direction of propagation. (b) Two-dimensional representation of the electric vector.

Wave Parameters

In Figure 22-1b, the *amplitude A* of the sinusoidal wave is shown as the length of the electric vector at a maximum in the wave. The time in seconds required for the passage of successive maxima or minima through a fixed point in space is called the *period, p,* of the radiation. The *frequency,* ν, is the number of oscillations of the field that occur per second and is equal to $1/p$. Another parameter of interest is the *wavelength,* λ, which is the linear distance between any two equivalent points on successive waves (for example, successive maxima or minima). Multiplication of the frequency in reciprocal seconds by the wavelength in meters gives the *velocity of propagation,* v_i, in meters per second:

$$v_i = \nu\lambda_i \tag{22-1}$$

It is important to realize that the frequency of a beam of radiation is determined by the source and *remains invariant.* In contrast, the velocity of radiation depends upon the composition of the medium through which it passes. Thus, it is apparent from Equation 22-1 that the wavelength of radiation is also dependent upon the medium. The subscript i in Equation 22-1 emphasizes these dependencies.

In a vacuum, the velocity of radiation becomes independent of wavelength and is at its maximum. This velocity, given the symbol c, has been determined to be 2.99792×10^8 m/s. It is noteworthy that the velocity of radiation in air differs only slightly from c (about 0.03% less); thus, for either air or vacuum, Equation 22-1 can be written to three significant figures as

$$c = \nu\lambda = 3.00 \times 10^8 \text{ m/s} = 3.00 \times 10^{10} \text{ cm/s} \tag{22-2}$$

In any medium containing matter, propagation of radiation is slowed by the interaction between the electromagnetic field of the radiation and the bound electrons in the atoms or molecules present. Since the radiant frequency is invariant and fixed by the source, the wavelength must decrease as radiation passes from a vacuum to some other medium (Equation 22-2). This effect is illustrated in Figure 22-2 for a beam of visible radiation. Note that the wavelength shortens by nearly 200 nm, or more than 30%, as it passes into glass; a reverse change occurs as the radiation again enters air.

The *wavenumber* $\bar{\nu}$, which is defined as the reciprocal of the wavelength in centimeters, is yet another way of describing electromagnetic radiation. The unit for $\bar{\nu}$ is cm^{-1}. Wavenumber is widely used in infrared vibrational spectroscopy. The wavenumber is a useful unit because, in contrast to wavelength, it is directly proportional to the frequency (and thus the energy) of radiation. Thus, we may write

$$\bar{\nu} = k\nu \tag{22-3}$$

where the proportionality constant k is dependent upon the medium and equal to the reciprocal of the velocity (Equation 22-1).

The **amplitude** of an electromagnetic wave is a vector quantity that provides a measure of the electrical or magnetic field strength at a maximum in the wave.

The **period** of an electromagnetic wave is the time in seconds for the passage of successive maxima or minima to pass a point in space.

The **frequency** of an electromagnetic wave is the number of oscillations that occur in one second. It has units of s^{-1}, or Hz.

The unit of frequency is the hertz (Hz), which corresponds to one oscillation per second.

$$1 \text{ Hz} = 1 \text{ s}^{-1}$$

The frequency of a beam of electromagnetic radiation is constant.

To three significant figures, the velocity of radiation in a vacuum and in air is 3.00×10^{10} cm/s.

In contrast to frequency, radiation velocity and wavelength both become smaller as the radiation passes from a vacuum or from air to a denser medium.

TABLE 22-1
Wavelength Units for Various Spectral Regions

Region	Unit	Definition
X-Ray	Angstrom, Å	10^{-10} m
Ultraviolet/ visible	Nanometer, nm	10^{-9} m
Infrared	Micrometer, μm	10^{-6} m

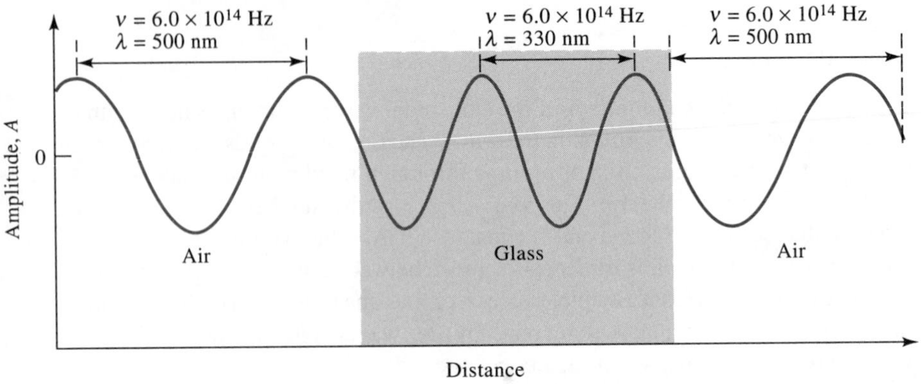

Figure 22-2
Change in wavelength as radiation passes from air into a dense glass and back to air.

The **refractive index** η of a medium is the ratio of the speed of light in a vacuum to the speed in the medium. That is, $\eta_i = c/v_i = 3.00 \times 10^{10}/v_i$, where v_i is the velocity of radiation of wavelength i. Both η_i and v_i are wavelength dependent.

For 510-nm yellow radiation, the refractive index of water at room temperature is 1.33, which means that yellow light passes through water at a rate of $v = c/1.33$, or
$$v = \frac{3.00 \times 10^{10}\ \text{cm/s}}{1.33} = 2.26 \times 10^{10}\ \text{cm/s}.$$

The **wavenumber** $\bar{\nu}$ of electromagnetic radiation is the reciprocal of its wavelength in centimeters. It bears units of cm^{-1}.

The wavenumber of $\bar{\nu}$ in cm^{-1} is generally used to describe infrared radiation. The most useful part of the infrared spectrum for the detection and determination of organic species is from 2.5 to 15 μm in wavelength, which corresponds to a wavenumber range of 4000 to 667 cm^{-1}.

A **photon** is a particle of electromagnetic radiation having zero mass and an energy of $h\nu$.

EXAMPLE 22-1

Calculate the wavenumber of a beam of infrared radiation with a wavelength of 5.00 μm.

$$\bar{\nu} = \frac{1}{5.00\ \mu\text{m} \times 10^{-4}\ \text{cm/}\mu\text{m}} = 2000\ \text{cm}^{-1}$$

Radiant Power and Intensity

The *power P* of radiation is the energy of the beam reaching a given area per second whereas the *intensity I* is the power per unit solid angle. These quantities are related to the square of the amplitude A (see Figure 22-1b). Although it is not strictly correct to do so, power and intensity are often used synonymously.

22A-2 The Particle Properties of Radiation

To understand many of the interactions between radiation and matter, it is necessary to postulate that electromagnetic radiation is made up of packets of energy called *photons* (or *quanta*). The energy of a photon depends upon the frequency of the radiation and is given by

$$E = h\nu \tag{22-4}$$

where h is Planck's constant (6.63×10^{-34} J$\cdot$s). In terms of wavelength and wavenumber,

$$E = \frac{hc}{\lambda} = hc\bar{\nu} \tag{22-5}$$

Note that wavenumber, like frequency, is directly proportional to energy.

EXAMPLE 22-2

Calculate the energy in joules of one photon of the radiation described in Example 22-1.

Applying Equation 22-5, we write

$$E = hc\bar{\nu} = 6.63 \times 10^{-34}(\text{J} \cdot \cancel{\text{s}}) \times 3.00 \times 10^{10} \frac{\cancel{\text{cm}}}{\cancel{\text{s}}} \times 2000 \ \cancel{\text{cm}^{-1}}$$

$$= 3.98 \times 10^{-20} \text{ J}$$

Both frequency and wavenumber are proportional to the energy of a photon.

On occasion we speak of "a mole of photons," meaning 6.02×10^{23} packets of radiation.

22B THE ELECTROMAGNETIC SPECTRUM

As shown in Figure 22-3, the electromagnetic spectrum encompasses an enormous range of wavelengths and frequencies (and thus energies). In fact, the range is so great that logarithmic scales are required. The figure also depicts qualitatively the major spectral regions. The divisions are based upon the methods required to generate and detect the various kinds of radiation. Several overlaps are evident. Note that the visible portion of the spectrum to which the human eye is sensitive is tiny when compared with other spectral regions. It should also be noted that spectrochemical methods that employ not only visible but also ultraviolet and infrared radiation are often called *optical methods* despite the fact that the human eye is sensitive to neither of the latter two types of radiation. This somewhat ambiguous terminology arises from the many common features of instruments for the three spectral regions and the similarities in how we view interactions of the three types of radiation with matter.

Table 22-2 lists the wavelength and wavenumber ranges for the regions of the spectrum that are important for analytical purposes and also gives the names of the various spectroscopic methods associated with each. The last column of the table lists the types of nuclear, atomic, or molecular quantum transitions that serve as the basis for the various spectroscopic techniques.

Optical methods are spectroscopic methods based upon ultraviolet, visible, and infrared radiation.

Recall the order of the colors in the spectrum by the mnemonic **ROY G BIV.**

Red
Orange
Yellow
Green
Blue
Indigo
Violet

The **visible region** of the spectrum extends from about 400 nm to 700 nm.

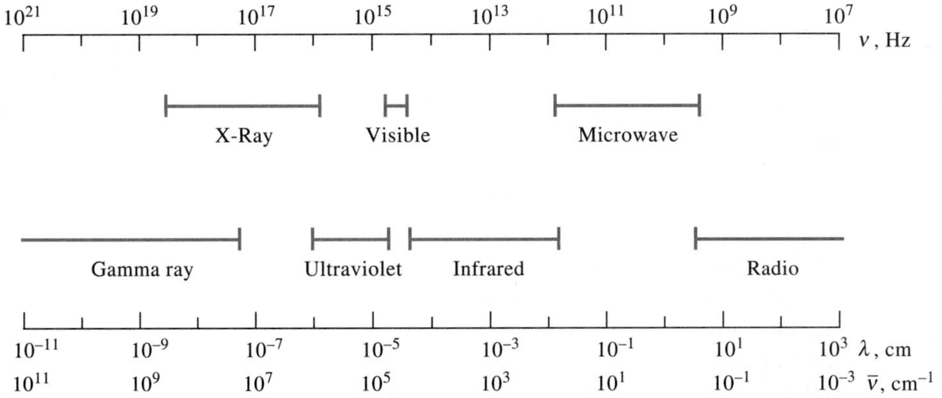

Figure 22-3
Regions of the electromagnetic spectrum.

TABLE 22-2 Common Spectroscopic Methods Based on Electromagnetic Radiation

Type Spectroscopy	Usual Wavelength Range*	Usual Wavenumber Range, cm^{-1}	Type of Quantum Transition
Gamma-ray emission	0.005–1.4 Å	—	Nuclear
X-Ray absorption, emission, fluorescence, and diffraction	0.1–100 Å	—	Inner electron
Vacuum ultraviolet absorption	10–180 nm	1×10^6 to 5×10^4	Bonding electrons
Ultraviolet visible absorption, emission, and fluorescence	180–780 nm	5×10^4 to 1.3×10^4	Bonding electrons
Infrared absorption and Raman scattering	0.78–300 μm	1.3×10^4 to 3.3×10^1	Rotation/vibration of molecules
Microwave absorption	0.75–3.75 mm	13–27	Rotation of molecules
Electron spin resonance	3 cm	0.33	Spin of electrons in a magnetic field
Nuclear magnetic resonance	0.6–10 m	1.7×10^{-2} to 1×10^3	Spin of nuclei in a magnetic field

*1 Å = 10^{-10} m = 10^{-8} cm
1 nm = 10^{-9} m = 10^{-7} cm
1 μm = 10^{-6} m = 10^{-4} cm

22C ABSORPTION OF RADIATION

Attenuate means to weaken.

The lowest energy state of an atom or molecule is called its **ground state.**

Excitation is a process in which a chemical species absorbs thermal, electrical, or radiant energy and is promoted to a higher energy state.

Relaxation is a process in which an excited species gives up its excess energy and returns to a lower energy state.

In spectroscopic nomenclature, *absorption* is a process in which a chemical species in a transparent medium selectively *attenuates* (decreases the intensity of) certain frequencies of electromagnetic radiation. According to quantum theory, every elementary particle (atom, ion, or molecule) has a unique set of energy states, the lowest of which is the *ground state;* at room temperature, most elementary particles are in their ground state. When a photon of radiation passes near an elementary particle, absorption becomes probable if (and only if) the energy of the photon matches *exactly* the energy difference between the ground state and one of the higher energy states of the particle. Under these circumstances, the energy of the photon is transferred to the atom, ion, or molecule, converting it to the higher energy state, which is termed an *excited state.* Excitation of a species M to its excited state M* can be depicted by the equation

$$M + h\nu \longrightarrow M^*$$

After a brief period (10^{-6} to 10^{-9} s), the excited species *relaxes* to its original, or ground, state, transferring its excess energy to other atoms or molecules in the medium. This process, which causes a small rise in temperature of the surroundings, is described by the equation

$$M^* \longrightarrow M + \text{heat}$$

Relaxation may also occur by *photochemical decomposition* of M* to form new species or by the *fluorescent* or *phosphorescent* reemission of radiation. It is important to note that the lifetime of M* is so very short that its concentration at any instant is ordinarily negligible. Furthermore, the amount of thermal energy released during relaxation is usually so small as to be undetectable. Thus, absorption measurements have the advantage of creating minimal disturbance of the system under study.

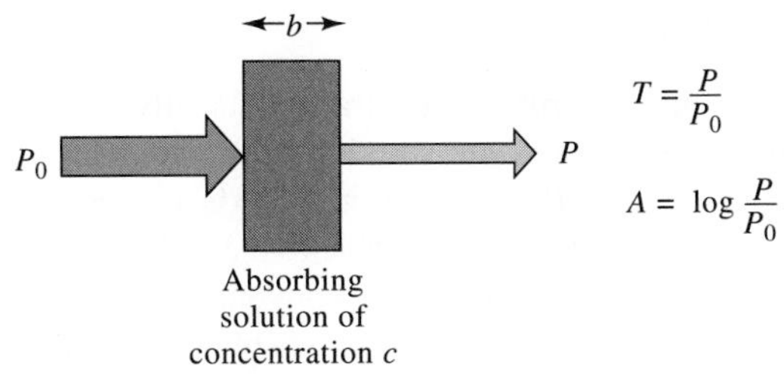

Figure 22-4
Attenuation of a beam of radiation by an absorbing solution.

22C-1 Absorption Spectra

The absorbing characteristics of a species are conveniently described by means of an *absorption spectrum,* which is a plot of some function of the attenuation of a beam of radiation versus wavelength, frequency, or wavenumber. Two terms are commonly employed as quantitative measures of beam attenuation: *transmittance* and *absorbance.*

The **transmittance** *T* of a solution is the fraction of the incident electromagnetic radiation that is transmitted by a sample. The **absorbance** *A* is then log(1/*T*). Note that the absorbance increases with an increase in attenuation of the beam or decreasing transmittance.

Transmittance

Figure 22-4 depicts a beam of parallel radiation before and after it has passed through a medium of thickness *b* cm and a concentration *c* of an absorbing species. As a consequence of interactions between the photons and absorbing particles, the power of the beam is attenuated from P_0 to *P*. The *transmittance T* of the medium is then the fraction of incident radiation transmitted by the medium

$$T = \frac{P}{P_0} \qquad (22\text{-}6)$$

Transmittance is often expressed as a percentage or

$$\%T = \frac{P}{P_0} \times 100\%$$

Absorbance

The absorbance *A* of a medium is defined by the equation

$$A = -\log_{10} T = \log \frac{P_0}{P} \qquad (22\text{-}7)$$

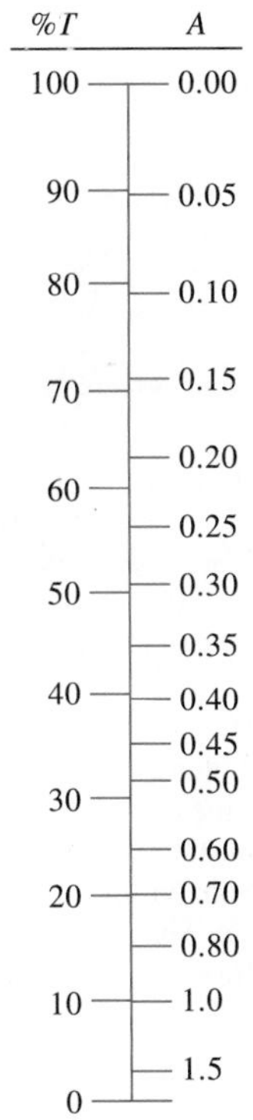

Figure 22-5
A spectrophotometer readout. The readout scales on some spectrophotometers are linear in percent transmittance. As shown below, the absorbance scales must then be logarithmic.

Note that, in contrast to transmittance, the absorbance of a medium increases as attenuation of the beam becomes greater. Figure 22-5 shows a typical readout scale of an instrument for measuring absorption. Note that this scale is linear in transmittance. As shown by Equation 22-7, the absorbance scale is then logarithmic.

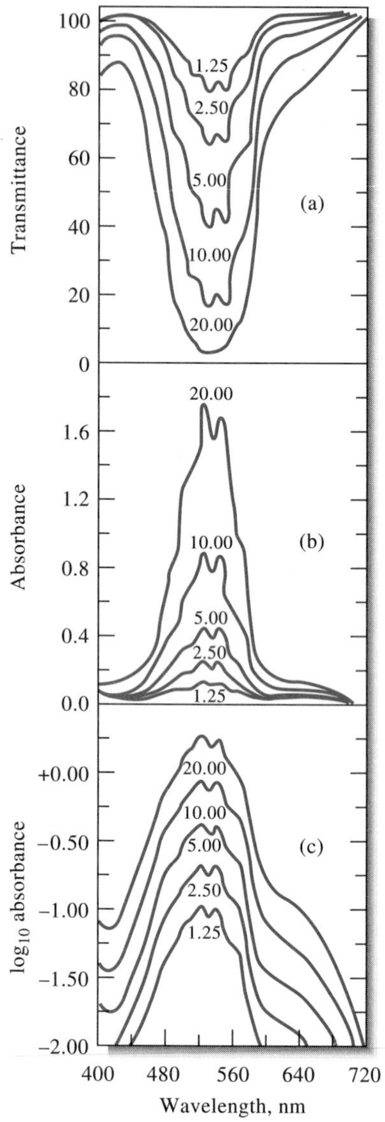

Figure 22-6
Methods for plotting spectral data. The numbers for the curves indicate ppm of $KMnO_4$ in the solution; $b =$ 2.00 cm. (From M. G. Mellon, *Analytical Absorption Spectroscopy,* pp. 104–106. New York: Wiley, 1950. With permission.)

At a higher sensitivity on the absorbance axis, a handful of additional lines would appear in Figure 22-8a. In addition, the three lines shown would split into pairs called *doublets* that differ in wavelength by a few tenths of a nanometer. The cause of this splitting is beyond the scope of this text.

Typical Spectral Plots

Ultraviolet and visible absorption spectra are usually obtained on a gaseous sample of the analyte or on a dilute solution of the analyte in a transparent solvent. As shown in Figure 22-6, the vertical axis of such plots may be percent transmittance, absorbance, or log absorbance. A plot with log A as ordinate leads to loss of spectral detail but is convenient for comparing samples of different concentration, since the curves are displaced equally along the vertical axis.

Figure 22-7 shows a typical infrared spectrum. In this case, the horizontal axis is most commonly wavenumber in cm^{-1}, and the vertical axis is percent transmittance.

22C-2 Atomic Absorption

When a beam of polychromatic ultraviolet or visible radiation passes through a medium containing gaseous atoms, only a few frequencies are attenuated by absorption, and the spectrum consists of a number of very narrow (about 0.005 nm) *absorption lines.* Figure 22-8a is an ultraviolet/visible absorption spectrum for gaseous sodium atoms. The ordinate is *absorbance.*

FEATURE 22-1
Why Is a Red Solution Red?

A solution such as aqueous $Fe(SCN)^{2+}$ is red not because the complex adds red radiation to the solvent, but because this complex absorbs the green component from the incoming white radiation and transmits the red component. Thus, in a colorimetric analysis of iron based upon its thiocyanate complex, the maximum change in absorbance with respect to concentration occurs with green radiation; the absorbance change with red radiation is negligible. In general, then, the radiation used for a colorimetric analysis should be the complementary color of the analyte solution. The table that follows shows this relationship for various parts of the visible spectrum.

The Visible Spectrum

Wavelength Region, nm	Color	Complementary Color
400–435	Violet	Yellow-green
435–480	Blue	Yellow
480–490	Blue-green	Orange
490–500	Green-blue	Red
500–560	Green	Purple
560–580	Yellow-green	Violet
580–595	Yellow	Blue
595–650	Orange	Blue-green
650–750	Red	Green-blue

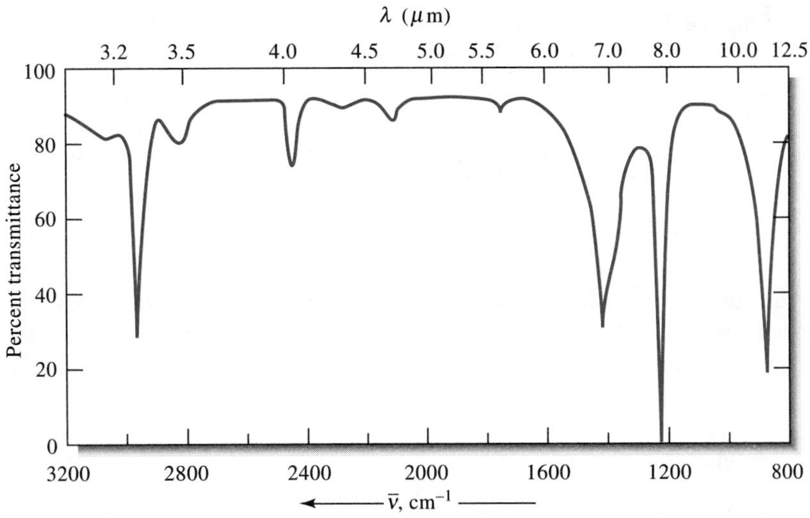

Figure 22-7
Infrared spectrum for CH_3I.

Figure 22-8b is a partial energy-level diagram for sodium that shows the transitions responsible for the three absorption lines in (a). These transitions involve excitation of the single outer electron of sodium from its room temperature or ground state 3*s* orbital to the 3*p*, 4*p*, and 5*p* orbitals. These excitations are brought on by absorption of photons of radiation whose energies *exactly* match the differences in energies between the excited states and the 3*s* ground state.

The **electron volt** (eV) is a unit of energy.

$$1\text{ eV} = 1.60 \times 10^{-19}\text{ J}$$
$$= 3.83 \times 10^{-20}\text{ calories}$$
$$= 1.58 \times 10^{-21}\text{ L} \cdot \text{atm}$$

EXAMPLE 22-3

The energy difference between the 3*p* and the 3*s* orbitals in Figure 22-8b is 2.107 eV. Calculate the wavelength of radiation that would be absorbed in

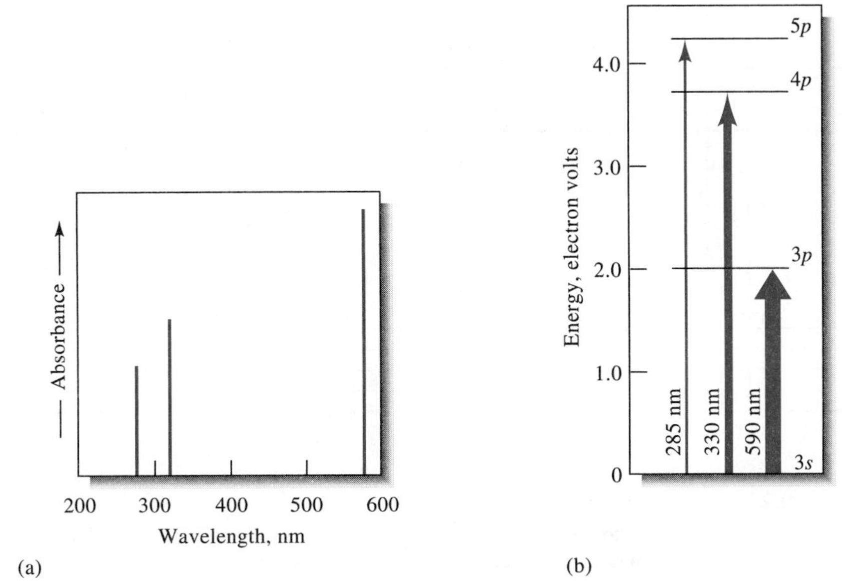

Figure 22-8
(a) Absorption spectrum for sodium vapor. (b) Partial energy-level diagram for sodium, showing the transitions resulting from absorption at 590, 330, and 285 nm.

exciting the 3s electron to the 3p state (1 eV = 1.60 × 10^{-19} J).
Rearranging Equation 22-5 gives

$$\lambda = hc/E$$

$$= \frac{6.63 \times 10^{-14}\ \cancel{J} \cdot \cancel{s} \times 3.00 \times 10^{10}\ \cancel{cm} \cdot \cancel{s^{-1}} \times 10^{7}\ nm \cdot \cancel{cm^{-1}}}{2.107\ \cancel{eV} \times 1.60 \times 10^{-9}\ \cancel{J}/\cancel{eV}}$$

$$= 590\ nm$$

Absorption spectra for the alkali metal atoms are much simpler than those of elements with additional outer electrons. Atomic spectra of the transition metals are particularly complex, with some elements exhibiting several thousand lines.

22C-3 Molecular Absorption

An **electronic transition** involves transfer of an electron from one electronic orbital to another. Both atoms and molecules can undergo this type of transition.

Molecules undergo three types of quantized transitions when excited by ultraviolet, visible, and infrared radiation. For ultraviolet and visible radiation, excitation involves promoting an electron residing in a low-energy molecular or atomic orbital to a higher-energy orbital. We have noted that the energy $h\nu$ of the photon must be exactly the same as the energy difference between the two orbital energies. The transition of an electron between two orbitals is called an *electronic transition,* and the absorption process is called *electronic absorption.*

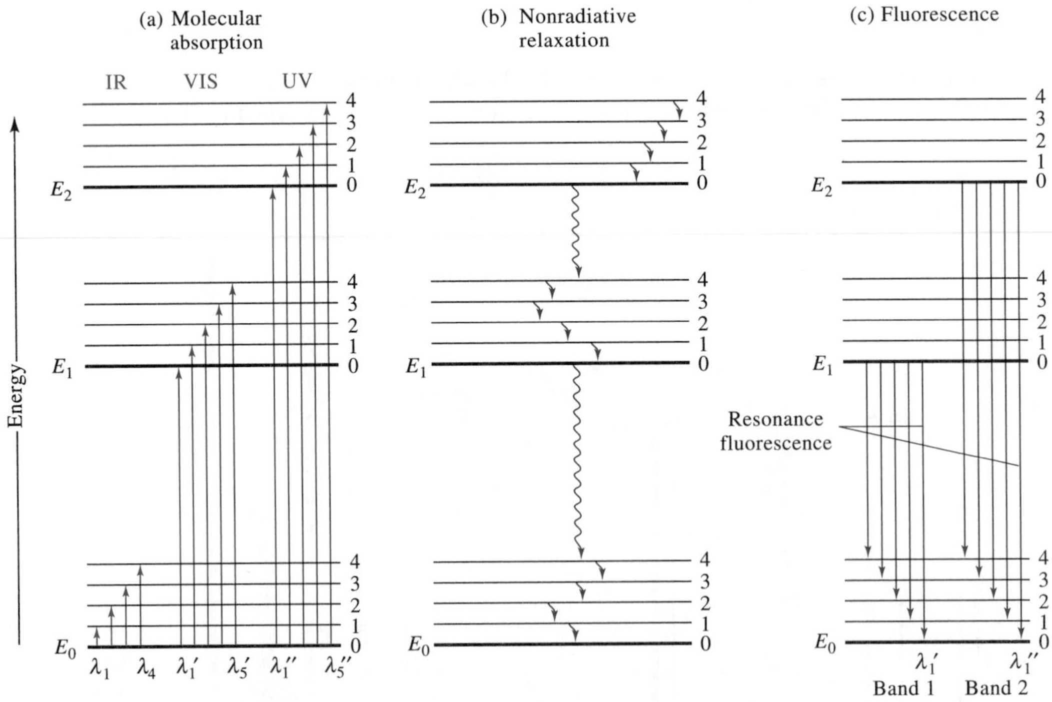

Figure 22-9
Energy-level diagram showing some of the energy changes that occur during absorption, nonradiative relaxation, and fluorescence by a molecular species.

In addition to electronic transitions, molecules exhibit two other types of radiation-induced transitions: *vibrational transitions* and *rotational transitions.* Vibrational transitions come about because a molecule has a multitude of quantized energy levels (or *vibrational states*) associated with the bonds that hold the molecule together.

Figure 22-9a is a partial energy-level diagram that depicts some of the processes that occur when a polyatomic species absorbs infrared, visible, and ultraviolet radiation. The energies E_1 and E_2, two of the several electronically excited states of a molecule, are shown relative to the energy of its ground state E_0. In addition, the relative energies of a few of the many vibrational states associated with each electronic state are indicated by the lighter horizontal lines.

An idea of the nature of vibrational states can be gained by picturing a bond in a molecule as a vibrating spring with atoms attached to both ends. In Figure 22-10a, two types of stretching vibration are shown. With each vibration, atoms first approach and then move away from one another. The potential energy of such a system at any instant depends upon the extent to which the spring is stretched or compressed. For an ordinary spring, the energy of the system varies continuously and reaches a maximum when the spring is fully stretched or fully compressed. In contrast, the energy of a spring system of atomic dimensions can assume only certain discrete energies called vibrational energy levels.

Figure 22-10b shows four other types of molecular vibrations. The energies associated with these vibrational states usually differ from one another and from the energies associated with stretching vibrations.

Vibrational and **rotational transitions** occur with polyatomic species because only this type of species has vibrational and rotational states with different energies.

The **ground state** of an atom or a molecular species is the minimum energy state of the species. At room temperature most species are in their ground state.

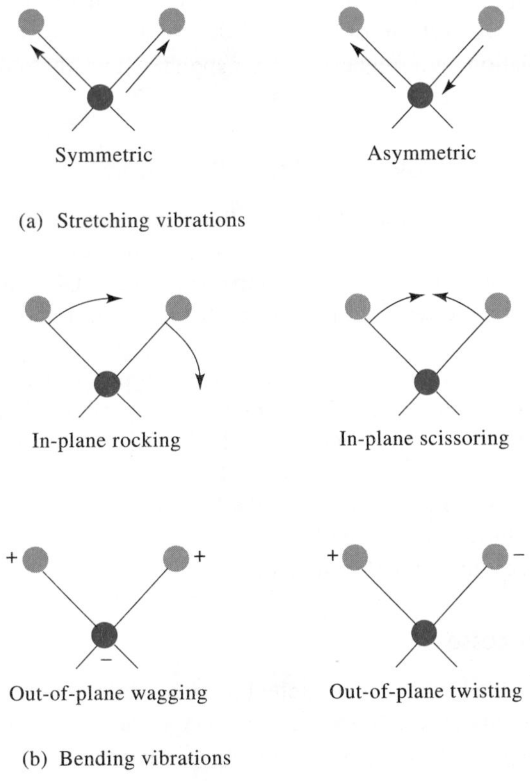

Figure 22-10

Types of molecular vibrations. Note: + indicates motion from the page toward the reader; − indicates motion away from the reader.

Some of the vibrational energy levels associated with each of the electronic states of a molecule are depicted by the lines labeled 1, 2, 3, and 4 in Figure 22-9a (the lowest vibrational levels are labeled 0). Note that the differences in energy among the vibrational states are significantly smaller than among energy levels of the electronic states (typically, an order of magnitude smaller).

Although they are not shown, a molecule has a host of quantized rotational states that are associated with the rotational motion of a molecule around its center of gravity. These rotational energy states are superimposed on each of the vibrational states shown in the energy diagram. The energy differences among these states are smaller than those among vibrational states by an order of magnitude.

The overall energy E associated with a molecule is then given by

$$E = E_{\text{electronic}} + E_{\text{vibrational}} + E_{\text{rotational}}$$

$\Delta E_{\text{electronic}} \approx 10\ \Delta E_{\text{vibrational}} \approx 100\ \Delta E_{\text{rotational}}$

where $E_{\text{electronic}}$ is the energy associated with the electrons in the various outer orbitals of the molecule, $E_{\text{vibrational}}$ is the energy of the molecule as a whole due to interatomic vibrations, and $E_{\text{rotational}}$ accounts for the energy associated with rotation of the molecule about its center of gravity.

Infrared Absorption

Infrared radiation is not sufficiently energetic to cause electronic transitions.

Infrared radiation generally is not sufficiently energetic to cause electronic transitions but can induce transitions in the vibrational and rotational states associated with *the ground electronic state* of the molecule. Four of these transitions are depicted in the lower left part of Figure 22-9a. For absorption to occur, the source has to emit radiation with frequencies corresponding exactly to the energies indicated by the lengths of the four arrows.

Absorption of Ultraviolet and Visible Radiation

The center arrows in Figure 22-9a suggest that the molecules under consideration absorb visible radiation of five wavelengths, thereby promoting electrons to the five vibrational levels of the excited electronic level E_1. Ultraviolet photons that are more energetic are required to produce the absorption indicated by the five arrows to the right.

As suggested by Figure 22-9a, molecular absorption in the ultraviolet and visible regions consists of absorption *bands* made up of closely spaced lines. (A real molecule has many more energy levels than shown here; thus, the typical absorption band consists of a multitude of lines.) In a solution the absorbing species are surrounded by solvent, and the band nature of molecular absorption often becomes blurred because collisions tend to spread the energies of the quantum states, thus giving smooth and continuous absorption peaks.

Relaxation Processes

As noted earlier, the lifetime of an excited species is brief because several mechanisms exist whereby an excited atom or molecule can give up its excess energy and relax to its ground state. Two of the most important of these mechanisms, nonradiative relaxation and fluorescent relaxation, are illustrated in Figures 22-9b and c.

Two types of nonradiative relaxation are shown in Figure 22-9b. *Vibrational deactivation,* or *relaxation,* depicted by the short wavy arrows between vibrational energy levels, takes place during collisions between excited molecules and molecules of the solvent. During the collisions, the excess vibrational energy is transferred to solvent molecules in a series of steps as indicated in the figure. The gain in vibrational energy of the solvent is reflected in a tiny increase in the temperature of the medium. Vibrational relaxation is such an efficient process that the average lifetime of an excited *vibrational* state is only about 10^{-15} s.

Nonradiative relaxation between the lowest vibrational level of an excited electronic state and the upper vibrational level of another electronic state can also occur. This type of relaxation, depicted by the two longer wavy arrows in Figure 22-9b, is much less efficient than vibrational relaxation, so the average lifetime of an electronic excited state is between 10^{-6} and 10^{-9} s. The mechanisms by which this type of relaxation occurs are not fully understood, but the net effect is again a rise in the temperature of the medium.

Figure 22-9c depicts another relaxation process: fluorescence. Molecular fluorescence is discussed in Chapter 25.

22C-4 Terms Employed in Absorption Spectrometry

Table 22-3 lists the common terms and symbols used in absorption spectrometry. This nomenclature is recommended by the American Society for Testing Materi-

TABLE 22-3 Important Terms and Symbols Employed in Absorption Measurement

Term and Symbol*	Definition	Alternative Name and Symbol
Radiant power P, P_0	Energy of radiation (in ergs) impinging on a 1-cm² area of a detector per second	Radiation intensity I, I_0
Absorbance A	$\log \frac{P_0}{P}$	Optical density D; extinction E
Transmittance T	$\frac{P}{P_0}$	Transmission T
Path length of radiation b	—	l, d
Concentration c	—	—
Absorptivity† a	$\frac{A}{bc}$	Extinction coefficient k
Molar absorptivity‡ ε	$\frac{A}{bc}$	Molar extinction coefficient

*Terminology recommended by the American Chemical Society (*Anal. Chem.*, **1990**, *62,* 91).

†c may be expressed in g/L or in other specified concentration units; b may be expressed in cm or in other units of length.

‡c is expressed in mol/L; b is expressed in cm.

als as well as the American Chemical Society. Column 3 contains alternate symbols found in the older literature. Throughout this book, we will use the recommended nomenclature and symbols and avoid using the symbols listed in column 3.

22C-5 Beer's Law

Beer's law is sometimes paraphrased:
The deeper the glass
The darker the blend
The smaller the amount
of light at the end.

For monochromatic radiation, absorbance is directly proportional to the path length b through the medium and the concentration c of the absorbing species. These relationships are given by

$$A = abc \tag{22-8}$$

where a is a proportionality constant called the *absorptivity*. The magnitude and dimensions of a will clearly depend upon the units used for b and c. For solutions of an absorbing species, b is often given in terms of centimeters and c in grams per liter. Absorptivity then has units of $L \cdot g^{-1} \cdot cm^{-1}$.

When the concentration in Equation 22-8 is expressed in moles per liter and the cell length is in centimeters, the absorptivity is called the *molar absorptivity* and given the special symbol ε. Thus, when b is in centimeters and c is in moles per liter,

$$A = \varepsilon bc \tag{22-9}$$

where ε has the units $L \cdot mol^{-1} \cdot cm^{-1}$.

Equations 22-8 and 22-9 are expressions of *Beer's law,* which is discussed in detail in the sections that follow.

22C-6 The Experimental Measurement of Transmittance and Absorbance

Ordinarily, transmittance and absorbance, as defined in Table 22-3, cannot be measured in the laboratory because the analyte solution must be held in some sort of a transparent container, or cell. As shown in Figure 22-11, reflection occurs at the two air/wall interfaces as well as at the two wall/solution interfaces. The resulting beam attenuation is substantial. For example, it can be shown that about 8.5% of a beam of yellow light is lost by reflection in passing through a glass cell containing water.[2] In addition, attenuation of a beam may occur as a result of scattering by large molecules and sometimes from absorption by the container walls. To compensate for these effects, the power of the beam transmitted by the analyte solution is ordinarily compared with the power of the beam transmitted by

[2]D. A. Skoog and J. J. Leary, *Principles of Instrumental Analysis,* 4th ed., p. 68. Philadelphia: Saunders College Publishing, 1992.

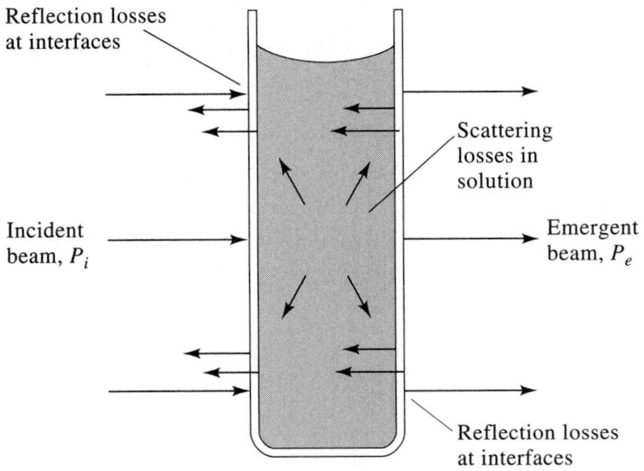

Figure 22-11
Reflection and scattering losses.

an identical cell containing only solvent. Experimental transmittances and absorbances that closely approximate the true transmittance and absorbance are then obtained with the equations

$$T = \frac{P_{\text{solution}}}{P_{\text{solvent}}} = \frac{P}{P_0} \tag{22-10}$$

$$A = \log \frac{P_{\text{solvent}}}{P_{\text{solution}}} \approx \log \frac{P_0}{P} \tag{22-11}$$

The terms P_0 and P, as used in the rest of this book, refer to the power of radiation after it has passed through cells containing the solvent and the analyte, respectively.

FEATURE 22-2
Derivation of Beer's Law

Beer's law can be derived as follows.[3] Consider the block of absorbing matter (solid, liquid, or gas) shown in Figure 22-12. A beam of parallel monochromatic radiation with power P_0 strikes the block perpendicular to a surface; after passing through a length b of the material, which contains n absorbing particles (atoms, ions, or molecules), its power is decreased to P as a result of absorption. Consider now a cross section of the block having an area S and an infinitesimal thickness dx. Within this section there are dn absorbing particles; associated with each particle, we can imagine a surface at which photon capture will occur. That is, if a photon reaches one of these areas by chance, absorption will follow immediately. The total projected area of these capture surfaces within the section is designated as dS; the

[3]The discussion that follows is based on a paper by F. C. Strong, *Anal. Chem.*, **1952**, *24*, 338.

ratio of the capture area to the total area, then, is dS/S. On a statistical average, this ratio represents the probability for the capture of photons within the section.

The power of the beam entering the section P_x is proportional to the number of photons per square centimeter per second, and dP_x represents the quantity removed per second within the section; the fraction absorbed is then $-dP_x/P_x$, and this ratio also equals the average probability for capture. The term is given a minus sign to indicate that P undergoes a decrease. Thus,

$$-\frac{dP_x}{P_x} = \frac{dS}{S} \tag{22-12}$$

Recall, now, that dS is the sum of the capture areas for particles within the section; it must therefore be proportional to the number of particles, or

$$dS = adn \tag{22-13}$$

where dn is the number of particles and a is a proportionality constant, which can be called the *capture cross section.* Combining Equations 22-12 and 22-13 and summing over the interval between 0 and n, we obtain

$$-\int_{P_0}^{P} \frac{dP_x}{P_x} = \int_0^n \frac{adn}{S}$$

which, upon integration, gives

$$-\ln \frac{P}{P_0} = \frac{an}{S}$$

Upon converting to base 10 logarithms and inverting the fraction to change the sign, we obtain

$$\log \frac{P_0}{P} = \frac{an}{2.303\ S} \tag{22-14}$$

where n is the total number of particles within the block shown in Figure 22-12. The cross-sectional area S can be expressed in terms of the volume of the block V in cm^3 and its length b in cm. Thus,

$$S = \frac{V}{b}\ \text{cm}^2$$

Substitution of this quantity into Equation 22-14 yields

$$\log \frac{P_0}{P} = \frac{anb}{2.303\ V} \tag{22-15}$$

Note that n/V has the units of concentration (that is, number of particles per cubic centimeter); we can readily convert n/V to moles per liter. Thus, the number of moles is given by

$$\text{number mol} = \frac{n \text{ particles}}{6.02 \times 10^{23} \text{ particles/mol}}$$

and c in mol/L is given by

$$c = \frac{n}{6.02 \times 10^{23}} \text{ mol} \times \frac{1000 \cancel{\text{cm}^3}/\text{L}}{V \cancel{\text{cm}^3}}$$

$$= \frac{1000\, n}{6.02 \times 10^{23}\, V} \text{ mol/L}$$

Combining this relationship with Equation 22-15 yields

$$\log \frac{P_0}{P} = \frac{6.02 \times 10^{23}\, abc}{2.303 \times 1000}$$

Finally, the constants in this equation can be collected into a single term ε to give

$$\log \frac{P_0}{P} = \varepsilon bc = A \tag{22-16}$$

which is a statement of Beer's law.

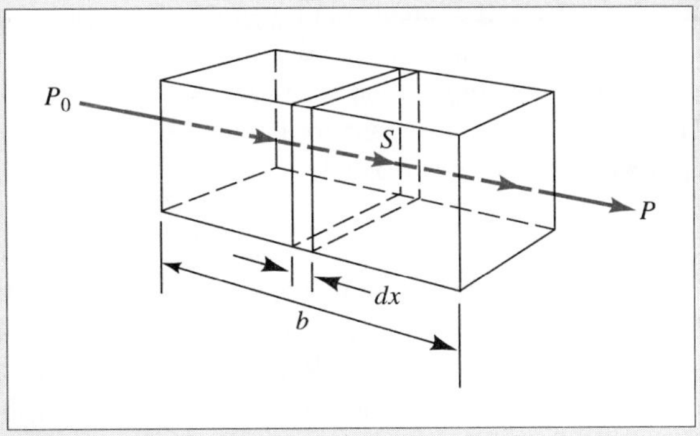

Figure 22-12 Attenuation of radiation with initial power P_0 by a solution containing c mol/L of absorbing solute and a path length of b cm ($P < P_0$).

22C-7 The Application of Beer's Law to Mixtures

Absorbances are additive.

Beer's law also applies to a medium containing more than one kind of absorbing substance. Provided there is no interaction among the various species, the total absorbance for a multicomponent system is given by

$$A_{total} = A_1 + A_2 + \cdots + A_n = \varepsilon_1 bc_1 + \varepsilon_2 bc_2 + \cdots + \varepsilon_n bc_n \quad (22\text{-}17)$$

where the subscripts refer to absorbing components 1, 2, . . . , n.

22C-8 Limitation to Beer's Law

Few if any exceptions are found to the generalization that absorbance is linearly related to path length. On the other hand, deviations from the direct proportionality between the measured absorbance and concentration when b is constant are frequently encountered. Some of these deviations are fundamental and represent real limitations of the law. Others occur as a consequence of the manner in which the absorbance measurements are made or as a result of chemical changes associated with concentration changes; the latter two are sometimes known as *instrumental deviations* and *chemical deviations.*

Real Limitations to Beer's Law

Real limitations to Beer's law are encountered only in relatively concentrated solutions of the analyte or in concentrated electrolyte solutions.

Beer's law is successful in describing the absorption behavior of media containing relatively low analyte concentrations; in this sense, it is a limiting law. At high concentrations (usually >0.01 M), the average distance between the species responsible for absorption is diminished to the point where each affects the charge distribution of its neighbors. This interaction, in turn, can alter their ability to absorb a given wavelength of radiation. Because the extent of interaction depends upon concentration, the occurrence of this phenomenon causes deviations from the linear relationship between absorbance and concentration. A similar effect is sometimes encountered in media containing low absorber concentrations but high concentrations of other species, particularly electrolytes. The close proximity of ions to the absorber alters the molar absorptivity of the latter as a result of electrostatic interactions; the effect is lessened by dilution.

While the effect of molecular interactions is ordinarily not significant at concentrations below 0.01 M, some exceptions are encountered among certain large organic ions or molecules. For example, the molar absorptivity at 436 nm for the cation of methylene blue in aqueous solutions is reported to increase by 88% as the dye concentration is increased from 10^{-5} to 10^{-2} M; even below 10^{-6} M, strict adherence to Beer's law is not observed.

Deviations from Beer's law also arise because ε is dependent upon the refractive index of the medium.[4] Thus, if concentration changes cause significant alterations in the refractive index η of a solution, departures from Beer's law are observed. A correction for this effect can be made by substitution of the quantity $\varepsilon\eta/(\eta^2 + 2)^2$ for ε in Equation 22-9. In general, this correction is never very large and is rarely significant at concentrations less than 0.01 M.

[4] G. Kortum and M. Seiler, *Angew. Chem.*, **1939**, *52*, 687.

Apparent Chemical Deviations

Apparent deviations from Beer's law arise when an analyte dissociates, associates, or reacts with a solvent to generate a product that has a different absorption spectrum from that of the analyte. A common example of this behavior is found with aqueous solutions of acid/base indicators. For example, the color change associated with a typical indicator HIn arises from shifts in the equilibrium

$$\underset{\text{color 1}}{\text{HIn}} \rightleftharpoons \text{H}^+ + \underset{\text{color 2}}{\text{In}^-}$$

Example 22-4 demonstrates how the shift in this equilibrium with dilution results in deviation from Beer's law.

EXAMPLE 22-4

The molar absorptivities of the weak acid HIn ($K_a = 1.42 \times 10^{-5}$) and its conjugate base In^- at 430 and 570 nm were determined by measurements of strongly acidic and strongly basic solutions of the indicator (where essentially all of the indicator was in the HIn and In^- form, respectively). The results were

	ε_{430}	ε_{570}
HIn	6.30×10^2	7.12×10^3
In^-	2.06×10^4	9.61×10^2

Derive absorbance data for unbuffered solutions having total indicator concentrations ranging from 2×10^{-5} to 16×10^{-5} M.

Let us calculate the molar concentrations of [HIn] and $[In^-]$ of the two species in a solution in which the total concentration of indicator is 2.00×10^{-5} M. Here,

$$\text{HIn} \rightleftharpoons \text{H}^+ + \text{In}^-$$

and

$$K_a = 1.42 \times 10^{-5} = \frac{[\text{H}^+][\text{In}^-]}{[\text{HIn}]}$$

From the equation for the dissociation process, we may write

$$[\text{H}^+] = [\text{In}^-]$$

Furthermore, the sum of the concentrations of the two indicator species must equal the total molar concentration of the indicator. Thus,

$$[\text{In}^-] + [\text{HIn}] = 2.00 \times 10^{-5}$$

Substitution of these relationships into the expression for K_a gives

$$\frac{[In^-]^2}{2.00 \times 10^{-5} - [In^-]} = 1.42 \times 10^{-5}$$

Rearrangement yields the quadratic expression

$$[In^-]^2 + 1.42 \times 10^{-5}\,[In^-] - 2.84 \times 10^{-10} = 0$$

The positive solution to this equation is

$$[In^-] = 1.12 \times 10^{-5}$$

$$[HIn] = 2.00 \times 10^{-5} - 1.12 \times 10^{-5} = 0.88 \times 10^{-5}$$

We are now able to calculate the absorbance at the two wavelengths. Thus, substituting into Equation 22-17 gives

$$A = \varepsilon_{In^-} b[In^-] + \varepsilon_{HIn} b[HIn]$$

$$\begin{aligned} A_{430} &= 2.06 \times 10^4 \times 1.00 \times 1.12 \times 10^{-5} + 6.30 \\ &\quad \times 10^2 \times 1.00 \times 0.88 \times 10^{-5} \\ &= 0.236 \end{aligned}$$

Similarly, at 570 nm,

$$\begin{aligned} A_{570} &= 9.61 \times 10^2 \times 1.00 \times 1.12 \times 10^{-5} + 7.12 \\ &\quad \times 10^3 \times 1.00 \times 0.88 \times 10^{-5} \\ &= 0.073 \end{aligned}$$

Additional data, obtained in the same way, are shown in Table 22-4.

Figure 22-13 is a plot of the data shown in Table 22-4, which illustrates the kinds of departures from Beer's law that arise when the absorbing system is capable of undergoing dissociation or association. Note that the direction of curvature is opposite at the two wavelengths.

Apparent Instrumental Deviations with Polychromatic Radiation

Strict adherence to Beer's law is observed only with truly monochromatic radiation; this observation is yet another manifestation of the limiting character of the law. Unfortunately, the use of radiation that is restricted to a single wavelength is

TABLE 22-4 Absorbance Data for Various Concentrations of the Indicators in Example 22-4

c_{HIn}, M	[HIn]	$[In^-]$	A_{430}	A_{570}
2.00×10^{-5}	0.88×10^{-5}	1.12×10^{-5}	0.236	0.073
4.00×10^{-5}	2.22×10^{-5}	1.78×10^{-5}	0.381	0.175
8.00×10^{-5}	5.27×10^{-5}	2.73×10^{-5}	0.596	0.401
12.00×10^{-5}	8.52×10^{-5}	3.48×10^{-5}	0.771	0.640
16.00×10^{-5}	11.9×10^{-5}	4.11×10^{-5}	0.922	0.887

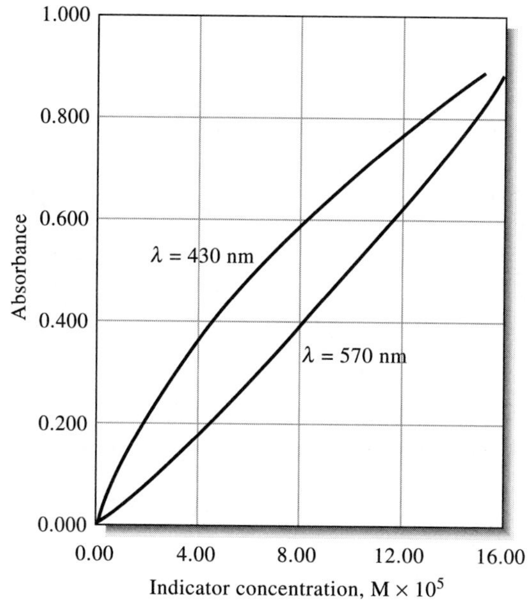

Figure 22-13
Chemical deviations from Beer's law for unbuffered solutions of the indicator HIn. For data, see Example 22-4.

seldom practical because devices that isolate portions of the output from a continuous source produce a more or less symmetric band of wavelengths around the desired one (see page 531 for example).

The following derivation shows the effect of polychromatic radiation on Beer's law.

Deviations from Beer's law often occur when polychromatic radiation is used to measure absorbance.

Consider a beam consisting of just two wavelengths λ' and λ''. Assuming that Beer's law applies strictly for each of these individually, we may write for radiation λ'

$$A' = \log \frac{P_0'}{P'} = \varepsilon' bc$$

or

$$P_0'/P' = 10^{\varepsilon' bc}$$

and

$$P' = P_0' 10^{-\varepsilon' bc}$$

Similarly, for λ''

$$P'' = P_0'' 10^{-\varepsilon'' bc}$$

When an absorbance measurement is made with radiation composed of both wavelengths, the power of the beam emerging from the solution is given by $(P' + P'')$ and that of the beam from the solvent by $(P_0' + P_0'')$. Therefore, the measured absorbance A_M is

$$A_M = \log \frac{(P_0' + P_0'')}{(P' + P'')}$$

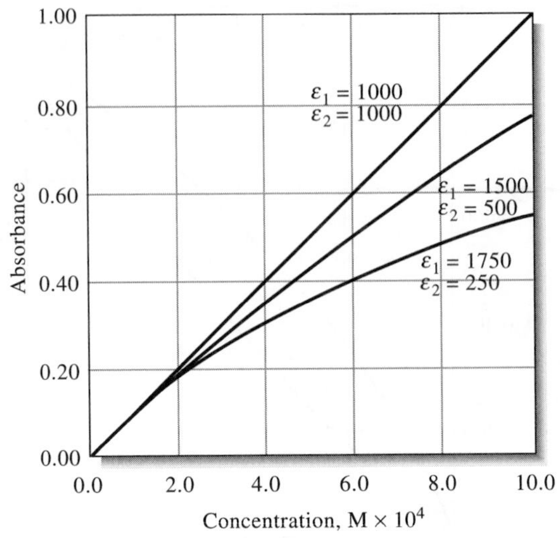

Figure 22-14
Deviations from Beer's law with polychromatic light. The absorber has the indicated molar absorptivities at the two wavelengths λ' and λ''.

Substituting for P' and P'' yields

$$A_M = \log \frac{(P'_0 + P''_0)}{(P'_0 10^{-\varepsilon' bc} + P''_0 10^{-\varepsilon'' bc})}$$

or

$$A_M = \log(P'_0 + P''_0) - \log(P'_0 10^{-\varepsilon' bc} + P''_0 10^{-\varepsilon'' bc})$$

Now, when $\varepsilon' = \varepsilon''$, this equation simplifies to

$$A_M = \varepsilon' bc$$

and Beer's law is followed. As shown in Figure 22-14, however, the relationship between A_M and concentration is no longer linear when the molar absorptivities differ; moreover, greater departures from linearity can be expected with increasing differences between ε' and ε''. This derivation can be expanded to include additional wavelengths; the effect remains the same.

It is an experimental fact that deviations from Beer's law resulting from the use of a polychromatic beam are not appreciable, provided the radiation used does not encompass a spectral region in which the absorber exhibits large changes in absorption as a function of wavelength. This observation is illustrated in Figure 22-15.

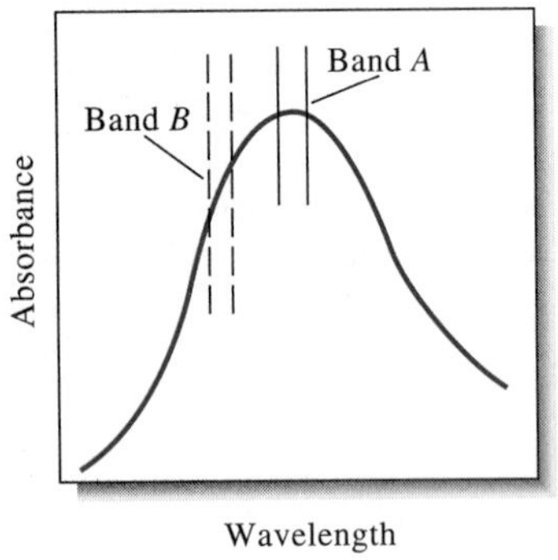

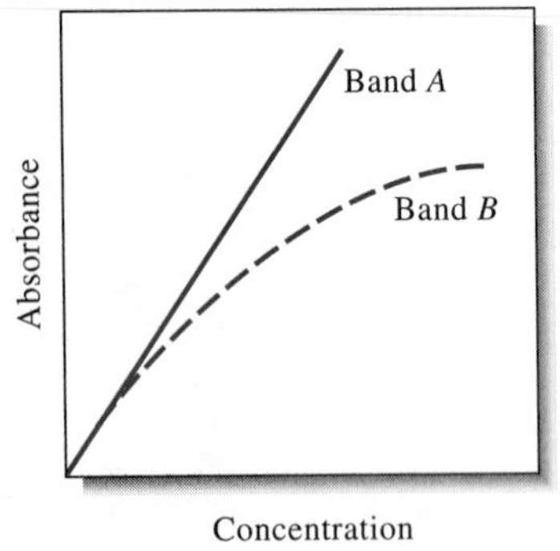

Figure 22-15
The effect of polychromatic radiation on Beer's law. Band A shows little deviation because ε does not change greatly throughout the band. Band B shows marked deviation because ε undergoes significant changes in this region.

Instrumental Deviations in the Presence of Stray Radiation

The radiation employed for absorbance measurements is usually contaminated with small amounts of *stray* radiation due to instrumental imperfections. Stray radiation is the result of scattering phenomena off the surfaces of prisms, lenses, filters, and windows (page 511). It often differs greatly in wavelength from the principal radiation and, in addition, may not have passed through the sample or solvent.

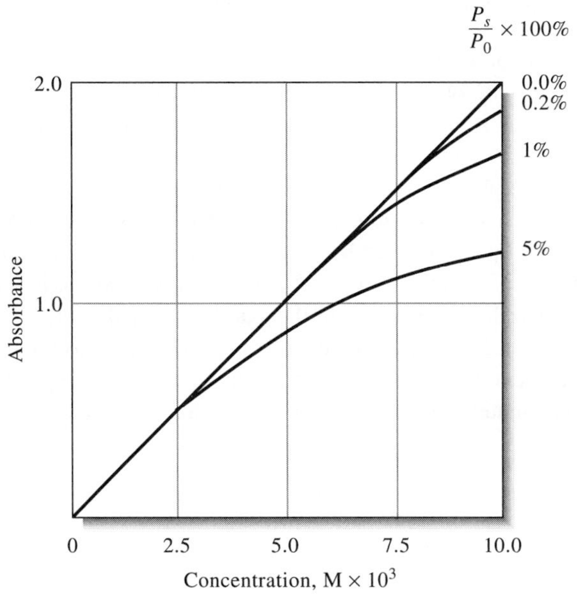

Figure 22-16
Apparent deviation from Beer's law caused by various amounts of stray radiation.

When measurements are made in the presence of stray radiation, the observed absorbance is given by

Generally, the better the instrument, the less likely are deviations from Beer's law due to polychromatic radiation.

$$A' = \log \frac{P_0 + P_s}{P + P_s}$$

where P_s is the power of nonabsorbed stray radiation. Figure 22-16 shows a plot of A' versus concentration for various ratios between P_s and P_0. It is noteworthy that at high concentrations and at longer path lengths, stray radiation can also cause deviations from the linear relationship between absorbance and path length.[5]

Note also that the instrumental deviations illustrated in Figures 22-14 and 22-16 result in absorbances that are smaller than theoretical. It can be shown that instrumental deviations always lead to negative absorbance errors.[6]

22D EMISSION OF ELECTROMAGNETIC RADIATION

Atoms, ions, and molecules can be excited to one or more higher energy levels by any of several processes, including bombardment with electrons or other elementary particles, exposure to a high-voltage ac spark, heat treatment in a flame or, arc, or exposure to a source of electromagnetic radiation. The lifetime of an excited species is generally transitory (10^{-6} to 10^{-9} s), and relaxation to a lower energy level or the ground state takes place with a release of the excess energy in the form of electromagnetic radiation, heat, or perhaps both.

Chemical species can be caused to emit light by (1) bombardment with electrons, (2) exposure to a high-voltage spark, (3) heating in a flame or an electric arc, or (4) irradiation with a beam of light.

[5]For a discussion of the effects of stray radiation, see M. R. Sharpe, *Anal. Chem.*, **1984**, *56*, 339A.

[6]E. J. Meehan, in *Treatise on Analytical Chemistry,* 2nd ed., P. J. Elving, E. J. Meehan, and I. M. Kolthoff, Eds., Part I, Vol. 7, p. 73. New York: Wiley, 1981.

22D-1 Emission Spectra

Radiation from a source is conveniently characterized by means of an *emission spectrum,* which usually takes the form of a plot of the relative power of the emitted radiation as a function of wavelength or frequency. Figure 22-17 illustrates a typical emission spectrum, which was obtained by aspirating a brine solution into an oxyhydrogen flame. Three types of spectra are evident in the figure: *line, band,* and *continuous.* The line spectrum is made up of a series of sharp, well-defined peaks caused by an excitation of individual atoms. The band spectrum consists of several groups of lines so closely spaced that they are not completely resolved. The source of the bands is small molecules or radicals. Finally, the continuous spectrum is responsible for the increase in the background that becomes evident above about 350 nm. The line and band spectra are superimposed on this continuum. The source of the continuum is described on page 523.

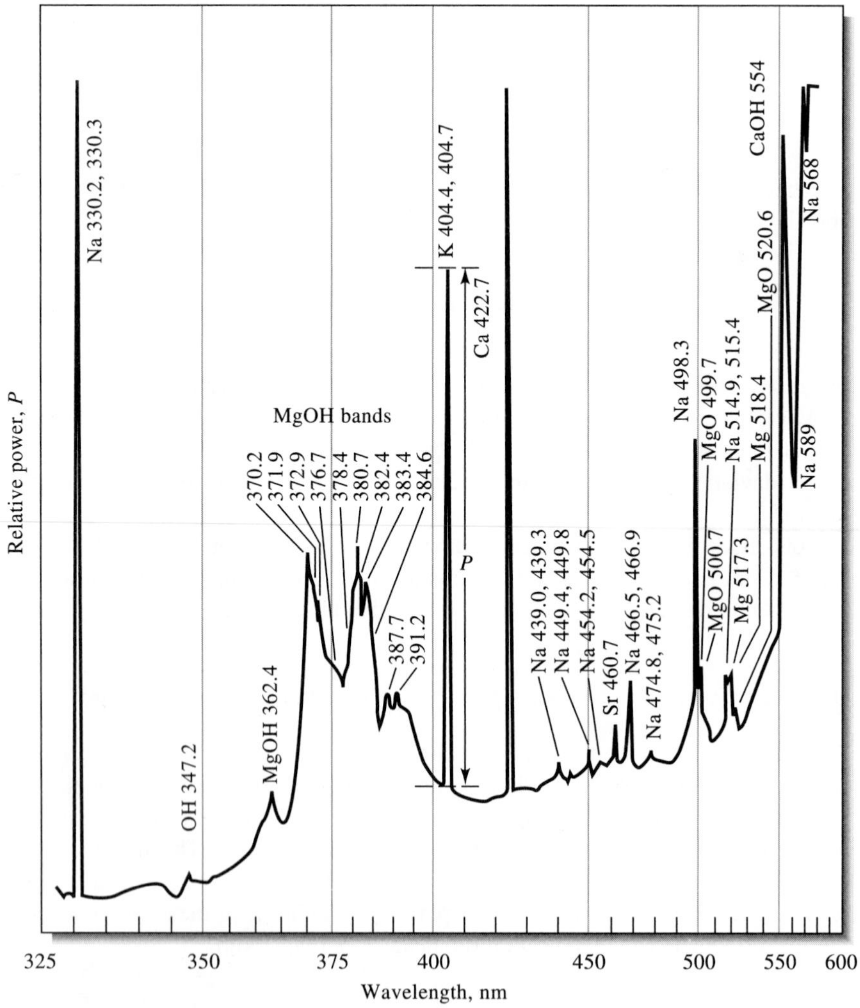

Figure 22-17

Emission spectrum of a brine obtained with an oxyhydrogen flame. (R. Hermann and C. T. J. Alkemade, *Chemical Analysis by Flame Photometry,* 2nd ed., p. 484. New York: Interscience, 1963. With permission.)

Line Spectra

Line spectra are encountered when the radiating species are individual atomic particles that are well separated, as in a gas. The individual particles in a gaseous medium behave independently, and the spectrum consists of a series of sharp lines with widths of about 10^{-2} Å. In Figure 22-17, lines for sodium, potassium, strontium, and calcium are identified.

The line widths in a typical atomic spectrum are about 0.01 Å. The wavelengths of atomic lines are unique for each element and are often used for qualitative analysis.

The energy-level diagram in Figure 22-18a shows the source of two of the lines in a typical emission spectrum of an element. The horizontal line labeled E_0 corresponds to the lowest, or ground-state, energy of the atom. The horizontal lines labeled E_1 and E_2 are two higher-energy electronic levels of the species. For example, the single outer electron in the ground state E_0 for a sodium atom is located in the $3s$ orbital. Energy level E_1 then represents the energy of the atom when this electron has been promoted to the $3p$ state by absorption of thermal, electrical, or radiant energy (see also Figure 22-8b). The promotion is depicted by the shorter wavy arrow on the left in Figure 22-18a. After perhaps 10^{-8} s, the atom returns to the ground state, emitting a photon whose frequency and wavelength are given by Equations 22-4 and 22-5

$$\nu_1 = (E_1 - E_0)/h$$
$$\lambda_1 = hc/(E_1 - E_0)$$

This emission process is illustrated by the shorter straight arrow on the right in Figure 22-18a.

For the sodium atom, E_2 in the figure corresponds to the more energetic $4p$ state; the resulting radiation λ_2 would then appear at a shorter wavelength. The line appearing at about 330 nm in Figure 22-17 results from this transition; the $3p$-to-$3s$ transition provides a line at about 590 nm. It is important to note that the emitted wavelengths *are identical to the wavelengths of the absorption peaks* for sodium (Figure 22-8) because the transitions involved are between the same two states.

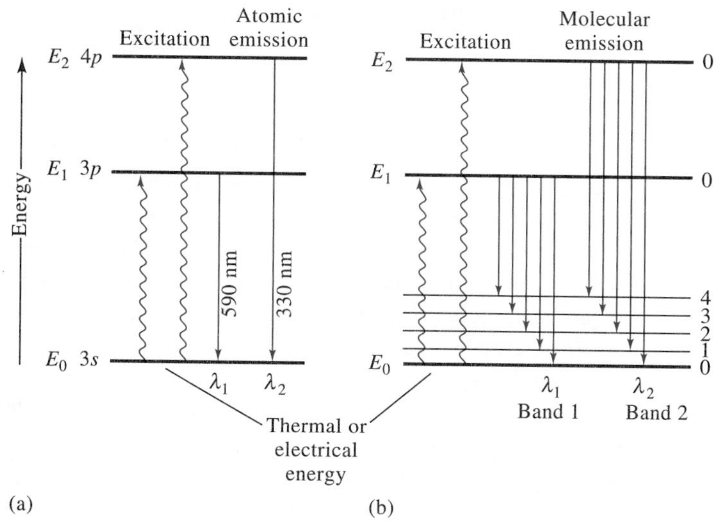

Figure 22-18
Energy-level diagrams for a sodium atom and a simple molecule, showing the source of (a) a line spectrum and (b) a band spectrum.

Band Spectra

An emission band spectrum is made up of many closely spaced lines that are difficult to resolve.

Band spectra are often encountered in spectral sources because of the presence of gaseous radicals or small molecules. For example, in Figure 22-17 bands for OH, MgOH, and MgO are labeled and consist of a series of closely spaced lines that are not fully resolved by the instrument used to obtain the spectrum. Bands arise from the numerous quantized vibrational levels that are superimposed on the ground-state electronic energy level of a molecule.

Figure 22-18b is a partial energy-level diagram for a molecule showing its ground state E_0 and two of its several excited electronic states, E_1 and E_2. A few of the many vibrational levels associated with the ground state are also shown. Vibrational levels associated with the two excited states have been omitted because the lifetime of an excited vibrational state is brief compared with that of an electronically excited state (about 10^{-15} s vs. 10^{-8} s). A consequence of this tremendous difference in lifetimes is that when an electron is excited to one of the higher vibrational levels of an electronic state, relaxation to the lowest vibrational level of that state occurs before an electronic transition to the ground state can occur. Therefore, the radiation produced by the electrical or thermal excitation of polyatomic species nearly always involves a transition from the *lowest vibrational level of an excited electronic state* to any of the several vibrational levels of the ground state.

The mechanism by which a vibrationally excited species relaxes to the nearest electronic state involves a transfer of its excess energy to other atoms in the system through a series of collisions. As noted, this process takes place at an enormous speed. Relaxation from one electronic state to another can also occur by collisional transfers of energy, but the rate of this process is slow enough that relaxation by photon release is favored.

The energy-level diagram in Figure 22-18b illustrates the mechanism by which two radiation bands consisting of five closely spaced lines are emitted by a molecule excited by thermal or electrical energy. For a real molecule, the number of individual lines is much larger because the ground state contains many more vibrational levels than are shown. In addition, a multitude of rotational states would be superimposed on each of the vibrational levels. The differences in energy among the rotational levels are perhaps an order of magnitude smaller than that for vibrational states. Thus, a real molecular band would be made up of many more lines than shown in Figure 22-17b, and these lines would be much more closely spaced.

Continuous Spectra

Continuous emission spectra have no line character and are generally produced by heating solids to a high temperature.

As shown in Figure 22-19, truly continuous radiation is produced when solids are heated to incandescence. Thermal radiation of this kind, which is called *blackbody radiation,* is more characteristic of the temperature of the emitting surface than of the material of which that surface is composed. Blackbody radiation is produced by the innumerable atomic and molecular oscillations excited in the condensed solid by the thermal energy. Note that the energy peaks in Figure 22-19 shift to shorter wavelengths with increasing temperature. It is clear that very high temperatures are needed to cause a thermally excited source to emit a substantial fraction of its energy as ultraviolet radiation.

Part of the continuous background radiation exhibited in the flame spectrum

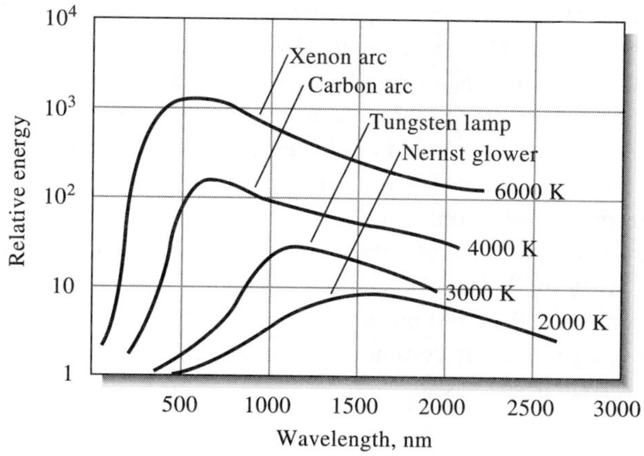

Figure 22-19
Blackbody radiation curves.

shown in Figure 22-17 is probably thermal emission from incandescent particles in the flame. Note that this background decreases rapidly as the ultraviolet region is approached.

Heated solids are important sources of infrared, visible, and longer wavelength ultraviolet radiation for analytical instruments.

The Effect of Concentration on Line and Band Spectra The radiant power P of a line or a band depends directly upon the number of excited atoms or molecules, which in turn is proportional to the total concentration c of the species present in the source. Thus we can write

$$P = kc \qquad (22\text{-}18)$$

where k is a proportionality constant. This relationship is the basis of quantitative emission spectroscopy.

22D-2 Emission by Fluorescence and Phosphorescence

Fluorescence and phosphorescence are analytically important emission processes in which atoms or molecules are excited by the absorption of a beam of electromagnetic radiation. The excited species then relax to the ground state, giving up their excess energy as photons. Fluorescence takes place much more rapidly than phosphorescence and is generally complete in about 10^{-5} s (or less) from the time of excitation. Phosphorescence emission may extend for minutes or even hours after irradiation has ceased.

Fluorescence is considerably more important than phosphorescence in analytical chemistry. Thus, our discussions focus largely on the former.

Atomic Fluorescence

Gaseous atoms may fluoresce when they are exposed to radiation that has a wavelength that exactly matches one of the absorption (or emission) lines of the element in question. For example, gaseous sodium atoms are promoted to the excited energy state E_1 shown in Figure 22-18a through absorption of 590-nm radiation. Relaxation may then take place by fluorescent reemission of radiation of the identical wavelength. Fluorescence in which the excitation and emission wavelengths are the same is termed *resonance fluorescence.* Sodium atoms could also exhibit resonance fluorescence when exposed to 330-nm radiation. In addition, however, the element could also produce nonresonance fluorescence by first relaxing to energy level E_1 through a series of nonradiative collisions with other species in the medium. Further relaxation to the ground state can then take place either by the emission of a 590-nm photon or by further collisional deactivation.

Resonance fluorescence is radiation that is identical in wavelength to radiation that excited the fluorescence.

Resonance fluorescence is commonly encountered with atoms and to a lesser extent with molecular species.

Molecular Fluorescence

The number of molecules that fluoresce is relatively small because fluorescence requires structural features that slow the rate of the nonradiative relaxation processes illustrated in Figure 22-9b and enhance the rate of fluorescence relaxation shown in Figure 22-9c. Most molecules lack these features and undergo nonradiative relaxation at a rate that is significantly greater than the radiative relaxation rate; thus, fluorescence is precluded.

As shown in Figure 22-9c, bands of radiation are produced when molecules fluoresce. Like molecular absorption bands, molecular fluorescence bands are made up of a multitude of closely spaced lines that are often difficult to resolve. Note that the lines that terminate the two fluorescence bands on the short-wavelength, or high-energy, side (λ_1' and λ_1'') are resonance lines. That is, molecular fluorescence bands consist largely of lines with wavelengths that are longer than the band of absorbed radiation responsible for their excitation. This shift in wavelength is sometimes called the *Stokes shift.* A more detailed discussion of the theory and applications of molecular fluorescence is given in Chapter 25.

The **Stokes shift** refers to fluorescence radiation that occurs at wavelengths that are longer than the wavelength of radiation used to excite the fluorescence.

22E QUESTIONS AND PROBLEMS

22-1. What kind of transitions are responsible for the
- *(a) absorption of ultraviolet radiation?
- (b) absorption of infrared radiation?
- *(c) emission of band spectra?
- (d) fluorescence of atoms?
- *(e) fluorescence of molecules?
- (f) emission of line spectra?

22-2. Define
- *(a) ground state.
- (b) excited electronic state.
- *(c) photon.
- (d) band spectra.
- *(e) continuous spectra.
- (f) line spectra.
- *(g) resonance fluorescence.
- (h) phosphorescence.
- *(i) percent transmittance.
- (j) absorbance.
- *(k) molar absorptivity.
- (l) absorptivity.
- *(m) stray radiation.
- *(n) wavenumber.
- *(o) relaxation.
- (p) Stokes shift.

22-3. Calculate the frequency in hertz of
- *(a) an X-ray beam with a wavelength of 2.65 Å.
- (b) an emission line for copper at 211.0 nm.
- *(c) the line at 694.3 nm produced by a ruby laser.
- (d) the output of a CO_2 gas laser at 10.6 μm.
- *(e) an infrared absorption peak at 19.6 μm.
- (f) a microwave beam at 1.86 cm.

22-4. Calculate the wavelength in centimeters of
- *(a) an airport tower transmitting at 118.6 MHz.
- (b) a VOR (radio navigation aid) transmitting at 114.10 kHz.
- *(c) an NMR signal at 105 MHz.
- (d) an infrared absorption peak having a wavenumber of 1210 cm^{-1}.

TABLE 22-5

	A	% T	ε	b, cm	c, M	c, ppm	a, cm^{-1} ppm^{-1}
*(a)	0.416			1.40	1.25×10^{-4}		
(b)		45.5		2.10	8.15×10^{-3}		
*(c)	1.424			0.996			0.137
(d)		19.6	5.42×10^{3}		2.50×10^{-4}		
*(e)			3.46×10^{3}	2.50		3.33	
(f)			1.214×10^{4}	1.25	7.77×10^{-4}		
*(g)		48.3		0.250		6.72	
(h)	0.842		7.73×10^{3}	2.00			
*(i)		76.3		1.10			0.0631
(j)		6.54	9.82×10^{2}		8.64×10^{-3}		

***22-5.** A typical simple infrared spectrophotometer covers a wavelength range from 3 to 15 μm. Express its range (a) in wavenumbers and (b) in hertz.

22-6. A sophisticated ultraviolet/visible/near-IR instrument has a wavelength range of 185 to 3000 nm. What are its wavenumber and frequency ranges?

22-7. Express the following absorbances in terms of percent transmittance:

*(a) 0.064. (d) 0.209.
*(b) 0.765. (e) 0.437.
*(c) 0.318. (f) 0.413.

22-8. Convert the accompanying transmittance data to absorbances:

*(a) 19.4%. (d) 4.51%.
*(b) 0.863. (e) 0.100.
*(c) 27.2%. (f) 79.8%.

***22-9.** Calculate the percent transmittance of solutions having twice the absorbance of the solutions in Problem 22-7.

***22-10.** Calculate the absorbances of solutions having half the percent transmittance of those in Problem 22-8.

22-11. Use the data in Table 22-5 to evaluate the missing quantities. Wherever necessary, assume that the molecular weight of the analyte is 250.

***22-12.** A solution containing 4.48 ppm $KMnO_4$ has a transmittance of 0.309 in a 1.00-cm cell at 520 nm. Calculate the molar absorptivity of $KMnO_4$.

22-13. A solution containing 3.75 mg/100 mL of A (220 g/mol) has a transmittance of 39.6% in a 1.50-cm cell at 480 nm. Calculate the molar absorptivity of A.

***22-14.** A solution containing the complex formed between Bi(III) and thiourea has a molar absorptivity of 9.32×10^3 L · cm^{-1} · mol^{-1} at 470 nm.
(a) What is the absorbance of a 6.24×10^{-5} M solution of the complex at 470 nm in a 1.00-cm cell?
(b) What is the percent transmittance of the solution described in (a)?
(c) What is the molar concentration of the complex in a solution that has the absorbance described in (a) when measured at 470 nm in a 5.00-cm cell?

22-15. At 580 nm, which is the wavelength of its maximum absorption, the complex $Fe(SCN)^{2+}$ has a molar absorptivity of 7.00×10^3 L · cm^{-1} · mol^{-1}. Calculate
(a) the absorbance of a 2.50×10^{-5} M solution of the complex at 580 nm in a 1.00-cm cell.
(b) the absorbance of a solution in which the concentration of the complex is twice that in (a).
(c) the transmittance of the solutions described in (a) and (b).
(d) the absorbance of a solution that has half the transmittance of that described in (a).

***22-16.** A 2.50-mL aliquot of a solution that contains 3.8 ppm iron(III) is treated with an appropriate excess of KSCN and diluted to 50.0 mL. What is the absorbance of the resulting solution at 580 nm in a 2.50-cm cell? See Problem 22-15 for absorptivity data.

22-17. Zinc(II) and the ligand L form a product that absorbs strongly at 600 nm. As long as the molar concentration of L exceeds that of zinc(II) by a factor of 5, the absorbance is dependent only on the cation concentration. Neither zinc(II) nor L absorbs at 600 nm. A solution that is 1.60×10^{-4} M in zinc(II) and 1.00×10^{-3} M in L has an absorbance of 0.464 in a 1.00-cm cell at 600 nm. Calculate
(a) the percent transmittance of this solution.
(b) the percent transmittance of this solution in a 2.50-cm cell.
(c) the molar absorptivity of the complex.

***22-18.** The equilibrium constant for the conjugate acid/base pair

$$HIn + H_2O \rightleftharpoons H_3O^+ + In^-$$

is 8.00×10^{-5}. From the additional information

Species	Absorption Maximum, nm	Molar Absorptivity 430 nm	Molar Absorptivity 600 nm
HIn	430	8.04×10^3	1.23×10^3
In^-	600	0.775×10^3	6.96×10^3

(a) calculate the absorbance at 430 nm and 600 nm for the following indicator concentrations: 3.00 × 10^{-4} M, 2.00 × 10^{-4} M, 1.00 × 10^{-4} M, 0.500 × 10^{-4} M, and 0.250 × 10^{-4} M.

(b) plot absorbance as a function of indicator concentration.

22-19. The equilibrium constant for the reaction

$$2CrO_4^{2-} + 2H^+ \rightleftharpoons Cr_2O_7^{2-} + H_2O$$

is 4.2 × 10^{14}. The molar absorptivities for the two principal species in a solution of $K_2Cr_2O_7$ are

λ	$\varepsilon_1(CrO_4^{2-})$	$\varepsilon_2(Cr_2O_7^{2-})$
345	1.84 × 10^3	10.7 × 10^2
370	4.81 × 10^3	7.28 × 10^2
400	1.88 × 10^3	1.89 × 10^2

Four solutions were prepared by dissolving 4.00 × 10^{-4}, 3.00 × 10^{-4}, 2.00 × 10^{-4}, and 1.00 × 10^{-4} moles of $K_2Cr_2O_7$ in water and diluting to 1.00 L with a pH 5.60 buffer. Derive theoretical absorbance values (1.00-cm cells) for each solution and plot the data for (a) 345 nm; (b) 370 nm; (c) 400 nm.

Color Plate 6
The Tyndall effect (see margin note, page 82).

Color Plate 7
Precipitation of 0.1041 g of Fe(III) with NH_3 (left), and homogeneous precipitation of the same amount of iron with urea (right) (see Section 5B-5, page 92).

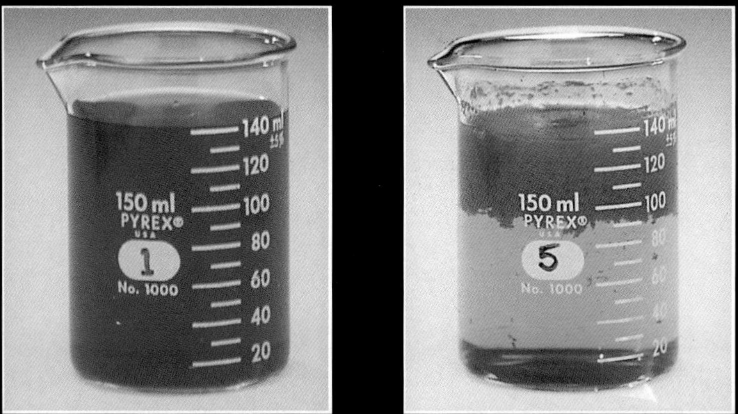

Color Plate 8
When dimethylglyoxime is added to a slightly basic solution of $Ni^{2+}(aq)$ shown on the left, a bright red precipitate of $Ni(C_4H_7N_2O_2)_2$ is formed as seen in the beaker on the right (Section 5D-3, page 95).

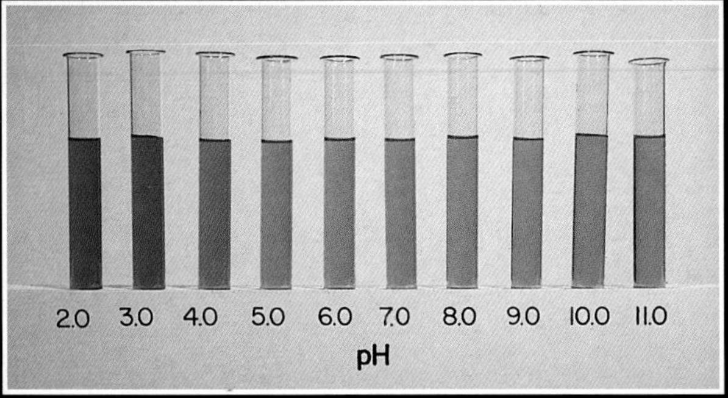

Methyl orange (3.1–4.4)

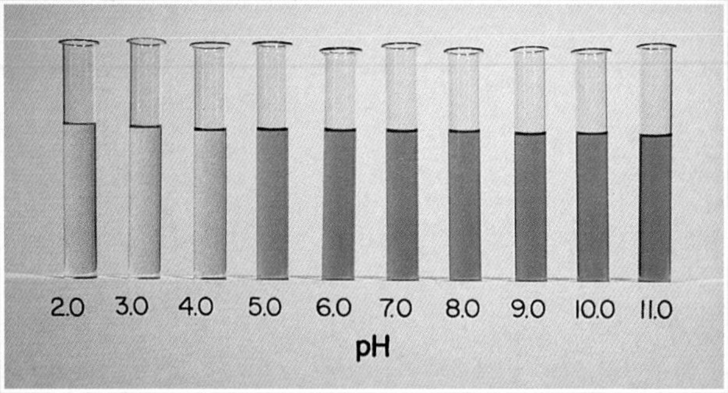

Bromocresol green (3.8–5.4)

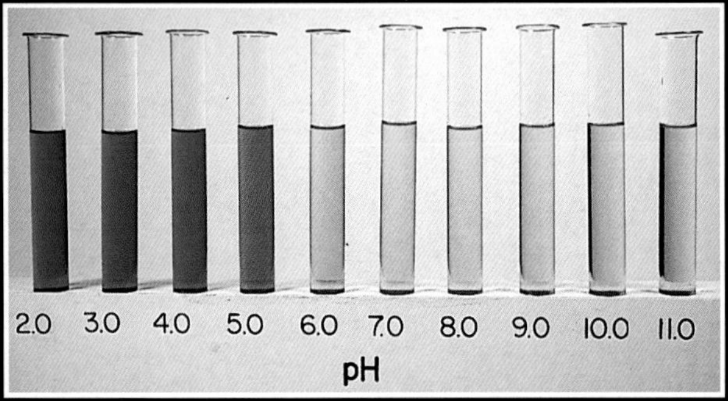

Methyl red (4.2–6.3)

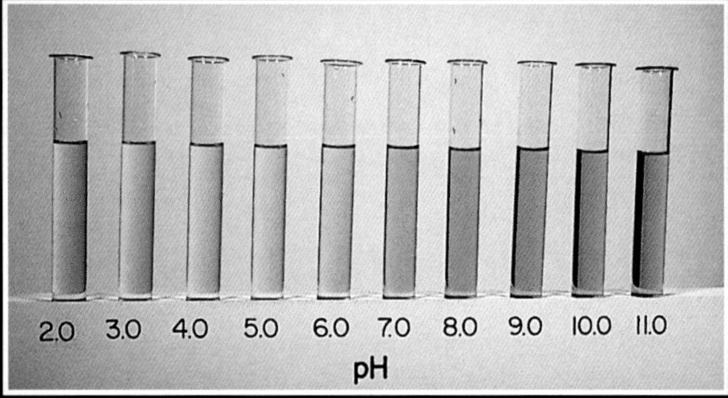

Chlorophenol red (5.2–8.8)

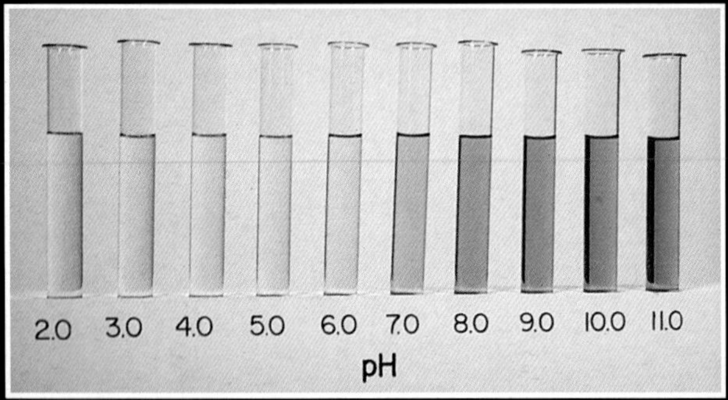

Bromothymol blue (6.0–7.6)

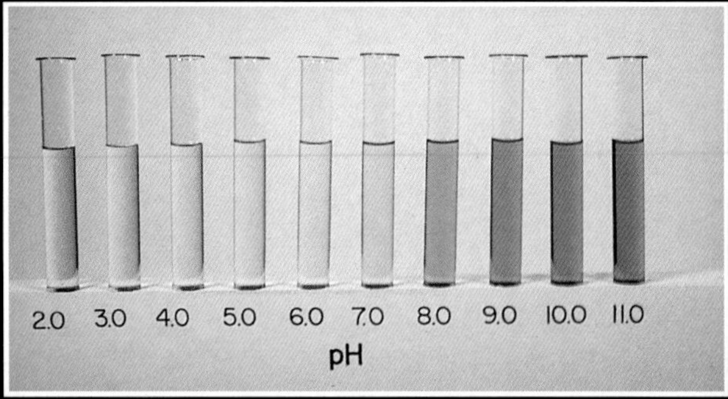

Cresol red (7.2–8.8)

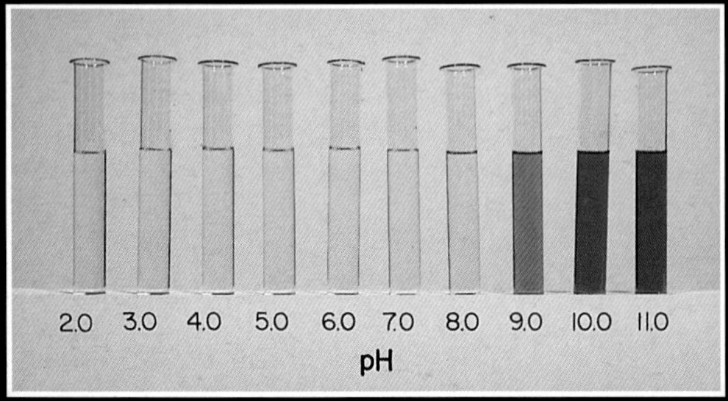

Phenolphthalein (8.3–10.0)

Color Plate 9

Acid-base indicators and their transition pH ranges (Section 10A-2).

a b c d

Color Plate 10
Argentometric determination of chloride: Fajans method (Section 13B-1).
(a) aqueous 2′,7′-dichlorofluorescein; (b) same, plus 1 mL 0.10 M Ag^+. Note the absence of a precipitate; (c) same, plus AgCl and an excess of Cl^-; (d) same, plus AgCl and the first slight excess of Ag^+.

a b c d

Color Plate 11
In (c) and (d) the AgCl has coagulated. Note that in (d) the dye has been carried down on the precipitate and that the supernatant solution is clear while in (c) the dye remains in solution. (Section 13B-1).

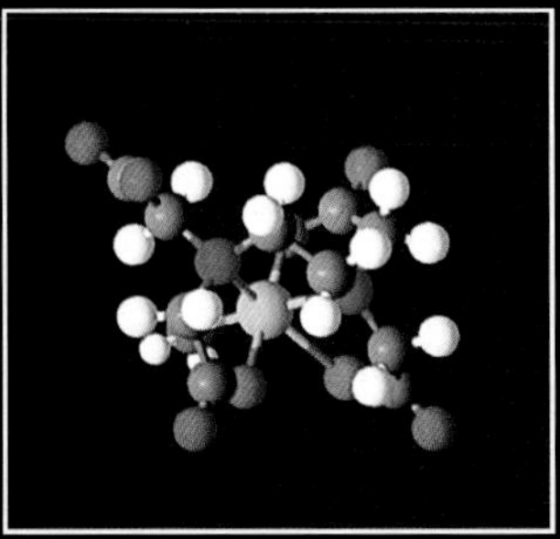

Color Plate 12
Model of an EDTA/cation chelate (Section 14B-2).

Color Plate 13
Reduction of silver(I) by direct reaction with copper: the "silver tree" (Section 15A-2).

Color Plate 14
Reduction of silver(I) by copper in an electrochemical cell (Section 15A-2).

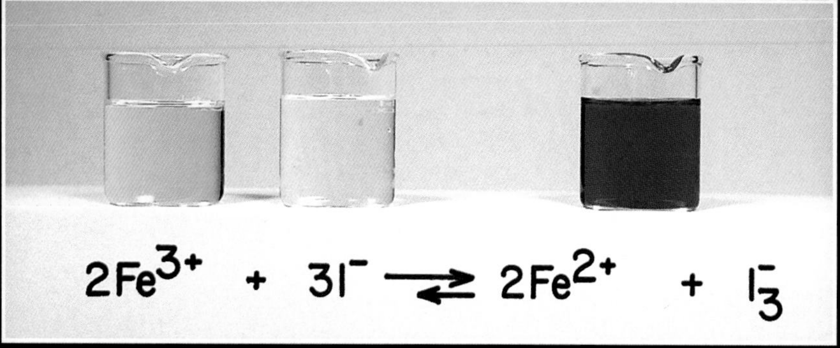

Color Plate 15
Reaction between iron(III) and iodide (see margin note, page 322).

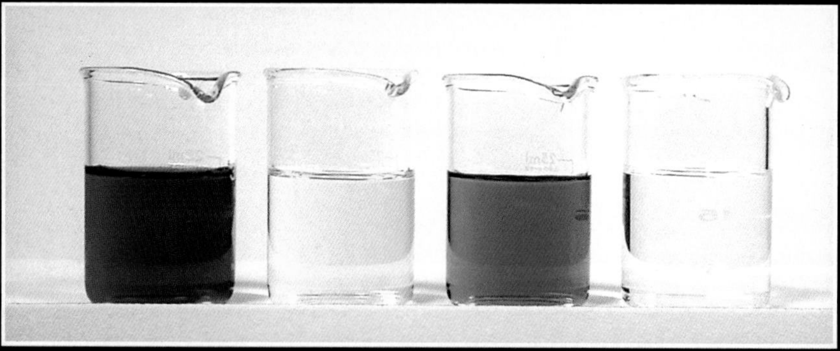

a b c d

Color Plate 16
Starch/iodine end point (Section 17B-2): (a) iodine solution; (b) same, within a few drops of the equivalence point; (c) same as (b), with starch added; (d) same as (c), at the equivalence point.

Color Plate 17
The time dependence of the reaction between permanganate and oxalate (Section 17C-1).

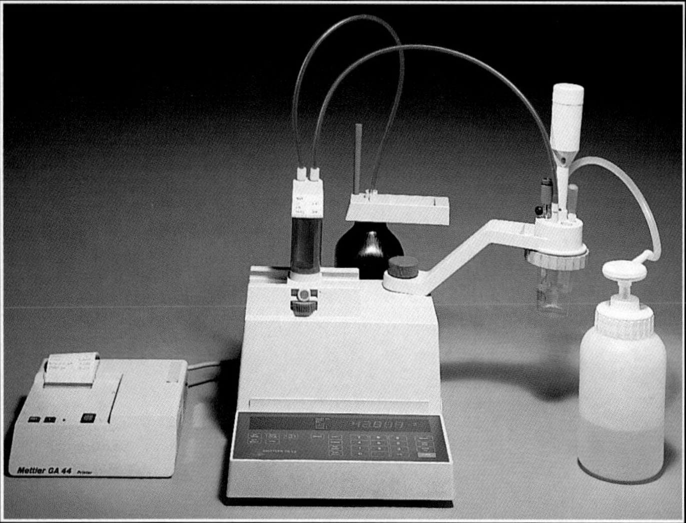

Color Plate 18

Computer-controlled automatic potentiometric titrator. The sample is pipetted into the sample cup. The titrator adds titrant, measures the pH of the titration mixture as titrant is added, and locates the equivalence point. The results of the titration are printed to provide a permanent record of the analysis.

Photo courtesy of Mettler-Toledo, Inc., Hightstown, NJ.

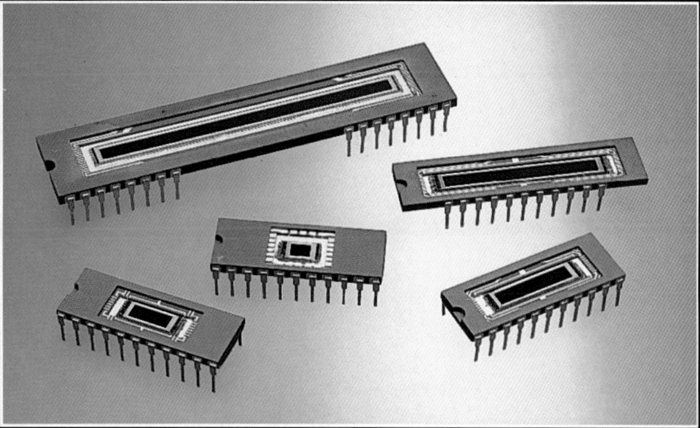

Color Plate 19

Photodiode arrays of various sizes. The arrays contain 256, 512, 1024, 2048, and 4096 diodes (Section 23A-4).

Photo courtesy of EG&G Reticon, Sunnyvale, CA.

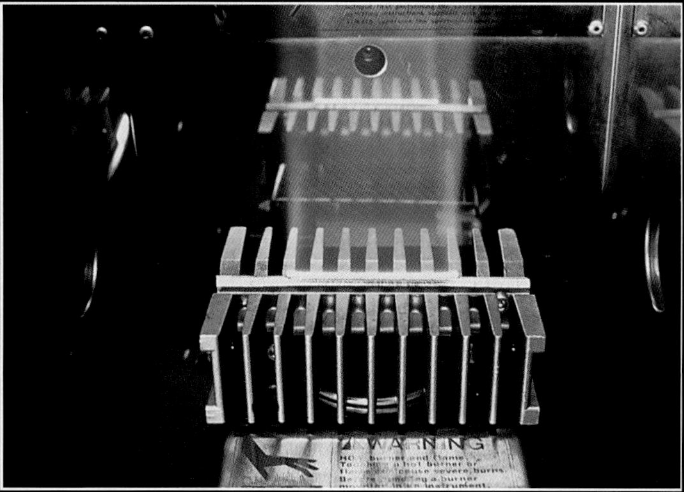

Color Plate 20

Laminar flow burner for atomic absorption spectroscopy (Section 26B-1).

Photo courtesy of Varian Instruments, Sunnyvale, CA.

Color Plate 21

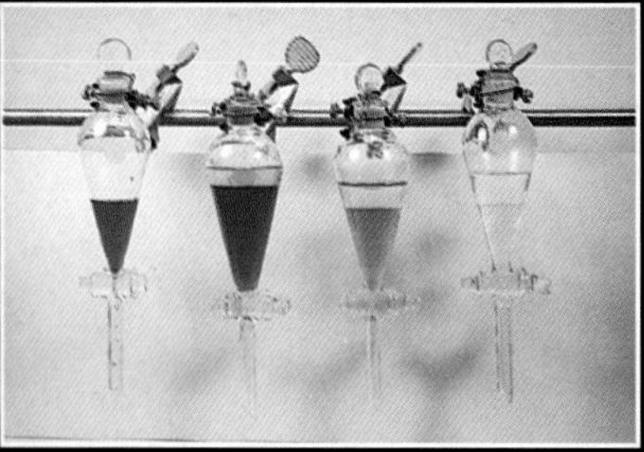
a

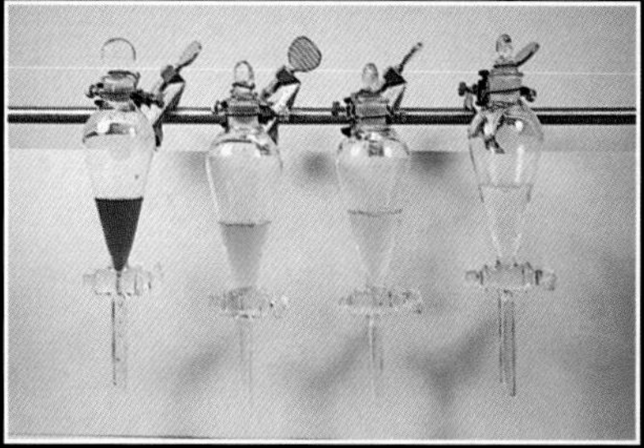
b

(a) Extraction of iodine with chloroform (Section 34C). Left to right: iodine solution; after single 40-mL extraction; after two 20-ml extractions; after four 10-mL extractions.

(b) Extraction of iodine with chloroform, aqueous layers only.

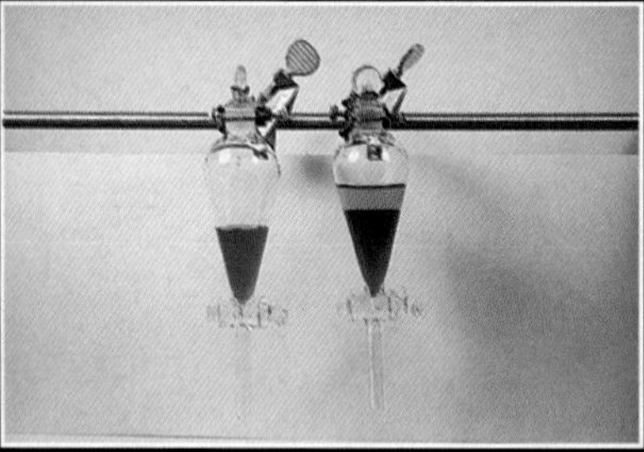
c

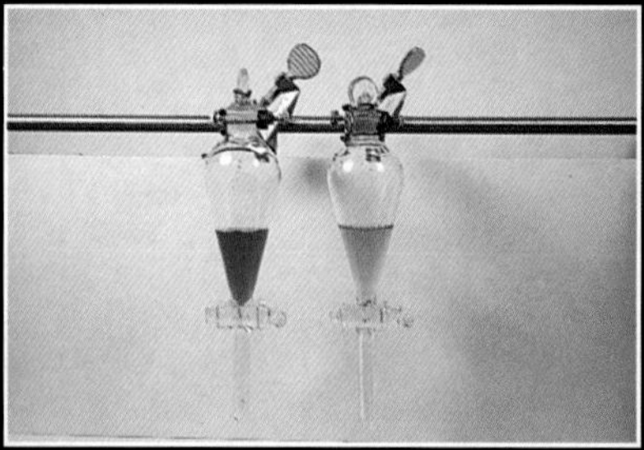
d

(c) Extraction of iodine with chloroform (Section 34C). Original solution (left); after single 40-mL extraction (right).

(d) Extraction of iodine with chloroform, aqueous phase only.

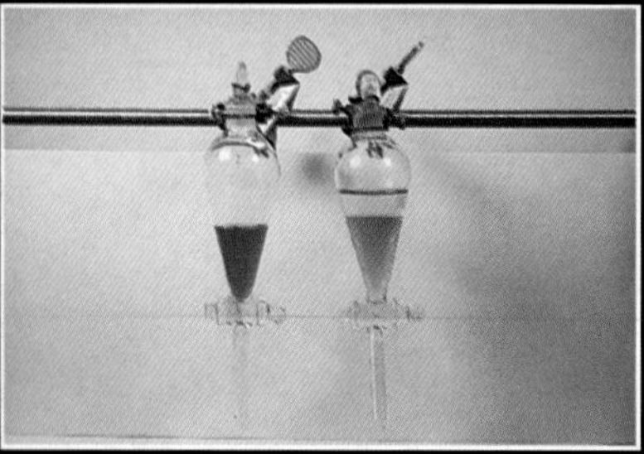
e

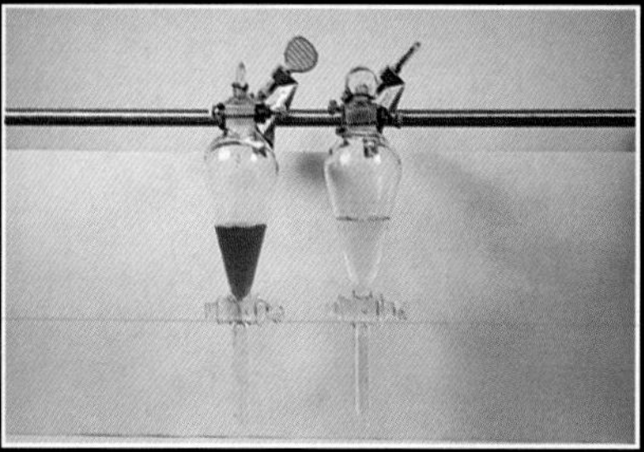
f

(e) Extraction of iodine with chloroform (Section 34C).

(f) Extraction of iodine with chloroform.

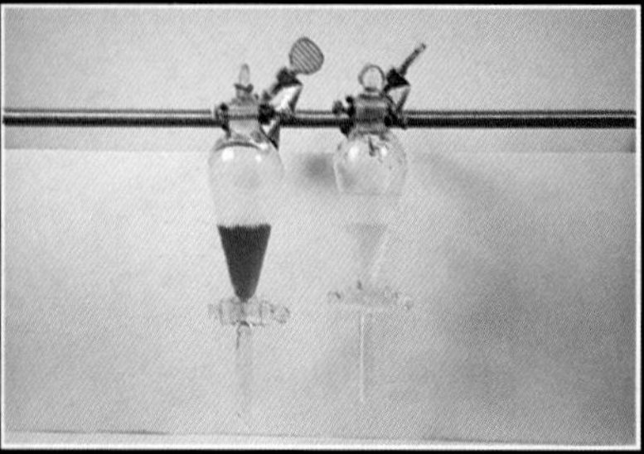
g

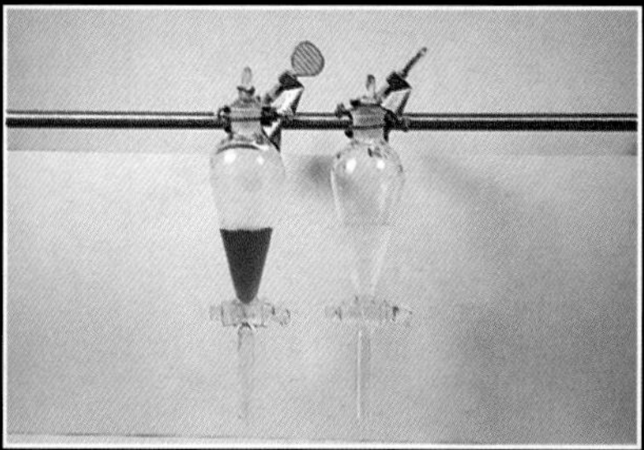
h

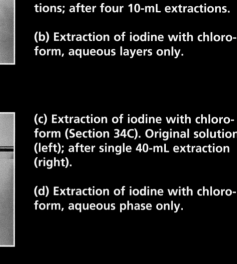

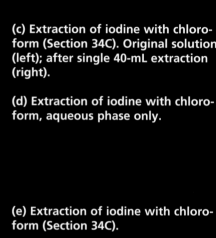

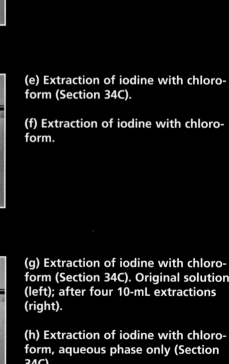

(g) Extraction of iodine with chloroform (Section 34C). Original solution (left); after four 10-mL extractions (right).

(h) Extraction of iodine with chloroform, aqueous phase only (Section 34C).

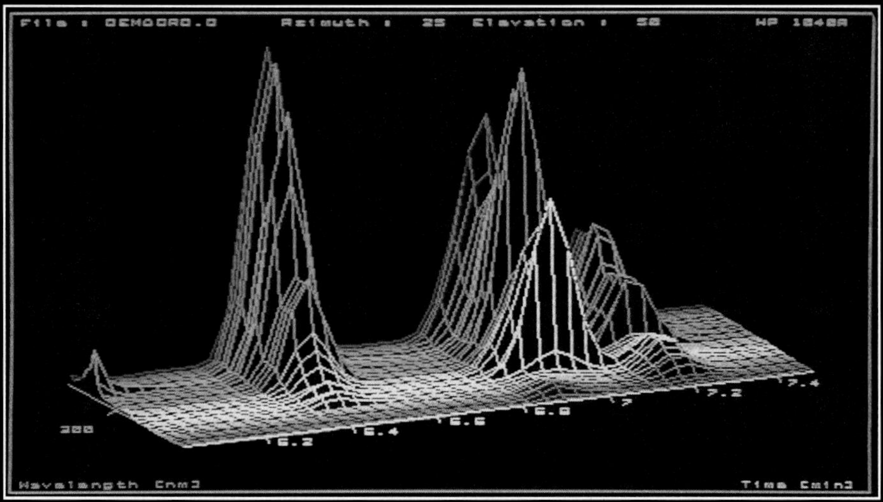

Color Plate 22

High performance liquid chromatogram (Section 30B) obtained with a multichannel diode array spectrometer (Section 24A-2) as detector. The chromatogram is displayed on the terminal of a computer, which controls the instrument. Retention time is displayed horizontally, and various detected wavelengths are presented in different colors.

Photo courtesy of Hewlett-Packard Company, Palo Alto, CA.

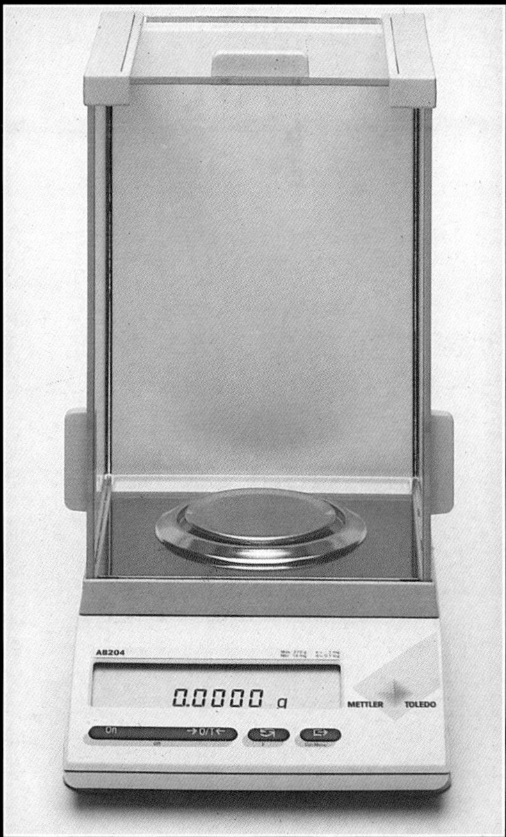

Color Plate 23

Electronic analytical balance with 0.1 mg precision (Section 35D-2).

Photo courtesy of Mettler-Toledo, Inc., Hightstown, NY.

Color Plate 24

Stamps from various countries celebrating scientists and events of importance to analytical chemistry. The numbers following each citation refer to the Scott Standard Postage Stamp Catalogue. (1) Stamp announcing an international spectroscopy colloquium in Madrid, Spain, #1570. (2) Stamp depicting the visible spectrum and atomic lines, Canada, #613. (3) Stamp honoring Archer J. P. Martin and Richard L. M. Synge for their work in chromatography, Great Britain, #808. Stamps commemorating (4) Cato Guldberg and Peter Waage, Norway, # 453; (5) Allessandro Volta, Italy, #527; (6) Svante Arrhenius, Sweden, #549; (7) Gustav Robert Kirchhoff, Germany, #9N345; (8) Jöns Jacob Berzelius, Sweden, #1293; (9) André Marie Ampére, Monaco, #1001; (10) T. W. Richards, Sweden, #1104; (11) the Centennial of the Metric Convention of 1875, France #1475; (12) Walther Nernst, Sweden, #655; (13) spectroscopic analysis, Vatican, #655. *These stamps are from the private collection of Professor C. M. Lang.*

CHAPTER

23

Instruments for Optical Spectrometry

The basic components of analytical instruments for emission, absorption, and fluorescence spectroscopy are remarkably alike in function and general performance requirements regardless of whether the instruments are designed for ultraviolet, visible, or infrared radiation. Because of their similarities, such instruments are frequently referred to as *optical instruments* even though they may be applied to spectral regions to which the eye is insensitive. In this chapter, we first examine the characteristics of the components common to all optical instruments and point out those features that are independent of the wavelength region being employed as well as those that are not. We then consider the general design characteristics of typical instruments, particularly those used for absorption spectroscopy.

Optical instruments are spectroscopic devices that employ ultraviolet, visible, or infrared radiation.

23A INSTRUMENT COMPONENTS

Most spectroscopic instruments are made up of five components: (1) a stable source of radiant energy, (2) a wavelength selector that permits the isolation of a restricted wavelength region, (3) one or more sample containers, (4) a radiation detector, or transducer, which converts radiant energy to a measurable signal (usually electrical), and (5) a signal processor and readout. Figure 23-1 shows how these components are assembled in instruments for emission, absorption, and fluorescence spectroscopy. Note that the configuration of components 4 and 5 is the same for the three types of instruments.

Emission instruments differ from the other two types in that components 1 and 3 are combined. That is, the sample container is an arc, a spark, a heated surface, or a flame that both holds the sample and causes it to emit characteristic radiation. In contrast, absorption and fluorescence spectroscopy require an external source of radiant energy and a cell to hold the sample. For absorption measurements, the beam from the source passes through the sample after leaving the wavelength selector (in some instruments, the positions of the sample and selector are reversed). In fluorescence, the source induces the sample to emit characteristic radiation, which is usually measured at a 90-deg angle with respect to the beam from the source.

In emission spectroscopy, the sample serves as the radiation source.

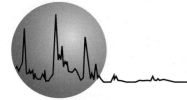

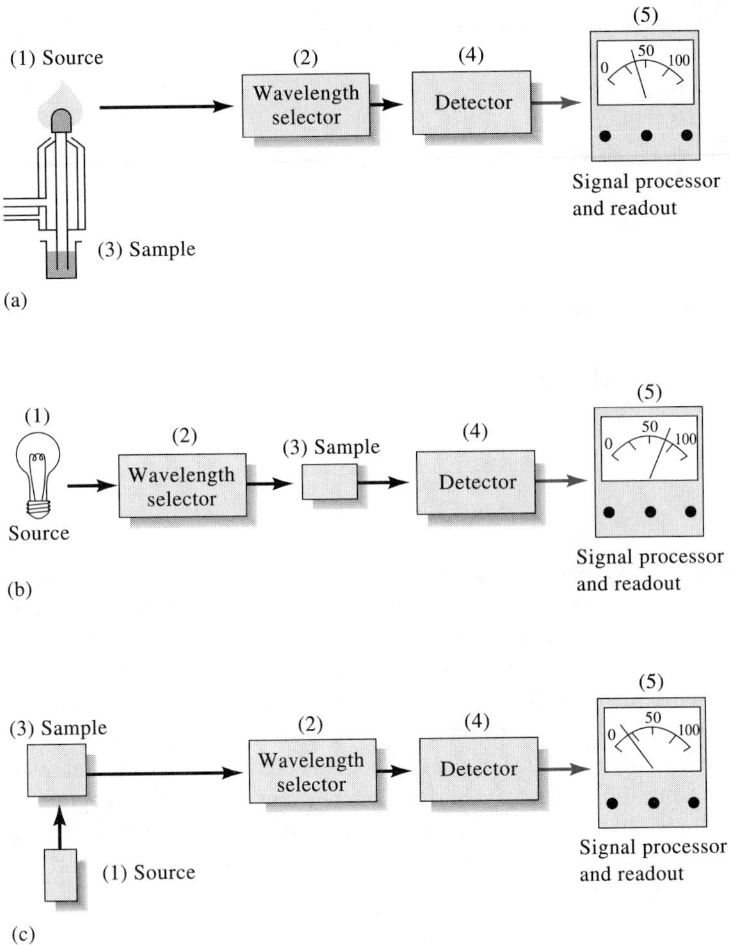

Figure 23-1
Components of various types of instruments for optical spectroscopy: (a) emission spectroscopy; (b) absorption spectroscopy; (c) fluorescence and scattering spectroscopy.

23A-1 Optical Materials

The cells, windows, lenses, and wavelength-dispersing elements in a spectrophotometer or photometer must be transparent in the wavelength region being employed for analyses. Figure 23-2 shows the usable wavelength ranges for several construction materials that find use for the ultraviolet, visible, and infrared regions of the spectrum. Ordinary silicate glass is completely adequate for the visible region and has the considerable advantage of low cost. Below about 380 nm, glass begins to absorb and fused silica or quartz must be substituted for glass.

Infrared spectroscopy requires the use of windows and cells fashioned from such materials as polished sodium chloride, potassium chloride, or silver chloride. All infrared transparent materials tend to become fogged as a result of moisture and must therefore be polished regularly until clear again.

23A-2 Spectroscopic Sources

A source for absorption and fluorescence measurements must generate a beam with sufficient power in the wavelength region of interest to permit ready detec-

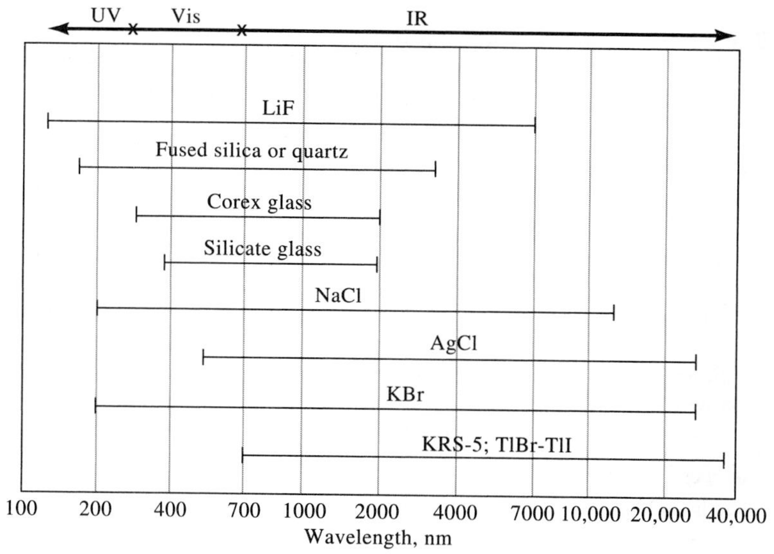

Figure 23-2
Transmittance range for various optical materials.

tion and measurement. In addition, the source must provide a stable output for the period needed to measure P_0 and P in Equations 22-6 and 22-7. Only under these conditions are reproducible absorbance measurements obtained. Some instruments are designed to measure P_0 and P simultaneously, or nearly so, thus making the stability requirement less stringent.

Spectroscopic sources are either *continuous sources* or *line sources.* The former are made up of radiation of all wavelengths within the spectral region for which they are to be used (see Figures 23-3b and 23-4b for examples of continuous radiation). A line source consists of one or more very narrow bands of radiation whose wavelengths are known exactly. Table 23-1 lists the common sources that are used in optical spectroscopy. In this section, we describe sources of continuous ultraviolet, visible, and infrared radiation. Details concerning line sources will be found in Section 26D.

A **line source** yields several very narrow (<0.01 mm) bands of radiation having wavelengths that are characteristic for the source.

A **continuous source** produces radiation of all wavelengths within a given spectral region. The radiant intensity changes only relatively slowly from one wavelength to the next.

Tungsten-Filament Lamps

The most common source of visible and near-infrared radiation is the tungsten-filament lamp, shown in Figure 23-3a. The energy distribution of this source approximates that of a blackbody and is thus temperature dependent. Figure 22-19 illustrates the output of the tungsten-filament lamp at 3000 K. In most absorption instruments, the operating filament temperature is about 2900 K; the bulk of the energy is thus emitted in the infrared region (Figure 23-3b). A tungsten-filament lamp is useful for the wavelength region between 320 and 2500 nm. The lower limit is imposed by absorption by the glass envelope that houses the filament.

The energy output of a tungsten lamp in the visible region varies approximately as the fourth power of the operating voltage, thus making close voltage control essential. For this reason, constant-voltage transformers or electronic voltage regulators are employed. As an alternative, the lamp can be operated from a 6V storage battery, which provides a remarkably stable voltage if it is maintained in good condition.

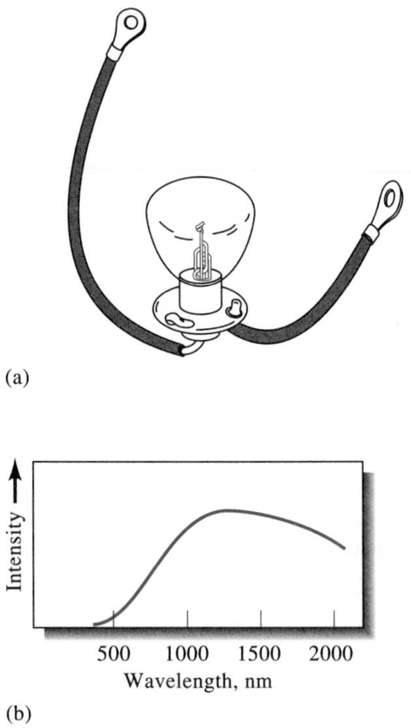

Figure 23-3
(a) A tungsten lamp. (b) Its emission spectrum.

TABLE 23-1 Sources for Spectroscopy

Source	Wavelength Region, nm	Type of Spectroscopy
Continuous Sources		
Xenon lamp	250–600	Molecular fluorescence; Raman
H_2 and D_2 lamps	160–380	UV molecular absorption
Tungsten/halogen lamp	240–2500	UV/vis/near-IR molecular absorption
Tungsten lamp	350–2200	Vis/near-IR molecular absorption
Nernst glower	400–20,000	IR molecular absorption
Nichrome wire	750–20,000	IR molecular absorption
Globar	1200–40,000	IR molecular absorption
Line Sources		
Hollow cathode lamp	UV/vis	Atomic absorption; atomic fluorescence
Electrodeless discharge lamp	UV/vis	Atomic absorption; atomic fluorescence
Metal vapor lamp	UV/vis	Atomic absorption; molecular fluorescence; Raman
Laser	UV/vis/IR	Raman; molecular absorption; molecular fluorescence

Tungsten-halogen lamps contain a small amount of iodine, which permits them to be operated at up to 3500 K and extends their range into the near-ultraviolet.

Tungsten-halogen lamps contain a small quantity of iodine within a quartz envelope that houses the filament. Quartz allows the filament to be operated at a temperature of about 3500 K, which leads to higher intensities and extends the range of the lamp well into the ultraviolet. The lifetime of a tungsten-halogen lamp is more than double that of an ordinary tungsten lamp because the life of the latter is limited by sublimation of the tungsten from the filament. In the presence of iodine, the sublimed tungsten reacts to give gaseous WI_2 molecules, which then diffuse back to the hot filament, where they decompose and redeposit as tungsten atoms. Tungsten/halogen lamps are finding ever-increasing use in modern spectroscopic instruments because of their extended wavelength range, greater intensity, and longer life.

Hydrogen and Deuterium Lamps

A truly continuous spectrum in the ultraviolet region is produced by the electrical excitation of deuterium or hydrogen at low pressure. The mechanism by which a continuum is produced involves the formation of an excited molecule (D_2^* or H_2^*) by absorption of electrical energy. This species then dissociates to give two hydrogen or deuterium atoms plus an ultraviolet photon. The reactions for hydrogen are

$$H_2 + E_e \longrightarrow H_2^* \longrightarrow H' + H'' + h\nu$$

where E_e is the electrical energy absorbed by the molecule. The energy for the overall process is

$$E_e = E_{H_2^*} = E_{H'} + E_{H''} + h\nu$$

where $E_{H_2^*}$ is the *fixed quantized energy* of H_2^*, and $E_{H'}$ and $E_{H''}$ are the *kinetic energies* of the two hydrogen atoms. The sum of the latter two energies can vary from zero to $E_{H_2^*}$. Thus, the energy and the frequency of the photon can also vary within this range of energies. That is, when the two kinetic energies are by chance small, $h\nu$ is large, and when the two energies are large, $h\nu$ is small. The consequence is a truly continuous spectrum from about 160 nm to the beginning of the visible region.

Most modern lamps for generating ultraviolet radiation contain deuterium and are of a low-voltage type in which an arc is formed between a heated, oxide-coated filament and a metal electrode (see Figure 23-4a). The heated filament provides electrons to maintain a direct current at a potential of about 40 V; a regulated power supply is required for constant intensities.

Both deuterium and hydrogen lamps provide a useful continuous spectrum in the region from 160 to 375 nm, as shown in Figure 23-4b. The intensity of the deuterium lamp is greater than that of the hydrogen lamp, however, which accounts for the more widespread use of the former. At longer wavelengths (>360 nm), the lamps generate emission lines, which are superimposed on the continuum. For many applications, these lines are a nuisance. They are useful for wavelength calibration of absorption instruments, however.

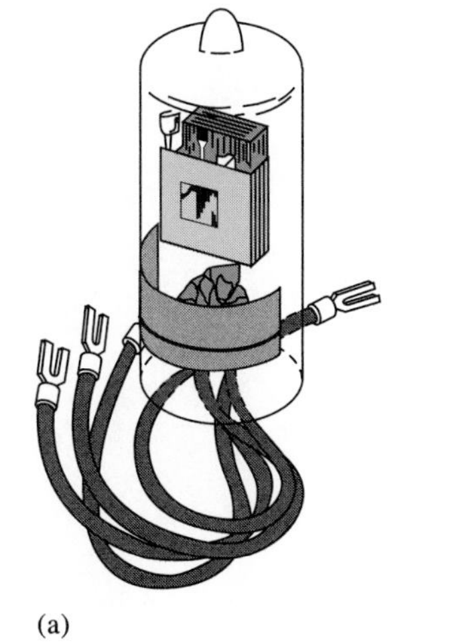

(a)

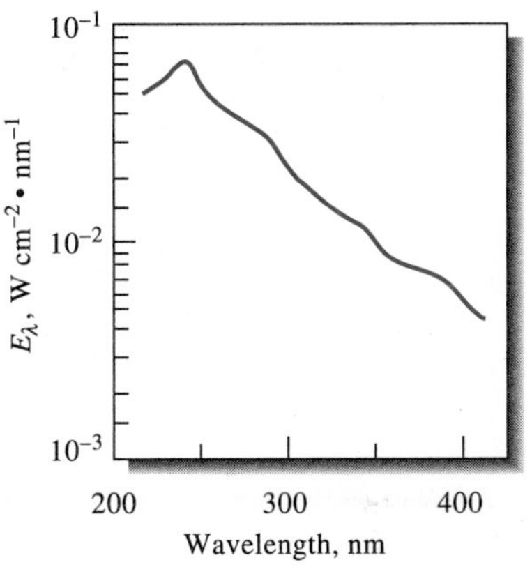

(b)

Figure 23-4
(a) A deuterium lamp. (b) Its emission spectrum.

Infrared Sources

Continuous infrared radiation is obtained from hot inert solids. A *Globar* source consists of a 5 by 50 mm silicon carbide rod. Radiation in the region from 1 to 40 μm is emitted when the Globar is heated to about 1500°C by the passage of electricity.

A *Nernst glower* is a cylinder of zirconium and yttrium oxides having typical dimensions of 2 by 20 mm; it emits infrared radiation when heated to a high temperature by an electric current. Electrically heated spirals of nichrome wire also serve as infrared sources.

23A-3 Wavelength Selectors

Spectroscopic instruments generally require a device that restricts the wavelength that is to be used to a narrow band that is absorbed or emitted by the analyte. Such devices greatly enhance both the selectivity and the sensitivity of an instrument. In addition, for absorption measurements, narrow bandwidths increase the likelihood of adherence to Beer's law.

At the outset, it must be understood that no selector is capable of producing radiation of a single wavelength. Instead, the output of such a device is a range of contiguous wavelengths called a band; these wavelengths are distributed more or less symmetrically about a central *nominal wavelength.* As shown in Figure 23-5, the *effective bandwidth,* or *bandwidth,* of a selector is defined as the width of the band in wavelength units at half peak height. Note that the ordinate in this plot is

The **effective bandwidth** of a wavelength selector is the width of the band of transmitted radiation in wavelength units at half peak height.

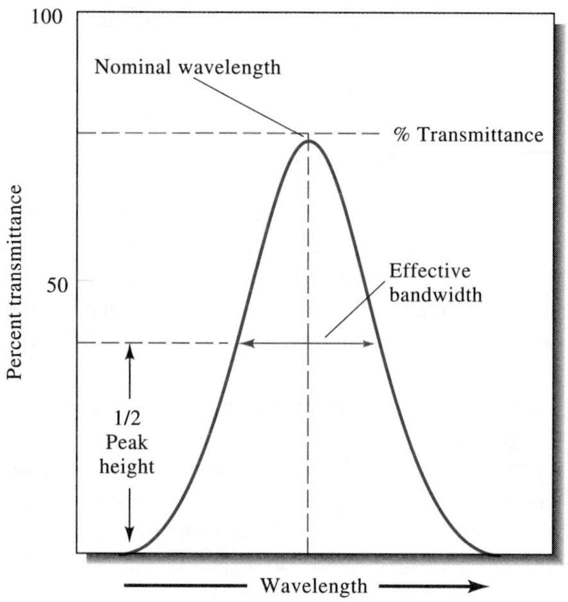

Figure 23-5
Output of a typical wavelength selector.

the percentage of incident radiation of a given wavelength that is transmitted. Bandwidths vary enormously from one wavelength selector to another. For example, a high-quality monochromator for the visible region may have an effective bandwidth of a few tenths of a nanometer or less, whereas an absorption filter in this same region may possess a bandwidth of 200 nm or more.

As shown in Table 23-2, two general types of wavelength selectors, *filters* and *monochromators,* are used to provide narrow bands of radiation. Monochromators have the advantage that the output wavelength can be varied continuously over a considerable spectral range. Filters have the advantages of simplicity, ruggedness, and low cost.

Radiation Filters

Filters operate by absorbing all but a restricted band of radiation from a continuous source. As shown in Figure 23-6, two types of filters are used in absorption

TABLE 23-2 Wavelength Selectors for Spectroscopy

Type	Wavelength Range, nm	Note
Continuously Variable		
Grating monochromator	100–40,000	3000 lines/mm for vacuum UV; 50 lines/mm for far IR
Prism monochromator	120–30,000	See Figure 23–2 for construction materials.
Discontinuous		
Interference filter	200–14,000	
Absorption filter	380–750	

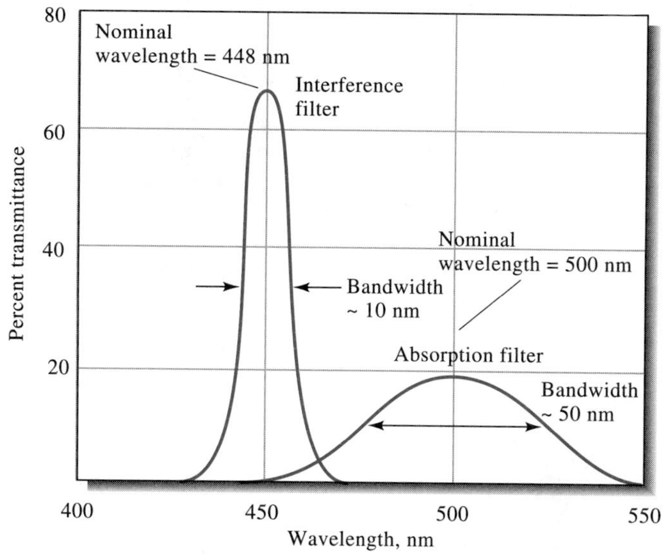

Figure 23-6
Bandwidths for two types of filters.

measurements: *interference filters* and *absorption filters.* Interference filters generally transmit a much greater percentage of radiation at their nominal wavelengths than do absorption filters.

Interference Filters Interference filters find use with ultraviolet and visible radiation, as well as with wavelengths up to about 14 μm in the infrared region. As the name implies, an interference filter relies on optical interference to provide a relatively narrow band of radiation.

As shown in Figure 23-7a, an interference filter consists of a very thin layer of a transparent *dielectric material* (frequently calcium fluoride or magnesium fluoride) coated on both sides with a film of metal that is thin enough so it transmits approximately half of the radiation striking it and reflects the other half. This array is sandwiched between two glass plates that protect it from the atmosphere. When radiation strikes the central array at a 90-deg angle, approximately half is transmitted by the first metallic layer and the other half reflected. The transmitted radiation undergoes a similar partition when it reaches the second layer of metal. If the reflected portion from the second layer is of the proper wavelength, it is partially reflected from the inner portion of the first layer in phase with the incoming light of the same wavelength. The result is constructive interference of the radiation of this wavelength and destructive removal of most other wavelengths. As shown in Feature 23-1, the nominal wavelength for an interference filter λ_{max} is given by the equation

A **dielectric** is a nonconducting substance or insulator. Such materials are usually optically transparent.

$$\lambda_{max} = 2t\eta/\mathbf{n} \qquad (23\text{-}1)$$

where t is the thickness of the central fluoride layer, η is its refractive index, and $\mathbf{n}$ is an integer called the *interference order.* The glass layers of the filter are often selected to absorb all but one of the wavelengths transmitted by the central layer, thus restricting the transmission of the filter to a single order.

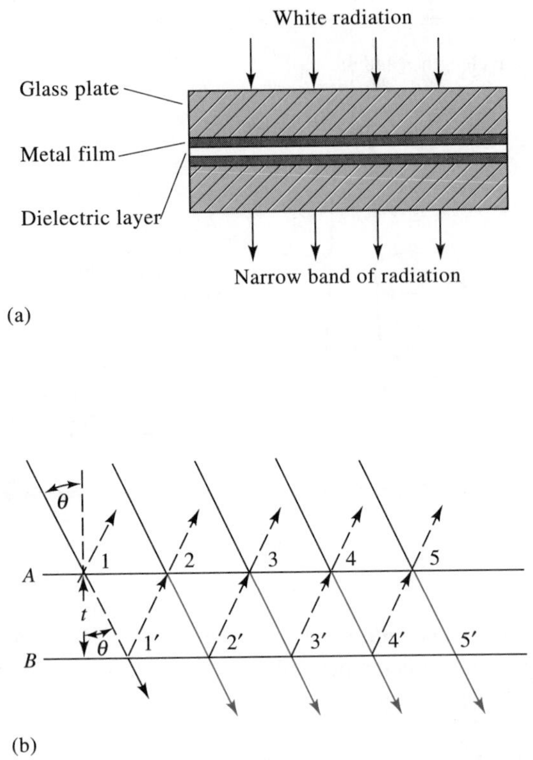

Figure 23-7
(a) Schematic cross section of an interference filter. Note that the drawing is not to scale and the three central bands are much narrower than shown.
(b) Schematic to show the conditions for constructive interference.

Figure 23-6 illustrates the performance characteristics of a typical interference filter. Most filters of this type have bandwidths of better than 1.5% of the nominal wavelength, although this figure is lowered to 0.15% in some narrow-band filters; the latter have a maximum transmittance of about 10%.

FEATURE 23-1
Derivation of Equation 23-1

The relationship between the thickness of the dielectric layer t and the transmitted wavelength λ can be found with the aid of Figure 23-7b. For purposes of clarity, the incident beam is shown as arriving at an angle θ from the perpendicular. At point 1, the radiation is partially reflected and partially transmitted to point 1′, where partial reflection and transmission again takes place. The same process occurs at 2, 2′, and so forth. For reinforcement to occur at point 2, the distance traveled by the beam reflected at 1′ must be some multiple of its wavelength in the medium λ'. Since the path length between surfaces can be expressed as $t/\cos\theta$, the condition for reinforcement is that $\mathbf{n}\lambda' = 2t/\cos\theta$ where **n** is a small whole number.

In ordinary use, θ approaches zero and $\cos\theta$ approaches unity so that the equation derived from Figure 23-7 simplifies to

$$\mathbf{n}\lambda' = 2t \qquad (23\text{-}2)$$

where λ' is the wavelength of radiation *in the dielectric* and t is the thickness of the dielectric. The corresponding wavelength in air is given by

$$\lambda = \lambda'\eta$$

where η is the refractive index of the dielectric medium. Thus, the wavelengths of radiation transmitted by the filter are

$$\boxed{\lambda = \frac{2t\eta}{\mathbf{n}}} \tag{23-3}$$

Absorption Filters Absorption filters, which are generally less expensive and more rugged than interference filters, are limited in application to the visible region. This type of filter usually consists of a colored glass plate that removes part of the incident radiation by absorption. Absorption filters have effective bandwidths that range from perhaps 30 to 250 nm. Filters that provide the narrowest bandwidths also absorb a significant fraction of the desired radiation and may have a transmittance of 1% or less at their band peaks. Figure 23-6 contrasts the performance characteristics of a typical absorption filter with its interference counterpart.

Glass filters with transmittance maxima throughout the entire visible region are available from commercial sources. While their performance characteristics are distinctly inferior to those of interference filters, their cost is appreciably less, and they are perfectly adequate for many routine applications.

Monochromators

As shown in Figure 23-8, monochromators are of two general types, one of which employs a grating to disperse radiation into its component wavelengths; the other uses a prism for this purpose. Early spectroscopic instruments were of the prism type. Currently, however, nearly all commercially available monochromators are based upon *reflection gratings* because they are less expensive than prisms and make possible the design of more compact and simpler instruments. We shall limit our discussion of monochromators to those based upon reflection gratings.

Figure 23-8a shows the design of a typical grating monochromator. Radiation from a source enters the monochromator via a narrow rectangular opening, or *slit.* The radiation is then collimated by a concave mirror, which produces a parallel beam that strikes the surface of a *reflection grating.* Angular dispersion results from diffraction, which occurs at the reflective surface. For illustrative purposes, the radiation entering the monochromator is shown as consisting of just two wavelengths, λ_1 and λ_2, where λ_1 is longer than λ_2. The pathway of the longer radiation after it is reflected from the grating is shown by the dashed lines; the solid lines show the path of the shorter wavelength. Note that the shorter wavelength radiation λ_2 is reflected off the grating at a sharper angle than is λ_1. That is, *angular dispersion* of the radiation takes place at the grating surface. The two wavelengths are focused by another concave mirror onto the *focal plane* of the

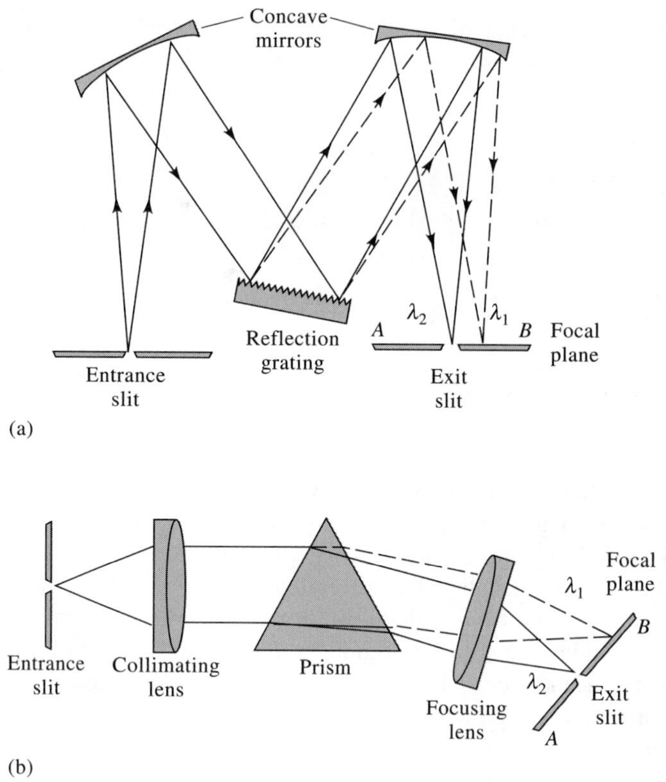

Figure 23-8
Two types of monochromators: (a) Czerney-Turner grating monochromator; (b) Bunsen prism monochromator. (In both instances, $\lambda_1 > \lambda_2$.)

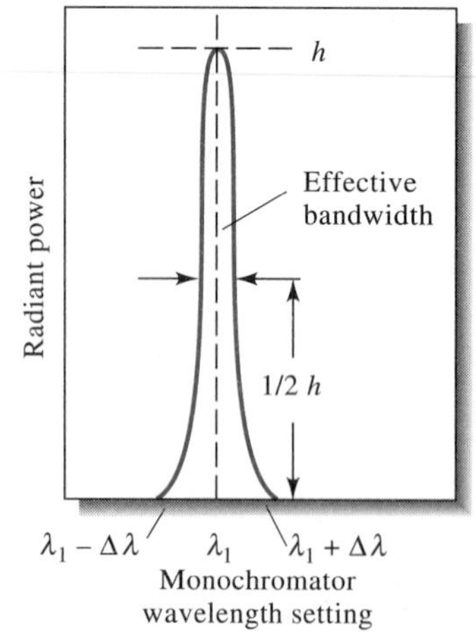

Figure 23-9
Output of an exit slit as monochromator is scanned from $\lambda_1 - \Delta\lambda$ to $\lambda_1 + \Delta\lambda$.

monochromator, where they appear as two images of the entrance slit, one for λ_1 and the other for λ_2. By rotating the grating, either one of these images can be focused on the exit slit.

If a detector is located at the exit slit of the monochromator shown in Figure 23-8a and the grating is rotated so that one of the lines shown (say, λ_1) is *scanned* across the slit from $\lambda_1 - \Delta\lambda$ to $\lambda_1 + \Delta\lambda$, where $\Delta\lambda$ is a small wavelength difference, the output of the detector takes the Gaussian shape shown in Figure 23-9. The effective bandwidth of the monochromator, which is defined in the figure, depends upon the size and quality of the dispersing element, the slit widths, and the focal length of the monochromator. A high-quality monochromator will exhibit an effective bandwidth of a few tenths of a nanometer or less in the ultraviolet/visible region. The effective bandwidth of a monochromator that is satisfactory for most quantitative applications is from about 1 to 20 nm.

Many monochromators are equipped with adjustable slits to permit some control over the bandwidth. A narrow slit decreases the effective bandwidth but also diminishes the power of the emergent beam. Thus, the minimum practical bandwidth may be limited by the sensitivity of the detector. For qualitative analysis, narrow slits and minimum effective bandwidths are required if a spectrum is made up of narrow peaks. For quantitative work, on the other hand, wider slits permit operation of the detector system at lower amplification, which in turn provides greater reproducibility of response.

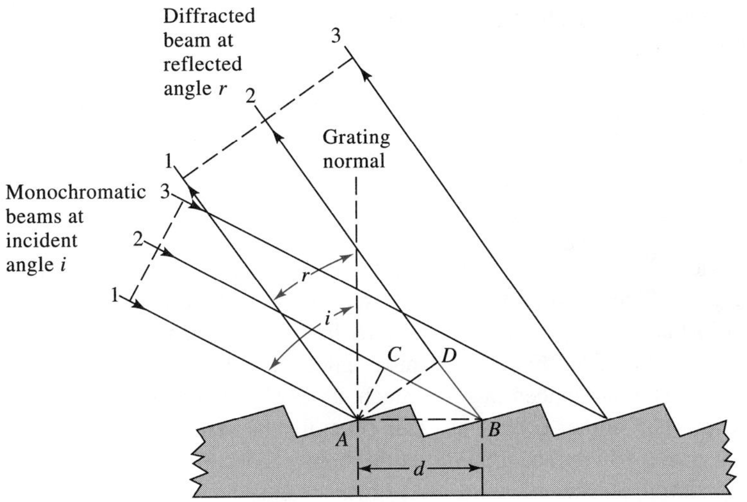

Figure 23-10
Mechanism of diffraction from an echellette-type grating.

Gratings

Most gratings in modern monochromators are *replica gratings,* which are obtained by making castings of a *master grating.*[1] The latter consists of a hard, optically flat, polished surface upon which a large number of parallel and closely spaced grooves have been ruled with a suitably shaped diamond tool. A magnified cross-sectional view of a few typical grooves is shown in Figure 23-10. A grating for the ultraviolet and visible region will typically contain from 300 to 2000 grooves/mm, with 1200 to 1400 being most common. For the infrared region, 10 to 200 grooves/mm are encountered; for spectrophotometers designed for the most widely used infrared range of 5 to 15 μm, a grating with about 100 grooves/mm is often employed. The construction of good master grating is tedious, time-consuming, and expensive because the grooves must be identical in size, exactly parallel, and equally spaced over the length of the grating (3 to 10 cm).

Replica gratings are formed from a master grating by a liquid resin casting process that preserves virtually perfectly the optical accuracy of the original master grating on a clear resin surface. This surface is ordinarily made reflective by a coating of aluminum, or sometimes gold or platinum.

The Echellette Grating One of the most common types of reflection gratings is the *echellette grating.* Figure 23-10 is a schematic representation of this type of grating, which is grooved or *blazed* such that it has relatively broad faces from which reflection occurs and narrow unused faces. This geometry provides highly efficient diffraction of radiation. In the figure, a parallel beam of monochromatic radiation is shown approaching the grating surface at an angle i relative to the grating normal. The incident radiation consists of three parallel beams that make up a wave front labeled 1, 2, 3. The diffracted beam is reflected at the angle r, which depends upon the wavelength of the radiation. In Feature 23-2 we show

---

[1]For an interesting and informative discussion of the manufacture, testing, and performance characteristics of gratings, see *Diffraction Grating Handbook,* Rochester, NY: Bausch and Lomb, Inc. (now Milton Roy Company), 1970.

that the angle of reflection, r, is related to the wavelength of the incoming radiation by the equation

$$\mathbf{n}\lambda = d(\sin i + \sin r) \tag{23-4}$$

Equation 23-4 suggests that several values of λ exist for a given diffraction angle r. Thus, if a first-order line ($\mathbf{n} = 1$) of 900 nm is found at r, second-order (450 nm) and third-order (300 nm) lines also appear at this angle. Ordinarily, the first-order line is the most intense; indeed, it is possible to design gratings that concentrate as much as 90% of the incident intensity in this order. The higher-order lines can generally be removed by filters. For example, glass, which absorbs radiation below 350 nm, eliminates the high-order spectra associated with first-order radiation in most of the visible region.

EXAMPLE 23-1

An echellette grating containing 1450 blazes per millimeter was irradiated with a polychromatic beam at an incident angle 48 deg to the grating normal. Calculate the wavelengths of radiation that would appear at an angle of reflection of +20, +10, and 0 deg (angle r, Figure 23-10).

To obtain d in Equation 23-4, we write

$$d = \frac{1 \text{ mm}}{1450 \text{ blazes}} \times 10^6 \frac{\text{nm}}{\text{mm}} = 689.7 \frac{\text{nm}}{\text{blaze}}$$

When r in Figure 23-10 equals +20 deg, λ can be obtained by substitutions into Equation 23-4. Thus,

$$\lambda = \frac{689.7}{\mathbf{n}} \text{ nm} (\sin 48 + \sin 20) = \frac{748.4}{\mathbf{n}} \text{ nm}$$

and the wavelengths for the first-, second-, and third-order reflections are 748, 374, and 249 nm, respectively. Further calculations of a similar kind yield the following data:

	Wavelength (nm) for		
r, deg	$\mathbf{n} = 1$	$\mathbf{n} = 2$	$\mathbf{n} = 3$
20	748	374	249
10	632	316	211
0	513	256	171

FEATURE 23-2
Derivation of Equation 23-4

In Figure 23-10, parallel beams of monochromatic radiation labeled 1 and 2 are shown striking two of the broad faces at an incident angle i to the *grating normal.* Maximum constructive interference is shown as occurring at the reflected angle r. It is evident that beam 2 travels a greater distance than beam 1 and that this difference is equal to $\overline{CB} + \overline{BD}$. For constructive interference to occur, this difference must equal $\mathbf{n}\lambda$:

$$\mathbf{n}\lambda = \overline{CB} + \overline{BD}$$

where **n**, a small whole number, is called the diffraction *order.* Note, however, that angle CAB is equal to angle i and that angle DAB is identical to angle r. Therefore, from trigonometry,

$$\overline{CB} = d \sin i$$

where d is the spacing between the reflecting surfaces. It is also seen that

$$\overline{BD} = d \sin r$$

Substitution of the last two expressions into the first gives Equation 23-4. That is,

$$\mathbf{n}\lambda = d(\sin i + \sin r)$$

One of the advantages of a monochromator with an echellette grating is that, in contrast to a prism monochromator, the dispersion of radiation along the focal plane is, for all practical purposes, *linear.* Figure 23-11 demonstrates this property. The linear dispersion of a grating greatly simplifies the design of monochromators.

Concave Gratings Gratings can be formed on a concave surface in much the same way as on a plane surface. A concave grating permits the design of a monochromator without auxiliary collimating and focusing mirrors or lenses because the concave surface both disperses the radiation and focuses it on the exit slit. Such an arrangement is advantageous in terms of cost; in addition, the reduction in number of optical surfaces increases the energy throughput of monochromators containing a concave grating.

Holographic Gratings[2] One of the products from the emergence of laser technology is an optical (rather than mechanical) technique for forming gratings on

[2]See J. Flamand, A. Grillo, and G. Hayat, *Amer. Lab.*, **1975**, *7*(5), 47; J. M. Lerner, et al., *Proc. Photo-Opt. Instrum. Eng.*, **1980**, *240,* 72, 82.

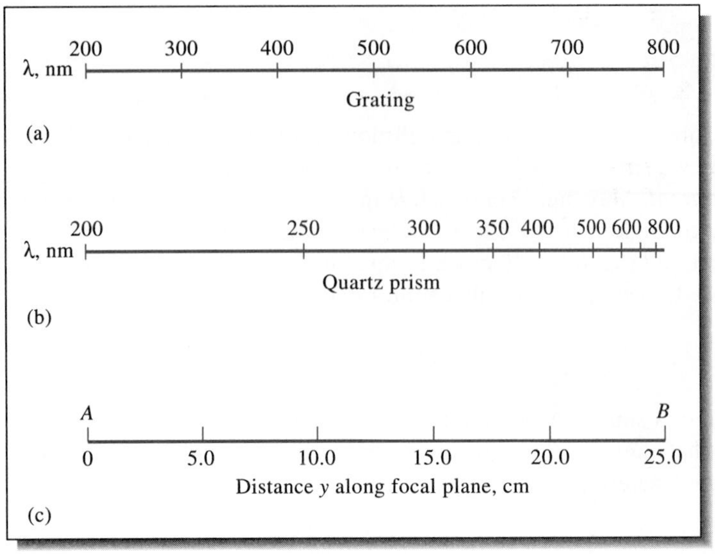

Figure 23-11
Dispersion of radiation along the focal plane *AB* of a typical (a) prism and (b) echellette grating. The positions of *A* and *B* in the scale in (c) are shown in Figure 23-8.

plane or concave glass surfaces. *Holographic gratings* produced in this way are appearing in ever-increasing numbers in modern optical instruments, even some of the less expensive ones. Because of their greater perfection with respect to line shape and dimensions, holographic gratings provide spectra that are freer from stray radiation and ghosts (double images).

In the preparation of holographic gratings, the beams from a pair of identical lasers are brought to bear at suitable angles upon a glass surface coated with photoresist. The resulting interference fringes from the two beams sensitize the photoresist so that it can be dissolved away, leaving a grooved structure that can be coated with aluminum or another reflecting substance to produce a reflection grating. The spacing of the grooves can be changed by changing the angle of the two laser beams with respect to one another. Nearly perfect, large (~ 50 cm) gratings having as many as 6000 lines/mm can be manufactured in this way at a relatively low cost. As with ruled gratings, replica gratings can be cast from a master holographic grating. It has been reported that no optical test exists that can distinguish between a master and a replica holographic grating.[3]

23A-4 Radiation Detectors and Transducers

A *detector* is a device that indicates the existence of some physical phenomenon. Familiar examples of detectors include photographic film for indicating the presence of electromagnetic or radioactive radiation, the pointer of a balance for detecting mass differences, and the mercury level in a thermometer for detecting temperature changes. The human eye is also a detector; it converts visible radiation into an electrical signal that is passed to the brain via the chain of neurons in the optic nerve.

[3] I. R. Altelmose, *J. Chem. Educ.*, **1986**, *63*, A221.

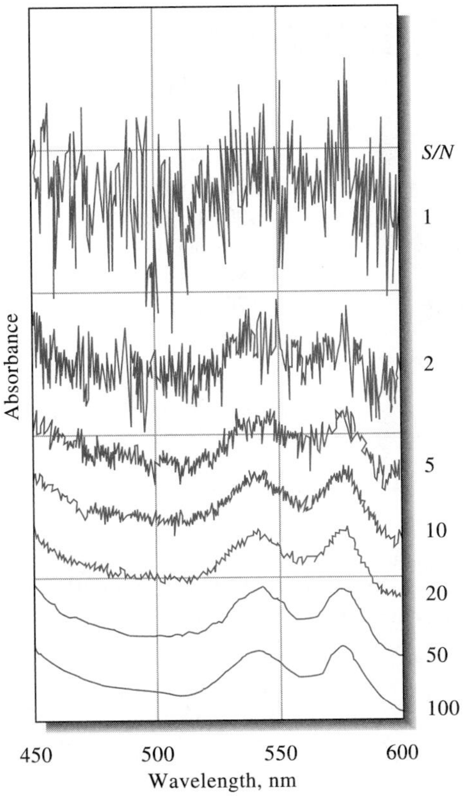

Figure 23-12
Absorption spectra of hemoglobin with identical signal levels but different amounts of noise. Note that the curves have been offset on the absorbance axis for clarity.

A *transducer* is a special type of detector that converts signals, such as light intensity, pH, mass, and temperature into *electrical* signals that can be subsequently amplified, manipulated, and finally converted into numbers that are related to the magnitude of the original signal.

A **transducer** is a type of detector that converts various types of chemical and physical quantities to voltage, charge, or current.

Properties of Transducers

The ideal electromagnetic radiation transducer responds rapidly to low levels of radiant energy over a broad wavelength range. In addition, it produces an electrical signal that is easily amplified and has a relatively low *noise* level (see Figure 23-12). Finally, it is essential that the electrical signal produced by the transducer be directly proportional to the beam power P:

Common causes for noise include vibration, pickup from 60-Hz lines, temperature variations, and frequency or voltage fluctuations in the power.

$$G = KP + K' \tag{23-5}$$

where G is the electrical response of the detector in units of current, resistance, or potential. The proportionality constant K measures the sensitivity of the detector in terms of electrical response per unit of radiant power. Many detectors exhibit a small constant response, known as a *dark current* K', even when no radiation impinges upon their surfaces. Instruments with detectors that have a significant dark-current response are ordinarily equipped with a compensating circuit that

A **dark current** is a current produced by a photoelectric detector in the absence of light.

TABLE 23-3 Detectors for Spectroscopy

Type	Wavelength Range, nm	Output
Photon Detectors		
Phototubes	150–1000	Current
Photomultiplier tubes	150–1000	Current
Silicon diodes	190–1100	Current
Photoconductive cells	750–6000	Resistance change
Photovoltaic cells	380–780	Current or voltage
Charge-transfer devices	170–800	Accumulated charge
Heat Detectors		
Thermocouples	600–20,000	Voltage
Bolometers	600–20,000	Resistance change
Pneumatic cells	600–40,000	Gas pressure
Pyroelectric cells	1000–20,000	Current

permits subtraction of a signal proportional to the dark current to reduce K' to zero. Thus, under ordinary circumstances, we can write

$$G = KP \tag{23-6}$$

FEATURE 23-3
Signals, Noise, and the Signal-to-Noise Ratio

Generally, the output from analytical instruments fluctuates in a random way as a consequence of the operation of a large number of uncontrolled variables. These fluctuations, which limit the sensitivity of an instrument, are called **noise.** The terminology is derived from radio engineering, where the presence of unwanted signal fluctuations is recognizable to the ear as static, or noise.

The output of an analytical instrument fluctuates in a random way. These fluctuations limit the precision of the instrument and are the net result of a large number of uncontrolled random variables in the instrument and in the chemical system under study. An example of such a variable is the random arrival of photons at the photocathode of a photomultiplier tube. The term *noise* is used to describe these fluctuations, and each uncontrolled variable is a noise source. The term comes from audio and electronic engineering, where undesirable signal fluctuations appear to the ear as static, or noise. The average value of the output of an electronic device is called the *signal,* and the standard deviation of the signal is a measure of the noise.

An important figure of merit for analytical instruments, stereos, compact-disk players, and many other types of electronic devices is the *signal-to-noise ratio* (S/N). The signal-to-noise ratio is usually defined as the ratio of the average value of the output signal to its standard deviation. The signal-to-noise behavior of an absorption spectrophotometer is illustrated in the spectra of hemoglobin shown in Figure 23-12. The spectrum at the bottom of the figure has $S/N = 100$, and you can easily pick out the peaks at 540 nm and 580 nm. As the S/N degrades to about two in the second spectrum from the top of the figure, the peaks are barely visible. Somewhere between $S/N = 2$ and $S/N = 1$, the peaks disappear altogether into the noise and are impossible to identify. As modern instruments have become computerized and controlled by sophisticated electronic circuits, various methods have been developed to increase the signal-to-noise ratio of instrument outputs. These methods include analog filtering, lock-in amplification, boxcar averaging, smoothing, and Fourier transformation.[4]

[4]D. A. Skoog and J. J. Leary, *Principles of Instrumental Analysis,* Chapter 4. Philadelphia: Saunders College Publishing, 1992.

Types of Transducers

As shown in Table 23-3, two general types of radiation transducers are encountered; one responds to *photons,* the other to *heat.* All photon detectors are based upon the interaction of radiation with a reactive surface to produce electrons (photoemission) or to promote electrons to energy states in which they can conduct electricity (photoconduction). Only ultraviolet, visible, and near-infrared radiation have sufficient energy to cause these processes to occur; thus, photon detectors are limited to wavelengths shorter than about 2 μm.

Generally, infrared radiation is detected by measuring the temperature rise of a blackened material located in the path of the beam. Because the temperature changes resulting from the absorption of the infrared energy are minute, close control of the ambient temperature is required if large errors are to be avoided. It is usually the detector system that limits the sensitivity and precision of an infrared instrument.

Photon Detectors

Six widely used types of photon detectors are described in the paragraphs that follow: (1) phototubes, (2) photomultiplier tubes, (3) silicon photodiodes, (4) photovoltaic cells, (5) photoconductive cells, and (6) charge-transfer devices.

Phototubes As shown in Figure 23-13, a phototube consists of a semicylindrical cathode and a wire anode sealed inside an evacuated transparent envelope.

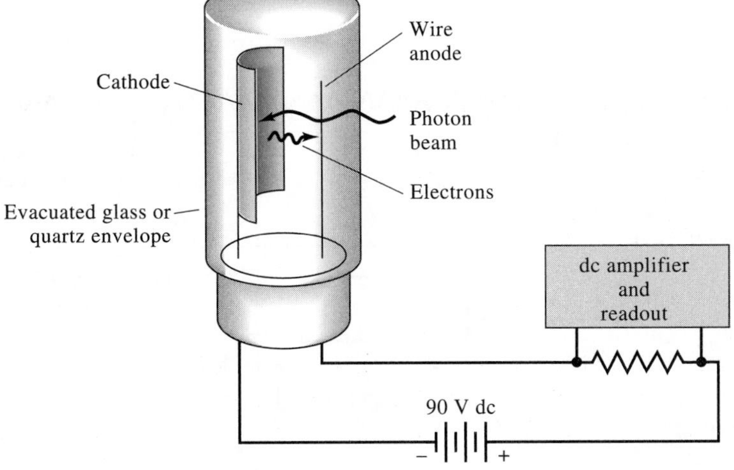

Figure 23-13
A phototube and accessory circuit. The photocurrent induced by the radiation causes a potential difference across the resistor; this potential is then amplified to drive a meter or recorder.

Photoelectrons are electrons that are ejected from a photosensitive surface as a result of absorption of electromagnetic radiation.

The concave surface of the cathode supports a layer of photoemissive material, such as an alkali metal or metal oxide, that tends to emit electrons upon being irradiated. When a potential is applied across the electrodes, the emitted *photoelectrons* flow to the wire anode, producing a current that is readily amplified and displayed or recorded.

The number of electrons ejected from a photoemissive surface is directly proportional to the radiant power of the beam striking that surface. With an applied potential of about 90 V, all these electrons reach the anode to give a current that is also proportional to radiant power. Phototubes frequently produce a small dark current in the absence of radiation (Equation 23-5) that results from thermally induced electron emission.

One of the major advantages of photomultipliers is their automatic internal amplification. About 10^6 to 10^7 electrons are produced at the anode for each photon that strikes the photocathode of a photomultiplier tube (PMT).

Photomultiplier Tubes The *photomultiplier tube* (PMT), shown schematically in Figure 23-14, is similar in construction to the phototube just described but is significantly more sensitive. Its cathode surface is similar in composition to that

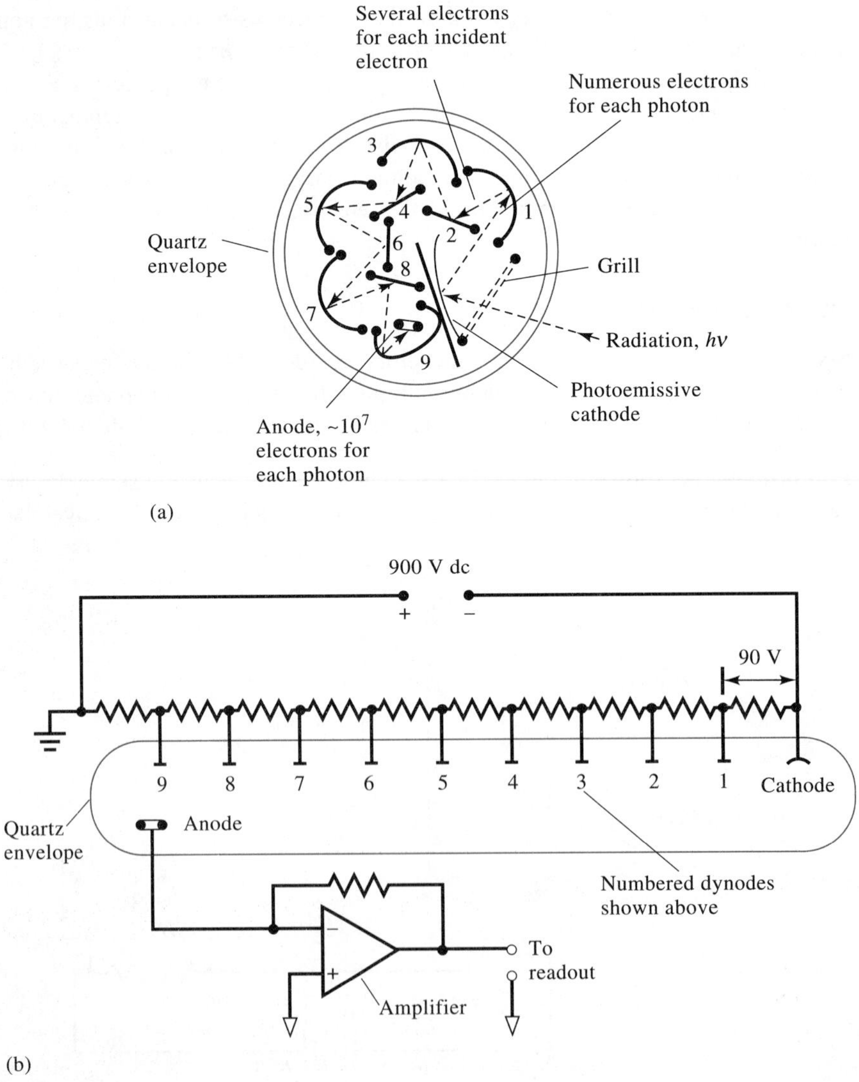

Figure 23-14
Diagram of a photomultiplier tube. (a) Cross section of the tube. (b) Electrical circuit.

of a phototube, with electrons being emitted upon exposure to radiation. The emitted electrons are accelerated toward a *dynode* (labeled 1 in the figure) maintained at a potential 90 V more positive than the cathode. Upon striking the dynode surface, each accelerated photoelectron produces several additional electrons, all of which are then accelerated to dynode 2, which is 90 V more positive than dynode 1. Here again, electron amplification occurs. By the time this process has been repeated at each of the remaining dynodes, 10^6 to 10^7 electrons have been produced for each photon; this cascade is finally collected at the anode. The resulting current is then further amplified electronically and measured.

Photomultiplier tubes are currently the most widely used types of transducers for detecting ultraviolet/visible radiation.

With modern electronic instrumentation, it is possible to detect the arrival of individual photons at the photocathode of a PMT. The photons are counted, and the accumulated count is a measure of the intensity of the electromagnetic radiation impinging on the PMT. Photon counting is advantageous when light intensity, or the frequency of arrival of photons at the photocathode, is low.

Silicon Photodiodes Crystalline silicon is a *semiconductor*—that is, a material whose electrical conductivity is less than that of a metal but greater than that of an electrical insulator. Silicon is a Group IV element and thus has four valence electrons. In a silicon crystal, each of these electrons is combined with electrons from four other silicon atoms to form four covalent bonds. At room temperature, sufficient thermal agitation occurs in this structure to liberate an occasional electron from its bonded state, leaving it free to move throughout the crystal. Thermal excitation of an electron leaves behind a positively charged region termed a *hole,* which, like the electron, is also mobile. The mechanism of hole movement is stepwise, with a bound electron from a neighboring silicon atom jumping into the electron-deficient region (the hole) and thereby creating another positive hole in its wake. Conduction in a semiconductor involves the movement of electrons and holes in opposite directions.

A **semiconductor** is a substance having a conductivity that lies between that of a metal and that of a dielectric (an insulator).

The conductivity of silicon can be greatly enhanced by *doping,* a process whereby a tiny, controlled amount (approximately 1 ppm) of a Group V or Group III element is distributed homogeneously throughout a silicon crystal. For example, when a crystal is doped with a Group V element, such as arsenic, four out of five of the valence electrons of the dopant form covalent bonds with four silicon atoms, leaving one electron free to conduct (see Figure 23-15). When the silicon is doped with a Group III element, such as gallium, which has only three valence electrons, an excess of holes develops, which also enhances conductivity (see Figure 23-16). A semiconductor containing unbonded electrons (*n*egative charges) is termed an *n-type* semiconductor, and one containing an excess of holes (*p*ositive charges) is a *p-type.* In an *n*-type semiconductor, electrons are the *majority carrier;* in a *p*-type, holes are the majority carrier.

In electronics, a **bias** is a dc voltage that is applied across a circuit element.

A silicon photodiode is a reverse-biased silicon diode that is used for measuring radiant power.

Present silicon technology makes it possible to fabricate what is called a *pn junction* or a *pn diode,* which is conductive in one direction and not in the other. Figure 23-17a is a schematic of a silicon diode. The *pn* junction is shown as a dashed line through the middle of the crystal. Electrical wires are attached to both ends of the device. Figure 23-17b shows the junction in its conduction mode, wherein the positive terminal of a dc source is connected to the *p* region and the negative terminal to the *n* region (the diode is said to be *forward-biased* under these conditions). The excess electrons in the *n* region and the positive holes in the *p* region move toward the junction, where they combine and annihilate each other. The negative terminal of the source injects new electrons into the *n* region, which can continue the conduction process. The positive terminal extracts electrons from the *p* region, thus creating new holes that are free to migrate toward the *pn* junction.

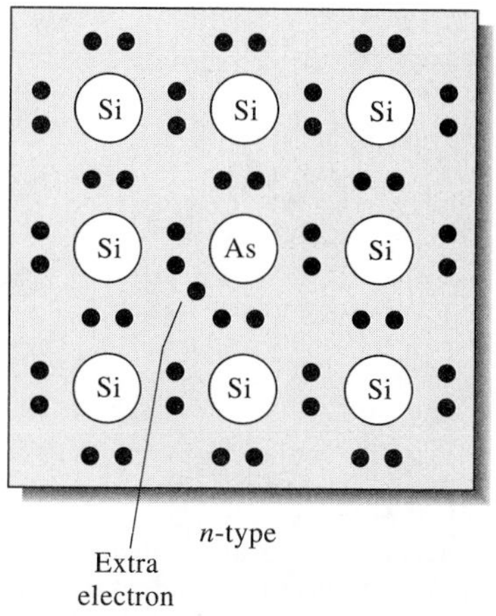

Figure 23-15
Two-dimensional representation of *n*-type silicon showing ''impurity'' atom.

Figure 23-17c illustrates the behavior of a silicon diode under *reverse biasing.* Here, the majority carriers are drawn away from the junction, leaving a nonconductive *depletion layer.* The conductance under reverse bias is only about

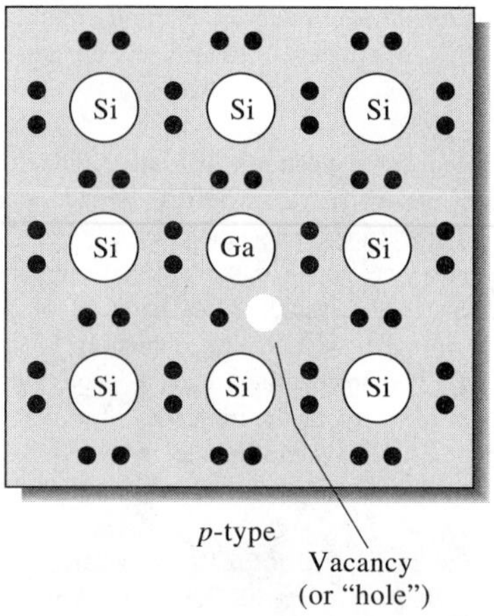

Figure 23-16
Two-dimensional representation of *p*-type silicon showing ''impurity'' atom.

Photovoltaic cells are sometimes called **barrier-layer cells.**

10^{-6} to 10^{-8} of that under forward biasing; thus, a silicon diode is a current rectifier.

A reverse-biased silicon diode can serve as a radiation detector because ultraviolet and visible photons are sufficiently energetic to create additional electrons and holes when they strike the depletion layer of a *pn* junction. The resulting increase in conductivity is readily measured and is directly proportional to radiant power. A silicon-diode detector is more sensitive than a simple vacuum phototube but less sensitive than a photomultiplier tube.

Diode-Array Detectors Silicon photodiodes have become of notable importance recently because 1000 or more can be fabricated side by side on a single small silicon chip; the width of the individual diodes is about 0.02 mm (see color plate 19). With one or two of the *diode-array detectors* placed along the length of the focal plane of a monochromator, all wavelengths can be monitored simultaneously, thus making high-speed spectroscopy possible. Multichannel instruments based upon diode arrays are discussed in Section 24A-2.

Photovoltaic Cells A photovoltaic cell (or photocell), the simplest of all radiation transducers, consists of a flat copper or iron electrode upon which is deposited a layer of a semiconducting material, such as selenium or copper(I) oxide (see Figure 23-18). The outer surface of the semiconductor is coated with a thin, transparent film of gold, silver, or lead, which serves as the second, or collector, electrode. When radiation is absorbed on the surface of the semiconductor, electrons and holes are formed and migrate in opposite directions, thus creating a current. If the two electrodes are connected through a low-resistance external circuit, the current produced is directly proportional to the power of the incident beam. The currents are large enough (10 to 100 μA) to be measured with a simple microammeter without amplification.

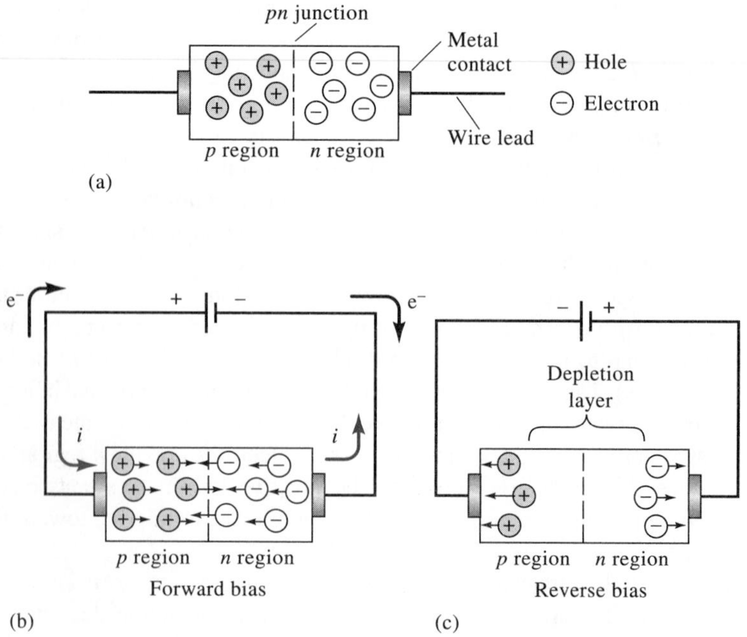

Figure 23-17
(a) Schematic of a silicon diode.
(b) Flow of electricity under forward bias. (c) Formation of depletion layer, which prevents flow of electricity under reverse bias.

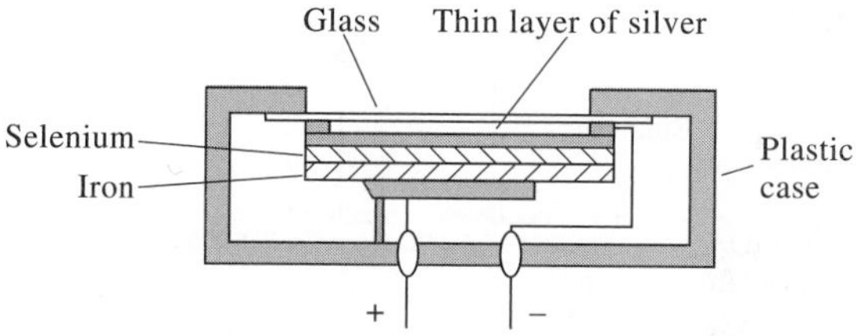

Figure 23-18
Schematic of a typical photovoltaic cell.

The typical photovoltaic cell has maximum sensitivity at about 550 nm, with the response falling off to perhaps 10% of maximum at 350 and 750 nm. The photocell constitutes a rugged, low-cost detector of visible radiation that has the advantage of not requiring an external power source. It is not as sensitive as other detectors, however, and in addition suffers from *fatigue,* which causes its current output to decrease gradually with continued illumination. Despite these disadvantages, photovoltaic cells are quite useful for simple, portable, low-cost filter instruments.

Photoconductivity Cells The most sensitive detectors for monitoring radiation in the near-infrared region (0.75 to 3 μm) are semiconductors whose resistances decrease when radiation within this range is absorbed. The useful range of photoconductors can be extended into the far-infrared region by cooling to suppress noise arising from thermally induced transitions among closely lying energy levels. This application of photoconductors is important in infrared Fourier transform instrumentation. Crystalline semiconductors are formed from the sulfides, selenides, and stibnides of such metals as lead, cadmium, gallium, and indium. Absorption of radiation by these materials promotes some of their bound electrons into an energy state in which they are free to conduct electricity. The resulting change in conductivity can then be measured.

Lead sulfide is the most widely used photoconductive material and offers the advantage that it can be used at room temperature. Lead sulfide detectors are sensitive in the region between 0.8 and 3 μm (12,500 to 3300 cm^{-1}). A thin layer of this compound is deposited on glass or quartz plates to form the cell. The entire assembly is then sealed in an evacuated container to protect the semiconductor from reaction with the atmosphere.

Charge-Transfer Device Detectors Photodiode arrays cannot match the performance of photomultiplier tubes in terms of sensitivity, dynamic range, and signal-to-noise ratio. Thus, their use has been limited to situations where the multichannel advantage outweighs their other shortcomings. In contrast, performance characteristics of charge-transfer device (CTD) detectors appear to approach those of photomultiplier tubes in addition to having the multichannel advantage. As a consequence, this type of detector is now appearing in ever-increasing numbers in modern spectroscopic instruments.[5] A further advantage of charge-transfer detectors is that they are two-dimensional in the sense that individual detector elements are arranged in rows and columns. For example, one detector that we describe in the next section consists of 244 rows of detector

[5]For details on charge-transfer detectors, see J. V. Sweedler, *Crit. Rev. Anal. Chem.,* **1993**, *24,* 59; J. V. Sweedler, R. B. Bilhorn, P. M. Epperson, G. R. Sims, and M. B. Denton, *Anal. Chem.,* **1988**, *60,* 282A, 327A.

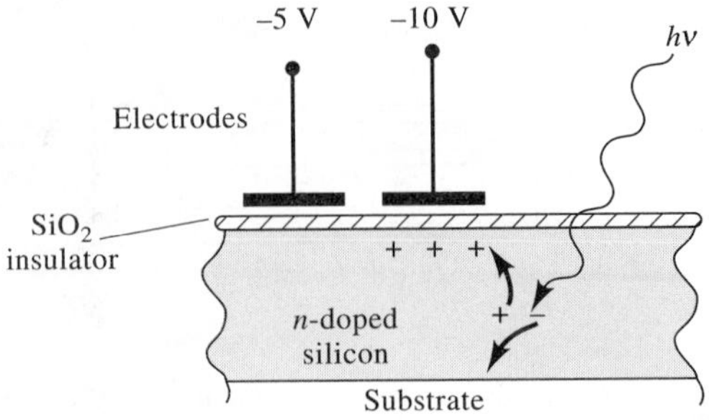

Figure 23-19
Cross section of one of the pixels of a charge-transfer device detector. The positive hole produced by the photon $h\nu$ is collected under the negative electrode.

elements, each row being made up of 388 detector elements, thus giving a two-dimensional array of 19,672 individual detectors, or *pixels,* contained on a silicon chip having dimensions of 6.5 mm by 8.7 mm. With this device, it becomes possible to record an entire two-dimensional spectrum.

Charge-transfer detectors operate much like a photographic film in the sense that they integrate signal information as radiation strikes them. Figure 23-19 is a cross-sectional depiction of one of the pixels making up a charge-transfer array. In this case, the pixel consists of two conductive electrodes overlying an insulating layer of silica (a pixel in some charge-transfer devices is made up of more than two electrodes). This silica layer separates the electrodes from a region of *n*-doped silicon. This assembly constitutes a metal oxide semiconductor capacitor that stores the charges formed when radiation strikes the doped silicon. When, as shown, a negative charge is applied to the electrodes, a charge inversion region is created under the electrodes, which is energetically favorable for the storage of positive holes. The mobile holes created by the absorption of photons by the silicon then migrate and collect in this region (typically, this region, which is called a *potential well,* is capable of holding as many as 10^5 to 10^6 charges before overflowing into an adjacent pixel). In the figure, one electrode is shown as being more negative than the other, making the accumulation of charge under this electrode more favorable. The amount of charge generated during exposure to radiation is measured in either of two ways. In a *charge-injection device* (CID) detector, the voltage change arising from movement of the charge from the region under one electrode to the region under the other is measured. In a *charge-coupled device* (CCD) detector, the charge is moved to a charge sensing amplifier for measurement.

Silica is silicon dioxide SiO_2, which is an electrical insulator.

Heat Detectors

The convenient photon detectors discussed in the previous section cannot be used to measure infrared radiation because photons of these frequencies lack the energy to cause photoemission of electrons; as a consequence, thermal detectors must be used. Unfortunately, the performance characteristics of thermal detectors are much inferior to those of phototubes, photomultiplier tubes, silicon diodes, or photovoltaic cells.

A thermal detector consists of a tiny blackened surface that absorbs infrared radiation and increases in temperature as a consequence. The temperature rise is converted to an electrical signal that is amplified and measured. Under the best of circumstances, the temperature changes involved are minuscule, amounting to a few thousandths of a degree Celsius. The difficulty of measurement is compounded by thermal radiation from the surroundings, which is always a potential source of uncertainty. To minimize the effects of this background radiation, or noise, thermal detectors are housed in a vacuum and are carefully shielded from their surroundings. To further minimize the effects of this external noise, the beam from the source is chopped by a rotating disk inserted between source and detector. Chopping produces a beam that fluctuates regularly from zero intensity to a maximum. The transducer converts this periodic radiation signal to an alternating electrical current that can be amplified and separated from the dc signal resulting from the background radiation. Despite all these measures, infrared measurements are significantly less precise than measurements of ultraviolet and visible radiation.

As shown in Table 23-3, four types of heat detectors are used for infrared spectroscopy. The most widely used is a tiny thermocouple or a group of thermocouples called a *thermopile.* These devices consist of one or more pairs of dissimilar metal junctions that develop a potential difference when their temperatures differ. The magnitude of the potential depends upon the temperature difference.

A *bolometer* consists of a conducting element whose electrical resistance changes as a function of temperature. Bolometers are fabricated from thin strips of metals, such as nickel or platinum, or from semiconductors consisting of oxides of nickel or cobalt; the latter are called *thermistors.*

A *pneumatic detector* consists of a small cylindrical chamber that is filled with xenon and contains a blackened membrane to absorb infrared radiation and heat the gas. One end of the cylinder is sealed with a window that is transparent to infrared radiation; the other is sealed with a flexible diaphragm that moves in and out as the gas pressure changes with cooling or heating. The temperature is determined from the position of the diaphragm.

Pyroelectric detectors are manufactured from crystals of a pyroelectric material, such as barium titanate or triglycine sulfate. When a crystal of either of these compounds is sandwiched between a pair of electrodes (one of which is transparent to infrared radiation), a temperature-dependent voltage develops that can be amplified and measured.

Signal Processors and Readouts

A signal processor is ordinarily an electronic device that amplifies the electrical signal from the detector; in addition, it may alter the signal from dc to ac (or the reverse), change the phase of the signal, and filter it to remove unwanted components. The signal processor may also be called upon to perform such mathematical operations on the signal as differentiation, integration, or conversion to a logarithm.

Several types of readout devices are found in modern instruments. Digital meters, scales of potentiometers, recorders, cathode-ray tubes, and monitors of microcomputers are some examples.

FEATURE 23-4
Measuring Photocurrents with Operational Amplifiers

The current produced by a reverse-biased silicon photodiode is typically 0.1 μA to 100 μA. These currents are so small that they must be converted to a voltage that is large enough to be measured with a digital voltmeter or other voltage-measuring device. We can perform such a conversion with the op amp circuit shown below. Light striking the reverse-biased photodiode causes a current I in the circuit. Because the op amp has a very large input resistance, essentially no current enters the op amp input designated by the minus sign. Thus, current in the photodiode must also pass through the resistor R. The current is conveniently calculated from Ohm's law: $E_{out} = -IR$. Since the current is proportional to the radiant power of the light striking the photodiode, $I = kP$, where k is a constant, and $E_{out} = -IR = -kPR = K'P$. A voltmeter is connected to the output of the op amp to give a direct readout that is proportional to the radiant power of the light. This same circuit can also be used with vacuum photodiodes or photomultipliers.

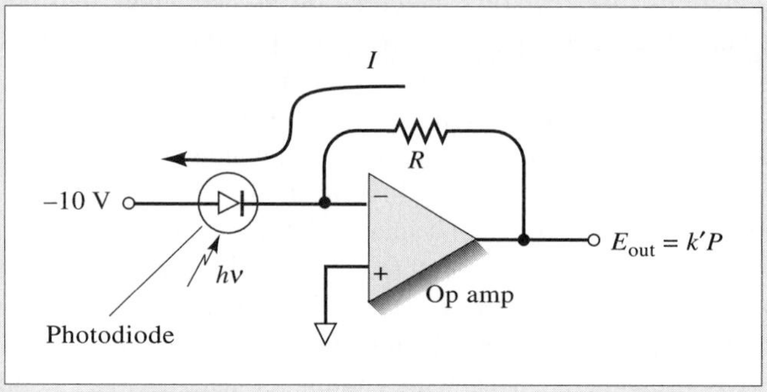

23B OPTICAL INSTRUMENT DESIGNS

The components discussed in the previous sections have been combined in various ways to produce dozens of optical instruments for absorption, emission, and fluorescence measurements. The designs of these instruments run the gamut from very simple to very sophisticated; it is not surprising that substantial differences also exist in their cost. No single instrument is best for all purposes. Selection must be determined by the type of work for which the instrument is intended.

Figure 23-20
One of Kirchhoff's early prism spectroscopes. From his paper in *Abhandl. Berlin Akad.*, **1862,** 227.

23B-1 Common Names of Some Optical Instruments

A *spectroscope* is an instrument for visually identifying the elements in a sample that have been excited in a flame or other hot medium. It consists of a modified monochromator, such as that shown in Figure 23-8b, in which the focal plane containing the exit slit is replaced by a movable eyepiece that permits visual detection of the emission lines. The wavelength of a line is determined from the angle between the incident beam and the path of the line to the eyepiece (Figure 23-20).

Strictly speaking, a *colorimeter* is an instrument for absorption measurements in which the human eye serves as the detector. One or more comparison standards are required each time the instrument is used. Feature 23-4 describes the *Duboscq* colorimeter, which was invented in the 1850s and is still being used today for simple colorimetric analyses.

A *photometer* is a photoelectric instrument that can be used for absorption, emission, or fluorescence measurements with ultraviolet, visible, or infrared radiation. A photometer is distinguished by its use of absorption or interference filters for wavelength selection and a photoelectric device for measuring radiant power. Instruments of this type that are used for absorption measurements with visible radiation are sometimes called *photoelectric colorimeters* or even simply *colorimeters.* Use of the latter term can lead to ambiguity. A photometer that is to be used for fluorescence measurements exclusively is often termed a *fluorometer.*

A *spectrograph* records spectra on a photographic plate or film located along the focal plane of a monochromator. Thus, the monochromators in Figure 23-8 could be converted to spectrographs by replacing the focal plane *AB* by a plate or film holder. The spectra would then appear as a series of black images of the entrance slit. Spectrographs are used primarily for qualitative elemental analysis based upon emission.

A *spectrometer* is a monochromator equipped with a fixed slit at the focal plane. The two monochromators shown in Figure 23-8 are examples of spectrometers. A spectrometer equipped with a phototransducer at the exit slit is called a *spectrophotometer.* Spectrometers can be used for absorption, emission, and fluorescence measurements. Fluorescence spectrometers are often called *spectrofluorometers.*

FEATURE 23-5
The Duboscq Colorimeter

Historically, the Duboscq colorimeter, or color comparator, is a remarkable analytical instrument in terms of its longevity. It was first described by Jules Duboscq in 1854, and only slightly modified versions of the original instrument are still being used and sold for colorimetric analyses 140 years later. A picture of the original instrument is shown in Figure 23-21. Here the sample and a standard solution of the analyte are contained in two similar flat-bottom tubes that are illuminated equally by reflected radiation from a lamp. The path lengths of the radiation through the standard and the sample (b_s and b_x, respectively) are varied by means of adjustable plungers that can be moved up and down in the solution. The beams from the two cylinders are combined by a prism and lens system so that they can be viewed as a split field in the eyepiece of the instrument. In use, the plungers are adjusted until the two fields appear identical. The position of the plungers is then read off from scales on either side of the instrument. If Beer's law is obeyed by the system, the concentration of the unknown c_x is obtained by multiplying the concentration of the standard by the ratio of the two path lengths. That is,

$$c_x = c_s(b_s/b_x)$$

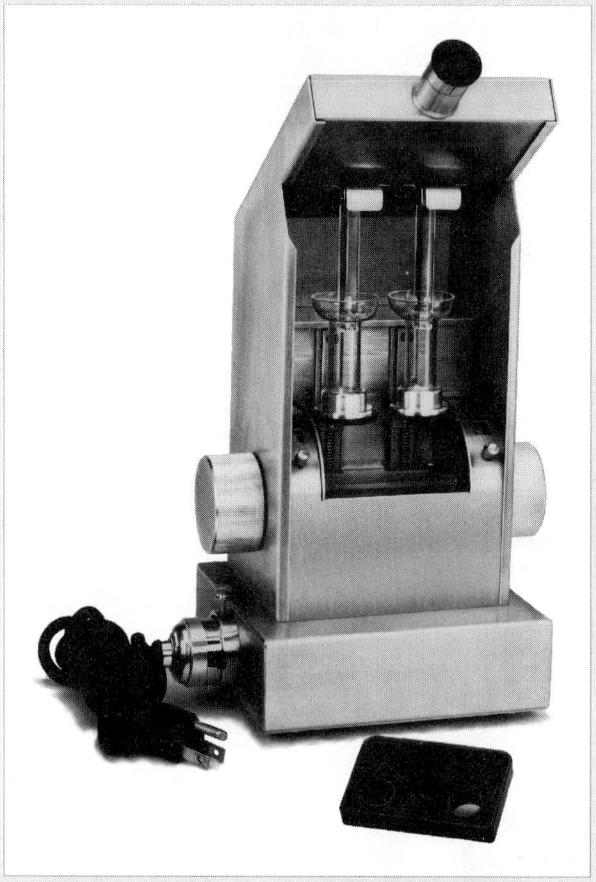

Figure 23-21 A Duboscq colorimeter.

23B-2 Instrument Types

In this section, we consider three general types of spectroscopic instruments: (1) single-beam, (2) double-beam in space, and (3) double-beam in time. All three types are found in instruments for absorption, emission, and fluorescence measurements. For illustrative purposes, we show how each of these types is applied for the measurement of transmittance and absorbance.

Single-Beam Instruments

Figure 23-22a is a schematic of a single-beam instrument for absorption measurements. It consists of one of the radiation sources shown in Table 23-1, a filter or a monochromator for wavelength selection (Table 23-2), matched cells that can be interposed alternately in the radiation beam, one of the detectors listed in Table 23-3, an amplifier, and a readout device.

The measurement of percent transmittance with a manual single-beam instrument involves three steps: (1) the *0% T adjustment,* (2) the *100% T adjustment,*

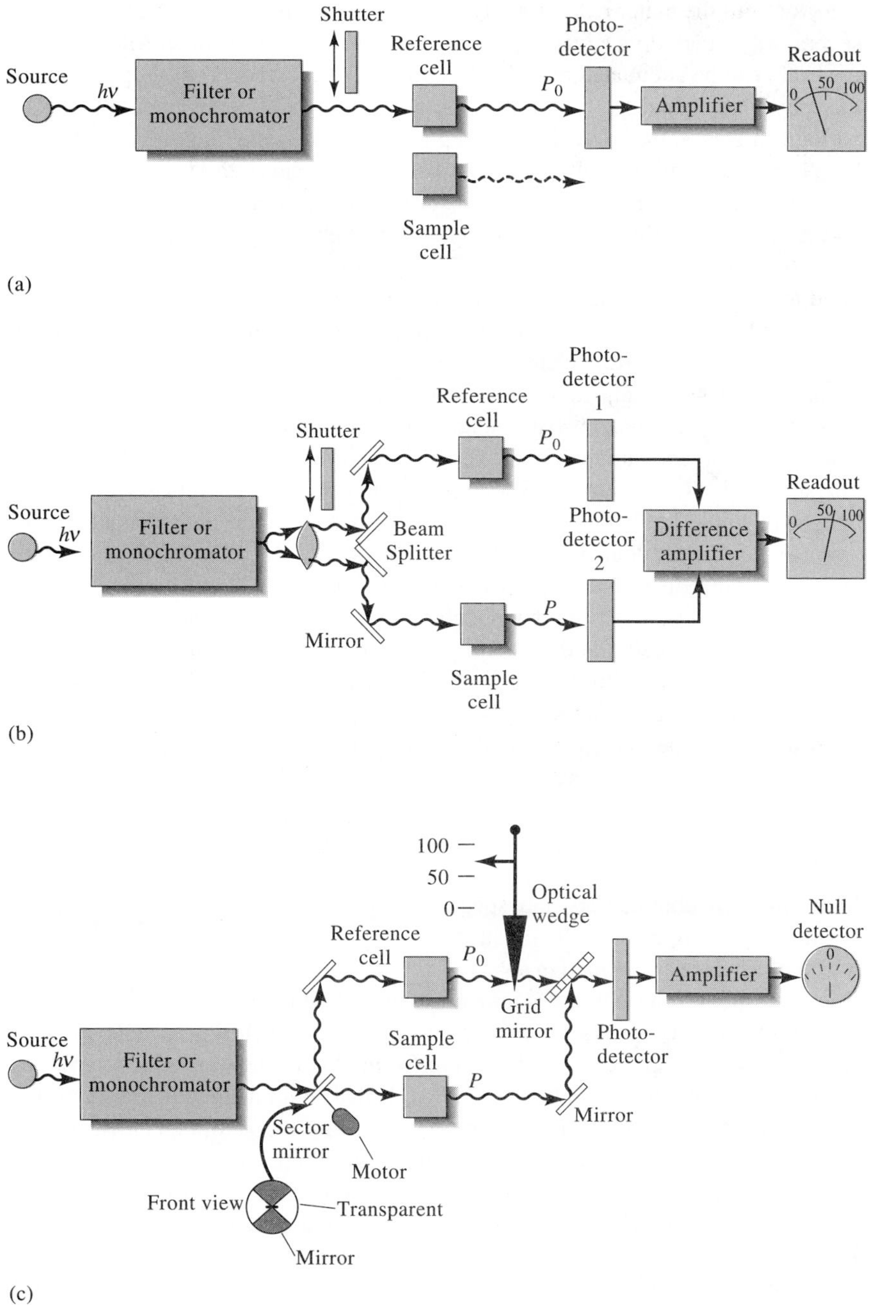

Figure 23-22
Instrument designs for photometers and spectrophotometers: (a) single-beam instrument; (b) double-beam instrument with beams separated in space; (c) double-beam instrument with beams separated in time.

and (3) the *determination of % T for the analyte.* The 0% *T* adjustment is carried out with the shutter imposed between the source and the photodetector. The meter is then mechanically or electrically adjusted until it reads 0. Step 2 is then carried out by placing the solvent in the light path, opening the shutter, and varying the intensity of the radiation or the amplification of the electrical signal from the

This calibration procedure should be performed before each absorbance measurement.

detector until the meter reads 100 (100% T). The beam intensity can be varied in several ways, including adjusting the electrical power to the source. Alternatively, the beam can be attenuated by a diaphragm, an optical wedge, or an optical comb that physically blocks a fraction of the radiation (the amount removed can be varied). In step 3, the solvent cell is replaced by one containing the analyte, and the percent transmittance is read from the scale. Because the transduced signal from a photodetector is linear with respect to the power of the radiation it receives, the scale reading with the sample in the light path is the percent transmittance (that is, the percent full scale). Clearly, a logarithmic scale can be substituted to give the absorbance of the solution directly (see Figure 22-5).

Normally, a single-beam instrument requires a stabilized voltage supply to avoid errors resulting from changes in the beam intensity during the time required to make the 100% T adjustment and determine % T for the analyte.

Single-beam instruments vary widely in their complexity and performance characteristics. The simplest and least expensive consists of a battery-operated tungsten bulb as the source, a set of glass filters for wavelength selection, test tubes as sample holders, a photovoltaic cell as a detector, and a small microammeter as the readout device. At the other extreme are sophisticated, computer-controlled instruments with a range of 200 to 1000 nm or more. These spectrophotometers have interchangeable tungsten/deuterium lamp sources, use rectangular silica cells, and are equipped with a high-resolution grating with variable slits. Photomultiplier tubes are used as detectors, and the output is often digitized and stored so that it can be printed out in several forms. Alternatively, the detector may be a multichannel diode array that permits the entire spectrum to be recorded in a matter of seconds.

Double-Beam Instruments

Many modern photometers and spectrophotometers are based upon a double-beam design. Figure 23-22b illustrates a double-beam-in-space instrument in which two beams are formed in space by a V-shaped mirror called a beam splitter. One beam passes through the reference solution to a photodetector, and the second simultaneously traverses the sample to a second, matched photodetector. The two outputs are amplified, and their ratio (or the log of their ratio) is determined electronically and displayed by the readout device. With manual instruments, the measurement is a two-step operation involving first the zero adjustment with a shutter in place between selector and beam splitter. In the second step, the shutter is opened and the transmittance or absorbance is read directly from the meter.

The second type of double-beam instrument is illustrated in Figure 23-22c. Here the beams are separated in time by a rotating sector mirror that directs the entire beam from the monochromator first through the reference cell and then through the sample cell. The pulses of radiation are recombined by another sector mirror, which transmits one pulse and reflects the other to the detector. As shown by the insert labeled "front view" in Figure 23-22c, the motor-driven sector mirror is made up of pie-shaped segments, half of which are mirrored and half of which are transparent. The mirrored sections are held in place by blackened metal frames that periodically interrupt the beam and prevent its reaching the detector. The detector circuit is programmed to use these periods to perform the dark-current adjustment.

The instrument shown in Figure 23-22c is a null type, in which the beam passing through the solvent is attenuated until its intensity just matches that of the beam passing through the sample. Attenuation is accomplished in this design with an optical wedge, the transmission of which decreases linearly along its length. Thus, the null point is reached by moving the wedge in the beam until the two electrical pulses are identical as indicated by the null detector. The transmittance (or absorbance) is then read directly from the pointer attached to the wedge.

Double-beam instruments offer the advantage that they compensate for all but the most short-term fluctuations in the radiant output of the source as well as for drift in the detector and amplifier. They also compensate for wide variations in source intensity with wavelength (see Figure 23-4b). Furthermore, the double-beam design lends itself well to the continuous recording of transmittance or absorbance spectra. Consequently, most modern ultraviolet and visible recording instruments are based on this design.

23C QUESTIONS AND PROBLEMS

23-1. Define the term *effective bandwidth of a filter.*

***23-2.** How many lines per millimeter are required in a grating if the first-order diffraction line at 500 nm is to be observed at a reflection angle of 10 deg when the angle of incidence is 60 deg?

23-3. An infrared grating has 72.0 lines/mm. Calculate the wavelengths of the first- and second-order diffraction spectra at reflection angles of (a) 0 deg and (b) + 15 deg. Assume the incident angle is 50 deg.

***23-4.** Describe how a spectroscope, a spectrograph, and a spectrophotometer differ from each other.

23-5. Why do quantitative and qualitative analyses often require different monochromator slit widths?

23-6. The Wien displacement law states that the wavelength maximum in micrometers for blackbody radiation is

$$\lambda_{max} T = 2.90 \times 10^3$$

where T is the temperature in kelvins. Calculate the wavelength maximum for a blackbody that has been heated to *(a) 4000 K, (b) 3000 K, *(c) 2000 K, and (d) 1000 K.

***23-7.** Stefan's law states that the total energy E_t emitted by a blackbody per unit time and per unit area is

$$E_t = \alpha T^4$$

where α is 5.69×10^{-8} W/m² · K⁴. Calculate the total energy output in W/m² for the blackbodies described in Problem 23-6.

***23-8.** The relationships described in Problems 23-6 and 23-7 may be of help in solving the following.

(a) Calculate the wavelength of maximum emission of a tungsten-filament bulb operated at 2870 K and at 3000 K.

(b) Calculate the total energy output of the bulb in W/cm².

23-9. Describe the differences between the following and list any particular advantages possessed by one over the other.

***(a)** hydrogen- and deuterium-discharge lamps as sources for ultraviolet radiation.

(b) filters and monochromators as wavelength selectors.

***(c)** photovoltaic cells and phototubes as detectors for electromagnetic radiation.

(d) phototubes and photomultiplier tubes.

***(e)** photometers and colorimeters.

(f) spectrophotometers and photometers.

***(g)** single-beam and double-beam instruments for absorbance measurements.

(h) conventional and diode-array spectrophotometers.

***23-10.** A portable photometer with a linear response to radiation registered 73.6 μA with a blank solution in the light path. Replacement of the blank with an absorbing solution yielded a response of 24.9 μA. Calculate

(a) the percent transmittance of the sample solution.

(b) the absorbance of the sample solution.

(c) the transmittance to be expected for a solution in which the concentration of the absorber is one third that of the original sample solution.

(d) the transmittance to be expected for a solution that has twice the concentration of the sample solution.

23-11. A photometer with a linear response to radiation gave a reading of 685 mV with a blank in the light path and 179 mV when the blank was replaced by an absorbing solution. Calculate

(a) the percent transmittance and absorbance of the absorbing solution.

(b) the expected transmittance if the concentration of absorber is one half that of the original solution.

(c) the transmittance to be expected if the light path through the original solution is doubled.

***23-12.** Why does a deuterium lamp produce a continuous rather than a line spectrum in the ultraviolet?

23-13. What are the differences between a photon detector and a heat detector?

***23-14.** How is the power of infrared radiation measured?

23-15. Why can photomultiplier tubes not be used with infrared radiation?

***23-16.** Why is iodine sometimes introduced into a tungsten lamp?

23-17. Describe how an absorption photometer and a fluorescence photometer differ from each other.

***23-18.** Describe the basic design difference between a spectrometer for absorption measurements and one for emission studies.

23-19. What data are needed to describe the performance characteristics of an interference filter?

23-20. Define

***(a)** dark current.

(b) transducer.

***(c)** scattered radiation (in a monochromator).

(d) *n*-type semiconductor.

***(e)** majority carrier.

(f) depletion layer.

***23-21.** An interference filter is to be constructed for isolation of the CS_2 absorption band at 4.54 μm.

(a) If the determination is to be based upon first-order interference, how thick should the dielectric layer be (refractive index 1.34)?

(b) What other wavelengths will be transmitted?

CHAPTER

24

Molecular Absorption Spectroscopy

Molecular spectroscopy based upon ultraviolet, visible, and infrared radiation is widely used for the identification and determination of a myriad of inorganic, organic, and biochemical species.[1] Molecular ultraviolet/visible absorption spectroscopy is employed primarily for quantitative analysis and is probably more widely used in chemical and clinical laboratories throughout the world than any other single method. Infrared absorption spectroscopy is one of the most powerful tools for determining the structure of both inorganic and organic compounds. In addition, it is now assuming a role in quantitative analysis, particularly for determining environmental pollutants.

24A ULTRAVIOLET AND VISIBLE MOLECULAR ABSORPTION SPECTROSCOPY

In this section, we first consider types of molecular species that absorb ultraviolet and visible radiation. We then describe examples of a few instruments that are currently marketed and used by chemists for molecular absorption studies. Finally, we discuss how these instruments are used to solve analytical problems.

24A-1 Absorbing Species

As noted in Section 22C-3, absorption of ultraviolet and visible radiation by molecules generally occurs in one or more electronic absorption bands, each of which is made up of numerous closely packed but discrete lines. Each line arises from the transition of an electron from the ground state to one of the many vibrational and rotational energy states associated with each excited electronic energy state. Because there are so many of these vibrational and rotational states and because their energies differ only slightly, the number of lines contained in the typical band is large and their displacement from one another minute.

A band consists of a large number of closely spaced vibrational and rotational lines. The energies associated with the lines differ little from one another.

[1]For more detailed treatment of absorption spectroscopy, see E. J. Meehan, in *Treatise on Analytical Chemistry,* 2nd ed., P. J. Elving, E. J. Meehan, and I. M. Kolthoff, Eds., Part I, Vol. 7, Chapter 2. New York: Wiley, 1981; *Techniques in Visible and Ultraviolet Spectrometry,* C. Burgess and A. Knowles, Eds., Vol. 1. New York: Chapman and Hall, 1981; J. D. Ingle Jr. and S. R. Crouch, *Spectrochemical Analysis,* Chapters 2, 3, and 13. Englewood Cliffs, NJ: Prentice-Hall, 1988.

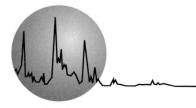

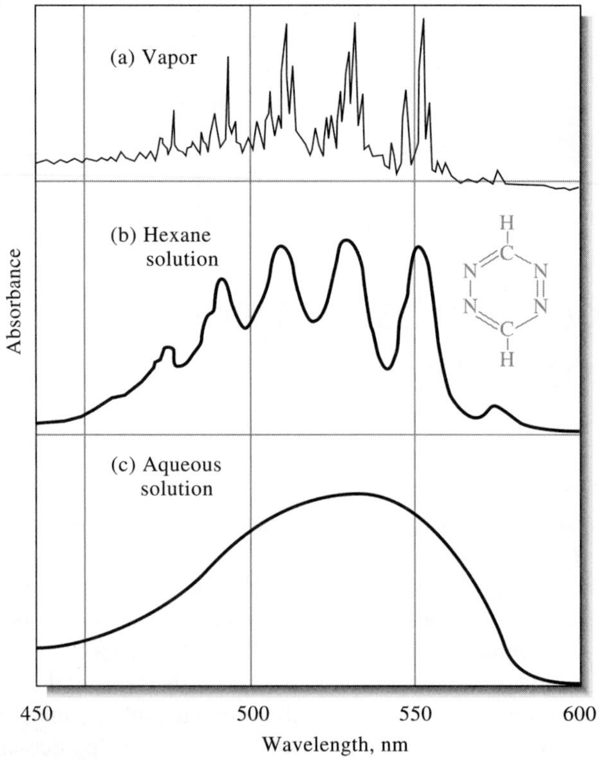

Figure 24-1
Typical ultraviolet absorption spectra. The compound is 1,2,4,5-tetrazine. (From S. F. Mason, *J. Chem. Soc.* **1959,** 1265. With permission.)

Figure 24-1a, which is part of the visible absorption spectrum for 1,2,4,5-tetrazine vapor, shows the fine structure that is due to the numerous rotational and vibrational levels associated with the excited electronic states of this aromatic molecule. In the gaseous state, the individual tetrazine molecules are sufficiently separated from one another to vibrate and rotate freely, and the many individual absorption lines resulting from the multitude of vibrational/rotational energy states are clearly evident. In the condensed state or in solution, however, freedom to rotate is largely lost, and lines due to differences in rotational energy levels are obliterated. Furthermore, in the presence of solvent molecules, energies of the various vibrational levels are modified in an irregular way. Thus, the energy of a given state in an assemblage of molecules takes on a Gaussian distribution; line broadening is the result. This effect is more pronounced in polar solvents, such as water, than in nonpolar hydrocarbon media. The solvent effect is illustrated in Figures 24-1b and 24-1c.

Absorption by Organic Compounds

Absorption of radiation by organic molecules in the wavelength region between 180 and 780 nm results from interactions between photons and those electrons that either participate directly in bond formation (and are thus associated with more than one atom) or are localized about such atoms as oxygen, sulfur, nitrogen, and the halogens.

The wavelength at which an organic molecule absorbs depends upon how tightly its several electrons are bound. The shared electrons in such single bonds as carbon/carbon or carbon/hydrogen are so firmly held that their excitation requires energies corresponding to wavelengths in the vacuum ultraviolet region (below 180 nm). Single-bond spectra have not been widely exploited for analytical purposes because of the experimental difficulties of working in this region. These difficulties arise because both quartz and atmospheric components absorb in this region, a circumstance that requires the use of evacuated spectrophotometers with lithium fluoride optics.

Electrons involved in double and triple bonds of organic molecules are not as strongly held and are therefore more easily excited by radiation; thus, species with unsaturated bonds generally exhibit useful absorption peaks. Unsaturated organic functional groups that absorb in the ultraviolet or visible regions are known as *chromophores.* Table 24-1 lists common chromophores and the approximate wavelengths at which they absorb. The data for position and peak intensity can only serve as a rough guide for identification purposes, since both are influenced by solvent effects as well as other structural details of the molecule. Moreover, conjugation between two (or more) chromophores tends to cause shifts in peak maxima to longer wavelengths. Finally, vibrational effects broaden absorption peaks in the ultraviolet and visible regions, which often makes precise determination of a maximum difficult. Typical spectra for organic compounds are shown in Figure 24-2.

Chromophores are unsaturated organic functional groups that absorb in the ultraviolet or visible region.

TABLE 24-1 Absorption Characteristics of Some Common Organic Chromophores

Chromophore	Example	Solvent	λ_{max}, nm	ε_{max}
Alkene	$C_6H_{13}CH{=}CH_2$	*n*-Heptane	177	13,000
Conjugated alkene	$CH_2{=}CHCH{=}CH_2$	*n*-Heptane	217	21,000
Alkyne	$C_5H_{11}C{\equiv}C{-}CH_3$	*n*-Heptane	178	10,000
			196	2,000
			225	160
Carbonyl	$CH_3\overset{O}{\overset{\parallel}{C}}CH_3$	*n*-Hexane	186	1,000
			280	16
	$CH_3\overset{O}{\overset{\parallel}{C}}H$	*n*-Hexane	180	Large
			293	12
Carboxyl	$CH_3\overset{O}{\overset{\parallel}{C}}OH$	Ethanol	204	41
Amido	$CH_3\overset{O}{\overset{\parallel}{C}}NH_2$	Water	214	60
Azo	$CH_3N{=}NCH_3$	Ethanol	339	5
Nitro	CH_3NO_2	Isooctane	280	22
Nitroso	C_4H_9NO	Ethyl ether	300	100
			665	20
Nitrate	$C_2H_5ONO_2$	Dioxane	270	12
Aromatic	Benzene	*n*-Hexane	204	7,900
			256	200

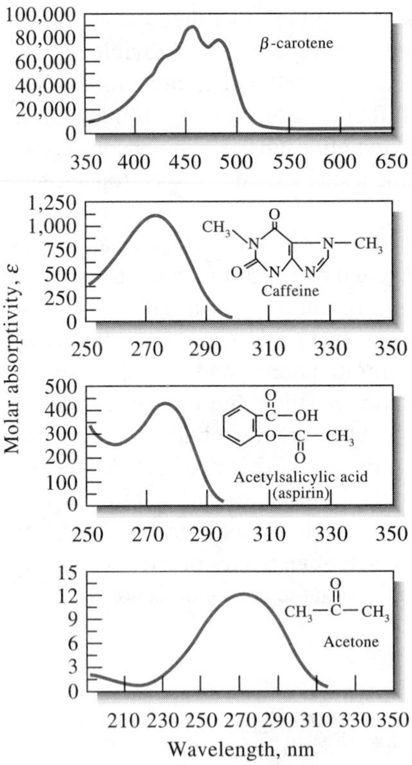

Figure 24-2
Spectra for typical organic compounds.

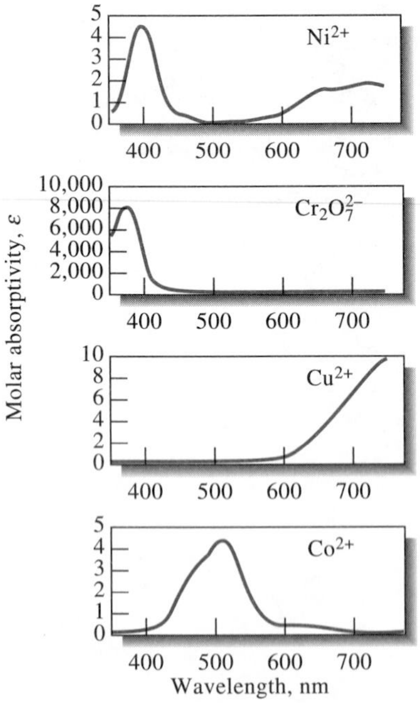

Figure 24-3
Absorption spectra of aqueous solutions of transition metal ions.

Saturated organic compounds containing heteroatoms—such as oxygen, nitrogen, sulfur, or halogens—contain nonbonding electrons that can be excited by radiation in the 170- to 250-nm range. Table 24-2 lists a few examples. Some of these compounds, such as alcohols and ethers, are used as solvents, and their absorption leads to lower cutoff spectral limits of 180 to 200 nm. Occasionally, this type of absorption is used for determining halogen- and sulfur-bearing compounds.

TABLE 24-2 Absorption by Organic Compounds Containing Unsaturated Heteroatoms

Compound	λ_{max}, nm	ε_{max}
CH_3OH	167	1480
$(CH_3)_2O$	184	2520
CH_3Cl	173	200
CH_3I	258	365
$(CH_3)_2S$	229	140
CH_3NH_2	215	600
$(CH_3)_3N$	227	900

Absorption by Inorganic Species

In general, the ions and complexes of elements in the first two transition series absorb broad bands of visible radiation in at least one of their oxidation states and are, as a consequence, colored (see, for example, Figure 24-3). Here, absorption involves transitions between filled and unfilled *d*-orbitals with energies that depend on the ligands bonded to the metal ions. The energy differences between these *d*-orbitals (and thus the positions of the corresponding absorption peaks) depend on the position of the element in the periodic table, its oxidation state, and the nature of the ligand bonded to it.

Absorption spectra of ions of the lanthanide and actinide transitions series differ substantially from those shown in Figure 24-3. The electrons responsible for absorption by these elements (4*f* and 5*f*, respectively) are shielded from external influences by electrons that occupy orbitals with larger principal quantum numbers. As a result, the bands tend to be narrow and relatively unaffected by the species involved in bonding with the outer electrons (see Figure 24-4).

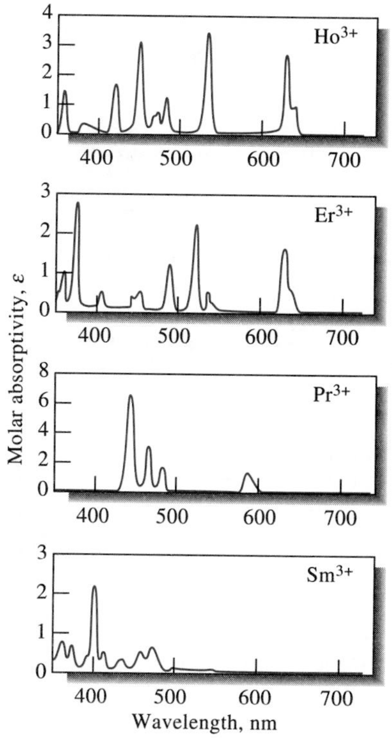

Figure 24-4
Absorption spectra of aqueous solutions of rare earth ions.

Charge-Transfer Absorption

For quantitative purposes, *charge-transfer absorption* is particularly important because molar absorptivities are unusually large ($\varepsilon_{max} > 10{,}000$), a circumstance that leads to high sensitivity. Many inorganic and organic complexes exhibit this type of absorption and are therefore called *charge-transfer complexes.*

A charge-transfer complex consists of an electron-donor group bonded to an electron acceptor. When this product absorbs radiation, an electron from the donor is transferred to an orbital that is largely associated with the acceptor. The excited state is thus the product of a kind of internal oxidation/reduction process. This behavior differs from that of an organic chromophore in which the excited electron is in a *molecular* orbital that is shared by two or more atoms.

A **charge-transfer complex** is a strongly absorbing species that is made up of an electron-donating species that is bonded to an electron-accepting species.

Familiar examples of charge-transfer complexes include the phenolic complex of iron(III), the 1,10-phenanthroline complex of iron(II), the iodide complex of molecular iodine, and the ferro/ferricyanide complex responsible for the color of Prussian blue. The red color of the iron(III)/thiocyanate complex is a further example of charge-transfer absorption. Absorption of a photon results in the transfer of an electron from the thiocyanate ion to an orbital that is largely associated with the iron(III) ion. The product is an excited species involving predominantly iron(II) and the thiocyanate radical SCN. As with other types of electronic excitation, the electron in this complex ordinarily returns to its original state after a brief period. Occasionally, however, an excited complex may dissociate and produce photochemical oxidation/reduction products. The spectra of three charge-transfer complexes are shown in Figure 24-5.

In most charge-transfer complexes involving a metal ion, the metal serves as the electron acceptor. Exceptions are the 1,10-phenanthroline complexes of iron(II) (Section 16D-1) and copper (I), where the ligand is the acceptor and the metal ion the donor. A few other examples of this type of complex are known.

24A-2 Typical Instruments for Absorption Measurements

The optical components described in Section 23A have been combined in various ways to produce two types of instruments that are used for absorption measure-

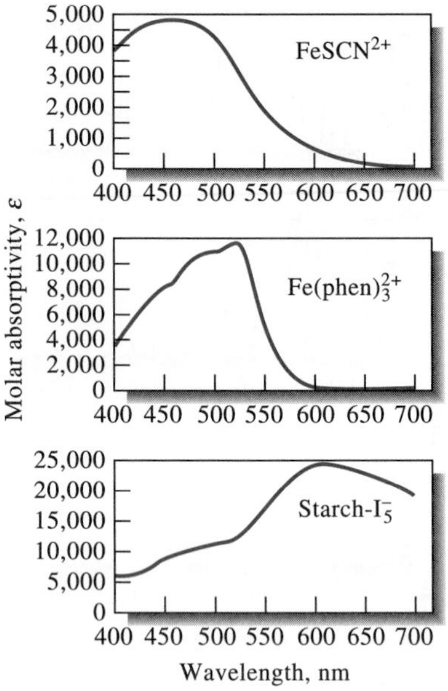

Figure 24-5
Charge-transfer spectra.

Photometers use filters to provide wavelength selection. Spectrophotometers use a grating or prism for this purpose.

ments with ultraviolet, visible, or infrared radiation: *spectrophotometers* and *photometers*. Spectrophotometers employ a grating or a prism monochromator to provide a narrow band of radiation for measurements. Photometers, in contrast, use an absorption filter or an interference filter for this purpose. Spectrophotometers offer the considerable advantage that the wavelength used can be varied continuously, thus making it possible to obtain entire absorption spectra.

Photometers

Photometers for absorption methods offer the advantages of low cost, simplicity, ruggedness, portability, and ease of maintenance. Moreover, where high spectral purity is not important (and often it is not), the accuracy and precision of measurements made with a photometer can approach those made with a spectrophotometer. The disadvantages of photometers are their lesser versatility, their inability to generate entire spectra, and their generally wider effective bandwidths.

Figure 24-6 is a diagram of a simple single-beam photometer used for quantitative measurements in the visible region. Photometers of this kind are ordinarily supplied with several filters, each of which transmits a different portion of the visible spectrum. Generally, a suitable filter is one whose color is the complement of the color of the analyte solution because it is this complementary color that is absorbed by the solution (see Feature 22-1). If several filters possessing the same general hue are available, the one that provides the greatest absorbance (or least transmittance) for a solution of the analyte should be used.

Ultraviolet photometers have become important detectors in high-performance liquid chromatography. In this application, a mercury-vapor lamp serves as the source and the emission line at 254 or 280 nm is isolated by an interference filter. This type of photometer is described briefly in Section 30A-5.

Ultraviolet/Visible Spectrophotometers

Several dozen models of spectrophotometers are marketed by various instrument manufacturers. Some are designed for the visible region alone. Others have ranges that extend from about 180 to 200 nm in the ultraviolet through the entire visible region to roughly 800 nm. A few cover the ultraviolet/visible region and into the near-infrared to approximately 3000 nm.

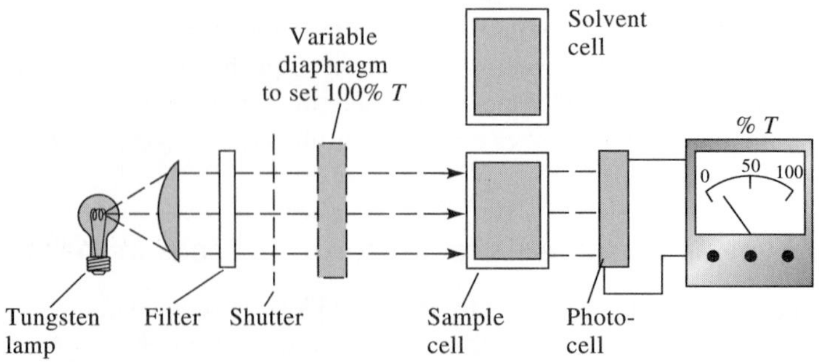

Figure 24-6
Single-beam photometer for absorption measurements in the visible region.

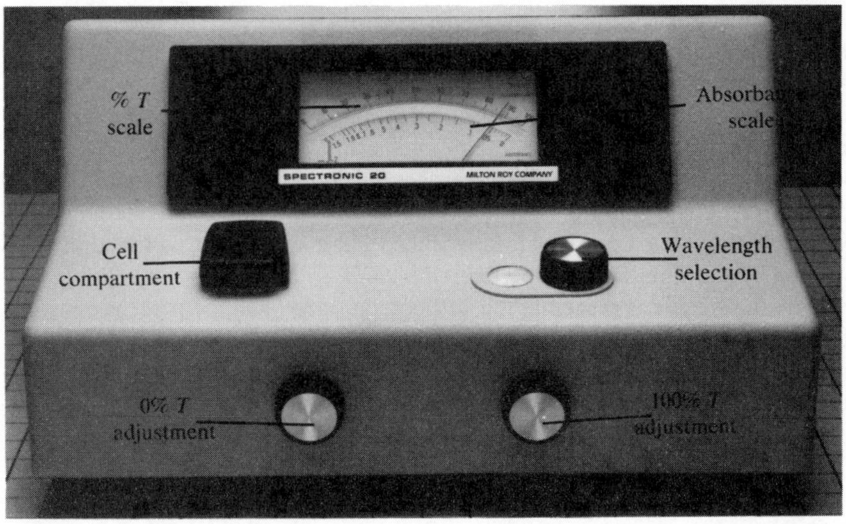

(a)

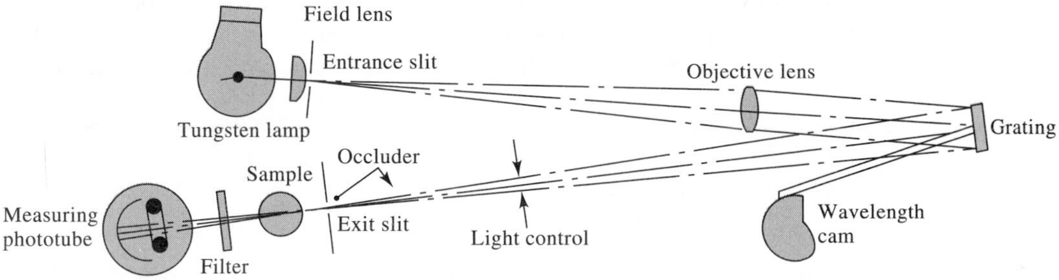

(b)

Figure 24-7

(a) The Spectronic 20 spectrophotometer. (b) Its optical diagram. (Courtesy of Spectronic Instruments, Inc., Analytical Products Division, Rochester, NY.)

Single-Beam Instruments Figure 24-7 shows the design of a simple and inexpensive spectrophotometer, the Spectronic 20. The original version of this instrument first appeared on the market in the mid-1950s, and the modified version shown in the figure is still being manufactured and widely sold. Undoubtedly, more of these instruments are currently in use throughout the world than any other single spectrophotometer model.

The Spectronic 20 employs a tungsten-filament light source operated by a stabilized power supply that provides radiation of constant intensity for sufficient time to provide good reproducibility for absorbance readings. The radiation from the source passes through a fixed slit to the surface of a diffraction grating. The diffracted radiation then passes through an exit slit to the sample or reference cell and thence to a phototube. The amplified electrical signal from the detector powers a meter with a $5\frac{1}{2}$-in. scale that is linear in percent transmittance and logarithmic in absorbance.

The Spectronic 20 is equipped with an occluder, which is a vane that automatically falls between the beam and the detector whenever the cuvette is removed

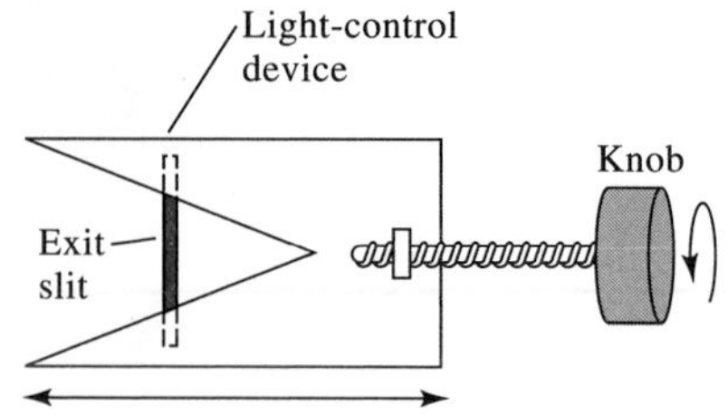

Figure 24-8
End view of the exit slit of the Spectronic 20 spectrophotometer pictured in Figure 24-7 showing the light-control device.

The 0% *T* and 100% *T* adjustments should be made immediately before each transmittance or absorbance measurement.

To obtain reproducible transmittance measurements, it is essential that the radiant power of the source remain constant during the time that the 100% *T* adjustment is made and the % *T* is read from the meter.

from its holder. The light-control device consists of a V-shaped aperture that is moved in and out of the beam in order to control the intensity of the beam falling on the photocell (see Figure 24-8).

To obtain a percent transmittance or an absorbance reading, the pointer of the meter is first zeroed with the sample compartment empty so that the occluder blocks off the beam and no radiation reaches the detector (the *0% T calibration,* or *adjustment*). The solvent is then inserted into the cell holder and the pointer is brought to the 100% *T* mark by means of the light-control knob, which adjusts the position of the light-control aperture (the *100% T calibration,* or *adjustment*). Finally, the sample is placed in the cell compartment, and the percent transmittance or the absorbance is read directly off the scale of the meter.

The spectral range of the Spectronic 20 is 340 to 625 nm (an accessory phototube extends the range to 950 nm). Other specifications for the instrument include an effective bandwidth of 20 nm and a wavelength accuracy of ±2.5 nm.

Single-beam instruments, such as the Spectronic 20, are well suited for quantitative absorption measurements at a single wavelength. Here, simplicity of instrumentation, low cost, and ease of maintenance offer distinct advantages.

Several instrument manufacturers offer single-beam instruments for both ultraviolet and visible measurements. The lower wavelength extremes for these instruments range from 190 to 210 nm and the upper from 800 to 1000 nm. All are equipped with interchangeable tungsten and deuterium or hydrogen lamps. Most employ photomultiplier tubes for detectors and gratings for dispersing radiation.

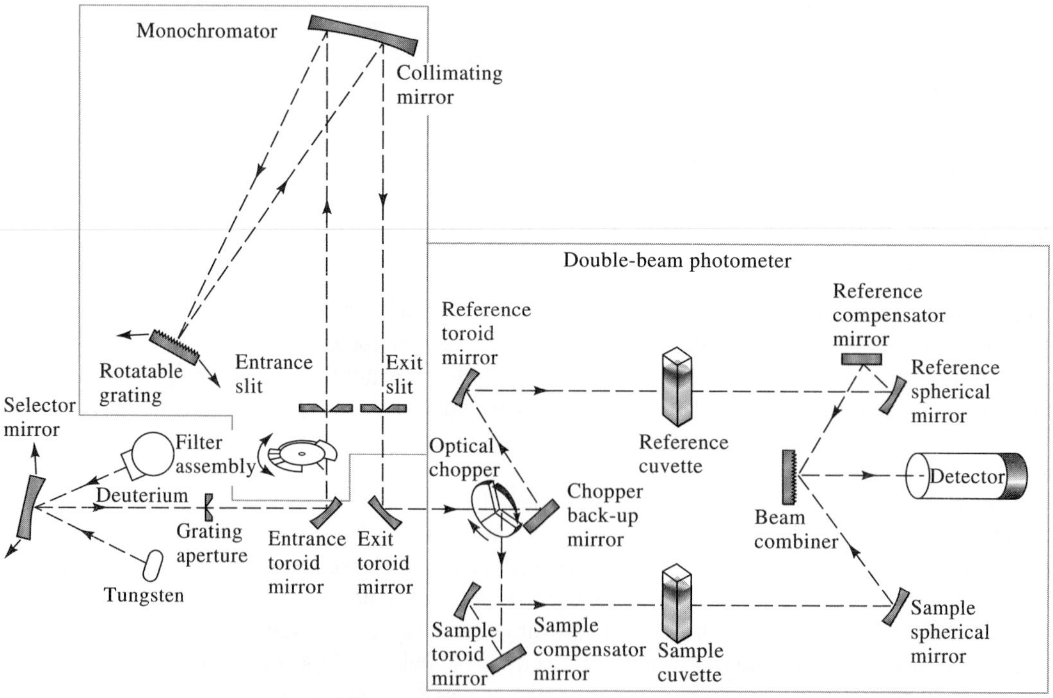

Figure 24-9
A double-beam recording spectrophotometer for the ultraviolet and visible regions; the Perkin-Elmer Series. (Courtesy of Coleman Instruments Division, Oak Brook, IL 50421.)

Some are equipped with digital readout devices; others employ large meters. Prices for these instruments range from \$2000 to \$8000.

Double-Beam Instruments Figure 24-9 shows the optics of a typical double-beam (in time) spectrophotometer designed to operate over a range from about 190 to 750 nm. The instrument is equipped with interchangeable deuterium/tungsten sources, a reflection grating monochromator, and a photomultiplier detector. The beam splitter is a motor-driven circular disk or chopper that is divided into three segments, one of which is transparent, the second reflecting, and the third opaque. With each rotation, the detector receives three signals, the first corresponding to P_0, the second to P, and the third to the dark current. The resulting electrical signals are then processed electronically to give the transmittance or absorbance on a readout device.

An instrument of this kind shown in Figure 24-9 is usually provided with a motor-driven grating that is synchronized with the paper drive of a recorder so that automatic scanning and recording of an entire spectrum becomes possible.

Multichannel Instruments Multichannel, or diode array, spectrometers are products of modern optoelectronic technology that makes it possible to record an entire ultraviolet or visible spectrum rapidly. The heart of these instruments is an array of several hundred silicon diode detectors (page 545) that are fabricated side by side on a single silicon chip as shown in Figure 24-10. Typically, the chips are 1 to 6 cm in length, and the widths of the individual diodes are 0.015 to 0.050 mm. The chip also contains a capacitor and an electronic switch for each diode. A computer-driven shift register sequentially closes each switch momentarily, which causes each capacitor to be charged to −5 V. Radiation impinging on the diode surface causes partial discharge of its capacitor. This lost charge is replaced during the next switching cycle. The resulting charging currents, which

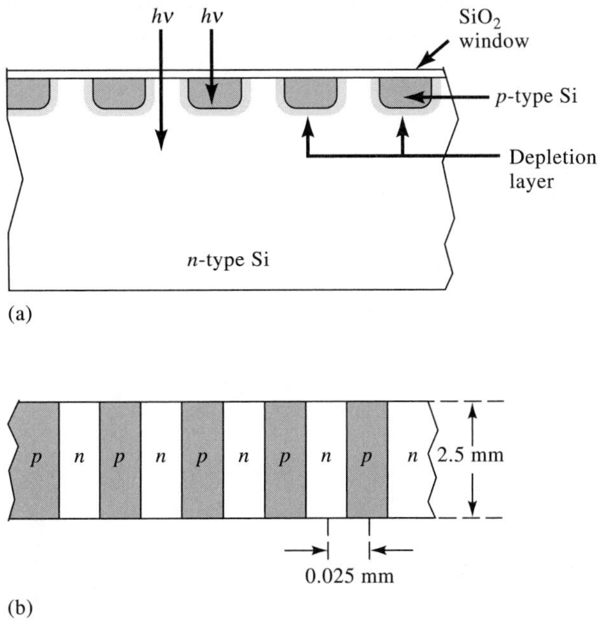

Figure 24-10
A reverse-biased linear diode array detector: (a) cross section; (b) top view.

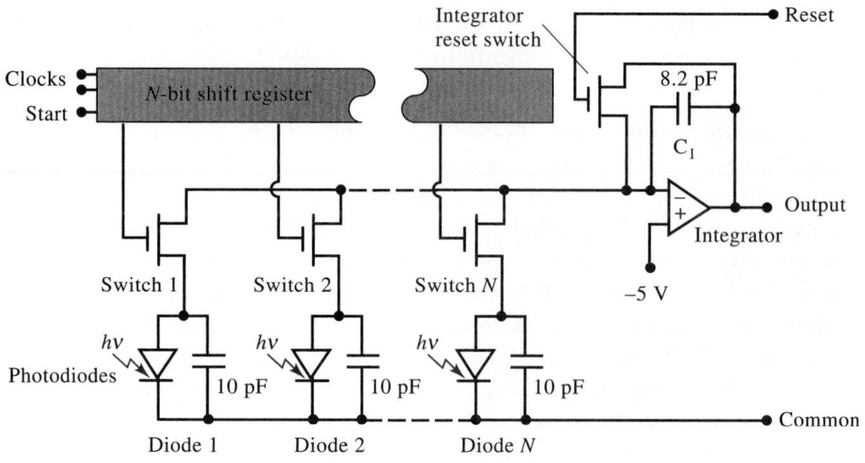

Figure 24-11
Block diagram of a photodiode-array detector chip.

Photos of diode-array detectors are shown in color plate 19.

are proportional to radiant power, are amplified, digitized, and stored in computer memory. With one or two of these diode arrays (Figure 24-11) placed along the length of the focal plane of a grating monochromator, all wavelengths can be monitored simultaneously and data for an entire spectrum collected and stored in a second or less.

Figure 24-12 shows an optical diagram of a typical multichannel ultraviolet/visible spectrophotometer. Because of the few optical components, the radiation throughput is much higher than that of traditional spectrophotometers. As a result, a single deuterium lamp can serve as a source for not only the ultraviolet but also the visible region (up to 820 nm). After passing through the solvent or analyte solution, the radiation is focused on an entrance slit and then passes onto the surface of a reflection grating. The detector is a diode array made up of 316 elements each with a dimension of 18 × 0.5 mm. The dispersion of the grating and the size of the diode elements are such that a resolution of 2 nm is realized throughout the entire spectral region. Because the system contains no moving parts, the wavelength reproducibility from scan to scan is exceedingly high.

The detector in the HP 8452A spectrometer shown in Figure 24-12 consists of 316 individual diode detectors, each coupled to its own capacitor and on-off switch, and all formed on a single silicon chip.

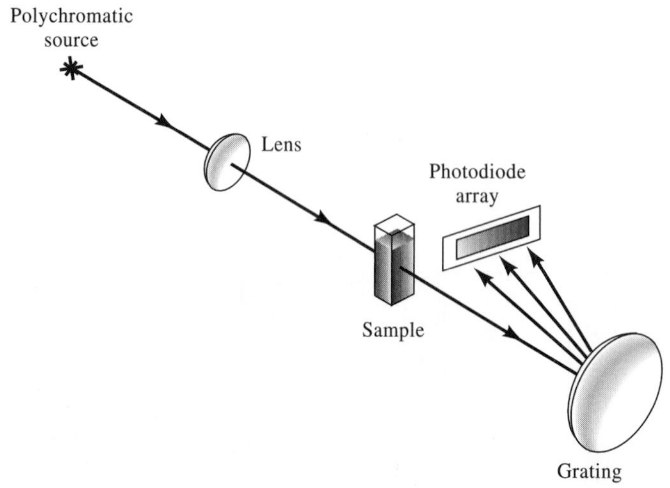

Figure 24-12
Diagram of a multichannel spectrophotometer based on a grating and photodiode detector.

A single scan from 200 to 820 nm with this instrument requires 0.1 s. In order to improve the precision of the measurements, however, spectra are generally scanned for a second or more with the data being collected in computer memory for subsequent averaging. With such short exposure times, photodecomposition of samples is minimized despite the location of the sample between the source and the monochromator. The stability of the source and electronic system is such that the solvent signal (100% *T*) needs to be observed and stored only every five to ten minutes.

The spectrophotometer shown in Figure 24-12 is designed to be interfaced with most personal computer systems. The instrument (without the computer) costs about \$7000 to \$9000.

24A-3 Qualitative Applications of Ultraviolet/Visible Spectroscopy

Spectrophotometric measurements with ultraviolet radiation are useful for detecting chromophoric groups, such as those shown in Table 24-1.[2] Because large parts of even the most complex organic molecules are transparent to radiation longer than 180 nm, the appearance of one or more peaks in the region from 200 to 400 nm is clear indication of the presence of unsaturated groups or of atoms such as sulfur or halogens. Often, an idea as to the identity of the absorbing groups can be gained by comparing the spectrum of an analyte with those of single molecules containing various chromophoric groups.[3] Ordinarily, however, ultraviolet spectra lack sufficient fine structure to allow unambiguous identification of an analyte. Thus, ultraviolet qualitative data must be supplemented with other physical or chemical evidence such as infrared, nuclear magnetic resonance, and mass spectra as well as solubility and melting- and boiling-point information.

Solvents

Ultraviolet spectra for qualitative analysis are most commonly derived for dilute solutions of the analyte. For volatile compounds, however, more useful spectra often result when the sample is examined as a gas (for example, compare Figures 24-1a and 24-1b). Such gas-phase spectra can often be obtained by allowing a drop or two of the pure compound to equilibrate with the atmosphere in a stoppered cuvette.

A solvent for ultraviolet/visible spectroscopy must be transparent throughout this region and should dissolve a sufficient quantity of the sample to give well-defined peaks. Moreover, consideration must be given to possible interactions with the absorbing species. For example, polar solvents, such as water, alcohols, esters, and ketones, tend to obliterate vibration spectra and should thus be avoided when spectral detail is desired. Nonpolar solvents, such as cyclohexane, often

---

[2]For a detailed discussion of ultraviolet absorption spectroscopy in the identification of organic functional groups, see R. M. Silverstein, G. C. Bassler, and T. C. Morrill, *Spectrometric Identification of Organic Compounds,* 5th ed., Chapter 7. New York: Wiley, 1991.

[3]Several organizations publish catalogs of spectra, including American Petroleum Institute, *Ultraviolet Spectral Data, A.P.I. Research Project 44.* Pittsburgh: Carnegie Institute of Technology; *Sadtler Ultraviolet Spectra.* Philadelphia: Sadtler Research Laboratories; American Society for Testing Materials, Committee E-13, Philadelphia.

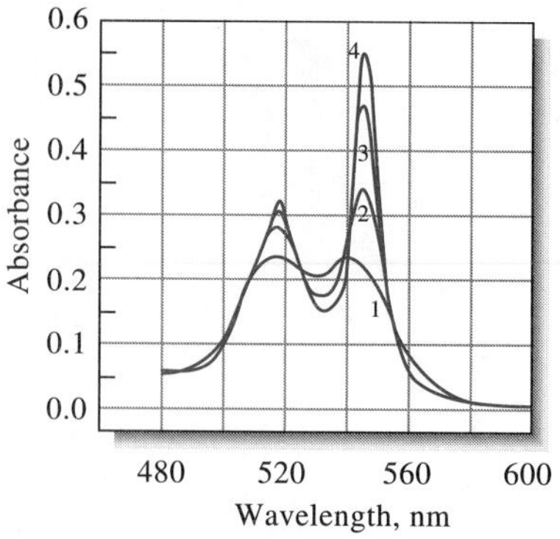

Figure 24-13

Spectra for reduced cytochrome *c* obtained with four spectral bandwidths: (1) 20 nm, (2) 10 nm, (3) 5 nm, and (4) 1 nm. At bandwidths < 1 nm, peak heights were the same but instrument noise became pronounced. (Courtesy of Varian Instrument Division, Palo Alto, CA.)

TABLE 24-3 Solvents for the Ultraviolet and Visible Regions

Solvent	Lower Wavelength Limit, nm	Solvent	Lower Wavelength Limit, nm
Water	180	Carbon tetrachloride	260
Ethanol	220	Diethyl ether	210
Hexane	200	Acetone	330
Cyclohexane	200	Dioxane	320
		Cellosolve	320

provide spectra that more closely approach that of a gas (compare, for example, the three spectra in Figure 24-1). In addition, the polarity of the solvent often influences the position of absorption maxima. Consequently a common solvent must be employed when comparing spectra for the purpose of identification.

Table 24-3 lists common solvents for studies in the ultraviolet and visible regions and their approximate lower wavelength limits. These limits are strongly dependent upon the purity of the solvent. For example, ethanol and the hydrocarbon solvents are frequently contaminated with benzene, which absorbs below 280 nm.[4]

The Effect of Slit Width

The effect of variation in slit width, and hence effective bandwidth, is illustrated by the spectra in Figure 24-13. Clearly, peak heights and separation are distorted at wider bandwidths. For this reason, spectra for qualitative applications are obtained at minimum slit widths.

Use minimum slit width for qualitative studies to preserve maximum detail in spectra.

The Effect of Scattered Radiation at the Wavelength Extremes of a Spectrophotometer

Earlier we demonstrated that scattered radiation may lead to instrumental deviations from Beer's law (page 518). Another undesirable effect of this type of radiation is that it occasionally causes false peaks to appear when a spectrophotometer is being operated at its wavelength extremes. Figure 24-14 shows an example of such behavior. Curve *B* is the true spectrum for a solution of cerium(IV) produced with a research-quality spectrophotometer responsive down to 200 nm or less. Curve *A* was obtained for the same solution with an inexpensive instrument operated with a tungsten source designed for work in the visible region only. The false peak at about 360 nm is directly attributable to scattered radiation, which was not absorbed because it was made up of wavelengths longer than 400 nm. Under most circumstances, such stray radiation has a negligible effect because its power is only a tiny fraction of the total power of the beam exiting from the monochromator. At wavelength settings below 380 nm, how-

[4]Most major suppliers of reagent chemicals in the United States offer spectrochemical grades of solvents. Spectral-grade solvents have been treated so as to remove absorbing impurities and meet or exceed the requirements set forth in *Reagent Chemicals, American Chemical Society Specifications,* 8th ed. Washington, DC: American Chemical Society, 1993.

ever, radiation from the monochromator is greatly attenuated as a result of absorption by glass optical components and cuvettes. In addition, both the output of the source and the photocell sensitivity fall off dramatically below 380 nm. These factors combine to cause a substantial fraction of the measured absorbance to be due to the scattered radiation of wavelengths to which cerium(IV) is transparent. A false peak results.

This same effect is sometimes observed with ultraviolet/visible instruments when attempts are made to measure absorbances at wavelengths lower than about 190 nm.

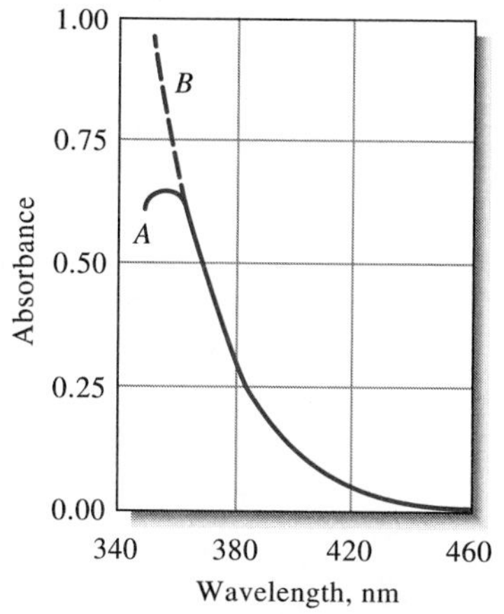

Figure 24-14
Spectra of cerium(IV) obtained with a spectrophotometer having glass optics (*A*) and quartz optics (*B*). The false peak in *A* arises from the transmission of stray radiation of longer wavelengths.

24A-4 Quantitative Applications

Absorption spectroscopy based upon ultraviolet and visible radiation is one of the most useful tools available to the chemist for quantitative analysis.[5] The important characteristics of spectrophotometric and photometric methods are

1. *Wide applicability.* Enormous numbers of inorganic, organic, and biochemical species absorb ultraviolet or visible radiation and are thus amenable to direct quantitative determination. Many nonabsorbing species can also be determined after chemical conversion to absorbing derivatives. It has been estimated that over 90% of the analyses performed in clinical laboratories are based upon ultraviolet and visible absorption spectroscopy.
2. *High sensitivity.* Typical detection limits for absorption spectroscopy range from 10^{-4} to 10^{-5} M. This range can often be extended to 10^{-6} or even 10^{-7} M with certain procedural modifications.
3. *Moderate to high selectivity.* If a wavelength can be found at which the analyte alone absorbs, preliminary separations become unnecessary. Furthermore, where overlapping absorption bands do occur, corrections based upon additional measurements at other wavelengths sometimes eliminate the need for a separation step.
4. *Good accuracy.* The relative errors in concentration encountered with a typical ultraviolet/visible spectrophotometric or photometric procedure lie in the range from 1% to 5%. Such errors can often be decreased to a few tenths of a percent with special precautions.
5. *Ease and convenience.* Spectrophotometric and photometric measurements are easily and rapidly performed with modern instruments. In addition, the methods readily lend themselves to automation.

Scope

The applications of absorption analysis are not only numerous but also touch upon every area in which quantitative information is sought. The reader can obtain a notion of the scope of spectrophotometry by consulting the series of review arti-

[5]For a wealth of detailed, practical information on spectrophotometric practices, see *Techniques in Visible and Ultraviolet Spectrometry,* Vol. I, *Standards in Absorption Spectroscopy,* C. Burgess and A. Knowles, Eds. London: Chapman and Hall, 1981; J. R. Edisbury, *Practical Hints on Absorption Spectrometry.* New York: Plenum Press, 1968.

cles published biennially in *Analytical Chemistry*[6] as well as monographs on the subject.[7]

Application to Absorbing Species Table 24-1 lists many common organic chromophoric groups. Spectrophotometric determination of organic compounds containing one or more of these groups is thus potentially feasible; many such applications can be found in the literature.

A number of inorganic species also absorb. We have noted that many ions of the transition metals are colored in solution and can thus be determined by spectrophotometric measurement. In addition, a number of other species—including nitrite, nitrate, and chromate ions, the oxides of nitrogen, the elemental halogens, and ozone—show characteristic absorption peaks.

Applications to Nonabsorbing Species Many nonabsorbing analytes can be determined photometrically by causing them to react with chromophoric reagents to produce products that absorb strongly in the ultraviolet and visible regions. The successful application of these color-forming reagents usually requires that their reaction with the analyte be forced to near completion.

Typical inorganic reagents include the following: thiocyanate ion for iron, cobalt, and molybdenum; the anion of hydrogen peroxide for titanium, vanadium, and chromium; and iodide ion for bismuth, palladium, and tellurium. Of even greater importance are organic chelating reagents that form stable colored complexes with cations. Common examples include diethyldithiocarbamate for the determination of copper, diphenylthiocarbazone for lead, 1,10-phenanthroline for iron, and dimethylglyoxime for nickel; Figure 24-15 shows the color-forming reaction for the first two of these reagents. The structure of the 1,10-phenanthroline complex of iron(II) is shown on page 355, and the reaction of nickel with dimethylglyoxime to form a red precipitate is described on page 95. In the application of the last reaction to the photometric determination of nickel, an aqueous solution of the cation is extracted with a solution of the chelating agent in an immiscible organic liquid. The absorbance of the resulting bright red organic layer serves as a measure of the concentration of the metal.

Other reagents are available that react with organic functional groups to produce colors that are useful for quantitative analysis. For example, the red color of the 1:1 complexes that form between low-molecular-weight aliphatic alcohols and cerium(IV) can be used for the quantitative estimation of such alcohols.

Procedural Details

A first step in any photometric or spectrophotometric analysis is the development of conditions that yield a reproducible relationship (preferably linear) between absorbance and analyte concentration.

---

[6]L. G. Hargis and J. A. Howell, *Anal. Chem.,* **1984,** *56*, 225R; **1986,** *58*, 108R; **1988,** *60*, 131R; **1990,** *62*, 155R; **1992,** *64*, 66R; **1994,** *66*, 445R.

[7]See, for example, E. B. Sandell and H. Onishi, *Photometric Determination of Traces of Metals,* 4th ed. New York: Wiley, 1978; H. Onishi, *Photometric Determination of Traces of Metals,* Part IIA, Part IIB, 4th ed. New York: Wiley, 1986, 1989; *Colorimetric Determination of Nonmetals,* 2nd ed., D. F. Boltz, Ed. New York: Interscience, 1978; F. D. Snell, *Photometric and Fluorometric Methods of Analysis.* New York: Wiley, 1978.

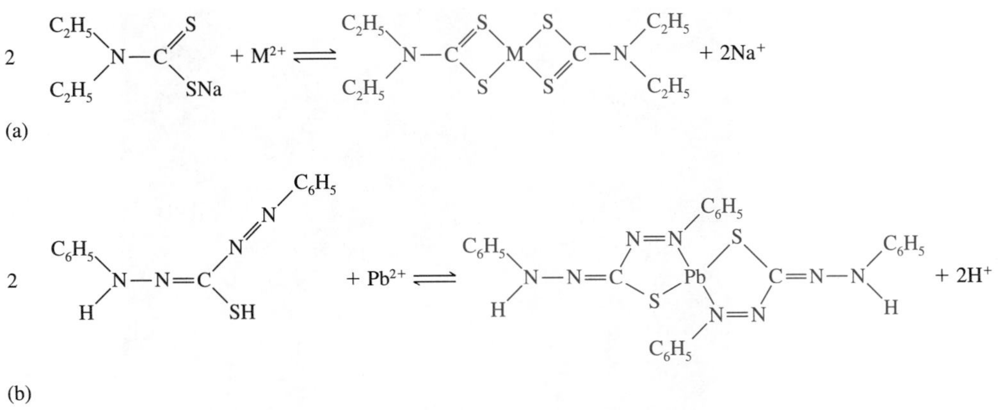

Figure 24-15
Typical chelating reagents for absorption. (a) Diethyldithiocarbamate. (b) Diphenylthiocarbazone.

Wavelength Selection In order to realize maximum sensitivity, spectrophotometric absorbance measurements are ordinarily made at a wavelength corresponding to an absorption peak because the change in absorbance per unit of concentration is greatest at this point. In addition, the absorption curve is often flat at a maximum, which leads to good adherence to Beer's law (see Figure 22-15) and less uncertainty from failure to reproduce precisely the wavelength setting of the instrument.

Variables That Influence Absorbance Common variables that influence the absorption spectrum of a substance include the nature of the solvent, the pH of the solution, the temperature, high electrolyte concentrations, and the presence of interfering substances. The effects of these variables must be known and conditions for the analysis chosen such that the absorbance will not be materially affected by small, uncontrolled variations in their magnitudes.

Cleaning and Handling of Cells Sample containers, which are usually called *cells* or *cuvettes,* must have windows fabricated from a material that is transparent in the spectral region of interest. Thus, as shown in Figure 23-2, quartz or fused silica is required for the ultraviolet region (below 350 nm) and may be used in the visible region and to about 3000 nm in the infrared. Because of its lower cost, silicate glass is ordinarily used for the region between 375 and 2000 nm. Plastic containers have also found application in the visible region. The most common window material for infrared studies is crystalline sodium chloride.

Avoid touching or scratching the windows of cuvettes.

The best cells have windows that are normal to the direction of the beam to minimize reflection losses. The most common cell length for studies in the ultraviolet/visible is 1 cm; matched, calibrated cells of this size are available from several commercial sources. Other path lengths, from shorter than 0.1 to 10 cm, can also be purchased. Transparent spacers for shortening the path length of 1-cm cells to 0.1 cm are also available. Some typical cells are shown in Figure 24-16.

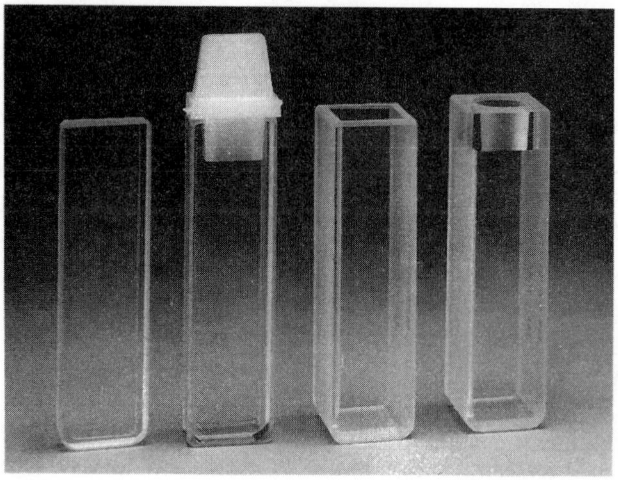

Figure 24-16
Typical commercially available cells.

For reasons of economy, cylindrical cells are sometimes encountered. Particular care must be taken to duplicate the position of such cells with respect to the beam; otherwise variations in path length and reflection loss at the curved surfaces can cause significant error.

The quality of spectroscopic data is critically dependent on the way the matched cells are used and maintained. Fingerprints, grease, or other deposits on the walls markedly alter the transmission characteristics of a cell. Thus, thorough cleaning before and after use is imperative, and care must be taken to avoid touching the windows after cleaning is complete. Matched cells should never be dried by heating in an oven or over a flame because this may cause physical damage or a change in path length. Matched cells should be calibrated against each other regularly with an absorbing solution.

Determination of the Relationship Between Absorbance and Concentration The calibration standards for a photometric or a spectrophotometric analysis should approximate as closely as possible the overall composition of the actual samples, and should encompass a reasonable range of analyte concentrations. Seldom, if ever, is it safe to assume adherence to Beer's law and use only a single standard to determine the molar absorptivity; it is even less prudent to base the results of an analysis on a literature value for the molar absorptivity.

See *Mathcad Applications for Analytical Chemistry,* **pp. 114–116.**

The Standard-Addition Method Ideally, calibration standards should approximate the composition of the samples to be analyzed not only with respect to the analyte concentration but also with regard to the concentrations of the other species in the sample matrix in order to minimize the effects of various components of the sample on the measured absorbance. For example, the absorbance of many colored complexes of metal ions is decreased to a varying degree in the presence of sulfate and phosphate ions as a consequence of the tendency of these anions to form colorless complexes with metal ions. The color formation reaction is often less complete as a consequence, and lowered absorbances are the result.

The matrix effect of sulfate and phosphate can often be counteracted by introducing into the standards amounts of the two species that approximate the amounts found in the samples. Unfortunately, when complex materials such as soils, minerals, and plant ash are being analyzed, preparation of standards that match the samples is often impossible or extremely difficult. When this is the case, the *standard addition method* is often helpful in counteracting matrix effects.

The standard addition method can take several forms.[8] The one most often chosen for photometric or spectrophotometric analyses, and the one that will be discussed here, involves adding one or more increments of a standard solution to sample aliquots of the same size. Each solution is then diluted to a fixed volume before measuring its absorbance. It should be noted that when the amount of sample is limited, standard additions can be carried out by successive introductions of increments of the standard to a single measured aliquot of the unknown. Measurements are made on the original and after each addition. This procedure is often more convenient for voltammetry.

Assume that several identical aliquots V_x of the unknown solution with a concentration c_x are transferred to volumetric flasks having a volume V_t. To each of these flasks is added a variable volume V_s mL of a standard solution of the analyte having a known concentration c_s. The color development reagents are then added, and each solution is diluted to volume. If Beer's law is followed, the absorbance of the solutions is described by

$$\begin{aligned} A_s &= \frac{\varepsilon b V_s c_s}{V_t} + \frac{\varepsilon b V_x c_x}{V_t} \\ &= kV_s c_s + kV_x c_x \end{aligned} \qquad (24\text{-}1)$$

where k is a constant equal to $\varepsilon b/V_t$. A plot of A_s as a function of V_s should yield a straight line of the form

$$A_s = mV_s + b$$

where the slope m and the intercept b are given by

$$m = kc_s$$

and

$$b = kV_x c_x$$

A least-squares analysis (Section 4E-2) can be used to determine m and b; c_x can then be obtained from the ratio of these two quantities and the known values of c_s, V_x, and V_s. Thus,

$$\frac{m}{b} = \frac{\cancel{k} c_s}{\cancel{k} V_x c_x}$$

[8]See M. Bader, *J. Chem. Educ.*, **1980,** *57*, 703.

which rearranges to

$$c_x = \frac{bc_s}{mV_x} \tag{24-2}$$

An approximate value for the standard deviation in c_x can then be obtained by assuming that the uncertainties in c_s, V_s, and V_t are negligible with respect to those in m and b. Then, the relative variance of the result $(s_c/c_x)^2$ is assumed to be the sum of the relative variances of m and b. That is,

$$\left(\frac{s_c}{c_x}\right)^2 = \left(\frac{s_m}{m}\right)^2 + \left(\frac{s_b}{b}\right)^2$$

where s_m and s_b are the standard deviations of the slope and intercept, respectively. Taking the square root of this equation gives

$$s_c = c_x\sqrt{\left(\frac{s_m}{m}\right)^2 + \left(\frac{s_b}{b}\right)^2} \tag{24-3}$$

EXAMPLE 24-1

Ten-millimeter aliquots of a natural water sample were pipetted into 50.00 mL volumetric flasks. Exactly 0.00, 5.00, 10.00, 15.00, and 20.00 mL of a standard solution containing 11.1 ppm of Fe^{3+} were added to each followed by an excess of thiocyanate ion to give the red complex $Fe(SCN)^{2+}$. After dilution to volume, absorbances for the five solutions, measured with a photometer equipped with a green filter, were found to be 0.240, 0.437, 0.621, 0.809, and 1.009, respectively (0.982-cm cells). (a) What was the concentration of Fe^{3+} in the water sample? (b) Calculate the standard deviation of the slope, the intercept, and the concentration of Fe.

(a) In this problem, c_s = 11.1 ppm, V_x = 10.00 mL, and V_t = 50.00 mL. A plot of the data, shown in Figure 24-17 demonstrates that Beer's law is obeyed.

To obtain the equation for the line in Figure 24-17 ($A = mV_s + b$), the procedure illustrated in Example 4-9 (page 63) is followed. The result is $m = 0.03820$ and $b = 0.2412$, and thus

$$A = 0.03820\ V_s + 0.2412$$

Substituting into Equation 24-2 gives

$$c_x = \frac{0.2412 \times 11.1}{0.03820 \times 10.00} = 7.01 \text{ ppm } Fe^{3+}$$

(b) Equations 4-16 and 4-17 give the standard deviation of the intercept and the slope. That is, $s_m = 3.07 \times 10^{-4}$ and $s_b = 3.76 \times 10^{-3}$.

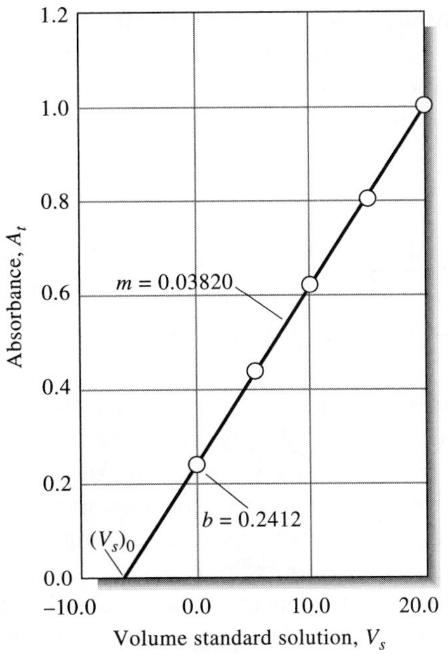

Figure 24-17
Data for standard addition method for the determination of Fe^{3+} as the SCN^- complex.

Substituting into Equation 24-3 gives

$$s_c = 7.01\sqrt{\left(\frac{3.07 \times 10^{-4}}{0.03820}\right)^2 + \left(\frac{3.76 \times 10^{-3}}{0.2412}\right)^2}$$
$$= 0.12 \text{ ppm } Fe^{3+}$$

In the interest of saving time or sample, it is possible to perform a standard addition analysis using only two increments of sample. Here, a single addition of V_s mL of standard would be added to one of the two samples and we can write

$$A_1 = \frac{\varepsilon b V_x c_x}{V_t}$$

$$A_2 = \frac{\varepsilon b V_x c_x}{V_t} + \frac{\varepsilon b V_s c_s}{V_t}$$

where A_1 and A_2 are absorbances of the diluted sample and the diluted sample plus standard, respectively. Dividing the second equation by the first gives upon rearrangement

$$c_x = \frac{A_1 c_s V_s}{(A_2 - A_1)V_x} \tag{24-4}$$

EXAMPLE 24-2

A 2.00-mL urine specimen was treated with reagent to generate a colored adduct with phosphate, and the sample was then diluted to 100 mL. Exactly

5.00 mL of a phosphate solution containing 0.0300 mg phosphate/mL were added to a second 2.00-mL sample, which was treated in the same way as the original sample. The absorbance of the first solution was 0.428; that of the second was 0.538. Calculate the milligrams of phosphate per millimeter of the specimen.

Here we substitute into Equation 24-4 and obtain

$$c_x = \frac{0.428 \times 0.0300\ (\text{mg PO}_4^{3-}/\text{mL}) \times 5.00\ \text{mL}}{(0.538 - 0.428) \times 2.00\ \text{mL sample}}$$

$$= 0.292\ \text{mg PO}_4^{3-}/\text{mL sample}$$

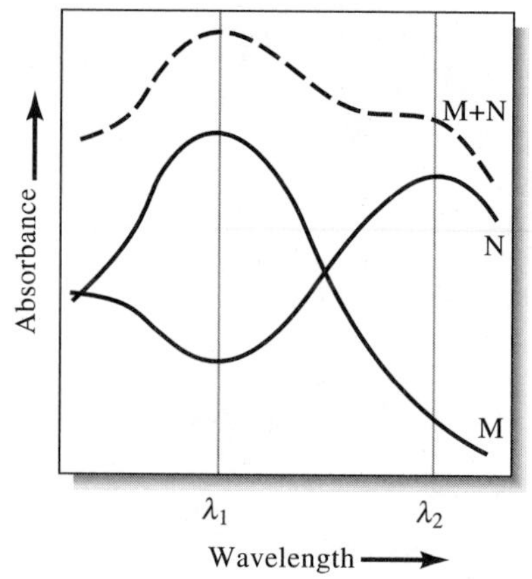

Figure 24-18
Absorption spectrum of a two-component mixture (M + N), with spectra of the individual components.

Analysis of Mixtures The total absorbance of a solution at any given wavelength is equal to the sum of the absorbances of the individual components in the solution (Equation 22-17). This relationship makes it possible in principle to determine the concentrations of the individual components of a mixture even if total overlap in their spectra exists. For example, Figure 24-18 shows the spectrum of a solution containing a mixture of species M and species N as well as absorption spectra for the individual components. Clearly, no wavelength exists at which the absorbance is due to just one of these components. To analyze the mixture, molar absorptivities for M and N are first determined at wavelengths λ_1 and λ_2 with enough standard to be sure that Beer's law is obeyed over an absorbance range that encompasses the absorbance of the sample. Note that the wavelengths selected are ones at which the two spectra differ significantly. Thus, at λ_1, the molar absorptivity of component M is much larger than that for component N. The reverse is true for λ_2. To complete the analysis, the absorbance of the mixture is determined at the same two wavelengths. Example 24-3 demonstrates how the composition of the mixture is derived from data of this kind.

EXAMPLE 24-3

Palladium(II) and gold(III) can be analyzed simultaneously through their reaction with methiomeprazine ($C_{19}H_{24}N_2S_2$). The absorption maximum for the Pd complex occurs at 480 nm, while that for Au complex is at 635 nm. Molar absorptivity data at these wavelengths are

	Molar Absorptivity, ε	
	480 nm	**635 nm**
Pd complex	3.55×10^3	5.64×10^2
Au complex	2.96×10^3	1.45×10^4

A 25.0-mL sample was treated with an excess of methiomeprazine and subsequently diluted to 50.0 mL. Calculate the molar concentrations of Pd(II), c_{Pd}, and Au(III), c_{Au}, in the sample if the diluted solution had an absorbance of 0.533 at 480 nm and 0.590 at 635 nm when measured in a 1.00-cm cell.

At 480 nm

$$0.533 = (3.55 \times 10^3)(1.00)c_{Pd} + (2.96 \times 10^3)(1.00)c_{Au}$$

or

$$c_{Pd} = \frac{0.533 - 2.96 \times 10^3 c_{Au}}{3.55 \times 10^3}$$

At 635 nm

$$0.590 = (5.64 \times 10^2)(1.00)c_{Pd} + (1.45 \times 10^4)(1.00)c_{Au}$$

Substitution for c_{Pd} in this expression gives

$$0.590 = \frac{5.64 \times 10^2(0.533 - 2.96 \times 10^3 c_{Au})}{3.55 \times 10^3} + 1.45 \times 10^4 c_{Au}$$

$$= 0.0847 - 4.70 \times 10^2 c_{Au} + 1.45 \times 10^4 c_{Au}$$

$$c_{Au} = (0.590 - 0.0847)/(1.403 \times 10^4) = 3.60 \times 10^{-5}\ \text{M}$$

and

$$c_{Pd} = \frac{0.533 - (2.96 \times 10^3)(3.60 \times 10^{-5})}{3.55 \times 10^3} = 1.20 \times 10^{-4}\ \text{M}$$

Since the analysis involved a twofold dilution, the concentrations of Pd(II) and Au(III) in the original sample were 7.20×10^{-5} and 2.40×10^{-4} M, respectively.

Mixtures containing more than two absorbing species can be analyzed, in principle at least, if one additional absorbance measurement is made for each added component. The uncertainties in the resulting data become greater, however, as the number of measurements increases. Some of the newer computerized spectrophotometers are capable of minimizing these uncertainties by overdetermining the system; that is, these instruments use many more data points than unknowns and effectively match the entire spectrum of the unknown as closely as possible by deriving synthetic spectra for various concentrations of the components. The derived spectra are then compared with that of the analyte until a close match is found. The spectrum for standard solutions of each component is required, of course.

The Effect of Instrumental Uncertainties[9]

The accuracy and precision of spectrophotometric analyses are often limited by the indeterminate error, or *noise,* associated with the instrument. As pointed out earlier, a spectrophotometric absorbance measurement entails three steps: a 0% T adjustment, a 100% T adjustment, and a measurement of % T. The random errors associated with each of these steps combine to give a net random error for the final value obtained for T. The relationship between the noise encountered in the

In the context of this discussion, **noise** refers to random variations in the instrument output due not only to electrical fluctuations but also to such other variables as the way the operator reads the meter, the position of the cell in the light beam, the temperature of the solution, and the output of the source.

[9]For further reading, see J. D. Ingle Jr. and S. R. Crouch, *Analytical Spectroscopy.* Chapter 5. Englewood Cliffs, NJ: Prentice-Hall, 1988.

measurement of T and the resulting *concentration uncertainty* can be derived by writing Beer's law in the form

$$c = -\frac{1}{\varepsilon b} \log T = \frac{-0.434}{\varepsilon b} \ln T$$

See *Mathcad Applications for Analytical Chemistry,* **pp. 124–127.**

Taking the partial derivative of this equation while holding εb constant leads to the expression

$$\partial c = \frac{-0.434}{\varepsilon b T} \partial T$$

where ∂c can be interpreted as the uncertainty in c that results from the noise (or uncertainty) ∂T in T. Dividing this equation by the previous one gives

$$\frac{\partial c}{c} = \frac{0.434}{\log T} \times \frac{\partial T}{T} \qquad \text{(24-5)}$$

where $\partial T/T$ is the *relative* random error in T attributable to the noise in the three measurement steps, and $\partial c/c$ is the resulting relative random concentration error.

The best and most useful measure of the random error ∂T is the standard deviation σ_T, which is easily measured for a given instrument by making 20 or more replicate transmittance measurements of an absorbing solution. Substituting σ_T and σ_c for the corresponding differential quantities in Equation 24-5 leads to

$$\frac{\sigma_c}{c} = \frac{0.434}{\log T} \times \frac{\sigma_T}{T} \qquad \text{(24-6)}$$

where σ_c/c and σ_T/T are relative standard deviations.

It is clear from an examination of Equation 24-6 that the uncertainty in a photometric concentration measurement varies in a complex way with the magnitude of the transmittance. The situation is even more complicated than suggested by the equation, however, because the uncertainty σ_T is, under many circumstances, also *dependent upon T.*

See *Mathcad Applications for Analytical Chemistry,* **pp. 147–149.**

In a detailed theoretical and experimental study, Rothman, Crouch, and Ingle[10] described several sources of instrumental random errors and showed the net effect of these errors on the precision of concentration measurements. The errors fall into three categories: those for which the magnitude of σ_T is (1) independent of T, (2) proportional to $\sqrt{T^2 + T}$, and (3) proportional to T. Table 24-4 summarizes information about these sources of uncertainty. When the three relationships for σ_T in the first column are substituted into Equation 24-6, three equations for the relative standard deviation in the concentration are obtained; these are shown in the third column.

Concentration Errors When $\sigma_T = k_1$ For many photometers and spectrophotometers, the standard deviation in the measurement of T is constant and inde-

[10] L. D. Rothman, S. R. Crouch, and J. D. Ingle Jr., *Anal. Chem.,* **1975,** *47,* 1226.

TABLE 24-4 Categories of Instrumental Indeterminate Errors in Transmittance Measurements

Category	Sources	Effect of T on Relative Standard Deviation of Concentration	
$\sigma_T = k_1$	Readout resolution; thermal detector noise; dark current and amplifier noise	$\dfrac{\sigma_c}{c} = \dfrac{0.434}{\log T} \times \dfrac{k_1}{T}$	(24-7)
$\sigma_T = k_2\sqrt{T^2 + T}$	Photon detector shot noise	$\dfrac{\sigma_c}{c} = \dfrac{0.434}{\log T} \times k_2\sqrt{1 + \dfrac{1}{T}}$	(24-8)
$\sigma_T = k_3 T$	Cell positioning uncertainty; fluctuation in source intensity	$\dfrac{\sigma_c}{c} = \dfrac{0.434}{\log T} \times k_3$	(24-9)

Note: σ_T is the standard deviation of the transmittance measurements; σ_c/c is the relative standard deviation of the concentration measurements; T is transmittance; and k_1, k_2, and k_3 are constants for a given instrument.

pendent of the magnitude of T. This type of random error is often encountered with direct-reading instruments and has its origin in the somewhat limited resolution of the meter scale. The size of a typical scale is such that a reading cannot be reproduced to better than a few tenths of a percent of the full-scale reading, and the magnitude of this uncertainty is the same from one end of the scale to the other. For typical inexpensive instruments, standard deviations of about 0.003 T ($\sigma_T = \pm 0.003\ T$) are observed.

EXAMPLE 24-4

A spectrophotometric analysis was performed with a manual instrument that exhibited an absolute standard deviation of $\pm 0.003\ T$ throughout its transmittance scale. Calculate the relative standard deviation in concentration that results from this uncertainty when the analyte solution has an absorbance of (a) 1.000 and (b) 2.000.

(a) To convert absorbance to transmittance, we write

$$\log T = -A = -1.00$$
$$T = \text{antilog}(-1.00) = 0.100$$

For this instrument, $\sigma_T = k_1 = \pm 0.003$ (see first entry in Table 24-4). Substituting this value and $T = 0.100$ into Equation 24-7 yields

$$\frac{\sigma_c}{c} = \frac{0.434}{\log 0.100} \times \frac{\pm 0.003}{0.1000} = \pm 0.013 \text{ or } \pm 1.3\%$$

(b) At $A = 2.000$, $T = \text{antilog}(-2.000) = 0.010$

$$\frac{\sigma_c}{c} = \frac{0.434}{\log 0.010} \times \frac{\pm 0.003}{0.010} = \pm 0.064 \text{ or } \pm 6.5\%$$

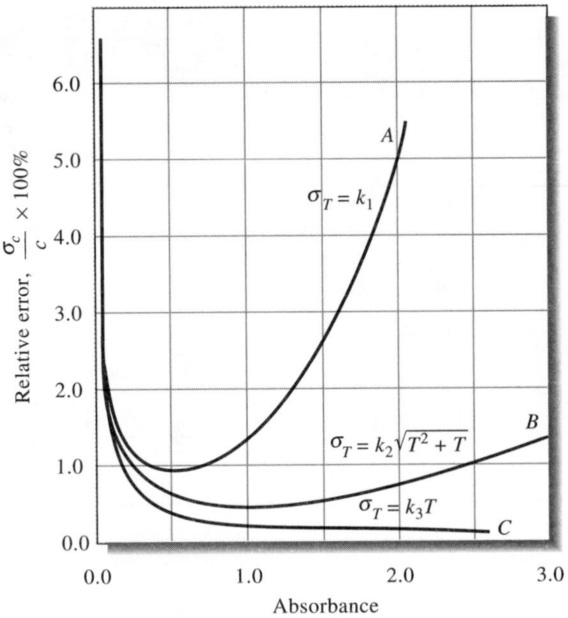

Figure 24-19
Error curves for various categories of instrumental uncertainties.

The data plotted as curve *A* in Figure 24-19 were derived from calculations similar to those in Example 24-4. Note that the relative standard deviation in the concentration passes through a minimum at an absorbance of about 0.5 and rises rapidly when the absorbance is less than about 0.1 or greater than approximately 1.5.

See *Mathcad Applications for Analytical Chemistry,* **p. 148.**

Figure 24-20a is a plot of the relative standard deviation for experimentally determined concentrations as a function of absorbance. It was obtained with the inexpensive spectrophotometer shown in Figure 24-7. The striking similarity between this curve and curve *A* in Figure 24-19 indicates that the instrument studied is affected by an absolute indeterminate error of about $\pm 0.003\ T$ and that this error is independent of transmittance. It is probable that the source of this uncertainty lies in the limited resolution of the transmittance scale.

Infrared spectrophotometers also exhibit an indeterminate error that is independent of transmittance. With such instruments, the source of this error lies in the thermal detector. Fluctuations in the output of this type of transducer are

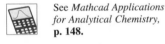

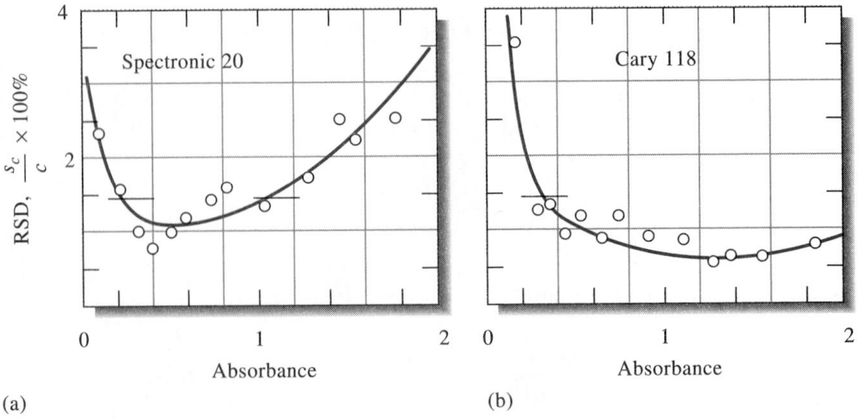

Figure 24-20
Experimental curves relating relative concentration uncertainties to absorbance for two spectrophotometers. Data obtained with (a) a Spectronic 20, a low-cost instrument (Figure 24-7), and (b) a Cary 118, a research-quality instrument. (From W. E. Harris and B. Kratochvil, *An Introduction to Chemical Analysis,* p. 384. Philadelphia: Saunders College Publishing, 1981. With permission.)

independent of the output; indeed, fluctuations are observed even in the absence of radiation. An experimental plot of data from an infrared spectrophotometer is similar in appearance to Figure 24-20a. The curve is displaced upward, however, because of the greater standard deviation associated with infrared measurements.

Concentration Errors When $\sigma_T = k_2\sqrt{T^2 + T}$ This type of random uncertainty is characteristic of the highest-quality spectrophotometers. It has its origin in the so-called *shot noise* that causes the output of photomultipliers and phototubes to fluctuate randomly about a mean value. Equation 24-8 in Table 24-4 describes the effect of shot noise on the relative standard deviation of concentration measurements. A plot of this relationship appears as curve B in Figure 24-19. In obtaining these data, k_2 was assumed to be ± 0.003, a typical value for high-quality spectrophotometers.

Figure 24-20b shows an analogous plot of experimental data obtained with an ultraviolet/visible spectrophotometer with performance characteristics similar to those of the instrument shown in Figure 24-9. Note that, in contrast to the less expensive instrument, absorbances of 2.0 or greater can be measured here without serious deterioration in the quality of the data.

Concentration Errors When $\sigma_T = k_3 T$ Substitution of $\sigma_T = k_3 T$ into Equation 24-6 reveals that the relative standard deviation in concentration from this type of uncertainty is inversely proportional to the logarithm of the transmittance (Equation 24-9 in Table 24-4). Curve C in Figure 24-19, which is a plot of Equation 24-9, reveals that this type of uncertainty is important at low absorbances (high transmittances) but approaches zero at high absorbances.

At low absorbances, the precision obtained with high-quality double-beam instruments is often described by Equation 24-9. The source of this behavior is failure to position cells reproducibly with respect to the beam during replicate measurements. This position dependence probably is the result of small imperfections in the cell windows, which cause reflective losses and transparency to differ from one area of the window to another.

Evaluation of k_3 in Equation 24-9 is possible by comparing the precision of absorbance measurements made in the usual way with measurements in which the cells are left undisturbed at all times with replicate solutions being introduced with a syringe. Experiments of this kind with a high-quality spectrophotometer yielded a value of 0.013 for k_3 (footnote 10, page 578). Curve C in Figure 24-19 was obtained by substituting this numerical value into Equation 24-9. Cell positioning errors affect all types of spectrophotometric measurements in which cells are repositioned between measurements.

Fluctuations in source intensity also yield standard deviations that are described by Equation 24-9. This type of behavior is sometimes encountered in inexpensive single-beam instruments that have unstable power supplies and in infrared instruments.

24A-5 Photometric and Spectrophotometric Titrations

Photometric and spectrophotometric measurements are useful for locating the equivalence points of titrations.[11] This application of absorption measurements

[11]For further information, see J. B. Headridge, *Photometric Titrations.* New York: Pergamon Press, 1961.

obviously requires that one or more of the reactants or products absorb radiation or that an absorbing indicator be present.

Titration Curves

A photometric titration curve is a plot of absorbance (corrected for volume change) as a function of titrant volume. If conditions are chosen properly, the curve consists of two straight-line regions with different slopes, one occurring at the outset of the titration and the other located well beyond the equivalence-point region; the end point is taken as the intersection of extrapolated linear portions of the two lines.

Figure 24-21 shows typical photometric titration curves. Figure 24-21a is the curve for the titration of a nonabsorbing species with an absorbing titrant that is decolorized by the reaction. An example is the titration of thiosulfate ion with triiodide ion. The titration curve for the formation of an absorbing product from colorless reactants is shown in Figure 24-21b; an example is the titration iodide ion with a standard solution of iodate ion to form triiodide. The remaining figures illustrate the curves obtained with various combinations of absorbing analytes, titrants, and products.

In order to obtain titration curves with linear portions that can be extrapolated, the absorbing system(s) must obey Beer's law. Furthermore, absorbances must be corrected for volume changes by multiplying the observed absorbance by $(V + v)/V$, where V is the original volume of the solution and v is the volume of added titrant.

Instrumentation

Photometric titrations are ordinarily performed with a spectrophotometer or a photometer that has been modified so that the titration vessel is held in the light path. After the instrument is set to a suitable wavelength (or an appropriate filter is inserted), the 0% T adjustment is made in the usual way. With radiation passing through the analyte solution to the detector, the instrument is then adjusted to a convenient absorbance reading by varying the source intensity or the detector sensitivity. Ordinarily, no attempt is made to measure the true absorbance since relative values are perfectly adequate for end-point detection. Titration data are then collected without alteration of the instrument settings. The power of the

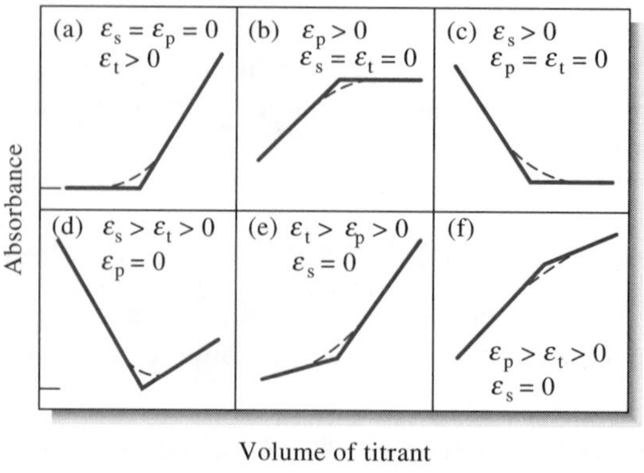

Figure 24-21
Typical photometric titration curves. Molar absorptivities of the substance titrated, the product, and the titrant are ε_s, ε_p, and ε_t.

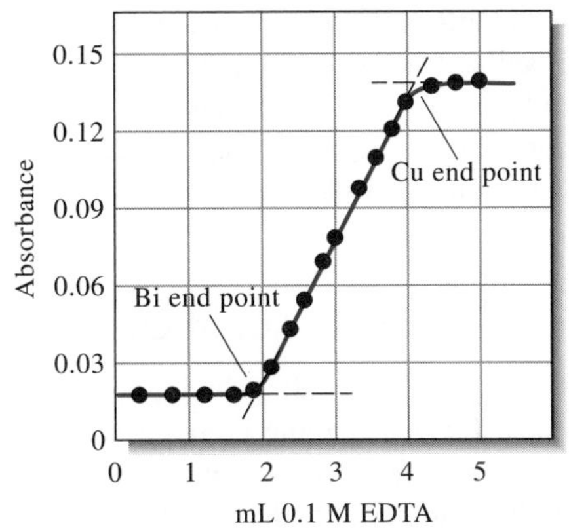

Figure 24-22
Photometric titration curve at 745 nm for 100 mL of a solution that was 2.0×10^{-3} M in Bi^{3+} and Cu^{2+}. (A. L. Underwood, *Anal. Chem.,* **1954,** *26,* 1322. With permission of the American Chemical Society.)

radiation source and the response of the detector must remain constant during a photometric titration. Cylindrical containers are ordinarily used, and care must be taken to avoid any movement of the vessel that might alter the length of the radiation path.

Both filter photometers and spectrophotometers have been employed for photometric titrations. The latter are preferred, however, because their narrower bandwidths enhance the probability of adherence to Beer's law.

Applications of Photometric Titrations

Photometric titrations often provide more accurate results than a direct photometric determination because the data from several measurements are pooled in determining the end point. Furthermore, the presence of other absorbing species may not interfere since only a change in absorbance is being measured.

One advantage of a photometric end point is that the experimental data are taken well away from the equivalence-point region. Consequently, the equilibrium constant for the reaction need not be as favorable as that required for a titration that depends upon observations near the equivalence point (for example, potentiometric or indicator end points). For the same reason, more dilute solutions may be titrated.

The photometric end point has been applied to all types of reactions.[12] For example, most standard oxidizing agents have characteristic absorption spectra and thus produce photometrically detectable end points. Although standard acids or bases do not absorb, the introduction of acid/base indicators permits photometric neutralization titrations. The photometric end point has also been used to great advantage in titrations with EDTA and other complexing agents. Figure 24-22 illustrates the application of this technique to the successive titration of bismuth(III) and copper(II). At 745 nm, the cations, the reagent, and the bismuth complex formed in the first part of the titration do not absorb, but the copper complex does. Thus, the solution exhibits no absorbance until essentially all the bismuth has been titrated. With the first formation of the copper complex, an increase in absorbance occurs. The increase continues until the copper equiva-

[12]See, for example, the review by A. L. Underwood, in *Advances in Analytical Chemistry and Instrumentation,* C. N. Reilley, Ed., Vol. 3, pp. 31–104. New York: Interscience, 1964.

lence point is reached. Further reagent additions cause no further absorbance change. Clearly, two well-defined end points result.

The photometric end point has also been adapted to precipitation titrations. The suspended solid product diminishes the radiant power by scattering; titrations are carried to a condition of constant turbidity.

24A-6 Spectrophotometric Studies of Complex Ions

This section shows how you can determine the composition of a complex in solution without actually isolating the complex as a pure compound.

Spectrophotometry is a valuable tool for elucidating the composition of complex ions in solution and for determining their formation constants. The power of the technique lies in the fact that quantitative absorption measurements can be performed without disturbing the equilibria under consideration. Although most spectrophotometric studies of complexes involve systems in which a reactant or a product absorbs, nonabsorbing systems can also be investigated successfully. For example, the composition and formation constant for a complex of iron(II) and a nonabsorbing ligand could probably be determined by measuring the color decreases that occur when solutions of the absorbing iron(II) complex of 1,10-phenanthroline are mixed with various amounts of the ligand. For this approach to be successful, the formation constant and the composition of the 1,10-phenanthroline complex would have to be known.

The three most common techniques employed for complex-ion studies are (1) the method of continuous variations, (2) the mole-ratio method, and (3) the slope-ratio method.

The Method of Continuous Variations[13]

In the method of continuous variations, cation and ligand solutions with identical analytical concentrations are mixed in such a way that the total volume (and hence the total moles) of reactants in each mixture is constant but the mole ratio of reactants varies systematically (for example, 1:9, 8:2, 7:3, and so forth). The absorbance of each solution is then measured at a suitable wavelength and corrected for any absorbance the mixture might exhibit if no reaction had occurred. The corrected absorbance is plotted against the volume fraction of one reactant, that is, $V_M/(V_M + V_L)$, where V_M is the volume of the cation solution and V_L that of the ligand. A typical plot is shown in Figure 24-23. A maximum (or minimum if the complex absorbs less than the reactants) occurs at a volume ratio V_M/V_L corresponding to the combining ratio of cation and ligand in the complex. In Figure 24-23 , $V_M/(V_M + V_L)$ is 0.33 and $V_L/(V_M + V_L)$ is 0.66; thus, V_M/V_L is 0.33/0.66, which suggests that the complex has the formula ML_2.

The curvature of the experimental lines in Figure 24-23 is the result of incompleteness of the complex-formation reaction. A formation constant for the complex can be evaluated from measurements of the deviations from the theoretical straight lines.

---

[13]See W. C. Vosburgh and G. R. Cooper, *J. Amer. Chem. Soc.*, **1941,** *63*, 437.

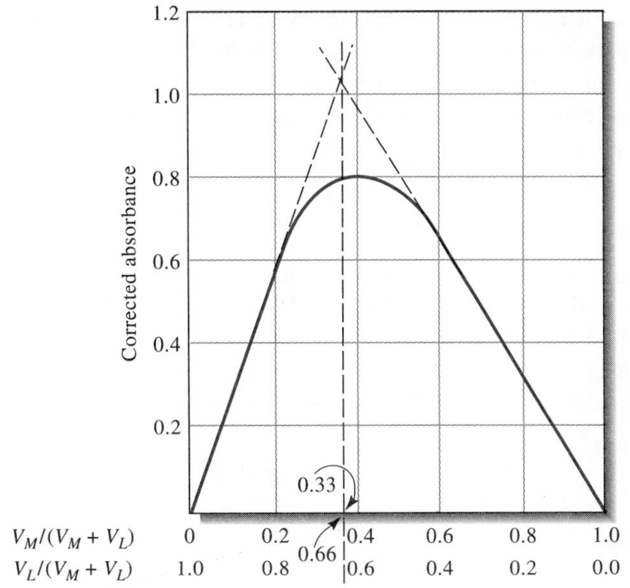

Figure 24-23
Continuous-variation plot for the 1 : 2 complex ML_2.

The Mole-Ratio Method

In the mole-ratio method, a series of solutions is prepared in which the analytical concentration of one reactant (usually the cation) is held constant while that of the other is varied. A plot of absorbance versus mole ratio of the reactants is then prepared. If the formation constant is reasonably favorable, two straight lines with different slopes are obtained. The two intersect at a mole ratio that corresponds to the combining ratio in the complex. Typical mole-ratio plots are shown in Figure 24-24. Note that the ligand of the 1 : 2 complex absorbs at the wavelength selected

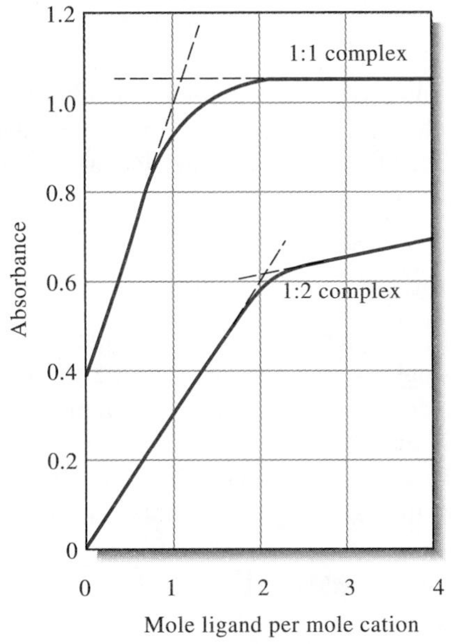

Figure 24-24
Mole-ratio plots for a 1 : 1 and a 1 : 2 complex. The 1 : 2 complex is the more stable, as indicated by less curvature near the stoichiometric ratio.

so that the slope beyond the equivalence point is greater than zero. We deduce that the uncomplexed cation involved in the 1 : 1 complex absorbs because the initial point has an absorbance greater than zero.

Formation constants can be evaluated from the data in the curved portion of mole-ratio plots.

EXAMPLE 24-5

Derive sufficient equations to permit calculation of the equilibrium concentrations of all the species involved in the 1 : 2 complex-formation reaction illustrated in Figure 24-24.

Two mass-balance expressions can be written that are based upon the preparatory data. Thus, for the reaction

$$M + 2L \rightleftharpoons ML_2$$

we can write

$$c_M = [M] + [ML_2]$$
$$c_L = [L] + 2[ML_2]$$

where c_M and c_L are the molar concentrations of M and L before reaction occurs. For 1-cm cells, the absorbance of the solution is

$$A = \varepsilon_M[M] + \varepsilon_L[L] + \varepsilon_{ML_2}[ML_2]$$

From the mole-ratio plot, we see that $\varepsilon_M = 0$. Values for ε_L and ε_{ML_2} can be obtained from the two straight-line portions of the curve. With one or more measurements of A in the curved region of the plot, sufficient data are available to calculate the three equilibrium concentrations and thus the formation constant.

A mole-ratio plot may reveal the stepwise formation of two or more complexes as successive slope changes, provided the complexes have different molar absorptivities and provided the formation constants are sufficiently different from each other.

The Slope-Ratio Method

This approach is particularly useful for weak complexes but is applicable only to systems in which a single complex is formed. The method assumes (1) that the complex-formation reaction can be forced to completion by a large excess of either reactant and (2) that Beer's law is followed under these circumstances.

Let us consider the reaction in which the complex M_xL_y is formed by the reaction of x moles of the cation M with y moles of a ligand L:

$$xM + yL \rightleftharpoons M_xL_y$$

Mass-balance expressions for this system are

$$c_M = [M] + x[M_xL_y]$$
$$c_L = [L] + y[M_xL_y]$$

where c_M and c_L are the molar analytical concentrations of the two reactants. We now assume that at very high analytical concentrations of L, the equilibrium is shifted far to the right and $[M] \ll x[M_xL_y]$. Under this circumstance, the first mass-balance expression simplifies to

$$c_M = x[M_xL_y]$$

If Beer's law obtains,

$$A_1 = \varepsilon b[M_xL_y] = \varepsilon bc_M/x$$

A plot of absorbance as a function of c_M becomes linear whenever sufficient L is present to satisfy the assumption that $[M] \ll x[M_xL_y]$. The slope of this plot is $\varepsilon b/x$.

When c_M is made very large, we assume that $[L] \ll y[M_xL_y]$, whereupon the second mass-balance equation reduces to

$$c_L = y[M_xL_y]$$

and

$$A_2 = \varepsilon b[M_xL_y] = \varepsilon bc_L/y$$

Again, if our assumptions are valid, a linear plot of A versus c_L is observed at high concentrations of M. The slope of this line is $\varepsilon b/y$.

The ratio of the slopes of the two straight lines gives the combining ratio between M and L:

$$\frac{\varepsilon b/x}{\varepsilon b/y} = \frac{y}{x}$$

24B AUTOMATED PHOTOMETRIC AND SPECTROPHOTOMETRIC METHODS

The first fully automated instrument for chemical analysis (the Technicon AutoAnalyzer®) appeared on the market in 1957. This instrument was designed to fulfill the needs of clinical laboratories, where blood and urine samples are routinely analyzed for a dozen or more chemical species. The number of such analyses demanded by modern medicine is enormous[14]; the need to keep their cost at a reasonable level is obvious. These two considerations motivated the development of analytical systems that perform several analyses simultaneously with a minimum input of human labor. The use of automatic instruments has spread from

[14]For example, in 1976, about 100 million samples were analyzed in domestic clinical laboratories. See L. Snyder *et al., Anal. Chem.*, **1976,** *48*, 942A.

clinical laboratories to laboratories for the control of industrial processes and the routine determination of a wide spectrum of species in air, water, soils, and pharmaceutical and agricultural products.[15] In the majority of these applications, the analyses are completed by photometric, spectrophotometric, or fluorometric measurements.

24B-1 Types of Automatic Analytical Systems

Automatic analytical instruments are of two general types, *discrete* and *continuous;* combinations of the two are sometimes encountered. In a discrete instrument, individual samples are maintained as separate entities and kept in separate vessels throughout such unit operations as sampling, definition of the sample (measurement of its mass or volume), dilution, reagent addition, mixing, centrifugation, and transportation to the measuring device. Discrete systems frequently require the use of robots.

In *continuous-flow systems,* the sample becomes a part of a flowing stream in which the several unit operations of the analysis, mentioned in the previous paragraph, take place as the sample is carried from the injection point to a flow-through measuring device (usually a photometer) and finally to waste.

This section is devoted to one type of continuous-flow method called a *flow-injection analysis* (FIA), which commonly employs photometric, spectrophotometric, or ion-selective measurements for completion of analyses.

24B-2 Flow-Injection Analysis

Flow-injection methods, in their present form, were first described by Ruzicka and Hansen in Denmark and Stewart and co-workers in the United States in the mid-1970s.[16] Flow-injection methods are an outgrowth of segmented-flow procedures, which were widely used in clinical laboratories in the 1960s and 1970s for automatic routine determination of a variety of species in blood and urine samples. In segmented-flow systems, which were manufactured by a single company in this country, samples were carried through the system to a detector by a flowing aqueous solution that contained closely spaced air bubbles. The purpose of the air bubbles was to prevent excess sample dispersion, to promote turbulent mixing of samples and reagents, and to scrub the walls of the conduit, thus preventing cross-contamination between successive samples. The discoverers of flow-injection analysis found, however, that excess dispersion and cross-contamination are nearly completely avoided in a properly designed system without air bubbles and that mixing of samples and reagent can be easily realized.[17]

---

[15]For an extensive treatment of automatic analysis, see J. K. Foreman and P. B. Stockwell, *Automatic Chemical Analysis.* New York: Wiley, 1975; V. Cerda and G. Ramis, *An Introduction to Laboratory Automation.* New York: Wiley, 1990.

[16]K. K. Stewart, G. R. Beecher, and P. E. Hare, *Anal. Biochem.,* **1976,** *70,* 167: J. Ruzicka and E. H. Hansen, *Anal. Chim. Acta,* **1975,** *78,* 145.

[17]For monographs on flow-injection analysis, see J. Ruzicka and E. H. Hansen. *Flow Injection Analysis,* 2nd ed. New York: Wiley, 1988; M. Valcarcel and M. D. Luque de Castro, *Flow Injection Analysis, Principles and Applications.* Chichester, England: Ellis Horwood, 1987; B. Karlberg and G. E. Pacey, *Flow Injection Analysis. A Practical Guide.* New York: Elsevier, 1989.

The absence of air bubbles imparts several important advantages to flow-injection measurements, including (1) higher analysis rates (typically 100 to 300 samples per hour), (2) enhanced response times (often less than 1 min between sample injection and detector response), (3) much more rapid start-up and shut-down times (less than 5 min for each), and (4) except for the injection system, simpler and more flexible equipment. The last two advantages are of particular importance because they make it feasible and economic to apply automated measurements to a relatively few samples of a nonroutine kind. That is, no longer are continuous-flow methods restricted to situations where the number of samples is large and the analytical method highly routine. As a consequence of these advantages, segmented-flow systems have been largely replaced by flow-injection methods (and also by discrete systems based upon robotics).

24B-3 Instrumentation

Figure 24-25a is a flow diagram of the simplest of all flow-injection systems. Here, a colorimetric reagent for chloride ion is pumped by a peristaltic pump directly into a valve that permits injection of samples into the flowing stream. The sample and reagent then pass through a 50-cm reactor coil where the reagent diffuses into the sample plug and produces a colored product by the sequence of reactions

$$Hg(SCN)_2(aq) + 2Cl^- \rightleftharpoons HgCl_2(aq) + 2SCN^-$$

$$Fe^{3+} + SCN^- \rightleftharpoons \underset{\text{red}}{Fe(SCN)^{2+}}$$

From the reactor coil, the solution passes into a flow-through photometer equipped with a 480-nm interference filter.

The recorder output from this system for a series of standards containing from 5 to 75 ppm of chloride is shown on the left of Figure 24-25b. Note that four

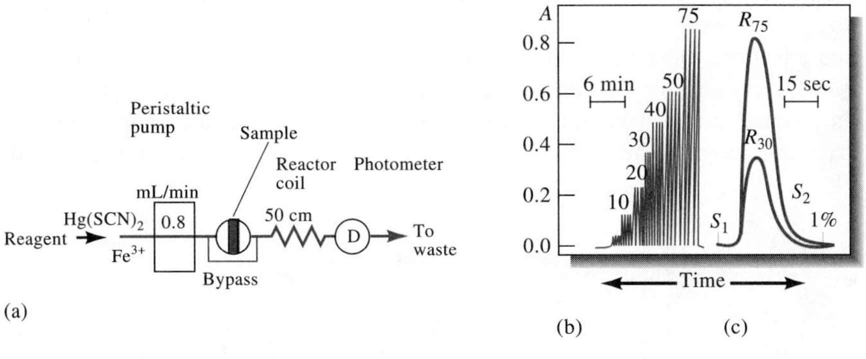

Figure 24-25
Flow-injection determination of chloride: (a) flow diagram; (b) recorder readout for quadruplicate runs on standards containing 5 to 75 ppm of chloride ion; (c) fast scan of two of the standards to demonstrate the low analyte carryover (less than 1%) from run to run. Note that the point marked 1% corresponds to where the response would just begin for a sample injected at time S_2. (From J. Ruzicka and E. H. Hansen, *Flow Injection Methods,* 2nd ed., p. 16. New York: Wiley, 1988. Reprinted by permission of John Wiley & Sons, Inc.)

injections of each standard were made to demonstrate the reproducibility of the system. The two curves to the right are high-speed recorder scans of one of the samples containing 30 ppm (R_{30}) and another containing 75 ppm (R_{75}) chloride. These curves demonstrate that cross-contamination is minimal in an unsegmented stream. Thus, less than 1% of the first analyte is present in the flow cell after 28 s, the time of the next injection (S_2). This system has been successfully used for the routine determination of chloride ion in brackish and waste waters as well as in serum samples.

Sample and Reagent Transport System

Ordinarily, the solution in a flow-injection analysis is pumped through flexible tubing in the system by a peristaltic pump, a device in which a fluid (liquid or gas) is squeezed through plastic tubing by rollers. Figure 24-26 illustrates the operating principle of the peristaltic pump. Here, the spring-loaded cam, or band, pinches the tubing against two or more of the rollers at all times, thus forcing a continuous flow of fluid through the tubing. Modern pumps generally have 8 to 10 rollers, arranged in a circular configuration so that half are squeezing the tube at any instant. This design leads to a flow that is relatively pulse free. The flow rate is controlled by the speed of the motor, which should be greater than 30 rpm, and by the inside diameter of the tube. A variety of tube sizes (i.d. = 0.25 to 4 mm) is available commercially that permit flow rates as small as 0.0005 mL/min and as great as 40 mL/min. The rollers of typical commercial peristaltic pumps are long enough that several tubes can be handled simultaneously.

As shown in Figure 24-25a, flow-injection systems often contain a coiled section of tubing (typical coil diameters are about 1 cm or less) whose purpose it is to enhance axial dispersion and to increase radial mixing of the sample and reagent, both of which lead to more symmetric peaks.

Sample Injectors and Detectors

Sample sizes for flow-injection analysis range from 5 to 200 μL, with 10 to 30 μL being typical for most applications. For a successful analysis, it is vital that the sample solution be injected rapidly as a pulse, or plug, of liquid; in addition, the injections must not disturb the flow of the carrier stream. To date, the most satisfactory injector systems are based upon sampling loops similar to those used in chromatography (see, for example, Figure 30-4). The method of operation of a sampling loop can be seen by reference to Figure 24-25a. With the valve of the

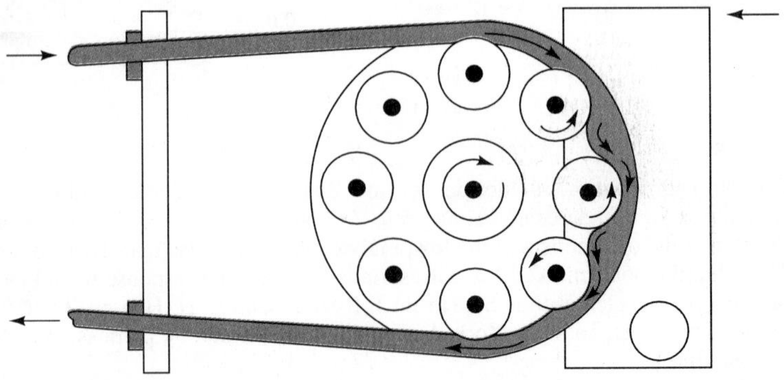

Figure 24-26
Diagram showing one channel of a peristaltic pump. Usually, several additional tubes may be located under the one shown (below the plane of the paper). (From B. Karlberg and G. E. Pacey, *Flow Injection Analysis. A Practical Guide,* p. 34. New York: Elsevier, 1989. With permission of Elsevier Science Publishers.)

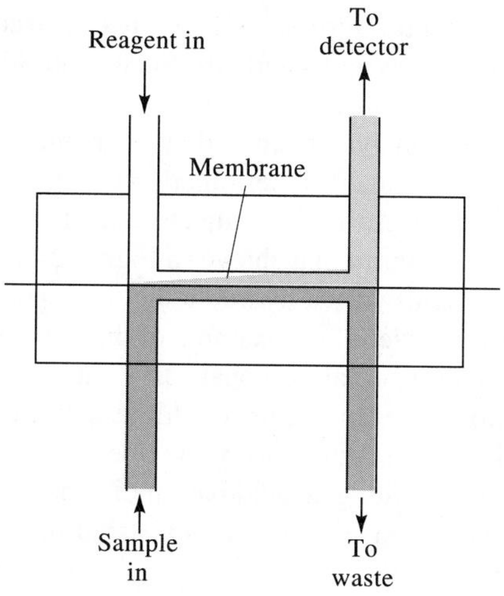

Figure 24-27
A dialysis module. The membrane is supported between two grooved Teflon blocks.

loop in the position shown, reagents flow through the bypass. When a sample has been injected into the loop and the valve is turned 90 deg, the sample enters the flow as a single, well-defined zone. For all practical purposes, flow through the bypass ceases with the valve in this position because the diameter of the sample loop is significantly greater than that of the bypass tubing.

Detection in flow-injection procedures has been carried out by atomic absorption and emission instruments, fluorometers, electrochemical systems, refractometers, spectrophotometers, and photometers. The last is perhaps the most common.

Separations in FIA

Separations by dialysis, by liquid/liquid extraction, and by gaseous diffusion are readily carried out automatically with flow-injection systems.[18]

Dialysis and Gas Diffusion Dialysis is often used in continuous-flow methods to separate inorganic ions, such as chloride or sodium, or small organic molecules, such as glucose, from high-molecular-weight species, such as proteins. Small ions and molecules diffuse relatively rapidly through hydrophilic membranes of cellulose acetate or nitrate, whereas large molecules do not. Dialysis usually precedes the determination of ions and small molecules in whole blood or serum.

Figure 24-27 is a diagram of a dialysis module in which analyte ions or small molecules diffuse from the sample solution through a membrane into a reagent stream, which often contains a species that reacts with the analyte to form a colored entity, which can then be determined photometrically. Large molecules, which interfere in the determination, remain in the original stream and are carried to waste. The membrane is supported between two Teflon plates in which congruent channels have been cut to accommodate the two stream flows. The transfer of smaller species through this membrane is usually incomplete (often less than

[18]For a review of applications of FIA for sample preparation and separations, see G. D. Clark, D. A. Whitman, G. D. Christian, and J. Ruzicka, *Crit. Rev. Anal. Chem.*, **1990,** *21*(5), 357.

50%). Thus, successful quantitative analysis requires close control of temperature and flow rates for both samples and standards. Such control is readily achieved in automated flow-injection systems.

Diffusion of a gaseous analyte from a donor stream to an acceptor stream containing a reagent that permits its determination is a highly selective technique that has found considerable use in flow-injection analysis. The separations are carried out in a module similar to that shown in Figure 24-27. In this application, however, the membrane is usually a hydrophobic microporous material, such as Teflon or isotactic polypropylene. An example of this type of separation is found in a method for determining total carbonate in an aqueous solution. Here the sample is injected into a carrier stream of dilute sulfuric acid, which is then directed into a gas-diffusion module, where the liberated carbon dioxide diffuses into an acceptor stream containing an acid/base indicator. This stream then passes through a photometric detector, which yields a signal that is proportional to the carbonate content of the sample.

Extraction Extraction is another separation method that is readily performed in a flow-injection apparatus. Most commonly, an aqueous solution of the analyte is mixed with an immiscible organic solvent, such as hexane or chloroform, which results in transfer of the analyte (or the interferents) into the organic layer. After passing the mixture through a coil of tubing in which the extraction is given time to occur, the more dense liquid is separated from the less dense and one or the other of the phases is passed into a detector for completion of the analysis. Figure 24-28 is a diagram of an apparatus in which the analyte is separated by extraction with chloroform.

24B-4 A Typical Application of Flow-Injection Analysis

Figure 24-28 illustrates a flow-injection system designed for the automatic spectrophotometric determination of caffeine in acetylsalicylic acid drug preparations after extraction of the caffeine into chloroform. After cooling in an ice bath to minimize evaporation, the chloroform solvent is mixed with the alkaline sample stream in a T-tube (see lower insert). After passing through the 2-m extraction coil, the mixture enters a T-tube separator, which is differentially pumped so that about 35% of the organic phase containing the caffeine passes into the flow cell while the other 65% accompanies the aqueous solution containing the rest of the sample to waste. In order to avoid contaminating the flow cell with water, Teflon fibers, which are not wetted by water, are twisted into a thread and inserted in the inlet to the T-tube in such a way as to form a smooth downward bend. The chloroform flow then follows this bend to the photometer cell where the caffeine concentration is determined based upon its absorption peak at 275 nm. The output of the photometer is similar in appearance to that shown in Figure 24-25b.

24C INFRARED ABSORPTION SPECTROSCOPY

Infrared spectrophotometry is one of the most powerful tools available to the chemist for identifying pure organic and inorganic compounds because, with the exception of a few homonuclear molecules, such as O_2, N_2, and Cl_2, all molecular species absorb infrared radiation. Furthermore, with the exception of chiral mole-

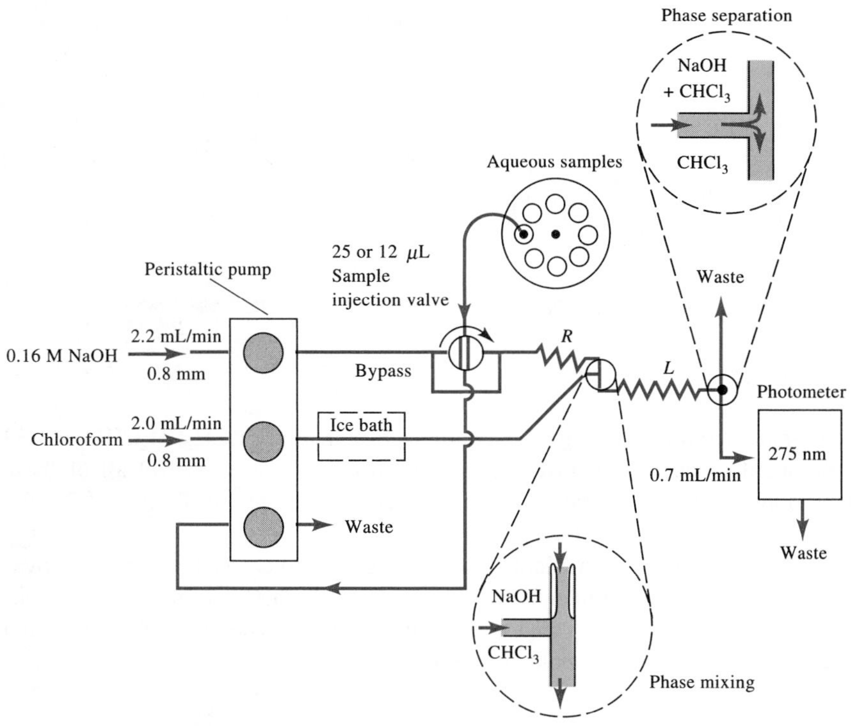

Figure 24-28
Flow-injection apparatus for the determination of caffeine in acetylsalicylic acid preparations. With the valve rotated at 90 deg, the flow in the bypass is essentially zero because of its small diameter. *R* and *L* are Teflon coils with 0.8-mm inside diameters; *L* has a length of 2 m, and the distance from the injection point through *R* to the mixing point is 0.15 m. (Adapted from B. Karlberg and S. Thelander, *Anal. Chim. Acta,* **1978,** *98,* 2. With permission.)

cules in the crystalline state, each molecular species has a unique infrared absorption spectrum. Thus, an exact match between the spectrum of a compound of known structure and that of an analyte unambiguously identifies the latter.

Infrared spectroscopy is a less satisfactory tool for quantitative analyses than its ultraviolet and visible counterparts because the narrow peaks that characterize infrared absorption usually lead to deviations from Beer's law. Furthermore, infrared absorbance measurements are considerably less precise. Nevertheless, where modest precision suffices, the unique nature of infrared spectra provides a degree of selectivity in a quantitative measurement that may offset these undesirable characteristics.[19]

24C-1 Infrared Absorption Spectra

Vibrational absorption occurs in the infrared region, where the energy of radiation is insufficient to excite electronic transitions. As shown in Figure 24-29, infrared spectra exhibit narrow, closely spaced absorption peaks resulting from transitions among the various vibrational quantum levels. Variations in rotational levels may also give rise to a series of peaks for each vibrational state; with liquid or solid samples, however, rotation is often hindered or prevented, and the effects of these small energy differences are not detected. Thus, a typical infrared spectrum for a liquid, such as that in Figure 24-29, consists of a series of vibrational peaks.

The number of ways a molecule can vibrate is related to the number of atoms, and thus the number of bonds, it contains. For even a simple molecule, the number

[19]For a detailed discussion of infrared spectroscopy, see N. B. Colthup, L. H. Daly, and S. E. Wiberley, *Introduction to Infrared and Raman Spectroscopy,* 3rd ed. New York: Academic Press, 1990.

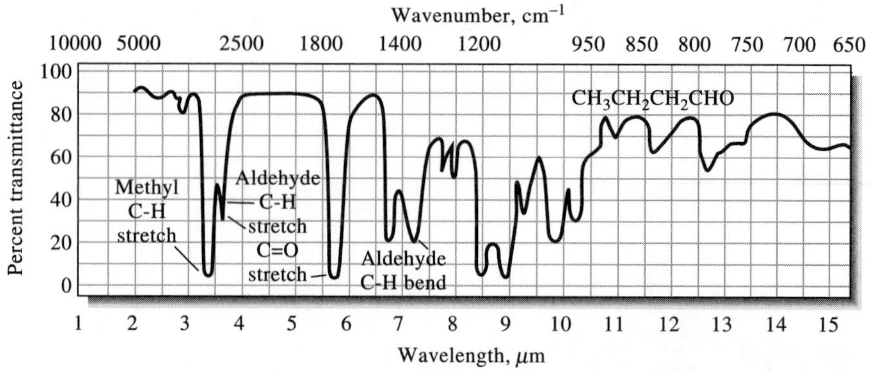

Figure 24-29
Infrared spectrum for *n*-butanal (*n*-butyraldehyde). Note that transmittance rather than absorbance is plotted.

of possible vibrations is large. For example, *n*-butanal ($CH_3CH_2CH_2CHO$) has 33 vibrational modes, most differing from each other in energy. Not all of these vibrations produce infrared peaks; nevertheless, as shown in Figure 24-29, the spectrum for *n*-butanal is relatively complex.

Infrared absorption occurs not only with organic molecules but also with covalently bonded metal complexes, which are generally active in the longer-wavelength infrared region. Infrared spectrophotometric studies have thus provided much useful information about complex metal ions.

24C-2 Instruments for Infrared Spectroscopy

Three types of infrared instruments are found in modern laboratories; dispersive spectrometers (or spectrophotometers), Fourier-transform (FTIR) spectrometers, and filter photometers. The first two are used for obtaining complete spectra for qualitative identification, whereas filter photometers are designed for quantitative work. Fourier-transform and filter instruments are nondispersive in the sense that neither employs a grating or prism to disperse radiation into its component wavelengths.

Dispersive Instruments

With one difference, dispersive infrared instruments are similar in general design to the double-beam (in time) spectrophotometers shown in Figures 23-22c and 24-9. The difference lies in the location of the cell compartment with respect to the monochromator. In ultraviolet/visible instruments, cells are always located between the monochromator and the detector in order to avoid photochemical decomposition, which may occur if samples are exposed to the full power of an ultraviolet or visible source. Infrared radiation, in contrast, is not sufficiently energetic to bring about photodecomposition; thus, the cell compartment can be located between the source and the monochromator. This arrangement is advantageous because any scattered radiation generated in the cell compartment is largely removed by the monochromator.

As shown in Section 23A, the components of infrared instruments differ considerably in detail from those in ultraviolet and visible instruments. Thus, infrared sources are heated solids rather than deuterium or tungsten lamps, infrared gratings are much coarser than those required for ultraviolet/visible radiation, and infrared detectors respond to heat rather than photons. Furthermore, the optical components of infrared instruments are constructed from polished solids, such as sodium chloride or potassium bromide.

Fourier-Transform Spectrometers

Fourier-transform infrared spectrometers offer the advantages of unusually high sensitivity, resolution, and speed of data acquisition (data for an entire spectrum can be obtained in 1 s or less). Offsetting these advantages are the complexity of the instruments and their high cost.

Fourier-transform instruments contain no dispersing element, and all wavelengths are detected and measured simultaneously. In order to separate wavelengths, it is necessary to modulate the source signal in such a way that it can subsequently be decoded by a Fourier transformation, a mathematical operation that requires a high-speed computer. The theory of Fourier-transform measurements is beyond the scope of this book.[20]

Filter Photometers

Infrared photometers designed to monitor the concentration of air pollutants, such as carbon monoxide, nitrobenzene, vinyl chloride, hydrogen cyanide, and pyridine, are now being marketed and used to ensure compliance with regulations established by the Occupational Safety and Health Administration (OSHA). Interference filters, each designed for the determination of a specific pollutant, are available. These transmit narrow bands of radiation in the range of 3 to 14 μm.

24C-3 Qualitative Applications of Infrared Spectrophotometry

An infrared absorption spectrum, even one for a relatively simple compound, often contains a bewildering array of sharp peaks and minima. Peaks useful for the identification of functional groups are located in the shorter-wavelength region of the infrared (from about 2.5 to 8.5 μm), where the positions of the maxima are only slightly affected by the carbon skeleton to which the groups are attached. Investigation of this region of the spectrum thus provides considerable information regarding the overall constitution of the molecule under investigation. Table 24-5 gives the positions of characteristic maxima for some common functional groups.[21]

Identification of the functional groups in a molecule is seldom sufficient to permit positive identification of the compound, and the entire spectrum from 2.5 to 15 μm must be compared with that of known compounds. Collections of spectra are available for this purpose.[22]

---

[20]For an elementary discussion of the principles of Fourier-transform spectroscopy, see D. A. Skoog and J. J. Leary, *Principles of Instrumental Analysis,* 4th ed., pp. 113–120 and 266–270. Philadelphia: Saunders College Publishing, 1992.

[21]For more detailed information, see N. B. Colthup, *J. Opt. Soc. Amer.,* **1950,** *40,* 397; R. M. Silverstein, G. W. Bassler, and T. C. Morrill, *Spectrometric Identification of Organic Compounds,* 5th ed., Chapter 3. New York: Wiley, 1991.

[22]American Petroleum Institute, *Infrared Spectral Data, A.P.I. Research Project 44.* Pittsburgh: Carnegie Institute of Technology; *Sadtler Standard Spectra.* Philadelphia: Sadtler Research Laboratories.

TABLE 24-5 Some Characteristic Infrared Absorption Peaks

Functional Group		Absorption Peaks: Wavenumber, cm^{-1}	Absorption Peaks: Wavelength, μm
O—H	Aliphatic and aromatic	3600–3000	2.8–3.3
NH_2	Also secondary and tertiary	3600–3100	2.8–3.2
C—H	Aromatic	3150–3000	3.2–3.3
C—H	Aliphatic	3000–2850	3.3–3.5
C≡N	Nitrile	2400–2200	4.2–4.6
C≡C—	Alkyne	2260–2100	4.4–4.8
COOR	Ester	1750–1700	5.7–5.9
COOH	Carboxylic acid	1740–1670	5.7–6.0
C=O	Aldehydes and ketones	1740–1660	5.7–6.0
$CONH_2$	Amides	1720–1640	5.8–6.1
C=C—	Alkene	1670–1610	6.0–6.2
ϕ—O—R	Aromatic	1300–1180	7.7–8.5
R—O—R	Aliphatic	1160–1060	8.6–9.4

24C-4 Quantitative Infrared Photometry and Spectrophotometry

Quantitative infrared absorption methods differ somewhat from their ultraviolet and visible counterparts because of the greater complexity of the spectra, the narrowness of the absorption bands, and the capabilities of the instruments available for measurements in this spectral region.[23]

Absorbance Measurements

The use of matched cuvettes for solvent and analyte is seldom practical for infrared measurements because of the difficulty in obtaining cells with identical transmission characteristics. Part of this difficulty stems from degradation of the transparency of infrared cell windows (typically polished sodium chloride) with use due to attack by traces of moisture in the atmosphere and in samples. Furthermore, path lengths are hard to reproduce because infrared cells are often less than 1 mm thick. Such narrow cells are required to permit the transmission of measurable intensities of radiation through pure samples or through very concentrated solutions of the analyte. Measurements of dilute analyte solutions, as is done in ultraviolet or visible spectroscopy, are frequently precluded by the lack of good solvents that transmit over appreciable regions of the infrared spectrum.

For these reasons, a reference absorber is often dispensed with entirely in qualitative infrared work, and the intensity of the radiation passing through the

---

[23]For an extensive discussion of quantitative infrared analysis, see A. L. Smith, in *Treatise on Analytical Chemistry,* 2nd ed., P. J. Elving, E. J. Meehan, and I. M. Kolthoff, Eds., Part I, Vol. 7, pp. 415–456. New York: Wiley, 1981.

sample is simply compared with that of the unobstructed beam; alternatively, a salt plate may be placed in the reference beam. Either way, the resulting transmittance is ordinarily less than 100%, even in regions of the spectrum where the sample is totally transparent. This effect is readily seen by examining the spectrum in Figure 24-29.

Applications of Quantitative Infrared Spectroscopy

Infrared spectrophotometry offers the potential for determining an unusually large number of substances because nearly all molecular species absorb in this region. Moreover, the uniqueness of an infrared spectrum provides a degree of specificity that is matched or exceeded by relatively few other analytical methods. This specificity has particular application to the analysis of mixtures of closely related organic compounds.

The recent proliferation of government regulations on atmospheric contaminants has demanded the development of sensitive, rapid, and highly specific methods for a variety of chemical compounds. Infrared absorption procedures appear to meet this need better than any other single analytical tool.

Table 24-6 illustrates the variety of atmospheric pollutants that can be determined with a simple, portable filter photometer equipped with a separate interference filter for each analyte species. Of the more than 400 chemicals for which maximum tolerable limits have been set by OSHA, half or more have absorption characteristics that make them amenable to determination by infrared photometry or spectrophotometry. Obviously, peak overlaps are to be expected with so many compounds absorbing; nevertheless, the method does provide a moderately high degree of selectivity.

TABLE 24-6 Examples of Infrared Vapor Analysis for OSHA Compliance*

Compound	Allowable Exposure, ppm†	Wavelength, μm	Minimum Detectable Concentration, ppm‡
Carbon disulfide	4	4.54	0.5
Chloroprene	10	11.4	4
Diborane	0.1	3.9	0.05
Ethylenediamine	10	13.0	0.4
Hydrogen cyanide	4.7§	3.04	0.4
Methyl mercaptan	0.5	3.38	0.4
Nitrobenzene	1	11.8	0.2
Pyridine	5	14.2	0.2
Sulfur dioxide	2	8.6	0.5
Vinyl chloride	1	10.9	0.3

*Courtesy of The Foxboro Company, Foxboro, MA 02035.

†1992–1993 OSHA exposure limits for 8-hr weighted average.

‡For 20.25-m cell.

§Short-term exposure limit: 15-min time-weighted average that shall not be exceeded at any time during the work day.

24D QUESTIONS AND PROBLEMS

24-1. Describe the differences between the following and list any particular advantages possessed by one over the other:
*(a) spectrophotometers and photometers.
(b) single-beam and double-beam instruments for absorbance measurements.
*(c) conventional and diode-array spectrophotometers.

*24-2. What minimum requirement is needed to obtain reproducible results with a single-beam spectrophotometer?

24-3. What is the purpose of (a) the 0% *T* adjustment and (b) the 100% *T* adjustment of a spectrophotometer?

*24-4. What experimental variables must be controlled to assure reproducible absorbance data?

24-5. What advantage can be claimed for the standard addition method? What minimum condition is needed for the successful application of this method?

*24-6. The molar absorptivity for the complex formed between bismuth(III) and thiourea is 9.32×10^3 L · cm^{-1} · mol^{-1} at 470 nm. Calculate the range of permissible concentrations for the complex if the absorbance is to be no less than 0.15 nor greater than 0.80 when the measurements are made in 1.00-cm cells.

24-7. The molar absorptivity for aqueous solutions of phenol at 211 nm is 6.17×10^3 L · cm^{-1} · mol^{-1}. Calculate the permissible range of phenol concentrations that can be used if the transmittance is to be less than 80% and greater than 5% when the measurements are made in 1.00-cm cells.

*24-8. The logarithm of the molar absorptivity for acetone in ethanol is 2.75 at 366 nm. Calculate the range of acetone concentrations that can be used if the transmittance is to be greater than 10.0% and less than 90.0% with a 1.50-cm cell.

24-9. The logarithm of the molar absorptivity of phenol in aqueous solution is 3.812 at 211 nm. Calculate the range of phenol concentrations that can be used if the absorbance is to be greater than 0.100 and less than 2.000 with a 1.25-cm cell.

*24-10. A photometer with a linear response to radiation gave a reading of 685 mV with a blank in the light path and 179 mV when the blank was replaced by an absorbing solution. Calculate
(a) the percent transmittance and absorbance of the absorbing solution.
(b) the expected transmittance if the concentration of absorber is one half that of the original solution.
(c) the transmittance to be expected if the light path through the original solution is doubled.

24-11. A portable photometer with a linear response to radiation registered 73.6 μA with a blank solution in the light path. Replacement of the blank with an absorbing solution yielded a response of 24.9 μA. Calculate
(a) the percent transmittance of the sample solution.
*(b) the absorbance of the sample solution.
(c) the transmittance to be expected for a solution in which the concentration of the absorber is one third that of the original sample solution.
*(d) the transmittance to be expected for a solution that has twice the concentration of the sample solution.

24-12. Sketch a photometric titration curve for the titration of Sn^{2+} with MnO_4^-. What color radiation should be used for this titration? Explain.

24-13. Iron(III) reacts with SCN^- to form the red complex, $Fe(SCN)^{2+}$. Sketch a photometric titration curve for Fe(III) with SCN^- ion when a photometer with a green filter is used to collect data. Why is a green filter used?

*24-14. Ethylenediaminetetraacetic acid abstracts bismuth(III) from its thiourea complex:

$$Bi(tu)_6^{3+} + H_2Y^{2-} \longrightarrow BiY^- + 6tu + 2H^+$$

where tu is the thiourea molecule, $(NH_2)_2CS$. Predict the shape of a photometric titration curve based on this process, given that the Bi(III)/thiourea complex is the only species in the system that absorbs at 465 nm, the wavelength selected for the analysis.

24-15. The accompanying data (1.00-cm cells) were obtained for the spectrophotometric titration of 10.00 mL of Pd(II) with 2.44×10^{-4} M Nitroso R (O. W. Rollins and M. M. Oldham, *Anal. Chem.*, **1971,** *43*, 262).

Volume of Nitroso R, mL	A_{500}
0	0
1.00	0.147
2.00	0.271
3.00	0.375
4.00	0.371
5.00	0.347
6.00	0.325
7.00	0.306
8.00	0.289

Calculate the concentration of the Pd(II) solution, given that the ligand-to-cation ratio in the colored product is 2 : 1.

*24-16. A 4.97-g petroleum specimen was decomposed by wet-ashing and subsequently diluted to 500 mL in a volumetric flask. Cobalt was determined by treating 25.00-mL aliquots of this diluted solution as follows:

Reagent Volume			
Co(II), 3.00 ppm	**Ligand**	**H_2O**	Absorbance
0.00	20.00	5.00	0.398
5.00	20.00	0.00	0.510

Assume that the Co(II)/ligand chelate obeys Beer's law, and calculate the percentage of cobalt in the original sample.

24-17. Iron(III) forms a complex with thiocyanate ion that has the formula $Fe(SCN)^{2+}$. The complex has an absorption maximum at 580 nm. A specimen of well water was assayed according to the following scheme:

	Volumes, mL					Absorbance,
Sample	**Sample Volume**	**Oxidizing Reagent**	**Fe(II) 2.75 ppm**	**KSCN 0.050 M**	**H_2O**	580 nm (1.00-cm cells)
1	50.00	5.00	5.00	20.00	20.00	0.549
2	50.00	5.00	0.00	20.00	25.00	0.231

Calculate the concentration of iron in parts per million.

24-18. A. J. Mukhedkar and N. V. Deshpande (*Anal. Chem.*, **1963,** *35*, 47) report on a simultaneous determination for cobalt and nickel based upon absorption by their 8-quinolinol complexes. Molar absorptivities are $\varepsilon_{Co} = 3529$ and $\varepsilon_{Ni} = 3228$ at 365 nm, and $\varepsilon_{Co} = 428.9$ and $\varepsilon_{Ni} = 0$ at 700 nm. Calculate the concentration of nickel and cobalt in each of the following solutions (1.00-cm cells):

Solution	A_{365}	A_{700}
1	0.0235	0.617
2	0.0714	0.755
3	0.0945	0.920
4	0.0147	0.592

24-19. Molar absorptivity data for the cobalt and nickel complexes with 2,3-quinoxalinedithiol are $\varepsilon_{Co} = 36{,}400$ and $\varepsilon_{Ni} = 5520$ at 510 nm, and $\varepsilon_{Co} = 1240$ and $\varepsilon_{Ni} = 17{,}500$ at 656 nm. A 0.425-g sample was dissolved and diluted to 50.0 mL. A 25.0-mL aliquot was treated to eliminate interferences; after addition of 2,3-quinoxalinedithiol, the volume was adjusted to 50.0 mL. This solution had an absorbance of 0.446 at 510 nm and 0.326 at 656 nm in a 1.00-cm cell. Calculate the parts per million of cobalt and nickel in the sample.

24-20. The indicator HIn has an acid dissociation constant of 4.80×10^{-6} at ordinary temperatures. The accompanying absorbance data are for 8.00×10^{-5} M solutions of the indicator measured in 1.00-cm cells in strongly acidic and strongly alkaline media.

	Absorbance	
λ, nm	**pH 1.00**	**pH 13.00**
420	0.535	0.050
445	0.657	0.068
450	0.658	0.076
455	0.656	0.085
470	0.614	0.116
510	0.353	0.223
550	0.119	0.324
570	0.068	0.352
585	0.044	0.360
595	0.032	0.361
610	0.019	0.355
650	0.014	0.284

Estimate the wavelength at which absorption by the indicator becomes independent of pH (that is, the isosbestic point).

24-21. Calculate the absorbance (1.00-cm cells) at 450 nm of a solution in which the total molar concentration of the indicator described in Problem 24-20 is 8.00×10^{-5} and the pH is *(a) 4.92, (b) 5.46, *(c) 5.93, (d) 6.16.

24-22. What is the absorbance at 595 nm (1.00-cm cells) of a solution that is 1.25×10^{-4} M in the indicator of Problem 24-20 and has a pH of *(a) 5.30, (b) 5.70, *(c) 6.10?

24-23. Several buffer solutions were made 1.00×10^{-4} M in the indicator of Problem 24-20. Absorbance data (1.00-cm cells) are

Solution	A_{450}	A_{595}
*A	0.344	0.310
B	0.508	0.212
*C	0.653	0.136
D	0.220	0.380

Calculate the pH of each solution.

24-24. Construct an absorption spectrum for an 8.00×10^{-5} M solution of the indicator of Problem 24-20 when measurements are made with 1.00-cm cells and

(a) $\dfrac{[HIn]}{[In^-]} = 3$

(b) $\dfrac{[HIn]}{[In^-]} = 1$

(c) $\dfrac{[HIn]}{[In^-]} = \dfrac{1}{3}$

24-25. Solutions of P and Q individually obey Beer's law over a large concentration range. Spectral data for these species in 1.00-cm cells are

	Absorbance	
λ, nm	**8.55 × 10⁻⁵ M P**	**2.37 × 10⁻⁴ M Q**
400	0.078	0.500
420	0.087	0.592
440	0.096	0.599
460	0.102	0.590
480	0.106	0.564
500	0.110	0.515
520	0.113	0.433
540	0.116	0.343
560	0.126	0.255
580	0.170	0.170
600	0.264	0.100
620	0.326	0.055
640	0.359	0.030
660	0.373	0.030
680	0.370	0.035
700	0.346	0.063

(a) Plot an absorption spectrum for a solution that is 8.55×10^{-5} M in P and 2.37×10^{-4} M in Q.

*(b) Calculate the absorbance (1.00-cm cells) at 440 nm of a solution that is 4.00×10^{-5} M in P and 3.60×10^{-4} M in Q.

*(c) Calculate the absorbance (1.00-cm cells) at 620 nm for a solution that is 1.61×10^{-4} M in P and 7.35×10^{-4} M in Q.

24-26. Use the data in Problem 24-25 to calculate the molar concentration of P and Q in each of the following solutions:

*(a)	0.357	0.803	(d)	0.910	0.338
(b)	0.830	0.448	*(e)	0.480	0.825
*(c)	0.248	0.333	(f)	0.194	0.315

24-27. A standard solution was put through appropriate dilutions to give the concentrations of iron shown below. The iron(II)-1,10-phenanthroline complex was then developed in 25.0-mL aliquots of these solutions, following which each was diluted to 50.0 mL. The following absorbances (1.00-cm cells) were recorded at 510 nm:

Fe(II) Concentration in Original Solutions, ppm	A_{510}
4.00	0.160
10.0	0.390
16.0	0.630
24.0	0.950
32.0	1.260
40.0	1.580

(a) Sketch a calibration curve from these data.

*(b) Use the method of least squares to derive an equation relating absorbance and the concentration of iron(II).

*(c) Calculate the standard deviation of the slope and intercept.

24-28. The method developed in Problem 24-27 was used for the routine determination of iron in 25.0-mL aliquots of ground water. Express the concentration (as ppm Fe) in samples that yielded the accompanying absorbance data (1.00-cm cell). Calculate the relative standard deviation of the result. Repeat the calculation assuming the absorbance data are means of three measurements.

*(a) 0.143	*(c) 0.068	*(e) 1.512
(b) 0.675	(d) 1.009	(f) 0.546

CHAPTER

25

Molecular Fluorescence Spectroscopy

Fluorescence is an analytically important emission process in which atoms or molecules are excited by the absorption of a beam of electromagnetic radiation. The excited species then relax to the ground state, giving up their excess energy as photons. In this chapter we consider applications of molecular fluorescence.

Fluorescence emission is over in 10^{-5} s or less. In contrast, phosphorescence may go on for several minutes or even hours. Fluorescence is much more widely used for chemical analysis than phosphorescence.

25A THEORY OF MOLECULAR FLUORESCENCE

Figure 25-1 is the partial energy diagram for a hypothetical molecular species. Three electronic energy states are shown: E_0, E_1, and E_2, where E_0 is the ground state and E_1 and E_2 are electronic excited states. Each of the electronic states is shown as having four excited vibrational states.

Irradiation of this species with a band of radiation made up of wavelengths λ_1 to λ_5 (Figure 25-1a) results in the momentary population of the five vibrational levels of the first excited electronic state E_1. Similarly, when the molecules are irradiated with a more energetic band radiation made up of shorter wavelengths λ_1' through λ_5', the five vibrational levels of the higher energy electronic state E_2 become briefly populated.

25A-1 Relaxation Processes

As noted in Section 22C, the lifetime of an excited species is brief because there are several ways an excited atom or molecule can give up its excess energy and relax to its ground state. Two of the most important of these mechanisms, nonradiative relaxation and fluorescent relaxation, are illustrated in Figures 25-1b and 25-1c.

Two types of nonradiative relaxation are shown in Figure 25-1b. *Vibrational deactivation,* or *relaxation,* depicted by the short wavy arrows between vibrational energy levels, takes place during collisions between excited molecules and molecules of the solvent. During these collisions, the excess vibrational energy is transferred to solvent molecules in a series of steps as indicated in the figure. The gain in vibrational energy of the solvent is reflected in a slight increase in the temperature of the medium. Vibrational relaxation is such an efficient process that the average lifetime of an excited *vibrational* state is only about 10^{-15} s.

Vibrational relaxation involves transfer of the excess energy of a vibrationally excited species to molecules of the solvent. This process takes place in less than 10^{-15} s and leaves the molecules in the lowest vibrational level of an electronic state.

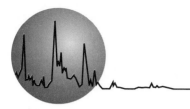

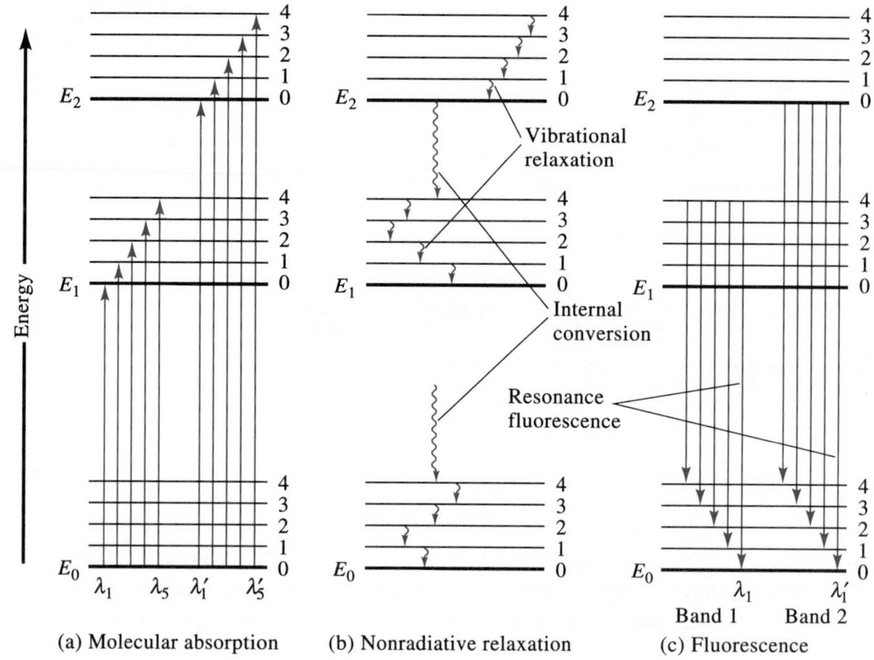

Figure 25-1
Energy-level diagram showing some of the energy changes that occur during (a) absorption, (b) nonradiative relaxation, and (c) fluorescence by a molecular species.

Internal conversion is a type of relaxation that involves transfer of the excess energy of a species in the lowest vibrational level of an excited electronic state to a lower electronic state.

Fluorescence bands consist of a host of closely spaced lines.

Nonradiative relaxation between the lowest vibrational level of an excited electronic state and the upper vibrational level of another electronic state can also occur. This type of relaxation, which is sometimes called *internal conversion,* is depicted by the two longer wavy arrows in Figure 25-1b. Internal conversion is much less efficient than vibrational relaxation, so the average lifetime of an electronic excited state is between 10^{-6} and 10^{-9} s. The mechanisms by which this type of relaxation occurs are not fully understood, but the net effect is again a tiny rise in the temperature of the medium.

Figure 25-1c depicts another relaxation process: fluorescence. Note that bands of radiation are produced when molecules fluoresce because the electronically excited molecules can relax to any of the several vibrational states of the ground electronic state. Like molecular absorption bands, molecular fluorescence bands are made up of a multitude of closely spaced lines that are often difficult to resolve.

Resonance Lines and the Stokes Shift

Resonance fluorescence has an identical wavelength to the radiation that caused the fluorescence.

Stokes shift fluorescence is longer in wavelength than the radiation that caused the fluorescence.

Note that the lines that terminate the two fluorescence bands on the short-wavelength, or high-energy, side (λ_1 and λ_1') are identical in energy to the two lines labeled λ_1 and λ_1' in the absorption diagram in Figure 25-1a. These lines are termed *resonance lines* because the fluorescence and the absorption wavelengths are identical. Note also that molecular fluorescence bands are made up largely of lines that are longer in wavelength or lower in energy than the band of absorbed radiation responsible for their excitation. This shift to longer wavelengths is sometimes called the *Stokes shift.*

To develop a better understanding of Stokes shifts, let us consider what occurs when the molecule under consideration is irradiated by a single wavelength λ_5. As shown in Figure 25-1a, absorption of this radiation promotes an electron into vibrational level 4 of the second excited electronic state E_2. In 10^{-15} s or less, vibrational relaxation to the zero vibrational level of E_2 occurs (Figure 25-1b). At this point, further relaxation can follow either the nonradiative route depicted in Figure 25-1b or the radiative route shown in Figure 25-1c. If the radiative route is followed, relaxation to any of the several vibrational levels of the ground state takes place, giving a band (band 2) of emitted wavelengths, as shown. All of these lines are lower in energy, or longer in wavelength, than the excitation line, λ_5.

Let us now turn to those molecules in excited state E_2 that undergo internal conversion to electronic state E_1. As before, further relaxation can take a nonradiative or a radiative route to the ground state. In the latter case, band 1 of fluorescence is produced. Note that in this case the Stokes shift is from ultraviolet radiation to visible. Note also that band 1 can be produced not only by the mechanism just described but also by the absorption of visible radiation of wavelengths λ_1 through λ_5 (Figure 25-2a).

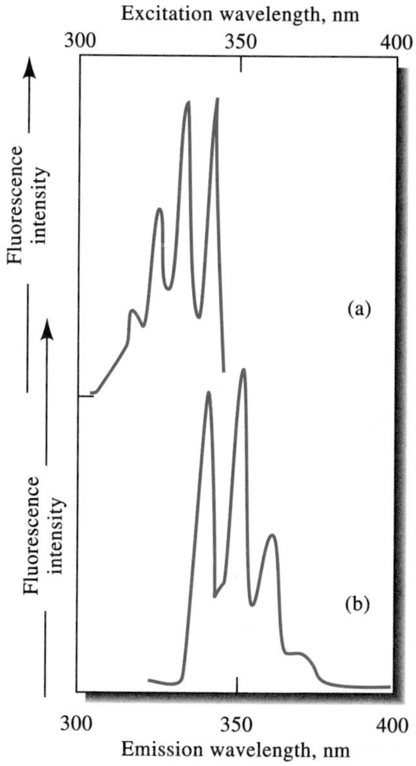

Figure 25-2
Fluorescence spectra for 1 ppm anthracene in alcohol: (a) excitation spectrum; (b) emission spectrum.

Relationship Between Excitation Spectra and Fluorescence Spectra

Because the energy differences between vibrational states is about the same for both ground and excited states, the absorption or *excitation spectrum* and the fluorescence spectrum for a compound often appear as approximate mirror images of one another with overlap occurring at the resonance line. This effect is demonstrated by the spectra shown in Figure 25-2.

25A-2 Fluorescent Species

As shown in Figure 25-1, fluorescence is one of several mechanisms by which a molecule can return to the ground state after it has been excited by absorption of radiation. Thus all absorbing molecules have the potential to fluoresce. Most compounds do not, however, because their structure provides radiationless pathways by which relaxation can occur *at greater rate* than fluorescent emission.

The *quantum yield* of molecular fluorescence is simply the ratio of the number of molecules that fluoresce to the total number of excited molecules (or the ratio of photons emitted to photons absorbed). Highly fluorescent molecules, such as fluorescein, have quantum efficiencies that approach unity under some conditions. Nonfluorescent species have efficiencies that are essentially zero.

Quantum efficiency is described in terms of **quantum yield,** which is equal to $\frac{r_f}{r_f + r_r}$ where r_f is the rate of fluorescent relaxation and r_r is the rate of radiationless relaxation.

Fluorescence and Structure

Compounds containing aromatic rings give the most intense and most useful molecular fluorescence emission. While certain aliphatic and alicyclic carbonyl compounds as well as highly conjugated double-bonded structures also fluoresce, their numbers are small in comparison with the number of fluorescent compounds containing aromatic systems.

Most unsubstituted aromatic hydrocarbons fluoresce in solution, with the quantum efficiency increasing with the number of rings and their degree of condensation. The simplest heterocyclics, such as pyridine, furan, thiophene, and pyrrole, do not exhibit molecular fluorescence (Figure 25-3), but fused-ring structures containing these rings often do (Figure 25-4).

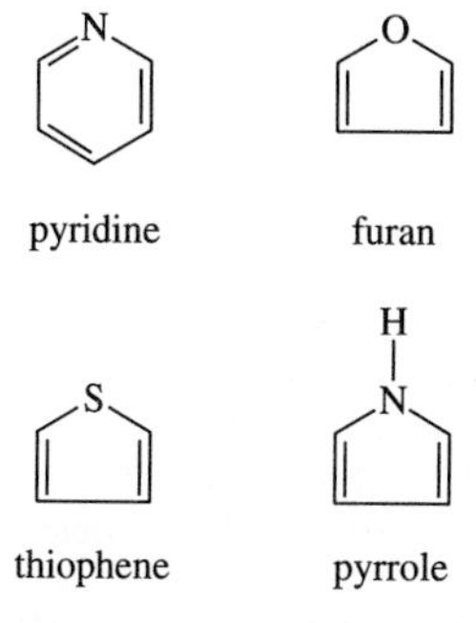

Figure 25-3
Typical aromatic molecules that do not fluoresce.

Substitution on an aromatic ring causes shifts in the wavelength of absorption maxima and corresponding changes in the fluorescence peaks. In addition, sub-

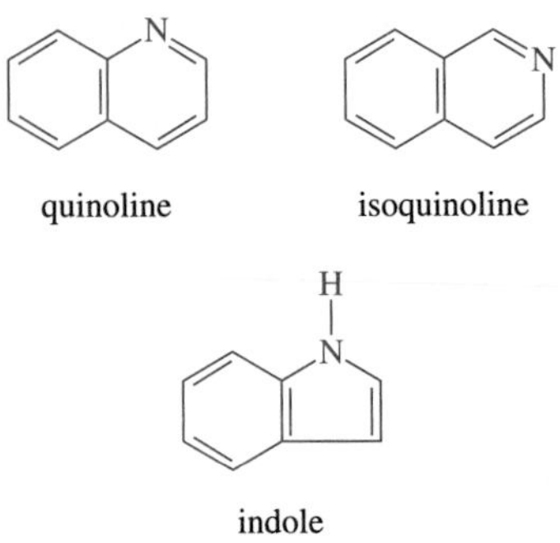

Figure 25-4
Typical aromatic molecules that fluoresce.

Many aromatic compounds fluoresce.

stitution frequently affects the fluorescence efficiency. These effects are demonstrated by the data in Table 25-1.

The Effect of Structural Rigidity

It is found experimentally that fluorescence is particularly favored in rigid molecules. For example, under similar conditions of measurement, the quantum efficiency of fluorene is nearly 1.0 whereas that of biphenyl is about 0.2 (Figure 25-5). The difference in behavior appears to be largely a result of the increased rigidity furnished by the bridging methylene group in fluorene. This rigidity lowers the rate of nonradiative relaxation to the point where relaxation by fluorescence has time to occur. Many similar examples can be cited. In addition, enhanced emission frequently results when fluorescing dyes are adsorbed on a solid surface; here again, the added rigidity provided by the solid may account for the observed effect.

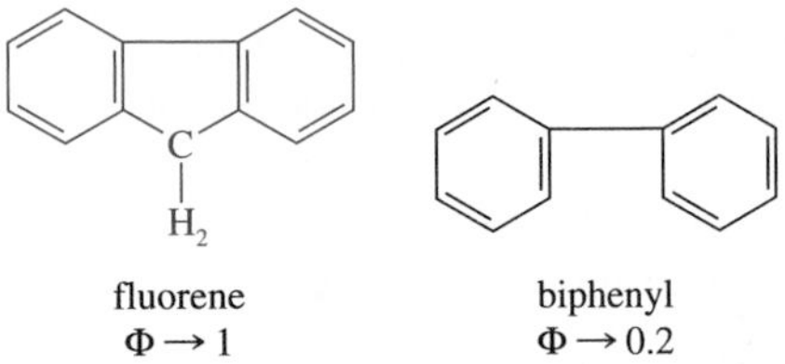

Figure 25-5
Effect of rigidity on quantum yield.

Rigid molecules or complexes tend to fluoresce.

TABLE 25-1
Effect of Substitution on the Fluorescence of Benzene Derivatives*

Compound	Relative Intensity of Fluorescence
Benzene	10
Toluene	17
Propylbenzene	17
Fluorobenzene	10
Chlorobenzene	7
Bromobenzene	5
Iodobenzene	0
Phenol	18
Phenolate ion	10
Anisole	20
Aniline	20
Anilinium ion	0
Benzoic acid	3
Benzonitrile	20
Nitrobenzene	0

*In ethanol solution. Taken from W. West, *Chemical Applications of Spectroscopy* (*Techniques of Organic Chemistry,* Vol. IX, p. 730). New York: Interscience, 1956. Reprinted by permission of John Wiley & Sons.

The influence of rigidity has also been invoked to account for the increase in fluorescence of certain organic chelating agents when they are complexed with a metal ion. For example, the fluorescence intensity of 8-hydroxyquinoline is much less than that of the zinc complex (Figure 25-6).

Temperature and Solvent Effects

In most molecules, the quantum efficiency of fluorescence decreases with increasing temperature because the increased frequency of collision at elevated temperatures improves the probability of collisional relaxation. A decrease in solvent viscosity leads to the same result.

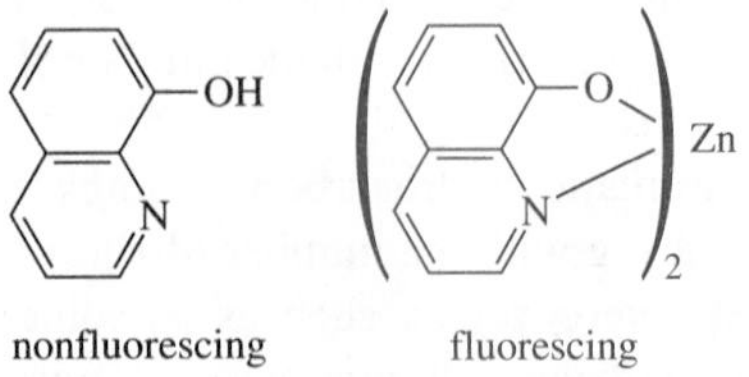

Figure 25-6
Effect of rigidity on fluorescence.

25B EFFECT OF CONCENTRATION ON FLUORESCENCE INTENSITY

The power of fluorescent radiation F is proportional to the radiant power of the excitation beam absorbed by the system:

$$F = K'(P_0 - P) \tag{25-1}$$

where P_0 is the power of the beam incident on the solution and P is its power after it traverses a length b of the medium. The constant K' depends upon the quantum efficiency of the fluorescence. In order to relate F to the concentration c of the fluorescing particle, we write Beer's law in the form

$$\frac{P}{P_0} = 10^{-\varepsilon bc} \tag{25-2}$$

where ε is the molar absorptivity of the fluorescing species and εbc is the absorbance A. By substituting Equation 25-2 into Equation 25-1, we obtain

$$F = K'P_0(1 - 10^{-\varepsilon bc}) \tag{25-3}$$

Expansion of the exponential term in Equation 25-3 leads to

$$F = K'P_0\left[2.3\varepsilon bc - \frac{(-2.3\varepsilon bc)^2}{2!} - \frac{(-2.3\varepsilon bc)^3}{3!} - \cdots\right] \tag{25-4}$$

Provided $\varepsilon bc = A < 0.05$, all the subsequent terms in the brackets are small with respect to the first and we can write

$$F = 2.3K'\varepsilon bc\, P_0 \tag{25-5}$$

or, at constant P_0

$$F = Kc \tag{25-6}$$

See *Mathcad Applications for Analytical Chemistry,* **pp. 111–114.**

Thus, a plot of the fluorescence power of a solution versus the concentration of the emitting species should be linear at low concentrations. When c becomes great enough that the absorbance is larger than about 0.05 (or the transmittance is smaller than about 90%), linearity is lost and F lies below an extrapolation of the straight-line plot. This effect is a result of *self-quenching* in which analyte molecules absorb the fluorescence produced by other analyte molecules. Indeed, at very high concentrations, F reaches a maximum and may even begin to decrease with increasing concentration. A typical plot of F versus concentration is shown in Figure 25-7.

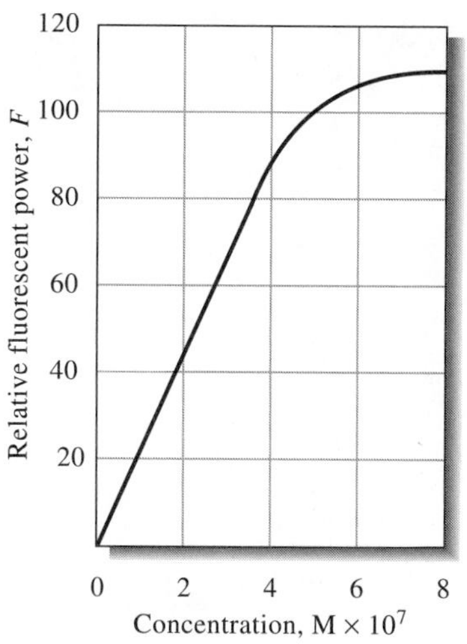

Figure 25-7
Calibration curve for the spectrofluorometric determination of tryptophan in soluble proteins from the lens of a mammalian eye.

25C FLUORESCENCE INSTRUMENTS

Figure 25-8 shows a typical configuration for the components of *fluorometers* and *spectrofluorometers.* These components are identical to the ones described in Section 23A for ultraviolet/visible spectroscopy. A fluorometer, like a photometer, employs filters for wavelength selection. Most spectrofluorometers, in contrast, employ a filter for limiting the excitation radiation and a grating monochromator for dispersing the fluorescence radiation from the sample. A few spectrofluorometers have two monochromator systems, one for the excitation radiation and one for the fluorescence. With such an instrument both *excitation* and *fluorescence* spectra can be obtained.

As shown in Figure 25-8, fluorescence instruments are usually double-beam in design in order to compensate for fluctuations in the power of the source. The beam to the sample first passes through a primary filter or a primary monochromator, which transmits radiation that excites fluorescence but excludes or limits radiation that corresponds to the fluorescence wavelengths. Fluorescence radiation is propagated from the sample in all directions but is most conveniently observed at right angles to the excitation beam; at other angles, increased scattering from the solution and the cell walls may cause large errors in the intensity measurement. The emitted radiation reaches a photoelectric detector after passing through the secondary filter or monochromator, which isolates a fluorescence peak for measurement.

The reference beam passes through an attenuator to decrease its power to approximately that of the fluorescence radiation (the power reduction is usually

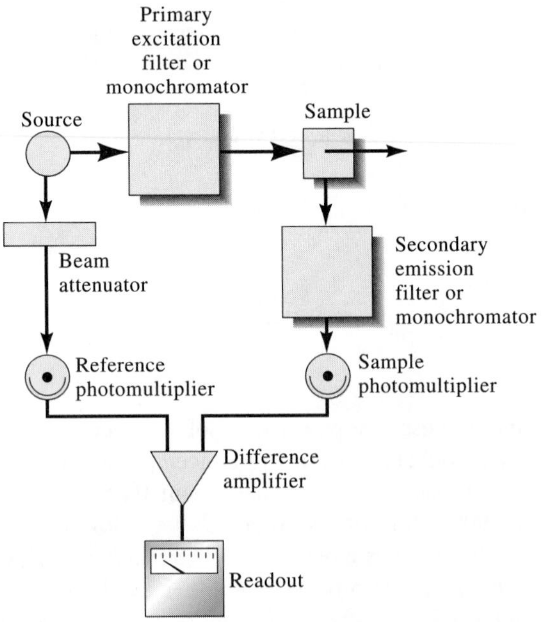

Figure 25-8
Components of a fluorometer or a spectrofluorometer.

by a factor of 100 or more). The signals from the reference and sample phototubes are then processed by a difference amplifier whose output is displayed on a meter or recorder. Many fluorescence instruments are of the null type, this state being achieved by optical or electrical attenuators.

The sophistication, performance characteristics, and cost of fluorometers and spectrofluorometers differ as widely as do the corresponding instruments for absorption measurements. In many regards, filter-type instruments are better suited for quantitative analytical work than are the more elaborate instruments based on monochromators. Generally, fluorometers are more sensitive than spectrofluorometers because filters have a higher radiation throughput than do monochromators. In addition, source and detector can be positioned closer to the sample in the simpler instrument, a factor that enhances sensitivity.

25D APPLICATIONS OF FLUORESCENCE METHODS

Fluorescence methods are generally one to three orders of magnitude more sensitive than methods based upon absorption because the sensitivity of the former can be enhanced either by increasing the power of the excitation beam (Equation 25-5) or amplifying the detector signal. Neither of these options improves the sensitivity of methods based upon absorption, however, because the concentration-related parameter, absorbance, is a ratio:

Fluorescence methods are 10 to 1000 times more sensitive than absorption methods.

$$c = k\text{A} = k \log(P_0/P)$$

Beer's law again.

where $k = 1/ab$ (Equation 22-8). Increasing the power P_0 increases P proportionately and thus has no effect on sensitivity. Similarly, increasing the amplification of the detector signal affects the two measured quantities in an identical way and leads to no improvement.

25D-1 Methods for Inorganic Species

Inorganic fluorometric methods are of two types. Direct methods are based upon the reaction of the analyte with a chelating agent to form a complex that fluoresces. In contrast, indirect methods depend upon the diminution, or *quenching,* of fluorescence of a reagent as a result of its reaction with the analyte. Quenching is used primarily for the determination of anions.

Quenching involves attenuation of fluorescence by some other species (often an anion) in a solution.

The most successful fluorometric reagents for the determination of cations are aromatic compounds with two or more donor functional groups that permit chelate formation with the metal ion. A typical example is 8-hydroxyquinoline, the structure of which is given in Section 5D-3. A few other fluorometric reagents and their applications are found in Table 25-2. With most of these reagents, the cation is extracted into a solution of the reagent in an immiscible organic solvent, such as chloroform. The fluorescence of the organic solvent is then measured. For a more

TABLE 25-2 Selected Fluorometric Methods for Inorganic Species*

Ion	Reagent	Wavelength, nm: Absorption	Wavelength, nm: Fluorescence	Sensitivity, μg/mL	Interference
Al^{3+}	Alizarin garnet R	470	500	0.007	Be, Co, Cr, Cu, F^-, NO_3^-, Ni, PO_4^{3-}, Th, Zr
F^-	Al complex of Alizarin garnet R (quenching)	470	500	0.001	Be, Co, Cr, Cu, Fe, Ni, PO_4^{3-}, Th, Zr
$B_4O_7^{2-}$	Benzoin	370	450	0.04	Be, Sb
Cd^{2+}	2-(*o*-Hydroxyphenyl)-benzoxazole	365	Blue	2	NH_3
Li^+	8-Hydroxyquinoline	370	580	0.2	Mg
Sn^{4+}	Flavanol	400	470	0.1	F^-, PO_4^{3-}, Zr
Zn^{2+}	Benzoin	—	Green	10	B, Be, Sb, colored ions

*From *Handbook of Analytical Chemistry,* L. Meites, Ed., pp. **6**-178 to **6**-181. New York: McGraw-Hill, 1963.

Numerous physiologically important compounds fluoresce.

complete summary of fluorescent chelating reagents, see the handbook by Meites.[1]

Nonradiative relaxation of transition-metal chelates is so efficient that fluorescence of these species is seldom encountered. It is noteworthy that most transition metals absorb in the ultraviolet or visible region, whereas nontransition metal ions do not. For this reason, fluorometry often complements ultraviolet/visible spectrophotometry as a method for the determination of cations (see Figure 25-9).

25D-2 Methods for Organic and Biochemical Species

The number of applications of fluorometric methods to organic problems is impressive. Weissler and White have summarized the most important of these in several tables.[2] More than 100 entries are found under the heading ''Organic and General Biochemical Substances,'' including such diverse compounds as adenine, anthranilic acid, aromatic polycyclic hydrocarbons, cysteine, guanidine, indole, naphthols, certain nerve gases, proteins, salicylic acid, skatole, tryptophan, uric acid, and warfarin. Some 50 medicinal agents that can be determined fluorometrically are listed. Included among these are adrenaline, alkylmorphine, chloroquine, digitalis principles, lysergic acid diethylamide (LSD), penicillin, phenobarbital, procaine, and reserpine. Methods for the analysis of ten steroids and an equal number of enzymes and coenzymes are also listed in these tables. Some of the plant products listed are chlorophyll, ergot alkaloids, rauwolfia serpentian alkaloids, flavonoids, and rotenone.

Without question, the most important application of fluorometry is in the analysis of food products, pharmaceuticals, clinical samples, and natural products. The sensitivity and selectivity of the method make it a particularly valuable tool in these fields.

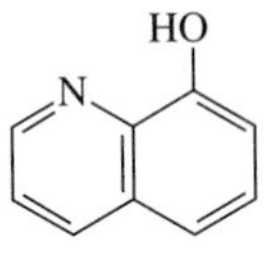

8-hydroxyquinoline
(reagent for Al, Be, and other metal ions)

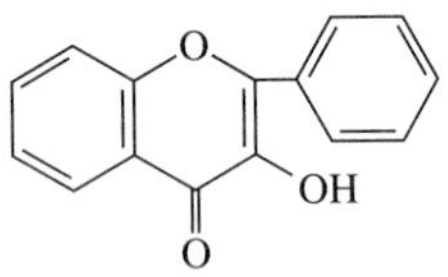

flavanol
(reagent for Zr and Sn)

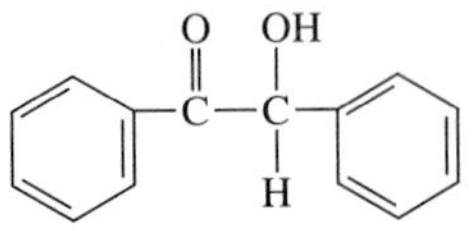

benzoin
(reagent for B, Zn, Ge, and Si)

Figure 25-9
Some fluorometric reagents for cations.

[1]*Handbook of Analytical Chemistry,* L. Meites, Ed., pp. **6**-178 to **6**-181. New York: McGraw-Hill, 1963.

[2]A. Weissler and C. E. White, in *Handbook of Analytical Chemistry,* L. Meites, Ed., pp. **6**-182 to **6**-196. New York: McGraw-Hill, 1963.

25E QUESTIONS AND PROBLEMS

25-1. Define the following terms:
 *(a) resonance fluorescence
 (b) vibrational relaxation
 *(c) internal conversion
 (d) quantum yield
 *(e) Stokes shift
 (f) self-quenching

*25-2. Why is spectrofluorometry potentially more sensitive than spectrophotometry?

25-3. Which compound below is expected to have a greater fluorescent quantum yield? Explain.

*(a)

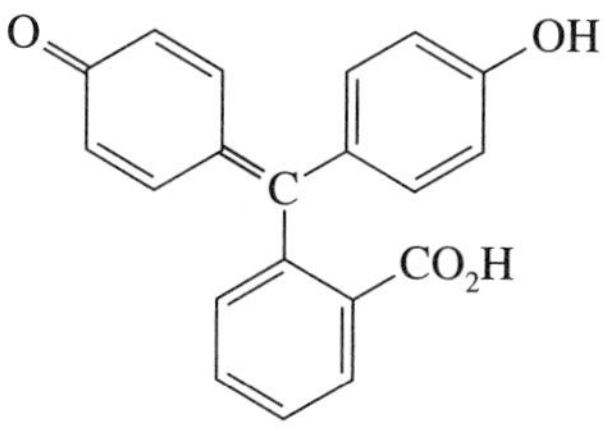

phenolphthalein

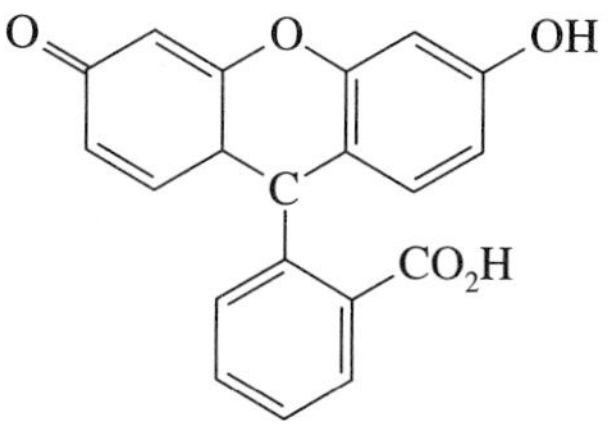

fluorescein

(b)

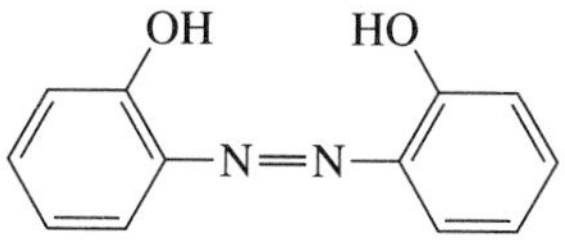

o,o'-dihydroxyazobenzene

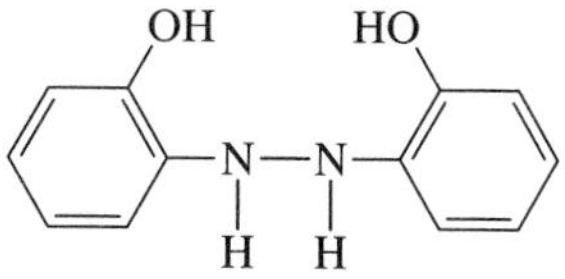

bis(*o*-hydroxyphenyl)hydrazine

*25-4. Why do some absorbing compounds fluoresce and others not?

25-5. Describe the characteristics of organic compounds that fluoresce.

*25-6. Explain why molecular fluorescence often occurs at a longer wavelength than the exciting radiation.

25-7. Describe the components making up a fluorometer.

*25-8. Why are most fluorescence instruments double beam in design?

25-9. Why are fluorometers often more useful than spectrofluorometers for quantitative analysis?

*25-10. The reduced form of nicotinamide adenine dinucleotide (NADH) is an important and highly fluorescent coenzyme. It has an absorption maximum of 340 nm and an emission maximum at 465 nm. Standard solutions of NADH gave the following fluorescence intensities:

Concn NADH, μmol/L	Relative Intensity
0.200	4.52
0.400	9.01
0.600	13.71
0.800	17.91

(a) Construct a calibration curve for NADH.
(b) Derive a least-squares equation for the plot in part (a).
(c) Calculate the standard deviation of the slope and the intercept for the plot.
(d) An unknown exhibits a relative fluorescence of 12.16. Calculate the concentration of NADH.
(e) Calculate the relative standard deviation for the result in part (d).
(f) Calculate the relative standard deviation for the result in part (d) if the reading of 12.16 was the mean of three measurements.

25-11. The following volumes of a solution containing 1.10 ppm of Zn^{2+} were pipetted into separatory funnels each containing 5.00 mL of an unknown zinc solution: 0.00, 1.00, 4.00, 7.00, and 11.00. Each was extracted with three 5-mL aliquots of CCl_4 containing an excess of 8-hydroxyquinoline. The extracts were then diluted to 25.0 mL and their fluorescence measured with a fluorometer. The results were

mL Std Zn^{2+}	Fluorometer Reading
0.000	6.12
4.00	11.16
8.00	15.68
12.00	20.64

(a) Plot the data.
(b) Derive by least squares an equation for the plot.
(c) Calculate the standard deviation of the slope and the intercept.
(d) Calculate the concentration of zinc in the sample.
(e) Calculate a standard deviation for the result in part (d).

*25-12. To four 10.0-mL aliquots of a water sample were added 0.00, 1.00, 2.00, 3.00 mL of a standard NaF solution containing 10.0 ppb F^-. Exactly 5.00 mL of a solution containing an excess of Al-acid Alizarin Garnet R complex, a strongly fluorescing complex, were added to each and the solutions were diluted to 50.0 mL. The fluorescent intensity of the four solutions plus a blank were

mL Sample	mL Std F^-	Meter Reading
5.00	0.00	68.2
5.00	1.00	55.3
5.00	2.00	41.3
5.00	3.00	28.8

(a) Explain the chemistry of the analytical method.
(b) Plot the data.
(c) By least squares derive an equation relating the decrease in fluorescence to the volume of standard reagent.
(d) Calculate the standard deviation of the slope and the intercept.
(e) Calculate the ppb F^- in the sample.
(f) Calculate the standard deviation of the result in part (e).

CHAPTER

26

Atomic Spectroscopy

Atomic spectroscopy is used for the qualitative and quantitative determination of perhaps 70 elements. Sensitivities of atomic methods lie typically in the parts-per-million to parts-per-billion range. Additional virtues of these methods are speed, convenience, unusually high selectivity, and moderate instrument costs.[1]

Spectroscopic determination of atomic species can only be performed on a gaseous medium in which the individual atoms (or sometimes, elementary ions, such as Fe^{+}, Mg^{+}, or Al^{+}) are well separated from one another. Consequently, the first step in all atomic spectroscopic procedures is *atomization,* a process in which the sample is volatilized and decomposed in such a way as to produce an atomic gas. The efficiency and reproducibility of the atomization step in large measure determine the method's sensitivity, precision, and accuracy; that is, atomization is by far the most critical step in atomic spectroscopy.

Atomization is a process in which a sample is converted into gaseous atoms or elementary ions.

As shown in Table 26-1, several methods are available to atomize samples for atomic spectroscopic studies. The most widely used of these is flame atomization, which is discussed in some detail in this chapter. Note that flame atomized samples produce atomic absorption, emission, and fluorescence spectra. We shall also consider briefly three other atomization methods listed in Table 26-1, namely electrothermal, inductively coupled plasma, and direct current plasma.

26A SOURCES OF ATOMIC SPECTRA

Without chemical bonding, there can be no vibrational or rotational energy states and transitions. Thus atomic emission, absorption, and fluorescence spectra are made up of a limited number of narrow peaks, or lines.

26A-1 Emission Spectra

Figure 26-1 is a partial energy-level diagram for atomic sodium that shows the source of three of the most prominent emission lines. These lines are generated by

[1] References that deal with the theory and applications of atomic spectroscopy include C. Th. J. Alkemade et al., *Metal Vapors in Flames.* Elmsford, NY: Pergamon Press, 1982; B. Magyar, *Guidelines to Planning Atomic Spectrometric Analysis.* New York: Elsevier, 1982; L. H. J. Lajunen, *Spectrochemical Analysis by Atomic Absorption and Emission,* Cambridge, England: Royal Society of Chemistry, 1992; J. D. Ingle Jr. and S. R. Crouch, *Spectrochemical Analysis,* Chapters 7–11. Englewood Cliffs, NJ: Prentice-Hall, 1988.

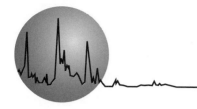

TABLE 26-1 Classification of Atomic Spectral Methods

Atomization Method	Typical Atomization Temperature, °C	Basis for Method	Common Name and Abbreviation of Method
Flame	1700–3150	Absorption	Atomic absorption spectroscopy, AAS
		Emission	Atomic emission spectroscopy, AES
		Fluorescence	Atomic fluorescence spectroscopy, AFS
Electrothermal	1200–3000	Absorption	Electrothermal atomic absorption spectroscopy
		Fluorescence	Electrothermal atomic fluorescence spectroscopy
Inductively coupled argon plasma	6000–8000	Emission	Inductively coupled plasma spectroscopy, ICP
		Fluorescence	Inductively coupled plasma fluorescence spectroscopy
Direct-current argon plasma	6000–10,000	Emission	DC plasma spectroscopy, DCP
Electric arc	4000–5000	Emission	Arc-source emission spectroscopy
Electric spark	40,000(?)	Emission	Spark-source emission spectroscopy

Atomic *p* orbitals are in fact split into two energy levels that differ only slightly in energy. The energy difference between the two levels is so small that the emission appears to be a single line as suggested by Figure 26-1. With a high-resolution spectrometer, each of the lines appears as a doublet.

Note that the wavelengths of the absorption and emission peaks for sodium are all identical.

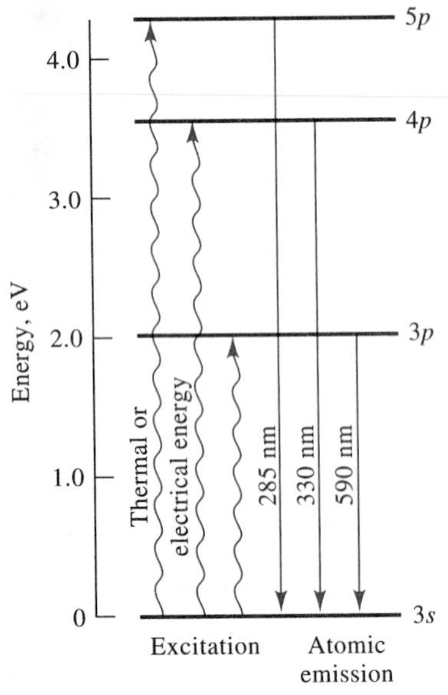

Figure 26-1
Source of three emission lines for sodium.

heating gaseous sodium to 2000°C to 3000°C in a flame. The heat promotes the single outer electron of the atoms from their ground state 3*s* orbitals to 3*p*, 4*p*, or 5*p* excited state orbitals. After a microsecond or less the excited atoms relax to the ground state, giving up their energy as photons of visible or ultraviolet radiation. As shown at the right of the figure, the wavelengths of the emitted radiation are 590, 330, and 285 nm.

26A-2 Absorption Spectra

Figure 26-2a shows three of several absorption peaks for sodium vapor. The source of these peaks is indicated in the partial energy diagram shown in Figure 26-2b. Here, absorption of radiation of 285, 330, and 590 nm excites the single outer electron of sodium from its ground-state 3*s* energy level to the excited 3*p*, 4*p*, and 5*p* orbitals, respectively. After a few microseconds, the excited atoms relax to their ground state by transferring their excess energy to other atoms or molecules in the medium. Alternatively, relaxation may take the form of fluorescence.

The absorption and emission spectra for sodium are relatively simple and consist of perhaps 40 peaks. For elements that have several outer electrons that can be excited, absorption spectra may be much more complex and consist of hundreds or even thousands of peaks.

26B ATOMIC SPECTROSCOPY BASED UPON FLAME ATOMIZATION

As shown in Table 26-1, three types of atomic spectroscopy are based upon flame atomization: (1) atomic absorption spectroscopy (AAS), (2) atomic emission spectroscopy (AES), and (3) atomic fluorescence spectroscopy (AFS). We shall consider the first two of these methods.

In flame atomization, a solution of the analyte (usually aqueous) is converted to a mist, or *nebulized,* and carried into the flame by a flow of gaseous oxidant or fuel. Emission and absorption spectra are generated in the resulting hot, gaseous medium.

26B-1 Flame Atomizers

A flame atomizer consists of a pneumatic nebulizer, which converts the sample solution into a mist, or *aerosol,* that is then fed into a burner. A common type of nebulizer is the concentric tube type, shown in Figure 26-3, in which the liquid sample is sucked through a capillary tube by a high-pressure stream of gas flowing around the tip of the tube (the *Bernoulli effect*). This process of liquid transport is called *aspiration.* The high-velocity gas breaks the liquid up into fine droplets of various sizes, which are then carried into the flame. Cross-flow nebulizers are also employed in which the high-pressure gas flows across a capillary tip at right angles. Often in this type of nebulizer, the liquid is pumped through the capillary. In most atomizers, the high-pressure gas is the oxidant, with the aerosol containing oxidant being mixed subsequently with the fuel.

Figure 26-4 is a diagram of a typical commercial laminar flow burner that employs a concentric tube nebulizer. The aerosol is mixed with fuel and flows past a series of baffles that remove all but the finest droplets. As a result of the baffles, the majority of the sample collects in the bottom of the mixing chamber, where it is drained to a waste container. The aerosol, oxidant, and fuel are then directed into a slotted burner, which provides a flame that is usually 5 or 10 cm in length.

Laminar flow burners provide a relatively quiet flame and a long path length. These properties tend to enhance sensitivity and reproducibility. The mixing chamber in this type of burner contains a potentially explosive mixture, which can be ignited by flashback if the flow rates are not sufficient. Note that the burner in Figure 26-4 is equipped with pressure relief vents for this reason.

26B-2 Properties of Flames

When a nebulized sample is carried into a flame, the solvent evaporates in the *primary combustion zone,* which is located just above the tip of the burner (Figure 26-5). The resulting finely divided solid particles are carried to a region in the center of the flame called the *interzonal region.* Here, in this hottest part of the flame, gaseous atoms and elementary ions are formed from the solid particles.

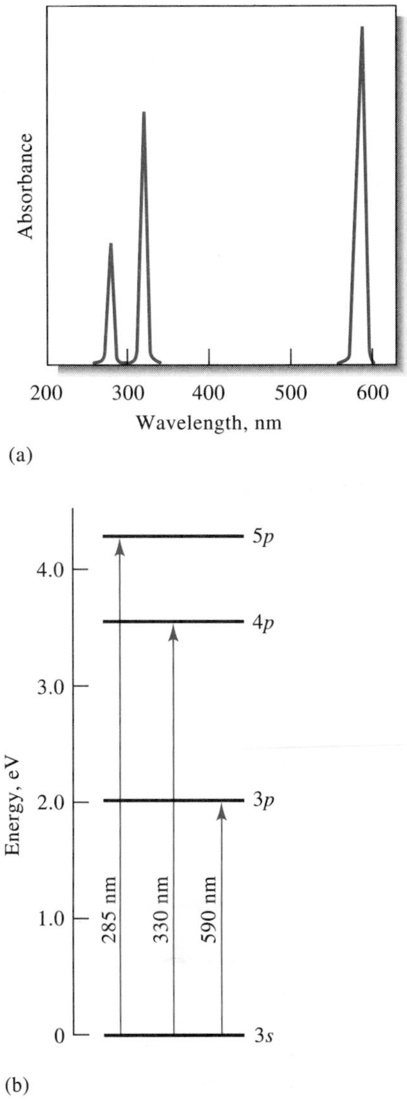

Figure 26-2
(a) Partial absorption spectrum for sodium vapor. (b) Electronic transitions, responsible for the lines in (a).

Nebulize means to reduce to a fine spray.

An **aerosol** is a suspension of finely divided liquid or solid particles in a gas.

Modern flame atomic absorption instruments use laminar flow burners almost exclusively.

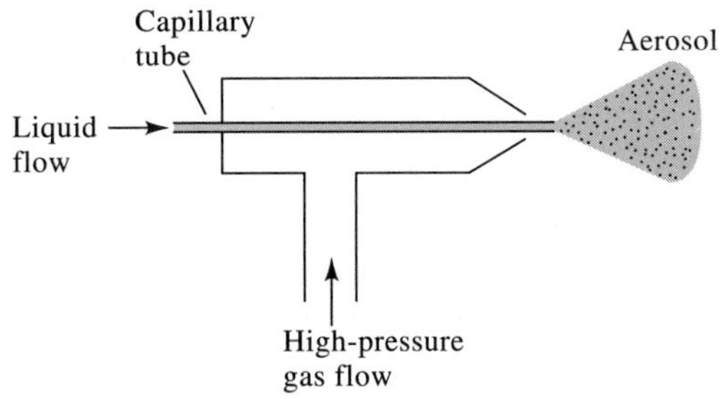

Figure 26-3
A concentric tube nebulizer.

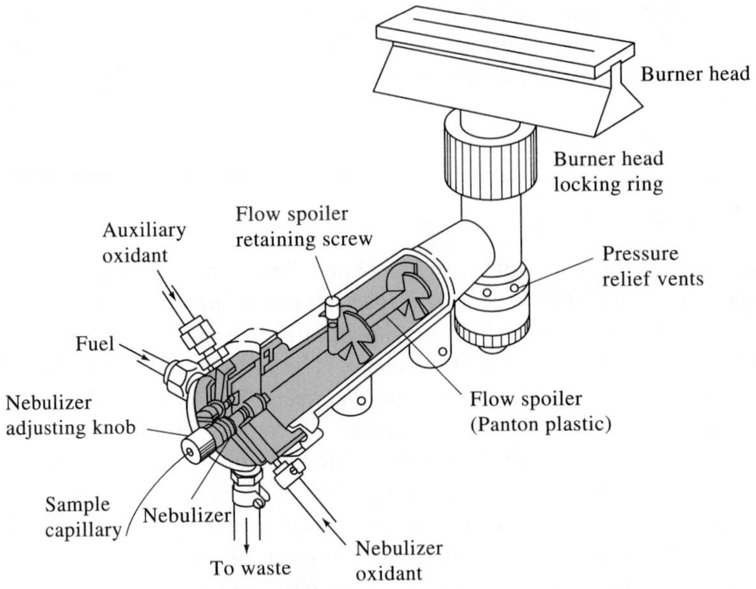

Figure 26-4
A laminar-flow burner. (Courtesy of Perkin-Elmer Corporation, Norwalk, CT.)

Excitation of atomic emission spectra also takes place in this region. Finally, the atoms and ions are carried to the outer edge, or *secondary combustion zone,* where oxidation may occur before the atomization products disperse into the atmosphere. Because the velocity of the fuel/oxidant mixture through the flame is high, only a fraction of the sample undergoes all these processes; indeed, a flame is not a very efficient atomizer.

Types of Flames Used in Atomic Spectroscopy

Table 26-2 lists the common fuels and oxidants employed in flame spectroscopy and the approximate range of temperatures realized with each of these mixtures.

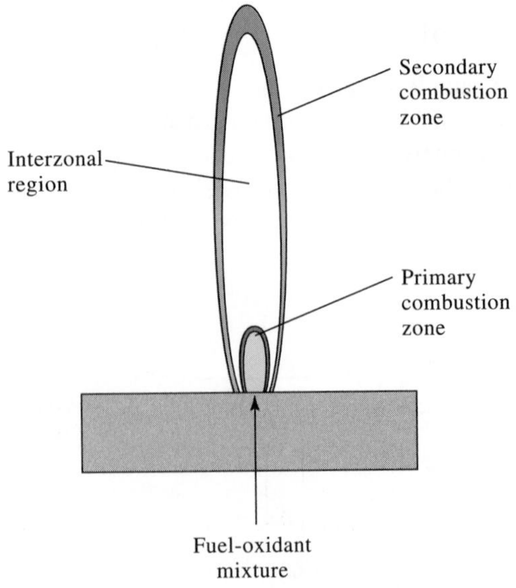

Figure 26-5
Regions in a flame.

Note that temperatures of 1700°C to 2400°C are obtained with the various fuels when air serves as the oxidant. At these temperatures, only easily excitable species such as the alkali and alkaline earth metals produce usable emission spectra. For heavy-metal species, which are less readily excited, oxygen or nitrous oxide must be employed as the oxidant. These oxidants produce temperatures of 2500°C to 3100°C with common fuels.

TABLE 26-2
Flames Used in Atomic Spectroscopy

Fuel and Oxidant	Temperature, °C
Gas/Air	1700–1900
Gas/O_2	2700–2800
H_2/Air	2000–2100
H_2/O_2	2550–2700
*C_2H_2/Air	2100–2400
*C_2H_2/O_2	3050–3150
*C_2H_2/N_2O	2600–2800

*Acetylene

The Effects of Flame Temperature

Both emission and absorption spectra are affected in a complex way by variations in flame temperature. Higher temperatures tend to increase the total atom population of the flame and thus the sensitivity. With certain elements, such as the alkali metals, however, this increase in atom population is more than offset by the loss of atoms by ionization.

Flame temperature also determines the relative number of excited and unexcited atoms in a flame. In an air/acetylene flame, for example, the ratio of excited to unexcited magnesium atoms can be computed to be about 10^{-8}, whereas in an oxygen/acetylene flame, which is about 700°C hotter, this ratio is about 10^{-6}. Control of temperature is thus of prime importance in flame emission methods. For example, with a 2500°C flame, a temperature increase of 10°C causes the number of sodium atoms in the excited $3p$ state to increase by about 3%. In contrast, the corresponding *decrease* in the much larger number of ground-state atoms is only about 0.002%. Therefore, emission methods, based as they are on the population of *excited atoms,* require much closer control of flame temperature than do absorption procedures, in which the analytical signal depends upon the number of *unexcited atoms.*

The number of unexcited atoms in a typical flame exceeds the number of excited ones by a factor of 10^3 to 10^{10} or more. This fact suggests that absorption methods should be significantly more sensitive than emission methods. In fact, however, several other variables also influence sensitivity, and the two methods tend to complement each other in this regard. Table 26-3 illustrates this point.

Absorption and Emission Spectra in Flames

Both atomic and molecular emission and absorption spectra are found when a sample is atomized in a flame. A typical flame-emission spectrum was shown in

TABLE 26-3 Comparison of Detection Limits for Various Elements by Flame Absorption Methods and Flame Emission Methods*

Flame Emission More Sensitive	Sensitivity About the Same	Flame Absorption More Sensitive
Al, Ba, Ca, Eu, Ga, Ho, In, K, La, Li, Lu, Na, Nd, Pr, Rb, Re, Ru, Sm, Sr, Tb, Tl, Tm, W, Yb	Cr, Cu, Dy, Er, Gd, Ge, Mn, Mo, Nb, Pd, Rh, Sc, Ta, Ti, V, Y, Zr	Ag, As, Au, B, Be, Bi, Cd, Co, Fe, Hg, Ir, Mg, Ni, Pb, Pt, Sb, Se, Si, Sn, Te, Zn

*Adapted with permission from E. E. Pickett and S. R. Koirtyohann, *Anal. Chem.*, **1969,** *41*(14), 42A. Copyright 1969 American Chemical Society.

The width of atomic emission peaks is about 10^{-3} nm.

Figure 22-17. Atomic emissions in this spectrum are made up of narrow peaks, or lines, such as that for sodium at about 330 nm, potassium at approximately 404 nm, and calcium at 423 nm. Also present are broad emission bands that result from excitation of molecular species such as MgOH, MgO, CaOH, and OH. Here, vibrational transitions superimposed upon electronic transitions produce closely spaced lines that are not completely resolved by the spectrometer.

Atomic absorption spectra are seldom recorded because to do so would require a sophisticated monochromator that produces radiation with unusually narrow bandwidths. Such spectra would have much the same general appearance as Figure 22-17 with both atomic and molecular absorption peaks being present. The vertical axis in this case would be absorbance rather than relative power.

Ionization in Flames

Ionization of an atomic species in a flame is an equilibrium process that can be treated by the law of mass action.

All elements ionize to some degree in a flame, which leads to a mixture of atoms, ions, and electrons in the hot medium. For example, when a sample containing barium is atomized, the equilibrium

$$Ba \rightleftharpoons Ba^+ + e^-$$

is established in the inner cone of the flame. The position of this equilibrium depends on the temperature of the flame and the total concentration of barium as well as on the concentration of the electrons produced from the ionization of *all elements* present in the sample. At the temperatures of the hottest flames (>3000 K), nearly half of the barium is ionized. The emission and absorption spectra of Ba and Ba^+ are, however, totally different from one another. Thus, in a high-temperature flame, two spectra for barium appear, one for the atom and one for its ion. For this reason (and others), control of the flame temperatures is important in flame spectroscopy.

The spectrum of an atom is entirely different from that of its ions.

26B-3 Flame Atomic Absorption Spectroscopy

Flame atomic absorption spectroscopy is currently the most widely used of all the atomic methods listed in Table 26-1 because of its simplicity, effectiveness, and relatively low cost. The general use of this technique by chemists for elemental analysis began in the early 1950s and grew explosively after that. The reason that atomic absorption methods were not widely used until that time was directly related to problems created by the very narrow widths of atomic absorption lines.

Line Widths

The natural width of an atomic absorption or an atomic emission line is on the order of 10^{-5} nm. Two effects, however, cause line widths to be broadened by a factor of 100 (or more). As will become apparent shortly, line broadening is an important consideration in the design of atomic absorption instruments.

Doppler Broadening Doppler broadening results from the rapid motion of atoms as they emit or absorb radiation. Atoms moving toward the detector emit wavelengths that are slightly shorter than the wavelengths emitted by atoms moving at right angles to the detector. This difference is a manifestation of the well-known Doppler shift; the effect is reversed for atoms moving away from the detector. The net effect is an increase in the width of the emission line. For precisely the same reason, the Doppler effect also causes broadening of absorption lines (see Figure 26-6). This type of broadening becomes more pronounced as the flame temperature increases because of the consequent increased rate of motion of the atoms.

Pressure Broadening Pressure broadening arises from collisions among atoms that cause slight variations in their ground-state energies and thus slight energy differences between ground and excited states. Like Doppler broadening, pressure broadening becomes greater with increases in temperature. Therefore, broader absorption and emission peaks are always encountered at elevated temperatures.

Both Doppler broadening and pressure broadening are temperature dependent.

The Effect of Narrow Line Widths on Absorbance Measurements Since transition energies for atomic absorption lines are unique for each element, analytical methods based upon atomic absorption are highly selective. The narrow lines do, however, create a problem in quantitative analysis that is not encountered in molecular absorption.

No ordinary monochromator is capable of yielding a band of radiation as narrow as the peak width of an atomic absorption line (0.002 to 0.005 nm). As a result, the use of radiation that has been isolated from a continuous source by a monochromator inevitably causes instrumental departures from Beer's law (see the discussion of instrument deviations from Beer's law in Section 22C-8). In addition, since the fraction of radiation absorbed from such a beam is small, the detector receives a signal that is less attenuated (that is, $P \rightarrow P_0$) and the sensitivity of the measurement is reduced. This effect is illustrated by the lower curve in Figure 22-15 (page 518).

The widths of atomic absorption peaks are much less than the effective bandwidths of most monochromators.

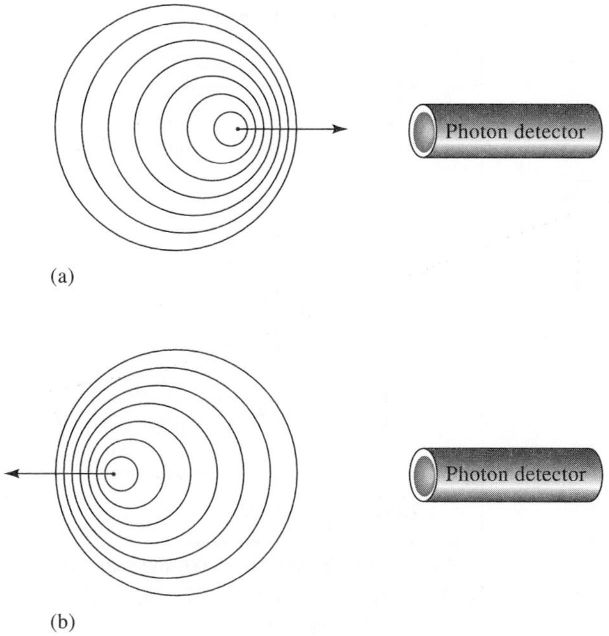

Figure 26-6
Cause of Doppler broadening. (a) When an atom moves toward a photon detector and emits radiation, the detector sees wave crests more often and detects radiation of higher frequency. (b) When an atom moves away from a photon detector and emits radiation, the detector sees crests less frequently and thus detects radiation of lower frequency.

The problem created by narrow absorption peaks was surmounted in mid-1950s by using radiation from a source that emits not only a *line of the same wavelength* as the one selected for absorption measurements but also one that is *narrower*. For example, a mercury vapor lamp is selected as the external radiation source for the determination of mercury. Gaseous mercury atoms electrically excited in such a lamp return to the ground state by *emitting* radiation with wavelengths that are identical to the wavelengths *absorbed* by the analyte mercury atoms in the flame. Since the lamp is operated at a temperature lower than that of the flame, the Doppler and pressure broadening of the mercury emission lines from the lamp are less than the corresponding broadening of the analyte absorption peaks in the hot flame that holds the sample. The effective bandwidths of the lines emitted by the lamp are therefore significantly less than the corresponding bandwidths of the absorption peaks for the analyte in the flame.

Figure 26-7 illustrates the strategy generally used in measuring absorbances in atomic absorption methods. Figure 26-7a shows four narrow *emission* lines from a typical atomic lamp source. Also shown is how one of these lines is isolated by a filter or monochromator. Figure 26-7b shows the flame *absorption spectrum* for the analyte between the wavelengths λ_1 and λ_2; note that the width of the absorption peak in the flame is significantly greater than the width of the emission line

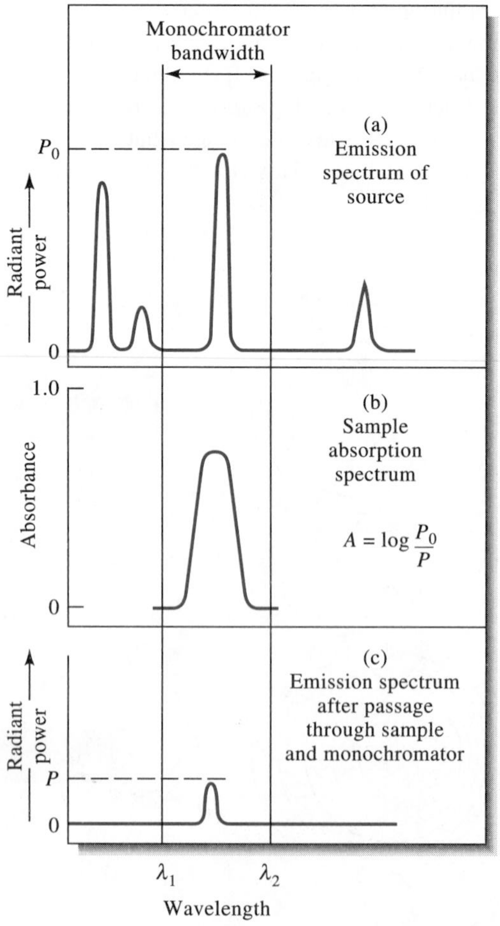

Figure 26-7
Atomic absorption of a resonance line.

from the lamp. As shown in Figure 26-7c, the intensity of the incident beam P_0 has been decreased to P by passage through the sample. Since the bandwidth of the emission line from the lamp is now significantly less than the bandwidth of the absorption peak in the flame, log P_0/P is likely to be linearly related to concentration.

Line Sources

Two types of lamps are used in atomic absorption instruments: *hollow-cathode lamps* and *electrodeless-discharge lamps.*

Hollow-Cathode Lamps The most useful radiation source for atomic absorption spectroscopy is the *hollow-cathode lamp,* shown schematically in Figure 26-8. It consists of a tungsten anode and a cylindrical cathode sealed in a glass tube containing an inert gas, such as argon, at a pressure of 1 to 5 torr. The cathode is either fabricated from the analyte metal or else serves as a support for a coating of that metal.

The application of a potential of about 300 V across the electrodes causes ionization of the argon and generation of a current of 5 to 10 mA as the argon cations and electrons migrate to the two electrodes. If the potential is sufficiently large, the argon cations strike the cathode with sufficient energy to dislodge some of the metal atoms and thereby produce an atomic cloud; this process is called *sputtering.* Some of the sputtered metal atoms are in an excited state and emit their characteristic wavelengths as they return to the ground state. It is important to recall that the atoms producing emission lines in the lamp are at a significantly lower temperature than the analyte atoms in the flame. Thus, the emission lines from the lamp are broadened less than the absorption peaks in the flame.

Sputtering is a process in which atoms or ions are ejected from a surface by a beam of charged particles.

The sputtered metal atoms in a lamp eventually diffuse back to the cathode surface (or to the walls of the lamp) and are deposited.

Hollow-cathode lamps for about 40 elements are available from commercial sources. Some are fitted with a cathode containing more than one element; such lamps provide spectral lines for the determination of several species. The development of the hollow-cathode lamp is widely regarded as the single most important event in the evolution of *atomic absorption* spectroscopy.

Hollow-cathode lamps made atomic absorption spectroscopy practical.

Electrodeless-Discharge Lamps Electrodeless-discharge lamps are useful sources of atomic line spectra and provide radiant intensities that are usually one to two orders of magnitude greater than their hollow-cathode counterparts. A typical lamp is constructed from a sealed quartz tube containing an inert gas, such

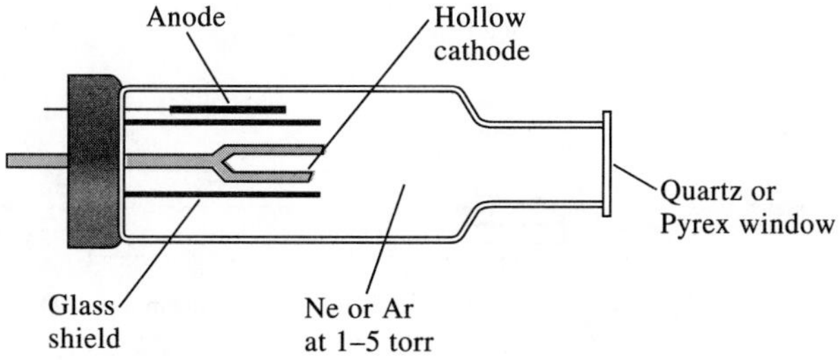

Figure 26-8
Diagram of a hollow-cathode lamp.

as argon, at a pressure of a few torr and a small quantity of the analyte metal (or its salt). The lamp contains no electrode but instead is energized by an intense field of radio-frequency or microwave radiation. The argon ionizes in this field, and the ions are accelerated by the high-frequency component of the field until they gain sufficient energy to excite (by collision) the atoms of the metal whose spectrum is sought.

Electrodeless-discharge lamps are available commercially for several elements. Their performance does not appear to be as reliable as that of the hollow-cathode lamp.

Source Modulation

In an atomic absorption measurement, it is necessary to discriminate between radiation from the source lamp and radiation from the flame. Much of the latter is eliminated by the monochromator, which is always located between the flame and the detector. The thermal excitation of a fraction of the analyte atoms in the flame, however, produces radiation of the wavelength at which the monochromator is set. Because such radiation is not removed, it acts as a potential source of interference.

Modulation is defined as changing some property of a signal such as frequency, amplitude, or wavelength. In AAS, the frequency of the source is modulated from continuous (dc) to intermittent (ac).

The effect of flame emission is overcome by *modulating* the output from the hollow-cathode lamp so that its intensity fluctuates at a constant frequency. The detector thus receives an alternating signal from the hollow-cathode lamp and a continuous signal from the flame and converts these signals into the corresponding types of electric current. A relatively simple electronic system then eliminates the unmodulated dc signal produced by the flame and passes the ac signal from the source to an amplifier and finally to the readout device.

Beam choppers are widely used in spectroscopy to produce pulsed beams of radiation.

Modulation is most often accomplished by interposing a motor-driven circular chopper *between the source and the flame* (Figure 26-9). Segments of the metal chopper have been removed so that radiation passes through the device half of the time and is reflected the other half. Rotation of the chopper at a constant speed causes the beam reaching the flame to vary periodically from zero intensity to some maximum intensity and then back to zero. As an alternative, the power supply for the source can be designed for intermittent (or ac) operation.

Instruments for Atomic Absorption Spectroscopy

An atomic absorption instrument contains the same basic components as an instrument designed for molecular absorption measurements: a source, a sample container (here, a flame reservoir), a wavelength selector, and a detector/readout system. Both single- and double-beam instruments are offered by numerous man-

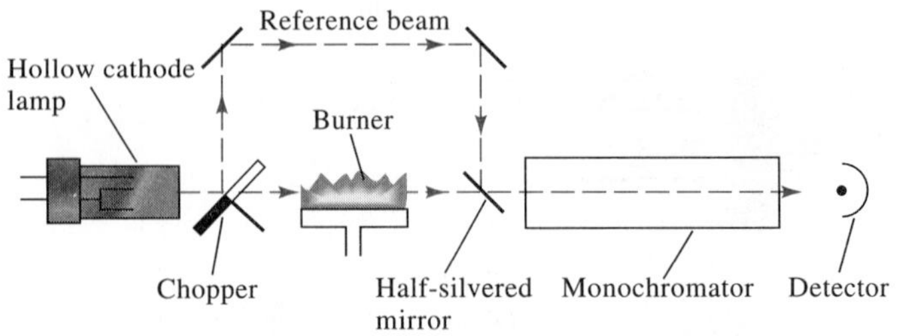

Figure 26-9
Optical paths in a double-beam atomic absorption spectrophotometer.

ufacturers. The range of sophistication and the cost (upward from a few thousand dollars) are both substantial.

Photometers As a minimum, an instrument for atomic absorption spectroscopy must be capable of providing a sufficiently narrow bandwidth to isolate the line chosen for a measurement from other lines that may interfere with or diminish the sensitivity of the method. A photometer equipped with a hollow-cathode source and filters is satisfactory for measuring concentrations of the alkali metals, which have only a few widely spaced resonance lines in the visible region. A more versatile photometer is sold with readily interchangeable interference filters and lamps. A separate filter and lamp are used for each element. Satisfactory results for the determination of 22 metals are claimed.

Spectrophotometers Most atomic absorption measurements are made with instruments equipped with an ultraviolet/visible grating monochromator. Figure 26-9 is a schematic of a typical double-beam instrument. Radiation from the hollow-cathode lamp is chopped and mechanically split into two beams, one of which passes through the flame and the other around the flame. A half-silvered mirror returns both beams to a single path, by which they pass alternately through the monochromator and to the detector. The signal processor then separates the ac signal generated by the chopped light source from the dc signal produced by the flame. The logarithm of the ratio of the reference and sample components of the ac signal is then computed and sent to the readout device for display as absorbance.

Interferences

Two types of interference are encountered in atomic absorption methods. *Spectral interferences* occur when particulate matter from the atomization process scatters the incident radiation from the source or when the absorption by an interfering species is so close to the analyte wavelength that overlap of absorption peaks occurs. *Chemical interferences* result from various chemical processes that occur during atomization and alter the absorption characteristics of the analyte.

Spectral Interferences Interference due to overlapping lines is rare because the emission lines of hollow-cathode sources are so very narrow. Nevertheless, such an interference can occur if the separation between two lines is on the order of 0.01 nm. For example, a vanadium line at 308.211 nm interferes in an analysis based upon the aluminum absorption line at 308.215 nm. The interference is readily avoided, however, by selecting a different aluminum line (309.27 nm, for example).

Scattering by particulate matter and absorption by molecular species in the flame cause interference in AAS.

Spectral interferences also result from the presence of either *molecular* combustion products that exhibit broadband absorption or particulate products that scatter radiation. Both diminish the power of the transmitted beam and lead to positive analytical errors. Where the source of these products is the fuel/oxidant mixture alone, corrections are readily obtained from absorbance measurements made with a blank aspirated into the flame.

A much more troublesome problem is encountered when the source of absorption or scattering originates in the sample matrix. In this type of interference, the power of the transmitted beam P is attenuated by the matrix components but the incident beam power P_0 is not; a positive error in absorbance, and thus concentration, results. For example, a potential matrix interference due to absorption occurs in the determination of barium in alkaline earth mixtures. The wavelength of the barium line used for atomic absorption analysis appears in the center of a broad absorption *band* for molecular CaOH; interference by calcium in a barium analysis results. The net effect is readily eliminated by substituting nitrous oxide for air as the oxidant; the higher temperature decomposes the CaOH and eliminates the absorption band.

Spectral interference due to scattering by products of atomization often occurs when concentrated solutions containing elements such as titanium, zirconium, and tungsten—which form stable oxides—are aspirated into the flame. Metal oxide particles with diameters greater than the wavelength of light appear to be formed and cause scattering of the incident beam.

Fortunately, spectral interferences by matrix products are not widely encountered with flame atomization and usually can be avoided by variations in such analytical parameters as temperature and fuel-to-oxidant ratio. Alternatively, if the source of interference is known, an excess of the interferent can be added to both sample and standards; provided the excess is large with respect to the concentration from the sample matrix, the contribution of the latter will become insignificant. The added substance is sometimes called a *radiation buffer.*

A **radiation buffer** is a substance that is added in large excess to both samples and standards in atomic spectroscopy to swamp out the effect of matrix species and thus minimize interference.

Chemical Interferences Chemical interferences can frequently be minimized by a suitable choice of operating conditions. Perhaps the most common type of chemical interference is by anions that form compounds of low volatility with the analyte and thus decrease the rate at which it is atomized. Low results are the consequence. An example is the decrease in calcium absorbance observed with increasing concentrations of sulfate or phosphate ions, both of which form nonvolatile compounds with calcium ion.

Chemical interferences in AAS are caused by reactions between the analyte and the interferents that decrease the analyte concentration in the flame.

Interferences due to the formation of species of low volatility can often be eliminated or moderated by use of higher temperatures. Alternatively, *releasing agents,* which are cations that react preferentially with the interference and prevent its interaction with the analyte, can be introduced. For example, the addition of excess strontium or lanthanum ion minimizes interference by phosphate in the determination of calcium. Here, the strontium or lanthanum replaces the analyte in the nonvolatile compound formed with the interferent.

Releasing agents are cations that react selectively with anions and thus prevent their interfering in the determination of a cationic analyte.

Protective agents prevent interference by preferentially forming stable but volatile species with the analyte. Three common reagents for this purpose are EDTA, 8-hydroxyquinoline, and APDC (the ammonium salt of 1-pyrrolidine-carbodithioic acid). For example, the presence of EDTA has been shown to eliminate interference by silicon, phosphate, and sulfate in the determination of calcium.

Protective agents are reagents that form stable volatile complexes with an analyte and thus prevent interference by anions that form nonvolatile compounds with the analyte.

Ionization Effects The ionization of atoms and molecules is usually inconsequential in combustion mixtures that involve air as the oxidant. In high-temperature flames, however, where oxygen or nitrous oxide serves as the oxidant, ionization becomes appreciable, and a significant concentration of free electrons exists as a consequence of the equilibrium

$$M \rightleftharpoons M^+ + e^-$$

where M represents a neutral atom or molecule and M^+ is its ion. Ordinarily, the spectrum of M^+ is quite different from that of M, so ionization of the analyte ions leads to low results. It is important to appreciate that treating the ionization process as an equilibrium—with free electrons as one of the products—implies that the degree of ionization of an analyte atom is strongly influenced by the presence of other ionizable metals in the flame. Thus, if the medium contains not only species M but species B as well, and if B ionizes according to the equation

Ionization of the analyte causes AA results based upon absorption by atoms to be low.

$$B \rightleftharpoons B^+ + e^-$$

then the degree of ionization of M is decreased by the mass-action effect of the electrons formed from B. The errors caused by analyte ionization can frequently be eliminated by addition of an *ionization suppressor,* which provides a relatively high concentration of electrons to the flame; suppression of analyte ionization results. Potassium salts are frequently used as ionization suppressors because of the low ionization energy of this element.

An **ionization suppressor** is a readily ionized species that represses ionization of an analyte by providing a high concentration of electrons in a flame.

Quantitative Analysis by Atomic Absorption Spectroscopy

Atomic absorption spectroscopy provides a sensitive means of determining more than 60 elements. The method is well suited for routine measurements by relatively unskilled operators.

Region of Flame for Quantitative Measurements Figure 26-10 shows the absorbance of three elements as a function of distance above the burner tip. For magnesium and silver, the initial rise in absorbance is a consequence of the longer exposure to heat, which leads to greater concentration of atoms in the radiation path. The absorbance for magnesium, however, reaches a maximum near the center of the flame and then falls off as oxidation of the magnesium to magnesium oxide takes place. This effect is not seen with silver because this element is much more resistant to oxidation. For chromium, which forms very stable oxides, maximum absorbance lies immediately above the burner tip. Here, oxide formation begins as soon as chromium atoms are formed.

The absorbance of an analyte varies from one part of the flame to another.

It is clear from Figure 26-10 that the part of a flame to be used in an analysis must vary from element to element and furthermore that the position of the flame with respect to the source must be reproduced closely during calibration and analysis. Generally, the flame position is adjusted to yield a maximum absorbance reading.

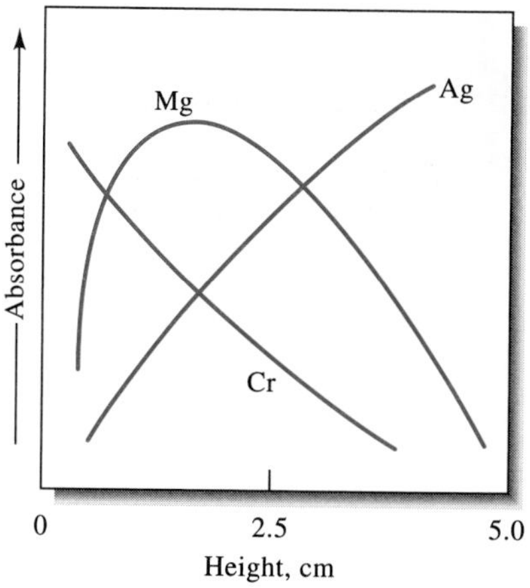

Figure 26-10
Flame-absorbance profile for three elements.

Quantitative Techniques Quantitative atomic absorption techniques are similar in principle to those described in Section 24A-4 for quantitative molecular absorption. Frequently analyses are based upon calibration curves involving a plot of absorbance versus concentration of standards that have been prepared to mimic solutions of the sample. In atomic absorption, departures from linearity are encountered more often than in molecular absorption, however, and analyses should *never* be based on the measurement of a single standard with the assumption that Beer's law is being followed. In addition, the production of an atomic vapor involves sufficient uncontrollable variables to warrant measuring the absorbance of at least one standard solution each time an analysis is performed. Often, two standards are employed whose absorbances bracket that of the unknown. Any deviation of the standard from its original calibration value can then be applied as a correction to the analytical results.

Standard addition methods, discussed on page 572, are also used extensively in atomic spectroscopy in order to try to compensate for differences between the composition of the standards and the unknowns.

Detection Limits and Accuracy Column 2 of Table 26-4 shows detection limits for a number of common elements by flame atomic absorption and also by flame emission methods. Under usual conditions, the relative error of flame absorption analysis is of the order of 1% to 2%. With special precautions, this figure can be lowered to a few tenths of one percent.

26B-4 Flame Emission Spectroscopy

Atomic emission spectroscopy can provide both qualitative and quantitative information about an analyte.

Atomic emission spectroscopy employing a flame (also called flame emission spectroscopy or flame photometry) has found widespread application in elemental analysis. Its most important uses are in the determination of sodium, potassium, lithium, and calcium, particularly in biological fluids and tissues. Because of its convenience, speed, and relative freedom from interferences, flame emission spectroscopy has become the method of choice for these elements, which are otherwise difficult to determine. The method has also been applied, with various

TABLE 26-4 Detection Limits of Atomic Spectroscopy Methods for Selected Elements*

Element	Absorption, Flame†	Absorption, Electrothermal‡	Emission, Flame†	Emission, ICP†,§
Al	30	0.005	5	2
As	100	0.02	0.0005	40
Ca	1	0.02	0.1	0.02
Cd	1	0.0001	800	2
Cr	3	0.01	4	0.3
Cu	2	0.002	10	0.1
Fe	5	0.005	30	0.3
Hg	500	0.1	0.0004	1
Mg	0.1	0.00002	5	0.05
Mn	2	0.0002	5	0.06
Mo	30	0.005	100	0.2
Na	2	0.0002	0.1	0.2
Ni	5	0.02	20	0.4
Pb	10	0.002	100	2
Sn	20	0.1	300	30
V	20	0.1	10	0.2
Zn	2	0.00005	0.0005	2

*All values in nanograms/milliliter = 10^{-3} μg/mL = 10^{-3} ppm.

†From V. A. Fassel and R. N. Kniseley, *Anal. Chem.*, **1974,** *46,* 1111A. With permission. Copyright 1974 American Chemical Society.

‡From C. W. Fuller, *Electrothermal Atomization for Atomic Absorption Spectroscopy,* pp. 65–83. London: The Chemical Society, 1977. With permission of The Royal Society of Chemistry.

§ICP = inductively coupled plasma.

degrees of success, to the determination of perhaps half the elements in the periodic table.

In addition to its quantitative applications, flame emission spectroscopy is also useful for qualitative analysis. Complete spectra are readily recorded; identification of the elements present is then based upon the peak wavelengths, which are unique for each element. In this respect, flame emission has a clear advantage over flame absorption, which does not provide complete absorption spectra because of the discontinuous nature of the radiation sources that must be used.

Instruments

Instruments for flame emission work are similar in design to flame absorption instruments except that in the former the flame now acts as the radiation source; a hollow-cathode lamp and chopper are therefore unnecessary.

Atomic emission and atomic absorption instruments are similar except that no lamp is required for emission measurement.

For nonroutine analysis, a recording ultraviolet/visible spectrophotometer with a resolution of perhaps 0.05 nm is desirable. Such an instrument can provide complete emission spectra that are useful for the identification of the elements present in a sample. Figure 22-17, page 520, is an example of a typical flame emission spectrum excited by an oxyhydrogen flame. The sample was a brine, and spectral lines and bands for several elements are identified. The figure illustrates how background corrections are made with a recorded spectrum.

Simple filter photometers often suffice for routine determinations of the alkali and alkaline earth metals. A low-temperature flame is employed to prevent excitation of most other metals. As a consequence, the spectra are simple, and interference filters can be used to isolate the desired emission line.

FEATURE 26-1
Automated Flame Photometers for Determining Sodium and Potassium in Blood Serum

Several instrument manufacturers supply partially or fully automated flame photometers designed specifically for the analysis of sodium and potassium in blood serum and other biological samples. Figure 26-11 is a schematic of one of these instruments. Note that it is made up of three separate photometers, one for sodium, the second for lithium, and the third for potassium. Each contains an interference filter that transmits a line of one of the three elements and removes all other radiation. Ordinarily lithium serves as an *internal standard* for the analysis. For this purpose, a fixed amount of lithium is introduced into each standard and sample. The output ratios between the sodium transducer and the lithium transducer as well as the potassium transducer and the lithium transducer then serve as analytical parameters. This system provides improved accuracy because the intensities of the three lines are affected in the same way by most analytical variables, such as flame temperature, fuel flow rate, and background radiation. Clearly, lithium must be absent from the sample.

In automated versions of these instruments, samples are withdrawn sequentially from a sample turntable, dialyzed to remove protein and particu-

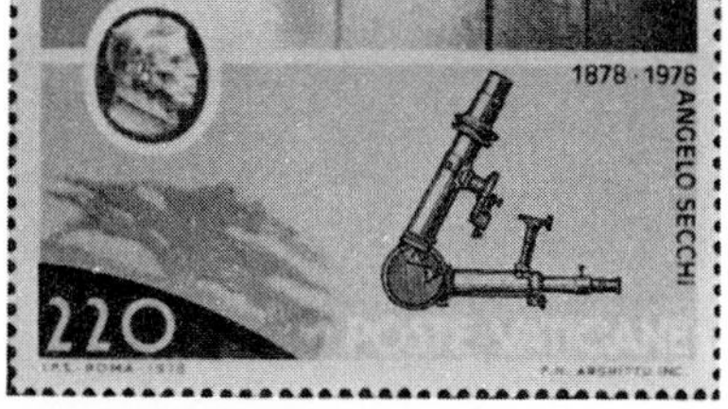

Gustav Robert Kirchhoff (1824–1877) was a German physicist who, along with his colleague, chemist Robert Wilhelm Bunsen (1811–1899), discovered spectroscopic analysis. Bunsen's burner was used to atomize samples of the elements, and Kirchhoff's prism spectroscope was used to analyze the light given off by the incandescent samples. Using this technique, they were able to identify several new elements and demonstrate the presence of a number of elements in the sun by analyzing sunlight. This discovery is depicted on this Vatican stamp. The sun's corona, one of Kirchhoff's spectroscopes, and the absorption spectrum of hydrogen are also illustrated.

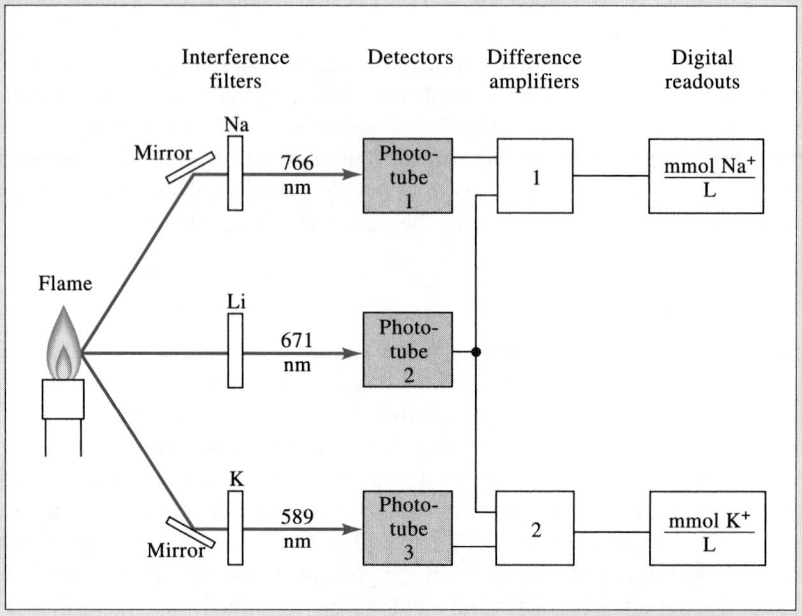

Figure 26-11 An automated flame photometer for Na and K in blood serum.

lates, diluted with a lithium internal standard, and aspirated into a flame. Calibration is performed automatically after every few samples. Simultaneous sodium and potassium determinations are made at a rate of over 100 samples per hour for urine, plasma, or serum samples.

Interferences

The interferences encountered in flame emission spectroscopy have the same sources as those encountered in atomic absorption (pages 621–623); the severity of any given interference often differs for the two procedures, however.

Analytical Techniques

The analytical techniques for flame emission spectroscopy are similar to those described for atomic absorption spectroscopy. Both calibration curves and the standard-addition method are employed. In addition, internal standards can be used to compensate for flame variables.

26C ATOMIC SPECTROSCOPY WITH ELECTROTHERMAL ATOMIZERS

Electrothermal atomizers, which first appeared on the market in about 1970, generally provide enhanced sensitivity because the entire sample is atomized in a short period, and the average residence time of the atoms in the optical path is a

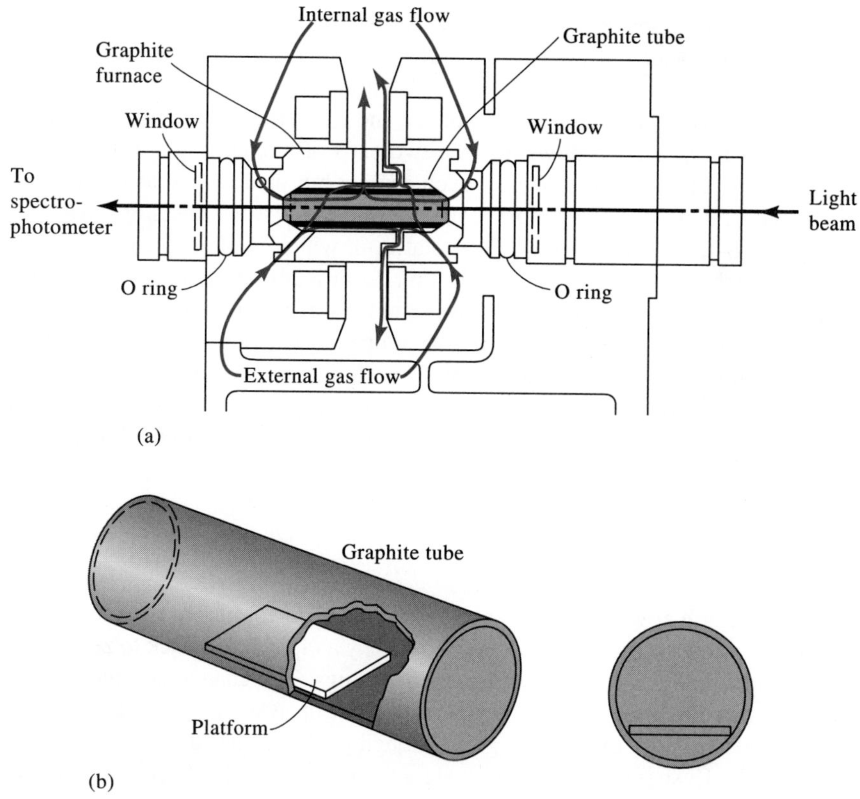

Figure 26-12
(a) Cross-sectional view of a graphite furnace. (b) The L'vov platform and its position in the graphite furnace. [(a) Courtesy of the Perkin-Elmer Corp., Norwalk, CT; (b) Reprinted with permission from: W. Slavin, *Anal. Chem.*, **1982,** *54*, 689A. Copyright 1982 American Chemical Society.]

second or more.[2] Electrothermal atomizers are used for atomic absorption and atomic fluorescence measurements, but have not been generally applied for emission work. They are, however, beginning to be used for vaporizing samples in inductively coupled emission spectroscopy.

In electrothermal atomizers, a few microliters of sample are first evaporated at a low temperature and then ashed at a somewhat higher temperature in an electrically heated graphite tube or cup. After ashing, the current is rapidly increased to several hundred amperes, which causes the temperature to soar to perhaps 2000°C to 3000°C; atomization of the sample occurs in a period of a few milliseconds to seconds. The absorption or fluorescence of the atomized analyte is then monitored in the region immediately above the heated surface.

26C-1 Atomizer Designs

Figure 26-12a is a cross-sectional view of a commercial electrothermal atomizer. Atomization occurs in a cylindrical graphite tube, which is open at both ends and has a central hole for introduction of sample by means of a micropipet. The tube is about 5 cm long and has an internal diameter of somewhat less than 1 cm. The

[2]For detailed discussions of electrothermal atomizers, see S. R. Koirtyohann and M. L. Kaiser, *Anal. Chem.*, **1982,** *54*, 1515A; C. W. Fuller, *Electrothermal Atomization for Atomic Absorption Spectroscopy*. London: The Chemical Society, 1978. A. Varma, *CRC Handbook of Furnace Atomic Absorption Spectroscopy*. Boca Raton, FL: CRC Press, 1989.

interchangeable graphite tube fits snugly into a pair of cylindrical graphite electrical contacts located at the two ends of the tube. These contacts are held in a water-cooled metal housing. Two inert gas streams are provided. The external stream prevents the entrance of outside air and a consequent incineration of the tube. The internal stream flows into the two ends of the tube and out the central sample port. This stream not only excludes air but also serves to carry away vapors generated from the sample matrix during the first two heating stages.

Figure 26-12b illustrates the L'vov platform, which is often used in graphite furnaces such as that shown in (a). The platform is also graphite and is located beneath the sample entrance port. The sample is evaporated and ashed on this platform in the usual way. When the tube temperature is raised rapidly, however, atomization is delayed since the sample is no longer located directly on the furnace wall. As a consequence, atomization occurs in an environment in which the temperature is not changing so rapidly. More reproducible peaks are obtained as a consequence.

26C-2 Output Signal

At a wavelength at which absorbance occurs, the detector output rises to a maximum after a few seconds of ignition followed by a rapid decay back to zero as the atomization products escape into the surroundings. The change is rapid enough (often < 1 s) to require a high-speed data acquisition system. Quantitative analyses are usually based on peak height, although peak area has also been used.

Figure 26-13 shows typical output signals from an atomic absorption spectrophotometer equipped with a carbon rod atomizer. The series of peaks on the right show the absorbance at the wavelength of a lead peak as a function of time when a 2 μL sample of a canned orange juice was atomized. During both drying and

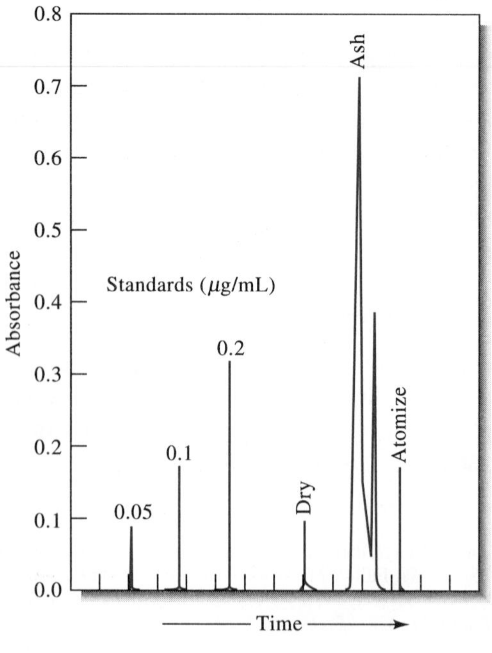

Figure 26-13
Typical output from a spectrophotometer equipped with an electrothermal atomizer. The time for drying and ashing are 20 s and 60 s, respectively. (Courtesy of Varian Instrument Division, Palo Alto, CA.)

ashing, peaks are produced probably due to particulate ignition products. The three peaks on the left are for lead standards employed for calibration. The sample peak on the far right indicates a lead concentration of 0.1 μg/mL of juice.

Applications of Electrothermal Atomizers

Electrothermal atomizers offer the advantage of unusually high sensitivity for small volumes of sample. Typically, sample volumes between 0.5 and 10 μL are employed; under these circumstances, absolute detection limits typically lie in the range of 10^{-10} to 10^{-13} g of analyte.

The relative precision of electrothermal methods is generally in the range of 5% to 10% compared with the 1% or better that can be expected for flame or plasma atomization. Furthermore, furnace methods are slow—typically requiring several minutes per element. Still another disadvantage is that chemical interference effects are often more severe with electrothermal atomization than with flame atomization. A final disadvantage is that the analytical range is low, being usually less than two orders of magnitude. Consequently, electrothermal atomization is ordinarily applied only when flame or plasma atomization provides inadequate detection limits.

FEATURE 26-2

Mercury and Its Determination by Cold-Vapor Atomic Absorption Spectrophotometry

People's fascination with mercury began when prehistoric cave dwellers discovered the mineral cinnabar (HgS) and used it as a red pigment. Our first written record of the element came from Aristotle who described it as ''liquid silver'' in the fourth century B.C. Today there are thousands of uses of mercury and its compounds in medicine, metallurgy, electronics, agriculture, and many other fields. Because it is a liquid metal at room temperature, mercury is used to make flexible and efficient electrical contacts in scientific, industrial, and household applications. Thermostats, silent light switches, and fluorescent light bulbs are but a few examples of electrical applications.

A useful property of metallic mercury is that it forms amalgams with other metals, which have a host of uses. For example, metallic sodium is produced as an amalgam by electrolysis of molten sodium chloride. Dentists use a 50% amalgam with an alloy of silver for fillings.

The toxicological effects of mercury have been known for many years. The bizarre behavior of the Mad Hatter in Lewis Carroll's *Alice in Wonderland* was a result of the effects of mercury and mercury compounds on the Hatter's brain. Mercury absorbed through the skin and lungs destroys brain cells, which are not regenerated. Hatters of the nineteenth century used mercury compounds in processing fur to make felt hats. These workers and workers in other industries have suffered the debilitating symptoms of mercurialism such as loosening of teeth, tremors, muscle spasms, personality changes, depression, irritability, and nervousness.

The toxicity of mercury is complicated by its tendency to form both inorganic and organic compounds. Inorganic mercury is relatively insoluble in body tissues and fluids, so it is expelled from the body about ten times faster than organic mercury. Organic mercury, usually in the form of alkyl compounds such as methyl mercury, is somewhat soluble in fatty tissues such as the liver. Methyl mercury accumulates to toxic levels and is expelled from the body quite slowly.

Mercury concentrates in the environment as illustrated in Figure 26-14. Inorganic mercury is converted to organic mercury by anaerobic bacteria in sludge deposited at the bottom of lakes, streams, and other bodies of water. Small aquatic animals consume the organic mercury and are in turn eaten by larger life forms. As the mercury moves up the food chain from microbes, to shrimp, to fish, and ultimately to larger animals such as swordfish, the mercury becomes ever more concentrated. Some sea creatures such as oysters may concentrate mercury by a factor of 100,000. At the top of the food chain, the concentration of mercury reaches levels as high as 20 ppm. The Food and Drug Administration has set a legal limit of one ppm in fish for human consumption. As a result, mercury levels in some areas threaten local fishing industries.

Analytical methods for the determination of mercury play an important role in monitoring the safety of food and water supplies. One of the most

Figure 26-14 Biological concentration of mercury in the environment.

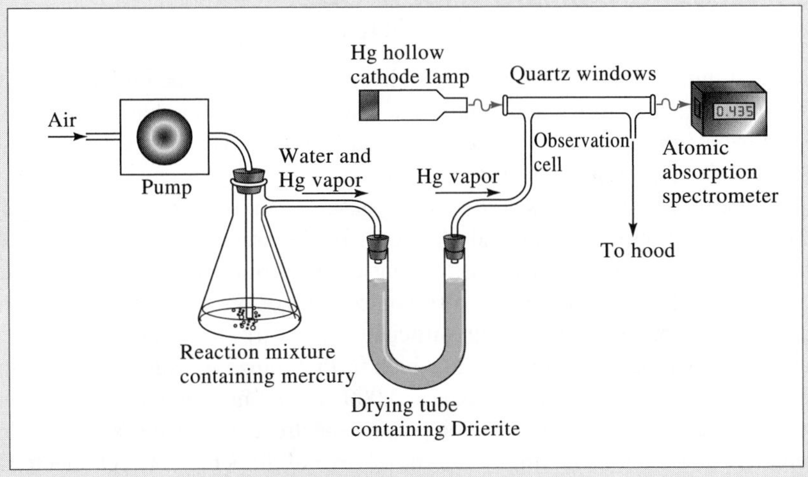

Figure 26-15 Apparatus for cold-vapor atomic absorption determination of mercury.

useful methods is based on the atomic absorption by mercury of 253.7 nm radiation. Figure 26-15 shows an apparatus that is used to determine mercury by atomic absorption at room temperature.[3]

A sample suspected of containing mercury is decomposed in a hot mixture of nitric acid and sulfuric acid, which converts the mercury to mercury(II) salts. These are reduced to the metal with a mixture of hydroxylamine sulfate and tin(II) sulfate. Air is then pumped through the solution to carry the resulting mercury-containing vapor through the drying tube and into the observation cell. Water vapor is trapped by Drierite in the drying tube so that only mercury vapor and air pass through the cell. The monochromator of the atomic absorption spectrophotometer is tuned to a band around 254 nm so that radiation from the 253.7 nm line of the mercury hollow-cathode lamp passes through the quartz windows of the observation cell, which is placed in the light path of the instrument. The absorbance is directly proportional to the concentration of mercury in the cell, which is in turn proportional to the concentration of mercury in the sample. Solutions of known mercury concentration are treated in a similar way to calibrate the apparatus. The method depends on the low solubility of mercury in the reaction mixture and its appreciable vapor pressure, which is 2×10^{-3} torr at 25°C. The sensitivity of the method is about 1 ppb, and it is used to determine mercury in foods, metals, ores, and environmental samples. The method has the advantages of sensitivity, simplicity, and room temperature operation.

---

[3]W. R. Hatch and W. L. Ott, *Anal. Chem.*, **1968,** *40*(14), 2085.

26D ATOMIC EMISSION METHODS BASED ON PLASMA SOURCES

Plasma atomizers, which became available commercially in the mid-1970s, offer several advantages over flame atomizers.[4] Plasma atomization has been used for both thermal emission and fluorescence spectroscopy. It has not been widely used as an atomizer for atomic absorption methods.

A **plasma** is a gas containing relatively high concentration of cations and electrons.

By definition, a plasma is a conducting gaseous mixture containing a significant concentration of cations and electrons. In the argon plasma employed for emission analyses, argon ions and electrons are the principal conducting species, although cations from the sample also contribute. Argon ions, once formed in a plasma, are capable of absorbing sufficient power from an external source to maintain the temperature at a level at which further ionization sustains the plasma indefinitely; temperatures as great as 10,000 K are encountered. Three power sources have been employed in argon plasma spectroscopy. One is a dc electrical source capable of maintaining a current of several amperes between electrodes immersed in the argon plasma. The second and third are powerful radio-frequency and microwave-frequency generators through which the argon flows. Of the three, the radio-frequency, or *inductively coupled plasma* (ICP), source appears to offer the greatest advantage in terms of sensitivity and freedom from interference. On the other hand, the *dc plasma source* (DCP) has the virtues of simplicity and lower cost.

26D-1 The Inductively Coupled Plasma Source

Figure 26-16a is a schematic drawing of an inductively coupled plasma torch. It consists of three concentric quartz tubes through which streams of argon flow at a total rate of between 11 and 17 L/min. The diameter of the largest tube is about 2.5 cm. Surrounding the top of this tube is a water-cooled induction coil powered by a radio-frequency generator capable of producing 2 kW of energy at about 27 MHz. Ionization of the flowing argon is initiated by a spark from a Tesla coil. The resulting ions, and their associated electrons, then interact with the fluctuating magnetic field (labeled H in Figure 26-16a) produced by the induction coil I. This interaction causes the ions and electrons within the coil to flow in the closed annular paths depicted in the figure; ohmic heating is the consequence of their resistance to this movement.

During the 1980s, low-flow, low-power torches appeared on the market. Currently, one manufacturer offers only this type of torch, whereas other companies offer them as an option. Typically, these torches require a total argon flow of lower than 10 L/min and require less than 800 W of radio-frequency power.

The temperature of this plasma is high enough to require that it be thermally isolated from the quartz cylinder. Isolation is achieved by flowing argon tangentially around the walls of the tube, as indicated by the arrows in Figure 26-16a. The tangential flow cools the inside walls of the central tube and centers the plasma radially.

---

[4]For a detailed discussion of the various plasma sources, see *Inductively Coupled Plasma Emission Spectroscopy,* Parts 1 and 2, P. W. J. M. Boumans, Ed., New York: Wiley 1987; *Inductively Coupled Plasma in Analytical Atomic Spectroscopy,* D. W. Golightly, Ed., New York: VCR Publishers, 1987; *Handbook of Inductively Coupled Plasma Spectroscopy,* M. Thompson and J. N. Walsh, Eds., New York: Chapman Hall, 1989.

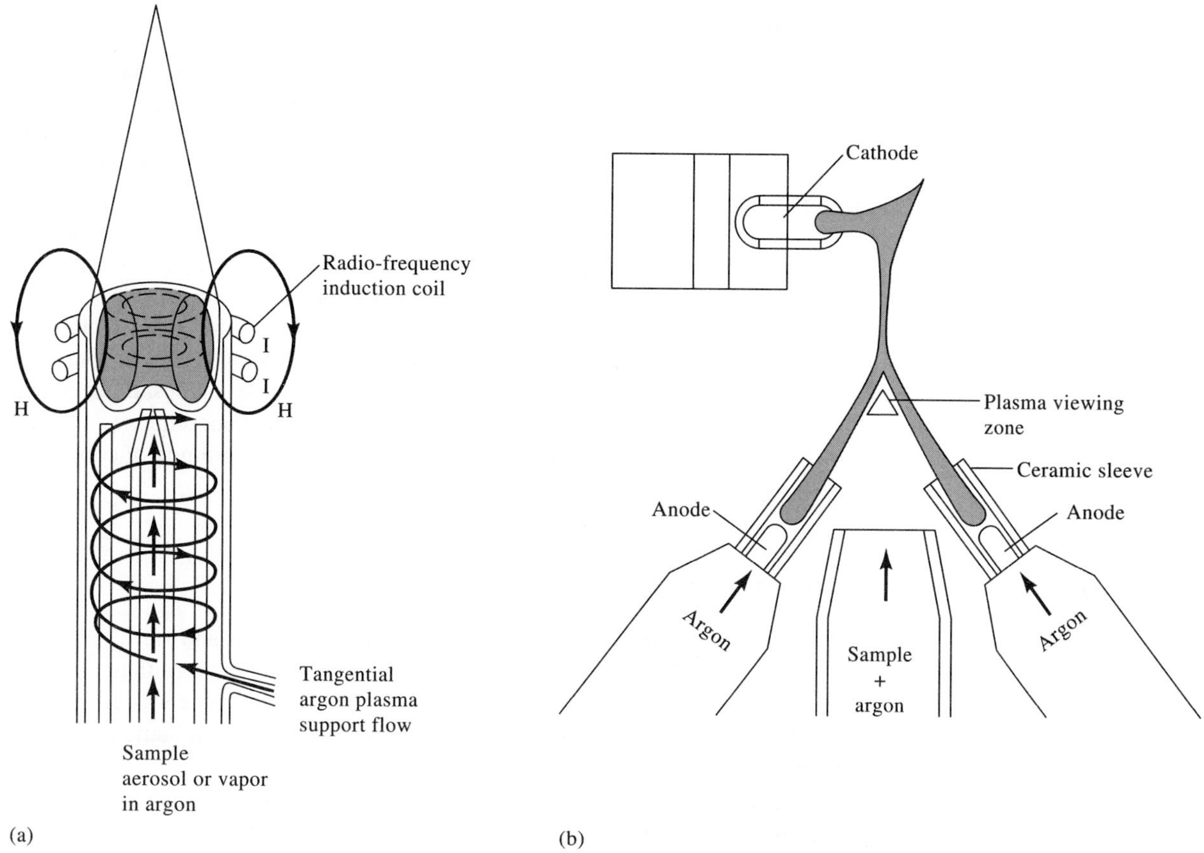

Figure 26-16

Plasma sources. (a) Inductively coupled plasma source. (b) A three-electrode dc plasma jet. [(a) From V. A. Fassel, *Science,* **1978,** *202,* 185. With permission. Copyright 1978 by the American Association for the Advancement of Science; (b) Courtesy of Applied Research Laboratories, Inc., Sunland, CA.]

Sample Injection

The sample is carried into the hot plasma at the head of the tubes by argon flowing at about 1 L/min through the central quartz tube. The sample can be an aerosol, a thermally generated vapor, or a fine powder.

Plasma Appearance and Spectra

The typical plasma has a very intense, brilliant white, nontransparent core topped by a flamelike tail. The core, which extends a few millimeters above the tube, produces a spectral continuum upon which is superimposed the atomic spectrum for argon. The continuum apparently results when argon and other ions recombine with electrons. In the region 10 to 30 mm above the core, the continuum fades and the plasma is optically transparent. Spectral observations are generally made 15 to 20 mm above the induction coil. Here, the background radiation is remarkably free of argon lines and is well suited to spectral measurements. Many of the most sensitive analyte lines in this region of the plasma are from ions such as Ca^+, Cd^+, Cr^+, and Mn^+.

Analyte Atomization and Ionization

Plasma temperatures are substantially higher than those in a flame.

By the time the sample atoms reach the observation point in the plasma, they have had a residence time of about 2 ms at temperatures ranging from 6000 to 8000 K. These times and temperatures are two to three times as great as those attainable in the hottest combustion flames (acetylene/nitrous oxide). As a consequence, atomization is more nearly complete and fewer chemical interferences are encountered. Surprisingly, ionization interference effects are small or nonexistent, perhaps because the large concentration of electrons from the ionization of the argon maintains a more or less constant electron concentration in the plasma.

Several other advantages are associated with the plasma source. First, atomization occurs in a chemically inert environment, which should also enhance the lifetime of the analyte. In addition, and in contrast to flame sources, the temperature cross section of the plasma is relatively uniform. As a consequence, calibration curves tend to remain linear over several orders of magnitude of concentration.

26D-2 The Direct-Current Argon Plasma Source

Direct-current plasma jets were first described in the 1920s and were systematically investigated as sources for emission spectroscopy for more than two decades. It was not until recently, however, that a source of this type was designed to provide data reproducible enough to successfully compete with flame and inductively coupled plasma sources.

Figure 26-16b is a diagram of a commercially available dc plasma source that is well suited to the excitation of emission spectra. This plasma-jet source consists of three electrodes arranged in an inverted Y configuration. A graphite anode is located in each arm of the Y, and a tungsten cathode is located at the inverted base. Argon flows from the two anode blocks toward the cathode. The plasma jet is formed when the cathode is momentarily brought into contact with the anodes. Ionization of the argon occurs, and the current that develops (~ 14 A) generates additional ions to sustain itself indefinitely. The temperature is perhaps 10,000 K in the arc core and 5000 K in the viewing region. The sample is aspirated into the area between the two arms of the Y, where it is atomized, excited, and its spectrum viewed.

Spectra produced by the plasma jet tend to have fewer lines than those produced by the inductively coupled plasma, and the lines formed in the former are largely from atoms rather than ions. Sensitivities achieved with the dc plasma jet appear to range from an order of magnitude lower to about the same as those obtainable with the inductively coupled plasma. The reproducibilities of the two systems are similar. Significantly less argon is required for the dc plasma, and the auxiliary power supply is simpler and less expensive. On the other hand, the graphite electrodes must be replaced every few hours, whereas the inductively coupled plasma source requires little or no maintenance.

26D-3 Instruments for Plasma Spectroscopy

Several manufacturers offer instruments for plasma emission spectroscopy. In general, these consist of a high-quality grating spectrophotometer for the ultraviolet and visible regions and a photomultiplier detector. Many are automated, so that an entire spectrum can be scanned sequentially. Others have several photomultiplier tubes located in the focal plane, so that the lines for several elements (two dozen or more) can be monitored simultaneously. Such instruments are very expensive.

26D-4 Quantitative Applications of Plasma Sources

Unquestionably, inductively coupled and dc plasma sources yield significantly better quantitative analytical data than other emission sources. The excellence of these results stems from the high stability, low noise, low background, and freedom from interferences of the sources when operated under appropriate experimental conditions. The performance of the inductively coupled plasma source is somewhat better than that of the dc plasma source in terms of detection limits. The latter, however, is less expensive to purchase and operate and is entirely adequate for many applications.

In general, detection limits with the inductively coupled plasma source appear comparable to or better than those of other atomic spectral procedures. Table 26-4 compares the sensitivity of several of these methods.

26E QUESTIONS AND PROBLEMS

***26-1.** Describe the basic differences between atomic emission and atomic absorption spectroscopy.

26-2. Define *(a) atomization, (b) pressure broadening, *(c) Doppler broadening, (d) nebulizer, *(e) plasma, (f) hollow-cathode lamp, *(g) sputtering, (h) ionization suppressor, *(i) spectral interference, (j) chemical interference, *(k) radiation buffer, (l) releasing agent, *(m) protective agent.

***26-3.** Why is atomic emission more sensitive to flame instability than atomic absorption or fluorescence?

26-4. Why is source modulation employed in atomic absorption spectroscopy?

***26-5.** In a hydrogen/oxygen flame, an atomic absorption peak for iron decreased in the presence of large concentrations of sulfate ion.

(a) Suggest an explanation for this observation.

(b) Suggest three possible methods for overcoming the potential interference of sulfate in a quantitative determination of iron.

26-6. Why are the lines from a hollow-cathode lamp generally narrower than the lines emitted by atoms in a flame?

***26-7.** In the concentration range from 500 to 2000 ppm of U, a linear relationship is found between absorbance at 351.5 nm and concentration. At lower concentrations, the relationship becomes nonlinear unless about 2000 ppm of an alkali metal salt is introduced. Explain.

26-8. What is the purpose of an internal standard in flame emission methods?

***26-9.** A 5.00-mL sample of blood was treated with trichloroacetic acid to precipitate proteins. After centrifugation, the resulting solution was brought to pH 3 and extracted with two 5-mL portions of methyl isobutyl ketone containing the organic lead-complexing agent APCD. The extract was aspirated directly into an air/acetylene flame and yielded an absorbance of 0.502 at 283.3 nm. Five-milliliter aliquots of standard solutions containing 0.400 and 0.600 ppm of lead were treated in the same way and yielded absorbances of 0.396 and 0.599. Calculate the parts per million of lead in the sample assuming that Beer's law is followed.

26-10. The sodium in a series of cement samples was determined by flame emission spectroscopy. The flame photometer was calibrated with a series of standards containing 0, 20.0, 40.0, 60.0, and 80.0 μg Na_2O per milliliter. The instrument readings for these solutions were 3.1, 21.5, 40.9, 57.1, and 77.3.

(a) Plot the data.

(b) Derive a least-squares line for the data.

(c) Calculate standard deviations for the slope and the intercept for the line in (b).

(d) The following data were obtained for replicate 1.00-g samples of cement dissolved in HCl and diluted to 100.0 mL after neutralization.

	Emission Reading			
	Blank	**Sample A**	**Sample B**	**Sample C**
Replicate 1	5.1	28.6	40.7	73.1
Replicate 2	4.8	28.2	41.2	72.1
Replicate 3	4.9	28.9	40.2	spilled

Calculate the % Na_2O in each sample. What are the absolute and relative standard deviations for the average of each determination?

***26-11.** The chromium in an aqueous sample was determined by pipetting 10.0 mL of the unknown into each of five 50.0-mL volumetric flasks. Various volumes of a standard containing 12.2 ppm Cr were added to the flasks, and the solutions were then diluted to volume.

Unknown, mL	Standard, mL	Absorbance
10.0	0.0	0.201
10.0	10.0	0.292
10.0	20.0	0.378
10.0	30.0	0.467
10.0	40.0	0.554

(a) Plot absorbance as a function of volume of standard V_s.

(b) Derive an expression relating absorbance to the concentrations of standard and unknown (c_s and c_x) and the volumes of the standards and unknown (V_s and V_x) as well as the volume to which the solutions were diluted (V_t).

(c) Derive expressions for the slope and the intercept of the straight line obtained in (a) in terms of the variables listed in (b).

(d) Show that the concentration of the analyte is given by the relationship $c_x = bc_s/mV_x$, where m and b are the slope and the intercept of the straight line in (a).

(e) Determine values for m and b by the method of least squares.

(f) Calculate the standard deviation for the slope and the intercept in (e).

(g) Calculate the ppm Cr in the sample using the relationship given in (d).

CHAPTER

27

Kinetic Methods of Analysis

Kinetic methods of analysis differ in a fundamental way from the equilibrium, or thermodynamic, methods we have dealt with so far. In kinetic methods, measurements are made under *dynamic* conditions in which the concentrations of reactants and products are changing continuously. Here, the rate of appearance of products or disappearance of reactants serves as the analytical parameter. The measurements in thermodynamic methods, in contrast, are performed upon systems that have come to equilibrium so that concentrations are *static*.

The distinction between the two types of methods is illustrated in Figure 27-1, which shows the progress over time of the reaction

$$A + R \rightleftharpoons P \quad (27\text{-}1)$$

where A represents the analyte, R the reagent, and P the product. Thermodynamic methods operate in the region beyond time t_e, when the bulk concentrations of reactants and product have become constant and the chemical system is at equilibrium. In contrast, kinetic methods are carried out during the time interval from 0 to t_e, when reactant and product concentrations are changing continuously and the rate of their disappearance or appearance is conveniently measured.

Selectivity in kinetic methods is achieved by choosing reagents and conditions that amplify differences in the *rates* at which the analyte and potential interferences react. Selectivity in thermodynamic methods is realized by choosing reagents and conditions that amplify differences in *equilibrium constants.*

In **kinetic methods,** we measure *changes* in concentration per unit time. In **equilibrium methods,** we measure absolute concentrations.

Kinetic methods greatly extend the number of chemical reactions that can be used for analytical purposes because they permit the use of reactions that are too slow or too incomplete for thermodynamic-based procedures.

Two types of kinetic methods are based upon catalyzed reactions. In the first, the analyte is the catalyst and is determined from its effect upon an *indicator reaction* involving reactants or products that are easily measured. Such methods are among the most sensitive in the chemist's repertoire. In the second type, a catalyst is introduced to hasten the reaction between analyte and reagent. This approach is often highly selective, or even specific, particularly when an enzyme serves as the catalyst.

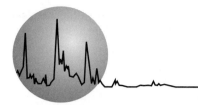

Undoubtedly, the most widespread use of kinetic methods is in biochemical and clinical laboratories, where more analyses are based upon kinetics than upon thermodynamics.[1]

27A RATES OF CHEMICAL REACTIONS; RATE LAWS

This section provides a brief introduction to the theory of chemical kinetics, which is needed to understand the basis for kinetic methods of analysis.

27A-1 Some Terms and Symbols Used in Chemical Kinetics

The *mechanism* for a chemical reaction consists of a series of chemical equations describing the individual elementary steps by which products are formed from reactants. Much of what chemists know about mechanisms has been gained from studies in which the rate at which reactants are consumed or products formed is measured as a function of such variables as reactant and product concentration, temperature, pressure, pH, and ionic strength. Such studies lead to an empirical *rate law* that relates the reaction rate to the concentrations of reactants, products, and intermediates at any instant. Mechanisms are derived by postulating a series of elementary steps that are chemically reasonable and consistent with the empirical rate law.

The rate law for a reaction is determined experimentally.

A **rate law** is an empirically determined relationship between the rate of a reaction and concentrations of reactants, products, and intermediates.

Concentration Terms in Rate Laws

Rate laws are algebraic expressions consisting of concentration terms and constants, which frequently resemble equilibrium constant expressions (see Equation 27-2). You should realize, however, that the square-bracketed terms in a rate expression represent molar concentrations *at a particular instant* rather than equilibrium molar concentrations (as in equilibrium constant expressions). This meaning is frequently emphasized by adding a subscript to show the time to which the concentration refers. Thus, $[A]_t$, $[A]_0$, and $[A]_\infty$ indicate the concentration of A at time t, time zero, and infinite time, respectively. Infinite time is regarded as any time greater than that required for the reaction under consideration to come to equilibrium. That is, $t_\infty > t_e$ in Figure 27-1.

Molar concentrations are still symbolized with square brackets. In the context of kinetic methods, however, their numerical values change with time.

Reaction Order

Let us assume that the empirical rate law for the general reaction shown as Equation 27-1 is found by experiment to take the form

$$\text{rate} = -\frac{d[A]}{dt} = -\frac{d[R]}{dt} = \frac{d[P]}{dt} = k[A]^m[R]^n \qquad (27\text{-}2)$$

where the rate is the derivative of the concentration of A, R, or P with respect to time. Note that the first two rates carry a negative sign because the concentrations of A and R decrease as the reaction proceeds. In this rate expression, k is the *rate*

Because A and R are being used up, the rates of change of [A] and [R] with respect to time are negative.

[1] H. O. Mottola, *Kinetic Aspects of Analytical Chemistry*. New York: Wiley, 1988.

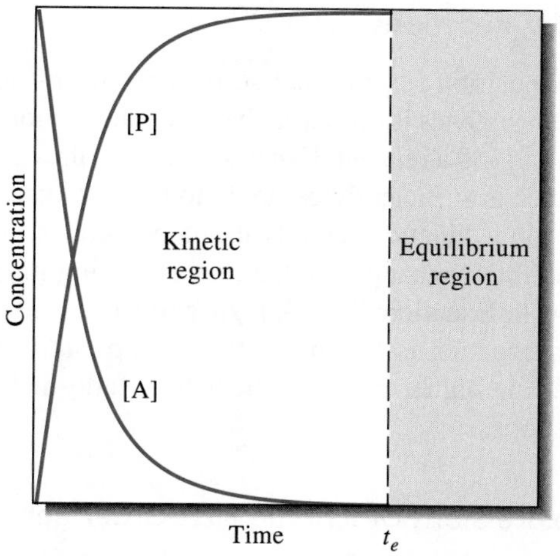

Figure 27-1
Change in concentration of analyte [A] and product [P] as a function of time.

constant, m is the *order of the reaction* with respect to A, and n is the order of the reaction with respect to R. The *overall order* of the reaction is $p = m + n$. Thus, if $m = 1$ and $n = 2$, the reaction is said to be first order in A, second order in R, and third order overall.

Units for Rate Constants

Since reaction rates are always expressed in terms of concentration per unit time, the units of the rate constant are determined by the overall order p of the reaction according to the relation

$$\frac{\text{concentration}}{\text{time}} = (\text{units of } k)(\text{concentration})^p$$

The units of the rate constant k for a first-order reaction are s^{-1}.

where $p = m + n$. Rearranging leads to

$$\text{units of } k = (\text{concentration})^{1-p}\ \text{time}^{-1}$$

Thus, the units for a first-order rate constant are s^{-1}, and the units for a second-order rate constant are $M^{-1} \cdot s^{-1}$.

27A-2 The Rate Law for First-Order Reactions

The simplest case in the mathematical analysis of reaction kinetics is that of the spontaneous irreversible decomposition of a species A:

Radioactive decay is an example of a spontaneous decomposition.

$$A \xrightarrow{k} P \tag{27-3}$$

The reaction is first order in A, and the rate is

$$\text{rate} = \frac{-d[A]}{dt} = k[A] \tag{27-4}$$

Pseudo-First-Order Reactions

Radioactive decay is an exception to this statement. The technique of neutron activation analysis is based on the measurement of the spontaneous decay of radionuclides created by irradiation of a sample in a nuclear reactor.

A first-order decomposition reaction per se is generally of no use in analytical chemistry because an analysis is ordinarily based upon reactions involving at least two species, an analyte and a reagent. Usually, however, the rate law for a reaction involving two species is sufficiently complex to be of no use for analytical purposes. In fact, the only kinetic methods that are useful are those that can be performed under conditions that permit the chemist to simplify complex rate laws to a form analogous to Equation 27-4. A higher-order reaction that is carried out so that such a simplification is feasible is termed a *pseudo-first-order reaction.* Methods for converting higher-order reactions to pseudo-first-order ones are described in later sections.

Mathematical Expressions Describing First-Order Behavior

Because essentially all kinetic analyses are performed under pseudo-first-order conditions, it is worthwhile to examine in detail some of the characteristics of reactions with rate laws that approximate Equation 27-4.

By rearranging Equation 27-4, we obtain

$$\frac{d[\mathrm{A}]}{[\mathrm{A}]} = -k\,dt \qquad (27\text{-}5)$$

The integral of this equation from time zero, when $[\mathrm{A}] = [\mathrm{A}]_0$, to time t, when $[\mathrm{A}] = [\mathrm{A}]_t$, is

$$\int_{[\mathrm{A}]_0}^{[\mathrm{A}]_t} \frac{d[\mathrm{A}]}{[\mathrm{A}]} = -k \int_0^t dt$$

Evaluation of the integrals gives

$$\ln \frac{[\mathrm{A}]_t}{[\mathrm{A}]_0} = -kt \qquad (27\text{-}6)$$

Finally, by taking the exponential of both sides of Equation 27-6, we obtain

See *Mathcad Applications for Analytical Chemistry,* **p. 141.**

$$\frac{[\mathrm{A}]_t}{[\mathrm{A}]_0} = e^{-kt} \quad \text{or} \quad [\mathrm{A}]_t = [\mathrm{A}]_0\, e^{-kt} \qquad (27\text{-}7)$$

This integrated form of the rate law gives the concentration of A as a function of the initial concentration $[\mathrm{A}]_0$, the rate constant k, and the time t. A plot of this relationship is depicted in Figure 27-1.

EXAMPLE 27-1

A reaction is first-order with $k = 0.0370\ \mathrm{s}^{-1}$. Calculate the concentration of reactant remaining 18.2 s after initiation of the reaction if its initial concentration is 0.0100 M.

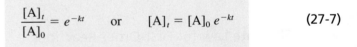

Substituting into Equation 27-7 gives

$$[A]_{18.2} = (0.0100\ M)e^{-(0.0370\ s^{-1})(18.2\ s)} = 0.00510\ M$$

When the rate of a reaction is being followed by the rate of appearance of a product P rather than the rate of disappearance of analyte A, it is useful to modify Equation 27-7 to relate the concentration of P at time t to the initial analyte concentration $[A]_0$. When 1 mol of product forms for 1 mol of analyte, the concentration of A at any time is equal to its original concentration minus the concentration of product. Thus,

$$[A]_t = [A]_0 - [P]_t \qquad (27\text{-}8)$$

Substitution of this expression for $[A]_t$ into Equation 27-7 and rearrangement give

$$[P]_t = [A]_0(1 - e^{-kt}) \qquad (27\text{-}9)$$

A plot of this relationship is also shown in Figure 27-1.

The form of Equations 27-7 and 27-9 is that of a pure exponential, which appears widely in science and engineering. A pure exponential in this case has the useful characteristic that equal elapsed times give equal fractional decreases in reactant concentration or increases in product concentration. As an example, consider a time interval $t = \tau = 1/k$, which upon substitution into Equation 27-7 gives

The fraction of reactant used (or product formed) in a first-order reaction is the same for any given period of time.

$$[A]_\tau = [A]_0 e^{-k\tau} = [A]_0 e^{-k/k} = (1/e)[A]_0$$

and likewise, for a period $t = 2\tau = 2/k$,

$$[A]_{2\tau} = (1/e)^2[A]_0$$

and so on for successive periods, as shown in Figure 27-2.

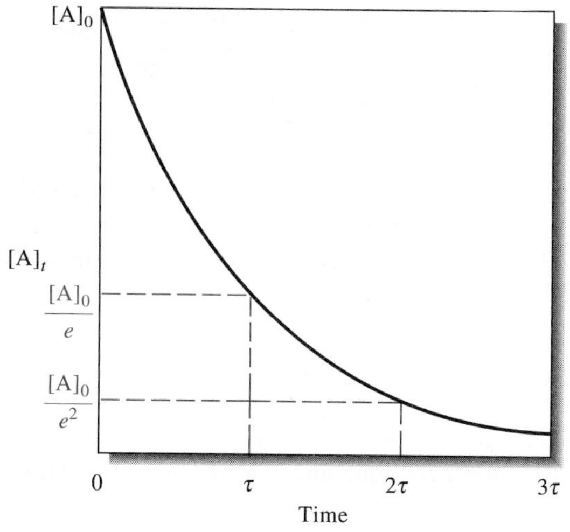

Figure 27-2
Rate curve for a first-order reaction showing that equal elapsed times produce equal fractional decreases in analyte concentration.

The period $\tau = 1/k$ is sometimes referred to as the *natural lifetime* of species A. During time τ, the concentration of A decreases to $1/e$ of its original value. A second period, from $t = \tau$ to $t = 2\tau$, produces an equivalent fractional decrease in concentration to $1/e$ of the value at the beginning of the second interval, which is $(1/e)^2$ of $[A]_0$. A more familiar example of this property of exponentials is found in the half-life $t_{1/2}$ of radionuclides. During a period $t_{1/2}$, half of the atoms in a sample of a radioactive element decay to products; a second period of $t_{1/2}$ decreases the amount of the element to one quarter of its original number, and so on for succeeding periods. Regardless of the time interval chosen, equal elapsed times produce equal fractional decreases in reactant concentration for a first-order process.

CHALLENGE: Derive an expression for $t_{1/2}$ in terms of τ.

EXAMPLE 27-2

Calculate the time required for a first-order reaction with $k = 0.0500\ s^{-1}$ to proceed to 99.0% completion.

For 99.0% completion, $[A]_t/[A]_0 = (100 - 99)/100 = 0.010$; substitution into Equation 27-6 then gives

$$\ln 0.010 = -kt = -(0.0500\ s^{-1})t$$

$$t = \frac{\ln 0.010}{0.0500\ s^{-1}} = 92\ s$$

27A-3 Rate Laws for Second-Order and Pseudo-First-Order Reactions

Let us consider a typical analytical reaction in which 1 mol of analyte A reacts with 1 mol of reagent R to give a single product P. For the present, we assume the reaction is irreversible and write

$$A + R \xrightarrow{k} P \qquad (27\text{-}10)$$

If the reaction occurs in a single elementary step, the rate is proportional to the concentration of each of the reactants and the rate law is

$$-\frac{d[A]}{dt} = k[A][R] \qquad (27\text{-}11)$$

The reaction is first order in each of the reactants and second order overall. If the concentration of R is chosen such that $[R] \gg [A]$, the concentration of R changes very little during the course of the reaction and we can write $k[R] = \text{constant} = k'$. Equation 27-11 is then rewritten as

$$-\frac{d[A]}{dt} = k'[A] \qquad (27\text{-}12)$$

which is identical in form to Equation 27-4. Hence, the reaction is said to be *pseudo-first-order* in A.

Second-order or other higher-order reactions can be performed under pseudo-first-order conditions through control of experimental conditions.

EXAMPLE 27-3

For a pseudo-first-order reaction in which the reagent is present in 100-fold excess, find the relative error resulting from the assumption that $k[\mathrm{R}] = k'$ is constant when the reaction is 40% complete.

The initial concentration of the reagent can be expressed as

$$[\mathrm{R}]_0 = 100[\mathrm{A}]_0$$

At 40% reaction, 60% of A remains. Thus,

$$[\mathrm{A}]_{40\%} = 0.60[\mathrm{A}]_0$$

$$[\mathrm{R}]_{40\%} = [\mathrm{R}]_0 - 0.40[\mathrm{A}]_0 = 100[\mathrm{A}]_0 - 0.40[\mathrm{A}]_0 = 99.6[\mathrm{A}]_0$$

Assuming pseudo-first-order behavior, the rate at 40% reaction is

$$-\frac{d[\mathrm{A}]_{40\%}}{dt} = k[\mathrm{R}]_0[\mathrm{A}]_{40\%} = k(100[\mathrm{A}]_0)(0.60[\mathrm{A}]_0)$$

The true rate at 40% reaction is $k(99.6[\mathrm{A}]_0)(0.60[\mathrm{A}]_0)$. Thus, the relative error is

$$\frac{k(100[\mathrm{A}]_0)(0.60[\mathrm{A}]_0) - k(99.6[\mathrm{A}]_0)(0.60[\mathrm{A}]_0)}{k(99.6[\mathrm{A}]_0)(0.60[\mathrm{A}]_0)} = 0.004 \qquad \text{(or 0.4\%)}$$

As Example 27-3 clearly illustrates, the error associated with the determination of the rate of a pseudo-first-order reaction with a 100-fold excess of reagent is negligible. A 50-fold reagent excess leads to a 1% error, and error of this magnitude is usually deemed acceptable in kinetic methods. Moreover, the error is even less significant at times when the reaction is less than 40% complete.

It may be necessary to account for the rate of the reverse reaction.

Reactions are seldom completely irreversible, and a rigorous description of the kinetics of a second-order reaction that occurs in a single step must take the reverse reaction into account. The rate of the reaction is the difference between the forward rate and the reverse rate:

$$-\frac{d[\mathrm{A}]}{dt} = k_1[\mathrm{A}][\mathrm{R}] - k_{-1}[\mathrm{P}]$$

where k_1 is the second-order rate constant for the forward reaction and k_{-1} is the first-order rate constant for the reverse reaction. In deriving this equation we have assumed for simplicity that a single product is formed, but more complex cases can be described as well.[2] As long as conditions are maintained such that k_{-1} and/or [P] are relatively small, the rate of the reverse reaction is negligible, and little error is introduced by the assumption of pseudo-first-order behavior.

---

[2]See J. H. Espenson, *Chemical Kinetics and Reaction Mechanisms*, pp. 442–448. New York: McGraw-Hill, 1981.

27A-4 Catalyzed Reactions

Enzymes are high-molecular-weight molecules that catalyze reactions in biological systems.

Catalyzed reactions, particularly those in which enzymes serve as catalysts, are widely used for the determination of a variety of biological and biochemical species as well as a number of inorganic cations and anions. We shall therefore use enzyme-catalyzed reactions to illustrate catalytic rate laws and to show how rate laws can be reduced to relatively simple algebraic relationships, such as the pseudo-first-order equation shown as Equation 27-12. These simplified relationships can then be used for analytical purposes.

Enzyme-Catalyzed Reactions

Enzymes serve as highly selective analytical reagents.

The species acted upon by an enzyme is called a **substrate.**

Enzymes are high-molecular-weight molecules that catalyze reactions of biological and biomedical importance. Enzymes are particularly useful as analytical reagents owing to their high degree of selectivity. Consequently, they are widely used in the determination of molecules with which they combine when acting as catalysts. Such molecules are usually designated as *substrates.* In addition to the determination of substrates, enzyme-catalyzed reactions are employed for the determination of activators, inhibitors, and, of course, enzymes.

The behavior of a large number of enzymes is consistent with the general mechanism

$$E + S \underset{k_{-1}}{\overset{k_1}{\rightleftharpoons}} ES \xrightarrow{k_2} P + E \qquad (27\text{-}13)$$

In this equation, the enzyme E is shown as reacting reversibly with the substrate S to form an enzyme-substrate complex ES; this complex then decomposes irreversibly to form the product(s) and the regenerated enzyme. The rate law for this mechanism assumes one of two forms, depending upon the relative rates of the two steps. If the second step is considerably slower than the first (case 1), the reactants and ES essentially achieve equilibrium. In this case, the second step determines the overall rate and is thus termed *the rate-determining step.* When the rates of the two steps are comparable in magnitude (case 2), however, ES decomposes as rapidly as it is formed, and its concentration can be assumed to be small and relatively constant throughout much of the reaction. In the two sections that follow, we show that in both cases, the reaction conditions can be arranged to yield simple relationships between rate and analyte concentration.

Case 1: Rate Determined by the Rate of Complex Decomposition

A rate law consistent with the mechanism of Equation 27-13 when the second step is rate determining is derived in the following way. The concentrations of S and E at time t are given by

$$[S]_t = [S]_0 - [ES]_t \qquad \text{and} \qquad [E]_t = [E]_0 - [ES]_t \qquad (27\text{-}14)$$

Since the equilibrium for the first step is achieved rapidly relative to the rate of the second step, we equate the forward and reverse rates of the first step and obtain

$$-\frac{d[E]}{dt} = -\frac{d[S]}{dt} \approx \frac{d[ES]}{dt}$$

At any time t, the rate of the forward reaction is proportional to the product of the concentration of the two reactants ($[E]_t[S]_t$); the rate of the reverse reaction is then proportional to the instantaneous concentration of the complex ($[ES]_t$). Thus,

$$k_1[E]_t[S]_t = k_{-1}[ES]_t$$

Rearranging this equation leads to

$$\frac{k_1}{k_{-1}} = \frac{[ES]_t}{[E]_t[S]_t} = K \tag{27-15}$$

When the rates of the forward and reverse reactions are equal, the system is *at chemical equilibrium.* Thus, the concentration terms in Equation 27-15 are *equilibrium concentrations,* and K in this equation is the *equilibrium constant* for the first step in the reaction. Substitution of $[S]_t$ and $[E]_t$ from Equation 27-14 into Equation 27-15 and rearrangement give

$$[ES]_t = K([S]_0 - [ES]_t)([E]_0 - [ES]_t) \tag{27-16}$$

In theory, either the substrate or the enzyme catalyst can be present in considerable excess; however, enzyme reagents are expensive and are thus nearly always maintained at low levels as a substrate molecule is determined. Furthermore, natural levels of enzymes are very low so that we generally assume that $[S]_0 \gg [ES]_t$. Equation 27-16 then becomes

$$[ES]_t = K[S]_0([E]_0 - [ES]_t) \tag{27-17}$$

When Equation 27-17 is solved for $[ES]_t$, we have

$$[ES]_t = \frac{K[S]_0[E]_0}{1 + K[S]_0} \tag{27-18}$$

and the rate of the reaction is

$$\frac{d[P]}{dt} = k_2[ES]_t = \frac{k_2K[S]_0[E]_0}{1 + K[S]_0} = \frac{k_2[S]_0[E]_0}{(1/K) + [S]_0} \tag{27-19}$$

It is usually possible to adjust conditions such that Equation 27-19 simplifies to a relationship that can be used for the determination of either the catalyst or the substrate. For enzyme determinations, the concentration of the substrate is made large enough so that $[S]_0 \gg 1/K$; the denominator of Equation 27-19 is then approximately equal to $[S]_0$, and the rate is

To determine an enzyme under case 1 conditions, $[S]_0 \gg 1/K$.

$$\frac{d[P]}{dt} = k_2[E]_0 \tag{27-20}$$

Thus, the rate of the reaction is directly proportional to the initial concentration of the enzyme catalyst and can be used for determining that species.

To determine a substrate under case 1 conditions, $[S]_0 \ll 1/K$.

If the substrate is the species to be determined, however, its initial concentration is made small enough that $[S]_0 \ll 1/K$. Equation 27-19 then reduces to

$$\frac{d[P]}{dt} = k_2K[S]_0[E]_0$$

In order to determine substrate concentrations, measurements are performed at constant enzyme concentration so that

$$\frac{d[P]}{dt} = k'[S]_0 \qquad (27\text{-}21)$$

where $k' = k_2K[E]_0$. Under these conditions, the rate of the reaction is clearly proportional to substrate concentration. Note that the catalyst is being regenerated continuously (Equation 27-13). Thus, $[E]_t \approx [E]_0$.

Case 2: Rate Determined by the Steady State Condition

In the event that the reaction rates of the two steps in the mechanism shown by Equation 27-13 are comparable, it is necessary to use what is called the *steady state approximation* to arrive at a rate law. In this treatment, which is described in many sources[3] and therefore is not reproduced here, the rate of change in the concentration of the ES complex is presumed to be negligible. This assumption leads to the rate law

$$\frac{d[P]}{dt} = \frac{k_2[E]_0[S]_t}{[S]_t + (k_{-1} + k_2)/k_1} = \frac{k_2[E]_0[S]_t}{[S]_t + K_m} \qquad (27\text{-}22)$$

Equation 27-22 is called the **Michaelis-Menten equation.**

The *Michaelis constant,* $K_m = (k_{-1} + k_2)/k_1$, is essentially an "equilibrium-like" constant analogous to $1/K$ in case 1; it expresses the ratio of the sum of the rates of the two reactions tending to destroy ES to the rate of the reaction favoring ES formation. The constant is usually expressed in units of millimoles/liter (mM) and ranges from 0.01 mM to 100 mM for common enzymes. Michaelis constants for some common enzymes are shown in Table 27-1.

TABLE 27-1
K_m for Some Enzymes

Enzyme and Substrate	K_m, mM
Catalase	
H_2O_2	25
Hexokinase	
Glucose	0.15
Fructose	1.5
Chymotrypsin	
n-Benzoyltyrosinamide	2.5
n-Formyltyrosinamide	12.0
n-Acetyltyrosinamide	32
Glycyltyrosinamide	122
Carbonic anhydrase	
HCO_3^-	9.0
Glucose oxidase	
Glucose (saturated with O_2)	0.013
Urease	
Urea	2.0

As in case 1, the rate equation can be simplified so that the reaction rate is proportional to either enzyme or substrate concentration. For example, if the concentration of substrate is made large enough that $[S] \gg K_m$, Equation 27-22 reduces to

$$\frac{d[P]}{dt} = k_2[E]_0 \qquad (27\text{-}23)$$

Under these conditions, when the rate is independent of substrate concentration, the reaction is said to be *pseudo-zero-order* in substrate, and the rate is directly proportional to the concentration of enzyme.

[3]For example, see J. H. Espenson, *Chemical Kinetics and Reaction Mechanisms,* pp. 72–80. New York: McGraw-Hill, 1981.

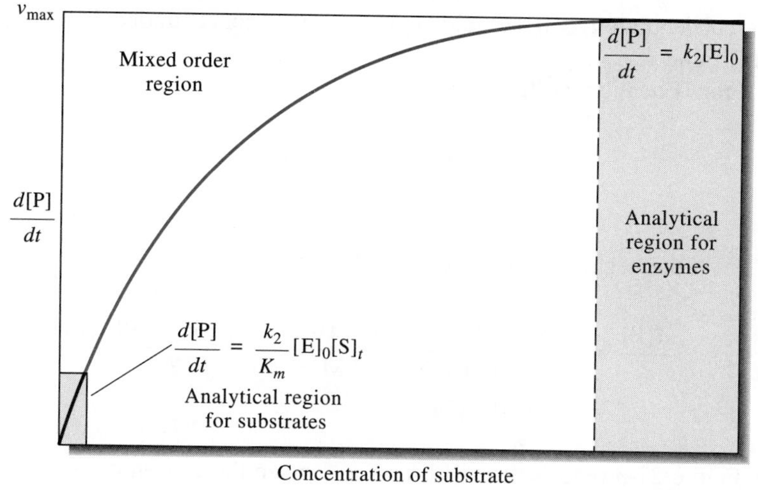

Figure 27-3
Change in rate of product formation as a function of substrate concentration, showing the parts of the curve useful for determination of substrate and enzyme.

When conditions are such that the concentration of S is small or when K_m is relatively large, then $[S]_t \ll K_m$, and Equation 27-22 simplifies to

$$\frac{d[P]}{dt} = \frac{k_2}{K_m}[E]_0[S]_t = k'[S]_t$$

where $k' = k_2[E]_0/K_m$. Hence, the kinetics are first order in substrate. In order to use this equation for determining analyte concentrations, it is necessary to measure $d[P]/dt$ at the beginning of the reaction, where $[S]_t \approx [S]_0$, so that

$$\frac{d[P]}{dt} \approx k'[S]_0 \qquad (27\text{-}24)$$

The regions where Equations 27-23 and 27-24 are applicable are illustrated in Figure 27-3, in which the initial rate of an enzyme-catalyzed reaction is plotted as a function of substrate concentration. When the substrate concentration is low, Equation 27-24, which is linear in substrate concentration, governs the shape of the curve, and it is this region that is used to determine the amount of substrate present.

To determine enzymes under case 2 conditions, $[S] \gg K_m$.

If it is desired to determine the amount of enzyme, the region of high substrate concentration, where Equation 27-23 applies, is employed, and the rate is independent of substrate concentration. The limiting rate of the reaction at large values of [S] is often referred to as v_{max}, as indicated in the figure. It can be shown that the value of the substrate concentration at exactly $v_{max}/2$ is equal to the Michaelis constant K_m.

To determine substrates under case 2 conditions, $[S] \ll K_m$.

EXAMPLE 27-4

The enzyme urease, which catalyzes the hydrolysis of urea, is widely used for determining urea in blood. Details of this application are given on page 656. The Michaelis constant for urease at room temperature is 2.0 mM, and $k_2 = 2.5 \times 10^4\ s^{-1}$ at pH 7.5. (a) Calculate the initial rate of the reaction when the

urea concentration is 0.030 mM and the urease concentration is 5.0 μM, and (b) find v_{max}.
(a) From Equation 27-22,

$$\frac{d[\mathrm{P}]}{dt} = \frac{k_2[\mathrm{E}]_0[\mathrm{S}]_t}{[\mathrm{S}]_t + K_m}$$

At the beginning of the reaction, $[\mathrm{S}]_t = [\mathrm{S}]_0$ and

$$\frac{d[\mathrm{P}]}{dt} = \frac{(2.5 \times 10^4\ \mathrm{s}^{-1})(5.0 \times 10^{-6}\ \mathrm{M})(0.030 \times 10^{-3}\ \mathrm{M})}{0.030 \times 10^{-3}\ \mathrm{M} + 2.0 \times 10^{-3}\ \mathrm{M}}$$
$$= 1.8 \times 10^{-3}\ \mathrm{M \cdot s^{-1}}$$

(b) Figure 27-3 reveals that $d[\mathrm{P}]/dt = v_{max}$ when the concentration of substrate is large and Equation 27-23 applies. Thus,

$$d[\mathrm{P}]/dt = v_{max} = k_2[\mathrm{E}]_0 = (2.5 \times 10^4\ \mathrm{s}^{-1})(5.0 \times 10^{-6}\ \mathrm{M})$$
$$= 0.125\ \mathrm{M \cdot s^{-1}}$$

Although our discussion thus far has been concerned with enzymatic methods, an analogous treatment for ordinary catalysis gives rate laws that are similar. These expressions often reduce to the first-order case for ease of data treatment, and many examples of kinetic-catalytic methods are found in the literature.[4]

27B THE DETERMINATION OF REACTION RATES

Several methods are used for the determination of reaction rates. In this section, we describe some of these methods and when in the course of a reaction they are used.

27B-1 Experimental Methods

A fast reaction is 50% complete in 10 s or less.

The way reaction rates are measured depends upon whether the reaction of interest is fast or slow. A reaction is generally regarded as fast if it proceeds to 50% of completion in 10 s or less. Analytical methods based upon fast reactions generally require special equipment that permits rapid mixing of reagents and fast recording of data.

If a reaction is sufficiently slow, conventional equilibrium methods of analysis can be used to determine the concentration of a reactant or product as a function of time. Often, however, the reaction of interest is too rapid for equilibrium measurements—that is, concentrations change appreciably during the measurement process. Under these circumstances, either the reaction must be stopped

[4]See K. B. Yatsimirskii, *Kinetic Methods of Analysis.* Oxford: Pergamon Press, 1966; H. B. Mark, G. A. Rechnitz, and R. A. Greinke, *Kinetics in Analytical Chemistry.* New York: Wiley-Interscience, 1968.

(quenched) while the measurement is made or an instrumental technique that records concentrations continuously as the reaction proceeds must be employed. In the former case, an aliquot is removed from the reaction mixture and rapidly quenched by mixing it with a reagent that combines with one of the reactants to stop the reaction. Alternatively, quenching is accomplished by lowering the temperature rapidly to slow the reaction to an acceptable level for the measurement step. Unfortunately, quenching techniques tend to be laborious and time-consuming, and are thus not widely used for analytical purposes.

The most convenient approach for obtaining kinetic data is to monitor the progress of the reaction continuously by spectrophotometry, conductometry, potentiometry, or some other instrumental technique. With the advent of inexpensive microcomputer technology, instrumental readings proportional to concentrations of reactants and/or products are often recorded directly as a function of time, stored in the computer's memory, and retrieved later for data processing.

In the following sections, we explore some strategies used to determine rates of reactions and thus concentrations from plots of analyte or product concentration as a function of time.

FEATURE 27-1
Fast Reactions and Stopped-Flow Mixing

One of the most popular and reliable methods for carrying out rapid reactions is stopped-flow mixing. In this technique, streams of reagent and sample are mixed rapidly, and the flow of the mixed solution is stopped suddenly. The reaction progress is then monitored at a position slightly downstream from the mixing point. The apparatus shown in the figure that follows is designed to perform stopped-flow mixing.

To illustrate the operation of the apparatus, we begin with the drive syringes filled with reagent and sample and with valves A, B, and C closed. The stop syringe is empty. The drive mechanism is then activated to move the drive syringe plungers forward rapidly. The reagent and sample pass into the mixer, where they are mixed, and immediately into the observation cell, as indicated by the colored arrows in the figure. The reaction mixture then passes into the stop syringe. Eventually, the stop syringe fills and the stop syringe plunger strikes the stop block. This event causes the flow to cease almost instantly with a recently mixed plug of solution in the observation cell. In this example, the observation cell is transparent so that a light beam can be passed through to make absorption measurements. In this way, the progress of the reaction can be recorded. All that is required is that the *dead time,* or the time between the mixing of reagents and the arrival of the sample in the observation cell, be short relative to the time required for the reaction to proceed to completion. For well-designed systems in which the turbulent flow of the mixer provides very rapid and efficient mixing, the dead time is on the order of 2 to 4 ms. Thus, first-order or pseudo-first-order reactions with $\tau \approx 25$ ms ($k \approx 40\ s^{-1}$) can be examined using the stopped-flow technique.

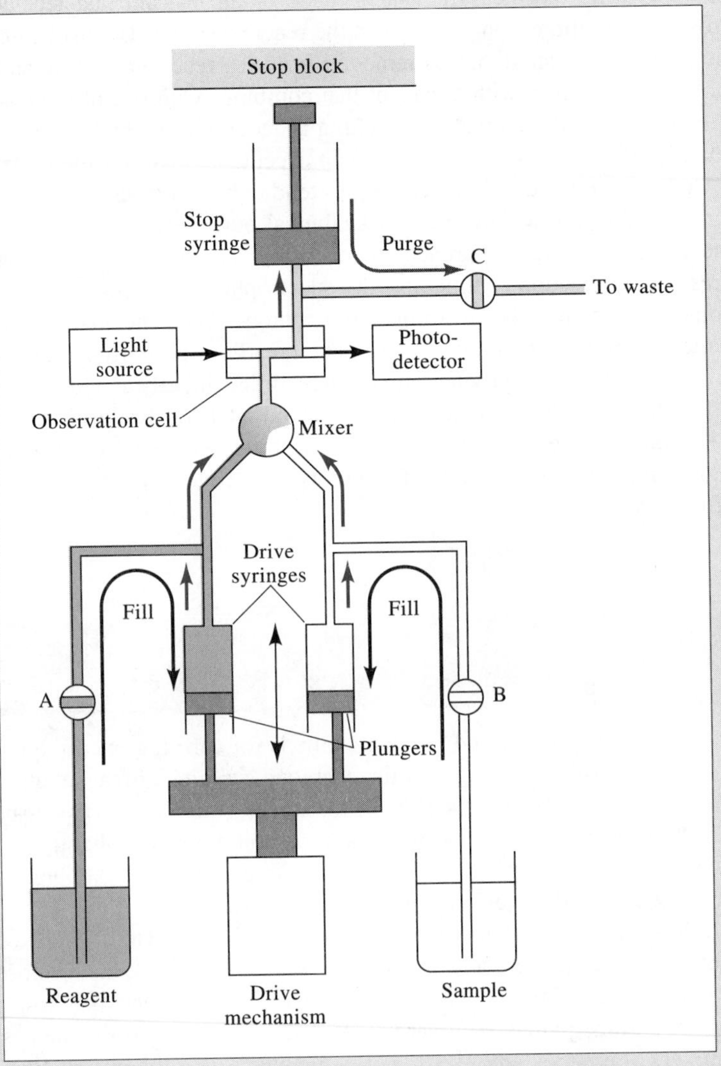

When the reaction is complete, valve C is opened, and the stop syringe plunger is then pushed down to purge the syringe of its contents (gray arrow). Valve C is then closed, valves A and B are opened, and the drive mechanism is moved down to fill the drive syringes with solution (black arrows). At this point, the apparatus is ready for another rapid mixing experiment. The entire apparatus can be placed under the control of a microcomputer, which can also collect and analyze the reaction rate data. Stopped-flow mixing has been used for fundamental studies of rapid reactions and for routine kinetic determinations of analytes involved in fast reactions. The principles of fluid dynamics that make stopped-flow mixing possible and the solution-handling capabilities of this and other similar devices are utilized in many contexts for automatically mixing solutions and measuring analyte concentrations in numerous industrial and clinical laboratories.

27B-2 Types of Kinetic Methods

Kinetic methods are classified according to the type of relationship that exists between the measured variable and the analyte concentration.

The Differential Method

In the *differential method,* concentrations are computed from reaction rates by means of a differential form of a rate expression. Rates are determined by measuring the slope of a curve relating analyte or product concentration to reaction time. To illustrate, let us substitute $[A]_t$ from Equation 27-7 for [A] in Equation 27-4:

$$\text{rate} = -\left(\frac{d[A]}{dt}\right)_t = k[A]_t = k[A]_0 e^{-kt} \quad (27\text{-}25)$$

As an alternative, the rate can be expressed in terms of the product concentration. That is,

$$\text{rate} = \left(\frac{d[P]}{dt}\right)_t = k[A]_0 e^{-kt} \quad (27\text{-}26)$$

Equations 27-25 and 27-26 give the dependence of the rate upon k, t, and, most important, $[A]_0$, the initial concentration of the analyte. At any fixed time t, the factor ke^{-kt} is a constant, and the rate is directly proportional to the initial analyte concentration.

EXAMPLE 27-5

The rate constant for a pseudo-first-order reaction is 0.156 s^{-1}. Find the initial concentration of the reactant if its rate of disappearance 10.00 s after the initiation of the reaction is 2.79×10^{-4} M · s^{-1}.

The proportionality constant ke^{-kt} is

$$ke^{-kt} = (0.156\ s^{-1})e^{-(0.156\ s^{-1})(10.00\ s)} = 3.28 \times 10^{-2}\ s^{-1}$$

Rearranging Equation 27-25 and substituting numerical values, we have

$$\begin{aligned}[A]_0 &= \text{rate}/ke^{-kt} \\ &= (2.79 \times 10^{-4}\ M \cdot s^{-1})/(3.28 \times 10^{-2}\ s^{-1}) \\ &= 8.51 \times 10^{-3}\ M\end{aligned}$$

Kinetic methods in which data are collected near the beginning of the reaction are called **initial rate methods.**

The choice of the time at which a reaction rate is measured is often based upon such factors as convenience, the existence of interfering side reactions, and the inherent precision of making the measurement at a particular time. It is often advantageous to make the measurement near $t = 0$ because this portion of the exponential curve is nearly linear (see, for example, the initial parts of the curves in Figure 27-1) and the slope is readily estimated from the tangent to the curve. Moreover, if the reaction is pseudo-first-order, such a small amount of excess

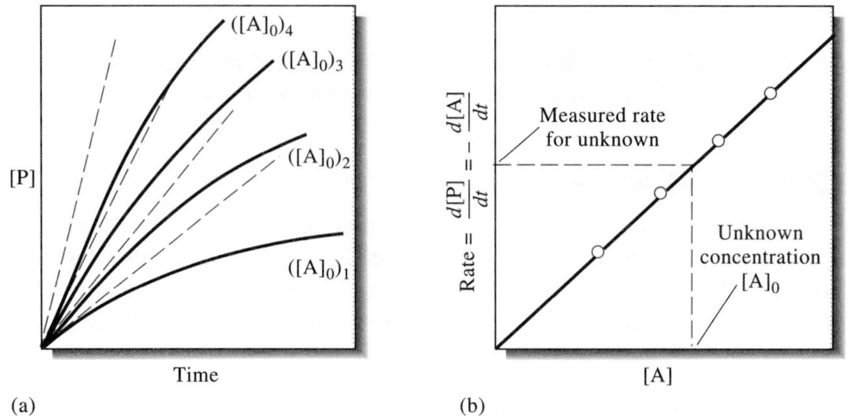

Figure 27-4
A plot of data for the determination of A by the differential method. (a) Solid lines are the experimental plots of product concentration as a function of time for four initial concentrations of A. Dashed lines are tangents to the curve at $t \rightarrow 0$. (b) A plot of the slopes obtained from the tangents in (a) as a function of analyte concentration.

reagent is consumed that no error arises from changes in k resulting from changes in reagent concentration. Finally, the relative error in determining the slope is minimal at the beginning of the reaction because the slope is a maximum in this region.

Figure 27-4 illustrates how the differential method is used to determine the concentration of an analyte $[A]_0$ from experimental rate measurements for the reaction shown as Equation 27-1. The solid curves in Figure 27-4a are plots of the experimentally measured product concentration [P] as a function of reaction time for four standard solutions of A. These curves are then used to prepare the differential calibration plot shown in Figure 27-4b. To obtain the rates, tangents are drawn to each of the curves in 27-4a at a time near zero (dashed lines in part a). The slopes of the tangents are then plotted as a function of [A], giving the straight line shown in 27-4b. Unknowns are treated in the same way, and analyte concentrations are determined from the calibration curve.

Of course, it is not necessary to record the entire rate curve, as has been done in Figure 27-4a, since only a small portion of the plot is utilized for measuring the slope. As long as sufficient data points are collected to determine the initial slope precisely, time is saved and the entire procedure is simplified. More sophisticated data-handling procedures and numerical analysis of the data make possible high-precision rate measurements at later times as well; under certain circumstances such measurements are more accurate and precise than those made near $t = 0$.

Integral Methods

Integral methods use integral forms of the first-order rate equation such as $[A] = [A]_0e^{-kt}$.

In contrast to the differential method, *integral methods* take advantage of integrated forms of rate laws, such as those shown by Equations 27-6, 27-7, and 27-9.

Graphical Methods Equation 27-6 may be rearranged to give

$$\ln[A]_t = -kt + \ln[A]_0 \tag{27-27}$$

See *Mathcad Applications for Analytical Chemistry,* **p. 141.**

Thus, a plot of the natural logarithm of experimentally measured concentrations of A (or P) as a function of time should yield a straight line with a slope of $-k$ and a y intercept of $\ln[A]_0$. Use of this procedure for the determination of nitromethane is illustrated in Example 27-6.

EXAMPLE 27-6

The data in the first two columns of Table 27-2 were recorded for the pseudo-first-order decomposition of nitromethane in the presence of excess base. Find the initial concentration of nitromethane and the pseudo-first-order rate constant for the reaction.

Computed values for the natural logarithms of nitromethane concentrations are shown in the third column of Table 27-2. The data are plotted in Figure 27-5. A least-squares analysis of the data (Section 4E-2) leads to an intercept *a* of

$$a = \ln[CH_3NO_2]_0 = -5.129$$

which upon exponentiation gives

$$[CH_3NO_2]_0 = 5.92 \times 10^{-3}\ M$$

The least-squares analysis also gives the slope of the line *b*, which turns out to be

$$b = -1.62 = -k$$

and thus

$$k = 1.62\ s^{-1}$$

TABLE 27-2
Data for the Decomposition of Nitromethane

Time, s	$[CH_3NO_2]$, M	$\ln[CH_3NO_2]$
0.25	3.86×10^{-3}	−5.557
0.50	2.59×10^{-3}	−5.956
0.75	1.84×10^{-3}	−6.298
1.00	1.21×10^{-3}	−6.717
1.25	0.742×10^{-3}	−7.206

Fixed-Time Methods Fixed-time methods are based upon Equation 27-7 or 27-9. The former can be rearranged to

$$[A]_0 = \frac{[A]_t}{e^{-kt}} \tag{27-28}$$

The simplest way of employing this relationship is to perform a calibration experiment with a standard solution that has a known concentration $[A]_0$. After a

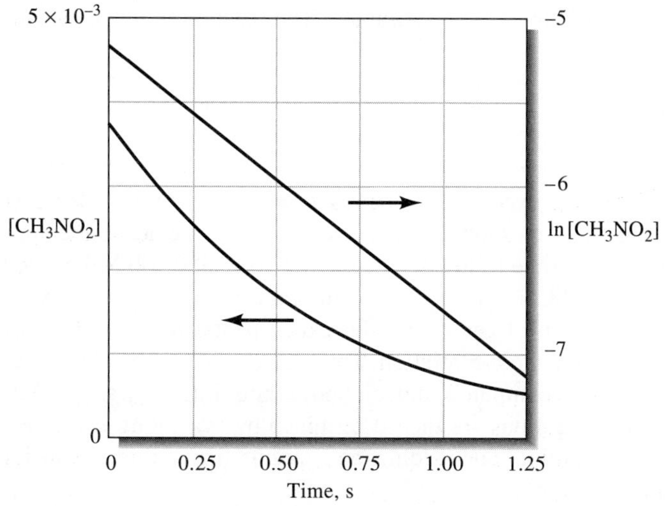

Figure 27-5
Plots of nitromethane concentration and the natural logarithm of nitromethane concentration as a function of time. The data are from Example 27-6.

carefully measured reaction time t, $[A]_t$ is determined and used to evaluate e^{-kt} by means of Equation 27-28. Unknowns are then analyzed by measuring $[A]_t$ after exactly the same reaction time and employing the calculated value for e^{-kt} to compute the analyte concentrations.

Equation 27-28 is easily modified for the situation where [P] is measured experimentally rather than [A]. Equation 27-9 may be rearranged to solve for $[A]_0$. That is,

$$[A]_0 = \frac{[P]_t}{1 - e^{-kt}} \tag{27-29}$$

A more desirable approach to the uses of Equation 27-28 or 27-29 is to measure [A] or [P] at two times t_1 and t_2. For example, if the product concentration is determined, we can write

$$[P]_{t_1} = [A]_0(1 - e^{-kt_1})$$
$$[P]_{t_2} = [A]_0(1 - e^{-kt_2})$$

Subtracting the first equation from the second yields, upon rearrangement,

$$[A]_0 = \frac{[P]_{t_2} - [P]_{t_1}}{e^{-kt_1} - e^{-kt_2}} = C([P]_{t_2} - [P]_{t_1}) \tag{27-30}$$

The reciprocal of the denominator is constant for constant t_1 and t_2 and is assigned the symbol C.

A major advantage of kinetic methods is their immunity to errors resulting from long-term drift of the measurement system.

The use of Equation 27-30 has the fundamental advantage common to most kinetic methods that the absolute determination of concentration or of a variable proportional to concentration is unnecessary. It is the *difference* between two concentrations that is proportional to the initial concentration of the analyte. This means that factors influencing the long-term stability of the measuring instrument are subtracted out.

An important example of uncatalyzed methods is the fixed-time method for the determination of thiocyanate ion based upon spectrophotometric measurements of the red iron(III) thiocyanate complex. The reaction in this application is

The stopped-flow mixing technique described in Feature 27-1 is often used to carry out the Fe^{3+}/SCN^- reaction.

$$Fe^{3+} + SCN^- \underset{k_{-1}}{\overset{k_1}{\rightleftharpoons}} \underset{\text{red}}{Fe(SCN)^{2+}}$$

Under conditions of excess Fe^{3+}, the reaction is pseudo-first-order in SCN^-. The curves in Figure 27-6a show the increase in absorbance due to the appearance of $Fe(SCN)^{2+}$ versus time following the rapid mixing of 0.100 M Fe^{3+} with various concentrations of SCN^- at pH 2. Since the concentration of $Fe(SCN)^{2+}$ is related to the absorbance by Beer's law, the experimental data can be used directly without conversion to concentration. Thus, the change in absorbance ΔA between times t_1 and t_2 is computed and plotted versus $[SCN^-]_0$ as in Figure 27-6b. Unknown concentrations are then determined by evaluating ΔA under the same experimental conditions and reading the concentration of thiocyanate ion from the linear working curve.

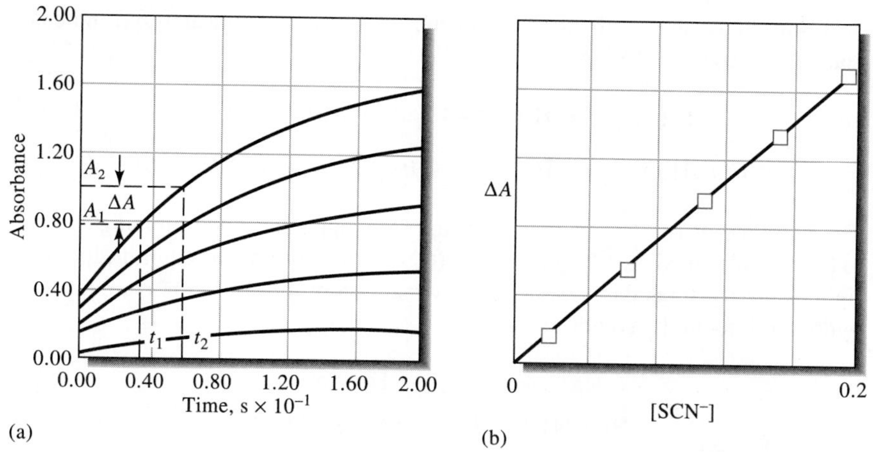

Figure 27-6
(a) Absorbance due to the formation of $Fe(SCN)^{2+}$ as a function of time for five concentrations of SCN^-. (b) A plot of the difference in absorbance ΔA at times t_2 and t_1 as a function of SCN^- concentration.

Fixed-time methods are advantageous because the measured quantity is directly proportional to the analyte concentration and because measurements can be made *at any time* during the progress of first-order reactions. When instrumental methods are used for monitoring reactions by fixed-time procedures, the precision of the analytical results approaches the precision of the instrument used.

27C APPLICATIONS OF KINETIC METHODS

Kinetic methods, which are used for the determination of both organic and inorganic species, fall into two categories, catalyzed and uncatalyzed. As noted earlier, uncatalyzed reactions are not nearly as widely used as catalyzed reactions because the selectivity and sensitivity of the former are often quite inferior to those of the latter. Uncatalyzed reactions are used to advantage when high-speed, automated measurements are required, however, or when the sensitivity of the detection method is great.[5]

27C-1 Catalytic Methods

The Determination of Inorganic Species

Many inorganic cations and anions catalyze *indicator reactions*—that is, reactions whose rates are readily measured by instrumental methods, such as spectrophotometry, fluorometry, or electroanalysis. Conditions are then employed such that the rate is proportional to the concentration of catalyst, and from the rate data, this concentration is determined. An example of this method involves the kinetic determination of traces of mercury(II).

The Catalytic Determination of Mercury The catalytic effect of mercury on the reaction of hexacyanoferrate(II) with nitrosobenzene has been employed to

[5]For reviews of applications of modern kinetic methods, see M. Kopanica and V. Stara, in *Comprehensive Analytical Chemistry,* G. Svehla, Ed., Vol. XVIII, pp. 11–227. New York: Elsevier, 1983; G. G. Guilbault, in *Treatise on Analytical Chemistry,* 2nd ed., I. M. Kolthoff and P. J. Elving, Eds., Part I, Vol. 1, Chapter 11. New York: Wiley, 1978.

determine trace levels of the metal in the +2 state. The uncatalyzed indicator reaction proceeds as follows:

$$Fe(CN)_6^{4-} + H_2O \underset{}{\overset{\text{slow}}{\rightleftharpoons}} Fe(CN)_5H_2O^{3-} + CN^- \qquad (27\text{-}31)$$

$$Fe(CN)_5H_2O^{3-} + C_6H_5NO \underset{}{\overset{\text{fast}}{\rightleftharpoons}} Fe(CN)_5C_6H_5NO^{3-} + H_2O \qquad (27\text{-}32)$$

The presence of Hg^{2+} accelerates the first step, apparently by removing cyanide to form an intermediate complex ion $Hg(CN)^+$, thus forcing the equilibrium to the right. Mercury(II) is then regenerated by reaction of the $Hg(CN)^+$ complex with hydrogen ions in the solution:

$$Hg^{2+} + CN^- \longrightarrow Hg(CN)^+$$

$$Hg(CN)^+ + H^+ \longrightarrow HCN + Hg^{2+}$$

Kinetic methods are particularly advantageous when reactions are so slow that it is impractical to wait until equilibrium is achieved.

Reaction 27-32 is relatively fast and serves as the indicator reaction for monitoring the formation of the product by spectrophotometric measurements at 528 nm. The formation of $Fe(CN)_5C_6H_5NO^{3-}$ is consistent with a general catalyst mechanism similar to that presented in Section 27A-4, so the rate of the reaction is given by

$$\frac{d[Fe(CN)_5C_6H_5NO^{3-}]}{dt} = k_{cat}[Hg^{2+}]$$

where k_{cat} is a collection of constants.

The reaction is slow enough that a fixed-time measurement of absorbance at $t = 1800$ s permits the determination of Hg^{2+} at levels as low as 3×10^{-7} M. The method has been used for the determination of mercury in biological materials, in industrial by-products, and in water.

The Scope of Catalyzed Methods in Inorganic Analysis Kinetic methods based on catalysis by inorganic analytes are widely applicable. For example, Kopanica and Stara[6] list 40 cations and 15 anions that have been determined by a variety of indicator reactions.

The Determination of Organic Species

Without question, the most important applications of catalyzed kinetic methods to organic analyses involve the use of enzymes as catalysts. These methods have been used for the determination of both enzymes and substrates, and serve as the basis for many of the routine and automated screening tests performed by the thousands in clinical laboratories throughout the world. One of these tests is for determining the quantity of urea in blood and is called the blood urea nitrogen (BUN) test. A brief description of this test follows.

The Enzymatic Determination of Urea The determination of urea in blood and urine is frequently carried out by measuring the rate of hydrolysis of urea in the presence of the enzyme urease. The equation for the reaction is

[6]M. Kopanica and V. Stara, in *Comprehensive Analytical Chemistry,* G. Svehla, Ed., Vol. XVIII, pp. 192–209. New York: Elsevier, 1983.

$$CO(NH_2)_2 + 2H_2O + H^+ \xrightarrow{\text{urease}} 2NH_4^+ + HCO_3^-$$

As suggested in Example 27-4, urea can be determined by measuring the initial rate of production of the products of this reaction. The high selectivity of the enzyme permits the use of nonselective detection methods, such as electrical conductance, for initial rate measurements. Commercial instruments operate on this principle. The sample is mixed with a small amount of an enzyme-buffer solution in a conductivity cell. The maximum rate of increase in conductance is measured within 10 s of mixing, and the concentration of urea is determined from a calibration curve consisting of a plot of maximum initial rate as a function of urea concentration. The precision of the instrument is on the order of 2% to 5% relative for concentrations in the physiological range of 2 to 10 mM.

Another method for following the rate of urea hydrolysis is based on an ion-selective electrode for ammonium ions (Section 18D-4).

The Scope of Methods Based on Enzyme-Catalyzed Reactions Enzyme-catalyzed reactions are used for the kinetic determination of a large and diverse group of substrate species. These methods have the twin virtues of remarkable selectivity and good sensitivity. In addition, they are generally readily automated.

A number of inorganic species can be determined by enzyme-catalyzed reactions, including ammonia, hydrogen peroxide, carbon dioxide, and hydroxylamine, as well as nitrate, phosphate, and pyrophosphate ions.

Methods have also been developed for the determination of individual components in mixtures of closely related organic compounds. For example, a procedure for the determination of the individual components in a mixture of 21 organic acids without preliminary separation has been described (several enzymes are used). Similarly, methods have been developed for the determination of components in mixtures of alcohols and hydroxy compounds, sugars, amines, and steroid alcohols.

Kinetic methods have been described for the quantitative determination of several hundred enzymes. In addition, some two dozen inorganic cations and anions are known to decrease the rates of certain enzyme-catalyzed indicator reactions. These *inhibitors* can thus be determined from the decrease in rate brought about by their presence.

Enzyme activators are substances, often inorganic ions, that are required for certain enzymes to become active as catalysts. Activators can be determined by their effect on the rates of enzyme-catalyzed reactions. For example, it has been reported that magnesium at concentrations as low as 10 ppb can be determined in blood plasma based on activation by this ion of the enzyme isocitrate dehydrogenase.

We may use enzymes for the determination of both inhibitors and activators.

27C-2 Noncatalytic Reactions

As noted earlier, kinetic methods based on uncatalyzed reactions are not nearly as widely used as those in which a catalyst is involved. We have already described two of these methods (pages 653 and 654).

Generally, uncatalyzed reactions can be rendered useful when selective reagents are employed in conjunction with sensitive detection methods. For example, the selectivity of complexing agents can be controlled by adjusting the pH of the medium in the determination of metal ions, as discussed in Section 14B-8. Sensitivity can be achieved through the use of spectrophotometric detection to monitor reagents that form complexes with large molar absorptivities. The determination of Cu^{2+} presented in Problem 27-13 is an example.

A highly sensitive alternative is to select complexes that fluoresce so that the rate of change of fluorescence can be used as a measure of analyte concentration (Problem 27-14).

The precision of both noncatalytic and catalytic kinetic methods depends upon such experimental conditions as pH, ionic strength, and temperature. With careful control of these variables, relative standard deviations of 1% to 10% are typical. Automation of kinetic methods and computerized data analysis can often improve the relative precision to 1% or less.

27C-3 The Kinetic Determination of Components in Mixtures

An important application of kinetic methods is in the determination of closely related species in mixtures, such as alkaline earth cations or organic compounds with the same functional groups. For example, suppose two species A and B react with a common excess reagent to form products under pseudo-first-order conditions:

$$\mathrm{A} + \mathrm{R} \xrightarrow{k_A} \mathrm{P}'$$

$$\mathrm{B} + \mathrm{R} \xrightarrow{k_B} \mathrm{P}''$$

Generally, k_A and k_B differ from each other. Thus, if $k_A > k_B$, A is depleted before B. It is possible to show that, if the ratio k_A/k_B is greater than about 500, the consumption of A is approximately 99% complete before 1% of B is used up. Thus, a differential determination of A with no significant interference from B is possible provided the rate is measured shortly after mixing.

When the ratio of the two rate constants is small, determination of both species is still possible by more complex methods of data treatment. These methods are beyond the scope of this text.[7]

[7]For examples of the use of kinetic methods for the analysis of multicomponent mixtures, see G. M. Ridder and D. W. Margerum, *Anal. Chem.*, **1977,** *49,* 2090.

27D QUESTIONS AND PROBLEMS

27-1. Define the following terms as they are used in reaction kinetics.
 *(a) order of a reaction
 (b) pseudo-first-order
 *(c) enzyme
 (d) substrate
 *(e) Michaelis constant
 (f) differential method
 *(g) integral method
 (h) indicator reaction

27-2. The analysis of multicomponent of mixtures by kinetic methods is sometimes referred to as ''kinetic separation.'' Explain the significance of this term.

***27-3.** Explain why pseudo-first-order conditions are utilized in most kinetic methods.

27-4. List three advantages of kinetic methods.

***27-5.** Develop an expression for the half-life of the reactant in a first-order process in terms of k.

27-6. Find the natural lifetime in seconds for first-order reactions corresponding to
 *(a) $k = 0.497\ \mathrm{s}^{-1}$.
 (b) $k = 6.62\ \mathrm{hr}^{-1}$.
 *(c) $[\mathrm{A}]_0 = 3.16$ M and $[\mathrm{A}]_t = 0.496$ M at $t = 3650$ s.
 (d) $[\mathrm{P}]_\infty = 0.176$ and $[\mathrm{P}]_t = 0.0423$ M at $t = 9.62$ s.
 *(e) half-life $t_{1/2} = 32.4$ years.
 (f) $t_{1/2} = 0.478$ s.

27-7. Find the first-order rate constant for a reaction that is 66.6% complete in

*(a) 0.0100 s. *(c) 1.00 s. *(e) 26.8 μs.
(b) 0.100 s. (d) 5280 s. (f) 8.86 ns.

27-8. Calculate the number of lifetimes τ required for a pseudo-first-order reaction to achieve the following levels of completion:
*(a) 10% *(c) 90% *(e) 99.9%
(b) 50% (d) 99% (f) 99.99%

*27-9. Find the number of half-lives $t_{1/2}$ required to reach the levels of completion listed in Problem 27-8.

27-10. Find the relative error associated with the assumption that k' is invariant during the course of a pseudo-first-order reaction under the following conditions.

	Extent of Reaction, %	Excess of Reagent
*(a)	1	5×
(b)	1	10×
*(c)	1	50×
(d)	1	100×
*(e)	5	5×
(f)	5	10×
*(g)	5	100×
(h)	63.2	5×
*(i)	63.2	10×
(j)	63.2	50×
*(k)	63.2	100×

*27-11. Show that for an enzyme reaction obeying Equation 27-22 the substrate concentration for which the rate equals $\nu_{max}/2$ is equal to K_m.

27-12. Equation 27-22 can be rearranged to produce the equation

$$\frac{1}{d[P]/dt} = \frac{K_m}{\nu_{max}[S]} + \frac{1}{\nu_{max}}$$

where $\nu_{max} = k_2[E]_0$ when [S] is large.
(a) Suggest a way to employ this equation in the construction of a working curve for the enzymatic determination of substrate.
(b) Describe how the resulting working curve can be used to find K_m and ν_{max}.

*27-13. Copper(II) forms a 1:1 complex with the organic complexing agent R in acidic medium. The formation of the complex can be monitored by spectrophotometry at 480 nm. Use the following data collected under pseudo-first-order conditions to construct a calibration curve of rate versus concentration of R. Find the concentration of copper(II) in an unknown whose rate under the same conditions was 6.2×10^{-3} $A \cdot s^{-1}$.

$c_{Cu^{2+}}$, ppm	Rate, $A \cdot s^{-1}$
3.0	3.6×10^{-3}
5.0	5.4×10^{-3}
7.0	7.9×10^{-3}
9.0	1.03×10^{-2}

27-14. Aluminum forms a 1:1 complex with 2-hydroxy-1-naphthaldehyde *p*-methoxybenzoylhydraxonal that exhibits fluorescence emission at 475 nm. Under pseudo-first-order conditions, a plot of the initial rate of the reaction (emission units per second) versus the concentration of aluminum (in μM) yields a straight line described by the equation

$$\text{rate} = 1.74c_{Al} - 0.225$$

Find the concentration of aluminum in a solution that exhibits a rate of 0.76 emission units per second under the same experimental conditions.

*27-15. The enzyme monoamine oxidase catalyzes the oxidation of amines to aldehydes. For tryptamine, K_m for the enzyme is 4.0×10^{-4} M and $\nu_{max} = k_2[E]_0 = 1.6 \times 10^{-3}$ μM/min at pH 8. Find the concentration of a solution of tryptamine that reacts at a rate of 0.18 μM/min in the presence of amine oxidase under the above conditions. Assume that [tryptamine] $\ll K_m$.

CHAPTER

28

An Introduction to Chromatographic Methods

Chromatography was invented by the Russian botanist Mikhail Tswett shortly after the turn of the century. He employed the technique to separate various plant pigments, such as chlorophylls and xanthophylls, by passing solutions of them through glass columns packed with finely divided calcium carbonate. The separated species appeared as colored bands on the column, which accounts for the name he chose for the method (Greek *chroma* meaning "color" and *graphein* meaning "to write").

The **stationary phase** in chromatography is a phase that is fixed in place either in a column or on a planar surface.

The **mobile phase** in chromatography is a phase that moves over or through the stationary phase, carrying the analyte with it.

Chromatography is a technique in which the components of a mixture are separated based upon the rates at which they are carried through a stationary phase by a gaseous or liquid mobile phase.

Planar and column chromatography are based upon the same types of equilibria.

Chromatography is an analytical method that is widely used for the separation, identification, and determination of the chemical components in complex mixtures. No other separation method is as powerful and generally applicable as is chromatography.[1]

28A A GENERAL DESCRIPTION OF CHROMATOGRAPHY

The term *chromatography* is difficult to define rigorously because the term has been applied to such a variety of systems and techniques. All of these methods, however, have in common the use of a *stationary phase* and a *mobile phase.* Components of a mixture are carried through the stationary phase by the flow of a gaseous or liquid mobile phase, separations being based on differences in migration rates among the sample components.

28A-1 Classification of Chromatographic Methods

Chromatographic methods are of two types. In *column chromatography,* the stationary phase is held in a narrow tube and the mobile phase is forced through the tube under pressure or by gravity. In *planar chromatography,* the stationary phase is supported on a flat plate or in the pores of a paper. Here the mobile phase moves through the stationary phase by capillary action or under the influence of gravity.

As shown in the first column of Table 28-1, chromatographic methods fall into three categories based upon the nature of the mobile phase: liquid, gas, and supercritical fluid. The second column of the table reveals that there are five types of liquid chromatography and three types of gas chromatography, which differ in the nature of the stationary phase and the types of equilibria between phases.

---

[1]General references on chromatography include *Chromatography: Fundamentals and Applications of Chromatography and Electrophoretic Methods,* Part A, *Fundamentals,* Part B, *Applications,* E. Heftmann, Ed. New York: Elsevier, 1983; P. Sewell and B. Clarke, *Chromatographic Separations.* New York: Wiley, 1988; *Chromatographic Theory and Basic Principles,* J. A. Jonsson, Ed. New York: Marcel Dekker, 1987; J. C. Giddings, *Unified Separation Science.* New York: Wiley, 1991.

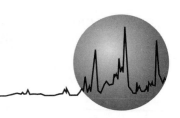

TABLE 28-1 Classification of Column Chromatographic Methods

General Classification	Specific Method	Stationary Phase	Type of Equilibrium
Liquid chromatography (LC) (mobile phase: liquid)	Liquid-liquid, or partition	Liquid adsorbed on a solid	Partition between immiscible liquids
	Liquid-bonded phase	Organic species bonded to a solid surface	Partition between liquid and bonded surface
	Liquid-solid, or adsorption	Solid	Adsorption
	Ion exchange	Ion-exchange resin	Ion exchange
	Size exclusion	Liquid in interstices of a polymeric solid	Partition/sieving
Gas chromatography (GC) (mobile phase: gas)	Gas-liquid	Liquid adsorbed on a solid	Partition between gas and liquid
	Gas-bonded phase	Organic species bonded to a solid surface	Partition between liquid and bonded surface
	Gas-solid	Solid	Adsorption
Supercritical-fluid chromatography (SFC) (mobile phase: supercritical fluid)		Organic species bonded to a solid surface	Partition between supercritical fluid and bonded surface

28A-2 Elution Chromatography

Figure 28-1 shows how two components A and B are resolved on a column by *elution chromatography.* Elution involves washing a solute through a column by additions of fresh solvent. A single portion of the sample dissolved in the mobile phase is introduced at the head of the column (at time t_0 in Figure 28-1), whereupon components A and B distribute themselves between the two phases. Introduction of additional mobile phase (the *eluent*) forces the dissolved portion of the sample down the column, where further partition between the mobile phase and fresh portions of the stationary phase occurs (time t_1). Partitioning between the fresh solvent and the stationary phase takes place simultaneously at the site of the original sample.

Liquid chromatography can be performed in columns and on planar surfaces, but gas chromatography is restricted to column procedures.

Elution is a process in which solutes are washed through a stationary phase by the movement of a mobile phase.

An **eluent** is a solvent used to carry the components of a mixture through a stationary phase.

Further additions of solvent carry solute molecules down the column in a continuous series of transfers between the two phases. Because solute movement can occur only in the mobile phase, the average *rate* at which a solute migrates *depends upon the fraction of time it spends in that phase.* This fraction is small for solutes that are strongly retained by the stationary phase (component B in Figure 28-1, for example) and large where retention in the mobile phase is more likely (component A). Ideally, the resulting differences in rates cause the components in a mixture to separate into *bands,* or *zones,* along the length of the column (see Figure 28-2). Isolation of the separated species is then accomplished by passing a sufficient quantity of mobile phase through the column to cause the individual bands to pass out the end (to be *eluted* from the column), where they can be collected (times t_3 and t_4 in Figure 28-1).

Chromatograms

If a detector that responds to solute concentration is placed at the end of the column and its signal is plotted as a function of time (or of volume of added

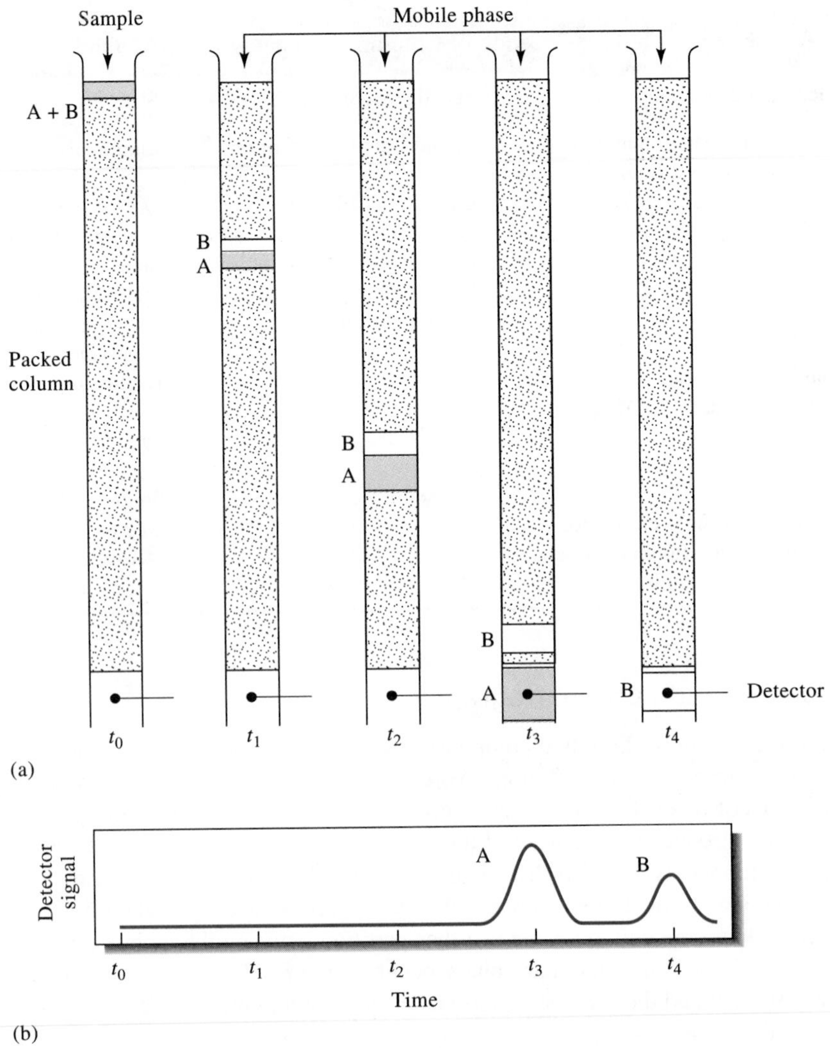

Figure 28-1
(a) Diagram showing the separation of a mixture of components A and B by column elution chromatography.
(b) The output of the signal detector at the various stages of elution shown in (a).

A **chromatogram** is a plot of some function of solute concentration versus elution time or elution volume.

mobile phase), a series of symmetric peaks is obtained, as shown in the lower part of Figure 28-1. Such a plot, called a *chromatogram,* is useful for both qualitative and quantitative analysis. The positions of the peaks on the time axis can be used to identify the components of the sample; the areas under the peaks provide a quantitative measure of the amount of each species.

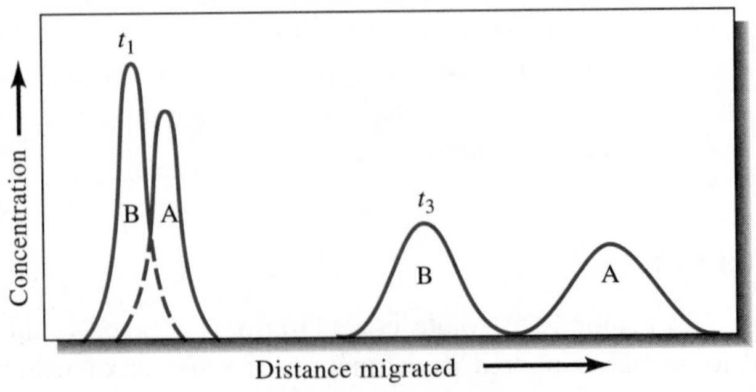

Figure 28-2
Concentration profiles of solute bands A and B at two different times in their migration down the column in Figure 28-1. The times t_1 and t_3 are indicated in Figure 28-1.

The Effects of Relative Migration Rates and Band Broadening on Resolution

Figure 28-2 shows concentration profiles for the bands containing solutes A and B on the column in Figure 28-1 at time t_1 and at a later time t_3.[2] Because B is more strongly retained by the stationary phase than is A, B lags during the migration. Clearly, the distance between the two increases as they move down the column. At the same time, however, broadening of both bands takes place, which lowers the efficiency of the column as a separating device. While band broadening is inevitable, conditions can often be found where it occurs more slowly than band separation. Thus, as shown in Figure 28-2, a clean separation of species is possible provided the column is sufficiently long.

Several chemical and physical variables influence the rates of band separation and band broadening. As a consequence, improved separations can often be realized by the control of variables that either (1) increase the rate of band separation or (2) decrease the rate of band spreading. These alternatives are illustrated in Figure 28-3.

The variables that influence the relative rates at which solutes migrate through a stationary phase are described in the next section. Following this discussion, we turn to those factors that play a part in zone broadening.

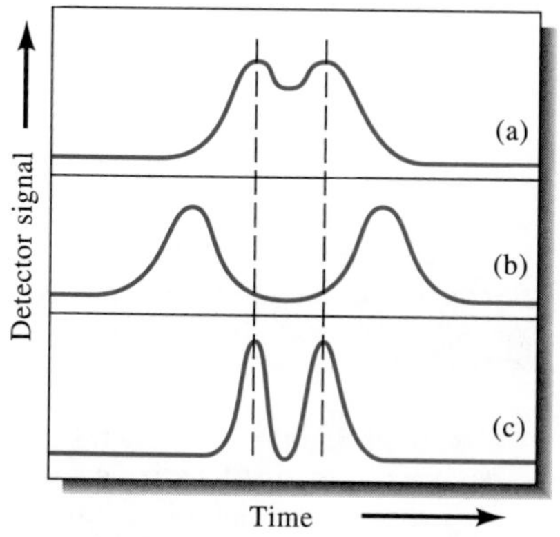

Figure 28-3
Two-component chromatograms illustrating two methods of improving separation: (a) original chromatogram with overlapping peaks; improvement brought about by (b) an increase in band separation; and (c) a decrease in bandwidth.

28B MIGRATION RATES OF SOLUTES

The effectiveness of a chromatographic column as a means of separating two solutes depends in part upon the relative rates at which the two species are eluted. These rates are in turn determined by the partition ratios of the solutes between the two phases.

28B-1 Partition Ratios in Chromatography

All chromatographic separations are based upon differences in the extent to which solutes are partitioned between the mobile and the stationary phase. For the solute species A, the equilibrium involved is described by the equation

$$A_{mobile} \rightleftharpoons A_{stationary}$$

The equilibrium constant K for this reaction is called a *partition ratio,* or *partition coefficient,* and is defined as

$$K = \frac{c_S}{c_M} \tag{28-1}$$

The **partition ratio,** or **partition coefficient,** for a solute in chromatography is equal to the ratio of its concentration in the stationary phase to its concentration in the mobile phase.

where c_S is the molar analytical concentration of a solute in the stationary phase and c_M is its analytical concentration in the mobile phase. Ideally, the partition

[2]Note that the relative positions of the bands for A and B in the concentration profile in Figure 28-2 appear to be reversed from their positions in the lower part of Figure 28-1. The difference is that the abscissa is distance along the column in Figure 28-2 but time in Figure 28-1. Thus, in Figure 28-1, the *front* of a peak lies to the left and the *tail* to the right; in Figure 28-2, the reverse is true.

ratio is constant over a wide range of solute concentrations; that is, c_S is directly proportional to c_M.

28B-2 Retention Time

The **dead time** t_M is the time required for an unretained species to pass through a column.

The **retention time** t_R is the time between injection of a sample and the appearance of a solute peak at the detector of a chromatographic column.

Figure 28-4 is a simple chromatogram that has just two peaks. The small peak on the left is for a species that is *not* retained by the stationary phase. The time t_M after sample injection for this peak to appear is sometimes called the *dead time.* The dead time provides a measure of the average rate of migration of the mobile phase and is an important parameter in identifying analyte peaks. Often the sample or the mobile phase will contain an unretained species. If they do not, such a species may be added to aid in peak identification. The larger peak on the right in Figure 28-4 is that of an analyte species. The time required for this peak to reach the detector after sample injection is called the *retention time* and is given the symbol t_R.

The average linear rate of solute migration, $\bar{v}$, is

$$\bar{v} = \frac{L}{t_R} \tag{28-2}$$

where L is the length of the column packing. Similarly, the average linear velocity, u, of the molecules of the mobile phase is

$$u = \frac{L}{t_M} \tag{28-3}$$

28B-3 The Relationship Between Migration Rate and Partition Ratio

In order to relate the rate of migration of a solute to its partition ratio, we express the rate as a fraction of the velocity of the mobile phase:

$$\bar{v} = u \times \text{fraction of time solute spends in mobile phase}$$

This fraction, however, equals the average number of moles of solute in the mobile phase at any instant divided by the total number of moles of solute in the column:

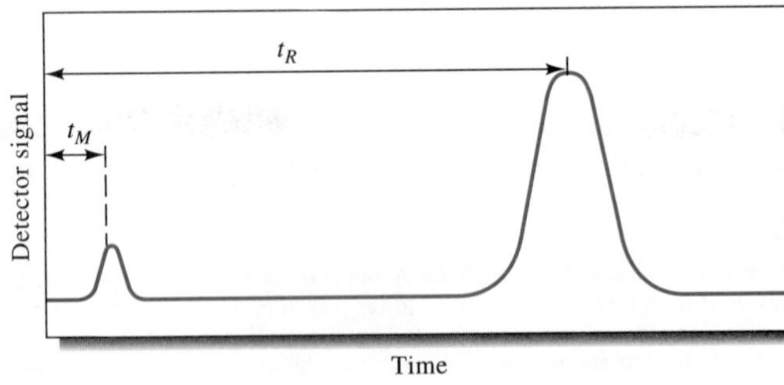

Figure 28-4
A typical chromatogram for a two-component mixture. The small peak on the left represents a solute that is not retained on the column and so reaches the detector almost immediately after elution is started. Thus, its retention time t_M is approximately equal to the time required for a molecule of the mobile phase to pass through the column.

$$\bar{v} = u \times \frac{\text{moles of solute in mobile phase}}{\text{total moles of solute}}$$

The total number of moles of solute in the mobile phase is equal to the molar concentration c_M of the solute in that phase multiplied by its volume V_M. Similarly, the number of moles of solute in the stationary phase is given by the product of the concentration c_S of the solute in the stationary phase and its volume V_S. Therefore,

$$\bar{v} = u \times \frac{c_M V_M}{c_M V_M + c_S V_S} = u \times \frac{1}{1 + c_S V_S / c_M V_M}$$

Substitution of Equation 28-1 into this equation gives an expression for the rate of solute migration as a function of its partition ratio as well as a function of the volumes of the stationary and mobile phases:

$$\bar{v} = u \times \frac{1}{1 + KV_S/V_M} \qquad \text{(28-4)}$$

The two volumes can be estimated from the method by which the column is prepared.

Michael Tswett (1872–1919), a Russian botanist, discovered the basic principles of column chromatography. He separated plant pigments by eluting a mixture of the pigments on a column of calcium carbonate. The various pigments separated into colored bands; hence the name chromatography.

28B-4 The Rate of Solute Migration: The Capacity Factor

The *capacity factor* is an important experimental parameter that is widely used to describe the migration rates of solutes on columns. For a solute A, the capacity factor k'_A is defined as

$$k'_A = \frac{K_A V_S}{V_M} \qquad \text{(28-5)}$$

where K_A is the partition ratio for the species A. Substitution of Equation 28-5 into 28-4 yields

$$\bar{v} = u \times \frac{1}{1 + k'_A} \qquad \text{(28-6)}$$

In order to show how k'_A can be derived from a chromatogram, we substitute Equations 28-2 and 28-3 into Equation 28-6:

$$\frac{L}{t_R} = \frac{L}{t_M} \times \frac{1}{1 + k'_A} \qquad \text{(28-7)}$$

This equation rearranges to

$$k'_A = \frac{t_R - t_M}{t_M} \qquad \text{(28-8)}$$

The capacity factor k'_A for solute A describes the rate at which A migrates through a column.

Ideally, the capacity factor for analytes in a sample is between 1 and 5.

As shown in Figure 28-4, t_R and t_M are readily obtained from a chromatogram. When the capacity factor for a solute is much less than unity, elution occurs so rapidly that accurate determination of the retention times is difficult. When the capacity factor is larger than perhaps 20 to 30, elution times become inordinately long. Ideally, separations are performed under conditions in which the capacity factors for the solutes in a mixture lie in the range between 1 and 5.

The capacity factors in gas chromatography can be varied by changing the temperature and the column packing. In liquid chromatography, capacity factors can often be manipulated to give better separations by varying the composition of the mobile phase and the stationary phase.

28B-5 Relative Migration Rates: The Selectivity Factor

The selectivity factor α for solutes A and B is defined as the ratio of the partition ratio of the more strongly retained solute (B) to the partition ratio of the less strongly held solute (A).

The selectivity factor for two analytes in a column provides a measure of how well the column will separate the two.

The *selectivity factor* α of a column for the two species A and B is defined as

$$\alpha = \frac{K_B}{K_A} \tag{28-9}$$

where K_B is the partition ratio for the more strongly retained species B and K_A is the constant for the less strongly held or more rapidly eluted species A. According to this definition, *α is always greater than unity.*

Substitution of Equation 28-5 and the analogous equation for solute B into Equation 28-9 provides after rearrangement a relationship between the selectivity factor for two solutes and their capacity factors:

$$\alpha = \frac{k'_B}{k'_A} \tag{28-10}$$

where k'_B and k'_A are the capacity factors for B and A, respectively. Substitution of Equation 28-8 for the two solutes into Equation 28-10 gives an expression that permits the determination of α from an experimental chromatogram:

$$\alpha = \frac{(t_R)_B - t_M}{(t_R)_A - t_M} \tag{28-11}$$

In Section 28D-1, we show how we use the selectivity factor to compute the resolving power of a column.

28C BAND BROADENING AND COLUMN EFFICIENCY

See *Mathcad Applications for Analytical Chemistry,* **pp. 129–133.**

Chromatographic efficiency is affected by the amount of band broadening that occurs when a compound passes through the column. Before defining column efficiency in more quantitative terms, let us examine the reasons that bands become broader as they move down a column.

28C-1 The Rate Theory of Chromatography

The *rate theory* of chromatography describes the shapes and breadths of elution peaks in quantitative terms based on a random-walk mechanism for the migration of molecules through a column. A detailed discussion of the rate theory is beyond the scope of this text. We can, however, give a qualitative picture of why bands broaden and what variables improve column efficiency.[3]

If you examine the chromatograms shown in this and the next chapter, you will see that the elution peaks look very much like the Gaussian or normal error curves that you encountered in Chapters 3 and 4. As shown in Section 3A-2, normal error curves are rationalized by assuming that the uncertainty associated with any single measurement is the summation of a much larger number of small, individually undetectable, and random uncertainties, each of which has an equal probability of being positive or negative. In a similar way, the typical Gaussian shape of a chromatographic band can be attributed to the additive combination of the random motions of the myriad molecules making up a band as it moves down the column.

Some chromatographic peaks are nonideal and exhibit *tailing* or *fronting*. In the former case the tail of the peak, appearing to the right on the chromatogram, is drawn out while the front is steepened. With fronting, the reverse is the case. A common cause of tailing and fronting is a nonlinear distribution coefficient. Fronting also arises when too large a sample is introduced onto a column. Distortions of this kind are undesirable because they lead to poorer separations and less reproducible elution times. In our discussion, tailing and fronting are assumed to be minimal.

It is instructive to consider a single solute molecule as it undergoes many thousands of transfers between the stationary and the mobile phase during elution. Residence time in either phase is highly irregular. Transfer from one phase to the other requires energy, and the molecule must acquire this energy from its surroundings. Thus, the residence time in a given phase may be transitory after some transfers and relatively long after others. Recall that movement down the column can occur *only while the molecule is in the mobile phase.* As a consequence, certain particles travel rapidly by virtue of their accidental inclusion in the mobile phase for a majority of the time whereas others lag because they happen to be incorporated in the stationary phase for a greater-than-average length of time. The result of these random individual processes is a symmetric spread of velocities around the mean value, which represents the behavior of the average analyte molecule.

28C-2 A Quantitative Description of Column Efficiency

Two related terms are widely used as quantitative measures of the efficiency of chromatographic columns: (1) *plate height H* and (2) *number of theoretical plates N*. The two are related by the equation

The plate height H is also known as the **height equivalent of a theoretical plate (HETP).**

$$N = L/H \tag{28-12}$$

where L is the length (usually in centimeters) of the column packing. Feature 28-1 describes how these measures of column efficiency got their names.

The efficiency of a chromatographic column increases as the number of plates becomes greater and as the plate height becomes smaller. Enormous differences in efficiencies are encountered in columns, depending upon their type and the kinds of mobile and stationary phases they contain. Efficiencies in terms of plate numbers can vary from a few hundred to several hundred thousand; plate heights from a few tenths to one thousandth of a centimeter or smaller are not uncommon.

The efficiency of a column is great when H is small and N is large.

[3]See S. J. Hawkes, *J. Chem. Educ.*, **1983,** *60*, 393.

FEATURE 28-1
What Is the Source of the Terms *Plate* and *Plate Height?*

The 1952 Nobel Prize was awarded to two Englishmen, A. J. P. Martin and R. L. M. Synge, for their work in the development of modern chromatography. In their theoretical studies, they adapted a model that was first developed in the early 1920s to describe separations on fractional distillation columns. Fractionating columns, which were first used in the petroleum industry for separating closely related hydrocarbons, consisted of numerous interconnected bubble-cap plates (see Figure 28-5) at which vapor-liquid equilibria were established when the column was operated under reflux conditions.

Martin and Synge treated a chromatographic column as if it were made up of a series of contiguous bubble-cap-like plates, within which equilibrium conditions always prevail. This plate model successfully accounts for the Gaussian shape of chromatographic peaks as well as for factors that influence differences in solute-migration rates. The plate model is totally incapa-

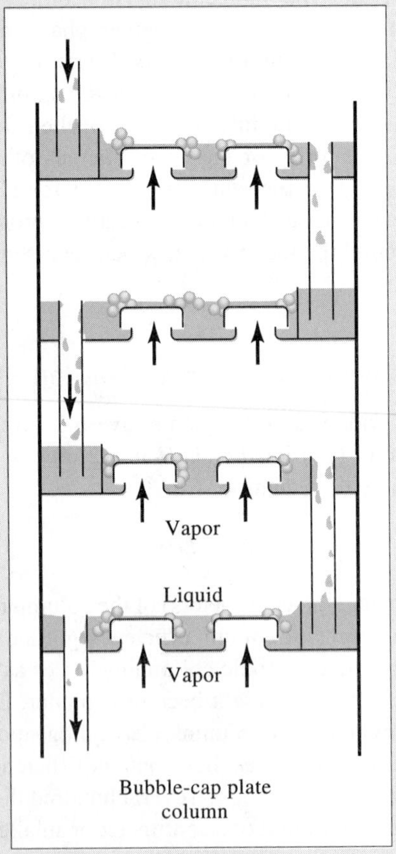

Figure 28-5
Plates in a fractionating column.

ble of accounting for zone broadening, however, because of its basic assumption that equilibrium conditions prevail throughout a column during elution. This assumption can never be valid in the dynamic state that exists in a chromatographic column, where phases are moving past one another at such a pace that sufficient time is not available for equilibration.

Because the plate model is an inadequate representation of a chromatographic column, you are strongly urged (1) to avoid attaching any real or imaginary significance to the terms *plate* and *plate height* and (2) to view these terms as designators of column efficiency that are retained for historic reasons only and not because they have physical significance. Unfortunately, the terms are so well entrenched in the chromatographic literature that their replacement by more appropriate designations seems unlikely, at least in the near future.

Definition of Plate Height, *H*

In Section 3B-2, we pointed out that the breadth of a Gaussian curve is described by the standard deviation σ and the variance σ^2. Because chromatographic bands are also Gaussian and because the efficiency of a column is reflected in the breadth of chromatographic peaks, the variance per unit length of column is used by chromatographers as a measure of column efficiency. That is, the column efficiency H is defined as

$$H = \frac{\sigma^2}{L} \qquad (28\text{-}13)$$

This definition of column efficiency is illustrated in Figure 28-6a, which shows a column having a packing L cm in length. Above this schematic (Figure 28-6b) is a plot showing the distribution of molecules along the length of the column at the moment the analyte peak reaches the end of the packing (that is, at the retention time t_R). The curve is Gaussian, and the locations of $L + 1\sigma$ and $L - 1\sigma$ are indicated as broken vertical lines. Note that L carries units of centimeters and σ^2 units of centimeters squared; thus, H represents a linear distance in centimeters as well (Equation 28-13). In fact, the plate height can be thought of as the length of column that contains a fraction of the analyte that lies between L and $L - \sigma$. Because the area under a normal error curve bounded by $\pm\sigma$ is about 68% of the total area (page 28), the plate height, as defined, contains 34% of the analyte.

Experimental Determination of the Number of Plates in a Column

The number of theoretical plates N and the plate height H are widely used in the literature and by instrument manufacturers as measures of column performance. Figure 28-7 shows how N can be determined from a chromatogram. Here, the retention time of a peak t_R and the width of the peak at its base W (in units of time) are measured. It can be shown (see Feature 28-2) that the number of plates can then be computed by the simple relationship

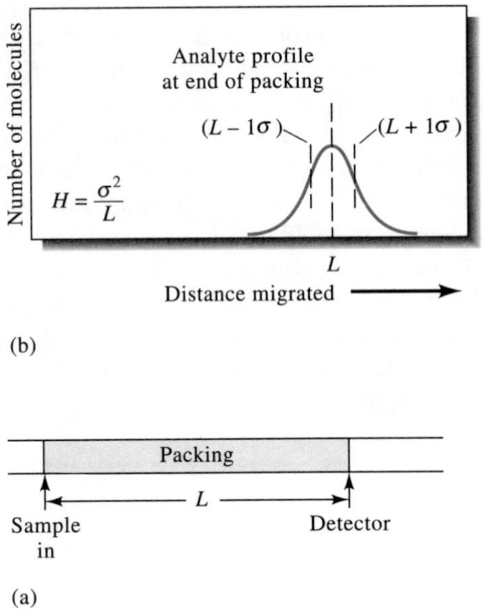

Figure 28-6
Definition of plate height $H = \sigma^2/L$.

See *Mathcad Applications for Analytical Chemistry,* **p. 130.**

$$N = 16\left(\frac{t_R}{W}\right)^2 \quad (28\text{-}14)$$

To obtain H, the length of the column L is measured and Equation 28-12 applied.

FEATURE 28-2
Derivation of Equation 28-14

The variance of the peak shown in Figure 28-7 has units of seconds squared because the abscissa is time in seconds (or sometimes in minutes). This time-based variance is usually designated as τ^2 to distinguish it from σ^2, which has units of centimeters squared. The two standard deviations τ and σ are related by

$$\tau = \frac{\sigma}{L/t_R} \quad (28\text{-}15)$$

where L/t_R is the average linear velocity of the solute in centimeters per second.

Figure 28-7 illustrates a simple means for approximating τ from an experimental chromatogram. Tangents at the inflection points on the two sides of the chromatographic peak are extended to form a triangle with the baseline. The area of this triangle can be shown to be approximately 96% of the total area under the peak. In Section 4A-1, it was shown that about 96% of the

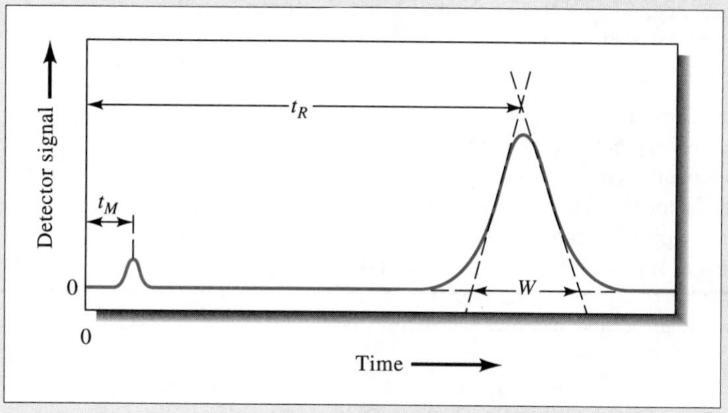

Figure 28-7
Determination of the standard deviation τ from a chromatographic peak: $W = 4\tau$.

area under a Gaussian peak is included within plus or minus two standard deviations ($\pm 2\sigma$) of its maximum. Thus, the intercepts shown in Figure 28-8 occur at approximately $\pm 2\tau$ from the maximum, and $W = 4\tau$, where W is the magnitude of the base of the triangle. Substituting this relationship into Equation 28-15 and rearranging yield

$$\sigma = \frac{LW}{4t_R}$$

Substitution of this equation for σ into Equation 28-13 gives

$$H = \frac{LW^2}{16t_R^2} \tag{28-16}$$

To obtain N, we substitute into Equation 28-12 and rearrange to get

$$N = 16\left(\frac{t_R}{W}\right)^2$$

Thus, N can be calculated from two time measurements, t_R and W; to obtain H, the length of the column packing L must also be known.

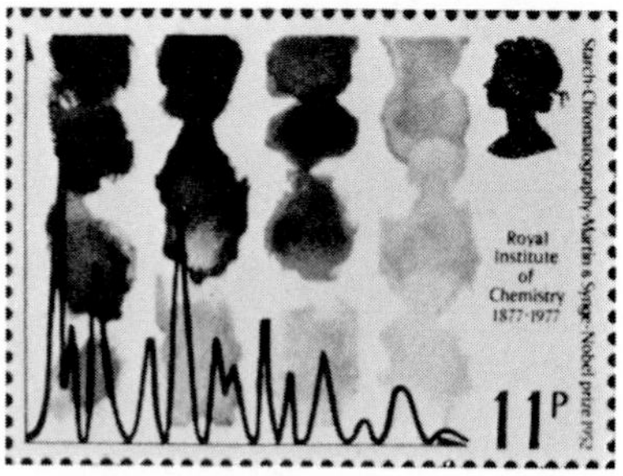

Stamp in honor of biochemists Archer J. P. Martin (1910–) and Richard L. M. Synge (1914–), who were awarded the 1952 Nobel Prize in chemistry for their contributions to the development of modern chromatography.

28C-3 Variables That Affect Column Efficiency

Band broadening, and thus loss of column efficiency, is the consequence of the finite rate at which several mass-transfer processes occur during migration of a solute down a column. Some of the variables that affect these rates are controllable and can be exploited to improve separations. Table 28-2 lists the most important of these variables.

TABLE 28-2 Variables That Affect Column Efficiency

Variable	Symbol	Usual Units
Linear velocity of mobile phase	u	$cm \cdot s^{-1}$
Diffusion coefficient in mobile phase*	D_M	$cm^2 \cdot s^{-1}$
Diffusion coefficient in stationary phase*	D_S	$cm^2 \cdot s^{-1}$
Capacity factor (Equation 28-8)	k'	unitless
Diameter of packing particle	d_p	cm
Thickness of liquid coating on stationary phase	d_f	cm

*Increases as temperature increases and viscosity decreases.

The Effect of Mobile-Phase Flow Rate

The magnitude of kinetic effects on column efficiency clearly depends upon the length of time the mobile phase is in contact with the stationary phase, which in turn depends upon the flow rate of the mobile phase. For this reason, efficiency studies have generally been carried out by determining H (using Equation 28-16) as a function of mobile-phase velocity. The data obtained from such studies are typified by the two plots shown in Figure 28-8, the one for liquid chromatography and the other for gas chromatography. While both show a minimum in H (or a maximum in efficiency) at low flow rates, the minimum for liquid chromatography usually occurs at flow rates that are well below those for gas chromatography and are often so low that they are not observed under normal operating conditions.

Generally, liquid chromatograms are obtained at lower flow rates than gas chromatograms. Furthermore, as shown in the figure, plate heights for liquid chromatographic columns are an order of magnitude or more smaller than those encountered with gas chromatographic columns. Offsetting this advantage is the fact that it is impractical to employ liquid columns that are longer than about 25 to 50 cm (because of high pressure drops), whereas gas-chromatographic columns may be 50 m or more in length. Consequently, the total number of plates is usually greater, and thus overall column efficiency usually superior, in gas-chromatographic columns.

Theory of Band Broadening

Over the last 40 years, an enormous amount of theoretical and experimental effort has been devoted to developing quantitative relationships that account for the effects of experimental variables listed in Table 28-2 on plate heights for various types of columns. Perhaps a dozen or more expressions for calculating plate height have been put forward and applied with various degrees of success. It is apparent that none of these is entirely adequate to explain the complex physical interactions and effects that lead to zone broadening and thus lower column efficiencies. Some of the equations, though imperfect, have been of considerable use, however, in pointing the way toward improved column performance. One of these is presented here.

The efficiency of most chromatographic columns can be approximated by the expression

$$H = B/u + C_S u + C_M u \qquad (28\text{-}17)$$

Theoretical studies of zone broadening in the 1950s by Dutch chemical engineers led to the *van Deemter equation,* which can be written in the form

$$H = A + B/u + Cu$$

where the constants A, B, and C are coefficients of eddy diffusion, longitudinal diffusion, and mass transfer, respectively. This equation is of considerable historic interest; its modernized version takes the form of Equation 28-17.

where H is the plate height in centimeters and u is the linear velocity of the mobile phase in centimeters per second.[4] The quantity B is the *longitudinal diffusion coefficient,* while C_S and C_M are *mass-transfer coefficients* for the stationary and mobile phases, respectively.

The Longitudinal Diffusion Term B/u Diffusion is a process by which species migrate from a more concentrated part of a medium to a more dilute. The rate of migration is proportional to the concentration difference between the regions as well as to the *diffusion coefficient* D_M of the species. The latter, which is a measure of the mobility of a substance in a given medium, is a constant equal to the velocity of migration under a unit concentration gradient.

See *Mathcad Applications for Analytical Chemistry,* **pp. 145–147.**

In chromatography, longitudinal diffusion results in the migration of a solute from the concentrated center of a band to the more dilute regions on either side (that is, toward and opposed to the direction of flow). Longitudinal diffusion is a common source of band broadening in gas chromatography where the rate at which molecules diffuse is high. The phenomenon is of little significance in liquid chromatography. The magnitude of the B term in Equation 28-17 is largely determined by the diffusion coefficient D_M of the analyte in the mobile phase and is directly proportional to this constant.

As shown by Equation 28-17, the contribution of longitudinal diffusion to plate height is inversely proportional to the linear velocity of the eluent. Such a relationship is not surprising inasmuch as the analyte is in the column for a briefer period when the flow rate is high. Thus, diffusion from the center of the band to the two edges has less time to occur.

The initial decreases in H shown in both curves in Figure 28-8 are a direct consequence of the longitudinal diffusion. Note that the effect is much less pronounced in liquid chromatography because of the much lower diffusion rates in a liquid mobile phase. The striking difference in plate heights shown by the two curves in Figure 28-8 can also be explained by considering the relative rates of longitudinal diffusion in the two mobile phases. That is, diffusion coefficients in gaseous media are orders of magnitude larger than in liquids. Thus, band broadening goes on to a much greater extent in gas chromatography than in liquid chromatography.

The Stationary Phase Mass-Transfer Coefficient $C_S u$ When the stationary phase is an immobilized liquid, the mass-transfer coefficient is directly proportional to the square of the thickness of the film on the support particles d_f^2 and inversely proportional to the diffusion coefficient D_S of the solute in the film. These effects can be understood by realizing that both reduce the average frequency at which analyte molecules reach the interface, where transfer to the mobile phase can occur. That is, with thick films, molecules must on the average travel farther to reach the surface, and with smaller diffusion coefficients, they

[4] S. J. Hawkes, *J. Chem. Educ.*, **1983,** *60*, 393.

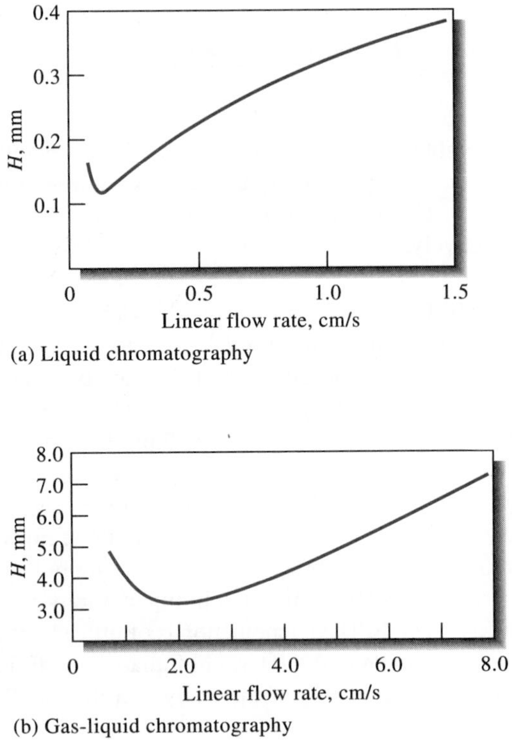

Figure 28-8
Effect of mobile phase flow rate on plate height for (a) liquid chromatography and (b) gas chromatography.

travel slower. The consequence is a slower rate of mass transfer and an increase in plate height.

When the stationary phase is a solid surface, the mass-transfer coefficient C_S is directly proportional to the time required for a species to be adsorbed or desorbed, which in turn is inversely proportional to the first-order rate constant for the processes.

The Mobile Phase Mass-Transfer Term $C_M u$ The mass-transfer processes that occur in the mobile phase are sufficiently complex to defy complete and rigorous analysis, at least to date. On the other hand, a good qualitative understanding of the variables affecting zone broadening from this cause exists, and this understanding has led to vast improvements in all types of chromatographic columns.

The mobile phase mass-transfer coefficient C_M is known to be inversely proportional to the diffusion coefficient of the analyte in the mobile phase D_M and also to be some function of the square of the particle diameter of the packing d_p^2, the square of the column diameter d_c^2, and the flow rate.

The contribution of mobile phase mass transfer to plate height is the product of the mass-transfer coefficient C_M (which is a function of solvent velocity) as well as the velocity of the solvent itself. Thus, the net contribution of $C_M u$ to plate height is not linear in u (see the curve labeled $C_M u$ in Figure 28-10) but bears a complex dependency of solvent velocity.

Zone broadening in the mobile phase is due in part to the multitude of pathways by which a molecule (or ion) can find its way through a packed column. As shown in Figure 28-9, the length of these pathways may differ significantly; thus, the residence time in the column for molecules of the same species is also variable. Solute molecules then reach the end of the column over a time interval, which leads to a broadened band. This effect, which is sometimes called *eddy diffusion,* would be independent of solvent velocity if it were not partially offset by ordinary diffusion, which results in molecules being transferred from a stream following one pathway to a stream following another. If the velocity of flow is very low, a large number of these transfers will occur, and each molecule in its movement down the column will sample numerous flow paths, spending a brief time in each. The rate at which each molecule moves down the column thus tends to approach that of the average. At low mobile phase velocities then, the molecules are not significantly dispersed by the multiple path nature of the packing. At moderate or high velocities, however, sufficient time is not available for diffusion averaging to occur, and band broadening due to the different path lengths is observed. At sufficiently high velocities, the effect of eddy diffusion becomes independent of flow rate.

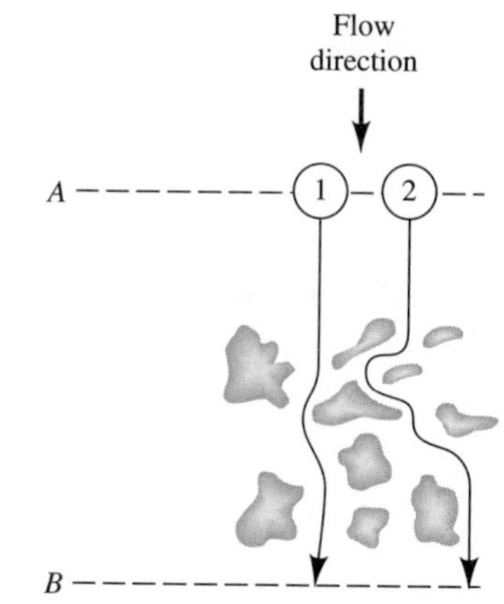

Figure 28-9
Typical pathways of two molecules during elution. Note that the distance traveled by molecule 2 is greater than that traveled by molecule 1. Thus, molecule 2 will arrive at B later than molecule 1.

Pathways for the mobile phase through the column are numerous and have different lengths.

Superimposed upon the eddy diffusion effect is one that arises from stagnant pools of the mobile phase retained in the stationary phase. Thus, when a solid serves as the stationary phase, its pores are filled with *static* volumes of mobile phase. Solute molecules must then diffuse through these stagnant pools before transfer can occur between the *moving* mobile phase and the stationary phase. This situation applies not only to solid stationary phases but also to liquid stationary phases immobilized on porous solids because the immobilized liquid does not usually fully fill the pores.

Static pools of solvent contribute to increases in H.

The presence of stagnant pools of mobile phase slows the exchange process and results in a contribution to the plate height that is directly proportional to the mobile phase velocity and inversely proportional to the diffusion coefficient for the solute in the mobile phase. An increase in internal volume accompanies increases in particle size.

Effect of Mobile Phase Velocity on Terms in Equation 28-17 Figure 28-10 shows the variation of the three terms in Equation 28-17 as a function of mobile phase velocity. The top curve is the summation of these various effects. Note that an optimum flow rate exists at which the plate height is a minimum and the separation efficiency is a maximum.

Summary of Methods for Reducing Band Broadening Two important controllable variables that affect column efficiency are the diameters of the particles

Band broadening is minimized by small packing diameter and small column diameter.

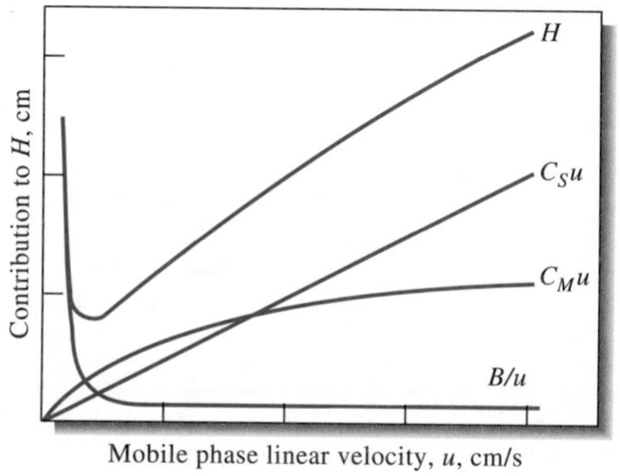

Figure 28-10
Contribution of various mass-transfer coefficients to column plate height. C_Su arises from the rate of mass transfer to and from the stationary phase, C_Mu comes from a limitation in the rate of mass transfer in the mobile phase, and B/u is associated with longitudinal diffusion.

making up the packing and the diameter of the column itself. The effect of particle diameter is demonstrated by the data shown in Figure 28-11. To take advantage of the effect of column diameter, narrower and narrower columns have been used in recent years.

The diffusion coefficient D_M has a greater effect on gas chromatography than on liquid chromatography.

With gaseous mobile phases, the rate of longitudinal diffusion can be reduced appreciably by lowering the temperature and thus the diffusion coefficient D_M. The consequence is significantly smaller plate heights at lower temperatures. This effect is usually not noticeable in liquid chromatography because diffusion is slow enough that the longitudinal diffusion term has little effect on overall plate height.

With liquid stationary phases, the thickness of the layer of adsorbed liquid should be minimized because C_S in Equation 28-17 is proportional to the square of this variable.

28D COLUMN RESOLUTION

The *resolution* R_s of a column provides a quantitative measure of its ability to separate two analytes. The significance of this term is illustrated in Figure 28-12, which consists of chromatograms for species A and B on three columns with different resolving powers. The resolution of each column is defined as

See *Mathcad Applications for Analytical Chemistry,* **p. 131.**

$$R_s = \frac{2\,\Delta Z}{W_A + W_B} = \frac{2[(t_R)_B - (t_R)_A]}{W_A + W_B} \tag{28-18}$$

where all of the terms on the right side are as defined in the figure.

The **resolution** of a chromatographic column is a quantitative measure of its ability to separate analytes A and B.

It is evident from Figure 28-12 that a resolution of 1.5 gives an essentially complete separation of A and B, whereas a resolution of 0.75 does not. At a resolution of 1.0, zone A contains about 4% B and zone B contains about 4% A. At a resolution of 1.5, the overlap is about 0.3%. The resolution for a given stationary phase can be improved by lengthening the column, thus increasing the number of plates. An adverse consequence of the added plates, however, is an increase in the time required for the resolution.

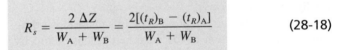

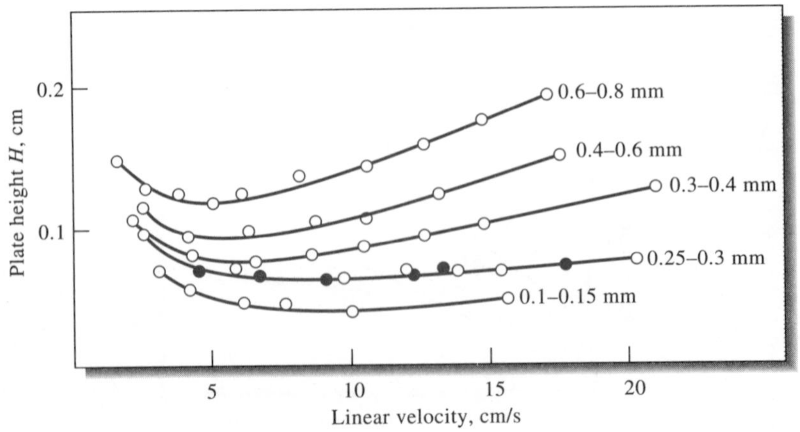

Figure 28-11
Effect of particle size on plate height. The numbers to the right are particle diameters. (From J. Boheman and J. H. Purnell, in *Gas Chromatography 1958,* D. H. Desty, Ed. New York: Academic Press, 1958. With permission of Butterworths, Stoneham, MA.)

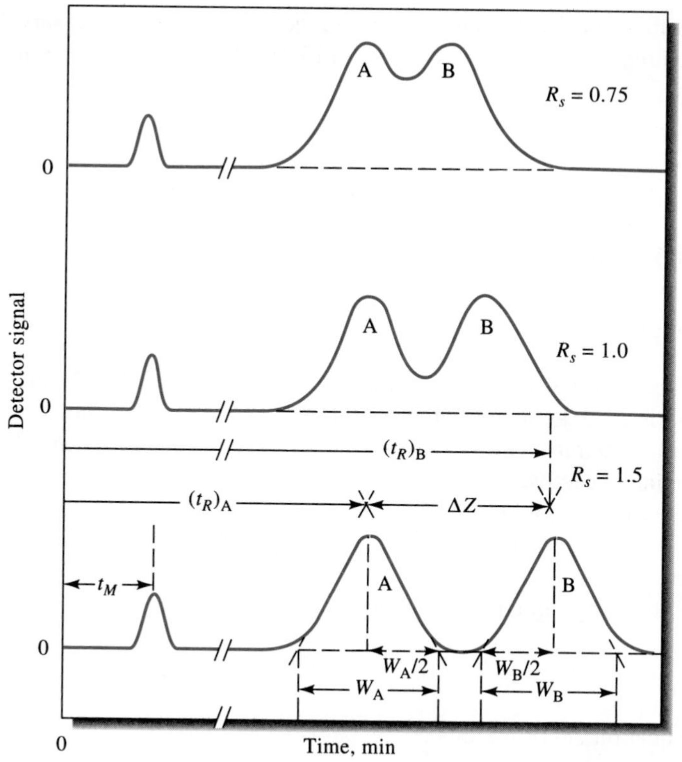

Figure 28-12
Separation at three resolutions: $R_s = 2\,\Delta Z/(W_A + W_B)$.

28D-1 Effect of Capacity and Selectivity Factors on Resolution

A useful equation is readily derived that relates the resolution of a column to the number of plates it contains as well as to the capacity and selectivity factors of a pair of solutes on the column. Thus, it can be shown[5] that for the two solutes A and B in Figure 28-12, the resolution is given by the equation

$$R_s = \frac{\sqrt{N}}{4}\left(\frac{\alpha - 1}{\alpha}\right)\left(\frac{k'_B}{1 + k'_B}\right) \tag{28-19}$$

where k'_B is the capacity factor of the slower-moving species and α is the selectivity factor. This equation can be rearranged to give the number of plates needed to realize a given resolution:

$$N = 16R_s^2\left(\frac{\alpha}{\alpha - 1}\right)^2\left(\frac{1 + k'_B}{k'_B}\right)^2 \tag{28-20}$$

28D-2 Effect of Resolution on Retention Time

As mentioned earlier, the goal in chromatography is the highest possible resolution in the shortest possible elapsed time. Unfortunately, these goals tend to be

---

[5]See D. A. Skoog and J. J. Leary, *Principles of Instrumental Analysis,* 4th ed., pp. 592–593. Philadelphia: Saunders College Publishing, 1992.

incompatible, and a compromise between the two is usually necessary. The time $(t_R)_B$ required to elute the two species in Figure 28-12 with a resolution of R_s is given by

$$(t_R)_B = \frac{16R_s^2H}{u}\left(\frac{\alpha}{\alpha - 1}\right)^2 \frac{(1 + k'_B)^3}{(k'_B)^2} \tag{28-21}$$

where u is the linear velocity of the mobile phase.

EXAMPLE 28-1

Substances A and B have retention times of 16.40 and 17.63 min, respectively, on a 30.0-cm column. An unretained species passes through the column in 1.30 min. The peak widths (at base) for A and B are 1.11 and 1.21 min, respectively. Calculate (a) column resolution, (b) average number of plates in the column, (c) plate height, (d) length of column required to achieve a resolution of 1.5, and (e) time required to elute substance B on the longer column.

(a) Employing Equation 28-18, we find

$$R_s = 2(17.63 - 16.40)/(1.11 + 1.21) = 1.06$$

(b) Equation 28-14 permits computation of N:

$$N = 16\left(\frac{16.40}{1.11}\right)^2 = 3493 \qquad \text{and} \qquad N = 16\left(\frac{17.63}{1.21}\right)^2 = 3397$$

$$N_{avg} = (3493 + 3397)/2 = 3445 = 3.4 \times 10^3$$

(c) $H = L/N = 30.0/3445 = 8.7 \times 10^{-3}$ cm

(d) k' and α do not change greatly with increasing N and L. Thus, substituting N_1 and N_2 into Equation 28-19 and dividing one of the resulting equations by the other yield

$$\frac{(R_s)_1}{(R_s)_2} = \frac{\sqrt{N_1}}{\sqrt{N_2}}$$

where the subscripts 1 and 2 refer to the original and longer columns, respectively. Substituting the appropriate values for N_1, $(R_s)_1$, and $(R_s)_2$ gives

$$\frac{1.06}{1.5} = \frac{\sqrt{3445}}{\sqrt{N_2}}$$

$$N_2 = 3445\left(\frac{1.5}{1.06}\right)^2 = 6.9 \times 10^3$$

But

$$L = NH = 6.9 \times 10^3 \times 8.7 \times 10^{-3} = 60 \text{ cm}$$

(e) Substituting $(R_s)_1$ and $(R_s)_2$ into Equation 28-20 and dividing yield

$$\frac{(t_R)_1}{(t_R)_2} = \frac{(R_s)_1^2}{(R_s)_2^2} = \frac{17.63}{(t_R)_2} = \frac{(1.06)^2}{(1.5)^2}$$

$$(t_R)_2 = 35 \text{ min}$$

Thus, to obtain the improved resolution, the separation time must be doubled.

28D-3 Optimization Techniques

Equations 28-19 and 28-21 serve as guides in choosing conditions that lead to a desired degree of resolution with a minimum expenditure of time. An examination of these equations reveals that each is made up of three parts. The first describes the efficiency of the column in terms of $\sqrt{N}$ or H. The second, which is the quotient containing α, is a selectivity term that depends on the properties of the two solutes. The third component is the capacity term, which is the quotient containing k'_B; the term depends on the properties of both the solute and the column.

Variation in Plate Height

As shown by Equation 28-19, the resolution of a column improves as the square root of the number of plates it contains increases. Example 28-1e reveals, however, that increasing the number of plates is expensive in terms of time unless the increase is achieved by reducing the plate height and not by increasing column length.

Methods for minimizing plate height, which are discussed in Section 28C, include reducing the particle size of the packing, the diameter of the column, the column temperature (gas chromatography), and the thickness of the liquid film (liquid chromatography). Optimizing the flow rate of the mobile phase is also helpful.

Variation in the Capacity Factor

Often, a separation can be improved significantly by manipulation of the capacity factor k'_B. Increases in k'_B generally enhance resolution (but at the expense of elution time). To determine the optimum range of values for k'_B, it is convenient to write Equation 28-19 in the form

$$R_s = Q\frac{k'_B}{1 + k'_B}$$

and Equation 28-21 as

$$(t_R)_B = Q'\frac{(1 + k'_B)^3}{(k'_B)^2}$$

where Q and Q' contain the rest of the terms in the two equations. Figure 28-13 is a plot of R_s/Q and $(t_R)_B/Q'$ as a function of k'_B, assuming Q and Q' remain

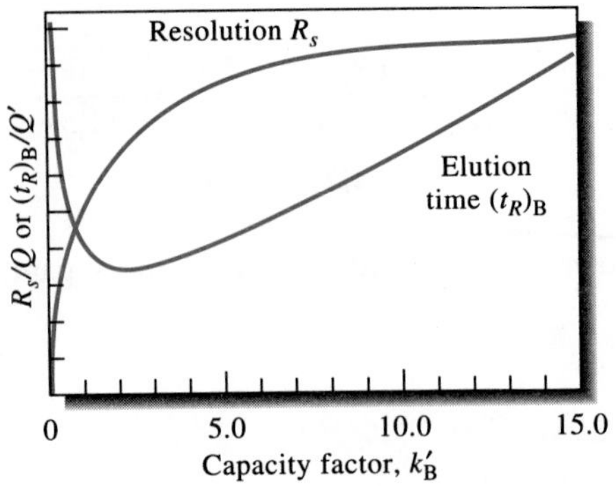

Figure 28-13
Effect of capacity factor k'_B on resolution R_s and elution time $(t_R)_B$. It is assumed that Q and Q' remain constant with variations in k'_B.

approximately constant. It is clear that values of k'_B greater than about 10 should be avoided because they provide little increase in resolution but markedly increase the time required for separations. The minimum in the elution-time curve occurs at $k'_B \approx 2$. Often, then, the optimal value of k'_B lies in the range from 1 to 5.

Usually, the easiest way to improve resolution is by optimizing k'. For gaseous mobile phases, k' can often be improved by temperature changes. For liquid mobile phases, changes in the solvent composition often permit manipulation of k' to yield better separations. An example of the dramatic effect that relatively simple solvent changes can bring about is demonstrated in Figure 28-14. Here, modest variations in the methanol/water ratio convert unsatisfactory chromatograms (a and b) to ones with well-separated peaks for each component (c and d). For most purposes, the chromatogram shown in (c) is best, since it shows adequate resolution in minimum time.

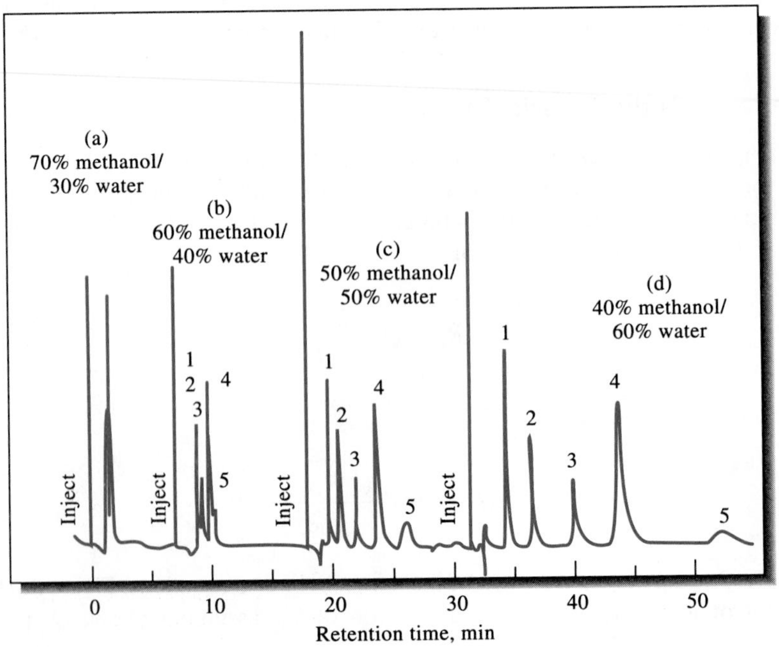

Figure 28-14
Effect of solvent variation on chromatograms. Analytes: (1) 9,10-anthraquinone; (2) 2-methyl-9,10-anthraquinone; (3) 2-ethyl-9,10-anthraquinone; (4) 1,4-dimethyl-9,10-anthraquinone; (5) 2-*t*-butyl-9,10-anthraquinone. (Courtesy of DuPont Biotechnology Systems, Wilmington, DE.)

Variation in the Selectivity Factor

Optimizing k' and increasing N are not sufficient to give a satisfactory separation of two solutes in a reasonable time when α approaches unity. Here, a means must be sought to increase α while maintaining k' in the range of 1 to 10. Several options are available; in decreasing order of their desirability based on promise and convenience, the options are (1) changing the composition of the mobile phase, (2) changing the column temperature, (3) changing the composition of the stationary phase, and (4) using special chemical effects.

An example of the use of option 1 has been reported for the separation of anisole ($C_6H_5OCH_3$) and benzene.[6] With a mobile phase that was a 50% mixture of water and methanol, k' was 4.5 for anisole and 4.7 for benzene, while α was only 1.04. Substitution of an aqueous mobile phase containing 37% tetrahydrofuran gave k' values of 3.9 and 4.7 and an α value of 1.20. Peak overlap was significant with the first solvent system and negligible with the second.

A less convenient but often highly effective method of improving α while maintaining values for k' in their optimal range is to alter the chemical composition of the stationary phase. To take advantage of this option, most laboratories that carry out chromatographic separations maintain several columns that can be interchanged with a minimum of effort.

Increases in temperature usually cause increases in k' but have little effect on α values in liquid-liquid and liquid-solid chromatography. In contrast, with ion-exchange chromatography, temperature effects can be large enough to make exploration of this option worthwhile before resorting to a change in column packing.

A final method for enhancing resolution is to incorporate into the stationary phase a species that complexes or otherwise interacts with one or more components of the sample. An example is the use of an adsorbent impregnated with silver ion to improve the separation of olefins as silver complexes.

28D-4 The General Elution Problem

Figure 28-15 shows hypothetical chromatograms for a six-component mixture made up of three pairs of components with widely different distribution coefficients and thus widely different capacity factors. In chromatogram (a), conditions have been adjusted so that the capacity factors for components 1 and 2 (k'_1 and k'_2) are in the optimal range of 1 to 5. The factors for the other components are far larger than the optimum, however. Thus, the peaks for components 5 and 6 appear only after an inordinate length of time has passed; furthermore, these peaks are so broad that they may be difficult to identify unambiguously.

As shown in chromatogram (b), changing conditions to optimize the separation of components 5 and 6 bunches the peaks for the first four components to the point where their resolution is unsatisfactory. Here, however, the total elution time is ideal.

The phenomenon illustrated in Figure 28-15 is encountered often enough to be given a name—the *general elution problem.* A common solution to this problem

---

[6]L. R. Snyder and J. J. Kirkland, *Introduction to Modern Liquid Chromatography,* 2nd ed., p. 75. New York: Wiley, 1979.

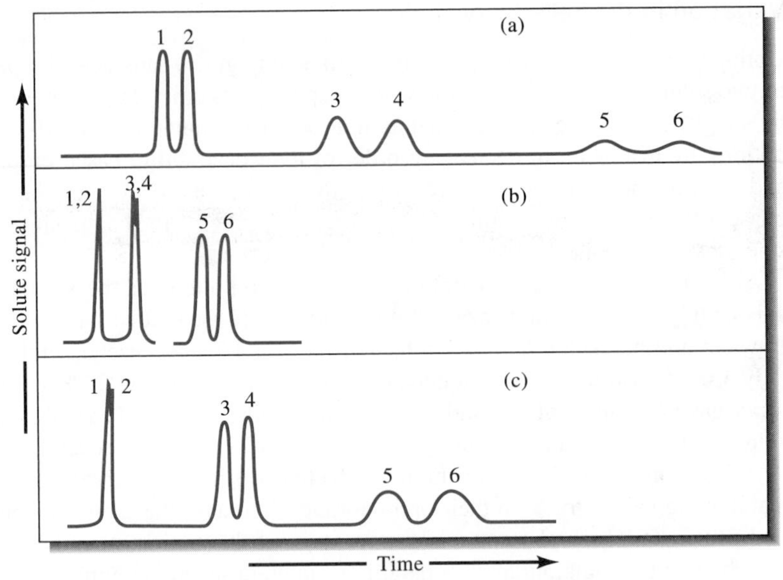

Figure 28-15
The general elution problem in chromatography.

is to change conditions that determine the values of k' as the separation proceeds. These changes can be performed in a stepwise manner or continuously. Thus, for the mixture shown in Figure 28-15, conditions at the outset could be those producing chromatogram (a). Immediately after the elution of components 1 and 2, conditions could be changed to those that are optimal for separating components 3 and 4 (as in chromatogram c). With the appearance of peaks for these components, the elution could be completed under the conditions used for producing chromatogram (b). Often such a procedure leads to satisfactory separation of all the components of a mixture in minimal time.

For liquid chromatography, variations in k' are brought about by varying the composition of the mobile phase during elution (*gradient elution* or *solvent programming*). For gas chromatography, temperature increases *(temperature programming)* achieve optimal conditions for separations.

28E APPLICATIONS OF CHROMATOGRAPHY

Chromatography is a powerful and versatile tool for separating closely related chemical species. In addition, it can be employed for the qualitative identification and quantitative determination of separated species.

28E-1 Qualitative Analysis

Chromatography is widely used for recognizing the presence or absence of components in mixtures that contain a limited number of species whose identities are known. For example, 30 or more amino acids in a protein hydrolysate can be detected with a reasonable degree of certainty with a chromatogram. On the other hand, because a chromatogram provides but a single piece of information about each species in a mixture (the retention time), the application of the technique to the qualitative analysis of complex samples of unknown composition is limited.

This limitation has been largely overcome by linking chromatographic columns directly with ultraviolet, infrared, and mass spectrometers. The resulting *hyphenated* instruments are powerful tools for identifying the components of complex mixtures.

It is important to note that while a chromatogram may not lead to positive identification of the species in a sample, it often provides sure evidence of the *absence* of species. Thus, failure of a sample to produce a peak at the same retention time as a standard obtained under identical conditions is strong evidence that the compound in question is absent (or present at a concentration below the detection limit of the procedure).

A **hyphenated** technique is an analytical method in which two instrumental techniques are coupled to produce a more powerful analytical procedure. Examples include a gas chromatography/mass spectrometry detector, a liquid chromatography/voltammetry detector, and an inductively coupled plasma source/mass spectrometry detector.

28E-2 Quantitative Analysis

Chromatography owes its enormous growth in part to its speed, simplicity, relatively low cost, and wide applicability as a tool for separations. It is doubtful, however, that its use would have become so widespread had it not been for the fact that it can provide quantitative information about separated species as well.

Quantitative chromatography is based upon a comparison of either the height or the area of an analyte peak with that of one or more standards. If conditions are properly controlled, both of these parameters vary linearly with concentration.

Analyses Based on Peak Height

The height of a chromatographic peak is obtained by connecting the baselines on the two sides of the peak by a straight line and measuring the perpendicular distance from this line to the peak. This measurement can ordinarily be made with reasonably high precision and yields accurate results, provided variations in column conditions do not alter peak width during the period required to obtain chromatograms for sample and standards. The variables that must be controlled closely are column temperature, eluent flow rate, and rate of sample injection. In addition, care must be taken to avoid overloading the column. The effect of sample-injection rate is particularly critical for the early peaks of a chromatogram. Relative errors of 5% to 10% due to this cause are not unusual with syringe injection.

Analyses Based on Peak Area

Peak area is independent of broadening effects caused by the variables mentioned in the previous paragraph. From this standpoint, therefore, area is a more satisfactory analytical parameter than peak height. On the other hand, peak heights are more easily measured and, for narrow peaks, more accurately determined.

Most modern chromatographic instruments are equipped with electronic integrators that provide precise measurements of relative peak areas. If such equipment is not available, a manual estimate must be made. A simple method that works well for symmetric peaks of reasonable widths is to multiply peak height by the width at one-half peak height.

Calibration with Standards

The most straightforward method for quantitative chromatographic analyses involves the preparation of a series of standard solutions that approximate the

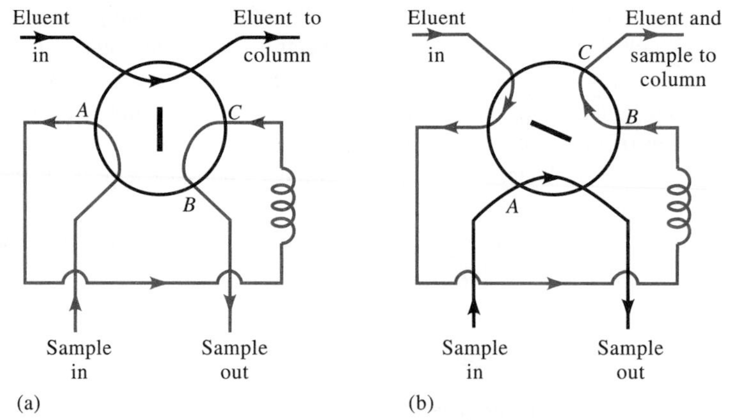

Figure 28-16

A rotary sample valve: (a) valve position for filling sample loop *ACB*; (b) valve position for introduction of sample into column.

composition of the unknown. Chromatograms for the standards are then obtained, and peak heights or areas are plotted as a function of concentration. A plot of the data should yield a straight line passing through the origin; analyses are based upon this plot. Frequent standardization is necessary for highest accuracy.

The most important source of error in analyses by the method based on calibration standards is usually the uncertainty in the volume of sample; occasionally, the rate of injection is also a factor. Samples are ordinarily small ($\approx 1\ \mu L$), and the uncertainties associated with the injection of a reproducible volume of this size with a microsyringe can amount to several percent relative. The situation is even worse in gas-liquid chromatography, where the sample must be injected into a heated sample port. Here, evaporation from the needle tip may lead to large variations in the volume injected.

Errors in sample volume can be reduced to perhaps 1% to 2% relative by means of a rotary sample valve such as that shown in Figure 28-16. The sample loop *ACB* in (a) is filled with liquid; rotation of the valve by 45 deg then introduces a reproducible volume of sample (the volume originally contained in *ACB*) into the mobile phase stream.

The Internal-Standard Method

The highest precision for quantitative chromatography is obtained by using internal standards because the uncertainties introduced by sample injection, flow rate, and variations in column conditions are minimized. In this procedure, a carefully measured quantity of an internal standard is introduced into each standard and sample, and the ratio of analyte peak area (or height) to internal-standard peak area (or height) is the analytical parameter. For this method to be successful, it is necessary that the internal-standard peak be well separated from the peaks of all other components in the sample, but it must appear close to the analyte peak. With a suitable internal standard, precisions of 0.5% to 1% relative are reported.

28F QUESTIONS AND PROBLEMS

28-1. Define
- *(a) elution.
- (b) mobile phase.
- *(c) stationary phase.
- (d) partition ratio.
- *(e) retention time.
- (f) capacity factor.
- *(g) selectivity factor.
- (h) plate height.

28-2. List the variables that lead to zone broadening.

*28-3. What is the difference between gas-liquid and liquid-liquid chromatography?

28-4. What is the difference between liquid-liquid and liquid-solid chromatography?

*28-5. Describe a method for determining the number of plates in a column.

28-6. Name two general methods for improving the resolution of two substances on a chromatographic column.

*28-7. The following data are for a liquid chromatographic column:

Length of Packing	24.7 cm
Flow rate	0.313 mL/min
V_M	1.37 mL
V_S	0.164 mL

A chromatogram of a mixture of species A, B, C, and D provided the following data:

	Retention Time, min	Width of Peak Base (W), min
Nonretained	3.1	—
A	5.4	0.41
B	13.3	1.07
C	14.1	1.16
D	21.6	1.72

Calculate
(a) the number of plates from each peak.
(b) the mean and the standard deviation for N.
(c) the plate height for the column.

*28-8. From the data in Problem 28-7, calculate for A, B, C, and D
(a) the capacity factor.
(b) the partition coefficient.

*28-9. From the data in Problem 28-7, calculate for species B and C
(a) the resolution.
(b) the selectivity factor.
(c) the length of column necessary to separate the two species with a resolution of 1.5.
(d) the time required to separate the two species on the column in part (c).

*28-10. From the data in Problem 28-7, calculate for species C and D
(a) the resolution.
(b) the length of column necessary to separate the two species with a resolution of 1.5.

28-11. The following data were obtained by gas-liquid chromatography on a 40-cm packed column:

Compound	t_R, min	$W_{1/2}$, min
Air	1.9	—
Methylcyclohexane	10.0	0.76
Methylcyclohexene	10.9	0.82
Toluene	13.4	1.06

Calculate
(a) an average number of plates from the data.
(b) the standard deviation for the average in (a).
(c) an average plate height for the column.

28-12. Referring to Problem 28-11, calculate the resolution for
(a) methylcyclohexene and methylcyclohexane.
(b) methylcyclohexene and toluene.
(c) methylcyclohexane and toluene.

28-13. If a resolution of 1.5 is desired in separating methylcyclohexane and methylcyclohexene in Problem 28-11,
(a) how many plates are required?
(b) how long must the column be if the same packing is employed?
(c) what is the retention time for methylcyclohexene on the column in Problem 28-11b?

*28-14. If V_S and V_M for the column in Problem 28-11 are 19.6 and 62.6 mL, respectively, and a nonretained peak appears after 1.9 min, calculate the
(a) capacity factor for each compound.
(b) partition coefficient for each compound.
(c) selectivity factor for methylcyclohexane and methylcyclohexene.

*28-15. From distribution studies, species M and N are known to have water/hexane partition coefficients of 6.01 and 6.20 ($K = [M]_{H_2O}/[M]_{hex}$). The two species are to be separated by elution with hexane in a column packed with silica gel containing adsorbed water. The ratio V_S/V_M for the packing is 0.422.
(a) Calculate the capacity factor for each solute.
(b) Calculate the selectivity factor.
(c) How many plates are needed to provide a resolution of 1.5?
(d) How long a column is needed if the plate height of the packing is 2.2×10^{-3} cm?
(e) If a flow rate of 7.10 cm/min is employed, how long will it take to elute the two species?

28-16. Repeat the calculations in Problem 28-15 assuming $K_M = 5.81$ and $K_N = 6.20$.

CHAPTER

29

Gas Chromatography

In gas chromatography, the sample is vaporized and injected onto the head of a chromatographic column. Elution is brought about by the flow of an inert gaseous mobile phase. In contrast to most other types of chromatography, the mobile phase does not interact with molecules of the analyte; its only function is to transport the analyte through the column. Two types of gas chromatography are encountered: *gas-solid chromatography* (GSC) and *gas-liquid chromatography* (GLC). Gas-liquid chromatography finds widespread use in all fields of science, where its name is usually shortened to *gas chromatography* (GC).

> In **gas-liquid chromatography,** the mobile phase is a gas and the stationary phase is a liquid that is retained on the surface of an inert solid by adsorption or chemical bonding.

> In **gas-solid chromatography,** the mobile phase is a gas and the stationary phase is a solid that retains the analytes by physical adsorption.

Gas-solid chromatography permits the separation and determination of low-molecular-mass gases, such as air components, hydrogen sulfide, carbon monoxide, and nitrogen oxides.

In all types of chromatography, liquid stationary phases are immobilized on solid surfaces by adsorption or by chemical bonding.

Gas-solid chromatography is based upon a solid stationary phase in which retention of analytes is the consequence of physical adsorption. Gas-solid chromatography has limited application owing to semipermanent retention of active or polar molecules and severe tailing of elution peaks (a consequence of the nonlinear character of the adsorption process). Thus, this technique has not found wide application except for the separation of certain low-molecular-weight gaseous species; we discuss the method briefly in Section 29D.

Gas-liquid chromatography is based upon the partition of the analyte between a gaseous mobile phase and a liquid phase immobilized on the surface of an inert solid. The concept of *gas-liquid chromatography* was first enunciated in 1941 by Martin and Synge, who were also responsible for the development of liquid-liquid partition chromatography. More than a decade was to elapse, however, before the value of gas-liquid chromatography was demonstrated experimentally.[1] Three years later, in 1955, the first commercial apparatus for gas-liquid chromatography appeared on the market. Since that time, the growth in applications of this technique has been phenomenal.[2] It has been estimated that as many as 200,000 gas chromatographs are currently in use throughout the world.[3]

---

[1] A. J. Jones and A. J. P. Martin, *Analyst,* **1952,** *77,* 915.

[2] For monographs on GLC, see J. Willet, *Gas Chromatography.* New York: Wiley, 1987; W. Jennings, *Analytical Gas Chromatography.* Orlando, FL: Academic Press, 1987; *Modern Practice of Gas Chromatography,* 2nd ed., R. L. Grob, Ed. New York: Wiley-Interscience, 1985; M. L. Lee, F. Yang, and K. Bartle, *Open Tubular Gas Chromatography: Theory and Practice.* New York: Wiley, 1984.

[3] R. Schill and R. R. Freeman, in *Modern Practice of Gas Chromatography,* 2nd ed., R. L. Grob, Ed., p. 294. New York: Wiley, 1985.

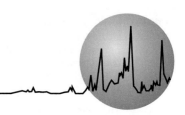

29A INSTRUMENTS FOR GAS-LIQUID CHROMATOGRAPHY

Today, well over 30 instrument manufacturers offer some 130 different models of gas-chromatographic equipment at costs that vary from perhaps $1500 to $40,000. The last two decades have been marked by many changes and improvements to gas chromatographic instruments. In the 1970s, electronic integrators and computer-based data processing equipment became common. The 1980s saw computers being used for automatic control of most instrument parameters, such as column temperature, flow rates, and sample injection; development of very high-performance instruments at moderate costs; and perhaps most important, the development of open tubular columns that are capable of separating a multitude of analytes in relatively short times.

The basic components of an instrument for gas chromatography are illustrated in Figure 29-1. A description of each component follows.

29A-1 Carrier Gas Supply

Carrier gases, which must be chemically inert, include helium, argon, nitrogen, carbon dioxide, and hydrogen. As will be shown later, the choice of gases is often dictated by the type of detector used. Associated with the gas supply are pressure regulators, gauges, and flow meters. In addition, the carrier gas system often contains a molecular sieve to remove water or other impurities.

Helium is the most common mobile phase in gas-liquid chromatography.

Flow rates are normally controlled by a two-stage pressure regulator at the gas cylinder and some sort of pressure regulator or flow regulator mounted in the chromatograph. Inlet pressures usually range from 10 to 50 psi (above room

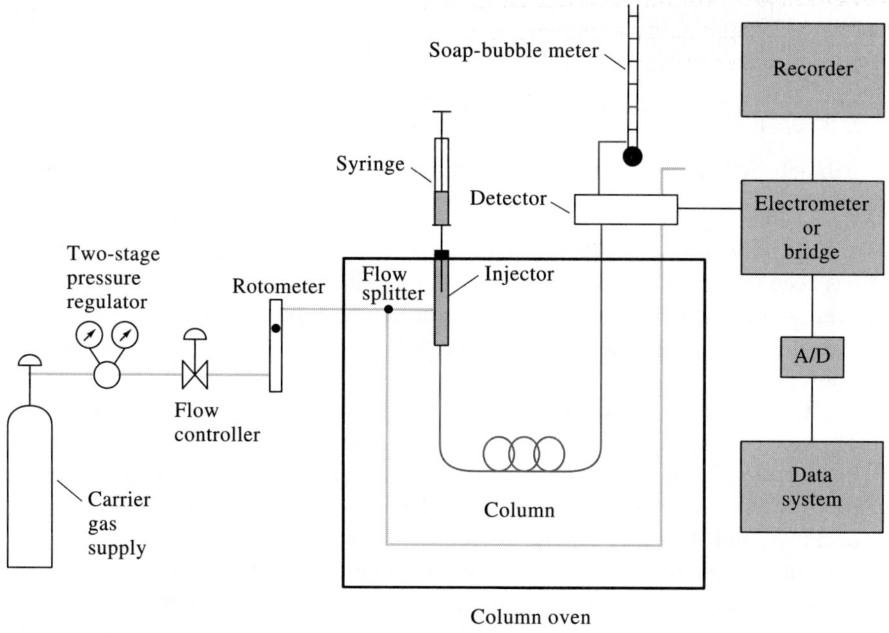

Figure 29-1
Schematic of a gas chromatograph.

pressure), which lead to flow rates of 29 to 150 mL/min with packed columns and 1 to 25 mL/min for open tubular capillary columns. Generally, it is assumed that flow rates will be constant if the inlet pressure remains constant. Flow rates can be established by a rotometer at the column head, but this device is not as accurate as a simple soap-bubble meter, which, as shown in Figure 29-1, is located at the end of the column. A soap film is formed in the path of the gas when a rubber bulb containing an aqueous solution of soap or detergent is squeezed; the time required for this film to move between two graduations on the buret is measured and converted to volumetric flow rate (see Figure 29-2).

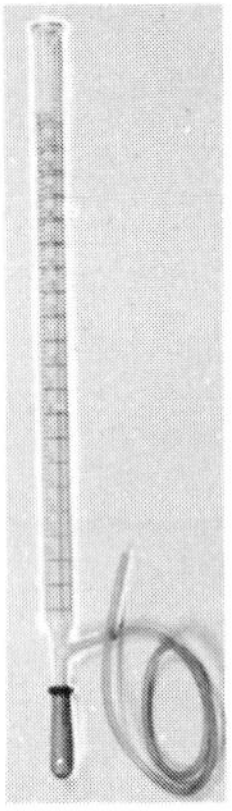

Figure 29-2
A soap-bubble flow meter. (Courtesy of Chrompack, Inc., Raritan, NJ.)

29A-2 Sample Injection System

Column efficiency requires that the sample be of suitable size and be introduced as a ''plug'' of vapor; slow injection of oversized samples causes band spreading and poor resolution. The most common method of sample injection involves the use of a microsyringe (Figure 29-3) to inject a liquid or gaseous sample through a silicone-rubber diaphragm or septum into a flash vaporizer port located at the head of the column (the sample port is ordinarily about 50°C above the boiling point of the least volatile component in the sample). Figure 29-4 is a schematic of a typical injection port. For ordinary analytical columns, sample sizes vary from a few tenths of a microliter to 20 μL. Capillary columns require much smaller samples ($\sim 10^{-3}$ μL); here, a sample splitter system is employed to deliver only a small fraction of the injected sample to the column head, with the remainder going to waste.

For quantitative work, more reproducible sample sizes for both liquids and gases are obtained by means of a sample valve such as that shown in Figure 28-16 (page 684). With such devices, sample sizes can be reproduced to better than 0.5% relative. Solid samples are introduced as solutions or, alternatively, are sealed into thin-walled vials that can be inserted at the head of the column and punctured or crushed from the outside.

29A-3 Column Configurations and Column Ovens

Two general types of columns are encountered in gas chromatography, *packed* and *open tubular,* or *capillary.* To date, the vast majority of gas chromatography has been carried out on packed columns. This situation is changing rapidly, however, and it seems probable that in the near future, packed columns will be replaced by the more efficient and faster open tubular columns for most applications.

Chromatographic columns vary in length from less than 2 m to 50 m or more. They are constructed of stainless steel, glass, fused silica, or Teflon. In order to fit into an oven for thermostating, they are usually formed as coils having diameters of 10 to 30 cm (Figure 29-5). A more detailed discussion of columns, column packings, and stationary phases is found in Section 29B.

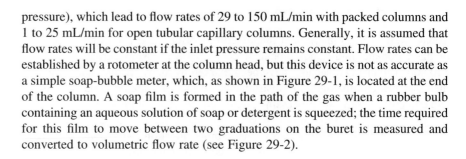

Figure 29-3
A set of microsyringes for sample injection. (Courtesy of Chrompack, Inc., Raritan, NJ.)

Column temperature is an important variable that must be controlled to a few tenths of a degree for precise work. Thus, the column is ordinarily housed in a thermostated oven. The optimum column temperature depends upon the boiling point of the sample and the degree of separation required. Roughly, a temperature equal to or slightly above the average boiling point of a sample results in a reasonable elution time (2 to 30 min). For samples with a broad boiling range, it is often desirable to employ *temperature programming* whereby the column temperature is increased either continuously or in steps as the separation proceeds. Figure 29-6 shows the improvement in a chromatogram brought about by temperature programming.

In general, optimum resolution is associated with minimum temperature; the cost of lowered temperature, however, is an increase in elution time and therefore the time required to complete an analysis. Figures 29-6a and 29-6b illustrate this principle.

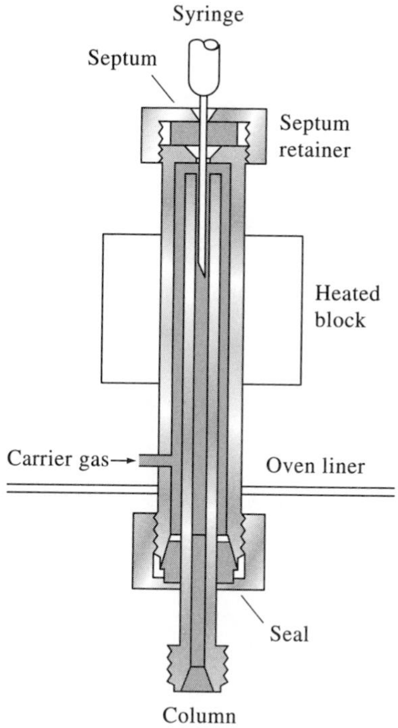

Figure 29-4
A heated sample port. (From H. H. Willard, L. L. Merritt, J. A. Dean, and F. A. Settle, *Instrumental Methods of Analysis,* 7th ed., p. 542. Belmont, CA: Wadsworth, 1988. With permission.)

Temperature programming in gas chromatography involves increasing the column temperature in a regular programmed way.

29A-4 Detectors

Dozens of detectors have been investigated and used during the development of gas chromatography. In the sections that follow immediately, we describe the most widely used of these.

Characteristics of the Ideal Detector

The ideal detector for gas chromatography has the following characteristics: (1) adequate sensitivity (in general, the sensitivities of present-day detectors lie in the analyte range of 10^{-8} to 10^{-15} g/s); (2) good stability and reproducibility; (3) a linear response to solutes that extends over several orders of magnitude; (4) a temperature range from room temperature to at least 400°C; (5) a quick response time that is independent of flow rate; (6) high reliability and ease of use (the detector should, to the extent possible, be foolproof in the hands of inexperienced operators); (7) similarity in response toward all solutes or alternatively a highly predictable and selective response toward one or more classes of solutes; and (8) nondestructive of sample.

Needless to say, no detector exhibits all of these characteristics, and it seems unlikely that such a detector will ever be developed.

Flame Ionization Detectors (FIDs)

The flame ionization detector is the most widely used and generally applicable detector for gas chromatography. With a burner such as that shown in Figure 29-7, the effluent from the column is mixed with hydrogen and air and then ignited electrically. Most organic compounds, when pyrolyzed at the temperature of a hydrogen/air flame, produce ions and electrons that can conduct electricity through the flame. A potential of a few hundred volts is applied across the burner tip and a collector electrode located above the flame. The resulting current ($\sim 10^{-12}$ A) is then directed into a high-impedance operational amplifier for measurement.

The ionization of carbon compounds in a flame is a poorly understood process, although it is observed that the number of ions produced is roughly proportional to the number of *reduced* carbon atoms in the flame. Because the flame ionization detector responds to the number of carbon atoms entering the detector per unit of time, it is a *mass-sensitive* rather than a concentration-sensitive device. As a consequence, this detector has the advantage that changes in flow rate of the mobile phase have little effect on detector response.

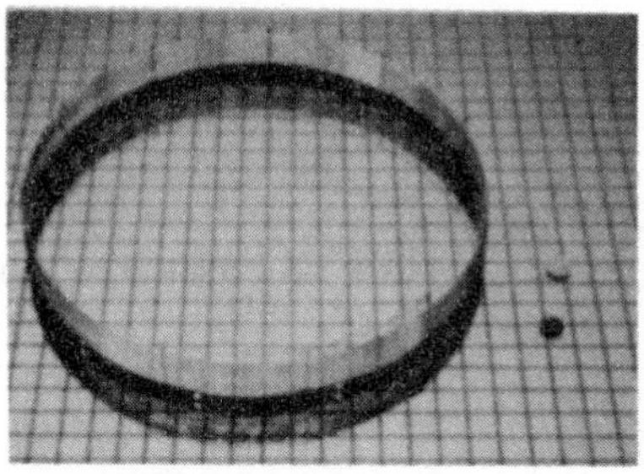

Figure 29-5
A 25-m fused silica capillary column. (Courtesy of Chrompack, Inc., Raritan, NJ.)

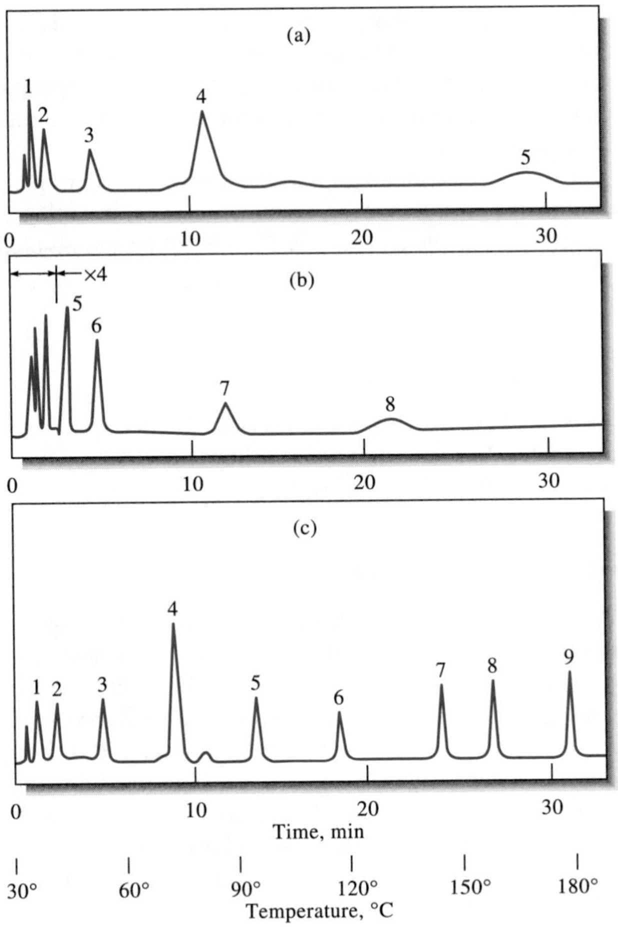

Figure 29-6
Effect of temperature on gas chromatograms. (a) Isothermal at 45°C; (b) isothermal at 145°C; (c) programmed at 30° to 180°C. (From W. E. Harris and H. W. Habgood, *Programmed Temperature Gas Chromatography,* p. 10. New York: Wiley, 1966. Reprinted with permission.)

The high sensitivity of the flame ionization detector comes at the expense of the sample, which is destroyed.

Functional groups, such as carbonyl, alcohol, halogen, and amine, yield fewer ions or none at all in a flame. In addition, the detector is insensitive toward noncombustible gases such as H_2O, CO_2, SO_2, and NO_x. These properties make the flame ionization detector a most useful general detector for the analysis of most organic samples including those that are contaminated with water and the oxides of nitrogen and sulfur.

The flame ionization detector exhibits a high sensitivity ($\sim 10^{-13}$ g/s), a large linear response range ($\sim 10^7$), and low noise. It is generally rugged and easy to use. A disadvantage is that it destroys the sample.

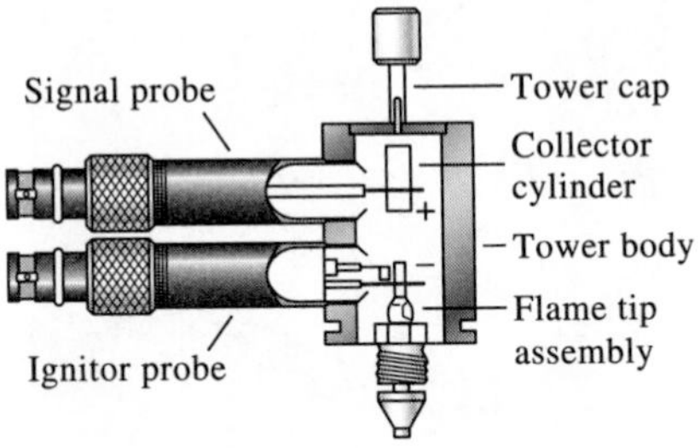

Figure 29-7
A flame ionization detector. (Courtesy of Varian Instrument Division, Palo Alto, CA.)

Thermal Conductivity Detectors (TCDs)

The thermal conductivity detector, or *katharometer,* was one of the earliest detectors employed in gas-chromatographic studies. This device, which still finds widespread use, is based upon changes in the thermal conductivity of the gas stream brought about by the presence of analyte molecules. The sensing element in the thermal conductivity detector is an electrically heated element whose temperature at constant electrical power depends upon the thermal conductivity of the surrounding gas. The heated element may be a fine platinum, gold, or tungsten wire, or a semiconducting thermistor. The resistance of the wire or thermistor is a measure of its temperature, which depends in part upon the rate at which the surrounding gas molecules conduct energy away from the detector element to the walls of a metal block in which it is housed. Figure 29-8 is a schematic drawing of a typical commercial thermal conductivity detector.

A **katharometer** is a thermal detector that depends upon a change in thermal conductivity of the mobile phase in the presence of analytes.

In chromatographic applications, a double detector system is usually employed, with one element being located in the gas stream *ahead* of the sample-injection chamber and the other immediately beyond the column. Alternatively, the gas stream may be split, as shown in Figure 29-1. In either case, the thermal conductivity of the carrier gas is canceled, and the effects of variation in flow rate, pressure, and electrical power are minimized. The resistances of the twin detectors are usually compared by incorporating them into two arms of a simple Wheatstone bridge circuit.

A twin detector system compensates for the thermal conductivity of the carrier gas as well as other column variables.

The thermal conductivities of hydrogen and helium are roughly six to ten times greater than those of most organic compounds. Thus, the presence of even small amounts of organic materials causes a relatively large decrease in the thermal conductivity of the column effluent; consequently, the detector undergoes a marked rise in temperature. The conductivities of other carrier gases more closely resemble those of organic constituents; therefore, a thermal conductivity detector dictates the use of hydrogen or helium.

The advantages of a thermal conductivity detector are its simplicity, its large linear dynamic range ($\sim 10^5$), its general response to both organic and inorganic species, and its nondestructive character, which permits the collection of solutes after detection. A limitation of the thermal conductivity detector to its relatively low sensitivity ($\sim 10^{-8}$ g solute/mL of carrier gas). Other detectors exceed this sensitivity by factors as large as 10^4 to 10^7. It should be noted that the low sensitivity of thermal detectors precludes their use in conjunction with capillary columns because of the very small samples that can be accommodated by such columns.

Thermionic Detectors (TIDs)

The *thermionic detector* is selective toward organic compounds containing phosphorus and nitrogen. Its response to a phosphorus atom is approximately 10 times greater than to a nitrogen atom and 10^4 to 10^6 larger than to a carbon atom. Compared with the flame ionization detector, the thermionic detector is approximately 500 times more sensitive for phosphorus containing compounds and 50 times more sensitive for nitrogen bearing species. These properties make ther-

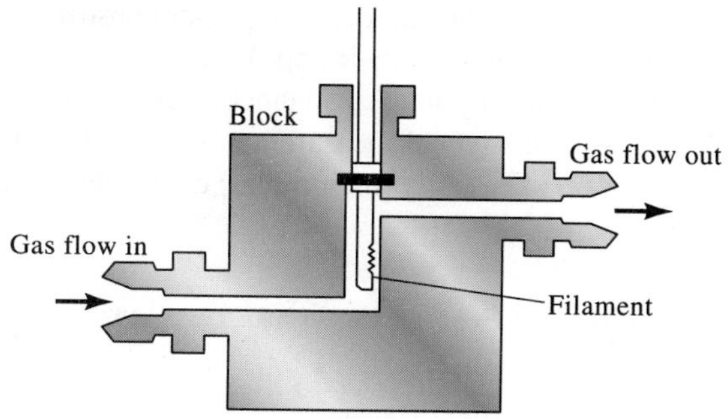

Figure 29-8
A typical thermal conductivity detector. (Courtesy of Varian Instrument Division, Palo Alto, CA.)

mionic detection particularly useful for detecting and determining the many phosphorus containing pesticides.

A thermionic detector is similar in structure to the flame detector shown in Figure 29-7. The column effluent is mixed with hydrogen, passes through the flame tip assembly, and is ignited. The hot gas then flows around an electrically heated rubidium silicate bead, which is maintained at about 180 V with respect to the collector. The heated bead forms a plasma that has a temperature of 600°C to 800°C. Exactly what occurs in the plasma to produce unusually large numbers of ions from phosphorus or nitrogen containing molecules is not understood, but large ion currents result, which are useful for determining compounds containing these two elements.

Other Types of Detectors

The *flame photometric detector* has been widely applied to the analysis of air and water pollutants, pesticides, and coal hydrogenation products. It is a selective detector that is primarily responsive to compounds containing sulfur and phosphorus. In this detector, the eluent is passed into a low-temperature hydrogen/air flame, which converts part of the phosphorus to an HPO species that emits bands of radiation centered about 510 and 526 nm. Sulfur in the sample is simultaneously converted to S_2, which emits a band centered at 394 nm. Suitable filters are employed to isolate these bands, and their intensities are recorded photometrically. Other elements that have been detected by flame photometry include the halogens, nitrogen, and several metals, such as tin, chromium, selenium, and germanium.

In the *photoionization detector,* the column eluent is irradiated with an intense beam of ultraviolet radiation varying in energy from 8.3 to 11.7 eV (λ = 149 to 106 nm), which causes ionization of the molecules. Application of a potential across a cell containing the ions leads to an ion current, which is amplified and recorded.

Selective Detectors

Hyphenated methods couple the separation capabilities of chromatography with the capacity for qualitative and quantitative detection of electrical or spectral methods.

Gas chromatography is often coupled with the selective techniques of spectroscopy and electrochemistry. The resulting so-called *hyphenated methods* (for instance, gas chromatography/mass spectroscopy (GC/MS) and gas chromatography/infrared spectroscopy (GC/IR) provide the chemist with powerful tools for identifying the components of complex mixtures.[4]

In early hyphenated methods, the eluates from the chromatographic column were collected as separate fractions in a cold trap, a nondestructive, nonselective detector being employed to indicate their appearance. The composition of each fraction was then investigated by nuclear magnetic resonance, infrared, or mass spectroscopy, or by electroanalytical measurements. A serious limitation to this approach was the very small (usually micromolar) quantities of solute contained

[4]For a review on hyphenated methods, see T. Hirschfeld, *Anal. Chem.,* **1980,** *52,* 297A; C. L. Wilkens, *Science,* **1983,** *222,* 251.

in a fraction; nonetheless, the general procedure proved useful for the qualitative analysis of many multicomponent mixtures.

A second general method, one that now finds widespread use, involves the application of spectroscopic or electroanalytical detectors to monitor the column effluent continuously. Generally, this procedure requires computerized instruments and a large computer memory for storing spectral or electrochemical data for subsequent display as spectra and chromatograms.

29B GAS-CHROMATOGRAPHIC COLUMNS AND THE STATIONARY PHASE

All of the pioneering gas-liquid chromatographic studies in the early 1950s were carried out on packed columns in which the stationary phase was a thin film of liquid retained by adsorption on the surface of a finely divided, inert solid support. From theoretical studies made during this early period, however, it became apparent that unpacked columns with inside diameters of a few tenths of a millimeter should provide separations that were much superior to packed columns in both speed and column efficiency. In such *capillary columns,* the stationary phase was a uniform film of liquid a few tenths of a micrometer thick that uniformly coated the interior of a capillary tubing. In the late 1950s such *open tubular columns* were constructed, and the predicted performance characteristics were experimentally confirmed in several laboratories, with open tubular columns having 300,000 plates or more being described.[5] Despite such spectacular performance characteristics, capillary columns did not gain widespread use until more than two decades after their invention. The reasons for the delay were several, including small sample capacities, fragility of columns, mechanical problems associated with sample introduction and connection of the column to the detector, difficulties in coating the column reproducibly, short lifetimes of poorly prepared columns, tendencies of columns to clog, and patents, which limited commercial development to a single manufacturer (the original patent expired in 1977). By the late 1970s these problems had become manageable, and several instrument companies began to offer open tubular columns at a reasonable cost. As a consequence, a major growth in the use of open tubular columns has occurred in the last few years.

29B-1 Packed Columns

Present-day packed columns are fabricated from glass, metal (stainless steel, copper, aluminum), or Teflon tubes that typically have lengths of 2 to 3 m and inside diameters of 2 to 4 mm. These tubes are densely packed with a uniform, finely divided packing material, or solid support, that is coated with a thin layer (0.05 to 1 μm) of the stationary liquid phase. In order to fit in a thermostating oven, the tubes are formed as coils having diameters of roughly 15 cm.

---

[5]In 1987, a world record for length of an open tubular column and number of theoretical plates was set, as attested in the *Guinness Book of Records,* by Chrompack International Corporation of the Netherlands. The column was fused silica drawn in one piece and having an internal diameter of 0.32 mm and a length of 2100 m or 1.3 miles. The column was coated with 0.1 μm film of polydimethyl siloxane. A 1300-m section of this column contained over 2 million plates.

Figure 29-9
A photomicrograph of a diatom. Magnification 5000×.

Solid Support Materials

The solid support in a packed column serves to hold the liquid stationary phase in place so that as large a surface area as possible is exposed to the mobile phase. The ideal support consists of small, uniform, spherical particles with good mechanical strength and a specific surface area of at least 1 m^2/g. In addition, the material should be inert at elevated temperatures and be uniformly wetted by the liquid phase. No substance that meets all of these criteria perfectly is yet available.

Today, the most widely used support material is prepared from naturally occurring diatomaceous earth, which is made up of the skeletons of thousands of species of single-celled plants that inhabited ancient lakes and seas (Figure 29-9). Such plants received their nutrients and disposed of their wastes via molecular diffusion through their pores. Their remains are suitable as support materials because gas chromatography is based upon the same kind of molecular diffusion.

Particle Size of Supports

As shown in Figure 28-11 (page 676), the efficiency of a gas-chromatographic column increases rapidly with decreasing particle diameter of the packing. The pressure difference required to maintain a given flow rate of carrier gas, however, varies inversely as the square of the particle diameter; the latter relationship has placed lower limits on the size of particles employed in gas chromatography because it is not convenient to use pressure differences that are greater than about 50 psi. As a result, the usual support particles are 60 to 80 mesh (0.25 to 0.17 mm) or 80 to 100 mesh (0.17 to 0.15 mm).

29B-2 Open Tubular Columns

Open tubular, or capillary, columns are two basic types, namely, *wall coated open tubular* (WCOT) and *support coated open tubular* (SCOT). Wall coated columns are simply capillary tubes coated with a thin layer of the stationary phase. In support coated open tubular columns, the inner surface of the capillary is lined with a thin film (~ 30 μm) of a support material, such as diatomaceous earth. This type of column holds several times as much stationary phase as does a wall coated column and thus has a greater sample capacity. Generally, the efficiency of a SCOT column is less than that of a WCOT column but significantly greater than that of a packed column.

Fused silica open tubular columns (FSOT columns) are currently the most widely used GLC columns.

Early WCOT columns were constructed of stainless steel, aluminum, copper, or plastic. Subsequently, glass was used. Often the glass was etched with gaseous hydrochloric acid, strong aqueous hydrochloric acid, or potassium hydrogen fluoride to give a rough surface, which bonded the stationary phase more tightly. The newest WCOT columns, which first appeared in 1979, are *fused silica open tubular columns* (FSOT columns). Fused silica capillaries are drawn from specially purified silica that contains minimal amounts of metal oxides. These capillaries have much thinner walls than their glass counterparts. The tubes are given added strength by an outside protective polyimide coating, which is applied as the capillary tubing is being drawn. The resulting columns are quite flexible and can be bent into coils having diameters of a few inches. Silica open tubular columns are available commercially and offer several important advantages such as physical strength, much lower reactivity toward sample components, and flexibility. For most applications, they have replaced the older WCOT glass columns.

The most widely used silica open tubular columns have inside diameters of 0.32 and 0.25 mm. Higher-resolution columns, with diameters of 0.20 and 0.15 mm, are also sold. Such columns are more troublesome to use and are more demanding upon the injection and detection systems. Thus, a sample splitter must be used to reduce the size of the sample injected onto the column, and a more sensitive detector system with a rapid response time is required. Recently, 530 μm capillaries, sometimes called *megabore* columns, which will tolerate sample sizes that are similar to those for packed columns, have appeared on the market. The performance characteristics of megabore open tubular columns are not as good as those of smaller diameter columns but significantly better than those of packed columns.

Table 29-1 compares the performance characteristics of fused silica capillary columns with other types of wall coated columns as well as with support coated and packed columns.

29B-3 The Stationary Liquid Phase

Desirable properties for the immobilized liquid phase in a gas-liquid chromatographic column include: (1) *low volatility* (ideally, the boiling point of the liquid should be at least 100°C higher than the maximum operating temperature for the column); (2) *thermal stability;* (3) *chemical inertness;* (4) *solvent characteristics* such that k' and α (see Sections 28B-4 and 28B-5) values for the solutes to be resolved fall within a suitable range.

A myriad of solvents have been proposed as stationary phases in the course of the development of gas-liquid chromatography. By now, only a handful—perhaps a dozen or less—suffice for most applications. The proper choice among these solvents is often critical to the success of a separation. Qualitative guide-

TABLE 29-1 Properties and Characteristics of Typical Gas-Chromatographic Columns

	Type of Column			
	FSOT*	**WCOT†**	**SCOT‡**	**Packed**
Length, m	10–100	10–100	10–100	1–6
Inside diameter, mm	0.1–0.3	0.25–0.75	0.5	2–4
Efficiency, plates/m	2000–4000	1000–4000	600–1200	500–1000
Sample size, ng	10–75	10–1000	10–1000	$10–10^6$
Relative pressure	Low	Low	Low	High
Relative speed	Fast	Fast	Fast	Slow
Chemical inertness	Best →			Poorest
Flexible?	Yes	No	No	No

*Fused silica open tubular column.

†Wall coated open tubular column.

‡Support coated open tubular column (also called porous layer open tubular, or PLOT).

TABLE 29-2 Some Common Liquid Stationary Phases for Gas-Liquid Chromatography

Stationary Phase	Common Trade Name	Maximum Temperature, °C	Common Applications
Polydimethyl siloxane	OV-1, SE-30	350	General purpose nonpolar phase; hydrocarbons; polynuclear aromatics; drugs; steroids; PCBs
5% Phenyl-polydimethyl siloxane	OV-3, SE-52	350	Fatty acid methyl esters; alkaloids; drugs; halogenated compounds
50% Phenyl-polydimethyl siloxane	OV-17	250	Drugs; steroids; pesticides; glycols
50% Trifluoropropyl-polydimethyl siloxane	OV-210	200	Chlorinated aromatics; nitroaromatic; alkyl substituted benzenes
Polyethylene glycol	Carbowax 20M	250	Free acids; alcohols; ethers; essential oils; glycols
50% Cyanopropyl-polydimethyl siloxane	OV-275	240	Polyunsaturated fatty acids; rosin acids; free acids; alcohols

lines exist for making this choice, but in the end, the best stationary phase can only be determined in the laboratory.

The retention time for a solute on a column depends upon its partition ratio (Equation 28-1), which in turn is related to the chemical nature of the stationary phase. Clearly, to be useful in gas-liquid chromatography, the immobilized liquid must generate different partition ratios for different solutes. In addition, however, these ratios must not be extremely large or extremely small because the former leads to prohibitively long retention times and the latter results in such short retention times that separations are incomplete.

To have a reasonable residence time in the column, a species must show some degree of compatibility (solubility) with the stationary phase. Here, the principle of "like dissolves like" applies, where "like" refers to the polarities of the solute and the immobilized liquid. Polarity is the electrical field effect in the immediate vicinity of a molecule and is measured by the dipole moment of the species. Polar stationary phases contain functional groups such as —CN, —CO, and —OH. Hydrocarbon-type stationary phases and dialkyl siloxanes are nonpolar, whereas polyester phases are highly polar. Polar solutes include alcohols, acids, and amines; solutes of medium polarity include ethers, ketones, and aldehydes. Saturated hydrocarbons are nonpolar. Generally, the polarity of the stationary phase should match that of the sample components. When the match is good, the order of elution is determined by the boiling point of the eluents.

Some Widely Used Stationary Phases

Table 29-2 lists the most widely used stationary phases for both packed and open tubular column gas chromatography in order of increasing polarity. These six liquids can probably provide satisfactory separations for 90% or more of the samples encountered by the scientist.

Five of the liquids listed in Table 29-2 are polydimethyl siloxanes that have the general structure

$$R-\underset{\underset{R}{|}}{\overset{\overset{R}{|}}{Si}}-O+\underset{\underset{R}{|}}{\overset{\overset{R}{|}}{Si}}-O+\underset{\underset{R}{|}}{\overset{\overset{R}{|}}{Si}}-R$$

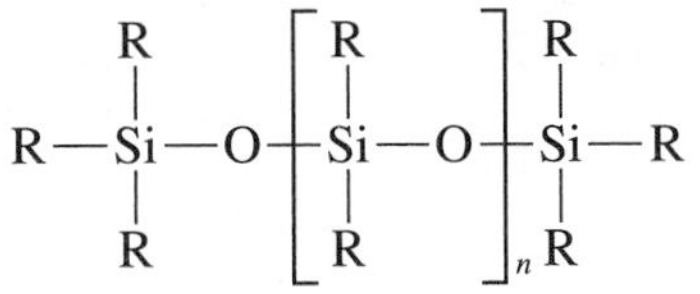

In the first of these, polydimethyl siloxane, the —R groups are all —CH_3, giving a liquid that is relatively nonpolar. In the other polysiloxanes shown in the table, a fraction of the methyl groups are replaced by functional groups such as phenyl (—C_6H_5), cyanopropyl (—C_3H_6CN), and trifluoropropyl (—$C_3H_6CF_3$). The percentage descriptions in each case give the amount of substitution of the named group for methyl groups on the polysiloxane backbone. Thus, for example, 5% phenyl polydimethyl siloxane has a phenyl ring bonded to 5% by number of the silicon atoms in the polymer. These substitutions increase the polarity of the liquids to various degrees.

The fifth entry in Table 29-2 is a polyethylene glycol with the structure

$$HO-CH_2-CH_2-(O-CH_2-CH_2)_n-OH$$

It finds widespread use for separating polar species.

Figure 29-10 illustrates applications of some of the liquid stationary phases listed in Table 29-2 for open tubular columns.

Bonded and Cross-Linked Stationary Phases

Commercial columns are advertised as having bonded and/or cross-linked stationary phases. The purpose of bonding and cross-linking is to provide a longer lasting stationary phase that can be rinsed with a solvent when the film becomes contaminated. With use, untreated columns slowly lose their stationary phase due to "bleeding," in which a small amount of immobilized liquid is carried out of the column during the elution process. Bleeding is exacerbated when a column must be rinsed with a solvent to remove contaminants. Chemical bonding and cross-linking inhibit bleeding.

Bonding involves attaching a monomolecular layer of the stationary phase to the silica surface of the column by a chemical reaction. For commercial columns, the nature of the reactions is ordinarily proprietary.

Cross-linking is carried out *in situ* after a column has been coated with one of the polymers listed in Table 29-2. One way of cross-linking is to incorporate a peroxide into the original liquid. When the film is heated, reaction between the methyl groups in the polymer chains is initiated by a free radical mechanism. The polymer molecules are then cross-linked through carbon-to-carbon bonds. The resulting films are less extractable and have considerably greater thermal stability than do untreated films. Cross-linking has also been initiated by exposing the coated columns to gamma radiation.

Film Thickness

Commercial columns are available having stationary phases that vary in thickness from 0.1 to 5 μm. Film thickness primarily affects the retentive character and the capacity of a column. Thick films are used with highly volatile analytes because

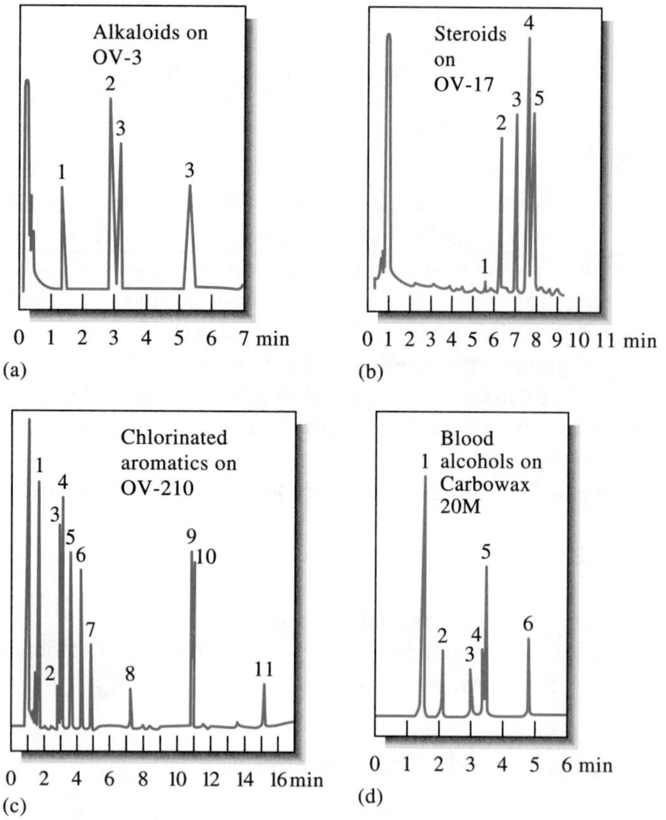

Figure 29-10
Typical chromatograms for open tubular columns coated with phases listed in Table 29-2. Compound identification: (a) 1. cocaine, 2. codeine, 3. morphine, 4. quinine; (b) 1. 17-α-estradiol, 2. dihydroequiline, 3. testosterone, 4. estrone, 5. equiline; (c) 1. chlorobenzene, 2. hexachloroethane, 3. 1,3-dichlorobenzene, 4. 1,4-dichlorobenzene, 5. 1,2-dichlorobenzene, 6. 4-chlorostyrene, 7. hexachlorobutadiene, 8. hexachlorocyclopentadiene, 9. 2-chloronaphthalene, 10. 1-chloronaphthalene, 11. hexachlorobenzene; (d) 1. acetaldehyde, 2. acetone, 3. methanol, 4. isopropanol, 5. ethanol, 6. n-propanol.

such films retain solutes for a longer time, thus providing a greater time for separation to take place. Thin films are useful for separating species of low volatility in a reasonable length of time. For most applications with 0.25- or 0.32-mm columns, a film thickness of 0.25 μm is recommended. With megabore columns, 1 to 1.5 μm films are often used. Today, columns with 8 μm films are also marketed.

29C APPLICATIONS OF GAS-LIQUID CHROMATOGRAPHY

In evaluating the importance of GLC, it is necessary to distinguish between the two roles the method plays. The first is as a tool for performing separations; in this capacity, it is unsurpassed when applied to complex organic, metal-organic, and biochemical systems. The second, and distinctly different, function is that of providing the means for completion of an analysis. Here, retention times or volumes are employed for qualitative identification, while peak heights or peak areas provide quantitative information. For qualitative purposes, GLC is much more limited than most of the spectroscopic methods considered in earlier chapters. As a consequence, an important trend in the field has been in the direction of combining the remarkable fractionation qualities of GLC with the superior identification properties of such instruments as mass, infrared, and NMR spectrometers.

29C-1 Qualitative Analysis

Gas chromatograms are widely used as criteria of purity for organic compounds. Contaminants, if present, are revealed by the appearance of additional peaks; the areas under these peaks provide rough estimates of the extent of contamination. The technique is also useful for evaluating the effectiveness of purification procedures.

In theory, retention times should be useful for the identification of components in mixtures. In fact, however, the applicability of such data is limited by the number of variables that must be controlled in order to obtain reproducible results. Nevertheless, gas chromatography provides an excellent means of confirming the presence or absence of a suspected compound in a mixture, provided an authentic sample of the substance is available. No new peaks in the chromatogram of the mixture should appear upon addition of the known compound, and enhancement of an existing peak should be observed. The evidence is particularly convincing if the effect can be duplicated on different columns and at different temperatures.

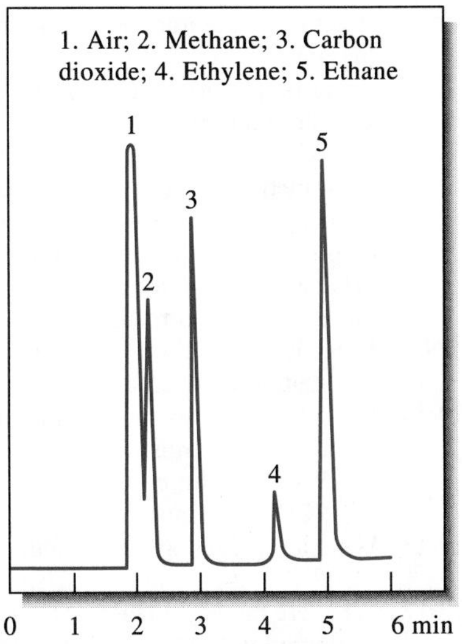

Figure 29-11
Typical gas-solid chromatogram on a PLOT column.

29C-2 Quantitative Analysis

The detector signal from a gas-liquid chromatographic column has had wide use for quantitative and semiquantitative analyses. An accuracy of 1% relative is attainable under carefully controlled conditions. As with most analytical tools, reliability is directly related to the control of variables; the nature of the sample also plays a part in determining the potential accuracy. The general discussion of quantitative chromatographic analysis given in Section 28E-2 applies to gas chromatography as well as to other types; therefore, no further consideration of this topic is given here.

29D GAS-SOLID CHROMATOGRAPHY

Gas-solid chromatography is based upon adsorption of gaseous substances on solid surfaces. Distribution coefficients are generally much larger than those for gas-liquid chromatography. Consequently, gas-solid chromatography is useful for the separation of species that are not retained by gas-liquid columns, such as the components of air, hydrogen sulfide, carbon disulfide, nitrogen oxides, carbon monoxide, carbon dioxide, and the rare gases.

Gas-solid chromatography is performed with both packed and open tubular columns. For the latter, a thin layer of the adsorbent is affixed to the inner walls of the capillary. Such columns are sometimes called *porous layer open tubular columns,* or PLOT columns. Figure 29-11 shows a typical application of a PLOT column.

29E QUESTIONS AND PROBLEMS

***29-1.** How do gas-liquid and gas-solid chromatography differ?

29-2. What kind of mixtures are separated by gas-solid chromatography?

***29-3.** Why is gas-solid chromatography not used nearly as extensively as gas-liquid chromatography?

29-4. How does a soap-bubble flow meter work?

*29-5. What is temperature programming as used in gas chromatography?

29-6. Describe the principle upon which each of the following gas chromatography detectors are based: **(a)** thermal conductivity, **(b)** flame ionization, **(c)** thermionic, **(d)** flame photometric, and **(e)** photoionization.

*29-7. What are the principal advantages and the principal limitations of each of the detectors listed in Problem 29-6?

29-8. What are *hyphenated* gas-chromatographic methods? List three hyphenated methods.

*29-9. What is the packing material used in most packed gas chromatographic columns?

29-10. How do the following open tubular columns differ?
(a) PLOT columns
(b) WCOT columns
(c) SCOT columns

*29-11. What are megabore open tubular columns? Why are they used?

29-12. What are the advantages of fused silica capillary columns compared with glass or metal columns?

*29-13. What properties should the stationary phase liquid for gas chromatography possess?

29-14. Why are gas-chromatographic stationary phases often bonded and cross-linked? What do these terms mean?

*29-15. What is the effect of stationary phase film thickness on gas chromatograms?

29-16. List the variables that lead to **(a)** band broadening and **(b)** band separation in gas-liquid chromatography.

*29-17. One method for the quantitative determination of the constituents in a sample analyzed by gas chromatography is the area-normalization method. In this technique, complete elution of all sample constituents is necessary; the area of each peak is then measured and corrected for differences in detector response to the different eluates. This correction involves multiplication or division of the area by an empirically determined correction factor. The concentration of the analyte is found from the ratio of its corrected area to the total corrected area of all peaks. For a chromatogram containing three peaks, the relative areas were 16.4, 45.2, and 30.2 in order of increasing retention time. The first peak was for methyl acetate, the second for methyl propionate, and the third for methyl butyrate. Calculate the percentage of each compound if the detector response correction requires division of the areas by 0.60, 0.78, and 0.88, respectively.

29-18. Peak areas and relative detector responses are to be used to determine the concentration of the five species in a sample. The area-normalization method described in Problem 29-17 is to be used. The relative areas for the five gas chromatographic peaks are given below. Also shown are the detector response correction factors. Calculate the percentage of each component in the mixture.

Compound	Relative Peak Area	Detector Response Divisor
A	32.5	0.70
B	20.7	0.72
C	60.1	0.75
D	30.2	0.73
E	18.3	0.78

CHAPTER

30

High-Performance Liquid Chromatography

Early liquid chromatography was performed in glass columns having inside diameters of perhaps 10 to 50 mm. The columns were packed with 50- to 500-cm lengths of solid particles coated with an adsorbed liquid that formed the stationary phase. To ensure reasonable flow rates through this type of stationary phase, the particle size of the solid was kept larger than 150 to 200 μm; even then, flow rates were a few tenths of a milliliter per minute, at best. Attempts to speed up this classic procedure by application of vacuum or pressure were not effective because increases in flow rates were accompanied by increases in plate heights and accompanying decreases in column efficiency.

Early in the development of the theory of liquid chromatography, it was postulated that large decreases in plate heights could be expected to accompany decreases in the particle size of packings. This effect is illustrated by the data in Figure 30-1. It is of interest to note that in none of these plots is the minimum shown in Figure 28-8a (page 674) reached. The reason for this difference is that diffusion in liquids is much slower than in gases; consequently, its effect on plate heights is observed only at extremely low flow rates.

It was not until the late 1960s, that the technology for producing and using packings with particle diameters as small as 5 to 10 μm was developed. This technology required sophisticated instruments that contrasted markedly with the simple devices that preceded them. The name *high-performance liquid chromatography* (HPLC) is often employed to distinguish these newer procedures from their predecessors, which still find considerable use for preparative purposes.[1]

High-performance liquid chromatography, HPLC, is a type of chromatography that employs a liquid mobile phase and a very finely divided stationary phase. In order to obtain satisfactory flow rates, the liquid must be pressurized to several hundred pounds per square inch or more.

Figure 30-2 shows the five most widely used types of high-performance liquid chromatography. These methods include: (1) *partition,* or *liquid-liquid, chromatography;* (2) *adsorption,* or *liquid-solid chromatography;* (3) *ion-exchange chromatography,* and two types of *size-exclusion chromatography;* (4) *gel-permeation chromatography;* and (5) *gel-filtration chromatography.* Figure 30-2 suggests that the various types of liquid chromatography tend to be complemen-

[1]References for HPLC include: L. R. Snyder and J. J. Kirkland, *Introduction to High-Performance Liquid Chromatography,* 2nd ed. New York: Chapman and Hall, 1982; P. Brown and R. A. Hartwick, *High-Performance Liquid Chromatography.* New York: Wiley, 1988; S. Lindsey, *High-Performance Liquid Chromatography.* New York: Wiley, 1987; V. Meyer, *Practical High-Performance Liquid Chromatography.* New York: Wiley, 1988.

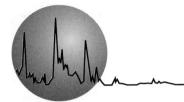

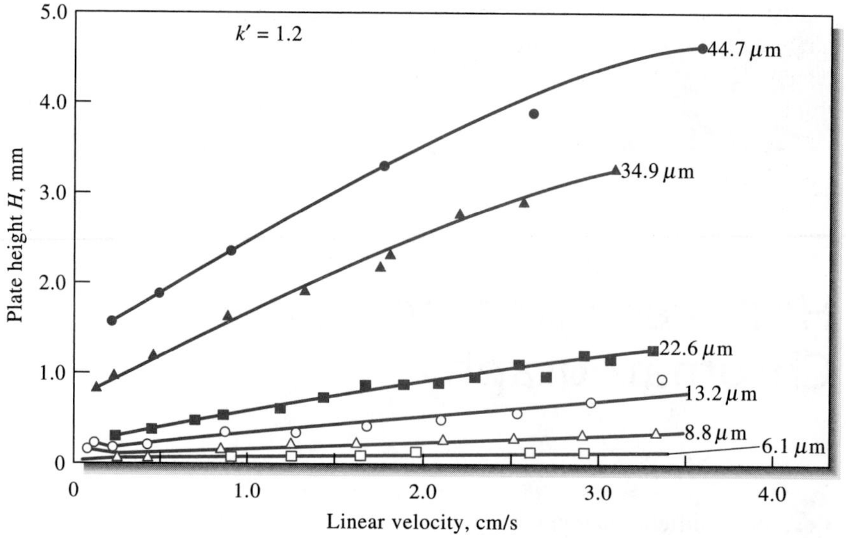

Figure 30-1
Effect of packing particle size and flow rate on plate height in liquid chromatography. Column dimensions: 30 cm × 2.4 mm. Solute: N,N-diethyl-*n*-aminoazobenzene. Mobile phase: mixture of hexane, methylene chloride, and isopropyl alcohol. (From R. E. Majors, *J. Chromatogr. Sci.*, **1973,** *11,* 92. With permission.)

tary insofar as applications are concerned. For example, for analytes having molecular masses greater than 10,000, one of the two size-exclusion methods is often used: gel permeation for nonpolar species and gel filtration for polar or ionic compounds. For ionic species having lower molar masses, ion-exchange chromatography is generally the method of choice. Smaller polar but nonionic species are best handled by partition methods.

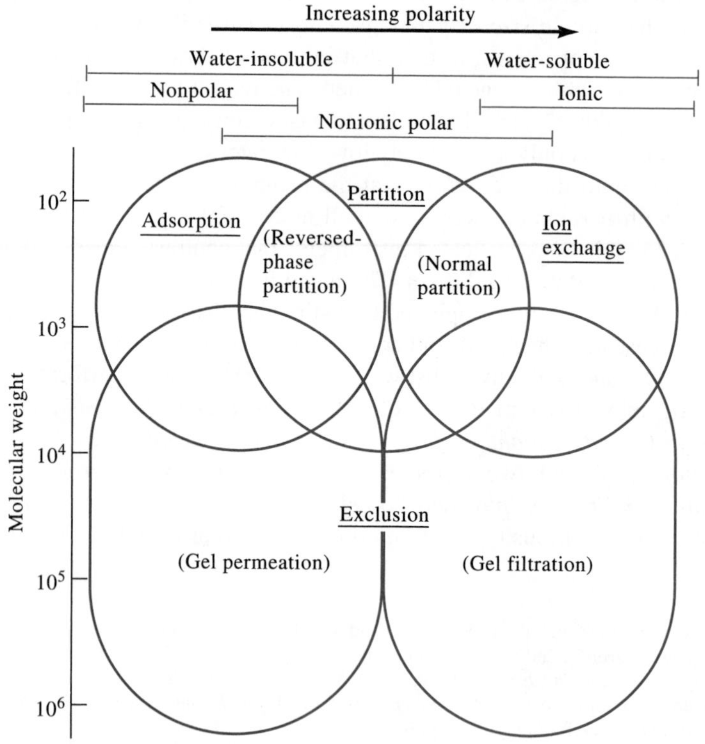

Figure 30-2
Applications of liquid chromatography. (From D. L. Saunders, in *Chromatography,* 3rd ed., E. Heftmann, Ed., p. 81. New York: Van Nostrand Reinhold, 1975. With permission.)

Because of its versatility and wide applicability, high-performance liquid chromatography is currently the most widely used of all separation techniques, with annual equipment sales approaching a billion dollar mark.

It is important to realize that the various types of liquid chromatography can be performed either in a column or on a flat surface. The latter type of chromatography is called *planar chromatography* and has the advantage of simplicity and inexpensive equipment. In this chapter we focus on column chromatography because of its more general use for quantitative analysis but have included a brief section on planar chromatography as well. Also included at the end of the chapter is a short section on *supercritical-fluid chromatography.* In this method a supercritical fluid rather than a liquid serves as the mobile phase.

30A INSTRUMENTS

Pumping pressures of several hundred atmospheres are required to achieve reasonable flow rates with packings in the 3 to 10 μm size range, which are common in modern liquid chromatography. As a consequence of these high pressures, the equipment for high-performance liquid chromatography tends to be considerably more elaborate and expensive than that encountered in other types of chromatography. Figure 30-3 is a diagram showing the important components of a typical high-performance chromatographic instrument.

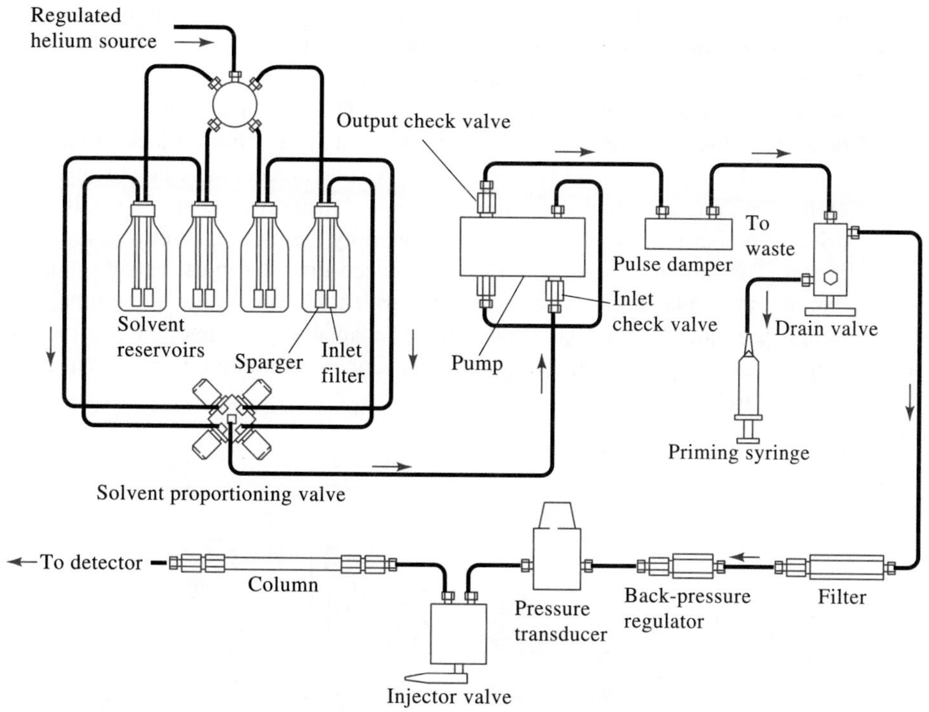

Figure 30-3
Schematic of an apparatus for high-performance liquid chromatography. (Courtesy of Perkin-Elmer, Norwalk, CT.)

30A-1 Mobile Phase Reservoirs and Solvent Treating Systems

Sparging is a process in which dissolved gases are swept out of a solvent by bubbles of an inert, insoluble gas.

An isocratic elution in HPLC is one in which the composition of the solvent remains constant.

A **gradient elution** in HPLC is one in which the composition of the solvent is changed continuously or in a series of steps.

A modern HPLC apparatus is equipped with one or more glass or stainless steel reservoirs, each of which contains 500 mL or more of a solvent. Provisions are often included to remove dissolved gases and dust from the liquids. The former produce bubbles in the column and thereby cause band spreading; in addition, both bubbles and dust interfere with the performance of detectors. Degassers may consist of a vacuum pumping system, a distillation system, a device for heating and stirring, or, as shown in Figure 30-3, a system for *sparging,* in which the dissolved gases are swept out of solution by fine bubbles of an inert gas that is not soluble in the mobile phase.

An elution with a single solvent of constant composition is termed *isocratic.* In *gradient elution,* two (and sometimes more) solvent systems that differ significantly in polarity are employed. The ratio of the two solvents is varied in a preprogrammed way, sometimes continuously and sometimes in a series of steps. Gradient elution frequently improves separation efficiency, just as temperature programming helps in gas chromatography. Modern high-performance liquid chromatography instruments are often equipped with proportionating valves that introduce liquids from two or more reservoirs at rates that vary continuously (Figure 30-3).

30A-2 Pumping Systems

The requirements for liquid chromatographic pumps are severe and include (1) the generation of pressures of up to 6000 psi (lb/in^2), (2) pulse-free output, (3) flow rates ranging from 0.1 to 10 mL/min, (4) flow reproducibilities of 0.5% relative or better, and (5) resistance to corrosion by a variety of solvents.

Two types of mechanical pumps are employed: a screw-driven syringe type and a reciprocating pump (see Figure 30-4). The former produces a pulse-free delivery whose flow rate is readily controlled; it suffers from a lack of capacity (~250 mL), however, and is inconvenient when solvents must be changed. Reciprocating pumps, which are more widely used, usually consist of a small cylindrical chamber that is filled and then emptied by the back-and-forth motion of a piston. The pumping motion produces a pulsed flow that must be subsequently damped. Advantages of reciprocating pumps include small internal volume, high output pressure (up to 10,000 psi), ready adaptability to gradient elution, and constant flow rates, which are largely independent of column back-pressure and solvent viscosity.

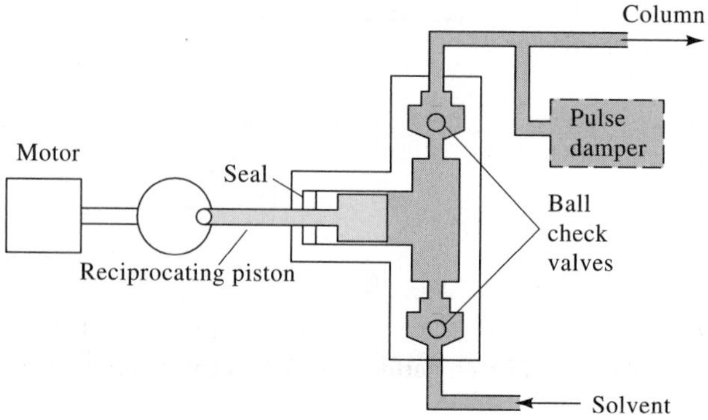

Figure 30-4
A reciprocating pump for HPLC.

Some instruments use a pneumatic pump, which in its simplest form consists of a collapsible solvent container housed in a vessel that can be pressurized by a compressed gas. Pumps of this type are simple, inexpensive, and pulse-free; they suffer from limited capacity and pressure output, however, and from pumping rates that depend upon solvent viscosity. In addition, they are not adaptable to gradient elution.

It should be noted that the high pressures generated by liquid-chromatographic pumps do not constitute an explosion hazard because liquids are not very compressible. Thus, rupture of a component results only in solvent leakage. To be sure, such leakage may constitute a fire hazard.

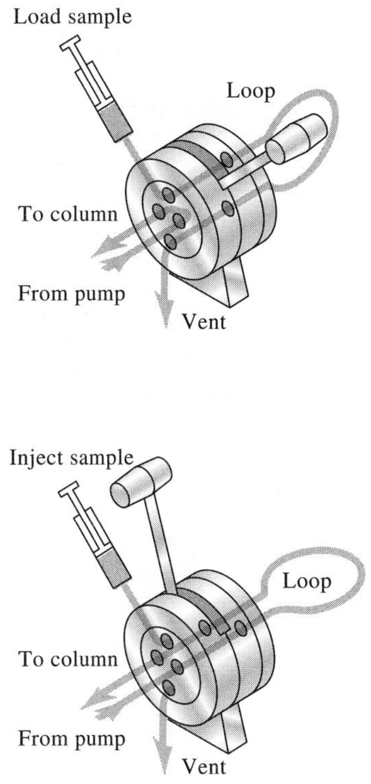

Figure 30-5
A sampling loop for liquid chromatography. (Courtesy of Beckman Instruments, Fullerton, CA.)

30A-3 Sample-Injection System

Although syringe injection through an elastomeric septum is often used in liquid chromatography, this procedure is not very reproducible and is limited to pressures less than about 1500 psi. In *stop-flow* injection, the solvent flow is stopped momentarily, a fitting at the column head is removed, and the sample is injected directly onto the head of the packing by means of a syringe. The most widely used method of sample introduction in liquid chromatography is based upon sampling loops such as that shown in Figure 30-5. These devices are often an integral part of modern liquid chromatography equipment and have interchangeable loops that provide a choice of sample sizes ranging from 5 to 500 μL. The reproducibility of injections with a typical sampling loop is a few tenths of a percent relative.

30A-4 Columns for High-Performance Liquid Chromatography

Liquid-chromatographic columns are usually constructed from stainless steel tubing, although heavy-walled glass tubing is sometimes employed for lower-pressure applications ($<$ 600 psi). Most columns range in length from 10 to 30 cm and have inside diameters of 4 to 10 mm. Column packings typically have particle sizes of 5 or 10 μm. Columns of this type often contain 40,000 to 60,000 plates/m. Recently, high-performance microcolumns with inside diameters of 1 to 4.6 mm and lengths of 3 to 7.5 cm have become available. These columns, which are packed with 3 or 5 μm particles, contain as many as 100,000 plates/m and have the advantage of speed and minimal solvent consumption. The latter property is of considerable importance because the high-purity solvents required for liquid chromatography are expensive to purchase and to dispose of after use. Figure 30-6 illustrates the speed with which a separation can be performed on this type of column. Here, eight diverse components are separated in about 15 s. The column is 4 cm in length and has an inside diameter of 4 mm; it is packed with 3 μm particles.

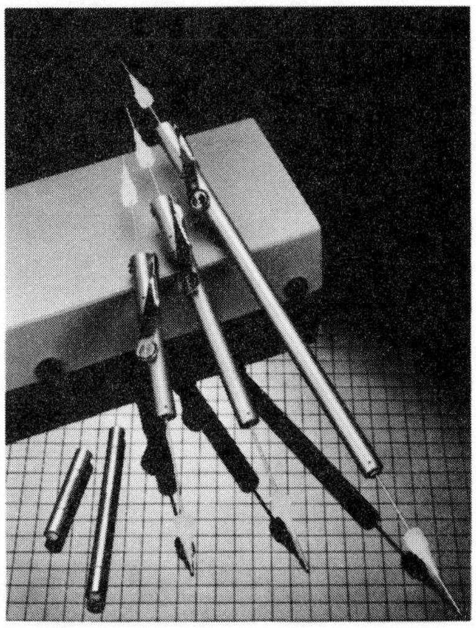

Three HPLC columns. (Courtesy of Chrompack, Inc., Raritan, NJ.)

The most common packing for liquid chromatography is from silica, prepared by agglomerating submicron silica particles under conditions that lead to larger particles with highly uniform diameters. The resulting particles are often coated with thin organic films, which are chemically or physically bonded to the surface. Other packing materials include alumina particles, porous polymer particles, and ion-exchange resins.

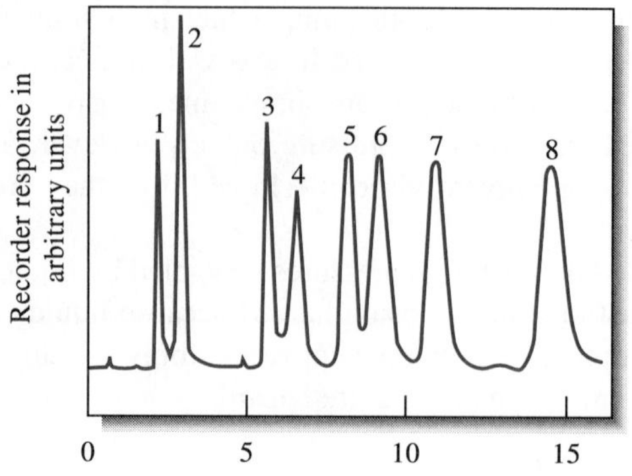

Figure 30-6
High-speed isocratic separation. Column dimensions: 4 cm length, 0.4 cm i.d.; Packing: 3-μm; Mobile phase: 4.1% ethyl acetate in *n*-hexane. Compounds: (1) *p*-xylene, (2) anisole, (3) benzyl acetate, (4) dioctyl phthalate, (5) dipentyl phthalate, (6) dibutyl phthalate, (7) dipropyl phthalate, (8) diethyl phthalate. (From R. P. W. Scott, *Small Bore Liquid Chromatography Columns: Their Properties and Uses,* p. 156. New York: Wiley, 1984. Reprinted with permission of John Wiley & Sons, Inc.)

Guard Columns

Often, a short guard column is introduced in front of the analytical column to increase the life of the analytical column by removing particulate matter and contaminants from the solvents. In addition, in liquid-liquid chromatography, the guard column serves to saturate the mobile phase with the stationary phase so that losses of the stationary phase from the analytical column are minimized. The composition of the guard-column packing should be similar to that of the analytical column; the particle size is usually larger, however, to minimize pressure drop.

Column Thermostats

For many applications, close control of column temperature is not necessary and columns are operated at room temperature. Often, however, better chromatograms are obtained by maintaining column temperatures constant to a few tenths of a degree Celsius. Most modern commercial instruments are now equipped with column heaters that control column temperatures to a few tenths of a degree from near-ambient to 150°C. Columns may also be fitted with water jackets fed from a constant temperature bath to give precise temperature control.

30A-5 Detectors

No highly sensitive, universal detector system, such as those for gas chromatography, is available for high-performance liquid chromatography. Thus, the system used will depend upon the nature of the sample. Table 30-1 lists some of the common detectors and their properties.

The most widely used detectors for liquid chromatography are based upon absorption of ultraviolet or visible radiation (see Figure 30-7). Photometers and spectrophotometers specifically designed for use with chromatographic columns are available from commercial sources. The former often makes use of the 254- and 280-nm lines from a mercury source because many organic functional groups absorb in the region. Deuterium or tungsten filament sources with interference filters also provide a simple means of detecting absorbing species. Some modern instruments are equipped with filter wheels that contain several interference filters, which can be rapidly switched into place. Spectrophotometric detectors are

TABLE 30-1 Performances of HPLC Detectors*

HPLC Detector	Commercially Available	Mass LOD (commercial detectors)†	Mass LOD (state of the art)‡
Absorbance	Yes§	100 pg–1 ng	1 pg
Fluorescence	Yes§	1–10 pg	10 fg
Electrochemical	Yes§	10 pg–1 ng	100 fg
Refractive index	Yes	100 ng–1 μg	10 ng
Conductivity	Yes	500 pg–1 ng	500 ng
Mass spectrometry	Yes	100 pg–1 ng	1 pg
FT–IR	Yes	1 μg	100 ng
Light scattering	Yes	10 μg	500 ng
Optical activity	No	—	1 ng
Element selective	No	—	10 ng
Photoionization	No	—	1 pg–1 ng

*From E. S. Yeung and R. E. Synovec, *Anal. Chem.,* **1986,** *58,* 1238. With permission. Copyright American Chemical Society.

†Mass LOD (limit of detection) is calculated for injected mass that yields a signal equal to five times the σ noise, using a molar mass of 200 g/mol, 10 μL injected for conventional or 1 μL injected for microbore HPLC.

‡Same definition as †, but the injected volume is generally smaller.

§Commercially available for microbore HPLC also.

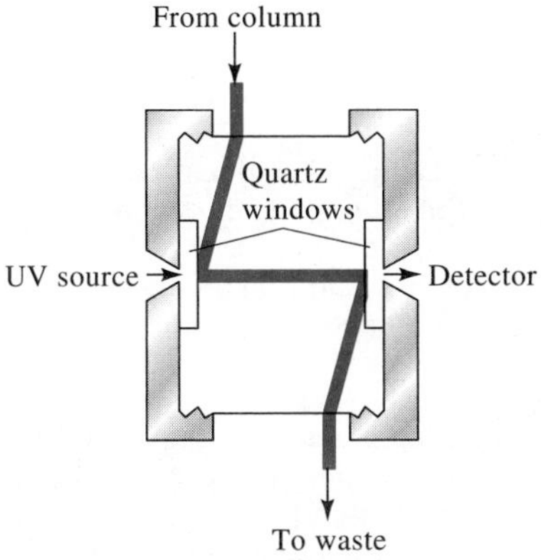

Figure 30-7
A UV detector for HPLC.

considerably more versatile than photometers and are also widely used in high-performance instruments. Often these are diode-array instruments that can display an entire spectrum as an analyte exits the column.

Another detector, which has found considerable application, is based upon the changes in the refractive index of the solvent that is caused by analyte molecules. In contrast to most of the other detectors listed in Table 30-1, the refractive index indicator is general rather than selective and responds to the presence of all solutes. The disadvantage of this detector is its somewhat limited sensitivity. Several electrochemical detectors have also been introduced that are based on potentiometric, conductometric, and voltammetric measurements. An example of the latter type of detector is shown in Figure 30-8.

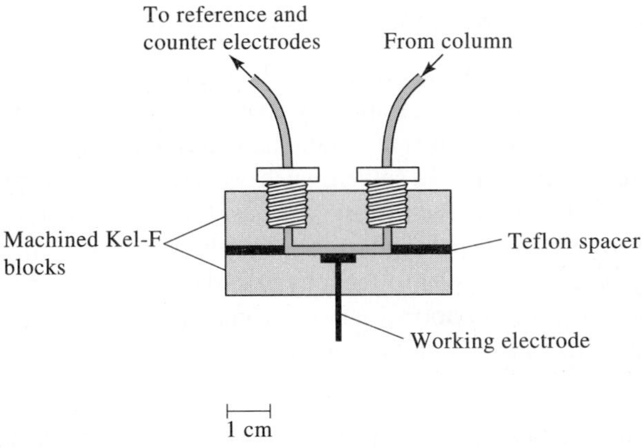

Figure 30-8
Amperometric detector for HPLC.

30B HIGH-PERFORMANCE PARTITION CHROMATOGRAPHY

Partition chromatography has become the most widely used of all liquid chromatographic procedures. This technique can be subdivided into *liquid-liquid* and *liquid bonded-phase* chromatography. The difference between the two lies in the method by which the stationary phase is held on the support particles of the packing. With liquid-liquid, retention is by physical adsorption, while with bonded-phase, covalent bonds are involved. Early partition chromatography was exclusively liquid-liquid; now, however, bonded-phase packings predominate because of their greater stability, with liquid-liquid packings being relegated to certain special applications.

> In **liquid-liquid partition chromatography,** the stationary phase is a solvent that is held in place by adsorption on the surface of packing particles.

> In **liquid bonded-phase partition chromatography,** the stationary phase is an organic species that is attached to the surface of the packing particles by chemical bonds.

30B-1 Bonded-Phase Packings

Most bonded-phase packings are prepared by reaction of an organochlorosilane with the —OH groups formed on the surface of silica particles by hydrolysis in hot, dilute hydrochloric acid. The product is an organosiloxane. The reaction for one such SiOH site on the surface of a particle can be written as

$$
-\underset{\diagdown}{\overset{\diagup}{Si}}-OH \;+\; Cl-\underset{\underset{CH_3}{|}}{\overset{\overset{CH_3}{|}}{Si}}-R \;\longrightarrow\; -\underset{\diagdown}{\overset{\diagup}{Si}}-O-\underset{\underset{CH_3}{|}}{\overset{\overset{CH_3}{|}}{Si}}-R
$$

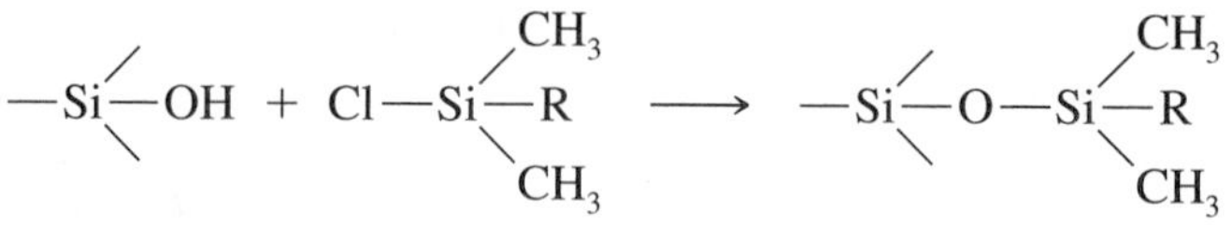

where R is often a straight-chain octyl- or octyldecyl-group. Other organic functional groups that have been bonded to silica surfaces include aliphatic amines, ethers, and nitriles, as well as aromatic hydrocarbons. Thus, a variety of polarities for the bonded stationary phase is available.

Bonded-phase packings have the advantage of markedly greater stability compared with physically held stationary phases. With the latter, periodic recoating of the solid surfaces is required because the stationary phase is gradually dissolved away in the mobile phase. Furthermore, gradient elution is not practical with liquid-liquid packings, again because of solubility losses in the mobile phase. The main disadvantage of bonded-phase packings is their somewhat limited sample capacity.

30B-2 Normal- and Reversed-Phase Packings

Two types of partition chromatography are distinguishable based upon the relative polarities of the mobile and stationary phases. Early work in liquid chromatography was based upon highly polar stationary phases such as triethylene glycol or water; a relatively nonpolar solvent such as hexane or *i*-propyl ether then served as the mobile phase. For historic reasons, this type of chromatography is now called *normal-phase chromatography*. In *reversed-phase chromatography*, the stationary phase is nonpolar, often a hydrocarbon, and the mobile phase is a relatively polar solvent.[2] In normal-phase chromatography, the *least* polar component is eluted first; *increasing* the polarity of the mobile phase then *decreases*

> In **normal-phase partition chromatography,** the stationary phase is polar and the mobile phase nonpolar. In **reversed-phase partition chromatography,** the polarity of these phases is reversed.

In normal-phase chromatography, the least polar analyte is eluted first. In reversed-phase chromatography, the least polar analyte is eluted last.

[2]For a detailed discussion of reversed-phase HPLC, see A. M. Krstulovic and P. R. Brown, *Reversed-Phase High-Performance Liquid Chromatography*. New York: Wiley, 1982.

the elution time. In contrast, in the reversed-phase method, the *most* polar component elutes first, and *increasing* the mobile phase polarity *increases* the elution time.

It has been estimated that more than three quarters of all HPLC separations are currently performed with reversed-phase, bonded, octyl- or octyldecyl siloxane packings. With such preparations, the long-chain hydrocarbon groups are aligned parallel to one another and perpendicular to the surface of the particle, giving a brushlike, nonpolar, hydrocarbon surface. The mobile phase used with these packings is often an aqueous solution containing various concentrations of such solvents as methanol, acetonitrile, or tetrahydrofuran.

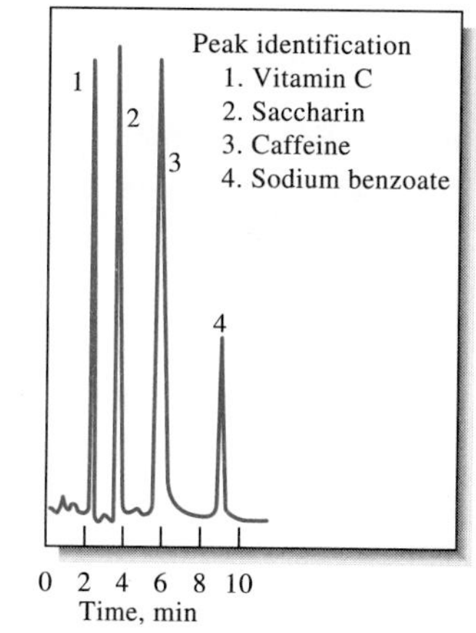

Figure 30-9
Liquid bonded-phase chromatogram of soft-drink additives. Packing: polar cyano packing. Isocratic elution with 6% HOAc/94% water. (Courtesy of DuPont Instrument Systems, Wilmington, DE.)

30B-3 Choice of Mobile and Stationary Phases

Successful partition chromatography requires a proper balance of intermolecular forces among the three participants in the separation process—the analyte, the mobile phase, and the stationary phase. These intermolecular forces are described qualitatively in terms of the relative polarity possessed by each of the three reactants. In general, the polarities of common organic functional groups in increasing order are aliphatic hydrocarbons $<$ olefins $<$ aromatic hydrocarbons $<$ halides $<$ sulfides $<$ ethers $<$ nitro compounds $<$ esters $\approx$ aldehydes $\approx$ ketones $<$ alcohols $\approx$ amines $<$ sulfones $<$ sulfoxides $<$ amides $<$ carboxylic acids $<$ water.

As a rule, most chromatographic separations are achieved by matching the polarity of the analyte to that of the stationary phase; a mobile phase of considerably different polarity is then used. This procedure is generally more successful than one in which the polarities of the analyte and the mobile phase are matched but are different from that of the stationary phase. Here, the stationary phase often cannot compete successfully for the sample components; retention times then become too short for practical application. At the other extreme is the situation where the polarities of the analyte and stationary phase are too much alike; here, retention times become inordinately long.

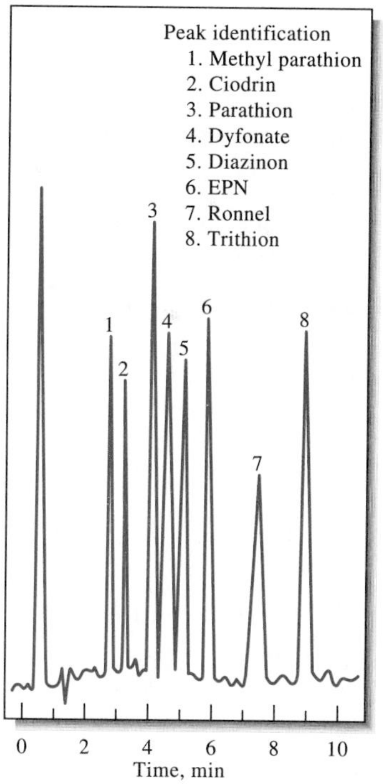

Figure 30-10
Liquid bonded-phase chromatogram of organic phosphate insecticides. Packing: nonpolar C_8. Gradient elution: 67% CH_3OH/33% H_2O to 80% CH_3OH/20% H_2O. (Courtesy of IBM Instruments, Inc., Danbury, CT.)

30B-4 Applications

Figures 30-9 and 30-10 illustrate typical applications of bonded-phase partition chromatography. Table 30-2 further illustrates the variety of samples to which the technique is applicable.

TABLE 30-2 Typical Applications of High-Performance Partition Chromatography

Field	Typical Mixtures
Pharmaceuticals	Antibiotics, sedatives, steroids, analgesics
Biochemicals	Amino acids, proteins, carbohydrates, lipids
Food products	Artificial sweeteners, antioxidants, aflatoxins, additives
Industrial chemicals	Condensed aromatics, surfactants, propellants, dyes
Pollutants	Pesticides, herbicides, phenols, PCBs
Forensic chemistry	Drugs, poisons, blood alcohol, narcotics
Clinical medicine	Bile acids, drug metabolites, urine extracts, estrogens

30C HIGH-PERFORMANCE ADSORPTION CHROMATOGRAPHY

In **adsorption chromatography,** analyte species are adsorbed onto the surface of a polar packing.

All of the pioneering work in chromatography was based upon adsorption of analyte species on a solid surface. Here, the stationary phase is the surface of a finely divided polar solid. With such a packing, the analyte competes with the mobile phase for sites on the surface of the packing, and retention is the result of adsorption forces.

30C-1 Stationary and Mobile Phases

Finely divided silica and alumina are the only stationary phases that find extensive use for adsorption chromatography. Silica is preferred for most (but not all) applications because of its higher sample capacity and its wider range of useful forms. The adsorption characteristics of the two substances parallel one another. For both, retention times become longer as the polarity of the analyte increases.

In adsorption chromatography, the mobile phase is usually an organic solvent or a mixture of organic solvents; the stationary phase is finely divided particles of silica or alumina.

In adsorption chromatography, the only variable that affects the partition coefficient of analytes is the composition of the mobile phase (in contrast to partition chromatography, where the polarity of the stationary phase can also be varied). Fortunately, enormous variations in retention and thus resolution accompany variations in the solvent system, and only rarely is a suitable mobile phase not available.

30C-2 Applications of Adsorption Chromatography

Currently, liquid-solid HPLC is used extensively for the separations of relatively nonpolar, water-insoluble organic compounds with molar masses that are less than about 5000. A particular strength of adsorption chromatography, which is not shared by other methods, is its ability to resolve isomeric mixtures such as meta and para substituted benzene derivatives.

Ion chromatography is a type of HPLC in which the analytes are ions dissolved in an aqueous mobile phase. The stationary phase is a finely ground ion-exchange resin.

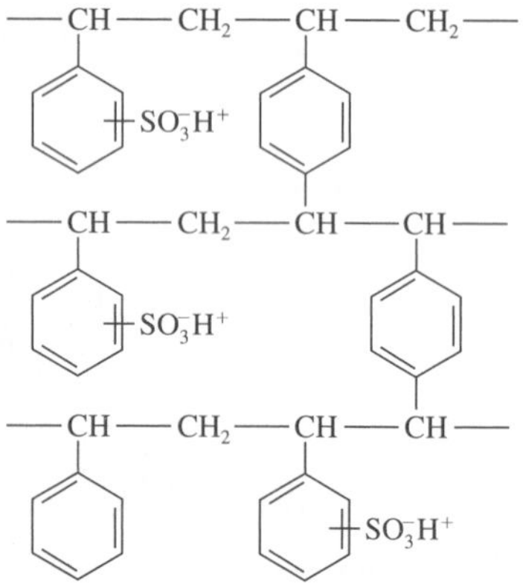

Figure 30-11

Structure of a cross-linked polystyrene ion-exchange resin. Similar resins are used in which the $—SO_3^-H^+$ group is replaced by $—COO^-H^+$, $—NH_3^+OH^-$, and $—N(CH_3)_3^+OH^-$ groups.

30D HIGH-PERFORMANCE ION-EXCHANGE CHROMATOGRAPHY

Ion-exchange resins serve as the stationary phase in *ion chromatography,* a separation procedure in which ions of like charge are separated by elution from a column packed with a finely divided resin.

30D-1 Ion-Exchange Resins

Synthetic ion-exchange resins are high-molecular-weight polymeric materials containing many ionic functional groups per molecule. Cation-exchange resins can be either a strong-acid type with sulfonic acid groups ($RSO_3^-H^+$) or a weak-acid type containing carboxylic acid groups (RCOOH) (see Figure 30-11); the former have wider application. Anion-exchange resins contain basic amine functional groups attached to the polymer molecule. Strong-base exchangers are quaternary amines [$RN(CH_3)_3^+OH^-$]; weak-base types contain secondary or tertiary amines.

An important property of all ion-exchange resins is that they are essentially insoluble in aqueous media. Thus, when a cation exchanger is immersed in an aqueous solution containing the M^{x+} cations, the following exchange equilibrium is quickly established between the solid and solution phases:

$$\underset{\text{solid}}{xRSO_3^-H^+} + \underset{\text{solution}}{M^{x+}} \rightleftharpoons \underset{\text{solid}}{(RSO_3^-)_xM^{x+}} + \underset{\text{solution}}{xH^+}$$

where M^{x+} is a cation and R represents *that part of a resin molecule containing one sulfonic acid group.* The analogous process involving a typical anion-exchange resin can be written as

$$\underset{\text{solid}}{xRN(CH_3)_3^+OH^-} + \underset{\text{solution}}{A^{x-}} \rightleftharpoons \underset{\text{solid}}{[RN(CN_3)_3^+]_xA^{x-}} + \underset{\text{solution}}{xOH^-}$$

where A^{x-} is an anion.

30D-2 Ion-Exchange Equilibria

Ion-exchange equilibria can be treated by the law of mass action. For example, when a dilute solution of calcium ions is brought into contact with a sulfonic acid resin, the following equilibrium develops.

$$Ca^{2+}(aq) + 2H^+(res) \rightleftharpoons Ca^{2+}(res) + 2H^+(aq)$$

Application of the mass law leads to

$$K' = \frac{[Ca^{2+}]_{res}[H^+]_{aq}^2}{[Ca^{2+}]_{aq}[H^+]_{res}^2}$$

where the bracketed terms are molar concentrations (strictly, activities). Note that $[Ca^{2+}]_{res}$ and $[H^+]_{res}$ are the concentrations of the two ions *in the solid phase.* In contrast to most solids, however, these concentrations can vary from zero to some maximum value at which all the available sites in the resin are occupied by one species only.

Ion-exchange separations are ordinarily performed under conditions in which one of the ions predominates in *both* phases. For example, in the removal of calcium ions from a dilute and somewhat acidic solution, the calcium ion concentration is much smaller than that of hydrogen ions in both the aqueous and the resin phase. Consequently, the concentration of hydrogen ions does not change significantly as a result of the exchange process. Under these circumstances, $[Ca^{2+}]_{res} \ll [H^+]_{res}$ and $[Ca^{2+}]_{aq} \ll [H^+]_{aq}$, and the foregoing equilibrium-constant expression can be rearranged to give

$$\frac{[Ca^{2+}]_{res}}{[Ca^{2+}]_{aq}} = K'\frac{[H^+]_{res}^2}{[H^+]_{aq}^2} = K \qquad (30\text{-}1)$$

where K is the distribution ratio as defined by Equation 28-1. Note that K in Equation 30-1 represents the affinity of the resin for calcium ion relative to another ion (here, H^+). In general, where K for an ion is large, a strong tendency for the stationary phase to retain that ion exists; where K is small, the opposite is true. Selection of a common reference ion (such as H^+) permits a comparison of distribution ratios for various ions on a given type of resin. Such experiments reveal that polyvalent ions are much more strongly retained than singly charged species. Within a given charge group, differences in K appear to be related to the size of the hydrated ion as well as other properties. Thus, for a typical sulfonated cation-exchange resin, values of K for univalent ions decrease in the order $Ag^+ > Cs^+ > Rb^+ > K^+ > NH_4^+ > Na^+ > H^+ > Li^+$. For divalent cations, the order is $Ba^{2+} > Pb^{2+} > Sr^{2+} > Ca^{2+} > Ni^{2+} > Cd^{2+} > Cu^{2+} > Co^{2+} > Zn^{2+} > Mg^{2+} > UO_2^{2+}$.

Polyvalent ions are more strongly held by ion-exchange resins than are univalent ions.

30D-3 Applications of Ion-Exchange Resins to Chromatography

In ion chromatography, analyte ions are introduced at the head of a column packed with a suitable ion-exchange resin. Elution is then carried out with a solution that contains an ion that competes with the analyte ions for the charged groups on the resin surface. For example, anions such as chloride, thiocyanate, sulfate, and phosphate can be separated on an anion-exchange resin in its basic form. In this application, the sample is first introduced onto the head of the column, where the anions are retained by the reaction

$$xRN(CH_3)_3^+OH^- + A^{x-} \rightleftharpoons [RN(CH_3)_3^+]_xA^{x-} + xOH^-$$

Elution is then carried out with a dilute solution of a base, which causes the foregoing reaction to be reversed and the anions released. Since the partition ratios for various anions differ from one another, fractionation occurs during the elution.

One of the attractive aspects of ion chromatography is that conductivity measurements provide a convenient general method for detecting and determining the concentration of the eluted species. Ion chromatography with conductometric detection of ions eluted from the column was first described in 1975 and by now has become a powerful and important technique for the quantitative determination of both inorganic and organic charged species.[3]

Two types of chromatography based on ion-exchange packings are currently in use: *suppressor-based* and *single-column* ion chromatography. They differ in the method used to prevent the conductivity of the eluting electrolyte from interfering with the measurement of the conductivity of the analytes.

30D-4 Ion Chromatography Based on Suppressors

The conductivity detector is well suited for ion chromatography.

Conductivity detectors have many of the properties of the ideal detector alluded to in Section 29A-4. They can be highly sensitive, they are universal for charged species, and as a general rule, they respond in a predictable way to concentration changes. Furthermore, such detectors are simple to operate, inexpensive to con-

[3]For a brief review of ion chromatography, see J. S. Fritz, *Anal. Chem.*, **1987,** *59,* 335A. For a detailed description of the method, see H. Small, *Ion Chromatography.* New York: Plenum Press, 1989; D. T. Gjerde and J. S. Fritz, *Ion Chromatography,* 2nd ed. New York: A. Heuthig, 1987.

struct and maintain, easy to miniaturize, and ordinarily give prolonged, trouble-free service. The only limitation to the use of conductivity detectors, which delayed their general application to ion chromatography until the mid-1970s, was due to the high electrolyte concentrations required to elute most analyte ions in a reasonable time. As a consequence, the conductivity from the mobile phase components tended to swamp that from the analyte ions, thus greatly reducing the detector sensitivity.

In 1975, the problem created by the high conductance of eluents was solved by the introduction of an *eluent suppressor column* immediately following the ion-exchange column.[4] The suppressor column is packed with a second ion-exchange resin that effectively converts the ions of the eluting solvent to a molecular species of limited ionization without affecting the conductivity due to the analyte ions. For example, when cations are being separated and determined, hydrochloric acid is chosen as the eluting reagent, and the suppressor column is an anion-exchange resin in the hydroxide form. The product of the reaction between the eluent and the suppressor is water:

In **suppressor-based ion chromatography,** the ion-exchange column is followed by a **suppressor column,** or a **suppressor membrane,** which converts an ionic eluent into a nonionic species that does not interfere with the conductometric detection of analyte ions.

$$H^+(aq) + Cl^-(aq) + resin^+OH^-(s) \longrightarrow resin^+Cl^-(s) + H_2O$$

The analyte cations are of course not retained by anionic groups in this second column.

For anion separations, the suppressor packing is the acid form of a cation-exchange resin and sodium bicarbonate or carbonate is the eluting agent. The reaction in the suppressor is

$$Na^+(aq) + HCO_3^-(aq) + resin^-H^+(s) \longrightarrow resin^-Na^+(s) + H_2CO_3(aq)$$

The largely undissociated carbonic acid does not contribute significantly to the conductivity.

An inconvenience associated with the original suppressor columns was the need to regenerate them periodically (typically, every 8 to 10 hr) in order to convert the packing back to the original acid or base form. Recently, however, micromembrane suppressors that operate continuously have become available.[5] For example, where sodium carbonate or bicarbonate is to be removed, the eluent is passed over a series of ultrathin cation-exchange membranes that separate it from a stream of acidic regenerating solution that flows continuously in the opposite direction. The sodium ions from the eluent undergo exchange with hydrogen ions on the inner surface of the exchanger membrane and then migrate to the other surface for exchange with hydrogen ions from the regenerating reagent. Hydrogen ions from the regeneration solution migrate in the reverse direction, thus preserving electrical neutrality.

Figures 30-12 and 30-13 show applications of ion chromatography based upon a suppressor column and conductometric detection. In each, the ions were present in the parts-per-million range; the sample size was 50 μL in one case and 20 μL in the other. The method is particularly important for anion analysis because no other rapid and convenient method for handling mixtures of this type exists.

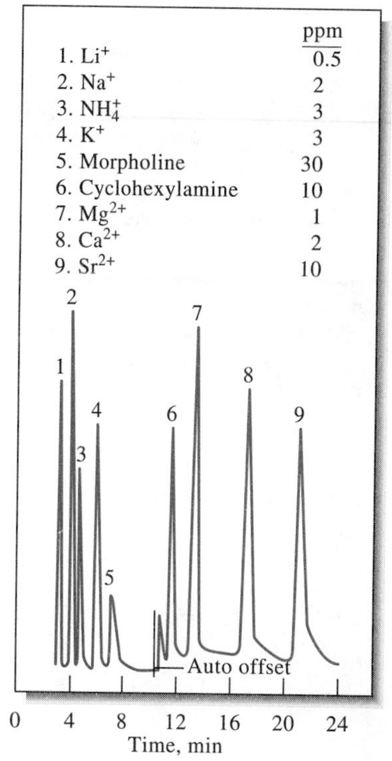

Figure 30-12
Ion chromatogram of a mixture of cations. (Courtesy of Dionex, Sunnyvale, CA.)

---

[4]H. Small, T. S. Stevens, and W. C. Bauman, *Anal. Chem.*, **1975,** *47,* 1801.

[5]For a description of this device, see G. O. Franklin, *Amer. Lab.*, **1985** (3), 71.

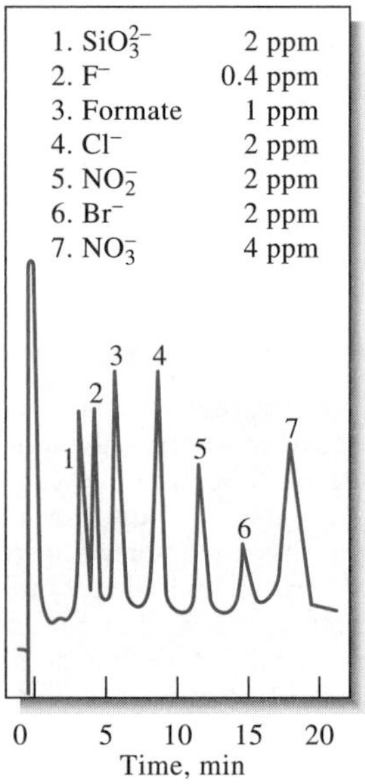

Figure 30-13
Ion chromatogram of a mixture of anions. (Courtesy of Dionex, Sunnyvale, CA.)

In **single-column ion-exchange chromatography,** analyte ions are separated on a low-capacity ion exchanger by means of a low-ionic strength eluent that does not interfere with the conductometric detection of analyte ions.

In **size-exclusion chromatography,** fractionation is based upon molecular size.

Gel filtration is a type of size-exclusion chromatography in which the packing is hydrophilic. It is used for separating polar species.

Gel permeation is a type of size-exclusion chromatography in which the packing is hydrophobic. It is used to separate nonpolar species.

30D-5 Single-Column Ion Chromatography

Recently, equipment has become available commercially for ion chromatography in which no suppressor column is used. This approach depends upon the small differences in conductivity between the eluted sample ions and the prevailing eluent ions. To amplify these differences, low-capacity exchangers that permit elution with dilute eluent solutions are used. Furthermore, eluents of low equivalent conductance are chosen.[6]

Single-column ion chromatography offers the advantage of not requiring special equipment for suppression. It is, however, a somewhat less sensitive method for determining anions than suppressor column methods.

30E HIGH-PERFORMANCE SIZE-EXCLUSION CHROMATOGRAPHY

Size-exclusion, or gel, chromatography is the newest of the liquid chromatographic procedures. It is a powerful technique that is particularly applicable to high-molecular-weight species.[7]

30E-1 Packings

Packings for size-exclusion chromatography consist of small (~10 μm) silica or polymer particles containing a network of uniform pores into which solute and solvent molecules can diffuse. While in the pores, molecules are effectively trapped and removed from the flow of the mobile phase. The average residence time of analyte molecules depends upon their effective size. Molecules that are significantly larger than the average pore size of the packing are excluded and thus suffer no retention; that is, they travel through the column at the rate of the mobile phase. Molecules that are appreciably smaller than the pores can penetrate throughout the pore maze and thus become entrapped for the greatest time; they are last to be eluted. Between these two extremes are intermediate-size molecules whose average penetration into the pores of the packing depends upon their diameters. Fractionation within this group is directly related to molecular size and, to some extent, molecular shape. Note that size-exclusion separations differ from the other chromatographic procedures in the respect that no chemical or physical interactions between analytes and the stationary phase are involved. Indeed, every effort is made to avoid such interactions because they lead to impaired column efficiencies.

Numerous size-exclusion packings are on the market. Some are hydrophilic for use with aqueous mobile phases; others are hydrophobic and are used with nonpolar organic solvents. Chromatography based on the hydrophilic packings is sometimes called *gel filtration,* while techniques based on hydrophobic packings are termed *gel permeation.* With both types of packings a wide variety of pore diameters are available. Ordinarily, a given packing will accommodate a 2- to

[6]See R. M. Becker, *Anal. Chem.,* **1980,** *52,* 1510; J. R. Benson, *Amer. Lab.,* **1985** (6), 30; T. Jupille, *Amer. Lab.,* **1985** (5), 114.

[7]See W. W. Yao, J. J. Kirkland, and D. D. Bly, *Modern Size Exclusion Liquid Chromatography.* New York: Wiley, 1979.

2.5-decade range of molecular weight. The average molecular weight suitable for a given packing may be as small as a few hundred or as large as several million.

30E-2 Applications

The chromatograms in Figures 30-14 and 30-15 illustrate typical applications of size-exclusion chromatography. In the first, a hydrophilic packing was used to exclude molecular weights greater than 1000. The second chromatogram was obtained with a hydrophobic packing in which the eluent was tetrahydrofuran. The sample was a commercial epoxy resin in which each monomer unit had a molecular weight of 280 (n = number of monomer units).

Another important application of size-exclusion chromatography involves the rapid determination of the molecular weight or the molecular weight distribution of large polymers or natural products. Here, the elution volumes of the sample are compared with elution volumes for a series of standard compounds that have the same chemical characteristics.

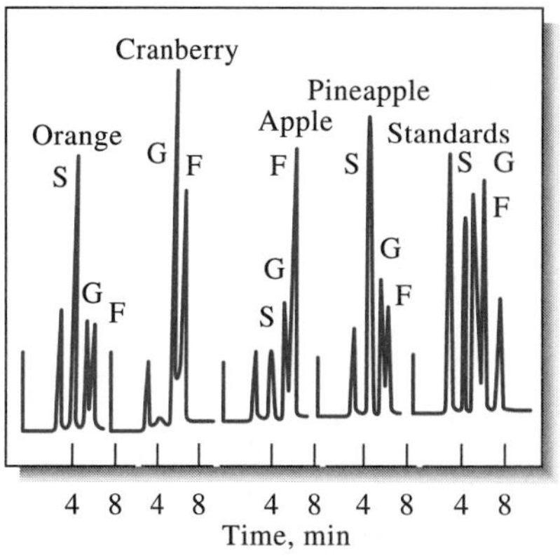

Figure 30-14
Gel-filtration chromatogram for glucose (G), fructose (F), and sucrose (S) in canned juices.

FEATURE 30-1
Buckyballs: The Chromatographic Separation of Fullerenes

Our ideas about the nature of matter are often profoundly influenced by chance discoveries. No event in recent memory has captured the imagination of both the scientific community and the public as did the serendipitous discovery in 1985 of the soccerball-shaped molecule C_{60}. This molecule, illustrated in Figure 30-16, and its cousin C_{70}, are more commonly called *buckyballs.*[8] The compounds are named in honor of the famous architect, R. Buckminster Fuller, who designed many geodesic dome buildings having the same hexagonal/pentagonal structure as buckyballs. Since their discovery, thousands of research groups throughout the world have studied various chemical and physical properties of these highly stable molecules. They represent a third allotropic form of carbon besides graphite and diamond.

The preparation of buckyballs is almost trivial. When an ac arc is established between two carbon electrodes in a flowing helium atmosphere, the soot that is collected is rich in C_{60} and C_{70}. Although the preparation is easy, the separation and purification of more than a few milligrams of C_{60} proved tedious and expensive. Recently, it was found that relatively large quantities of buckyballs could be separated using size-exclusion chromatography.[9] Fullerenes are extracted from soot prepared as mentioned above and injected on a 199 mm × 30 cm 500 Å Ultrastyragel column (Waters Chromatography Division of Millipore Corporation), using toluene as the mobile phase and UV/visible detection following separation. A typical chromatogram is shown in Figure 30-17. The peaks in the chromatogram are labeled with their identities and retention times.

[8]R. F. Curl and R. E. Smalley, *Scientific American,* **1991,** *265* (4), 54.

[9]M. S. Meier and J. P. Selegue, *J. Org. Chem.,* **1992,** *57,* 1924; A. Gügel and K. Müllen, *J. Chromatogr.,* **1993,** *628,* 23.

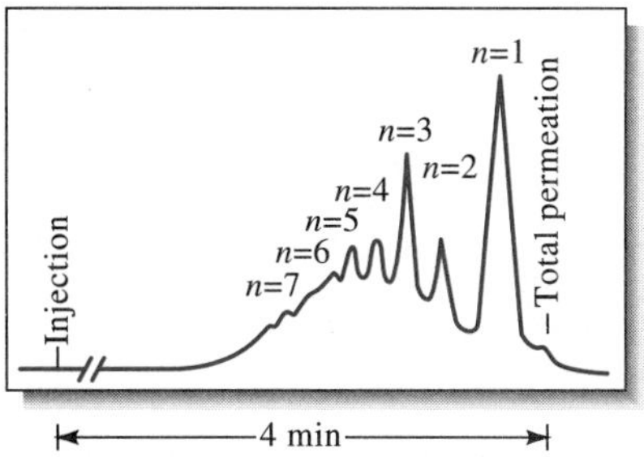

Figure 30-15
Gel-permeation separation of components in an epoxy resin.

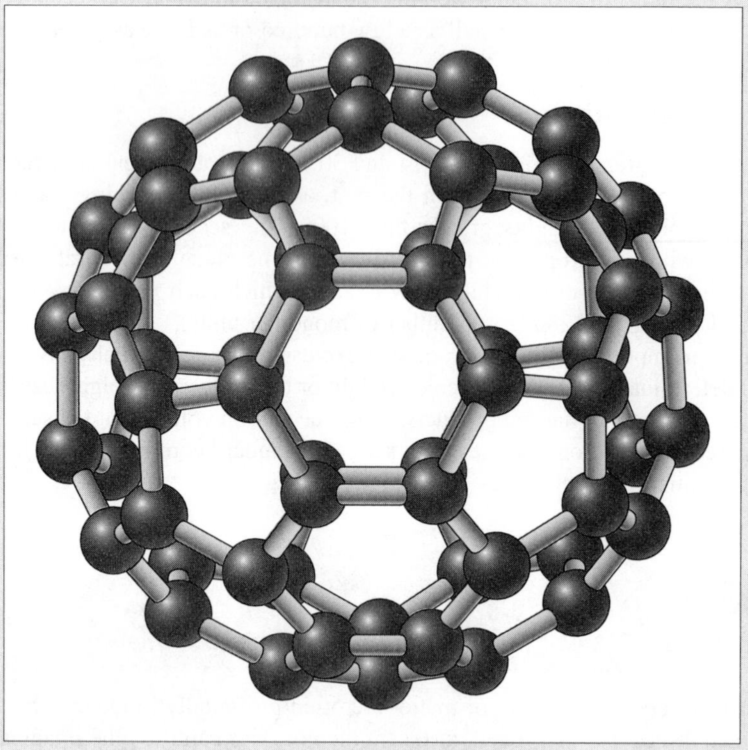

Figure 30-16 Buckminster fullerene, C_{60}.

Note that C_{60} elutes before C_{70} and the higher fullerenes. This is contrary to what we expect; the smallest molecule, C_{60}, should be retained more strongly than C_{70} and the higher fullerenes. It has been suggested that the interaction between the solute molecules and the gel is on the surface of the

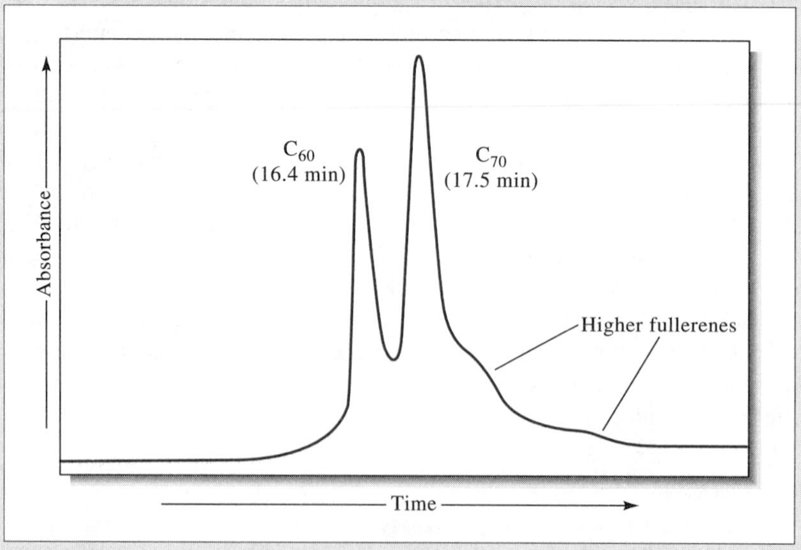

Figure 30-17 Separation of fullerenes.

gel rather than in its pores. Since C_{70} and the higher fullerenes have larger surface areas than C_{60}, the higher fullerenes are retained more strongly on the surface of the gel and thus elute after C_{60}. Using automated apparatus, this method of separation may be used to prepare several grams of 99.8% pure C_{60} from 5 to 10 g of a C_{60} to C_{70} mixture in a 24-hr period. These quantities of C_{60} can then be used to prepare and study the chemistry and physics of derivatives of this interesting and unusual form of carbon.

As an example, C_{60} forms stoichiometric compounds with the alkali metals that have the general formula M_3C_{60}, where M is potassium, rubidium, or cesium. Researchers have found that these compounds are superconductors at temperatures below 30 K. In the future these or similar compounds may help lead the way to practical high-temperature superconductors, which would revolutionize the electrical, electronic, and telecommunication industries and conserve vast amounts of energy.

30F COMPARISON OF HIGH-PERFORMANCE LIQUID CHROMATOGRAPHY WITH GAS-LIQUID CHROMATOGRAPHY

Table 30-3 provides a comparison between high-performance liquid chromatography and gas-liquid chromatography. When either is applicable, gas-liquid chromatography offers the advantage of speed and simplicity of equipment. On the other hand, high-performance liquid chromatography is applicable to nonvolatile substances (including inorganic ions) and thermally unstable materials, whereas gas-liquid chromatography is not. Often, the two methods are complementary.

TABLE 30-3 Comparison of High-Performance Liquid Chromatography and Gas-Liquid Chromatography

Characteristics of both methods

- Efficient, highly selective, widely applicable
- Only small sample required
- May be nondestructive of sample
- Readily adapted to quantitative analysis

Advantages of HPLC

- Can accommodate nonvolatile and thermally unstable samples
- Generally applicable to inorganic ions

Advantages of GLC

- Simple and inexpensive equipment
- Rapid
- Unparalleled resolution (with capillary columns)
- Easily interfaced with mass spectroscopy

30G SUPERCRITICAL-FLUID CHROMATOGRAPHY

Supercritical-fluid chromatography (SFC) is a hybrid of gas and liquid chromatography that combines some of the best features of each. For certain applications, it appears to be clearly superior to both gas-liquid and high-performance liquid chromatography.[10]

30G-1 Important Properties of Supercritical Fluids

A **supercritical fluid** is the physical state of a substance when it is held above its critical temperature.

The **critical temperature** is the temperature above which a distinct liquid phase for a substance cannot exist.

The density of a supercritical fluid is 200 to 400 times that of its gaseous state, and it is nearly as dense as its liquid state.

Supercritical fluids tend to dissolve large, nonvolatile molecules.

A *supercritical fluid* is formed whenever a substance is heated above its *critical temperature.* At the critical temperature, a substance can no longer be condensed into its liquid state through the application of pressure. For example, carbon dioxide becomes a supercritical fluid at temperatures above 31°C. In this state, the molecules of carbon dioxide act independently, just as they do in a gas.

As shown by the data in Table 30-4, the physical properties of a substance in the supercritical-fluid state can be remarkably different from the same properties in either the liquid or the gaseous state. For example, the density of a supercritical fluid is typically 200 to 400 times greater than that of the corresponding gas and approaches that of the substance in its liquid state. The properties compared in Table 30-4 are those that are of importance in gas, liquid, and supercritical-fluid chromatography.

An important property of supercritical fluids and one that is related to their high densities (0.2 to 0.5 g/cm^3), is their ability to dissolve large nonvolatile molecules. For example, supercritical carbon dioxide readily dissolves *n*-alkanes containing from 5 to 22 carbon atoms, di-*n*-alkylphthalates in which the alkyl groups contain 4 to 16 carbon atoms, and various polycyclic aromatic hydrocarbons consisting of several rings.[11]

Critical temperatures for fluids used in chromatography vary widely, from about 30°C to above 200°C. Lower critical temperatures are advantageous from several standpoints. For this reason, much of the work to date has focused on such supercritical fluids as carbon dioxide (31°C), ethane (32°C), and nitrous oxide (37°C). Note that these temperatures, and the pressures at these temperatures, are well within the operating conditions of ordinary high-performance liquid chromatography.

30G-2 Instrumentation and Operating Variables

Instruments for supercritical-fluid chromatography are similar in design to high-performance liquid chromatographs except that provision is made in the former for controlling and measuring the column pressure. Several manufacturers began to offer apparatus for supercritical-fluid chromatography in the mid-1980s.[12]

[10]M. D. Palmieri, *J. Chem. Educ.*, **1988,** *65,* A254; **1989,** *66,* A141; P. R. Griffiths, *Anal. Chem.*, **1988,** *60,* 593A; R. D. Smith, B. W. Wright, and C. R. Yonker, *Anal. Chem.*, **1988,** *60,* 1323A.

[11]Certain important industrial processes are based upon the high solubility of organic species in supercritical carbon dioxide. For example, this medium has been employed in extracting caffeine from coffee beans to give decaffeinated coffee and in extracting nicotine from cigarette tobacco.

[12]For descriptions of several commercial instruments for SFC, see F. Wach, *Anal. Chem.*, **1994,** *66,* 372A.

TABLE 30-4 Comparison of Properties of Supercritical Fluids, Liquids, and Gases*

	Gas (STP)	Supercritical Fluid	Liquid
Density, g/cm^3	$(0.6-2) \times 10^{-3}$	0.2–0.5	0.6–1.6
Diffusion coefficient, $cm^2 \cdot s^{-1}$	$(1-4) \times 10^{-1}$	$10^{-3}-10^{-4}$	$(0.2-2) \times 10^{-5}$
Viscosity, $g \cdot cm^{-1} \cdot s^{-1}$	$(1-3) \times 10^{-4}$	$(1-3) \times 10^{-4}$	$(0.2-3) \times 10^{-2}$

*All data order of magnitude only.

The Effect of Pressure

The density of a supercritical fluid increases rapidly and nonlinearly with pressure increases. Density increases also alter capacity factors (k') and thus elution times. For example, the elution time for hexadecane is reported to decrease from 25 to 5 min as the pressure of carbon dioxide is raised from 70 to 90 atm. Gradient elution can thus be achieved through linear increases in column pressure or through regulation of pressure to obtain linear density increases. A clear analogy exists between gradient elution through adjustment of pressure or density of a supercritical fluid and by temperature-gradient elution in gas-liquid chromatography and solvent-gradient elution in liquid chromatography.

Gradient elution can be achieved in SFC by systematically changing the column pressure or the density of the supercritical fluid.

Columns

Both packed columns and open tubular columns are used in supercritical fluid chromatography. Packed columns have the advantages of greater efficiency per unit time and the capability of handling larger sample volumes. Packed columns for SFC can be much longer than those for HPLC, providing well over 100,000 plates. Open tubular columns are similar to the fused silica open tubular (FSOT) columns described in Table 29-1. Because of the low viscosity of supercritical media, columns can be much longer than those used in liquid chromatography, and column lengths of 10 to 20 m and inside diameters of 50 or 100 μm are common. For difficult separations, columns 60 m in length and longer have been used.

Very long columns can be used in SFC because the viscosity of supercritical fluids is so low.

Many of the column coatings used in liquid chromatography have been applied to supercritical-fluid chromatography as well. Typically, these are polysiloxanes (see Table 29-2) that are chemically bonded to the surface of silica particles or to the inner silica wall of capillary tubing. Film thicknesses are 0.05 to 0.4 μm.

Mobile Phases

The most widely used mobile phase for supercritical-fluid chromatography is carbon dioxide. It is an excellent solvent for a variety of nonpolar organic molecules. In addition, it transmits in the ultraviolet and is odorless, nontoxic, readily available, and remarkably inexpensive relative to other chromatographic solvents. Its critical temperature of 31°C and its pressure of 73 atm at the critical temperature permit a wide selection of temperatures and pressures without exceeding the operating limits of modern high-performance liquid chromatography equipment.

In some applications, polar organic modifiers, such as methanol, are introduced in small concentrations (~1%) to modify α values for analytes.

A number of other substances have served as mobile phases in supercritical chromatography, including ethane, pentane, dichlorodifluoromethane, diethyl ether, and tetrahydrofuran.

Detectors

A major advantage of supercritical-fluid chromatography is that the sensitive and universal detectors of gas-liquid chromatography are applicable to this technique as well. For example, the convenient flame ionization detector of gas-liquid chromatography can be applied by simply allowing the supercritical carrier to expand through a restrictor and into a hydrogen flame, where ions formed from the analytes cause variations in the electrical conductivity of the medium.

30G-3 Supercritical-Fluid Chromatography Versus Other Column Methods

The information in Table 30-4, and other data as well, reveals that several physical properties of supercritical fluids are intermediate between the properties of gases and liquids. As a consequence, this new type of chromatography combines some of the characteristics of both gas and liquid chromatography. Thus, like gas chromatography, supercritical-fluid chromatography is inherently faster than liquid chromatography because of the lower viscosity and higher diffusion rates in the mobile phase. High diffusivity, however, leads to longitudinal band spreading (page 673), which is a significant factor with gas but not with liquid chromatography. Thus, the intermediate diffusivities and viscosities of supercritical fluids result in faster separations than are achieved with liquid chromatography and are accompanied by less zone spreading than is encountered in gas chromatography.

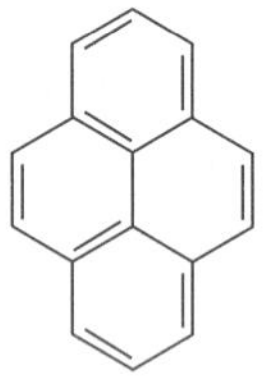
pyrene

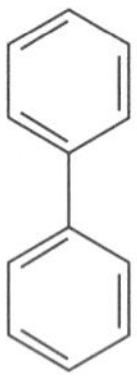
biphenyl

Figure 30-18 shows plots of plate heights (H) as a function of flow rate ($\bar{u}$) for high-performance liquid chromatography and supercritical-fluid chromatography. In both cases, the solute was pyrene and the stationary phase a reversed-phase octyldecyl silane maintained at 40°C. An acetonitrile/water solution served as the mobile phase for HPLC, and carbon dioxide for SFC. These conditions

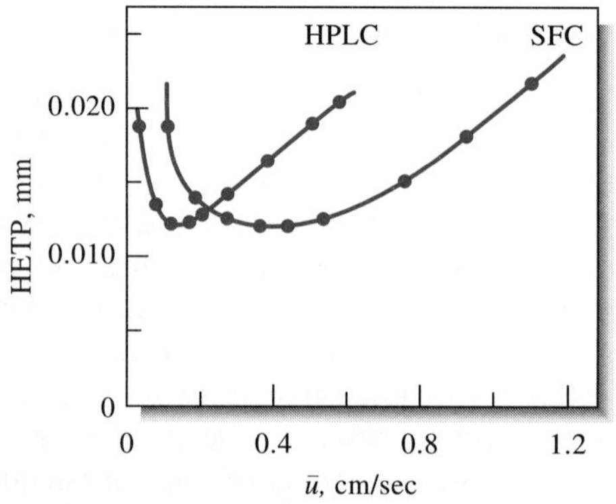

Figure 30-18
Comparison of the efficiency of an HPLC column with that of an SFC column. (From D. R. Gere, *Science,* **1983,** *222,* 254. With permission.)

yielded about the same capacity factor (k') for both mobile phases. Note that the minimum in plate height occurred at a flow rate of 0.13 cm/s with the HPLC and 0.40 cm/s for the SFC. The consequence of this difference is shown in Figure 30-19 where these same conditions are used for the separation of pyrene from biphenyl. Note that the HPLC separation required more than twice the time of the SFC separation.

30G-4 Applications

Supercritical-fluid chromatography, particularly of the open-tubular type, appears to have a potential niche in the spectrum of column-chromatographic methods because it is applicable to a class of compounds that is not readily amenable to either gas-liquid or liquid chromatography. These compounds include species that are nonvolatile or thermally unstable and, in addition, contain no chromophoric groups that can be used for photometric detection. Separation of these compounds is possible with supercritical-fluid chromatography at temperatures below 100°C; furthermore, detection is readily carried out by means of the highly sensitive flame ionization detector.

SFC with flame ionization detection works very well for nonvolatile or thermally unstable compounds that have no chromophores for photometric detection.

It is also noteworthy that supercritical columns have the added advantage of being much easier to interface with mass spectrometers than liquid chromatographic columns.

30H PLANAR CHROMATOGRAPHY

Planar chromatographic methods include *thin-layer chromatography* (TLC), *paper chromatography* (PC), and *electrochromatography*. Each makes use of a flat, relatively thin layer of material that is either self-supporting or is coated on a glass, plastic, or metal surface. The mobile phase moves through the stationary phase by capillary action, sometimes assisted by gravity or an electrical potential. Planar chromatography is sometimes called two-dimensional chromatography, although this description is not strictly correct inasmuch as the stationary phase does have a finite thickness.

Currently, most planar chromatography is based upon the thin-layer technique, which is faster, has better resolution, and is more sensitive than its paper counterpart. This section is devoted to thin-layer methods.

30H-1 The Scope of Thin-Layer Chromatography

In terms of theory, the types of stationary and mobile phases, and applications, thin-layer and liquid chromatography are remarkably similar. In fact, thin-layer plates can be profitably used to develop optimal conditions for separations by column liquid chromatography. The advantages of following this procedure are the speed and low cost of the exploratory thin-layer experiments. Some chromatographers take the position that thin-layer experiments should always precede column experiments.

Preliminary thin-layer studies provide valuable clues for successful liquid chromatography.

Thin-layer chromatography has become the workhorse of the drug industry for the all-important determination of product purity. It has also found widespread use in clinical laboratories and is the backbone of many biochemical and biologi-

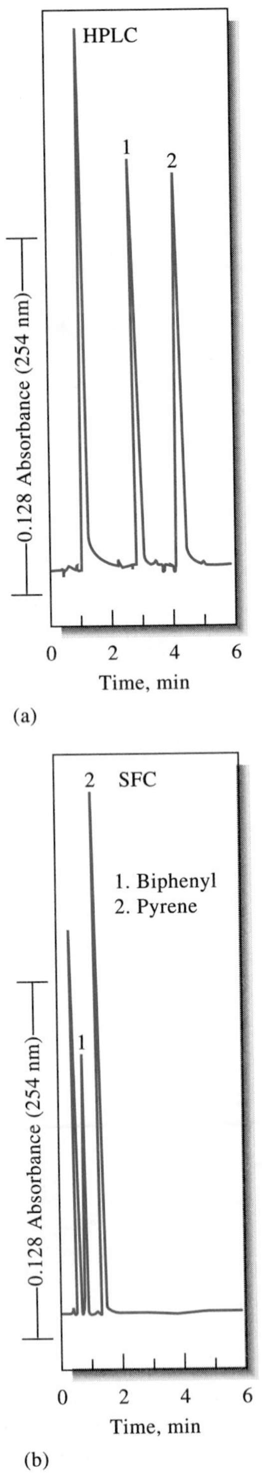

Figure 30-19

Separation of pyrene and biphenyl by (a) HPLC and (b) SFC. (From D. R. Gere, *Science,* **1983,** *222,* 255. With permission.)

cal studies. Finally, it finds widespread use in the industrial laboratories.[13] As a consequence of these many areas of application, it has been estimated that at least as many analyses are performed by thin-layer chromatography as by high-performance liquid chromatography.[14]

30H-2 The Principles of Thin-Layer Chromatography

Typical thin-layer separations are performed on a glass plate that is coated with a thin and adherent layer of finely divided particles; this layer constitutes the stationary phase. The particles are similar to those described in the discussion of adsorption, normal- and reversed-phase partition, ion-exchange, and size-exclusion column chromatography. Mobile phases are also similar to those employed in high-performance liquid chromatography.

The Preparation of Thin-Layer Plates

A thin-layer plate is prepared by spreading an aqueous slurry of the finely ground solid onto the clean surface of a glass or plastic plate or microscope slide. Often a binder is incorporated into the slurry to enhance adhesion of the solid particles to the glass and to one another. The plate is then allowed to stand until the layer has set and adheres tightly to the surface; for some purposes, it may be heated in an oven for several hours. Several chemical supply houses offer precoated plates of various kinds.

Plate Development

Plate development is the process in which a sample is carried through the stationary phase by a mobile phase; the process is analogous to elution in liquid chromatography. The most common way of developing a plate is to place a drop of the sample near one edge (most plates have dimensions of 5 × 20 or 20 × 20 cm) and mark its position with a pencil. After the sample solvent has evaporated, the plate is placed in a closed container that is saturated with vapors of the developing solvent. One end of the plate is immersed in the developing solvent, with care being taken to avoid direct contact between the sample and the developer (Figure 30-20). After the developer has traversed one half or two thirds of the length of the plate, the plate is removed from the container and dried. The positions of the components are then determined in any of several ways.

Figure 30-21 illustrates the separation of amino acids in a mixture by development in two directions *(two-dimensional planar chromatography).* The sample was placed in one corner of a square plate, and the plate was developed in the ascending direction with solvent A. This solvent was then removed by evaporation, and the plate was rotated 90 deg, following which ascending development

[13]Two monographs devoted to the principles and applications of thin-layer chromatography are R. Hamilton and S. Hamilton, *Thin-Layer Chromatography.* New York: Wiley, 1987; J. C. Touchstone, *Practice of Thin-Layer Chromatography,* 2nd ed. New York: Wiley, 1983. For briefer reviews, see D. C. Fenimore and C. M. Davis, *Anal. Chem.,* **1981,** *53,* 253A; C. F. Poole and S. K. Poole, *Anal. Chem.,* **1989,** *61,* 1257A.

[14]T. H. Mauch II, *Science,* **1982,** *216,* 161.

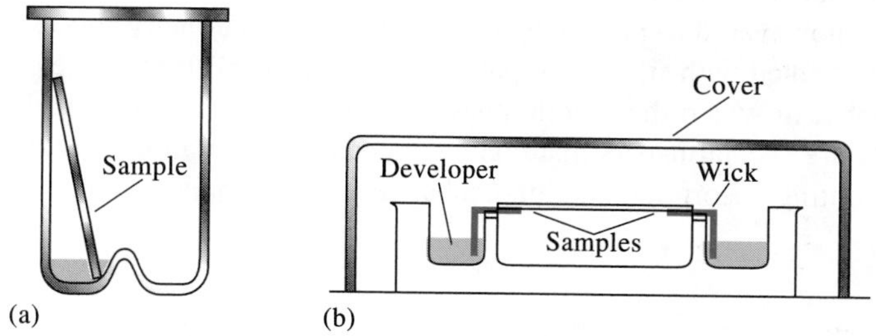

Figure 30-20

(a) Ascending-flow developing chamber. (b) Horizontal-flow developing chamber, in which samples are placed on both ends of the plate and developed toward the middle, thus doubling the number of samples that can be accommodated.

was performed with solvent B. After solvent removal, the positions of the amino acids were determined by spraying with ninhydrin, a reagent that forms a pink to purple product with amino acids. The spots were identified by comparison of their positions with those of standards.

Locating Analytes on the Plate

Several methods are employed to locate sample components after separation. Two common methods, which can be applied to most organic mixtures, involve spraying with a solution of iodine or sulfuric acid, both of which react with organic compounds to yield dark products. Several specific reagents (such as ninhydrin) are also useful for locating separated species.

The process of locating analytes on a thin-layer plate is often termed **visualization.**

Another method of detection is based upon incorporating a fluorescent material into the stationary phase. After development, the plate is examined under ultraviolet light. The sample components quench the fluorescence of the material so that all of the plate fluoresces except areas where nonfluorescing sample components are located.

30H-3 Paper Chromatography

Separations by paper chromatography are performed in the same way as those on thin-layer plates. The papers are manufactured from highly purified cellulose with close control over porosity and thickness. Such papers contain sufficient adsorbed

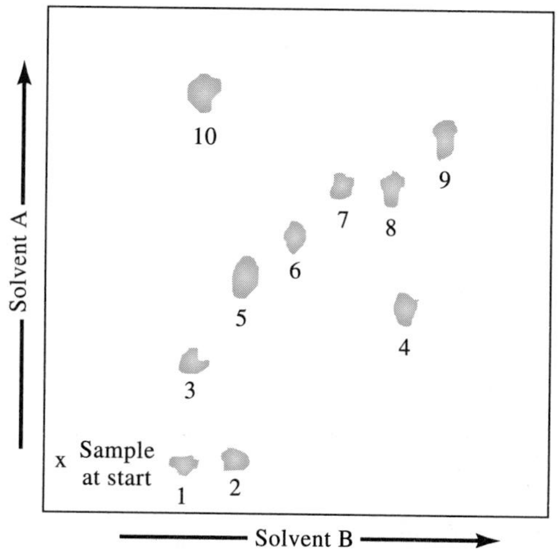

Figure 30-21

Two-dimensional thin-layer chromatogram (silica gel) of some amino acids. Solvent A: toluene/2-chloroethanol/pyridine. Solvent B: chloroform/benzyl alcohol/acetic acid. Amino acids: (1) aspartic acid, (2) glutamic acid, (3) serine, (4) β-alanine, (5) glycine, (6) alanine, (7) methionine, (8) valine, (9) isoleucine, (10) cysteine.

water to make the stationary phase aqueous. Other liquids can be made to displace the water, however, thus providing a different type of stationary phase. For example, paper treated with silicone or paraffin oil permits reversed-phase paper chromatography, in which the mobile phase is a polar solvent. Also available commercially are special papers that contain an adsorbent or an ion-exchange resin, thus permitting adsorption and ion-exchange paper chromatography.

30I QUESTIONS AND PROBLEMS

30-1. List the types of substances to which each of the following chromatographic methods is most applicable:
- *(a) gas-liquid
- (b) liquid partition
- *(c) ion-exchange
- (d) liquid adsorption
- *(e) gel permeation
- (f) gel filtration
- (g) supercritical fluid

30-2. Define:
- *(a) isocratic elution.
- (b) gradient elution.
- *(c) stop-flow injection.
- (d) reversed-phase packing.
- *(e) normal-phase packing.
- (f) ion chromatography.
- *(g) eluent-suppressor column.
- (h) gel filtration.
- *(i) gel permeation.
- (j) critical temperature.
- *(k) FSOT column.
- (l) two-dimensional thin-layer chromatography.
- *(m) supercritical fluid.

30-3. List the differences in properties and roles of the mobile phase in gas, liquid, and supercritical-fluid chromatography. How do these differences influence the characteristics of the three methods?

30-4. For a normal-phase separation, predict the order of elution of
- *(a) *n*-hexane, *n*-hexanol, benzene.
- (b) ethyl acetate, diethyl ether, nitrobutane.

***30-5.** For a reversed-phase separation, predict the order of elution of the solutes in Problem 30-4.

30-6. Describe the physical differences between open tubular and packed columns. What are the advantages and disadvantages of each?

***30-7.** Describe the fundamental difference between adsorption and partition chromatography.

30-8. Describe the fundamental difference between ion-exchange and size-exclusion chromatography.

***30-9.** What types of species can be separated by high-performance liquid chromatography but not by gas-liquid chromatography?

***30-10.** To what types of compounds is supercritical-fluid chromatography particularly applicable?

***30-11.** Describe the various kinds of pumps used in high-performance liquid chromatography. What are the advantages and disadvantages of each?

30-12. Describe the differences between single-column and suppressor-column ion chromatography.

***30-13.** List the advantages of supercritical fluid chromatography over (a) high-performance liquid chromatography and (b) gas-liquid chromatography.

30-14. Why is there more zone broadening with gas chromatography than with supercritical fluid chromatography? Why is there less zone broadening with high-performance liquid chromatography than with supercritical-fluid chromatography?

CHAPTER

31

The Analysis of Real Samples

Very early in this text (Section 1C), we pointed out that a quantitative analysis involves a sequence of steps: (1) selecting a method, (2) sampling, (3) preparing a laboratory sample, (4) defining replicate samples (by weight or volume measurements), (5) preparing solutions of the samples, (6) eliminating interferences, (7) completing the analysis by performing measurements that are related in a known way to analyte concentration, and (8) computing the results and estimating their reliability.

Thus far we have focused largely on steps 7 and 8 and to a lesser extent on steps 4 and 6. This emphasis is *not* because the earlier steps are unimportant or easy—quite the opposite. The preliminary steps may not only be more difficult and time-consuming than the two final steps of an analysis but also a greater source of error.

The reasons for postponing a discussion of the preliminary steps to this point are pedagogical. Experience has shown that it is easier to introduce students to analytical techniques by having them first perform measurements on simple materials for which no method selection is required and for which problems with sampling, sample preparation, and sample dissolution are either nonexistent or easily solved. Thus, we have been largely concerned so far with measuring the concentration of analytes in simple aqueous solutions that are uncluttered by interfering species.

The determination of an analyte in a simple solution is usually not particularly difficult once learned because the number of variables that must be controlled is small and the tools available are numerous and easy to use. Furthermore, with simple systems, our theoretical knowledge suffices to allow us to anticipate problems and correct for them. Thus, if chemical analysis involved only determining the concentration of a single species in a simple and readily soluble homogeneous mixture, analytical chemistry could profitably be entrusted to the hands of a skilled technician; certainly, a professional chemist can find more useful and challenging work for mind and hands.

In fact, the materials in which academic and industrial chemists are interested are not as a rule simple. To the contrary, most substances that require analysis are complex, consisting of several species or several tens of species. Such materials are frequently far from ideal in matters of solubility, volatility, stability, and homogeneity, and several steps must precede the final measurement step. Indeed, the final measurement is often anticlimactic in the sense that it is far easier and less time-consuming than any of the several steps that must precede it.

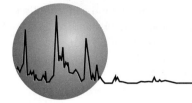

For example, we showed in earlier chapters that the calcium ion concentration of an aqueous solution is readily determined by titration with a standard EDTA solution or by potential measurements with an ion-selective electrode. Alternatively, the calcium content of a solution can be determined from flame absorption or emission measurements or by the precipitation of calcium oxalate followed by weighing or by titrating with a standard solution of potassium permanganate.

All the aforementioned methods provide a straightforward means of determining the calcium content of a simple salt, such as the carbonate. The chemist is seldom interested in the calcium content of calcium carbonate, however. More likely, what is needed is the percentage of this element in a sample of animal tissue, a silicate rock, or a piece of glass. The analysis thereby acquires new and formidable complexities. For example, none of these materials is soluble in water or dilute aqueous reagents. Before calcium can be determined, therefore, the sample must be decomposed by high-temperature treatment with concentrated reagents. Unless care is taken, this step may cause losses of some of the calcium from the sample or, equally bad, the introduction of that element as a contaminant in the relatively large quantities of reagent usually required to decompose a sample.

Even after the sample has been decomposed to give a solution of calcium ion, the excellent procedures mentioned in the two previous paragraphs cannot ordinarily be applied immediately to complete the analysis, for they are all based upon reactions or properties shared by several elements in addition to calcium. Thus, a sample of animal tissue, silicate rock, or glass almost surely contain one or more components that also react with EDTA, act as a chemical interference in a flame-absorption measurement, or form a precipitate with oxalate ion; furthermore, the high ionic strength resulting from the reagents used for sample decomposition would complicate a direct potentiometric measurement. As a consequence, several additional operations to free the solution of interferences are required before the final measurement is made.

Real samples are far more complex than those that you encounter in the instructional laboratory.

We have chosen the term *real samples* to describe materials such as those in the preceding illustration. In this context, most of the samples encountered in an elementary quantitative analysis laboratory course definitely are not real, for they are generally homogeneous, usually stable even with rough handling, readily soluble, and, above all, chemically simple. Moreover, well-established and thoroughly tested methods exist for their analysis. There is unquestioned value in introducing analytical techniques with such substances, for they do allow the student to concentrate on the mechanical aspects of an analysis. Once these mechanics have been mastered, however, there is little point in the continued analysis of unreal samples; to do so creates the impression that any chemical analysis involves nothing more than the slavish adherence to a well-defined and narrow regimen, the end result of which is a number that is accurate to one or two parts per thousand. Unfortunately, all too many chemists retain this view far into their professional lives.

In truth, the pathway leading to knowledge of the compositions of real samples is frequently more demanding of intellectual skills and chemical intuition than of mechanical aptitude. Furthermore, a compromise must often be struck between the available time and the accuracy that is believed necessary. The chemist is frequently happy to settle for an accuracy of one or two parts per hundred instead of one or two parts per thousand, knowing that the latter may require several hours or even days of additional effort. In fact, with complex materials, even parts-per-hundred accuracy may be unrealistic.

The difficulties encountered in the analysis of real samples stem from their complexity. As a consequence, the literature may not contain a well-tested analytical route for the kind of sample under consideration, and an existing procedure must then be modified to take into account compositional differences between the sample in hand and the samples for which the original method was designed. Alternatively, a new analytical method must be developed. In either case, the number of variables that have to be taken into account usually increases exponentially with the number of species contained in the sample.

As an example, let us contrast the problems associated with the flame photometric analysis of calcium carbonate with those for a real sample containing calcium. In the former, the number of components is small and those variables likely to affect the results are reasonably few. Principal among the variables are the physical losses of analyte during the evolution of carbon dioxide when the sample is dissolved in acid, the effect of the anion of the acid and of the fuel-to-air ratio on the intensity of the calcium emission line, the position of the inner cone of the flame with respect to the entrance slit of the photometer, and the quality of the standard calcium solutions used for calibration.

The determination of calcium in a sample, such as a silicate rock, which contains a dozen or more other elements, is far more complex. First of all, the sample can be dissolved only by fusing it at a high temperature with a large excess of a reagent such as sodium carbonate. Physical loss of the analyte is possible during this treatment unless suitable precautions are taken. Furthermore, the introduction of calcium from the excess sodium carbonate or the fusion vessel is of real concern. Following fusion, the sample and reagent are dissolved in acid. With this step, all the variables affecting the calcium carbonate sample are operating, but in addition, a host of new variables are introduced because of the dozens of components in the sample matrix. Now, measures must be taken to minimize instrumental and chemical interference brought about by the presence of various anions and cations in the solution being atomized.

The analysis of a real substance is often a challenging problem requiring knowledge, intuition, and experience. The development of a procedure for such materials is not to be taken lightly even by the experienced chemist.

31A CHOICE OF METHOD FOR THE ANALYSIS OF REAL SAMPLES

The choice of a method for the analysis of a complex substance requires good judgment and a sound knowledge of the advantages and limitations of the various analytical tools available. In addition, a familiarity with the literature of analytical chemistry is essential. We cannot be very explicit concerning how an analytical method is selected because there is no single best way that applies under all circumstances. We can, however, suggest a systematic approach to the problem and present some generalities that can aid in making intelligent decisions.

31A-1 Definition of the Problem

A first step, which must precede any choice of method, involves a clear definition of the analytical problem. The method of approach selected will be largely governed by the answers to the following questions:

At the outset, the objectives sought in an analysis must be clearly defined.

What is the concentration range of the species to be determined?
What degree of accuracy is demanded?
What other components are present in the sample?
What are the physical and chemical properties of the laboratory sample?
How many samples are to be analyzed?

The concentration range of the analyte may well limit the number of feasible methods. If, for example, the analyst is interested in an element present to the extent of a few parts per million, gravimetric or volumetric methods can generally be eliminated, and spectrometric, potentiometric, and other more sensitive methods become likely candidates. For components in the parts-per-million range, the chemist must guard against even small losses resulting from coprecipitation or volatility, and contamination from reagents and apparatus becomes a major concern. In contrast, if the analyte is a major component of the sample, these considerations become less important, and a classic analytical method may well be preferable.

The time required to carry out an analysis increases exponentially with the desired level of accuracy.

The answer to the question of required accuracy is of vital importance in the choice of method and in the way it is performed because the time required to complete an analysis increases, often exponentially, with demands for greater accuracy. Thus, to improve the reliability of analytical results from, say, 2% to 0.2% relative may require the operator time to increase by a factor of 100 or more. Consequently, the chemist should always spend a few minutes before undertaking an analysis in careful consideration of the degree of accuracy really needed. It is usually foolish to produce physical or chemical data with accuracies that are significantly greater than what is demanded by the use to which the data are to be put.

The demands of accuracy frequently dictate the procedure chosen for an analysis. For example, if the allowable error in the determination of aluminum is only a few parts per thousand, a gravimetric procedure is probably required. On the other hand, if an error of, say, 50 ppt can be tolerated, a spectroscopic or electroanalytical approach may be preferable.

The way in which an analysis is carried out is also affected by accuracy requirements. If precipitation with ammonia is chosen for the analysis of a sample containing 20% aluminum, the presence of 0.2% iron is of serious concern where accuracy in the parts-per-thousand range is demanded and a preliminary separation of the two elements is necessary. If an error of 50 ppt is tolerable, however, the separation of iron is not necessary. Furthermore, this tolerance governs other aspects of the method. For example, 1-g samples can be weighed to perhaps 10 mg and certainly no closer than 1 mg. In addition, less care is needed in transferring and washing the precipitate and in other time-consuming operations of the gravimetric method. The intelligent use of shortcuts is not a sign of carelessness but is instead a recognition of realities in matters of time and effort. The question of accuracy, then, must be settled in clear terms at the onset.

Often you can save considerable time by the use of permissible shortcuts in an analytical procedure.

In order to choose a method for the determination of one or more species in a sample, it is necessary to know what other elements and/or compounds are present. For lack of such information, a qualitative analysis must be undertaken in order to identify components that are likely to interfere in the various methods under consideration. As we have noted repeatedly, most analytical methods are based on reactions and physical properties that are shared by several elements or compounds. Thus, measurement of the concentration of a given element by a method that is simple and straightforward in the presence of one group of elements or compounds may require many tedious and time-consuming separations in the presence of others. A solvent suitable for one combination of compounds may be totally unsatisfactory when applied to another. Clearly, a knowledge of the qualitative chemical composition of the sample is a prerequisite for selecting a method for the quantitative determination of one or more of its components.

Identify the components of a sample before undertaking a quantitative analysis.

The chemist must also consider the physical state of the sample in order to determine whether it must be homogenized, whether volatility losses are likely, and whether its composition may change under laboratory conditions due to the absorption of water or to efflorescence. It is also important to determine what treatment is sufficient to decompose or dissolve the sample without loss of analyte. Preliminary tests of one sort or another may be needed to provide this type of information.

Finally, the number of samples to be analyzed is an important criterion in selecting a method. If there are many samples, considerable time can be expended in calibrating instruments, preparing reagents, assembling equipment, and investigating shortcuts, since the cost of these operations can be spread over the large number of samples. If a few samples at most are to be analyzed, however, a longer and more tedious procedure involving a minimum of these preparatory operations may prove to be the wiser choice from the economic standpoint.

Having answered the preliminary questions, the chemist is now in a position to consider possible approaches to the problem. Sometimes, based upon past experience, the route to be followed is obvious. In other instances, it is not, and the chemist must speculate on problems that are likely to be encountered in the analysis and how they can be solved. By now, some methods probably have been eliminated from consideration and others put on the doubtful list. Ordinarily, however, the chemist turns to the analytical literature in order to profit from the experience of others. This, then, is the next logical step in choosing an analytical method.

A little extra time spent in the library can save a tremendous amount of time and effort in the laboratory.

31A-2 Investigation of the Literature

A list of reference books and journals concerned with various aspects of analytical chemistry appears in Appendix 1. This list is not exhaustive but rather one that is adequate for most work. It is divided into several categories. In many instances, the division is arbitrary since some works could be logically placed in more than one category.

We usually begin a search of the literature by referring to one or more of the treatises on analytical chemistry or to those devoted to the analysis of specific types of materials. In addition, it is often helpful to consult a general reference to work relating to the compound or element of interest. From this survey, a clearer picture of the problem at hand may develop—what steps are likely to be difficult, what separations must be made, what pitfalls must be avoided. Occasionally, all the answers needed or even a set of specific instructions for the analysis may be found. Alternatively, journal references that lead directly to this information may be discovered. On other occasions, the chemist will acquire only a general notion of how to proceed. Several possible methods may appear suitable; others may be eliminated. At this point, it is often helpful to consider reference works concerned with specific substances or specific techniques. Alternatively, the various analytical journals may be consulted. Monographs on methods for completing the analysis are valuable in deciding among several possible techniques.

The technology for computer-based scientific information retrieval provides an efficient means for surveying the analytical literature.

A major problem in using analytical journals is locating articles pertinent to the problem at hand. The various reference books are useful since most are liberally annotated with references to the original journals. The key to a thorough search of the literature, however, is *Chemical Abstracts* and its *Index Guide.* Such a survey involves the expenditure of a great deal of time, however, and is often made unnecessary by consulting reliable reference works. The advent of computer-aided literature searches promises to minimize the time required for a careful literature survey.

31A-3 Choosing or Devising a Method

Having defined the problem and investigated the literature for possible approaches, the chemist must next decide upon the route to be followed in the laboratory. If the choice is simple and obvious, analysis can be undertaken directly. Frequently, however, the decision requires the exercise of considerable judgment and ingenuity; experience, an understanding of chemical principles, and perhaps intuition all come into play.

If the substance to be analyzed occurs widely, the literature survey usually yields several alternative methods for the analysis. Economic considerations may dictate a method that will yield the desired reliability with the least expenditure of time and effort. As mentioned earlier, the number of samples to be analyzed is often a determining factor in the choice.

Investigation of the literature does not invariably reveal a method designed specifically for the type of sample in question. Ordinarily, however, the chemist will encounter procedures for materials that are at least analogous in composition to the one in question; the decision then has to be made as to whether the variables introduced by differences in composition are likely to have any influence on the results. This judgment is often difficult and fraught with uncertainty; recourse to the laboratory may be the only way of obtaining an unequivocal answer.

Preliminary laboratory testing may be needed to validate proposed changes to established methods.

If it is decided that existing procedures are not applicable, consideration must be given to modifications that may overcome the problems imposed by the variation in composition. Again, it may be possible to propose only tentative alterations, owing to the complexity of the system; whether these modifications will accomplish their purpose without introducing new difficulties can be determined only in the laboratory.

After giving due consideration to existing methods and their modifications, the chemist may decide that none fits the problem, and an entirely new procedure must be developed. In doing so, all the facts on the chemical and physical properties of the analyte must be marshaled and given consideration. Several possible ways of performing the desired measurements may become evident from this information. Each possibility must then be examined critically, with consideration given to the influence of the other components in the sample as well as to the reagents that must be used for solution or decomposition. At this point, the chemist must try to anticipate sources of error and possible interferences attributable to interactions among sample components and reagents; it may be necessary to devise strategies to circumvent such problems. In the end, it is to be hoped that one or more tentative methods worth testing will have been located. It is probable that the feasibility of some of the steps in the procedure cannot be determined on the basis of theoretical considerations alone; recourse must be made to preliminary laboratory testing of such steps. Certainly, critical evaluation of the entire procedure can come only from careful laboratory work.

31A-4 Testing the Procedure

Once a procedure for an analysis has been selected, a decision must be made as to whether it can be employed directly, without testing, to the problem at hand. The answer to this question is not simple and depends upon a number of considerations. If the method chosen is the subject of a single, or at most a few, literature references, there may be a real point to preliminary laboratory evaluation. With experience, the chemist becomes more and more cautious about accepting claims regarding the accuracy and applicability of a new method. All too often, statements found in the literature tend to be optimistic; a few hours spent in testing the procedure in the laboratory may be enlightening.

Whenever a major modification of a standard procedure is undertaken or an attempt is made to apply it to a type of sample different from that for which it was designed, a preliminary laboratory test is advisable. The effects of such alterations simply cannot be predicted with certainty, and the chemist who dispenses with such precautions is optimistic indeed.

Finally, of course, a newly devised procedure must be extensively tested before it is adopted for general use. We must now consider the means by which a new method or a modification of an existing method can be tested for reliability.

The Analysis of Standard Samples

Unquestionably, the best technique for evaluating an analytical method involves the analysis of one or more standard samples whose analyte composition is reliably known. For this technique to be of value, however, it is essential that the standards closely resemble the samples to be analyzed with respect to both the analyte concentration range and the overall composition.

Occasionally, standards suitable for method testing can be synthesized by thoroughly homogenizing weighed quantities of pure compounds. Such a procedure is generally inapplicable, however, when the samples to be analyzed are complex, naturally occurring substances, such as rocks, animal tissue, and soils.

The National Institute of Standards and Technology has for sale a variety of standard reference materials (SRM) that have been specifically prepared for validation of analytical methods. Most standard reference materials are substances commonly encountered in commerce or in environmental, pollution, clinical, biological, and forensic studies. The concentration of one or more components in these materials is certified by the Institute based upon measurements using (1) a previously validated reference method, (2) two or more independent, reliable measurement methods, or (3) results from a network of cooperating laboratories, technically competent and thoroughly familiar with the material being tested. More than 800 of these materials are available, including such substances as ferrous and nonferrous metals; ores, ceramics, and cements; environmental gases, liquids, and solids; primary and secondary chemical standards; clinical, biological, and botanical samples; fertilizers; and glasses. Several industrial concerns also offer various kinds of standard materials designed for validating analytical procedures.

The National Institute of Standards and Technology is an important source for standard reference materials.

For literature describing standard reference materials, see the references in footnotes 2 and 3 in Chapter 2 on page 19.

When standard reference materials are not available (and often they are not), the best the chemist can do is to prepare a solution of known concentration whose composition approximates that of the sample after it has been decomposed and dissolved. Obviously, such a standard gives no information at all concerning the fate of the substance being determined during the important decomposition and solution steps.

Analysis by Other Methods

The results of an analytical method can sometimes be evaluated by comparison with data obtained from an entirely different method. Clearly, a second method must exist and in addition should be based on chemical principles that differ as much as possible from the one under examination. Comparable results from the two serve as presumptive evidence that both are yielding satisfactory results, inasmuch as it is unlikely that the same systematic errors would affect each. Such a conclusion does not apply to those aspects of the two methods that are similar.

Standard Addition to the Sample

Applications of standard-addition methods are presented in Chapters 19, 24, and 26.

When the foregoing approaches are inapplicable, the standard-addition method may prove useful. Here, in addition to being used to analyze the sample, the proposed procedure is tested against portions of the sample to which known amounts of the analyte have been added. The effectiveness of the method can then be established by evaluating the extent of recovery of the added quantity. The standard-addition method may reveal errors arising from the way the sample was treated or from the presence of the other elements and/or compounds in the matrix.

31B THE ACCURACY OBTAINABLE IN THE ANALYSIS OF COMPLEX MATERIALS

To provide a clear idea of the accuracy that can be expected for the analysis of a complex material, data on the determination of four elements in a variety of materials are presented in Tables 31-1 to 31-4. These data were taken from a

TABLE 31-1 Determination of Iron in Various Materials*

Materials	Iron, %	Number of Analysts	Average Absolute Error	Average Relative Error, %
Soda-lime glass	0.064 (Fe_2O_3)	13	0.01	15.6
Cast bronze	0.12	14	0.02	16.7
Chromel	0.45	6	0.03	6.7
Refractory	0.90 (Fe_2O_3)	7	0.07	7.8
Manganese bronze	1.13	12	0.02	1.8
Refractory	2.38 (Fe_2O_3)	7	0.07	2.9
Bauxite	5.66	5	0.06	1.1
Chromel	22.8	5	0.17	0.75
Iron ore	68.57	19	0.05	0.07

*From W. F. Hillebrand and G. E. F. Lundell, *Applied Inorganic Analysis,* p. 878. New York: Wiley, 1929.

TABLE 31-2 Determination of Manganese in Various Materials*

Material	Manganese, %	Number of Analysts	Average Absolute Error	Average Relative Error, %
Ferro-chromium	0.225	4	0.013	5.8
Cast iron	0.478	8	0.006	1.3
	0.897	10	0.005	0.56
Manganese bronze	1.59	12	0.02	1.3
Ferro-vanadium	3.57	12	0.06	1.7
Spiegeleisen	19.93	11	0.06	0.30
Manganese ore	58.35	3	0.06	0.10
Ferro-manganese	80.67	11	0.11	0.14

*From W. F. Hillebrand and G. E. F. Lundell, *Applied Inorganic Analysis,* p. 880. New York: Wiley, 1929.

TABLE 31-3 Determination of Phosphorus in Various Materials*

Material	Phosphorus, %	Number of Analysts	Average Absolute Error	Average Relative Error, %
Ferro-tungsten	0.015	9	0.003	20.
Iron ore	0.040	31	0.001	2.5
Refractory	0.069 (P_2O_5)	5	0.011	16.
Ferro-vanadium	0.243	11	0.013	5.4
Refractory	0.45	4	0.10	22.
Cast iron	0.88	7	0.01	1.1
Phosphate rock	43.77 (P_2O_5)	11	0.5	1.1
Synthetic mixtures	52.18 (P_2O_5)	11	0.14	0.27
Phosphate rock	77.56 ($Ca_3(PO_4)_2$)	30	0.85	1.1

*From W. F. Hillebrand and G. E. F. Lundell, *Applied Inorganic Analysis,* p. 882. New York: Wiley, 1929.

TABLE 31-4 Determination of Potassium in Various Materials*

Material	Potassium Oxide, %	Number of Analysts	Average Absolute Error	Average Relative Error, %
Soda-lime glass	0.04	8	0.02	50.
Limestone	1.15	15	0.11	9.6
Refractory	1.37	6	0.09	6.6
	2.11	6	0.04	1.9
	2.83	6	0.10	3.5
Lead-barium glass	8.38	6	0.16	1.9

*From W. F. Hillebrand and G. E. F. Lundell, *Applied Inorganic Analysis,* p. 883. New York: Wiley, 1929.

much larger set of results collected by W. F. Hillebrand and G. E. F. Lundell of the National Bureau of Standards and published in the first edition of their classic book on inorganic analysis.[1]

The materials analyzed were naturally occurring substances and items of commerce; they were especially prepared to give uniform and homogeneous samples and were distributed among chemists who were, for the most part, actively engaged in the analysis of similar materials. The analysts were allowed to use the methods they considered most reliable and best suited for the problem at hand. In most instances, special precautions were taken, and the results are consequently better than can be expected from the average routine analysis.

The numbers in the second column of Tables 31-1 to 31-4 are best values, obtained by the most painstaking analysis for the measured quantity. Each is considered to be the true value for calculation of the absolute and relative errors shown in the fourth and fifth columns. The fourth column was obtained by discarding extremely divergent results, determining the deviation of the remaining individual data from the best value (second column), and averaging these deviations. The fifth column was obtained by dividing the data in the fourth column by the best value (second column) and multiplying by 100%.

The results shown in these tables are typical of the data for 26 elements reported in the original publication. It is clear that analyses reliable to a few tenths of a percent relative are the exception rather than the rule in the analysis of complex mixtures by ordinary methods and that, unless we are willing to invest an inordinate amount of time in the analysis, errors on the order of 1% or 2% must be accepted. If the sample contains less than 1% of the analyte, even larger relative errors are to be expected.

It is clear from these data that the accuracy obtainable in the determination of an element is greatly dependent upon the nature and complexity of the substrate. Thus, the relative error in the determination of phosphorus in two phosphate rocks was 1.1%; in a synthetic mixture, it was only 0.27%. The relative error in an iron determination in a refractory was 7.8%; in a manganese bronze having about the

TABLE 31-5 Standard Deviation of Silica Results*

Year Reported	Sample Type	Number of Results	Standard Deviation (%, absolute)
1931	Glass	5	0.28†
1951	Granite	34	0.37
1963	Tonalite	14	0.26
1970	Feldspar	9	0.10
1972	Granite	30	0.18
1972	Syenite	36	1.06
1974	Granodiorite	35	0.46

*From S. Abbey, *Anal. Chem.*, **1981,** *53* (4), 529A. With permission of the American Chemical Society.

†0.09 after eliminating one result.

[1]W. F. Hillebrand and G. E. F. Lundell, *Applied Inorganic Analysis,* pp. 874–887. New York: Wiley, 1929.

same iron content, it was only 1.8%. Here, the limiting factor in the accuracy is not in the completion step but rather in the dissolution of the samples and the elimination of interferences.

The data in these four tables are more than 65 years old, and it is tempting to think that analyses carried out with more modern tools and additional experience are likely to be significantly better in terms of accuracy and precision. A study by S. Abbey suggests that this assumption is not valid, however.[2] For example, the data in Table 31-5, which were taken from his paper, reveal no significant improvement in silicate analyses of standard reference glass and rock samples in the 43-year period from 1931 to 1974. Indeed, the standard deviation among participating laboratories appears to be larger in later years.

Fundamental sources of uncertainty and error that were with us 65 years ago are still with us today.

From the data in the five tables in this chapter, it is clear that the chemist is well advised to adopt a pessimistic viewpoint regarding the accuracy of an analysis, be it one's own or one performed by someone else.

[2]S. Abbey, *Anal. Chem.*, **1981,** *53,* 529A.

CHAPTER

32

Preparing Samples for Analysis

This chapter deals with two preliminary steps that are important parts of most chemical analyses: (1) sampling and (2) either drying the sample or determining its moisture content. Under some circumstances, neither of these steps is important or necessary, but more commonly, one or both are vital and may limit the accuracy and significance of the analytical data.

32A SAMPLING

Generally, a chemical analysis is performed on only a small fraction of the material whose composition is of interest. Clearly, the composition of this fraction must reflect as closely as possible the average composition of the bulk of the material if the results are to have value. The process by which a representative fraction is acquired is termed *sampling.* Often, sampling is the most difficult step in the entire analytical process and the step that limits the accuracy of the analysis. This statement is particularly true when the material to be analyzed is a large and inhomogeneous liquid, such as a lake, or an inhomogeneous solid, such as an ore, a soil, or a piece of animal tissue.

Sampling is often the most difficult aspect of an analysis.

The end product of the sampling step is a quantity of homogeneous material weighing a few grams, or at most a few hundred grams, that may constitute as little as one part in 10^7 or 10^8 of the material whose composition is sought. Yet the composition of this minute fraction, which makes up the *laboratory sample,* must, as closely as possible, be identical to the average composition of the total mass. Where, as with the examples just mentioned, the quantity of material to be sampled is large and its composition inherently inhomogeneous, the task of producing a representative sample is indeed formidable. Clearly, the reliability of the analysis cannot exceed that of the sampling step, and the painstaking analysis of a poor sample is wasted effort.

The composition of the sample must closely resemble the average composition of the total mass of material to be analyzed.

Sampling is the process by which a sample population is reduced in size to an amount of homogeneous material that can be conveniently handled in the laboratory and whose composition is representative of the population.

The literature on sampling is extensive[1]; at best we can provide only a brief outline of sampling methods here.

[1]See, for example, B. W. Woodget and D. Cooper, *Samples and Standards.* London: Wiley, 1987; F. F. Pitard, *Pierre Gy's Sampling Theory and Sampling Practice.* Boca Raton, FL: CRC Press, 1989; *Standard Methods of Chemical Analysis,* 6th ed., F. J. Welcher, Ed., Vol. 2, Part A, pp. 21–55. Princeton, NJ: Van Nostrand, 1963. An extensive bibliography of specific sampling information has been compiled by C. A. Bicking, in *Treatise on Analytical Chemistry,* 2nd ed., I. M. Kolthoff and P. J. Elving, Eds., Part I, Vol. 1, p. 299. New York: Interscience, 1978. For a brief review of sampling problems, see B. Kratochvil and J. K. Taylor, *Anal. Chem.,* **1981,** *53,* 924A.

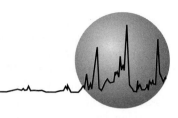

As shown in Figure 32-1, three steps are involved in sampling bulk materials: (1) identification of the population from which the sample is to be obtained, (2) collection of a *gross sample* that is truly representative of the population being sampled, and (3) reduction of the gross sample to a few hundred grams of a homogeneous *laboratory sample* that is suitable for analysis.

Ordinarily, Step 1 is straightforward, with the population being as diverse as a carload of ore, a carton of bottles containing vitamin tablets, a field of wheat, the brain of a rat, or the mud from a stretch of river bottom. Steps 2 and 3 are seldom simple and may require more effort and ingenuity than any other step in the entire analytical scheme.

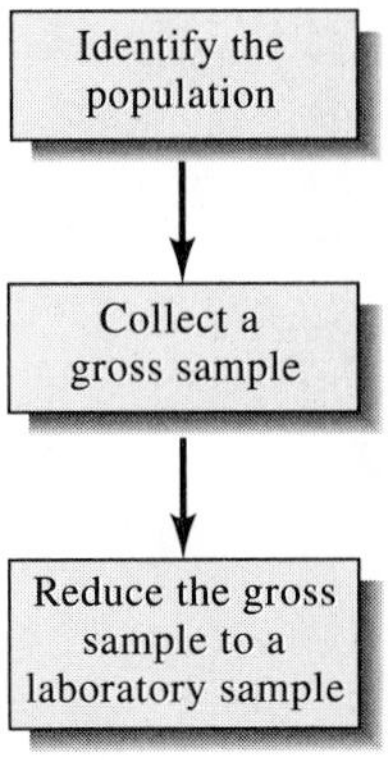

Figure 32-1
Steps in obtaining a laboratory sample.

32A-1 The Effects of Sampling Uncertainties

In Chapter 2, we concluded that both systematic and random errors in analytical data can be traced to instrument, method, and personal causes. Most systematic errors can be eliminated by exercising care, by calibration, and by the proper use of standards, blanks, and reference materials. Random errors, which are reflected in the precision of data, can generally be kept at an acceptable level by close control of the variables that influence the measurements. Errors due to invalid sampling are unique in the sense that they are not controllable by the use of blanks and standards or by closer control of experimental variables. For this reason, sampling errors are ordinarily treated separately from the other uncertainties associated with an analysis.

For random uncertainties, the overall standard deviation s_o for an analytical measurement is related to the standard deviation of the sampling process s_s, and to the standard deviation of the method s_m by the relationship

$$s_o^2 = s_s^2 + s_m^2 \tag{32-1}$$

In many cases, the method variance will be known from replicate measurements of a single laboratory sample. Under this circumstance, s_s can be computed from measurements of s_o for a series of laboratory samples, each of which is obtained from several gross samples.

Youden has shown that, once the measurement uncertainty has been reduced to one third or less of the sampling uncertainty (that is, $s_m < s_s/3$), further improvement in the measurement uncertainty is fruitless.[2] As a consequence, if the sampling uncertainty is large and cannot be improved, it is often wise to switch to a less precise but rapid method of analysis so that more samples can be analyzed in a given length of time, thus improving the precision of the average for the lot.

When $s_m < s_s/3$, there is no point in trying to improve the measurement precision.

32A-2 The Gross Sample

Ideally, the gross sample is a miniature replica of the entire mass of material to be analyzed. It corresponds to the whole not only in chemical composition but, *equally important,* in particle-size distribution.

The **gross sample** must be representative of the whole not only in the matter of chemical composition but also in its particle-size distribution.

[2]W. J. Youden, *J. Assoc. Off. Anal. Chem.*, **1981,** *50,* 1007.

The Size of the Gross Sample

From the standpoint of convenience and economy, it is desirable that the gross sample weigh no more than absolutely necessary. Basically, gross sample weight is determined by (1) the uncertainty that can be tolerated between the composition of the gross sample and that of the whole, (2) the degree of heterogeneity of the whole, and (3) the level of particle size at which heterogeneity begins.[3]

The last point warrants amplification. A well-mixed, homogeneous solution of a gas or liquid is heterogeneous only on the molecular scale, and the weight of the molecules themselves governs the minimum weight of the gross sample. A particulate solid, such as an ore or a soil, represents the opposite situation. In such materials, the individual pieces of solid differ from each other in composition. Here, heterogeneity develops in particles that may have dimensions on the order of a centimeter or more and may weigh several grams. Intermediate between these extremes are colloidal materials and solidified metals. Heterogeneity in a colloid is first encountered in the range of 10^{-5} cm or less. In an alloy, heterogeneity first occurs in the crystal grains.

The number of particles required in a gross sample ranges from a few particles to 10^{12} particles.

In order to obtain a truly representative gross sample, a certain number **n** of the particles referred to in (3) must be taken. The magnitude of this number depends upon (1) and (2) and may involve only a relatively few particles, several millions, or even several millions of millions. The need for large numbers of particles is of no great concern for homogeneous gases and liquids, since heterogeneity among particles first occurs at the molecular level. Thus, even a very small weight of sample will contain more than the requisite number of particles. However, the individual particles of a particulate solid may weigh a gram or more, which sometimes leads to a gross sample that weighs several tons. Sampling of such material is a costly, time-consuming procedure at best; determination of the smallest weight of material required to provide the needed information minimizes this expense.

The composition of a gross sample removed randomly from a bulk of material is governed by the law of chance; thus, by suitable statistical manipulations, it is possible to predict the probability that a given fraction is similar to the whole. A simple, idealized case serves as an example. A carload of lead ore is made up of just two kinds of particles: galena (lead sulfide) and a gangue containing no lead. All particles have the same size. The car contains, let us say, 100 million particles, and we wish to know how many of these are galena. Since the two components differ in appearance, the composition of the carload could be obtained exactly by counting all the galena particles as well as the remaining particles. This approach would probably involve several lifetimes of work, however. We must therefore settle for the lesser certainty involved in counting some reasonable fraction of the total number of particles. The number of particles contained in this fraction depends, of course, upon the error we are willing to tolerate in the measurement.

The relationship between the allowable error and the number of particles **n** to be counted can be stated as[4]

---

[3]For a paper on sample weight as a function of particle size, see G. H. Fricke, P. G. Mischler, F. P. Staffieri, and C. L. Housmyer, *Anal. Chem.*, **1987,** *59,* 1213.

[4]For a discussion of the derivation and significance of Equations 32-2 and 32-3, see A. A. Benedetti-Pichler, in *Physical Methods in Chemical Analysis,* W. G. Berl, Ed., Vol. 3, pp. 183–194. New York: Academic Press, 1956; A. A. Benedetti-Pichler, *Essentials of Quantitative Analysis,* Chapter 19. New York: Ronald Press, 1956.

$$\mathbf{n} = \frac{1 - p}{p\sigma_r^2} \tag{32-2}$$

where p is the fraction of galena particles, $1 - p$ is the fraction of gangue particles, and σ_r is the allowable relative standard deviation in the count of the galena particles. Thus, for example, if 80% of the particles are galena ($p = 0.8$) and the tolerable standard deviation is 1% ($\sigma_r = 0.01$), the number of particles making up the gross sample should be

$$\mathbf{n} = \frac{1 - 0.8}{0.8(0.01)^2} = 2500$$

Thus, a random sample of 2500 particles should be collected. A standard deviation of 0.1% requires a gross sample containing 250,000 particles.

Let us now make the problem more realistic and assume that one of the components in the car contains a higher percentage of lead P_1 and the other component contains a lesser amount P_2. Furthermore, the average density d of the shipment differs from the densities d_1 and d_2 of these components. We are now interested in deciding what number of particles and thus what weight we should take to ensure a sample possessing the overall average percent of lead P with a sampling relative standard deviation of σ_r. Equation 32-2 can be extended to include these stipulations:

$$\mathbf{n} = p\,(1 - p)\left(\frac{d_1 d_2}{d^2}\right)^2\left(\frac{P_1 - P_2}{\sigma_r P}\right)^2 \tag{32-3}$$

From this equation, we see that the demands of accuracy are costly, in terms of the sample size required, because of the inverse-square relationship between the allowable standard deviation and the number of particles taken. Furthermore, a greater number of particles must be taken as the average percentage P of the element of interest becomes smaller.

The degree of heterogeneity as measured by $P_1 - P_2$ has a profound effect on the number of particles required, with the number increasing as the square of the difference in composition of the two components of the mixture.

The problem of deciding upon the weight of the gross sample for a solid material is ordinarily more difficult than this example because most samples not only contain more than two components but also consist of a range of particle sizes. In most instances, the first of these problems can be met by dividing the sample into an imaginary two-component system. Thus, with an actual lead ore, one component selected might be all the various lead-bearing minerals of the ore and the other all the residual components containing little or no lead. After average densities and percentages of lead are assigned to each part, the system is treated as if it has only two components.

To simplify the problem of defining the weight of a gross sample for a multicomponent mixture, assume that the sample is a hypothetical two-component mixture.

The problem of variable particle size can be handled by calculating the number of particles that would be needed if the sample consisted of particles of a single size. The gross sample weight is then determined by taking into account the particle-size distribution. One approach is to calculate the needed weight by assuming that all particles are the size of the largest. This procedure is not very efficient, however, for it usually calls for removal of a larger weight of material

The gross sample of a particulate solid can weigh several tons.

than necessary. Benedetti-Pichler gives alternative methods for computing the weight of gross sample to be chosen.[5]

An interesting conclusion from Equation 32-3 is that the number of particles in the gross sample is *independent* of particle size. The weight of the sample, of course, increases directly as the volume (or as the cube of the particle diameter), so reduction in the particle size of a given material has a large effect on the weight required in the gross sample.

Clearly, a great deal of information must be known about a substance in order to make use of Equation 32-3. Fortunately, reasonable estimates of the various parameters in the equation can often be made. These estimates can be based upon a qualitative analysis of the substance, visual inspection, and information from the literature on substances of similar origin. Crude measurements of the density of the various sample components may also be necessary.

EXAMPLE 32-1

Assume that the average particle in the carload of lead ore just considered is judged to be approximately spherical with a radius of about 5 mm. Roughly 4% of the particles appear to be galena (~70% Pb), which has a density of 7.6 g/cm³; the remaining particles have a density of about 3.5 g/cm³ and contain little or no lead. How many pounds of ore should the gross sample contain if the sampling uncertainty is to be kept below 0.5% relative?

We first compute values for the average density and percent lead:

$$d = 0.04 \times 7.6 + 0.96 \times 3.5 = 3.7 \text{ g/cm}^3$$

$$P = \frac{(0.04 \times 7.6 \times 0.70) \text{ g Pb/cm}^3}{3.7 \text{ g sample/cm}^3} \times 100\% = 5.8\% \text{ Pb}$$

Then, substituting into Equation 32-3 gives

$$\mathbf{n} = 0.04\,(1 - 0.04)\left[\frac{7.6 \times 3.5}{(3.7)^2}\right]^2\left(\frac{70 - 0}{0.005 \times 5.8}\right)^2$$

$$= 8.45 \times 10^5 \text{ particles required}$$

$$\text{wt of sample} = 8.45 \times 10^5 \text{ particles} \times \frac{4}{3}\pi(0.05)^3 \frac{\text{cm}^3}{\text{particle}} \times \frac{3.7 \text{ g}}{\text{cm}^3} \times \frac{1}{454 \text{ g/lb}}$$

$$= 3.61 \times 10^3 \text{ lb or about 1.8 ton}$$

Sampling Homogeneous Solutions of Liquids and Gases

Solutions of liquids and gases require only a very small sample because they are homogeneous down to the molecular level.

For solutions of liquids or gases, the gross sample can be relatively small, since ordinarily nonhomogeneity first occurs at the molecular level, and even small volumes of sample will contain many more particles than the number computed

[5] A. A. Benedetti-Pichler, in *Physical Methods in Chemical Analysis,* W. G. Berl, Ed., Vol. 3, p. 192. New York: Academic Press, 1956.

from Equation 32-3. Whenever possible, the liquid or gas to be analyzed should be stirred well prior to sampling to make sure that the gross sample is homogeneous. With large volumes of solutions, mixing may be impossible; it is then best to sample several portions of the container with a ''sample thief,'' a bottle that can be opened and filled at any desired location in the solution. This type of sampling is important, for example, in determining the constituents of liquids exposed to the atmosphere. Thus, the oxygen content of lake water may vary by a factor of 1000 or more over a depth difference of a few feet.

Industrial gases or liquids are often sampled continuously as they flow through pipes, with care being taken to ensure that the sample collected represents a constant fraction of the total flow and that all portions of the stream are sampled.

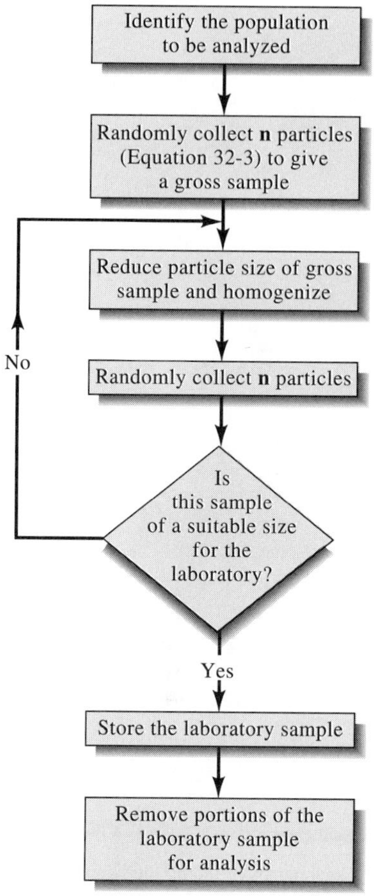

Figure 32-2
Steps in sampling a particulate solid.

Sampling Particulate Solids

Obtaining a random sample from a bulky particulate material is often difficult. It can best be accomplished while the material is being transferred. For example, randomly chosen shovelfuls or wheelbarrow loads may be consigned to a sample pile, or portions of the material may be intermittently removed from a conveyor belt. Alternatively, the material may be forced through a mechanical device called a *riffle* or a series of riffles that continuously isolate a fraction of the stream. Mechanical devices of this sort have been developed for handling coals and ores. Details regarding sampling of these materials are beyond the scope of this book.

Sampling Metals and Alloys

Obtain samples of metals and alloys by machining a small sample of chips from a large piece of the material.

Samples of metals and alloys are obtained by sawing, milling, or drilling. In general, it is not safe to assume that chips of the metal removed from the surface are representative of the entire bulk, and solid from the interior must be sampled as well. With billets or ingots of metal, a representative sample can be obtained by sawing across the piece at random intervals and collecting the ''sawdust'' as the sample. Alternatively, the specimen may be drilled, again at various randomly spaced intervals, and the drillings collected as the sample; the drill should pass entirely through the block or halfway through from opposite sides. The drillings can then be broken up and mixed or melted together in a graphite crucible. A granular sample can often then be produced by pouring the melt into distilled water.

32A-3 Preparation of a Laboratory Sample

The laboratory sample should have the same number of particles as the gross sample.

For inhomogeneous solids, the gross sample may weigh several hundred pounds or more (see Example 32-1), so reduction of the gross sample to a finely ground and homogeneous laboratory sample, weighing at the most a few pounds, is necessary. As shown in Figure 32-2, this process involves a cycle of operations that includes crushing and grinding, sieving, mixing, and dividing the sample (often into halves) to decrease its weight. During each division, a weight of sample that contains the number of particles computed from Equation 32-3 is retained.

EXAMPLE 32-2

It is desired to reduce the gross sample in Example 32-1 to a laboratory sample that weighs about one pound. How can this be done?

The laboratory sample should contain the same number of particles as the gross sample, or 8.45×10^5. Each particle will weigh on the average

$$\text{avg wt of particle} = 1 \text{ lb} \times 454 \frac{\text{g}}{\text{lb}} \times \frac{1}{8.45 \times 10^5 \text{ particles}}$$
$$= 5.37 \times 10^{-4} \text{ g/particle}$$

The average weight of a particle is related to its radius by the equation

$$\text{avg wt of particle} = \frac{4}{3}\pi[r(\text{cm})]^3 \times \frac{3.7 \text{ g}}{\text{cm}^3}$$

Equating these two relationships and solving for r give

$$r = \left(5.37 \times 10^{-4} \text{ g} \times \frac{3}{4\pi} \times \frac{\text{cm}^3}{3.7 \text{ g}}\right)^{1/3} = 3.3 \times 10^{-2} \text{ cm or } 0.3 \text{ mm}$$

Thus, the sample should be repeatedly ground, mixed, and divided until the particles are about 0.3 mm in diameter.

Crushing and Grinding of Laboratory Samples

Crushing and grinding the sample can sometimes change its composition.

A certain amount of crushing and grinding is ordinarily required to decrease the particle size of solid samples. Because these operations tend to alter the composition of the sample, the particle size should be reduced no more than is required for homogeneity and ready attack by reagents.

Several factors can cause appreciable changes in sample composition as a result of grinding. The heat inevitably generated can cause losses of volatile components. In addition, grinding increases the surface area of the solid and thus increases its susceptibility to reaction with the atmosphere. For example, it has been observed that the iron(II) content of a rock may decrease by as much as 40% during grinding—apparently, a direct result of the iron being oxidized to the +3 state.

Frequently, the water content of a sample is altered substantially during grinding. Increases are observed as a consequence of the increased specific surface area that accompanies a decrease in particle size (page 88). The increased surface area leads to greater amounts of adsorbed water. For example, it has been found that the water content of a piece of porcelain changed from 0% to 0.6% when it was ground to a fine powder.

In contrast, decreases in the water content of hydrates often take place during grinding as a result of localized frictional heating. For example, the water content of a gypsum ($CaSO_4 \cdot 2H_2O$) sample decreased from about 21% to 5% when the compound was ground to a fine powder.

Differences in hardness of the component can also introduce errors during crushing and grinding. Softer materials are ground to fine particles more rapidly than are hard ones and may be lost as dust as the grinding proceeds. In addition, flying fragments tend to contain a higher fraction of the harder components.

Intermittent screening often increases the efficiency of grinding. Screening involves shaking the ground sample on a wire or cloth sieve that will pass particles of a desired size. The residual particles are then returned for further grinding; the operation is repeated until the entire sample passes through the screen. The hardest materials, which often differ in composition from the bulk of the sample, are last to be reduced in particle size and are thus last through the screen. Therefore, grinding must be continued until every particle has been passed if the screened sample is to have the same composition as it had before grinding and screening.

Crushing and grinding must be continued until the entire sample passes through a screen of the desired mesh size.

A serious contamination error can arise during grinding and crushing due to mechanical wear and abrasion of the grinding surfaces. Even though these surfaces are fabricated from hardened steel, agate, or boron carbide, contamination of the sample is nevertheless occasionally encountered. The problem is particularly acute in analyses for minor constituents.

Mechanical abrasion of the surfaces of the grinding device can contaminate the sample.

A variety of tools are employed for reducing the particle size of solids, including jaw crushers and disk pulverizers for large samples containing large lumps, ball mills for medium-size samples and particles, and various types of mortars for small amounts of material.

The *ball mill* is a useful device for grinding solids that are not too hard. It consists of a porcelain crock with a capacity of perhaps two liters that can be sealed and rotated mechanically. The container is charged with approximately equal volumes of the sample and flint or porcelain balls with a diameter of 20 to 50 mm. Grinding and crushing occur as the balls tumble in the rotating container. A finely ground and well-mixed powder can be produced in this way.

The *Plattner diamond mortar* (Figure 32-3) is used for crushing hard, brittle materials. It is constructed of hardened tool steel and consists of a base plate, a removable collar, and a pestle. The sample is placed on the base plate inside the collar. The pestle is then fitted into place and struck several blows with a hammer; this reduces the solid to a fine powder that is collected on glazed paper after the apparatus has been disassembled.

Mixing Solid Laboratory Samples

It is essential that solid materials be thoroughly mixed to ensure a random distribution of the components in the analytical samples. A common method for mixing powders involves rolling the sample on a sheet of glazed paper. A pile of the substance is placed in the center and mixed by lifting one corner of the paper enough to roll the particles of the sample to the opposite corner. This operation is repeated many times, with the four corners of the sheet being lifted alternately.

Finely ground materials may segregate after standing for a long time.

Effective mixing of solids is also accomplished by rotating the sample for some time in a ball mill or a twin-shell V-blender. The latter consists of two connected cylinders that form a V-shaped container for the sample. As the blender is rotated, the sample is split and recombined with each rotation, leading to highly efficient mixing.

It is worthwhile noting that, with long standing, finely ground homogeneous materials may segregate on the basis of particle size and density. For example, analyses of layers of a set of student unknowns that had not been used for several years revealed a regular variation in the analyte concentration from top to bottom

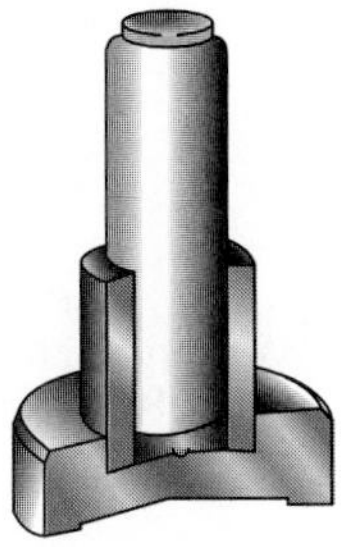

Figure 32-3
A Plattner diamond mortar.

of the container. Segregation apparently occurred as a consequence of vibrations and of density differences in the sample components.

32B MOISTURE IN SAMPLES

Laboratory samples of solids often contain water that is in equilibrium with the atmosphere. As a consequence, unless special precautions are taken, the composition of the sample depends upon the relative humidity and ambient temperature at the time it is analyzed. To cope with this variability in composition, it is common practice to remove moisture from solid samples prior to weighing or, if this is not possible, to bring the water content to some reproducible level that can be duplicated later if necessary. Traditionally, drying was accomplished by heating the sample in a conventional oven or a vacuum oven or by storing in a desiccator at a fixed humidity. These processes were continued until the material had become constant in weight. These treatments were time-consuming, often requiring several hours or even several days. In order to speed up sample drying, microwave ovens or infrared lamps are being used for sample preparation.[6] Several companies now offer equipment for this type of sample treatment.

An alternative to drying samples before beginning an analysis involves determining the water content at the time the samples are weighed for analysis so that the results can be corrected to a dry basis. In any event, many analyses are preceded by some sort of preliminary treatment designed to take into account the presence of water.

32B-1 Forms of Water in Solids

Essential Water

Essential water is the water that is an integral part of a solid chemical compound in a stoichiometric amount in a stable solid hydrate such as $BaCl_2 \cdot 2H_2O$.

Water of constitution is water that is formed when a pure solid is decomposed by heat or other chemical treatment.

Essential water forms an integral part of the molecular or crystalline structure of a compound in its solid state. Thus, the *water of crystallization* in a stable solid hydrate (for example, $CaC_2O_4 \cdot 2H_2O$ and $BaCl_2 \cdot 2H_2O$) qualifies as a type of essential water. *Water of constitution* is a second type of essential water; it is found in compounds that yield stoichiometric amounts of water when heated or otherwise decomposed. Examples of this type of water are found in potassium hydrogen sulfate and calcium hydroxide, which when heated come to equilibrium with the moisture in the atmosphere, as shown by the reactions

$$2KHSO_4(s) \rightleftharpoons K_2S_2O_7(s) + H_2O(g)$$
$$Ca(OH)_2(s) \rightleftharpoons CaO(s) + H_2O(g)$$

Nonessential Water

Nonessential water is the water that is physically retained by a solid.

Nonessential water is retained by the solid as a consequence of physical forces. It is not necessary for characterization of the chemical constitution of the sample and therefore does not occur in any sort of stoichiometric proportion.

---

[6]For a comparison of the reproducibility of these various methods of drying, see E. S. Berry, *Anal. Chem.*, **1988,** *60,* 742.

Adsorbed water is a type of nonessential water that is retained on the surface of solids. The amount adsorbed is dependent upon humidity, temperature, and the specific surface area of the solid. Adsorption of water occurs to some degree on all solids.

A second type of nonessential water is called *sorbed water* and is encountered with many colloidal substances, such as starch, protein, charcoal, zeolite minerals, and silica gel. In contrast to adsorption, the quantity of sorbed water is often large, amounting to as much as 20% or more of the total weight of the solid. Interestingly enough, solids containing even this amount of water may appear as perfectly dry powders. Sorbed water is held as a condensed phase in the interstices or capillaries of the colloidal solid. The quantity contained in the solid is greatly dependent upon temperature and humidity.

A third type of nonessential moisture is *occluded water,* liquid water entrapped in microscopic pockets spaced irregularly throughout solid crystals. Such cavities often occur in minerals and rocks (and in gravimetric precipitates).

32B-2 The Effect of Temperature and Humidity on the Water Content of Solids

In general, the concentration of water in a solid tends to decrease with increasing temperature and decreasing humidity. The magnitude of these effects and the rate at which they manifest themselves differ considerably according to the manner in which the water is retained.

Compounds Containing Essential Water

The chemical composition of a compound containing essential water is dependent upon temperature and relative humidity. For example, anhydrous barium chloride tends to take up atmospheric moisture to give one of two stable hydrates, depending upon temperature and relative humidity. The equilibria involved are

$$BaCl_2(s) + H_2O(g) \rightleftharpoons BaCl_2 \cdot H_2O(s)$$

$$BaCl_2 \cdot H_2O(s) + H_2O(g) \rightleftharpoons BaCl_2 \cdot 2H_2O(s)$$

At room temperature and at a relative humidity between 25% and 90%, $BaCl_2 \cdot 2H_2O$ is the stable species. Since the relative humidity in most laboratories is well within these limits, the essential water content of the dihydrate is ordinarily independent of atmospheric conditions. Exposure of either $BaCl_2$ or $BaCl_2 \cdot H_2O$ to these conditions causes compositional changes that ultimately lead to formation of the dihydrate. On a very dry winter day (relative humidity $< 25\%$), however, the situation changes; the dihydrate becomes unstable with respect to the atmosphere, and a molecule of water is lost from the new stable species $BaCl_2 \cdot H_2O$. At relative humidities less than about 8%, both hydrates lose water and the anhydrous compound is the stable species. From these remarks, it is apparent that the composition of a sample containing essential water is greatly dependent upon the relative humidity of its environment.

Many hydrated compounds can be converted to the anhydrous condition by oven-drying at 100°C to 120°C for an hour or two. Such treatment often precedes an analysis of samples containing hydrated compounds.

Relative humidity is the ratio of the vapor pressure of water in the atmosphere to its vapor pressure in air that is saturated with moisture. At 25°C, the partial pressure of water in saturated air is 23.76 torr. Thus, when air contains water at a partial pressure of 6 torr, the relative humidity is

$$\frac{6.00}{23.76} = 0.253 \text{ (or 25.3\%)}$$

The essential water content of a compound depends upon the temperature and relative humidity of its surroundings.

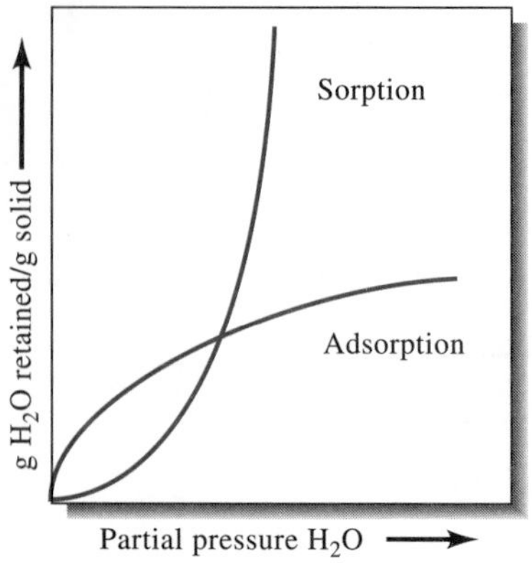

Figure 32-4

Typical adsorption and sorption isotherms.

Adsorbed water resides on the surface of the particles of a material.

Sorbed water is contained within the interstices of the molecular structure of a colloidal compound.

Compounds Containing Adsorbed Water

Figure 32-4 shows an *adsorption isotherm,* in which the weight of the water adsorbed on a typical solid is plotted against the partial pressure of water in the surrounding atmosphere. It is apparent from the diagram that the extent of adsorption is particularly sensitive to changes in water-vapor pressure at low partial pressures.

The amount of water adsorbed on a solid decreases as the temperature of the solid increases and generally approaches zero when the solid is heated above 100°C. Adsorption or desorption of moisture usually occurs rapidly, with equilibrium often being reached after 5 or 10 min. The speed of the process is often apparent during the weighing of finely divided anhydrous solids, where a continuous increase in weight will occur unless the solid is contained in a tightly stoppered vessel.

Compounds Containing Sorbed Water

The quantity of moisture sorbed by a colloidal solid varies tremendously with atmospheric conditions, as shown in Figure 32-5. In contrast to the behavior of adsorbed water, however, equilibrium may require days or even weeks for attainment, particularly at room temperature. Moreover, the amounts of water retained by the two processes are often quite different from each other; typically, adsorbed moisture amounts to a few tenths of a percent of the weight of the solid, whereas sorbed water can amount to 10% to 20%.

The amount of water sorbed in a solid also decreases as the solid is heated. Complete removal of this type of moisture at 100°C is by no means a certainty, however, as indicated by the drying curves for an organic compound shown in Figure 32-5. After this material was dried for about 70 min at 105°C, constant weight was apparently reached. It is clear, however, that additional moisture was removed by elevating the temperature. Even at 230°C, dehydration was probably not complete.

Compounds Containing Occluded Water

Occluded water is not in equilibrium with the atmosphere and is therefore insensitive to changes in humidity. Heating a solid containing occluded water may

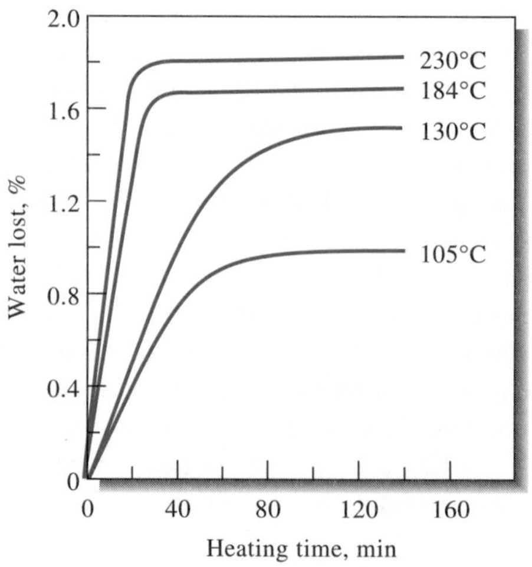

Figure 32-5

Removal of sorbed water from an organic compound at various temperatures. (Data from C. O. Willits, *Anal. Chem.,* **1951,** *23,* 1058. With permission of the American Chemical Society. Copyright 1951.)

cause a gradual diffusion of the moisture to the surface, where evaporation occurs. Frequently, heating is accompanied by *decrepitation,* in which the crystals of the solid are suddenly shattered by the steam pressure created from moisture contained in the internal cavities.

Occluded water is trapped in random microscopic pockets of solids, particularly minerals and rocks.

Decrepitation is a process in which a crystalline material containing occluded water suddenly explodes during heating as a consequence of a buildup in internal pressure resulting from steam formation.

32B-3 Drying the Analytical Sample

The methods for dealing with moisture in solid samples depend upon the information desired. When the composition of the material on an as-received basis is needed, the principal concern is that the moisture content not be altered as a consequence of grinding or other preliminary treatment and storage. Where such changes are unavoidable or probable, it is often advantageous to determine the weight loss upon drying by some reproducible procedure (say, heating to constant weight at 105°C) immediately upon receipt of the sample. Then, when the analysis is to be performed, the sample is again dried at this temperature so that the data can be corrected back to the original basis.

We have already noted that the moisture content of some substances is markedly changed by variations in humidity and temperature. Colloidal materials containing large amounts of sorbed moisture are particularly susceptible to the effects of these variables. For example, the moisture content of a potato starch has been found to vary from 10% to 21% as a consequence of an increase in relative humidity from 20% to 70%. With substances of this sort, comparable analytical data from one laboratory to another or even within the same laboratory can be achieved only by carefully specifying a procedure for taking the moisture content into consideration. For example, samples are frequently dried to constant weight at 105°C or at some other specified temperature. Analyses are then performed and results reported on this dry basis. While such a procedure may not render the solid completely free of water, it usually lowers the moisture content to a reproducible level.

32C THE DETERMINATION OF WATER IN SAMPLES

Often the only sure way to obtain a result on a dry basis requires a separate determination of moisture in a set of samples taken concurrently with the samples that are to be analyzed. Several methods of determining water in solid samples are available. The simplest involves determining the weight loss after the sample has been heated at 100°C to 110°C (or some other specified temperature) until the weight of the dried sample becomes constant. Unfortunately, this simple procedure is not at all specific for water, and large positive systematic errors occur in samples that yield volatile decomposition products (in addition to water) upon being heated. This method can also yield negative errors when applied to samples containing sorbed moisture (for example, see Figure 32-5).

Several highly selective methods have been developed for the determination of water in solid and liquid samples. One of these, the Karl Fischer method, is described in Section 17D-3. Several others are described in a monograph by Mitchell and Smith.[7]

[7]J. J. Mitchell Jr. and D. M. Smith, *Aquametry,* 2nd ed., 3 volumes. New York: Wiley, 1977–1980.

32D QUESTIONS AND PROBLEMS

*32-1. Describe the steps in a sampling operation.

32-2. What is the object of the sampling step in an analysis?

32-3. Differentiate between
*(a) sorbed water, adsorbed water, and occluded water.
(b) water of crystallization and water of constitution.
*(c) essential water and nonessential water.
(d) gross sample and laboratory sample.

32-4. What factors determine the weight of a gross sample?

*32-5. The following results were obtained for the determination of calcium in a NIST limestone sample: % CaO = 50.38, 50.20, 50.31, 50.22, and 50.41. Five gross samples were then obtained for a carload of limestone. The average percent CaO values for the gross samples were found to be 49.53, 50.12, 49.60, 49.87, and 50.49. Calculate the relative standard deviation associated with the sampling step.

*32-6. A coating that weighs at least 3.00 mg is needed to impart adequate shelf life to a pharmaceutical tablet. A random sampling of 250 tablets revealed that 14 failed to meet this requirement.
(a) Use this information to estimate the relative standard deviation for the measurement.
(b) What is the 90% confidence interval for the number of unsatisfactory tablets?
(c) Assuming that the fraction of rejects remains unchanged, how many tablets should be taken to ensure a relative standard deviation of 10% in this measurement?

32-7. Changes in the method used to coat the tablets lowered the percentage of rejects from 5.6% (Problem 32-6) to 2.0%. How many tablets should be taken for inspection if the permissible relative standard deviation in the measurement is to be
*(a) 25%? (b) 10%? *(c) 5%? (d) 1%?

32-8. The mishandling of a shipping container loaded with 750 cases of wine caused some of the bottles to break. An insurance adjuster proposed to settle the claim at 20.8% of the value of the shipment, based upon a random 250-bottle sample in which 52 were cracked or broken. Calculate
(a) the relative standard deviation of the adjuster's evaluation.
(b) the absolute standard deviation for the 750 cases (12 bottles per case).
(c) the 90% confidence interval for the total number of bottles.
(d) the size of a random sampling needed for a relative standard deviation of 5.0%, assuming a breakage rate of about 21%.

*32-9. Approximately 15% of the particles in a shipment of silver-bearing ore are judged to be argentite, Ag_2S ($d = 7.3\ g \cdot cm^{-3}$, 87% Ag); the remainder are siliceous ($d = 2.6\ g \cdot cm^{-3}$) and contain essentially no silver.
(a) Calculate the number of particles that should be taken for the gross sample if the relative standard deviation due to sampling is to be 1% or less.
(b) Estimate the mass of the gross sample, assuming that the particles are spherical and have an average diameter of 4.0 mm.
(c) The sample taken for analysis is to weigh 0.600 g and contain the same number of particles as the gross sample. To what diameter must the particles be ground to satisfy these criteria?

32-10. The average diameter of the particles in a shipment of copper appears to be 5.0 mm. Approximately 5% of the particles are cuprite (80% Cu) with a density of $6\ g \cdot cm^{-3}$; the remainder is estimated to have a density of $4\ g \cdot cm^{-3}$ and contain 3% Cu.
(a) How many particles of the ore should be sampled if the relative standard deviation due to sampling is to be 4% or less?
(b) What should the weight of the gross sample be?
(c) To what diameter must the particles be ground in order to yield a sample for analysis that weighs 0.500 g and has the same number of particles as the gross sample?

*32-11. The average particle diameter of an ore sample is 2.0 mm. It is estimated that the stibnite content ($d_{Sb_2S_3}$ = $4.5\ g \cdot cm^{-3}$, 71.7% Sb) is approximately 2.0%; the remainder has a density of $3.0\ g \cdot cm^{-3}$ and contains about 1% Sb.
(a) How many particles of the ore should be taken if the relative standard deviation due to sampling is to be 1% or less?
(b) What should the weight of the gross sample be?
(c) To what diameter must the particles be ground in order to yield a sample for analysis that weighs 0.750 g and has the same number of particles as the gross sample?

32-12. The seller of a mining claim took a random ore sample that weighed approximately 5 lb and had an average particle diameter of 5.0 mm. Inspection revealed that about 1% of the sample was argentite (Problem 32-9), and the remainder had a density of about $2.6\ g \cdot cm^{-3}$ and contained no silver. The prospective buyer insisted upon knowing the silver content of the claim with a relative error no greater than 5%. Establish whether the seller provided a sufficiently large sample to permit such an evaluation.

CHAPTER

33

Decomposing and Dissolving the Sample

Most analytical measurements are performed on solutions (usually aqueous) of the analyte. While some samples dissolve readily in water or aqueous solutions of the common acids or bases, others require powerful reagents and rigorous treatment. For example, when sulfur or halogens are to be determined in an organic compound, the sample must be subjected to high temperatures and potent reagents in order to rupture the strong bonds between these elements and carbon. Similarly, drastic conditions are usually required to destroy the silicate structure of a siliceous mineral, thus rendering its cations free for analysis.

The proper choice among the various reagents and techniques for decomposing and dissolving analytical samples can be critical to the success of an analysis, particularly where refractory substances are involved or where the analyte is present in trace amounts. In this chapter, we first consider the types of errors that can occur during decomposing and dissolving an analytical sample. We then describe four general methods of decomposing solid and liquid samples to obtain aqueous solutions of analytes. These methods differ in the temperature at which they are carried out and the strengths of the reagents employed and include (1) heating with aqueous strong acids (or occasionally bases) in open vessels; (2) microwave heating with acids in closed vessels; (3) high-temperature ignition in air or oxygen; and (4) fusion in molten salt media.[1]

A **refractory substance** is a material that is resistant to heat and attack by strong chemical reagents.

33A SOURCES OF ERRORS IN DECOMPOSITION AND DISSOLUTION

Several sources of error are encountered in the sample decomposition step. In fact, such errors often limit the accuracy that can be realized in an analysis. The sources of these errors include the following:

1. **Incomplete dissolution of the analytes.** Ideally the sample treatment should dissolve the sample completely because attempts to leach analytes quantita-

Ideally, the reagent selected should dissolve the entire sample, not just the analyte.

[1]For an extensive discussion of this topic, see R. Bock, *A Handbook of Decomposition Methods in Analytical Chemistry.* New York: Wiley, 1979; Z. Sulcek and P. Povondra, *Methods of Decomposition in Inorganic Analysis.* Boca Raton, FL: CRC Press, 1989.

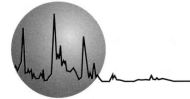

tively from an insoluble residue are usually not successful, with portions of the analyte being retained within the residue.

2. **Losses of the analyte by volatilization.** An important concern when dissolving samples is the possibility that some portion of the analyte may volatilize. For example, carbon dioxide, sulfur dioxide, hydrogen sulfide, hydrogen selenide, and hydrogen telluride are generally volatilized when a sample is dissolved in strong acid, whereas ammonia is commonly lost when a basic reagent is employed. Similarly, hydrofluoric acid reacts with silicates and boron containing compounds to produce volatile fluorides. Strong oxidizing solvents often cause the evolution of chlorine, bromine, or iodine; reducing solvents may lead to the volatilization of such compounds as arsine, phosphine, and stibine.

   A number of elements form volatile chlorides that are partially or completely lost from hot hydrochloric acid solutions. Among these are the chlorides of tin(IV), germanium(IV), antimony(III), arsenic(III), and mercury(II). The oxychlorides of selenium and tellurium also volatilize to some extent from hot hydrochloric acid. The presence of chloride ion in hot concentrated sulfuric or perchloric acid solutions can cause volatilization losses of bismuth, manganese, molybdenum, thallium, vanadium, and chromium.

   Boric acid, nitric acid, and the halogen acids are lost from boiling aqueous solutions. Certain volatile oxides can also be lost from hot acidic solutions, including the tetroxides of osmium and ruthenium and the heptoxide of rhenium.

3. **Introduction of analyte as a solvent contaminant.** Ordinarily, the mass of solvent required to dissolve a sample exceeds the mass of the sample by one or two orders of magnitude. As a consequence, the presence of analyte species in the solvent even at small concentrations may lead to significant error, particularly when the analyte is present in trace amounts in the sample.

4. **Introduction of contaminants from reaction of the solvent with vessel walls.** This source of error is often encountered in decompositions that involve high-temperature fusions. Again, this source of error becomes of particular concern in trace analyses.

33B DECOMPOSING SAMPLES BY INORGANIC ACIDS IN OPEN VESSELS

The most common reagents for open-vessel decomposition of inorganic analytical samples are the mineral acids (much less frequently, ammonia and aqueous solutions of the alkali metal hydroxides are employed). Ordinarily, a suspension of the sample in the acid is heated by flame or a hot plate until the dissolution is judged to be complete by the disappearance of a solid phase. The temperature of the decomposition is the boiling (or decomposition) point of the acid reagent.

33B-1 Hydrochloric Acid

Concentrated hydrochloric acid is an excellent solvent for inorganic samples but finds limited application for decomposing organic materials. It finds widespread use for dissolving many metal oxides as well as metals more easily oxidized than hydrogen; often, it is a better solvent for oxides than the oxidizing acids. Concentrated hydrochloric acid is about 12 M, but upon heating, hydrogen chloride is lost until a constant-boiling 6 M solution remains (boiling point about 110°C).

33B-2 Nitric Acid

Hot concentrated nitric acid is a strong oxidant that dissolves all common metals with the exception of aluminum and chromium, which become passive to the reagent as a consequence of surface oxide formation. When alloys containing tin, tungsten, or antimony are treated with the hot reagent, slightly soluble hydrated oxides, such as $SnO_2 \cdot 4H_2O$, form. After coagulation, these colloidal materials can be separated from other metallic species by filtration.

Hot nitric acid alone or in combination with other acids and oxidizing agents, such as hydrogen peroxide and bromine, is widely used for decomposing organic samples prior to determining their trace metal content. This decomposition process, which is called *wet ashing,* converts the organic sample to carbon dioxide and water. Unless it is carried out in a closed vessel, nonmetallic elements, such as the halogens, sulfur, and nitrogen, are wholly or partially lost by volatilization.

Wet ashing is a process of oxidation decomposition of organic samples by liquid oxidizing reagents, such as HNO_3, H_2SO_4, $HClO_4$, or mixtures of these acids.

33B-3 Sulfuric Acid

Many materials are decomposed and dissolved by hot concentrated sulfuric acid, which owes part of its effectiveness as a solvent to its high boiling point (about 340°C). Most organic compounds are dehydrated and oxidized at this temperature and are thus eliminated from samples as carbon dioxide and water by this wet ashing treatment. Most metals and many alloys are attacked by the hot acid.

33B-4 Perchloric Acid

Hot concentrated perchloric acid, a potent oxidizing agent, attacks a number of iron alloys and stainless steels that are intractable to other mineral acids. Care must be taken in using the reagent, however, because of *its potentially explosive nature.* The cold concentrated acid is not explosive, nor are heated dilute solutions. *Violent explosions occur, however, when hot concentrated perchloric acid comes into contact with organic materials or easily oxidized inorganic substances.* Because of this property, the concentrated reagent should be heated only in special hoods, which are lined with glass or stainless steel, are seamless, and have a fog system for washing down the walls with water. A perchloric acid hood should always have its own fan system, one that is independent of all other systems.[2]

Perchloric acid is marketed as the 60% to 72% acid. A constant-boiling mixture (72.4% $HClO_4$) is obtained at 203°C.

33B-5 Oxidizing Mixtures

More rapid wet ashing can sometimes be obtained by the use of mixtures of acids or by the addition of oxidizing agents to a mineral acid. *Aqua regia,* a mixture

[2]See A. A. Schilt, *Perchloric Acid and Perchlorates.* Columbus, OH: G. Frederick Smith Chemical Company, 1979.

containing three volumes of concentrated hydrochloric acid and one of nitric acid, is well known. The addition of bromine or hydrogen peroxide to mineral acids often increases their solvent action and hastens the oxidation of organic materials in the sample. Mixtures of nitric and perchloric acid are also useful for this purpose and less dangerous than perchloric acid alone. With this mixture, however, care must be taken to avoid evaporation of all the nitric acid before oxidation of the organic material is complete. *Severe explosions and injuries have resulted from failure to observe this precaution.*

33B-6 Hydrofluoric Acid

The primary use of hydrofluoric acid is for the decomposition of silicate rocks and minerals in the determination of species other than silica. In this treatment, silicon is evolved as the tetrafluoride. After decomposition is complete, the excess hydrofluoric acid is driven off by evaporation with sulfuric acid or perchloric acid. Complete removal is often essential to the success of an analysis because fluoride ion reacts with several cations to form extraordinarily stable complexes that then interfere with the determination of these cations. For example, precipitation of aluminum (as $Al_2O_3 \cdot xH_2O$) with ammonia is quite incomplete if fluoride is present even in small amounts. Frequently, removal of the last traces of fluoride ion from a sample is so difficult and time-consuming as to negate the attractive features of the parent acid as a solvent for silicates.

Hydrofluoric acid finds occasional use in conjunction with other acids in attacking steels that dissolve with difficulty in other solvents.

Because hydrofluoric acid is extremely toxic, dissolution of samples and evaporation to remove excess reagent *should always be carried out in a well-ventilated hood.* Hydrofluoric acid *causes serious damage and painful injury* when brought into contact with the skin. Its effects may not become evident until hours after exposure. If the acid comes into contact with the skin, the affected area should be immediately washed with copious quantities of water. Treatment with a dilute solution of calcium ion, which precipitates fluoride ion, may also be of help.

33C MICROWAVE DECOMPOSITIONS

The use of microwave ovens for the decomposition of both inorganic and organic samples was first proposed in the mid-1970s and by now has become an important method for sample preparation.[3] Microwave digestions can be carried out in either closed or open vessels, but the former are much preferred because of the higher pressures and thus higher temperatures that are realized. Most of our discussion will focus on closed-vessel microwave decompositions.

One of the main advantages of microwave decompositions compared with conventional methods using a flame or hot plate is speed. Typically, microwave decompositions of even difficult samples can be accomplished in five to ten minutes. In contrast, the same results require several hours when carried out by

[3]For more detailed discussions of microwave decompositions see *Anal. Chem.,* **1986,** *58,* 1424A; P. Aysola, P. Anderson, and C. H. Langford, *Anal. Chem.,* **1987,** *59,* 1582; H. M. Kingston and L. B. Jassie, *Anal. Chem.,* **1986,** *58,* 2534; *Introduction to Microwave Sample Preparation: Theory and Practice,* H. M. Kingston and L. B. Jassie, Eds. Washington, DC: American Chemical Society, 1988.

heating over a flame or hot plate. The difference is due to the different mechanism by which energy is transferred to the molecules of the solution in the two methods. Heat transfer is by conduction in the conventional method. Because the vessels used in conductive heating are usually poor conductors, time is required to heat the vessel and then transfer the heat to the solution by conduction. Furthermore, because of convection within the solution, only a small fraction of the liquid is maintained at the temperature of the vessel and thus at its boiling point. In contrast, microwave energy is transferred directly to all of the molecules of the solution nearly simultaneously without heating the vessel. Thus, boiling temperatures are reached throughout the entire solution very quickly.

As noted earlier, an advantage of employing closed vessels for microwave decompositions is the higher temperatures that develop as a consequence of the increased pressure. In addition, because evaporative losses are avoided, significantly smaller amounts of reagent are required, thus reducing interference by reagent contaminants. A further advantage of decompositions of this type is that loss of volatile components of samples is virtually eliminated. Finally, closed-vessel microwave decompositions are often easy to automate, thus reducing the operator time required to prepare samples for analysis.

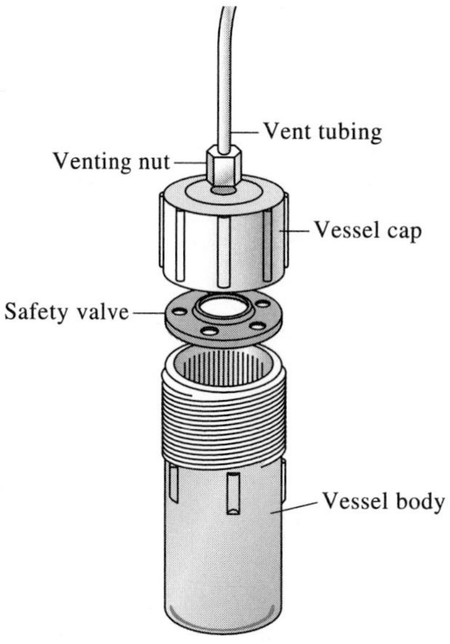

Figure 33-1
A moderate-pressure vessel for microwave decomposition. (Courtesy of CEM Corporation, Indian Trail, NC.)

33C-1 Vessels for Moderate-Pressure Digestions

Microwave digestion vessels are constructed from the low-loss materials that are transparent to microwaves. These materials must also be thermally stable and resistant to chemical attack by the various acids used for decompositions. Teflon is a nearly ideal material for many of the acids commonly used for dissolutions. It is transparent to microwaves, has a melting point of about 300°C, and is not attacked by any of the common acids. Sulfuric and phosphoric acids, however, have boiling points above the melting point of Teflon, which means that care must be exercised to control the temperature during decompositions. For these acids, quartz or borosilicate glass vessels are sometimes used in place of Teflon containers. Silicate vessels have the disadvantage, however, that they are attacked by hydrofluoric acid, a reagent that is often used for decomposing silicates and refractory alloys.

Figure 33-1 is a schematic of a commercially available closed digestion vessel designed for use in a microwave oven. It consists of a Teflon body, a cap, and a safety relief valve designed to operate at 120 $\pm$ 10 psi. At this pressure the safety valve opens and then reseals.

33C-2 High-Pressure Microwave Vessels

Figure 33-2 is a schematic of a commercial microwave bomb designed to operate at 80 atm, or about 10 times the pressure that can be tolerated by the moderate-pressure vessels described in the previous section. The maximum recommended temperature with this device is 250°C. The heavy-wall bomb body is constructed of a polymeric material that is transparent to microwaves. The decomposition is carried out in a Teflon cup supported in the bomb body. The microwave bomb incorporates a Teflon O-ring in the liner cap that seats against a narrow rim on the exterior of the liner and its cap when the retaining jacket is screwed into place. If overpressurization occurs, the O-ring distorts, and the excess pressure then compresses the sealer disk that allows the gases to escape into the surroundings. The

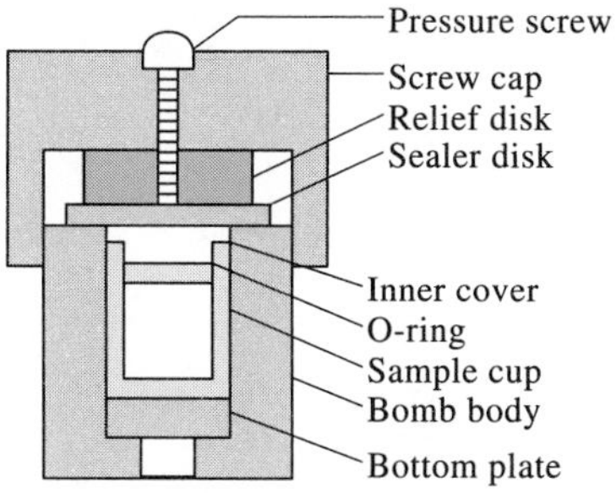

Figure 33-2
A bomb for high-pressure microwave decomposition. (Courtesy of Parr Instrument Company, Moline, IL.)

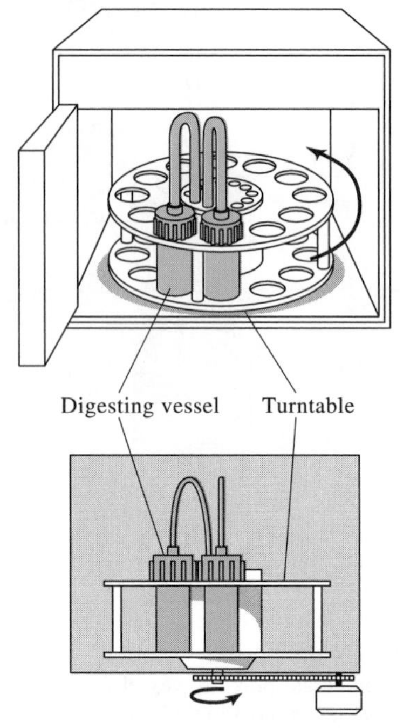

Figure 33-3
A microwave oven designed for use with 12 vessels of the type shown in Figure 33-1. (Courtesy of CEM Corporation, Indian Trail, NC.)

sample is compromised when this occurs. The internal pressure in the bomb can be judged roughly by the distance the pressure screw protrudes from the cap. This microwave bomb is particularly useful for dissolving highly refractory materials that are incompletely decomposed in the moderate-pressure vessel described in the earlier section.

33C-3 Microwave Ovens

Figure 33-3 is a schematic of a microwave oven designed to heat simultaneously 12 of the moderate-pressure vessels described in Section 33C-1. The vessels are held on a turntable that rotates continuously through 360 deg so that the average energy received by each of the vessels is approximately the same.

33C-4 Microwave Furnaces

Recently, microwave furnaces have been developed for performing fusions and for dry ashing samples containing large amounts of organic materials. These furnaces consist of a small chamber constructed of silicon carbide that is surrounded by quartz insulation. When microwaves are focused on this chamber, temperatures of 1000°C are reached in two minutes. The advantage of this type of furnace relative to a conventional muffle furnace is the speed at which high temperatures are attained. In contrast, conventional muffle furnaces are usually operated continuously because of the elapsed time required to get them up to temperature. Furthermore, with a microwave furnace there are no burned-out heating coils such as are frequently encountered with conventional furnaces. Finally, the operator is not exposed to high temperatures when samples are introduced or removed from the furnace. A disadvantage of the microwave furnace is the small volume of the heating cavity, which only accommodates an ordinary size crucible.

33C-5 Applications of Closed-Vessel Microwave Decompositions

During the last decade and a half, hundreds of reports have appeared in the literature regarding the use of closed-vessel decompositions carried out in microwave ovens with the reagents described in Section 33B. These applications fall into two categories: (1) oxidative decompositions of organic and biological samples (wet ashing), and (2) decomposition of refractory inorganic materials encountered in industry. In both cases this new technique is replacing older conventional methods because of the large economic gains that result from significant savings in time.

33D COMBUSTION METHODS FOR DECOMPOSING ORGANIC SAMPLES[4]

33D-1 Combustion Over an Open Flame (Dry Ashing)

The simplest method for decomposing an organic sample prior to determining the cations it contains is to heat the sample over a flame in an open dish or crucible

[4]For a thorough treatment of this topic, see T. S. Ma and R. C. Rittner, *Modern Organic Elemental Analysis*. New York: Marcel Dekker, 1979.

until all carbonaceous material has been oxidized to carbon dioxide. Red heat is often required to complete the oxidation. Analysis of the nonvolatile components follows dissolution of the residual solid. Unfortunately, there is always substantial uncertainty about the completeness of recovery of supposedly nonvolatile elements from a dry-ashed sample. Some losses probably result from the entrainment of finely divided particulate matter in the convection currents around the crucible. In addition, volatile metallic compounds may be lost during the ignition. For example, copper, iron, and vanadium are appreciably volatilized when samples containing porphyrin compounds are ashed.

Although dry ashing is the simplest method for decomposing organic compounds, it is often the least reliable. It should not be employed unless tests have demonstrated its applicability to a given type of sample.

Dry ashing is a process of oxidizing an organic sample with oxygen or air at high temperature, leaving the inorganic component for analysis.

33D-2 Combustion-Tube Methods

Several common and important elemental components of organic compounds are converted to gaseous products as a sample is pyrolyzed in the presence of oxygen. With suitable apparatus, it is possible to trap these volatile compounds quantitatively, thus making them available for the analysis of the element of interest. The heating is commonly performed in a glass or quartz combustion tube through which a stream of carrier gas is passed. The stream transports the volatile products to parts of the apparatus where they are separated and retained for the measurement; the gas may also serve as the oxidizing agent. Elements susceptible to this type of treatment are carbon, hydrogen, nitrogen, the halogens, sulfur, and oxygen.

Automated combustion-tube analyzers are now available for the determination of either carbon, hydrogen, and nitrogen or carbon, hydrogen, and oxygen in a single sample.[5] The apparatus requires essentially no attention by the operator, and the analysis is complete in less than 15 min. In one such analyzer, the sample is ignited in a stream of helium and oxygen and passes over an oxidation catalyst consisting of a mixture of silver vanadate and silver tungstate. Halogens and sulfur are removed with a packing of silver salts. A packing of hot copper is located at the end of the combustion train to remove oxygen and convert nitrogen oxides to nitrogen. The exit gas, consisting of a mixture of water, carbon dioxide, nitrogen, and helium, is collected in a glass bulb. The analysis of this mixture is accomplished with three thermal-conductivity measurements (Section 29B-5). The first is made on the intact mixture, the second is made on the mixture after water has been removed by passage of the gas through a dehydrating agent, and the third is made on the mixture after carbon dioxide has been removed by an absorbent. The relationship between thermal conductivity and concentration is linear, and the slope of the curve for each constituent is established by calibration with a pure compound such as acetanilide.

33D-3 Combustion with Oxygen in a Sealed Container

A relatively straightforward method for the decomposition of many organic substances involves combustion with oxygen in a sealed container. The reaction

---

[5]For a description of these instruments, see Chapters 2, 3, and 4 of the reference in footnote 4.

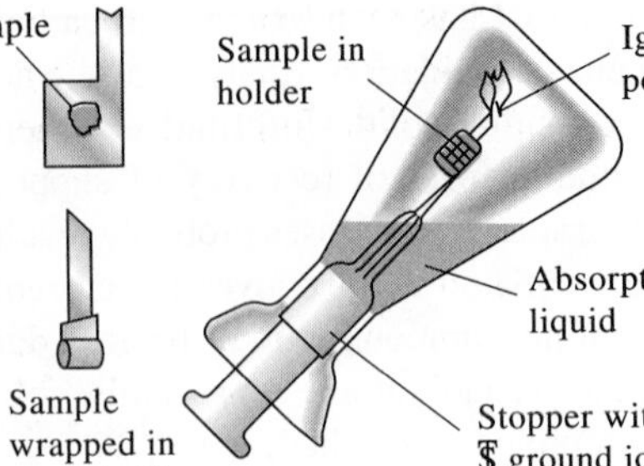

Figure 33-4
Schöniger combustion apparatus. (Courtesy Thomas Scientific, Swedesboro, NJ 08085.)

products are absorbed in a suitable solvent before the reaction vessel is opened; they are subsequently analyzed by ordinary methods.

A remarkably simple apparatus for performing such oxidations has been suggested by Schöniger (Figure 33-4).[6] It consists of a heavy-walled flask of 300- to 1000-mL capacity fitted with a ground-glass stopper. Attached to the stopper is a platinum gauze basket that holds from 2 to 200 mg of sample. If the substance to be analyzed is a solid, it is wrapped in a piece of low-ash filter paper cut in the shape shown in Figure 33-4. Liquid samples are weighed into gelatin capsules, which are then wrapped in a similar fashion. The paper tail serves as the ignition point.

A small volume of an absorbing solution (often sodium carbonate) is placed in the flask, and the air in the flask is displaced by oxygen. The tail of the paper is ignited, the stopper is quickly fitted into the flask, and the flask is inverted to prevent the escape of the volatile oxidation products. The reaction ordinarily proceeds rapidly, being catalyzed by the platinum gauze surrounding the sample. During the combustion step, the flask is shielded to minimize damage in case of explosion.

After cooling, the flask is shaken thoroughly and disassembled, and the inner surfaces are carefully rinsed. The analysis is then performed on the resulting solution. This procedure has been applied to the determination of halogens, sulfur, phosphorus, fluorine, arsenic, boron, carbon, and various metals in organic compounds.

33E DECOMPOSITION OF INORGANIC MATERIALS BY FLUXES

Many common substances—notably silicates, some mineral oxides, and a few iron alloys—are attacked slowly, if at all, by the methods just considered. In such cases, recourse to a fused-salt medium is indicated. Here, the sample is mixed with an alkali metal salt, called the *flux,* and the combination is then fused to form a water-soluble product called the *melt.* Fluxes decompose most substances by virtue of the high temperature required for their use (300°C to 1000°C) and the high concentration of reagent brought in contact with the sample.

While they are very effective solvents, fluxes introduce high concentrations of ionic species to aqueous solutions of the melt.

[6]W. Schöniger, *Mikrochim. Acta,* **1955,** *123;* **1956,** *869.* See also the review articles by A. M. G. MacDonald, in *Advances in Analytical Chemistry and Instrumentation,* C. N. Reilley, Ed., Vol. 4, p. 75. New York: Interscience, 1965.

Where possible, the employment of a flux is avoided, for several dangers and disadvantages attend its use. Among these is the possible contamination of the sample by impurities in the flux. This possibility is exacerbated by the relatively large amount of flux (typically at least ten times the sample weight) required for a successful fusion. Moreover, the aqueous solution that results when the melt from a fusion is dissolved has a high salt content, which may cause difficulties in the subsequent steps of the analysis. In addition, the high temperatures required for a fusion increase the danger of volatilization losses. Finally, the container in which the fusion is performed is almost inevitably attacked to some extent by the flux; again, contamination of the sample is the result.

For a sample containing only a small fraction of material that dissolves with difficulty, it is common practice to employ a liquid reagent first; the undecomposed residue is then isolated by filtration and fused with a relatively small quantity of a suitable flux. After cooling, the melt is dissolved and combined with the major portion of the sample.

33E-1 Carrying out a Fusion

The sample in the form of a very fine powder is mixed intimately with perhaps a tenfold excess of the flux. Mixing is usually carried out in the crucible in which the fusion is to be performed. The time required for fusion can range from a few minutes to hours. The production of a clear melt signals completion of the decomposition, although often this condition is not always obvious.

When the fusion is complete, the mass is allowed to cool slowly; just before solidification, the crucible is rotated to distribute the solid around the walls to produce a thin layer of melt that is easy to dislodge.

33E-2 Types of Fluxes

With few exceptions, the common fluxes used in analysis are compounds of the alkali metals. Alkali metal carbonates, hydroxides, peroxides, and borates are basic fluxes that attack acidic materials. The acidic fluxes are pyrosulfates, acid fluorides, and boric oxide. If an oxidizing flux is required, sodium peroxide can be used. As an alternative, small quantities of the alkali nitrates or chlorates can be mixed with sodium carbonate.

The properties of the common fluxes are summarized in Table 33-1.

Sodium Carbonate

Silicates and certain other refractory materials can be decomposed by heating to 1000°C to 1200°C with sodium carbonate. This treatment generally converts the cationic constituents of the sample to acid-soluble carbonates or oxides; the nonmetallic constituents are converted to soluble sodium salts.

Carbonate fusions are normally carried out in platinum crucibles.

Potassium Pyrosulfate

Potassium pyrosulfate is a potent acidic flux that is particularly useful for attacking the more intractable metal oxides. Fusions with this reagent are performed at

TABLE 33-1 Common Fluxes

Flux	Melting Point, °C	Type of Crucible for Fusion	Type of Substance Decomposed
Na_2CO_3	851	Pt	Silicates and silica-containing samples, alumina-containing samples, sparingly soluble phosphates and sulfates
Na_2CO_3 + an oxidizing agent, such as KNO_3, $KClO_3$, or Na_2O_2	—	Pt (not with Na_2O_2), Ni	Samples requiring an oxidizing environment; that is, samples containing S, As, Sb, Cr, etc.
$LiBO_2$	849	Pt, Au, Glassy carbon	Powerful basic flux for silicates, most minerals, slags, ceramics
NaOH or KOH	318 380	Au, Ag, Ni	Powerful basic fluxes for silicates, silicon carbide, and certain minerals (main limitation is purity of reagents)
Na_2O_2	Decomposes	Fe, Ni	Powerful basic oxidizing flux for sulfides; acid-insoluble alloys of Fe, Ni, Cr, Mo, W, and Li; platinum alloys; Cr, Sn, Zr minerals
$K_2S_2O_7$	300	Pt, porcelain	Acidic flux for slightly soluble oxides and oxide-containing samples
B_2O_3	577	Pt	Acidic flux for silicates and oxides where alkali metals are to be determined
$CaCO_3 + NH_4Cl$	—	Ni	Upon heating the flux, a mixture of CaO and $CaCl_2$ is produced; used to decompose silicates for the determination of the alkali metals

about 400°C; at this temperature, the slow evolution of the highly acidic sulfur trioxide takes place:

$$K_2S_2O_7 \longrightarrow K_2SO_4 + SO_3(g)$$

Potassium pyrosulfate can be prepared by heating potassium hydrogen sulfate:

$$2KHSO_4 \longrightarrow K_2S_2O_7 + H_2O$$

Lithium Metaborate

Lithium metaborate $LiBO_2$ by itself or mixed with lithium tetraborate finds considerable use for attacking refractory silicate and alumina minerals, particularly for AAS, ICP, and X-ray absorption or emission analyses. These fusions are generally carried out in graphite or platinum crucibles at about 900°C. The glass that results upon cooling the melt can be used directly for X-ray fluorescence measurements. It is also readily soluble in mineral acids. Boric oxide is removed after solution of the melt by evaporation to dryness with methyl alcohol; methyl borate, $B(OCH_3)_3$, distills.

33F QUESTIONS AND PROBLEMS

***33-1.** Differentiate between wet ashing and dry ashing.

33-2. What is a flux? When is its use called for?

***33-3.** What fluxes are suitable for the determination of alkali metals in silicates?

33-4. What flux is commonly used for the decomposition of certain refractory oxides?

***33-5.** Under what conditions is the use of perchloric acid likely to be dangerous?

33-6. How are organic compounds decomposed for the determination of

***(a)** halogens?

(b) sulfur?

***(c)** nitrogen?

(d) heavy-metal species?

CHAPTER

34

Elimination of Interferences

An interference in a chemical analysis arises whenever a species in the sample matrix either produces a signal that is indistinguishable from that of the analyte or, alternatively, attenuates the analyte signal. Few if any analytical signals are so specific as to be free of interference. As a consequence, most analytical methods require one or more preliminary steps to eliminate the effects of interferences.

A **masking agent** is a reagent that chemically binds an interference and prevents it from causing errors in an analysis.

Two general methods are available for dealing with interferences. The first makes use of a *masking agent* to immobilize or chemically bind the interfering species in a form in which it no longer contributes to or attenuates the signal from the analyte.[1] Clearly, a masking agent must not affect the behavior of the analyte significantly. In earlier chapters, we have encountered several masking agents. An example is the use of fluoride ion to prevent iron(III) from interfering in the iodometric determination of copper(II). Here, masking results from the strong tendency of fluoride ions to complex iron(III) but not copper(II). The consequence is a decrease in the electrode potential of the iron(III) system to the point where only copper(II) ions from the sample oxidize iodide to iodine.

Chemical species are generally separated by converting them to different phases that can then be mechanically isolated.

The second approach to dealing with an interference involves converting either the analyte or the interferences into a separate phase that can be separated mechanically. Currently, the most common way of accomplishing this type of separation is by chromatography, which was treated in detail in Chapters 28 through 30. In this chapter we consider the classical methods of separation that were employed prior to the widespread use of chromatography and still find use on occasion.

34A THE NATURE OF THE SEPARATION PROCESS

The **distribution ratio** is the ratio of the amount of a substance in one phase to the amount in a second phase.

All separation processes have in common the distribution of the components in a mixture between two phases that subsequently can be separated mechanically. If the ratio of the amount of one particular component in each phase (the *distribution ratio*) differs significantly from that of another, a separation of the two is potentially feasible. To be sure, the complexity of the separation depends upon the magnitude of this difference in distribution ratios. Where the difference is large, a single-stage process suffices. For example, a single precipitation with silver ion is

[1]For a monograph on masking agents, see D. D. Perrin, *Masking and Demasking Reactions.* New York: Wiley-Interscience, 1970.

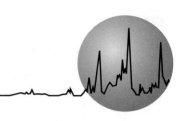

adequate for the isolation of chloride from many other anions because in the presence of excess silver ions, the ratio of the amount of chloride ion in the solid phase to that in the aqueous phase is immense while comparable ratios for, say, nitrate or perchlorate ion approach zero.

A more complex situation prevails when the distribution ratio for one component is essentially zero, as in the foregoing example, but the ratio for the other is not very large. Here, a multistage process is required. For example, uranium(VI) can be extracted into ether from an aqueous nitric acid solution. Although the distribution ratio approximates unity for a single extraction, uranium(VI) can nevertheless be isolated quantitatively by repeated, or *exhaustive,* extraction of the aqueous solution with fresh portions of ether.

The most complex procedures are required when the distribution ratios of the species to be separated are both greater than zero and approach one another in magnitude; here, multistage *fractionation* techniques, such as chromatography, are necessary. Like its simpler counterparts, fractionation is based upon differences in the distribution ratios of solutes. Two factors, however, account for the gain in separation efficiency with fractionation. First, the number of times that partitioning occurs between phases is increased enormously; second, distribution occurs between fresh portions of both phases. An exhaustive extraction differs from a fractionation in that fresh portions of only one phase are involved in the former.

34B SEPARATION BY PRECIPITATION

Separations by precipitation require large solubility differences between analyte and potential interferences. The theoretical feasibility of this type of separation can be determined by solubility calculations such as those shown in Section 9C. Unfortunately, several other factors may preclude the use of precipitation to achieve a separation. For example, the various coprecipitation phenomena described in Section 5B-4 may cause extensive contamination of a precipitate by an unwanted component, even though the solubility product of the contaminant has not been exceeded. Likewise, the rate of an otherwise feasible precipitation may be too slow to be useful for a separation. Finally, when precipitates form as colloidal suspensions, coagulation may be difficult and slow, particularly when the isolation of a small quantity of a solid phase is attempted.

Many precipitating agents have been employed for quantitative inorganic separations. Some of the most generally useful are described in the sections that follow.

34B-1 Separations Based upon Control of Acidity

Enormous differences exist in the solubilities of the hydroxides, hydrous oxides, and acids of various elements. Moreover, the concentration of hydrogen or hydroxide ions in a solution can be varied by a factor of 10^{15} or more and can be readily controlled by the use of buffers. As a consequence, many separations based on pH control are, in theory, available to the chemist. In practice, these separations can be grouped in three categories: (1) those made in relatively concentrated solutions of strong acids, (2) those made in buffered solutions at intermediate pH values, and (3) those made in concentrated solutions of sodium or

TABLE 34-1 Separations Based upon Control of Acidity

Reagent	Species Forming Precipitates	Species Not Precipitated
Hot concd HNO_3	Oxides of W(VI), Ta(V), Nb(V), Si(IV), Sn(IV), Sb(V)	Most other metal ions
NH_3/NH_4Cl buffer	Fe(III), Cr(III), Al(III)	Alkali and alkaline earths, Mn(II), Cu(II), Zn(II), Ni(II), Co(II)
$HOAc/NH_4OAc$ buffer	Fe(III), Cr(III), Al(III)	Common dipositive ions
$NaOH/Na_2O_2$	Fe(III), most dipositive ions, rare earths	Zn(II), Al(III), Cr(VI), V(V), U(VI)

potassium hydroxide. Table 34-1 lists common separations that can be achieved by control of acidity.

34B-2 Sulfide Separations

Recall (Equation 9-57) that

$$[S^{2-}] = \frac{1.2 \times 10^{-22}}{[H_3O^+]^2}$$

With the exception of the alkali and alkaline-earth metals, most cations form sparingly soluble sulfides whose solubilities differ greatly from one another. Because it is relatively easy to control the sulfide ion concentration of an aqueous solution by adjustment of pH (Section 9C-2), separations based on the formation of sulfides have found extensive use. Sulfides can be conveniently precipitated from homogeneous solution, with the anion being generated by the hydrolysis of thioacetamide (Table 5-3).

A theoretical treatment of the ionic equilibria that influence the solubility of sulfide precipitates was considered in Section 9C-2. Such a treatment may fail to provide realistic conclusions regarding the feasibility of separations, however, because of coprecipitation and the slow rate at which some sulfides form. As a consequence, resort must be made to empirical observations.

Table 34-2 shows some common separations that can be accomplished with hydrogen sulfide through control of pH.

TABLE 34-2 Precipitation of Sulfides

Elements	Conditions for Precipitation*	Conditions for No Precipitation*
Hg(II), Cu(II), Ag(I)	1, 2, 3, 4	
As(V), As(III), Sb(V), Sb(III)	1, 2, 3	4
Bi(III), Cd(II), Pb(II), Sn(II)	2, 3, 4	1
Sn(IV)	2, 3	1, 4
Zn(II), Co(II), Ni(II)	3, 4	1, 2
Fe(II), Mn(II)	4	1, 2, 3

*1 = 3 M HCl; 2 = 0.3 M HCl; 3 = buffered to pH 6 with acetate; 4 = buffered to pH 9 with $NH_3/(NH_4)_2S$.

34B-3 Separations by Other Inorganic Precipitants

No other inorganic ion is as generally useful for separations as hydroxide and sulfide ions. Phosphate, carbonate, and oxalate ions are often employed as precipitants for cations, but their behavior is nonselective; therefore, separations must ordinarily precede their use.

Chloride and sulfate are useful because of their highly selective behavior. The former is used to separate silver from most other metals, and the latter is frequently employed to isolate a group of metals that includes lead, barium, and strontium.

34B-4 Separations by Organic Precipitants

Selected organic reagents for the isolation of various inorganic ions were discussed in Section 5D-3. Some of these organic precipitants, such as dimethylglyoxime, are useful because of their remarkable selectivity in forming precipitates with a few ions. Others, such as 8-hydroxyquinoline, yield slightly soluble compounds with a host of cations. The selectivity of this sort of reagent is due to the wide range of solubility products among its reaction products and also to the fact that the precipitating reagent is ordinarily an anion that is the conjugate base of a weak acid. Thus, separations based on pH control can be realized just as with hydrogen sulfide.

34B-5 Separation of Species Present in Trace Amounts by Precipitation

A common problem in trace analysis is that of isolating microgram quantities of the species of interest from the major components of the sample. Although such a separation is sometimes based on a precipitation, the techniques required differ from those used when the analyte is present in generous amounts.

Several problems attend the quantitative isolation of a trace element by precipitation even when solubility losses are not important. Supersaturation often delays formation of the precipitate, and coagulation of small amounts of a colloidally dispersed substance is often difficult. In addition, it is likely that an appreciable fraction of the solid will be lost during transfer and filtration. To minimize these difficulties, a quantity of some other ion that also forms a precipitate with the reagent is often added to the solution. The precipitate from the added ion is called a *collector* and carries the desired minor species out of solution. For example, in isolating manganese as the sparingly soluble manganese dioxide, a small amount of iron(III) is frequently added to the analyte solution before the introduction of ammonia as the precipitating reagent. The basic iron(III) oxide carries down even the smallest traces of the manganese dioxide. Other examples are basic aluminum oxide as a collector of trace amounts of titanium and copper sulfide for collection of traces of zinc and lead. Many other collectors are described by Sandell and Onishi.[2]

A collector may entrain a trace constituent as a result of similarities in their solubilities. Others involve coprecipitation, in which the minor component is

A **collector** is used to remove trace constituents from solution.

[2]E. B. Sandell and H. Onishi, *Colorimetric Determination of Traces of Metals,* 4th ed., pp. 709–721. New York: Interscience, 1978.

adsorbed on or incorporated into the collector precipitate as the result of mixed-crystal formation. Clearly, the collector must not interfere with the method selected for determining the trace component.

34B-6 Separation by Electrolytic Precipitation

Electrolytic precipitation is a highly useful method for accomplishing separations. In this process, the more easily reduced species, be it the wanted or the unwanted component of the mixture, is isolated as a separate phase. The method becomes particularly effective when the potential of the working electrode is controlled at a predetermined level (Section 20C-2).

The mercury cathode (page 445) has found wide application in the removal of many metal ions prior to the analysis of the residual solution. In general, metals more easily reduced than zinc are conveniently deposited in the mercury, leaving such ions as aluminum, beryllium, the alkaline earths, and the alkali metals in solution. The potential required to decrease the concentration of a metal ion to any desired level is readily calculated from polarographic data.

34C SEPARATION BY EXTRACTION

The extent to which solutes, both inorganic and organic, distribute themselves between two immiscible liquids differs enormously, and these differences have been used for decades to accomplish analytical separations. This section considers applications of the distribution phenomenon to such separations.

34C-1 Theory

Two terms are employed to describe the distribution of a solute between two immiscible solvents: *distribution coefficients* and *distribution ratio.* It is important to have a clear understanding of the distinction between the two.

The Distribution Coefficient

The distribution coefficient is an equilibrium constant that describes the distribution of a solute species between two immiscible solvents. For example, when an aqueous solution of an organic solute A is shaken with an organic solvent, such as hexane, there is quickly established an equilibrium that is described by the equation

$$\mathrm{A}(aq) \rightleftharpoons \mathrm{A}(org) \qquad (34\text{-}1)$$

Be sure you understand the difference between the distribution coefficient and the distribution ratio.

where (*aq*) and (*org*) refer to the aqueous and organic phases. Ideally, the ratio of the activities of species A in the two phases is constant and independent of the total quantity of A. That is, at any given temperature

$$K_{\mathrm{d}} = \frac{[\mathrm{A}]_{org}}{[\mathrm{A}]_{aq}} \qquad (34\text{-}2)$$

where the equilibrium constant K_d is the distribution coefficient. The terms in brackets are strictly the activities of species A in the two solvents, but molar concentrations can frequently be substituted without serious error. Often, K_d is approximately equal to the ratio of the solubility of A in the two solvents.

When the solute exists in different states of aggregation in the two solvents, the equilibrium becomes

$$xA_y(aq) \rightleftharpoons yA_x(org)$$

and the distribution coefficient takes the form

$$K_d = \frac{[A_x]^y_{org}}{[A_y]^x_{aq}}$$

The Distribution Ratio

The distribution ratio D for an analyte is defined as the ratio of its *analytical* concentration in two immiscible solvents. For a simple system, such as that described by Equation 34-1, the distribution ratio is identical to the distribution coefficient. For more complex systems, however, the two can be quite different. For example, for the distribution of a fatty acid HA between water and diethyl ether, we can write

$$D = \frac{c_{org}}{c_{aq}} \tag{34-3}$$

where c_{org} and c_{aq} are the molar *analytical* concentrations of HA in the two phases. In the aqueous medium, the analytical concentration of the acid is equal to the sum of the equilibrium concentrations of the weak acid and its conjugate base:

$$c_{aq} = [HA]_{aq} + [A^-]_{aq}$$

In contrast, no significant dissociation of the acid occurs in the nonpolar organic layer, so the analytical and equilibrium concentrations of HA are identical, and we can write

$$c_{org} = [HA]_{org}$$

Substituting the last two relationships into Equation 34-3 gives

$$D = \frac{[HA]_{org}}{[HA]_{aq} + [A^-]_{aq}} \tag{34-4}$$

In order to relate D to K_a for *species* [HA] we write the acid dissociation expression for HA_{aq} as

$$K_a = \frac{[H_3O^+]_{aq}[A^-]_{aq}}{[HA]_{aq}} \quad \text{or} \quad [A^-]_{aq} = \frac{[HA]_{aq}K_a}{[H_3O^+]_{aq}}$$

Substituting into Equation 34-4 gives

$$D = \frac{[\text{HA}]_{org}}{[\text{HA}]_{aq} + [\text{HA}]_{aq}K_a/[\text{H}_3\text{O}^+]_{aq}} = \frac{[\text{HA}]_{org}}{[\text{HA}]_{aq}} \times \frac{1}{1 + K_a/[\text{H}_3\text{O}^+]_{aq}}$$

Substituting Equation 34-2 gives

$$D = K_d \times \frac{1}{1 + K_a/[\text{H}_3\text{O}^+]_{aq}}$$

Rearranging demonstrates the relationship between D and K_d for HA. That is,

$$D = \frac{c_{org}}{c_{aq}} = K_d \times \frac{[\text{H}_3\text{O}^+]_{aq}}{[\text{H}_3\text{O}^+]_{org} + K_a} \quad (34\text{-}5)$$

A **distribution coefficient** is a ratio of **species** molar concentrations. A **distribution ratio** is a ratio of **analytical** molar concentrations (Equation 34-3).

Equation 34-5 can be used to compute the extent of extraction of HA from buffered aqueous solutions. (Note the difference between the distribution coefficient K_d and the distribution ratio D.)

EXAMPLE 34-1

The distribution coefficient for a weak acid between diethyl ether and water is found to be 800, and its acid dissociation constant in water is 1.50×10^{-5}. Calculate the analytical concentration of HA, c_{HA}, remaining in an aqueous solution after the extraction of 50.0 mL of 0.0500 M HA with 25.0 mL of ether, assuming the aqueous solution is buffered to a pH of (a) 2.00 and (b) 8.00.

(a) Substituting $[\text{H}_3\text{O}^+] = 1.00 \times 10^{-2}$ and the two equilibrium constants into Equation 34-5 yields

$$D = \frac{c_{org}}{c_{aq}} = \frac{800 \times 1.00 \times 10^{-2}}{1.00 \times 10^{-2} + 1.50 \times 10^{-5}} = 799 \quad (34\text{-}6)$$

The total number of millimoles of acid contained in the two solvents is equal to the original number of millimoles in the aqueous solution:

$$\text{total no. mmol HA} = 50.0 \times 0.0500 = 2.50$$

After extraction, the 2.50 mmol of HA is distributed between the two solvents, so

$$50.0\ c_{aq} + 25.0\ c_{org} = 2.50$$

where c_{aq} and c_{org} are the analytical concentrations in the two solvents. Rearranging Equation 34-6 shows that

$$c_{org} = 799\ c_{aq}$$

and so we can write

$$50.0\ c_{aq} + (25.0)(799\ c_{aq}) = 2.50$$

$$c_{aq} = 1.25 \times 10^{-4}\ \text{M}$$

(b) When $[H_3O^+] = 1.00 \times 10^{-8}$,

$$D = \frac{800 \times 1.00 \times 10^{-8}}{1.50 \times 10^{-5} + 1.00 \times 10^{-8}} = 0.533$$

Proceeding as in part (a), we have

$$50.0 \times c_{aq} + (25.0)(0.533\, c_{aq}) = 2.50$$
$$c_{aq} = 3.95 \times 10^{-2}\ \text{M}$$

The Completeness of Multiple Extractions

Distribution coefficients and distribution ratios are useful because they provide guidance as to the most efficient way to perform extractive separations. For example, consider again the extraction of a species HA from an aqueous solution with a pH of 2.00 where Equation 34-3 applies. Suppose that V_{aq} milliliters of water containing a_0 millimoles of HA is extracted with V_{org} milliliters of diethyl ether. At equilibrium, a_1 millimoles of HA remain in the aqueous layer and $a_0 - a_1$ millimoles have been transferred to the organic layer. The analytical concentration of HA in each layer is

$$c_{aq1} = \frac{a_1}{V_{aq}}$$

$$c_{org1} = \frac{a_0 - a_1}{V_{org}}$$

Substitution of these quantities into Equation 34-3 and rearrangement give

$$a_1 = \left(\frac{V_{aq}}{V_{org}D + V_{aq}}\right) a_0 \qquad \textbf{(34-7)}$$

The number of millimoles a_2 remaining in the aqueous layer after a second extraction with an identical volume of solvent is, by the same reasoning,

$$a_2 = \left(\frac{V_{aq}}{V_{org}D + V_{aq}}\right) a_1$$

When this expression is multiplied by Equation 34-7, a_1 cancels and we obtain

$$a_2 = \left(\frac{V_{aq}}{V_{org}D + V_{aq}}\right)^2 a_0$$

By the same reasoning, after n extractions, the number of millimoles of HA remaining in the aqueous layer is

$$a_n = \left(\frac{V_{aq}}{V_{org}D + V_{aq}}\right)^n a_0 \qquad \textbf{(34-8)}$$

Equation 34-8 can be rewritten in terms of the initial and final analytical concentrations of HA in the water by substituting the relationships

$$a_n = (c_{aq})_n V_{aq} \qquad \text{and} \qquad a_0 = (c_{aq})_0 V_{aq}$$

where $(c_{aq})_n$ is the analytical concentration of HA in the aqueous phase after n extractions. Substitution of these relationships into Equation 34-8 gives

$$(c_{aq})_n = \left(\frac{V_{aq}}{V_{org}D + V_{aq}}\right)^n (c_{aq})_0 \qquad (34\text{-}9)$$

As shown in the example that follows, a more efficient extraction is achieved with several small volumes of solvent than with a single large one.

EXAMPLE 34-2

The distribution coefficient for iodine between CCl_4 and H_2O is 85. Calculate the concentration of I_2 remaining in the aqueous layer after extraction of 50.0 mL of 1.00×10^{-3} M I_2 with the following quantities of CCl_4: (a) 50.0 mL; (b) two 25.0-mL portions; (c) five 10.0-mL portions.

Substitution into Equation 34-9 gives

$$\text{(a)}\ (c_{aq})_1 = \left(\frac{50.0}{50.0 \times 85 + 50.0}\right)^1 \times 1.00 \times 10^{-3} = 1.16 \times 10^{-5}\ \text{M}$$

$$\text{(b)}\ (c_{aq})_2 = \left(\frac{50.0}{25.0 \times 85 + 50.0}\right)^2 \times 1.00 \times 10^{-3} = 5.28 \times 10^{-7}\ \text{M}$$

$$\text{(c)}\ (c_{aq})_3 = \left(\frac{50.0}{10.0 \times 85 + 50.0}\right)^5 \times 1.00 \times 10^{-3} = 5.29 \times 10^{-10}\ \text{M}$$

It is always better to use several small portions of solvent to extract a sample than to extract with one large portion.

Figure 34-1 shows that the improved efficiency of multiple extractions falls off rapidly as a total fixed volume is subdivided into smaller and smaller portions. Clearly, little is to be gained by dividing the extracting solvent into more than five or six portions.

34C-2 Types of Extraction Procedures

Three types of separation procedures are based upon distribution equilibria between immiscible solvents: *simple, exhaustive,* and *countercurrent extractions.*

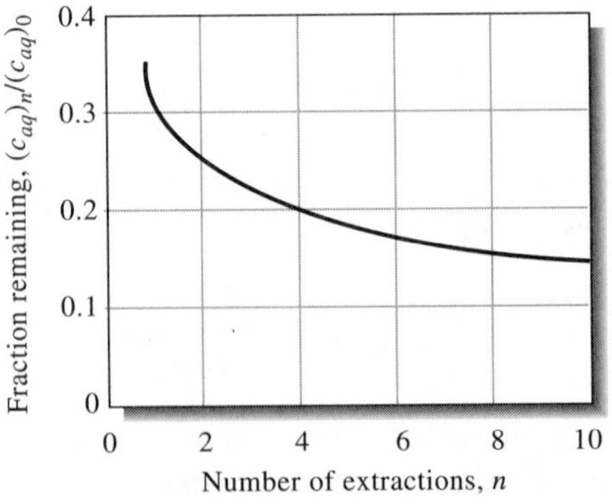

Figure 34-1
Plot of Equation 34-8, assuming $K_d = 2$ and $V_{aq} = 100$. The total volume of the organic solvent was also assumed to be 100, so that $V_{org} = 100/n$.

Simple Extractions

When the distribution ratio for one species in a mixture is reasonably favorable (on the order of 5 to 10 or greater) and that for the others is unfavorable (< 0.001), an extractive separation can be simple, rapid, and quantitative. The solution containing the analyte is extracted successively with up to six portions of fresh solvent. An ordinary separatory funnel is used, with either the original solution or the extract being retained for completing the analysis.

Exhaustive Extractions

Exhaustive extraction permits the separation of components of a mixture that have relatively unfavorable distribution ratios (< 1) from those with ratios that approach zero. An apparatus is used in which the organic solvent is automatically distilled, condensed, and caused to pass continuously through the aqueous layer (see Figure 34-2). Thus, the equivalent of several hundred extractions with fresh solvent is accomplished in one hour or less with equipment that requires no attention.

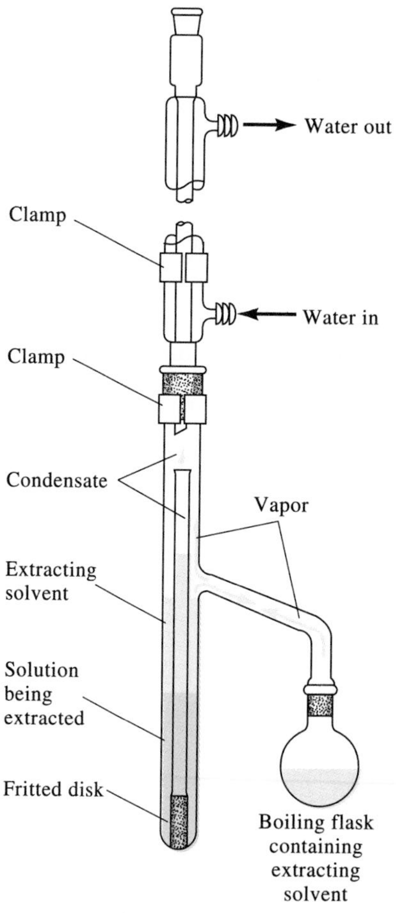

Figure 34-2
A Soxhlet extractor.

Countercurrent Fractionation

Automated devices that permit hundreds of automatic successive extractions have been developed. With these instruments, fractionation occurs by a *countercurrent* scheme in which distribution between *fresh* portions of the two phases occurs in a series of discrete steps. An exhaustive extraction differs from the countercurrent technique in that fresh portions of only one phase are introduced in the former.

The countercurrent method permits the separation of components with nearly identical partition ratios. For example, Craig[3] has demonstrated that ten amino acids can be separated by countercurrent extraction even though their partition coefficients differ by less than 0.1.

34D APPLICATION OF EXTRACTION PROCEDURES

Extraction is often more attractive than classical precipitation for separating inorganic species because equilibration and separation of phases in a separatory funnel are less tedious and time-consuming than precipitation, filtration, and washing. In addition, the problems of coprecipitation and postprecipitation are avoided. Finally, extractions are ideally suited for isolation of trace quantities of a species.

34D-1 The Extractive Separation of Metal Ions as Chelates

Many organic chelating agents are weak acids that react with metal ions to give uncharged complexes that are highly soluble in organic solvents such as ethers, hydrocarbons, ketones, and chlorinated species (including chloroform and carbon tetrachloride). The metal chelates, on the other hand, are usually nearly insoluble in water. Similarly, the chelating agents themselves are often quite soluble in organic solvents but have limited solubility in water.

[3]L. C. Craig, *Anal. Chem.*, **1950,** *22,* 1346.

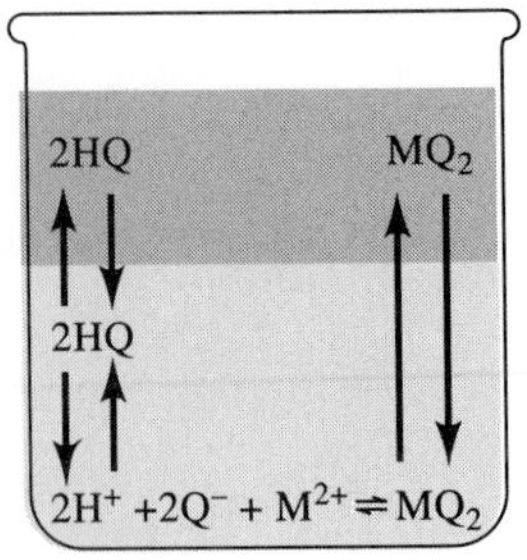

Figure 34-3
Equilibria in the extraction of an aqueous cation M^{2+} into an immiscible organic solvent containing 8-hydroxyquinoline.

The Effect of pH and Reagent Concentration on Distribution Ratios

Several chelating agents have proved useful for separations based upon the selective extraction of metal ions from a buffered aqueous solution into a nonaqueous solvent containing these agents. As shown in Figure 34-3, such a process involves several equilibria and several species. Among the latter are the undissociated ligand HQ, its conjugate base Q^-, the metal-ligand complex MQ_n, and metal and hydronium ions. The important equilibria are

$$HQ(aq) \rightleftharpoons HQ(org) \qquad K_{d1} = \frac{[HQ]_{org}}{[HQ]_{aq}} \tag{34-10}$$

$$HQ(aq) + H_2O \rightleftharpoons H_3O^+(aq) + Q^-(aq) \qquad K_a = \frac{[H_3O^+]_{aq}[Q^-]_{aq}}{[HQ]_{aq}} \tag{34-11}$$

$$M^{n+}(aq) + nQ^-(aq) \rightleftharpoons MQ_n(aq) \qquad K_f = \frac{[MQ_n]_{aq}}{[M^{n+}]_{aq}[Q^-]^n_{aq}} \tag{34-12}$$

$$MQ_n(aq) \rightleftharpoons MQ_n(org) \qquad K_{d2} = \frac{[MQ_n]_{org}}{[MQ_n]_{aq}} \tag{34-13}$$

Organic chelating agents as well as neutral metal chelates are usually highly soluble in organic liquids, so the distribution coefficients K_{d1} and K_{d2} are generally large numerically. Furthermore, the concentration of M^{n+} in the nonpolar organic layer ordinarily approaches zero in most cases. The selectivity of the reagent is determined by the relative magnitudes of the formation constants K_f for various cations. As shown by Equation 34-11, the concentration of the active reagent Q^- is pH-dependent. Thus, by controlling pH, one can control the concentration of Q^- and thus which cations are extracted and which are not.

In order to derive an expression relating the amount of a cation extracted to pH and concentration of chelating agent, we employ the distribution ratio, which for the system shown in Figure 34-3 takes the form

$$D = \frac{c_{org}}{c_{aq}} = \frac{[MQ_n]_{org}}{[M^{n+}]_{aq} + [MQ_n]_{aq}} \approx \frac{[MQ_n]_{org}}{[M^{n+}]_{aq}} \tag{34-14}$$

where c_{org} and c_{aq} are the molar analytical concentrations of M^{n+} in the organic and aqueous phases. Ordinarily, the assumption that $[MQ_n]_{aq} \ll [M^{n+}]_{aq}$ is valid because (1) the chelate is generally not very soluble in water and (2) that which is in solution is largely dissociated. As shown by the following derivation, D is independent of the total amount of metal in the two phases but dependent upon both the concentration of HQ in the organic layer and the hydronium ion concentration in the aqueous solution.

If c_Q is the original molar concentration of HQ in the organic phase, mass balance requires that

$$c_Q = [HQ]_{org} + [HQ]_{aq} + [Q^-]_{aq} + n[MQ_n]_{aq} + n[MQ_n]_{org}$$

Ordinarily, extractions are carried out with such a large excess of chelating agent that the concentration of the *species* HQ in the organic layer far exceeds the concentration of all other species containing Q. Thus, the foregoing mass-balance expression simplifies to

$$c_Q \approx [HQ]_{org} \tag{34-15}$$

In order to arrive at an expression that relates D for this system to the original concentration of the chelating agent in the organic solution and to the pH of the aqueous solution, let us multiply Equation 34-12 by 34-13 and rearrange, which leads to

$$[MQ_n]_{org} = K_f K_{d2}[M^{n+}]_{aq}[Q^-]^n_{aq}$$

Substituting into Equation 34-14 gives

$$D = \frac{c_{org}}{c_{aq}} = K_f K_{d2}[Q]^n_{aq} \tag{34-16}$$

Dividing Equation 34-11 by 34-10 allows us to express $[Q^-]_{aq}$ in terms of $[H_3O^+]$ and $[HQ]_{org}$:

$$[Q^-]_{aq} = \frac{K_a}{K_{d1}} \frac{[HQ]_{org}}{[H_3O^+]_{aq}}$$

Substitution of this equation and Equation 34-15 into Equation 34-16 leads to the desired relationship:

$$D = \frac{c_{org}}{c_{aq}} = \frac{K_f K_{d2} K^n_a}{K^n_{d1}} \times \frac{c^n_Q}{[H_3O^+]^n_{aq}} \tag{34-17}$$

Combining the four equilibrium constants into a single constant K_{ex} yields

$$D = \frac{c_{org}}{c_{aq}} = \frac{K_{ex} c^n_Q}{[H_3O^+]^n_{aq}} \tag{34-18}$$

EXAMPLE 34-3

Lead forms a neutral complex PbQ_2 with the ligand Q^-. The constant K_{ex} for the distribution of this complex between water and CCl_4 has been found by experiment to be 2.0×10^4. A 25.0-mL aliquot of an aqueous solution that is 5.00×10^{-4} M in Pb^{2+} and 0.500 M in $HClO_4$ is extracted with two 10.0-mL portions of CCl_4 that are 0.0250 M in HQ. Calculate the percentage of unrecovered Pb^{2+} in the aqueous solution.

Substituting into Equation 34-18 gives

$$D = \frac{(2.0 \times 10^4)(0.0250)^2}{(0.500)^2} = 50.0$$

Substituting into Equation 34-9 for two extractions gives

$$(c_{aq})_2 = \left(\frac{25.0}{10.0 \times 50.0 + 2.50}\right)^2 (5.00 \times 10^{-4}) = 1.13 \times 10^{-6}$$

$$\%\text{ unextracted Pb}^{2+} = \frac{1.13 \times 10^{-6}}{5.0 \times 10^{-4}} \times 100\% = 0.23$$

Some Separations Based on the Extraction of Metal Chelates

Extractions with Diphenylthiocarbazone Diphenylthiocarbazone, or dithizone, is a useful reagent for separating minute quantities of a dozen or more metal ions. Its reaction with a divalent cation, such as Pb^{2+}, can be written as

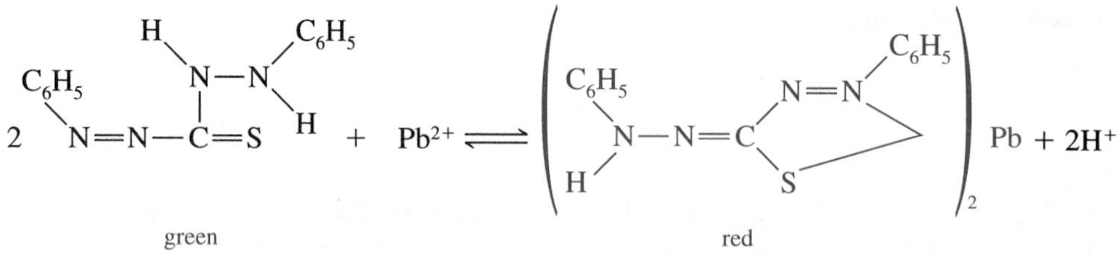

Both dithizone and its metal chelates are essentially insoluble in water but dissolve readily in solvents such as carbon tetrachloride and chloroform. Solutions of the reagent are deep green, whereas solutions of the metal chelates are intensely red, violet, orange, or yellow and provide a sensitive means for the photometric determination of the separated ions.

The distribution ratio of the various metal dithizonates is described by Equations 34-18 and 34-14. Thus, for lead,

$$D = \frac{c_{org}}{c_{aq}} = \frac{[\text{PbDz}_2]_{org}}{[\text{Pb}^{2+}]_{aq}} = \frac{K_{ex}c_{\text{Dz}}^2}{[\text{H}_3\text{O}^+]_{aq}^2}$$

Figure 34-4 shows how the percentages of lead and other metal ions extracted into a chloroform solution of dithizone change as the pH of the aqueous medium

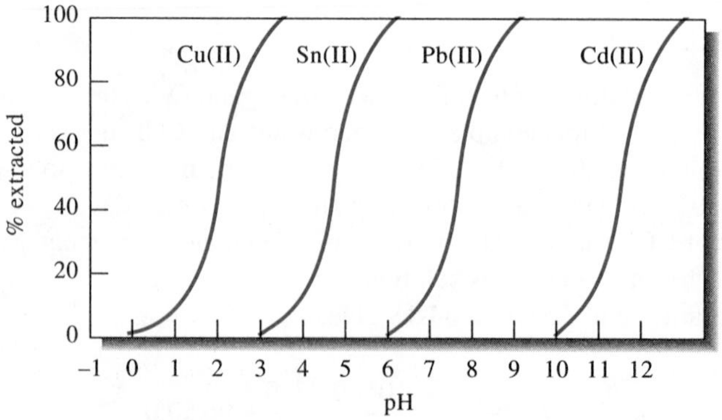

Figure 34-4
Effect of pH on the extraction of cations with CCl_4 solutions of dithizone.

TABLE 34-3 Diethyl Ether Extraction of Various Chlorides from 6 M Hydrochloric Acid*

Percent Extracted	Elements and Oxidation State
90–100	Fe(III), 99%; Sb(V), 99%†; Ga(III), 97%; Ti(III), 95%†; Au(III), 95%
50–90	Mo(VI), 80–90%; As(III), 80%†‡; Ge(IV), 40–60%
1–50	Te(IV), 34%; Sn(II), 15–30%; Sn(IV), 17%; Ir(IV), 5%; Sb(III), 2.5%†
<1, >0	As(V)†, Cu(II), In(III), Hg(II), Pt(IV), Se(IV), V(V), V(IV), Zn(II)
0	Al(III), Bi(III), Cd(II), Cr(III), Co(II), Be(II), Fe(II), Pb(II), Mn(II), Ni(II), Os(VIII), Pd(II), Rh(III), Ag(I), Th(IV), Ti(IV), W(VI), Zr(IV)

*From E. H. Swift, *Introductory Quantitative Analysis,* p. 431. Englewood Cliffs, NJ: Prentice-Hall, 1950. With permission.

†Isopropyl ether employed rather than diethyl ether.

‡8 M HCl rather than 6 M.

changes. The data for these curves were derived in a way analogous to the calculation shown in Example 34-3. It is apparent from these curves that clean separations of copper, lead, and cadmium can be achieved by pH control. Depending upon its absorption spectrum, tin(II) may interfere slightly in the determination of copper and lead.

Extractions with 8-Hydroxyquinoline Many of the chelates of 8-hydroxyquinoline (Section 5D-3) are readily extracted into various organic solvents. The extraction process can be formulated as

$$2HQ(org) + M^{2+}(aq) \rightleftharpoons MQ_2(org) + 2H^+(aq)$$

where HQ symbolizes the chelating agent. The equilibrium is clearly pH-dependent, which permits the separation of metals through control of the aqueous phase pH. The method has proved particularly useful for the separation of trace amounts of metals.

Extractions with Other Chelating Agents Separations that make use of other organic chelating agents are described in several reference works.[4]

34D-2 The Extraction of Metal Chlorides

The data in Table 34-3 indicate that a substantial number of metal chlorides can be extracted into diethyl ether from 6 M hydrochloric acid solution; equally important, a large number of metal ions are either unaffected or extracted only slightly

[4]G. H. Morrison and H. Freiser, *Solvent Extraction in Analytical Chemistry.* New York: Wiley, 1957; A. K. De, S. M. Khophar, and R. A. Chalmers, *Solvent Extraction of Metals.* New York: Van Nostrand, 1970; E. B. Sandell and H. Onishi, *Colorimetric Determination of Traces of Metals,* 4th ed. New York: Interscience, 1978.

under these conditions. Thus, many useful separations are possible. One of the most important is the separation of iron(III) (99% extracted) from a host of other cations. For example, the greater part of the iron from steel or iron ore samples can be removed by extraction prior to analysis for such trace elements as chromium, aluminum, titanium, and nickel. The species extracted has been shown to be the ion pair $H_3O^+FeCl_4^-$. It has also been shown that the percentage of iron transferred to the organic phase depends upon the hydrochloric acid content of the aqueous phase (little is removed from solutions that are below 3 M and above 9 M HCl) and, to some extent, upon the iron content. Unless special precautions are taken, extraction of the last traces of iron is incomplete.

Methyl isobutyl ketone also extracts iron from hydrochloric acid solutions. It is reported to have a somewhat more favorable distribution ratio than diethyl ether and has the added advantage of being less flammable.

34D-3 The Extraction of Nitrates

Certain nitrate salts are selectively extracted by diethyl ether as well as other organic solvents. For example, uranium(VI) is conveniently separated from such elements as lead and thorium by ether extraction of an aqueous solution that is saturated with ammonium nitrate and has a nitric acid concentration of about 1.5 M. Bismuth and iron(III) nitrates are also extracted to some extent under these conditions.

34E ION-EXCHANGE SEPARATIONS

Ion-exchange resins have several applications in analytical chemistry. The most important is in high-performance liquid chromatography, described in Section 30D. A brief description of other analytical applications of these useful materials follows.

34E-1 The Separation of Interfering Ions of Opposite Charge

Ion exchange is particularly effective where the charge on the analyte is opposite that of the interfering ion(s).

Ion-exchange resins are useful for the removal of interfering ions, particularly where these ions have a charge opposite that of the analyte. For example, iron(II), aluminum(III), and other cations interfere in the gravimetric determination of sulfate by virtue of their tendency to coprecipitate with barium sulfate. Passage of a solution to be analyzed through a column containing a cation-exchange resin results in the retention of all the cations and the liberation of a corresponding number of protons. The sulfate ion passes freely through such a column, and the analysis can then be performed on the effluent. In a similar manner, phosphate ion, which interferes in the determination of barium and calcium ions, can be removed by passing the sample through an anion-exchange resin.

34E-2 The Concentration of Traces of an Electrolyte

A useful application of ion exchangers is the concentration of traces of an ion from a very dilute solution. Cation-exchange resins, for example, have been employed to collect traces of metallic elements from large volumes of natural waters. The ions are then liberated by treating the resin with acid; the result is a considerably more concentrated solution for analysis.

34E-3 The Conversion of Salts to Acids or Bases

The total salt content of a sample can be determined by titrating the hydrogen ions released as an aliquot of sample is washed through a cation exchanger in its acidic form. Similarly, a standard hydrochloric acid solution can be prepared by passing a solution containing a known weight of sodium chloride through a cation-exchange resin in its acid form. The effluent and washings are collected in a volumetric flask and diluted to volume. In an analogous way, standard sodium hydroxide solutions can be prepared by treatment of an anion-exchange resin with a known quantity of sodium chloride.

34F THE SEPARATION OF INORGANIC SPECIES BY DISTILLATION

Distillation permits the separation of components whose solution/vapor phase distribution ratios differ significantly from one another. If one species has a distribution ratio that is large compared with those of the other components of the mixture, the separation is simple. Table 34-4 lists some inorganic species that can be conveniently separated by simple distillation.

TABLE 34-4 Separation of Some Inorganic Species by Distillation

Analyte	Sample Treatment	Volatile Species	Method of Collection
CO_3^{2-}	Acidification	CO_2	$Ba(OH)_2(aq) + CO_2(g) \rightarrow BaCO_3(s) + H_2O$ or on Ascarite
SO_3^{2-}	Acidification	SO_2	$SO_2(g) + H_2O_2(aq) \rightarrow H_2SO_4(aq)$
S^{2-}	Acidification	H_2S	$Cd^{2+}(aq) + H_2S(g) \rightarrow CdS(s) + 2\,H^+$
F^-	Addition of SiO_2 and acidification	H_2SiF_6	Basic solution
Si	Addition of HF	SiF_4	Basic solution
H_3BO_3	Addition of H_2SO_4 and methanol	$B(OCH_3)_3$	Basic solution
$Cr_2O_7^{2-}$	Addition of concd HCl	CrO_2Cl_2	Basic solution
NH_4^+	Addition of NaOH	NH_3	Acidic solution
As, Sb	Addition of concd HCl and H_2SO_4	$AsCl_3$ $SbCl_3$	Water
Sn	Addition of HBr	$SnBr_4$	Water

34G QUESTIONS AND PROBLEMS

*34-1. What is a masking agent and how does it function?

34-2. Differentiate between
(a) an exhaustive extraction and a countercurrent extraction.
(b) a distribution coefficient and a distribution ratio.

*34-3. The distribution coefficient for X between chloroform and water is 9.6. Calculate the concentration of X remaining in the aqueous phase after 50.0 mL of 0.150 M X is treated by extraction with the following quantities of chloroform: (a) one 40.0-mL portion, (b) two 20.0-mL portions, (c) four 10.0-mL portions, and (d) eight 5.00-mL portions.

34-4. The distribution coefficient for Z between *n*-hexane and water is 6.25. Calculate the percentage of Z remaining in 25.0 mL of water that was originally 0.0600 M in Z after extraction with the following volumes of *n*-hexane: (a) one 25.0-mL portion, (b) two 12.5-mL portions, (c) five 5.00-mL portions, and (d) ten 2.50-mL portions.

*34-5. What volume of $CHCl_3$ is required to decrease the concentration of X in Problem 34-3 to 1.00×10^{-4} M if 25.0 mL of 0.0500 M X is extracted with (a) 25.0-mL portions of $CHCl_3$, (b) 10.0-mL portions of $CHCl_3$, and (c) 2.0-mL portions of $CHCl_3$?

34-6. What volume of *n*-hexane is required to decrease the concentration of Z in Problem 34-4 to 1.00×10^{-5} M if 40.0 mL of 0.0200 M Z is extracted with (a) 50.0-mL portions of *n*-hexane, (b) 25.0-mL portions, (c) 10.0-mL portions?

*34-7. What is the minimum distribution coefficient that permits removal of 99% of a solute from 50.0 mL of water with (a) two 25.0-mL extractions with cyclohexane and (b) five 10.0-mL extractions with cyclohexane?

34-8. If 30.0 mL of water that is 0.0500 M in Q is to be extracted with four 10.0-mL portions of an immiscible organic solvent, what is the minimum distribution coefficient that allows transfer of all but the following percentages of the solute to the organic layer: *(a) 1.00×10^{-4}, (b) 1.00×10^{-2}, (c) 1.00×10^{-3}?

*34-9. A 0.150 M aqueous solution of the weak organic acid HA was prepared from the pure compound, and three 50.0-mL aliquots were transferred to 100-mL volumetric flasks. Solution 1 was diluted to 100 mL with 1.0 M $HClO_4$, solution 2 was diluted to the mark with 1.0 M NaOH, and solution 3 was diluted to the mark with water. A 25.0-mL aliquot of each was extracted with 25.0-mL of *n*-hexane. The extract from solution 2 contained no detectable trace of A-containing species, indicating that A^- is not soluble in the organic solvent. The extract from solution 1 contained no ClO_4^- or $HClO_4$ but was found to be 0.0454 M in HA (by extraction with standard NaOH and back-titration with standard HCl). The extract from solution 3 was found to be 0.0225 M in HA. Assume that HA does not associate or dissociate in the organic solvent, and calculate
(a) the distribution ratio for HA between the two solvents.
(b) the concentration of the *species* HA and A^- in aqueous solution 3 after extraction.
(c) the dissociation constant of HA in water.

34-10. To determine the equilibrium constant for the reaction

$$I_2 + 2SCN^- \rightleftharpoons I(SCN)_2^- + I^-$$

25.0 mL of a 0.0100 M aqueous solution of I_2 were extracted with 10.0 mL of CCl_4. After extraction, spectrophotometric measurements revealed that the I_2 concentration *of the aqueous layer* was 1.12×10^{-4} M. An aqueous solution that was 0.0100 M in I_2 and 0.100 M in KSCN was then prepared. After extraction of 25.0 mL of this solution with 10.0 mL of CCl_4, the concentration of I_2 *in the CCl_4 layer* was found from spectrophotometric measurement to be 1.02×10^{-3} M.
(a) What is the distribution coefficient for I_2 between CCl_4 and H_2O?
(b) What is the formation constant for $I(SCN)_2^-$?

*34-11. An organic acid was isolated and purified by recrystallization of its barium salt. To determine the equivalent weight of the acid, a 0.393-g sample of the salt was dissolved in about 100 mL of water. The solution was passed through a strong-acid ion-exchange resin, and the column was then washed with water; the eluate and washings were titrated with 18.1 mL of 0.1006 M NaOH to a phenolphthalein end point.
(a) Calculate the equivalent weight of the organic acid (the mass of the acid that contains one titratable proton).
(b) A potentiometric titration curve of the solution resulting when a second sample was treated in the same way revealed two end points, one at pH 5 and the other at pH 9. What is the molecular mass of the acid?

*34-12. Copper(II) reacts with the chelating agent HL to give a complex CuL_2 that is readily soluble in $CHCl_3$. A spectrophotometric study revealed that when a 1.00×10^{-4} aqueous solution of copper(II) was extracted with $CHCl_3$ that was 0.0100 M in H_2L, the analytical concentration of copper in the two phases was identical at pH 5.65.
(a) Write equations describing the equilibria in the system, assuming that dissociation of CuL_2 in the organic phase is negligible.
(b) Calculate K_{ex}.
(c) Calculate the distribution ratio for the system at pH 6.00.
(d) If 50.0 mL of 5.00×10^{-5} M Cu^{2+} in a pH 6.00 buffer were to be extracted with 25.0 mL portions of 0.0100 M H_2L in $CHCl_3$, how many extractions would be required to remove 99% of the copper from the aqueous phase?
(e) Repeat the calculations in part (d) for 99.9% removal.

34-13. Exactly 25.0 mL of a standard solution that was 2.24×10^{-4} in Ag^+ and buffered to pH 4.30 was extracted with 10.0 mL of a *n*-hexane solution that was 0.025 M in the chelating agent H_2Z (the extracted complex has the formula Ag_2Z). After the phases were separated, the silver concentration of the aqueous phase was found to be 5.41×10^{-6} M.

(a) Calculate the distribution ratio for the system.

(b) Calculate K_{ex} for the system, assuming that the organic phase contained no uncomplexed Ag^+.

(c) How many 10.0-mL extractions would be required to extract 99.5% of the original Ag^+ at pH 4.30?

(d) Repeat the calculations in part (c) for a solution buffered to pH 3.00 and having a silver ion concentration of 5.15×10^{-4} M.

CHAPTER

35

The Chemicals, Apparatus, and Unit Operations of Analytical Chemistry

This chapter is concerned with the practical aspects of the unit operations encountered in an analytical laboratory as well as with the apparatus and chemicals used in these operations.

35A THE SELECTION AND HANDLING OF REAGENTS AND OTHER CHEMICALS

The purity of reagents has an important bearing upon the accuracy attained in any analysis. It is therefore essential that the quality of a reagent be consistent with the use for which it is intended.

35A-1 The Classification of Chemicals

Reagent Grade

Reagent-grade chemicals conform to the minimum standards set forth by the Reagent Chemical Committee of the American Chemical Society[1] and are used wherever possible in analytical work. Some suppliers label their products with the maximum limits of impurity allowed by the ACS specifications; others print actual assay for the various impurities.

Primary-Standard Grade

The qualities required of a *primary standard*—in addition to extraordinary purity—are set forth in Section 6A-3. Primary-standard reagents have been carefully analyzed by the supplier, and the assay is printed on the container label. The National Institute of Standards and Technology is an excellent source for primary

The National Institute of Standards and Technology (NIST) is the current name of what was formerly the National Bureau of Standards.

[1]Committee on Analytical Reagents, *Reagent Chemicals,* 8th ed. Washington, DC: American Chemical Society, 1993.

standards. This agency also provides *reference standards,* which are complex substances that have been exhaustively analyzed.[2]

Special-Purpose Reagent Chemicals

Chemicals that have been prepared for specific applications are also available. Included among these are solvents for spectrophotometry and high-performance liquid chromatography. Information pertinent to the intended use is supplied with these reagents. Data provided with a spectrophotometric solvent, for example, might include its absorbance at selected wavelengths and its ultraviolet cutoff wavelength.

35A-2 Rules for Handling Reagents and Solutions

High quality in a chemical analysis requires reagents and solutions of established purity. A freshly opened bottle of a reagent-grade chemical can ordinarily be used with confidence; whether this same confidence is justified when the bottle is half empty depends entirely on the way it has been handled after being opened. The following rules should be observed to prevent the accidental contamination of reagents and solutions.

1. Select the best grade of chemical available for analytical work. Whenever possible, pick the smallest bottle that will supply the desired quantity.
2. Replace the top of every container *immediately* after removal of the reagent; do not rely on someone else to do this.
3. Hold the stoppers of reagent bottles between your fingers; never set a stopper on a desktop.
4. *Unless specifically directed otherwise, never return any excess reagent to a bottle.* The money saved by returning excesses is seldom worth the risk of contaminating the entire bottle.
5. Unless directed otherwise, never insert spatulas, spoons, or knives into a bottle that contains a solid chemical. Instead, shake the capped bottle vigorously or tap it gently against a wooden table to break up an encrustation; then pour out the desired quantity. These measures are occasionally ineffective, and in such cases a clean porcelain spoon should be used.
6. Keep the reagent shelf and the laboratory balance clean and neat. Clean up any spills immediately, even though someone else is waiting to use the same chemical or reagent.
7. Observe local regulations concerning the disposal of surplus reagents and solutions.

35B THE CLEANING AND MARKING OF LABORATORY WARE

A chemical analysis is ordinarily performed in duplicate or triplicate. Thus, each vessel that holds a sample must be marked so that its contents can be positively

---

[2]United States Department of Commerce, *NIST Standard Reference Materials Catalog, 1995–96, NIST Special Publication 260.* Washington, DC 20234: Government Printing Office, 1995.

Figure 35-1
Arrangement for the evaporation of a liquid.

Unless you are directed otherwise, do not dry the interior surfaces of glassware or porcelain ware.

identified. Flasks, beakers, and some crucibles have small etched areas on which semipermanent markings can be made with a pencil.

Special marking inks are available for porcelain surfaces. The marking is baked permanently into the glaze by heating at a high temperature. A saturated solution of iron(III) chloride, while not as satisfactory as the commercial preparation, can also be used for marking.

Every beaker, flask, or crucible that will contain the sample must be thoroughly cleaned before being used. The apparatus should be washed with a hot detergent solution and then rinsed—initially with copious amounts of tap water and finally with several small portions of deionized water.[3] Properly cleaned glassware will be coated with a uniform and unbroken film of water. *It is seldom necessary to dry the interior surface of glassware before use;* drying is ordinarily a waste of time at best and a potential source of contamination at worst.

An organic solvent may be effective in removing grease films. Chemical suppliers also market preparations for the elimination of such films.

35C THE EVAPORATION OF LIQUIDS

It is frequently necessary to decrease the volume of a solution that contains a nonvolatile solute. Figure 35-1 illustrates how this operation is performed. The ribbed cover glass permits vapors to escape and protects the remaining solution from accidental contamination. Less satisfactory is the use of glass hooks to provide space between the rim of the beaker and a conventional cover glass.

Bumping is sudden, often violent boiling that tends to spatter solution out of its container.

Evaporation is frequently difficult to control, owing to the tendency of some solutions to overheat locally. The *bumping* that results can be sufficiently vigorous to cause partial loss of the solution. Careful and gentle heating will minimize the danger of such loss. Where their use is permissible, glass beads will also minimize bumping.

Some unwanted species can be eliminated during evaporation. For example, chloride and nitrate can be removed from a solution by adding sulfuric acid and evaporating until copious white fumes of sulfur trioxide are observed (this operation must be performed in a hood). Urea is effective in removing nitrate ion and nitrogen oxides from acidic solutions. The removal of ammonium chloride is best accomplished by adding concentrated nitric acid and evaporating the solution to a small volume. Ammonium ion is rapidly oxidized upon heating; the solution is then evaporated to dryness.

Organic constituents can frequently be eliminated from a solution by adding sulfuric acid and heating to the appearance of sulfur trioxide fumes (hood); this process is known as *wet ashing*. Nitric acid can be added toward the end of heating to hasten oxidation of the last traces of organic matter.

35D THE MEASUREMENT OF MASS

In most analyses, an *analytical balance* must be used to obtain highly accurate masses. Less accurate *laboratory balances* are also employed in the analytical laboratory for mass measurements where the demands for reliability are not critical.

[3]References to deionized water in this chapter and the one that follows apply equally to distilled water.

35D-1 Types of Analytical Balances

By definition, an *analytical balance* is a weighing instrument with a maximum capacity that ranges from a gram to a few kilograms with a precision of at least one part in 10^5 at maximum capacity. The precision and accuracy of many modern analytical balances exceed one part in 10^6 at full capacity.

An **analytical balance** has a maximum capacity that ranges from one gram to several kilograms and a precision at maximum capacity of at least one part in 10^5.

The most commonly encountered analytical balances *(macrobalances)* have a maximum capacity ranging between 160 and 200 g; measurements can be made with a standard deviation of ± 0.1 mg. *Semimicroanalytical balances* have a maximum loading of 10 to 30 g with a precision of ± 0.01 mg. A typical *microanalytical balance* has a capacity of 1 to 3 g and a precision of ± 0.001 mg.

A **macrobalance** is the most common type of analytical balance; it has a maximum load of 160 to 200 g and a precision of 0.1 mg.

A **semimicroanalytical balance** has a maximum load of 10 to 30 g and a precision of 0.01 mg.

A **microanalytical balance** has a maximum load of 1 to 3 g and a precision of 0.001 mg or 1 μg.

The analytical balance has undergone a dramatic evolution over the past several decades. The traditional analytical balance had two pans attached to either end of a light-weight beam that pivoted about a knife edge located in the center of the beam. The object to be weighed was placed on one pan; sufficient standard weights were then added to the other pan to restore the beam to its original position. Weighing with such an *equal-arm* balance was tedious and time-consuming.

The first *single-pan analytical balance* appeared on the market in 1946. The speed and convenience of weighing with this balance were vastly superior to what could be realized with the traditional equal-arm balance. Consequently, this balance rapidly replaced the latter in most laboratories. The single-pan balance is still extensively used. The design and operation of a single-pan balance are discussed briefly in Section 35D-3.

The single-pan balance is currently being replaced by the *electronic analytical balance,* which has neither a beam nor a knife edge; this type of balance is discussed in Section 35D-2.

35D-2 The Electronic Analytical Balance[4]

Figure 35-2 is a diagram of an electronic analytical balance. The pan rides above a hollow metal cylinder that is surrounded by a coil and fits over the inner pole of a cylindrical permanent magnet. An electric current in the coil creates a magnetic field that supports or *levitates* the cylinder, the pan and indicator arm, and whatever load is on the pan. The current is adjusted so that the level of the indicator arm is in the null position when the pan is empty. Placing an object on the pan causes the pan and indicator arm to move downward, which increases the amount of light striking the photocell of the null detector. The increased current from the photocell is amplified and fed into the coil, creating a larger magnetic field, which returns the pan to its original null position. A device such as this, in which a small electric current causes a mechanical system to maintain a null position, is called a *servo system.* The current required to keep the pan and object in the null position is directly proportional to the mass of the object and is readily measured, digitized, and displayed. The calibration of an electronic balance involves the use of a standard mass and adjustment of the current so that the mass of the standard is exhibited on the display.

To **levitate** means to cause an object to float in air.

A **servo system** is a device in which a small electric signal causes a mechanical system to return to a null position.

[4]For a more detailed discussion, see R. M. Schoonover, *Anal. Chem.,* **1982,** *54,* 973A; K. M. Lang, *Amer. Lab.,* **1983,** *15* (3), 72.

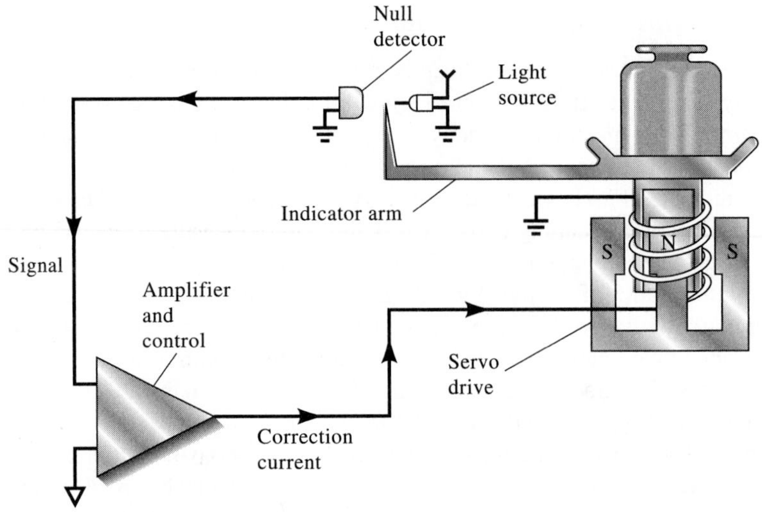

Figure 35-2
Electronic analytical balance. (From R. M. Schoonover, *Anal. Chem.*, **1982,** *54,* 974A. Published 1982 American Chemical Society.)

Figure 35-3 shows the configurations for two electronic analytical balances. In each, the pan is tethered to a system of constraints known collectively as a *cell.* The cell incorporates several *flexures* that permit limited movement of the pan and prevent torsional forces (resulting from off-center loading) from disturbing the alignment of the balance mechanism. At null, the beam is parallel to the gravitational horizon and each flexure pivot is in a relaxed position.

Figure 35-3a shows an electronic balance with the pan located below the cell. Higher precision is achieved with this arrangement than with the top-loading

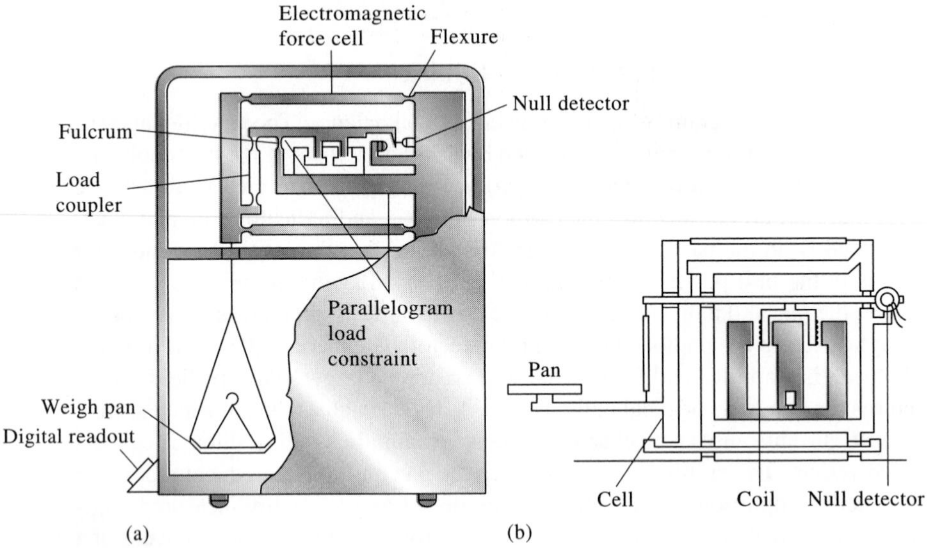

Figure 35-3
Electronic analytical balances. (a) Classical configuration with pan beneath the cell. (From R. M. Schoonover, *Anal. Chem.*, **1982,** *54,* 976A. Published 1982 American Chemical Society.) (b) A top-loading design. Note that the mechanism is enclosed in a windowed case. (Reprinted with permission *Amer. Lab.*, **1983,** *15* (3), 72. Copyright 1983 by International Scientific Communications, Inc.)

design shown in Figure 35-3b. Even so, top-loading electronic balances have a precision that equals or exceeds that of the best mechanical balances and additionally provide unencumbered access to the pan.

Electronic balances generally feature an automatic *taring control* that causes the display to read zero with a container (such as a boat or weighing bottle) on the pan. Most balances permit taring to 100% of capacity.

A **tare** is the mass of an empty sample container. Taring is the process of setting a balance to read zero in the presence of the tare on the balance pan.

Some electronic balances have dual capacities and dual precisions. These features permit the capacity to be decreased from that of a macrobalance to that of a semimicrobalance (30 g) with a concomitant gain in precision to 0.01 mg. Thus, the chemist has effectively two balances in one.

A modern electronic analytical balance provides unprecedented speed and ease of use. For example, one instrument is controlled by touching a single bar at various positions along its length. One position on the bar turns the instrument on or off, another automatically calibrates the balance against a standard weight, and a third zeros the display, either with or without an object on the pan. Reliable weighing data are obtainable with little or no instruction or practice.

A photograph of a modern electronic balance is shown in color plate 23.

35D-3 The Single-Pan Mechanical Analytical Balance

Components

Although they differ considerably in appearance and performance characteristics, all mechanical balances—equal-arm as well as single-pan—have several common components. Figure 35-4 is a diagram of a typical single-pan balance. Fundamental to this instrument is a lightweight *beam* that is supported on a planar surface by a prism-shaped *knife edge (A)*. Attached to the left end of the beam is a pan for holding the object to be weighed and a full set of weights held in place by hangers. These weights can be lifted from the beam one at a time by a mechanical arrangement that is controlled by a set of knobs on the exterior of the balance case.

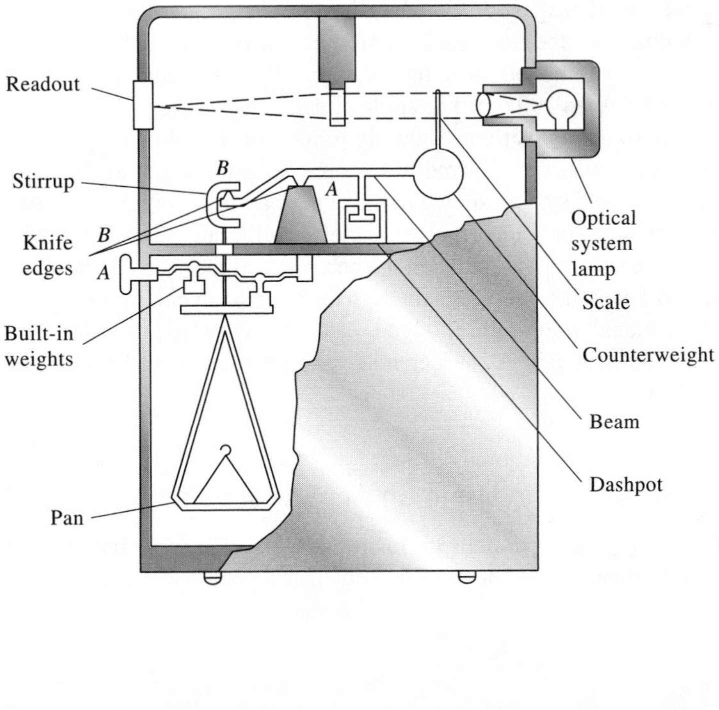

Figure 35-4
Modern single-pan analytical balance. (From R. M. Schoonover, *Anal. Chem.*, **1982,** *54,* 973A. Published 1982 American Chemical Society.)

The right end of the beam holds a counterweight of such a size as to just balance the pan and weights on the left end of the beam.

The two **knife edges** in a mechanical balance are prism-shaped agate or sapphire devices that form low-friction bearings with two planar surfaces contained in **stirrups,** also of agate or sapphire.

A second knife edge *(B)* is located near the left end of the beam and serves to support a second planar surface, which is located in the inner side of a *stirrup* that couples the pan to the beam. The two knife edges and their planar surfaces are fabricated from extraordinarily hard materials (agate or synthetic sapphire) and form two bearings that permit motion of the beam and pan with a very minimum of friction. The performance of a mechanical balance is critically dependent upon the perfection of these two bearings.

To avoid damage to the knife edges and bearing surfaces, the arrest system for a mechanical balance should be engaged at all times other than during actual weighing.

Single-pan balances are also equipped with a *beam arrest* and a *pan arrest.* The beam arrest is a mechanical device that raises the beam so that the central knife edge no longer touches its bearing surface and simultaneously frees the stirrup from contact with the outer knife edge. The purpose of both arrest mechanisms is to prevent damage to the bearings while objects are being placed upon or removed from the pan. When engaged, the pan arrest supports most of the mass of the pan and its contents and thus prevents oscillation. Both arrests are controlled by a lever mounted on the outside of the balance case and should be engaged whenever the balance is not in use.

An *air damper* (also known as a *dashpot*) is mounted near the end of the beam opposite the pan. This device consists of a piston that moves within a concentric cylinder attached to the balance case. Air in the cylinder undergoes expansion and contraction as the beam is set in motion; the beam rapidly comes to rest as a result of this opposition to motion.

Protection from air currents is needed to permit discrimination between small differences in mass (< 1 mg). An analytical balance is thus always enclosed in a case equipped with doors to permit the introduction or removal of objects.

Weighing with a Single-Pan Balance

The beam of a properly adjusted balance assumes an essentially horizontal position with no object on the pan and all of the weights in place. When the pan and beam arrests are disengaged, the beam is free to rotate around the knife edge. Placing an object on the pan causes the left end of the beam to move downward. Weights are then removed systematically one by one from the beam until the imbalance is less than 100 mg. The angle of deflection of the beam with respect to its original horizontal position is directly proportional to the milligrams of additional mass that must be removed to restore the beam to its original horizontal position. The optical system shown in the upper part of Figure 35-4 measures this angle of deflection and converts this angle to milligrams. A *reticle,* which is a small transparent screen mounted on the beam, is scribed with a scale that reads 0 to 100 mg. A beam of light passes through the scale to an enlarging lens, which in turn focuses a small part of the enlarged scale onto a frosted glass plate located on the front of the balance. A vernier makes it possible to read this scale to the nearest 0.1 mg.

35D-4 Precautions in Using an Analytical Balance

An analytical balance is a delicate instrument that you must handle with care. Consult with your instructor for detailed instructions for weighing with your particular model of balance. Observe the following general rules for working with an analytical balance regardless of make or model.

1. Center the load on the pan as well as possible.
2. Protect the balance from corrosion. Objects to be placed on the pan should be limited to nonreactive metals, nonreactive plastics, and vitreous materials.
3. Observe special precautions (Section 35E-6) for the weighing of liquids.
4. Consult the instructor if the balance appears to need adjustment.
5. Keep the balance and its case scrupulously clean. A camel's-hair brush is useful for the removal of spilled material or dust.
6. Always allow an object that has been heated to return to room temperature before weighing it.
7. Use tongs or finger pads to prevent the uptake of moisture by dried objects.

35D-5 Sources of Error in Weighing

Correction for Buoyancy[5]

A *buoyancy error* will affect data if the density of the object being weighed differs significantly from that of the standard weights. This error has its origin in the difference in the buoyant force exerted by the medium (air) upon the object and upon the weights. Correction for buoyancy is accomplished with the equation

> A **buoyancy error** is the weighing error that develops when the object being weighed has a significantly different density than the weights.

$$W_1 = W_2 + W_2\left(\frac{d_{air}}{d_{obj}} - \frac{d_{air}}{d_{wts}}\right) \tag{35-1}$$

where W_1 is the corrected mass of the object, W_2 is the mass of the standard weights, d_{obj} is the density of the object, d_{wts} is the density of the weights, and d_{air} is the density of the air displaced by them; d_{air} has a value of 0.0012 $g \cdot cm^{-3}$.

The consequences of Equation 35-1 are shown in Figure 35-5, in which the relative error due to buoyancy is plotted against the density of objects weighed in air against stainless steel weights. Note that this error is less than 0.1% for objects that have a density of 2 $g \cdot cm^{-3}$ or greater. It is thus seldom necessary to apply a correction to the mass of most solids. The same cannot be said for low-density solids, liquids, or gases, however; for these, the effects of buoyancy are significant and a correction must be applied.

The density of weights used in single-pan balances and in calibrating electronic balances ranges from 7.8 to 8.4 $g \cdot cm^{-3}$, depending upon the manufacturer. Use of 8 $g \cdot cm^{-3}$ is adequate for most purposes. If a greater accuracy is required, the specifications for the balance to be used should be consulted for the necessary density data.

EXAMPLE 35-1

A bottle weighed 7.6500 g empty and 9.9700 g after introduction of an organic liquid with a density of 0.92 $g \cdot cm^{-3}$. The balance was equipped

[5]For further information, see R. Battino and A. G. Williamson, *J. Chem. Educ.*, **1984,** *64,* 51.

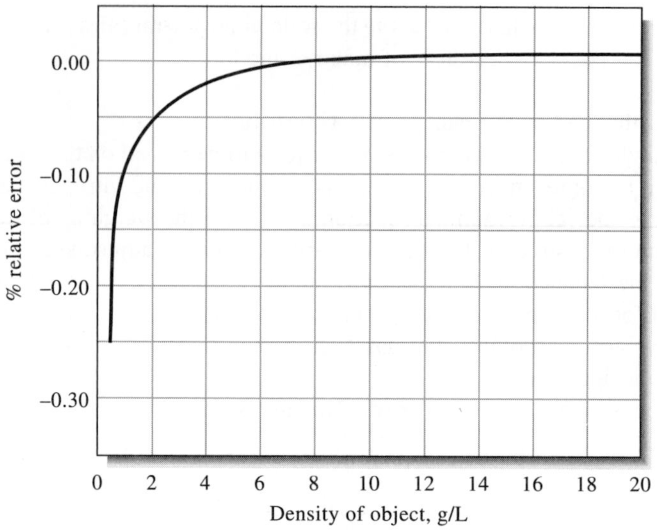

Figure 35-5
Effect of buoyancy on weighing data (density of weights = 8 g · cm^{-3}). Plot of relative error as a function of the density of the object weighed.

with stainless steel weights (d = 8.0 g · cm^{-3}). Correct the mass of the sample for the effects of buoyancy.

The apparent mass of the liquid is 9.9700 − 7.6500 = 2.3200 g. The same buoyant force acts on the container during both weighings; thus, we need to consider only the force that acts on the 2.3200 g of liquid. Substitution of 0.0012 g · cm^{-3} for d_{air}, 0.92 g · cm^{-3} for d_{obj}, and 8.0 g · cm^{-3} for d_{wts} in Equation 35-1 gives

$$W_1 = 2.3200 + 2.3200\left(\frac{0.0012}{0.92} - \frac{0.0012}{8.0}\right) = 2.3227 \text{ g}$$

Temperature Effects

Always allow heated objects to return to room temperature before you attempt to weigh them.

Attempts to weigh an object whose temperature is different from that of its surroundings will result in a significant error. Failure to allow sufficient time for a heated object to return to room temperature is the commonest source of this problem. Errors due to a difference in temperature have two origins. First, convection currents within the balance case exert a buoyant effect on the pan and object. Second, warm air trapped in a closed container weighs less than the same volume at a lower temperature. Both effects cause the apparent mass of the object to be low. This error can amount to as much as 10 to 15 mg for a typical porcelain filtering crucible or a weighing bottle (Figure 35-6). Heated objects must always be cooled to room temperature before being weighed.

Other Sources of Error

A porcelain or glass object will occasionally acquire a static charge that is sufficient to cause a balance to perform erratically; this problem is particularly serious when the relative humidity is low. Spontaneous discharge frequently occurs after a short period. A low-level source of radioactivity (such as a photographer's brush) in the balance case will provide sufficient ions to relieve the charge. Alternatively, the object can be wiped with a faintly damp chamois.

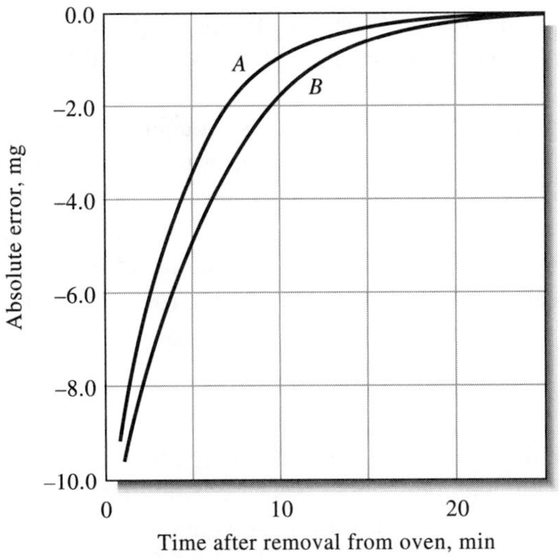

Figure 35-6
Effect of temperature on weighing data. Absolute error in weight as a function of time after the object was removed from a 110°C drying oven. *A*: porcelain filtering crucible. *B*: weighing bottle containing about 7.5 g of KCl.

The optical scale of a single-pan balance should be checked regularly for accuracy, particularly under loading conditions that require the full scale range. A standard 100-mg weight is used for this check.

35D-6 Auxiliary Balances

Balances that are less precise than analytical balances find extensive use in the analytical laboratory. These offer the advantages of speed, ruggedness, large capacity, and convenience; they should be used whenever high sensitivity is not required.

Use auxiliary laboratory balances for weighings that do not require great accuracy.

Top-loading auxiliary balances are particularly convenient. A sensitive top-loading balance will accommodate 150 to 200 g with a precision of about 1 mg —an order of magnitude less than a macroanalytical balance. Some balances of this type tolerate loads as great as 25,000 g with a precision of $\pm$ 0.05 g. Most are equipped with a taring device that brings the balance reading to zero with an empty container on the pan. Some are fully automatic, require no manual dialing or weight handling, and provide a digital readout of the mass. Modern top-loading balances are electronic.

A triple-beam balance with a sensitivity less than that of a typical top-loading auxiliary balance is also useful. This is a single-pan balance with three decades of weights that slide along individual calibrated scales. The precision of a triple-beam balance may be one or two orders of magnitude less than that of a top-loading instrument but is adequate for many weighing operations. This type of balance offers the advantages of simplicity, durability, and low cost.

35E THE EQUIPMENT AND MANIPULATIONS ASSOCIATED WITH WEIGHING

The mass of many solids changes with humidity, owing to their tendency to absorb weighable amounts of moisture. This effect is especially pronounced when a large surface area is exposed, as with a reagent chemical or a sample that has

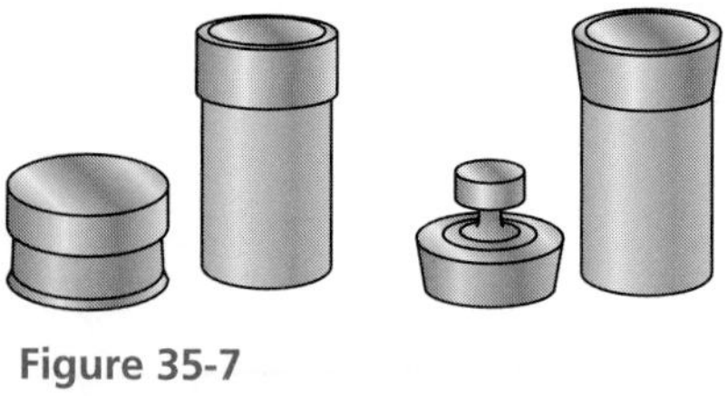

Figure 35-7
Typical weighing bottles.

Drying or ignition to constant mass is a process in which a solid is cycled through heating, cooling, and weighing steps until its mass becomes constant to within 0.2 to 0.3 mg.

been ground to a fine powder. The first step in a typical analysis, then, involves drying the sample so that the results will not be affected by the humidity of the surrounding atmosphere.

A sample, a precipitate, or a container is brought to *constant mass* by a cycle that involves heating (ordinarily for an hour or more) at an appropriate temperature, cooling, and weighing. This cycle is repeated as many times as needed to obtain successive masses that agree within 0.2 to 0.3 mg of one another. The establishment of constant mass provides some assurance that the chemical or physical processes that occur during the heating (or ignition) are complete.

35E-1 Weighing Bottles

Solids are conveniently dried and stored in *weighing bottles,* two common varieties of which are shown in Figure 35-7. The ground-glass portion of the cap-style bottle shown on the left is on the outside and does not come into contact with the contents; this design eliminates the possibility of some of the sample becoming entrained upon and subsequently lost from the ground-glass surface.

Plastic weighing bottles are available; ruggedness is the principal advantage of these bottles over their glass counterparts.

35E-2 Desiccators and Desiccants

A **desiccator** is a device for drying substances or objects.

Oven drying is the most common way of removing moisture from solids. This approach is not appropriate for substances that decompose or for those from which water is not removed at the temperature of the oven.

Dried materials are stored in *desiccators* while they cool in order to minimize the uptake of moisture. Figure 35-8 shows the components of a typical desiccator.

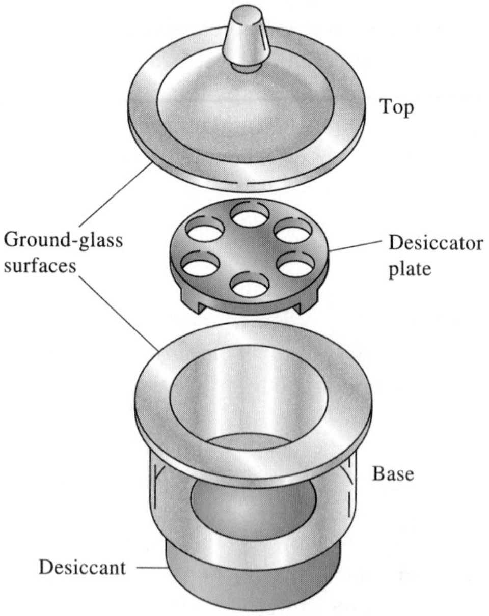

Figure 35-8
Components of a typical desiccator.

The base section contains a chemical drying agent, such as anhydrous calcium chloride, calcium sulfate (Drierite), anhydrous magnesium perchlorate (Anhydrone or Dehydrite), or phosphorus pentoxide. The ground-glass surfaces are lightly coated with grease.

When you remove or replace the lid of a desiccator, use a sliding motion to minimize the likelihood of disturbing the sample. An airtight seal is achieved by slight rotation and downward pressure upon the positioned lid.

When you place a heated object in a desiccator, the increase in pressure as the enclosed air is warmed may be sufficient to break the seal between lid and base. Conversely, if the seal is not broken, the cooling of heated objects can cause development of a partial vacuum. Both of these conditions can cause the contents of the desiccator to be physically lost or contaminated. Although it defeats the purpose of the desiccator somewhat, you should allow some cooling to occur before the lid is seated. It is also helpful to break the seal once or twice during cooling to relieve any excessive vacuum that develops. Finally, you should lock the lid in place with your thumbs while moving the desiccator from one place to another.

Very hygroscopic materials should be stored in containers equipped with snug covers, such as weighing bottles; the covers remain in place while in the desiccator. Most other solids can be safely stored uncovered.

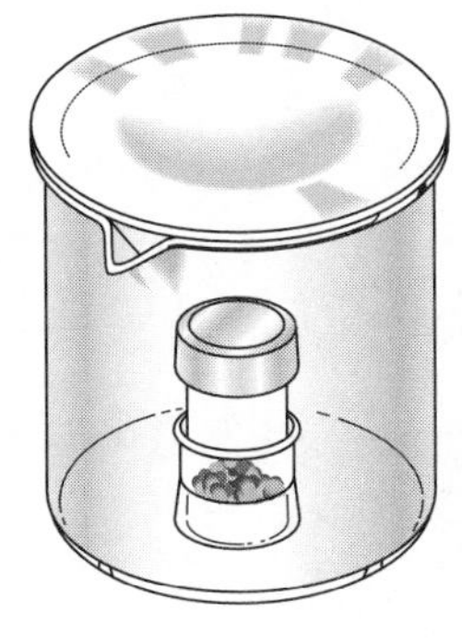

Figure 35-9
Arrangement for the drying of samples.

35E-3 The Manipulation of Weighing Bottles

Heating at 105°C to 110°C is sufficient to remove the moisture from the surface of most solids. Figure 35-9 depicts the arrangement recommended for drying a sample. The weighing bottle is contained in a labeled beaker with a ribbed cover glass. This arrangement protects the sample from accidental contamination and also allows for the free access of air. Crucibles containing a precipitate that can be freed of moisture by simple drying can be treated similarly. The beaker containing the weighing bottle or crucible to be dried must be carefully marked to permit identification.

You should avoid handling dried objects with your fingers because detectable amounts of water or oil from your skin may be transferred to the object. This problem is avoided by using tongs, chamois finger cots, clean cotton gloves, or strips of paper to handle dried objects for weighing. Figure 35-10 shows how a weighing bottle is manipulated with strips of paper.

Figure 35-10
Method for quantitative transfer of a solid sample. Note the use of paper strips to avoid contact between glass and skin.

35E-4 Weighing by Difference

Weighing by difference is a simple method for determining a series of sample weights. First, the bottle and its contents are weighed. One sample is then transferred from the bottle to a container; gentle tapping of the bottle with its top and slight rotation of the bottle provide control over the amount of sample removed. Following transfer, the bottle and its residual contents are weighed. The mass of the sample is the difference between these two weighings. It is essential that all the solid removed from the weighing bottle be transferred without loss to the container.

35E-5 Weighing Hygroscopic Solids

Hygroscopic substances rapidly absorb moisture from the atmosphere and therefore require special handling. You need a weighing bottle for each sample to be weighed. Place the approximate amount of sample needed in the individual bottles and heat for an appropriate time. When heating is complete, quickly cap the bottles and cool in a desiccator. Weigh one of the bottles after opening it momentarily to relieve any vacuum. Quickly empty the contents of the bottle into its receiving vessel, cap immediately, and weigh the bottle again (along with any solid that did not get transferred). Repeat for each sample and determine the sample masses by difference.

35E-6 Weighing Liquids

The mass of a liquid is always obtained by difference. Liquids that are noncorrosive and relatively nonvolatile can be transferred to previously weighed containers with snugly fitting covers (such as weighing bottles); the mass of the container is subtracted from the total mass.

A volatile or corrosive liquid should be sealed in a weighed glass ampoule. The ampoule is heated, and the neck is then immersed in the sample; as cooling occurs, the liquid is drawn into the bulb. The ampoule is then inverted and the neck sealed off with a small flame. The ampoule and its contents, along with any glass removed during sealing, are cooled to room temperature and weighed. The ampoule is then transferred to an appropriate container and broken. A volume correction for the glass of the ampoule may be needed if the receiving vessel is a volumetric flask.

35F WEIGHT TITRATIONS

Weight (or gravimetric) titrimetry is discussed in Section 6D. Directions for the weight titration of chloride ion by the Mohr method are found in Section 36C-3.

A convenient reagent dispenser for a weight titration is a small polyethylene bottle equipped with a fine delivery tip (Figure 35-11). Weighings are normally

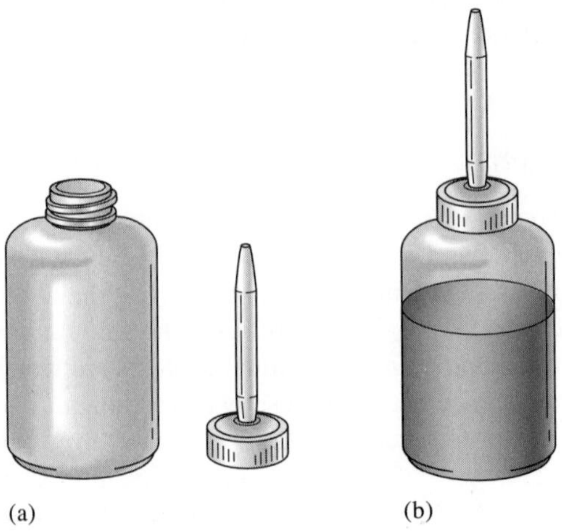

Figure 35-11
Reagent dispenser for weight titrations.

performed with a top-loading balance that has a sensitivity of 1 mg or 0.001 mL. It has been found that the mass of such a bottle does not change significantly when it is handled with bare hands.

35F-1 Directions for Preparing a Reagent Dispenser

A dispenser is readily fashioned from a 60-mL polyethylene bottle with a screw cap [such as the Nalge 2-oz (~ 60 mL) or 4-oz (~ 125 mL) Boston Round Bottle]. With a cork borer, make a hole in the cap that is slightly smaller than the outside diameter of the tip (Note). Carefully force the tip through the hole; apply a bead of epoxy cement to seal the tip to the cap. Use a pressure-sensitive label to identify the contents of the bottle.

Note

A tip can be prepared by constricting the opening of an ordinary medicine dropper in a flame. Equally satisfactory are glass tips for conventional Kimax® burets.

35F-2 Directions for Performing a Weight Titration

Fill the reagent dispenser with a quantity of the standard titrant, wipe away any liquid spilled on the outside of the container with an absorbing tissue, and tighten the screw cap firmly. Weigh the bottle and its contents to the nearest milligram. Introduce a suitable indicator into the solution of the analyte. Grasp the dispenser so that its tip is below the lip of the flask and deliver several increments of the reagent by squeezing the bottle while rotating the flask with your other hand. When it is judged that only a few more drops of reagent are needed, ease the pressure on the bottle so that the flow ceases; then touch the tip to the inside of the flask and further reduce the pressure on the dispenser so that the liquid in the tip is drawn back into the bottle as the tip is removed from the flask. Set the dispenser on a piece of clean, dry glazed paper and rinse down the inner walls of the flask with a stream of distilled or deionized water (Note 1). Add reagent a drop at a time until the end point is reached (Note 2). Weigh the dispenser and record the data.

Notes

1. Instead of rinsing the walls, the titration flask can be tilted and rotated so that the bulk of the liquid picks up droplets that adhere to the inner surface.
2. Increments smaller than an ordinary drop can be added by forming a partial drop on the tip and then touching the tip to the wall. This partial drop is then combined with the bulk of the solution by rinsing the walls with wash water or tilting the flask until the droplet is incorporated into the bulk.

35G THE EQUIPMENT AND MANIPULATIONS FOR FILTRATION AND IGNITION

35G-1 Apparatus

Simple Crucibles

Simple crucibles serve only as containers. Porcelain, aluminum oxide, silica, and platinum crucibles maintain constant mass—within the limits of experimental error—and are used principally to convert a precipitate into a suitable weighing form. The solid is first collected on a filter paper. The filter and contents are then transferred to a weighed crucible, and the paper is ignited.

Simple crucibles of nickel, iron, silver, and gold are used as containers for the high-temperature fusion of samples that are not soluble in aqueous reagents. Attack by both the atmosphere and the contents may cause these crucibles to suffer mass changes. Moreover, such attack will contaminate the sample with species derived from the crucible. The chemist selects the crucible whose products will offer the least interference in subsequent steps of the analysis.

Filtering Crucibles

Filtering crucibles serve not only as containers but also as filters. A vacuum is used to hasten the filtration; a tight seal between crucible and filtering flask is accomplished with any of several types of rubber adapters (see Figure 35-12; a complete filtration train is shown in Figure 35-17). Collection of a precipitate with a filtering crucible is frequently less time-consuming than with paper.

Sintered-glass (also called *fritted-glass*) crucibles are manufactured in fine, medium, and coarse porosities (marked *f*, *m*, and *c*). The upper temperature limit for a sintered-glass crucible is ordinarily about 200°C. Filtering crucibles made entirely of quartz can tolerate substantially higher temperatures without damage. The same is true for crucibles with unglazed porcelain or aluminum oxide frits. The latter are not as costly as quartz. Appendix 6 lists the porosities of various kinds of commercially available filtering crucibles.

A *Gooch crucible* has a perforated bottom that supports a fibrous mat. Asbestos was at one time the filtering medium of choice for a Gooch crucible; current regulations concerning this material have virtually eliminated its use. Small circles of glass matting have now replaced asbestos; they are used in pairs to protect against disintegration during the filtration. Glass mats can tolerate temperatures in excess of 500°C and are substantially less hygroscopic than asbestos.

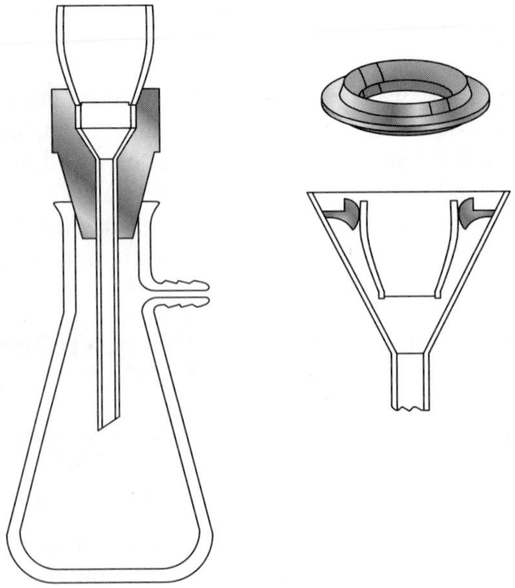

Figure 35-12
Adapters for filtering crucibles.

Filter Paper

Paper is an important filtering medium. Ashless paper is manufactured from cellulose fibers that have been treated with hydrochloric and hydrofluoric acids to remove metallic impurities and silica; ammonia is then used to neutralize the acids. The residual ammonium salts in many filter papers may be sufficient to affect the analysis for nitrogen by the Kjeldahl method (Section 36B-11).

All papers tend to pick up moisture from the atmosphere, and ashless paper is no exception. It is thus necessary to destroy the paper by ignition if the precipitate collected on it is to be weighed. Typically, 9- or 11-cm circles of ashless paper leave a residue that weighs less than 0.1 mg, an amount that is ordinarily negligible. Ashless paper can be obtained in several porosities (Appendix 6).

Gelatinous precipitates, such as hydrous iron(III) oxide, clog the pores of any filtering medium. A coarse-porosity ashless paper is most effective for the filtration of such solids, but even here clogging occurs. This problem can be minimized by mixing a dispersion of ashless filter paper with the precipitate prior to filtration. Filter paper pulp is available in tablet form from chemical suppliers; if necessary, the pulp can be prepared by treating a piece of ashless paper with concentrated hydrochloric acid and washing the disintegrated mass free of acid.

Table 35-1 summarizes the characteristics of common filtering media. None satisfies all requirements.

Heating Equipment

Many precipitates can be weighed directly after being brought to constant mass in a low-temperature drying oven. Such an oven is electrically heated and capable of maintaining a constant temperature to within 1°C (or better). The maximum attainable temperature ranges from 140°C to 260°C, depending upon make and model; for many precipitates, 110°C is a satisfactory drying temperature. The efficiency of a drying oven is greatly increased by the forced circulation of air. The passage of predried air through an oven designed to operate under a partial vacuum represents an additional improvement.

TABLE 35-1 Comparison of Filtering Media for Gravimetric Analyses

Characteristic	Paper	Gooch Crucible, Glass Mat	Glass Crucible	Porcelain Crucible	Aluminum Oxide Crucible
Speed of filtration	Slow	Rapid	Rapid	Rapid	Rapid
Convenience and ease of preparation	Troublesome, inconvenient	Convenient	Convenient	Convenient	Convenient
Maximum ignition temperature, °C	None	>500	200–500	1100	1450
Chemical reactivity	Carbon has reducing properties	Inert	Inert	Inert	Inert
Porosity	Many available	Several available	Several available	Several available	Several available
Convenience with gelatinous precipitates	Satisfactory	Unsuitable; filter tends to clog	Unsuitable; filter tends to clog	Unsuitable; filter tends to clog	Unsuitable; filter tends to clog
Cost	Low	Low	High	High	High

Microwave laboratory ovens are currently appearing on the market. Where applicable, these greatly shorten drying cycles. For example, slurry samples that require 12 to 16 hr for drying in a conventional oven are reported to be dried within 5 to 6 min in a microwave oven.[6]

An ordinary heat lamp can be used to dry a precipitate that has been collected on ashless paper and to char the paper as well. The process is conveniently completed by ignition at an elevated temperature in a muffle furnace.

Burners are convenient sources of intense heat. The maximum attainable temperature depends upon the design of the burner and the combustion properties of the fuel. Of the three common laboratory burners, the Meker provides the highest temperatures, followed by the Tirrill and Bunsen types.

A heavy-duty electric furnace *(muffle furnace)* is capable of maintaining controlled temperatures of 1100°C or higher. Long-handled tongs and heat-resistant gloves are needed for protection when transferring objects to or from such a furnace.

35G-2 The Manipulations Associated with Filtration and Ignition

Preparation of Crucibles

Backwashing a filtering crucible is done by turning the crucible upside down in the adapter (Figure 35-12) and drawing water through the inverted crucible.

We have already noted that a crucible used to convert a precipitate to a form suitable for weighing must maintain a constant mass throughout the drying or ignition. The crucible is first cleaned thoroughly (filtering crucibles are conveniently cleaned by backwashing on a filtration train) and subjected to the same regimen of heating and cooling as that required for the precipitate. This process is repeated until constant mass (page 788) has been achieved; that is, until consecutive weighings differ by 0.3 mg or less.

Filtration and Washing of Precipitates

Decantation is the process of pouring a liquid gently so as to not disturb a solid in the bottom of the container.

The steps involved in filtering an analytical precipitate are *decantation, washing,* and *transfer.* In decantation, as much supernatant liquid as possible is passed through the filter while the precipitated solid is kept essentially undisturbed in the beaker where it was formed. This procedure speeds the overall filtration rate by delaying the time at which the pores of the filtering medium become clogged with precipitate. A stirring rod is used to direct the flow of decantate (Figure 35-13). When flow ceases, the drop of liquid at the end of the pouring spout is collected with the stirring rod and returned to the beaker. Wash liquid is next added to the beaker and thoroughly mixed with precipitate. The solid is allowed to settle, following which this liquid is also decanted through the filter. Several such washings may be required, depending upon the precipitate. Most washing is thus carried out *before* the solid is transferred; this results in a more thoroughly washed precipitate and a more rapid filtration.

---

[6]D. G. Kuehn, R. L. Brandvig, D. C. Lundean, and R. H. Jefferson, *Amer. Lab.,* **1986,** *18* (7), 31. See also *Anal. Chem.,* **1986,** *58,* 1424A; E. S. Beary, *Anal. Chem.,* **1988,** *60,* 742.

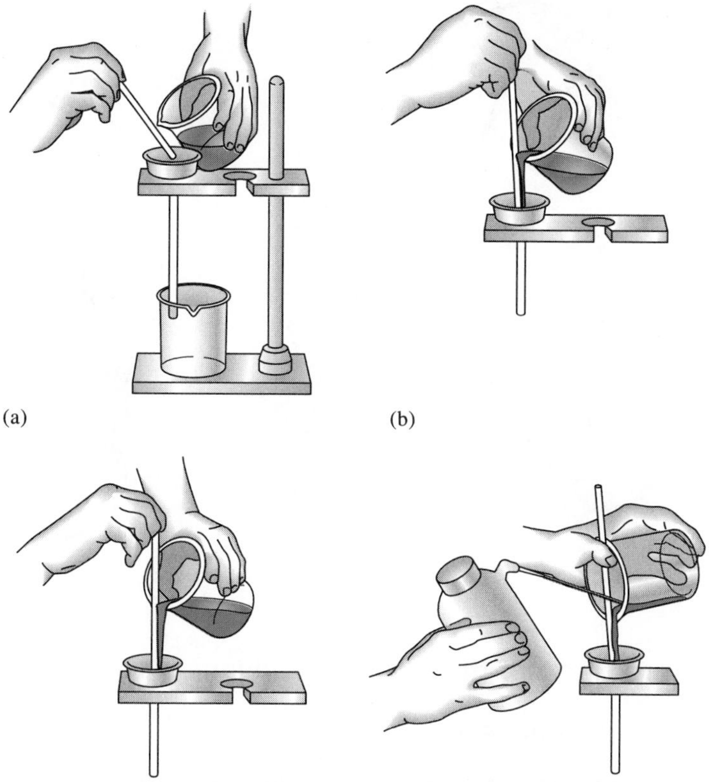

Figure 35-13
Steps in filtering: (a and b) washing by decantation; (c and d) transfer of the precipitate.

The transfer process is illustrated in Figures 35-13c and d. The bulk of the precipitate is moved from beaker to filter by suitably directed streams of wash liquid. As in decantation and washing, a stirring rod provides direction for the flow of material to the filtering medium.

The last traces of precipitate that cling to the inside of the beaker are dislodged with a *rubber policeman,* which is a small section of rubber tubing that has been crimped on one end. The open end of the tubing is fitted onto the end of a stirring rod and is wetted with wash liquid before use. Any solid collected with it is combined with the main portion on the filter. Small pieces of ashless paper can be used to wipe the last traces of hydrous oxide precipitates from the wall of the beaker; these papers are ignited along with the paper that holds the bulk of the precipitate.

Many precipitates possess the exasperating property of *creeping,* or spreading over a wetted surface against the force of gravity. Filters are never filled to more than three quarters of capacity, owing to the possibility that some of the precipitate could be lost as the result of creeping. The addition of a small amount of nonionic detergent, such as Triton X-100, to the supernatant liquid or to the wash liquid can be helpful in minimizing creeping.

Creeping is a process in which a solid moves up the side of a wetted container or filter paper.

A gelatinous precipitate must be completely washed before it is allowed to dry. These precipitates shrink and develop cracks as they dry. Further additions of wash liquid simply pass through these cracks and accomplish little or no washing.

Do not permit a gelatinous precipitate to dry until it has been washed completely.

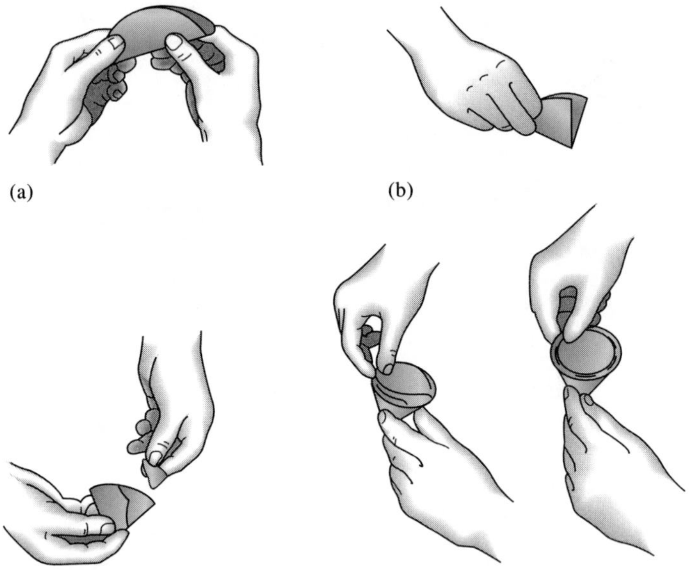

Figure 35-14
Folding and seating a filter paper.

35G-3 Directions for the Filtration and Ignition of Precipitates

Preparation of a Filter Paper

Figure 35-14 shows the sequence followed in folding a filter paper and seating it in a 60-deg funnel. The paper is folded exactly in half (a), firmly creased, and folded again (b). A triangular piece from one of the corners is torn off parallel to the second fold (c). The paper is then opened so that the untorn half forms a cone (d). The cone is fitted into the funnel, and the second fold is creased (e). Seating is completed by dampening the cone with water from a wash bottle and *gently* patting it with a finger. There will be no leakage of air between the funnel and a properly seated cone; in addition, the stem of the funnel will be filled with an unbroken column of liquid.

The Transfer of Paper and Precipitate to a Crucible

After filtration and washing have been completed, the filter and its contents must be transferred from the funnel to a crucible that has been brought to constant mass. Ashless paper has very low wet strength and must be handled with care during the transfer. The danger of tearing is lessened considerably if the paper is allowed to dry somewhat before it is removed from the funnel.

Figure 35-15 illustrates the transfer process. The triple-thick portion of the filter paper is drawn across the funnel to flatten the cone along its upper edge (a); the corners are next folded inward (b); the top edge is then folded over (c). Finally, the paper and its contents are eased into the crucible (d) so that the bulk of the precipitate is near the bottom.

Ashing of a Filter Paper

You should have a burner for each crucible. You can tend to the ashing of several filter papers at the same time.

If a heat lamp is to be used, the crucible is placed on a clean, nonreactive surface, such as a wire screen covered with aluminum foil. The lamp is then positioned

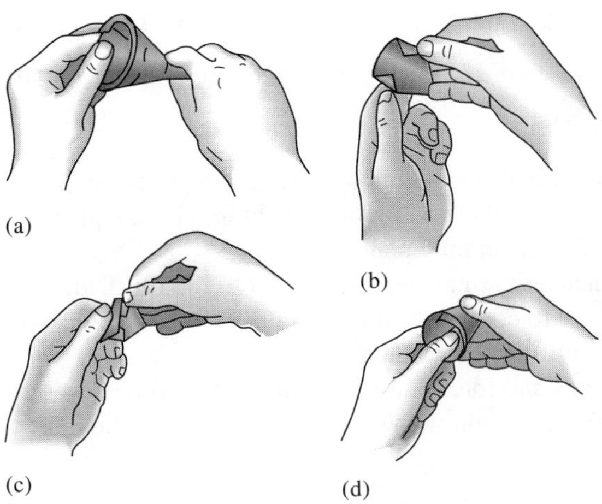

Figure 35-15
Transferring a filter paper and precipitate to a crucible.

about 1 cm above the rim of the crucible and turned on. Charring takes place without further attention. The process is considerably accelerated if the paper is moistened with no more than one drop of concentrated ammonium nitrate solution. Elimination of the residual carbon is accomplished with a burner, as described in the next paragraph.

Considerably more attention must be paid if a burner is used to ash a filter paper. The burner produces much higher temperatures than a heat lamp. The possibility thus exists for the mechanical loss of precipitate if moisture is expelled too rapidly in the initial stages of heating or if the paper bursts into flame. Also, partial reduction of some precipitates can occur through reaction with the hot carbon of the charring paper; such reduction is a serious problem if reoxidation following ashing is inconvenient. These difficulties can be minimized by positioning the crucible as illustrated in Figure 35-16. The tilted position allows for the ready access of air; a clean crucible cover should be available to extinguish any flame that might develop.

Heating is commenced with a small flame. The temperature is gradually increased as moisture is evolved and the paper begins to char. The intensity of heating that can be tolerated can be gauged by the amount of smoke given off. Thin wisps are normal. A significant increase in the amount of smoke indicates that the paper is about to flash and that heating should be temporarily discontinued. Any flame that does appear should be immediately extinguished with a crucible cover. (The cover may become discolored, owing to the condensation of carbonaceous products; these products must ultimately be removed from the cover by ignition to confirm the absence of entrained particles of precipitate.) When no further smoking can be detected, heating is increased to eliminate the residual carbon. Strong heating, as necessary, can then be undertaken.

This sequence ordinarily precedes the final ignition of a precipitate in a muffle furnace, where a reducing atmosphere is equally undesirable.

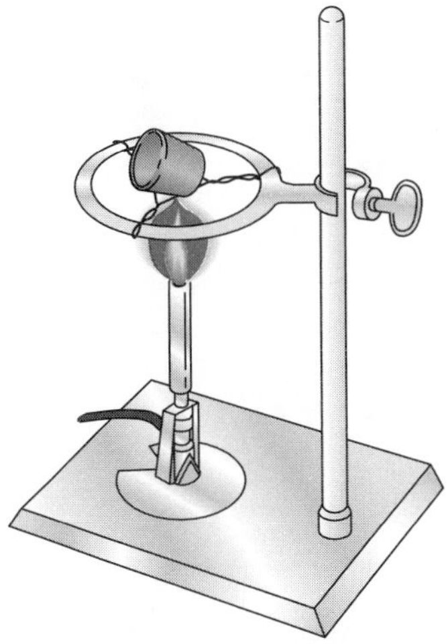

Figure 35-16
Ignition of a precipitate. Proper crucible position for preliminary charring.

The Use of Filtering Crucibles

A vacuum filtration train (Figure 35-17) is used where a filtering crucible can be used instead of paper. The trap isolates the filter flask from the source of vacuum.

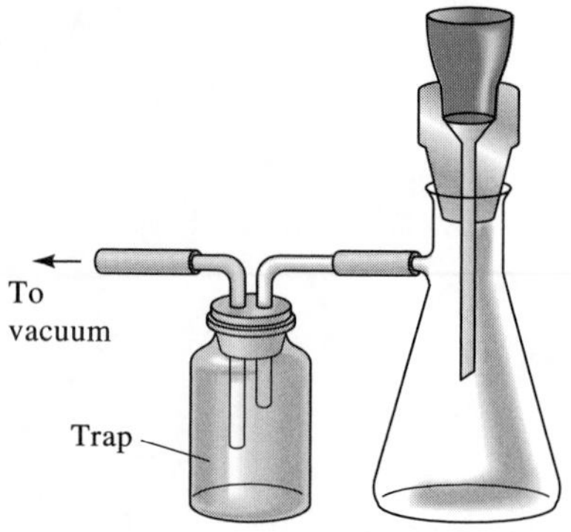

Figure 35-17
Train for vacuum filtration.

35G-4 Rules for the Manipulation of Heated Objects

Careful adherence to the following rules will minimize the possibility of accidental loss of a precipitate.

1. Practice unfamiliar manipulations before you put them to use.
2. *Never* place a heated object on the benchtop; instead, place it on a wire gauze or a heat-resistant ceramic plate.
3. Allow a crucible that has been subjected to the full flame of a burner or to a muffle furnace to cool momentarily (on a wire gauze or ceramic plate) before you transfer it to the desiccator.
4. Keep the tongs and forceps used to handle heated objects scrupulously clean. In particular, do not allow the tips to touch the benchtop.

35H THE MEASUREMENT OF VOLUME

The precise measurement of volume is as important to many analytical methods as is the precise measurement of mass.

35H-1 Units of Volume

The unit of volume is the *liter* (L), defined as one cubic decimeter. The *milliliter* (mL) is 1/1000 L and is used where the liter represents an inconveniently large volume unit.

35H-2 The Effect of Temperature on Volume Measurements

The volume occupied by a given mass of liquid varies with temperature, as does the device that holds the liquid during measurement. Most volumetric measuring devices are made of glass, however, which fortunately has a small coefficient of expansion. Consequently, variations in the volume of a glass container with temperature need not be considered in ordinary analytical work.

The coefficient of expansion for dilute aqueous solutions (approximately 0.025%/°C) is such that a 5°C change has a measurable effect upon the reliability of ordinary volumetric measurements.

EXAMPLE 35-2

A 40.00-mL sample is taken from an aqueous solution at 5°C; what volume does it occupy at 20°C?

$$V_{20^\circ} = V_{5^\circ} + 0.00025(20 - 5)(40.00) = 40.00 + 0.15 = 40.15 \text{ mL}$$

Volumetric measurements must be referred to some standard temperature; this reference point is ordinarily 20°C. The ambient temperature of most laboratories is sufficiently close to 20°C to eliminate the need for temperature corrections in volume measurements for aqueous solutions. In contrast, the coefficient of expansion for organic liquids may require corrections for temperature differences of 1°C or less.

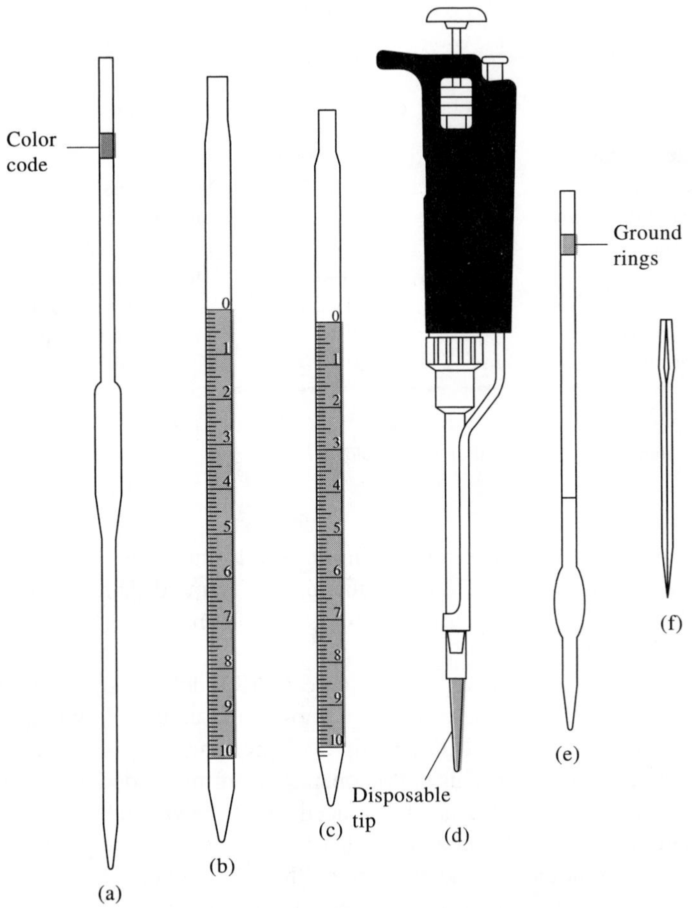

Figure 35-18
Typical pipets: (a) volumetric, (b) Mohr, (c) serological, (d) Eppendorf micropipet, (e) Ostwald-Folin, (f) lambda.

35H-3 Apparatus for the Precise Measurement of Volume

The reliable measurement of volume is performed with a *pipet,* a *buret,* and a *volumetric flask.*

Volumetric equipment is marked by the manufacturer to indicate not only the manner of calibration (usually TD for ''to deliver'' or TC for ''to contain'') but also the temperature at which the calibration strictly applies. Pipets and burets are ordinarily calibrated to deliver specified volumes, whereas volumetric flasks are calibrated on a to-contain basis.

Tolerances, Class A Transfer Pipets

Capacity, mL	Tolerances, mL
0.5	±0.006
1	±0.006
2	±0.006
5	±0.01
10	±0.02
20	±0.03
25	±0.03
50	±0.05
100	±0.08

Pipets

Pipets permit the transfer of accurately known volumes from one container to another. Common types are shown in Figure 35-18; information concerning their use is given in Table 35-2. A *volumetric,* or *transfer,* pipet (Figure 35-18a) delivers a single, fixed volume between 0.5 and 200 mL. Many such pipets are color-coded by volume for convenience in identification and sorting. *Measuring* pipets (Figures 35-18b and c) are calibrated in convenient units to permit delivery of any volume up to a maximum capacity ranging from 0.1 to 25 mL.

Volumetric and measuring pipets are filled to a calibration mark at the outset; the manner in which the transfer is completed depends upon the particular type.

TABLE 35-2 Characteristics of Pipets

Name	Type of Calibration*	Function	Available Capacity, mL	Type of Drainage
Volumetric	TD	Delivery of fixed volume	1–200	Free
Mohr	TD	Delivery of variable volume	1–25	To lower calibration line
Serological	TD	Delivery of variable volume	0.1–10	Blow out last drop†
Serological	TD	Delivery of variable volume	0.1–10	To lower calibration line
Ostwald-Folin	TD	Delivery of fixed volume	0.5–10	Blow out last drop†
Lambda	TC	Containment of fixed volume	0.001–2	Wash out with suitable solvent
Lambda	TD	Delivery of fixed volume	0.001–2	Blow out last drop†
Eppendorf	TD	Delivery of variable or fixed volume	0.001–1	Tip emptied by air displacement

*TD: to deliver; TC: to contain.

†A frosted ring near the top of recently manufactured pipets indicates that the last drop is to be blown out.

Because an attraction exists between most liquids and glass, a small amount of liquid tends to remain in the tip after the pipet is emptied. This residual liquid is never blown out of a volumetric pipet or from some measuring pipets; it is blown out of other types of pipets (Table 35-2).

Handheld Eppendorf micropipets (Figure 35-18d) deliver adjustable microliter volumes of liquid. With these pipets, a known and adjustable volume of air is displaced from the plastic disposable tip by depressing the pushbutton on the top of the pipet to a first stop. This button operates a spring-loaded piston that forces air out of the pipet. The volume of displaced air can be varied by a locking digital micrometer adjustment located on the front of the device. The plastic tip is then inserted into the liquid and the pressure on the button released, causing liquid to be drawn into the tip. The tip is then placed against the walls of the receiving vessel, and the pushbutton is again depressed to the first stop. After one second, the pushbutton is depressed further to a second stop, which completely empties the tip. The range of volumes and precision of typical pipets of this type are shown in the margin.

Range and Precision of Typical Eppendorf Micropipets

Volume Range, μL	Standard Deviation, μL
1–20	<0.04 @ 2 μL
	<0.06 @ 20 μL
10–100	<0.10 @ 15 μL
	<0.15 @ 100 μL
20–200	<0.15 @ 25 μL
	<0.30 @ 200 μL
100–1000	<0.6 @ 250 μL
	<1.3 @ 1000 μL
500–5000	<3 @ 1.0 mL
	<8 @ 5.0 mL

Numerous *automatic* pipets are available for situations that call for the repeated delivery of a particular volume. In addition, a motorized, computer-controlled microliter pipet is now available (see Figure 35-19). This device is programmed to function as a pipet, a dispenser of multiple volumes, a buret, and a means for diluting samples. The volume desired is entered on a keyboard and is displayed on a panel. A motor-driven piston dispenses the liquid. Maximum volumes range from 10 to 2500 μL.

Burets

Tolerances, Class A Burets

Volume, mL	Tolerances, mL
5	±0.01
10	±0.02
25	±0.03
50	±0.05
100	±0.20

Like measuring pipets, burets enable the analyst to deliver any volume up to their maximum capacity. The precision attainable with a buret is substantially greater than with a pipet.

A buret consists of a calibrated tube to hold titrant plus a valve arrangement by which the flow of titrant is controlled. This valve is the principal source of difference among burets. The simplest pinchcock valve consists of a close-fitting glass bead inside a short length of rubber tubing that connects the buret and its tip (Figure 35-20a); only when the tubing is deformed does liquid flow past the bead.

A buret equipped with a glass stopcock for a valve relies upon a lubricant between the ground-glass surfaces of stopcock and barrel for a liquid-tight seal. Some solutions, notably bases, cause a glass stopcock to freeze upon long contact; therefore thorough cleaning is needed after each use. Valves made of Teflon are commonly encountered; these are unaffected by most common reagents and require no lubricant (Figure 35-20b).

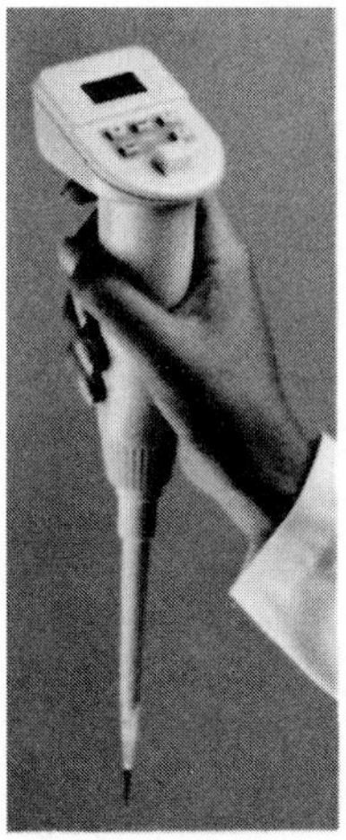

Figure 35-19
A handheld, battery-operated, motorized pipet. (Courtesy of Rainin Instrument Co., Inc., Woburn, MA.)

Volumetric Flasks

Volumetric flasks are manufactured with capacities ranging from 5 mL to 5 L and are usually calibrated to contain a specified volume when filled to a line etched on the neck (Figure 35-21). They are used for the preparation of standard solutions and for the dilution of samples to a fixed volume prior to taking aliquots with a pipet. Some are also calibrated on a to-deliver basis; these are readily distinguished by two reference lines on the neck. If delivery of the stated volume is desired, the flask is filled to the upper line.

Tolerances, Class A Volumetric Flasks

Capacity, mL	Tolerances, mL
5	±0.02
10	±0.02
25	±0.03
50	±0.05
100	±0.08
250	±0.12
500	±0.20
1000	±0.30
2000	±0.50

35H-4 General Considerations Concerning the Use of Volumetric Equipment

Volume markings are blazed upon clean volumetric equipment by the manufacturer. An equal degree of cleanliness is needed in the laboratory if these markings are to have their stated meanings. Only clean glass surfaces support a uniform film of liquid. Dirt or oil causes breaks in this film; the existence of breaks is a certain indication of an unclean surface.

Cleaning

A brief soaking in a warm detergent solution is usually sufficient to remove the grease and dirt responsible for water breaks. Prolonged soaking should be avoided because a rough area or ring is likely to develop at a detergent/air interface. This ring cannot be removed and causes a film break that destroys the usefulness of the equipment.

After being cleaned, the apparatus must be thoroughly rinsed with tap water and then with three or four portions of distilled water. It is seldom necessary to dry volumetric ware.

A **meniscus** is the curved surface of a liquid at its interface with the atmosphere.

Avoiding Parallax

The top surface of a liquid confined in a narrow tube exhibits a marked curvature, or *meniscus*. It is common practice to use the bottom of the meniscus as the point of reference in calibrating and using volumetric equipment. This minimum can be established more exactly by holding an opaque card or piece of paper behind the graduations (Figure 35-22).

Parallax is the apparent displacement of a liquid level or of a pointer as an observer changes position. It occurs when an object is viewed from a position that is not at a right angle to the object.

In reading volumes, your eye must be at the level of the liquid surface to avoid an error due to *parallax*. Parallax is a phenomenon that causes the volume to appear smaller than its actual value if the meniscus is viewed from above and larger if the meniscus is viewed from below (Figure 35-22).

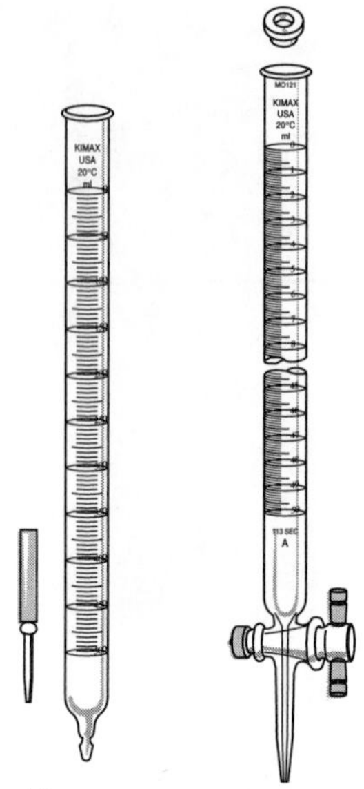

(a) (b)

Figure 35-20
Burets: (a) Pinchcock valve; (b) Teflon valve.

35H-5 Directions for the Use of a Pipet

The following directions pertain specifically to volumetric pipets but can be modified for the use of other types as well.

Liquid is drawn into a pipet through the application of a slight vacuum. *Your mouth should never be used for suction because of the possibility of accidentally ingesting the liquid being pipetted.* Instead, a rubber suction bulb (Figure 35-23a) or a rubber tube connected to a vacuum source should be used.

Cleaning

Use a rubber bulb to draw detergent solution to a level 2 to 3 cm above the calibration mark of the pipet. Drain this solution and then rinse the pipet with several portions of tap water. Inspect for film breaks; repeat this portion of the cleaning cycle if necessary. Finally, fill the pipet with distilled water to perhaps one third of its capacity and carefully rotate it so that the entire interior surface is wetted. Repeat this rinsing step at least twice.

Measurement of an Aliquot

Use a rubber bulb to draw a small volume of the liquid to be sampled into the pipet and thoroughly wet the entire interior surface. Repeat with *at least* two additional portions. Then carefully fill the pipet to a level somewhat above the graduation mark (Figure 35-23a). Quickly replace the bulb with a *forefinger* to arrest the outflow of liquid (Figure 35-23b). Make certain there are no bubbles in the bulk of the liquid or foam at the surface. Tilt the pipet slightly from the vertical and wipe the exterior free of adhering liquid (Figure 35-23c). Touch the tip of the pipet to the wall of a glass vessel (*not* the container into which the aliquot is to be transferred), and slowly allow the liquid level to drop by partially releasing the forefinger (Note 1). Halt further flow as the bottom of the meniscus coincides exactly with the graduation mark. Then place the pipet tip well within the receiving vessel, and allow the liquid to drain. When free flow ceases, rest the tip against the inner wall of the receiver for a full ten seconds (Figure 35-23d). Finally, withdraw the pipet with a rotating motion to remove any liquid adhering to the tip. *The small volume remaining inside the tip of a volumetric pipet should not be blown or rinsed into the receiving vessel* (Note 2).

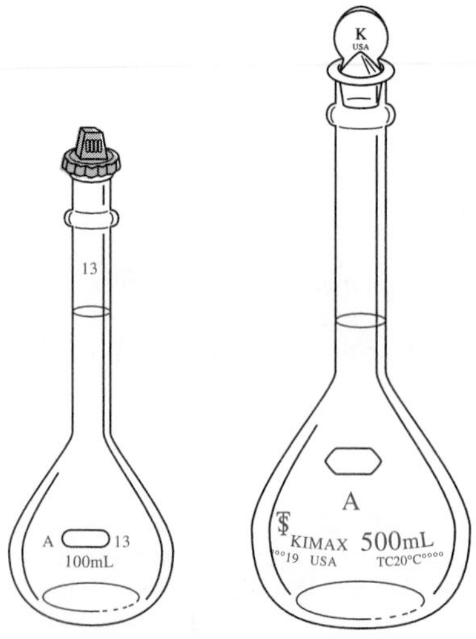

Figure 35-21
Typical volumetric flasks.

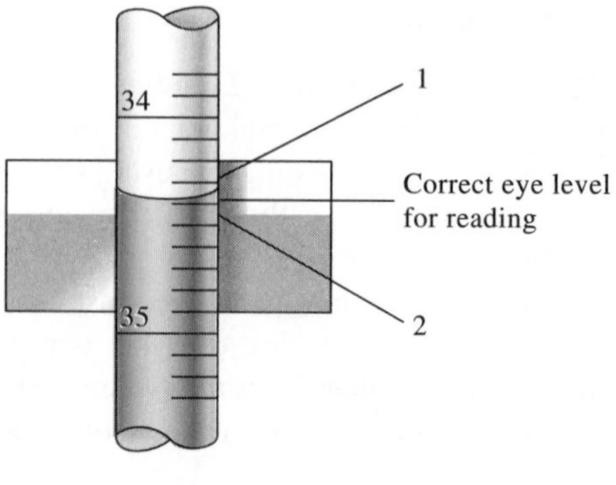

Figure 35-22
Method for reading a buret. The eye should be level with the meniscus. The reading shown is 34.39 mL. If viewed from position 1, the reading appears smaller than 34.39 mL; from position 2, it appears larger.

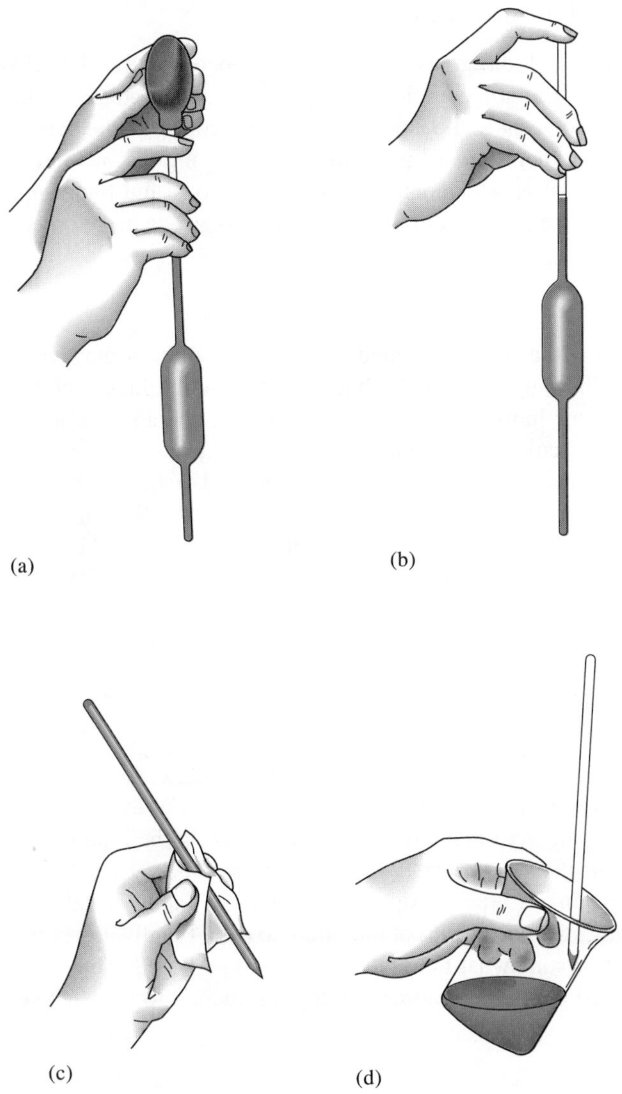

Figure 35-23
Steps in dispensing an aliquot.

Notes

1. The liquid can best be held at a constant level if your forefinger is *faintly* moist. Too much moisture makes control impossible.
2. Rinse the pipet thoroughly after use.

35H-6 Directions for the Use of a Buret

Before it is placed in service, a buret must be scrupulously clean; in addition, its valve must be liquid-tight.

Cleaning

Thoroughly clean the tube of the buret with detergent and a long brush. Rinse thoroughly with tap water and then with distilled water. Inspect for water breaks. Repeat the treatment if necessary.

Lubrication of a Glass Stopcock

Carefully remove all old grease from a glass stopcock and its barrel with a paper towel and dry both parts completely. Lightly grease the stopcock, taking care to avoid the area adjacent to the hole. Insert the stopcock into the barrel and rotate it vigorously with slight inward pressure. A proper amount of lubricant has been used when (1) the area of contact between stopcock and barrel appears nearly transparent, (2) the seal is liquid-tight, and (3) no grease has worked its way into the tip.

Notes

1. Grease films that are unaffected by cleaning solution may yield to organic solvents. Thorough washing with detergent should follow such treatment. The use of silicone lubricants is not recommended; contamination by such preparations is difficult—if not impossible—to remove.
2. So long as the flow of liquid is not impeded, fouling of a buret tip with stopcock grease is not a serious matter. Removal is best accomplished with organic solvents. A stoppage during a titration can be freed by *gentle* warming of the tip with a lighted match.
3. Before a buret is returned to service after reassembly, it is advisable to test for leakage. Simply fill the buret with water and establish that the volume reading does not change with time.

Buret readings should be estimated to the nearest 0.01 mL.

Filling

Make certain the stopcock is closed. Add 5 to 10 mL of the titrant, and carefully rotate the buret to wet the interior completely. Allow the liquid to drain through the tip. *Repeat this procedure at least two more times.* Then fill the buret well above the zero mark. Free the tip of air bubbles by rapidly rotating the stopcock and permitting small quantities of the titrant to pass. Finally, lower the level of the liquid just to or somewhat below the zero mark. Allow for drainage ($\sim$ 1 min), and then record the initial volume reading, estimating to the nearest 0.01 mL.

Titration

Figure 35-24 illustrates the preferred method for the manipulation of a stopcock; when the hand is held as shown, any tendency for lateral movement by the stopcock will be in the direction of firmer seating. Be sure the tip of the buret is well within the titration vessel (ordinarily a flask). Introduce the titrant in increments of about 1 mL. Swirl (or stir) constantly to ensure thorough mixing. Decrease the size of the increments as the titration progresses; add titrant dropwise in the immediate vicinity of the end point (Note 2). When it is judged that only a few more drops are needed, rinse the walls of the container (Note 3). Allow for drainage (at least 30 seconds) at the completion of the titration. Then record the final volume, again to the nearest 0.01 mL.

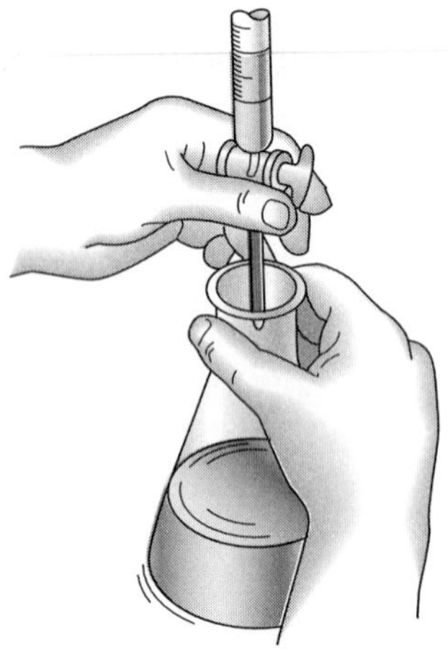

Figure 35-24
Recommended method for manipulation of a buret stopcock.

Notes

1. When unfamiliar with a particular titration, many chemists prepare an extra sample. No care is lavished on its titration, since its functions are to reveal the nature of the end point and to provide a rough estimate of titrant requirements.

This deliberate sacrifice of one sample frequently results in an overall saving of time.

2. Increments smaller than one drop can be taken by allowing a small volume of titrant to form on the tip of the buret and then touching the tip to the wall of the flask. This partial drop is then combined with the bulk of the liquid as in Note 3.
3. Instead of being rinsed toward the end of a titration, the flask can be tilted and rotated so that the bulk of the liquid picks up any drops that adhere to the inner surface.

35H-7 Directions for the Use of a Volumetric Flask

Before being put into use, volumetric flasks should be washed with detergent and thoroughly rinsed. Only rarely do they need to be dried. If required, however, drying is best accomplished by clamping the flask in an inverted position. Insertion of a glass tube connected to a vacuum line hastens the process.

Direct Weighing into a Volumetric Flask

The direct preparation of a standard solution requires the introduction of a known mass of solute to a volumetric flask. Use of a powder funnel minimizes the possibility of loss of solid during the transfer. Rinse the funnel thoroughly; collect the washings in the flask.

The foregoing procedure may be inappropriate if heating is needed to dissolve the solute. Instead, weigh the solid into a beaker or flask, add solvent, heat to dissolve the solute, and allow the solution to cool to room temperature. Transfer this solution quantitatively to the volumetric flask, as described in the next section.

The Quantitative Transfer of Liquid to a Volumetric Flask

Insert a funnel into the neck of the volumetric flask; use a stirring rod to direct the flow of liquid from the beaker into the funnel. Tip off the last drop of liquid on the spout of the beaker with the stirring rod. Rinse both the stirring rod and the interior of the beaker with distilled water and transfer the washings to the volumetric flask, as before. Repeat the rinsing process *at least* two more times.

Dilution to the Mark

After the solute has been transferred, fill the flask about half-full and swirl the contents to hasten solution. Add more solvent and again mix well. Bring the liquid level almost to the mark, and allow time for drainage (≈ 1 min); then use a medicine dropper to make such final additions of solvent as are necessary (Note). Firmly stopper the flask, and invert it repeatedly to ensure thorough mixing. Transfer the contents to a storage bottle that either is dry or has been thoroughly rinsed with several small portions of the solution from the flask.

The solute should be completely dissolved *before* you dilute to the mark.

Note

If, as sometimes happens, the liquid level accidentally exceeds the calibration mark, the solution can be saved by correcting for the excess volume. Use a gummed label to mark the location of the meniscus. After the flask has been emptied, carefully refill to the manufacturer's etched mark with water. Then use a buret to determine the additional volume needed to fill the flask so that the meniscus is at the gummed-label mark. This volume must be added to the nominal volume of the flask when calculating the concentration of the solution.

35I THE CALIBRATION OF VOLUMETRIC WARE

Volumetric glassware is calibrated by measuring the mass of a liquid (usually distilled water) of known density and temperature that is contained in (or delivered by) the volumetric ware. In carrying out a calibration, a buoyancy correction must be made (Section 35D-4), since the density of water is quite different from that of the weights.

The calculations associated with calibration, while not difficult, are somewhat involved. The raw weighing data are first corrected for buoyancy with Equation 35-1. Next, the volume of the apparatus at the temperature of calibration (T) is obtained by dividing the density of the liquid at that temperature into the corrected mass. Finally, this volume is corrected to the standard temperature of 20°C as in Example 35-2.

Table 35-3 is provided to ease the computational burden of calibration. Corrections for buoyancy with respect to stainless steel or brass weights (the density difference between the two is small enough to be neglected) and for the volume change of water and of glass containers have been incorporated into these data. Multiplication by the appropriate factor from Table 35-3 converts the mass of water at temperature T to (1) the corresponding volume at that temperature or (2) the volume at 20°C.

EXAMPLE 35-3

A 25-mL pipet delivers 24.976 g of water weighed against stainless steel weights at 25°C. Use the data in Table 35-3 to calculate the volume delivered by this pipet at 25°C and 20°C.

At 25°C: $V = 24.976 \text{ g} \times 1.0040 \text{ mL/g} = 25.08 \text{ mL}$

At 20°C: $V = 24.976 \text{ g} \times 1.0037 \text{ mL/g} = 25.07 \text{ mL}$

35I-1 General Directions for Calibration Work

All volumetric ware should be painstakingly freed of water breaks before being calibrated. Burets and pipets need not be dry; volumetric flasks should be thoroughly drained and dried at room temperature. The water used for calibration should be in thermal equilibrium with its surroundings. This condition is best established by drawing the water well in advance, noting its temperature at frequent intervals, and waiting until no further changes occur.

Although an analytical balance can be used for calibration, weighings to the nearest milligram are perfectly satisfactory for all but the very smallest volumes.

TABLE 35-3 Volume Occupied by 1.000 g of Water Weighed in Air Against Stainless Steel Weights*

Temperature, *T* °C	Volume, mL At *T*	Volume, mL Corrected to 20°C
10	1.0013	1.0016
11	1.0014	1.0016
12	1.0015	1.0017
13	1.0016	1.0018
14	1.0018	1.0019
15	1.0019	1.0020
16	1.0021	1.0022
17	1.0022	1.0023
18	1.0024	1.0025
19	1.0026	1.0026
20	1.0028	1.0028
21	1.0030	1.0030
22	1.0033	1.0032
23	1.0035	1.0034
24	1.0037	1.0036
25	1.0040	1.0037
26	1.0043	1.0041
27	1.0045	1.0043
28	1.0048	1.0046
29	1.0051	1.0048
30	1.0054	1.0052

*Corrections for buoyancy (stainless steel weights) and change in container volume have been applied.

Thus, a top-loading balance is more conveniently used. Weighing bottles or small, well-stoppered conical flasks can serve as receivers for the calibration liquid.

Calibration of a Volumetric Pipet

Determine the empty mass of the stoppered receiver to the nearest milligram. Transfer a portion of temperature-equilibrated water to the receiver with the pipet, weigh the receiver and its contents (again, to the nearest milligram), and calculate the mass of water delivered from the difference in these masses. Calculate the volume delivered with the aid of Table 35-3. Repeat the calibration several times; calculate the mean volume delivered and its standard deviation.

Calibration of a Buret

Fill the buret with temperature-equilibrated water and make sure that no air bubbles are trapped in the tip. Allow about 1 min for drainage; then lower the liquid level to bring the bottom of the meniscus to the 0.00-mL mark. Touch the tip to the wall of a beaker to remove any adhering drop. Wait 10 min and recheck the volume; if the stopcock is tight, there should be no perceptible change. During this interval, weigh (to the nearest milligram) a 125-mL conical flask fitted with a rubber stopper.

Once tightness of the stopcock has been established, slowly transfer (at about 10 mL/min) approximately 10 mL of water to the flask. Touch the tip to the wall of the flask. Wait 1 min, record the volume that was apparently delivered, and refill the buret. Weigh the flask and its contents to the nearest milligram; the difference between this mass and the initial value gives the mass of water delivered. Use Table 35-3 to convert this mass to the true volume. Subtract the apparent volume from the true volume. This difference is the correction that should be applied to the apparent volume to give the true volume. Repeat the calibration until agreement within ± 0.02 mL is achieved.

Starting again from the zero mark, repeat the calibration, this time delivering about 20 mL to the receiver. Test the buret at 10-mL intervals over its entire volume. Prepare a plot of the correction to be applied as a function of volume delivered. The correction associated with any interval can be determined from this plot.

Calibration of a Volumetric Flask

Weigh the clean, dry flask to the nearest milligram. Then fill to the mark with equilibrated water and reweigh. Calculate the volume contained with the aid of Table 35-3.

Calibration of a Volumetric Flask Relative to a Pipet

An **aliquot** is a measured fraction of the volume of a liquid sample.

The calibration of a volumetric flask relative to a pipet provides an excellent method for partitioning a sample into aliquots. These directions pertain to a 50-mL pipet and a 500-mL volumetric flask; other combinations are equally convenient.

Carefully transfer ten 50-mL aliquots from the pipet to a dry 500-mL volumetric flask. Mark the location of the meniscus with a gummed label. Cover with a label varnish to ensure permanence. Dilution to the label permits the same pipet to deliver precisely a one-tenth aliquot of the solution in the flask. Note that recalibration is necessary if another pipet is to be used.

35J THE LABORATORY NOTEBOOK

A laboratory notebook is needed to record measurements and observations concerning an analysis. The book should be permanently bound with consecutively numbered pages (if necessary, the pages should be hand-numbered before any entries are made). Most notebooks have more than ample room; there is no need to crowd entries.

The first few pages should be saved for a table of contents that is updated as entries are made.

35J-1 Rules for the Maintenance of a Laboratory Notebook

1. *Record all data and observations directly into the notebook in ink.* Neatness is desirable, but you should not realize neatness by transcribing data from a sheet of paper to the notebook or from one notebook to another. The risk of misplacing—or incorrectly transcribing—crucial data and thereby ruining an experiment is unacceptable.
2. Supply each entry or series of entries with a heading or label. A series of weighing data for a set of empty crucibles should carry the heading ''empty crucible mass'' (or something similar), for example, and the mass of each crucible should be identified by the same number or letter used to label the crucible.
3. Date each page of the notebook as it is used.
4. *Never* attempt to erase or obliterate an incorrect entry. Instead, cross it out with a single horizontal line and locate the correct entry as nearby as possible. Do not write over incorrect numbers; with time, it may become impossible to distinguish the correct entry from the incorrect one.
5. Never remove a page from the notebook. Draw diagonal lines across any page that is to be disregarded. Provide a brief rationale for disregarding the page.

Remember that you can discard an experimental measurement *only if you have certain knowledge that you made an experimental error.* Thus, you must carefully record experimental observations in your notebook as soon as they occur.

An entry in a laboratory notebook should never be erased but should be crossed out instead.

35J-2 Format

The instructor should be consulted concerning the format to be used in keeping the laboratory notebook.[7] One convention involves using each page consecutively for the recording of data and observations as they occur. The completed analysis is then summarized on the next available page spread (that is, left and right facing pages). As shown in Figure 35-25, the first of these two facing pages should contain the following entries:

1. The title of the experiment (''The Gravimetric Determination of Chloride'').
2. A brief statement of the principles upon which the analysis is based.
3. A complete summary of the weighing, volumetric, and/or instrument response data needed to calculate the results.
4. A report of the best value for the set and a statement of its precision.

The second page should contain the following items:

1. Equations for the principal reactions in the analysis.
2. An equation showing how the results were calculated.
3. A summary of observations that appear to bear upon the validity of a particular result or the analysis as a whole. *Any such entry must have been originally recorded in the notebook at the time the observation was made.*

---

[7]See also Howard M. Kanare, *Writing the Laboratory Notebook.* Washington, DC 20036: The American Chemical Society, 1985.

08

Gravimetric Determination of Chloride

The chloride in a soluble sample was precipitated as AgCl and weighed as such.

Sample masses	1	2	3
Mass bottle plus sample, g	27.6115	27.2185	26.8105
-less bottle, g	27.2185	26.8105	26.4517
mass sample, g	0.3930	0.4080	0.3588
Crucible masses, empty	~~20.7925~~	~~22.8311~~	~~21.2488~~
	20.7926	22.8311	~~21.2482~~
			21.2483
Crucible masses, with AgCl, g	~~21.4294~~	~~23.4920~~	~~21.8327~~
	~~21.4297~~	~~23.4914~~	21.8323
	21.4296	23.4915	
Mass of AgCl, g	0.6370	0.6604	0.5840
Percent Cl-	40.10	40.04	40.27
Average percent Cl-		40.12	

Relative standard deviation 3.0 *parts per thousand*

Date Started 1-9-96

Date Completed 1-16-96

Figure 35-25
Laboratory notebook data page.

35K SAFETY IN THE LABORATORY

Work in a chemical laboratory necessarily involves a degree of risk; accidents can and do happen. Strict adherence to the following rules will go far toward preventing (or minimizing the effect of) accidents.

1. At the outset, learn the location of the nearest eye fountain, fire blanket, shower, and fire extinguisher. Learn the proper use of each, and do not hesitate to use this equipment should the need arise.
2. ***Wear eye protection at all times.*** The potential for serious and perhaps permanent eye injury makes it mandatory that adequate eye protection be worn at all times by students, instructors, and visitors. Eye protection should be donned before entering the laboratory and should be used continuously until it is time to leave. Serious eye injuries have occurred to people performing such innocuous tasks as computing or writing in a laboratory notebook; such incidents are usually the result of someone else's losing control of an experiment. Regular prescription glasses are not adequate substitutes for eye pro-

tection approved by the Occupational Safety and Health Administration (OSHA). Contact lenses should *never* be used in the laboratory because laboratory fumes may react with them and have a harmful effect on the eyes.

3. Most of the chemicals in a laboratory are toxic; some are very toxic, and some—such as concentrated solutions of acids and bases—are highly corrosive. Avoid contact between these liquids and your skin. In the event of such contact, *immediately* flood the affected area with copious quantities of water. If a corrosive solution is spilled on clothing, remove the garment immediately. Time is of the essence; modesty cannot be a matter of concern.
4. ***Never*** perform an unauthorized experiment. Such activity is grounds for disqualification at many institutions.
5. Never work alone in the laboratory; be certain that someone is always within earshot.
6. Never bring food or beverages into the laboratory. Do not drink from laboratory glassware. Do not smoke in the laboratory.
7. Always use a bulb to draw liquids into a pipet; ***never*** use your mouth to provide suction.
8. Wear adequate foot covering (no sandals). Confine long hair with a net. A laboratory coat or apron will provide some protection and may be required.
9. Be extremely tentative in touching objects that have been heated; hot glass looks just like cold glass.
10. Always fire-polish the ends of freshly cut glass tubing. ***Never*** attempt to force glass tubing through the hole of a stopper. Instead, make sure that both tubing and hole are wet with soapy water. Protect hands with several layers of towel while inserting glass into a stopper.
11. Use fume hoods whenever toxic or noxious gases are likely to be evolved. Be cautious in testing for odors; use your hand to waft vapors above containers toward your nose.
12. Notify the instructor in the event of an injury.
13. Dispose of solutions and chemicals as instructed. It is illegal to flush solutions containing heavy metal ions or organic liquids down the drain in many localities; alternative arrangements are required for the disposal of such liquids.

CHAPTER

36

Selected Methods of Analysis

This chapter contains detailed directions for performing a variety of chemical analyses. The methods have been chosen to introduce you to analytical techniques that are widely used by chemists. For most of these analyses, the composition of the samples is known to the instructor. Thus, you will be able to judge how well you are mastering these techniques.

Your chances of success in the laboratory will greatly improve if you take time before beginning any analysis to read carefully and *understand* each step in the method and to develop a plan for how and when you will perform each step. For greatest efficiency, such study and planning should take place *before you enter the laboratory.*

The discussion in this section is aimed at helping you develop efficient work habits in the laboratory and also at providing you with some general information about an analytical chemistry laboratory.

Background Information

The better you understand the reasons for the steps in an analysis, the greater are your chances for success.

Before you start an analysis, you should understand the significance of each step in the procedure in order to avoid the pitfalls and potential sources of error that exist in all analytical methods. Information about these steps can usually be found in (1) preliminary discussion sections, (2) earlier chapters that are referred to in the discussion section, and (3) the ''Notes'' that follow many of the procedures. If, after reading these materials, you still don't understand the reason for one or more of the steps in the method, consult your instructor *before you begin laboratory work.*

The Accuracy of Measurements

In looking over an analytical procedure, you should decide which measurements must be made with maximum precision, and thus with maximum care, as opposed to those that should be carried out rapidly with little concern for precision. Generally, measurements that appear in the equation used to compute the results must be performed with maximum precision. The remaining measurements can *and should* be made less carefully to conserve time. The words *about* and *approximately* are frequently used to indicate that a measurement does not have to be done carefully. You should not waste time and effort to measure, let us say, a volume to $\pm$ 0.02 mL when an uncertainty of $\pm$ 0.5 mL or even $\pm$ 5 mL will have no discernible effect on the results.

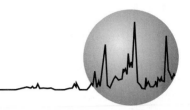

In some procedures, a statement such as "weigh three 0.5-g samples to the nearest 0.1 mg" is encountered. Here, samples of perhaps 0.4 to 0.6 g are acceptable, but their masses must be known to the nearest 0.1 mg. The number of significant figures in the specification of a volume or a mass is also a guide as to the care that should be taken in making a measurement. For example, the statement "add 10.00 mL of a solution to the beaker" indicates that you should measure the volume carefully with a buret or a pipet, with the aim of limiting the uncertainty to perhaps ± 0.02 mL. In contrast, if the directions read "add 10 mL," the measurement can be made with a graduated cylinder.

Time Utilization

You should study carefully the time requirements of the several unit operations involved in an analysis *before work is started.* Such study will reveal operations that require considerable elapsed, or clock, time but little or no operator time—for example, when a sample is dried in an oven, cooled in a desiccator, or evaporated on a hot plate. The experienced chemist plans to use such periods of waiting to perform other operations or perhaps to begin a new analysis. Some people find it worthwhile preparing a written time schedule for each laboratory period to avoid periods when no work can be done.

Time planning is also needed to identify places where an analysis can be interrupted for overnight or longer, as well as those operations that must be completed without a break.

Reagents

Directions for the preparation of reagents accompany many of the procedures. Before preparing such reagents, be sure to check to see if they are already prepared and available on a side shelf for general use.

If a reagent is known to pose a hazard, you should plan *in advance of the laboratory period* the steps that you should take to minimize injury or damage. Furthermore, you must acquaint yourself with the rules that apply in your laboratory for the disposal of waste liquids and solids. These rules vary from one part of the country to another and even among laboratories in the same locale.

Water

Some laboratories use deionizers to purify water; others employ stills for this purpose. The terms "distilled water" and "deionized water" are used interchangeably in the directions that follow. Either type is satisfactory for analytical work.

You should use tap water only for preliminary cleaning of glassware. The cleaned glassware is then rinsed with at least three small portions of distilled or deionized water.

36A GRAVIMETRIC METHODS OF ANALYSIS

General aspects, calculations, and typical applications of gravimetric analysis are discussed in Chapter 5.

36A-1 The Gravimetric Determination of Chloride in a Soluble Sample

Discussion

The chloride content of a soluble salt can be determined by precipitation as silver chloride:

$$Ag^+ + Cl^- \longrightarrow AgCl(s)$$

The precipitate is collected in a weighed filtering crucible and washed. Its mass is then determined after it has been dried to constant mass at 110°C.

The solution containing the sample is kept slightly acidic during the precipitation to eliminate possible interference from anions of weak acids (such as CO_3^{2-}) that form sparingly soluble silver salts in a neutral environment. A moderate excess of silver ion is needed to diminish the solubility of silver chloride, but a large excess is avoided to minimize coprecipitation of silver nitrate.

Silver chloride forms first as a colloid and is subsequently coagulated with heat. Nitric acid and the small excess of silver nitrate promote coagulation by providing a moderately high electrolyte concentration. Nitric acid in the wash solution maintains the electrolyte concentration and eliminates the possibility of peptization during the washing step; the acid subsequently decomposes to give volatile products when the precipitate is dried. See Section 5B-2 for additional information concerning the properties and treatment of colloidal precipitates.

In common with other silver halides, finely divided silver chloride undergoes photodecomposition:

$$2AgCl(s) \xrightarrow{h\nu} 2Ag(s) + Cl_2(g)$$

The elemental silver produced in this reaction is responsible for the violet color that develops in the precipitate. In principle, this reaction leads to low results for chloride ion. In practice, however, its effect is negligible provided you avoid direct and prolonged exposure of the precipitate to sunlight.

If photodecomposition of silver chloride occurs before filtration, the additional reaction

$$3Cl_2(aq) + 3H_2O + 5Ag^+ \longrightarrow 5AgCl(s) + ClO_3^- + 6H^+$$

tends to cause high results.

As the analysis is ordinarily performed, some photodecomposition of silver chloride is inevitable. It is worthwhile to minimize exposure of the solid to intense sources of light as far as possible.

Because silver nitrate is expensive, any unused reagent should be collected in a storage container; similarly, precipitated silver chloride should be retained after the analysis is complete.[1]

---

[1]Silver can be recovered from silver chloride and from surplus reagent by reduction with ascorbic acid; see J. W. Hill and L. Bellows, *J. Chem. Educ.*, **1986,** *63* (4), 357; see also J. P. Rawat and S. Iqbal M. Kamoonpuri, *ibid.*, **1986,** *63* (6), 537 for recovery (as $AgNO_3$) based on ion exchange. See D. D. Perrin, W. L. F. Armarego, and D. R. Perrin, *Chemistry International,* **1987,** *9* (1), 3 concerning a potential hazard in the recovery of silver nitrate.

Procedure

Clean three medium-porosity sintered-glass or porcelain filtering crucibles by allowing about 5 mL of concentrated HNO_3 to stand in each for about 5 min. Use a vacuum (Figure 35-17) to draw the acid through the crucible. Rinse each crucible with three portions of tap water, and then discontinue the vacuum. Next, add about 5 mL of 6 M NH_3 and wait for about 5 min before drawing it through the filter. Finally, rinse each crucible with six to eight portions of distilled or deionized water. Provide each crucible with an identifying mark. Dry the crucibles to constant mass by heating at 110°C while the other steps in the analysis are being carried out. The first drying should be for at least 1 hr; subsequent heating periods can be somewhat shorter (30 to 40 min). This process of heating and drying should be repeated until the mass becomes constant to within 0.2 to 0.3 mg.

Be sure to label your beakers and crucibles.

Transfer the unknown to a weighing bottle and dry it at 110°C (Figure 35-9) for 1 to 2 hr; allow the bottle and contents to cool to room temperature in a desiccator. Weigh (to the nearest 0.1 mg) individual samples by difference into 400-mL beakers (Note 1). Dissolve each sample in about 100 mL of distilled water to which 2 to 3 mL of 6 M HNO_3 has been added.

Slowly, and with good stirring, add 0.2 M $AgNO_3$ to each of the cold sample solutions until AgCl is observed to coagulate (Notes 2 and 3); then introduce an additional 3 to 5 mL. Heat almost to boiling, and digest the solids for about 10 min. Add a few drops of $AgNO_3$ to confirm that precipitation is complete. If more precipitate forms, add about 3 mL of $AgNO_3$, digest, and again test for completeness of precipitation. Pour any unused $AgNO_3$ into a waste container (*not* into the original reagent bottle). Cover each beaker, and store in a dark place for at least 2 hr and preferably until the next laboratory period.

Digest means to heat an unstirred precipitate in the **mother liquid,** i.e., the solution from which it is formed.

Read the instructions for filtration in Section 35G-2. Decant the supernatant liquids through weighed filtering crucibles. Wash the precipitates several times (while they are still in the beaker) with a solution consisting of 2 to 5 mL of 6 M HNO_3 per liter of distilled water; decant these washings through the filters. Quantitatively transfer the AgCl from the beakers to the individual crucibles with fine streams of wash solution; use rubber policemen to dislodge any particles that adhere to the walls of the beakers. Continue washing until the filtrates are essentially free of Ag^+ ion (Note 4).

Dry the precipitates at 110°C for at least 1 hr. Store the crucibles in a desiccator while they cool. Determine the mass of crucibles and their contents. Repeat the cycle of heating, cooling, and weighing until consecutive weighings agree to within 0.2 mg. Calculate the percentage of Cl^- in the sample.

Upon completion of the analysis, remove the precipitates by gently tapping the crucibles over a piece of glazed paper. Transfer the collected AgCl to a container for silver wastes. Remove the last traces of AgCl by filling the crucibles with 6 M NH_3 and allowing them to stand.

Notes

1. Consult with the instructor concerning an appropriate sample size.
2. Determine the approximate amount of $AgNO_3$ needed by calculating the volume that would be required if the unknown were pure NaCl.
3. Use a separate stirring rod for each sample and leave it in its beaker throughout the determination.

4. To test the washings for Ag^+, collect a small volume in a test tube and add a few drops of HCl. Washing is judged complete when little or no turbidity develops.

36A-2 The Gravimetric Determination of Tin in Brass

Discussion

Brasses are important alloys. Copper is ordinarily the principal constituent, with lesser amounts of lead, zinc, tin, and possibly other elements as well. Treatment of a brass with nitric acid results in the formation of the sparingly soluble "metastannic acid" $H_2SnO_3 \cdot xH_2O$; all other constituents are dissolved. The solid is filtered, washed, and ignited to SnO_2.

The gravimetric determination of tin provides experience in the use of ashless filter paper and is frequently performed in conjunction with a more inclusive analysis of a brass sample.

Procedure

Provide identifying marks on three porcelain crucibles and their covers. During waiting periods in the experiment, bring these to constant mass by ignition at 900°C in a muffle furnace.

Do not dry the unknown. If so instructed, rinse it with acetone to remove any oil or grease. Weigh (to the nearest 0.1 mg) approximately 1-g samples of the unknown into 250-mL beakers. Cover the beakers with watch glasses. Place the beakers in the hood, and cautiously introduce a mixture containing about 15 mL of concentrated HNO_3 and 10 mL of H_2O. Digest the samples for at least 30 min; add more HNO_3 if necessary. Rinse the watch glasses; then evaporate the solutions to about 5 mL but not to dryness (Note 1).

Add about 5 mL of 3 M HNO_3, 25 mL of distilled water, and one quarter of a tablet of filter pulp to each sample; heat without boiling for about 45 min. Collect the precipitated $H_2SnO_3 \cdot xH_2O$ on fine-porosity ashless filter papers (Section 35G-3 and Notes 2 and 3). Use many small volumes of hot 0.3 M HNO_3 to wash the last traces of copper from the precipitate. Test for completeness of washing with a drop of NH_3 on the top of the precipitate; wash further if the precipitate turns blue.

Remove the filter paper and its contents from the funnels, fold, and place in crucibles that, with their covers, have been brought to constant weight (Figure 35-15). Ash the filter paper at as low a temperature as possible. There must be free access of air throughout the charring (Section 35G-3 and Figure 35-16). Gradually increase the temperature until all the carbon has been removed. Then bring the covered crucibles and their contents to constant weight in a 900°C furnace (Note 4). Calculate the percentage of tin in the unknown.

Notes

1. It is often time-consuming and difficult to redissolve the soluble components of the residue after a sample has been evaporated to dryness.

2. The filtration step can be quite time-consuming and once started cannot be interrupted.
3. If the unknown is to be analyzed electrolytically for its lead and copper content (Section 36J-1), collect the filtrates in tall-form beakers. The final volume should be about 125 mL; evaporate to that volume if necessary. If the analysis is for tin only, the volume of washings is not important.
4. Partial reduction of SnO_2 may cause the ignited precipitate to appear gray. In this case, add a drop of nitric acid, cautiously evaporate, and ignite again.

36A-3 The Gravimetric Determination of Nickel in Steel

Discussion

The nickel in a steel sample can be precipitated from a slightly alkaline medium with an alcoholic solution of dimethylglyoxime (Section 5D-3). Interference from iron(III) is eliminated by masking with tartaric acid. The product is freed of moisture by drying at 110°C.

Iron(III) forms a highly stable complex with tartrate ion, which prevents it from precipitating as $Fe_2O_3 \cdot xH_2O$ in slightly alkaline solutions.

The bulky character of nickel dimethylglyoxime limits the mass of nickel that can be accommodated conveniently and thus the sample mass. Care must also be taken to control the excess of alcoholic dimethylglyoxime used. If too much is added, the alcohol concentration becomes sufficient to dissolve appreciable amounts of the nickel dimethylglyoxime, which leads to low results. If the alcohol concentration becomes too low, however, some of the reagent may precipitate and cause a positive error.

Preparation of Solutions

(a) *Dimethylglyoxime, 1% (w/v)* (sufficient for about 50 precipitations). Dissolve 10 g of dimethylglyoxime in 1 L of ethanol.
(b) *Tartaric acid, 15% (w/v)* (sufficient for about 50 precipitations). Dissolve 225 g of tartaric acid in sufficient water to give 1500 mL of solution. Filter before use if the solution is not clear.

Procedure

Clean and mark three medium-porosity sintered-glass crucibles (Note 1); bring them to constant mass by drying at 110°C for at least 1 hr.

Weigh (to the nearest 0.1 mg) samples containing between 30 and 35 mg of nickel into individual 400-mL beakers (Note 2). Dissolve each sample in about 50 mL of 6 M HCl with gentle warming *(hood)*. Carefully add approximately 15 mL of 6 M HNO_3, and boil gently to expel any oxides of nitrogen that may have been produced. Dilute to about 200 mL and heat to boiling. Introduce about 30 mL of 15% tartaric acid and sufficient concentrated NH_3 to produce a faint odor of NH_3 in the vapors over the solutions (Note 3); then add another 1 to 2 mL of NH_3. If the solutions are not clear at this stage, proceed as directed in Note 4. Make the solutions acidic with HCl (no odor of NH_3), heat to 60°C to

80°C, and add about 20 mL of the 1% dimethylglyoxime solution. With good stirring, add 6 M NH_3 until a slight excess exists (faint odor of NH_3) plus an additional 1 to 2 mL. Digest the precipitates for 30 to 60 min, cool for at least 1 hr, and filter.

Wash the solids with water until the washings are free of Cl^- (Note 5). Bring the crucibles and their contents to constant mass at 110°C. Report the percentage of nickel in the sample. The dried precipitate has the composition $Ni(C_4H_7O_2N_2)_2$ (288.92 g/mol).

Notes

1. Medium-porosity porcelain filtering crucibles or Gooch crucibles with glass pads can be substituted for sintered-glass crucibles in this determination.
2. Use a separate stirring rod for each sample and leave it in its beaker throughout.
3. The presence or absence of excess NH_3 is readily established by odor; use a waving motion with your hand to waft the vapors toward your nose.
4. If $Fe_2O_3 \cdot xH_2O$ forms upon addition of NH_3, acidify the solution with HCl, introduce additional tartaric acid, and neutralize again. Alternatively, remove the solid by filtration. Thorough washing with a hot NH_3/NH_4Cl solution is required; the washings are combined with the solution containing the bulk of the sample.
5. Test the washings for Cl^- by collecting a small portion in a test tube, acidifying with HNO_3, and adding a drop or two of 0.1 M $AgNO_3$. Washing is judged complete when little or no turbidity develops.

36B NEUTRALIZATION TITRATIONS

Neutralization titrations are performed with standard solutions of strong acids or bases. While a single solution (of either acid or base) is sufficient for the titration of a given type of analyte, it is convenient to have standard solutions of both acid and base available in the event a back-titration is needed to locate end points more exactly. The concentration of one solution is established by titration against a primary standard; the concentration of the other is then determined from the acid/base ratio (that is, the volume of acid needed to neutralize 1.000 mL of the base).

36B-1 The Effect of Atmospheric Carbon Dioxide on Neutralization Titrations

Water in equilibrium with the atmosphere is about 1×10^{-5} M in carbonic acid as a consequence of the equilibrium

$$CO_2(g) + H_2O \rightleftharpoons H_2CO_3(aq)$$

At this concentration level, the amount of 0.1 M base consumed by the carbonic acid in a typical titration is negligible. With more dilute reagents (<0.05 M), however, the water used as a solvent for the analyte and in the preparation of reagents must be freed of carbonic acid by boiling for a brief period.

Water that has been purified by distillation rather than by deionization is often supersaturated with carbon dioxide and may thus contain sufficient acid to affect the results of an analysis.[2] The instructions that follow are based upon the assumption that the amount of carbon dioxide in the water supply can be neglected without causing serious error. For further discussion on the effects of carbon dioxide in neutralization titrations, see Section 12A-3.

36B-2 Preparation of Indicator Solutions for Neutralization Titrations

Discussion

The theory of acid/base indicators is discussed in Section 10A-2. An indicator exists for virtually any pH range between 1 and 13.[3] Directions follow for the preparation of indicator solutions suitable for most neutralization titrations.

Procedure

Stock solutions ordinarily contain between 0.5 and 1.0 g of indicator per liter. (One liter of indicator is sufficient for hundreds of titrations.)

(a) *Bromocresol green.* Dissolve the sodium salt directly in distilled water.
(b) *Phenolphthalein, thymolphthalein.* Dissolve the solid indicator in a solution consisting of 800 mL ethanol and 200 mL of distilled or deionized water.

36B-3 Preparation of Dilute Hydrochloric Acid Solutions

Discussion

The preparation and standardization of acids are considered in Sections 12A-1 and 12A-2.

Procedure

For a 0.1 M solution, add about 8 mL of concentrated HCl to about 1 L of distilled water (Note). Mix thoroughly, and store in a glass-stoppered bottle.

Note

It is advisable to eliminate CO_2 from the water by a preliminary boiling if very dilute solutions ($<$0.05 M) are being prepared.

---

[2]Water that is to be used for neutralization titrations can be tested by adding 5 drops of phenolphthalein to a 500-mL portion. Less than 0.2 to 0.3 mL of 0.1 M OH^- should suffice to produce the first faint pink color of the indicator. If a larger volume is needed, the water should be boiled and cooled before it is used to prepare standard solutions or to dissolve samples.

[3]See, for example, J. Beukenkemp and W. Rieman III, in *Treatise in Analytical Chemistry,* I. M. Kolthoff and P. J. Elving, Eds., Part I, Vol. 11, pp. 6987–7001. New York: Wiley, 1974.

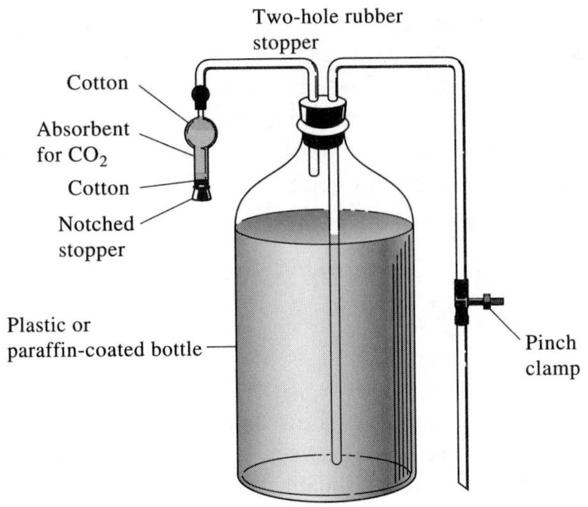

Figure 36-1
Arrangement for the storage of standard base solutions.

36B-4 Preparation of Carbonate-Free Sodium Hydroxide

Discussion

See Sections 12A-3 and 12A-4 for information concerning the preparation and standardization of bases.

Standard solutions of base are reasonably stable as long as they are protected from contact with the atmosphere. Figure 36-1 shows an arrangement for preventing the uptake of atmospheric carbon dioxide during storage and when the reagent is dispensed. Air entering the vessel is passed over a solid absorbent for CO_2, such as soda lime or Ascarite II®.[4] The contamination that occurs as the solution is transferred from this storage bottle to the buret is ordinarily negligible.

Solutions of bases should be stored in polyethylene bottles rather than glass because of the reaction between bases and glass. Such solutions should never be stored in glass-stoppered bottles; after standing for a period, removal of the stopper often becomes impossible.

As an alternative to the storage system in Figure 36-1, a tightly capped polyethylene bottle usually provides sufficient short-term protection against the uptake of atmospheric carbon dioxide. Before capping, the bottle is squeezed to minimize the interior air space. Care should also be taken to keep the bottle closed except during the brief periods when the contents are being transferred to a buret.

The concentration of solutions of sodium hydroxide decreases slowly (0.1% to 0.3% per week) when the base is stored in glass bottles. The loss in strength is caused by the reaction of the base with the glass to form sodium silicates. For this reason, standard solutions of base should not be stored for extended periods (longer than 1 or 2 weeks) in glass containers. In addition, bases should never be kept in glass-stoppered containers because the reaction between the base and the stopper may cause the latter to "freeze" after a brief period. Finally, to avoid the same type of freezing, burets with glass stopcocks should be promptly drained and thoroughly rinsed with water after use with standard base solutions. This problem is avoided with burets equipped with Teflon stopcocks.

[4]Thomas Scientific, Swedesboro, NJ. Ascarite II consists of sodium hydroxide deposited on a nonfibrous silicate structure.

Procedure

If so directed by the instructor, prepare a bottle for protected storage (Figure 36-1; also Note 1). Transfer 1 L of distilled water to the storage bottle (see the Note in Section 36B-3). Decant 4 to 5 mL of 50% NaOH into a small container (Note 2), add it to the water, and *mix thoroughly*. ***Use extreme care in handling 50% NaOH,*** which is highly corrosive. If the reagent comes into contact with skin, ***immediately*** flush the area with *copious* amounts of water.

Protect the solution from unnecessary contact with the atmosphere.

Notes

1. A solution of base that will be used up within two weeks can be stored in a tightly capped polyethylene bottle. After each removal of base, squeeze the bottle while tightening the cap to minimize the air space above the reagent. The bottle will become brittle after extensive use as a container for bases.
2. Be certain that any solid Na_2CO_3 in the 50% NaOH has settled to the bottom of the container and that the decanted liquid is absolutely clear. If necessary, filter the base through a glass mat in a Gooch crucible; collect the clear filtrate in a test tube inserted in the filter flask.

36B-5 The Determination of the Acid/Base Ratio

Discussion

If both acid and base solutions have been prepared, it is useful to determine their volumetric combining ratio. Knowledge of this ratio and the concentration of one solution permits calculation of the molarity of the other.

Procedure

Instructions for placing a buret into service are given in Sections 35H-4 and 35H-6; consult these instructions if necessary. Place a test tube or small beaker over the top of the buret that holds the NaOH solution to minimize contact between the solution and the atmosphere.

Record the initial volumes of acid and base in the burets to the nearest 0.01 mL. Deliver 35 to 40 mL of the acid into a 250-mL conical flask. Touch the tip of the buret to the inside wall of the flask, and rinse down with a little distilled water. Add two drops of phenolphthalein (Note 1) and then sufficient base to render the solution a definite pink. Introduce acid dropwise to discharge the color, and again rinse down the walls of the flask. Carefully add base until the solution again acquires a faint pink hue that persists for at least 30 s (Notes 2 and 3). Record the final buret volumes (again, to the nearest 0.01 mL). Repeat the titration. Calculate the acid/base volume ratio. The ratios for duplicate titrations should agree to within 1 to 2 ppt. Perform additional titrations, if necessary, to achieve this order of precision.

Notes

1. The volume ratio can also be determined with an indicator that has an acidic transition range, such as bromocresol green. If the NaOH is contaminated with

carbonate, the ratio obtained with this indicator will differ significantly from the value obtained with phenolphthalein. In general, the acid/base ratio should be evaluated with the indicator that is to be used in subsequent titrations.
2. Fractional drops can be formed on the buret tip, touched to the wall of the flask, and then rinsed down with a small amount of water.
3. The phenolphthalein end point fades as CO_2 is absorbed from the atmosphere.

36B-6 Standardization of Hydrochloric Acid Against Sodium Carbonate

Discussion

See Section 12A-2.

Procedure

Dry a quantity of primary-standard Na_2CO_3 for about 2 hr at 110°C (Figure 35-9), and cool in a desiccator. Weigh individual 0.20- to 0.25-g samples (to the nearest 0.1 mg) into 250-mL conical flasks, and dissolve each in about 50 mL of distilled water. Introduce 3 drops of bromocresol green, and titrate with HCl until the solution just begins to change from blue to green. Boil the solution for 2 to 3 min, cool to room temperature (Note 1), and complete the titration (Note 2).

Determine an indicator correction by titrating approximately 100 mL of 0.05 M NaCl and 3 drops of indicator. Boil briefly, cool, and complete the titration. Subtract any volume needed for the blank from the titration volumes. Calculate the concentration of the HCl solution.

Notes

1. The indicator should change from green to blue as CO_2 is removed during heating. If no color change occurs, an excess of acid was added originally. This excess can be back-titrated with base, provided the acid/base combining ratio is known; otherwise, the sample must be discarded.
2. It is permissible to back-titrate with base to establish the end point with greater certainty.

36B-7 Standardization of Sodium Hydroxide Against Potassium Hydrogen Phthalate

Discussion

See Section 12A-4.

Procedure

Dry a quantity of primary-standard potassium hydrogen phthalate (KHP) for about 2 hr at 110°C (Figure 35-9), and cool in a desiccator. Weigh individual 0.7-

to 0.8-g samples (to the nearest 0.1 mg) into 250-mL conical flasks, and dissolve each in 50 to 75 mL of distilled water. Add 2 drops of phenolphthalein; titrate with base until the pink color of the indicator persists for 30 s (Note). Calculate the concentration of the NaOH solution.

Note

It is permissible to back-titrate with acid to establish the end point more precisely. Record the volume used in the back-titration. Use the acid/base ratio to calculate the net volume of base used in the standardization.

36B-8 The Determination of Potassium Hydrogen Phthalate in an Impure Sample

Discussion

The unknown is a mixture of KHP and a neutral salt. This analysis is conveniently performed concurrently with the standardization of the base.

Procedure

Consult with the instructor concerning an appropriate sample size. Then follow the directions in Section 36B-7.

36B-9 The Determination of the Acid Content of Vinegars and Wines

Discussion

The total acid content of a vinegar or a wine is readily determined by titration with a standard base. It is customary to report the acid content of vinegar in terms of acetic acid, the principal acidic constituent, even though other acids are present. Similarly, the acid content of a wine is expressed as percent tartaric acid, even though there are other acids in the sample. Most vinegars contain about 5% acid (w/v) expressed as acetic acid; wines ordinarily contain somewhat under 1% acid (w/v) expressed as tartaric acid.

Procedure

(a) *If the unknown is a vinegar* (Note 1), pipet 25.00 mL into a 250-mL volumetric flask and dilute to the mark with distilled water. Mix thoroughly, and pipet 50.00-mL aliquots into 250-mL conical flasks. Add about 50 mL of water and 2 drops of phenolphthalein (Note 2) to each, and titrate with standard 0.1 M NaOH to the first permanent ($\approx$ 30 s) pink color.

Report the acidity of the vinegar as percent (w/v) CH_3COOH (60.053 g/mol).

(b) *If the unknown is a wine,* pipet 50.00-mL aliquots into 250-mL conical flasks, add about 50 mL of distilled water and 2 drops of phenolphthalein to each

(Note 2), and titrate to the first permanent (≈ 30 s) pink color.

Express the acidity of the sample as percent (w/v) tartaric acid $C_2H_4O_2(COOH)_2$ (150.09 g/mol). (Note 3)

Notes

1. The acidity of bottled vinegar tends to decrease on exposure to air. It is recommended that unknowns be stored in individual vials with snug covers.
2. The amount of indicator used should be increased as necessary to make the color change visible in colored samples.
3. Tartaric acid has two acidic hydrogens, both of which are titrated at a phenolphthalein end point.

36B-10 The Determination of Sodium Carbonate in an Impure Sample

Discussion

The titration of sodium carbonate is discussed in Section 12A-2 in connection with its use as a primary standard; the same considerations apply for the determination of carbonate in an unknown that has no interfering contaminants.

Procedure

Dry the unknown at 110°C for 2 hr, and then cool in a desiccator. Consult with the instructor on an appropriate sample size. Then follow the instructions in Section 36B-6.

Report the percentage of Na_2CO_3 in the sample.

36B-11 The Determination of Amine Nitrogen by the Kjeldahl Method

Discussion

These directions are suitable for the Kjeldahl determination of protein in materials such as blood meal, wheat flour, pasta products, dry cereals, and pet foods. A simple modification permits the analysis of unknowns that contain more highly oxidized forms of nitrogen.[5]

In the Kjeldahl method (see Section 12B-1), the organic sample is digested in hot concentrated sulfuric acid, which converts amine nitrogen in the sample to ammonium sulfate. After cooling, the sulfuric acid is neutralized by the addition of an excess of concentrated sodium hydroxide. The ammonia liberated by this treatment is then distilled into a measured excess of a standard solution of acid; the excess is determined by back-titration with standard base.

---

[5]See *Official Methods of Analysis,* 14th ed., p. 16. Washington, DC: Association of Official Analytical Chemists, 1984.

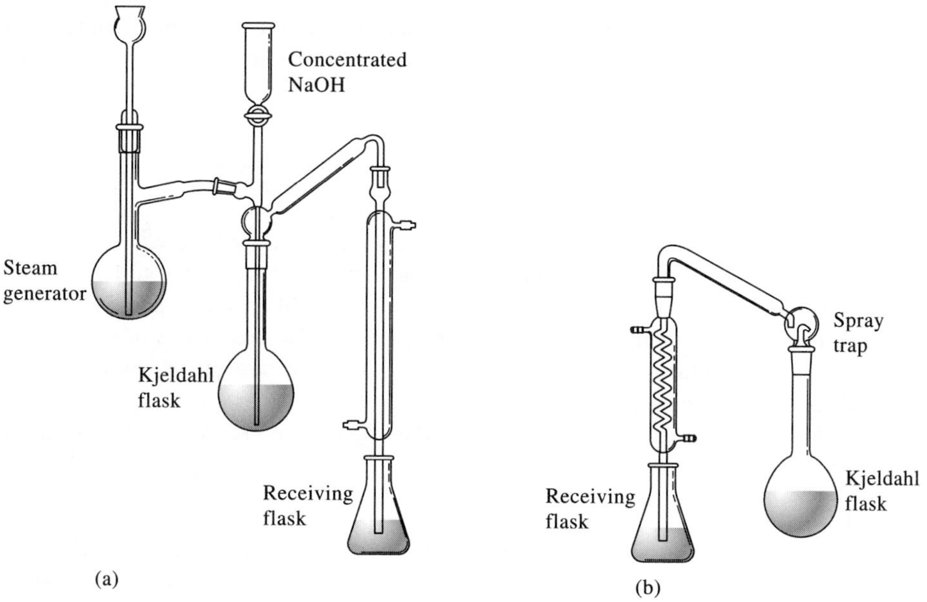

Figure 36-2
Kjeldahl distillation apparatus.

Figure 36-2 illustrates typical equipment for a Kjeldahl distillation. The long-necked container, which is used for both digestion and distillation, is called a *Kjeldahl flask.* In the apparatus in Figure 36-2a, the base is added slowly by partially opening the stopcock from the NaOH storage vessel; the liberated ammonia is then carried to the receiving flask by steam distillation.

In an alternative method (Figure 36-2b), a dense, concentrated sodium hydroxide solution is carefully poured down the side of the Kjeldahl flask to form a second, lower layer. The flask is then quickly connected to a spray trap and an ordinary condenser before loss of ammonia can occur. Only then are the two layers mixed by gentle swirling of the flask.

Quantitative collection of ammonia requires the tip of the condenser to extend into the liquid in the receiving flask throughout the distillation step. The tip must be removed before heating is discontinued, however. Otherwise, the liquid will be drawn back into the apparatus.

Two methods are commonly used for collecting and determining the ammonia liberated from the sample. In one, the ammonia is distilled into a measured volume of standard acid. After the distillation is complete, the excess acid is back-titrated with standard base. An indicator with an acidic transition range (such as bromocresol green) is required because of the acidity of the ammonium ions present at equivalence. A convenient alternative, which requires only one standard solution, involves the collection of the ammonia in an unmeasured excess of boric acid, which retains the ammonia by the reaction

$$H_3BO_3 + NH_3 \longrightarrow NH_4^+ + H_2BO_3^-$$

The dihydrogen borate ion produced is a reasonably strong base that can be titrated with a standard solution of hydrochloric acid:

$$H_2BO_3^- + H_3O^+ \longrightarrow H_3BO_3 + H_2O$$

At the equivalence point, the solution contains boric acid and ammonium ions; an indicator with an acidic transition interval is again required.

Procedure

Preparation of Samples

Consult with the instructor on sample size. *If the unknown is powdered* (such as blood meal), weigh samples onto individual 9-cm filter papers (Note 1). Fold the paper around the sample and drop each into a Kjeldahl flask (the paper keeps the samples from clinging to the neck of the flask). *If the unknown is not powdered* (such as breakfast cereals or pasta), the samples can be weighed by difference directly into the Kjeldahl flasks.

Add 25 mL of concentrated H_2SO_4, 10 g of powdered K_2SO_4, and the catalyst (Note 2) to each flask.

Digestion

Clamp the flasks in a slanted position in a hood or vented digestion rack. Heat carefully to boiling. Discontinue heating briefly if foaming becomes excessive; never allow the foam to reach the neck of the flask. Once foaming ceases and the acid is boiling vigorously, the samples can be left unattended; prepare the distillation apparatus during this time. Continue digestion until the solution becomes colorless or faint yellow; 2 to 3 hr may be needed for some materials. If necessary, *cautiously* replace the acid lost by evaporation.

When digestion is complete, discontinue heating, and allow the flasks to cool to room temperature; swirl the flasks if the contents show signs of solidifying. Cautiously add 250 mL of water to each flask and again allow the solution to cool to room temperature.

Distillation of Ammonia

Arrange a distillation apparatus similar to that shown in Figure 36-2b. Pipet 50.00 mL of standard 0.1 M HCl into the receiver flask (Note 3). Clamp the flask so that the tip of the adapter extends below the surface of the standard acid. Circulate water through the condenser jacket.

Hold the Kjeldahl flask at an angle and gently introduce about 60 mL of 50% (w/v) NaOH solution, taking care to minimize mixing with the solution in the flask. *The concentrated caustic solution is highly corrosive and should be handled with great care* (Note 4). Add several pieces of granulated zinc (Note 5) and a small piece of litmus paper. *Immediately* connect the Kjeldahl flask to the spray trap. Cautiously mix the contents by gentle swirling. The litmus paper should be blue after mixing is complete, indicating that the solution is basic.

Bring the solution to a boil, and distill at a steady rate until one half to one third of the original volume remains. Control the rate of heating to prevent the liquid in the receiver flask from being drawn back into the Kjeldahl flask. After distillation is judged complete, lower the receiver flask to bring the adapter well clear of the liquid. Discontinue heating, disconnect the apparatus, and rinse the inside of the condenser with small portions of distilled water, collecting the washings in the receiver flask. Add 2 drops of bromocresol green to the receiver flask, and titrate the residual HCl with standard 0.1 M NaOH to the color change of the indicator.

Report the percentage of nitrogen and the percentage of protein (Note 6) in the unknown.

Notes

1. If filter paper is used to hold the sample, carry a similar piece through the analysis as a blank. Acid-washed filter paper is frequently contaminated with measurable amounts of ammonium ion and should be avoided if possible.
2. Any of the following catalyze the digestion: a crystal of $CuSO_4$, 0.1 g of selenium, 0.2 g of $CuSeO_3$. The catalyst can be omitted, if desired.
3. A modification of this procedure uses about 50 mL of 4% boric acid solution in lieu of the standard HCl in the receiver flask. After distillation is complete, the ammonium borate produced is titrated with standard 0.1 M HCl, with 2 to 3 drops of bromocresol green as indicator.
4. If any sodium hydroxide solution comes into contact with your skin, wash the affected area *immediately* with copious amounts of water.
5. Granulated zinc (10 to 20 mesh) is added to minimize bumping during the distillation; it reacts slowly with the base to give small bubbles of hydrogen that prevent superheating of the liquid.
6. The percentage of protein in the unknown is calculated by multiplying the % N by an appropriate factor: 5.70 for cereals, 6.25 for meats, and 6.38 for dairy products.

36C PRECIPITATION TITRATIONS

As noted in Chapter 13, most precipitation titrations make use of a standard silver nitrate solution as titrant. Directions follow for the volumetric titration of chloride ion using an adsorption indicator and the gravimetric titration of chloride with a precipitation indicator.

36C-1 Preparation of a Standard Silver Nitrate Solution

Procedure

Use a top-loading balance to transfer the approximate mass of $AgNO_3$ to a weighing bottle (Note 1). Dry at 110°C for about 1 hr but not much longer (Note 2), and then cool to room temperature in a desiccator. Weigh the bottle and contents (to the nearest 0.1 mg). Transfer the bulk of the $AgNO_3$ to a volumetric flask using a powder funnel. Cap the weighing bottle, and reweigh it and any solid that remains. Rinse the powder funnel thoroughly. Dissolve the $AgNO_3$, dilute to the mark with water, and mix well (Note 3). Calculate the molar concentration of this solution.

Notes

1. Consult with the instructor concerning the volume and concentration of $AgNO_3$ to be prepared. The mass of $AgNO_3$ to be taken is as follows:

Silver Ion Concentration, M	Approximate Mass (g) of $AgNO_3$ Needed to Prepare		
	1000 mL	**500 mL**	**250 mL**
0.10	16.9	8.5	4.2
0.05	8.5	4.2	2.1
0.02	3.4	1.8	1.0

2. Prolonged heating causes partial decomposition of $AgNO_3$. Some discoloration may occur, even after only 1 hr at 110°C; the effect of this decomposition on the purity of the reagent is ordinarily imperceptible.
3. Silver nitrate solutions should be stored in a dark place when not in use.

36C-2 The Determination of Chloride by Titration with an Adsorption Indicator

Discussion

In this titration, the anionic adsorption indicator dichlorofluorescein is used to locate the end point. With the first excess of titrant, the indicator becomes incorporated in the counter-ion layer surrounding the silver chloride and imparts color to the solid (Section 13B-1). In order to obtain a satisfactory color change, it is desirable to maintain the particles of silver chloride in the colloidal state. Dextrin is added to the solution to stabilize the colloid and prevent its coagulation.

Preparation of Solutions

Dichlorofluorescein indicator (sufficient for several hundred titrations). Dissolve 0.2 g of dichlorofluorescein in a solution prepared by mixing 75 mL of ethanol and 25 mL of water.

Procedure

Dry the unknown at 110°C for about 1 hr; allow it to return to room temperature in a desiccator. Weigh replicate samples (to the nearest 0.1 mg) into individual conical flasks, and dissolve them in appropriate volumes of distilled water (Note 1). To each, add about 0.1 g of dextrin and 5 drops of indicator. Titrate (Note 2) with $AgNO_3$ to the first permanent pink color of silver dichlorofluoresceinate. Report the percentage of Cl^- in the unknown.

Notes

1. Use 0.25-g samples for 0.1 M $AgNO_3$ and about half that amount for 0.05 M reagent. Dissolve the former in about 200 mL of distilled water and the latter in about 100 mL. If 0.02 M $AgNO_3$ is to be used, weigh a 0.4-g sample into a 500-mL volumetric flask, and take 50-mL aliquots for titration.
2. Colloidal AgCl is sensitive to photodecomposition, particularly in the presence of the indicator; attempts to perform the titration in direct sunlight will fail. If photodecomposition appears to be a problem, establish the approximate end point with a rough preliminary titration, and use this information to estimate the volumes of $AgNO_3$ needed for the other samples. For each subsequent sample, add the indicator and dextrin only after most of the $AgNO_3$ has been added, and then complete the titration without delay.

36C-3 The Determination of Chloride by a Weight Titration

Discussion

The Mohr method uses CrO_4^{2-} ion as an indicator in the titration of chloride ion with silver nitrate. The first excess of titrant results in the formation of a red silver chromate precipitate, which signals the end point.

Instead of a buret, a balance is employed in this procedure to determine the weight of silver nitrate solution needed to reach the end point. The concentration of the silver nitrate is most conveniently determined by standardization against primary-standard sodium chloride, although direct preparation by weight is also feasible. The reagent concentration is expressed as weight molarity (mmol $AgNO_3$/g of solution). See Section 6D-1 for additional details.

Preparation of Solutions

(a) *Silver nitrate, approximately 0.1 mmol/g of solution* (sufficient for about ten titrations). Dissolve about 4.5 g of $AgNO_3$ in about 500 mL of distilled water. Standardize the solution against weighed quantities of reagent-grade NaCl as directed in Note 1 of the procedure. Express the concentration as weight molarity (mmol $AgNO_3$/g of solution). When it is not in use, store the solution in a dark place.

(b) *Potassium chromate, 5%* (sufficient for about ten titrations). Dissolve about 1.0 g of K_2CrO_4 in about 20 mL of distilled water.

Note

Alternatively, standard $AgNO_3$ can be prepared directly by weight. To do so, follow the directions in Section 36C-1 for weighing out a known amount of primary-standard $AgNO_3$. Use a powder funnel to transfer the weighed $AgNO_3$ to a 500-mL polyethylene bottle that has been previously weighed to the nearest 10 mg. Add about 500 mL of water and weigh again. Calculate the weight molarity.

Procedure

Dry the unknown at 110°C for at least 1 hr (Note 1). Cool in a desiccator. Consult with your instructor for a suitable sample size. Weigh (to the nearest 0.1 mg) individual samples into 250-mL conical flasks, and dissolve in about 100 mL of distilled water. Add small quantities of $NaHCO_3$ until effervescence ceases. Introduce about 2 mL of K_2CrO_4 solution, and titrate (Note 2) to the first permanent appearance of red Ag_2CrO_4.

Determine an indicator blank by suspending a small amount of chloride-free $CaCO_3$ in 100 mL of distilled water containing 2 mL of K_2CrO_4.

Correct reagent weights for the blank. Report the percentage of Cl^- in the unknown.

Dispose of AgCl and reagents as directed by the instructor.

Notes

1. The $AgNO_3$ is conveniently standardized concurrently with the analysis. Dry reagent-grade NaCl for about 1 hr. Cool; then weigh (to the nearest 0.1 mg) 0.25-g portions into conical flasks and titrate as above.

2. Directions for performing a weight titration are given in Section 35F-2.

36D COMPLEX-FORMATION TITRATIONS WITH EDTA

See Chapter 14 for a discussion of the analytical uses of EDTA as a chelating reagent. Directions follow for a direct titration of magnesium and a determination of the hardness of a natural water.

36D-1 Preparation of Solutions

Procedure

A pH-10 buffer and an indicator solution are needed for these titrations.

(a) *Buffer solution, pH 10* (sufficient for 80 to 100 titrations). Dilute 57 mL of concentrated NH_3 and 7 g of NH_4Cl in sufficient distilled water to give 100 mL of solution.

(b) *Eriochrome Black T indicator* (sufficient for about 100 titrations). Dissolve 100 mg of the solid in a solution containing 15 mL of ethanolamine and 5 mL of absolute ethanol. This solution should be freshly prepared every two weeks; refrigeration slows its deterioration.

36D-2 Preparation of Standard 0.01 M EDTA Solution

Discussion

See Section 14B-1 for a description of the properties of reagent-grade $Na_2H_2Y \cdot 2H_2O$ and its use in the direct preparation of standard EDTA solutions.

Procedure

Dry about 4 g of the purified dihydrate $Na_2H_2Y \cdot 2H_2O$ (Note 1) for 1 hr at 80°C to remove superficial moisture. Cool to room temperature in a desiccator. Weigh (to the nearest milligram) about 3.8 g into a 1-L volumetric flask (Note 2). Use a powder funnel to ensure quantitative transfer; rinse the funnel well with water before removing it from the flask. Add 600 to 800 mL of water (Note 3) and swirl periodically. Dissolution may take 15 min or longer. When all the solid has dissolved, dilute to the mark with water and mix well (Note 4). In calculating the molarity of the solution, correct the weight of the salt for the 0.3% moisture it ordinarily retains after drying at 80°C.

Notes

1. Directions for the purification of the disodium salt are described by W. J. Blaedel and H. T. Knight, *Anal. Chem.,* **1954,** *26* (4), 741.

2. The solution can be prepared from the anhydrous disodium salt, if desired. The weight taken should be about 3.6 g.

3. Water used in the preparation of standard EDTA solutions must be totally free of polyvalent cations. If any doubt exists concerning its quality, pass the water through a cation-exchange resin before use.
4. As an alternative, an EDTA solution that is approximately 0.01 M can be prepared and standardized by direct titration against a Mg^{2+} solution of known concentration (using the directions in Section 36D-3).

36D-3 The Determination of Magnesium by Direct Titration

Discussion

See Section 14B-7.

Procedure

Submit a clean 500-mL volumetric flask to receive the unknown, dilute to the mark with water, and mix thoroughly. Transfer 50.00-mL aliquots to 250-mL conical flasks, add 1 to 2 mL of pH-10 buffer and 3 to 4 drops of Eriochrome Black T indicator to each. Titrate with 0.01 M EDTA until the color changes from red to pure blue (Notes 1 and 2).

Express the results as parts per million of Mg^{2+} in the sample.

Notes

1. The color change tends to be slow in the vicinity of the end point. Care must be taken to avoid overtitration.
2. Other alkaline earths, if present, are titrated along with the Mg^{2+}; removal of Ca^{2+} and Ba^{2+} can be accomplished with $(NH_4)_2CO_3$. Most polyvalent cations are also titrated. Precipitation as hydroxides or the use of a masking reagent may be needed to eliminate this source of interference.

36D-4 The Determination of Calcium by Displacement Titration

Discussion

A solution of the magnesium/EDTA complex is useful for the titration of cations that form more stable complexes than the magnesium complex but for which no indicator is available. Magnesium ions in the complex are displaced by a chemically equivalent quantity of analyte cations. The remaining uncomplexed analyte and the liberated magnesium ions are then titrated with Eriochrome Black T as indicator. Note that the concentration of the magnesium solution is not important; all that is necessary is that the molar ratio between Mg^{2+} and EDTA in the reagent be exactly unity.

Procedure

Preparation of the Magnesium/EDTA Complex, 0.1 M (sufficient for 90 to 100 titrations)

To 3.72 g of $Na_2H_2Y_2 \cdot 2H_2O$ in 50 mL of distilled water, add an equivalent quantity (2.46 g) of $MgSO_4 \cdot 7H_2O$. Add a few drops of phenolphthalein, followed by sufficient 0.1 M NaOH to turn the solution faintly pink. Dilute to about 100 mL with water. The addition of a few drops of Eriochrome Black T to a portion of this solution buffered to pH 10 should cause development of a dull violet color. Moreover, a single drop of 0.01 M Na_2H_2Y solution added to the violet solution should cause a color change to blue, and an equal quantity of 0.01 M Mg^{2+} should cause a change to red. The composition of the original solution should be adjusted with additional Mg^{2+} or H_2Y^{2-} until these criteria are met.

Titration

Weigh a sample of the unknown (to the nearest 0.1 mg) into a 500-mL beaker (Note 1). Cover with a watch glass, and carefully add 5 to 10 mL of 6 M HCl. After the sample has dissolved, remove CO_2 by adding about 50 mL of deionized water and boiling gently for a few minutes. Cool, add a drop or two of methyl red, and neutralize with 6 M NaOH until the red color is discharged. Quantitatively transfer the solution to a 500-mL volumetric flask, and dilute to the mark. Take 50.00-mL aliquots of the diluted solution for titration, treating each as follows: Add about 2 mL of pH-10 buffer, 1 mL of Mg/EDTA solution, and 3 to 4 drops of Eriochrome Black T or Calmagite indicator. Titrate (Note 2) with standard 0.01 M Na_2H_2Y to a color change from red to blue.

Report the number of milligrams of CaO in the sample.

Notes

1. The sample taken should contain 150 to 160 mg of Ca^{2+}.
2. Interferences with this titration are substantially the same as those encountered in the direct titration of Mg^{2+} and are eliminated in the same way.

36D-5 The Determination of Hardness in Water

Discussion

See Section 14B-9.

Procedure

Acidify 100.0-mL aliquots of the sample with a few drops of HCl, and boil gently for a few minutes to eliminate CO_2. Cool, add 3 to 4 drops of methyl red, and neutralize with 0.1 M NaOH. Introduce 2 mL of pH-10 buffer, 3 to 4 drops of Eriochrome Black T, and titrate with standard 0.01 M Na_2H_2Y to a color change from red to pure blue (Note).

Report the results in terms of milligrams of $CaCO_3$ per liter of water.

Note

The color change is sluggish if Mg^{2+} is absent. In this event, add 1 to 2 mL of 0.1 M MgY^{2-} before starting the titration (see Section 36D-4 for preparation of this solution).

36E TITRATIONS WITH POTASSIUM PERMANGANATE

The properties and uses of potassium permanganate are described in Section 17C-1. Directions follow for the determination of iron in an ore and calcium in a limestone.

36E-1 Preparation of 0.02 M Potassium Permanganate

Discussion

See page 367 for a discussion of the precautions needed in the preparation and storage of permanganate solutions.

Procedure

Dissolve about 3.2 g of $KMnO_4$ in 1 L of distilled water. Keep the solution at a gentle boil for about 1 hr. Cover and let stand overnight. Remove MnO_2 by filtration (Note 1) through a fine-porosity filtering crucible (Note 2) or through a Gooch crucible fitted with glass mats. Transfer the solution to a clean glass-stoppered bottle; store in the dark when not in use.

Notes

1. The heating and filtering can be omitted if the permanganate solution is standardized and used on the same day.
2. Remove the MnO_2 that collects on the fritted plate with 1 M H_2SO_4 containing a few milliliters of 3% H_2O_2, followed by a rinse with copious quantities of water.

36E-2 Standardization of Potassium Permanganate Solutions

Discussion

See Section 17C-1 for a discussion of primary standards for permanganate solutions. Directions follow for standardization with sodium oxalate.

Procedure

Dry about 1.5 g of primary-standard $Na_2C_2O_4$ at 110°C for at least 1 hr. Cool in a desiccator; weigh (to the nearest 0.1 mg) individual 0.2- to 0.3-g samples into 400-mL beakers. Dissolve each in about 250 mL of 1 M H_2SO_4. Heat each solution to 80°C to 90°C, and titrate with $KMnO_4$ while stirring with a thermometer. The pink color imparted by one addition should be permitted to disappear before any further titrant is introduced (Notes 1 and 2). Reheat if the temperature drops below 60°C. Take the first persistent ($\approx$ 30 s) pink color as the end point (Notes 3 and 4). Determine a blank by titrating an equal volume of the 1 M H_2SO_4.

Correct the titration data for the blank, and calculate the concentration of the permanganate solution (Note 5).

Notes

1. Promptly wash any $KMnO_4$ that spatters on the walls of the beaker into the bulk of the liquid with a stream of water.
2. Finely divided MnO_2 will form along with Mn^{2+} if the $KMnO_4$ is added too rapidly and will cause the solution to acquire a faint brown discoloration. Precipitate formation is not a serious problem as long as sufficient oxalate remains to reduce the MnO_2 to Mn^{2+}; the titration is simply discontinued until the brown color disappears. The solution must be free of MnO_2 at the end point.
3. The surface of the permanganate solution rather than the bottom of the meniscus can be used to measure titrant volumes. Alternatively, backlighting with a flashlight or a match will permit reading of the meniscus in the conventional manner.
4. A permanganate solution should not be allowed to stand in a buret any longer than necessary because partial decomposition to MnO_2 may occur. Freshly formed MnO_2 can be removed from a glass surface with 1 M H_2SO_4 containing a small amount of 3% H_2O_2.
5. As noted on page 369, this procedure yields molarities that are a few tenths of a percent low. For more accurate results, introduce from a buret sufficient permanganate to react with 90% to 95% of the oxalate (about 40 mL of 0.02 M $KMnO_4$ for a 0.3-g sample). Let the solution stand until the permanganate color disappears. Then warm to about 60°C and complete the titration, taking the first permanent pink ($\approx$ 30 s) as the end point (Notes 3 and 4). Determine a blank by titrating an equal volume of the 1 M H_2SO_4.

36E-3 The Determination of Calcium in a Limestone

Discussion

In common with a number of other cations, calcium is conveniently determined by precipitation with oxalate ion. The solid calcium oxalate is filtered, washed free of excess precipitating reagent, and dissolved in dilute acid. The oxalic acid liberated in this step is then titrated with standard permanganate or some other oxidizing reagent. This method is applicable to samples that contain magnesium and the alkali metals. Most other cations must be absent since they either precipitate or coprecipitate as oxalates and cause positive errors in the analysis.

Factors Affecting the Composition of Calcium Oxalate Precipitates It is essential that the mole ratio between calcium and oxalate be exactly unity in the precipitate and thus in solution at the time of titration. A number of precautions are needed to ensure this condition. For example, the calcium oxalate formed in a neutral or ammoniacal solution is likely to be contaminated with calcium hydroxide or a basic calcium oxalate, either of which will cause low results. The formation of these compounds is prevented by adding the oxalate to an acidic solution of the sample and slowly forming the desired precipitate by the dropwise addition of ammonia. The coarsely crystalline calcium oxalate that is produced under these conditions is readily filtered. Losses resulting from the solubility of calcium oxalate are negligible above pH 4, provided washing is limited to freeing the precipitate of excess oxalate.

Coprecipitation of sodium oxalate becomes a source of positive error in the determination of calcium whenever the concentration of sodium in the sample exceeds that of calcium. The error from this source can be eliminated by reprecipitation.

Reprecipitation is a method for minimizing coprecipitation errors by dissolving the initial precipitate and then reforming the solid.

Magnesium, if present in high concentration, may also be a source of contamination. An excess of oxalate ion helps prevent this interference through the formation of soluble oxalate complexes of magnesium. Prompt filtration of the calcium oxalate can also help prevent interference because of the pronounced tendency of magnesium oxalate to form supersaturated solutions from which precipitate formation occurs only after an hour or more. These measures do not suffice for samples that contain more magnesium than calcium. Here, reprecipitation of the calcium oxalate becomes necessary.

The Composition of Limestones Limestones are composed principally of calcium carbonate; dolomitic limestones contain large amounts of magnesium carbonate as well. Calcium and magnesium silicates are also present in smaller amounts, along with the carbonates and silicates of iron, aluminum, manganese, titanium, sodium, and other metals.

Hydrochloric acid is an effective solvent for most limestones. Only silica, which does not interfere with the analysis, remains undissolved. Some limestones are more readily decomposed after they have been ignited; a few yield only to a carbonate fusion.

The method that follows is remarkably effective for determining calcium in most limestones. Iron and aluminum, in amounts equivalent to that of calcium, do not interfere. Small amounts of manganese and titanium can also be tolerated.

Procedure

Sample Preparation

Dry the unknown for 1 to 2 hr at 110°C, and cool in a desiccator. If the material is readily decomposed in acid, weigh 0.25- to 0.30-g samples (to the nearest 0.1 mg) into 250-mL beakers. Add 10 mL of water to each sample and cover with a watch glass. Add 10 mL of concentrated HCl dropwise, taking care to avoid losses due to spattering as the acid is introduced.

Precipitation of Calcium Oxalate

Add 5 drops of saturated bromine water to oxidize any iron in the samples and boil gently *(hood)* for 5 min to remove the excess Br_2. Dilute each sample solution to about 50 mL, heat to boiling, and add 100 mL of hot 6% (w/v) $(NH_4)_2C_2O_4$ solution. Add 3 or 4 drops of methyl red, and precipitate CaC_2O_4 by slowly adding 6 M NH_3. As the indicator starts to change color, add the NH_3 at a rate of one drop every 3 to 4 s. Continue until the solutions turn to the intermediate yellow-orange color of the indicator (pH 4.5 to 5.5). Allow the solutions to stand for no more than 30 min (Note) and filter; medium-porosity filtering crucibles or Gooch crucibles with glass mats are satisfactory. Wash the precipitates with several 10-mL portions of cold water. Rinse the outside of the crucibles to remove residual $(NH_4)_2C_2O_4$, and return them to the beakers in which the CaC_2O_4 was formed.

Titration

Add 100 mL of water and 50 mL of 3 M H_2SO_4 to each of the beakers containing the precipitated calcium oxalate and the crucible. Heat to 80°C to 90°C, and titrate with 0.02 M permanganate. The temperature should be above 60°C throughout the titration; reheat if necessary.

Report the percentage of CaO in the unknown.

Note

The period of standing can be longer if the unknown contains no Mg^{2+}.

36E-4 The Determination of Iron in an Ore

Discussion

The common ores of iron are hematite (Fe_2O_3), magnetite (Fe_3O_4), and limonite ($2Fe_2O_3 \cdot 3H_2O$). Steps in the analysis of these ores are (1) dissolution of the sample, (2) reduction of iron to the divalent state, and (3) titration of iron(II) with a standard oxidant.

The Decomposition of Iron Ores Iron ores often decompose completely in hot concentrated hydrochloric acid. The rate of attack by this reagent is increased by the presence of a small amount of tin(II) chloride. The tendency of iron(II) and iron(III) to form chloro complexes accounts for the effectiveness of hydrochloric acid over nitric or sulfuric acid as a solvent for iron ores.

Many iron ores contain silicates that may not be entirely decomposed by treatment with hydrochloric acid. Incomplete decomposition is indicated by a dark residue that remains after prolonged treatment with the acid. A white residue of hydrated silica, which does not interfere in any way, is indicative of complete decomposition.

The Prereduction of Iron Because part or all of the iron is in the trivalent state after decomposition of the sample, prereduction to iron(II) must precede titration with the oxidant. Any of the methods described in Section 17A-1 can be used. Perhaps the most satisfactory prereductant for iron is tin(II) chloride:

$$2Fe^{3+} + Sn^{2+} \longrightarrow 2Fe^{2+} + Sn^{4+}$$

The only other common species reduced by this reagent are the high oxidation states of arsenic, copper, mercury, molybdenum, tungsten, and vanadium.

The excess reducing agent is eliminated by the addition of mercury(II) chloride:

$$Sn^{2+} + 2HgCl_2 \longrightarrow Hg_2Cl_2(s) + Sn^{4+} + 2Cl^-$$

The slightly soluble mercury(I) chloride (Hg_2Cl_2) does not reduce permanganate, nor does the excess mercury(II) chloride ($HgCl_2$) reoxidize iron(II). Care must be taken, however, to prevent the occurrence of the alternative reaction

$$Sn^{2+} + HgCl_2 \longrightarrow Hg(l) + Sn^{4+} + 2Cl^-$$

Elemental mercury reacts with permanganate and causes the results of the analysis to be high. The formation of mercury, which is favored by an appreciable excess of tin(II), is prevented by careful control of this excess and by the rapid addition of excess mercury(II) chloride. A proper reduction is indicated by the appearance of a small amount of a silky white precipitate after the addition of mercury(II). Formation of a gray precipitate at this juncture indicates the presence of metallic mercury; the total absence of a precipitate indicates that an insufficient amount of tin(II) chloride was used. In either event, the sample must be discarded.

The Titration of Iron(II) The reaction of iron(II) with permanganate is smooth and rapid. The presence of iron(II) in the reaction mixture, however, *induces* the oxidation of chloride ion by permanganate, a reaction that does not ordinarily proceed rapidly enough to cause serious error. High results are obtained if this parasitic reaction is not controlled. Its effects can be eliminated through removal of the hydrochloric acid by evaporation with sulfuric acid or by introduction of *Zimmermann-Reinhardt reagent,* which contains manganese(II) in a fairly concentrated mixture of sulfuric and phosphoric acids.

The oxidation of chloride ion during a titration is believed to involve a direct reaction between this species and the manganese(III) ions that form as an intermediate in the reduction of permanganate ion by iron(II). The presence of manganese(II) in the Zimmermann-Reinhardt reagent is believed to inhibit the formation of chlorine by decreasing the potential of the manganese(III)/manganese(II) couple. Phosphate ion is believed to exert a similar effect by forming stable manganese(III) complexes. Moreover, phosphate ions react with iron(III) to form nearly colorless complexes so that the yellow color of the iron(II)/chloro complexes does not interfere with the end point.[6]

Preparation of Reagents

The following solutions suffice for about 100 titrations.

(a) *Tin(II) chloride, 0.25 M.* Dissolve 60 g of iron-free $SnCl_2 \cdot 2H_2O$ in 100 mL of concentrated HCl; warm if necessary. After the solid has dissolved, dilute to 1 L with distilled water and store in a well-stoppered bottle. Add a few pieces of mossy tin to help preserve the solution.

(b) *Mercury(II) chloride, 5% (w/v).* Dissolve 50 g of $HgCl_2$ in 1 L of distilled water.

(c) *Zimmermann-Reinhardt reagent.* Dissolve 300 g of $MnSO_4 \cdot 4H_2O$ in 1 L of water. Cautiously add 400 mL of concentrated H_2SO_4, 400 mL of 85% H_3PO_4, and dilute to 3 L.

Procedure

Sample Preparation

Dry the ore at 110°C for at least 3 hr, and then allow it to cool to room temperature in a desiccator. Consult with the instructor for a sample size that will require from 25 to 40 mL of standard 0.02 M $KMnO_4$. Weigh samples into 500-mL conical flasks. To each, add 10 mL of concentrated HCl and about 3 mL of 0.25 M $SnCl_2$

[6]The mechanism by which Zimmermann-Reinhardt reagent acts has been the subject of much study. For a discussion of this work, see H. A. Laitinen, *Chemical Analysis,* pp. 369–372. New York: McGraw-Hill, 1960.

(Note 1). Cover each flask with a small watch glass or Tuttle flask cover. Heat the flasks in a hood at just below boiling until the samples are decomposed and the undissolved solid—if any—is pure white (Note 2). Use another 1 to 2 mL of $SnCl_2$ to eliminate any yellow color that may develop as the solutions are heated. Heat a blank consisting of 10 mL of HCl and 3 mL of $SnCl_2$ for the same amount of time.

After the ore has been decomposed, remove the excess Sn(II) by the dropwise addition of 0.02 M $KMnO_4$ until the solutions become faintly yellow. Dilute to about 15 mL. Add sufficient $KMnO_4$ solution to impart a faint pink color to the blank; then decolorize with one drop of the $SnCl_2$ solution.

Take samples and blank individually through subsequent steps to minimize air-oxidation of iron(II).

Reduction of Iron

Heat the sample solution nearly to boiling, and make dropwise additions of 0.25 M $SnCl_2$ until the yellow color just disappears; then add two more drops (Note 3). Cool to room temperature, and *rapidly* add 10 mL of 5% $HgCl_2$ solution. A small amount of silky white Hg_2Cl_2 should precipitate (Note 4).

Titration

Following addition of the $HgCl_2$, wait 2 to 3 min. Then add 25 mL of Zimmermann-Reinhardt reagent and 300 mL of water. Titrate *immediately* with standard 0.02 M $KMnO_4$ to the first faint pink that persists for 15 to 20 s. Do not add the $KMnO_4$ rapidly at any time. Correct the titrant volume for the blank.

Report the percentage of Fe_2O_3 in the sample.

Notes

1. The $SnCl_2$ hastens decomposition of the ore by reducing iron(III) oxides to iron(II). Insufficient $SnCl_2$ is indicated by the appearance of yellow iron(III)/chloride complexes.
2. If dark particles persist after the sample has been heated with acid for several hours, filter the solution through ashless paper, wash the residue with 5 to 10 mL of 6 M HCl, and retain the filtrate and washings. Ignite the paper and its contents in a small platinum crucible. Mix 0.5 to 0.7 g of Na_2CO_3 with the residue and heat until a clear melt is obtained. Cool, add 5 mL of water, and then cautiously add a few milliliters of 6 M HCl. Warm the crucible until the melt has dissolved, and combine the contents with the original filtrate. Evaporate the solution to 15 mL and continue the analysis.
3. The solution may not become entirely colorless but instead may acquire a faint yellow-green hue. Further additions of $SnCl_2$ will not alter this color. If too much $SnCl_2$ is added, it can be removed by adding 0.02 M $KMnO_4$ and repeating the reduction.
4. The absence of precipitate indicates that insufficient $SnCl_2$ was used and that the reduction of iron(III) was incomplete. A gray residue indicates the presence of elementary mercury, which reacts with $KMnO_4$. The sample must be discarded in either event.
5. These directions can be used to standardize a permanganate solution against primary-standard iron. Weigh (to the nearest 0.1 mg) 0.2-g lengths of electrolytic iron wire into 250-mL conical flasks and dissolve in about 10 mL of concentrated HCl. Dilute each sample to about 75 mL. Then take each individually through the reduction and titration steps.

36F TITRATIONS WITH IODINE

The oxidizing properties of iodine, the composition and stability of triiodide solutions, and the applications of this reagent in volumetric analysis are discussed in Section 17C-3. Starch is ordinarily employed as an indicator for iodimetric titrations.

36F-1 Preparation of Reagents

Procedure

(a) *Iodine approximately 0.05 M.* Weigh about 40 g of KI into a 100-mL beaker. Add 12.7 g of I_2 and 10 mL of water. Stir for several minutes (Note 1). Introduce an additional 20 mL of water, and stir again for several minutes. Carefully decant the bulk of the liquid into a storage bottle containing 1 L of distilled water. It is essential that any undissolved iodine remain in the beaker (Note 2).

Notes

1. Iodine dissolves slowly in the KI solution. Thorough stirring is needed to hasten the process.
2. Any solid I_2 inadvertently transferred to the storage bottle will cause the concentration of the solution to increase gradually. Filtration through a sintered-glass crucible eliminates this potential source of difficulty.

(b) *Starch indicator* (sufficient for about 100 titrations). Rub 1 g of soluble starch and 15 mL of water into a paste. Dilute to about 500 mL with boiling water, and heat until the mixture is clear. Cool; store in a tightly stoppered bottle. For most titrations, 3 to 5 mL of the indicator are used.

The indicator is readily attacked by airborne organisms and should be freshly prepared every few days.

36F-2 Standardization of Iodine Solutions

Discussion

Arsenic(III) oxide, long a favored primary standard for iodine solutions, is now seldom used because of the elaborate federal regulations governing the use of even small amounts of arsenic containing compounds. Barium thiosulfate monohydrate and anhydrous sodium thiosulfate have been proposed as alternative standards.[7,8] Perhaps the most convenient method for determining the concentration of an iodine solution is the titration of aliquots with a sodium thiosulfate solution that has been standardized against pure potassium iodate. Instructions for this method follow.

[7]W. M. McNevin and O. H. Kriege, *Anal. Chem.*, **1953,** *25* (5), 767.

[8]A. A. Woolf, *Anal. Chem.*, **1982,** *54* (12), 2134.

Preparation of Reagents

(a) *Sodium thiosulfate, 0.1 M.* Follow the directions in Sections 36G-1 and 36G-2 for the preparation and standardization of this solution.
(b) *Starch indicator.* See Section 36F-1(b).

Procedure

Transfer 25.00-mL aliquots of the iodine solution to 250-mL conical flasks, and dilute to about 50 mL. *Take each aliquot individually through subsequent steps.* Introduce approximately 1 mL of 3 M H_2SO_4, and titrate immediately with standard sodium thiosulfate until the solution becomes a faint straw yellow. Add about 5 mL of starch indicator, and complete the titration, taking as the end point the change in color from blue to colorless (Note).

Note
The blue color of the starch/iodine complex may reappear after the titration has been completed, owing to the air-oxidation of iodide ion.

36F-3 The Determination of Antimony in Stibnite

Discussion

The analysis of stibnite, a common antimony ore, is a typical application of iodimetry and is based upon the oxidation of Sb(III) to Sb(V):

$$SbO_3^{3-} + I_2 + H_2O \rightleftharpoons SbO_4^{3-} + 2I^- + 2H^+$$

The position of this equilibrium is strongly dependent upon the hydrogen ion concentration. In order to force the reaction to the right, it is common practice to carry out the titration in the presence of an excess of sodium hydrogen carbonate, which consumes the hydrogen ions as they are produced.

Stibnite is an antimony sulfide ore containing silica and other contaminants. Provided the material is free of iron and arsenic, the analysis of stibnite for its antimony content is straightforward. Samples are decomposed in hot concentrated hydrochloric acid to eliminate sulfide as gaseous hydrogen sulfide. Care is needed to prevent loss of volatile antimony(III) chloride during this step. The addition of potassium chloride helps by favoring formation of nonvolatile chloro complexes such as $SbCl_4^-$ and $SbCl_6^{3-}$.

Sparingly soluble basic antimony salts, such as SbOCl, often form when the excess hydrochloric acid is neutralized; these react incompletely with iodine and cause low results. The difficulty is overcome by adding tartaric acid, which forms a soluble complex ($SbOC_4H_4O_6^-$) from which antimony is rapidly oxidized by the reagent.

Procedure

Dry the unknown at 110°C for 1 hr, and allow it to cool in a desiccator. Weigh individual samples (Note 1) into 500-mL conical flasks. Introduce about 0.3 g of

KCl and 10 mL of concentrated HCl to each flask. Heat the mixtures *(hood)* just below boiling until only white or slightly gray residues of SiO_2 remain.

Add 3 g of tartaric acid to each sample and heat for an additional 10 to 15 min. Then, with good swirling, add water (Note 2) from a pipet or buret until the volume is about 100 mL. If reddish Sb_2S_3 forms, discontinue dilution and heat further to eliminate H_2S; add more HCl if necessary.

Add 3 drops of phenolphthalein, and neutralize with 6 M NaOH to the first faint pink of the indicator. Discharge the color by the dropwise addition of 6 M HCl, and then add 1 mL in excess. Introduce 4 to 5 g of $NaHCO_3$, taking care to avoid losses of solution by spattering during the addition. Add 5 mL of starch indicator, rinse down the inside of the flask, and titrate with standard 0.05 M I_2 to the first blue color that persists for 30 s.

Report the percentage of Sb_2S_3 in the unknown.

Notes

1. Samples should contain between 1.5 and 2 mmol of antimony; consult with the instructor for an appropriate sample size. Weighings to the nearest milligram are adequate for samples larger than 1 g.
2. The slow addition of water, with efficient stirring, is essential to prevent the formation of SbOCl.

36G TITRATIONS WITH SODIUM THIOSULFATE

Numerous methods are based upon the reducing properties of iodide ion:

$$2I^- \longrightarrow I_2 + 2e^-$$

Iodine, the reaction product, is ordinarily titrated with a standard sodium thiosulfate solution, with starch serving as the indicator:

$$I_2 + 2S_2O_3^{2-} \longrightarrow 2I^- + S_4O_6^{2-}$$

A discussion of thiosulfate methods is found in Section 17B-2.

36G-1 Preparation of 0.1 M Sodium Thiosulfate

Procedure

Boil about 1 L of distilled water for 10 to 15 min. Allow the water to cool to room temperature; then add about 25 g of $Na_2S_2O_3 \cdot 5H_2O$ and 0.1 g of Na_2CO_3. Stir until the solid has dissolved. Transfer the solution to a clean glass or plastic bottle, and store in a dark place.

36G-2 Standardization of Sodium Thiosulfate Against Potassium Iodate

Discussion

Solutions of sodium thiosulfate are conveniently standardized by titration of the iodine produced when an unmeasured excess of potassium iodide is added to a known volume of an acidified standard potassium iodate solution. The reaction is

$$IO_3^- + 5I^- + 6H^+ \longrightarrow 3I_2 + 3H_2O$$

Note that each mole of iodate results in the production of three moles of iodine. The procedure that follows is based upon this reaction.

Preparation of Solutions

(a) *Potassium iodate, 0.0100 M.* Dry about 1.2 g of primary-standard KIO_3 at 110°C for at least 1 hr and cool in a desiccator. Weigh (to the nearest 0.1 mg) about 1.1 g into a 500-mL volumetric flask; use a powder funnel to ensure quantitative transfer of the solid. Rinse the funnel well, dissolve the KIO_3 in about 200 mL of distilled water, dilute to the mark, and mix thoroughly.
(b) *Starch indicator.* See Section 36F-1(b).

Procedure

Pipet 50.00-mL aliquots of standard iodate solution into 250-mL conical flasks. *Treat each sample individually from this point to minimize error resulting from the air-oxidation of iodide ion.* Introduce 2 g of iodate-free KI, and swirl the flask to hasten solution. Add 2 mL of 6 M HCl, and immediately titrate with thiosulfate until the solution becomes pale yellow. Introduce 5 mL of starch indicator, and titrate with constant stirring to the disappearance of the blue color. Calculate the molarity of the iodine solution.

36G-3 Standardization of Sodium Thiosulfate Against Copper

Discussion

Thiosulfate solutions can also be standardized against pure copper wire or foil. This procedure is advantageous when the solution is to be used for the determination of copper because any systematic error in the method tends to be canceled.

Copper(II) is reduced quantitatively to copper(I) by iodide ion:

$$2Cu^{2+} + 4I^- \longrightarrow 2CuI(s) + I_2$$

The importance of CuI formation in forcing this reaction to completion can be seen from the following standard electrode potentials:

$$Cu^{2+} + e^- \rightleftharpoons Cu^+ \qquad E^0 = 0.15V$$

$$I_2 + 2e^- \rightleftharpoons 2I^- \qquad E^0 = 0.54V$$

$$Cu^{2+} + I^- + e^- \rightleftharpoons CuI(s) \qquad E^0 = 0.86V$$

The first two potentials suggest that iodide should have no tendency to reduce copper(II); the formation of CuI, however, favors the reduction. The solution must contain at least 4% excess iodide to force the reaction to completion. Moreover, the pH must be less than 4 to prevent the formation of basic copper species that react slowly and incompletely with iodide ion. The acidity of the solution cannot be greater than about 0.3 M, however, because of the tendency of iodide ion to undergo air-oxidation, a process catalyzed by copper salts. Nitrogen oxides also catalyze the air-oxidation of iodide ion. A common source of these oxides is the nitric acid ordinarily used to dissolve metallic copper and other copper-containing solids. Urea is used to scavenge nitrogen oxides from these solutions:

$$(NH_2)_2CO + 2HNO_2 \longrightarrow 2N_2(g) + CO_2(g) + 3H_2O$$

The titration of iodine by thiosulfate tends to yield slightly low results owing to the adsorption of small but measurable quantities of iodine upon solid CuI. The adsorbed iodine is released only slowly, even when thiosulfide is in excess; transient and premature end points result. This difficulty is largely overcome by the addition of thiocyanate ion. The sparingly soluble copper(I) thiocyanate replaces part of the copper iodide at the surface of the solid:

$$CuI(s) + SCN^- \longrightarrow CuSCN(s) + I^-$$

Accompanying this reaction is the release of the adsorbed iodine, which thus becomes available for titration. The addition of thiocyanate must be delayed until most of the iodine has been titrated to prevent interference from a slow reaction between the two species, possibly

$$2SCN^- + I_2 \longrightarrow 2I^- + (SCN)_2$$

Preparation of Solutions

(a) *Urea, 5% (w/v).* Dissolve about 5 g of urea in sufficient water to give 100 mL of solution. Approximately 10 mL will be needed for each titration.
(b) *Starch indicator.* See Section 36F-1(b).

Procedure

Use scissors to cut copper wire or foil into 0.20- to 0.25-g portions. Wipe the metal free of dust and grease with a filter paper; do not dry it. The pieces of copper should be handled with paper strips, cotton gloves, or tweezers to prevent contamination by contact with the skin.

Use a weighed watch glass or weighing bottle to obtain the mass of individual copper samples (to the nearest 0.1 mg) by difference. Transfer each sample to a 250-mL conical flask. Add 5 mL of 6 M HNO_3, cover with a small watch glass, and warm gently *(hood)* until the metal has dissolved. Dilute with about 25 mL of distilled water, add 10 mL of 5% (w/v) urea, and boil briefly to eliminate nitrogen oxides. Rinse the watch glass, collecting the rinsings in the flask. Cool.

Add concentrated NH_3 dropwise and with thorough mixing to produce the intensely blue $Cu(NH_3)_4^{2+}$; the solution should smell faintly of ammonia (Note). Make dropwise additions of 3 M H_2SO_4 until the color of the complex just disappears, and then add 2.0 mL of 85% H_3PO_4. Cool to room temperature.

Treat each sample individually from this point on to minimize the air-oxidation of iodide ion. Add 4.0 g of KI to the sample, and titrate immediately with $Na_2S_2O_3$ until the solution becomes pale yellow. Add 5 mL of starch indicator, and continue the titration until the blue color becomes faint. Add 2 g of KSCN; swirl vigorously for 30 s. Complete the titration, using the disappearance of the blue starch/I_2 color as the end point.

Calculate the molarity of the $Na_2S_2O_3$ solution.

Note

Do not sniff vapors directly from the flask; instead, waft them toward your nose with a waving motion of your hand.

36G-4 The Determination of Copper in Brass

Discussion

The standardization procedure described in Section 36G-3 is readily adapted to the determination of copper in brass, an alloy that also contains appreciable amounts of tin, lead, and zinc (and perhaps minor amounts of nickel and iron). The method is relatively simple and applicable to brasses with less than 2% iron. A weighed sample is treated with nitric acid, which causes the tin to precipitate as a hydrated oxide of uncertain composition. Evaporation with sulfuric acid to the appearance of sulfur trioxide eliminates the excess nitrate, redissolves the tin compound, and possibly causes the formation of lead sulfate. The pH is adjusted through the addition of ammonia, followed by acidification with a measured amount of phosphoric acid. An excess of potassium iodide is added, and the liberated iodine is titrated with standard thiosulfate. See Section 36G-3 for additional discussion.

Procedure

If so directed, free the metal of oils by treatment with an organic solvent; briefly heat in an oven to drive off the solvent. Weigh (to the nearest 0.1 mg) 0.3-g samples into 250-mL conical flasks, and introduce 5 mL of 6 M HNO_3 into each; warm *(hood)* until solution is complete. Add 10 mL of concentrated H_2SO_4, and evaporate *(hood)* until copious white fumes of SO_3 are given off. Allow the mixture to cool. Cautiously add 30 mL of distilled water, boil for 1 to 2 min, and again cool.

Follow the instructions in the third and fourth paragraphs of the procedure in Section 36G-3.

Report the percentage of Cu in the sample.

36H TITRATIONS WITH POTASSIUM BROMATE

Applications of standard bromate solutions to the determination of organic functional groups are described in Section 17D-1. Directions follow for the determination of ascorbic acid in vitamin C tablets.

36H-1 Preparation of Solutions

Procedure

(a) *Potassium bromate, 0.015 M.* Transfer about 1.5 g of reagent-grade potassium bromate to a weighing bottle, and dry at 110°C for at least 1 hr. Cool in a desiccator. Weigh approximately 1.3 g (to the nearest 0.1 mg) into a 500-mL volumetric flask; use a powder funnel to ensure quantitative transfer of the solid. Rinse the funnel well, and dissolve the $KBrO_3$ in about 200 mL of distilled water. Dilute to the mark, and mix thoroughly.

Solid potassium bromate can cause a fire if it comes into contact with damp organic material (such as paper toweling in a waste container). Consult with the instructor concerning the disposal of any excess.

(b) *Sodium thiosulfate, 0.05 M.* Follow the directions in Section 36G-1; use about 12.5 g of $Na_2S_2O_3 \cdot 5H_2O$ per liter of solution.

(c) *Starch indicator.* See Section 36F-1(b).

36H-2 Standardization of Sodium Thiosulfate Against Potassium Bromate

Discussion

Iodine is generated by the reaction between a known volume of standard potassium bromate and an unmeasured excess of potassium iodide:

$$BrO_3^- + 6I^- + 6H^+ \longrightarrow Br^- + 3I_2 + 3H_2O$$

The iodine produced is titrated with the sodium thiosulfate solution.

Procedure

Pipet 25.00-mL aliquots of the $KBrO_3$ solution into 250-mL conical flasks and rinse the interior wall with distilled water. *Treat each sample individually beyond this point.* Introduce 2 to 3 g of KI and about 5 mL of 3 M H_2SO_4. Immediately titrate with $Na_2S_2O_3$ until the solution is pale yellow. Add 5 mL of starch indicator, and titrate to the disappearance of the blue color.

Calculate the concentration of the thiosulfate solution.

36H-3 The Determination of Ascorbic Acid in Vitamin C Tablets by Titration with Potassium Bromate

Discussion

Ascorbic acid, $C_6H_8O_6$, is cleanly oxidized to dehydroascorbic acid by bromine:

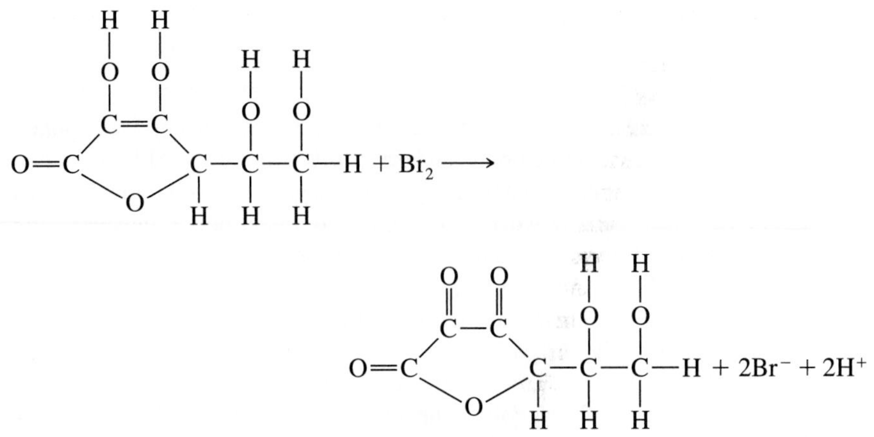

An unmeasured excess of potassium bromide is added to an acidified solution of the sample. The solution is titrated with standard potassium bromate to the first permanent appearance of excess bromine; this excess is then determined iodometrically with standard sodium thiosulfate. The entire titration must be performed without delay to prevent air-oxidation of the ascorbic acid.

Procedure

Weigh (to the nearest milligram) 3 to 5 vitamin C tablets (Note 1). Pulverize them thoroughly in a mortar, and transfer the powder to a dry weighing bottle. Weigh individual 0.40- to 0.50-g samples (to the nearest 0.1 mg) into dry 250-mL conical flasks. *Treat each sample individually beyond this point.* Dissolve the sample (Note 2) in 50 mL of 1.5 M H_2SO_4; then add about 5 g of KBr. Titrate immediately with standard $KBrO_3$ to the first faint yellow due to excess Br_2. Record the volume of $KBrO_3$ used. Add 3 g of KI and 5 mL of starch indicator; back-titrate (Note 3) with standard 0.05 M $Na_2S_2O_3$.

Calculate the average mass (in milligrams) of ascorbic acid (176.13 g/mol) in each tablet.

Notes

1. This method is not applicable to chewable vitamin C tablets.
2. The binder in many vitamin C tablets remains in suspension throughout the analysis. If the binder is starch, the characteristic color of the complex with iodine appears upon the addition of KI.
3. The volume of thiosulfate needed for the back-titration seldom exceeds a few milliliters.

36I POTENTIOMETRIC METHODS

Potentiometric measurements provide a highly selective method for the quantitative determination of numerous cations and anions. A discussion of the principles and applications of potentiometric measurements is found in Chapters 18 and 19. Detailed instructions are given in this section on the use of potentiometric measurements to locate end points in volumetric titrations. In addition, a procedure for the direct potentiometric determination of fluoride ion in drinking water and in toothpaste is described.

36I-1 General Directions for Performing a Potentiometric Titration

The procedure that follows is applicable to the titrimetric methods described in this section. With the proper choice of indicator electrode, it can also be applied to most of the volumetric methods given in Sections 36B through 36H.

1. Dissolve the sample in 50 to 250 mL of water. Rinse a suitable pair of electrodes with deionized water, and immerse them in the sample solution. Provide magnetic (or mechanical) stirring. Position the buret so that reagent can be delivered without splashing.
2. Connect the electrodes to the meter, commence stirring, and record the initial buret volume and the initial potential (or pH).
3. Record the meter reading and buret volume after each addition of titrant. Introduce fairly large volumes (about 5 mL) at the outset. Withhold a succeeding addition until the meter reading remains constant within 1 to 2 mV (or 0.05 pH unit) for at least 30 s (Note). Judge the volume of reagent to be added by estimating a value for $\Delta E/\Delta V$ after each addition. In the immediate vicinity of the equivalence point, introduce the reagent in 0.1-mL increments. Continue the titration 2 to 3 mL beyond the equivalence point, increasing the volume increments as $\Delta E/\Delta V$ again becomes smaller.

Note

Stirring motors occasionally cause erratic meter readings; it may be advisable to turn off the motor while meter readings are being made.

36I-2 The Potentiometric Titration of Chloride and Iodide in a Mixture

Discussion

Figure 13-3 (page 269) is a theoretical argentometric titration curve for a mixture of iodide and chloride ions. Initial additions of silver nitrate result in formation of silver iodide exclusively because the solubility of that salt is only about 5×10^{-7} that of silver chloride. It can be shown that this solubility difference is great enough that formation of silver chloride is delayed until all but 7×10^{-5}% of the iodide has precipitated. Thus, short of the equivalence point, the curve is essentially indistinguishable from that for iodide alone (Figure 13-2). Just beyond the iodide equivalence point, the silver ion concentration is determined by the concentration of chloride ion in the solution, and the titration curve becomes essentially identical to that for chloride ion by itself.

Curves resembling Figure 13-3 can be obtained experimentally by measuring the potential of a silver electrode immersed in the analyte solution. Hence, a chloride/iodide mixture can be analyzed for each of its components. This technique is not as applicable to analyzing iodide/bromide or bromide/chloride mixtures, however, because the solubility differences between the silver salts are not great enough. Thus, the more soluble salt begins to form in significant amounts before precipitation of the less soluble salt is complete. The silver indicator electrode can be a commercial billet type or simply a polished wire. A calomel electrode can be used as reference, although diffusion of chloride ion from the salt bridge may cause the results of the titration to be measurably high. This source of error can be eliminated by placing the calomel electrode in a potassium nitrate solution that is in contact with the analyte solution by means of a KNO_3 salt bridge. Alternatively, the analyte solution can be made slightly acidic with several drops of nitric acid; a glass electrode can then serve as the reference electrode because the pH of the solution and thus its potential will remain essentially constant throughout the titration.

The titration of I^-/Cl^- mixtures demonstrates how a potentiometric titration can have multiple end points. The potential of the silver electrode is proportional to pAg. Thus, a plot of E_{Ag} against titrant volume yields an experimental curve with the same shape as the theoretical curve shown in Figure 13-3 (the ordinate units will be different, of course).

Experimental curves for the titration of I^-/Cl^- mixtures do not show the sharp discontinuity that occurs at the first equivalence point of the theoretical curve (Figure 13-3). More important, the volume of silver nitrate needed to reach the I^- end point is generally somewhat greater than theoretical. This effect is the result of coprecipitation of the more soluble AgCl during formation of the less soluble AgI. An overconsumption of reagent thus occurs in the first part of the titration. The total volume closely approaches the correct amount.

Despite this coprecipitation error, the potentiometric method is useful for the analysis of halide mixtures. With approximately equal quantities of iodide and chloride, relative errors can be kept within 2% relative.

Preparation of Reagents

(a) *Silver nitrate, 0.05 M.* Follow the instructions in Section 36C-1.

(b) *Potassium nitrate salt bridge.* Bend an 8-mm glass tube into a U-shape with arms that are long enough to extend nearly to the bottom of two 100-mL beakers. Heat 50 mL of water to boiling, and stir in 1.8 g of powdered agar; continue to heat and stir until a uniform suspension is formed. Dissolve 12 g of KNO_3 in the hot suspension. Allow the mixture to cool somewhat. Clamp the U-tube with the openings facing up, and use a medicine dropper to fill it with the warm agar suspension. Cool the tube under a cold-water tap to form the gel. When the bridge is not in use, immerse the ends in 2.5 M KNO_3.

Procedure

Obtain the unknown in a clean 250-mL volumetric flask; dilute to the mark with water, and mix well.

Transfer 50.00 mL of the sample to a clean 100-mL beaker, and add a drop or two of concentrated HNO_3. Place about 25 mL of 2.5 M KNO_3 in a second 100-mL beaker, and make contact between the two solutions with the agar salt bridge. Immerse a silver electrode in the analyte solution and a calomel reference electrode in the second beaker. Titrate with $AgNO_3$ as described in Section 36I-1. Use small increments of titrant in the vicinity of the two end points.

Plot the data, and establish end points for the two analyte ions. Plot a theoretical titration curve, assuming the measured concentrations of the two constituents to be correct.

Report the number of milligrams of I^- and Cl^- in the sample or as otherwise instructed.

36I-3 The Potentiometric Determination of Solute Species in a Carbonate Mixture

Discussion

A glass/calomel electrode system can be used to locate end points in neutralization titrations and to estimate dissociation constants. As a preliminary step to the titrations, the electrode system is standardized against a buffer of known pH.

The unknown is issued as an aqueous solution prepared from one or perhaps two adjacent members of the following series: $NaHCO_3$, Na_2CO_3, and NaOH (see Section 12B-2). The object is to determine which of these components were used to prepare the unknown as well as the weight percent of each solute.

Most unknowns require a titration with either standard acid or standard base. A few may require separate titrations, one with acid and one with base. The initial pH of the unknown provides guidance concerning the appropriate titrant(s); a study of Figure 12-2 and Table 12-2 may be helpful in interpreting the data.

Preparation of Solutions

Standardized 0.1 M HCl and/or 0.1 M NaOH. Follow the directions in Sections 36B-3 through 36B-7.

Procedure

Obtain the unknown in a clean 250-mL volumetric flask. Dilute to the mark and mix well. Transfer a small amount of the diluted unknown to a beaker, and determine its pH. Titrate a 50.00-mL aliquot with standard acid or standard base (or perhaps both). Use the resulting titration curves to select indicator(s) suitable for end-point detection, and perform duplicate titrations with these.

Identify the solute species in the unknown, and report the mass/volume percentage of each. Calculate the approximate dissociation constant that can be obtained for any carbonate containing species from the titration data. Estimate the ionic strength of the solution, and correct the calculated constant to give an approximate thermodynamic dissociation constant.

36I-4 The Potentiometric Titration of Copper with EDTA

Discussion

Mercury serves as an electrode of the second kind (Section 18D-1) for the titration of many cations with EDTA. A small volume of the mercury(II)/EDTA complex is added to a solution of the sample. The formation constant of HgY^{2-} is so favorable that $[HgY^{2-}]$ remains essentially constant throughout the titration; the electrode is thus responsive to $[Y^{4-}]$ and indirectly to the concentration of the cation being titrated (in this experiment, Cu^{2+}). See Section 18D-1 for additional information.

Preparation of Solutions

(a) *Copper(II), 0.02 M.* Weigh (to the nearest 0.1 mg) about 0.32 g of copper into a 150-mL beaker. Cover with a watch glass, and dissolve in a minimum volume of dilute HNO_3 *(hood);* warm gently, if necessary, to hasten solution. Rinse the underside of the watch glass and the inner wall of the beaker. Transfer the solution quantitatively to a 250-mL volumetric flask, dilute to the mark, and mix well.

(b) *EDTA, about 0.02 M.* Weigh about 1.9 g (to the nearest 0.1 mg) of purified, dried $Na_2H_2Y \cdot 2H_2O$ into a 250-mL volumetric flask and dissolve in water, following the instructions given in Section 36D-2.

Recall that the solid should be dried at 80°C.

(c) *Acetate buffer* (sufficient for 40 to 50 titrations). Dilute about 7 mL of glacial acetic acid to about 100 mL. With a pH meter, adjust the pH to 4.6 through the addition of 6 M NaOH. Dilute to about 250 mL with water.

(d) *Mercury(II), 0.02 M* (sufficient for several hundred titrations). Dissolve 0.40 g of Hg in a minimum volume of HNO_3 *(hood).* Transfer quantitatively to a 100-mL volumetric flask, dilute to the mark, and mix well.

Procedure

Adjust the meter to read in millivolts. Connect the mercury electrode and a calomel electrode to the meter. Transfer a 25.00-mL aliquot of the copper solution to a 250-mL beaker. Add about 5 mL of acetate buffer and one drop of a solution prepared by mixing equal volumes of the EDTA and Hg(II) solutions. Add water, if needed, to cover the ends of the electrodes (Note 1). Titrate (Note 2) with EDTA according to Section 36I-1.

Compare the milligrams of copper found with the amount taken. Calculate the relative error in the titration.

Notes

1. Sufficient chloride ion may leak from the calomel electrode to interfere with the functioning of the mercury indicator electrode. If necessary, place the reference electrode in a beaker that is connected by a KNO_3 bridge to the analyte container; see Section 36I-2.
2. The titration must be performed slowly in the vicinity of the end point.

36I-5 The Direct Potentiometric Determination of Fluoride Ion

Discussion

The solid state fluoride electrode (Section 18D-6) has found extensive use in the determination of fluoride in a variety of materials. Directions follow for the determination of this ion in drinking water and in toothpaste. A total ionic strength adjustment buffer (TISAB) is used to adjust all unknowns and standards to essentially the same ionic strength; when this reagent is used, the *concentration* of fluoride, rather than its activity, is measured. The pH of the buffer is about 5, a level at which F^- is the predominant fluorine containing species. The buffer also contains cyclohexylaminedinitrilotetraacetic acid, which forms stable chelates with iron(III) and aluminum(III), thus freeing fluoride ion from its complexes with these cations.

Review Sections 18D-6 and 19B before undertaking these experiments.

Preparation of Solutions

(a) *Total ionic strength adjustment buffer (TISAB).* This solution is marketed commercially under the trade name TISAB.[9] Sufficient buffer for 15 to 20 determinations can be prepared by mixing (with stirring) 57 mL of glacial acetic acid, 58 g of NaCl, 4 g of cyclohexylaminedinitrilotetraacetic acid, and 500 mL of distilled water in a 1-L beaker. Cool the contents in a water or ice bath, and carefully add 6 M NaOH to a pH of 5.0 to 5.5. Dilute to 1 L with water, and store in a plastic bottle.

(b) *Standard fluoride solution, 100 ppm.* Dry a quantity of NaF at 110°C for 2 hr. Cool in a desiccator; then weigh (to the nearest milligram) 0.22 g into a 1-L volumetric flask. (***Caution!*** *NaF is highly toxic. Immediately* wash any skin touched by this compound with copious quantities of water.) Dissolve in water, dilute to the mark, mix well, and store in a plastic bottle. Calculate the exact concentration of fluoride in parts per million.

A standard F^- solution can be purchased from commercial sources.

Procedure

The apparatus for this experiment consists of a solid state fluoride electrode, a saturated calomel electrode, and a pH meter. A sleeve-type calomel electrode is needed for the toothpaste determination because the measurement is made on a suspension that tends to clog the liquid junction. The sleeve must be loosened momentarily to renew the interface after each series of measurements.

Determination of Fluoride in Drinking Water

Transfer 50.00-mL portions of the water to 100-mL volumetric flasks, and dilute to the mark with TISAB solution.

Prepare a 5-ppm F^- solution by diluting 25.0 mL of the 100-ppm standard to 500 mL in a volumetric flask. Transfer 5.00-, 10.0-, 25.0-, and 50.0-mL aliquots of the 5-ppm solution to 100-mL volumetric flasks, add 50 mL of TISAB solution, and dilute to the mark. (These solutions correspond to 0.5, 1.0, 2.5, and 5.0 ppm F^- in the sample.)

After thorough rinsing and drying with paper tissue, immerse the electrodes in the 0.5-ppm standard. Stir mechanically for 3 min; then record the potential. Repeat with the remaining standards and samples.

[9]Orion Research, Boston, MA.

Plot the measured potential against the log of the concentration of the standards. Use this plot to determine the concentration in parts per million of fluoride in the unknown.

Determination of Fluoride in Toothpaste[10]

Weigh (to the nearest milligram) 0.2 g of toothpaste into a 250-mL beaker. Add 50 mL of TISAB solution, and boil for 2 min with good mixing. Cool and then transfer the suspension quantitatively to a 100-mL volumetric flask, dilute to the mark with distilled water, and mix well. Follow the directions for the analysis of drinking water, beginning with the second paragraph.

Report the parts per million of F^- in the sample.

36J ELECTROGRAVIMETRIC METHODS

A convenient example of an electrogravimetric method of analysis is the simultaneous determination of copper and lead in a sample of brass. Additional information concerning electrogravimetric methods is found in Section 20C.

36J-1 The Electrogravimetric Determination of Copper and Lead in Brass

Discussion

This procedure is based upon the deposition of metallic copper on a cathode and of lead as PbO_2 on an anode. As a first step, the hydrous oxide of tin ($SnO_2 \cdot xH_2O$) that forms when the sample is treated with nitric acid must be removed by filtration. Lead dioxide is deposited quantitatively at the anode from a solution with a high nitrate ion concentration; copper is only partially deposited on the cathode under these conditions. It is therefore necessary to eliminate the excess nitrate after deposition of the PbO_2 is complete. Removal is accomplished through the addition of urea:

$$6NO_3^- + 6H^+ + 5(NH_2)_2CO \longrightarrow 8N_2(g) + 5CO_2(g) + 13H_2O$$

Copper then deposits quantitatively from the solution after the nitrate ion concentration has been decreased.

Procedure

Preparation of Electrodes

Immerse the platinum electrodes in hot 6 M HNO_3 for about 5 min (Note 1). Wash them thoroughly with distilled water, rinse with several small portions of ethanol, and dry in an oven at 110°C for 2 to 3 min. Cool and weigh both anodes and cathodes to the nearest 0.1 mg.

[10]From T. S. Light and C. C. Cappuccino, *J. Chem. Educ.*, **1975**, *52*, 247.

Preparation of Samples

It is not necessary to dry the unknown. Weigh (to the nearest 0.1 mg) 1-g samples into 250-mL beakers. Cover the beakers with watch glasses. Cautiously add about 35 mL of 6 M HNO_3 *(hood).* Digest for at least 30 min; add more acid if necessary. Evaporate to about 5 mL but never to dryness (Note 2).

To each sample, add 5 mL of 3 M HNO_3, 25 mL of water, and one quarter of a tablet of filter paper pulp; digest without boiling for about 45 min. Filter off the $SnO_2 \cdot xH_2O$, using a fine-porosity filter paper (Note 3); collect the filtrates in tall-form electrolysis beakers. Use many small washes with hot 0.3 M HNO_3 to remove the last traces of copper; test for completeness with a few drops of NH_3. The final volume of filtrate and washings should be between 100 and 125 mL; either add water or evaporate to attain this volume.

Electrolysis

With the current switch off, attach the cathode to the negative terminal and the anode to the positive terminal of the electrolysis apparatus. Briefly turn on the stirring motor to be sure the electrodes do not touch. Cover the beakers with split watch glasses and commence the electrolysis. Maintain a current of 1.3 A for 35 min.

Rinse the cover glasses and add 10 mL of 3 M H_2SO_4 followed by 5 g of urea to each beaker. Maintain a current of 2 A until the solutions are colorless. To test for completeness of the electrolysis, remove one drop of the solution with a medicine dropper, and mix it with a few drops of NH_3 in a small test tube. If the mixture turns blue, rinse the contents of the tube back into the electrolysis vessel and continue the electrolysis for an additional 10 min. Repeat the test until no blue $Cu(NH_3)_4^{2+}$ is produced.

When electrolysis is complete, discontinue stirring but leave the current on. Rinse the electrodes thoroughly with water as they are removed from the solution. After rinsing is complete, turn off the electrolysis apparatus (Note 4), disconnect the electrodes, and dip them in acetone. Dry the cathodes for about 3 min and the anodes for about 15 min at 110°C. Allow the electrodes to cool in air, and then weigh them.

Report the percentages of lead (Note 5) and copper in the brass.

Notes

1. Alternatively, grease and organic materials can be removed by heating platinum electrodes to redness in a flame. Electrode surfaces should not be touched with your fingers after cleaning because grease and oil cause nonadherent deposits that can flake off during washing and weighing.
2. Chloride ion must be totally excluded from this determination because it attacks the platinum anode during electrolysis. This reaction is not only destructive but also causes positive errors in the analysis by codepositing platinum with copper on the cathode.
3. If desired, the tin content can be determined gravimetrically by ignition of the $SnO_2 \cdot xH_2O$ to SnO_2.
4. It is important to maintain a potential between the electrodes until they have been removed from the solution and washed. Some copper may redissolve if this precaution is not observed.
5. Experience has shown that a small amount of moisture is retained by the PbO_2 and that better results are obtained if 0.8643 is used instead of 0.8662, the stoichiometric factor.

36K COULOMETRIC TITRATIONS

In a coulometric titration, the ''reagent'' is a constant direct current of exactly known magnitude. The time required for this current to oxidize or reduce the analyte quantitatively (directly or indirectly) is measured. See Section 20D-5 for a discussion of this electroanalytical method.

36K-1 The Coulometric Titration of Cyclohexene

Discussion[11]

Many olefins react sufficiently rapidly with bromine to permit their direct titration. The reaction is carried out in a largely nonaqueous environment with mercury(II) as a catalyst. A convenient way of performing this titration is to add excess bromide ion to a solution of the sample and generate the bromine at an anode that is connected to a constant-current source. The electrode processes are

$$2Br^- \longrightarrow 2Br_2 + 2e^- \qquad \text{anode}$$
$$2H^+ + 2e^- \longrightarrow H_2(g) \qquad \text{cathode}$$

The hydrogen produced does not react with bromine rapidly enough to interfere. The bromine reacts with an olefin, such as cyclohexene, to give the addition product:

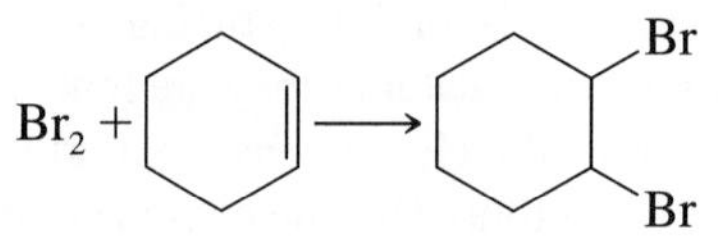

The amperometric method with twin-polarized electrodes (page 480) provides a convenient way to detect the end point in this titration. A potential difference of 0.2 to 0.3 V is maintained between two small electrodes. This potential is not sufficient to cause the generation of hydrogen at the cathode. Thus, short of the end point, the cathode is polarized and no current is observed. At the end point, the first excess of bromine depolarizes the cathode and produces a current. The electrode reactions at the twin indicator electrodes are

$$2Br^- \longrightarrow Br_2 + 2e^- \qquad \text{anode}$$
$$Br_2 + 2e^- \longrightarrow 2Br^- \qquad \text{cathode}$$

The current is proportional to the bromine concentration and is readily measured with a microammeter.

A convenient way to perform several analyses is to initially generate sufficient bromine in the solvent to give a readily measured current, say, 20 μA. An aliquot of the sample is then introduced, whereupon the current immediately decreases and approaches zero. Generation of bromine is again commenced, and the time needed to regain a current of 20 μA is measured. A second aliquot of the sample is

[11]This procedure was described by D. H. Evans in *J. Chem. Educ.*, **1968,** *45* (1), 88.

added to the same solution, and the process is repeated. Several samples can thus be analyzed without changing the solvent.

The procedure that follows is for the determination of cyclohexene in a methanol solution. Other olefins can be determined as well.

Preparation of Solvent

Dissolve about 9 g of KBr and 0.5 g of mercury(II) acetate (Note 1) in a mixture consisting of 300 mL of glacial acetic acid, 130 mL of methanol, and 65 mL of water (sufficient for about 35 mmol of Br_2). (***Caution!*** Mercury compounds are highly toxic and the solvent is a skin irritant. If inadvertent contact occurs, flood the affected area with copious quantities of water.)

Procedure

Obtain the unknown in a 100-mL volumetric flask; dilute to the mark with methanol, and mix well. The temperature of the methanol should be between 18°C and 20°C (Note 2).

Add sufficient acetic acid/methanol solvent to cover the indicator and generator electrodes in the electrolysis vessel. Apply about 0.2 V to the indicator electrodes. Activate the generator electrode system, and generate bromine until a current of about 20 μA is indicated on the microammeter. Stop the generation of bromine, record the indicator current to the nearest 0.1 μA, and set the timer to zero. Transfer 10.00 mL of the unknown to the solvent; the indicator current should decrease to almost zero. Resume bromine generation. Produce bromine in smaller and smaller increments by activating the generator for shorter and shorter periods as the indicator current rises and approaches the previously recorded value. Read and record the time needed to reach the original indicator current. Reset the timer to zero, introduce a second aliquot of sample (make the volume larger if the time needed for the first titration was too short, and conversely), and repeat the process. Titrate several aliquots.

Report the number of milligrams of cyclohexene in the unknown.

Notes

1. Mercury(II) ions catalyze the addition of bromine to olefinic double bonds.
2. The coefficient of expansion for methanol is 0.11%/°C; thus, significant volumetric errors result if the temperature is not controlled.

36L VOLTAMMETRY

Various aspects of polarographic and amperometric methods are considered in Chapter 21. Two examples that illustrate these methods are described in this section. Enormous diversity exists in the instrumentation available for these determinations. It will thus be necessary for you to consult the manufacturer's operating instructions concerning the details of operation for the particular instrument you will use.

36L-1 The Polarographic Determination of Copper and Zinc in Brass

Discussion

The percentage of copper and zinc in a sample of brass can be determined from polarographic measurements. The method is particularly useful for rapid, routine analyses; in return for speed, however, the accuracy is considerably lower than that obtained with volumetric or gravimetric methods.

The sample is dissolved in a minimum amount of nitric acid. It is not necessary to remove the $SnO_2 \cdot xH_2O$ produced. Addition of an ammonia/ammonium chloride buffer causes the precipitation of lead as a basic oxide. A polarogram of the supernatant liquid has two copper waves. The one at about -0.2 V (versus SCE) corresponds to the reduction of copper(II) to copper(I), and the one at about -0.5 V represents further reduction to the metal. The analysis is based upon the total diffusion current of the two waves. The zinc concentration is determined from its wave at -1.3 V. For instruments that permit current offset, the copper waves are measured at the highest feasible sensitivity. These waves are then suppressed by the offset control of the instrument, and the zinc wave is obtained, again at the highest possible sensitivity setting.

Preparation of Solutions

(a) *Copper(II) solution, 2.5×10^{-2} M.* Weigh (to the nearest milligram) 0.4 g of copper wire. Dissolve in 5 mL of concentrated HNO_3 *(hood).* Boil briefly to remove oxides of nitrogen; then cool, dilute with water, transfer quantitatively to a 250-mL volumetric flask, dilute to the mark with water, and mix thoroughly.

(b) *Zinc(II) solution, 2.5×10^{-2} M.* Dry reagent-grade ZnO for 1 hr at 110°C, cool in a desiccator, and weigh 0.5 g (to the nearest milligram) into a small beaker. Dissolve in a mixture of 25 mL of water and 5 mL of concentrated HNO_3. Transfer to a 250-mL volumetric flask, and dilute to the mark with water.

(c) *Gelatin, 0.1%.* Add about 0.1 g of gelatin to 100 mL of boiling water.

(d) *Ammonia/ammonium chloride buffer* (sufficient for about 15 polarograms). Mix 27 g of NH_4Cl and 35 mL of concentrated ammonia in sufficient distilled water to give about 500 mL. This solution is about 1 M in NH_3 and 1 M in NH_4^+.

Procedure

Calibration

Use a buret to transfer 0-, 1-, 8-, and 15-mL portions of standard Cu(II) solution to 50-mL volumetric flasks. Add 5 mL of gelatin solution and 30 mL of buffer to each. Dilute to the mark, and mix well. Prepare an identical series of Zn(II) solutions.

Rinse the polarographic cell three times with small portions of a copper(II) solution; then fill the cell. Bubble nitrogen through the solution for 10 to 15 min to remove oxygen. Apply a potential of about -1.6 V, and adjust the sensitivity to cause the detector to give a response that is essentially full scale. Obtain a polarogram, scanning from 0 to − 1.5 V (versus SCE). Measure the limiting current at a potential just beyond the second wave. Obtain a diffusion current by subtracting the current for the blank (Note) at this same potential. Calculate i_d/c.

Repeat the foregoing with the other two copper solutions and with the zinc solutions.

Sample Preparation

Weigh 0.10- to 0.15-g (to the nearest 0.5 mg) samples of brass into 50-mL beakers, and dissolve in 2 mL of concentrated HNO_3 *(hood).* Boil briefly to eliminate oxides of nitrogen. Cool, add 10 mL of distilled water, and transfer each quantitatively to a 50-mL volumetric flask; dilute to the mark with water, and mix well.

Transfer 10.00 mL of the diluted sample to another 50.0-mL volumetric flask, add 5 mL of gelatin and 30 mL of buffer, dilute to the mark with water, and mix well.

Analysis

Follow the calibration directions in the second paragraph. Evaluate the diffusion currents for copper and zinc.

Calculate the percentage of Cu and Zn in the brass sample.

Note

The polarogram for the blank (that is, 0 mL of standard) should be obtained at the sensitivity setting used for the standard with the lowest metal ion concentration.

36L-2 The Amperometric Titration of Lead

Discussion

Amperometric titrations are discussed at the end of Section 21B-4. In the procedure that follows, the lead concentration of an aqueous solution is determined by titration with a standard potassium dichromate solution. The reaction is

$$Cr_2O_7^{2-} + 2Pb^{2+} + H_2O \longrightarrow 2PbCrO_4(s) + 2H^+$$

The titration can be performed with a dropping mercury electrode maintained at either 0 or − 1.0 V (versus SCE). At 0 V, the current remains near zero short of the end point but rises rapidly immediately thereafter, owing to the reduction of dichromate, which is now in excess. At − 1.0 V, both lead ion and dichromate ion are reduced. The current thus decreases in the region short of the end point (reflecting the decreases in the lead ion concentration as titrant is added), passes through a minimum at the end point, and rises as dichromate becomes available. The end point at − 1.0 V should be the easier of the two to locate exactly. Removal of oxygen is unnecessary, however, for the titration at 0 V.

Preparation of Solutions

(a) *Supporting electrolyte.* Dissolve 10 g of KNO_3 and 8.2 g of sodium acetate in about 500 mL of distilled water. Add glacial acetic acid to bring the pH to 4.2

(pH meter); about 20 mL of the acid will be required (sufficient for about 20 titrations).

(b) *Gelatin, 0.1%.* Add 0.1 g of gelatin to 100 mL of boiling water.

(c) *Potassium dichromate, 0.01 M.* Use a powder funnel to weigh about 0.75 g (to the nearest 0.5 mg) of primary-standard $K_2Cr_2O_7$ into a 250-mL volumetric flask. Dilute to the mark with distilled water, and mix well.

(d) *Potassium nitrate salt bridge.* See Section 36I-2.

Procedure

Obtain the unknown (Note) in a clean 100-mL volumetric flask; dilute to the mark with distilled water, and mix well. Perform the titration in a 100-mL beaker. Locate a saturated calomel electrode in a second 100-mL beaker, and provide contact between the solutions in the two containers with the KNO_3 salt bridge.

Transfer a 10.00-mL aliquot of the unknown to the titration beaker. Add 25 mL of the supporting electrolyte and 5 mL of gelatin. Insert a dropping mercury electrode in the sample solution. Connect both electrodes to the polarograph. Measure the current at an applied potential of 0 V. Add 0.01 M $K_2Cr_2O_7$ in 1-mL increments; record the current and volume after each addition. Continue the titration to about 5 mL beyond the end point. Correct the currents for volume change and plot the data. Evaluate the end-point volume.

The corrected current is given by

$$(i_d)_{corr} = i_d\left(\frac{V + v}{V}\right)$$

where V is the original volume, v is the volume of reagent added, and i_d is the uncorrected current.

Repeat the titrations at $-$1.0 V. Here, you must bubble nitrogen through the solution for 10 to 15 min before the titration and briefly after each addition of reagent. The flow of nitrogen must, of course, be interrupted during the current measurements. Again, correct the currents for volume change, plot the data, and determine the end-point volume.

Report the number of milligrams of Pb in the unknown.

Note

A stock 0.0400 M solution can be prepared by dissolving 13.5 g of reagent-grade $Pb(NO_3)_2$ in 10 mL of 6 M HNO_3 and diluting to 1 L with water. Unknowns should contain 15 to 25 mL of this solution.

36M METHODS BASED ON THE ABSORPTION OF RADIATION

Molecular absorption methods are discussed in Chapter 24. Directions follow for (1) the use of a calibration curve for the determination of iron in water, (2) the use of a standard-addition procedure for the determination of manganese in steel, and (3) a spectrophotometric determination of the pH of a buffer solution.

36M-1 The Cleaning and Handling of Cells

The accuracy of spectrophotometric measurements is critically dependent upon the availability of good-quality matched cells. These should be calibrated against one another at regular intervals to detect differences resulting from scratches, etching, and wear. Equally important is the proper cleaning of the exterior sides (the *windows*) just before the cells are inserted into a photometer or spectrophotometer. The preferred method is to wipe the windows with a lens paper soaked in methanol; the methanol is then allowed to evaporate, leaving the windows free of contaminants. It has been shown that this method is far superior to the usual procedure of wiping the windows with a dry lens paper, which tends to leave a residue of lint and a film on the window.[12]

36M-2 The Determination of Iron in a Natural Water

Discussion

The red-orange complex that forms between iron(II) and 1,10-phenanthroline (orthophenanthroline) is useful for determining iron in water supplies. The reagent is a weak base that reacts to form phenanthrolinium ion, $phenH^+$, in acidic media. Complex formation with iron is thus best described by the equation

$$Fe^{2+} + 3phenH^+ \rightleftharpoons Fe(phen)_3^{2+} + 3H^+$$

The structure of the complex is shown in Section 16D-1. The formation constant for this equilibrium is 2.5×10^6 at 25°C. Iron(II) is quantitatively complexed in the pH range between 3 and 9. A pH of about 3.5 is ordinarily recommended to prevent precipitation of iron salts, such as phosphates.

An excess of a reducing agent, such as hydroxylamine or hydroquinone, is needed to maintain iron in the +2 state. The complex, once formed, is very stable.

This determination can be performed with a spectrophotometer set at 508 nm or with a photometer equipped with a green filter.

Preparation of Solutions

(a) *Standard iron solution, 0.01 mg/mL.* Weigh (to the nearest 0.2 mg) 0.0702 g of reagent-grade $Fe(NH_4)_2(SO_4)_2 \cdot 6H_2O$ into a 1-L volumetric flask. Dissolve in 50 mL of water that contains 1 to 2 mL of concentrated sulfuric acid; dilute to the mark, and mix well.
(b) *Hydroxylamine hydrochloride* (sufficient for 80 to 90 measurements). Dissolve 10 g of $H_2NOH \cdot HCl$ in about 100 mL of distilled water.
(c) *Orthophenanthroline solution* (sufficient for 80 to 90 measurements). Dissolve 1.0 g of orthophenanthroline monohydrate in about 1 L of water. Warm slightly if necessary. Each milliliter is sufficient for no more than about 0.09 mg of Fe. Prepare no more reagent than needed; it darkens on standing and must then be discarded.
(d) *Sodium acetate, 1.2 M* (sufficient for 80 to 90 measurements). Dissolve 166 g of $NaOAc \cdot 3H_2O$ in 1 L of distilled water.

Procedure

Preparation of a Calibration Curve

Transfer 25.00 mL of the standard iron solution to a 100-mL volumetric flask and 25 mL of distilled water to a second 100-mL volumetric flask. Add 1 mL of

[12]For further information, see J. O. Erickson and T. Surles, *Amer. Lab.*, **1976,** *8* (6), 50.

hydroxylamine, 10 mL of sodium acetate, and 10 mL of orthophenanthroline to each flask. Allow the mixtures to stand for 5 min; dilute to the mark and mix well.

Clean a pair of matched cells for the instrument. Rinse each cell with at least three portions of the solution it is to contain. Determine the absorbance of the standard with respect to the blank.

Repeat the above procedure with at least three other volumes of the standard iron solution; attempt to encompass an absorbance range between 0.1 and 1.0. Plot a calibration curve.

Determination of Iron

Transfer 10.00 mL of the unknown to a 100-mL volumetric flask; treat in exactly the same way as the standards, measuring the absorbance with respect to the blank. Alter the volume of unknown taken to obtain absorbance measurements for replicate samples that are within the range of the calibration curve.

Report the parts per million of iron in the unknown.

36M-3 The Determination of Manganese in Steel

Discussion

Small quantities of manganese are readily determined photometrically by the oxidation of Mn(II) to the intensely colored permanganate ion. Potassium periodate is an effective oxidizing reagent for this purpose. The reaction is

$$5IO_4^- + 2Mn^{2+} + 3H_2O \longrightarrow 5IO_3^- + 2MnO_4^- + 6H^+$$

Permanganate solutions that contain an excess of periodate are quite stable.

Interferences to the method are few. The presence of most colored ions can be compensated for with a blank. Cerium(III) and chromium(III) are exceptions; these yield oxidation products with periodate that absorb to some extent at the wavelength used for the measurement of permanganate.

The method given here is applicable to steels that do not contain large amounts of chromium. The sample is dissolved in nitric acid. Any carbon in the steel is oxidized with peroxodisulfate. Iron(III) is eliminated as a source of interference by complexation with phosphoric acid. The standard-addition method (page 572) is used to establish the relationship between absorbance and amount of manganese in the sample.

A spectrophotometer set at 525 nm or a photometer with a green filter can be used for the absorbance measurements.

Preparation of Solutions

Standard manganese(II) solution (sufficient for several hundred analyses). Weigh 0.1 g (to the nearest 0.1 mg) of manganese into a 50-mL beaker, and dissolve in about 10 mL of 6 M HNO_3 *(hood)*. Boil gently to eliminate oxides of nitrogen. Cool; then transfer the solution quantitatively to a 1-L volumetric flask. Dilute to the mark with water, and mix thoroughly. The manganese in 1 mL of the standard solution, after being converted to permanganate, causes a volume of 50 mL to increase in absorbance by about 0.09.

Procedure

The unknown does not require drying. If there is evidence of oil, rinse with acetone and dry briefly. Weigh (to the nearest 0.1 mg) duplicate samples (Note 1) into 150-mL beakers. Add about 50 mL of 6 M HNO_3 and boil gently *(hood);* heating for 15 to 30 min should suffice. Cautiously add about 1 g of ammonium peroxidisulfate, and boil gently for an additional 10 to 15 min. If the solution is pink or has a deposit of MnO_2, add 1 mL of NH_4HSO_3 (or 0.1 g of $NaHSO_3$) and heat for 5 min. Cool; transfer quantitatively (Note 2) to 250.0-mL volumetric flasks. Dilute to the mark with water, and mix well. Use a 20.00-mL pipet to transfer three aliquots of each sample to individual beakers. Treat as follows:

Aliquot	Volume of 85% H_3PO_4, mL	Volume of Standard Mn, mL	Weight of KIO_4, g
1	5	0.00	0.4
2	5	5.00 (Note 3)	0.4
3	5	0.00	0.0

Boil each solution gently for 5 min, cool, and transfer it quantitatively to a 50-mL volumetric flask. Mix well. Measure the absorbance of aliquots 1 and 2 using aliquot 3 as the blank (Note 4).

Report the percentage of manganese in the unknown.

Notes

1. The sample size depends upon the manganese content of the unknown; consult with the instructor.
2. If there is evidence of turbidity, filter the solutions as they are transferred to the volumetric flasks.
3. The volume of the standard addition may be dictated by the absorbance of the sample. It is useful to obtain a rough estimate by generating permanganate in about 20 mL of sample, diluting to about 50 mL, and measuring the absorbance.
4. A single blank can be used for all measurements, provided the samples weigh within 50 mg of one another.

36M-4 The Spectrophotometric Determination of pH

Discussion

The pH of an unknown buffer is determined by addition of an acid/base indicator and spectrophotometric measurement of the absorbance of the resulting solution. Because overlap exists between the spectra for the acid and base forms of the indicator, it is necessary to evaluate individual molar absorptivities for each form at two wavelengths. See page 576 for further discussion.

The relationship between the two forms of bromocresol green in an aqueous solution is described by the equilibrium

$$HIn + H_2O \rightleftharpoons H_3O^+ + In^-$$

for which

$$K_a = \frac{[H_3O^+][In^-]}{[HIn]} = 1.6 \times 10^{-5}$$

The spectrophotometric evaluation of $[In^-]$ and $[HIn]$ permits the calculation of $[H_3O^+]$.

Preparation of Solutions

(a) *Bromocresol green, 1.0×10^{-4} M* (sufficient for about five determinations). Dissolve 40.0 mg (to the nearest 0.1 mg) of the sodium salt of bromocresol green (720 g/mol) in water, and dilute to 500 mL in a volumetric flask.
(b) *HCl, 0.5 M.* Dilute about 4 mL of concentrated HCl to approximately 100 mL with water.
(c) *NaOH, 0.4 M.* Dilute about 7 mL of 6 M NaOH to about 100 mL with water.

Procedure

Determination of Individual Absorption Spectra

Transfer 25.00-mL aliquots of the bromocresol green indicator solution to two 100-mL volumetric flasks. To one add 25 mL of 0.5 M HCl; to the other add 25 mL of 0.4 M NaOH. Dilute to the mark and mix well.

Obtain the absorption spectra for the acid and conjugate-base forms of the indicator between 400 and 600 nm, using water as a blank. Record absorbance values at 10-nm intervals routinely and at closer intervals as needed to define maxima and minima. Evaluate the molar absorptivity for HIn and In^- at wavelengths corresponding to their absorption maxima.

Determination of the pH of an Unknown Buffer

Transfer 25.00 mL of the stock bromocresol green indicator to a 100-mL volumetric flask. Add 50.0 mL of the unknown buffer, dilute to the mark, and mix well. Measure the absorbance of the diluted solution at the wavelengths for which absorptivity data were calculated.

Report the pH of the buffer.

36N MOLECULAR FLUORESCENCE

The phenomenon of fluorescence and its analytical applications are discussed in Chapter 25. Directions follow for the determination of quinine in beverages, in which the quinine concentration is typically between 25 and 60 ppm.

36N-1 The Determination of Quinine in Beverages[13]

Discussion

Solutions of quinine fluoresce strongly when excited by radiation at 350 nm. The relative intensity of the fluorescent peak at 450 nm provides a sensitive method

[13]These directions were adapted from J. E. O'Reilly, *J. Chem. Educ.*, **1975,** *52* (9), 610.

for the determination of quinine in beverages. Preliminary measurements are needed to define a concentration region in which fluorescent intensity is either linear or nearly so. The unknown is then diluted as necessary to produce readings within this range.

Preparation of Reagents

(a) *Sulfuric acid, 0.05 M.* Add about 17 mL of 6 M H_2SO_4 to 2 L of distilled water.
(b) *Quinine sulfate standard, 1 ppm.* Weigh (to the nearest 0.5 mg) 0.100 g of quinine sulfate into a 1-L volumetric flask, and dilute to the mark with 0.05 M H_2SO_4 (sufficient for 60 to 70 analyses). Transfer 10.00 mL of this solution to another 1-L volumetric flask, and again dilute to the mark with 0.05 M H_2SO_4. This latter solution contains 1 ppm of quinine; it should be prepared daily and stored in the dark when not in use.

Procedure

Determination of a Suitable Concentration Range

To find a suitable working range, measure the relative fluorescent intensity of the 1 ppm standard at 450 nm (or with a suitable filter that transmits in this range). Use a graduated cylinder to dilute 10 mL of the 1 ppm solution with 10 mL of 0.05 M H_2SO_4; again measure the relative fluorescence. Repeat this dilution and measurement process until the relative intensity approaches that of a blank consisting of 0.05 M H_2SO_4. Make a plot of the data, and select a suitable range for the analysis (that is, a region within which the plot is linear).

Preparation of a Calibration Curve

Use volumetric glassware to prepare three or four standards that span the linear region; measure the fluorescence intensity for each. Plot the data.

Analysis

Obtain an unknown. Make suitable dilutions with 0.05 M H_2SO_4 to bring its fluorescence intensity within the calibration range.

Calculate the parts per million of quinine in the unknown.

36O ATOMIC SPECTROSCOPY

Several methods of analysis based upon atomic spectroscopy are discussed in Chapter 26. One such application is atomic absorption, which is demonstrated in the experiment that follows.

36O-1 The Determination of Lead in Brass by Atomic Absorption Spectroscopy

Discussion

Brasses and other copper-based alloys contain from 0% to about 10% lead as well as tin and zinc. Atomic absorption spectroscopy permits the quantitative estimation of these elements. The accuracy of this procedure is not as great as that obtainable with gravimetric or volumetric measurements, but the time needed to acquire the analytical information is considerably less.

A weighed sample is dissolved in a mixture of nitric and hydrochloric acids, the latter being needed to prevent the precipitation of tin as metastannic acid, $SnO_2 \cdot xH_2O$. After suitable dilution, the sample is aspirated into a flame, and the absorption of radiation from a hollow cathode lamp is measured.

Preparation of Solutions

Standard lead solution, 100 mg/L. Dry a quantity of reagent-grade $Pb(NO_3)_2$ for about 1 hr at 110°C. Cool; weigh (to the nearest 0.1 mg) 0.17 g into a 1-L volumetric flask. Dissolve in a solution of 5 mL water and 1 to 3 mL of concentrated HNO_3. Dilute to the mark with distilled water, and mix well.

Procedure

Weigh duplicate samples of the unknown (Note 1) into 150-mL beakers. Cover with watch glasses, and then dissolve *(hood)* in a mixture consisting of about 4 mL of concentrated HNO_3 and 4 mL of concentrated HCl (Note 2). Boil gently to remove oxides of nitrogen. Cool; transfer the solutions quantitatively to 250-mL volumetric flasks, dilute to the mark with water, and mix well.

Use a buret to deliver 0-, 5-, 10-, 15-, and 20-mL portions of the standard lead solution to individual 50-mL volumetric flasks. Add 4 mL of concentrated HNO_3 and 4 mL of concentrated HCl to each, and dilute to the mark with water.

Transfer 10.00-mL aliquots of each sample to 50-mL volumetric flasks, add 4 mL of concentrated HCl and 4 mL of concentrated HNO_3, and dilute to the mark with water.

Set the monochromator at 283.3 nm, and measure the absorbance for each standard and the sample at that wavelength. Take at least three—and preferably more—readings for each measurement.

Plot the calibration data. Report the percentage of lead in the brass.

Notes

1. The mass of sample depends on the lead content of the brass and on the sensitivity of the instrument used for the absorption measurements. A sample containing 6 to 10 mg of lead is reasonable. Consult with the instructor.
2. Brasses that contain a large percentage of tin require additional HCl to prevent the formation of metastannic acid. The diluted samples may develop some turbidity on prolonged standing; a slight turbidity has no effect on the determination of lead.

36O-2 The Determination of Sodium, Potassium, and Calcium in Mineral Waters by Atomic Emission Spectroscopy

Discussion

A convenient method for the determination of alkali and alkaline earth metals in water and in blood serum is based upon the characteristic spectra these elements emit upon being aspirated into a natural gas/air flame. The accompanying directions are suitable for the analysis of the three elements in water samples. Radiation buffers (page 622) are used to minimize the effect of each element upon the emission intensity of the others.

Preparation of Solutions

(a) *Standard calcium solution, approximately 500 ppm.* Dry a quantity of $CaCO_3$ for about 1 hr at 110°C. Cool in a desiccator; weigh (to the nearest milligram) 1.25 g into a 600-mL beaker. Add about 200 mL of distilled water and about 10 mL of concentrated HCl; cover the beaker with a watch glass during the addition of the acid to avoid loss due to spattering. After reaction is complete, transfer the solution quantitatively to a 1-L volumetric flask, dilute to the mark and mix well.

(b) *Standard potassium solution, approximately 500 ppm.* Dry a quantity of KCl for about 1 hr at 110°C. Cool; weigh (to the nearest milligram) about 0.95 g into a 1-L volumetric flask. Dissolve in distilled water, and dilute to the mark.

(c) *Standard sodium solution, approximately 500 ppm.* Proceed as in (b), using 1.25 g (to the nearest milligram) of dried NaCl.

(d) *Radiation buffer for the determination of calcium.* Prepare about 100 mL of a solution that has been saturated with NaCl, KCl, and $MgCl_2$, in that order.

(e) *Radiation buffer for the determination of potassium.* Prepare about 100 mL of a solution that has been saturated with NaCl, $CaCl_2$, and $MgCl_2$, in that order.

(f) *Radiation buffer for the determination of sodium.* Prepare about 100 mL of a solution that has been saturated with $CaCl_2$, KCl, and $MgCl_2$, in that order.

Procedure

Preparation of Working Curves

Add 5.00 mL of the appropriate radiation buffer to each of a series of 100-mL volumetric flasks. Add volumes of standard that will produce solutions that range from 0 to 10 ppm in the cation to be determined. Dilute to the mark with water, and mix well.

Measure the emission intensity for each solution, taking at least three readings for each. Aspirate distilled water between each set of measurements. Correct the average values for background luminosity, and prepare a working curve from the data.

Repeat for the other two cations.

Analysis of a Water Sample

Prepare duplicate aliquots of the unknown as directed for preparation of working curves. If necessary, use a standard to calibrate the response of the instrument to the working curve; then measure the emission intensity of the unknown. Correct the data for background. Determine the cation concentration in the unknown by comparison with the working curve.

36P APPLICATIONS OF ION-EXCHANGE RESINS

36P-1 The Separation of Cations

The application of ion-exchange resins to the separation of ionic species of opposite charge is discussed in Section 34E-1. Directions follow for the ion-exchange separation of nickel(II) from zinc(II) based upon converting the zinc ions to negatively charged chloro complexes. After separation, each of the cations is determined by EDTA titration.

Discussion

The separation of the two cations is based upon differences in their tendency to form anionic complexes. Stable chlorozincate(II) complexes (such as $ZnCl_3^-$ and $ZnCl_4^{2-}$) are formed in 2 M hydrochloric acid and retained on an anion-exchange resin. In contrast, nickel(II) is not complexed appreciably in this medium and passes rapidly through such a column. After separation is complete, elution with water effectively decomposes the chloro complexes and permits removal of the zinc.

Both nickel and zinc are determined by titration with standard EDTA at pH 10. Eriochrome Black T is the indicator for the zinc titration. Bromopyrogallol Red or murexide is used for the nickel titration.

Preparation of Solutions

(a) *Standard EDTA.* See Section 36D-2.
(b) *pH-10 buffer.* See Section 36D-1.
(c) *Eriochrome Black T indicator.* See Section 36D-1.
(d) *Bromopyrogallol Red indicator* (sufficient for 100 titrations). Dissolve 0.5 g of the solid indicator in 100 mL of 50% (v/v) ethanol.
(e) *Murexide indicator.* The solid is approximately 0.2% indicator by weight in NaCl. Approximately 0.2 g is needed for each titration. The solid preparation is used because solutions of the indicator are quite unstable.

Preparation of Ion-Exchange Columns

A typical ion-exchange column is a cylinder 25 to 40 cm in length and 1 to 1.5 cm in diameter. A stopcock at the lower end permits adjustment of liquid flow through the column. A buret makes a convenient column. It is recommended that two columns be prepared to permit the simultaneous treatment of duplicate samples.

Insert a plug of glass wool to retain the resin particles. Then introduce sufficient strong-base anion-exchange resin (Note) to give a 10- to 15-cm column. Wash the column with about 50 mL of 6 M NH_3, followed by 100 mL of water and 100 mL of 2 M HCl. At the end of this cycle, the flow should be stopped so that the liquid level remains about 1 cm above the resin column. *At no time should the liquid level be allowed to drop below the top of the resin.*

Note
Amberlite® CG 400 or its equivalent can be used.

Procedure

Obtain the unknown, which should contain between 2 and 4 mmol of Ni^{2+} and Zn^{2+}, in a clean 100-mL volumetric flask. Add 16 mL of 12 M HCl, dilute to the mark with distilled water, and mix well. The resulting solution is approximately 2 M in acid. Transfer 10.00 mL of the diluted unknown onto the column. Place a 250-mL conical flask beneath the column, and slowly drain until the liquid level is barely above the resin. Rinse the interior of the column with several 2- to 3-mL portions of the 2 M HCl; lower the liquid level to just above the resin surface after each washing. Elute the nickel with about 50 mL of 2 M HCl at a flow rate of 2 to 3 mL/min.

Elute the Zn(II) by passing about 100 mL of water through the column, using the same flow rate; collect the liquid in a 500-mL conical flask.

Titration of Nickel

Evaporate the solution containing the nickel to dryness to eliminate excess HCl. Avoid overheating; the residual $NiCl_2$ must not be permitted to decompose to NiO. Dissolve the residue in 100 mL of distilled water, and add 10 to 20 mL of pH-10 buffer. Add 15 drops of bromopyrogallol red indicator or 0.2 g of murexide. Titrate to the color change (blue to purple for bromopyrogallol red, yellow to purple for murexide).

Calculate the milligrams of nickel in the unknown.

Titration of Zinc

Add 10 to 20 mL of pH-10 buffer and 1 to 2 drops of Eriochrome Black T to the eluate. Titrate with standard EDTA solution to a color change from red to blue.

Calculate the milligrams of zinc in the unknown.

36P-2 Determination of Magnesium by Ion-Exchange Chromatography

Discussion

Magnesium hydroxide in milk of magnesia tablets can be determined by dissolving the sample in a minimum of acid and diluting to a known volume. The excess acid is established by titrating several aliquots of the diluted sample with standard base. Other aliquots are passed through a cation exchange column, where magnesium ion is retained and replaced in solution by a chemically equivalent quantity of hydrogen ion:

$$Mg^{2+} + 2H^+_{res} \longrightarrow Mg^{2+}_{res} + 2H^+$$

The acid in the eluate is then titrated; the difference between the volumes of base needed for the two series of titrations is proportional to the magnesium ion in the sample. Thus,

$$\text{net mmol } H^+ = 2 \times \text{mmol } Mg^+$$

Procedure

Step 1. Prepare approximately 1 L of 0.05 M NaOH. Standardize this solution against weighed portions of dried primary-standard potassium hydrogen phthalate (KHP); use about 0.4-g (to the nearest 0.1 mg) samples of KHP for each standardization.

Step 2. Record the number of tablets in your sample. Transfer the tablets to a clean 500-mL volumetric flask (Note 1). Dissolve them in a minimum volume of 3 M HCl (4 to 5 mL/tablet should be sufficient). Then dilute to the mark with distilled water.

Step 3. Titrate the free acid in several 15.00-mL aliquots of the dissolved sample; phenolphthalein is a satisfactory indicator.

Step 4. Condition the column with about 15 mL of 3 M HCl (Note 2), followed by three 15-mL portions of distilled water. **Never permit the liquid level to drop below the top of the column packing.**

Step 5. Charge each column with a 15.00-mL aliquot of the sample. Elute at a rate of about 2 to 3 mL/min. Wash the column with three 15-mL portions of water. Collect eluate and washings in a conical flask. Repeat this step with additional aliquots of the sample (Note 3).

Step 6. Titrate the eluted samples (and washings) with standard base. Correct the total volume for that needed to titrate the free acid, and calculate the milligrams of $Mg(OH)_2$ in each tablet.

Notes

1. The sample will dissolve more quickly if the tablets are first ground in a mortar. If you choose this alternative, you will need to know (a) the total mass of your tablets as well as (b) the mass of ground sample that you transfer to the volumetric flask.
2. The volume of the sample aliquots and the volume of base used in the several titrations must be measured with the utmost care, i.e., to the nearest 0.01 mL. **All other volumes can and should be approximations only.**
3. A 25-mL buret packed to a depth of about 15 cm with Dowex-50 cation exchange resin makes a satisfactory column. Re-conditioning after 4 or 5 elutions is recommended.

36Q GAS-LIQUID CHROMATOGRAPHY

As noted in Chapter 29, gas-liquid chromatography enables the analyst to resolve the components of complex mixtures. The accompanying directions are for the determination of ethanol in beverages.

36Q-1 The Gas-Chromatographic Determination of Ethanol in Beverages[14]

Discussion

Ethanol is conveniently determined in aqueous solutions by means of gas chromatography. The method is readily extended to measurement of the *proof* of

[14]Adapted from J. J. Leary, *J. Chem. Educ.*, **1983,** *60,* 675.

alcoholic beverages. By definition, the proof of a beverage is two times its volume percent of ethanol at 60°F.

The operating instructions pertain to a 1/4-in. (o.d.) × 0.5-m Poropack column containing 80- to 100-mesh packing. A thermal conductivity detector is needed. (Flame ionization is not satisfactory because of its insensitivity to water.)

The analysis is based upon a calibration curve in which the ratio of the area under the ethanol peak to the area under the ethanol-plus-water peak is plotted as a function of the volume percent of ethanol:

$$\text{vol \% EtOH} = \frac{\text{vol EtOH}}{\text{vol soln}}$$

This relationship is not strictly linear. At least two reasons can be cited to account for the curvature. First, the thermal conductivity detector responds linearly to mass ratios rather than volume ratios. Second, at the high concentrations involved, the volumes of ethanol and water are not strictly additive, as would be required for linearity. That is,

$$\text{vol EtOH} + \text{vol H}_2\text{O} \neq \text{vol soln}$$

Preparation of Standards

Use a buret to measure 10.00, 20.00, 30.00, and 40.00 mL of absolute ethanol into separate 50-mL volumetric flasks (Note). Dilute to volume with distilled water, and mix well.

Note

The coefficient of thermal expansion for ethanol is approximately five times that for water. It is thus necessary to keep the temperature of the solutions used in this experiment to ±1°C during volume measurements.

Procedure

The following operating conditions have yielded satisfactory chromatograms for this analysis:

Column temperature	100°C
Detector temperature	130°C
Injection-port temperature	120°C
Bridge current	100 mA
Flow rate	60 mL/min

Inject a 1 μL sample of the 20% (v/v) standard, and record the chromatogram. Obtain additional chromatograms, adjusting the recorder speed until the water peak has a width of about 2 mm at half-height. Then vary the volume of sample injected and the attenuation until peaks with a height of at least 40 mm are produced. Obtain chromatograms for the remainder of the standards (including pure water and pure ethanol) in the same way. Measure the area under each peak, and plot $\text{area}_{\text{EtOH}}/(\text{area}_{\text{EtOH}} + \text{area}_{\text{H}_2\text{O}})$ as a function of the volume percentage of ethanol.

Obtain chromatograms for the unknown. Report the volume percentage of ethanol.

Glossary

A

Absolute error The difference between an experimental measurement and its true (or accepted) value.

Absolute standard deviation A precision estimate based on the deviations between individual members in a set and the mean of that set.

Absorbance, A The logarithm of the ratio between the initial power of a beam of radiation and its power after it has traversed an absorbing medium.

Absorption A process in which one substance becomes incorporated within another substance; also used to describe interactions between electromagnetic radiation and matter.

Absorption of electromagnetic radiation Processes in which radiation causes transitions in atoms and molecules to excited states; the absorbed energy is lost, ordinarily as heat, as the excited species return to their *ground states.*

Absorption filter A colored medium (usually glass) that transmits a relatively narrow portion of the electromagnetic spectrum.

Absorption spectrum A plot of absorbance as a function of wavelength.

Absorptivity, a A proportionality constant in the Beer's law equation; since A is dimensionless, b has units of centimeters, and c has units of concentration, then a has units of

$$a = \text{length}^{-1}\ \text{concentration}^{-1}$$

Accuracy A measure of the agreement between an analytical result and the accepted value for the quantity measured; this agreement is measured in terms of *error.*

Acid/base indicators Weak organic acids (or bases), solutions of which are intensely colored and whose colors are pH dependent.

Acid dissociation constant, K_{HA} An expression that describes the equilibrium that exists between a weak acid HA and its conjugate base A^- in dilute solution.

Acid error The tendency of a glass electrode to register anomalously high pH response in highly acidic media.

Acidic flux A salt that exhibits acidic properties in the molten state; used to convert refractory substances into water-soluble products.

Acid rain Rainwater that has been rendered acidic from absorption of airborne nitrogen and sulfur oxides produced mainly by humans.

Acids Species that are capable of donating protons to other species that in turn are capable of accepting these protons.

Acid salt A conjugate base that possesses an acidic hydrogen.

Activator A species that is needed in order for an enzyme to show catalytic activity.

Activity, a The effective concentration of a participant in a chemical equilibrium; given by the product of the species and its activity coefficient.

Activity coefficient, γ_X A unitless quantity whose numerical value depends on the ionic strength of a solution; the activity for a species is equal to the product of its molar equilibrium concentration and its activity coefficient.

Addition reactions Reactions that open olefinic double bonds.

Adsorbed water Nonessential water that is held on the surface of solids.

Adsorption A process in which a substance becomes attached to the surface of another substance.

Adsorption chromatography A separation technique in which the stationary phase is a finely divided solid; partition is accomplished by altering the composition of the mobile phase.

Agar A polysaccharide that forms a conducting gel with electrolyte solutions; used in salt bridges to provide electric contact between dissimilar solutions without mixing.

Air damper A device that hastens achievement of equilibrium by the beam of a mechanical analytical balance; also called a *dashpot.*

Aliquot A volume of liquid that is a known fraction of a larger volume.

Alkaline error The tendency of many glass electrodes to provide an anomalously low pH response in highly alkaline environments.

α-Values The ratio between the concentration of a particular species and the analytical concentration of the solute from which it is derived.

Alumina The common name for aluminum oxide. In a finely divided state, used as a stationary phase in adsorption chromatography; also finds application as support for a liquid stationary phase, HPLC.

Amines Derivatives of ammonia with one or more organic groups replacing hydrogen.

Amino acids Weak organic acids that also contain basic amine groups; the amine group is α to the carboxylic acid group in amino acids derived from proteins.

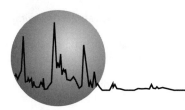

Ammonium-1-pyrrolidinecarbodithioate (APDC) A *protective agent,* atomic spectroscopy.
Amperometric titration A method based on the application of a constant potential to a working electrode and measurement of the resulting current; a linear segment curve is obtained.
Amperostat An instrument that maintains a constant current; used for coulometric titrations.
Amperostatic coulometry Constant current coulometry; synonymous with coulometric titrimetry.
Amphiprotic substances Species that can either donate protons or accept protons, depending on the environment.
Amylose A component of starch, the β-form of which is a specific indicator for iodine.
Analyte The species in the sample about which analytical information is sought.
Analytical balance An instrument for the accurate determination of mass.
Analytical molarity, c_X The moles of solute, X, that have been dissolved in sufficient solvent to give 1.000 L of solution; also numerically equal to the number of millimoles of solute per milliliter of solution. Compare with *species molarity.*
Analytical sample That carefully measured mass (or volume) of the sample matrix upon which the analysis is performed; not to be confused with the *statistical sample.*
Analytical separations A general term for preliminary steps needed to isolate an analyte from other species that might affect the outcome of an analysis.
Ångstrom unit A unit of length, convenient for the description of X-rays; 1 Å $= 1 \times 10^{-10}$ meter.
Angular dispersion $dr/d\lambda$ Measures the change in the angle of reflection or refraction of radiation by a prism or grating as a function of wavelength.
Anhydrone® Trade name for magnesium perchlorate, a drying agent.
Anion-exchange resins High-molecular-weight polymers to which amine groups are bonded; permit the exchange of anions in solution for hydroxide ions from the exchanger.
Anode The electrode of an electrochemical cell at which *oxidation* occurs.
Aqua regia A mixture of concentrated hydrochloric and nitric acids; a potent oxidizing solution.
Arithmetic mean Synonymous with *mean, average.*
Arnd's alloy A copper/magnesium alloy used to reduce nitrates and nitrites to ammonia in a nearly neutral solution.
Asbestos A fibrous mineral, some varieties of which are carcinogenic; once used as a filtering medium in a *Gooch crucible* but currently subject to stringent regulation.
Ashing The process whereby ashless filter paper is destroyed. See also *dry ashing, wet ashing.*
Ashless filter paper Paper produced from cellulose fibers that have been treated to eliminate nonvolatile inorganic species.
Aspiration The process by which a sample solution is drawn by suction into the fuel/oxidant mixture, atomic spectroscopy.
Assay The process of determining how much of a given sample is the material indicated by its name.
Asymmetry potential A small potential that results from slight differences between the two surfaces of a glass membrane electrode.
Atomic absorption The process by which unexcited atoms absorb characteristic radiation from a source. See also *hollow cathode lamp.*
Atomic emission The emission of radiation by atoms that have been promoted to excited states by an arc, a spark, or a flame.
Atomic fluorescence Radiant emission from atoms that have been excited by electromagnetic radiation, the energy of which exactly matches differences in levels of the atomic energy states.
Atomic mass unit A unit of mass based upon $\frac{1}{12}$ of the mass of the most abundant isotope of carbon, ^{12}C.
Atomic spectroscopy Methods that make use of absorption (AAS), emission (AES), and fluorescence (AFS) characteristic of the analyte.
Atomization The process of generating an atomic gas by applying heat to a sample.
Attenuation A decrease in the power of radiant energy.
Attenuator A device for diminishing the radiant power in the beam of an optical instrument.
Autocatalysis A condition in which the product of a reaction catalyzes the reaction itself.
Autoprotolysis A process in which a solvent undergoes self-dissociation.
Auxiliary balance A generic term for a balance that is less sensitive but more rugged than an analytical balance; synonymous with *laboratory balance.*
Average Synonymous with *mean, arithmetic mean.*
Average current Polarographic current determined by dividing the total charge accumulated by a mercury drop by its lifetime.
Average linear velocity, u The length L, of a chromatographic column divided by the time, t_M, required for an unretained species to pass through the column.
Azo indicators A group of acid/base indicators that have in common the structure R—N=N—R′.

B

Back-titration The amount of a second titrant needed to react with a first titrant that is in stoichiometric excess of the amount needed to react with the analyte.
Backwashing A method for cleaning filtering crucibles; wash liquid is drawn through the crucible in the opposite direction.
Ball mill A device for decreasing the particle size of the laboratory sample.
Band The distribution, ideally Gaussian, of an analyte as it exits a chromatographic column.
Band broadening The tendency of analytes to spread as they pass through a chromatographic column; caused by various diffusion and mass transfer processes.
Band spectra Regions in which spectral lines are so numerous and close together that they cannot be resolved.
Bandwidth The Gaussian distribution of wavelengths about the nominal setting of a wavelength selector.
Barrier-layer cell Synonymous with *photovoltaic cell.*
Base dissociation constant, K_B An expression that accounts for the equilibrium that exists between a weak base BH^+ and its conjugate acid in dilute solution.
Base region, of flame The region in which the solvent is evaporated, leaving the analyte as a very finely divided solid.
Bases Species that are capable of accepting protons from donor (acid) species.

Basic flux A substance with basic characteristics in the molten state; used to solubilize refractory samples, principally silicates.
Beam The principal moving part of a mechanical analytical balance.
Beam arrest A mechanism that lifts the beam from its bearing surface when an analytical balance is not in use, or when the load is being changed.
Beam splitter A device for dividing radiation from a monochromator such that one portion passes through the sample while the other passes through the blank.
Beer's law The fundamental relationship for the absorption of radiation by matter.
Bernoulli effect In atomic spectroscopy, the mechanism by which analyte droplets are aspirated into the fuel/oxidant mixture.
β-Amylose That component of starch that serves as a specific indicator for iodine.
Bias The tendency to skew estimates in the direction that favors the anticipated result. Also used to describe the effect of a *systematic error* on a set of measurements; also, a dc voltage that is in series with a circuit element.
Blackbody radiation Continuous radiation produced by an incandescent solid.
Blank determination The process of performing all steps of an analysis in the absence of sample; used to detect and compensate for systematic errors in an analysis.
Bolometer A detector for infrared radiation based on changes in resistance with changes in temperature.
Bonded-phase packings In HPLC, a support medium to which a liquid stationary phase is chemically bonded. Compare with *liquid-liquid chromatography.*
Bonded stationary phase A liquid stationary phase that is chemically bonded to the support medium.
Boundary potential, E_b The resultant of the two potentials that develop at the surfaces of a glass membrane electrode.
Brønsted-Lowry acids and bases A concept of acid/base behavior in which an acid is defined as a proton donor and a base as a proton acceptor; the loss of a proton by an acid results in the formation of a species that is a potential proton acceptor, or *conjugate base,* of the parent acid.
Buffer capacity The number of moles of strong acid (or strong base) needed to alter the pH of 1.00 L of a buffer solution by 1.00 unit.
Buffer solutions Solutions that tend to resist changes in pH as the result of dilution or the addition of small amounts of acids or bases.
Bumping The result of local overheating of a liquid.
Buoyancy The displacement of the medium (ordinarily air) by an object, resulting in an apparent loss of mass; an appreciable source of error when the densities of the object and the comparison standards (masses) are significantly different.
Buret A graduated tube from which accurately known volumes can be dispensed.
Burners Sources of heat for laboratory operations.

C

Calibration The process whereby the analytical signal is caused to bear a known relation to the amount of analyte in the sample; also, a process whereby volumes contained by (or delivered from) graduated glassware are verified.
Calomel electrode A versatile reference electrode based on the half-reaction

$$Hg_2Cl_2(s) + 2e^- \rightleftharpoons 2Hg(l) + 2Cl^-$$

Capacity factor, k'_A A term used to describe the migration of species through a chromatographic column; the difference between the retention time t_R and the dead time t_M, divided by t_M.
Capillary column A small-diameter chromatographic column for GLC, fabricated of metal, glass, or fused silica; the stationary phase is a thin coating of liquid on the interior wall of the tube.
Carbonate error A systematic error caused by absorption of carbon dioxide by standard solutions of base that will be used in the titration of weak acids.
Carbonyl group R—CO—R′ where R′ can be hydrogen (an aldehyde) or an organic group (a ketone).
Carboxylic acid group RCOOH.
Carrier gas The mobile phase for gas chromatography (GLC).
Catalytic methods Analytical methods based on pseudo-first-order kinetics; particularly useful for enzyme-catalyzed reactions.
Catalytic reaction A reaction whose progress toward equilibrium is hastened by a substance that is not consumed in the overall process.
Cathode The electrode in an electrochemical cell at which *reduction* takes place; also, the *indicator electrode* in potentiometric measurements.
Cathode depolarizer A substance that is more easily reduced than hydrogen ion; used to prevent codeposition of hydrogen during an electrolysis.
Cathodic stripping analysis A method by which the analyte is concentrated on an anodic microelectrode, and subsequently stripped cathodically.
Cation-exchange resins High-molecular-weight polymers to which acidic groups are bonded; permit the replacement of cations in solution by hydrogen ions from the exchanger.
Cell A term with several meanings. In statistics, the combination of adjacent data for display in a histogram. In electrochemistry, an array consisting of a pair of electrodes immersed in solutions that are in electrical contact through a salt bridge; the electrodes are connected externally by a metallic conductor. In *spectroscopy,* the container that holds the sample in the light path of an optical instrument. In an *electronic balance,* a system of constraints that ensure alignment of the pan.
Cells without liquid junction Electrochemical cells in which both anode and cathode are immersed in a common electrolyte.
Charge-balance equation The equality of positive and negative charge in a solution expressed in terms of the molar concentrations of the species that carry the charge.
Charge transfer A process involving formation of a complex containing an electron acceptor group and an electron donor group; associated with large molar absorptivities.
Charge-transfer device detectors (CTD) Semiconductor detectors with multichannel capability; the charge generated by exposure to radiation is accomplished by *charge coupling* (CCD) or with *charge injection* (CID).
Charging current A positive or a negative nonfaradaic current resulting from a surplus or a deficiency of electrons in a mercury droplet at the instant of detachment.
Chelating agents Substances with multiple sites available for coordination bonding with metal ions; such bonding typically results in the formation of five- or six-membered rings.

Chelation The reaction between a metal ion and a chelating reagent.
Chemical Abstracts A major source of chemical information, worldwide; entered through an extensive system of indexes or a computer database.
Chemical deviations In absorption spectroscopy, departures from Beer's law resulting from association or dissociation of the analyte for which account has not been taken; in *atomic spectroscopy,* chemical interactions of the analyte with interferents that thus affect the absorption or emission properties of the analyte.
Chemical equilibrium A dynamic state in which the rates of forward and reverse reactions are identical; a system in equilibrium will not spontaneously depart from this condition.
Chemical formula Specifies the number and the identity of atoms in a molecule or polyatomic ion.
Chemiluminescence The emission of energy as electromagnetic radiation during a chemical reaction.
Chopper A mechanical device that alternately transmits and blocks radiation from a source.
Chromatogram A plot of analyte signal as a function of elution time or elution volume.
Chromatographic band The distribution (ideally Gaussian) of eluted species about a central value; the result of variations in the time that the analyte species resides in the mobile phase.
Chromatographic zones Synonymous with *chromatographic bands.*
Chromatography A term for methods of separation based upon the partition of analyte species between a *stationary phase* and a *mobile phase.*
Clark oxygen sensor A voltammetric sensor for dissolved oxygen.
Coagulation The process whereby particles with colloidal dimensions are caused to form larger aggregates.
Coefficient of variation, CV The relative standard deviation, expressed as a percentage.
Collector A species used to scavenge a trace analyte by precipitation from solution.
Colloidal suspension A mixture (commonly of a solid in a liquid) in which the particles are so finely divided that they have no tendency to settle.
Colorimeter An optical instrument for the measurement of electromagnetic radiation that requires the human eye to function as the detector.
Column chromatography As the name implies, chromatographic methods in which the stationary phase is held within or on the surface of a column. Compare with *planar chromatography.*
Column efficiency A measure of the degree of broadening of a chromatographic band; often expressed in terms of *plate height, H,* or the *number of theoretical plates, N.* Insofar as the distribution of analyte is Gaussian within the band, the plate height is given by the variance, σ^2, divided by the length, L, of the column packing.
Column resolution, *R* Measures the capability of a column to separate two analyte bands.
Column thermostat A system that maintains the temperature of a chromatographic column at an optimum value.
Combustion-tube methods Any of several closed-tube methods for freeing analytes from chemical compounds or other intractable matrices.
Common-ion effect The shift in the position of equilibrium caused by the addition of a participating ion.
Complex formation The process whereby a species with one or more unshared electron pairs forms coordinate bonds with metal ions.
Concentration-based equilibrium constant, K' The equilibrium constant based on molar equilibrium concentrations; the numerical value of K' depends on the ionic strength of the medium.
Concentration polarization A condition in an electrochemical cell in which the transport of species to and from electrode surfaces is insufficient to maintain current levels.
Concentration profile The Gaussian distribution displayed by analytes as they emerge from a chromatographic column.
Condenser current Synonymous with *charging current.*
Conduction of electricity The movement of charge by ions in solution, by electrochemical reaction at the surface of electrodes, and by metals.
Conductometric detector A detector for charged species; finds use in ion chromatography.
Confidence interval Defines bounds about the experimental mean within which—with a given probability—the true mean will be located.
Confidence limits The values that define the confidence interval.
Conjugate acid/base pairs Species that differ from one another by one proton.
Constant-boiling HCl Solutions of hydrochloric acid with concentrations that depend on the atmospheric pressure.
Constant error A systematic error whose relative magnitude increases as the size of the sample taken for analysis decreases. Contrast with *proportional error.*
Constant mass The condition in which the mass of an object is no longer altered by heating.
Constructive interference Increase in the amplitude of a wave in regions where two or more wave fronts are in phase with one another.
Continuous source, absorption spectroscopy Lamps that emit a continuum of wavelengths; examples include the *tungsten filament lamp* and the *deuterium lamp.*
Continuous spectrum Radiation consisting of a continuous band of wavelengths as opposed to discrete lines. Incandescent solids provide continuous output *(blackbody radiation)* in the visible and infrared regions; the *deuterium lamp* and the *hydrogen lamp* yield continuous spectra in the ultraviolet region.
Control chart A plot that demonstrates statistical control of a product or service as a function of time.
Control circuit A three-electrode electrochemical apparatus that maintains a constant potential between the working electrode and the reference electrode.
Controlled potential methods Coulometric and electrogravimetric methods that involve the continuous monitoring of the potential applied to the cell in order to maintain a constant potential between the working electrode and a reference electrode.
Convection A mechanism that contributes to the mass transport of species through a solution; stirring, mechanical agitation, and temperature gradients may be involved.
Coordination compounds Species formed between metal ions and electron-pair donating groups; the product may be anionic, neutral, or cationic.

Coprecipitation The carrying down of otherwise soluble species either on the surface or in the interior of a solid as it precipitates.
Corning 015 glass The formulation used in the membrane of many glass pH electrodes.
Coulomb, C The quantity of charge provided by a constant current of one ampere in one second.
Coulometer A device that permits measurement of the quantity of charge. Electronic coulometers evaluate the integral of the current/time curve; chemical coulometers are based on the extent of reaction in an auxiliary cell.
Coulometric titration A variety of coulometric analysis that involves measurement of the time needed for a constant current to complete a chemical reaction.
Counter electrode The electrode that with the working electrode forms the electrolysis circuit in an three-electrode arrangement.
Counter-ion layer A region of solution surrounding a colloidal particle within which there exists a quantity of charge sufficient to balance the charge on the surface of the particle.
Creeping The tendency of some precipitates to spread over a wetted surface.
Critical temperature The temperature above which a substance can no longer exist in the liquid state, regardless of pressure.
Cross-linked stationary phase The result of an *in situ* reaction to stabilize the stationary phase on the wall of a GLC column.
Crucibles Containers for the isolation of precipitates.
Crystalline membrane electrode Electrodes in which the sensing element is a solid that responds to the activity of an ionic analyte.
Crystalline precipitates Solids that tend to form as large-particulate, easily filtered solids.
Crystalline suspensions Particles with greater-than-colloidal dimensions that are temporarily dispersed in a liquid.
Current, *i* The amount of electric charge that passes through an electrical circuit per unit time.
Current density The current (A) per unit area (cm^2) of an electrode; many electrochemical processes are sensitive to current density.
Current efficiency Measures of the effectiveness of a quantity of electricity to bring about an equivalent amount of chemical change in an analyte; coulometric methods require 100% current efficiency.
Current maxima Anomalous peaks in the current of a polarographic cell; eliminated by the introduction of surface active agents.
Current-sampled polarography Synonymous with *TAST polarography.*
Current-to-voltage converter A circuit in which the current of an electrochemical cell is proportional to the output voltage of an operational amplifier.
Cuvette The container that holds the analyte in the light path, absorption spectrocopy.

D

Dalton Synonymous with *atomic mass unit.*
Dark currents Small currents that occur even when no radiation is reaching a photometric transducer.
Dashpot Synonymous with *air damper.*
dc Plasma (DCP) spectroscopy A method that makes use of an electrically induced argon plasma to excite the emission spectra of analyte species.
Dead time A term with two meanings. In *column chromatography,* it is the time, t_M, required for an unretained species to traverse the column; in stopped-flow kinetics, it is the time between the mixing of reactants and the arrival of the mixture at the observation cell.
Debye-Hückel equation An expression that permits calculation of activity coefficients in media with ionic strengths up to 0.1.
Debye-Hückel Limiting Law A simplified form of the Debye-Hückel equation, applicable to solutions in which the ionic strength is less than 0.01.
Decantation The transfer of supernatant liquid and washings from a container to a filter without disturbing the precipitated solid in the container.
Decrepitation The shattering of a crystalline solid as it is heated; caused by vaporization of occluded water.
Degrees of freedom The number of members in a statistical sample that provide an independent measure of the precision of the set.
Dehydration The loss of water; a glass electrode that has become dehydrated responds erratically as a pH sensor.
Dehydrite® Trade name for magnesium perchlorate, a drying agent.
Density The ratio between the mass of an object and its volume. Compare with *specific gravity.*
Depletion layer A nonconductive region in a reverse-biased semiconductor.
Depolarizer An additive that undergoes reaction at an electrode in preference to an otherwise undesirable process. See *cathode depolarizer.*
Derivative titration curve A plot of the change in the quantity measured per unit volume against the volume of titrant added; a derivative curve displays a maximum where there is a point of inflection in a conventional titration curve. See *second derivative curve.*
Desiccants Drying agents.
Desiccator A container that provides a dry atmosphere for the storage of samples, crucibles, and precipitates.
Destructive interference The decrease in amplitude of radiation resulting from the superposition of two or more wave fronts that are not in phase with one another.
Detection limit The minimum amount of analyte that a system or method is capable of measuring.
Detector A device that responds to some characteristic of the system under observation and converts that response into a measurable signal.
Determinate error A class of errors that—in principle, at least—have a known cause, affect results in one and only one way, and can be compensated for. Synonymous with *systematic error.*
Deuterium lamp A source that provides continuous radiation in the ultraviolet region of the spectrum; successor to the *hydrogen lamp.*
Devarda's alloy An alloy of copper, aluminum, and zinc; used to reduce nitrates and nitrites to ammonia in a basic medium.
Deviation The difference between an individual measurement and the mean (or median) value for a set of replicate experimental data.

Diatomaceous earth The siliceous skeletons of unicellular algae; used as a solid support, GLC.
Differential methods, kinetic Methods for evaluating the concentrations of two or more components in a mixture of analytes by measuring the differences in their rates of reaction with an analytical reagent.
Differential pulse polarography A variety of pulse polarography.
Differentiating solvents Solvents in which differences in the strength of solute acids or bases are enhanced. Compare with *leveling solvents.*
Diffraction order, n Integer multiples of a wavelength at which constructive interference occurs.
Diffusion The migration of species from a region of high concentration in a solution to a more dilute region; one of the mechanisms (along with convection and migration) responsible for the transport of charge in a solution.
Diffusion coefficient *(polarographic, D; chromatographic,* D_M*)* A measure of analyte mobility in a medium; has units of cm^2/sec.
Diffusion current, i_d The difference between the limiting current and the residual current in a polarographic wave.
Digestion The practice of maintaining an unstirred mixture of freshly formed precipitate and solution from which it was formed at temperatures just below boiling; results in improved purity, particle size.
Dimethylglyoxime A precipitating reagent that is specific for nickel(II).
Diode array detector A silicon chip that accommodates numerous photodiodes; provides the capability to scan entire spectral regions simultaneously.
Diphenylthiocarbazide A chelating reagent, also known as *dithizone;* adducts with cations are sparingly soluble in water but are readily extracted with organic solvents.
Direct potentiometric measurement Relates the potential of an indicator electrode to the activity of an analyte in solution.
Dissociation The splitting of molecules of a substance into two simpler entities.
Distillation A separation technique that involves vaporization, followed by condensation.
Distribution coefficient, K_d The ratio of the molar equilibrium concentrations of an analyte between two immiscible solvents.
Distribution ratio, D The ratio of the molar analytical concentrations of an analyte between two immiscible solvents.
Dithizone Synonymous with *diphenylthiocarbazide.*
Doping The intentional introduction of traces of group III or group V elements to increase the semiconductor properties of a silicon or germanium crystal.
Doppler broadening Absorption or emission of radiation by a species in rapid motion, resulting in a broadening of spectral lines; a wavelength that is slightly shorter or longer than nominal is received by the detector, depending on the direction of motion of the species in the light path.
Double-beam instrument An optical instrument design that eliminates the need to alternate blank and analyte solutions manually in the light path. A *beam splitter* partitions the radiation in a *double beam in space* spectrometer; a *chopper* directs the beam alternately between blank and analyte in a *double beam in time* instrument.
Double precipitation Synonymous with *reprecipitation.*
Drierite® Trade name for calcium sulfate, a drying agent.
Dry ashing The elimination of organic matter from a sample by direct heating.
Duboscq colorimeter An instrument that permits variation in light paths through an analyte solution and an optical standard. Light from each is displayed in an eyepiece; the paths are adjusted until a visual match is achieved.
Dumas method A method of analysis based on the decomposition of bound nitrogen in analytes to the elemental state; the analysis is completed by volumetric measurement of the collected nitrogen.
Dynamic methods Synonymous with *kinetic methods;* concerned with the changes that occur in chemical systems. Contrast with *static methods.*
Dynode An intermediate electrode in a photomultiplier tube.

E

Echellette grating A grating that is blazed with reflecting surfaces that are larger than the nonreflecting faces.
Eddy diffusion Diffusion of solutes that contributes to broadening of chromatographic bands, the result of differences in the pathways for solutes as they traverse a column.
EDTA An abbreviation of *ethylenediaminetetraacetic acid.*
Effective bandwidth The bandwidth of a monochromator or interference filter at which the transmittance is 50% that at the nominal wavelength.
Electric double layer Refers to the charge on the surface of a colloidal particle and the counter-ion layer that balances this charge; also, the charged layer on the surface of a working voltammetric electrode.
Electroanalytical methods A large group of methods that have in common the measurement of an electrical property of the system that is related to the amount of analyte in the sample.
Electrochemical cell An array consisting of two electrodes in contact with solutions that ordinarily have dissimilar composition and are separated from one another. A *salt bridge* provides electrical contact between the solutions; an external conductor connects the electrodes.
Electrochemical methods A large group of methods for which the analytical signal is derived from the electrochemical properties of the analyte.
Electrochemical reversibility Refers to the property of some cell processes to reverse themselves when the direction of the current is reversed; in an irreversible cell, reversal causes a different reaction at one or both electrodes.
Electrode A conductor at the surface of which electron transfer to or from the surrounding solution takes place.
Electrode of the first kind A metallic electrode whose potential is proportional to the logarithm of the concentration (strictly, activity) of a cation that is derived from the electrode metal.
Electrode of the second kind A metallic electrode whose response is proportional to the logarithm of the concentration (strictly, activity) of an anion that either forms a sparingly soluble species or complexes with a cation that is derived from the electrode metal.
Electrodeless discharge tube A source of atomic line spectra; powered by radio frequencies or microwaves.

Electrode potential The potential of an electrochemical cell in which the electrode of interest is the cathode and the standard hydrogen electrode (E_{SHE} = 0.000 V) is the anode.

Electrode potential of the system, E_{system} The potential assumed by all oxidation/reduction couples in a solution at equilibrium.

Electrogravimetric analysis A branch of gravimetric analysis that involves measuring the mass of species deposited on an electrode of an electrochemical cell.

Electrolysis circuit In a three-electrode arrangement, a dc source and a potential divider to permit regulation of the potential between the working electrode and the counter electrode.

Electrolyte effect The dependence of numerical values for equilibrium constants upon the total electrolyte concentration of the solution.

Electrolytes Solute species whose solutions conduct electricity.

Electrolytic cell An electrochemical cell that requires an external source of energy to drive the cell reaction. Compare with *galvanic cell.*

Electrolytic separations Separations that take advantage of differences in the electrochemical behavior of species in an electrolytic cell.

Electromagnetic radiation A form of energy with properties that can be described in terms of waves, or alternatively as particulate photons, depending on the method of observation.

Electromagnetic spectrum The wavelengths (or energies) of electromagnetic radiation, ranging from gamma rays to radio waves.

Electronic balance A balance in which an electromagnetic field supports the position of the pan; the current needed to restore the loaded pan to its original position is proportional to the mass on the pan.

Electronic transitions The promotion of an electron from an electronic state to an excited electronic state, and conversely.

Electrothermal analyzer Any of several devices that make use of electrical heating to produce an atomized gas containing the analyte; used to increase the sensitivity of atomic absorbance and atomic fluorescence measurements.

Eluent Synonymous with *mobile phase.*

Eluent suppressor column In ion chromatography, a column downstream from the analytical column where ionic eluents are converted to molecular species, while analyte ions remain unaffected.

Elution chromatography Describes processes in which analytes are separated from one another on a column owing to differences in the time they reside in the mobile phase.

Emission spectrum The collection of spectral lines that are observed when species in excited states relax by giving off their excess energy as electromagnetic radiation.

Empirical formula The simplest whole-number combination of atoms in a molecule.

End point An observable change which signals that the amount of titrant added is chemically equivalent to the amount of analyte in the sample.

Enzymatic sensor A membrane electrode that has been coated with an immobilized enzyme; the electrode responds to the product formed in the enzyme/analyte reaction.

Enzyme substrate complex (ES) The intermediate formed in the process

$$\text{substrate} + \text{enzyme} \rightleftharpoons ES \longrightarrow \text{product} + \text{enzyme}$$

Eppendorf pipet A device for metering small, reproducible volumes of liquid.

Equal-arm balance An analytical balance equipped with two pans equidistant from the fulcrum—the one for the load, the other to accommodate an equal mass of known *weights.*

Equilibrium-constant expression An algebraic statement that describes the equilibrium relationship among the participants of a chemical reaction.

Equilibrium molarity The concentration of a solute species (in mol/L or mmol/mL); synonymous with *species molarity.*

Equivalence point That point in a titration at which the amount of standard titrant added is chemically equivalent to the amount of analyte in the sample.

Equivalence-point potential The electrode potential of the system in an oxidation/reduction titration when the amount of titrant that has been added is chemically equivalent to the amount of analyte in the sample.

Equivalent of chemical change The mass of a species that is directly or indirectly equivalent to one faraday (6.02×10^{23} electrons).

Equivalent weight A specialized term for expressing mass in chemical terms; similar to, but different from, *molar mass.* As a consequence of definition, one equivalent of an analyte reacts with one equivalent of a reagent, even if the stoichiometry of their reaction is not one to one.

Error The difference between an experimental measurement and its accepted value. See also *accuracy.*

Essential water Water in a solid that exists in a fixed amount, either within the molecular structure *(water of constitution)* or within the crystalline structure *(water of crystallization).*

Ester group A product, along with water, of the reaction between an alcohol (ROH) and a carboxylic acid (R′COOH).

Ethylenediaminetetraacetic acid Probably the most versatile reagent for complex-formation titrations; forms chelates with most cations.

Excitation The promotion of an atom, an ion, or a molecule to a state that is more energetic than the ground state.

Excitation signal In *pulse polarography,* the resultant of superimposing a pulsed potential on an increasingly incremental (staircase) potential.

Excitation spectrum In fluorescence spectroscopy, a plot of fluorescence intensity as a function of wavelength; wavelengths that are responsible for the promotion of species to an excited state appear as peaks.

Exhaustive extraction A cycle in which an organic solvent, after percolation through an aqueous phase containing the solute of interest, is distilled, condensed, and again passed through the aqueous phase.

F

Faradaic current Transport of charge resulting from oxidation/reduction processes in an electrochemical cell.

Faraday, *F* The quantity of electricity associated with 6.022×10^{23} electrons.

Fast reactions Reactions that are half-complete in ten seconds (or less).
Fatigue The dependence of a photocell's response upon time of exposure to radiation.
Ferroin A common name for the 1,10-phenanthroline-iron(II) complex, which is a versatile redox indicator.
Filtering crucible A crucible that serves as both filtering medium and container for a precipitate.
Fixed-time methods Kinetic methods based on the extent of reaction that occurs in a specified time interval.
Flame emission spectroscopy Methods that utilize a flame to cause an analyte to emit its characteristic emission spectrum; also known as flame photometry.
Flame ionization detector (FID) A detector for GLC based on *conduction* by ions produced during the pyrolysis of organic analytes.
Flame photometric detector A detector for GLC based on absorbance by pyrolysis products of sulfur- and phosphorus-containing analytes.
Flexure A mechanical system within an electronic balance that maintains alignment of the pan.
Fluorescence A pathway by which an excited atom or molecule relaxes to its ground state; characterized by emission of radiant energy in all directions.
Fluorescence bands Groups of fluorescence lines that are so closely grouped that they cannot be resolved.
Fluorescence spectrum A plot of fluorescence intensity vs. wavelength in which the excitation wavelength is held constant.
Fluorometer An instrument for quantitative fluorescence measurements.
Fluxes Substances that in the molten state possess acidic or basic properties; used to solubilize the analyte in refractory samples.
Focal plane A surface upon which dispersed radiation from a prism or a grating is focused.
Formality, F The number of molar masses of solute contained in each liter of solution. Synonymous with *analytical molarity.*
Formal potential, E^{f} The electrode potentital for a couple when the analytical concentrations of all participants are unity and the concentrations of other species in the solution are defined.
Formula weight The summation of atomic weights in the chemical formula of a substance; synonymous with *molar mass.*
Forward biasing, of silicon diode A circuit in which *p-n* junction is conductive in one direction and not in the other.
Fractionation A continuous partitioning of an analyte between fresh portions of both phases in a separation. Compare with *exhaustive extraction.*
Free diffusion junction An arrangement that replenishes the interface between a calomel electrode and its surroundings with fresh electrolyte.
Frequency, ν, of electromagnetic radiation The number of oscillations per second; has units of hertz (Hz), which is one oscillation per second.
Fritted glass crucible A filtering crucible equipped with a porous glass mat; also called a *sintered glass crucible.*
Fronting Describes a nonideal chromatographic peak in which the early portions tend to be drawn out. Compare with *tailing.*
***F*-test** A statistical method that permits comparison of the precision of two sets of measurement.
Fused silica open tubular (FSOT) column A wall-coated GLC column that has been fabricated from purified silica.

G

Galvanic cell An electrochemical cell that provides energy during its operation; synonymous with *voltaic cell.*
Galvanostat Synonymous with *amperostat.*
Gas chromatography Methods that make use of a gaseous mobile phase and a liquid (GLC) or a solid (GSC) stationary phase.
Gas electrode An electrode that involves the formation or consumption of a gas during its operation.
Gas-sensing probe An indicator/reference electrode system that is isolated from the analyte solution by a hydrophobic membrane. The membrane is permeable to a gas; the composition of the internal solution, and thus the potential, is proportional to the gas content of the analyte solution.
Gaussian distribution A distribution in which small departures from a central result occur more frequently than large ones and in which positive and negative variations occur with equal frequency; encountered in connection with *random error* in analytical results, the output of a *detector* on the focal plane of a monochromator, and the transport of an *analyte* through a chromatographic column.
Gel filtration chromatography A type of *size exclusion chromatography* that makes use of a hydrophilic packing.
Gel permeation chromatography A type of *size exclusion chromatography* that makes use of a hydrophobic packing.
General elution problem The compromise between elution time and resolution; addressed through *gradient elution* (for liquid chromatography) or *temperature programming* (for gas chromatography).
General redox indicators Indicators that respond to changes in E_{system}.
Ghosts Double images in the output of a grating, the result of imperfections in the ruling engine used in its preparation.
Glass electrode An electrode within which a potential develops across a thin glass membrane that separates solutions with different hydronium ion compositions.
Gooch crucible A porcelain filtering crucible; filtration is accomplished by a glass fiber mat.
Gradient elution The systematic alteration of mobile phase composition to optimize the chromatographic resolution of the components in a mixture; also known as *solvent programming.*
Gram 1×10^{-3} kg.
Graphical kinetic methods Methods of determining reaction rates from plots of reactant or product concentrations as a function of time.
Grating A device used to disperse polychromatic radiation into its component wavelengths.
Gravimetric analysis A group of analytical methods in which the amount of analyte is established through the measurement of a mass.
Gravimetric factor, GF The stoichiometric ratio between the analyte and the precipitate in a gravimetric analysis.
Gravimetric titrimetry Titrations in which the concentration of titrant is expressed in millimoles per gram of solution (rather than the more familiar mmol/ml). Compare with *volumetric titrimetry.*

Gross error An occasional error, neither random nor systematic, that results in the occurrence of a questionable outlier.
Gross sample A representative portion of the material that is being sampled; with further treatment, a portion of the gross sample becomes the laboratory sample.
Ground state The lowest energy state of an atom or molecule.
Guard column A precolumn located ahead of an HPLC column; composition of the packing in the guard column is selected to extend the useful lifetime of the analytical column.

H

H_4Y An abbreviation for *ethylenediaminetetraacetic acid.*
Half-cell potential The potential of an electrochemical half-cell with respect to the standard hydrogen electrode.
Half-life, $t_{1/2}$ The time interval during which the amount of reactant has decreased by one half.
Half-reaction A method of portraying the oxidation or the reduction of a species; a balanced oxidation/reduction equation consists of two half-reactions, the one supplying electrons to the other.
Half-wave potential $E_{1/2}$ The potential (ordinarily vs. SCE) at which the current of a polarographic wave is one half that of the diffusion current, i_d.
Hanging mercury drop electrode (HMDE) A microelectrode that is operated as a cathode to concentrate traces of cations; the analysis is completed by measuring the currents when the drop is subsequently made the anode.
Heat detector A device that is sensitive to changes in the temperature of its surroundings; used to monitor infrared radiation.
Height equivalent of a theoretical plate (HETP) A measure of chromatographic column efficiency. See *column efficiency.*
Henderson-Hasselbalch equation A dissociation-constant expression expressed in logarithmic form.
Heterogeneity, and sampling A determinant in selecting the mass of the gross sample.
Heyrovsky, J. A Czechoslovakian chemist responsible for the discovery and much of the development of polarographic analysis; he received the 1959 Nobel Prize for his efforts.
High-performance adsorption chromatography Synonymous with *liquid-solid chromatography.*
High-performance ion-exchange chromatography Synonymous with *ion chromatography.*
High-performance liquid chromatography (HPLC) A term for column methods in which the mobile phase is a liquid, often under pressure.
High-performance size exclusion chromatography Synonymous with *size exclusion chromatography.*
Histogram A way of summarizing data. Replicate results are grouped according to ranges of magnitude along the horizontal axis and by frequency of occurrence on the vertical axis.
Hollow cathode lamp The principal radiation source for atomic absorption spectroscopy.
Holographic grating A grating that has been produced by optical interference on a coated plate rather than by mechanical means.
Homogeneous precipitation The chemical generation of a precipitating agent within the solution that contains the analyte; also known as *precipitation from homogeneous solution.*
Hundred percent T adjustment Adjustment of an optical instrument to register 100% T with a suitable blank in the light path.
Hydrodynamic voltammetry Voltammetry performed with the analyte solution in constant motion relative to the electrode surface; motion is produced by moving the solution past a stationary electrode or by rotating the electrode through the solution.
Hydrofluoric acid A very corrosive acid; forms a volatile product with silica.
Hydrogen lamp A source of continuous radiation in the ultraviolet region.
Hydronium ion The symbol used for the hydrated proton H_3O^+.
Hydroxyl group The group (—OH) characteristic of alcohols.
8-Hydroxyquinoline A versatile chelating reagent; used in gravimetric analysis, volumetric analysis, as a protective reagent in atomic spectroscopy, and as an extracting reagent; also known as *oxine.*
Hygroscopic glass A glass that absorbs minute amounts of water on its surface; hygroscopicity is an essential property in the membrane of a glass electrode.
Hyphenated methods Methods involving the combination of two or more types of instrumentation; the product has greater capabilities than either instrument alone.
Hypothesis testing A tentative assertion against which statistical tests are applied.

I

Ilkovic equation An equation that relates the diffusion current to the variables that affect it, i.e., the number of electrons involved (n) in the reaction with the analyte, the square root of the diffusion current ($D^{1/2}$), and the capillary constant ($m^{2/3}t^{1/6}$) of the dropping mercury electrode.
Immobilized enzyme detector Synonymous with enzymatic detector.
Indeterminate error Synonymous with *random error.*
Index Guide to Chemical Analysis An indispensable tool for the efficient retrieval of information from *Chemical Abstracts.*
Indicator electrode An electrode whose potential is proportional to the logarithm of the analyte concentration (strictly, activity).
Indicator reaction, kinetic A reaction that is influenced by the catalytic activity of the analyte.
Inductively coupled plasma (ICP) spectroscopy A method that makes use of an argon plasma to excite the emission spectrum of analyte species.
Inert electrode An electrode that responds to the potential of the system, E_{system}, and is not otherwise involved in the cell reaction.
Infrared radiation Electromagnetic radiation in the range of about 1.0 to 300 μm.
Inhibitor, catalytic A species that decreases the rate of an enzyme-catalyzed reaction.
Initial rate methods Kinetic methods based on measurements made near the onset of a reaction.
Inner cone In atomic spectroscopy, the region of the flame in which particles from an aspirated sample are decomposed to atoms.
Instrumental deviations to Beer's law Departures from linearity between absorbance and concentration that are attributable to the measuring device.
Integral methods Kinetic methods based on the integrated form of the first-order rate expression.

Intensity, *I*, of electromagnetic radiation The power per unit of solid angle; often used synonymously with power, *P*.
Intercept, *b*, of a calibration line The analytical signal obtained from a linear working curve when the concentration of the species responsible for the signal is zero.
Interference filter An optical filter that provides narrow bandwidths owing to constructive interference.
Interference order, n An integer that, along with the thickness and the refractive index of the dielectric material, determines the wavelength transmitted by an interference filter.
Interferences Species that affect the signal on which an analysis is based.
Internal conversion Nonradiative relaxation of an atom or molecule to the lowest vibrational level of an excited electronic state.
Internal standard A known quantity of a species with properties similar to the analyte species that is introduced into a sample to provide a known response in addition to that of the analyte. The unknown concentration of the analyte is then calculated from the ratio of the two responses, the known ratio of the responses of the analyte and the internal standard, and the known concentration of the internal standard. The method is useful in atomic emission spectroscopy, chromatography, and other fields.
International Union of Pure and Applied Chemistry (IUPAC) An international organization devoted to the worldwide acceptance of chemical usages.
Ion chromatography A separation technique based on the partition of ionic species between a liquid mobile phase and solid polymeric *ion exchanger;* synonymous with *ion-exchange chromatography.*
Ion exchanger A high-molecular-weight polymer to which myriad acidic or basic groups have been bonded. Cathodic resins permit the exchange of hydrogen ion for cations in solution; anodic resins substitute hydroxide ion for anions.
Ionic strength, μ A property of a solution that depends on the total concentration of ions in the solution as well as on the charge carried by each of these ions.
Ionization suppressor In atomic spectroscopy, an easily ionized species (such as potassium) that is introduced to suppress ionization by the analyte.
***IR* drop** The potentital drop across a cell due to resistance to the movement of charge; also known as the *ohmic potential.*
Irreversible cell, voltammetry An electrochemical cell in which the chemical reaction as a galvanic cell is different from that which occurs when the current is reversed. Compare with a *reversible cell.*
Irreversible reaction A reaction that yields a poorly defined polarogram; caused by the irreversibility of electron transfer at the microelectrode.
Isocratic elution Elution with a single solvent. Compare with *gradient elution.*
Isoelectric point The pH at which an amino acid has no tendency to migrate under the influence of an electric field.
IUPAC convention A set of definitions relating to electrochemical cells and their potentials; also known as the *Stockholm convention.*

J

Jones reductor A column packed with amalgamated zinc; used for the prereduction of analytes.
Joule A unit of work equal to a newton-meter.
Junction potential The potential that develops at the interface between solutions with dissimilar composition; synonymous with liquid junction potential.

K

Karl Fischer reagent A reagent for the titrimetric analysis of water.
Katharometer A detector for GLC; based on differences in the thermal conductivity of analytes.
Kilogram The SI base unit of mass.
Kinetic methods Analytical methods based on measurement of systems as they undergo change. Compare with equilibrium methods, in which measurements are made on systems under conditions in which there is no discernible change.
Kinetic polarization Nonlinear behavior of an electrochemical cell caused by the slowness of reaction at the surface of one or both electrodes.
Kjeldahl flask A long-necked flask used for the digestion of samples with hot, concentrated sulfuric acid.
Kjeldahl method A method for the determination of amine nitrogen.
Knife edge The nearly friction-free contact that forms part of the bearing of a mechanical balance.

L

Laboratory balance Synonymous with *auxiliary balance.*
Laboratory notebook A book into which all pertinent data and observations are recorded.
Laboratory sample A quantity of material taken from—and representative of—the gross sample.
Laminar flow Flow in a region near the surface of an electrode within which layers of liquid slide by one another; laminar flow occurs between a region of *turbulent flow* in the bulk of the solution and the *Nernst diffusion layer* at the surface of the electrode.
Laminar flow burner A burner for atomic spectroscopy in which the nebulized sample is dispersed in the fuel and oxidant in a mixing chamber.
Least-squares method A statistical approach that permits generation of a linear calibration curve that minimizes random departures of plot points from strict linearity.
Le Châtelier principle Application of a stress to an equilibrium system will cause the position of the equilibrium to shift in the direction that tends to oppose the stress.
Leveling solvents Solvents in which the strength of solute acids or bases tend to be the same. Compare with *differentiating solvents.*
Levitation The suspension of the pan of an electronic balance by a magnetic field.
Ligand A molecule or ion with at least one pair of unshared electrons available for coordinate bonding with cations.
Limiting current, i_l Synonymous with *diffusion current* in linear scan polarography.
Linear scan voltammetry Methods that involve measurement of the current in a cell as the potential is linearly increased (or decreased); the basis for *hydrodynamic voltammetry* and *polarography.*

Linear segment curve A titration curve in which the end point is extrapolated from measurements well removed from the equivalence point; useful for reactions that do not strongly favor the formation of products.
Line source In atomic absorption spectroscopy, a radiation source that emits lines characteristic of the analyte. An important line source is the *hollow cathode lamp;* also encountered is the *electrodeless discharge tube.*
Liquid bonded-phase chromatography Partition chromatography that makes use of a stationary phase that is chemically bonded to the column packing.
Liquid junction The interface between two liquids with different composition.
Liquid-liquid chromatography Partition chromatography in which the stationary phase is adsorbed on the surface of the column packing.
Liquid-solid chromatography Partition chromatography in which the stationary phase is a polar solid; synonymous with *adsorption chromatography.*
Liter One cubic decimeter.
Lithium metaborate A basic *flux.*
Loading error An error in the measurement of a cell potential when the resistance of the measuring instrument is comparable with that of the cell.
Longitudinal diffusion coefficient, *B* A measure of the tendency for analyte species to migrate from regions of high concentration to those of lesser concentration; a contributor to *band broadening* in a chromatographic column.
Longitudinal diffusion term, *B/u* The ratio of the longitudinal diffusion term to the velocity of the mobile phase; one of three terms in the van Deemter equation to account for band broadening.
Lower control limit (LCL) The lower criterion that has been set for satisfactory performance of a process or measurement.
L'vov platform A device for the electrochemical atomization of samples in atomic absorption spectroscopy.

M

Macrobalance An analytical balance with a capacity of 160 to 200 g and a precision of 0.1 mg.
Majority carrier The species principally responsible for the transport of electricity in a semiconductor.
Martin, A.J.P. With R.L.M. Synge, pioneered the development of modern chromatography.
Masking agent A reagent that combines with a species that would otherwise influence the results of an analysis.
Mass An invariant measure of the amount of matter in an object.
Mass action effect The shift in the position of equilibrium caused by the addition or removal of a participant in the equilibrium. See also *Le Châtelier principle.*
Mass balance equation An expression that accounts for the molecular analytical concentration of a substance as the sum of molar equilibrium concentrations of species derived from that substance in a solution.
Mass-sensitive detector, GLC Synonymous with *flame ionization detector.*
Mass-transfer coefficients, C_S, C_M Terms needed to account for band broadening attributable to the lack of equilibrium between the analyte and the two phases in a chromatographic column.
Mass transport In electrochemical cells, the movement of species caused by diffusion, convection, and electrostatic forces.
Matrix The medium that contains the analyte.
Mean Synonymous with *arithmetic mean, average;* a way of reporting what is considered the most representative value for a set of replicate measurements.
Mean activity coefficient, $\gamma_\pm$ An experimentally measured activity coefficient for an ionic compound; it is not possible to resolve the mean activity coefficient into values for the individual participants.
Measuring pipet A pipet calibrated to deliver any volume up to its maximum capacity. Compare with *volumetric pipet.*
Mechanical entrapment The incorporation of impurities within a growing crystal.
Mechanism of reaction The elementary steps involved in the formation of product from reactant(s).
Median The central value in a set of replicate measurements; for an even number of data, the median is the average of the central pair.
Megabore column An open tubular column that can accommodate samples that are larger than those in an ordinary packed column.
Melt The fused mass produced by the action of a flux.
Membrane electrode An indicator electrode whose response is due to ion-exchange processes on each side of a thin membrane.
Meniscus The curved surface displayed by a liquid.
Mercury electrode An electrode often used to eliminate easily reduced species from solution prior to the analysis for other species that are not reduced.
Mercury film electrode An electrode that has been coated with a thin layer of elemental mercury; used in place of the hanging mercury drop electrode for anodic stripping analysis.
Metal oxide field effect transistor (MOSFET) A semiconductor device; when suitably coated, can be used in lieu of a traditional ion-selective electrode.
Method uncertainty, s_m The standard deviation associated with a method of measurement; a factor—with the standard deviation in sampling—in determining the overall standard deviation of an analysis.
Michaelis constant A collection of constants in the rate equation for enzyme kinetics that are characteristic of the enzyme under study.
Michaelis-Menten equation A rate law that describes the kinetic behavior of many enzyme-catalyzed reactions.
Microanalytical balance An analytical balance with a capacity of 1 to 3 g with precision of 1 μg.
Microelectrode An electrode with small physical dimensions; used in voltammetric studies.
Microporous membrane A hydrophobic membrane with pore size that permits the passage of gases and is impermeable to other species; the sensing element of a *gas-sensing probe.*
Migration Synonymous with *mass transport.*
Migration rate, $\bar{\nu}$ The rate at which an analyte traverses a chromatographic column.
Milligram 1×10^{-6} kg
Milliliter 1×10^{-3} L
Millimole, mmol 1×10^{-3} mol
Mixed anodic/cathodic voltammetry A relatively uncommon situation in which an analyte in two oxidation states yields an nodic wave at the outset and a cathodic wave at more negative potentials.

Mixed crystal formation A variety of coprecipitation; impurities with molecular dimensions similar to those of a crystalline precipitate are incorporated into the lattice of the growing crystal.
Mobile phase A liquid or a gas that carries analytes through a liquid or solid stationary phase.
Mobile phase mass-transfer coefficient, $C_M u$ A quantity that affects plate height; nonlinear in solvent velocity u and influenced by the diffusion coefficient of the analyte, the particle size of the stationary phase, and the inside diameter of the column.
Modulation Process of causing the analytical signal from a spectrometer to fall on the photodetector at a fixed frequency that differs from the frequency of the background signal. The analytical signal is then distinguished from the background signal electronically.
Mohr's salt A common name for iron(II) ammonium sulfate hexahydrate.
Molar absorptivity, ε The proportionality constant in the Beer's law expression when b is in cm and c is in mol/L; characteristic of the absorbing species.
Molarity A concentration term that expresses the moles of solute contained in one liter of solution or the millimoles of solution in one milliliter of solution.
Molar mass, $\mathcal{M}$ The mass, in grams, of one mole of a chemical substance.
Mole The SI base unit for the amount of substance.
Molecular absorption The absorption of ultraviolet, visible, and infrared radiation brought about by quantized transitions in molecules.
Molecular fluorescence The process whereby excited-state electrons in molecules return to the ground electronic state, with the excess energy being given off as electromagnetic radiation.
Molecular formula A formula that includes structural information in addition to the number and identity of atoms in a molecule.
Molecular weight Synonymous with molecular mass.
Monochromatic radiation *Ideally,* electromagnetic radiation that consists of a single wavelength; *in practice,* a very narrow band of wavelengths. Compare with *polychromatic radiation.*
Monochromator A device for resolving polychromatic radiation into its component wavelengths.
Mother liquor The solution that remains following the precipitation of a solid.
Muffle furnace A heavy-duty oven, capable of maintaining temperatures in excess of 1100°C.

N

Nanometer, nm 1×10^{-9} m
Natural lifetime, τ The time period during which the concentration of the reactant in a first-order reaction decreases to $1/e$ of its original value.
National Institute of Standards and Technology (NIST) An agency of the U.S. Department of Commerce; formerly the *National Bureau of Standards* (NBS); a major source for primary standards and exhaustively analyzed standard reference materials (SRM).
Nebulization The transformation of a liquid into myriad droplets.
Nernst diffusion layer, δ A thin layer of stagnant liquid at the surface of an electrode; caused by friction between the surface and the liquid that flows past the surface.
Nernst equation A mathematical expression that relates the potential of an electrode to the activities of those species in solution that are responsible for the potential.
Nernst glower A source of infrared radiation.
Nernstian behavior Behavior of potentiometric sensors whose response is described by the Nernst equation.
Nichrome A nickel/chromium alloy; when incandescent, a source of infrared radiation.
Noise Any signal that interferes with detection of the analyte signal.
Nominal wavelength The principal wavelength provided by a wavelength selector.
Nonessential water Water that is retained in (or on) a solid by physical, rather than chemical, forces.
Nonfaradaic current A current that results from processes other than oxidation or reduction at the surface of an electrode.
Normal error curve A plot of a Gaussian distribution.
Normal hydrogen electrode, NHE Synonymous with *standard hydrogen electrode, SHE.*
Normality, c_N A method of expressing concentration based on the equivalent, a specialized unit of chemical mass; normality is equal to the number of equivalents of solute per liter of solution (or the number of meq solute/mL of solution).
Normal-phase partition chromatography Liquid-liquid chromatography involving a polar stationary phase and a nonpolar mobile phase. Compare with *reversed-phase partition chromatography.*
Nucleation A process involving formation of very small aggregates of a precipitating solid.
Null hypothesis The assumption that measurements being compared are, in fact, identical; statistical tests are devised to validate or invalidate—with a specified level of probability—the null hypothesis.
Number of theoretical plates, N A characteristic of a chromatographic column used to describe efficiency.

O

Occluded water Nonessential water that has been entrained in a growing crystal.
Occlusion The physical entrainment of impurities in pockets in a growing crystal.
Occupational Safety and Health Administration (OSHA) A federal agency charged with assuring safety in the laboratory and the workplace.
Oesper's salt Common name for iron(II) ethylenediamine sulfate tetrahydrate.
Ohmic potential Synonymous with *IR drop.*
Open tubular column, GLC A *capillary column* within which the stationary phase is coated directly on the wall or on a film of support material.
Operational amplifier A versatile electronic device for performing mathematical tasks and for the precise measurement of such instrumental outputs as voltage, current, and resistance.
Optical instruments A broad term for instruments that measure the absorption, emission, or fluorescence by analyte species.
Optical methods Synonymous with *spectrochemical methods.*

Optical wedge An adjustable optical body used to adjust the intensity of radiation in the light path of an optical instrument.
Order The exponent associated with the concentration of a species that appears in the rate law for a reaction.
Organic polarographic analysis Voltammetry of organic functional groups that are reducible at a microelectrode; waves tend to be drawn out and pH dependent but are still useful for analysis.
OSHA Acronym for Occupational Safety and Health Administration.
Outlier A result that appears at odds with the other members in a set of replicate measurements.
Overall order A summation of the exponents for the species whose concentrations appear in the rate law for a chemical reaction.
Overall standard deviation, s_o The square root of the sum of the variance of the measurement process and the variance of the sampling step.
Overpotential, overvoltage, Π The excess voltage needed to maintain the current in an electrochemical cell that is affected by kinetic polarization.
Oxidant Synonymous with *oxidizing agent.*
Oxidation The loss of electrons by a species in an oxidation/reduction reaction.
Oxidation potential The potential of an electrode process that is written as an oxidation.
Oxidizing agent A substance that acquires electrons in an oxidation/reduction reaction.
Oxine A common name for 8-hydroxyquinoline.
Oxygen wave At the dropping mercury electrode, oxygen provides two waves, the first due to the formation of peroxide, the second to further reduction to water; although a source of interference in the analysis of other species, these waves are useful for the determination of dissolved oxygen.

P

Packed columns Chromatographic columns packed with porous materials to provide a large surface area for transfer of analytes to and from the mobile phase.
Pan arrest A device to protect the knife edges of a mechanical balance from damage while loading is changed and when the balance is not in use.
Parallax The apparent change in position of an object as a result of movement by the observer; results in systematic errors in volume readings with burets and pipets, as well as the position of pointers of meters.
Particle growth The formation of larger particles of a precipitating solid; an alternative to nucleation during the precipitation process.
Particle properties of electromagnetic radiation Behavior that is consistent with radiation acting as small particles, or *quanta,* of energy.
Particle size, and sampling To retain a constant relative error in the sampling process, a constant number of particles must be retained in each step of sample reduction as the particle size is decreased from that of the gross sample to that of the laboratory sample, and finally to that of the analytical sample.
Partition chromatography A class of separation methods that includes liquid-liquid chromatography, and bonded-phase chromatography.
Partition coefficient An equilibrium constant for the distribution of a solute between two immiscible phases.
Partition ratio Synonymous with *partition coefficient.*
Parts per million, ppm A convenient method for expressing the concentration of a solute species that exists in trace amounts; for dilute aqueous solutions, ppm is synonymous with milligrams of solute per liter of solution.
Peak area, peak height Properties of chromatographic peaks that can be used, in comparison with standards, for quantitative measurements.
Peptization A process in which a coagulated colloid returns to its dispersed state.
Perchloric acid A strong mineral acid. Dilute solutions do not require special care; hot, concentrated solutions may react explosively with organic matter.
Period, p, of electromagnetic radiation The time required for successive peaks of an electromagnetic wave to pass a fixed point in space.
pH A logarithmic scale for expressing the hydrogen ion activity of a solution.
pH effects on polarograms Of particular concern for organic analysis; strong buffering is needed to assure reproducible waves.
pH effects on redox indicators Hydrogen ions are frequent participants in the transition of redox indicators; as a result, the color change of the indicator will be pH dependent.
Phosphorescence Similar to fluorescence; excited species that relax via phosphorescence typically have significantly longer lifetimes than those that relax via fluorescence.
Phosphorus pentoxide A drying agent.
Photocell A rugged detector of electromagnetic radiation; output suffers from *fatigue.*
Photoconduction Conductivity resulting from the generation of electrons and holes in a semiconductor.
Photodecomposition The formation of new species from excited molecules; one of several ways by which excitation energy is dissipated.
Photodiode, vacuum A photoelectric cell consisting of a wire anode and a photoemissive cathode encapsulated in an evacuated glass or quartz envelope and maintained at a potential of about 90 V; a current is produced when the cathode is illuminated.
Photoelectric colorimeter Synonymous with *photometer.*
Photoelectron An electron released by the action of a photon as a result of striking an emissive surface.
Photoionization detector A chromatographic detector that uses intense ultraviolet radiation to ionize analyte species; the resulting currents that are amplified and recorded are proportional to analyte concentrations.
Photometer An instrument for the measurement of absorbance; incorporates a filter for wavelength selection and a photoelectric detector.
Photomultiplier tube A sensitive detector for electromagnetic radiation; amplification is accomplished by a series of *dynodes,* which produce a cascade of electrons for each photon received by the tube.
Photon detector A generic term for transducers that convert an optical signal to an electrical signal.

Photons Packets of electromagnetic radiation; also known as *quanta.*
Phototube Synonymous with *photodiode.*
Photovoltaic cell Synonymous with *photocell.*
Phthalein indicators Acid/base indicators derived from phthalic anhydride, the most common of which is phenolphthalein.
pIon meter An instrument that directly measures the present concentration (strictly, activity) of an analyte in terms of its negative logarithm; consists of an indicator electrode, a reference electrode, and a potential-measuring device.
Pipet A device to permit the transfer of known volumes of solution from one container to another.
Pixel A single detector element on a diode array detector or a charge-transfer detector.
Planar chromatography The term used to describe chromatographic methods that make use of a flat stationary phase; the mobile phase migrates across the surface by gravity or capillary action.
Plasma A gaseous medium that owes its conductivity to appreciable amounts of ions and electrons.
Plate height, H A term that describes the efficiency of a chromatographic column. See *column efficiency.*
Platinum electrode Used extensively in electrochemical systems in which an inert metallic electrode is required.
Plattner diamond mortar A device for crushing small amounts of brittle materials.
***p-n* junction diode** A junction between electron-rich and electron-deficient regions of a semiconductor; permits currents in one direction only.
Pneumatic detector A heat detector that is based on changes in the pressure that a gas exerts on a flexible diaphragm.
Polarization A condition in which the potential of an electrochemical cell differs from that predicted by the Nernst equation and *IR* drop. See *concentration polarization, kinetic polarization.*
Polarogram The current/voltage plot obtained from polarographic measurements.
Polarography A type of voltammetry based on the diffusion-controlled migration of an analyte to the surface of a dropping mercury electrode.
Policeman Synonymous with *rubber policeman.*
Polychromatic radiation Electromagnetic radiation consisting of more than one wavelength. Compare with *monochromatic radiation.*
Polyfunctional acids and bases Species that contain more than one acidic or basic functional group.
Population of data A theoretical aggregate of the infinite number of values that a measurement can take; also referred to as a *universe of data.*
Population mean, μ The mean value for an infinite set of measurements. The true value for a measurement that is free of systematic error.
Population standard deviation, σ A precision estimate based in principle on a population containing an infinite number of measurements.
Porous layer open tube (PLOT) column A capillary column for gas-solid chromatography in which a thin layer of the stationary phase is adsorbed on the walls of the column.
Potassium pyrosulfate An acidic *flux.*
Potential divider Synonymous with *voltage divider.*
Potential of the system, E_{system} Synonymous with *electrode potential of the system.*
Potentiometric titration A titrimetric method involving measurement of the potential between two electrodes (a *reference electrode* and an *indicator electrode*) as a function of titrant volume.
Potentiometry That branch of electrochemistry concerned with the relation between potential and analyte concentration.
Potentiostat An electronic device that alters the applied potential so that the potential between the working electrode and the reference electrode is maintained at a fixed value.
Potentiostatic methods Coulometric and gravimetric methods that utilize a constant potential between the working electrode and a reference electrode.
Power, P, of electromagnetic radiation The energy that reaches a given area per second; often used synonymously with intensity, although the two are different.
Precipitation from homogeneous solution Synonymous with *homogeneous precipitation.*
Precipitation methods of analysis Gravimetric and titrimetric methods involving the formation (or less frequently, the disappearance) of a precipitate.
Precision A measure of internal agreement among a set of replicate observations.
Premix burner Synonymous with *laminar flow burner.*
Pressure broadening An effect that increases the width of an atomic absorption line; caused by collisions among atoms that result in slight variations in their energy states.
Primary adsorption layer A charged layer of ions on the surface of a solid, resulting from the attraction of lattice ions for ions of opposite charge in the solution.
Primary monochromator A device for dispersing source radiation in a spectrofluorometer. Compare with *secondary monochromator.*
Prism An optical body that disperses polychromatic radiation into its component wavelengths.
Primary standard A chemical that possesses such properties as extraordinary purity, large combining weight, and straightforward stoichiometry. Most titrimetric methods require a calibration step involving a primary standard.
Proof A method of specifying the alcoholic content of distilled beverages: 100% alcohol = 200 proof.
Proportional error An error whose magnitude depends on the size of the sample that is taken for analysis. Contrast with *constant error.*
Protective agent In atomic spectroscopy, species that form soluble complexes with the analyte and thereby prevent the formation of compounds that have low volatility.
Pseudo-order reactions Chemical systems in which the concentration of a reactant (or reactants) is large—and essentially invariant—with respect to that of the component (or components) of interest.
Pulse polarography Voltammetric methods that periodically impose a pulse upon the linearly increasing excitation voltage; the difference in measured current, Δi, yields a peak whose height is proportional to the analyte concentration.
p-Values The concentration of a solute species expressed as its negative logarithm; the use of p-values permits expression of

enormous concentration ranges in terms of relatively small numbers.
Pyroelectric detector A heat detector based on the temperature-dependent potential that develops between electrodes that are separated by a pyroelectric material.

Q

***Q*-test** A statistical test that indicates, with a specified level of probability, whether an outlying measurement in a set of replicate data is a member of a given Gaussian distribution.
Quality assessment A protocol to assure that quality control methods are providing the information needed to evaluate satisfactory performance of a product or service.
Quality assurance A protocol designed to demonstrate that a product or service is meeting criteria that have been established for satisfactory performance.
Quanta Synonymous with *photons.*
Quantum yield of fluorescence The fraction of radiation emitted during relaxation of excited-state molecules that appears as fluorescence.
Quenching (1) A process by which molecules in an excited state lose energy to other species without fluorescing, (2) the diminution in the intensity of a fluorescing reagent as a result of reaction with the analyte, and (3) an action that brings about the cessation of a reaction.

R

Radiation buffers Potential interferents that are intentionally added in large amounts to samples and standards to swamp out their effects on atomic emission measurements.
Random error Uncertainties resulting from the operation of small uncontrolled variables that are inevitable as measurement systems are extended to and beyond their limits; also referred to as *indeterminate error.*
Range, *w*, of data The difference between extreme values in a set of data; synonymous with *spread.*
Rate constant, *k* A proportionality constant in the rate expression.
Rate-determining step The slow step in the sequence of reactions of a mechanism.
Rate law The empirical relationship describing the rate of a reaction in terms of the concentrations of participating species.
Rate theory A theory that accounts for the shapes of chromatographic peaks.
Reaction order Synonymous with *order.*
Reagent grade chemicals Chemicals that meet the standards of the Reagent Chemical Committee of the American Chemical Society.
Redox Synonymous with *oxidation/reduction.*
Redox electrode An inert electrode that responds to the electrode potential of the system.
Reducing agent The species that supplies electrons in an oxidation/reduction reaction.
Reductant Synonymous with *reducing agent.*
Reduction The process whereby a species acquires electrons.
Reduction potential The potential of an electrode process expressed as a reduction; synonymous with *electrode potential.*
Reductor A column packed with a granular metal through which a sample is passed to prereduce an analyte.
Reference electrode An electrode whose potential is known and against which potentials of other electrodes can be measured; the potential of a reference electrode is completely independent of the analyte concentration.
Reference standards Complex materials that have been extensively analyzed; a prime source for these standards is the *National Institute of Standards and Technology* (NIST).
Reflection The return of radiation from a surface.
Reflection grating An optical body that disperses polychromatic radiation into its component wavelengths. Consists of lines ruled on a reflecting surface; dispersion is the result of constructive and destructive interference.
Refractive index The ratio of the velocity of electromagnetic radiation *in vacuo* and the velocity in some other medium.
Refractory materials Substances that resist attack by ordinary laboratory acids or bases; brought into solution by high-temperature fusion with a flux.
Regression analysis A statistical technique for derivation of the "best" fit line for a set of calibration data.
Relative electrode potential The potential of an electrode with respect to another (ordinarily the standard hydrogen electrode).
Relative error The error in a measurement divided by the true (or accepted) value for the measurement; often expressed as a percentage.
Relative humidity The ratio, often expressed as a percentage, between the ambient vapor pressure of water and its saturated vapor pressure at a given temperature.
Relative standard deviation (RSD) The standard deviation divided by the mean value for a set of data; when expressed as a percentage, the relative standard deviation is called the *coefficient of variation.*
Relative supersaturation The difference between the instantaneous (Q) and the equilibrium (S) solubility of a solid, divided by S; provides general guidance for the generation of filterable solids.
Relaxation The return of excited species to their *ground state;* the process is accompanied by the release of excitation energy as heat (absorption) or light (fluorescence, radiant emission).
Releasing agent In atomic absorption spectroscopy, species introduced to combine with sample components that would otherwise interfere by forming compounds of low volatility with the analyte.
Replica grating An impression of a master grating; used as the dispersing element in most grating instruments, owing to the high cost of a master.
Replicate samples Samples upon which a chemical analysis for the analyte is performed; agreement among the results of replicate assays is taken as validation of the analysis.
Reprecipitation A method for improving the purity of precipitates, involving formation and filtration of the solid, followed by redissolving and again forming the precipitate.
Residual The deviation of a plot point from a fitted straight line or curve.
Residual current Small parasitic nonfaradaic currents associated with the dropping mercury electrode; trace impurities may also contribute.
Resolution, R_s Measures the ability of a chromatographic column to separate two analytes.

Resonance fluorescence The fluorescence emission of a wavelength that is identical with the excitation wavelength.
Resonance line The wavelength of radiation responsible for both the excitation and the fluorescent emission of an atom or molecule.
Retention time, t_R The time between sample injection and the arrival of the analyte peak at the detector.
Reticle An optical scale that permits weighing to the nearest 0.1 mg with a single-pan analytical balance.
Reverse biasing Describes a circuit in which electrons and holes are drawn away from a *p-n* junction, thus rendering the junction nonconducting.
Reversed-phase partition chromatography Liquid-liquid chromatography that makes use of a nonpolar stationary phase and a polar liquid phase. Compare with *normal-phase partition chromatography.*
Reversible cell An electrochemical cell in which the oxidation/reduction process is reversed when the direction of current is reversed. Compare with *irreversible cell.*
Rheostat A type of *voltage divider.*
Riffle A mechanical device that isolates a fraction of material from the bulk; that which is retained becomes the *gross sample.*
Rotational states Quantized states associated with the rotation of a molecule about its center of mass.
Rubber policeman A small length of rubber tubing that has been crimped on one end; used to dislodge adherent particles of precipitate from beaker walls; also called *policeman.*

S

Salt A generic term for the species formed, in addition to water, as the result of neutralization.
Salt bridge A device that permits electrical contact between two solutions while preventing their becoming mixed.
Salt effect Synonymous with *electrolyte effect.*
Sample of data A group of replicate measurements; not to be confused with an *analytical sample.*
Sample matrix Refers to the medium that contains the analyte.
Sample mean, $\bar{x}$ The mean of a finite set of measurements.
Sample splitter A device that permits the introduction of small and reproducible portions of sample to a chromatographic column.
Sample standard deviation, s A precision estimate based on the mean, $\bar{x}$, of a finite set; also referred to as *standard deviation.*
Sample thief A device for the removal of samples from the interior of a quantity of matter.
Sampling The process of collecting a small portion whose composition is representative of the bulk of material from which it was taken.
Sampling loop A type of *sample splitter.*
Sampling uncertainty, s_s The standard deviation associated with the taking of a sample; a factor—with the method uncertainty—that contributes to the overall standard deviation of an analysis.
Sampling valve A type of *sample splitter.*
Saponification The cleavage of an ester group to regenerate the alcohol and the acid from which the ester was derived.
Saturated calomel electrode (SCE) A *calomel electrode* in which the inner solution is saturated potassium chloride; widely used as a reference electrode for potentiometry, coulometry, and voltammetry.
Schöniger apparatus A device for the combustion of samples in an oxygen-rich environment.
Screening, of samples An aid in bringing about the complete physical decomposition of a sample.
Secondary monochromator A dispersing device in spectrofluorometers that disperses the radiation produced by the fluorescing sample.
Secondary standard A substance whose purity has been established and verified by chemical analysis.
Second derivative curve A plot of $\Delta^2E/\Delta V^2$ for a potentiometric titration; function undergoes a change in sign at the inflection point in a conventional titration curve.
Sector mirror A disk with portions that are partially mirrored and partially nonreflecting; when rotated, directs radiation from the monochromator of a double-beam spectrophotometer alternately through the sample and the blank.
Selectivity The tendency for a reagent or instrumental method to respond similarly with several species; any one species thus represents a potential interference in the analysis of any of the others of the group.
Selectivity coefficient, $k_{H,B}$ A term that accounts for interferences in the equation for direct potentiometry.
Selectivity factor, α A fraction consisting of the partition ratios (K_A, K_B) for two retained species on a chromatographic column; by convention, the ratio for the more strongly held species is in the numerator.
Self-quenching The absorption of fluorescent energy by unexcited molecules of the analyte; causes a decrease in overall fluorescence intensity.
Semiconductor A material with electrical conductivity that is intermediate between a metal and an insulator.
Semimicroanalytical balance A balance with a capacity of about 30 g and a precision of 0.01 mg.
Servo system An arrangement that produces a response that is proportional to the difference between the measured quantity and the reference quantity.
SHE Abbreviation for *standard hydrogen electrode.*
Sigmoid curve An S-shaped curve that is typical of titrations.
Signal-to-noise *(S/N)* ratio The ratio of the mean analyte output signal to the standard deviation of the inherently random fluctuations of the electronic device used to measure the signal.
Significant figure convention A system of imparting to the reader information concerning the reliability of numerical data; in general, all digits known with certainty, plus the first uncertain digit, are significant.
Silica Common name for silicon dioxide; used in the manufacture of crucibles, the cells for optical analysis, and as a chromatographic support medium.
Silicon photodiode A photon detector based on a reverse-biased silicon diode; exposure to radiation creates new holes and electrons, thereby increasing conductivity.
Silver-silver chloride electrode A widely encountered reference electrode based on the half-reaction

$$AgCl(s) + e^- \rightleftharpoons Ag(s) + Cl^-$$

Simple extraction The process whereby extraction is accomplished with successive portions of fresh solvent.

Single-beam instruments Photometric instruments that require the sample and the blank to be alternately positioned in the light path.
Single column ion exchange Use is made of a low-capacity ion exchanger and an eluent with low ionic strength to permit conductometric detection with just a single column.
Single electrode potential Synonymous with *relative electrode potential.*
Single pan balance An unequal beam with the pan and weights on one side of the fulcrum and an air damper on the other; the weighing operation involves removal of standard weights in an amount equal to the mass of the object on the pan.
Sintered glass crucible Synonymous with *fritted glass crucible.*
SI units A system of measurement that makes use of seven base units; all other units are derived from these seven units.
Size exclusion chromatography Partition is based on molecular size and polarity. See *gel filtration chromatography, gel permeation chromatography.*
Slope, *m*, of a calibration line The ratio between values for the ordinate and abscissa of a straight line.
Soap-bubble meter A device for measuring gas flow, GLC.
Soda lime A mixture of calcium oxide and sodium hydroxide; an absorbent for carbon dioxide.
Sodium carbonate A basic *flux.*
Solubility-product constant, K_{sp} A numerical constant that describes the condition of equilibrium in a saturated solution of a sparingly soluble solute.
Soluble starch β-Amylose, an aqueous suspension of which is a specific indicator for iodine.
Solvent programming The systematic alteration of mobile-phase composition to optimize migration rates of solutes in a chromatographic column.
Sorbed water Nonessential water that is retained in the interstices of polymeric materials.
Sparging The removal of an unwanted dissolved gas by aeration with another gas.
Special-purpose chemicals Reagents that have been specially purified for a particular end use.
Species molarity The equilibrium concentration of a species expressed in moles per liter and symbolized with square brackets [ ]; synonymous with *equilibrium molarity.*
Specific gravity, sp gr The ratio between the density of a substance and that of water at a specified temperature (ordinarily 4°C).
Specific indicator A species that functions as an indicator for the presence of another particular species. Compare with *general redox indicators.*
Specificity Refers to methods or reagents that react with one and only one analyte.
Specific surface area The ratio between the surface area of a solid and its mass.
Spectra Plots of absorbance, transmittance, or emission as a function of wavelength (or wavenumber).
Spectral interference The obscuring of an atomic absorption line due to broad band absorption by combustion products; less commonly, interference resulting from overlapping lines.
Spectrochemical methods Synonymous with *spectrometric methods.*
Spectrofluorometer An instrument that incorporates a monochromator to disperse and isolate fluorescent radiation; some also have a monochromator to provide control over the wavelength of the excitation radiation.
Spectrograph An optical instrument with a dispersing element such as a grating or prism that provides a photographic record of spectra.
Spectrometer An instrument equipped with a monochromator or polychromator, a photodetector, and an electronic readout that displays a number that is proportional to the intensity of an isolated band.
Spectrometric methods Methods based on the absorption, the emission, or the fluorescence of electromagnetic radiation that is proportional to the amount of analyte in the sample.
Spectrophotometer A spectrometer designed for the measurement of absorbance.
Spectroscope An optical instrument similar to a spectrometer except that a movable eyepiece is used in place of an electronic detector; spectral lines are observed visually.
Spectroscopy A general term to describe analytical methods based on absorbance, chemiluminescence, emission, or fluorescence.
Spread, *w*, of data A precision estimate; synonymous with *range.*
Sputtering The process whereby an atomic cloud is produced by collisions between excited argon ions and the element on the surface of the cathode of a hollow-cathode lamp.
Square-wave polarography A variety of *pulse polarography.*
Standard addition method An alternative to production of a calibration curve; known incremental portions of the analyte are added to the sample, and instrumental readings are recorded after each addition. The method compensates for matrix interferences.
Standard deviation, σ (or *s*) Defines the bounds about the mean μ (or $\bar{x}$) within which approximately 67% of the data in a Gaussian distribution can be expected.
Standard deviation about regression, s_r The standard deviation based on deviations from a least square straight line.
Standard electrode potential, E^0 The potential (relative to the standard hydrogen electrode) of an oxidation/reduction process when the activities of all reactants and products are unity.
Standard error of the mean, σ_m The standard deviation divided by the square root of the number of measurements in the set.
Standard hydrogen electrode, SHE A gas electrode consisting of a platinized platinum electrode immersed in a solution that has a hydrogen ion activity of 1.00 and is kept saturated with hydrogen at a pressure of 1.00 atm.
Standardization Determination of the concentration of a solution through reaction—directly or indirectly—with a primary standard.
Standard reference materials (SRM) Samples of various materials in which the concentration of one or more species is known.
Standard solution A solution in which the concentration of a solute is known with high reliability.
Static methods Methods based on observation of systems in equilibrium; compare with *kinetic methods.*
Stationary phase A solid or immobilized liquid upon which analyte species are partitioned; compare with *mobile phase.*
Stationary phase mass-transfer term, $C_S u$ A measure of the ability of analyte molecules to traverse the stationary phase.

Statistical control The condition in which performance by a product or service is deemed within bounds that have been set for quality assurance; defined by the upper and lower control limits.
Statistical sample A finite set of measurements, considered to have been drawn from an infinite number of possible measurements.
Steady-state approximation The assumption that the concentration of the intermediate in a multistep reaction remains essentially constant.
Stirrup The link between the beam of a mechanical balance and its pan (or pans).
Stockholm convention A set of conventions relating to electrochemical cells and their potentials; formulated in 1953.
Stoichiometry Refers to the combining ratios among molar quantities of species in a chemical reaction.
Stokes shifts Differences in wavelengths absorbed and subsequently emitted by a fluorescent species.
Stop-flow injection In HPLC, introduction of the sample at the head of the column while solvent flow is temporarily discontinued.
Stopped-flow mixing A technique in which reactants are rapidly mixed, the course of the reaction being observed downstream after the flow is stopped.
Stray radiation Radiation of wavelength other than the wavelength selected for optical measurement.
Strong acids and bases Acids and bases that are completely dissociated in a particular solvent. See *leveling solvents.*
Strong electrolytes Solutes that are completely dissociated into ions in a particular solvent.
Student's *t* See *t-test.*
Substitution reactions Organic reactions in which existing atoms or groups are replaced by others.
Substrate A substance acted upon, usually by an enzyme.
Successive approximation method A procedure for solving higher order equations through the use of intermediate estimates of the answer.
Sulfide separations The use of sulfide precipitation for the selective separation of cations.
Supercritical fluid A substance that has been heated above its critical temperature; has properties intermediate between those of a liquid and a gas.
Supercritical fluid chromatography A type of chromatography involving a supercritical fluid as the eluent.
Supersaturation An unstable condition that exists when the solubility of a solute exceeds its equilibrium solubility.
Support coated open tubular (SCOT) columns Capillary GLC columns, the interior walls of which are lined with a solid support.
Supporting electrolyte A salt added to the solution in a polarographic cell to cause the rate of migration of the analyte to the electrode surface to be essentially independent of the applied potential.
Suppressor-based chromatography A chromatographic system involving a column or a membrane located between the analytical column and a conductivity detector, the purpose of which is to convert ions of the eluting solvent into nonconducting species while passing ions of the sample.
Surface adsorption The entrainment of foreign species on the surface of a solid.
Swamping The introduction of a potential interferent to both calibration standards and the solution of the analyte in order to minimize the effect of the interferent in the sample matrix.
Synge, R.L.M. With A.J.P. Martin, pioneered the development of modern chromatography.
Systematic error Errors that have a known source, affect measurements in one and only one way, and can, in principle, be accounted for; also called *determinate error.*

T

0% *T* adjustment A calibration step to eliminate *dark current* from the response of a spectrophotometer.
100% *T* adjustment Adjustment of a photometric instrument to register 100% transmittance with a blank in the light path.
Tailing A nonideal condition in a chromatogram where the latter portions of peaks are drawn out. Compare with *fronting.*
Tare Compensation for the mass of a container on an analytical balance.
Temperature programming In GLC, the systematic adjustment of column temperature to optimize migration rates for solutes.
THAM *tris*-(Hydroxymethyl)aminomethane, a primary standard for bases.
Thermal conductivity detector Synonymous with *katharometer.*
Thermionic detector (TID) A detector for GLC, similar to a flame ionization detector; particularly sensitive for analytes that contain nitrogen or phosphorus.
Thermistor A temperature-sensing semiconductor; used in some *bolometers.*
Thermodynamic equilibrium constant, *K* The equilibrium constant expressed in terms of the activities for all reactants and products.
Thermodynamic potential, *E* The potential of an electrochemical cell in which there is no current.
TISAB (Total Ion Strength Adjustment Buffer) A solution used to swamp the effect of electrolytes on the activity of the analyte in the sample matrix.
Titration The process of systematically adding a stoichiometrically equivalent quantity of standard solution to a known quantity of sample.
Titration error The difference between the titrant volumes needed to reach the equivalence point and the end point, respectively.
Titrator An instrument that performs titrations automatically.
Titrimetry The process of systematically introducing an amount of titrant that is chemically equivalent to the quantity of analyte in a sample.
Total ionic strength adjustment buffer See *TISAB.*
Transducer A device that converts a chemical or physical phenomenon into an electrical signal.
Transfer pipet Synonymous with *volumetric pipet.*
Transition pH range The span of acidities (frequently about 2 pH units) over which an acid/base indicator changes from its pure acid color to that of its conjugate base.
Transition potential The range in E_{system} over which an oxidation/reduction indicator changes from the color of its reduced form to that of its oxidized form.
Transmittance, *T* The ratio of the power of a beam after it has traversed an absorbing medium to its original power; often expressed as a percentage

$$\%T = \log P/P_0 \times 100\%$$

Transverse wave Wave motion in which the direction of displacement is perpendicular to the direction of propagation.
Triple-beam balance A rugged laboratory balance that is used to weigh approximate amounts.
TRIS Synonymous with *THAM.*
Tswett, M. The inventor of chromatography.
***t*-test** A statistical test used to decide whether a measurement belongs to a population with a given level of confidence; used with s and $\bar{x}$ when estimated σ and μ are not available.
Tungsten filament lamp A convenient source of visible and near-infrared radiation.
Tungsten-halogen lamp An improved version of the tungsten filament lamp.
Turbulent flow Describes the random motion of liquid in the bulk of a flowing solution. Compare with *laminar flow.*
Twin detector system In GLC, the location of one detector ahead of the column and another detector at its terminus; designed to minimize the effects of column variables.
Tyndall effect The scattering of radiation by particles that have colloidal dimensions.

U

Ultramicro voltammetric electrode A very small electrode; because *IR* effects are so small, there is no need for a third reference electrode.
Ultraviolet/visible detector, HPLC Use of radiation from a UV/visible source to monitor eluted species as they exit a chromatographic column.
Ultraviolet/visible region The region of the electromagnetic spectrum between 180 and 780 nm; associated with electronic transitions in atoms and molecules.
Universe of data Synonymous with a *population of data.*

V

Valinomycin An antibiotic that has also found application in a membrane electrode for potassium.
van Deemter equation An equation that relates the effects of eddy diffusion, longitudinal diffusion, and mass transport to plate height.
Variance, s^2 A precision estimate consisting of the square of the standard deviation. Also a measure of column performance; given the symbol τ^2 where the abscissa of the chromatogram has units of time. Variances are additive.
V-blender A device that is used to thoroughly mix dry samples.
Velocity, v_i, of electromagnetic radiation *In vacuo,* 3×10^{10} cm/sec.
Vernier An aid for making estimates between graduation marks on a scale.
Vibrational deactivation A very efficient process in which excited molecules relax to the lowest vibrational level of an electronic state.
Vibrational relaxation Synonymous with *vibrational deactivation.*
Vibrational transition Transitions between vibrational states of the ground electronic state that are responsible for infrared absorption.
Visible radiation That portion of the electromagnetic spectrum (340–780 nm) to which the human eye is perceptive.
Volatilization The process of converting a liquid (or a solid) to the vapor state.
Volatilization methods of analysis A variant of the gravimetric method based on mass loss caused by heating or ignition; the analysis is completed by determining either the mass of the volatilized matter or the mass of residue following volatilization.
Voltage divider A device that provides voltages ranging from zero to the maximum of the power supply.
Voltage follower An operational amplifier that isolates voltage sources from their measurement circuits.
Voltaic cell Synonymous with *galvanic cell.*
Voltammetric wave Synonymous with *voltammogram.*
Voltammetry A collective term for a large group of instrumental techniques that are based on the current resulting from the application of a continuously varying voltage to small electrodes. *Hydrodynamic voltammetry* is performed in stirred solutions; *polarography* involves diffusion-controlled reaction of analytes at a dropping mercury electrode.
Voltammogram A plot of current as a function of the potential applied to a working electrode.
Volume change correction Compensation of measurements where differences in volume are involved.
Volume percent (v/v) The ratio between the volume of solute and the volume of its solution, multiplied by 100.
Volumetric flask A device for the preparation of precise volumes of solution.
Volumetric methods Methods of analysis in which the final measurement is a volume, commonly that of the standard titrant needed to react with the analyte in a known quantity of sample *(volumetric titrimetry);* the measurement of a gas is occasionally involved.
Volumetric pipet A device that will deliver a precise volume from one container to another; also called a *measuring pipet.*

W

Walden reductor A column packed with finely divided silver granules; used to prereduce analytes.
Wall coated open tubular (WCOT) column A capillary column coated with a thin layer of stationary phase.
Water of constitution Essential water that is derived from the molecular composition of the species.
Water of crystallization Essential water that is an integral part of the crystal structure of a solid.
Wavelength, λ, of electromagnetic radiation The distance between successive maxima (or minima) of a wave.
Wavenumber, $\bar{\nu}$ The reciprocal of wavelength; has units of cm^{-1}.
Wavelength selector A device that limits the range of wavelengths used for an optical measurement.
Wave properties, electromagnetic radiation Behavior of radiation best described in terms of wave properties.
Weak acid/conjugate base pairs In the Brønsted-Lowry view, solute pairs that differ from one another by one proton.
Weak acids and bases Acids and bases that are only partially dissociated in a particular solvent. Compare with *differentiating solvents.*

Weak electrolytes Solutes that are incompletely dissociated into ions in a particular solvent.
Weighing bottle A lightweight container for the storage and weighing of analytical samples.
Weighing by difference The process of weighing a container plus the sample, followed by weighing the container after the sample has been removed.
Weighing form In gravimetric analysis, the species collected whose mass is proportional to the amount of analyte in the sample.
Weight The attraction between an object and its surroundings, terrestrially, the earth.
Weight molarity, M_w The concentration of titrant expressed as millimoles per gram. See *gravimetric titrimetry.*
Weight percent (w/w) The ratio between the mass of a solute and the mass of its solution, multiplied by 100.
Weight titrimetry Synonymous with *gravimetric titrimetry.*
Weight/volume percent (w/v) The ratio between the mass of a solute and the volume of solution in which it is dissolved, multiplied by 100.
Wet ashing The use of strong oxidizing reagents to decompose the organic matter in a sample.
Windows, of cells Surfaces of cells through which radiation passes.

Z

Zero percent *T* adjustment A calibration step to eliminate *dark current* from the response of a photometric instrument.
Zimmermann-Reinhardt reagent A preparation that prevents the induced oxidation of chloride ion by permanganate during the titration of iron(II).
Zones, chromatographic Synonymous with *chromatographic bands.*
Zwitterion The species that results from the transfer of a proton from an acidic group to an acceptor site on the same molecule.

APPENDIX

1

Selected References to the Literature of Analytical Chemistry

Treatises

As used here, the term *treatise* means a comprehensive presentation of one or more broad areas of analytical chemistry.

N. H. Furman and F. J. Welcher, Eds., *Standard Methods of Chemical Analysis,* 6th ed. New York: Van Nostrand, 1962–1966. In five parts; largely devoted to specific applications.

I. M. Kolthoff and P. J. Elving, Eds., *Treatise on Analytical Chemistry.* New York: Wiley, 1961–1986. Part I, 2nd ed. (14 volumes) is devoted to theory; Part II (17 volumes) deals with analytical methods for inorganic and organic compounds; Part III (4 volumes) treats industrial analytical chemistry.

B. W. Rossitor and R. C. Baetzold, Eds., *Physical Methods of Chemistry,* 2nd ed. New York: Wiley, 1986–1993. This series consists of 12 volumes devoted to various types of physical and chemical measurements performed by chemists.

G. Svehla, C. L. Wilson, and D. W. Wilson, Eds., *Comprehensive Analytical Chemistry.* New York: Elsevier, 1959–1992. To 1992, 29 volumes of this work have appeared.

Official Methods of Analysis

These publications are often single volumes that provide a useful source of analytical methods for the determination of specific substances in articles of commerce. The methods have been developed by various scientific societies and serve as standards in arbitration as well as in the courts.

Standard Methods for the Examination of Water and Wastewater, 18th ed., A. E. Greenberg, Ed. New York: American Public Health Association, 1992.

ASTM Book of Standards. Philadelphia: American Society for Testing Materials. This 48-volume work is revised annually and contains methods for both physical testing and chemical analysis. Volume 3.06, *Chemical Analysis of Metals and Related Materials,* is one of the useful volumes for the chemist.

C. A. Watson, *Official and Standardized Methods of Analysis,* 3rd ed. London: Royal Society of Chemistry, 1994.

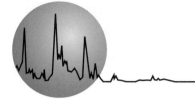

Official Methods of Analysis, 15th ed. Washington, DC: Association of Official Analytical Chemists, 1990. This is a very useful source of methods for the analysis of such materials as drugs, food, pesticides, agricultural materials, cosmetics, vitamins, and nutrients. It is revised every five years.

Review Serials

Analytical Chemistry, Fundamental Reviews. These reviews appear biennially as a supplement to the April issue of *Analytical Chemistry* in even-numbered years. Most of the significant developments occurring in the past two years in 30 or more areas of analytical chemistry are covered.

Analytical Chemistry, Application Reviews. These reviews appear as part of *Analytical Chemistry* biennially in odd-numbered years. The articles are devoted to recent analytical work in 18 specific areas, such as water analysis, clinical chemistry, petroleum products, and air pollution.

Critical Reviews in Analytical Chemistry. This publication appears quarterly and provides in-depth articles covering the newest developments in the analysis of biochemical substances.

D. Glick, Ed., *Methods of Biochemical Analysis.* This annual publication consists of a series of review articles covering the newest developments in the analysis of biochemical substances.

Tabular Compilations

A. E. Martell and R. M. Smith, *Critical Stability Constants.* New York: Plenum Press, 1974–1989. Six volumes.

L. Meites, Ed., *Handbook of Analytical Chemistry.* New York: McGraw-Hill, 1963.

G. Milazzo, S. Caroli, and V. K. Sharma, *Tables of Standard Electrode Potentials.* New York: Wiley, 1978.

A. J. Bard, R. Parsons, and T. Jordan, Eds., *Standard Potentials in Aqueous Solution.* New York: Marcel Dekker, 1985.

Advanced Analytical and Instrumental Textbooks

J. N. Butler, *Ionic Equilibrium: A Mathematical Approach.* Reading, MA: Addison-Wesley, 1964.

G. D. Christian and J. E. O'Reilly, *Instrumental Analysis,* 2nd ed. Boston: Allyn and Bacon, 1986.

H. A. Laitinen and W. E. Harris, *Chemical Analysis,* 2nd ed. New York: McGraw-Hill, 1975.

E. D. Olsen, *Modern Optical Methods of Analysis.* New York: McGraw-Hill, 1975.

D. A. Skoog and J. J. Leary, *Principles of Instrumental Analysis,* 4th ed. Philadelphia: Saunders College Publishing, 1992.

H. Strobel and W. R. Heineman, *Chemical Instrumentation: A Systematic Approach,* 3rd ed. Boston: Addison-Wesley, 1989.

H. H. Willard, L. L. Merritt Jr., J. A. Dean, and F. A. Settle, *Instrumental Methods of Analysis,* 7th ed. New York: Wadsworth, 1988.

Monographs

Hundreds of monographs devoted to limited areas of analytical chemistry are available. In general, these are authored by experts and are excellent sources of information. Representative monographs in various areas are listed here.

Gravimetric and Titrimetric Methods

M. R. F. Ashworth, *Titrimetric Organic Analysis.* New York: Interscience, 1965. Two volumes.

L. Erdey, *Gravimetric Analysis.* Oxford: Pergamon, 1965.

J. S. Fritz, *Acid-Base Titration in Nonaqueous Solvents.* Boston: Allyn and Bacon, 1973.

W. F. Hillebrand, G. E. F. Lundell, H. A. Bright, and J. I. Hoffman, *Applied Inorganic Analysis,* 2nd ed. New York: Wiley, 1953, reissued 1980.

I. M. Kolthoff, V. A. Stenger, and R. Belcher, *Volumetric Analysis.* New York: Interscience, 1942–1957. Three volumes.

T. S. Ma and R. C. Ritner, *Modern Organic Elemental Analysis.* New York: Marcel Dekker, 1979.

W. Wagner and C. J. Hull, *Inorganic Titrimetric Analysis.* New York: Marcel Dekker, 1971.

Organic Analysis

S. Siggia and J. G. Hanna, *Quantitative Organic Analysis via Functional Groups,* 4th ed. New York: Wiley, 1979.

F. T. Weiss, *Determination of Organic Compounds: Methods and Procedures.* New York: Wiley-Interscience, 1970.

Spectrometric Methods

D. F. Boltz and J. A. Howell, *Colorimetric Determination of Nonmetals,* 2nd ed. New York: Wiley-Interscience, 1978.

J. D. Ingle and S. R. Crouch, *Analytical Spectroscopy.* Englewood Cliffs, NJ: Prentice-Hall, 1988.

L. H. Lajunen, *Spectrochemical Analysis by Atomic Absorption and Emission.* Boca Raton, FL: CRC Press, 1992.

W. J. Price, *Spectrochemical Analysis by Atomic Absorption.* London: Heyden, 1979.

E. B. Sandell and H. Onishi, *Colorimetric Determination of Traces of Metals,* 4th ed. New York: Interscience, 1978–1989. Two volumes.

F. D. Snell, *Photometric and Fluorometric Methods of Analysis.* New York: Wiley, 1978–1981. Two volumes.

Electroanalytical Methods

A. J. Bard and L. R. Faulkner, *Electrochemical Methods.* New York: Wiley, 1980.

P. T. Kissinger and W. R. Heineman, Eds., *Laboratory Techniques in Electroanalytical Chemistry.* New York: Marcel Dekker, 1984.

J. J. Lingane, *Electroanalytical Chemistry,* 2nd ed. New York: Interscience, 1954.

D. T. Sawyer and J. L. Roberts Jr., *Experimental Electrochemistry for Chemists.* New York: Wiley, 1974.

Analytical Separations

E. Heftmann, Ed., *Chromatography: Fundamentals and Applications of Chromatography and Electrophoretic Methods,* Part A: *Fundamentals,* Part B: *Applications.* New York: Elsevier, 1983.

P. Sewell and B. Clarke, *Chromatographic Separations.* New York: Wiley, 1988.

J. A. Jonsson, Ed., *Chromatographic Theory and Basic Principles.* New York: Marcel Dekker, 1987.

R. M. Smith, *Gas and Liquid Chromatography in Analytical Chemistry.* New York: Wiley, 1988.

E. Katz, *Quantitative Analysis Using Chromatographic Techniques.* New York: Wiley, 1987.

J. C. Giddings, *Unified Separation Science.* New York: Wiley, 1991.

Miscellaneous

R. G. Bates, *Determination of pH: Theory and Practice,* 2nd ed. New York: Wiley, 1973.

R. Bock, *Decomposition Methods in Analytical Chemistry.* New York: Wiley, 1979.

G. H. Morrison, Ed., *Trace Analysis.* New York: Interscience, 1965.

D. D. Perrin, *Masking and Demasking Chemical Reactions.* New York: Wiley, 1970.

M. Pinta, *Modern Methods for Trace Element Analysis.* Ann Arbor, MI: Ann Arbor Science, 1978.

W. Rieman and H. F. Walton, *Ion Exchange in Analytical Chemistry.* Oxford: Pergamon, 1970.

W. J. Williams, *Handbook of Anion Determination.* London: Butterworths, 1979.

Periodicals

Numerous journals are devoted to analytical chemistry; these are primary sources of information in the field. Some of the best-known titles are listed here. (The boldface portion of the title is the *Chemical Abstracts* abbreviation for the journal.)

American **Lab**oratory

Analyst, The

Analytical **Biochem**istry

Analytical **Chem**istry

Analytica **Chim**ica **Acta**

Analytical **Instrum**entation

Analytical **Lett**ers

Applied **Spectros**copy

Clinical **Chem**istry

Fresenius' Journal of **Anal**ytical **Chem**istry

Journal of the **Assoc**iation of **Off**icial **Anal**ytical **Chem**ists

Journal of **Chromatogr**aphic **Sci**ence

Journal of Chromatography
Journal of Electroanalytical Chemistry and Interfacial Electrochemistry
Microchemical Journal
Mikrochimica Acta
Separation Science
Spectrochimica Acta
Talanta
Zeitschrift für Analytische Chemie

APPENDIX

2

Solubility Product Constants at 25°C

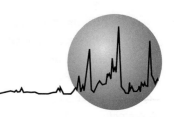

Compound	Formula	K_{sp}	Notes
Aluminum hydroxide	$Al(OH)_3$	3×10^{-34}	
Barium carbonate	$BaCO_3$	5.0×10^{-9}	
Barium chromate	$BaCrO_4$	2.1×10^{-10}	
Barium hydroxide	$Ba(OH)_2 \cdot 8H_2O$	3×10^{-4}	
Barium iodate	$Ba(IO_3)_2$	1.57×10^{-9}	
Barium oxalate	BaC_2O_4	1×10^{-6}	
Barium sulfate	$BaSO_4$	1.1×10^{-10}	
Cadmium carbonate	$CdCO_3$	1.8×10^{-14}	
Cadmium hydroxide	$Cd(OH)_2$	4.5×10^{-15}	
Cadmium oxalate	CdC_2O_4	9×10^{-8}	
Cadmium sulfide	CdS	1×10^{-27}	
Calcium carbonate	$CaCO_3$	4.5×10^{-9}	Calcite
	$CaCO_3$	6.0×10^{-9}	Aragonite
Calcium fluoride	CaF_2	3.9×10^{-11}	
Calcium hydroxide	$Ca(OH)_2$	6.5×10^{-6}	
Calcium oxalate	$CaC_2O_4 \cdot H_2O$	1.7×10^{-9}	
Calcium sulfate	$CaSO_4$	2.4×10^{-5}	
Cobalt(II) carbonate	$CoCO_3$	1.0×10^{-10}	
Cobalt(II) hydroxide	$Co(OH)_2$	1.3×10^{-15}	
Cobalt(II) sulfide	CoS	5×10^{-22}	α
	CoS	3×10^{-26}	β
Copper(I) bromide	$CuBr$	5×10^{-9}	
Copper(I) chloride	$CuCl$	1.9×10^{-7}	
Copper(I) hydroxide*	Cu_2O	2×10^{-15}	
Copper(I) iodide	CuI	1×10^{-12}	
Copper(I) thiocyanate	$CuSCN$	4.0×10^{-14}	
Copper(II) hydroxide	$Cu(OH)_2$	4.8×10^{-20}	
Copper(II) sulfide	CuS	8×10^{-37}	
Iron(II) carbonate	$FeCO_3$	2.1×10^{-11}	
Iron(II) hydroxide	$Fe(OH)_2$	4.1×10^{-15}	
Iron(II) sulfide	FeS	8×10^{-19}	
Iron(III) hydroxide	$Fe(OH)_3$	2×10^{-39}	
Lanthanum iodate	$La(IO_3)_3$	1.0×10^{-11}	
Lead carbonate	$PbCO_3$	7.4×10^{-14}	
Lead chloride	$PbCl_2$	1.7×10^{-5}	
Lead chromate	$PbCrO_4$	3×10^{-13}	

Compound	Formula	K_{sp}	Notes
Lead hydroxide	PbO†	8×10^{-16}	Yellow
	PbO†	5×10^{-16}	Red
Lead iodide	PbI_2	7.9×10^{-9}	
Lead oxalate	PbC_2O_4	8.5×10^{-9}	$\mu = 0.05$
Lead sulfate	$PbSO_4$	1.6×10^{-8}	
Lead sulfide	PbS	3×10^{-28}	
Magnesium ammonium phosphate	$MgNH_4PO_4$	3×10^{-13}	
Magnesium carbonate	$MgCO_3$	3.5×10^{-8}	
Magnesium hydroxide	$Mg(OH)_2$	7.1×10^{-12}	
Manganese carbonate	$MnCO_3$	5.0×10^{-10}	
Manganese hydroxide	$Mn(OH)_2$	2×10^{-13}	
Manganese sulfide	MnS	3×10^{-11}	Pink
	MnS	3×10^{-14}	Green
Mercury(I) bromide	Hg_2Br_2	5.6×10^{-23}	
Mercury(I) carbonate	Hg_2CO_3	8.9×10^{-17}	
Mercury(I) chloride	Hg_2Cl_2	1.2×10^{-18}	
Mercury(I) iodide	Hg_2I_2	4.7×10^{-29}	
Mercury(I) thiocyanate	$Hg_2(SCN)_2$	3.0×10^{-20}	
Mercury(II) hydroxide	HgO‡	3.6×10^{-26}	
Mercury(II) sulfide	HgS	2×10^{-53}	Black
	HgS	5×10^{-54}	Red
Nickel carbonate	$NiCO_3$	1.3×10^{-7}	
Nickel hydroxide	$Ni(OH)_2$	6×10^{-16}	
Nickel sulfide	NiS	4×10^{-20}	α
	NiS	1.3×10^{-25}	β
Silver arsenate	Ag_3AsO_4	6×10^{-23}	
Silver bromide	$AgBr$	5.0×10^{-13}	
Silver carbonate	Ag_2CO_3	8.1×10^{-12}	
Silver chloride	$AgCl$	1.82×10^{-10}	
Silver chromate	$AgCrO_4$	1.2×10^{-12}	
Silver cyanide	$AgCN$	2.2×10^{-16}	
Silver iodate	$AgIO_3$	3.1×10^{-8}	
Silver iodide	AgI	8.3×10^{-17}	
Silver oxalate	$Ag_2C_2O_4$	3.5×10^{-11}	
Silver sulfide	Ag_2S	8×10^{-51}	
Silver thiocyanate	$AgSCN$	1.1×10^{-12}	
Strontium carbonate	$SrCO_3$	9.3×10^{-10}	
Strontium oxalate	SrC_2O_4	5×10^{-8}	
Strontium sulfate	$SrSO_4$	3.2×10^{-7}	
Thallium(I) chloride	$TlCl$	1.8×10^{-4}	
Thallium(I) sulfide	Tl_2S	6×10^{-22}	
Zinc carbonate	$ZnCO_3$	1.0×10^{-10}	
Zinc hydroxide	$Zn(OH)_2$	3.0×10^{-16}	Amorphous
Zinc oxalate	ZnC_2O_4	8×10^{-9}	
Zinc sulfide	ZnS	2×10^{-25}	α
	ZnS	3×10^{-23}	β

Most of these data were taken from A. E. Martell and R. M. Smith, *Critical Stability Constants,* Vol. 3–6. New York: Plenum, 1976–1989. In most cases, the ionic strength was 0.0 and the temperature 25°C.

*$Cu_2O(s) + H_2O \rightleftharpoons 2Cu^+ + 2OH^-$

†$PbO(s) + H_2O \rightleftharpoons Pb^{2+} + 2OH^-$

‡$HgO(s) + H_2O \rightleftharpoons Hg^{2+} + 2OH^-$

APPENDIX

3

Acid Dissociation Constants at 25°C

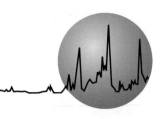

Acid	Formula	K_1	K_2	K_3
Acetic acid	CH_3COOH	1.75×10^{-5}		
Ammonium ion	NH_4^+	5.70×10^{-10}		
Anilinium ion	$C_6H_5NH_3^+$	2.51×10^{-5}		
Arsenic acid	H_3AsO_4	5.8×10^{-3}	1.1×10^{-7}	3.2×10^{-12}
Arsenous acid	H_3AsO_3	5.1×10^{-10}		
Benzoic acid	C_6H_5COOH	6.28×10^{-5}		
Boric acid	H_3BO_3	5.81×10^{-10}		
1-Butanoic acid	$CH_3CH_2CH_2COOH$	1.52×10^{-5}		
Carbonic acid	H_2CO_3	4.45×10^{-7}	4.69×10^{-11}	
Chloroacetic acid	$ClCH_2COOH$	1.36×10^{-3}		
Citric acid	$HOOC(OH)C(CH_2COOH)_2$	7.45×10^{-4}	1.73×10^{-5}	4.02×10^{-7}
Dimethyl ammonium ion	$(CH_3)_2NH_2^+$	1.68×10^{-11}		
Ethanol ammonium ion	$HOC_2H_4NH_3^+$	3.18×10^{-10}		
Ethyl ammonium ion	$C_2H_5NH_3^+$	2.31×10^{-11}		
Ethylene diammonium ion	$^+H_3NCH_2CH_2NH_3^+$	1.42×10^{-7}	1.18×10^{-10}	
Formic acid	$HCOOH$	1.80×10^{-4}		
Fumaric acid	*trans*-$HOOCCH{:}CHCOOH$	8.85×10^{-4}	3.21×10^{-5}	
Glycolic acid	$HOCH_2COOH$	1.47×10^{-4}		
Hydrazinium ion	$H_2NNH_3^+$	1.05×10^{-8}		
Hydrazoic acid	HN_3	2.2×10^{-5}		
Hydrogen cyanide	HCN	6.2×10^{-10}		
Hydrogen fluoride	HF	6.8×10^{-4}		
Hydrogen peroxide	H_2O_2	2.2×10^{-12}		
Hydrogen sulfide	H_2S	9.6×10^{-8}	1.3×10^{-14}	
Hydroxyl ammonium ion	$HONH_3^+$	1.10×10^{-6}		
Hypochlorous acid	$HOCl$	3.0×10^{-8}		
Iodic acid	HIO_3	1.7×10^{-1}		
Lactic acid	$CH_3CHOHCOOH$	1.38×10^{-4}		
Maleic acid	*cis*-$HOOCCH{:}CHCOOH$	1.3×10^{-2}	5.9×10^{-7}	
Malic acid	$HOOCCHOHCH_2COOH$	3.48×10^{-4}	8.00×10^{-6}	
Malonic acid	$HOOCCH_2COOH$	1.42×10^{-3}	2.01×10^{-6}	
Mandelic acid	$C_6H_5CHOHCOOH$	4.0×10^{-4}		
Methyl ammonium ion	$CH_3NH_3^+$	2.3×10^{-11}		
Nitrous acid	HNO_2	7.1×10^{-4}		
Oxalic acid	$HOOCCOOH$	5.60×10^{-2}	5.42×10^{-5}	
Periodic acid	H_5IO_6	2×10^{-2}	5×10^{-9}	

Acid	Formula	K_1	K_2	K_3
Phenol	C_6H_5OH	1.00×10^{-10}		
Phosphoric acid	H_3PO_4	7.11×10^{-3}	6.32×10^{-8}	4.5×10^{-13}
Phosphorous acid	H_3PO_3	3×10^{-2}	1.62×10^{-7}	
o-Phthalic acid	$C_6H_4(COOH)_2$	1.12×10^{-3}	3.91×10^{-6}	
Picric acid	$(NO_2)_3C_6H_2OH$	4.3×10^{-1}		
Piperidinium ion	$C_5H_{11}NH^+$	7.50×10^{-12}		
Propanoic acid	CH_3CH_2COOH	1.34×10^{-5}		
Pyridinium ion	$C_5H_5NH^+$	5.90×10^{-6}		
Pyruvic acid	$CH_3COCOOH$	3.2×10^{-3}		
Salicylic acid	$C_6H_4(OH)COOH$	1.06×10^{-3}		
Sulfamic acid	H_2NSO_3H	1.03×10^{-1}		
Succinic acid	$HOOCCH_2CH_2COOH$	6.21×10^{-5}	2.31×10^{-6}	
Sulfuric acid	H_2SO_4	Strong	1.02×10^{-2}	
Sulfurous acid	H_2SO_3	1.23×10^{-2}	6.6×10^{-8}	
Tartaric acid	$HOOC(CHOH)_2COOH$	9.20×10^{-4}	4.31×10^{-5}	
Thiocyanic acid	HSCN	0.13		
Thiosulfuric acid	$H_2S_2O_3$	0.3	2.5×10^{-2}	
Trichloroacetic acid	Cl_3CCOOH	3		
Trimethyl ammonium ion	$(CH_3)_3NH^+$	1.58×10^{-10}		

Most data are for zero ionic strength. (From A. E. Martell and R. M. Smith, *Critical Stability Constants,* Vol. 1–6. New York: Plenum Press, 1974–1989.)

APPENDIX 4

Formation Constants at 25°C

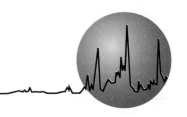

Ligand	Cation	log K_1	log K_2	log K_3	log K_4	Ionic Strength
Acetate (CH_3COO^-)	Ag^+	0.73	−0.9			0.0
	Ca^{2+}	1.18				0.0
	Cd^{2+}	1.93	1.22			0.0
	Cu^{2+}	2.21	1.42			0.0
	Fe^{3+}	3.38*	3.1*	1.8*		0.1
	Hg^{2+}	$\log K_1K_2 = 8.45$				0.0
	Mg^{2+}	1.27				0.0
	Pb^{2+}	2.68	1.40			0.0
Ammonia (NH_3)	Ag^+	3.31	3.91			0.0
	Cd^{2+}	2.55	2.01	1.34	0.84	0.0
	Co^{2+}	1.99*	1.51	0.93	0.64	0.0
			$\log K_5 = 0.06$	$\log K_6 = -0.74$		0.0
	Cu^{2+}	4.04	3.43	2.80	1.48	0.0
	Hg^{2+}	8.8	8.6	1.0	0.7	0.5
	Ni^{2+}	2.72	2.17	1.66	1.12	0.0
			$\log K_5 = 0.67$	$\log K_6 = -0.03$		0.0
	Zn^{2+}	2.21	2.29	2.36	2.03	0.0
Bromide (Br^-)	Ag^+		$Ag^+ + 2Br^- \rightleftharpoons AgBr_2^-$	$\log K_1K_2 = 7.5$		0.0
	Hg^{2+}	9.00	8.1	2.3	1.6	0.5
	Pb^{2+}	1.77				0.0
Chloride (Cl^-)	Ag^+		$Ag^+ + 2Cl^- \rightleftharpoons AgCl_2^-$	$\log K_1K_2 = 5.25$		0.0
			$AgCl_2^- + Cl^- \rightleftharpoons AgCl_3^{2-}$	$\log K_3 = 0.37$		0.0
	Cu^+		$Cu^+ + 2Cl^- \rightleftharpoons CuCl_2^-$	log = 5.5*		0.0
	Fe^{3+}	1.48	0.65			0.0
	Hg^{2+}	7.30	6.70	1.0	0.6	0.0
	Pb^{2+}		$Pb^{2+} + 3Cl^- \rightleftharpoons PbCl_3^-$	$\log K_1K_2K_3 = 1.8$		0.0
	Sn^{2+}	1.51	0.74	−0.3	−0.5	0.0
Cyanide (CN^-)	Ag^+		$Ag^+ + 2CN^- \rightleftharpoons Ag(CN)_2^-$	$\log K_1K_2 = 20.48$		0.0
	Cd^{2+}	6.01	5.11	4.53	2.27	0.0
	Hg^{2+}	17.00	15.75	3.56	2.66	0.0
	Ni^{2+}		$Ni^{2+} + 4CN^- \rightleftharpoons Ni(CN)_4^-$	$\log K_1K_2K_3K_4 = 30.22$		0.0
	Zn^{2+}	$\log K_1K_2 = 11.07$		4.98	3.57	0.0
EDTA	See Table 14-1, page 282.					
Fluoride (F^-)	Al^{3+}	7.0	5.6	4.1	2.4	0.0
	Fe^{3+}	5.18	3.89	3.03		0.0

Ligand	Cation	log K_1	log K_2	log K_3	log K_4	Ionic Strength
Hydroxide (OH^-)	Al^{3+}		$Al^{3+} + 4OH^- \rightleftharpoons Al(OH)_4^-$	$\log K_1K_2K_3K_4 = 33.4$		0.0
	Cd^{2+}	3.9	3.8			0.0
	Cu^{2+}	6.5				0.0
	Fe^{2+}	4.6				0.0
	Fe^{3+}	11.81	11.5			0.0
	Hg^{2+}	10.60	11.2			0.0
	Ni^{2+}	4.1	4.9	3		0.0
	Pb^{2+}	6.4	$Pb^{2+} + 3OH^- \rightleftharpoons Pb(OH)_3^-$	$\log K_1K_2K_3 = 13.9$		0.0
	Zn^{2+}	5.0	$Zn^{2+} + 4OH^- \rightleftharpoons Zn(OH)_4^{2-}$	$\log K_1K_2K_3K_4 = 15.5$		0.0
Iodide (I^-)	Cd^{2+}	2.28	1.64	1.0	1.0	0.0
	Cu^+		$Cu^+ + 2I^- \rightleftharpoons CuI_2^-$	$\log K_1K_2 = 8.9$		0.0
	Hg^{2+}	12.87	10.95	3.8	2.2	0.5
	Pb^{2+}		$Pb^{2+} + 3I^- \rightleftharpoons PbI_3^-$	$\log K_1K_2K_3 = 3.9$		0.0
			$Pb^{2+} + 4I^- \rightleftharpoons PbI_4^{2-}$	$\log K_1K_2K_3K_4 = 4.5$		0.0
Oxalate ($C_2O_4^{2-}$)	Al^{3+}	5.97	4.96	5.04		0.1
	Ca^{2+}	3.19				0.0
	Cd^{2+}	2.73	1.4	1.0		1.0
	Fe^{3+}	7.58	6.23	4.8		1.0
	Mg^{2+}	3.42(18°C)				
	Pb^{2+}	4.20	2.11			1.0
Sulfate (SO_4^{2-})	Al^{3+}	3.89				0.0
	Ca^{2+}	2.13				0.0
	Cu^{2+}	2.34				0.0
	Fe^{3+}	4.04	1.34			0.0
	Mg^{2+}	2.23				0.0
Thiocyanate (SCN^-)	Cd^{2+}	1.89	0.89	0.1		0.0
	Cu^+		$Cu^+ + 3SCN^- \rightleftharpoons Cu(SCN)_3^{2-}$	$\log K_1K_2K_3 = 11.60$		0.0
	Fe^{3+}	3.02	0.62*			0.0
	Hg^{2+}		$\log K_1K_2 = 17.26$	2.7	1.8	0.0
	Ni^{2+}	1.76				0.0
Thiosulfate ($S_2O_3^{2-}$)	Ag^+	8.82*	4.7	0.7		0.0
	Cu^{2+}		$\log K_1K_2 = 6.3$			0.0
	Hg^{2+}		$\log K_1K_2 = 29.23$	1.4		0.0

Data from A. E. Martell and R. M. Smith, *Critical Stability Constants,* Vol. 3–6. New York: Plenum Press, 1974–1989.

*20°C

APPENDIX 5

Some Standard and Formal Electrode Potentials

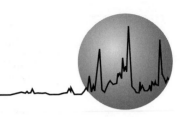

Half-Reaction	E^0, V*	Formal Potential, V†
Aluminum		
$Al^{3+} + 3e^- \rightleftharpoons Al(s)$	−1.662	
Antimony		
$Sb_2O_5(s) + 6H^+ + 4e^- \rightleftharpoons 2SbO^+ + 3H_2O$	+0.581	
Arsenic		
$H_3AsO_4 + 2H^+ + 2e^- \rightleftharpoons H_3AsO_3 + H_2O$	+0.559	0.577 in 1 M HCl, $HClO_4$
Barium		
$Ba^{2+} + 2e^- \rightleftharpoons Ba(s)$	−2.906	
Bismuth		
$BiO^+ + 2H^+ + 3e^- \rightleftharpoons Bi(s) + H_2O$	+0.320	
$BiCl_4^- + 3e^- \rightleftharpoons Bi(s) + 4Cl^-$	+0.16	
Bromine		
$Br_2(l) + 2e^- \rightleftharpoons 2Br^-$	+1.065	1.05 in 4 M HCl
$Br_2(aq) + 2e^- \rightleftharpoons 2Br^-$	+1.087‡	
$BrO_3^- + 6H^+ + 5e^- \rightleftharpoons \frac{1}{2}Br_2(l) + 3H_2O$	+1.52	
$BrO_3^- + 6H^+ + 6e^- \rightleftharpoons Br^- + 3H_2O$	+1.44	
Cadmium		
$Cd^{2+} + 2e^- \rightleftharpoons Cd(s)$	−0.403	
Calcium		
$Ca^{2+} + 2e^- \rightleftharpoons Ca(s)$	−2.866	
Carbon		
$C_6H_4O_2$ (quinone) $+ 2H^+ + 2e^- \rightleftharpoons C_6H_4(OH)_2$	+0.699	0.696 in 1 M HCl, $HClO_4$, H_2SO_4
$2CO_2(g) + 2H^+ + 2e^- \rightleftharpoons H_2C_2O_4$	−0.49	
Cerium		
$Ce^{4+} + e^- \rightleftharpoons Ce^{3+}$		+1.70 in 1 M $HClO_4$; +1.61 in 1 M HNO_3; +1.44 in 1 M H_2SO_4
Chlorine		
$Cl_2(g) + 2e^- \rightleftharpoons 2Cl^-$	+1.359	
$HClO + H^+ + e^- \rightleftharpoons \frac{1}{2}Cl_2(g) + H_2O$	+1.63	
$ClO_3^- + 6H^+ + 5e^- \rightleftharpoons \frac{1}{2}Cl_2(g) + 3H_2O$	+1.47	
Chromium		
$Cr^{3+} + e^- \rightleftharpoons Cr^{2+}$	−0.408	
$Cr^{3+} + 3e^- \rightleftharpoons Cr(s)$	−0.744	
$Cr_2O_7^{2-} + 14H^+ + 6e^- \rightleftharpoons 2Cr^{3+} + 7H_2O$	+1.33	

Half-Reaction	E^0, V*	Formal Potential, V†
Cobalt		
$Co^{2+} + 2e^- \rightleftharpoons Co(s)$	−0.277	
$Co^{3+} + e^- \rightleftharpoons Co^{2+}$	+1.808	
Copper		
$Cu^{2+} + 2e^- \rightleftharpoons Cu(s)$	+0.337	
$Cu^{2+} + e^- \rightleftharpoons Cu^+$	+0.153	
$Cu^+ + e^- \rightleftharpoons Cu(s)$	+0.521	
$Cu^{2+} + I^- + e^- \rightleftharpoons CuI(s)$	+0.86	
$CuI(s) + e^- \rightleftharpoons Cu(s) + I^-$	−0.185	
Fluorine		
$F_2(g) + 2H^+ + 2e^- \rightleftharpoons 2HF(aq)$	+3.06	
Hydrogen		
$2H^+ + 2e^- \rightleftharpoons H_2(g)$	0.000	−0.005 in 1 M HCl, $HClO_4$
Iodine		
$I_2(s) + 2e^- \rightleftharpoons 2I^-$	+0.5355	
$I_2(aq) + 2e^- \rightleftharpoons 2I^-$	+0.615‡	
$I_3^- + 2e^- \rightleftharpoons 3I^-$	+0.536	
$ICl_2^- + e^- \rightleftharpoons \frac{1}{2}I_2(s) + 2Cl^-$	+1.056	
$IO_3^- + 6H^+ + 5e^- \rightleftharpoons \frac{1}{2}I_2(s) + 3H_2O$	+1.196	
$IO_3^- + 6H^+ + 5e^- \rightleftharpoons \frac{1}{2}I_2(aq) + 3H_2O$	+1.178‡	
$IO_3^- + 2Cl^- + 6H^+ + 4e^- \rightleftharpoons ICl_2^- + 3H_2O$	+1.24	
$H_5IO_6 + H^+ + 2e^- \rightleftharpoons IO_3^- + 3H_2O$	+1.601	
Iron		
$Fe^{2+} + 2e^- \rightleftharpoons Fe(s)$	−0.440	
$Fe^{3+} + e^- \rightleftharpoons Fe^{2+}$	+0.771	0.700 in 1 M HCl; 0.732 in 1 M $HClO_4$; 0.68 in 1 M H_2SO_4
$Fe(CN)_6^{3-} + e^- \rightleftharpoons Fe(CN)_6^{4-}$	+0.36	0.71 in 1 M HCl; 0.72 in 1 M $HClO_4$, H_2SO_4
Lead		
$Pb^{2+} + 2e^- \rightleftharpoons Pb(s)$	−0.126	−0.14 in 1 M $HClO_4$; −0.29 in 1 M H_2SO_4
$PbO_2(s) + 4H^+ + 2e^- \rightleftharpoons Pb^{2+} + 2H_2O$	+1.455	
$PbSO_4(s) + 2e^- \rightleftharpoons Pb(s) + SO_4^{2-}$	−0.350	
Lithium		
$Li^+ + e^- \rightleftharpoons Li(s)$	−3.045	
Magnesium		
$Mg^{2+} + 2e^- \rightleftharpoons Mg(s)$	−2.363	
Manganese		
$Mn^{2+} + 2e^- \rightleftharpoons Mn(s)$	−1.180	
$Mn^{3+} + e^- \rightleftharpoons Mn^{2+}$		1.51 in 7.5 M H_2SO_4
$MnO_2(s) + 4H^+ + 2e^- \rightleftharpoons Mn^{2+} + 2H_2O$	+1.23	
$MnO_4^- + 8H^+ + 5e^- \rightleftharpoons Mn^{2+} + 4H_2O$	+1.51	
$MnO_4^- + 4H^+ + 3e^- \rightleftharpoons MnO_2(s) + 2H_2O$	+1.695	
$MnO_4^- + e^- \rightleftharpoons MnO_4^{2-}$	+0.564	
Mercury		
$Hg_2^{2+} + 2e^- \rightleftharpoons 2Hg(l)$	+0.788	0.274 in 1 M HCl; 0.776 in 1 M $HClO_4$; 0.674 in 1 M H_2SO_4
$2Hg^{2+} + 2e^- \rightleftharpoons Hg_2^{2+}$	+0.920	0.907 in 1 M $HClO_4$
$Hg^{2+} + 2e^- \rightleftharpoons Hg(l)$	+0.854	
$Hg_2Cl_2(s) + 2e^- \rightleftharpoons 2Hg(l) + 2Cl^-$	+0.268	0.244 in sat'd KCl; 0.282 in 1 M KCl; 0.334 in 0.1 M KCl
$Hg_2SO_4(s) + 2e^- \rightleftharpoons 2Hg(l) + SO_4^{2-}$	+0.615	
Nickel		
$Ni^{2+} + 2e^- \rightleftharpoons Ni(s)$	−0.250	

Half-Reaction	E^0, V*	Formal Potential, V†
Nitrogen		
$N_2(g) + 5H^+ + 4e^- \rightleftharpoons N_2H_5^+$	−0.23	
$HNO_2 + H^+ + e^- \rightleftharpoons NO(g) + H_2O$	+1.00	
$NO_3^- + 3H^+ + 2e^- \rightleftharpoons HNO_2 + H_2O$	+0.94	0.92 in 1 M HNO_3
Oxygen		
$H_2O_2 + 2H^+ + 2e^- \rightleftharpoons 2H_2O$	+1.776	
$HO_2^- + H_2O + 2e^- \rightleftharpoons 3OH^-$	+0.88	
$O_2(g) + 4H^+ + 4e^- \rightleftharpoons 2H_2O$	+1.229	
$O_2(g) + 2H^+ + 2e^- \rightleftharpoons H_2O_2$	+0.682	
$O_3(g) + 2H^+ + 2e^- \rightleftharpoons O_2(g) + H_2O$	+2.07	
Palladium		
$Pd^{2+} + 2e^- \rightleftharpoons Pd(s)$	+0.987	
Platinum		
$PtCl_4^{2-} + 2e^- \rightleftharpoons Pt(s) + 4Cl^-$	+0.73	
$PtCl_6^{2-} + 2e^- \rightleftharpoons PtCl_4^{2-} + 2Cl^-$	+0.68	
Potassium		
$K^+ + e^- \rightleftharpoons K(s)$	−2.925	
Selenium		
$H_2SeO_3 + 4H^+ + 4e^- \rightleftharpoons Se(s) + 3H_2O$	+0.740	
$SeO_4^{2-} + 4H^+ + 2e^- \rightleftharpoons H_2SeO_3 + H_2O$	+1.15	
Silver		
$Ag^+ + e^- \rightleftharpoons Ag(s)$	+0.799	0.228 in 1 M HCl; 0.792 in 1 M $HClO_4$; 0.77 in 1 M H_2SO_4
$AgBr(s) + e^- \rightleftharpoons Ag(s) + Br^-$	+0.073	
$AgCl(s) + e^- \rightleftharpoons Ag(s) + Cl^-$	+0.222	0.228 in 1 M KCl
$Ag(CN)_2^- + e^- \rightleftharpoons Ag(s) + 2CN^-$	−0.31	
$Ag_2CrO_4(s) + 2e^- \rightleftharpoons 2Ag(s) + CrO_4^{2-}$	+0.446	
$AgI(s) + e^- \rightleftharpoons Ag(s) + I^-$	−0.151	
$Ag(S_2O_3)_2^{3-} + e^- \rightleftharpoons Ag(s) + 2S_2O_3^{2-}$	+0.017	
Sodium		
$Na^+ + e^- \rightleftharpoons Na(s)$	−2.714	
Sulfur		
$S(s) + 2H^+ + 2e^- \rightleftharpoons H_2S(g)$	+0.141	
$H_2SO_3 + 4H^+ + 4e^- \rightleftharpoons S(s) + 3H_2O$	+0.450	
$SO_4^{2-} + 4H^+ + 2e^- \rightleftharpoons H_2SO_3 + H_2O$	+0.172	
$S_4O_6^{2-} + 2e^- \rightleftharpoons 2S_2O_3^{2-}$	+0.08	
$S_2O_8^{2-} + 2e^- \rightleftharpoons 2SO_4^{2-}$	+2.01	
Thallium		
$Tl^+ + e^- \rightleftharpoons Tl(s)$	−0.336	−0.551 in 1 M HCl; −0.33 in 1 M $HClO_4$, H_2SO_4
$Tl^{3+} + 2e^- \rightleftharpoons Tl^+$	+1.25	0.77 in 1 M HCl
Tin		
$Sn^{2+} + 2e^- \rightleftharpoons Sn(s)$	−0.136	−0.16 in 1 M $HClO_4$
$Sn^{4+} + 2e^- \rightleftharpoons Sn^{2+}$	+0.154	0.14 in 1 M HCl
Titanium		
$Ti^{3+} + e^- \rightleftharpoons Ti^{2+}$	−0.369	
$TiO^{2+} + 2H^+ + e^- \rightleftharpoons Ti^{3+} + H_2O$	+0.099	0.04 in 1 M H_2SO_4
Uranium		
$UO_2^{2+} + 4H^+ + 2e^- \rightleftharpoons U^{4+} + 2H_2O$	+0.334	
Vanadium		
$V^{3+} + e^- \rightleftharpoons V^{2+}$	−0.256	0.21 in 1 M $HClO_4$
$VO^{2+} + 2H^+ + e^- \rightleftharpoons V^{3+} + H_2O$	+0.359	
$V(OH)_4^+ + 2H^+ + e^- \rightleftharpoons VO^{2+} + 3H_2O$	+1.00	1.02 in 1 M HCl, $HClO_4$

Half-Reaction	E^0, V*	Formal Potential, V†
Zinc		
$Zn^{2+} + 2e^- \rightleftharpoons Zn(s)$	−0.763	

*G. Milazzo, S. Caroli, and V. K. Sharma, *Tables of Standard Electrode Potentials.* London: Wiley, 1978.

†E. H. Swift and E. A. Butler, *Quantitative Measurements and Chemical Equilibria.* New York: Freeman, 1972.

‡These potentials are hypothetical because they correspond to solutions that are 1.00 M in Br_2 or I_2. The solubilities of these two compounds at 25°C are 0.18 M and 0.0020 M, respectively. In saturated solutions containing an excess of $Br_2(l)$ or $I_2(s)$, the standard potentials for the half-reaction $Br_2(l) + 2e^- \rightleftharpoons 2Br^-$ or $I_2(s) + 2e^- \rightleftharpoons 2I^-$ should be used. In contrast, at Br_2 and I_2 concentrations less than saturation, these hypothetical electrode potentials should be employed.

APPENDIX

6

Designations Carried by Ashless Filter Papers

Manufacturer	Fine Crystals	Moderately Fine Crystals	Coarse Crystals		Gelatinous Precipitates	
Schleicher and Schuell*	507, 590 589 blue ribbon	589 white ribbon	589 green ribbon	589 black ribbon	589 black ribbon	589-1H
Munktell†	OOH	OK OO	OOR		OOR	
Whatman‡	42	44, 40	41		41	41H
Eaton-Dikeman§	90	80	60		50	

Manufacturers' literature should be consulted for more complete specifications. Tabulated are manufacturer designations of papers suitable for filtration of the indicated type of precipitate.

*Schleicher and Schuell, Inc., Keene, NH.

†E. H. Sargent and Company, Chicago, IL, agents.

‡H. Reeve Angel and Company, Inc., Clifton, NJ, agents.

§Eaton-Dikeman Company, Mount Holly Springs, PA.

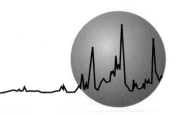

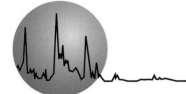

APPENDIX

7

Designations and Porosities for Filtering Crucibles

Crucible Type	Designation				
Glass, Pyrex†		C (60)*	M (15)	F (5.5)	
Glass, Kimax‡	EC (170–220)	C (40–60)	M (10–15)	F (4–4.5)	VF (2–2.5)
Porcelain, Coors U.S.A.§	Medium (40)	Fine (5)	VF (1.2)		
Porcelain, Selas‖	XF (100)	XFF (40)	#10 (8.8)	#01 (6)	
Aluminum oxide, ALUNDUM¶	Extra Coarse (30)	Coarse (20)	Medium (5)	Fine (0.1)	

*The numbers in parentheses are nominal maximum pore diameters in micrometers.

†Corning Glass Works, Corning, NY.

‡Owens-Illinois, Toledo, OH.

§Coors Porcelain, Golden, CO.

‖Selas Corporation of America, Dresher, PA.

¶Norton, Worcester, MA.

APPENDIX 8

Use of Exponential Numbers and Logarithms

Scientists frequently find it necessary (or convenient) to use exponential notation to express numerical data. A brief review of this notation follows.

Exponential Notation

An exponent is used to describe the process of repeated multiplication or division. For example, 3^5 means

$$3 \times 3 \times 3 \times 3 \times 3 = 3^5 = 243$$

The power 5 is the exponent of the number (or base) 3; thus, 3 raised to the fifth power is equal to 243.

A negative exponent represents repeated division. For example, 3^{-5} means

$$\frac{1}{3} \times \frac{1}{3} \times \frac{1}{3} \times \frac{1}{3} \times \frac{1}{3} = \frac{1}{3^5} = 3^{-5} = 0.00412$$

Note that changing the sign of the exponent yields the *reciprocal* of the number; that is,

$$3^{-5} = \frac{1}{3^5} = \frac{1}{243} = 0.00412$$

It is important to note that a number raised to the first power is the number itself, and any number raised to the zero power has a value of 1. For example,

$$4^1 = 4$$

$$4^0 = 1$$

$$67^0 = 1$$

Fractional Exponents

A fractional exponent symbolizes the process of extracting the root of a number. The fifth root of 243 is 3; this process is expressed exponentially as

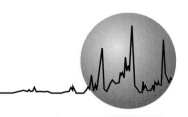

$$(243)^{1/5} = 3$$

Other examples are

$$25^{1/2} = 5$$

$$25^{-1/2} = \frac{1}{25^{1/2}} = \frac{1}{5}$$

The Combination of Exponential Numbers in Multiplication and Division

Multiplication and division of exponential numbers having the same base are accomplished by adding and subtracting the exponents. For example,

$$3^3 \times 3^2 = (3 \times 3 \times 3)(3 \times 3) = 3^{(3+2)} = 3^5 = 243$$

$$3^4 \times 3^{-2} \times 3^0 = (3 \times 3 \times 3 \times 3)\left(\frac{1}{3} \times \frac{1}{3}\right) \times 1 = 3^{(4-2+0)} = 3^2 = 9$$

$$\frac{5^4}{5^2} = \frac{5 \times 5 \times 5 \times 5}{5 \times 5} = 5^{(4-2)} = 5^2 = 25$$

$$\frac{2^3}{2^{-1}} = \frac{(2 \times 2 \times 2)}{1/2} = 2^4 = 16$$

Note that in the last equation the exponent is given by the relationship

$$3 - (-1) = 3 + 1 = 4$$

Extraction of the Root of an Exponential Number

To obtain the root of an exponential number, the exponent is divided by the desired root. Thus,

$$(5^4)^{1/2} = (5 \times 5 \times 5 \times 5)^{1/2} = 5^{(4/2)} = 5^2 = 25$$

$$(10^{-8})^{1/4} = 10^{(-8/4)} = 10^{-2}$$

$$(10^9)^{1/2} = 10^{(9/2)} = 10^{4.5}$$

The Use of Exponents in Scientific Notation

Scientists and engineers are frequently called upon to use very large or very small numbers for which ordinary decimal notation is either awkward or impossible. For example, to express Avogadro's number in decimal notation would require 21 zeros following the number 602. In scientific notation the number is written as a multiple of two numbers, the one number in decimal notation and the other expressed as a power of 10. Thus, Avogadro's number is written as 6.02×10^{23}. Other examples are

$$4.32 \times 10^3 = 4.32 \times 10 \times 10 \times 10 = 4320$$

$$4.32 \times 10^{-3} = 4.32 \times \frac{1}{10} \times \frac{1}{10} \times \frac{1}{10} = 0.00432$$

$$0.002002 = 2.002 \times \frac{1}{10} \times \frac{1}{10} \times \frac{1}{10} = 2.002 \times 10^{-3}$$

$$375 = 3.75 \times 10 \times 10 = 3.75 \times 10^{2}$$

It should be noted that the scientific notation for a number can be expressed in any of several equivalent forms. Thus,

$$4.32 \times 10^{3} = 43.2 \times 10^{2} = 432 \times 10^{1} = 0.432 \times 10^{4} = 0.0432 \times 10^{5}$$

The number in the exponent is equal to the number of places the decimal must be shifted to convert a number from scientific to purely decimal notation. The shift is to the right if the exponent is positive and to the left if it is negative. The process is reversed when decimal numbers are converted to scientific notation.

Arithmetic Operations with Scientific Notation

The use of scientific notation is helpful in preventing decimal errors in arithmetic calculations. Some examples follow.

Multiplication

Here, the decimal parts of the numbers are multiplied and the exponents are added; thus,

$$\begin{aligned} 420{,}000 \times 0.0300 &= (4.20 \times 10^{5})(3.00 \times 10^{-2}) \\ &= 12.60 \times 10^{3} = 1.26 \times 10^{4} \\ 0.0060 \times 0.000020 &= 6.0 \times 10^{-3} \times 2.0 \times 10^{-5} \\ &= 12 \times 10^{-8} = 1.2 \times 10^{-7} \end{aligned}$$

Division

Here, the decimal parts of the numbers are divided; the exponent in the denominator is subtracted from that in the numerator. For example,

$$\frac{0.015}{5000} = \frac{15 \times 10^{-3}}{5.0 \times 10^{3}} = 3.0 \times 10^{-6}$$

Addition and Subtraction

Addition or subtraction in scientific notation requires that all numbers be expressed to a common power of 10. The decimal parts are then added or subtracted, as appropriate. Thus,

$$\begin{aligned} &2.00 \times 10^{-11} + 4.00 \times 10^{-12} - 3.00 \times 10^{-10} \\ &\quad = 2.00 \times 10^{-11} + 0.400 \times 10^{-11} - 30.0 \times 10^{-11} \\ &\quad = -2.76 \times 10^{-10} = -27.6 \times 10^{-11} \end{aligned}$$

Raising to a Power a Number Written in Exponential Notation

Here, each part of the number is raised to the power separately. For example,

$$(2 \times 10^{-3})^4 = (2.0)^4 \times (10^{-3})^4 = 16 \times 10^{-(3\times4)}$$
$$= 16 \times 10^{-12} = 1.6 \times 10^{-11}$$

Extraction of the Root of a Number Written in Exponential Notation

Here, the number is written in such a way that the exponent of 10 is evenly divisible by the root. Thus,

$$(4.0 \times 10^{-5})^{1/3} = \sqrt[3]{40 \times 10^{-6}} = \sqrt[3]{40} \times \sqrt[3]{10^{-6}}$$
$$= 3.4 \times 10^{-2}$$

Logarithms

In this discussion, we will assume that the reader has available an electronic calculator for obtaining logarithms and antilogarithms of numbers. (The key for the antilogarithm function on most calculators is designated as 10^x.) It is desirable, however, to understand what a logarithm is as well as some of its properties. The discussion that follows provides this information.

A logarithm (or log) of a number is the power to which some base number (usually 10) must be raised in order to give the desired number. Thus, a logarithm is an exponent of the base 10. From the discussion in the previous paragraphs about exponential numbers, we can draw the following conclusions with respect to logs:

1. The logarithm of a product is the sum of the logarithms of the individual numbers in the product.

$$\log (100 \times 1000) = \log 10^2 + 10^3 = 2 + 3 = 5$$

2. The logarithm of a quotient is the difference between the logarithms of the individual numbers.

$$\log (100/1000) = \log 10^2 - \log 10^3 = 2 - 3 = -1$$

3. The logarithm of a number raised to some power is the logarithm of the number multiplied by that power.

$$\log (1000)^2 = 2 \times \log 10^3 = 2 \times 3 = 6$$
$$\log (0.01)^6 = 6 \times \log 10^{-2} = 6 \times (-2) = -12$$

4. The logarithm of a root of a number is the logarithm of that number divided by the root.

$$\log (1000)^{1/3} = \frac{1}{3} \times \log 10^3 = \frac{1}{3} \times 3 = 1$$

The following examples illustrate these statements:

$$\log 40 \times 10^{20} = \log 4.0 \times 10^{21} = \log 4.0 + \log 10^{21}$$
$$= 0.60 + 21 = 21.60$$
$$\log 2.0 \times 10^{-6} = \log 2.0 + \log 10^{-6} = 0.30 + (-6) = -5.70$$

For some purposes it is helpful to dispense with the subtraction step shown in the last example and report the log as a *negative* integer and a *positive* decimal number; that is,

$$\log 2.0 \times 10^{-6} = \log 2.0 + \log 10^{-6} = \bar{6}.30$$

The last two examples demonstrate that the logarithm of a number is the sum of two parts, a *characteristic* located to the left of the decimal point and a *mantissa* that lies to the right. The characteristic is the logarithm of 10 raised to a power and serves to indicate the location of the decimal point in the original number when that number is expressed in decimal notation. The mantissa is the logarithm of a number in the range between 0.00 and 9.99. . . . Note that the mantissa is *always positive.* As a consequence, the characteristic in the last example is -6 and the mantissa is $+0.30$.

APPENDIX

9

Volumetric Calculations Using Normality and Equivalent Weight

The *normality* of a solution expresses the number of equivalents of solute contained in 1 L of solution or the number of milliequivalents in 1 mL. The equivalent and milliequivalent, like the mole and millimole, are units for describing the amount of a chemical species. The former two, however, are defined so that we may state that, at the equivalence point in *any* titration,

$$\text{no. meq analyte present} = \text{no. meq standard reagent added} \qquad \text{(A9-1)}$$

or

$$\text{no. eq analyte present} = \text{no. eq standard reagent added} \qquad \text{(A9-2)}$$

As a consequence, stoichiometric ratios such as those described in Section 6C-5 (page 111) need not be derived every time a volumetric calculation is performed. Instead, the stoichiometry is taken into account by how equivalent or milliequivalent weight is defined.

9A-1 The Definitions of Equivalent and Milliequivalent

In contrast to the mole, the amount of a substance contained in one equivalent can vary from reaction to reaction. Consequently, the weight of one equivalent of a compound can never be computed *without reference to a chemical reaction* in which that compound is, directly or indirectly, a participant. Similarly, the normality of a solution can never be specified *without knowledge about how the solution will be used.*

Equivalent Weights in Neutralization Reactions

One equivalent weight of a substance participating in a neutralization reaction is that amount of substance (molecule, ion, or paired ion such as NaOH) that either

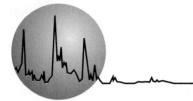

Once again we find ourselves using the term *weight* when we really mean *mass.* The term *equivalent weight* is so firmly engrained in the literature and vocabulary of chemistry that we retain it in this discussion.

reacts with or supplies 1 mol of hydrogen ions *in that reaction.*[1] A milliequivalent is simply 1/1000 of an equivalent.

The relationship between equivalent weight (eqw) and the molar mass ($\mathcal{M}$) is straightforward for strong acids or bases and for other acids or bases that contain a single reaction hydrogen or hydroxide ion. For example, the equivalent weights of potassium hydroxide, hydrochloric acid, and acetic acid are equal to their molar masses because each has but a single reactive hydrogen ion or hydroxide ion. Barium hydroxide, which contains two identical hydroxide ions, reacts with two hydrogen ions in any acid/base reaction, and so its equivalent weight is one half its molar mass:

$$\text{eqw Ba(OH)}_2 = \frac{\mathcal{M}_{\text{Ba(OH)}_2}}{2}$$

The situation becomes more complex for acids or bases that contain two or more reactive hydrogen or hydroxide ions with different tendencies to dissociate. With certain indicators, for example, only the first of the three protons in phosphoric acid is titrated:

$$H_3PO_4 + OH^- \longrightarrow H_2PO_4^- + H_2O$$

With certain other indicators, a color change occurs only after two hydrogen ions have reacted:

$$H_3PO_4 + 2OH^- \longrightarrow HPO_4^{2-} + 2H_2O$$

For a titration involving the first reaction, the equivalent weight of phosphoric acid is equal to the molar mass; for the second, the equivalent weight is one half the molar mass. (Because it is not practical to titrate the third proton, an equivalent weight that is one third the molar mass is not generally encountered for H_3PO_4.) If it is not known which of these reactions is involved, an unambiguous definition of the equivalent weight for phosphoric acid *cannot be made.*

Equivalent Weights in Oxidation/Reduction Reactions

The equivalent weight of a participant in an oxidation/reduction reaction is that amount that directly or indirectly produces or consumes 1 mol of electrons. The numerical value for the equivalent weight is conveniently established by dividing the molar mass of the substance of interest by the change in oxidation number associated with its reaction. As an example, consider the oxidation of oxalate ion by permanganate ion:

$$5C_2O_4^{2-} + 2MnO_4^- + 16H^+ \longrightarrow 10CO_2 + 2Mn^+ + 8H_2O \quad \textbf{(A9-3)}$$

---

[1]An alternative definition, proposed by the International Union of Pure and Applied Chemistry is as follows: An equivalent is "that amount of substance, which, in a specified reaction, releases or replaces that amount of hydrogen that is combined with 3 g of carbon-12 in methane $^{12}CH_4$" (see *Information Bulletin* No. 36, International Union of Pure and Applied Chemistry, August 1974). This definition applies to acids. For other types of reactions and reagents, the amount of hydrogen referred to may be replaced by the equivalent amount of hydroxide ions, electrons, or cations. The reaction to which the definition is applied must be specified.

In this reaction, the change in oxidation number of manganese is 5 because the element passes from the +7 to the +2 state; the equivalent weights for MnO_4^- and Mn^{2+} are therefore one fifth their molar masses. Each carbon atom in the oxalate ion is oxidized from the +3 to the +4 state, leading to the production of two electrons by that species. Therefore, the equivalent weight of sodium oxalate is one half its molar mass. It is also possible to assign an equivalent weight to the carbon dioxide produced by the reaction. Since this molecule contains but a single carbon atom and since that carbon undergoes a change in oxidation number of 1, the molar mass and equivalent weight of the two are identical.

It is important to note that in evaluating the equivalent weight of a substance, *only* its change in oxidation number during the titration is considered. For example, suppose the manganese content of a sample containing Mn_2O_3 is to be determined by a titration based upon the reaction given in Equation A9-3. The fact that each manganese in the Mn_2O_3 has an oxidation number of +3 plays no part in determining equivalent weight. That is, we must assume that by suitable treatment, all the manganese is oxidized to the +7 state before the titration is begun. Each manganese from the Mn_2O_3 is then reduced from the +7 to the +2 state in the titration step. The equivalent weight is thus the molar mass of Mn_2O_3 divided by $2 \times 5 = 10$.

As in neutralization reactions, the equivalent weight for a given oxidizing or reducing agent is not invariant. Potassium permanganate, for example, reacts under some conditions to give MnO_2:

$$MnO_4^- + 3e^- + 2H_2O \longrightarrow MnO_2(s) + OH^-$$

The change in the oxidation state of manganese in this reaction is from +7 to +4, and the equivalent weight of potassium permanganate is now equal to its molar mass divided by 3 (instead of 5 as in the earlier example).

Equivalent Weights in Precipitation and Complex-Formation Reactions

The equivalent weight of a participant in a precipitation or a complex-formation reaction is that weight which reacts with or provides one mole of the *reacting* cation if it is univalent, one half mole if it is divalent, one third mole if it is trivalent, and so on. It is important to note that the cation referred to in this definition is always *the cation directly involved in the analytical reaction* and not necessarily the cation contained in the compound whose equivalent weight is being defined.

EXAMPLE A9-1

Define equivalent weights for $AlCl_3$ and BiOCl if the two compounds are determined by a precipitation titration with $AgNO_3$:

$$Ag^+ + Cl^- \longrightarrow AgCl(s)$$

In this instance, the equivalent weight is based upon the number of moles of *silver ions* involved in the titration of each compound. Since 1 mol of Ag^+ reacts with 1 mol of Cl^- provided by one third mole of $AlCl_3$, we can write

$$\text{eqw AlCl}_3 = \frac{\mathcal{M}_{\text{AlCl}_3}}{3}$$

Because each mole of BiOCl reacts with only 1 Ag^+ ion,

$$\text{eqw BiOCl} = \frac{\mathcal{M}_{\text{BiOCl}}}{1}$$

Note that the fact that Bi^{3+} (or Al^{3+}) is trivalent has no bearing because the definition is based *upon the cation involved in the titration:* Ag^+.

A9-2 The Definition of Normality

The normality c_N of a solution expresses the number of milliequivalents of solute contained in 1 mL of solution or the number of equivalents contained in 1 L. Thus, a 0.20 N hydrochloric acid solution contains 0.20 meq of HCl in each milliliter of solution or 0.20 eq in each liter.

The normal concentration of a solution is defined by equations analogous to Equation 6-1. Thus, for a solution of the species A, the normality $c_{N(A)}$ is given by the equations

$$c_{N(A)} = \frac{\text{no. meq A}}{\text{no. mL solution}} \qquad \text{(A9-4)}$$

$$c_{N(A)} = \frac{\text{no. eq A}}{\text{no. L solution}} \qquad \text{(A9-5)}$$

A9-3 Some Useful Algebraic Relationships

Two pairs of algebraic equations, analogous to Equations 6-4 and 6-5 as well as 6-6 and 6-7 in Chapter 6, apply when normal concentrations are used:

$$\text{amount A} = \text{no. meq A} = \frac{\text{mass A (g)}}{\text{meqw A (g/meq)}} \qquad \text{(A9-6)}$$

$$\text{amount A} = \text{no. eq A} = \frac{\text{mass A (g)}}{\text{eqw A (g/eq)}} \qquad \text{(A9-7)}$$

$$\text{amount A} = \text{no. meq A} = V\,(\text{mL}) \times c_{N(A)}(\text{meq/mL}) \qquad \text{(A9-8)}$$

$$\text{amount A} = \text{no. eq A} = V\,(\text{L}) \times c_{N(A)}(\text{eq/L}) \qquad \text{(A9-9)}$$

A9-4 Calculation of the Normality of Standard Solutions

Example A9-2 shows how the normality of a standard solution is computed from preparatory data. Note the similarity between this example and Example 6-8 (Chapter 6).

EXAMPLE A9-2

Describe the preparation of 5.000 L of 0.1000 N Na_2CO_3 (105.99 g/mol) from the primary-standard solid, assuming the solution is to be used for titrations in which the reaction is

$$CO_3^{2-} + 2H^+ \longrightarrow H_2O + CO_2$$

Applying Equation A9-9 gives

$$\begin{aligned}\text{amount } Na_2CO_3 &= V \text{ soln (L)} \times c_{N(Na_2CO_3)}(\text{eq/L}) \\ &= 5.000 \text{ L} \times 0.1000 \text{ eq/L} = 0.5000 \text{ eq } Na_2CO_3\end{aligned}$$

Rearranging Equation A9-7 gives

$$\text{mass } Na_2CO_3 = \text{no. eq } Na_2CO_3 \times \text{eqw } Na_2CO_3$$

But 2 eq of Na_2CO_3 are contained in each mole of the compound; therefore,

$$\text{mass } Na_2CO_3 = 0.5000 \text{ eq } Na_2CO_3 \times \frac{105.99 \text{ g } Na_2CO_3}{2 \text{ eq } Na_2CO_3} = 26.50 \text{ g}$$

Therefore, dissolve 26.50 g in water and dilute to 5.000 L.

It is worth noting that, when the carbonate ion reacts with two protons, the weight of sodium carbonate required to prepare a 0.10 N solution is just one half that required to prepare a 0.10 M solution.

A9-5 The Treatment of Titration Data With Normalities

Calculation of Normalities from Titration Data

Examples A9-3 and A9-4 illustrate how normality is computed from standardization data. Note that these examples are similar to Examples 6-11 and 6-12 in Chapter 6.

EXAMPLE A9-3

Exactly 50.00 mL of an HCl solution required 29.71 mL of 0.03926 N $Ba(OH)_2$ to give an end point with bromocresol green indicator. Calculate the normality of the HCl.

Note that the molarity of $Ba(OH)_2$ is one half its normality. That is,

$$c_{Ba(OH)_2} = 0.03926 \frac{\cancel{\text{meq}}}{\text{mL}} \times \frac{1 \text{ mmol}}{2 \cancel{\text{meq}}} = 0.01963 \text{ M}$$

Because we are basing our calculations on the milliequivalent, we write

$$\text{no. meq HCl} = \text{no. meq } Ba(OH)_2$$

The number of milliequivalents of standard is obtained by substituting into Equation A9-8:

$$\text{amount Ba(OH)}_2 = 29.71\ \cancel{\text{mL Ba(OH)}_2} \times 0.03926\ \frac{\text{meq Ba(OH)}_2}{\cancel{\text{mL Ba(OH)}_2}}$$

To obtain the number of milliequivalents of HCl, we write

$$\text{amount HCl} = (29.71 \times 0.03926)\ \cancel{\text{meq Ba(OH)}_2} \times \frac{1\ \text{meq HCl}}{1\ \cancel{\text{meq Ba(OH)}_2}}$$

Equating this result to Equation A9-8 yields

$$\begin{aligned} \text{amount HCl} &= 50.00\ \text{mL} \times c_{\text{N(HCl)}} \\ &= (29.71 \times 0.03926 \times 1)\ \text{meq HCl} \\ c_{\text{N(HCl)}} &= \frac{(29.71 \times 0.03926 \times 1)\ \text{meq HCl}}{50.00\ \text{mL HCl}} = 0.02333\ \text{N} \end{aligned}$$

EXAMPLE A9-4

A 0.2121-g sample of pure $Na_2C_2O_4$(134.00 g/mol) was titrated with 43.31 mL of $KMnO_4$. What is the normality of the $KMnO_4$ solution? The chemical reaction is

$$2MnO_4^- + 5C_2O_4^{2-} + 16H^+ \longrightarrow 2Mn^{2+} + 10CO_2 + 8H_2O$$

By definition, at the equivalence point in the titration,

$$\text{no. meq Na}_2\text{C}_2\text{O}_4 = \text{no. meq KMnO}_4$$

Substituting Equations A9-8 and A9-6 into this relationship gives

$$\begin{aligned} V_{\text{KMnO}_4} \times c_{\text{N(KMnO}_4)} &= \frac{\text{mass Na}_2\text{C}_2\text{O}_4(\cancel{\text{g}})}{\text{meqw Na}_2\text{C}_2\text{O}_4(\cancel{\text{g}}/\text{meq})} \\ 43.31\ \text{mL KMnO}_4 \times c_{\text{N(KMnO}_4)} &= \frac{0.2121\ \cancel{\text{g Na}_2\text{C}_2\text{O}_4}}{0.13400\ \cancel{\text{g Na}_2\text{C}_2\text{O}_4}/2\ \text{meq}} \\ c_{\text{N(KMnO}_4)} &= \frac{0.2121\ \cancel{\text{g Na}_2\text{C}_2\text{O}_4}}{43.31\ \text{mL KMnO}_4 \times 0.1340\ \cancel{\text{g Na}_2\text{C}_2\text{O}_4}/2\ \text{meq}} \\ &= 0.073093\ \text{meq/mL KMnO}_4 = 0.07309\ \text{N} \end{aligned}$$

Note that the normality found here is five times the molarity computed in Example 6-12.

Calculation of the Quantity of Analyte from Titration Data

The examples that follow illustrate how analyte concentrations are computed when normalities are involved. Note that these examples are similar to Examples 6-13 and 6-15 in Chapter 6.

EXAMPLE A9-5

A 0.8040-g sample of an iron ore was dissolved in acid. The iron was then reduced to Fe^{2+} and titrated with 47.22 mL of 0.1121 N (0.02242 M) $KMnO_4$ solution. Calculate the results of this analysis in terms of (a) percent Fe (55.847 g/mol) and (b) percent Fe_3O_4 (231.54 g/mol). The reaction of the analyte with the reagent is described by the equation

$$MnO_4^- + 5Fe^{2+} + 8H^+ \longrightarrow Mn^{2+} + 5Fe^{3+} + 4H_2O$$

(a) At the equivalence point, we know that

$$\text{no. meq } KMnO_4 = \text{no. meq } Fe^{2+} = \text{no. meq } Fe_3O_4$$

Substituting Equations A9-8 and A9-6 leads to

$$V_{KMnO_4}(\cancel{mL}) \times c_{N(KMnO_4)}(\text{meq}/\cancel{mL}) = \frac{\text{mass } Fe^{2+}(\cancel{g})}{\text{meqw } Fe^{2+}(\cancel{g}/\text{meq})}$$

Substituting numerical data into this equation gives, after rearranging,

$$\text{mass } Fe^{2+} = 47.22 \cancel{\text{mL } KMnO_4} \times 0.1121 \frac{\cancel{\text{meq}}}{\cancel{\text{mL } KMnO_4}} \times \frac{0.055847 \text{ g}}{1 \cancel{\text{meq}}}$$

Note that the milliequivalent weight of the Fe^{2+} is equal to its millimolar mass. The percentage of iron is

$$\text{percent } Fe^{2+} = \frac{(47.22 \times 0.1121 \times 0.055847) \text{ g } Fe^{2+}}{0.8040 \text{ g sample}} \times 100\%$$

$$= 36.77\%$$

(b) Here,

$$\text{no. meq } KMnO_4 = \text{no. meq } Fe_3O_4$$

and

$$V_{KMnO_4}(\cancel{mL}) \times c_{N(KMnO_4)}(\text{meq}/\cancel{mL}) = \frac{\text{mass } Fe_3O_4(\cancel{g})}{\text{meqw } Fe_3O_4(\cancel{g}/\text{meq})}$$

Substituting numerical data and rearranging give

$$\text{mass } Fe_3O_4 = 47.22 \cancel{mL} \times 0.1121 \frac{\cancel{\text{meq}}}{\cancel{mL}} \times 0.23154 \frac{\text{g } Fe_3O_4}{3 \cancel{\text{meq}}}$$

Note that the milliequivalent weight of Fe_3O_4 is one third its millimolar mass because each Fe^{2+} undergoes a one-electron change and the compound is converted to $3Fe^{2+}$ before titration. The percentage of Fe_3O_4 is then

$$\text{percent } Fe_3O_4 = \frac{(47.22 \times 0.1121 \times 0.23154/3) \text{ g } Fe_3O_4}{0.8040 \text{ g sample}} \times 100\%$$

$$= 50.81\%$$

Note that the answers to this example are identical to those in Example 6-13.

EXAMPLE A9-6

A 0.4755-g sample containing $(NH_4)_2C_2O_4$ and inert compounds was dissolved in water and made alkaline with KOH. The liberated NH_3 was distilled into 50.00 mL of 0.1007 N (0.05035 M) H_2SO_4. The excess H_2SO_4 was back-titrated with 11.13 mL of 0.1214 N NaOH. Calculate the percentage of N (14.007 g/mol) and of $(NH_4)_2C_2O_4$ (124.10 g/mol) in the sample.

At the equivalence point, the number of milliequivalents of acid and base are equal. In this titration, however, two bases are involved: NaOH and NH_3. Thus,

$$\text{no. meq } H_2SO_4 = \text{no. meq } NH_3 + \text{no. meq NaOH}$$

After rearranging,

$$\text{no. meq } NH_3 = \text{no. meq N} = \text{no. meq } H_2SO_4 - \text{no. meq NaOH}$$

Substituting Equations A9-6 and A9-8 for the number of milliequivalents of N and H_2SO_4, respectively, yields

$$\frac{\text{mass N } (\cancel{\text{g}})}{\text{meqw N } (\cancel{\text{g}}/\text{meq})} = 50.00 \cancel{\text{mL } H_2SO_4} \times 0.1007 \frac{\text{meq}}{\cancel{\text{mL } H_2SO_4}} - 11.13 \cancel{\text{mL NaOH}} \times 0.1214 \frac{\text{meq}}{\cancel{\text{mL NaOH}}}$$

$$\text{mass N} = (50.00 \times 0.1007 - 11.13 \times 0.1214) \cancel{\text{meq}} \times 0.014007 \text{ g N}/\cancel{\text{meq}}$$

$$\text{percent N} = \frac{(50.00 \times 0.1007 - 11.13 \times 0.1214) \times 0.014007 \text{ g N}}{0.4755 \text{ g sample}} \times 100\%$$

$$= 10.85\%$$

The number of milliequivalents of $(NH_4)_2C_2O_4$ is equal to the number of milliequivalents of NH_3 and N, but the milliequivalent weight of the $(NH_4)_2C_2O_4$ is equal to one half its molar mass. Thus,

$$\text{mass } (NH_4)_2C_2O_4 = (50.00 \times 0.1007 - 11.13 \times 0.1214) \cancel{\text{meq}} \times 0.12410 \text{ g}/2 \cancel{\text{meq}}$$

percent $(NH_4)_2C_2O_4$

$$= \frac{(50.00 \times 0.1007 - 11.13 \times 0.1214) \times 0.06205 \ \cancel{g\,(NH_4)_2C_2O_4}}{0.4755 \ \cancel{g\ sample}} \times 100\%$$

$$= 48.07\%$$

Note that the results obtained here are identical to those obtained in Example 6-15.

APPENDIX

10

Compounds Recommended for the Preparation of Standard Solutions of Some Common Elements*

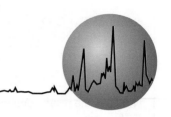

Element	Compound	FW	Solvent†	Notes
Aluminum	Al metal	26.98	Hot dil HCl	a
Antimony	$KSbOC_4H_4O_6 \cdot \frac{1}{2}H_2O$	333.93	H_2O	c
Arsenic	As_2O_3	197.84	dil HCl	i,b,d
Barium	$BaCO_3$	197.35	dil HCl	
Bismuth	Bi_2O_3	465.96	HNO_3	
Boron	H_3BO_3	61.83	H_2O	d,e
Bromine	KBr	119.01	H_2O	a
Cadmium	CdO	128.40	HNO_3	
Calcium	$CaCO_3$	100.09	dil HCl	i
Cerium	$(NH_4)_2Ce(NO_3)_6$	548.23	H_2SO_4	
Chromium	$K_2Cr_2O_7$	294.19	H_2O	i,d
Cobalt	Co metal	58.93	HNO_3	a
Copper	Cu metal	63.55	dil HNO_3	a
Fluorine	NaF	41.99	H_2O	b
Iodine	KIO_3	214.00	H_2O	i
Iron	Fe metal	55.85	HCl, hot	a
Lanthanum	La_2O_3	325.82	HCl, hot	f
Lead	$Pb(NO_3)_2$	331.20	H_2O	a
Lithium	Li_2CO_3	73.89	HCl	a
Magnesium	MgO	40.31	HCl	
Manganese	$MnSO_4 \cdot H_2O$	169.01	H_2O	g
Mercury	$HgCl_2$	271.50	H_2O	b
Molybdenum	MoO_3	143.94	1 M NaOH	
Nickel	Ni metal	58.70	HNO_3, hot	a
Phosphorus	KH_2PO_4	136.09	H_2O	
Potassium	KCl	74.56	H_2O	a
	$KHC_8H_4O_4$	204.23	H_2O	i,d
	$K_2Cr_2O_7$	294.19	H_2O	i,d
Silicon	Si metal	28.09	NaOH, concd	
	SiO_2	60.08	HF	j

Element	Compound	FW	Solvent†	Notes
Silver	$AgNO_3$	169.87	H_2O	a
Sodium	NaCl	58.44	H_2O	i
	$Na_2C_2O_4$	134.00	H_2O	i,d
Strontium	$SrCO_3$	147.63	HCl	a
Sulfur	K_2SO_4	174.27	H_2O	
Tin	Sn metal	118.69	HCl	
Titanium	Ti metal	47.90	H_2SO_4, 1 : 1	a
Tungsten	$Na_2WO_4 \cdot 2H_2O$	329.86	H_2O	h
Uranium	U_3O_8	842.09	HNO_3	d
Vanadium	V_2O_5	181.88	HCl, hot	
Zinc	ZnO	81.37	HCl	a

*The data in this table were taken from a more complete list assembled by B. W. Smith and M. L. Parsons, *J. Chem. Educ.*, **1973,** *50,* 679. Unless otherwise specified, compounds should be dried to constant weight at 110°C.

†Unless otherwise specified, acids are concentrated analytical grade.

[a]Conforms well to the criteria listed in Section 5B-1 and approaches primary standard quality.

[b]Highly toxic.

[c]Loses $\frac{1}{2}H_2O$ at 110°C. After drying, molar mass = 324.92. The dried compound should be weighed quickly after removal from the desiccator.

[d]Available as a primary standard from the National Institute of Standards and Technology.

[e]H_3BO_3 should be weighed directly from the bottle. It loses 1 mole H_2O at 100°C and is difficult to dry to constant weight.

[f]Absorbs CO_2 and H_2O. Should be ignited just before use.

[g]May be dried at 110°C without loss of water.

[h]Loses both waters at 110°C. Molar mass = 293.82. Keep in desiccator after drying.

[i]Primary standard.

[j]HF is highly toxic and dissolves glass.

APPENDIX

11

Derivation of Error Propagation Equations

See *Mathcad Applications for Analytical Chemistry*, **p. 121.**

In this appendix we derive several equations that permit the calculation of the standard deviation for the results from various types of arithmetical computations.

A11-A Propagation of Measurement Uncertainties

The calculated result for a typical analysis ordinarily requires data from several independent experimental measurements, each of which is subject to a random uncertainty and each of which contributes to the net random error of the final result. For the purpose of showing how such random uncertainties affect the outcome of an analysis, let us assume that a result y is dependent upon the experimental variables, $a, b, c, \ldots$, each of which fluctuates in a random and independent way. That is, y is a function of $a, b, c, \ldots$, so we may write

$$y = f(a, b, c, \ldots) \qquad \text{(A11-1)}$$

The uncertainty dy_i is generally given in terms of the deviation from the mean or $(y_i - \bar{y})$, which will depend upon the size and sign of the corresponding uncertainties $da_i, db_i, dc_i, \ldots$. That is,

$$dy_i = (y_i - \bar{y}) = f(da_i, db_i, dc_i, \ldots)$$

The variable in dy as a function of the uncertainties in $a, b, c, \ldots$ can be derived by taking the total differential of Equation A11-1. That is,

$$dy = \left(\frac{\partial y}{\partial a}\right)_{b,c,\ldots} da + \left(\frac{\partial y}{\partial b}\right)_{a,c,\ldots} db + \left(\frac{\partial y}{\partial c}\right)_{a,b,\ldots} dc + \ldots \qquad \text{(A11-2)}$$

In order to develop a relationship between the standard deviation of y and the standard deviations of a, b, and c for N replicate measurements, we employ Equation 3-3 (p. 28), which requires that we square Equation A11-2, sum between $i = 0$ and $i = N$, divide by $N - 1$, and take the square root of the result. The square of Equation A11-2 takes the form

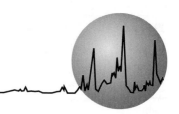

$$(dy)^2 = \left[\left(\frac{\partial y}{\partial a}\right)_{b,c,\,\ldots} da + \left(\frac{\partial y}{\partial b}\right)_{a,c,\,\ldots} db + \left(\frac{\partial y}{\partial c}\right)_{a,b,\,\ldots} dc + \;\ldots\right]^2 \qquad \text{(A11-3)}$$

This equation must then be summed between the limits of $i = 1$ to $i = N$.

In squaring Equation A11-2, two types of terms emerge from the right-hand side of the equation: (1) square terms and (2) cross terms. Square terms take the form

$$\left(\frac{\partial y}{\partial a}\right)^2 da^2,\ \left(\frac{\partial y}{\partial b}\right)^2 db^2,\ \left(\frac{\partial y}{\partial c}\right)^2 dc^2,\ \ldots$$

Square terms are always positive and can, therefore, *never* cancel when summed. In contrast, cross terms may be either positive or negative in sign. Examples are

$$\left(\frac{\partial y}{\partial a}\right)\left(\frac{\partial y}{\partial b}\right) dadb,\ \left(\frac{\partial y}{\partial a}\right)\left(\frac{\partial y}{\partial c}\right) dadc,\ \ldots$$

If da, db, and dc represent *independent* and *random uncertainties,* some of the cross terms will be negative and others positive. Thus, the *sum of all such terms should approach zero,* particularly when N is large.

As a consequence of the tendency of cross terms to cancel, the sum of Equation A11-3 from $i = 1$ to $i = N$ can be assumed to be made up exclusively of square terms. This sum then takes the form

$$\Sigma\,(dy_i)^2 = \left(\frac{\partial y}{\partial a}\right)^2 \Sigma\,(da_i)^2 + \left(\frac{\partial y}{\partial b}\right)^2 \Sigma\,(db_i)^2 + \left(\frac{\partial y}{\partial c}\right)^2 \Sigma\,(dc_i)^2 + \cdots \qquad \text{(A11-4)}$$

Dividing through by $N - 1$ gives

$$\frac{\Sigma\,(dy_i)^2}{N-1} = \left(\frac{\partial y}{\partial a}\right)^2 \frac{\Sigma\,(da_i)^2}{N-1} + \left(\frac{\partial y}{\partial b}\right)^2 \frac{\Sigma\,(db_i)^2}{N-1} + \left(\frac{\partial y}{\partial c}\right)^2 \frac{\Sigma\,(dc_i)^2}{N-1} + \cdots \qquad \text{(A11-5)}$$

From Equation 3-3, however, we see that

$$\frac{\Sigma\,(dy_i)^2}{N-1} = \frac{\Sigma\,(y_i - \bar{y})^2}{N-1} = s_y^2$$

where s_y^2 is the variance of y. Similarly,

$$\frac{\Sigma\,(da_i)^2}{N-1} = \frac{\Sigma\,(a_i - \bar{a})^2}{N-1} = s_a^2$$

and so forth. Thus, Equation A11-5 can be written in terms of the variances of the variables; that is,

$$s_y^2 = \left(\frac{\partial y}{\partial a}\right)^2 s_a^2 + \left(\frac{\partial y}{\partial b}\right)^2 s_b^2 + \left(\frac{\partial y}{\partial c}\right)^2 s_c^2 + \cdots \qquad \text{(A11-6)}$$

A11-B The Standard Deviation of Computed Results

In this section, we employ Equation A11-6 to derive relationships that permit calculation of standard deviations for the results produced by five types of arithmetic operations.

A11-B.1 Addition and Subtraction

Consider the case where we wish to compute the quantity y from the three experimental quantities a, b, and c by means of the equation

$$y = a + b - c$$

We assume that the standard deviations for these quantities are s_y, s_a, s_b, and s_c. Applying Equation A11-6 leads to

$$s_y^2 = \left(\frac{\partial y}{\partial a}\right)^2_{b,c} s_a^2 + \left(\frac{\partial y}{\partial b}\right)^2_{a,c} s_b^2 + \left(\frac{\partial y}{\partial c}\right)^2_{a,b} s_c^2$$

The partial derivatives of y with respect to the three experimental quantities are

$$\left(\frac{\partial y}{\partial a}\right)_{b,c} = 1; \quad \left(\frac{\partial y}{\partial b}\right)_{a,c} = 1; \quad \left(\frac{\partial y}{\partial c}\right)_{a,b} = -1$$

Therefore, the variance of y is given by

$$s_y^2 = (1)^2 s_a^2 + (1)^2 s_b^2 + (-1)^2 s_c^2 = s_a^2 + s_b^2 + s_c^2$$

or the standard deviation of the result is given by

$$s_y = \sqrt{s_a^2 + s_b^2 + s_c^2} \qquad \text{(A11-7)}$$

Thus, the *absolute* standard deviation of a sum or difference is equal to the square root of the sum of the squares of the *absolute* standard deviation of the numbers making up the sum or difference.

A11-B.2 Multiplication and Division

Let us now consider the case where

$$y = \frac{ab}{c}$$

The partial derivatives of y with respect to a, b, and c are

$$\left(\frac{\partial y}{\partial a}\right)_{b,c} = \frac{b}{c}; \quad \left(\frac{\partial y}{\partial b}\right)_{a,c} = \frac{a}{c}; \quad \left(\frac{\partial y}{\partial c}\right) = -\frac{ab}{c^2}$$

Substituting into Equation A11-6 gives

$$s_y^2 = \left(\frac{b}{c}\right)^2 s_a^2 + \left(\frac{a}{c}\right)^2 s_b^2 + \left(-\frac{ab}{c^2}\right)^2 s_c^2$$

Dividing this equation by the square of the original equation ($y^2 = a^2b^2/c^2$) gives

$$\frac{s_y^2}{y^2} = \frac{s_a^2}{a^2} + \frac{s_b^2}{b^2} + \frac{s_c^2}{c^2}$$

or

$$\frac{s_y}{y} = \sqrt{\left(\frac{s_a}{a}\right)^2 + \left(\frac{s_b}{b}\right)^2 + \left(\frac{s_c}{c}\right)^2} \qquad \text{(A11-8)}$$

Thus, for products and quotients, the *relative* standard deviation of the result is equal to the sum of the squares of the *relative* standard deviation of the number making up the product or quotient.

A11-B.3 Exponential Calculations

Consider the following computation

$$y = a^x$$

Here, Equation A11-6 takes the form

$$s_y^2 = \left(\frac{\partial a^x}{\partial y}\right)^2 s_a^2$$

or

$$s_y = \frac{\partial a^x}{\partial y} s_a$$

But

$$\frac{\partial a^x}{\partial y} = x a^{(x-1)}$$

Thus

$$s_y = x a^{(x-1)} s_a$$

and dividing by the original equation ($y = a^x$) gives

$$\frac{s_y}{y} = \frac{x a^{(x-1)} s_a}{a^x} = x\,\frac{s_a}{a} \qquad \text{(A11-9)}$$

Thus, the relative error of the result is equal to the relative error of numbers to be exponentiated, multiplied by the exponent.

It is important to note that the error propagated in taking a number to a power is different from the error propagated in multiplication. For example, consider the uncertainty in the square of 4.0(±0.2). Here the relative error in the result (16.0) is given by Equation A11-9

$$s_y/y = 2 \times (0.2/4) = 0.1 \qquad \text{or} \qquad 10\%$$

Consider now the case when y is the product of two *independently measured* numbers that by chance happen to have values of $a = 4.0(\pm 0.2)$ and $b = 4.0(\pm 0.2)$. In this case the relative error of the product $ab = 16.0$ is given by Equation A11-8:

$$s_y/y = \sqrt{(0.2/4)^2 + (0.2/4)^2} = 0.07 \qquad \text{or} \qquad 7\%$$

The reason for this apparent anomaly is that in the second case the sign associated with one error can be the same or different from that of the other. If they happen to be the same, the error is identical to that encountered in the first case, where the signs *must* be the same. In contrast, the possibility exists that one sign could be positive and the other negative, in which case the relative errors tend to cancel one another. Thus, the probable error lies between the maximum (10%) and zero.

A11-B.4 Calculation of Logarithms

Consider the computation

$$y = \log_{10} a$$

In this case, we can write Equation A11-6 as

$$s_y^2 = \left(\frac{\partial \log_{10} a}{\partial y}\right)^2 s_a^2$$

But

$$\frac{\partial \log_{10} a}{\partial y} = \frac{0.434}{a}$$

and

$$s_y = 0.434 \frac{s_a}{a} \qquad \text{(A11-10)}$$

Thus, the absolute standard deviation of a logarithm is determined by the *relative* standard deviation of the number.

A11-B.5 Calculation of Antilogarithms

Consider the relationship

$$y = \text{antilog}_{10}\, a = 10^a$$

$$\left(\frac{\partial y}{\partial a}\right) = 10^a \log_e 10 = 10^a \ln 10 = 2.303 \times 10^a$$

$$s_y^2 = \left(\frac{\partial y}{\partial a}\right)^2 s_a^2$$

or

$$s_y = \frac{\partial y}{\partial a} s_a = 2.303 \times 10^a s_a$$

Dividing by the original relationship gives

$$\frac{s_y}{y} = 2.303 s_a \qquad \text{(A11-11)}$$

Thus, the *relative* standard deviation of the antilog of a number is determined by the absolute standard deviation of the number.

Answers to Selected Questions and Problems

Chapter 2

2-1. (a) *Constant errors* are the same magnitude regardless of the sample size. *Proportional errors* are proportional in size to the sample size.
(c) The *mean* is the sum of the measurements in a set divided by the number of measurements. The *median* is the central value for a set of data; half of the measurements are larger and half are smaller than the median.

2-2. (1) Random temperature fluctuations causing random changes in the length of the metal rule; (2) uncertainties arising from having to move and position the rule twice; (3) personal judgment in reading the rule; (4) vibrations in the table and/or rule; (5) uncertainty in locating the rule perpendicular to the edge of the table.

2-3. The three types of systematic error are *instrumental error, method error,* and *personal error.*

2-5. Constant errors.

2-6. (a) -0.04% (c) -0.3%

2-7. (a) 13 g (c) 3 g

2-8. (a) $+0.06\%$ (c) $+0.12\%$

2-9. (a) -1.0% (c) -0.10%

Chapter 3

3-1. (a) The *spread* or *range* for a set of replicate data is the numerical difference between the highest and lowest value.
(c) *Significant figures* include all of the digits in a number that are known with certainty plus the first uncertain digit.

3-2. (a) The *sample variance, s^2*, is given by the expression

$$s^2 = \frac{\sum_{i=1}^{N}(x_i - \bar{x})^2}{N-1}$$

where $\bar{x}$ is the sample mean.

The *sample standard deviation* is given by

$$s = \sqrt{\frac{\sum_{i=1}^{N}(x_i - \bar{x})^2}{N-1}}$$

(c) *Accuracy* represents the agreement between an experimentally measured value and the true value. *Precision* describes the agreement among measurements that have been performed in exactly the same way.

3-3. (a) In statistics, a sample is a small set of replicate measurements. In chemistry, a sample is a portion of a material that is used for analysis.

3-5.

	Mean	Median	Spread	Standard Deviation	CV
A	2.08	2.1	0.9	0.35	17%
C	0.0918	0.0894	0.0116	0.0055	6.0%
E	69.53	69.635	0.44	0.22	0.31%

3-6.

	Absolute Error	Relative Error, ppt
A	$+0.08$	$+40$
C	-0.0012	-13
E	$+0.48$	$+7.0$

3-7.

	s_y	CV	y
(a)	0.030	5.2%	$0.57(\pm 0.03)$
(c)	0.14×10^{-16}	2.0%	$6.9(\pm 0.1) \times 10^{-16}$
(e)	5.1×10^{-3}	8.5%	$6.0(\pm 0.5) \times 10^{-2}$

3-8.

	s_y	CV	y
(a)	0.03×10^{-7}	2%	$-1.37(\pm 0.03) \times 10^{-7}$
(c)	1.8	50%	4 ± 2
(e)	2×10^1	50%	$5(\pm 2) \times 10^1$

3-9.

	s_y	CV	y
(a)	0.0065	0.18%	$-3.70(\pm 0.01)$
(c)	0.11	0.7%	$15.8(\pm 0.1)$

3-10. (a) $s_y = 2.0$; CV = 1.9%; $y = 105(\pm 2)$
3-11. $2.2(\pm 0.1) \times 10^5$ L
3-13. $s_1 = 0.095$; $s_2 = 0.12$; $s_3 = 0.11$; $s_4 = 0.10$; $s_5 = 0.10$; $s_{\text{pooled}} = 0.11\%$ K
3-15. $s_{\text{pooled}} = 0.29\%$

Chapter 4

4-1. Set A 2.1 ± 0.4; Set C 0.092 ± 0.009; Set E 69.5 ± 0.3
4-2. Set A 2.1 ± 0.2; Set C 0.092 ± 0.007; Set E 69.5 ± 0.2
4-3. Set A Retain; Set C Reject; Set E Reject
4-4. (a) 80% CL, 18 ± 3 μg/mL; 95% CL, 18 ± 5 μg/mL
(b) 80% CL, 18 ± 2 μg/mL; 95% CL, 18 ± 3 μg/mL
(c) 80% CL, 18 ± 2 μg/mL; 95% CL, 18 ± 2 μg/mL
4-6. 95%, 10 measurements; 99%, 17 measurements
4-8. (a) 3.22 ± 0.15 mmol/L (b) 3.22 ± 0.06 mmol/L
4-10. (a) 12 measurements
4-11. (a) Systematic error is indicated at 95% confidence.
(b) No systematic error is demonstrated.
4-13. At the 99% confidence level, the difference between the cobalt results should be no greater than ±0.053 ppm. The actual difference was −0.07 ppm. Similarly, the difference in the thorium results should be no greater than ±0.09 ppm. The actual difference was −0.12 ppm. Therefore, reasonable doubt exists as to the driver's guilt.
4-15. (a) 0.58 mg (b) 0.41 mg (c) 0.35 mg
4-16. (a) No difference demonstrated.
(c) No difference demonstrated.
4-17. (a) Outlier is retained.
(b) Outlier is rejected.
4-19. (a) $R = 0.162 + 0.232\, C_x$
(d) $C_x = 15.1$ mg SO_4^{2-}/L; $s = 1.4$ mg SO_4^{2-}/L; CV = 9.3%
(e) $s = 0.81$ mg SO_4^{2-}/L; CV = 5.4%

Chapter 5

5-1. (a) *Mass, m,* is an invariant measure of the amount of matter in an object. *Weight, w,* is the force of attraction between an object and earth.
(c) The individual particles of a *colloid* are smaller than about 10^{-5} mm in diameter, while those of a *crystalline precipitate* are larger. As a consequence, crystalline precipitates settle out of solution relatively rapidly, whereas colloidal particles do not unless they can be caused to agglomerate.
(e) *Precipitation* is the process by which a solid phase forms and is carried out of solution when the solubility product of a species is exceeded. *Coprecipitation* is the process in which a normally soluble species is carried out of solution during the formation of a precipitate.
(g) *Occlusion* is a type of coprecipitation in which an impurity is entrapped in a pocket formed by a rapidly growing crystal. *Mixed-crystal formation* is a type of coprecipitation in which a foreign ion is incorporated into a growing crystal in a lattice position that is ordinarily occupied by one of the ions of the precipitate.
5-2. (a) The dalton, which is synonymous with the atomic mass unit, is a relative mass unit that is equal to 1/12 of the mass of one neutral ^{12}C atom.
(c) Stoichiometry is the mass relationship among reacting chemical species.
(e) *Digestion* is a process for improving the purity and filterability of a precipitate by heating the solid in contact with the solution from which it is formed (the *mother liquor*).
(g) In *reprecipitation,* a precipitate is filtered, washed, redissolved, and then re-formed from the new solution. Because the concentration of contaminant is lower in this new solution than in the original, the second precipitate contains less coprecipitated impurity.
(i) The *electric double layer* consists of lattice ions adsorbed on the surface of a solid (the primary adsorption layer) and a volume of solution surrounding the particle (the counter-ion layer) in which an excess of ions of opposite charge exists.
(k) *Relative supersaturation* is given by the expression

$$\text{relative supersaturation} = \frac{Q - S}{S}$$

where Q is the concentration of a solute in a solution at any instant and S is its equilibrium solubility ($Q > S$).
5-3. A *chelating agent* is an organic compound or ion that contains two or more electron-donor groups located in such a way that five- or six-membered rings are formed when the donor groups complex a cation.
5-5. (a) positive charge (b) adsorbed Ag^+ (c) NO_3^-
5-7. *Peptization* is the process in which a coagulated colloid returns to its original dispersed state as a consequence of a decrease in the electrolyte concentration of the solution in contact with the precipitate. Peptization of a coagulated colloid can be avoided by washing with an electrolyte solution rather than with pure water.
5-9. (a) Generate hydroxide ions from urea.
(c) Generate hydrogen sulfide from thioacetamide.
5-10. The atomic mass of iron is 55.85 daltons, or 55.85 amu. Its molar mass is 55.85 grams.
5-12. (a) 0.0982 mol B_2O_3
(b) 7.76×10^{-4} mol $Na_2B_4O_7 \cdot 10H_2O$
(c) 0.0382 mol Mn_3O_4
(d) 5.26×10^{-4} mol CaC_2O_4
5-14. (a) 4.20×10^4 mg HNO_3
(b) 1.21×10^4 mg MgO
(c) 1.52×10^6 mg NH_4NO_3
(d) 2.92×10^6 mg $(NH_4)_2Ce(NO_3)_6$
5-16. (a) $\dfrac{1 \text{ mol } CO_2}{\text{mol } BaCO_3}$
(c) $\dfrac{1 \text{ mol } K_2O}{2 \times \text{mol } (C_6H_5)_4BK}$
(e) $\dfrac{1 \text{ mol } H_2S}{\text{mol } CdSO_4}$
(g) $\dfrac{1 \text{ mol } C_8H_6O_3Cl_2}{2 \times \text{mol AgCl}}$
(i) $\dfrac{1 \text{ mol } CoSiF_6 \cdot 6H_2O}{6 \times \text{mol } H_2O}$

5-18. (a) 1.172 g Ag_2CrO_4 (b) 2.050 g Ag_2CrO_4
5-20. (a) 0.318 g AgCl (c) 0.438 g AgCl
5-21. (a) 0.369 g $Ba(IO_3)_2$ (b) 0.0149 g $BaCl_2 \cdot 2H_2O$
5-24. 95.35% KCl
5-26. 18.99% C
5-29. 46.40% NH_3
5-30. 0.03219 g $C_{21}H_{29}NH_2$/tablet
5-31. 1.867% P_2O_5
5-33. 96.12% $C_6H_5NO_2$
5-35. (a) 0.239 g sample (b) 0.494 g AgCl
(c) 0.4065 g sample
5-37. 4.72% Cl^- 27.05% I^-
5-39. 1.80% KI
5-40. (a) 80.00% Ag and 20.00% Cu
(c) 90.00% Ag and 10.00% Cu
(e) 50.00% Ag and 50.00% Cu

Chapter 6

6-2. (a) The *millimole* is the amount of an elementary species, such as an atom, an ion, a molecule, or an electron. A millimole contains

$$6.02 \times 10^{23} \frac{\text{particles}}{\cancel{\text{mole}}} \times 10^{-3} \frac{\cancel{\text{mole}}}{\text{millimole}} = 6.02 \times 10^{20} \frac{\text{particles}}{\text{millimole}}$$

(c) The *stoichiometric factor* is the molar ratio of two species that appear in a balanced chemical equation.

6-3. (a) The *equivalence point* in a titration is that point at which sufficient titrant has been added so that stoichiometrically equivalent amounts of analyte and titrant are present. The *end point* in a titration is the point at which an observable physical change signals the equivalence point.

(c) A *primary standard* is a highly purified substance that serves as the basis for a titrimetric method. It is used either (1) to prepare a standard solution directly by mass or (2) to standardize a solution to be used in a titration.

A *secondary standard* is a material or solution whose concentration is determined from the stoichiometry of its reaction with a primary-standard material. Secondary standards are employed when a reagent is not available in primary-standard quality. For example, solid sodium hydroxide is hygroscopic and cannot be used to prepare a standard solution directly. A secondary-standard solution of the reagent is readily prepared, however, by standardizing a solution of sodium hydroxide against a primary-standard reagent such as potassium hydrogen phthalate.

6-4. $\frac{\text{mg}}{\text{L}} = \frac{10^{-3}\text{ g solute}}{10^{3}\text{ g solution}} = \frac{1\text{ g solute}}{10^{6}\text{ g solution}} = 1\text{ ppm}$

6-5. (a) $\frac{1\text{ mol } H_2NNH_2}{2\text{ mol } I_2}$

(c) $\frac{1\text{ mol } Na_2B_4O_7 \cdot 10H_2O}{2\text{ mol } H^+}$

6-6. (a) 5.52 mmol
(b) 31.2 mmol
(c) 6.58×10^{-3} mmol
(d) 966 mmol

6-8. (a) 1.33×10^3 mg (b) 520 mg
6-9. (a) 2.51 g (b) 2.88×10^{-3} g
6-10. 19.0 M
6-12. (a) 6.161 M (c) 4.669 M
6-13. (a) Dilute 80 g ethanol to 500 mL with water.
(b) Dilute 80.0 mL ethanol to 500 mL with water.
(c) Dilute 80.0 g ethanol with 420 g water.
6-15. Dilute 26 mL of the concentrated reagent to 2.0 L.
6-17. (a) Dissolve 6.37 g of $AgNO_3$ in water and dilute to 500 mL.
(b) Dilute 52.5 mL of 6.00 M HCl to 1.00 L.
(c) Dissolve 4.56 g $K_4Fe(CN)_6$ in water and dilute to 600 mL.
(d) Dilute 144 mL of 0.400 M $BaCl_2$ to a volume of 400 mL.
(e) Dilute 25 mL of the commercial reagent to a volume of 2.0 L.
(f) Dissolve 1.67 g Na_2SO_4 in water and dilute to 9.00 L.
6-19. 9.151×10^{-2} M
6-21. 0.06114 M $BaCl_2$
6-22. 0.2970 M $HClO_4$; 0.3259 M NaOH
6-24. 0.08411 M
6-25. 345.8 ppm S
6-27. 5.471% As_2O_3
6-28. 7.317% $(NH_2)_2CS$
6-30. (a) 9.36×10^{-3} M $Ba(OH)_2$
(b) 1.9×10^{-5} M
(c) −3 ppt
6-32. (a) $\frac{0.02966\text{ mmol } KMnO_4}{\text{g soln}}$
(b) 31.68% Fe_2O_3
6-34. (a) 1.821×10^{-2} M
(b) 1.821×10^{-2} M
(c) 5.463×10^{-2} M
(d) 0.506% (w/v)
(e) 1.366 mmol Cl^-
(f) 712 ppm K^+

Chapter 7

7-1. (a) A weak electrolyte is a substance that ionizes only partially in a solvent.
(c) The conjugate base of a Brønsted-Lowry acid is the species formed when the acid has donated a proton.
(e) An amphiprotic solute is one that can act either as an acid or as a base when dissolved in a solvent.
(g) Autoprotolysis is self-ionization of a solvent to give a conjugate acid and a conjugate base.
(i) The Le Châtelier principle states that the position of equilibrium in a system always shifts in a direction that tends to relieve an applied stress to the system.

7-2. (a) An amphiprotic solvent is a solvent that acts as a base with acidic solutes and as an acid with basic solutes.

(c) A leveling solvent is one in which a series of acids (or bases) all dissociate completely.

7-3. For an aqueous equilibrium in which water is a reactant or a product, the concentration of water is normally so much larger than the concentrations of the reactants and products that its concentration can be assumed to be constant and independent of the position of the equilibrium. Thus, its concentration is assumed to be constant and is included in the equilibrium constant. For a solid reactant or product, it is the concentration of that reactant in the solid phase that would influence the position of equilibrium. However, the concentration of a species in the solid phase is constant. Thus, as long as some solid exists as a second phase, its effect on the equilibrium is constant, and its concentration is included in the equilibrium constant.

7-4.

	Acid	Conjugate Base
(a)	HCN	CN^-
(c)	NH_4^+	NH_3
(e)	$H_2PO_4^-$	HPO_4^{2-}

7-6. (a) $2H_2O \rightleftharpoons H_3O^+ + OH^-$
(c) $2CH_3NH_2 \rightleftharpoons CH_3NH_3^+ + CH_3NH^-$

7-7. (a) $K_{sp} = [Ag^+][I^-] = 8.3 \times 10^{-17}$
(c) $K_{sp} = [Ag^+]^2[CrO_4^{2-}] = 1.2 \times 10^{-12}$

7-8. (a) $K_b = \dfrac{K_w}{K_a} = \dfrac{1.00 \times 10^{-14}}{2.31 \times 10^{-11}} = 4.33 \times 10^{-4}$
$= \dfrac{[C_2H_5NH_3^+][OH^-]}{[C_2H_5NH_2]}$
(c) $K_a = 5.90 \times 10^{-6} = \dfrac{[H_3O^+][C_5H_5N]}{[C_5H_5NH^+]}$
(e) $\beta_3 = K_1K_2K_3 = 2 \times 10^{-21} = \dfrac{[H_3O^+]^3[AsO_4^{3-}]}{[H_3AsO_4]}$

7-9. (a) $K_{sp} = [Ag^+][IO_3^-] = 3.1 \times 10^{-8}$
(b) $K_{sp} = [Ag^+]^2[SO_3^{2-}]$
(c) $K_{sp} = [Ag^+]^3[AsO_4] = 6 \times 10^{-23}$
(d) $K_{sp} = [Pb^{2+}][Cl^-][F^-]$

7-10. (a) $K_{sp} = S^2$ (b) $K_{sp} = 4S^3$ (c) $K_{sp} = 27S^4$
(d) $K_{sp} = S^3$

7-11. (a) 4.0×10^{-16}
(c) 3.1×10^{-6}
(e) 3.5×10^{-10}

7-12. (a) 8.0×10^{-15} M (c) 3.9×10^{-3} M
(e) 6.4×10^{-4} M

7-13. (a) 8.0×10^{-15} M (c) 1.2×10^{-3} M
(e) 2.8×10^{-6} M

7-14. (a) 2.18×10^{-7} M (b) 0.98 M

7-16. (a) 0.0250 M Ce^{3+} (b) 1.7×10^{-2} M
(c) 1.9×10^{-3} M (d) 7.6×10^{-7} M

7-18. (a) $PbI_2 > TlI > BiI_3 > AgI$
(b) $PbI_2 > TlI > AgI > BiI_3$
(c) $PbI_2 > TlI > BiI_3 > AgI$

7-21. (a) $[H_3O^+] = 2.4 \times 10^{-5}$ M; $[OH^-] = 4.1 \times 10^{-10}$ M
(c) $[H_3O^+] = 1.1 \times 10^{-12}$ M; $[OH^-] = 9.3 \times 10^{-3}$ M
(e) $[H_3O^+] = 5.0 \times 10^{-11}$ M; $[OH^-] = 2.0 \times 10^{-4}$ M
(g) $[H_3O^+] = 3.32 \times 10^{-4}$ M; $[OH^-] = 3.02 \times 10^{-11}$ M

7-22. (a) $[H_3O^+] = 1.10 \times 10^{-2}$ M
(b) $[H_3O^+] = 1.17 \times 10^{-8}$ M
(e) $[H_3O^+] = 1.46 \times 10^{-4}$ M

Chapter 8

8-1. (a) *Activity, a,* is the effective concentration of a species A in solution. The *activity coefficient,* γ_A, is the factor needed to convert a molar concentration to activity:

$$a_A = \gamma_A[A]$$

(b) The *thermodynamic equilibrium constant* refers to an ideal system within which each species is unaffected by any others. A *concentration equilibrium constant* takes account of the influence exerted by solute species upon one another. A thermodynamic constant is based upon activities of reactants and products; a concentration constant is based upon molar concentrations of reactants and products.

8-3. (a) Ionic strength should decrease.
(b) Ionic strength should be unchanged.
(c) Ionic strength should increase.

8-5. For a given ionic strength, activity coefficients for ions with multiple charge show greater departures from ideality.

8-7. (a) 0.16 (c) 1.2

8-8. (a) 0.20 (c) 0.073

8-9. (a) 0.21 (c) 0.079

8-10. (a) 1.7×10^{-12} (c) 7.6×10^{-11}

8-11. (a) 5.2×10^{-6} M (b) 6.2×10^{-6} M
(c) 9.5×10^{-12} M (d) 1.5×10^{-7} M

8-12. (a) (1) 1.4×10^{-6} M (2) 1.0×10^{-6} M
(b) (1) 2.1×10^{-3} M (2) 1.3×10^{-3} M
(c) (1) 2.8×10^{-5} M (2) 1.0×10^{-5} M
(d) (1) 1.4×10^{-5} M (2) 2.0×10^{-6} M

Chapter 9

9-1. In the calculation of the molar solubility of $Fe(OH)_2$, the hydronium ion concentration may be assumed to be negligibly small. This assumption cannot be made in the case of $Fe(OH)_3$. In the latter case we assume $[H_3O^+] = [OH^-] \approx 1.00 \times 10^{-7}$.

9-4. A charge-balance equation is derived by relating the concentration of cations and anions in such a way that no. mol/L positive charge = no. mol/L negative charge. For a doubly charged ion such as Ba^{2+}, the concentration of electrons for each mole is twice the molar *concentration* of the Ba^{2+}. That is, mol/L positive charge = $2[Ba^{2+}]$. Thus, the molar concentration of all multiply charged species is always multiplied by the charge in a charge-balance equation.

9-5. (a) $0.10 = [H_3PO_4] + [H_2PO_4^-] + [HPO_4^{2-}] + [PO_4^{3-}]$
(c) $0.100 + 0.0500 = [HNO_2] + [NO_2^-]$
(e) $0.100 = [Na^+] = [OH^-] + 2[Zn(OH)_4^{2-}]$
(g) $[Ca^{2+}] = \frac{1}{2}([F^-] + [HF])$

9-7. (a) 5.2×10^{-3} M (c) 3.6×10^{-4} M

9-8. (a) 1.47×10^{-4} M (c) 7.42×10^{-5} M

9-9. (a) 2.5×10^{-9} M (b) 2.5×10^{-12} M

9-11. (a) 5×10^{-2} M

9-12. (a) 1.4×10^{-5} M (b) 2.53×10^{-6} M

9-14. (a) $Cu(OH)_2$ forms first
(b) 9.8×10^{-10} M (c) 9.6×10^{-9} M

9-16. (a) 8.3×10^{-11} M (b) 1.6×10^{-11} M (c) 5.2×10^{6} M (d) 1.3×10^{4}
9-17. (a) not feasible (c) feasible
9-18. 1.877 g
9-20. (a) 0.0095 M; 48% (b) 6.6×10^{-3} M; 70%

Chapter 10

In the answers to this chapter, (Q) indicates that the answer was obtained by solving a quadratic equation.

10-1. (a) The initial pH of the NH_3 solution will be less than that for the solution containing NaOH. With the first addition of titrant, the pH of the NH_3 solution will decrease rapidly and then level off and become nearly constant throughout the middle part of the titration. In contrast, additions of standard acid to the NaOH solution will cause the pH of the NaOH solution to decrease gradually and nearly linearly until the equivalence point is approached. The equivalence-point pH for the NH_3 solution will be well below 7, whereas for the NaOH solution it will be exactly 7.
(b) Beyond the equivalence point, the pH is determined by the excess titrant. Thus, the curves become identical in this region.

10-3. The limited sensitivity of the eye to small color differences requires that there be a roughly tenfold excess of one or the other form of the indicator to be present in order for the color change to be seen. This change corresponds to a pH range of ± 1 pH unit about the pK of the indicator.

10-5. The standard reagents in neutralization titrations are always strong acids or strong bases because the reactions with this type of reagent are more complete than with those of their weaker counterparts. Sharper end points are the consequence of this difference.

10-7. The *buffer capacity* of a solution is the number of moles of hydronium ion or hydroxide ion needed to cause 1.00 L of the buffer to undergo a unit change in pH.

10-9. The three solutions will have the same pH, since the ratios of the amounts of weak acid to conjugate base are identical. They will differ in buffer capacity, however, with (a) having the greatest and (c) the least.

10-10. (a) Malic acid/sodium hydrogen malate (c) NH_4Cl/NH_3
10-11. (a) NaOCl (c) CH_3NH_2
10-12. (a) HIO_3 (c) $CH_3COCOOH$
10-14. 3.24
10-16. (a) 14.94
10-17. (a) 12.94
10-18. −0.607
10-20. 7.04 (Q)
10-22. (a) 1.05 (b) 1.05 (c) 1.81 (d) 1.81 (e) 12.60
10-24. (a) 1.30 (b) 1.37
10-26. (a) 4.26 (b) 4.76 (c) 5.76
10-28. (a) 11.12 (b) 10.62 (c) 9.53 (Q)
10-30. (a) 12.04 (Q) (b) 11.48 (Q) (c) 9.97 (Q)
10-32. (a) 1.94 (b) 2.45 (c) 3.52
10-34. (a) 2.41 (Q) (b) 8.35 (c) 12.35 (d) 3.84
10-37. (a) 3.85 (b) 4.06 (c) 2.63 (Q) (d) 2.10 (Q)
10-39. (a) 0.00 (c) −1.000 (e) −0.500 (g) 0.000
10-40. (a) −5.00 (c) −0.097 (e) −3.369 (g) −0.017
10-41. (a) 5.00 (c) 0.079 (e) 3.272 (g) 0.017
10-42. (b) −0.141
10-43. 15.5 g sodium formate
10-45. 194 mL HCl

10-47.

V_{HCl}	pH	V_{HCl}	pH
0.00	13.00	49.00	11.00
10.00	12.82	50.00	7.00
25.00	12.52	51.00	3.00
40.00	12.05	55.00	2.32
45.00	11.72	60.00	2.04

10-48. The theoretical pH at 24.95 mL is 6.44; at 25.05 mL it is 9.82. Thus, the indicator should change color in the range of pH 6.5 to 9.8. Cresol purple (range: 7.6 to 9.2) (Table 10-1) would be quite suitable.

10-50.

Vol, mL	(a) pH	(c) pH
0.00	2.09 (Q)	3.12
5.00	2.38 (Q)	4.28
15.00	2.82 (Q)	4.86
25.00	3.17 (Q)	5.23
40.00	3.76 (Q)	5.83
45.00	4.11 (Q)	6.18
49.00	4.85 (Q)	6.92
50.00	7.92	8.96
51.00	11.00	11.00
55.00	11.68	11.68
60.00	11.96	11.96

10-51.

Vol HCl, mL	(a) pH
0.00	11.12
5.00	10.20
15.00	9.61
25.00	9.24
40.00	8.64
45.00	8.29
49.00	7.55
50.00	5.27
51.00	3.00
55.00	2.32
60.00	2.04

10-52.

Vol, mL	(a) pH	(c) pH
0.00	2.80	4.26
5.00	3.65	6.57
15.00	4.23	7.15
25.00	4.60	7.52
40.00	5.20	8.12
49.00	6.29	9.21
50.00	8.65	10.11
51.00	11.00	11.00
55.00	11.68	11.68
60.00	11.96	11.96

10-53. (a) $\alpha_0 = 0.215$; $\alpha_1 = 0.785$
(c) $\alpha_0 = 0.769$; $\alpha_1 = 0.231$
(e) $\alpha_0 = 0.917$; $\alpha_1 = 0.083$

10-54. 6.61×10^{-2} M

10-56. (a) Lactic; pH = 3.61; [HA] = 0.0768; $[A^-] = 0.0432$; $\alpha_1 = 0.360$
(b) Sulfamic; [HA] = 0.095; $[A^-] = 0.155$; $\alpha_0 = 0.380$; $\alpha_1 = 0.620$

Chapter 11

11-1. Not only is NaHA a proton donor, it is also the conjugate base of the parent acid H_2A.

$$HA^- + H_2O \rightleftharpoons H_3O^+ + A^-$$
$$HA^- + H_2O \rightleftharpoons H_2A + OH^-$$

Solutions of acid salts are acidic or alkaline, depending upon which of these equilibria predominates. In order to compute the pH of this type, it is necessary to take both equilibria into account.

11-3. The HPO_4^{2-} is such a weak acid ($K_a = 4.5 \times 10^{-13}$) that the change in pH in the vicinity of the third equivalence point is too small to be observable.

11-4. (a) Since the *K*s are essentially identical, the solution should be approximately neutral. (c) neutral (e) basic (g) acidic

11-6. phenolphthalein

11-8. (a) cresol purple (c) cresol purple (e) bromocresol green (g) phenolphthalein

11-9. (a) 1.86 (Q) (c) 1.64 (Q) (e) 4.21

11-10. (a) 4.71 (c) 4.28 (e) 9.80

11-11. (a) 12.32 (Q) (c) 9.70 (e) 12.58 (Q)

11-12. (a) 2.07 (Q) (b) 7.18 (c) 10.63 (d) 2.55 (Q) (e) 2.06

11-14. (a) 1.54 (b) 1.99 (c) 12.07 (d) 12.01

11-16. (a) $[SO_3^{2-}]/[HSO_3^-] = 15.2$
(b) $[HCit^{2-}][Cit^{3-}] = 2.5$
(c) $[HM^-][M^{2-}] = 0.498$
(d) $[HT^-][T^{2-}] = 0.0232$

11-18. 50.2 g

11-20. (a) 2.11 (Q) (b) 7.38

11-22. Mix 442 mL of 0.300 M Na_2CO_3 with 558 mL of 0.200 M HCl.

11-24. Mix 704 mL of the HCl with 296 mL of the Na_3AsO_4.

11-28.

Vol reagent, mL	(a) pH	(c) pH
0.00	11.66	0.96
12.50	10.33	1.28
20.00	9.73	1.50
24.00	8.95	1.63
25.00	8.34	1.67
26.00	7.73	1.70
37.50	6.35	2.19
45.00	5.75	2.70
49.00	4.97	3.46
50.00	3.83	7.35
51.00	2.70	11.30
60.00	1.74	12.26

11-29. 0.00 mL, pH = 13.00; 10.00 mL, pH = 12.70; 20.00 mL, pH = 12.15; 24.00 mL, pH = 11.43; 25.00 mL, pH = 10.35; 26.00 mL, pH = 9.26; 35.00 mL, pH = 7.98; 44.00 mL, pH = 6.70; 45.00 mL, pH = 4.68; 46.00 mL, pH = 2.68; 50.00 mL, pH = 2.00

11-31. (a) $\dfrac{[H_3AsO_4][HAsO_4^{2-}]}{[H_2AsO_4^-]^2} = 1.9 \times 10^{-5}$

11-32. $\dfrac{[NH_3][HOAc]}{[NH_4^+][OAc^-]} = 3.26 \times 10^{-5}$

11-33.

	pH	*D*	α_0	α_1	α_2	α_3
(a)	2.00	1.112×10^{-4}	0.899	0.101	3.94×10^{-5}	
	6.00	5.500×10^{-9}	1.82×10^{-4}	0.204	0.796	
	10.00	4.379×10^{-9}	2.28×10^{-12}	2.56×10^{-5}	1.000	
(c)	2.00	1.075×10^{-6}	0.931	6.93×10^{-2}	1.20×10^{-4}	4.82×10^{-9}
	6.00	1.882×10^{-14}	5.31×10^{-5}	3.96×10^{-2}	0.685	0.275
	10.00	5.182×10^{-15}	1.93×10^{-16}	1.44×10^{-9}	2.49×10^{-4}	1.000
(e)	2.00	4.000×10^{-4}	0.250	0.750	1.22×10^{-5}	
	6.00	3.486×10^{-9}	2.87×10^{-5}	0.861	0.139	
	10.00	4.863×10^{-9}	2.06×10^{-12}	6.17×10^{-4}	0.999	

Chapter 12

12-1. Carbon dioxide is not strongly bonded by water molecules, and thus is readily volatilized from aqueous media. Gaseous HCl molecules, on the other hand, are fully dissociated into H_3O^+ and Cl^- when dissolved in water; neither of these species is volatile.

12-3. Primary-standard Na_2CO_3 can be obtained by heating primary-standard-grade $NaHCO_3$ for about an hour at 270° to 300°C. The reaction is

$$2NaHCO_3(s) \longrightarrow Na_2CO_3(s) + H_2O(g) + CO_2(g)$$

12-5. For, let us say, a 40-mL titration

mass $KH(IO_3)_2$ required = 0.16 g

mass HBz required = 0.049 g

The $KH(IO_3)_2$ is preferable because the relative weighing error would be less with a 0.16-g sample than with a 0.049-g sample. A second advantage of $KH(IO_3)_2$ is that it acts as a strong acid, which makes the choice of indicator simpler.

12-8. (a) Dissolve 17 g KOH in water and dilute to 2.0 L.
(b) Dissolve 9.5 g of $Ba(OH)_2 \cdot 2H_2O$ in water and dilute to 2.0 L.
(c) Dilute about 120 mL of the reagent HCl to 2.0 L.

12-10. (a) 0.1026 M
(b) s = 0.00039 and CV = 0.38%

12-12. (a) 0.1388 M (b) 0.1500 M

12-14. (a) 0.08387 M (b) 0.1007 M (c) 0.1311 M

12-16. (a) 0.28 to 0.36 g Na_2CO_3 (c) 0.85 to 1.1 g HBz
(e) 0.17 to 0.22 g TRIS

12-17. (a) 0.067% (b) 0.16% (c) 0.043%

12-19. (a) 0.1217 g H_2T/100 mL

12-21. (a) 46.25% $Na_2B_4O_7$
(b) 87.67% $Na_2B_4O_7 \cdot 10H_2O$
(c) 32.01% B_2O_3
(d) 9.94% B

12-23. 24.4% HCHO

12-25. 7.079% RS_4

12-27. $MgCO_3$ with a molar mass of 84.31 seems a likely candidate.

12-29. 3.35×10^3 ppm CO_2

12-31. 6.333% P

12-32. 13.86% analyte

12-33. 22.08% RN_4

12-35. 3.885% N

12-37. (a) 10.09% N (c) 47.61% $(NH_4)_2SO_4$

12-39. 15.23% $(NH_4)_2SO_4$ and 24.39% NH_4NO_3

12-40. 69.84% KOH; 21.04% K_2CO_3; 9.12% H_2O

12-42. (a) 18.15 mL HCl (b) 45.37 mL HCl
(c) 38.28 mL HCl (d) 12.27 mL HCl

12-44. (a) 4.314 mg NaOH/mL
(b) 7.985 mg Na_2CO_3/mL and 4.358 mg $NaHCO_3$/mL
(c) 3.455 mg Na_2CO_3/mL and 4.396 mg NaOH/mL
(d) 8.215 mg Na_2CO_3/mL
(e) 13.46 mg $NaHCO_3$/mL

12-46. The equivalent weight of an acid is that weight of the pure material that contains one mole of titratable protons in a specified reaction. The equivalent weight of base is that weight of a pure compound that consumes one mole of protons in a specified reaction.

Chapter 13

13-1. The Fajan determination of chloride involves a direct titration, while the Volhard approach requires two standard solutions and a filtration step to eliminate AgCl.

13-3. In contrast to Ag_2CO_3 and AgCN, the solubility of AgI is unaffected by the acidity and additionally is less soluble than AgSCN. The filtration step is thus unnecessary, whereas it is with the other two compounds.

13-5. Potassium is determined by precipitation with an excess of a standard solution of sodium tetraphenylboron. An excess of standard $AgNO_3$ is then added, which precipitates the excess tetraphenylboron ion. The excess $AgNO_3$ is then titrated with a standard solution of SCN^-. The reactions are

$$K^+ + B(C_6H_5)_4^- \rightleftharpoons KB(C_6H_5)_4(s)$$

[measured excess $B(C_6H_5)_4^-$]

$$Ag^+ + B(C_6H_5)_4^- \rightleftharpoons AgB(C_6H_5)_4(s)$$

[measured excess $AgNO_3$]

The excess $AgNO_3$ is then determined by a Volhard titration with KSCN.

13-7. (a) 51.78 mL $AgNO_3$ (c) 10.64 mL $AgNO_3$
(e) 46.24 mL $AgNO_3$

13-9. (a) 44.70 mL $AgNO_3$ (c) 14.87 mL $AgNO_3$

13-11. 28.5% Cl

13-13. 18.9 ppm H_2S

13-15. 1.472% P_2O_5

13-16. 116.7 mg analyte

13-19. 0.07052 M

13-20. Only one of the chlorines in the heptachlor reacts with $AgNO_3$.

13-23. 15.60 mg saccharin/tablet

13-24. 21.5% CH_2O

13-25. 0.4348% $C_{19}H_{16}O_4$

13-28. 10.60% Cl^-; 55.65% ClO_4^-

13-29. (a)

V_{NH_4SCN} (mL)	$[Ag^+]$	$[SCN^-]$	pAg
30.00	9.09×10^{-3}	1.2×10^{-10}	2.04
40.00	3.85×10^{-3}	2.9×10^{-10}	2.42
49.00	3.38×10^{-4}	3.3×10^{-9}	3.47
50.00	1.05×10^{-6}	1.05×10^{-6}	5.98
51.00	3.3×10^{-9}	3.3×10^{-4}	8.48
60.00	3.7×10^{-10}	2.94×10^{-3}	9.43
70.00	2.1×10^{-10}	5.26×10^{-3}	9.68

(c)

V_{NaCl} (mL)	$[Ag^+]$	$[Cl^-]$	pAg
10.00	3.75×10^{-2}	4.85×10^{-9}	1.43
20.00	1.50×10^{-2}	1.21×10^{-8}	1.82
29.00	1.27×10^{-3}	1.43×10^{-7}	2.90
30.00	1.35×10^{-5}	1.35×10^{-5}	4.87
31.00	1.48×10^{-7}	1.23×10^{-3}	6.83
40.00	1.70×10^{-8}	1.07×10^{-2}	7.77
50.00	9.71×10^{-9}	1.88×10^{-2}	8.01

(e)

$V_{Na_2SO_4}$ (mL)	$[Ba^{2+}]$	$[SO_4^-]$	pBa
0.00	2.50×10^{-2}	0.0	1.60
10.00	1.00×10^{-2}	1.1×10^{-8}	2.00
19.00	8.48×10^{-4}	1.3×10^{-7}	3.07
20.00	1.05×10^{-5}	1.05×10^{-5}	4.98
21.00	1.3×10^{-7}	8.20×10^{-4}	6.87
30.00	1.5×10^{-8}	7.14×10^{-3}	7.81
40.00	8.8×10^{-9}	1.25×10^{-2}	8.06

13-30.

V_{AgNO_3} (mL)	$[Ag^+]$
5.00	1.57×10^{-11}
40.00	7.07×10^{-7}
45.00	2.63×10^{-3}

Chapter 14

14-1. (a) A *chelate* is a cyclic complex consisting of a metal ion and a reagent that contains two or more electron donor groups located in such a position that they can bond with the metal ion to form a heterocyclic ring structure.

(c) A *ligand* is a species that contains one or more electron pair donor groups that tend to form bonds with metal ions.

(e) A *conditional formation constant* is an equilibrium constant for the reaction between a metal ion and a complexing agent that applies only when the pH and/or the concentration of other complexing ions are carefully specified.

(g) *Water hardness* is the concentration of calcium carbonate that is equivalent to the total concentration of all of the multivalent metal carbonates in the water.

14-2. Three general methods for performing EDTA titrations are (1) direct titration, (2) back-titration, and (3) displacement titration. Method (1) is simple, rapid, and requires but one standard reagent. Method (2) is advantageous for those metals that react so slowly with EDTA as to make direct titration inconvenient. In addition, the procedure is useful for cations for which satisfactory indicators are not available. Finally, it is useful for analyzing samples that contain anions that form sparingly soluble precipitates with the analyte under the analytical conditions. Method (3) is particularly useful in situations where no satisfactory indicators are available for direct titration.

14-4. (a) $Ag^+ + S_2O_3^{2-} \rightleftharpoons AgS_2O_3^-$

$$K_1 = \frac{[AgS_2O^-]}{[Ag^+][S_2O_3^{2-}]}$$

$$AgS_2O_3^- + S_2O_3^{2-} \rightleftharpoons Ag(S_2O_3)_2^{3-}$$

$$K_2 = \frac{[Ag(S_2O_3)_2^{3-}]}{[AgS_2O_3^-][S_2O_3^{2-}]}$$

14-5. The overall formation constant β_n is equal to the product of the individual stepwise constants. Thus, the overall constant for formation of $Ni(SCN)_3^-$ in Problem 14-4(b) is

$$\beta_3 = K_1K_2K_3 = \frac{[Ni(SCN)_3^-]}{[Ni^{2+}][SCN^-]^3}$$

which is the equilibrium constant for the reaction

$$Ni^{2+} + 3SCN^- \rightleftharpoons Ni(SCN)_3^-$$

and

$$\beta_2 = K_1K_2 = \frac{Ni(SCN)_2}{[Ni^{2+}][SCN^-]^2}$$

where the overall constant β_2 is for the reaction

$$Ni^{2+} + 2SCN^- \rightleftharpoons Ni(SCN)_2$$

14-8. 0.01032 M EDTA

14-10. (a) 39.1 mL EDTA (c) 41.6 mL EDTA (e) 31.2 mL EDTA

14-12. 3.028% Zn

14-14. 1.228% Tl_2SO_4

14-16. 184.0 ppm Fe^{3+}; 213.0 ppm Fe^{2+}

14-18. 55.16% Pb; 44.86% Cd

14-20. 99.7% ZnO; 0.256% Fe_2O_3

14-22. 64.68 ppm K^+

14-24. 8.518% Pb; 24.86% Zn: 64.08% Cu; 2.54% Sn

14-25. (a) 4.6×10^9; (b) 1.1×10^{12}; (c) 7.4×10^{13}

14.27.

Vol, mL	pSr	Vol, mL	pSr
0.00	2.00	25.00	5.37
10.00	2.30	25.10	6.16
24.00	3.57	26.00	7.16
24.90	4.57	30.00	7.86

Chapter 15

15-1. (a) *Oxidation* is a process in which a species loses one or more electrons.

(c) A *salt bridge* is a device that provides electrical contact but prevents mixing of dissimilar solutions in an electrochemical cell.

(e) The *Nernst equation* relates the potential to the concentrations (strictly, activities) of the participants in an electrochemical reaction.

15-2. (a) The *electrode potential* is the potential of an electrochemical cell in which a standard hydrogen electrode acts as anode and the half-cell of interest is the cathode.

(c) The *standard electrode potential* for a half-reaction is the potential of a *cell* consisting of a cathode at which that half-reaction is occurring and a standard hydrogen electrode behaving as the anode. The activities of all of the participants in the half-reaction are specified as having a value of unity. The additional specification that the standard hydrogen electrode is the anode implies that the standard potential for a half-reaction is always a *reduction potential.*

(e) An *oxidation potential* is the potential of an electrochemical cell in which the cathode is a standard hydrogen electrode and the half-cell of interest acts as anode.

15-3. (a) *Reduction* is the process whereby a substance acquires electrons; a *reducing agent* is a supplier of electrons.
(c) The *anode* of an electrochemical cell is the electrode at which oxidation occurs. The *cathode* is the electrode at which reduction occurs.
(e) The *standard electrode potential* is the potential of an electrochemical cell in which the standard hydrogen electrode acts as an anode and all participants in the cathode process have unit activity. The formal potential differs in that the molar *concentrations* of the reactants and products are unity and the concentration of other species in the solution are carefully specified.

15-4. The first standard potential is for a solution that is saturated with I_2, which has an $I_2(aq)$ activity significantly less than one. The second potential is for a *hypothetical* half-cell in which the $I_2(aq)$ activity is unity. Such a half-cell, if it existed, would have a greater potential, since the driving force for the reduction would be greater at the higher I_2 concentration. The second half-cell potential, although hypothetical, is nevertheless useful for calculating electrode potentials for solutions that are undersaturated in I_2.

15-5. It is necessary to bubble hydrogen through the electrolyte in a hydrogen electrode in order to keep the solution saturated with the gas. Only under these circumstances is the hydrogen activity constant so that the electrode potential is constant and reproducible.

15-7. (a) $2Fe^{3+} + Sn^{2+} \rightarrow 2Fe^{2+} + Sn^{4+}$
(c) $2NO_3^- + Cu(s) + 4H^+ \rightarrow 2NO_2(g) + 2H_2O + Cu^{2+}$
(e) $Ti^{3+} + Fe(CN)_6^{3-} + H_2O \rightarrow TiO^{2+} + Fe(CN)_6^{4-} + 2H^+$
(g) $2Ag(s) + 2I^- + Sn^{4+} \rightarrow 2AgI(s) + Sn^{2+}$
(i) $5HNO_2 + 2MnO_4^- + H^+ \rightarrow 5NO_3^- + 2Mn^{2+} + 3H_2O$

15-8. (a) Oxidizing agent Fe^{3+}; $Fe^{3+} + e^- \rightleftharpoons Fe^{2+}$
Reducing agent Sn^{2+}; $Sn^{2+} \rightleftharpoons Sn^{4+} + 2e^-$
(c) Oxidizing agent NO_3^-;
$NO_3^- + 2H^+ + e^- \rightleftharpoons NO_2(g) + H_2O$
Reducing agent Cu; $Cu(s) \rightleftharpoons Cu^{2+} + 2e^-$
(e) Oxidizing agent $Fe(CN)_6^{3-}$;
$Fe(CN)_6^{3-} + e^- \rightleftharpoons Fe(CN)_6^{4-}$
Reducing agent Ti^{3+};
$Ti^{3+} + H_2O \rightleftharpoons TiO^{2+} + 2H^+ + e^-$
(g) Oxidizing agent Sn^{4+}; $Sn^{4+} + 2e^- \rightleftharpoons Sn^{2+}$
Reducing agent Ag;
$Ag(s) + I^- \rightleftharpoons AgI(s) + e^-$
(i) Oxidizing agent MnO_4^-;
$MnO_4^- + 8H^+ + 5e^- \rightleftharpoons Mn^{2+} + 4H_2O$
Reducing agent HNO_2;
$HNO_2 + H_2O \rightleftharpoons NO_3^- + 3H^+ + 2e^-$

15-9. (a) $MnO_4^- + 5VO^{2+} + 11H_2O \rightarrow Mn^{2+} + 5V(OH)_4^+ + 2H^+$
(c) $Cr_2O_7^{2-} + 3U^{4+} + 2H^+ \rightarrow 2Cr^{3+} + 3UO_2^{2+} + H_2O$
(e) $IO_3^- + 5I^- + 6H^+ \rightarrow 3I_2 + H_2O$
(g) $HPO_3^{2-} + 2MnO_4^- + 3OH^- \rightarrow PO_4^{3-} + 2MnO_4^{2-} + 2H_2O$
(i) $V^{2+} + 2V(OH)_4^+ + 2H^+ \rightarrow 3VO^{2+} + 5H_2O$

15-10. (a) Oxidizing agent MnO_4^-;
$MnO_4^- + 5e^- + 8H^+ \rightleftharpoons Mn^{2+} + 4H_2O$
Reducing agent VO^{2+};
$VO^{2+} + 3H_2O \rightleftharpoons V(OH)_4^+ + 2H^+ + e^-$
(c) Oxidizing agent $Cr_2O_7^{2-}$;
$Cr_2O_7^{2-} + 6e^- + 14H^+ \rightleftharpoons 2Cr^{3+} + 7H_2O$
Reducing agent U^{4+};
$U^{4+} + 2H_2O \rightleftharpoons UO_2^{2+} + 2e^- + 4H^+$
(e) Oxidizing agent IO_3^-;
$IO_3^- + 5e^- + 6H^+ \rightleftharpoons \frac{1}{2}I_2 + 3H_2O$
Reducing agent I^-; $I^- \rightleftharpoons \frac{1}{2}I_2 + e^-$
(g) Oxidizing agent MnO_4^-;
$MnO_4^- + e^- \rightleftharpoons MnO_4^{2-}$
Reducing agent HPO_3^{2-};
$HPO_3^{2-} + H_2O \rightleftharpoons PO_4^{3-} + 2e^- + 3H^+$
(i) Oxidizing agent $V(OH)_4^+$;
$V(OH)_4^+ + e^- + 2H^+ \rightleftharpoons VO^{2+} + 3H_2O$
Reducing agent V^{2+};
$V^{2+} + H_2O \rightleftharpoons VO^{2+} + 2e^- + 2H^+$

15-11. (a) $AgBr(s) + e^- \rightleftharpoons Ag(s) + Br^-$; $V^{2+} \rightleftharpoons V^{3+} + e^-$
$Tl^{3+} + 2e^- \rightleftharpoons Tl^+$; $Fe(CN)_6^{4-} \rightleftharpoons Fe(CN)_6^{3-} + e^-$
$V^{3+} + e^- \rightleftharpoons V^{2+}$; $Zn \rightleftharpoons Zn^{2+} + 2e^-$
$Fe(CN)_6^{3-} + e^- \rightleftharpoons Fe(CN)_6^{4-}$;
$Ag(s) + Br^- \rightleftharpoons AgBr(s) + e^-$
$S_2O_8^{2-} + 2e^- \rightleftharpoons 2SO_4^{2-}$; $Tl^+ \rightleftharpoons Tl^{3+} + 2e^-$

(b), (c)	E^0
$S_2O_8^{2-} \rightleftharpoons 2SO_4^{2-}$	2.01
$Tl^{3+} + 2e^- \rightleftharpoons Tl^+$	1.25
$Fe(CN)_6^{3-} + e^- \rightleftharpoons Fe(CN)_6^{4-}$	0.36
$AgBr(s) + e^- \rightleftharpoons Ag(s) + Br^-$	0.073
$V^{3+} + e^- \rightleftharpoons V^{2+}$	−0.256
$Zn^{2+} + 2e^- \rightleftharpoons Zn(s)$	−0.763

15-13. (a) 0.297 V (b) 0.190 V (c) −0.152 V
(d) 0.047 V (e) 0.007 V

15-15. −0.121 V

15-16. (a) 0.78 V (b) 0.198 V (c) −0.355 V
(d) 0.210 V (e) 0.177 V (f) 0.86 V

15-18. (a) −0.280 anode (b) −0.090 anode
(c) 1.003 cathode
(d) 0.171 V cathode (e) −0.009 V anode

15-20. 0.390 V

15-22. −0.964 V

15-24. −1.25 V

Chapter 16

16-1. The electrode potential of a system is the electrode potential of all half-cell processes at equilibrium in the system.

16-2. (a) *Equilibrium* is the state that a system assumes after each addition of reagent. *Equivalence* refers to a particular equilibrium state when a stoichiometric amount of titrant has been added.

16-4. For points before equivalence, potential data are computed from the analyte standard potential and the analytical concentrations of the analyte and its product. Postequivalence point data are based upon the standard potential for the titrant and its analytical concentrations. The equivalence point potential is computed from the two standard potentials and the stoichiometric relation between the analyte and titrant.

16-6. An asymmetric titration curve will be encountered whenever the titrant and the analyte react in a ratio that is not 1:1.

16-8. (a) 0.452 V, galvanic (d) −0.401 V, electrolytic
(b) 0.031 V, galvanic (e) −0.208 V, electrolytic
(c) 0.414 V, galvanic (f) 0.724 V, galvanic

16-9. (a) 0.631 V (c) 0.331 V (e) −0.620 V

16-11. (a) 2.2×10^{17} (c) 3.2×10^{22}
(e) 9×10^{37} (g) 2.4×10^{10}

16-13. (a) Phenosafranine (e) Erioglaucin A
(c) Indigo tetrasulfonate or Methylene blue (g) None

16-14.

	E, V		
Vol, mL	(a)	(c)	(e)
10.00	−0.292	0.32	0.316
25.00	−0.256	0.36	0.334
49.00	−0.156	0.46	0.384
49.90	−0.097	0.52	0.414
50.00	0.017	0.95	1.17
50.10	0.074	1.17	1.48
51.00	0.104	1.20	1.49
60.00	0.133	1.23	1.50

Chapter 17

17-1. (a) $2Mn^{2+} + 5S_2O_8^{2-} + 8H_2O \rightarrow 10SO_4^{2-} + 2MnO_4^- + 16H^+$
(b) $NaBiO_3(s) + 2Ce^{3+} + 4H^+ \rightarrow BiO^+ + 2Ce^{4+} + 2H_2O + Na^+$
(c) $H_2O_2 + U^{4+} \rightarrow UO_2^{2+} + 2H^+$
(d) $V(OH)_4^+ + Ag(s) + Cl^- + 2H^+ \rightarrow VO^{2+} + AgCl(s) + 3H_2O$
(e) $2MnO_4^- + 5H_2O_2 + 6H^+ \rightarrow 5O_2 + 2Mn^{2+} + 8H_2O$
(f) $ClO_3^- + 6I^- + 6H^+ \rightarrow 3I_2 + Cl^- + 3H_2O$

17-3. Only in the presence of Cl^- ion is Ag a sufficiently good reducing agent to be very useful for prereductions. In the presence of Cl^- ion, the half-reaction occurring in the Walden reductor is

$$Ag(s) + Cl^- \longrightarrow AgCl(s) + e^-$$

The excess HCl increases the tendency of this reaction to occur by the common ion effect.

17-5. $UO_2^{2+} + 2Ag(s) + 4H^+ + 2Cl^- \rightleftharpoons U^{4+} + 2AgCl(s) + H_2O$

17-7. Standard solutions of reductants find somewhat limited use because of their susceptibility to air oxidation.

17-8. Standard $KMnO_4$ solutions are seldom used to titrate solutions containing HCl because of the tendency of MnO_4^- to oxidize Cl^- to Cl_2, thus causing an overconsumption of MnO_4^-.

17-10. $2MnO_4^- + 3Mn^{2+} + 2H_2O \rightarrow 5MnO_2(s) + 4H^+$

17-13. $4MnO_4^- + 2H_2O \rightarrow \underset{\text{brown}}{4MnO_2(s)} + 3O_2 + 4OH^-$

17-15. Iodine is not sufficiently soluble in water to produce a useful standard reagent. It is quite soluble in solutions that contain an excess of iodide, however, as a consequence of the formation of the triiodide complex. The rate at which iodine dissolves in iodide solutions increases as the concentration of iodide ion becomes greater. For this reason, iodine is always dissolved in a very concentrated solution of potassium iodide and diluted only after solution is complete.

17-17. $S_2O_3^{2-} + H^+ \rightarrow HSO_3^- + S(s)$

17-19. $BrO_3^- + \underset{\text{excess}}{6I^-} + 6H^+ \rightarrow Br^- + 3I_2 + 3H_2O$
$I_2 + 2S_2O_3^{2-} \rightarrow 2I^- + S_4O_6^{2-}$

17-21. $2I_2 + N_2H_4 \rightarrow N_2 + 4H^+ + 4I^-$

17-23. (a) 0.1122 M Ce^{4+} (c) 0.02245 M MnO_4^-
(e) 0.02806 M IO_3^-

17-24. Dissolve 2.574 g $K_2Cr_2O_7$ in sufficient water to give 250.0 mL of solution.

17-26. Dissolve about 24 g of $KMnO_4$ in 1.5 L of water.

17-28. 0.01518 M $KMnO_4$

17-30. 0.06711 M $Na_2S_2O_3$

17-32. (a) 14.72% Sb (b) 20.54% Sb_2S_3

17-34. 9.38% analyte

17-35. (a) 32.08% Fe (b) 45.86% Fe_2O_3

17-37. 0.03074 M

17-39. 56.53% $KClO_3$

17-41. 0.701% As_2O_3

17-43. 3.64% C_2H_5SH

17-45. 2.635% KI

17-46. 69.07% Fe and 21.07% Cr

17-48. 0.5554 g Tl

17-49. 10.4 ppm SO_2

17-51. 19.5 ppm H_2S

Chapter 18

18-1. (a) An indicator electrode is an electrode used in potentiometry that responds to variations in the activity of an analyte ion or molecule.
(c) An electrode of the first kind is a metal electrode that is used to determine the concentration of the cation of that metal in a solution.

18-2. (a) A liquid-junction potential is the potential that develops across the interface between two solutions having different electrolyte compositions.

18-3. (a) An electrode of the first kind for Hg(II) would take the form

$$\|Hg^{2+}(xM)|Hg$$

$$E_{Hg} = E^0_{Hg} - \frac{0.0592}{2}\log\frac{1}{[Hg^{2+}]}$$

$$= E^0_{Hg} + \frac{0.0592}{2}\text{pHg}$$

(b) An electrode of the second kind for EDTA would take the form

$$\|HgY^{2-}(yM), Y^{4-}(xM)|Hg$$

where a small and fixed amount of HgY^{2-} is introduced into the analyte solution so that its concentration is xM. Here the potential of the mercury electrode is given by

$$E_{Hg} = K - \frac{0.0592}{2}\log [Y^{4-}] = K + \frac{0.0592}{2}\text{pY}$$

where

$$K = E^0_{HgY^{2-}} - \frac{0.0592}{2}\log\frac{1}{a_{HgY^{2-}}}$$
$$\approx 0.21 - \frac{0.0592}{2}\log\frac{1}{[HgY^{2-}]}$$

18-5. The pH-dependent potential that develops across a glass membrane arises from the difference in positions of dissociation equilibria that arise on each of the two surfaces. These equilibria are described by the equation

$$\underset{\text{membrane}}{H^+Gl^-} \rightleftharpoons \underset{\text{soln}}{H^+} + \underset{\text{membrane}}{Gl^-}$$

The surface exposed to the solution having the higher hydrogen ion activity then becomes positive with respect to the other surface. This charge difference, or potential, serves as the analytical parameter when the pH of the solution on one side of the membrane is held constant.

18-7. Uncertainties that may be encountered in pH measurements include (1) the acid error in highly acidic solutions, (2) the alkaline error in strongly basic solutions, (3) the error that arises when the ionic strength of the calibration standards differ from that of the analyte solution, (4) uncertainties in the pH of the standard buffers, (5) nonreproducible junction potentials when samples of low ionic strength are measured, and (6) dehydration of the working surface.

18-9. The alkaline error arises when a glass electrode is employed to measure the pH of solutions having pH values in the 10 to 12 range or greater. In the presence of alkali ions, the glass surface becomes responsive to both hydrogen and alkali ions. Low pH values arise as a consequence.

18-11. (b) The *boundary potential* for a membrane electrode is a potential that develops when the membrane separates two solutions that have different concentrations of a cation or an anion that the membrane binds selectively. For an aqueous solution, the following equilibria develop when the membrane is positioned between two solutions of A^+:

$$\underset{\text{membrane}_1}{A^+M^-} \rightleftharpoons \underset{\text{soln}_1}{A^+} + \underset{\text{membrane}_1}{M^-}$$

$$\underset{\text{membrane}_2}{A^+M^-} \rightleftharpoons \underset{\text{soln}_2}{A^+} + \underset{\text{membrane}_2}{M^-}$$

where the subscripts refer to the two sides of the membrane. A potential develops across this membrane if one of these equilibria proceeds further to the right than the other, and this potential is the boundary potential. For example if the concentration of A^+ is greater in solution 1 than in solution 2, the negative charge on side 1 of the membrane will be less than that on side 2 because the equilibrium on side 1 will lie further to the left. Thus, a greater fraction of the negative charge on side 1 will be neutralized by A^+.

(d) The membrane in a solid state electrode for F^- is crystalline LaF_3, which when immersed in aqueous solution LaF_3 dissociates according to the equation

$$LaF_3(s) \rightleftharpoons La^{3+} + 3F^-$$

Thus, a boundary potential develops across this membrane when it separates two solutions of different F^- ion concentration. The source of this potential is described in part (b) of this answer.

Chapter 19

19-1. The direct potentiometric measurement of pH provides a measure of the equilibrium activity of hydronium ions present in a solution of the sample. A potentiometric titration provides information on the amount of reactive protons, both ionized and nonionized, that are present in a sample.

19-2. (a) 0.354 V
(b) $SCE\|IO_3^-(xM), AgIO_3(\text{sat'd})|Ag$
(c) $pIO_3 = \dfrac{E_{cell} - 0.110}{0.0592}$
(d) 3.11

19-4. (a) $SCE\|SCN^-(xM), AgSCN(\text{sat'd})|Ag$
(c) $SCE\|SO_3^{2-}(xM), Ag_2SO_3(\text{sat'd})|Ag$

19-5. (a) $pSCN = \dfrac{E_{cell} + 0.153}{0.0592}$
(c) $pSO_3 = \dfrac{2(E_{cell} - 0.146)}{0.0592}$

19-6. (a) 4.65 (c) 5.20

19-7. 6.76

19-9. (a) 0.157 V (b) −0.026 V

19-11. (a) pH = 12.629; $a_{H^+} = 2.35 \times 10^{-13}$
(b) pH = 5.579; $a_{H^+} = 2.64 \times 10^{-6}$
(c) pH = 12.596; $a_{H^+} = 2.54 \times 10^{-13}$
pH = 12.663; $a_{H^+} = 2.17 \times 10^{-13}$
and
pH = 5.545 and 5.612
$a_{H^+} = 2.85 \times 10^{-6}$ and 2.44×10^{-6}

19-13. 2.0×10^{-6}

19-15. 136 g HA/mol

19-17.
(a)

mL Reagent	*E* vs. SCE, V	mL Reagent	*E* vs. SCE, V
5.00	0.58	49.00	0.66
10.00	0.59	50.00	0.80
15.00	0.60	51.00	1.10
25.00	0.61	55.00	1.14
40.00	0.63	60.00	1.15

19-19. An operational definition of pH is a definition describing in detail how a pH is determined experimentally. To obtain an operational pH_U the potential E_S of an electrode system immersed in buffers prepared by a carefully prescribed method and having a pH of pH_S, is measured. The E_U for an unknown is then measured with the same electrode system. The operation of H_U of the unknown is then defined by

$$pH_U = pH_S - \frac{(E_U - E_S)}{0.0592}$$

19-20. (a) 90% (c) 31%

Chapter 20

20-1. (a) *Concentration polarization* is a condition in which the current in an electrochemical cell is limited by the rate at which reactants are brought to or removed from the surface of one or both electrodes. *Kinetic polarization* is a condition in which the current in an electrochemical cell is limited by the rate at which electrons are transferred between the electrode surfaces and reactants in solution. For either type of polarization, the current is no longer proportional to the cell potential.

(c) Both the *coulomb* and the *Faraday* are units of quantity of charge, or electricity. The former is the quantity transported by one ampere of current in one second; the latter is equal to 96,495 coulombs or one mole of electrons.

(e) The *electrolysis circuit* consists of a working electrode and a counter electrode. The *control circuit* regulates the applied potential such that the potential between the working electrode and a reference electrode in the control circuit is constant and at a desired level.

20-2. (a) *Current density* is the current at an electrode divided by the surface area of that electrode. Ordinarily, it has units of amperes per square centimeter.

(c) A *coulometric titration* is an electroanalytical method in which a constant current of known magnitude generates a reagent that reacts with the analyte. The time required to generate enough reagent to complete the reaction is measured.

(e) *Current efficiency* is a measure of agreement between the number of faradays of current and the number of moles of reactant oxidized or reduced at a working electrode.

20-3. Mass transport in an electrochemical cell results from one or more of the following: (1) *diffusion,* which arises from the concentration difference between the solution immediately adjacent to the electrode surface and the bulk of the solution; (2) *migration,* which results from electrostatic attraction or repulsion between a species and an electrode; and (3) *convection,* which results from stirring, vibration, or temperature difference.

20-5. Both kinetic and concentration polarization cause the potential of a cell to be more negative than the thermodynamic potential. Concentration polarization arises from the slow rate at which reactants or products are transported to or away from the electrode surfaces. Kinetic polarization arises from the slow rate of the electrochemical reactions at the electrode surfaces.

20-7. Kinetic polarization is often encountered when the product of a reaction is a gas, particularly when the electrode is a soft metal such as mercury, zinc, or copper. It is likely to occur at low temperatures and high current densities.

20-9. Temperature, current density, complexation of the analyte, and codeposition of a gas influence the physical properties of an electrogravimetric deposit.

20-11. (a) An *amperostat* is an instrument that provides a constant current to an electrolysis cell.

(b) A *potentiostat* controls the applied potential to maintain a constant potential between the working electrode and a reference electrode.

20-12. In *amperostatic coulometry,* the cell is operated so that the current in the cell is held constant. In *potentiostatic coulometry,* the potential of the working electrode is maintained constant.

20-13. The species produced at the counter electrode is a potential interference by reacting with the product at the working electrode. Isolation of the one from the other is ordinarily necessary.

20-15. (b) 6.2×10^{16} cations

20-16. (a) -0.738 V (c) -0.337 V

20-17. -0.913 V

20-18. -0.493 V

20-21. (a) -0.94 V (b) -0.35 V (c) -2.09 V (d) -2.37 V

20-22. (a) 4.2×10^{-6} M (b) -0.425 V

20-25. (a) Separation is impossible. (b) Separation is feasible. (c) Separation would be feasible.

20-26. (a) 0.237 V (c) 0.0395 V (e) 0.118 V (g) 0.276 V (i) 0.0789 V

20-27. (a) 28.4 min (b) 9.47 min

20-29. 112.6 g/eq

20-31. 79.5 ppm $CaCO_3$

20-33. 4.06% $C_6H_5NO_2$

20-35. 2.471% CCl_4 and 1.854% $CHCl_3$

20-37. 50.9 μg $C_6H_5NH_2$

Chapter 21

21-1. (a) *Voltammetry* is an analytical technique that is based upon measuring the current that develops in a microelectrode as the applied potential is varied. *Polarography* is a particular type of voltammetry in which the microelectrode is a dropping mercury electrode.

(c) As shown in Figures 21-19 (page 486) and 21-22 (page 488) *differential pulse polarography* and *square-wave polarography* differ in the type of pulse sequence used.

(e) A *residual current* in polarography is a nonfaradaic charging current that arises from the flow of electrons required to charge individual drops of mercury as they form and fall. A *limiting current* is a constant faradaic current that is limited in magnitude by the rate at which a reactant is brought to the surface of a microelectrode.

(g) *Turbulent flow* is a type of liquid flow that has no regular pattern. *Laminar flow* is a type of liquid flow in which layers of liquid slide by one another in a direction that is parallel to a solid surface.

21-2. (a) A *voltammogram* is a plot of current as a function of voltage applied to a microelectrode.

(c) The *Nernst diffusion layer* is the thin layer of stagnant solution that is immediately adjacent to the surface of an electrode in hydrodynamic voltammetry.

(e) The *half-wave potential* is the potential on a voltammetric wave at which the current is one half the limiting current.

21-3. Most organic electrode processes involve hydrogen ions. A typical reaction is

$$R + nH^+ + ne^- \longrightarrow RH_n$$

where R and RH_n are the oxidized and reduced forms of the organic molecule. Unless buffered solutions are used, marked pH changes will occur at the electrode surface. Because the reduction potential (and sometimes the product) is affected by pH, drawn out and poorly defined waves are observed in the absence of a buffer.

21-5. A plot of $E_{applied}$ versus $\log i/(i_d - i)$ will yield a straight line having a slope of $-0.0592/n$. Thus, n is readily computed from the slope.

21-6. (a) −0.059 V

21-8. (a) 0.067% (b) 0.13% (c) 0.40%

21-9. (a) 0.369 mg Cd/mL (c) 0.144 mg Cd/mL

21-11. (a) −0.260 V

21-12. Stripping methods are generally more sensitive than other voltammetric procedures because the analyte can be removed from a relatively large volume of solution and concentrated on a tiny electrode for a long period (often many minutes). After concentration, the potential is reversed and all of the analyte that has been stored on the electrode is rapidly oxidized or reduced, producing a large current.

21-14. The advantages of voltammetry at ultramicroelectrodes include (1) their low *IR* loss, which permits their use in low-dielectric-constant solvents, such as hydrocarbons; (2) their small size, which permits electrochemical studies on exceedingly small samples, such as the inside of living organs; and (3) their great speed of equilibration even in unstirred solution.

Chapter 22

22-1. (a) Transitions among various electronic energy levels.

(c) Transitions from the lowest vibrational levels of the various excited electronic states to the various vibrational states of the ground state.

(e) Transitions from molecules in excited electronic states to lower-energy electronic states.

22-2. (a) The *ground state* is the lowest energy state of an atom, ion, or molecule of a species at room temperature.

(c) A *photon* is a bundle or a particle of radiant energy whose magnitude is given by $h\nu$, where h is Planck's constant and ν is the frequency of the radiation.

(e) A *continuous spectrum* is one that is produced by heated bodies. Even at highest resolution, a continuous spectrum cannot be dispersed into lines.

(g) *Resonance fluorescence* is a type of fluorescence in which the emitted radiation is identical in frequency to the excitation frequency.

(i) $\% T = (P/P_0) \times 100\%$, where T is the transmittance, P_0 is the power of a beam of radiation that is incident upon an absorbing medium, and P is the power of the beam after having passed through a layer of the medium.

(k) The *molar absorptivity* ε is defined by the equation

$$\varepsilon = A/bc$$

where A is the absorbance of a medium having a path length of b cm and an analyte concentration of c molar.

(m) *Stray radiation* is unwanted radiation that appears at the exit slit of a monochromator as a result of scattering by dust particles and reflection off of interior surfaces of the monochromator. Its wavelength usually differs from the wavelength to which the monochromator is set.

(o) *Relaxation* is a process whereby an excited species loses energy and returns to a lower energy state.

22-3. (a) 1.13×10^{18} Hz (c) 4.318×10^{14} Hz

(e) 1.53×10^{13} Hz

22-4. (a) 252.8 cm

(c) 286 cm

22-5. (a) 3.33×10^3 cm^{-1} to 6.67×10^2 cm^{-1}

(b) 1.00×10^{14} Hz to 2.00×10^{13} Hz

22-7. (a) 86.3% (b) 17.2% (c) 48.1%

22-8. (a) 0.712 (b) 0.064 (c) 0.565

22-9. (a) 74.5% (b) 2.95% (c) 23.1%

22-10. (a) 1.013 (b) 0.365 (c) 0.866

22-11. (a) $\% T = 38.4$; $\varepsilon = 2.38 \times 10^3$; $c = 31.2$ ppm; $a = 1.33 \times 10^{-2}$

(c) $\% T = 3.77$; $c = 10.4$ ppm; $c = 4.18 \times 10^{-5}$ M; $\varepsilon = 3.42 \times 10^4$

(e) $c = 1.33 \times 10^{-5}$ M; $A = 0.115$; $\% T = 76.7$; $a = 0.0138$

(g) $A = 0.316$; $c = 2.69 \times 10^{-5}$ M; $\varepsilon = 4.70 \times 10^4$; $a = 0.188$

(i) $A = 0.117$; $c = 1.69$ ppm; $c = 6.77 \times 10^{-6}$ M; $\varepsilon = 1.58 \times 10^4$

22-12. $\varepsilon = 1.80 \times 10^4$ L cm^{-1} mol^{-1}

22-14. (a) 0.582 (b) 26.2% (c) 1.25×10^{-5} M

22-16. $A = 0.0595$

22-18. (a)

c_{ind}, M	[In$^-$]	[HIn]	A_{430}	A_{600}
3.00×10^{-4}	1.20×10^{-4}	1.80×10^{-4}	1.54	1.06
2.00×10^{-4}	9.27×10^{-5}	1.07×10^{-4}	0.935	0.777
1.00×10^{-4}	5.80×10^{-5}	4.20×10^{-5}	0.383	0.455
0.500×10^{-4}	3.48×10^{-5}	1.52×10^{-5}	0.149	0.261
0.250×10^{-4}	2.00×10^{-5}	5.00×10^{-6}	0.056	0.145

Chapter 23

23-2. 2079 lines/mm

23-4. A *spectroscope* consists of a monochromator that has been modified so that the focal plane contains a movable eyepiece that permits visual detection of the various emission lines of the elements. A *spectrograph* is a monochromator equipped with a photographic film or plate holder located along its focal plane; spectra are recorded photographically. A *spectrophotometer* is a monochromator equipped with a photoelectric detector that is located behind an exit slit located on the focal plane of the device.

23-6. **(a)** 725 nm **(c)** 1.45 μm

23-7. **(a)** 1.46×10^7 W/m^2 **(c)** 9.10×10^5 W/m^2

23-8. **(a)** $\lambda_{max} = 1.01$ μm; $\lambda'_{max} = 0.967$ μm
(b) $E_t = 3.86 \times 10^6$ W/m^2; $E'_t = 4.61 \times 10^6$ W/m^2

23-9. **(a)** *Hydrogen* and *deuterium lamps* differ only in the gases they contain. The latter produces radiation of somewhat higher intensity.
(c) A *phototube* is a vacuum tube equipped with a photoemissive cathode. It has a high electrical resistance and requires a potential of 90 V or more to produce a photocurrent. The currents are generally small enough to require considerable amplification before they can be measured. A *photovoltaic cell* consists of a photosensitive semiconductor sandwiched between two electrodes. A current is generated between the electrodes when radiation is absorbed by the semiconducting layer. The current is generally large enough to be measured directly with a microammeter. The advantages of a phototube are greater sensitivity and wavelength range as well as better reproducibility. The advantages of the photocell are its simplicity, low cost, and general ruggedness. In addition, it does not require an external power supply or elaborate electric circuitry. Its use is limited to visible radiation, however, and in addition it suffers from fatigue whereby its electrical output decreases gradually with time.
(e) A *photometer* is an instrument for absorption measurements that consists of a source, a filter, and a photoelectric detector. A *colorimeter* differs from a photometer in the respect that the human eye serves as the detector in the former. The photometer offers the advantage of greater precision and the ability to discriminate between colors provided they are not too much alike. The main advantages of a colorimeter are simplicity, low cost, and the fact that no power supply is needed.
(g) A *single-beam spectrophotometer* employs a fixed beam of radiation that irradiates first the solvent and then the analyte solution. In a *double-beam instrument,* the solvent and solution are irradiated simultaneously or nearly so. The advantages of the double-beam instruments are freedom from problems arising from fluctuations in the source intensity from drift in electronic circuits; in addition, they are more easily adapted to automatic spectral recording. The single-beam instrument offers the advantages of simplicity and lower cost.

23-10. **(a)** % $T = 33.8\%$ **(b)** $A = 0.471$
(c) $T = 0.697$ **(d)** $T = 0.114$

23-12. In a *deuterium lamp,* the input energy from the power source produces an excited deuterium molecule that dissociates into two atoms in the ground state and a photon of radiation. As the excited deuterium molecule relaxes, its quantized energy is distributed between the energy of the photon and the kinetic energies of the two deuterium atoms. The latter can vary from nearly zero to the original energy of the excited molecule. Therefore, the energy of the radiation, which is the difference between the quantized energy of the excited molecule and the kinetic energies of the atoms, can also vary continuously over the same range. Consequently, the emission spectrum is continuous.

23-14. The power of an infrared beam is measured with a heat detector, which consists of a tiny blackened surface that is warmed as a consequence of radiation absorption. This surface is attached to a transducer, which converts the heat signal to an electrical one. Heat transducers are of three types: (1) a *thermopile,* which consists of several dissimilar metal junctions that develop a potential that depends upon the difference in temperature of the junctions; (2) a *bolometer,* which is fashioned from a conductor whose electrical resistance depends upon temperature; and (3) a *pneumatic detector,* whose internal pressure is temperature dependent.

23-16. *Tungsten/halogen lamps* contain a small amount of iodine in the evacuated quartz envelope that contains the tungsten filament. The iodine prolongs the life of the lamp and permits it to operate at a higher temperature. The iodine combines with gaseous tungsten that sublimes from the filament and causes the metal to be redeposited, thus adding to the life of the lamp.

23-18. The basic difference between an *absorption spectrometer* and an *emission spectrometer* is that the former requires a separate radiation source and a sample compartment that holds containers for the sample and its solvent. With an emission spectrometer, the sample container is a hot flame, a heated surface, or an electric arc or spark that also serves as the radiation source.

23-20. **(a)** The *dark current* is the small current that develops in a radiation transducer in the absence of radiation.
(c) *Scattered radiation* in a monochromator is unwanted radiation that reaches the exit slit as a result of reflection and scattering. Its wavelength is usually different from that of the radiation reaching the slit directly from the dispersing element.
(e) The *majority carrier* in a semiconductor is the mobile charge carrier in either n-type or p-type semiconductors. For n-type, the majority carrier is the electron; in p-type, the majority carrier is a positively charged hole.

23-21. **(a)** $t = 1.69$ μm
(b) $\lambda_1 = 4.54$ μm; $\lambda_2 = 2.27$ μm; $\lambda_3 = 1.51$ μm

Chapter 24

24-1. **(a)** *Spectrophotometers* use a grating or a prism to provide narrow bands of radiation; *photometers* utilize filters for this purpose. The advantages of spectrophotometers are greater versatility and the ability to procure entire spectra. The advantages of photometers are simplicity, ruggedness, and low cost.

(c) *Diode-array spectrophotometers* detect the entire spectral range essentially simultaneously and can produce a spectrum in less than a second. *Conventional spectrophotometers* require several minutes to perform the same task. Accordingly, diode-array instruments can be used to monitor processes that occur on very fast time scales. Their resolution is usually lower than a conventional spectrophotometer, however.

24-2. As a minimum requirement, the radiation emitted by the source of a single-beam instrument must be stable for however long it takes to make the 0% T adjustment, the 100% T adjustment, and the measurement of T for the sample.

24-4. Electrolyte concentration, pH, temperature.

24-6. 1.6×10^{-5} M to 8.6×10^{-5} M

24-8. $c_{10\%} = 1.2 \times 10^{-3}$ M; $c_{90\%} = 5.4 \times 10^{-5}$ M

24-10. **(a)** $T = 26.1\%$; $A = 0.583$ **(b)** $A = 0.291$; $T = 51.1\%$
(c) $A = 1.166$; $T = 0.0682$

24-11. **(b)** $A = 0.471$
(d) $T = 0.114$

24-14.

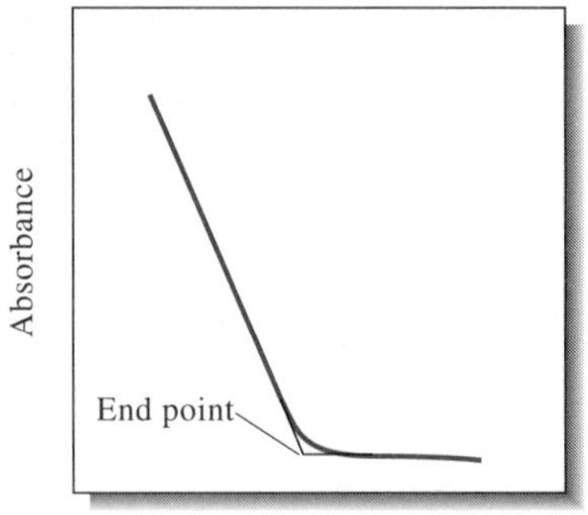

24-16. %Co = 0.0214

24-18.

	c_{Co}, M	c_{Ni}, M
(a)	5.48×10^{-5}	1.31×10^{-4}
(b)	1.66×10^{-4}	5.19×10^{-5}
(c)	2.20×10^{-4}	4.41×10^{-5}
(d)	3.43×10^{-5}	1.46×10^{-4}

24-21. **(a)** 0.492 **(c)** 0.190

24-22. **(a)** 0.301 **(c)** 0.491

24-23. For solution A, pH = 5.60
For solution C, pH = 4.80

24-25. **(b)** 0.956 **(c)** 0.749

24-26.

	c_P, M	c_Q, M
(a)	2.08×10^{-4}	4.91×10^{-5}
(c)	8.37×10^{-5}	6.10×10^{-5}
(e)	2.11×10^{-4}	9.65×10^{-5}

24-27. **(b)** $A_{510} = 0.03949\ c_{Fe} - 0.001008$
(c) $s_m = 1.1 \times 10^{-4}$; $s_b = 2.7 \times 10^{-3}$

24-28.

		s_c, rel %	
	$c_{Fe(II)}$, ppm	**1 Result**	**3 Results**
(a)	3.65	2.8%	2.1
(c)	1.75	6.1%	4.6
(e)	38.3	0.27%	0.20

Chapter 25

25-1. **(a)** *Resonance fluorescence* is observed when excited atoms emit radiation of the same wavelength as that used to excite them.

(c) *Internal conversion* is the nonradiative relaxation of a molecule from the lowest vibrational level of an excited electronic state to a high vibrational level of a lower electronic state.

(e) The *Stokes shift* is the difference in wavelength between the radiation used to excite fluorescence and the wavelength of the emitted radiation.

25-2. For spectrofluorometry, the analytical signal F is given by $F = 2.3\ K'\epsilon bcP_0$. The magnitude of F, and thus sensitivity, can be enhanced by increasing the source intensity P_0 or the transducer sensitivity.

For spectrophotometry, the analytical signal A is given by $A = \log P_0/P$. Increasing P_0 or the detector's response to P_0 is accompanied by a corresponding increase in P. Thus, the ratio does not change nor does the analytical signal. Consequently, no improvement in sensitivity accompanies such changes.

25-3. **(a)** Fluorescein because of its greater structural rigidity due to the bridging —O— groups.

25-4. Compounds that fluoresce have structures that slow the rate of nonradiative relaxation to the point where there is time for fluorescence to occur. Compounds that do not fluoresce have structures that permit rapid relaxation by nonradiative processes.

25-6. Excitation of fluorescence usually involves transfer of an electron to a high vibrational state of an upper electronic state. Relaxation to a lower vibrational state of this electronic state goes on much more rapidly than fluorescence relaxation. When fluorescence relaxation occurs it is to a high vibrational state of the ground state or to a high vibrational state of an electronic state that is above the ground state. Such transitions involve less energy than the excitation wavelength.

25-8. Most fluorescence instruments are double beamed in order to compensate for fluctuations in the analytical signal arising from variations in the power supply to the source.

25-10. **(a)** Plotting the data gives a straight line that passes through the origin. The slope is estimated to be 22.3.
(b) $y = mx + b = 22.4x + 0.700$
(c) $s_m = 0.340$ and $s_b = 0.186$
(d) 0.540 μmol NADH/L
(e) 1.4%
(f) 0.96%

25-12. (a) The fluoride ion must react with and decompose the fluorescent complex to give nonfluorescing products. Thus, if we symbolize the complex as AlG_x^{3+}, we may write

$$\underset{\text{fluorescent}}{AlG_x^{3+}} + 6F^- \rightleftharpoons \underset{\text{nonfluorescent}}{AlF_6^{3-}} + \underset{\text{nonfluorescent}}{xG}$$

(b) A plot of the meter reading R versus the volume of standard V_s is a straight line with a negative slope.
(c) meter reading = − 13.22 volume + 68.23
(d) $S_{slope} = 0.19$; $S_{intercept} = 0.36$
(e) 10.3 ppb F^-
(f) $s = 0.16$ ppb F^-

Chapter 26

26-1. In *atomic emission spectroscopy* the radiation source is the sample itself. The energy for excitation of analyte atoms is supplied by a flame, an oven, a plasma, or an electric arc or spark. The signal is the measured intensity of the source at the wavelength of interest. In *atomic absorption spectroscopy* the radiation source is usually a line source such as a hollow cathode lamp, and the output signal is the absorbance calculated from the incident power of the source and the resulting power after the light has passed through the atomized sample in the source.

26-2. (a) *Atomization* is a process in which a sample, usually in solution is volatilized and decomposed to produce an atomic vapor.
(c) *Doppler broadening* is an increase in the width of atomic absorption or emission lines caused by the Doppler effect in which atoms moving toward a detector absorb or emit wavelengths that are slightly shorter than those absorbed or emitted by atoms moving at right angles to the detector. The effect is reversed for atoms moving away from the detector.
(e) A *plasma* is a conducting gas that contains a large concentration of ions and/or electrons.
(g) *Sputtering* is a process in which atoms of an element are dislodged from the surface of a cathode by bombardment by a stream of inert gas ions that have been accelerated toward the cathode by a high electrical potential.
(i) A *spectral interference* in atomic spectroscopy occurs when an absorption or emission peak of an element in the sample matrix overlaps that of the analyte.
(k) A *radiation buffer* is a substance that is added in large excess to both standards and samples in atomic spectroscopy to prevent the presence of that substance in the sample matrix from having an appreciable effect on the results.
(m) A *protective agent* in atomic spectroscopy is a substance, such as EDTA or 8-hydroxyquinoline, that forms stable but volatile complexes with analyte cation and thus prevents interference by anions that form nonvolatile species with the analyte.

26-3. In atomic emission spectroscopy, the analytical signal is produced by *excited* atoms or ions, whereas in atomic absorption and fluorescence the signal results from absorption by *unexcited* species. Typically, the number of unexcited species exceeds that of the excited ones by several orders of magnitude. The ratio of unexcited to excited atoms in a hot medium varies exponentially with temperature. Thus, a small change in temperature brings about a large change in the number of excited atoms. The number of unexcited atoms changes very little, however, because they are present in an enormous excess. Therefore, emission spectroscopy is more sensitive to temperature changes than is absorption or fluorescence spectroscopy.

26-5. (a) Sulfate ion forms complexes with Fe(III) that are not readily atomized. Thus, the concentration of iron atoms in a flame is lower in the presence of sulfate ions.
(b) Sulfate interference could be overcome (1) by adding a releasing agent that forms complexes with sulfate that are more stable than the iron complexes, (2) by adding a protective agent, such as EDTA, that forms a highly stable but volatile complex with the Fe(III), and (3) by employing a high-temperature flame (oxygen/acetylene or nitrous oxide/acetylene).

26-7. The lack of linearity is a consequence of ionization of the uranium, which decreases the relative concentration of uranium atoms. Alkali metal atoms ionize readily and thus give a high concentration of electrons, which represses the uranium ionization.

26-9. 0.504 ppm Pb

26-11. (b) $A = kc_s\dfrac{V_s}{V_t} + kc_x\dfrac{V_x}{V_t}$

(c) slope $= \dfrac{kc_s}{V_t}$; intercept $= \dfrac{kc_sV_x}{V_t}$

(d) Dividing the intercept by the slope from part (c) gives

$$c_x = \frac{b}{m} \times \frac{c_s}{V_x}$$

(e) slope $= 8.81 \times 10^{-3}$ and intercept $= 0.202$
(f) $s_m = 4.11 \times 10^{-5}$ and $s_b = 1.01 \times 10^{-3}$
(g) 28.0 ppm Cr

Chapter 27

27-1. (a) The *order of a reaction* is the numerical sum of the exponents of the concentration terms in the rate law for the reaction.
(c) *Enzymes* are high-molecular-weight organic molecules that catalyze reactions of biochemical importance.
(e) The *Michaelis constant* is an equilibrium-like constant defined by the equation $K_m = (k_{-1} + k_2)/k_1$, where k_1 and k_{-1} are the rate constants for the forward and reverse reactions in the formation of an enzyme/substrate complex, which is the intermediate in an enzyme-catalyzed reaction. The term k_2 is the

rate constant for the decomposition of the complex to give products.

(g) *Integral methods* use integrated forms of rate equations to compute concentrations from kinetic data.

27-3. Pseudo-first-order conditions are used in kinetic methods because under these conditions the reaction rate is directly proportional to the concentration of the analyte.

27-5. $t_{1/2} = 0.693/k$

27-6. (a) $\tau = 2.01$ s (c) $\tau = 1.97 \times 10^3$ s (e) $\tau = 1.47 \times 10^9$ s

27-7. (a) $k = 40.5\ \text{s}^{-1}$ (c) $k = 0.405\ \text{s}^{-1}$ (e) $k = 1.51 \times 10^4\ \text{s}^{-1}$

27-8. (a) $n = 0.105$ (c) $n = 2.3$ (e) $n = 6.9$

27-10. (a) 0.2% (c) 0.02% (e) 1.0% (g) 0.05% (i) 6.8% (k) 0.64%

27-11. $\frac{d[\text{P}]}{dt} = \frac{k_2[\text{E}]_0[\text{S}]_t}{[\text{S}]_t + K_m}$ (Equation 27-22)

At v_{max}, $\frac{dP}{dt} = k_2[\text{E}]_0$. Thus, at $v_{max}/2$, we can write

$$\frac{dP}{dt} = \frac{k_2[\text{E}]_0}{2} = \frac{k_2[\text{E}]_0[\text{S}]_t}{[\text{S}]_t + K_m}$$

$$[\text{S}]_t + K_m = 2[\text{S}]_t$$

$$K_m = [\text{S}]_t$$

27-13. $c_{Cu^{2+}} = 5.5$ ppm

27-15. $[\text{tryp}]_t = 4.5 \times 10^{-2}\ \mu\text{mol/L}$

Chapter 28

28-1. (a) *Elution* is a process in which species are washed through a chromatographic column by additions of fresh mobile phase.

(c) The *stationary phase* in a chromatographic column is a solid or liquid that is fixed in place. A mobile phase then passes over or through the stationary phase.

(e) The *retention time* for an analyte is the time interval between its injection onto a column and the appearance of its peak at the other end of the column.

(g) The *selectivity factor* α of a column toward two species is given by the equation $\alpha = K_B/K_A$, where K_B is the partition ratio of the more strongly held species B and K_A is the corresponding ratio for the less strongly held solute A.

28-3. In gas-liquid chromatography, the mobile phase is a gas, whereas in liquid-liquid chromatography it is a liquid.

28-5. The number of plates in a column can be determined by measuring the retention time t_R and width of a peak at its base W. The number of plates N is then given by the equation $N = 16(t_R/W)^2$.

28-7. (a) A 2775; B 2474; C 2363; D 2523 (b) $N = 2.5 \times 10^3$ and $s = 0.2 \times 10^3$ (c) $H = 0.0097$ cm

28-8. (a) A 0.74; B 3.3; C 3.5; D 6.0 (b) $K_A = 6.2$; $K_B = 27$; $K_C = 30$; $K_D = 50$

28-9. (a) $R_S = 0.72$ (b) $\alpha_{C,B} = 1.1$ (c) $L = 108$ cm (d) $(t_R)_2 = 62$ min

28-10. (a) $R_S = 5.2$ (b) $L = 2.0$ cm

28-14. (a) $k_1' = 4.3$; $k_2' = 4.7$; $k_3' = 6.1$ (b) $K_1 = 14$; $K_2 = 15$; $K_3 = 19$ (c) $\alpha_{2,1} = 1.11$

28-15. (a) $k_M' = 2.54$; $k_N' = 2.62$ (b) $\alpha = 1.03$ (c) $N = 8.1 \times 10^4$ plates (d) $L = 1.8 \times 10^2$ cm (e) $(t_R)_N = 91$ min

Chapter 29

29-1. In *gas-liquid chromatography,* the stationary phase is a liquid that is immobilized on a solid. Retention of sample constituents involves equilibria between a gaseous and a liquid phase. In gas-solid chromatography, the stationary phase is a solid surface that retains analytes by physical adsorption. Here separations involve adsorption equilibria.

29-3. Gas-solid chromatography has limited application because active or polar compounds are retained more or less permanently on the packings. In addition, severe tailing is often observed owing to the nonlinear character of the physical adsorption process.

29-5. *Temperature programming* involves increasing the temperature of a gas-chromatographic column as a function of time. This technique is particularly useful for samples that contain constituents whose boiling points differ significantly. Low-boiling-point constituents are separated initially at temperatures that provide good resolution. As the separation proceeds, the column temperature is increased so that the higher boiling constituents come off the column with good resolution and at reasonable lengths of time.

29-7. (a) Advantages, thermal conductivity: general applicability, large linear range simplicity, nondestructive. Disadvantage: low sensitivity.

(b) Advantages, flame ionization: high sensitivity, large linear response range, low noise, rugged, ease of use, and response that is largely independent of flow rate. Disadvantage: destructive of sample.

(c) Advantage, thermionic: high sensitivity for compounds containing nitrogen and phosphorus, good linear range. Disadvantages: destructive, not applicable for many types of analytes.

(d) Advantages, flame photometric: selective toward sulfur and phosphorus containing analytes, good sensitivity. Disadvantages: destructive, limited applicability.

(e) Advantages, photoionization: versatility, nondestructive, large linear range.

29-9. The typical column packing is made up of diatomaceous earth particles having diameters ranging from 250 to 170 μm or 170 to 149 μm.

29-11. Megapore columns are open-tubular columns that have a greater inside diameter (530 μm) than typical open-tubular columns, which range in diameter from 150 to 320 μm. The typical megabore column has a diameter of 530 μm.

29-13. Desirable properties of a stationary phase for GLC include low volatility, thermal stability, chemical inertness, and solvent characteristics that provide suitable k' and α values for the various analytes to be separated.

29-15. Film thickness of stationary phases influences the rate at which analytes are carried through the column with the rate increasing as the thickness is decreased. Less band broadening is encountered with thin films.

29-17. % methyl acetate = 22.9%
% methyl propionate = 48.5%
% methyl *n*-butyrate = 28.7%

Chapter 30

30-1. (a) Substances that are somewhat volatile and are thermally stable.
(c) Substances that are ionic.
(e) High-molecular-weight compounds that are soluble in nonpolar solvents.
(g) Nonvolatile or thermally unstable species that contain no chromophoric groups.

30-2. (a) In an *isocratic elution,* the solvent composition is held constant throughout the elution.
(c) In a *stop-flow injection,* the flow of solvent is stopped, a fitting at the head of the column is removed, and the sample is injected directly onto the head of the column. The fitting is then replaced and pumping is resumed.
(e) In a normal-phase packing, the stationary phase is quite polar and the mobile phase is relatively nonpolar.
(g) An *eluent-suppressor column* is located after the ion-exchange column in ion chromatography. It converts the ionized species used to elute analyte ions to largely undissociated molecules that do not interfere with conductometric detection.
(i) *Gel permeation chromatography* is a type of size-exclusion chromatography in which the packings are hydrophobic and the eluents are nonaqueous. It is used for separating high-molecular-weight nonpolar species.
(k) *FSOT columns* are fused silica open-tubular columns used in gas chromatography.
(m) A *supercritical fluid* is a substance that is maintained above its critical temperature so that it cannot be condensed into a liquid no matter how great the pressure.

30-4. (a) *n*-hexane, benzene, *n*-hexanol

30-5. (a) *n*-hexanol, benzene, *n*-hexane

30-7. In *adsorption chromatography,* the sample components are selectively retained on the surface of a solid stationary phase by adsorption. In partition chromatography, selective retention occurs in a liquid or liquid-like stationary phase.

30-9. Nonvolatile and thermally unstable compounds.

30-10. Supercritical-fluid chromatography is particularly applicable to nonvolatile or thermally unstable species that are difficult to detect because they have no chromophoric groups.

30-11. The simplest type of pump for liquid chromatography is a *pneumatic pump,* which consists of a collapsible solvent container housed in a vessel that can be pressurized by a compressed gas. This type of pump is simple, inexpensive, and pulse-free. It has limited capacity and pressure output, it is not adaptable to gradient elution, and its pumping rate depends upon the viscosity of the solvent.

A *screw-driven syringe pump* consists of a large syringe in which the piston is moved in or out by means of a motor-driven screw. It also is pulse-free, and the rate of delivery is easily varied. It suffers from lack of capacity and is inconvenient to use when solvents must be changed.

The most versatile and widely used pump is the *reciprocating pump* that usually consists of a small cylindrical chamber that is filled and then emptied by the back-and-forth motion of a piston. Advantages of the reciprocating pump include small internal volume, high output pressures, adaptability to gradient elution, and flow rates that are constant and independent of viscosity and back pressure. The main disadvantage is pulsed output, which must be damped.

30-13. (a) The advantages over HPLC are (1) that it can be used with general detectors, such as the flame ionization detector, and (2) that it is inherently faster or gives better resolution for the same speed.
(b) The advantage over GLC is that it can be used for nonvolatile and thermally unstable samples.

Chapter 32

32-1. The steps in sampling are (1) identification of the population from which the sample is to be drawn, (2) collection of a gross sample, and (3) reduction of the gross sample to a small quantity of homogeneous material for analysis.

32-3. (a) *Sorbed water* is that held as a condensed liquid phase in the capillaries of a colloid. *Adsorbed water* is that retained by adsorption on the surface of a finely ground solid. *Occluded water* is that held in cavities distributed irregularly throughout a crystalline solid.
(c) *Essential water* is chemically bound water that occurs as an integral part of the molecular or crystalline structure of a compound in its solid state. *Nonessential water* is that retained by a solid as a consequence of physical forces.

32-5. $(s_s/\bar{x}_i) = 0.76\%$

32-6. (a) 26% (b) 14 ± 6 (c) 1.69×10^3

32-7. (a) $n = 784$ (c) $n = 2.0 \times 10^4$

32-9. (a) $n = 3.5 \times 10^4$ particles
(b) mass = 3.9×10^3 g or 8.5 lb
(c) diameter = 0.22 mm

32-11. (a) $n = 2.2 \times 10^5$ particles
(b) mass = 2.8×10^3 g or 6.1 lb
(c) diameter = 0.13 mm

Chapter 33

33-1. *Dry ashing* is carried out by igniting the sample in air or sometimes oxygen. *Wet ashing* is carried out by heating the sample in an aqueous medium containing such oxidizing agents as H_2SO_4, $HClO_4$, HNO_3, H_2O_2, or some combination of these.

33-3. Ba_2O_3 or $CaCO_3/NH_4Cl$

33-5. When hot concentrated $HClO_4$ comes in contact with organic materials or other oxidizable species, explosions are highly probable.

33-6. (a) Samples for halogen determination may be decomposed in a Schöniger combustion flask, combusted in a tube furnace in a stream of oxygen, or fused in a peroxide bomb.
(c) Samples for nitrogen determination are decomposed in hot concentrated H_2SO_4 in a Kjeldahl flask or oxidized by CuO in a tube furnace in the Dumas method.

Chapter 34

34-1. A *masking agent* is a complexing reagent that reacts selectively with one or more components of a solution to prevent them from interfering in an analysis.

34-3. (a) 1.73×10^{-2} M (b) 6.40×10^{-3} M
(c) 2.06×10^{-5} M (d) 6.89×10^{-4} M

34-5. (a) 75 mL (b) 40 mL (c) 22 mL

34-7. (a) 18.0 (b) 7.56

34-8. (a) 91.9

34-9. (a) $K_D = 1.53$
(b) $[HA]_{aq} = 0.0147$; $[A^-]_{aq} = 0.0378$
(c) $K_a = 9.7 \times 10^{-2}$

34-11. (a) 148 g/eq (b) 296 g/mol

34-12. (a) $HL(aq) \rightleftharpoons HL(org)$
$HL(aq) + H_2O \rightleftharpoons H_3O^+ + L^-(aq)$
$Cu^{2+}(aq) + 2L^-(aq) \rightleftharpoons CuL_2(aq)$
$CuL_2(aq) \rightleftharpoons CuL_2(org)$
(b) $K_{ex} = 5.0 \times 10^{-8}$
(c) $D = 5.01$
(d) 4 extractions
(e) 6 extractions

INDEX

Page numbers in **boldface** refer to specific laboratory directions. Page numbers followed by *t* refer to tabular entries. Entries preceded by A refer to pages in the Appendices.

A

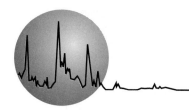

B

D

E

F

G

H

I

J

K

L

N

Q

R

T

U

V

MATHCAD® APPLICATIONS FOR ANALYTICAL CHEMISTRY

Mathcad Applications for Analytical Chemistry is a student workbook that is designed to introduce students to *Mathcad,* provide a guide for the application of *Mathcad* to analytical chemistry, and furnish students and instructors with useful templates for creating their own *Mathcad* documents. This workbook is keyed to *Fundamentals of Analytical Chemistry, 7th Edition* and *Analytical Chemistry: An Introduction, 6th Edition,* both books by Douglas A. Skoog, Donald M. West, and F. James Holler.

Cross-references between *Mathcad Applications for Analytical Chemistry* and *Fundamentals of Analytical Chemistry, 7th Edition* are presented on the page opposite this page, and cross-references to *Analytical Chemistry: An Introduction, 6th Edition* are provided in the end pages of *Mathcad Applications for Analytical Chemistry.*

Mathcad is a powerful and versatile program that is available for the IBM-PC™ running under *Windows*™ and MS-DOS® and for the Macintosh®. Particularly attractive are *Mathcad's* ability to present mathematical expressions nearly exactly as we would write them down on paper and its capability for creating complex graphs with just a few keystrokes or mouse clicks. *Mathcad* is a mathematical notebook, computational tool, and equation solver that allows students to perform many calculations of analytical chemistry quickly and accurately. *Mathcad* is available from MathSoft, Inc., Cambridge, MA.

Table of Contents

Mathcad Applications for Analytical Chemistry
Mathcad Applications for Analytical Chemistry is available from Saunders College Publishing, ISBN: 0-03-076017-8.

CROSS-REFERENCES FOR *MATHCAD®* APPLICATIONS FOR ANALYTICAL CHEMISTRY (MAAC)

Text Reference	Text Page	*MAAC* Chapter	*MAAC* Page
Example 2-1	13	2	15–18
Equation 3-1	27	2	19
Example 3-1	29	2	19–21
Example 3-2	31	2	32
Section 3C-2	35	9	114–116
Section 3D-1	40	1	4
Figure 4-1	48	2	28–30
Equation 4-7	54	2	32
Table 4-4	58	2	30–31
Section 4E	60	9	111–114
Example 4-9	63	9	111–114
Example 5-5	79	1	4–8
Example 5-7	81	1	8–12
Example 7-4	136	3	35–37
Feature 7-5	142	4	50–51
Example 7-10	144	3	35–36
Table 8-1	154	3	38, 40
Section 8A	148	3	37
Example 9-5	164	4	44–46
Example 9-7	168	4	46–50
Example 9-8	172	4	50–51
Example 9-11	179	4	51–56
Section 10D-2	205	5	68
Equation 10-17	217	5	64
Section 11F	234	6	79–85
Section 11G	240	6	85–88
Section 11H	241	6	88–90
Section 11I	243	6	77–79
Example 13-1	266	5	61–64
Figure 14-2	280	7	96–98
Example 14-4	286	7	98–100
Figure 14-6	289	7	100
Section 16C	343	8	107–109
Section 16C-2	346	8	109–110
Section 24A-4	572	9	114–116
Section 24A-4	578	10	124–127
Section 24A-4	578	12	147–149
Figures 24-19, 24-20	580	12	148
Section 25B	605	9	111–114
Equation 27-7	640	12	141
Section 27B-2	652	12	141
Section 28C	666	11	129–133
Equation 28-14	670	11	130
Section 28C-3	673	12	145–147
Equation 28-18	676	11	131
Appendix 11	A-34	10	121

INTERNATIONAL ATOMIC MASSES

Element	Symbol	Atomic Number	Atomic Mass
Actinium	Ac	89	227
Aluminum	Al	13	26.981539
Americium	Am	95	243
Antimony	Sb	51	121.757
Argon	Ar	18	39.948
Arsenic	As	33	74.92159
Astatine	At	85	210
Barium	Ba	56	137.327
Berkelium	Bk	97	247
Beryllium	Be	4	9.012182
Bismuth	Bi	83	208.98037
Boron	B	5	10.811
Bromine	Br	35	79.904
Cadmium	Cd	48	112.411
Calcium	Ca	20	40.078
Californium	Cf	98	251
Carbon	C	6	12.011
Cerium	Ce	58	140.115
Cesium	Cs	55	132.90543
Chlorine	Cl	17	35.4527
Chromium	Cr	24	51.9961
Cobalt	Co	27	58.93320
Copper	Cu	29	63.546
Curium	Cm	96	247
Dysprosium	Dy	66	162.50
Einsteinium	Es	99	252
Erbium	Er	68	167.26
Europium	Eu	63	151.965
Fermium	Fm	100	257
Fluorine	F	9	18.9984032
Francium	Fr	87	223
Gadolinium	Gd	64	157.25
Gallium	Ga	31	69.723
Germanium	Ge	32	72.61
Gold	Au	79	196.96654
Hafnium	Hf	72	178.49
Helium	He	2	4.002602
Holmium	Ho	67	164.93032
Hydrogen	H	1	1.00794
Indium	In	49	114.82
Iodine	I	53	126.90447
Iridium	Ir	77	192.22
Iron	Fe	26	55.847
Krypton	Kr	36	83.80
Lanthanum	La	57	138.9055
Lawrencium	Lr	103	262
Lead	Pb	82	207.2
Lithium	Li	3	6.941
Lutetium	Lu	71	174.967
Magnesium	Mg	12	24.305
Manganese	Mn	25	54.93805
Mendelevium	Md	101	258
Mercury	Hg	80	200.59
Molybdenum	Mo	42	95.94
Neodymium	Nd	60	144.24
Neon	Ne	10	20.1797
Neptunium	Np	93	237
Nickel	Ni	28	58.6934
Niobium	Nb	41	92.90638
Nitrogen	N	7	14.00674
Nobelium	No	102	259
Osmium	Os	76	190.2
Oxygen	O	8	15.9994
Palladium	Pd	46	106.42
Phosphorus	P	15	30.973762
Platinum	Pt	78	195.08
Plutonium	Pu	94	244
Polonium	Po	84	210
Potassium	K	19	39.0983
Praseodymium	Pr	59	140.90765
Promethium	Pm	61	145
Protactinium	Pa	91	231.03588
Radium	Ra	88	226
Radon	Rn	86	221
Rhenium	Re	75	186.207
Rhodium	Rh	45	102.90550
Rubidium	Rb	37	85.4678
Ruthenium	Ru	44	101.07
Samarium	Sm	62	150.36
Scandium	Sc	21	44.955910
Selenium	Se	34	78.96
Silicon	Si	14	28.0855
Silver	Ag	47	107.8682
Sodium	Na	11	22.989768
Strontium	Sr	38	87.62
Sulfur	S	16	32.066
Tantalum	Ta	73	180.9479
Technetium	Tc	43	98
Tellurium	Te	52	127.60
Terbium	Tb	65	158.92534
Thallium	Tl	81	204.3833
Thorium	Th	90	232.0381
Thulium	Tm	69	168.93421
Tin	Sn	50	118.710
Titanium	Ti	22	47.88
Tungsten	W	74	183.85
Uranium	U	92	238.0289
Vanadium	V	23	50.9415
Xenon	Xe	54	131.29
Ytterbium	Yb	70	173.04
Yttrium	Y	39	88.90585
Zinc	Zn	30	65.39
Zirconium	Zr	40	91.224